# The Microbiomes of Humans, Animals, Plants, and the Environment

Volume 3

This series covers microbiome topics from all natural habitats. Microbiome research is a vibrant field of science that offers a new perspective on Microbiology with a more comprehensive view on different microorganisms (microbiota) living and working together as a community (microbiome). Even though microbial communities in the environment have long been examined, this scientific movement also follows the increasing interest in microbiomes from humans, animals and plants. First and foremost, microbiome research tries to unravel how individual species within the community influence and communicate with each other. Additionally, scientists explore the delicate relationship between a microbiome and its habitat, as small changes in either, can have a profound impact on the other. With individual research volumes, this series reflects the vast diversity of Microbiomes and highlights the impact of this field in Microbiology.

More information about this series at https://link.springer.com/bookseries/16462

Lucas J. Stal • Mariana Silvia Cretoiu
Editors

# The Marine Microbiome

Second Edition

*Editors*
Lucas J. Stal
Department of Freshwater and Marine Ecology – IBED
University of Amsterdam
Amsterdam, The Netherlands

Mariana Silvia Cretoiu
University of Utrecht
Utrecht, The Netherlands

ISSN 2662-611X ISSN 2662-6128 (electronic)
The Microbiomes of Humans, Animals, Plants, and the Environment
ISBN 978-3-030-90382-4 ISBN 978-3-030-90383-1 (eBook)
https://doi.org/10.1007/978-3-030-90383-1



This Springer imprint is published by the registered company Springer Nature Switzerland AG.
The registered company address is: Gewerbestrasse 11, 6330 Cham, Switzerland

*Dedicated to*
*Prof. dr. Wolfgang Elisabeth Krumbein*
*Geomicrobiologist*
*1937–2021*
*In loving memory*

# Foreword

Marine microbiology has been a latecomer to the field of microbial ecology. But the last three decades have been enough to revert the situation. The ocean is now arguably the best known major microbial ecosystem on Earth. It was to be expected that its physical structure at first sight is much more amenable to sampling and interpretation than the richly micro-structured habitats like sediments or soil, not to mention animal microbiomes. The ocean is a single gigantic lake that is homogenized by the equally colossal global circulation. It has relatively constant and largely moderate conditions (apart from extreme oligotrophy, for exceptions see Chap. 14). On the other hand, it is also the largest, oldest and likely most critical ecosystem for the environmental health of the planet. It has also been the least impacted by the arrival of multicellular plants and animals ca. 500 million years ago. It is possible to envisage vast marine areas in which the conditions have changed little after the great oxygenation event ca. 2.2 billion years ago. Only in the last hundred years has anthropic impact started to be noticed in the Pacific and Atlantic central gyres that remained nearly pristine until the arrival of intensive whaling by the mid-nineteenth century.

Microbiology is now at a crossroads or, if you wish, a new beginning. It turns out that the fathers of microbiology were lucky to be able to discover the causative agents of the major infectious diseases of their time. Human pathogens tend to be copiotrophs well suited for growth in laboratory pure cultures, but most microbes are not, and this includes most bacteria, archaea and protists that live in the ocean. We know that because now we can sequence their genomes directly from the environment. However, the new microbiology that arose from nucleic acid sequencing is not without limitations and drawbacks. First and foremost, the capability of annotating genes and genomes (that is inferring function from sequence) is very unsatisfactory and largely based on a few model organisms very distant from their wild counterparts. Too little effort is invested in detecting new functions and too many mystifying sequences added to the humongous databases. What is worse, we still miss an evolutionary model that would consider the major impact of the pangenome of prokaryotic species. The key role played by the pangenome dynamics in the evolution of microbes and their ecology is yet to be fully incorporated in models of evolutionary biology or ecosystem functioning.

There is also an overall lack of finesse in microbiome studies that often gravitate over scale (big data) rather than detailed analysis of individual depth profiles, time series or specific microbes. To compound it all, the classification (nomenclature) of microbes has suffered from indiscriminate abuse by major rearrangements based on relationships, at the sequence level, of ribosomal components without contemplating that a microbe is much more than the pedigree of its ribosomes (and that is assuming that ribosomal RNAs or proteins really provide a reliable relationship). It is often unclear what the names mean in terms of the genome make-up and the biology of the microbes.

At this point in time, Lucas Stal and Silvia Cretoiu have undertaken the task of publishing a book (this) that updates the field and aggregates a large amount of relevant information; for that they have managed to gather a remarkable set of distinguished authors who have collectively done an amazing job of using the 'tsunami' of data generated by high-throughput sequencing technologies to describe most of relevant topics in the field, covering from population genomics to biogeography. They have introduced novel perspectives like the micro-seascape 'vast expanses of extremely dilute background seawater punctuated by rich hotspots of dissolved and particulate nutrient resources' (from Chap. 2). Everyone knows I am not an enthusiast of review books, although I believe they are certainly useful for newcomers such as PhD or master's students. The speed at which science is progressing these days coupled with the time required to collect and review chapters detracts from their usefulness. However, some books have the power to change a field. I hope this book will help to move marine microbiology into the next frontier in which a better understanding of microbes (largely, but not only, derived from genomics) will allow a more profound understanding of the oldest and largest microbiome on Earth.

Universitat Miguel Hernández,
Sant Joan d'Alacant, Spain

Francisco Rodriguez Valera

# Preface

In 2015, Springer invited us to edit a book on marine microbiology. At the time, we were leading the project 'MaCuMBA' ('Marine Microorganisms: Cultivation Methods for Improving their Biotechnological Applications') with a large consortium of European marine microbiologists and with funding of the European Commission. We decided to produce the book with contributions of MaCuMBA consortium members and entitled it *The Marine Microbiome* (with the subtitle *An Untapped Source of Biodiversity and Biotechnological Potential*). We thought that with this title we emphasize the importance of considering the total of microorganisms in the ocean and its adjacent seas, bays and estuaries as an entity, and that this entity plays a critical role in the functioning of the marine ecosystem, and that it would stimulate its research. The book appeared in 2016 (Stal and Cretoiu 2016) and was successful, and indeed, the research on the marine microbiome enjoyed more attention since (see below). The success of *The Marine Microbiome* was the reason for Springer to inquire with us already in 2019, whether we would be willing to edit a second edition. Meanwhile, Springer launched a book series *The Microbiomes of Humans, Animals, Plants, and the Environment* and the second edition of *The Marine Microbiome* fits perfectly in that series. While in our opinion the first edition is still recent and quite up-to-date, we decided that the second edition would need substantial new material, and therefore, only a few of the chapters of the first edition remained and were rewritten and updated. Therefore, the second edition of *The Marine Microbiome* might as well be considered *The Marine Microbiome* Part 2 being an important addition to the first.

Most of the authors of this volume agreed to their contributions in late 2019 and early 2020, not knowing the impact that the COVID-19 pandemic would have on their work routine. Many authors with young children were faced with closed schools and while mainly working from home had to divide their attention to their work as well as to the assistance of home education of their children. Also, their work and work load changed and from one day to another routine teaching, research and administration had to be completely reorganized in a way never done before. Meanwhile we, the editors, were pressing for delivery of the book chapters. We are therefore really grateful to all contributors for the great chapters that they wrote for this book, and we apologize to those authors who met the original deadline and had to wait much longer to see their work in print than originally anticipated.

We would also like to take this preface as an opportunity to exchange our thoughts about the title of this book (and of the previous edition) and about the use of the concept ‘microbiome’. We used it in the first edition more or less to replace the term ‘marine microbiology’, and it seemed to become fashionable to use the term ‘microbiome’ in a variety of situations in which microbiologists want to describe the whole of all microorganisms in a certain ‘biome’. Now, although we kept the term ‘microbiome’ as it seems that it has become commonplace in microbiology, we feel that it is semantically incorrectly used.

Let us consider the term ‘microorganism’. ‘Organism’ is defined as ‘a living being or entity adapted for living by means of organs separate in function but dependent on one another; any living being or its material structure; any complete whole which by the integration, interaction and mutual dependence of its parts is comparable to a living being’. And ‘life’ is the state of an organism characterized by certain processes or abilities that include metabolism, growth, reproduction and response. ‘Micro-‘ as a prefix means small or minute, or a millionth of (e.g. a metre). Therefore, ‘microorganism’ is a small, minute organism. These definitions were obtained from the Webster’s Dictionary of the English language.

Now, ‘biome’ is defined as ‘a large community of organisms having a peculiar form of vegetation and characteristic animals’ and usually covers large areas. This term was coined by Frederic E. Clements in his lecture ‘*The development and structure of biotic communities*’ at the Ecological Society of America in 1916 and published in 2017 in the *Journal of Ecology* (p. 120–121). He wrote:

> The biotic community is regarded as an organic unit comprising all the species of plants and animals at home in a particular habitat. While plants are regarded as exerting the dominant influence in the community, it is recognized that this rôle may sometimes be taken by the animals. The biotic community, or biome, is fundamentally controlled by the habitat, and exhibits a corresponding development and structure. In its development the biotic formation reacts upon the habitat, and thus produces a succession of biomes, comparable in practically all essentials to the succession of plant communities. Every such succession, or biosere, terminates regularly in a climax. The bioseres of each climax are either primary or secondary, and these may be further distinguished as hydroseres, xeroseres, etc.

A ‘microbiome’ would therefore mean a small or minute form of a biome which is in fact a contradiction in terms. This is not what has become the meaning of it. It has unfortunately introduced more and unnecessary ‘jargon’ in scientific language—jargon that is rarely precisely defined and increases confusion among ourselves but even more to society for which our scientific research should be relevant. Now that we seem to be stuck with ‘microbiome’, it may be good to briefly review its definition.

Berg et al. (2020) adopted the definition of Whipps and his colleagues (1988) as ‘a characteristic microbial community’ in a ‘reasonably well-defined habitat which has distinct physicochemical properties’ as their ‘theatre of activity’ (this book chapter of Whipps et al. (1988) is cited in Berg et al. (2020), and these authors made some explanatory comments to it).

BioConcepts (www.biological-concepts.com) gives two definitions of 'microbiome':

1. A population of microorganisms inhabiting a specific environment: a microbial community or ecosystem. This definition has been attributed to Mohr (1952) who mentioned the term microbiome without, however, giving an explicit definition.
2. The collective genomes of all the microorganisms inhabiting a specific environment.
   The latter was proposed by the 1958 Nobel laureate Joshua Lederberg (2001) (cited in Hooper and Gordon 2001) (citation: *'It includes Lederberg's own recent coinage of* microbiome, *to signify the ecological community of commensal, symbiotic, and pathogenic microorganisms that literally share our body space and have all but ignored as determinants of health and disease.'*) Hooper and Gordon (2001) citing Lederberg: *'microbiome' to describe the collective genome of our indigenous microbes (microflora).*

Shade and Handelsman (2012) introduced the term 'core microbiome' in the hope to find microorganisms common to any biome and hypothesized that it would fulfil the basic and fundamental roles required for any ecosystem to function. This is an interesting idea that awaits the discovery and description of examples that fit this hypothesis. Shade and Handelsman (2012) defined microbiome as '*an assemblage of microorganisms existing in or associated with a habitat; includes active and interacting member as well transient or inactive members*', definition that was based on Lederberg and McCray (2001). They gave as examples the human microbiome, earth microbiome, Lake Erie microbiome and soil microbiome (the marine or ocean microbiome had not yet been discovered; see Fig. 2). Remarkably, Shade and Handelsman (2012) define biome as '*the world's major ecosystems, defined by temperature gradients in latitude and altitude, precipitation and seasonality*' and cite Walter and Box (1976). Examples given by them are subtropical, Mediterranean and polar. It is difficult to see a microbiome with this definition.

The Earth microbiome in fact comprises all microorganisms on Earth and all three domains of life (bacteria, archaea and eukarya), and we also like to include viruses and other forms of genetic information that interact with cells by keeping some sort of functional equilibrium between the different organisms, exchanging genetic information and generating diversity.

A search in the Web of Science (WoS) using the term 'microbiome' in the title of scientific publications tells us that it appeared for the first time in 2006 (3 hits) culminating to more than 3500 hits in 2020 (Fig. 1). The combination 'marine' and 'microbiome' gave the first 2 hits in 2012 and 12 in 2020 (Fig. 2). Just to give a little more insight in the popularity of microbiome research, a WoS search on 19 May 2021 with 'microbiome' in the title returns 16,138 hits. 28% and 10% of those are with respectively 'gut' or 'human' in the title and 3.3 and 0.3% with the term 'soil' or 'rhizosphere' and 'marine', respectively. These numbers show that microbiome research becomes increasingly popular but that most of it focuses on humans and occasionally other animal model systems and that the second-most investigated

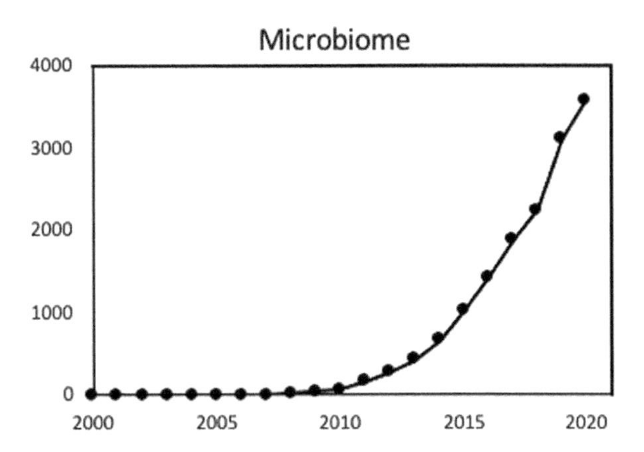


**Fig. 1** Hits in the Web of Science with keyword 'microbiome' in the title of publications by year of publication

microbiomes are connected to (crop) plants and food production. The microbiome of the ocean, arguably the largest ecosystem and crucial for the Earth's climate and food production, has still received little attention.

The Human Microbiome Project (2008–2012) was dominating microbiome research and until 2010 only human (gut) microbiome and few other animal gut microbiomes returned as hits in the title of a WoS search. In 2010, the Earth Microbiome was launched (Gilbert et al. 2010). Microbiome research stays focused on the human microbiome up to today. Other microbiomes studied are mostly in other animal models and to some extent plants or a rare environment or habitat turned up in the title together with the term microbiome. In 2013, the scientific journal *Microbiome* appeared next to a few other even more specialized journals as well as *Environmental Microbiome* (2019).

This book uses the term 'microbiome' as a description of all microorganisms in a certain biome. We consider the ocean and its adjacent seas, bays and estuaries as the 'marine biome' and refer to the total of microorganisms in it as the 'marine microbiome'.

We hope and expect that this book along with the first edition (Part 1) will inspire microbiologists to speed up the study of marine microorganisms and increase the knowledge of their diversity and function because it is urgently needed to understand the changes our planet is facing today and in the coming decennia. Fortunately, since 2016 important new initiatives have been launched such as the Atlantic Ocean Research Alliance (AORA) that published the 'Marine Microbiome Roadmap 2020', the United Nations 'Decade of Ocean Science for Sustainable Development' and the EU Ocean Literacy projects 'Sea Change' and 'ResponSEAble' (see Chap. 18). We are confident that this book will crank up marine microbiological research.

One of us (LJS) retired in 2017 but remained as emeritus professor 'Marine Microbiology' affiliated to the University of Amsterdam. This book may be his last product of an almost 45-year career in marine microbiology, and Chap. 1 of this book will be his last scientific paper. LJS would like to take this opportunity (again) to thank all his friends and colleagues in science and particularly in marine

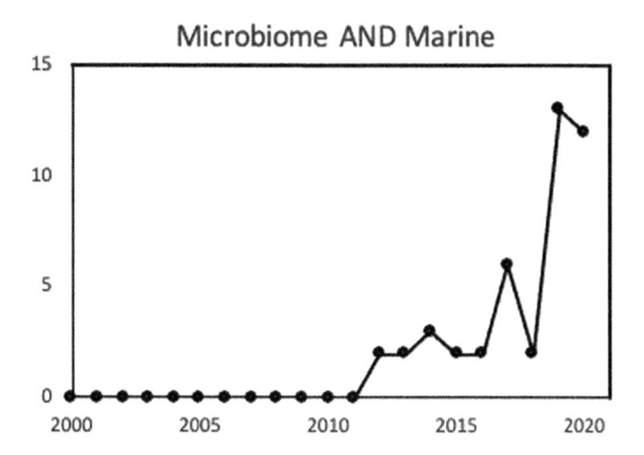


**Fig. 2** Hits in the Web of Science with keywords 'microbiome' AND 'marine' in the title of publications by year of publication

microbiology for the wonderful time he had, accomplishing scientific work and contributing to the world's knowledge on the unseen majority.

Finally, we would like to thank the project coordinators of Springer for their help and encouragement and for giving us the opportunity to publish this second edition of *The Marine Microbiome*. We particularly thank Mr. Bharat Sabnani, Mrs. Andrea Schlitzberger and Mr. Markus Spaeth. We also like to thank Professor Francisco Eduardo Rodriguez Valera for his kind foreword to this book.

Amsterdam, the Netherlands — Lucas J. Stal
Utrecht, the Netherlands — Mariana Silvia Cretoiu

## References

Berg G, Rybakova D, Fischer D, Cernava T, Champomier Vergès M-C, Charles T, Chen X, Cocolin L, Eversole K, Herrero Coral G, Kazou M, Kinkel L, Lange L, Lima N, Loy A, Macklin JA, Maguin E, Mauchline T, McClure R, Mitter B, Ryan M, Sarand I, Smidt H, Schelke B, Roume H, Kiran GS, Selvin J, Soares Correa de Souza R, van Overbeek L, Singh BK, Wagner M, Walsh A, Sessitsch A, Schloter M (2020) Microbiome definition re-visited: old concepts and new challenges. Microbiome 8:103

Gilbert JA, Meyer F, Jansson J, Gordon J, Pace N, Tiedje J, Ley R, Fierer N, Field D, Kyrpides N, Glöckner F-O, Klenk H-P, Wommack KE, Glass E, Docherty K, Gallery R, Stevens R, Knight R (2010) The Earth microbiome project: meeting report of the "1st EMP meeting on sample selection and acquisition" at Argonne National Laboratory October 6th 2010. Stand Genome Sci 3:249–253

Hooper LV, Gordon JI (2001) Commensal host-bacterial relationships in the gut. Science 292:1115–1118

Lederberg J, McCray AT (2001) 'Ome sweet 'omics - A genealogical treasury of words. The Scientist 8:April 2

Mohr JL (1952) Protozoa as indicators of pollution. Sci Month 74:7–9

Shade A, Handelsman J (2012) Beyond the Venn diagram: the hunt for a core microbiome. Environ Microbiol 14:4–12

Stal LJ, Cretoiu MS (2016) The marine microbiome. An untapped source of biodiversity and biotechnological potential. Springer International Publishing, Cham, 498 p

# Contents

# A Sea of Microbes: What's So Special about Marine Microbiology

# 1

Lucas J. Stal

**Abstract**

This chapter investigates what justifies marine microbiology as a discipline in its own right. Do marine microorganisms really exist? And if so, what distinguishes them from freshwater- or terrestrial microorganisms, or from microorganisms living in any other specialized habitat? The marine environment—particularly the ocean—is the largest continuous habitat on Earth. This makes the ocean different in terms of scale and sharing space and nutrients, as well as in terms of the distribution, dispersal, and encounters of microorganisms that inhabit it. The ocean comprises a large variety of confined sub-habitats. It has an impact on the function of the planet Earth as a whole. A critical property of seawater is that it contains a large amount of salt and that this requires microorganisms that live in it are able to adjust their osmotic pressure. The marine microbiome is composed of the three domains of life: bacteria, archaea, and eukarya, as well as viruses, all of which in dazzling numbers and diversity. All of the known microbial lineages are represented and many are exclusively found in the ocean and there is little doubt that life originated in the ocean.

**Keywords**

Marine microbiology · Milestones of marine microbiology · Nitrogen fixation · Ocean · Redfield ratio · Salinity

L. J. Stal (✉)
Department of Freshwater and Marine Ecology – IBED, University of Amsterdam, Amsterdam, The Netherlands



L. J. Stal, M. S. Cretoiu (eds.), *The Marine Microbiome*, The Microbiomes of Humans, Animals, Plants, and the Environment 3,
https://doi.org/10.1007/978-3-030-90383-1_1

## 1.1 Introduction

> *Nowhere else where life is possible, probably in no other place in the Universe except another ocean, are so many conditions so stable and so enduring.*

Baas Becking (1934) put this quote from Henderson's book "*The Fitness of the Environment*" (Henderson 1913) above Chap. 9 "De Zee" ("The Sea") of his book "*Geobiologie of inleiding tot de milieukunde*" (Fig. 1.1). But immediately, Baas Becking put this quote in perspective by stating that it might refer only to the physicochemical characteristics, because the ocean is considered to be stable because it is composed mainly of water. He continues by stating that when viewing it from a biological perspective no environment has such a diversity as the ocean with its adjacent seas, bays, estuaries, and coasts, with their currents and variations in illumination, temperature, salinity, hydrostatic pressure, and a variety of chemical components and organisms. Baas Becking's book, although written in Dutch (an English translation appeared in Canfield 2015, published by Wiley, and edited by Don Canfield), became famous by the quote: "*. . .alles is overal:* maar *het milieu*

GEOBIOLOGIE
OF
INLEIDING TOT DE
MILIEUKUNDE
DOOR
Dr. L. G. M. BAAS BECKING
HOOGLEERAAR AAN DE UNIVERSITEIT TE LEIDEN
MET VELE ILLUSTRATIE[illegible]
LITERATUURLIJST EN INDE[illegible]
DEN HAAG
W. P. VAN STOCKUM & ZOON N.V.
1934

**Fig. 1.1** Scan of the title page of Baas Becking's book "*Geobiologie of inleiding tot de mileukunde*" with a photo of the author on board of the RV "*Max Weber*" of the Zoological Station Den Helder (predecessor of the Netherlands Institute for Sea Research, NIOZ) while doing research in the waters around the island of Texel in the Netherlands in 1935. The photo is an original copy which was made by Dr. H. Oomen, who also was the original owner of this copy of the book which he obtained in 1935 during the same cruise on board of the RV "*Max Weber.*" The book is now in possession of L.J. Stal

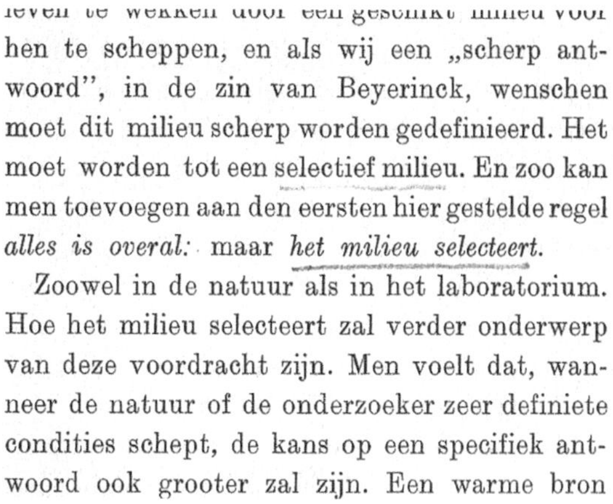

leven te wekken door een geschikt milieu voor
hen te scheppen, en als wij een „scherp antwoord", in de zin van Beyerinck, wenschen moet dit milieu scherp worden gedefinieerd. Het moet worden tot een selectief milieu. En zoo kan men toevoegen aan den eersten hier gestelde regel *alles is overal:* maar *het milieu selecteert.*

Zoowel in de natuur als in het laboratorium. Hoe het milieu selecteert zal verder onderwerp van deze voordracht zijn. Men voelt dat, wanneer de natuur of de onderzoeker zeer definiete condities schept, de kans op een specifiek antwoord ook grooter zal zijn. Een warme bron

**Fig. 1.2** Scan of the text of Baas Becking's book "*Geobiologie of inleiding tot de mileukunde*" with his famous quote: "…everything is everywhere: *but* the environment selects." Underlining presumably by Dr. H. Oomen, the first owner of this copy of the book

*selecteert*" (p. 15; italics and roman are from Baas Becking) (Fig. 1.2) ("… *everything is everywhere:* but *the environment selects*"). This quote has been cited often wrong, e.g., by leaving out "but" or replacing it by "and," both of which fundamentally change the meaning as nicely explained by De Wit and Bouvier (2006). Does this physicochemically stable ocean select for microorganisms that can be considered as truly marine?

"A Sea of Microbes" was the title of a special issue of "Oceanography" published in 2007 and edited by Proctor and Karl (2007). It was also the title of a book review (Karl 2001) of Kirchman's edited publication "Microbial Ecology of the Oceans" (2000) (second edition published in 2008) (Kirchman 2000, 2008). I took the translation in Dutch as the title for my inaugural lecture for celebrating the (first) chair of "Marine Microbiology" in the Netherlands at the University of Amsterdam on 11 April 2008 in Amsterdam: "*Een Zee van Bacteriën*." It is a double entendre in both languages meaning "a lot of microbes" as well as "a sea owned by the microbes." Here, I borrowed it again as the title for this introductory chapter, which tries to examine the question why marine microbiology would be a discipline in its own right.

## 1.2 Planet Ocean

The planet Earth is covered for more than 70% of its surface by the ocean and because water preferentially absorbs red light it leaves the complementary blue color. Therefore, "Ocean" would be a more appropriate name for this planet. The ocean has an average depth of 3700 m and its deepest point is the Challenger Deep in the Mariana Trench with ~11,000 m. The ocean contains 1.35 billion cubic kilometers of water (97% of Earth's water, Table 1.1) and is distributed over 5 named ocean basins: the Pacific Ocean, the Indian Ocean, the Atlantic Ocean, the Arctic Ocean, and the Southern (Antarctic) Ocean (Table 1.2). These oceans are all interconnected and altogether there is one world ocean. Five main currents called

**Table 1.1** Distribution of water on Earth

| Water body | % |
|---|---|
| Ocean, seas, bays | 96.5 |
| Ice caps, glaciers, permanent snow | 1.74 |
| Saline ground water | 0.94 |
| Fresh ground water | 0.76 |
| Ground ice and permafrost | 0.022 |
| Fresh water lakes | 0.007 |
| Saline lakes | 0.006 |
| Soil moisture | 0.001 |
| Atmosphere | 0.001 |
| Swamp water | 0.0008 |
| Rivers | 0.0002 |
| Biological water | 0.0001 |

https://en.wikipedia.org/wiki/Water_distribution_on_Earth

**Table 1.2** Distribution of water between the main oceanic basins (% of total water on Earth)

| Ocean | % |
|---|---|
| Pacific | 48.30 |
| Atlantic | 22.40 |
| Indian | 19.00 |
| Southern | 5.18 |
| Arctic | 1.35 |

https://en.wikipedia.org/wiki/Water_distribution_on_Earth

gyres connect the oceanic basins: the North Pacific Gyre, the South Pacific Gyre, the Indian Ocean Gyre, the North Atlantic Gyre, and the South Atlantic Gyre. These currents are formed by wind and the rotation of the Earth (Coriolis effect). It is estimated that to complete one circulation may take up to 1000 years. This means that oceanic microorganisms are globally distributed by these currents and that this one ocean, therefore, represents the largest habitat in the biosphere. Coastal areas where surface water is replaced by cold, nutrient-rich, deep water are called upwelling zones that are characterized by high productivity. The surface water is heated in the tropics and cooled down in the artic. Freezing rises the salinity and together with the decrease in temperature the density of the seawater increases and seawater sinks and flows to the equatorial regions where it emerges to the surface. These gyres altogether form the thermohaline circulation, a global water transport system that is also known as "conveyor belt."

### 1.2.1 Salinity

A remarkable property of seawater is that it contains salt. More precisely, it contains on average 35 ppt (parts per thousand) or 35 g salts per kilogram of water (‰) (Millero et al. 2008). Note that salinity is dimensionless. The quantitative most important components (solutes) of seawater are listed in Table 1.3. The salinity is

**Table 1.3** Composition of standard seawater compared to a hypersaline sea, a freshwater lake, and a brackish sea

| | Standard Seawater[a] (35‰) | Dead Sea[b] (230‰) | Lake Michigan[c] (0.1‰) | Baltic Sea (Bornholm Sea)[d] (~10‰) |
|---|---|---|---|---|
| Solute | $g\ kg^{-1}$ | | | |
| $Na^+$ | 10.781 | 30.156 | 0.016 | ~3.080 |
| $Mg^{2+}$ | 1.284 | 31.156 | 0.015 | ~0.367 |
| $Ca^{2+}$ | 0.412 | 13.001 | 0.016 | ~0.118 |
| $K^+$ | 0.399 | 5.616 | 0.013 | ~0.114 |
| $Sr^{2+}$ | 0.008 | 0.231 | n.r. | ~0.002 |
| $Li^+$ | n.r.[e] | 0.015 | n.r. | |
| $Cl^-$ | 19.353 | 163.397 | 0.025 | ~5.529 |
| $SO_4^{2-}$ | 2.712 | 0.385 | 0.013 | ~0.775 |
| $HCO_3^-$ | 0.105 | 0.154 | 0.009 | |
| $Br^-$ | 0.067 | 3.923 | n.r. | |
| $CO_3^{2-}$ | 0.014 | n.r. | n.r. | |
| $B(OH)_4^-$ | 0.008 | n.r. | n.r. | |
| $F^-$ | 0.00004 | n.r. | n.r. | |
| $B(OH)_3$ | 0.0006 | n.r. | n.r. | |
| $CO_2$ | 0.0004 | n.r. | n.r. | |

The salinity and precise composition of seawater depend on the exact location (geographic; depth) from which the seawater is considered. Here, the standardized seawater as defined by Millero et al. (2008) is listed as a good average. The Dead Sea is taken as just one example. Depending on the hypersaline system considered, the composition will be totally different. The same is true for freshwater systems, which may have greatly different composition in different lakes or rivers. The Baltic Sea is the largest brackish sea but characterized by a salinity gradient from the North Sea (almost full seawater salinity) to the Gulf of Bothnia (almost freshwater). Also, there is a pycnocline because the deeper waters are more saline (North Sea water entering the Baltic Sea dives underneath the lighter less saline Baltic Sea water). Here, the brackish Bornholm Sea is taken as an example. The salinity of the Baltic Sea does not show anomalies and the composition shown is just derived from standard seawater by dividing by 3.5

[a]Millero et al. (2008)
[b]Nissenbaum (1975)
[c]Chapra et al. (2012)
[d]Grasshoff and Voipio (1981)
[e]Not reported

not everywhere the same and not homogeneously distributed in the ocean and its adjacent waters. For instance, the Mediterranean Sea salinity is higher (38–40 ppt) because of its limited exchange with the Atlantic Ocean and higher evaporation. The (coastal) North Sea has a slightly lower salinity (30–34 ppt) because of river discharge. The Baltic Sea has a salinity gradient from the North Sea to almost freshwater in the Bothnian Gulf and an average of 10 ppt in the Bornholm Sea. This sea has also a stable halocline because the saltier and therefore denser North Sea water dives underneath the less dense surface water and does not mix. Deep hypersaline anoxic basins (DHABs) are found on the bottom of the Mediterranean Sea, Red Sea, the Black Sea, and the Gulf of Mexico. They are formed by dissolution of ancient evaporites and due to their high salinity do not mix with the overlying seawater (Merlino et al. 2018; Wallmann et al. 1997). Also, partial enclosed bays or lagunas in tropical areas where high evaporation occurs and no freshwater discharges, may exhibit above seawater salinity such as is the case in Hamelin Pool (Shark Bay, Australia) (Edgcomb et al. 2014). Estuaries are typically exhibiting a salinity gradient decreasing from the sea to the river. In some cases, the denser seawater dives underneath the discharging fresh river water (salt wedge estuary) (Heip et al. 1995; Watanabe et al. 2014). Upstream some tidal estuaries even experience tides of freshwater. Solar lakes in tropical regions are separated from the sea by a sand bar through which seawater seeps through. The water in this lake evaporates due to the high temperature and is heated by the high solar irradiation. It becomes denser due to the increasing salinity and the hot surface water sinks producing a thermocline with hot (65 °C) water below (Cohen et al. 1977). There is a variety of hypersaline systems. Natural or man-made solar salterns that evaporate seawater to produce sodium chloride (table salt) or magnesium salts (bitters) are made of a series of evaporation ponds with increasing salinity until saturation (Oren 2009). There are many examples of salt lakes in the world that are not connected to the sea such as the Dead Sea in the middle east (Table 1.3) or Great Salt Lake in the USA, which are different in their chemical composition. They are formed by evaporation of the river water that is discharged in them and their chemical composition is a function of their watersheds.

These hypersaline lakes, ponds, and lagoons are considered as extreme environments that are the habitats of specialized microorganisms. Microbiologists working in those hypersaline environments laugh about regular seawater salinity, which they consider not too much different from freshwater (see Table 1.3). In fact, seawater is probably less extreme than freshwater as it more closely equilibrates the concentrated cytoplasm. Most likely, seawater was the medium in which life evolved.

The sea receives salts via river discharge and run-off. This water contains minerals from rock dissolution by the acidic rain in the riverine catchment area. Rain dilutes seawater and this dilution is counteracted by evaporation keeping salinity in equilibrium. Also, melting sea ice contributes to the freshwater input into the ocean, which is enhanced by global warming. Whether or not this will cause a different equilibrium of the seawater salinity is an open question. Because the chemical composition of seawater did not change during the last 100 million years minerals must have been removed, which among other may have been accomplished by processes such as mineral precipitation and deposition (e.g., calcification), sulfur emission to the atmosphere (dimethylsulfide, DMS), organic matter decomposition, -deposition, -burial, and mineral removal. These are fast processes and without any of these mineral removal components the ocean could have reached its present salinity in only a few million years. It would also mean that seawater salinity is not in equilibrium and that the ocean would become more saline. However, there is evidence that the ocean became saline during its formation more than 4 billion years ago, i.e. before life evolved.

### 1.2.2 Origin of Salinity and Early Ocean

Knauth (2005) estimated that the salinity of the Archean ocean was 1.5–2 times the present value. This high salinity remained until continental cratons developed that sequestered halite and when brines formed that derived from evaporating seawater. Knauth (2005) also estimated that the surface of the Archaean ocean had a temperature of 55–85 °C and that only 1.2 Ga bp ocean surface temperature cooled down to present levels. The high salinity and temperature of the seawater prevented the dissolution of oxygen ($O_2$) and as a consequence the ocean stayed anoxic even when the atmosphere contained $O_2$ to ~70% of its modern value and hence the ocean remained a domain for anaerobic microorganisms (Knauth 2005). Yet, modern hypersaline environments have largely an aerobic microbiota. Multicellular eukaryotic $O_2$-dependent organisms may have, therefore, evolved on land and moved to the sea after its surface became oxygenated. Gaucher et al. (2003) predicted on the basis of the phylogeny of resurrected proteins that deep-routed bacteria must have been thermophiles with a temperature optimum of 55–65 °C and not meso- or hyperthermophiles. Hence, that would have allowed bacterial life emerging globally rather than in isolated spots around hyperthermal vents as have been hypothesized previously (Weiss et al. 2016) (but see Schreiber and Mayer 2020).

While chlorine ($Cl^-$) is the most abundant anion in seawater, sulfate ($SO_4^{2-}$) is quantitatively the second-most important anion with about 28 mM. This high concentration guarantees that sulfur is never a limiting nutrient for microorganisms thriving in the sea (which may be sometimes the case in freshwater systems) and that when oxygen is in short supply sulfate may take over as electron acceptor and reduced sulfur compounds (such as sulfide: $H_2S$, $HS^-$, $S^{2-}$) may dominate as, for instance, is the case in the so-called sulfureta (Overmann and van Gemerden 2000). Assimilatory sulfate reduction is used by (aerobic) microorganisms to satisfy their sulfur demand. Dissimilatory sulfate reduction is a form of anaerobic respiration that becomes important when oxygen is at low supply such as in marine aggregates (marine snow), oxygen minimum zones, intertidal sediment, or other anoxic marine or saline environments. Sulfate is also the main oxidant of deep-sea cold seeps of methane, which is achieved by aggregates of a methanogenic archaea and a sulfate-reducing bacteria, in which the former reversed its metabolism to methane oxidation rather than methanogenesis (Boetius et al. 2000).

### 1.2.3 Microorganisms in the Ocean

Bacteria have been around for almost 4 billion years and probably evolved in the ocean. Marine microorganisms comprise all three domains of life: bacteria, archaea, and eukarya, as well as the viruses as another important biological entity. They are the engines and engineers of marine ecosystems and are at the basis of the marine foodweb (Kirchman 2000, 2008). Marine microorganisms are responsible for the primary and secondary production and recycle nutrients and organic matter. Many biogeochemical processes are carried out exclusively by microorganisms and without them an ecosystem would not function. With their vast numbers, untold diversity, and high reproduction rates they truly represent the unseen majority (Whitman et al. 1998). Microorganisms represent 90% of the biomass in the oceans. There are $10^6$ bacterial and archaeal cells and $10^7$ viral particles in 1 ml of seawater. The total of microorganisms in the open ocean amounts to $>10^{29}$ (Whitman et al. 1998). These incredibly large numbers of individual cells are statistically separated by 100–200 cell-lengths (Moran 2015; Zehr et al. 2017).

Smith and Baker (1982) showed that ocean surface chlorophyll concentrations could be obtained by using satellite data and that this opened the possibility to obtain such data on a scale that is impossible to obtain from shipboard measurements. These satellite data were subsequently used for the estimation of primary production at the same (global) scale (Behrenfeld and Falkowski 1997).

The primary production in the ocean by marine phytoplankton accounts for roughly half of the global primary production (Field et al. 1998). A quarter of this production has been attributed to only two genera of cyanobacteria, the picocyanobacteria *Prochlorococcus* and *Synechococcus* (Flombaum et al. 2013). *Prochlorococcus* is exclusively found in the ocean, while *Synechococcus* has also brackish and freshwater lineages.

Marine phytoplankton and to a substantial extent also chemoautotrophic bacterioplankton fix an enormous amount of $CO_2$ of which a small but important amount escapes degradation and is transported to the deep-sea where it is buried and turned over on geological time-scales (carbon pump). This is a crucial process because it serves as a sink for the increasing emission of $CO_2$ because of anthropogenic activities, namely the burning of fossil fuel. The carbon pump counteracts global warming. The bulk of the autotrophically fixed $CO_2$ is immediately respired by heterotrophic microorganisms although there are seasonal variations in the ratio of production to respiration (Sherr and Sherr 1996). These authors noted a ratio $> 1$ during 3–5 months in winter and spring while during the rest of the year it was $<1$. It is generally assumed that during its voyage to the deep-sea the most labile dissolved organic carbon (DOC) is respired first leaving the more refractory DOC for the greater depths. Arrieta et al. (2015) proposed an alternative explanation. They hypothesized that despite its lability the concentration of organic matter in the deep ocean is too low to yield net energy upon consumption by heterotrophic bacteria.

### 1.2.4 The Oceanic Habitat

The ocean is probably the largest habitat and cohesive ecosystem in the biosphere. It is not just the vast amount of water but it is also a habitat that is composed of a vast diversity of sub-habitats. Just some of them are mentioned here. The sea surface–

atmosphere boundary is an important but often ignored habitat. The pelagic habitat is subdivided into five major realms. The epipelagic extends to a depth of 200 m and is defined as the layer that receives enough light for photosynthesis. This is followed by the mesopelagic that extends to 1000 m and is also known as the twilight or disphotic zone and the bathypelagic covers the depth between 1000–4000 m. Everything deeper is called the abyssal or abyssopelagic. The ocean floor represents a vast surface because of a complex landscape of ridges, mountains, and valleys that covers an area much larger than the 70% of the surface of the Earth that is covered by the ocean. The ocean floor also contains brine lakes and hydrothermal vents and cold seeps each with their special environmental conditions. Thousands of meters below the ocean floor in what is called the deep subsurface is a habitat of vast numbers ($3.5 \times 10^{30}$; Whitman et al. 1998) of specialized microorganisms, barophilic species that grow extremely slow. Then there are the adjacent seas, bays, lagunas, coral reefs, intertidal areas, and estuaries. Last but not least are epiphytic habitats on macroalgae and seagrasses as well as associations with all marine animals (both inside and outside). The atmosphere is another important component that drives the transport of microorganisms. It has been estimated that $4.3 \times 10^{21}$ microbes are airborne of which 33–68% have a marine origin and are transported over large distances (Mayol et al. 2017). Dust is also transported through the atmosphere and may supply the ocean with nutrients such as iron, which is thought to limit primary production in large areas (Rijkenberg et al. 2008). Diazotrophs are known to have a high iron demand and may depend on such dust depositions (Langlois et al. 2012).

Temperature has often been assigned a critical role in determining the composition and activity of microbial communities (Logares et al. 2020; Sunagawa et al. 2015). However, Gilbert et al. (2012) on the contrary concluded that day-length is a major force for the composition of microbial communities. This conclusion was also reached by Stal and Walsby (2000) who showed that blooms of diazotrophic cyanobacteria in the Baltic Sea would develop independent on temperature but determined by day-length (i.e., the total irradiance received by the system). Temperature may affect the size of the total standing stock biomass because it affects respiration and production differently. Another important effect of temperature is that it determines the density of the water and together with wind force may or may not lead to stratification of the water column. A stratified or mixed water column in the euphotic layer determines the amount of light that is received by the phytoplankton and hence their production.

## 1.3 What Is a Marine Microorganism?

### 1.3.1 What Is a Microorganism?

A microorganism is by definition small, too small to be seen by the naked eye, which means smaller than 0.1 mm. Microorganisms comprise the three domains of life we know: bacteria, archaea, and eukarya. The former two are also jointly known as "prokaryotes" because they do not possess a nucleus (karyon; from Greek káruon, which means "nut" or "kernel," while the prefix "pro" means "before") (or a cell organelle such as a mitochondrium or chloroplast), while eukaryotes (means with a true ("eu-") nucleus) do possess a nucleus (and cell organelles). The opposed of the Greek "eu-" would be "dys-," but introducing the term "dyskaryote" would not make much sense. Grouping two domains of life based on lacking a property is probably not a good idea (Pace 2006, 2009). Moreover, phylogenetically, archaea and eukarya are more closely related than each of them to bacteria (Imachi et al. 2020) and this is another argument against the term "prokaryote" for binning the two domains. The term "prokaryote" has been used from the days that it was synonymous to bacteria as being microorganisms without possessing a nucleus and that that property distinguished them from those cells that did possess a nucleus (Stanier and Van Niel 1962). There were only two domains of life: prokaryotes and eukaryotes. With the discovery of archaea and their phylogenetic position in the tree of life, the term "prokaryote" is in fact obsolete. Moreover, the meaning "before the nucleus" is also confusing because neither bacteria nor archaea possess or developed one. Nevertheless, it is still commonly used; not seldom in a sloppy way when only bacteria are referred to and excluding archaea. When archaea were first recognized as a domain, they were initially termed "archaeabacteria" because they were assumed to have evolved first in the extreme conditions of the early Earth. The bacteria were then termed "eubacteria" (meaning "true" bacteria as opposed to the untrue bacteria, the archaeabacteria). Obviously, the term "eubacteria" is also obsolete and should not be used.

Microorganisms are unicellular although some may form multicellular forms but with few exceptions these are all composed of the same undifferentiated cell. Macroorganisms all belong to the eukarya but by far most eukarya are in fact microorganisms. A few bacteria are big, i.e. bigger than 0.1 mm (Schulz and Jørgensen 2001). They are nevertheless considered as microorganisms just because they are bacteria. Viruses are not a domain of life because these biological entities do not multiply independently, which is said to be a property of life. Nevertheless, viruses are important because they are agents that help exchanging genetic material between cells of the same species and even interdomain exchanges have been reported. They also keep in check too successful organisms (killing the winner) and aid the recycling of nutrients in the ecosystem (Thingstad and Lignell 1997). With the discovery of giant viruses with large genomes and with the size of small bacteria as well as with the introduction of advanced sequencing techniques, the border between viruses and cells is fading away (Harris and Hill 2021). Hence, viruses fulfill essential roles in any ecosystem including in the marine environment and in the ocean where they occur in a dazzling high number of up to 10 billion viruses per liter. The marine microbiome cannot be understood without considering viruses.

### 1.3.2 Do Marine Microorganisms Exist?

Throughout the history of (non-medical) microbiological studies terrestrial (soil) and aquatic microbiology have been treated almost as different disciplines. This has been largely due to historical determined developments in the scientific approach, methodology, and the problem focus (e.g., agriculture). Such a difference did not develop between freshwater- and marine microbiology and therefore it is less obvious to define each of them as a discipline. Microorganisms growing in freshwater may also be found and grow in sea water and vice versa. Many microorganisms are found and may grow in a wide range of aquatic environments. Even terrestrial microorganisms may end up in the ocean through run-off and through air transport (Mayol et al. 2017). For a long time, it was thought that marine fungi do not exist and those that were found were considered to be accidently deposited there from a terrestrial source and non-growing in the marine environment. There is no doubt that fungi inhabit and grow in the marine environment and fulfill the same ecological role as terrestrial fungi do in terrestrial environments (see Chap. 5). Hence, what is a marine microorganism? MacLeod (1965) stated ".... the only one which clearly distinguishes them (marine bacteria, sic) from bacteria in other habitats is a capacity to survive and grow in the sea." Both ZoBell and Rittenberg (1938) and Stanier (1941) considered bacteria marine when obtained from a marine environment and degrade typical marine derived substrates such as chitin and agar-agar, respectively. The former is a major compound of Arthropoda and other marine animals and the latter is derived from marine red algae.

Many marine bacteria require elevated $Na^+$ for transporting substrates into their cells or to retain intracellular solutes (osmotic balance). However, this seems not to be uniquely the case for marine bacteria because all cells require $Na^+$ and marine bacteria exhibit a wide variety of tolerances for elevated salt concentrations (MacLeod 1985, 1986). ZoBell (1946) writes: "*Although there are no infallible criteria for the differentiation of marine from non-marine bacteria, most of the bacteria which occur in the open ocean differ in certain respects from those found in non-marine habitats. That is probably because adventitious organisms either fail to perpetuate themselves in the marine environment or else lose their identity in becoming acclimatized thereto.*" Can we agree simply on any microorganism that occurs and thrives in a marine or saline environment and their adjacent environments such as brackish seas, estuaries, bays, and lagunas as MacLeod (1985) proposed.

Most of the ocean is an extremely oligotrophic environment. The concentration of inorganic nutrients as well as of organic matter is low. Many marine pelagic microorganisms (plankton) have adapted to this environment by their small size (Ghai et al. 2013; Schut et al. 1995). The small size provides these microorganisms with a large cell surface area to cell volume ratio which enhances their affinity for the uptake of nutrients and substrate. The most abundant bacteria in the ocean are *Prochlorococcus* and *Pelagibacter* and are among the smallest microbes that exist. These microorganisms encounter their substrate by Brownian diffusion and that explains why these marine microorganisms are in general non-motile because theory predicts that swimming to a nutrient hotspot would be inefficient (Zehr et al. 2017).

However, when motile, marine bacteria swim much faster than freshwater species which can be attributed to the shallow gradients in the sea (see Chap. 2). During diatom blooms and their subsequent collapse chemotactic motile bacteria outcompete the non-motile species (Smriga et al. 2016).

### 1.3.3 How Many Species of Marine Microorganisms Exist?

In average there may be a billion of microorganisms in a liter of seawater but how many species are there? The number of species in seawater has been estimated from ~1500 (Hagström et al. 2002) to ~20,000 (Sogin et al. 2006) to one million (Curtis et al. 2002) or even tenths or hundreds of millions (Dykhuizen 1998). These large numbers may refer to sequence diversity but that does not make them species (Riley and Lizotte-Waniewski 2009). This makes clear that the number of species is in fact not known nor much is known about the role of this microbial diversity in the ocean. The species concept in microbiology is still a matter of controversy. It is usually based on the level of sequence identity of the 16S rRNA gene as well as a number of phenotypic characteristics. Rosselló-Mora and Amann (2001) wrote in the abstract of their review on the species concept for prokaryotes (sic) "*a monophyletic and genomically coherent cluster of individual organisms that show a high degree of overall similarity with respect to many independent characteristics, and is diagnosable by a discriminative phenotypic property.*" They refer to it as the "phylo-phenetic" species concept. However, what to say of 2 phenotypically different strains of picocyanobacteria (*Synechococcus*) of which one is red because of the possession of phycoerythrin and the other is blue-green because it does not produce the red pigment (Haverkamp et al. 2009). The red strain is several times bigger than the green strain. Also, the green strain grows much faster and there are more differences. Would you consider these as two different species? But these two strains have 100% identical 16S ribosomal RNA gene sequences and differ only 2 base pairs in ~1000 bp long internally transcribed spacer (ITS). And there are several other similar cases known from the literature. Obviously, phenotypic characteristics are important and should not be ignored (Margulis 1992). It shows the problem of defining a microbial species only by their 16S rRNA gene sequence, even though it is true that organisms with <98.7% 16S rRNA gene sequence identity can be considered to be different species (Achtman and Wagner 2008). An alternative for delineating microbial species is to compare average nucleotide identity (ANI) and a cut-off of 95% ANI to define a species. Using this criterion would require the whole genome of an organism. Riley and Lizotte-Waniewski (2009) adopted the core genome which encodes the essential housekeeping genes as the basis to define bacterial species. This would still not solve the question of the concept of a microbial species nor allows to answer the question whether it exists or not (Achtman and Wagner 2008). Wilkins (2006) argues that there is a continuum of recombination of genetic information from viruses and bacteria to the sexual recombination of 50% of the two parents as in metazoans and metaphytes. However, the question is whether there is such a continuum because evolutionary forces would prevent this (Achtman and Wagner 2008). Wilkins (2006), therefore, proposes to combine the recombination and the ecological species concept. Whether this will solve the problem of recognition of a microbial species remains to be seen.

Approximately 10,000 "species" of marine microorganisms have been isolated in culture and described to some extent. It is said that the vast majority ("90–99%") of species cannot be cultured using standard media and growth conditions. For many years it was known that counting marine microorganisms on agar plates yielded only a fraction of the number that was directly microscopically counted (Jannasch and Jones 1959) or by using fluorescent microscopy of stained bacteria fixed on nucleopore filters (Hobbie et al. 1977). This difference between "culturable" and total bacterial count has become known as "The Great Plate Count Anomaly" (Staley and Konopka 1985). Hence, this majority is only known by their DNA sequence or at best as microscopic images using techniques such as Fluorescent In Situ Hybridization (FISH) and its derived methods. The description, identification, and the growth and metabolic characteristics of these uncultured microorganisms will remain difficult if not impossible unless brought into culture. Similarly, the circumnavigating cruises such as the Global Ocean Sampling (GOS) and TARA Oceans Expedition have revealed many sequences of open reading frames (ORF) that could not be linked to any known gene or function. This again points to a huge hidden diversity and metabolic potential.

## 1.4 (Some) Milestones of Marine Microbiology

Karl and Proctor (2007) wrote a nice overview on the "Foundations of Microbial Oceanography" and I refer those who would like to have a more extended account on the subject to that paper and the references therein.

Fischer (1894) wrote one of the several monographs that appeared at the end of nineteenth century in several countries and in their own languages that focused specifically on marine microorganisms. This author made the following remark in the introduction of his book written in German "Die Bakterien des Meeres" (Fig. 1.3) (The Bacteria of the Sea) "*Nun war aber, als die Expedition unternommen wurde, über die Bakterien des Meeres noch so gut wie gar nichts bekannt geworden, wenigstens fehlte es damals und fehlt es auch heute noch in der Literatur an systematischen Untersuchungen über die Bakterien des Meeres, ja es finden sich selbst bis auf den heutigen Tag noch nicht einmal Angaben darüber, ob auf hoher See Bakterien überhaupt vorkommen.*" (Fischer 1894). Translated by me: "However, almost nothing was known about marine bacteria when the expedition (*1889 plankton expedition in the Atlantic Ocean of the Humboldt Foundation*, added by me) took place, at least at that time as well as of today systematic investigations about marine bacteria were lacking in the literature, and yes even of today no reports exist whether or not bacteria are present on the high seas."

Fischer (1894) reported counts of bacterial colonies on gelatin (supplemented with agar-agar when high temperatures prevented the use of gelatin). The medium was supplemented with NaCl and fish extract. Fischer concluded that salt

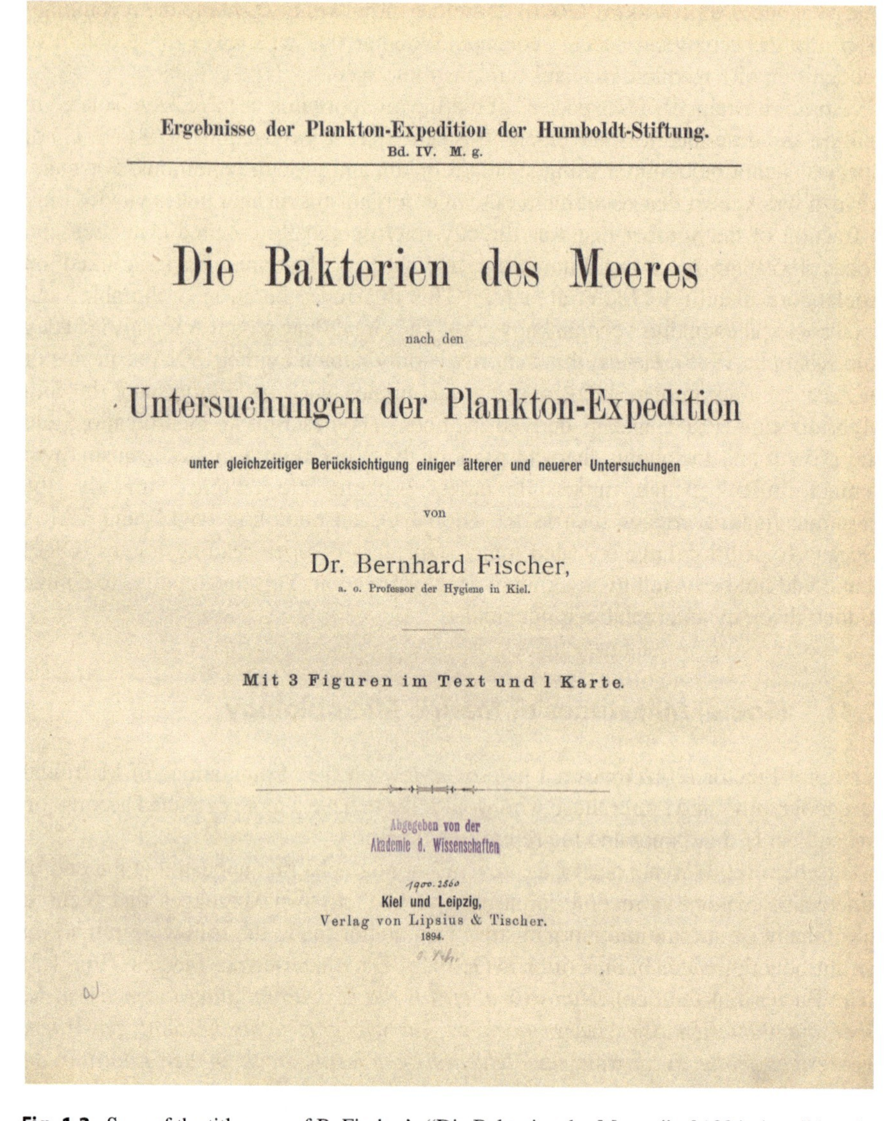
Ergebnisse der Plankton-Expedition der Humboldt-Stiftung.
Bd. IV. M. g.

Die Bakterien des Meeres

nach den

Untersuchungen der Plankton-Expedition

unter gleichzeitiger Berücksichtigung einiger älterer und neuerer Untersuchungen

von

Dr. Bernhard Fischer,
a. o. Professor der Hygiene in Kiel.

Mit 3 Figuren im Text und 1 Karte.

Abgegeben von der
Akademie d. Wissenschaften

Kiel und Leipzig,
Verlag von Lipsius & Tischer.
1894.

**Fig. 1.3** Scan of the title page of B. Fischer's "Die Bakterien des Meeres" of 1894, describing the results of a Plankton expedition in the Atlantic Ocean of the Humboldt foundation in 1889

requirement was a distinguishing characteristic for growth of marine bacteria. Water samples were obtained from the surface, deep water, and from different geographical locations and seasons. The numbers he found were low and rarely exceeded 1000 $ml^{-1}$. Colony counts from deep water samples were rarely higher than 10 $ml^{-1}$ and not seldom zero, which Fischer attributed to the hydrostatic pressure. He also isolated some marine bacteria into pure culture and frequently reported specifically bioluminescent bacteria. He concluded that marine bacteria exist everywhere on high seas but that much is unknown and advocated further studies on this subject.

In 1934 A.C. Redfield published a remarkable finding of a constant ratio of N:P of 20 in a wide range of samples taken from different geographical regions of the Atlantic Ocean and from different depths as well as from other oceans (Redfield 1934). It reflected the stoichiometry in marine plankton and its degradation would consequently also be recovered from seawater. Redfield concluded that oceanic plankton was remarkably uniform in their chemical composition. He noted that this composition may be different, for instance, in coastal regions. Redfield also noted the possibility of $N_2$-binding plankton introducing new nitrogen but only at the expense of phosphorus: "*... the quantity of nitrate in the sea may be regulated by biological agencies and its absolute value determined by the quantity of phosphate present.*" This number of N:P was later adjusted to 16 and became known as the Redfield ratio. The constant stoichiometry was subsequently extended to other major elements and adjustments have been made for specific cases. But ever since the Redfield ratio remained intact and continues to stay an important keystone for biological oceanography.

While "Die Bakterien des Meeres" may be considered as the hesitantly beginning of "Marine Microbiology," it became the subject of many studies and part of (biological) oceanographic cruises and research institutes during the years to follow and culminating in Claude ZoBell's seminal monograph "Marine Microbiology" published in 1946 (ZoBell 1946) (Fig. 1.4). As the subtitle of the book indicates, ZoBell's monograph focused on bacteria and left out all other microorganisms. Between the lines ZoBell was criticized for this even in the foreword of his book written by Selman Waksman and by other contemporaries. Nevertheless, his book is still remarkably actual as a basic text on the subject of marine microorganisms, even though much new knowledge has been obtained since, the basics as presented by Claude ZoBell did not see many fundamental changes and so it stands as an important keystone of marine microbiology.

The binding ("fixation") of atmospheric dinitrogen ($N_2$) in the marine cyanobacterium *Trichodesmium thiebautii* in the Sargasso Sea was reported in 1961 by Richard Dugdale and colleagues (Dugdale et al. 1961). Until then, $N_2$ was considered unimportant in the ocean, even though Redfield (1934) pointed out that the binding of $N_2$ could contribute to maintain the N:P ratio in marine plankton. Later, this author, therefore, concluded that oceanic primary production could not be limited by nitrogen (Redfield 1958) (which was re-emphasized by Tyrrell in 1999). Dugdale's report showed for the first time that *Trichodesmium* was able to incorporate the stable isotope $^{15}N$ from $^{15}N_2$ and that this process depended on light. Now it is known that the ocean is responsible for ~50% of the global $N_2$ binding (excluding the contribution of the chemical Haber–Bosch process for the production of fertilizer). Richard Dugdale, David Menzel, and John Ryther's discovery was the start of an extensive worldwide research of oceanic $N_2$ binding, which is still ongoing today and that resulted in many exciting discoveries (see Zehr and Capone 2021).

MARINE

MICROBIOLOGY

*A MONOGRAPH ON HYDROBACTERIOLOGY*

*BY*

CLAUDE E. ZOBELL, Ph.D.

*Associate Professor of Marine Microbiology,*
*Scripps Institution of Oceanography,*
*University of California, La Jolla*

Foreword by SELMAN A. WAKSMAN
*Professor of Microbiology, Rutgers University, etc.*

1946

*WALTHAM, MASS., U.S.A.*

Published by the Chronica Botanica Company

**Fig. 1.4** Scan of the title page of C. ZoBell's "Marine Microbiology." (first print 1946 possessed by L.J. Stal)

In 1974, Pomeroy published a keystone publication in which data from more than a decade of plankton studies were put together and a synthesis was made. Pomeroy (1974) understood that the classical picture of the oceanic food web needed revision. Marine microorganisms produce and recycle organic matter many times before it enters higher trophic levels. This part of the oceanic food web that recycles energy has later been named the "microbial loop" (Azam et al. 1983).

Waterbury et al. (1979) reported the widespread occurrence of small *Synechococcus* spp. containing the red pigment phycoerythrin. Due to the small size (<2 μm) of these chroococcalean cells they were later addressed as picoplankton or picocyanobacteria in case it concerned cyanobacteria. These tiny organisms carry out most of the primary production in the ocean (Platt et al. 1983).

A symposium series on marine microbiology was initiated in Europe since the beginning of the 1980s. The seventh European Marine Microbiology Symposium (EMMS) took place from 17–22 September 2000 in Noordwijkerhout in the Netherlands and was combined with the seventh International Workshop on the Measurement of Microbial Activities in the Carbon Cycle in Aquatic Environments. During this joint meeting it was decided to continue as a combined meeting under the new name "Symposium on Aquatic Microbial Ecology" (SAME) (but keeping the count as by coincidence both meetings were at the SAME sequence). An important reason for this merge was that both meeting series attracted largely the SAME group of scientists. This emphasized that the marine aspect was not a distinguishing characteristic for this group of scientists. They use largely the SAME methods and focus on the SAME fundamental research questions.

Ducklow (1983) argued that small heterotrophic protists actively predate on bacteria and microalgae in the ocean. Subsequently, Hagström et al. (1988) quantified the predation of bacteria in the oligotrophic marine environment by these nanoflagellates. These authors also showed that the carbon that is excreted by the nanoflagellates is returned to the heterotrophic bacteria thereby completing the "microbial loop."

Waterbury and colleagues discovered in 1985 a marine unicellular cyanobacterium belonging to *Synechococcus* that is capable of swimming (Waterbury et al. 1985). Until then the only motility known among cyanobacteria was gliding, which was used mainly by filamentous species on a solid substrate or along other cyanobacterial trichomes. Swimming appeared to be in the absence of flagella and therefore unique among bacteria. Although the precise mechanism of motility has not yet been elucidated, several genes involved in it have been identified in the genome of this organism and particularly in modification of the cell surface that might play a role in motility of this cyanobacterium (Palenik et al. 2003). This research also revealed marked difference with typical freshwater representatives of the same genus, particularly with regard to transporters and nutrient sources (N, P, Fe) and the authors considered this *Synechococcus* an ecological generalist for which strategy motility may be advantageous.

Shipboard flow cytometry discovered in 1988 the highly abundant cyanobacterium *Prochlorococcus* (Chisholm et al. 1988). This organism was later considered to be the most abundant photosynthetic organism on Earth. Originally, it was thought to be related to *Prochloron* based on its remarkable pigment composition, which is unusual for cyanobacteria. It contained a divinyl derivative of chlorophyll *a* as well as chlorophyll *b*, the latter typical for the chloroplasts of higher plants and a few other cyanobacteria which were at the time collectively assigned to as prochlorophytes. Another remarkable property of *Prochlorococcus* is the absence of phycoerythrin, which is characteristic for the oceanic picoplanktonic *Synechococcus*, which was discovered a decade earlier.

Since 1989, as the result of a new methodology, it became clear that ocean water contained high numbers of viral particles, while until then viruses were considered to be low in number and importance in aquatic environments (Bergh et al. 1989). Subsequent research revealed that viral abundances could reach as high as ten million per milliliter of seawater and established their role in the recycling of nutrients, in controlling successful microorganisms ("killing the winner"), and generate diversity by horizontal gene transfer (Suttle 2005).

Analysis of the sequences of clone libraries of 16S rRNA was introduced in 1990 in order to investigate the diversity of marine bacterioplankton allowing the culturing-independent study of marine microbial communities (Fuhrman et al. 1992; Giovannoni et al. 1990; Schmidt et al. 1991). Using this approach, Giovannoni and colleagues discovered the highly abundant bacterium SAR11, which became later known as *Pelagibacter* while Fuhrman and colleagues discovered meanwhile abundant marine archaea as component of the oceanic plankton even in cold and well-oxygenated waters. Until then archaea were rather considered to be associated with more extreme environments. Very soon after these important discoveries, phylogenetic stains were developed and used to microscopically visualize marine bacteria in natural seawater samples (Lee et al. 1993).

In 1992, the German Max Planck Society founded the Institute for Marine Microbiology (MPIMM) in Bremen, Germany (founding director was Prof. Bo Barker Jørgensen and co-director Prof. Friedrich Widdel). It may have been the first research institute that is fully dedicated to the study of marine microorganisms. Remarkably, MPIMM is located in the city of Bremen (Germany) rather than directly at the coast. This recalls MacLeod who said in 1985 that marine microbiology can be done far from the sea (MacLeod 1985). At MPIMM marine microbiological research is strongly internationally oriented. Since its foundation, research at MPIMM contributed importantly to new knowledge in the field of marine microbiology.

The European Commission launched the Marine Science and Technology (MAST) Program in 1997. Under the umbrella of MAST a wide variety of research projects were funded many of which focused on the biotechnological use of marine microorganisms.

In 1997, it was recognized that the oligotrophic unproductive ocean behaved as a source of $CO_2$ because bacterial respiration seemed to exceed primary production in such areas emphasizing the importance of bacterial activity (del Giorgio et al. 1997). However, this conclusion has been challenged by Richard Geider (Geider 1997 and replied by del Giorgio and Cole).

In 2000, metagenomic techniques such as the cloning and sequencing of bacterial artificial chromosomes (BACs) revealed the existence of proteorhodopsin hinting to the widespread existence of photoheterotrophy in the sea (Béjà et al. 2000). Kolber et al. (2001) discovered the widespread presence of bacteriochlorophyll *a* and demonstrated aerobic anoxygenic photosynthesis in the ocean. Subsequently, a variety of aerobic anoxygenic phototrophic bacteria were discovered and isolated and their ecology investigated (Rathgeber et al. 2004).

New investigations in 2002 and sequencing of 18S ribosomal genes of tiny protists (eukaryotic microorganisms) in the ocean revealed a much larger diversity among this neglected group of organisms (Massana et al. 2002).

The Global Ocean Sampling (GOS) expedition (Fig.1.5) took place from 2004 to 2006, after a successful pilot in the Sargasso Sea in 2003 of which the metagenome was constructed from shotgun sequencing. This revealed a large number of unknown microbial functions in the ocean (Venter et al. 2004). The new high-throughput sequencing technologies that were applied to the GOS samples revealed a much higher microbial diversity than expected at the time and that the ocean contains an extreme large number of microbial genes of which most of them with unknown function. Another Sorcerer II expedition took place from 2007 to 2008 sampling special environments such as polar ice, deep-sea thermal vents, and high saline ponds. In 2009–2010, a Sorcerer expedition took place focusing on the Baltic Sea, the Black Sea, and the Mediterranean Sea, large seas that are mostly disconnected from the ocean. These expeditions circumnavigated the globe and took water samples that were size-fractionated by filtration in order to obtain DNA from different size classes and eventually from viruses and from free DNA in the filtrate. The filters were shipped to the laboratory where DNA was extracted and sequenced and the genetic diversity of the microorganisms was analyzed. Genes and genomes were partly assembled from the fragments using advanced computer algorithms. The genomic databases were enriched with millions of novel genes and genomes most of which from uncultured and hitherto unknown microorganisms of each of three domains of life as well as of marine viruses. This was unprecedented at the time and boosted the knowledge of marine microbiological diversity and its global distribution (Kannan et al. 2007; Rusch et al. 2007; Yooseph et al. 2007). Most of the sequences could not be assigned. A considerable part (15%) of the recruited sequence reads belong to only three groups of widely abundant marine microorganisms: *Pelagibacter*, *Prochlorococcus*, and *Synechococcus*. Much of the discoveries from the GOS databases hint to life strategies and adaptations typical for the oceanic and marine environment. It should be noted that GOS sampled only surface waters and it only unveiled the tip of the iceberg of marine microbial diversity.

The Census of Marine Life (CoML) was a 10-year initiative involving a global network of scientists. It started in 2000 and received initial funding from the Sloan Foundation. Although there were several "field projects" within CoML that focused in part on marine microorganisms, the International Census of Marine Microbes (ICoMM) that was launched as a CoML field project in 2004 was the only one that aimed at making an inventory of all diversity of bacteria, archaea, protists, and

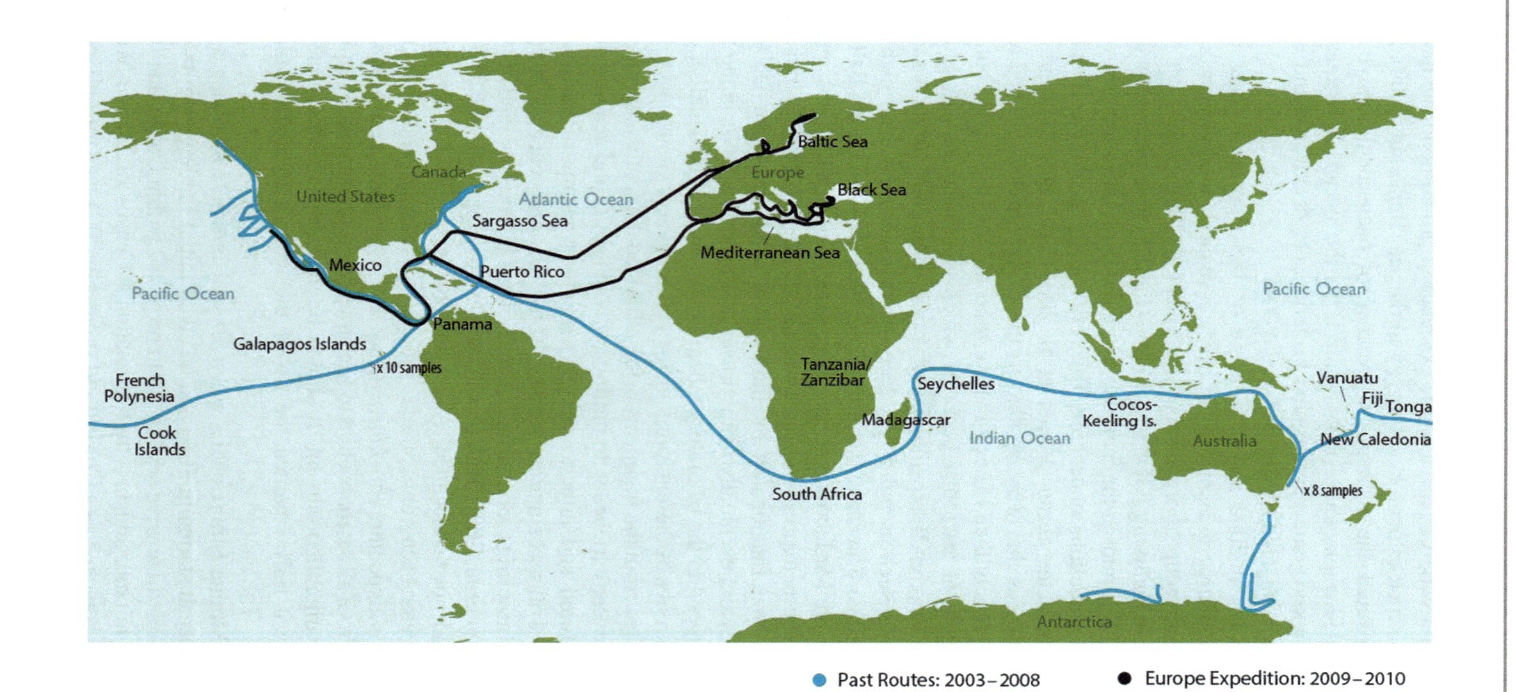


**Fig. 1.5** Global Ocean Sampling (GOS) Sorcerer II Expeditions during the years 2003–2010 by the J. Craig Venter Institute. Graph taken from https://www.jcvi.org/research/gos#past-voyages (public domain)

viruses in the ocean. It was the first attempt to sequence massively the microorganisms in a wide variety of marine environments using at the time novel parallel tag sequencing (Sogin et al. 2006). This work revealed a far higher diversity than was anticipated and shed light at the so-called rare biosphere, microorganisms that are present in low abundance (Pedrós-Alió 2006). The technique was subsequently applied on a wide range of samples from a variety of environments (Amaral-Zettler et al. 2010).

Until 2001, it was thought that *Trichodesmium* was the only quantitatively important diazotroph ($N_2$ binding; living on $N_2$ as source of nitrogen) in the ocean. This changed with the discovery of Zehr and colleagues that unicellular cyanobacteria were equally important as diazotrophs (Zehr et al. 2001). This discovery was subsequently confirmed and extended by many reports of phylogenetically different groups of unicellular diazotrophs, including obligate symbiotic organisms.

In 2002, it became clear that SAR11 (*Pelagibacter ubique*) dominates the ocean surface waters (Morris et al. 2002) and this organism was brought in culture using a dilution to extinction method, which was an important breakthrough (Rappé et al. 2002). At the same time innovative methods became available to isolate marine microorganisms among others by encapsulating them in gel microdroplets followed by flow cytometric detection and sorting (Zengler et al. 2002). During the years following 2002, many exciting discoveries were made of new marine microorganisms and metabolic pathways. Anaerobic ammonium-oxidizing (Anammox) bacteria belonging to the Planctomycetes were discovered in the Black Sea and from marine sediments (Jetten et al. 2003; Kuypers et al. 2003; Thamdrup and Dalsgaard 2002) and a meso- to psychrophilic marine aerobic ammonium-oxidizing archaea was isolated (Könneke et al. 2005), which subsequently became known as Thaumarchaeota and are widespread in the ocean (Brochier-Armanet et al. 2008; Wuchter et al. 2006).

The Gordon and Betty Moore Foundation launched in 2004 the Marine Microbiology Initiative, which supported an enormous amount of scientific research. It ended in 2021 and has been one of the longest research programs in marine microbiology with an investment of more than USD250 million. This program also supported the sequencing of the genomes of a large number of marine microorganisms that were isolated and brought into culture in the laboratory, including several cyanobacteria.

The Tara Oceans Expedition project was launched in 2009 and collected 35,000 plankton samples from all major oceans and at depths up to 2000 m (Sunagawa et al. 2020; Zhang and Ning 2015). The project was a true milestone in marine microbiology. Tara Oceans was led by the European Molecular Biology Laboratory (EMBL) and ended in 2013. The analysis of the genomics data resulted in a wealth of new knowledge on the functioning of the marine microbiome. Among many other discoveries Tara Oceans Expedition data revealed that marine microbial community composition seemed to be driven by temperature and the remarkable finding that three quarters of the ocean microbial core functionality is shared with that of the human gut (Logares et al. 2020; Sunagawa et al. 2015, 2020) (see Chap. 8). How special is the marine microbiome?

The 8 months lasting Spanish Malaspina expedition took place from December 2010 to July 2011 and circumvented the globe with the *RV* Hespérides. This expedition was named after the famous late eighteenth century scientific expedition led by Alejandro Malaspina. The 2010 Malaspina expedition included also microorganisms and particularly focused on deep water.

The European Commission funded the project "Marine Microorganisms: Cultivation Methods for Improving their Biotechnological Applications" (MaCuMBA) (FP7 grant agreement 311,975), which ran from 2012 to 2016 and comprised a total budget of more than 12 million euros and joined 23 participants from 11 countries. This project was designed by the awareness that marine microorganisms comprise an incredible diversity that is mostly unknown. These microorganisms form an almost untapped resource of biotechnological potential that may help mitigate climate change, control disease, and generate alternative energy sources. However, to access this resource culturing of these microorganisms is essential. The challenge of this project was to increase the success rate of isolating and culturing novel marine microorganisms. The project developed automated high-throughput culturing techniques (robotics) and increased the efficiency of growth by introducing co-culturing and nature-mimicking techniques as well as by genetic strategies. MaCuMBA noted the importance of culture collections in which isolated microorganisms are identified, stored, and maintained, and searchable by the public. This is still an underestimated challenge. The German Collection of Microorganisms and Cell Cultures GmbH (DSMZ) plays an important role in the collection of marine microorganisms. MaCuMBA led the initiative for the first edition of "The Marine Microbiome" (Stal and Cretoiu 2016). MaCuMBA formed a strategic alliance with two other large EU projects PharmaSea (which used the marine microbiome to discover new pharmaceuticals; see Chap. 17) and MicroB3.

Ocean sampling day (OSD) (Kopf et al. 2015) was an initiative of the European Union project Micro B3 (Marine Microbial Biodiversity, Bioinformatics, Biotechnology) and started in 2014. In 2016, after the Micro B3 project ended OSD was taken over by the Institute of Marine Biology, Biotechnology, and Aquaculture of the Hellenic Center for Marine Research in Heraklion, Crete. The idea of this sampling day was to obtain a worldwide inventory of the marine microbial biodiversity by extracting and sequencing nucleic acids from the samples taken at one single day (solstice, 21st July) together with their metadata. A worldwide consortium was put into place to coordinate this sampling which is still going on. An important aspect on this sampling effort was that sampling, sample handling, metadata reporting, and data processing were done according to a compulsory protocol resulting in a true standardized dataset.

## 1.5 Selected Aspects of the Marine Microbial System

### 1.5.1 The Redfield Ratio

Redfield observed a constant average ratio of C:N:P of 106:16:1 in the ocean (Redfield 1934, 1958). The Redfield ratio works for a large system as the ocean but less well for small lakes (Hecky et al. 1993). This has been attributed to the low water residence times and a greater contribution of continental watershed (They et al. 2017). The water residence time of the ocean is ~1000 years and exceeds the residence times of nitrate and phosphate which are an order of magnitude shorter (Falkowski and Davis 2004). This explains that the ratio of nitrate and phosphate in the ocean (Fig. 1.6) equals more or less the average composition of the

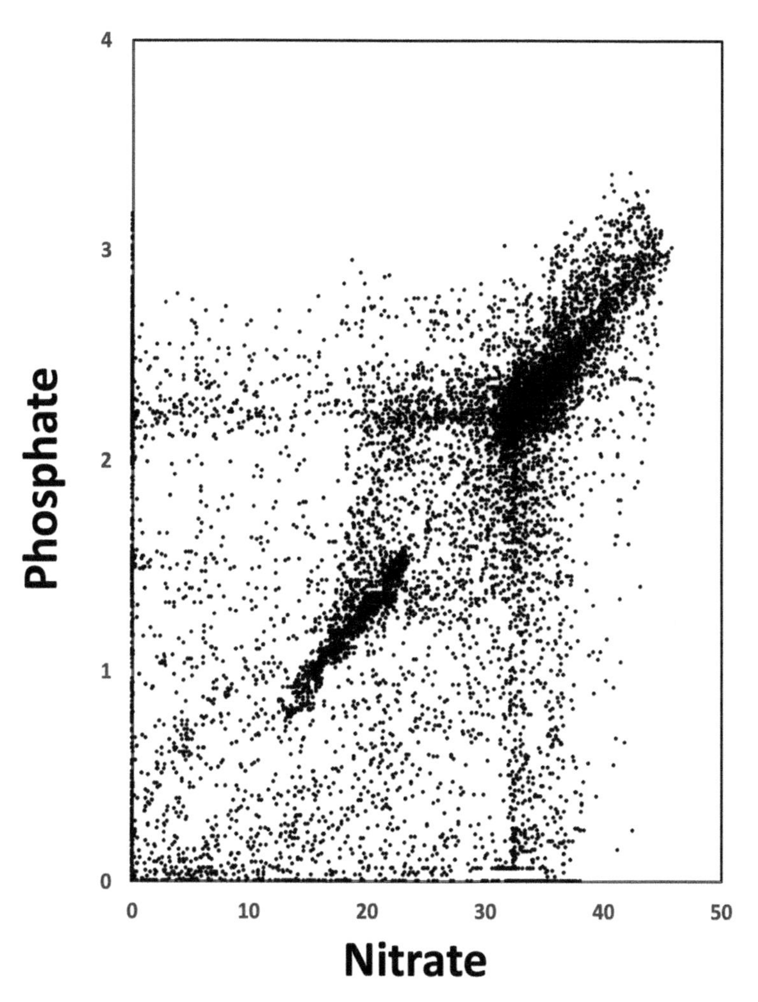


**Fig. 1.6** [$NO_3^-$] versus [$PO_4^{3-}$] (μmol $kg^{-1}$) scatter plot using data from GEOSECS global data set (https://iridl.ldeo.columbia.edu/SOURCES/.GEOSECS/)

phytoplankton. The smaller the system the greater is the deviation from the Redfield ratio. Therefore, an important aspect of marine microbiology is a matter of scale: the ocean is the largest continuous (eco)system. Also, in the ocean there is variability of the elemental ratios depending on the latitude. Martiny et al. (2013) reported C:N:P ratios varying from 195:28:1, 137:18:1, to 78:13:1 in the warm nutrient-depleted low-latitude gyres, warm nutrient-rich upwelling zones, and cold nutrient-rich high latitudes, respectively. A low N:P ratio has been taken as an indicator for the loss of nitrogen through denitrification, while a high ratio represents the binding of dinitrogen ($N_2$) ("nitrogen fixation") (Gruber and Sarmiento 1997). These processes can be quantified on an ocean-based scale by the tracer $N^*$ (Gruber and Sarmiento 1997). Using a modeling study, Tyrrell (1999) found that phosphate would eventually be the ultimate limiting factor for primary production in the ocean because the binding of dinitrogen by diazotrophic plankton would always alleviate nitrogen limitation. Falkowski (1997) noted that on a geological scale, phosphate could not be responsible for the variations of primary production due to its more or less constant supply by continental weathering and fluvial discharge. He argues that iron availability controls nitrogen fixation and that small changes in the ratio of $N_2$ fixation to denitrification may control primary production and carbon export which is limited by nitrate.

There are various processes that disobey Redfield. It is known that fast-growing organisms have lower N:P ratios than slow-growing organisms (Koeve and Kähler 2010). This is because fast-growing organisms need a higher number of the phosphate-rich ribosomes. Using an ecosystem model, Mills and Arrigo (2010) showed that fast-growing organisms depleted the phosphate thereby decreasing the phosphate available for diazotrophs. For the same reason, slow-growing organisms stimulated nitrogen fixation. If these non-Redfield uptake processes are ignored, estimates of oceanic nitrogen fixation may be wrong. Mills and Arrigo (2010) conceive that the relative composition of fast- and slow-growing phytoplankton controls nitrogen fixation in the ocean. Moreover, while, on the one hand, cyanobacteria may exhibit N:P ratios that vary over one order of magnitude, diatoms, on the other hand, are characterized by low N:P ratios (Planavsky 2014). Hence, the phytoplankton composition controls the global ratio of the nitrate to phosphate concentration in the ocean explaining that locally and temporally large deviations from the Redfield ratio are observed.

Staal et al. (2007) measured $N_2$ fixation activity along a transect following the West-African coastline in the southern hemisphere from latitude N4.698; longitude E-18.497 to latitude N-19.200; longitude E5.744. North of latitude N-10.972; longitude E1.321 nitrogenase activity was detected and south of it not. Along the whole transect N:P ratios were low and decreasing from north to south. The average N:P in the area where nitrogen fixation took place was 5.5 (with the highest value 14.1) and in the area negative for nitrogen fixation this ratio was on average 1. The phototrophic biomass (measured as chlorophyll *a*) in the area positive for nitrogen fixation (average 0.187 mg $L^{-1}$) was twice as high as in the area without nitrogen fixation (average 0.098 mg $L^{-1}$). The phosphate concentration in that area was on average 196 nmol $L^{-1}$ and about 4 times higher than in the area positive for nitrogen fixation (47 nmol $L^{-1}$). The concentration of silicate behaved opposite and was twice as high in the area positive for than nitrogen fixation compared to the area negative for it (1209 vs 707 nmol $L^{-1}$). The concentrations of combined nitrogen were equally low throughout the transect. These data were interpreted as that primary production was nitrogen limited in the southern part of the transect where no nitrogen fixation took place. The draw-down of silicate hints to diatoms as an important component of the phytoplankton. It is known that areas dominated by diatoms are characterized by low N:P ratios (Planavsky 2014). Also, the low levels of combined nitrogen made that a substantial part of the phosphate could not be used resulting in the observed low N:P ratios. The limitation in the northern part of the transect where nitrogen fixation took place cannot be determined with certainty. Most likely it was phosphate but it might also have been the availability of iron, which could have limited nitrogen fixation (and thus primary production still nitrogen limited) or the lower temperature might have limited nitrogen fixation (Stal 2009). The average surface water temperature in the area positive for nitrogen fixation was 27.9 °C, while it was 24.9 °C in the area negative for nitrogen fixation. Temperature may be a crucial factor for nitrogen fixation in the ocean (Stal 2009).

### 1.5.2 Nitrogen Fixation

Nitrogen is after carbon quantitatively the second-most abundant element in structural cell material where it occurs in its reduced (amino) form and represents about 10% of the dry weight biomass. Organisms obtain nitrogen mostly as organic nitrogen, ammonium, or nitrate and nitrite, which are commonly referred to as "bioavailable" nitrogen. However, the largest source of nitrogen is dinitrogen gas ($N_2$), which is in fact "bioavailable" as well, but for a limited group of selected bacteria and a few archaea (and not eukarya!, which can use it only in symbiosis with a specialized bacterium, when ignoring *Homo sapiens*, who with its industrial production of fertilizer—by the Haber–Bosch process—is the main nitrogen fixer and responsible for about 50% of the total global nitrogen fixation!). All these microorganisms capable of using $N_2$ as a source of nitrogen possess nitrogenase, a highly oxygen-sensitive enzyme, which is capable of breaking and reducing the strong double bond between the two nitrogen atoms. Leaving out the Haber–Bosch process, about half of the global nitrogen binding occurs in the ocean (Canfield et al. 2010), where it supplies "new" nitrogen to drive primary production. It counteracts the loss of combined "bioavailable" nitrogen by denitrification although it is unclear whether these two processes are in balance. Estimates indicate that the loss may exceed the gain, which has led to the idea that the modern ocean might be losing nitrogen (ocean is a "black hole" for nitrogen). However, the application of the $N^*$ tracer (Gruber and Sarmiento 1997) indicated that the fixation of $N_2$ had to be higher than estimated based on measurements of the known diazotrophs. Subsequently, many new diazotrophs were discovered in the ocean and estimates of "new" nitrogen input are going up (Zehr and Capone 2021).

Until 1961, $N_2$ fixation was thought to be unimportant in the ocean. Dugdale et al. (1961) discovered that *Trichodesmium thiebautii* fixed dinitrogen in the Sargasso Sea. *Trichodesmium* is a filamentous non-heterocystous cyanobacterium that forms surface blooms in the tropical and subtropical ocean. Subsequently, a variety of unicellular diazotrophic cyanobacteria were found (Moisander et al. 2010; Zehr et al. 2001). A wide range of diazotrophic cyanobacteria live in symbiosis with microalgae, seagrasses, and corals, including the remarkable anoxygenic, photosystem-2-lacking, and non-$CO_2$ fixing *Candidatus* Atelocyanobacterium thalassa (Foster and Zehr 2019; Thompson et al. 2012; Zehr et al. 2008, 2016). There are several reports of marine diazotrophs other than cyanobacteria but it is uncertain to what extent they contribute to the import of "new" nitrogen in the ocean (Farnelid et al. 2011; Langlois et al. 2015; Moisander et al. 2014). The detection of *nif* genes belonging to chemotrophic microorganisms and even their expression does not necessarily translate into $N_2$ fixation. Moreover, such genes have even been found in a cyanobacterium (Bolhuis et al. 2010)!

Cyanobacteria are responsible for the import of the "new" nitrogen into the ocean and thereby driving primary production. Hence, cyanobacteria support the ocean food web and are also key to $CO_2$ fixation and crucial for climate and global change. Cyanobacteria are oxygenic photoautotrophs and therefore the diazotrophs among them must have ways to prevent the inactivation of the oxygen-sensitive nitrogenase. Terrestrial and freshwater diazotrophic cyanobacteria are filamentous and differentiate special cells, the heterocyst, which is the site of nitrogen fixation in those organisms. In the ocean there are no free-living heterocystous cyanobacteria. They appear in some symbiotic associations or on sandy or rocky shores and beaches, which will not be considered here. The heterocyst lacks the oxygenic photosystem 2 (quite like *Candidatus* Atelocyanobacterium thalassa) and possesses a special cell wall that serves as a gas diffusion barrier and limits the rate by which oxygen enters the cell. This oxygen is scavenged by the respiratory system and keeps the intracellular oxygen partial pressure sufficient low to prevent nitrogenase inactivation. The gas diffusion is a trade-off between sufficient entry of $N_2$ to satisfy nitrogen fixation and the highest entry of $O_2$ that still can be scavenged by respiration. The solubility of $O_2$ in water decreases with increasing temperature with approximately the same quotient as gas diffusion rate increases, which therefore counteracts each other. However, metabolic processes vary with temperature according to a $Q_{10} \sim 2$, which means a doubling or halving of the rate with each 10 °C increase or decrease, respectively. That means that in cold water the heterocyst cell wall must be a stronger diffusion barrier ("thicker" cell wall) because respiration will be slower and less $O_2$ can be scavenged while in warm water the opposite is the case. Staal et al. (2003) showed that this was the case and demonstrated that *Trichodesmium* would not need such a special cell wall, also aided by the lower solubility of $O_2$ in the warm (sub)tropical ocean. Because *Trichodesmium* fixes $N_2$ during the daytime it still needs anoxygenic cells. These cells exist and have been termed diazocytes (Bergman et al. 2013). The unicellular diazotrophic cyanobacteria solved this problem of the incompatibility of oxygenic photosynthesis and nitrogen fixation by carrying out the latter during the night. The gas diffusion and respiratory capacity still requires a minimum cell size (~5 μm) in order to keep the intracellular level of $O_2$ sufficient low. *Candidatus* Atelocyanobacterium thalassa is smaller and fixes $N_2$ during the day but does so in symbiosis which apparently makes the situation much different. Also, benthic systems such as marine cyanobacterial mats that occur worldwide in coastal intertidal areas and all of which are diazotrophic behave different from the oceanic planktonic system (Severin and Stal 2008). While chemotrophic bacteria are all too small to be able to keep their cells close to anoxic in a fully oxygenated environment, they are, therefore, unlikely to be able to fix $N_2$. However, this might be different in oxygen minimum zones (OMZs) or in aggregates (e.g., marine snow) where chemotrophic diazotrophs may indeed contribute to the binding of $N_2$ (Paulmier and Ruiz-Pino 2009; Schunck et al. 2013; Shanks and Reeder 1993).

While there is a reasonably good explanation why nitrogen fixation seems to be restricted to the euphotic surface waters of the warmer areas of the ocean it is an enigma why heterocystous cyanobacteria do not seem to have this function in the temperate or cold ocean. Heterocystous cyanobacteria form blooms in freshwater lakes irrespective of their geographic location. The brackish Baltic Sea is exhibiting massive blooms of heterocystous cyanobacteria, notably *Nodularia*, *Aphanizomenon*, and *Dolichospermum* (formerly *Anabaena*). The Baltic Sea exhibits a salinity gradient from freshwater in the northern Bothnian Gulf to full seawater salinity in the Kattegat and Skagerrak toward the North Sea. Diazotrophic blooms of the heterocystous cyanobacteria suddenly stop in the Bornholm Sea at a salinity of ~9‰ (Stal et al. 1999). These authors speculated that the high sulfate concentration may prevent nitrogen fixation rather than NaCl, e.g., by depleting reducing equivalents through sulfate reduction. Although this may be part of the explanation the absence of heterocystous cyanobacteria in the temperate and cold ocean is still an unresolved question.

### 1.5.3 Adaptation to Salt

While NaCl is the most important component that contributes to the salinity of seawater, it might not be the most important one that determines the microbiology of seawater compared to freshwater. In terms of salinity, freshwater and seawater are not too much different (compared to hypersaline waters) and a whole range of salinities exists between freshwater and full-salinity seawater (or even a bit higher such as in the Mediterranean) but the relative ionic composition is the same in that range. The difference of salinity would probably not pose a serious problem for many microorganisms. Seawater may even be closer to a physiological solution than freshwater. Environments with salinities far beyond seawater are called hypersaline and are really challenging environments for any organism. Culturing hypersaline microorganisms is challenging. It is not just the concentration of NaCl. The composition varies greatly among the different hypersaline ecosystems worldwide and the actual ionic composition of a particular saline water may be critical for the growth of a microorganism.

Cells need a certain turgor (osmotic or hydrostatic pressure) to be able to carry out their cellular functions. Water can freely pass the cytoplasmic membrane and the cell has no means to actively control this water transport in order to regulate the turgor pressure. The only way is to regulate the cytoplasmic solute concentration in response to hyper- or hypoosmotic stress (Bremer and Krämer 2019). When the salt concentration in the surrounding environment increases the intracellular solute concentration must go up. In order to counteract the osmotic value of the surroundings, the cell will accumulate the so-called osmo-protectants or osmolytes (and several other synonyms). These are small molecules that need to be highly soluble in order to reach sufficient high concentrations. Another important requirement of an osmolyte is that they do not interfere with the cellular metabolism, i.e. they must be compatible. Therefore, they are also called "compatible solutes." There is a whole range of such compounds in microorganisms and they are generally involved in any situation of low water potential (high salinity, drought) and (low) temperature. An important role is to help folding of proteins and improve their stability and hydration level to allow their function. Microorganisms may employ a "salt-in" or "salt-out" strategy. The first is used by some halophilic microorganisms, who are adapted to high intracellular concentrations of salt (KCl). Most microorganisms pump the salt out of the cell while synthesizing or taking up from the environment organic compatible solutes. In the constant pelagic oceanic environment microorganisms are not exposed to great salinity changes and therefore do not need to respond to them. This is different, for instance, in coastal tidal mudflats or sandy beaches, where microorganisms are exposed to large fluctuations of salinity. For instance, while osmotically adjusted to seawater a benthic microorganism inhabiting a tidal environment needs to respond quickly to a rain shower by releasing or degrading its osmoprotectant otherwise the cell may lyse due to the influx of water (Kirsch et al. 2019). It also needs to take up or synthesize osmoprotectant if the opposite situation takes place. Upon a hyperosmotic shock, $Na^+$ that enters the cell is actively pumped out by an antiporter transporter in the cytoplasmic membrane and subsequently replaced by $K^+$, which is subsequently replaced by the organic osmolyte that is synthesized in response of the salt shock. Organic osmolytes in many microorganisms may also be used as a storage compound and metabolized in times of energy shortage. They may temporarily be exchanged for inorganic ions.

Compatible solutes that serve as osmoprotectant can be subdivided into three major groups: sugars (such as the disaccharides sucrose and trehalose), polyols (e.g., glycerol), heterosides (e.g., glucosyl glycerol), and amino acids and derivatives thereof (e.g., ectoine, betaine, proline, glutamate) (Bremer and Krämer 2019; Kirsch et al. 2019). Sucrose is often found as a compatible solute at low salinities while another disaccharide, trehalose, is more frequently found at intermediate salinities and glucosyl glycerol is typical for full marine salinity. At salinities above normal seawater glycine betaine, ectoine, and glycerol are found. There are many other derivatives of these compounds and amino acids (proline) that are used as compatible solutes in microorganisms (Bremer and Krämer 2019; Kirsch et al. 2019; Kumar et al. 2020). Certain marine microalgae produce dimethyl sulfoniopropionate

(DMSP), which upon release is converted to dimethylsulfide (DMS) that has been thought to play a role in a climate feedback.

### 1.5.4 Sulfate

An important component of seawater is sulfate ($SO_4^{2-}$) that is present in a concentration of 28 mM. This high concentration has several consequences. Even though sulfur is not a major constituent of cellular biomass, it is a component for some essential amino acids without which proteins cannot be synthesized. In the marine environment sulfur will not become limiting as is sometimes the case in freshwater. Under low oxygen conditions such as in oxygen minimum zones (OMZs), marine aggregates (marine snow), marine sediments, and cold seeps, sulfate may become the preferred terminal electron acceptor by sulfate-reducing bacteria. These bacteria reduce sulfate to the extremely toxic sulfide that also reacts with oxygen and leads to hypoxia, which has led to the so-called dead-zones in coastal waters and sediments (Middelburg and Levin 2009). They are not really "dead" because they are teeming with anaerobic microorganisms. The anaerobic respiration by sulfate-reducing bacteria is especially important in marine systems whereas methanogenic archaea fulfill this role in anoxic freshwater systems. Nevertheless, the latter are certainly also present in selected marine systems thriving on substrates not used by sulfate-reducing bacteria. The oxidation of methane emitted from cold seeps on the ocean floor is carried out by aggregates of methanogenic archaea and sulfate-reducing bacteria (Boetius et al. 2000).

Sulfur may also be incorporated in certain phytoplankton species in the form of dimethyl sulfoniopropionate (DMSP) where it serves a role as osmoprotectant. Upon release (e.g., through lysis or grazing), DMSP is converted into the gas dimethyl sulfide (DMS) which may be transported to the atmosphere and conceived to play a role in climate feedback but also transports sulfur to the continents.

The molybdate ion is a structural analog of sulfate and a co-factor in certain enzymes such as nitrogenase and nitrate reductase. It has been hypothesized that the high sulfate concentration in seawater would complicate the uptake of molybdate and thereby the processes depending on it. However, in the light of the abundant presence of diazotrophs and phytoplankton depending on nitrate as the source of nitrogen, this scenario does not seem plausible. Also, molybdate may be taken up by designated transporters that do not compete with sulfate. The high sulfate concentration may nevertheless prevent the proliferation of free-living heterocystous cyanobacteria in the ocean (Howarth and Cole 1985; Stal 2009; Stal et al. 1999).

### 1.5.5 Freshwater- and Marine Microbiomes: What Are the Boundaries?

The limiting nutrient in temperate marine waters is often nitrogen, while in freshwater lakes is often phosphate. Blomqvist et al. (2005) attributed this difference to salinity, more precisely to the high sulfate concentration of seawater (28 mM). Sulfide produced by sulfate-reducing bacteria scavenges iron, which will precipitate as FeS. Upon oxygenation the ferric iron will combine with phosphate. While in anoxic marine systems the ratio of dissolved iron and phosphate is such that there will be phosphate left after the iron phosphate precipitated, this is not the case in freshwater. The residual phosphate in the marine system results in a nitrogen limitation while in a freshwater system phosphate will be the limiting nutrient for primary production.

Dupont et al. (2014) concluded that the differences between the freshwater- and marine microbiomes can be found in the core metabolic functions and pathways of microorganisms. These authors observed large differences in a variety of central metabolic processes. The genomic information of these central metabolic pathways hint at an early differentiation of freshwater and marine microorganisms. These adaptations determined whether or not a microorganism is able to thrive at a certain salinity. Salinity is, therefore, the main driver of microbial community composition. Other environmental factors determining the microbial community composition and distinguishing freshwater- and marine microbiomes were N:P ratio and total phosphorus. The boundary between freshwater- and marine environments appears to be strict and difficult to cross despite the high rates of dispersal and large population sizes of their microbial inhabitants (Logares et al. 2009). These observations describe the boundary between freshwater- and marine environments and the phylogenetic split between their microbiomes but do not explain them. Why does it seem so difficult to migrate and adapt in either direction?

## 1.6 On a Personal Note: How Did I Become a Marine Microbiologist

At the time of writing this I have been retired from science for more than 3 years after working in the field for approximately 40 years. Thus, I thought it might be a good idea to also add a bit of a personal account on what I think is so special about this discipline of Marine Microbiology and how I got involved in it.

I studied Biology at the University of Groningen in The Netherlands from 1972 to 1978 and specialized in Microbiology and Molecular Genetics. I was taught microbiology by Prof. Hans Veldkamp (1923–2002) who was the founder and first Head of the Department of Microbiology at the University of Groningen in 1963. He belonged to the Delft School of Microbiology where he studied with Prof. Albert Jan Kluyver and later worked with Prof. Cornelis Bernardus van Niel at Hopkins Marine Station at Pacific Grove in California. Prof. Veldkamp established the field of general microbiology at Groningen University in the spirit of these founding fathers of microbial ecology. During those days the Department of Microbiology at Groningen University had three specialized working groups: Microbial Ecology (led by Prof. Veldkamp), Microbial Physiology (led by Prof. Wim Harder (1939–2008)), and Molecular Microbiology (led by Prof. Wil Konings (1937–2014)). In the second year of my biology studies, I decided to take the course

General Microbiology with a series of morning lectures that covered the field of Microbiology into great depth and were backed up by Veldkamp's lecture notes and two text books: Prof. Hans Schlegel's Allgemeine Mikrobiologie (in German; later translated in English: General Microbiology) and Prof. Roger Stanier's General Microbiology (together with Doudoroff and Adelberg). In the afternoon, there were hands-on practical lessons. These were focused on the enrichment of various microorganisms from natural samples, their isolation and purification to pure cultures as well as carrying out a range of physiological tests in order to identify the organism. This was well before the time of DNA-based identification of microorganisms. In the third year, I decided to take an extended course in Microbial Ecology which was offered by Prof. Veldkamp and two other staff: Dr. Gijs Kuenen (became later professor microbiology at Delft University) and Dr. Hans van Gemerden (1936–2006). This course was offered for the first time and was strongly modeled to the scientific research of the Microbial Ecology group. Their research was focused on the growth and behavior of microorganisms under sub-maximum growth rates and their competition with other microorganisms. At the time most of the microbiological research was done using the so-called batch cultures. In batch cultures all media components are present in excess. Microorganisms inoculated in this medium would, after a so-called lag phase, in which growth is sub-optimal, grow exponentially, i.e., at its maximum growth rate under the given conditions, until it runs out of its growth substrate and/or when the accumulation of metabolites exuded into the medium would inhibit further growth. The culture is said to enter its stationary phase, which eventually will be followed by a collapse of the culture and the death of the microorganisms in it. The research of the Microbial Ecology group was based on the idea that growth as it happens in a batch culture would rarely occur under natural conditions. Under natural conditions there would be a growth-limiting substrate and its concentration would be low and not supporting the maximum specific growth rate of the organism (compare the Michaelis–Menten model). In order to overcome the limitation of batch cultivation, the Microbial Ecology group adopted the chemostat as their culturing device and even as a device to enrich for particular microorganisms. This culturing technique was at the time quite revolutionary, even though its theory was already described in the fifties of the previous century by the scientists Monod (Monod 1949) and its application by Herbert (Herbert et al. 1956). Instead of a closed culture system such as batch culturing, the chemostat is an open continuous culture system. Sterile growth medium in which one component is the growth-limiting substrate (e.g., the source of energy or a nutrient) is pumped at a constant rate in the growth vessel. The concentration of the growth-limiting substrate will determine the eventual cell number or microbial biomass in the growth container. The volume of the growth container is kept constant by means of an overflow of culture into a waste container. The culture in the growth container needs to be well mixed. Often the pH has to be maintained at a specified value by means of an automatic titration system and depending on the cultured organism and required conditions, other conditions (e.g., the level of dissolved oxygen) need to be controlled and maintained at certain set of values. Theory predicts that at steady state the growth rate of the microorganisms equals the dilution rate imposed by the medium pump and all conditions (chemical composition of the culture; biomass/cell number) stay constant (chemostat). Hence, by simply changing the pumping rate of the medium supply, the growth rate of the cultured microorganism could be imposed from very low growth rates to close to the maximum specific growth rate. When using the chemostat as a device to enrich and isolate a microorganism it is inoculated with a natural sample. The theory predicts that the organism with the best affinity for the limiting growth substrate of the medium will become dominant and outcompete all other microorganisms in the chemostat at steady state (usually considered to be reached after ~5 volume changes). Of course, the reality was often less simple but nevertheless many microorganisms have been isolated into pure culture using this technique, while batch enrichments would always have led to the isolation of the organism with the highest growth rate in that medium and set of conditions. It was said that the continuous culture was named after the continuous attention that was required by its operator and that this has scared off many and kept the batch culture as a popular technique in microbiology.

The technique of the continuous culture was also used for the study of phototrophic microorganisms. Dr. Hans van Gemerden used it for the growth of anoxygenic purple and green sulfur bacteria. In those cultures, the source of energy (light) was supplied independently from the medium pump but the growth-limiting substrate could be a component of the growth medium such as the electron donor sulfide. Prof. Luuc Mur (University of Amsterdam) introduced the continuous culture of oxygenic phototrophs such as the eukaryotic microalgae and cyanobacteria. By choosing light as the growth-limiting factor, the chemostat was turned into a photostat (in which the average photon flux density in the culture vessel was kept constant), while the growth rate was still imposed by the medium pump rate. In steady state, the density of the culture is determined by the average photon density in the culture. However, these oxygenic phototrophs could also be grown as a chemostat by choosing nutrients such as phosphate or nitrate as the growth-limiting substrate. Apart from studying the behavior of these microorganisms at sub-maximum growth rates and at different growth limitations, these continuous cultures were also excellent tools to study the competition of different species of microorganisms. Prof. Jef Huisman (the successor of Prof. Mur at the University of Amsterdam) has brought this type of investigations to a higher level and increased our understanding of competition, collaboration, and co-existence of different microorganisms tremendously. In particular, these experiments shed light into the "paradox of the plankton" as was formulated by Hutchinson (1961). The paradox is that so many phytoplankton species co-exist in seemingly homogeneous aquatic environments freshwater as well as marine, while under such conditions one would expect that only one or a few would outcompete all others.

After finishing my Master's in Biology, I was offered the possibility to move to marine microbiology by joining the geomicrobiologist Prof. Wolfgang E. Krumbein (1937–2021) at the Carl von Ossietzky University in Oldenburg, Germany. With him I started a project on dinitrogen fixation by marine benthic cyanobacteria that built microbial mats in coastal intertidal sediments. Through his inspiring enthusiasm I became interested in geology, geomicrobiology, and biogeochemistry and learned that microorganisms and their activities are instrumental in many geological processes throughout the history of planet Earth. Sadly, Wolfgang Krumbein passed away on 4th April 2021.

While working on my research project on dinitrogen fixation in microbial mats of the desert island Mellum, belonging to the barrier islands of the southern North Sea, I was attracted to the wonderful layered structure of those microbial mats that developed on the intertidal sand flats, characterized by a sediment of fine sand and some silt. They showed a nicely lamination of the green cyanobacteria and purple sulfur bacteria underneath. This was known from the German literature as "Farbstreifen-Sandwatt," which means so much as colored striped sand. ("Watt" is not translated in English because it is a proper name of that sea (Wattenmeer in German, Waddenzee in Dutch, Vadehavet in Danish) and the German word "Watt" refers to the intertidal sand- or mudflats that are exposed at low tide. Despite several efforts, the German concept of "Farbstreifen-Sandwatt" did not make it in the English literature, nor did any English synonym such as color-striped sand and this and similar systems are generally addressed to as microbial mats.

I went to my former Microbiology Department at Groningen University and told them enthusiastically about this unique microbial ecosystem. Dr. Hans van Gemerden was a microbial ecologist studying purple sulfur bacteria and other anoxygenic phototrophs doing competition experiments in chemostats. His model organisms until then were derived from freshwater lakes. He became enthusiastic about the microbial mat as a model system and I invited him on our next field trip to the desert North Sea island Mellum (while we would normally organize a boat trip or get a lift of the research vessel "Senckenberg" of the same-named institute in Wilhelmshaven, on that particular expedition we could reach the island only by helicopter). Hans van Gemerden had prepared lots of tubes with sterile enrichment medium in order to isolate anoxygenic phototrophs. When I asked him about his medium and whether he realizes that we are working in a marine environment he admitted that he completely forgot about that. While we had to work with what was available in the only house on the island, we took a letter scale and kitchen salt and amended every single tube with salt to seawater salinity. It worked very well!! Unfortunately, we forgot about a control to check whether the salt addition made any difference.

In 1988, I joined the Department of Aquatic Microbiology of Luuc Mur at the University of Amsterdam and developed marine microbiology as a sub-discipline, with a continued focus on phototrophic microorganisms, mainly cyanobacteria and the eukaryotic diatoms.

In 1996, I founded the Department of Marine Microbiology in the Netherlands Institute of Ecology (NIOO). It was the first scientific department in the Netherlands devoted to this research field. Ten years later, the University of Amsterdam founded a chair in Marine Microbiology for which I was appointed until my retirement in 2017, when I was followed up by Prof. Linda Amarall Zettler who is currently holding the chair. The chair of Marine Microbiology at the University of Amsterdam was the first and still the only one in the Netherlands. This is remarkable considering the long history of and internationally highly recognized field of microbiology as well as the long tradition of marine and maritime research in the Netherlands.

For a long time, pure (axenic) cultures were the golden standard in microbiology. Anything that could not be isolated and grown in the laboratory as a pure culture was basically ignored and could not be formally described as a (new) species (genus, or other taxonomic level). Culturing meant in the first place that it demanded that the organism could be grown in the laboratory on more or less defined growth media and secondly that it could be separated from other types of organisms (the other microorganisms were referred to as contaminants). Isolation was usually done by starting an enrichment culture. The growth medium composition and the culture conditions were chosen such that it would specifically benefit the target organism. This implicated that the microbiologist already knew what the main properties of the target organism would be. After one or several rounds of enriching the culture for the target microbe, it was separated by dilution in growth medium supplemented with agar-agar gel that was cooled down just before it would solidify or streaked by an inoculation loop on the surface of agar-agar solidified growth medium. The idea was that the spatially isolated cell of the target organism would grow out to a colony visible by the naked eye. Such colony was picked up and re-diluted or re-streaked in or on agar-agar and this procedure was repeated at least one more time in order to be sure that the colony was formed by only cells of the target organism, which was in the past checked among other by microscopy and by now more regularly by DNA sequencing. Many microbes do not tolerate agar-agar while others, particularly marine microorganisms, degrade it. Even when highly purified agar-agar is used or in some cases agarose, certain microbes refuse to grow on it. In some cases, other gelling agents (gelatin, silica gel) could be applied as an alternative but they usually do not have the properties of agar-agar that makes it so suitable for microbiological purposes. Moreover, liquid enrichment media are often not suitable for microbes which live attached to surfaces such as in biofilms.

Although this technique has been used throughout the history of microbiology and is still used, there are many limitations. It starts with the design of the growth medium and culture conditions which is a prejudication of the scientist, mainly a precedent of what the scientist already knows, rather than aiming at the unknown. It selects for organisms that grow in that specific medium and conditions. Moreover, using batch enrichment cultures it selects for the fastest growing organism and in chemostats for the organism with the highest affinity for the growth-limiting substrate at the chosen dilution (= specific growth rate). Another technique that allows to isolate the most abundant organism(s) in a sample is by dilution to extinction. This prevents that abundant species are overgrown by rare species in the sample.

If a single species is brought into culture, it means that this particular microorganism is capable of growing in isolation. However, many species of microorganisms require one or more other types of organisms in order to be able to carry out the trophic reaction that allows them to proliferate. They will only form a colony when in very close proximity. The classical case of *Methanobacillus omelianskii* is what was thought to be an axenic culture that could ferment ethanol to methane and that later was found to be in fact a co-culture of two different organisms that could perform the fermentation only jointly (Barker 1939; Bryant et al. 1967). This way of life is known as syntrophy (feeding together) (Morris et al. 2013). Another good example is the methane-oxidizing consortium of a methanogenic archaea and a sulfate-reducing bacterium (Boetius et al. 2000).

Pure cultures were critical for the classical work on the cause of tuberculosis and other infectious diseases by R. Koch in the late nineteenth century (Koch 1884). Koch's postulates required the isolation of the bacteria (or viruses, fungi/yeasts, parasites, as well as bacteria contaminated with viruses that are the cause of disease) from a person suffering the disease. Koch's postulates have been used in a simplified form to solve scientific questions in microbial ecology and biogeochemistry (Grimes 2006). These simplified Koch's postulates can also be applied to mixed cultures, which overcomes the problem of the isolation of pure cultures, but still requires culturability of environmental microorganisms or more or less defined mixtures of such microorganisms.

The stringent procedures that were advocated by many traditional and hard-core microbiologists have led to the ignorance of an enormous diversity of microorganisms that could not be isolated and grown in axenic culture and led to experimentation with "lab rats." Those studies have revealed an enormous amount of knowledge on the physiology of microorganisms and their application in some major biotechnological processes. However, it has put back the development of microbial ecology enormously (see, e.g., Dubos 1974). Many environmental microbiologists made important observations by microscopy and laboratory and field experiments but as they were unable to isolate the microbes they observed in pure culture, their observations were treated lightly and considered by many as soft science. This has changed completely after nucleic acid-based studies were introduced in microbial ecology. The concept of pure cultures as the only standard in microbiology has been abandoned. Now, microbial ecologists use defined mixed cultures with 2 or more different types of microorganisms or even undefined natural communities as their study objects and investigate the interactions or communication between the microbial components of the system. This discipline has also become known as microbial systems biology.

One of those hard-core molecular microbiologists who always had looked down on microbial ecology entrusted me shortly before his retirement that he would have become a microbial ecologist if he could do it all over again.

## 1.7 Concluding Remarks

Much of the modern research on marine microbiology can be summarized as taking samples as many as possible from as many as possible oceanic and marine areas, including from the deep-sea, measure the physicochemical and other oceanographic parameters for metadata, filter the (size-fractionated) water samples, extract the nucleic acids, sequence them, and put the sequences in public databases. Bioinformatic tools are used to assemble genes and (partial) genomes and analyze them. What is often missing are (1) extensive metadata from the samples and the areas and times they were taken, (2) rate measurements of relevant microbiological processes, (3) biogeochemical processes, and (4) biochemical identification of predicted genes. To go out at sea, filter water, extract nucleic acids and have them sequenced is nowadays easy and does not require much scientific originality. Much of this research is not scientific research question driven and trusts that, once the sequence reads are in the data bases, interesting questions may come up and may be answered. This may be the case and in fact many interesting discoveries have been made this way. But the approach is taken just because it can be done and perhaps not in the last place because of its public outreach impact. This aspect is not unimportant, for instance, because of fund raising and public awareness. When the microbial and biogeochemical processes become linked with the genetic information it will become possible to have a system's view of the oceanic ecosystem and understand how it is regulated. Such approach is urgently needed to address the questions of global change (increasing temperature, acidification, sea level rise, plastisphere).

In the near future marine microbiology will rely less on expensive and inefficient research vessels. Many data may be obtained from automatic systems and sensors, either operating autonomous or carried by ships of opportunity (ferries, merchant ships) (Paul et al. 2007). Examples are among others the autonomous Argo Floats (Claustre et al. 2020), remote sensing (Coles et al. 2004; De Monte et al. 2013; Wurl et al. 2018), autonomous measurement of gene expression using an environmental sample processor connected to a drifting buoy for Lagrangian sampling (Birch 2018; Ottensen et al. 2013), autonomous tracking instruments (Zhang et al. 2020), and in situ flow cytometry and plankton analyzers. Other important projects are the so-called time series, where at a certain location sampling is done at regular time intervals and a large number of biological, chemical, and physical parameters are measured using strict and standardized analytical protocols. Some of these time series are running for several decades. While many marine research institutes run their own long-term samplings often off their own docks, others are only reached by ship. Among them are the Hawaiian Ocean Time Series (HOT) in the Pacific Ocean, the Bermuda Atlantic Time Series (BATS) in the western Atlantic Ocean, and the Cape Verde Ocean Observatory in the east Atlantic Ocean.

There may be many answers to the question of what is so special about marine microbiology but it depends in the first place on the devotion with which scientists that might call themselves marine microbiologists try to understand the life and role of microorganisms in the marine environment. And with the words of Louis Pasteur:

> Messieurs, c'est les microbes qui auront le dernier mot.
> (Gentlemen, it is the microbes who will have the last word)

## References

Achtman M, Wagner M (2008) Microbial diversity and the genetic nature of microbial species. Nature Rev Microbiol 6:431–440

Amaral-Zettler L, Artigas LF, Baross J, Bharathi PAL, Boetius A, Chandramohan D, Herndl G, Kogure K, Neal P, Pedrós-Alió C, Ramette A, Schouten S, Stal L, Thessen A, de Leeuw J, Sogin

M (2010) A global census of marine microbes. In: McIntyre AD (ed) Life in the World's Oceans. Blackwell, Oxford, pp 223–245
Arrieta JM, Mayol E, Hansman RL, Herndl GJ, Dittmar T, Duarte CM (2015) Dilution limits dissolved organic carbon utilization in the deep ocean. Science 348:331–333
Azam F, Fenchel T, Field JG, Gray JS, Meyer-Reil LA, Thingstad F (1983) The ecological role of water-column microbes in the sea. Mar Ecol Progr Ser 10:257–263
Baas Becking LGM (1934) Geobiologie of inleiding tot de milieukunde. WP van Stockum & Zoon NV, Den Haag, p 263
Barker HA (1939) Studies upon the methane fermentation. IV. The isolation and culture of *Methanobacterium omelianskii*. Ant Leeuwenhoek 6:201–220
Behrenfeld MJ, Falkowski PG (1997) Photosynthetic rates derived from satellite-based chlorophyll concentration. Limnol Oceanogr 42:1–20
Béjà O, Aravind L, Koonin EV, Suzuki MT, Hadd A, Nguyen LP, Jovanovich SB, Gates CM, Feldman RA, Spudich JL, Spudich EN, DeLong EF (2000) Bacterial rhodopsin: evidence for a new type of phototrophy in the sea. Science 289:1902–1906
Bergh Ø, Børsheim KY, Bratbak G, Heldal M (1989) High abundance of viruses found in aquatic environments. Nature 340:467–468
Bergman B, Sandh G, Lin S, Larsson J, Carpenter EJ (2013) *Trichodesmium* – a widespread marine cyanobacterium with unusual nitrogen fixation properties. FEMS Microbiol Rev 37:286–302
Birch J (2018) Collecting and processing samples in remote and dangerous places: the environmental sample processor as a case study. Pure Appl Chem 90:1625–1630
Blomqvist S, Gunnars A, Elmgren R (2005) Why the limiting nutrient differs between temperate coastal seas and freshwater lakes: a matter of salt. Limnol Oceanogr 49:2236–2241
Boetius A, Ravenschlag K, Schubert CJ, Rickert D, Widdel F, Gieseke A, Amann R, Jørgensen BB, Witte U, Pfannkuche O (2000) Microbial interactions involving sulfur bacteria: implications for the ecology and evolution of bacterial communities. Nature 407:623–626
Bolhuis H, Severin I, Confurius-Guns V, Wollenzien UIA, Stal LJ (2010) Horizontal transfer of the nitrogen fixation gene cluster in the cyanobacterium *Microcoleus chthonoplastes*. ISME J 4: 121–130
Bremer E, Krämer R (2019) Responses of microorganisms to osmotic stress. Ann Rev Microbiol 73:313–334
Brochier-Armanet C, Boussau B, Gribaldo S, Forterre P (2008) Mesophilic crenarchaeota: proposal for a third archaeal phylum, the Thaumarchaeota. Nature Rev Microbiol 6:245–252
Bryant MP, Wolin EA, Wolin MJ, Wolfe RS (1967) *Methanobacillus omelianskii*, a symbiotic association of two species of bacteria. Arch Mikrobiol 59:20–31
Canfield DE (2015) Baas Becking's Geobiology – or introduction to environmental science. Wiley Blackwell, Oxford, p 152
Canfield DE, Glazer AN, Falkowski PG (2010) The evolution and future of Earth's nitrogen cycle. Science 330:192–196
Chapra SC, Dove A, Warren GJ (2012) Long-term trends of Great Lakes major ion chemistry. J Great Lakes Res 38:550–560
Chisholm SW, Olson RJ, Zettler ER, Goericke R, Waterbury JB, Welschmeyer NA (1988) A novel free-living prochlorophyte abundant in the oceanic euphotic zone. Nature 334:340–343
Claustre H, Johnson KS, Takeshita Y (2020) Observing the global ocean with biogeochemical-Argo. Annu Rev Mar Sci 12:23–48
Cohen Y, Krumbein WE, Goldberg M, Shilo M (1977) Solar Lake (Sinai). 1. Physical and chemical limnology. Limnol Oceanogr 22:597–608
Coles VJ, Wilson C, Hood RR (2004) Remote sensing of new production fuelled by nitrogen fixation. Geophys Res Lett 31:L06301
Curtis TP, Sloan WT, Scannell JW (2002) Estimating prokaryotic diversity and its limits. Proc Natl Acad Sci U S A 99:10494–10499
De Monte S, Soccodato A, Alvain S, d'Ovidio F (2013) Can we detect oceanic biodiversity hotspots from space? ISME J 7:2054–2056

De Wit R, Bouvier T (2006) 'Everything is everywhere, but, the environment selects'; what did Baas Becking and Beijerinck really say? Env Microbiol 8:755–758
del Giorgio PA, Cole JJ, Cimbleris A (1997) Respiration rates in bacteria exceed phytoplankton production in unproductive aquatic systems. Nature 385:148–151
Dubos R (1974) Pasteur's dilemma – the road not taken. ASM News 40:703–709
Ducklow HW (1983) Production and fate of bacteria in the oceans. BioSciences 33:494–501
Dugdale RC, Menzel DW, Ryther JH (1961) Nitrogen fixation in the Sargasso Sea. Deep-Sea Res 7: 297–300
Dupont CL, Larsson J, Yooseph S, Ininbergs K, Goll J, Asplund-Samuelsson J, McCrow JP, Celepli N, Zeigler Allen L, Ekman M, Lucas AJ, Hagström A, Thiagarajan M, Brindefalk B, Richter AR, Andersson AF, Tenney A, Lundin D, Tovchigrechko A, Nylander JAA, Brami D, Badger JH, Allen AE, Rusch DB, Hoffman J, Norrby E, Friedman R, Pinhassi J, Venter JC, Bergman B (2014) Functional tradeoffs underpin salinity-driven divergence in microbial community composition. PLoS One 9(2):e89549
Dykhuizen DE (1998) Santa Rosalia revisited: why are there so many species of bacteria? Antonie Van Leeuwenhoek 73:25–33
Edgcomb VP, Bernhard JM, Summons RE, Orsi W, Beaudoin D, Visscher PT (2014) Active eukaryotes in microbialites from Highborne Cay, Bahamas, and Hamelin Pool (Shark Bay), Australia. ISME J 8:418–429
Falkowski PG (1997) Evolution of the nitrogen cycle and its influence on the biological sequestration of $CO_2$ in the ocean. Nature 387:272–274
Falkowski PG, Davis CS (2004) Natural proportions. Redfield ratios: the uniformity of elemental ratios in the oceans and the life they contain underpins our understanding of marine biogeochemistry. Nature 431:131
Farnelid H, Andersson AF, Bertilsson S, Al-Soud WA, Hansen LH, Sørensen S, Steward GF, Hagström A, Riemann L (2011) Nitrogenase gene amplicons from global marine surface waters are dominated by genes of non-cyanobacteria. PLoS One 6(4):e19223
Field CB, Behrenfeld MJ, Randerson JT, Falkowski P (1998) Primary production of the biosphere: integrating terrestrial and oceanic components. Science 281:237–240
Fischer B (1894) Die Bakterien des Meeres. Verlag von Lipsius & Tischer, Kiel und Leipzig, p 82
Flombaum P, Gallegos JL, Gordillo RA, Rincón J, Zabala LL, Jiao N, Karl DM, Li WKW, Lomas MW, Veneziano D, Vera CS, Vrugt JA, Martiny AC (2013) Present and future global distributions of the marine cyanobacteria *Prochlorococcus* and *Synechococcus*. Proc Natl Acad Sci U S A 110:9824–9829
Foster RA, Zehr JP (2019) Diversity, genomics, and distribution of phytoplankton-cyanobacterium single-cell symbiotic associations. Ann Rev Microbiol 73:435–456
Fuhrman JA, McCallum K, Davis AA (1992) Novel major archaebacterial group from marine plankton. Nature 356:148–149
Gaucher EA, Thomson JM, Burgan MF, Benner SA (2003) Inferring the palaeoenvironment of ancient bacteria on the basis of resurrected proteins. Nature 425:285–288
Geider RJ (1997) Photosynthesis or planktonic respiration? Nature 388:132–133
Ghai R, Megumi Mizuno C, Picazo A, Camacho A, Rodriguez-Valera F (2013) Metagenomics uncovers a new group of low GC and ultra-small marine Actinobacteria. Sci Rep 3:2471
Gilbert JA, Steele JA, Caporaso JG, Steinbrück L, Reeder J, Temperton B, Huse S, McHardy AC, Knight R, Joint I, Somerfield P, Fuhrman JA, Field D (2012) Defining seasonal marine microbial community dynamics. ISME J 6:298–308
Giovannoni SJ, Britschgi TB, Moyer CL, Field KG (1990) Genetic diversity in Sargasso Sea bacterioplankton. Nature 345:60–63
Grasshoff K, Voipio A (1981) Chemical oceanography. In: Voipio A (ed) The Baltic Sea. Elsevier Scientific Publishing Company, Amsterdam, pp 183–218. 418 pp
Grimes DJ (2006) Koch's postulates – then and now. Microbe 1:223–228
Gruber N, Sarmiento JL (1997) Global patterns of marine nitrogen fixation and denitrification. Glob Biogeochem Cyc 11:235–266

Hagström Å, Azam F, Wikner J, Rassoulzadegan F (1988) Microbial loop in an oligotrophic pelagic marine ecosystem: possible roles of cyanobacteria and nanoflagellates in the organic fluxes. Mar Ecol Progr Ser 49:171–178

Hagström Å, Pommier T, Rohwer F, Simu K, Stolte W, Svensson D, Zweifel UL (2002) Use of 16S ribosomal DNA for delineation of marine bacterioplankton species. Appl Env Microbiol 68: 3628–3633

Harris HMB, Hill C (2021) A place for viruses on the tree of life. Front Microbiol 11:604048

Haverkamp THA, Schouten D, Doeleman M, Wollenzien U, Huisman J, Stal LJ (2009) Colorful microdiversity of *Synechococcus* strains (picocyanobacteria) isolated from the Baltic Sea. The ISME J 3:397–408

Hecky RE, Campbell P, Hendzel LL (1993) The stoichiometry of carbon, nitrogen, and phosphorus in particulate matter of lakes and oceans. Limnol Oceanogr 38:709–724

Heip CHR, Goosen NK, Herman PMJ, Kromkamp J, Middelburg JJ, Soetaert K (1995) Production and consumption of biological particles in temperate tidal estuaries. Oceanogr Mar Biol Ann Rev 33:1–149

Henderson LJ (1913) The fitness of the environment. An inquiry into the biological significance of the properties of matter. MacMillan, New York, p 317

Herbert D, Elsworth R, Telling RC (1956) The continuous culture of bacteria; a theoretical and experimental study. J Gen Microbiol 14:601–622

Hobbie JE, Daley RJ, Jasper S (1977) Use of nucleopore filters for counting bacteria by fluorescence microscopy. Appl Environ Microbiol 33:1225–1228

Howarth RW, Cole JJ (1985) Molybdenum availability, nitrogen limitation and phytoplankton growth in natural waters. Science 229:653–655

Hutchinson GE (1961) The paradox of the plankton. Am Nat 95:137–145

Imachi H, Nobu MK, Nakahara N, Morono Y, Ogawara M, Takaki Y, Takano Y, Uematsu K, Ikuta T, Ito M, Matsui Y, Miyazaki M, Murata K, Saito Y, Sakai S, Song C, Tasumi E, Yamanaka Y, Yamaguchi T, Kamagata Y, Tamaki H, Takai K (2020) Isolation of an archaeon at the prokaryote-eukaryote interface. Nature 577:519–525

Jannasch HW, Jones A (1959) Bacterial populations in sea water as determined by different methods of enumeration. Limnol Oceanogr 4:128–139

Jetten MSM, Sliekers O, Kuypers M, Dalsgaard T, van Niftrik L, Cirpus I, van de Pas-Schoonen K, Lavik G, Thamdrup B, Le Paslier D, Op den Camp HJM, Hulth S, Nielsen LP, Abma W, Third K, Engström J, Kuenen JG, Jørgensen BB, Canfield DE, Sinninghe Damsté JS, Revsbech NP, Fuerst J, Weissenbach J, Wagner M, Schmidt I, Schmid M, Strous M (2003) Anaerobic ammonium oxidation by marine and freshwater planctomycete-like bacteria. Appl Microbiol Biotechnol 63:107–114

Kannan N, Taylor SS, Zhai Y, Venter JC, Manning G (2007) Structural and functional diversity of the microbial kinome. PLOS Biol 5(3):e17

Karl DM (2001) A sea of microbes. Trends Microbiol 9:44–45

Karl DM, Proctor LM (2007) Foundations of microbial oceanography. Oceanography 20:16–27

Kirchman DL (2000) Microbial ecology of the oceans. Wiley-Liss, Hoboken, NJ, p 542

Kirchman DL (2008) Microbial ecology of the oceans. Wiley-Liss, Hoboken, NJ, p 593

Kirsch F, Klähn S, Hagemann M (2019) Salt-regulated accumulation of the compatible solutes sucrose and glucosylglycerol in cyanobacteria and its biotechnological potential. Front Microbiol 10:2139

Knauth LP (2005) Temperature and salinity history of the Precambrian Ocean: implications for the course of microbial evolution. Palaeogr Palaeoclimat Palaeoecol 219:53–69

Koch R (1884) Die Aetiologie der Tuberkulose. Mitt Kaiserl Ges 2:1–88

Koeve W, Kähler P (2010) Balancing Ocean nitrogen. Nat Geosci 3:383–384

Kolber ZS, Plumley FG, Lang AS, Beatty JT, Blankenship RE, VanDover CL, Vetriani C, Koblizek M, Rathgeber C, Falkowski PG (2001) Contribution of aerobic photoheterotrophic bacteria to the carbon cycle in the ocean. Science 292:2492–2495

Könneke M, Bernhard AE, de la Torre JR, Walker CB, Waterbury JB, Stahl DA (2005) Isolation of an autotrophic ammonia-oxidizing marine archaeon. Nature 437:543–546
Kopf A et al (2015) The ocean sampling day consortium. GigaScience 4:27
Kumar S, Paul D, Bhushan B, Wakchaure GC, Meena KK, Shouche Y (2020) Traversing the "Omic" landscape of microbial halotolerance for key molecular processes and new insights. Crit Rev Microbiol 46:631–653
Kuypers MMM, Sliekers AO, Lavik G, Schmid M, Jørgensen BB, Kuenen JG, Sinninghe Damste JS, Strous M, Jetten MSM (2003) Anaerobic ammonium oxidation by anammox bacteria in the Black Sea. Nature 422:608–611
Langlois RJ, Mills MM, Ridame C, Croot P, LaRoche J (2012) Diazotrophic bacteria respond to Saharan dust additions. Mar Ecol Progr Ser 470:1–14
Langlois RJ, Großkopf T, Mills M, Takeda S, LaRoche J (2015) Widespread distribution and expression of gamma a (UMB), an uncultured, diazotrophic, γ-proteobacterial *nifH* phylotype. PLoS One 10(6):e0128912
Lee SH, Malone C, Kemp PF (1993) Use of multiple 16S rRNA-targeted fluorescent probes to increase signal strength and measure cellular RNA from natural planktonic bacteria. Mar Ecol Progr Ser 101:193–201
Logares R, Bråte J, Bertilsson S, Clasen JL, Shalchian-Tabrizi K, Rengefors K (2009) Infrequent marine–freshwater transitions in the microbial world. Trends Microbiol 17:414–422
Logares R, Deutschmann IM, Junger PC, Giner CR, Krabberød AK, Schmidt TSB, Rubinat-Ripoll L, Mestre M, Salazar G, Ruiz-González C, Sebastián M, de Vargas C, Acinas SG, Duarte CM, Gasol JM, Massana R (2020) Disentangling the mechanisms shaping the surface ocean microbiota. Microbiome 8:55
MacLeod RA (1965) The question of the existence of specific marine bacteria. Bact Rev 29:9–23
MacLeod RA (1985) Marine microbiology far from the sea. Ann Rev Microbiol 39:1–20
MacLeod RA (1986) Salt requirements for membrane transport and solute retention in some moderate halophiles. FEMS Microbiol Rev 39:109–113
Margulis L (1992) Biodiversity: molecular biological domains, symbiosis and kingdom origins. Biosystems 27:39–51
Martiny AC, Pham CTA, Primeau FW, Vrugt JA, Moore JK, Levin SA, Lomas MW (2013) Strong latitudinal patterns in the elemental ratios of marine plankton and organic matter. Nat Geosci 6: 279–283
Massana R, Guillou L, Díez B, Pedrós-Alió C (2002) Unveiling the organisms behind novel eukaryotic ribosomal DNA sequences from the ocean. Appl Env Microbiol 68:4554–4558
Mayol E, Arrieta JM, Jiménez MA, Martínez-Asensio A, Garcias-Bonet N, Dachs J, González-Gaya B, Royer S-J, Benítez-Barrios VM, Fraile-Nuez E, Duarte CM (2017) Long-range transport of airborne microbes over the global tropical and subtropical ocean. Nature Comm 8:201
Merlino G, Barozzi A, Michoud G, Ngugi DK, Daffonchio D (2018) Microbial ecology of deep-sea hypersaline anoxic basins. FEMS Microbiol Ecol 94:fiy085
Middelburg JJ, Levin LA (2009) Coastal hypoxia and sediment biogeochemistry. Biohgeosciences 6:1273–1293
Millero FJ, Feistel R, Wright DG, McDougall TJ (2008) The composition of standard seawater and the definition of the reference-composition salinity scale. Deep-Sea Res I 55:50–72
Mills MM, Arrigo KR (2010) Magnitude of oceanic nitrogen fixation influenced by the nutrient uptake ratio of phytoplankton. Nat Geosci 3:412–416
Moisander PH, Beinart RA, Hewson I, White AE, Johnson KS, Carlson DJ, Montoya JP, Zehr JP (2010) Unicellular cyanobacterial distributions broaden the oceanic $N_2$ fixation domain. Science 327:1512–1514
Moisander PH, Serros TRC, Paerl RW, Beinart RA, Zehr JP (2014) Gammaproteobacterial diazotrophs and nifH gene expression in surface waters of the South Pacific Ocean. ISME J 8: 1962–1973
Monod J (1949) The growth of bacterial cultures. Annu Rev Microbiol 3:371–394

Moran MA (2015) The global ocean microbiome. Science 350:aac8455
Morris RM, Rappé MS, Connon SA, Vergin KL, Siebold WA, Carlson CA, Giovannoni SJ (2002) SAR11 clade dominates ocean surface bacterioplankton communities. Nature 420:806–810
Morris BEL, Henneberger R, Huber H, Moissl-Eichinger C (2013) Microbial syntrophy: interaction for the common good. FEMS Microbiol Rev 37:384–406
Nissenbaum A (1975) The microbiology and biogeochemistry of the Dead Sea. Microb Ecol 2:139–161
Oren A (2009) Saltern evaporation ponds as model systems for the study of primary production processes under hypersaline conditions. Aq Microb Ecol 56:193–204
Ottensen EA, Young CR, Eppley JM, Ryan JP, Chavez FP, Scholin CA, DeLong EF (2013) Pattern and synchrony of gene expression among sympatric marine microbial populations. Proc Natl Acad Sci 110:E488–E497
Overmann J, van Gemerden H (2000) Microbial interactions involving sulfur bacteria: implications for the ecology and evolution of bacterial communities. FEMS Microbiol Rev 24:591–599
Pace NR (2006) Time for a change. Nature 441:289–289
Pace NR (2009) It's time to retire the prokaryote. Microbiol Today 5:85–87
Palenik B, Brahamsha B, Larimer FW, Land M, Hauser L, Chain P, Lamerdin J, Regala W, Allen EE, McCarren J, Paulsen I, Dufresne A, Partensky F, Webb EA, Waterbury J (2003) The genome of a motile marine *Synechococcus*. Nature 424:1037–1042
Paul J, Scholin C, van den Engh G, Perry MJ (2007) In situ instrumentation. Oceanography 20:70–78
Paulmier A, Ruiz-Pino D (2009) Oxygen minimum zones (OMZs) in the modern ocean. Progr Oceanogr 80:113–128
Pedrós-Alió C (2006) Marine microbial diversity: can it be determined? Trends Microbiol 14:257–263
Planavsky NJ (2014) The elements of marine life. Nat Geosci 7:855–856
Platt T, Subba Rao DV, Irwin B (1983) Photosynthesis of picoplankton in the oligotrophic ocean. Nature 301:702–704
Pomeroy LR (1974) The ocean's food web, a changing paradigm. Bioscience 24:499–504
Proctor LM, Karl DM (2007) A sea of microbes. Introduction Oceanography 20:14–15
Rappé MS, Connon SA, Vergin KL, Giovannoni SJ (2002) Cultivation of the ubiquitous SAR11 marine bacterioplankton clade. Nature 418:630–633
Rathgeber C, Beatty JT, Yurkov V (2004) Aerobic phototrophic bacteria: new evidence for the diversity, ecological importance and applied potential of this previously overlooked group. Photosynthesis Res 81:113–128
Redfield AC (1934) On the proportions of organic derivatives in sea water and their relation to the composition of plankton. In: James Johnstone memorial volume. University Press of Liverpool, Liverpool, pp 176–192
Redfield AC (1958) The biological control of chemical factors in the environment. Am Sci 46:205–221
Rijkenberg MJA, Powell CF, Dall'Osto D, Nielsdottir MC, Patey MD, Hill PG, Baker AR, Jickells TD, Harrison RM, Achterberg EP (2008) Changes in iron speciation following a Saharan dust event in the tropical North Atlantic Ocean. Mar Chem 110:56–67
Riley MA, Lizotte-Waniewski M (2009) Population genomics and the bacterial species concept. Meth Mol Biol 532:367–377
Rosselló-Mora R, Amann R (2001) The species concept for prokaryotes. FEMS Microbiol Rev 25: 39–67
Rusch DB, Halpern AL, Sutton G, Heidelberg KB, Williamson S, Yooseph S, Wu D, Eisen JA, Hoffman JM, Remington K, Beeson K, Tran B, Smith H, Baden-Tillson H, Stewart C, Thorpe J, Freeman J, Andrews-Pfannkoch C, Venter JE, Li K, Kravitz S, Heidelberg JF, Utterback T, Rogers Y-H, Falcón LI, Souza V, Bonilla-Rosso G, Eguiarte LE, Karl DM, Sathyendranath S, Platt T, Bermingham E, Gallardo V, Tamayo-Castillo G, Ferrari MR, Strausberg RL,

Nealson K, Friedman R, Frazier M, Venter JC (2007) The *sorcerer II* global ocean sampling expedition: Northwest Atlantic through eastern tropical Pacific. PLOS Biol 5(3):e77
Schmidt TM, DeLong EF, Pace NR (1991) Analysis of a marine picoplankton community by 16S rRNA gene cloning and sequencing. J Bact 173:4371–4378
Schreiber UC, Mayer C (2020) The first cell. Springer, Cham, Switzerland, p 178
Schulz HN, Jørgensen BB (2001) Big bacteria. Ann Rev Microbiol 55:105–137
Schunck H, Lavik G, Desai DK, Großkopf T, Kalvelage T, Löscher CR, Paulmier A, Contreras S, Siegel H, Holtappels M, Rosenstiel P, Schilhabel MB, Graco M, Schmitz RA, Kuypers MMM, LaRoche J (2013) Giant hydrogen sulfide plume in the oxygen minimum zone off Peru supports chemolithoautotrophy. PLoS One 8(8):e68661
Schut F, Jansen M, Gomes TMP, Gottschal JC, Harder W, Prins RA (1995) Substrate uptake and utilization by a marine ultramicrobacterium. Microbiology 141:351–361
Severin I, Stal LJ (2008) Light dependency of nitrogen fixation in a coastal cyanobacterial mat. ISME J 2:1077–1088
Shanks AL, Reeder ML (1993) Reducing microzones and sulfide production in marine snow. Mar Ecol Progr Ser 96:43–47
Sherr EB, Sherr BF (1996) Temporal offset in oceanic production and respiration processes implied by seasonal changes in atmospheric oxygen: the role of heterotrophic microbes. Aq Microb Ecol 11:91–100
Smith RC, Baker KS (1982) Oceanic chlorophyll concentrations as determined by satellite (Nimbus-7 coastal zone color scanner). Mar Biol 66:269–279
Smriga S, Fernandez VI, Mitchell JG, Stocker R (2016) Chemotaxis toward phytoplankton drives organic matter partitioning among marine bacteria. Proc Natl Acad Sci U S A 113:1576–1581
Sogin ML, Morrison HG, Huber JA, Welch DM, Huse SM, Neal PR, Arrieta JM, Herndl GJ (2006) Microbial diversity in the deep sea and the underexplored "rare biosphere". Proc Natl Acad Sci U S A 193:12115–12120
Staal M, Meysman FJR, Stal LJ (2003) Temperature excludes $N_2$-fixing heterocystous cyanobacteria in the tropical oceans. Nature 425:504–507
Staal M, te Lintel HS, Brummer GJ, Veldhuis M, Sikkens C, Persijn S, Stal LJ (2007) Nitrogen fixation along a north-south transect in the eastern Atlantic Ocean. Limnol Oceanogr 52:1305–1316
Stal LJ (2009) Is the distribution of nitrogen-fixing cyanobacteria in the oceans related to temperature? Env Microbiol 11:1632–1645
Stal LJ, Cretoiu MS (eds) (2016) The marine microbiome. An untapped source of biodiversity and biotechnological potential. Springer, Switzerland, p 498
Stal LJ, Walsby AE (2000) Photosynthesis and nitrogen fixation in a cyanobacterial bloom in the Baltic Sea. Eur J Phycol 35:97–108
Stal LJ, Staal M, Villbrandt M (1999) Nutrient control of cyanobacterial blooms in the Baltic Sea. Aq Microb Ecol 18:165–173
Staley JT, Konopka A (1985) Measurements of *in situ* activities of nonphotosynthetic microorganisms in aquatic and terrestrial habitats. Ann Rev Microbiol 39:321–346
Stanier RY (1941) Studies on marine agar-digesting bacteria. J Bact 42:527–559
Stanier RY, van Niel CB (1962) The concept of a bacterium. Arch Mikrobiol 42:17–35
Sunagawa S, Coelho LP, Chaffron S, Kultima JR, Labadie K, Salazar G, Djahanschiri B, Zeller G, Mende DR, Alberti A, Cornejo-Castillo FM, Costea PI, Cruaud C, d'Ovidio F, Engelen S, Ferrera I, Gasol JM, Guidi L, Hildebrand F, Kokoszka F, Lepoivre C, Lima-Mendez G, Poulain J, Poulos BT, Royo-Llonch M, Sarmento H, Vieira-Silva S, Dimier C, Picheral M, Searson S, Kandels-Lewis S, Bowler C, de Vargas C, Gorsky G, Grimsley N, Hingamp P, Iudicone D, Jaillon O, Not F, Ogata H, Pesant S, Speich S, Stemmann L, Sullivan MB, Weissenbach J, Wincker P, Karsenti E, Raes J, Acinas SG, Bork P (2015) Structure and function of the global ocean microbiome. Science 348(6237):1261359
Sunagawa S, Acinas SG, Bork P, Bowler C, Coordinators TO, Eveillard D, Gorsky G, Guidi L, Iudicone D, Karsenti E, Lombard F, Ogata H, Pesant S, Sullivan MB, Wincker P, de Vargas C

(2020) Tara oceans: towards global ocean ecosystems biology. Nature Rev Microbiol 18:428–445
Suttle CA (2005) Viruses in the sea. Nature 437:356–361
Thamdrup B, Dalsgaard T (2002) Production of $N_2$ through anaerobic ammonium oxidation coupled to nitrate reduction in marine sediments. Appl Env Microbiol 68:1312–1318
They NH, Amado AM, Cotner JB (2017) Redfield ratios in inland waters: higher biological control of C:N:P ratios in tropical semi-arid high water residence time lakes. Front Microbiol 8:1505
Thingstad TF, Lignell R (1997) Theoretical models for the control of bacterial growth rate, abundance, diversity and carbon demand. Aq Microb Ecol 13:19–27
Thompson AW, Foster RA, Krupke A, Carter BJ, Musat N, Vaulot D, Kuypers MMM, Zehr JP (2012) Unicellular cyanobacterium symbiotic with a single-celled eukaryotic alga. Science 337: 1546–1550
Tyrrell T (1999) The relative influences of nitrogen and phosphorus on oceanic primary production. Nature 400:525–531
Venter JC, Remington K, Heidelberg JF, Halpern AL, Rusch D, Eisen JA, Wu D, Paulsen I, Nelson KE, Nelson W, Fouts DE, Levy S, Knap AH, Lomas MW, Nealson K, White O, Peterson W, Hoffman J, Parsons R, Baden-Tillson H, Pfannkoch C, Rogers Y-H, Smith HO (2004) Environmental genome shotgun sequencing of the Sargasso Sea. Science 304:66–74
Wallmann K, Suess E, Westbrook GH, Winckler G, Cita MB (1997) Salt brines on the Mediterranean Sea floor. Nature 387:31–32
Watanabe K, Kasai A, Antonio ES, Suzuki K, Ueno M, Yamashita Y (2014) Influence of salt-wedge intrusion on ecological processes at lower trophic levels in the Yura estuary, Japan. Est Coast Shelf Sci 139:67–77
Waterbury JB, Watson SW, Guillard RRL, Brand LE (1979) Widespread occurrence of a unicellular marine planktonic cyanobacterium. Nature 277:293–293
Waterbury JB, Willey JM, Franks DG, Valois FW, Watson SW (1985) A cyanobacterium capable of swimming motility. Science 230:74–76
Weiss MC, Sousa FL, Mrnjavac N, Neukirchen S, Roettger M, Nelson-Sathi S, Martin WF (2016) The physiology and habitat of the last universal common ancestor. Nat Microbiol 1:116
Whitman WB, Coleman DC, Wiebe WJ (1998) Prokaryotes: the unseen majority. Proc Nat Acad Sci 95:6578–6583
Wilkins JS (2006) The concept and causes of microbial species. Hist Philos Life Sci 28:329–348
Wuchter C, Abbas B, Coolen MJL, Herfort L, van Bleijswijk J, Timmers P, Strous M, Teira E, Herndl GJ, Middelburg JJ, Schouten S, Sinninghe Damsté JS (2006) Archaeal nitrification in the ocean. Proc Natl Acad Sci U S A 33:12317–12322
Wurl O, Bird K, Cunliffe M, Landing WM, Miller U, Mustaffa NIH, Ribas-Ribas M, Witte C, Zappa CJ (2018) Warming and inhibition of salinization at the ocean's surface by cyanobacteria. Geophys Res Lett 45:4230–4237
Yooseph S, Sutton G, Rusch DB, Halpern AL, Williamson SJ, Remington K, Eisen JA, Heidelberg KB, Manning G, Li W, Jaroszewski L, Cieplak P, Miller CS, Li H, Mashiyama ST, Joachimiak MP, van Belle C, Chandonia J-M, Soergel DA, Zhai Y, Natarajan K, Lee S, Raphael BJ, Bafna V, Friedman R, Brenner SE, Godzik A, Eisenberg D, Dixon JE, Taylor SS, Strausberg RL, Frazier M, Venter JC (2007) The *sorcerer II* global ocean sampling expedition: expanding the universe of protein families. PLOS Biol 5(3):e16
Zehr JP, Capone DG (2021) Marine nitrogen fixation. Springer, Cham, Switzerland, p 186
Zehr JP, Waterbury JB, Turner PJ, Montoya JP, Omoregie E, Steward GF, Hansen A, Karl DM (2001) Unicellular cyanobacteria fix $N_2$ in the subtropical North Pacific Ocean. Nature 412:635–638
Zehr JP, Bench SR, Carter BJ, Hewson I, Niazi F, Shi T, Tripp HJ, Affourtit JP (2008) Globally distributed uncultivated oceanic $N_2$-fixing cyanobacteria lack oxygenic photosystem II. Science 322:1110–1112

Zehr JP, Shilova IN, Farnelid HM, del Carmen M-MM, Turk-Kubo K (2016) Unusual marine unicellular symbiosis with the nitrogen-fixing cyanobacterium UCYN-A. Nature Microbiol 2: 16214

Zehr JP, Weitz JS, Joint I (2017) How microbes survive in the open ocean. Science 357:646–647

Zengler K, Toledo G, Rappé M, Elkins J, Mathur EJ, Short JM, Keller M (2002) Cultivating the uncultured. Proc Natl Acad Sci U S A 99:15681–15686

Zhang H, Ning K (2015) The Tara oceans project: new opportunities and greater challenges ahead. Genom Proteom Bioinform 13:275–277

Zhang Y, Kieft B, Hobson BW, Ryan JP, Barone B, Preston CM, Roman B, Raanan B-Y, Marin R III, O'Reilly TC, Rueda CA, Pargett D, Yamahara KM, Poulos S, Romano AE, Foreman G, Ramm H, Wilson ST, DeLong EF, Karl DM, Birch JM, Bellingham JG, Scholin CA (2020) Autonomous tracking and sampling of the deep chlorophyll maximum layer in an open-ocean eddy by a long-range autonomous underwater vehicle. IEEE J Ocean Eng 45:1308–1321

ZoBell CE (1946) Marine microbiology, a monograph on hydrobacteriology. Mass, Waltham, p 240

ZoBell CE, Rittenberg SC (1938) The occurrence and characteristics of chitinoclastic bacteria in the sea. J Bact 35:275–287

# Part I

# Diversity and Evolution of Marine Microorganisms

# 2 Survival in a Sea of Gradients: Bacterial and Archaeal Foraging in a Heterogeneous Ocean

Estelle E. Clerc, Jean-Baptiste Raina, François J. Peaudecerf, Justin R. Seymour, and Roman Stocker

**Abstract**

Marine microbial ecology is usually investigated over large spatial and temporal scales, under the assumption that planktonic organisms and solutes are homogeneously mixed. However, it is becoming increasingly apparent that the seascape experienced by marine microorganisms is in fact rather heterogeneous and punctuated by chemical hotspots derived from planktonic organisms as well as sinking and suspended organic particles. Motile bacteria and archaea can exploit these hotspots to enhance their growth. Within these chemically rich microscale environments shaped by diffusion and flow, individual cells can also interact with other microorganisms, inducing ripple effects that have consequences across the entire marine food web. Here we describe the physical and biological processes that structure the ocean at the scale of marine microbes, the adaptations enabling them to navigate this patchy seascape, and the way these microscale behaviors can scale-up to influence large-scale biogeochemical processes.

**Keywords**

Chemotaxis · Microscale diffusion · Motility · Particles · Symbiosis · Turbulence

E. E. Clerc · F. J. Peaudecerf · R. Stocker (✉)
Institute of Environmental Engineering, Department of Civil, Environmental and Geomatic Engineering, ETH, Zurich, Switzerland
e-mail: clerc@ifu.baug.ethz.ch; peaudecerf@ifu.baug.ethz.ch

J.-B. Raina · J. R. Seymour
Climate Change Cluster, University of Technology Sydney, Sydney, Australia
e-mail: Jean-Baptiste.Raina@uts.edu.au; Justin.Seymour@uts.edu.au



L. J. Stal, M. S. Cretoiu (eds.), *The Marine Microbiome*, The Microbiomes of Humans, Animals, Plants, and the Environment 3,
https://doi.org/10.1007/978-3-030-90383-1_2

## 2.1 Introduction

The marine environment is one of the largest reservoirs of bacteria and archaea on Earth, with each liter of seawater containing approximately 1 billion cells. Latest estimates suggest that a total of $10^{29}$ bacteria and archaea populates the world's ocean (Kallmeyer et al. 2012), accounting for ~70% of the total marine biomass (Bar-On et al. 2018). This abundance encompasses a wide phylogenetic diversity and a broad range of trophic strategies (Lauro et al. 2009). At one end of the trophic spectrum, oligotrophs are adapted to environments with low levels of nutrients and these microorganisms are characterized by slow growth, low metabolic rates, small cell sizes, streamlined genomes, and a lack of motility (Overmann and Lepleux 2016). Some, such as *Pelagibacter ubique*, numerically dominate open ocean communities (Giovannoni 2017) and need only a few specific nutrients to grow (Carini et al. 2013; Tripp et al. 2008). In the nutrient-poor waters they inhabit, oligotrophs rely on molecular diffusion, which brings enough nutrients to their contact to sustain growth (Zehr et al. 2017). At the other end of the trophic spectrum, copiotrophs thrive in nutrient-rich environments. They grow rapidly, have high metabolic rates, large cell sizes, possess larger genome sizes, and are often motile. In addition, they are well equipped to sense, integrate, and respond to extracellular stimuli (Lauro et al. 2009), allowing them to find and exploit nutrient patches and hotspots. Although copiotrophs represent a small percentage of the free-living microorganisms in the open ocean, they account for most of the organisms on sinking particles (Lambert et al. 2019). While oligotrophy and copiotrophy are often represented as a dichotomy, there is in fact a continuum of trophic strategies between these extremes (Lauro et al. 2009), which enables a wide diversity of microorganisms to exploit hotspots of nutrients in the ocean.

The environment experienced by individual microbial cells in the water column is surprisingly heterogeneous, punctuated by chemical patches and pulses, as well as sinking and suspended organic particles (Azam 1998; Stocker 2012). This microscale heterogeneity influences the behavior, physiology, and trophic interactions of microorganisms and ultimately impacts their contribution to biogeochemical cycles (Azam and Long 2001; Smriga et al. 2016; Stocker et al. 2008). Yet, recognition of this heterogeneity and its importance is only recent. As a result, the ecology of marine bacterial and archaeal communities has only rarely been studied at scales that reflect the microscale environments experienced by these organisms. Indeed, microbial processes within the pelagic ocean are traditionally investigated over large spatial and temporal scales, under the assumption that planktonic organisms and solutes are homogeneously mixed by turbulence. Consequently, patterns in microbial abundance, activity, and diversity have most commonly been examined in the context of mesoscale oceanographic features (e.g., currents, eddies, and gyres) or large-scale gradients (e.g., temperature and salinity) using bulk sampling techniques such as Niskin bottles that collect liters of seawater. However, multi-liter seawater samples exceed the scales of important microscale features and microbial interactions by over one million-fold in volume. To draw a parallel, this would be equivalent to studying the foraging behavior of coral reef fishes by sampling them with a device 100 times larger than an oil supertanker.

Evidence has confirmed that many marine bacteria and archaea are well equipped to navigate and exploit the heterogeneous seascape they inhabit (Blackburn et al. 1998; Brumley et al. 2019; Son et al. 2016; Stocker et al. 2008), emphasizing the importance of studying these organisms at the appropriate scale and the need to integrate the effects of microscale gradients into studies of marine microbial ecology. The few studies that have examined the distribution and diversity of marine microbes at sub-centimeter scales have revealed that bacterial abundances in localized hotspots can be an order of magnitude higher than background (Seymour et al. 2000), that microscale patchiness exists in species richness (Long and Azam 2001), and that marine bacteria and archaea can use motility and chemotaxis to aggregate in microscale nutrient hotspots (Fenchel 2001; Lambert et al. 2017; Mitchell et al. 1996). Although these fine-scale field results are consistent with theoretical predictions (Azam 1998; Kiørboe and Jackson 2001) and laboratory studies (Blackburn et al. 1998; Smriga et al. 2016; Stocker et al. 2008), they are rare, and our perception of the life of a microbe in the ocean is only beginning to emerge. In this chapter, we consider the pelagic ocean from the perspective of a planktonic microorganism, by describing the microscale physics of seawater, the chemical and biological phenomena that define microscale seascapes, and the behavioral and physiological adaptations that permit marine microbes to succeed in this patchy and dynamic world.

## 2.2 The Physics of Marine Microenvironments

While the ocean represents an incredibly complex environment at the microscopic scale, rich with a multitude of nutrient sources and microbial species, our understanding of the physics at play enables us to establish some general principles. In this section, we outline how two key physical drivers, namely molecular diffusion and fluid flow, define marine microenvironments in terms of nutrient concentration dynamics. We then present some general considerations of the challenges and opportunities for bacteria in the resulting resource seascape, with the overall goal of providing an intuition about microbial processes at the microscale.

### 2.2.1 Diffusion and Flow Shape Microscale Nutrient Seascapes

Molecular diffusion is the key physical phenomenon that shapes the chemical seascape at the microscale. In the absence of flow, even an initially localized release of nutrients, for example, from a lysing cell, will result in a slowly extending patch, as thermal agitation at the molecular scale disperses the resource. This patch of nutrient will thus be smoothed out progressively over time until it reaches the level of background concentration, at which point it becomes difficult for copiotrophic bacteria to exploit. Typically, for a molecule with diffusivity $D$, a point source spreads to a distance $L = (6Dt)^{1/2}$ after a time $t$. As this scaling law shows, the rate of expansion ($dL/dt = (3D/2\,t)^{1/2}$) will decrease with time, and is set by the compound's diffusivity $D$. Considering a typical diffusivity of $D = 0.5 \times 10^{-9}$ m$^2$/s for small molecules, a point source will become a patch of 250 μm diameter in 20 s and of 2 mm diameter in 20 min. As the patch expands with time, the gradients of concentration at its edges become smaller, as do the peak and mean concentrations of the patch which scale as $t^{-3/2}$ (Fig. 2.1a, b) (Berg 1993; Blackburn et al. 1997; Jackson 2012). For example, the concentration of a compound originating from a lysing event from a cell of radius $R$ will be diluted 1000-fold relative to the intracellular concentration when the pulse has expanded to a distance $L = 10R$, which occurs over a typical time $t = (10R)^2/6D$, or ~ 20 s for $R = 25$ μm and $D = 0.5 \times 10^{-9}$ m$^2$/s. The time to dilution will thus vary strongly between lysing cells of different sizes (Fig. 2.1c).

The dependence of the timescale for diffusive spreading on the diffusivity of the solute means that different compounds will not diffuse at the same rate: high-molecular-weight compounds diffuse more slowly than their smaller dissolved counterparts, and will generate more persistent gradients (Stocker and Seymour 2012). For instance, the diffusivity of the small dissolved monomer glucose is $D = 0.5 \times 10^{-9}$ m$^2$/s (Ziegler et al. 1987), whereas the large polysaccharide laminarin is $D = 1.5 \times 10^{-10}$ m$^2$/s (Elyakova et al. 1994), demonstrating the variability of molecular diffusivities found in organic compounds. As the content of a cell is a rich cocktail of many substances (Hellebust 1965) that diffuse at different rates, the lifetime and dynamics of a nutrient pulse from a cell lysing event depend on the molecular composition of the cell's cytoplasm. Overall, diffusion will smooth out any localized and transient source of nutrient into the background concentration. As we discuss below, active foraging of copiotrophic bacteria can, among other benefits, provide a way to cope with this need for timely exploitation of resources before they disappear.

Diffusion also governs the nutrient profiles around sources of nutrients with a steady release, such as live phytoplankton cells with a constant leakage of photosynthates (Fig. 2.1d; Bell and Mitchell 1972; Blackburn et al. 1998; Cirri and Pohnert 2019; Seymour et al. 2017; Smriga et al. 2016). In this case, in the absence of flow and without strong uptake by other cells, a steady decreasing nutrient profile is established by diffusion around the algal cell (Fig. 2.1e), with concentration inversely proportional to the distance from the center of the cell (Kiørboe et al. 2002). The microenvironment immediately surrounding phytoplankton cells is known as the phycosphere, it is characterized by a concentration of nutrients higher than the bulk seawater but also by its bacterial accumulation potential (Bell and Mitchell 1972; Cole 1982; Seymour et al. 2017; Stocker 2012) and represents one of the most well-studied nutrient hotspots in the ocean (Azam and Malfatti 2007; Mühlenbruch et al. 2018; Thornton 2014). The extent of the phycosphere increases with the size of the phytoplankton cell and typically extends over a few cell radii. For example, for a phytoplankton of radius 10 μm, the above-background nutrient concentration will typically establish over 20–50 μm (Fig. 2.1f).

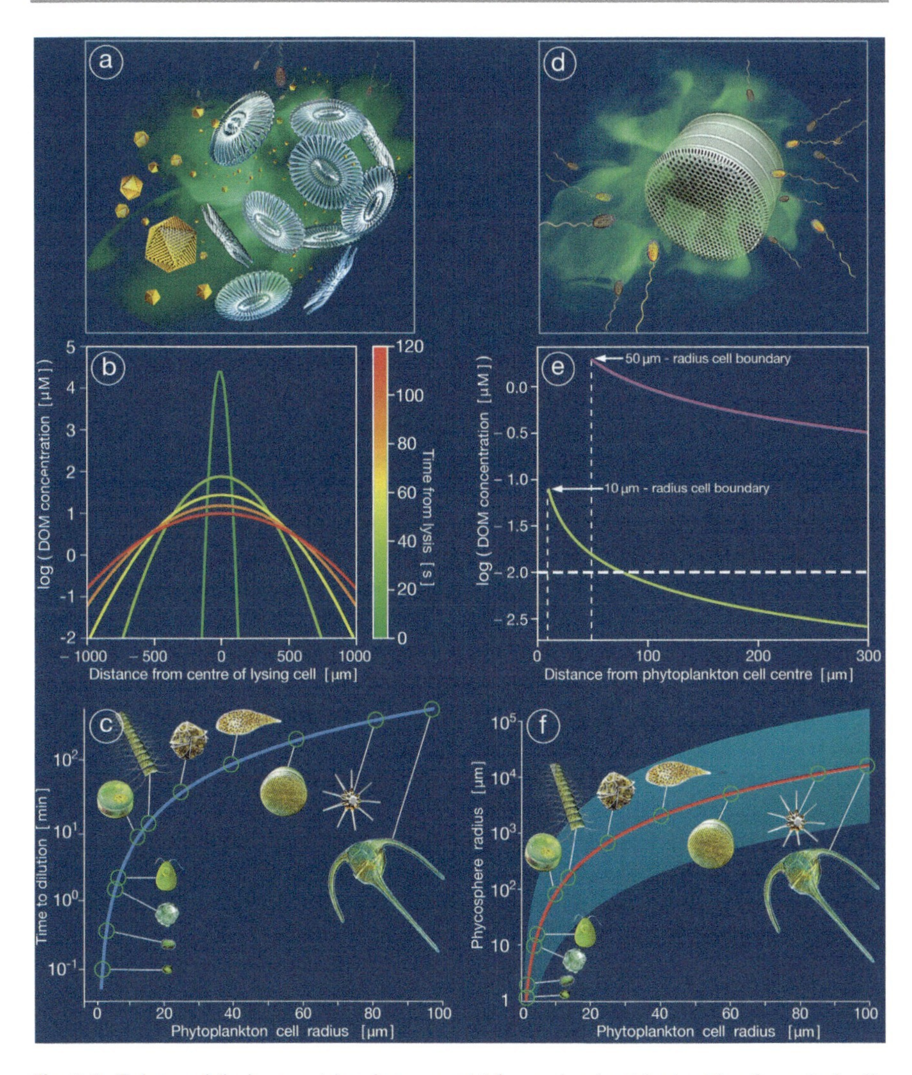


**Fig. 2.1** Ephemeral (lysing events) and permanent (phycosphere) nutrient patches from algal cells. (**a**) An artist's impression of the lysis of a phytoplankton cell, resulting in a strong yet ephemeral pulse of dissolved organic matter (DOM). (**b**) The DOM concentration field as it evolves over time after a lysis event. Concentration was computed using a mathematical model of diffusion from a pulse source, following Seymour et al. (2010a). The horizontal axis shows the distance from the center of a lysing cell of radius 25 μm, which bursts open at time 0, releasing an intracellular concentration of 100 mM of a small-molecule compound (diffusivity $0.5 \times 10^{-9}$ m$^2$/s). (**c**) Scaling of the time to dilution of a lysis patch for different phytoplankton sizes. The intracellular concentration was set to 100 mM with diffusivity $D = 0.5 \times 10^{-9}$ m$^2$/s. For each cell size, the size of the patch grows as $L = (6Dt)^{1/2}$ and is considered diluted when its average concentration has been reduced to 10 times the background concentration of 10 nM. (**d**) An artist's impression of the diffusion boundary layer around an individual phytoplankton cell, which incorporates the phycosphere where concentrations of DOM are enhanced over background level. (**e**) The decay of DOM concentration with distance from the center of a DOM-exuding phytoplankton. Concentrations are shown for phytoplankton of two different radii: 10 μm (bottom light green curve) and 50 μm (top magenta curve). The black dashed line shows a bulk background concentration of 10 nM (typical of many organic solutes in the ocean). The DOM concentration fields were

Steady diffusive profiles can also be found emanating from the surface of large organisms leaking nutrients, such as corals or sponges, and from the sediment–water interface. At these interfaces, the nutrient concentration decays with distance from the surface. Specifically, assuming a constant and uniform release rate (for example, of hydrogen sulfide from the sediment surface), molecular diffusion will spread the nutrient away from the surface according to a linear concentration gradient extending a fraction of a millimeter (0.1–1 mm) into the water (Schulz and Jørgensen 2001).

These characteristics of nutrient sources, both the transient hotspots linked to the sudden release of nutrients and the more stable nutrient profiles around steady sources, were established considering only diffusive transport. The resulting size of these hotspots and the resulting strength of nutrient gradients within them directly influence microbial foraging, for example, by determining growth of microbes able to localize within phycospheres and the gradients that chemotactic microbes can exploit to seek phytoplankton cells. One would intuitively think that fluid flow and turbulence in the ocean significantly modify these nutrient profiles. It turns out that at the microscale, the mixing effect of turbulence remains subordinate to diffusion in governing the concentration of nutrients. If we consider an initial nutrient patch on the scale of millimeters to centimeters (Fig. 2.2), turbulence will stir, stretch, and fold the patch into thin sheets and filaments (Taylor and Stocker 2012). As a consequence, turbulence initially enhances heterogeneity at the microscale. These fine structures become progressively smaller, down to the Batchelor scale (Box 2.1), which typically ranges from 30 to 300 μm in the ocean (Guasto et al. 2012). As even very large sources of solute are ultimately stirred into Batchelor-scale filaments and sheets, the Batchelor scale provides a universal scaling for microbial oceanography (Stocker 2015). For any patch smaller than this scale, turbulence will not fragment the profile formed by diffusion, but will simply stretch it. The importance of this deformation relative to pure diffusion is captured by the turbulent Péclet number (Box 2.2; Guasto et al. 2012).

---

**Fig. 2.1** (continued) obtained by solving the steady diffusion equation for a constant source, following Seymour et al. (2010a). The phytoplankton cell was assumed to have an intracellular concentration of 100 mM, a 1-day typical doubling time, and to exude 100% of its daily production of the solute. This upper limit for the exudation rate is most applicable to stressed or senescent cells. The diffusivity for the solute was $D = 0.5 \times 10^{-9}$ m$^2$/s. (**f**) Phycosphere radius as a function of cell radius. The red line corresponds to the size of the region around a cell where the concentration of a specific compound is >50% above background, shown here for a compound with diffusivity of $D = 0.5 \times 10^{-9}$ m$^2$/s, a leakage fraction of 5%, a phytoplankton growth rate of one per day, and a background concentration of this compound of 10 nM. The light-blue shaded region shows the variation of the phycosphere size when exudation rate is 10 times lower to 10 times higher. The cells presented on panels (**c**) and (**f**) are, from smaller to larger, *Prochlorochoccus*, *Synechococcus*, *Emiliania*, *Chlamydomonas*, *Thalassiosira*, *Chaetoceros*, *Alexandrium*, *Chattonella*, *Coscinodiscus*, *Asterionellopsis*, and *Ceratium*

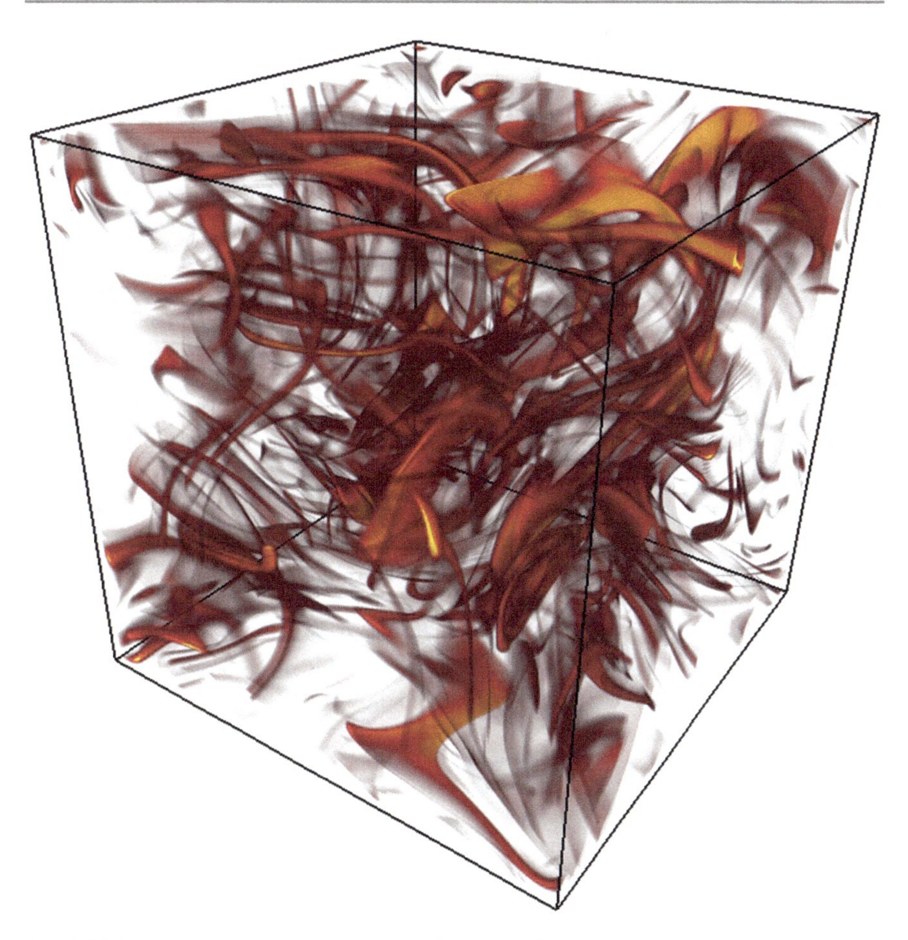

**Fig. 2.2** Turbulence can contribute to patchiness and heterogeneity. The cube represents a numerical simulation of the effect of turbulence on a patch of dissolved organic matter (DOM), with shading indicating the DOM concentration. Turbulence stretches, folds, and stirs the initial DOM patch to create a tangled web of sheets and filaments as small as the Batchelor scale (30–300 μm in the ocean). The characteristic timescale of this process, for a 2.5-mm patch in moderately strong turbulence (turbulent dissipation rate = $10^{-6}$ W/kg), is in the order of 1 min. The computational domain size is 5.65 cm (Taylor and Stocker unpublished)

**Box 2.1 The Batchelor Scale**

The Batchelor scale, $L_B = (\nu D^2/\varepsilon)^{1/4}$, is the lowest scale at which turbulence can generate variance in the distribution of nutrients. Below this size, molecular diffusion dissipates gradients, thereby truly mixing solutes. Here, $\nu = 10^{-6}$ m$^2$/s is the kinematic viscosity of water, $D$ is the solute's diffusivity, and $\varepsilon$ is the turbulent dissipation rate, characterizing the intensity of

(continued)

**Box 2.1** (continued)
turbulence. The Batchelor scale $L_B$ increases with diffusivity, and decreases with increasing dissipation rate. For typical marine conditions, the turbulent dissipation rate $\varepsilon$ varies between $10^{-6}$ and $10^{-10}$ W/kg, which for small molecules ($D \sim 10^{-9}$ m$^2$/s) corresponds to $L_B = 30$ μm to 300 μm (Guasto et al. 2012).

**Box 2.2 The Péclet Number**
The Péclet number Pe is a nondimensional parameter estimating the ratio of the magnitude of transport by both flow and molecular diffusion, characterizing how important each mechanism is at moving nutrients around an object. If Pe < 1, diffusion is the dominant mechanism of transport and flow plays a lesser role in the formation of nutrient concentration profiles. Its general expression $Pe = UR/D$ depends on a typical speed $U$, a typical size $R$, and the diffusivity of the solute $D$. For example, for a small patch of size $R$ in turbulence with dissipation rate $\varepsilon$, the typical speed will be $R(\varepsilon/\nu)^{1/2}$ and thus $Pe = R^2(\varepsilon/\nu)^{1/2}/D$ will quantify how much turbulence deforms this patch from its purely diffusive shape. Strong turbulence ($\varepsilon = 10^{-6}$ W/kg) acting on a patch of small molecules ($D \sim 10^{-9}$ m$^2$/s) of radius $R = 50$ μm thus results in Pe = 2.5, characteristic of a strong deformation of the patch by the turbulent flows. Alternatively, a particle of size $R$ sinking at speed $U$ will have an associated $Pe = UR/D$ that will determine the shape of its plume (Fig. 2.3; Guasto et al. 2012; Kiørboe and Jackson 2001; Stocker et al. 2008).

Turbulence also impacts the nutrient profiles originating from solid surfaces. In this situation, turbulence does not mix freely, but is damped by the presence of the surface. The region close to the surface where turbulence is quenched and diffusion dominates transport is called the "diffusion boundary layer" (DBL) (Schulz and Jørgensen 2001; Thar and Kühl 2002). DBLs can be found around all solid surfaces in the ocean, from corals to sediments, and they also surround marine snow particles and phytoplankton cells. The diffusive transport results in mostly steady gradients of solute, providing a robust cue to the location of the source to chemotactic microorganisms. The stronger the surrounding flow, the thinner the DBL, and hence the greater the solute transport to and from the surface (for example, over corals in flow) (Kühl et al. 1995). For phytoplankton cells, the phycosphere as described above corresponds generally with the DBL (Seymour et al. 2017).

Finally, the shape of the nutrient seascape in the ocean is also strongly determined by the sinking motion of leaking objects, such as marine snow particles and fecal pellets (Kiørboe and Jackson 2001). As they move through the water column while releasing solutes, marine snow particles and fecal pellets generate a quite different

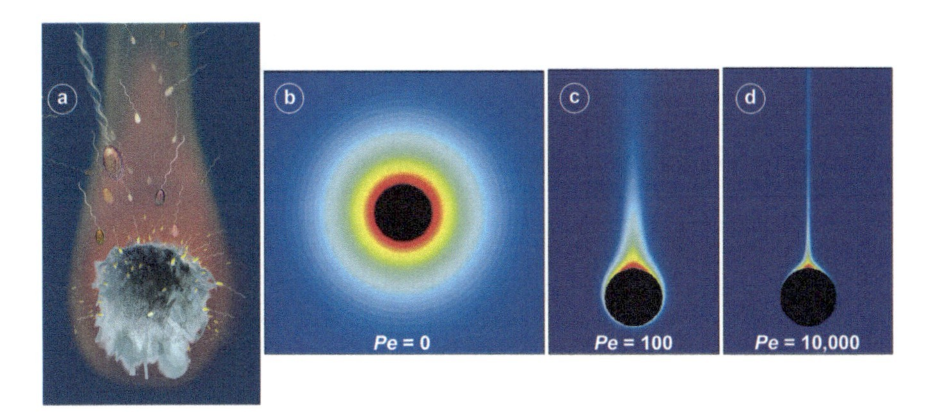


**Fig. 2.3** Marine particles. (**a**) An artist's impression of the plume of dissolved organic matter emanating from a sinking marine snow particle (**b–d**). The shape of the plume for sinking particles of different Péclet numbers. The Péclet number $Pe = UR/D$ for a particle of size $R$ releasing a solute of diffusivity $D$ and sinking at speed $U$ characterizes the importance of flow in shaping the plume with respect to pure diffusion ($Pe = 0$). Shown are plumes for Péclet numbers of (**b**) $Pe = 0$, (**c**) $Pe = 100$, and (**d**) $Pe = 10,000$, corresponding to a static particle, a slow sinking particle, and a fast-sinking particle, respectively. Reproduced from Kiørboe et al. (2001), with permission

nutrient signature from the static diffusive sources described above, as what would be a spherical DBL is deformed into a comet-like plume. The concentration seascape they generate is strongly asymmetric with a thin layer of higher concentration characterized by strong gradients preceding these particles and a long solute tail with concentration higher than background in their wake (Kiørboe and Jackson 2001; Stocker et al. 2008). This asymmetry, and thus the slenderness of the plume, increases with increasing Péclet number, and thus, for example, with increasing sinking speed (Box 2.2 and Fig. 2.3). This plume is itself subject to diffusion and turbulence and ultimately becomes diluted in the background (Kiørboe and Jackson 2001; Visser and Jackson 2004).

## 2.2.2 A Bacterial View of the Microscale Ocean

Physical phenomena define the chemical seascape at the microscale, resulting in a heterogeneous mosaic of transient hotspots amidst otherwise nutrient-poor waters. To understand the value of microbial behaviors, such as motility and chemotaxis, to navigate this seascape of resources, it is useful to picture the typical distances between cells in the ocean, as these distances have direct implications for the rates at which bacteria might expect to encounter, for example, a phytoplankton cell. If we take a typical concentration of bacteria of $10^6$ cells/mL—a relatively conserved value across the world's ocean—and distribute these cells uniformly in space, then the distance between a bacterium and its nearest neighbor would be 100 μm, i.e., ~100 body lengths. This separation does not change much with small changes in cell concentration, as it varies with the cubic root of the cell density (for example, $10^5$ cells/mL corresponds to a nearest neighbor distance of roughly 200 μm). Regarding the typical distance of these bacteria from potential sources of nutrients at the microscale, consider again a bacteria concentration of $10^6$ cells/mL together with a phytoplankton population at a typical cell density of $10^3$ cells/mL, both uniformly distributed. Then, for each individual bacterium, the nearest phytoplankton cell is at a distance of the order of 1 mm. Once again, this distance hardly varies with small variations in cell density. Overall, these considerations paint a picture of the ocean as a dilute suspension of microorganisms, with bacteria separated by many cell diameters from other bacteria and potential nutrient sources such as phytoplankton cells.

How can bacteria then find their way to nutrient hotspots? In the absence of motility, bacterial cells are subjected to Brownian motion, the small fluctuations in position driven by random collisions with water molecules. Brownian motion of a bacterium can be quantified by the diffusivity $D_B = kT/(6\pi\mu R)$ which is inversely proportional to bacterial radius $R$ and proportional to temperature $T$, with other parameters $k$ Boltzmann's constant and $\mu$ the dynamic viscosity of seawater. For typical seawater conditions at 10 °C, $D_B$ is of the order of $3.5 \times 10^{-13}$ m$^2$/s for a bacterium of radius 0.4 μm. As a result of the random path they follow by Brownian motion, nonmotile bacteria will cover a typical distance $L = (6D_B t)^{1/2}$ in a time $t$. These distances are small: for example, $L \sim 35$ μm in 10 minutes and $L \sim 450$ μm in one day. These values suggest that over the timescale of a day, a bacterium might encounter another bacterium. However, if we ask how long it would take to cover the typical separation with a phytoplankton cell $L \sim 1$ mm, the estimated time rises to $t \sim 6$ days (Smriga et al. 2016). This timescale is not only large compared to the doubling time of the bacterium, which could thus begin to starve during its random search, but it is also large compared to the lifetime of a transient hotspot of nutrients. For example, the mean concentration of an algal lysis patch with initial size $R = 25$ μm for a small nutrient with $D = 0.5 \times 10^{-9}$ m$^2$/s decreases from an intracellular concentration of 100 mM (ten million times the background concentration of 10 nM) to 100 nM (just 10 times the background concentration) after around 30 min. Therefore, in most cases, random motion by Brownian diffusion will not increase the chances of nonmotile bacteria encountering transient nutrient sources beyond the rare case of a hotspot arising near them by chance (Fig. 2.4a) (Smriga et al. 2016).

We can actually estimate the encounter rate between microbes more precisely, based on their respective sizes and diffusivities. If we consider one bacterial cell (having radius $r$ and Brownian diffusivity $D_B$) and ask how often it is expected to encounter an algal cell (of radius $R$, Brownian diffusivity $D_{B,a}$, and cell concentration $C_2$), the average encounter rate will be given by $4\pi(D_B + D_{B,a})(r + R)C_2$, in cells encountered per day. If we consider nonmotile bacteria of radius $r = 0.4$ μm, and algal cells of radius $R = 25$ μm and cell concentration $C_2 = 10^3$ cells/mL, on average one bacterium will encounter 0.01 algal cells over the course of a day, making the probability of arriving at the time of a lysis event very small.

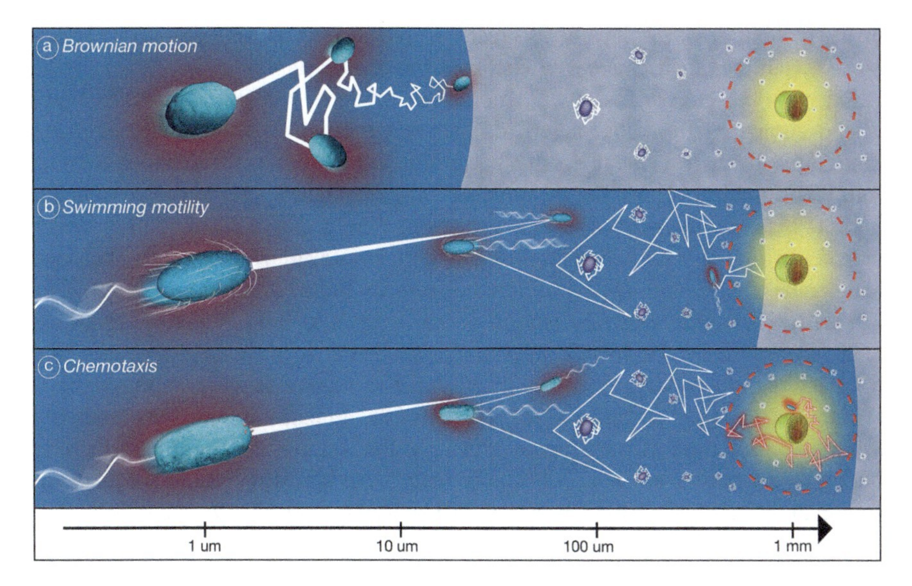


**Fig. 2.4** Different motility strategies result in different probabilities of encounter with other cells and nutrient hotspots. (**a**) Nonmotile bacteria diffuse randomly driven by Brownian motion, with small domains explored in an example timescale of 10 min, giving a low probability of encountering other bacterial cells, and an even lower probability of entering a phycosphere (dashed line around phytoplankton cell). (**b**) Motile bacteria swim with a random pattern alternating straight runs and random reorientation. This more rapid random walk allows them to explore much larger domains, with the potential for many encounters with other bacteria and the potential "lucky" encounter with a phycosphere. (**c**) Chemotactic bacteria swim with the same random motion in the absence of a nutrient gradient. However, as soon as they detect a patch of higher nutrient concentration (e.g., entering the phycosphere marked by a dashed line), their random walk becomes biased toward the source, thus enabling them to rapidly navigate to the center of a nutrient patch and retain position there (red path)

Given the typical length scales and timescales that characterize the heterogeneous seascape of nutrients at the microscale, random Brownian motion is thus not an effective strategy to exploit transient nutrient hotspots. In contrast, bacterial motility, a feature of most copiotrophic bacteria, represents a game changer. Swimming bacteria possess one or several corkscrew-shaped flagella that they rotate to move through fluids (the characteristics and distribution of motility in marine bacteria are described in Sect. 2.4). The resulting motion achieves typical speeds of 50 μm/s, with some species measured at speeds as fast as several hundred micrometers per second (e.g., large sulfur bacteria living above sediment such as *Thiovolum majus* (Fenchel 1994)), which represents several hundreds of body lengths per second. This fast-swimming motion does not follow a straight line: similarly to the run-and-tumble motion of the enteric bacterium *Escherichia coli*, marine bacteria often alternate straight "runs" with random reorientations (described in Sect. 2.4). This motility pattern, like Brownian motion, results in a random walk in space, but the greater magnitude of displacements greatly increases the volume explored by bacterial cells and thus their chances of encountering resources.

Indeed, a bacterial diffusivity $D_b$ (not to be confused with Brownian diffusivity) can be computed from the pattern of bacterial motion. When tracking in three dimensions the displacement $r(t)$ of a bacterium with time (i.e., the distance covered from its original position), the mean square displacement $<r(t)^2>$ (where angular brackets denote a mean over several choices of time origin) for a randomly swimming cell will evolve linearly with time after a timescale $t$ of a few seconds, the slope being equal to $6D_b$, with $D_b$ the bacterial diffusivity linked to random motility. This diffusivity is ranging typically from $5 \times 10^{-10}$ m$^2$/s to $8 \times 10^{-9}$ m$^2$/s (Kiørboe 2008), so varying at most by one order of magnitude between bacterial species. From this bacterial diffusivity, one can determine the typical lengths explored by a swimming bacterium. During a time $t$, the size of the domain explored by a randomly swimming bacterium is once again given by the scaling $L = (6D_b t)^{1/2}$ (similar scaling as for Brownian motion but now with bacterial diffusivity in the place of Brownian diffusivity). Using a bacterial diffusivity $D_b = 10^{-9}$ m$^2$/s, we thus deduce that a motile bacterium explores a domain of typical size $L \sim 2$ mm in 10 min and size $L \sim 2$ cm in a day! This is almost 50 times larger than the distance that could be reached by purely passive Brownian diffusion, and the volume explored is correspondingly greater by a factor of $50^3$. Moreover, using the diffusive encounter rate formula above replacing Brownian diffusivity $D_\mathrm{B}$ by swimming diffusivity $D_b = 10^{-9}$ m$^2$/s reveals that these motile bacteria would now on average encounter around 25 algal cells (of radius $R = 25$ μm and at a cell concentration of $10^3$ cells/mL) over the course of one day (note that more precise estimates would require a fuller model of encounter than the simplified diffusive process used here as a first approximation). Motility thus greatly increases the chances of bacteria encountering other cells or nutrient hotspots (Fig. 2.4b) (Lambert et al. 2019). The possibility to reach a transient resource such as a lysing algal cell, or a sinking particle, is significantly enhanced by motility, and the rewards associated with these rich nutrient sources could explain the preservation of this behavior in the oceans, where the background concentration of nutrients can be very low (as described below).

It should be noted that this description of random encounters, while providing an idea of the time and length scales at play, represents a simplification with regards to natural systems, in which other parameters such as cell shape and flow could also influence successful encounters. For example, it has been shown that elongated motile bacterial cells can be reoriented by the flow around a sinking particle. This interaction with flow can reorient cells so that their initially random swimming direction ends up facing the particle, favoring their arrival onto the particle and thus increasing the number of successful encounters (Słomka et al. 2020). The characterization of bacterial encounters in the ocean, in general, has many facets that await study, and we foresee many potential developments in this area.

We have up to here considered only random encounters, based on either Brownian motion or random bacterial motility. However, a large number of motile bacteria are also chemotactic, which means that they can sense gradients of certain nutrients and move in the direction of higher concentrations (behavior described in detail in Sect. 2.4). This chemotactic behavior is achieved by incorporating a bias into the random swimming pattern described above. Bacterial cell bodies are generally too small to be able to sense a gradient of nutrients over their cell length, with only few known exceptions for larger bacteria (Thar and Kühl 2003). Therefore, chemotaxis occurs by sensing how the local nutrient concentration varies during a straight run. If the bacterium senses an increasing concentration, the run lasts longer; in contrast, if it senses a decreasing concentration, it tends to "tumble" and change direction earlier. The net result of these biased runs is a general drift of the bacterium toward higher concentrations, such as the center of a nutrient patch or a phycosphere (Fig. 2.4c). Therefore, a chemotactic bacterium does not rely upon randomly moving to the center of a nutrient patch, as above for the random motility case. As bacteria can sense small concentration differences, a bacterium simply needs to randomly encounter the gradients of concentrations at the edge of a nutrient patch and then chemotaxis allows it to quickly swim toward the center of the patch to reach the highest nutrient concentrations. As diffusion disperses nutrients over large distances, this random encounter with the edge of a diffusing patch leading to quick motion to its nutrient-rich center is much more frequent than random swimming leading a cell by chance to the center of a patch. For example, let us consider again phytoplankton cells of radius 25 μm at a cell density of $10^3$ cells/mL surrounded by chemotactic bacteria with random swimming diffusivity $D_b = 10^{-9}$ $m^2$/s in the absence of gradients. If we assume that these bacteria can detect the edge of a phycosphere at a distance ~250 μm—so 10 times cell radius—away from an exudating phytoplankton cell, the average number of phycospheres encountered by random motility will be 10 times more than encounters with the phytoplankton cells themselves. Using the diffusive encounter rate estimate presented above, we can indeed estimate that a chemotactic bacterium will encounter on average 250 phycospheres per day. Each random encounter with the edge of a phycosphere can lead afterward to chemotactic behavior to reach the phytoplankton cell. Moreover, chemotactic bacteria will be better able to retain position in a patch that they have encountered without dispersing away as would bacteria that swim randomly.

These simple estimates suggest that chemotaxis could significantly enhance the access to transient nutrient patches for copiotrophic bacteria before they diffuse away, and thus promote the survival of these strains. Indeed, intense aggregations of bacterial cells at the microscopic scale in seawater have been observed, and it has been proposed that they arise from chemotaxis to microscale pulses of nutrients (Blackburn et al. 1998). Video microscopy observations have revealed a fast and strong chemotactic response of the marine bacterium *Pseudoalteromonas haloplanktis* to nutrient pulses, resulting in up to 87% increase in potential nutrient uptake of the population (Stocker et al. 2008). While earlier theoretical studies of these behaviors predicted only modest gains from chemotaxis (Jackson 1987; Mitchell et al. 1985), they were based, in the absence of specific information for marine bacteria, on the behavioral parameters of *E. coli*. Marine bacteria have since been shown to possess chemotactic abilities that can be much higher than that of

*E. coli* allowing cells to efficiently exploit localized nutrient patches (Brumley et al. 2019; Mitchell et al. 1995a, b, 1996; Stocker et al. 2008).

Imaging of the chemotaxis of bacteria from seawater enrichments toward lysing diatoms, combined with modeling of nutrient uptake, suggests that uptake is heterogeneous, for both chemotactic and nonmotile bacterial populations alike (Smriga et al. 2016). This heterogeneity stems from the short duration of the nutrient pulses, which gives advantage to cells that were already close to the source at the time of lysis. Scaling up of these results to the typical phytoplankton concentrations in the ocean predicts that chemotactic, copiotrophic strains can outcompete nonmotile, oligotrophic strains during phytoplankton blooms and bloom collapse conditions, periods characterized by abundant lysing events (Smriga et al. 2016).

Chemotaxis is also important for the exploitation of particles by bacteria, where motility and chemotaxis can be beneficial in two ways. First, by increasing the encounter rate with particles, as has been demonstrated experimentally using model particles made of agar for which colonization could be quantified (Kiørboe et al. 2002). Motile bacteria have much higher rates of colonization than nonmotile bacteria with chemotaxis further enhancing encounter by a factor of 5–10 with respect to random motility. Similarly, mathematical modeling predicts that chemotaxis enhances encounter rates with particles by two- to fivefold for particles ranging from 200 μm to 1.5 cm diameter (Kiørboe and Jackson 2001). This increased encounter rate could explain how motile, chemotactic species—which represent a minority of bacterial cells found in the water column—are often dominant on marine particles (Fontanez et al. 2015; Ganesh et al. 2014; Guidi et al. 2016; Lambert et al. 2019; Lauro and Bartlett 2008). Second, chemotactic bacteria can exploit the plume of nutrients leaking out of particles as they sink (Fig. 2.3). Assuming optimal chemotactic behavior, this use of marine snow plumes by chemotactic bacteria increases the growth rate of free-living bacteria by twofold for large particles (1.5 cm radius) and by 20-fold for small particles (200 μm radius) (Kiørboe and Jackson 2001). Based on established particle size spectra in the ocean, this suggests that chemotactic bacteria would achieve a growth rate 10 times that of a non-chemotactic motile population. These predictions are supported by observations of strong bacterial accumulation within DOM plumes created in a microfluidic device, resulting in a predicted fourfold nutrient gain of chemotactic bacteria compared to nonmotile ones (Stocker et al. 2008). Together, results from these modeling and experimental studies indicate that motility and chemotaxis can greatly enhance the ability of marine bacteria to use particles and their plumes as resources.

Finally, recent work has started to reveal the impact of motility and chemotaxis on bacteria population diversity (Gude et al. 2020) and fitness (Cremer et al. 2019; Liu et al. 2019). While these works used the model enteric bacterium *E. coli* growing in soft agar plates, their findings have potential implications for the marine environment. Indeed, chemotaxis could provide a fitness advantage by driving population colonization of unexplored substrates ahead of complete nutrient depletion and starvation, as established by observing *E. coli* colony expansion on soft-agar plates (Cremer et al. 2019). This ability to colonize new resource islands could play a role in particles or sediments. Similarly, motility can promote bacterial diversity on a structured patch of resources, allowing coexistence between slow-growing motile strains and fast-growing nonmotile populations (Gude et al. 2020). Indeed, while motility and chemotaxis provide fitness benefits in terms of access to nutrients, these traits are costly behaviors. It has recently been suggested that *E. coli* invests in motility and chemotactic behavior in proportion to the fitness benefit obtained by chemotaxis (Ni et al. 2020), thus demonstrating the delicate optimization of this behavior.

## 2.3 Sources and Nature of Microscale Gradients in the Ocean

In the ocean, the chemical seascape that bacteria experience is highly heterogeneous, characterized by vast expanses of extremely dilute background seawater punctuated by rich hotspots of dissolved and particulate nutrient resources. These chemical microenvironments are likely to be tremendously important for the growth, abundance, and diversity of bacteria (especially copiotrophs), and are derived from a diverse assortment of ecological processes, such as digestion, exudation, lysis, and excretion by the numerous members of the microbial community, but also from the diverse macroorganisms inhabiting the water column. Here, we review the multiple sources of chemical gradients in the ocean that form important nutrient hotspots for heterotrophic bacteria (Fig. 2.5).

### 2.3.1 The Phycosphere

Phytoplankton cells can release up to 50% of the carbon that they fix through photosynthesis into the surrounding seawater (Thornton 2014), either in the form of dissolved organic molecules such as carbohydrates (monosaccharides, oligosaccharides, and polysaccharides), nitrogenous compounds (amino acids, polypeptides, and proteins), fatty acids, and organic acids (glycolate, tricarboxylic acids, hydroxamate, and vitamins), or as matrices made of complex polysaccharides and lipids (Aaronson 1978; Fogg 1977; Fossing et al. 1995; Hellebust 1965; Jenkinson et al. 2015; Jones and Cannon 1986; Lancelot 1984). The exuded DOM is broadly representative of the molecular composition of phytoplankton cells, which contain approximately 25–50% proteins, 5–50% polysaccharides, 5–20% lipids, 3–20% pigments, and 20% nucleic acids (Emerson and Hedges 2008). The monosaccharide composition of the surface ocean is similar to phytoplankton exudates, suggesting that phytoplankton is a significant source of carbohydrates in the ocean (Aluwihare et al. 1997; Biersmith and Benner 1998).

The phycosphere is the microenvironment directly surrounding phytoplankton cells that are characterized by locally elevated concentrations of organic matter arising from the exudation of photosynthates by the phytoplankton (Bell and Mitchell 1972; Cirri and Pohnert 2019; Seymour et al. 2017; Smriga et al. 2016). It is considered as the aquatic analog of the rhizosphere in soil ecosystems (Philippot et al. 2013; Trolldenier 1987) and has direct implications for nutrient fluxes to and

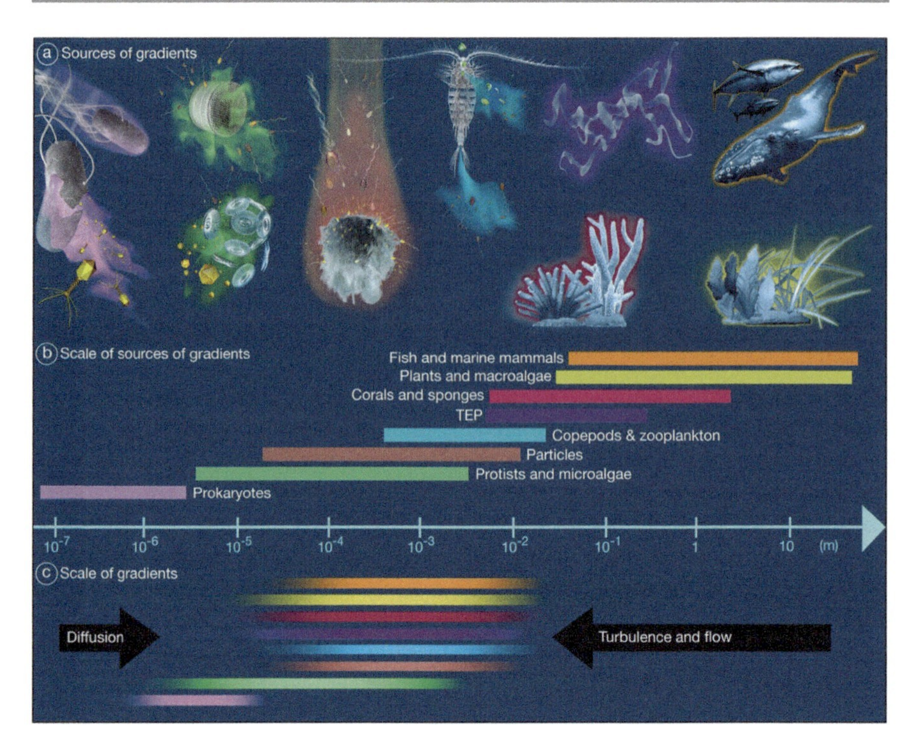


**Fig. 2.5** Sources of gradients in marine environments and their size ranges. (**a**) Sources of gradients. From left to right: bacterial lysis due to viral infection; phycosphere and phytoplankton lysis; sinking particle; zooplankton excretion and sloppy feeding; transparent exopolymeric polymers (TEPs); benthic organisms such as corals and sponges; fishes and marine mammals; marine plants and macroalgae. (**b**) The scale of the sources of gradients, spanning eight orders of magnitude. (**c**) The scale of the gradients themselves, spanning four orders of magnitude. The scale of a gradient depends on the initial size of a resource patch, modulated by two processes, diffusion and turbulence/flow (as presented in Sect. 2.2)

from algal cells (Amin et al. 2012; Seymour et al. 2017). Bell and Mitchell (1972) coined the term "phycosphere" based on a series of experiments where they demonstrated that filtrates from lysed phytoplankton cultures elicited significant chemotaxis, and that the release of dissolved organic matter by phytoplankton cultures contributes to the structure of their bacterial communities (Bell et al. 1974).

The phycosphere extends to a distance of a few cell diameters (Azam and Ammerman 1984; Bell and Mitchell 1972) and, hence, display a large variation in sizes across species which parallels the two orders of magnitude variation in size among phytoplankton taxa (Figs 2.1f and 2.5). Furthermore, the size of the phycosphere will vary depending on phytoplankton growth and exudation rate, as well as the diffusivity of the exuded compounds and their background concentration in bulk water (Seymour et al. 2017). For example, older cells tend to release high molecular weight DOM by secretion or cell lysis (Buchan et al. 2014; Passow 2002), and the diverse sizes and lability of these molecules directly impact the physical characteristics of the phycosphere as well as the metabolism of surrounding bacteria. Phycospheres are also present around phytoplankton cells in motion. Indeed, the chemical plume left in the wake of swimming or sinking phytoplankton cells offers a rich nutrient microenvironment that is exploitable by chemotactic bacteria (Barbara and Mitchell 2003; Jackson 1987, 1989).

### 2.3.2 Zooplankton Excretion and Sloppy Feeding

Zooplankton ingestion, digestion, excretion, and exudation of dissolved organic carbon also contribute to the patchiness of the microscale seascape (Möller et al. 2012; Stocker and Seymour 2012; Tang 2005), which can also impact the growth of bacteria and phytoplankton (Birtel and Matthews 2016; Goldman et al. 1979; Jackson 1980; Lehman and Scavia 1982a, b). For example, zooplankton cells release organic nutrients into the water column through a process called "sloppy feeding" in which they consume their prey only partially (Blackburn et al. 1997; Lampert 1978) and the remains, therefore, form a nutrient hotspot available to bacteria (Fig. 2.5) (Möller et al. 2012; Møller 2005; Peduzzi and Herndl 1992; Saba et al. 2011).

Zooplankton excretion events increase the concentrations of organic and inorganic substrates in the surrounding water (Lampert 1978; Peduzzi and Herndl 1992). Quantification of zooplankton-mediated DOM release has shown that *Daphnia pulex* and *Calanus hyperboreus* release up to 20% of ingested algal-derived carbon (Copping and Lorenzen 1980; Lampert 1978). While the size and composition of the chemical patch created by zooplankton emission will vary based on the identity of the organism releasing it, excretion events by zooplankton have been modeled as ~100 μm wide pulses of inorganic substrates such as ammonium with initial concentrations in the range 0.2–5 μM (Jackson 1980; McCarthy and Goldman 1979).

Copepod activity may also play a key role in providing organic substrates for bacterial growth. It was recently revealed that copepods potentially benefit from influencing the composition of microbial communities by attracting and "farming" specific bacterial species in their "zoosphere" (Shoemaker et al. 2019). Indeed, copepods may attract and support the growth of bacterial species of *Vibrionaceae, Oceanospirillales*, and *Rhodobacteraceae* in waters surrounding them but also appear to support the growth of specific groups of bacteria in or on the copepod body, particularly *Flavobacteriaceae* and *Pseudoalteromonadaceae* (Shoemaker et al. 2019).

### 2.3.3 Cell Lysis Events

Viruses also contribute to the formation of microscale chemical gradients in the water column (Blackburn et al. 1998; Ma et al. 2018; Moran et al. 2016; Riemann and Middelboe 2002). Viral infection of phytoplankton, bacterioplankton, protozoa, and other microorganisms can result in cell lysis (Middelboe 2000; Riemann and

Middelboe 2002; Suttle 1994; Suttle and Chan 1994), whereby the host cell's internal content is discharged into the water column (Fig. 2.5). It has been estimated that 20–50% of all microbial biomass is killed daily by viruses at a rate of $10^{23}$ lysis events per second, introducing each year as much as 3 Gt of carbon (Suttle 2007) into the organic carbon pool of the ocean (45–50 Gt) (Granum 2002; Granum et al. 2002).

As the internal content of a microbial cell contains concentrations of organic compounds that are up to six orders of magnitude higher than the bulk seawater (Flynn et al. 2008), nutrient-rich micropatches can be suddenly created upon lysis (Figs. 2.1a–c and 2.5), and can persist for several minutes (Stocker and Seymour 2012). Virus infected cultures of the phytoplankton *Micromonas pusilla* release dissolved organic carbon (DOC) enriched in peptides 4.5 times faster than noninfected cultures, resulting in local enrichment of seawater (Lønborg et al. 2013). Studies with laboratory-based virus–host systems have shown that lysis can alter the composition of DOM as well as its concentration (Middelboe and Jørgensen 2006; Weinbauer and Peduzzi 1995). In addition, the DOM released from virally infected cells is enriched in nitrogen, amino acids, and cell wall compounds, relative to the metabolites of noninfected cells (Ankrah et al. 2014; Middelboe and Jørgensen 2006). Estimates have also revealed a stochiometric mismatch between phages and their bacterial hosts with the former being enriched in phosphorus (Jover et al. 2014), and a subsequent modelling approach predicted that most of the phosphorus and a large fraction of the iron contained in bacterial cells might in fact be sequestered in phage particles during an infection (Bonnain et al. 2016). These studies have important implications for the composition and bioavailability of bacterial lysates.

### 2.3.4 Particles

Suspended and sinking particles provide a rich source of organic and inorganic nutrients for heterotrophic bacteria (Figs. 2.3 and 2.5). Concentrations of substrates associated with particles can exceed those present in the bulk seawater by more than two orders of magnitude (Alldredge and Gotschalk 1990). Particle dimensions range from sub-micrometer-sized colloids (Isao et al. 1990) to millimeter-sized aggregates called "marine snow" (Alldredge and Silver 1988), which are mostly made of coagulated dead phytoplankton cells (Alldredge and Silver 1988; Jackson 1990; Simon et al. 2002), zooplankton fecal pellets (Jacobsen and Azam 1984), aggregated microbial cells, and extracellular polymers (Passow 2002).

Marine snow and other particles represent key resource hotspots prone to colonization by heterotrophic bacteria (Kiørboe et al. 2002; Ploug and Grossart 2000; Simon et al. 2002). These microorganisms use surface-bound enzymes, including proteases, lipases, chitinases, and phosphatases to dissolve particulate organic matter (POM) contained in the aggregates through rapid hydrolysis (Smith et al. 1992). This process interrupts carbon export, converting POM into DOM, which can subsequently remain in the upper ocean. There is rapid turnover (0.2–2.1 days) of particulate amino acids into the dissolved phase with bacteria producing DOM much faster than they can use it (Smith et al. 1992). Consequently, the resources made available by this hydrolysis go beyond the particle-attached community because a substantial fraction of the solubilized organic matter leaks into the surrounding water where it becomes available to free-living bacteria (Alldredge and Cohen 1987; Kiørboe and Jackson 2001; Long and Azam 2001; Stocker et al. 2008). This DOM forms a comet-like plume in the wake of particles as they sink (Figs. 2.3 and 2.5) (Grossart and Simon 1998; Kiørboe et al. 2001; Smith et al. 1992; Ya et al. 1998) and it has been predicted that free-living bacteria can exploit the DOM plume to support growth rates that are up to 10 times higher than would be possible in the surrounding seawater (Kiørboe and Jackson 2001).

The community composition on particles is taxonomically distinct from free-living bacterial communities (DeLong et al. 1993), likely as a result of the selective pressure involved in the colonization and degradation of these nutrient hotspots. The use of synthetic polysaccharide particles has enabled to unravel the complexity of the microbial interactions occurring within these nutrient hotspots. Bacterial communities attached to these particles undergo rapid and reproducible successions under laboratory conditions (Datta et al. 2016). Indeed, a shift occurs from early colonizers that are motile and degrade organic matter derived from the particles to secondary consumers that fully rely on the metabolic byproducts of the primary degraders and who cannot directly consume particle-derived carbon (Datta et al. 2016). In addition, metabolic interactions alongside the spatial organization of bacteria on these particles influence the uptake of particle-derived carbon (Ebrahimi et al. 2019).

### 2.3.5 Transparent Exopolymer Particles

The traditional view of the microbial seascape as a simple dichotomy between particulate matter and dissolved organic matter may be an oversimplification. Diverse compounds known as TEP (transparent exopolymer particles), including organic gels, colloids, and matrices (Alldredge et al. 1993; Long and Azam 2001), bridge the two ends of the spectrum of the resource seascape, which has been described instead as an "organic matter continuum" (Azam 1998). Diatoms and bacteria release large quantities of exopolymeric materials (Alldredge et al. 1993; Gärdes et al. 2011; Jenkinson et al. 2015; Long and Azam 2001) that are characteristic of the transition phase between particulate and dissolved organic matter. The sticky nature of exopolymers promotes the aggregation of organic matter and microbes and therefore promotes carbon export from the euphotic zone to the deep ocean (Engel et al. 2004; Mari et al. 2017). However, exopolymers also facilitate the attachment of particle-degrading bacteria for which they provide an additional carbon source (Passow 2002; Taylor and Cunliffe 2017) (Fig. 2.5). For example, by using DNA stable-isotope probing, members of the *Alteromonadaceae* have been shown to assimilate $^{13}$C-TEP carbon (Taylor and Cunliffe 2017), which is consistent with their capability to produce a suite of polysaccharide-degrading enzymes (Teeling et al. 2016).

### 2.3.6 Larger Organisms

Chemical gradients in the water column also emanate from larger organisms such as fish and mammals (Fig. 2.5). As a rich source of organic molecules, the skin of fish can be colonized by numerous bacterial taxa (Sar and Rosenberg 1987; Shotts et al. 1990), such as the mucus-colonizing community dominated by the phylum *Proteobacteria* found on the skin of Atlantic salmon (Minniti et al. 2017). Benthic organisms, such as coral, seaweed, sponges, and bivalves, represent other sources of strong chemical gradients (Fig. 2.5). Surprisingly, surface metabolites released by the coral colonies can form concentration gradients extending up to 5 cm away from the coral surface (Ochsenkühn et al. 2018), a distance much greater than typical diffusion boundary layers. Corals excrete a mucus layer that can be hundreds of microns thick (Paul et al. 1986; Rohwer et al. 2001, 2002). This mucus contains organic and inorganic compounds at concentrations that are 3–4 orders of magnitude higher than the background seawater (Broadbent and Jones 2004; Wild et al. 2004) and it has been proposed that microbial colonization of this coral surface layer plays an important role in coral–microbe interactions and even in symbioses (Blackall et al. 2015; Pogoreutz et al. 2021; Pollock et al. 2018). Marine sponges also exude chemicals and are known "microbial hotspots" because they harbor dense and diverse microbial communities, which can account for up to 40% of a sponge's biomass (Taylor et al. 2007; Webster and Taylor 2012; Webster and Thomas 2016). Marine plants and algae represent another important source of chemical gradients (Haas and Wild 2010; Moriarty et al. 1986). For instance, the benthic alga *Halimeda opuntia* releases a large amount of carbohydrates and proteins into the surrounding seawater (up to 2 mg $m^{-2}$ $h^{-1}$) sustaining the growth of bacteria in their vicinity (Haas and Wild 2010).

The solid–liquid interface at the surface of marine sediments is a site of intense microbial activity due to its high concentration of organic material, which originates from deposited marine particulate organic matter that sank from the surface waters (Burdige and Komada 2015; Cai et al. 2019; Rossel et al. 2016; Zhang et al. 2018).

The concentration of dissolved organic carbon present in coastal sediments can be more than an order of magnitude higher than in the water directly above this interface (Burdige and Gardner 1998) indicating net production of DOM in sediments resulting from degradation processes (Burdige and Komada 2015). Experiments have demonstrated that in certain cases bacteria and archaea can sustain these gradients on timescales of hours to tens of hours in response to substrate addition by producing extracellular enzymes triggering extremely rapid hydrolysis of high molecular weight organic matter to low molecular weight DOM (Arnosti 2004; Burdige and Komada 2015), resulting in the production of small organic molecules and inorganic compounds ($H_2S$, $NH_4^+$, $Fe^{2+}$) (Jørgensen and Revsbech 1983; Ramsing et al. 1993; Schulz and Jørgensen 2001).

The active biological degradation of organic matter in sediments results in steep vertical gradients of nutrients and counter-gradients of oxygen or hydrogen sulfide (Jørgensen and Revsbech 1983; Jørgensen et al. 2019), which generates chemical habitats that are remarkably different from the pelagic environment (Schulz and

Jørgensen 2001). Indeed, the water–sediment interface is best described as a one-dimensional collection of chemical gradients that are more stable through time than the complex three-dimensional and often short-lived chemical gradients present in the water column (Fig. 2.5). The surface sediments are often dominated by sulfur oxidizers, such as *Thiovulum majus*, one of the fastest swimming bacteria recorded (swimming at up to 600 μm $s^{-1}$; Jørgensen and Revsbech 1983), which uses chemotaxis to form dense aggregations in the narrow region where optimal concentrations of oxygen and hydrogen sulfide coexist (Petroff and Libchaber 2014; Petroff et al. 2015). Furthermore, the large size of *T. majus* of up to 25 μm in diameter makes this bacterium immune to Brownian rotational diffusion and therefore considerably more effective at controlling its swimming direction than the small bacteria of the water column (Fenchel 1994).

### 2.3.7 Molecular Diversity of Chemoattractants

Oceanic DOM is extremely diverse in its chemical composition as might be expected from the wide range of organisms contributing to this pool, and recent estimates revealed that hundreds of thousands of different organic molecules might be present (Amon et al. 2001; Kim et al. 2003; Kujawinski et al. 2016) amounting to almost as much carbon as $CO_2$ in the atmosphere (Moran et al. 2016). Recent advances in DNA sequencing (DeLong and Karl 2005), mass spectrometry (Hartmann et al. 2017), and bioinformatics (Dührkop et al. 2015; Watrous et al. 2012) have enabled a giant step forwards in the identification and characterization of the key chemical currencies in the ocean, each of which has the potential to induce behavioral responses, generate interactions among organisms, and sustain the growth of specific marine bacteria. However, not all compounds have the same nutritive value to bacteria, as foraging strategies and chemotactic preferences are strain-specific (Amin et al. 2012; Seymour et al. 2010a).

DOM can be categorized along a gradient of reactivity, from labile to semi-labile to refractory, based on the persistence of these compounds in the water column (Hansell 2013). Compounds referred to as labile DOM are typically consumed within hours to days of production, although their half-lives have been estimated to be in the order of minutes at picomolar concentrations (Azam 1998), which complicates an accurate quantification of their abundance and lifetime in the water column. For example, the bulk concentrations of amino acids or sugars are usually just above the detection limit of most analytical instruments (a few nM per liter) (Kaiser and Benner 2012; Mopper et al. 1992). Yet, they represent a large proportion of the DOM taken up by bacteria (Hollibaugh and Azam 1983) and their rapid turnover is likely to keep their concentrations low. Most of the labile DOM consists of highly diverse compounds derived from phytoplankton primary production and is dominated by proteins and carbohydrates (Ferguson and Sunda 1984; Hodson et al. 1981; Vorobev et al. 2018) but also contains mono- and dicarboxylic acids (Gifford et al. 2013; Poretsky et al. 2010), glycerols and fatty acids (Gifford et al. 2013; McCarren et al. 2010), single-carbon compounds such as methanol (Gifford et al.

2013; Lidbury et al. 2014; McCarren et al. 2010), sulfonates (Durham et al. 2015), as well as the nitrogen-containing metabolites taurine, choline, polyamines, and ectoine (Gifford et al. 2013; Lidbury et al. 2014; Liu et al. 2015).

Compounds referred to as semi-labile are less reactive and persist longer in the surface ocean, from weeks to years (Hansell 2013), but might ultimately be exported to depth and buried in marine sediments for millennia (Hansell and Carlson 1998). Examples of semi-labile DOM include large polysaccharides and dissolved combined neutral sugars (Panagiotopoulos et al. 2019). Finally, the term refractory DOM is used to characterize the least reactive and most persistent fraction potentially stored in ocean basins for millennia (Follett et al. 2014; Williams and Druffel 1987). These refractory molecules might account for 95% of the dissolved organic carbon found in the ocean (~624 Gt of C) and contribute to long-term carbon storage (Jiao et al. 2010; Ogawa et al. 2001). Despite its ubiquitous presence in the ocean, the pool of refractory DOM is poorly characterized and the role that specific microbial species play in producing or partially degrading these molecules has not been elucidated (Osterholz et al. 2015). In addition, the distribution of specific molecules between these three categories is not well established, mostly because of the vast chemical diversity of the DOM pool. Our understanding of the "chemical preferences" of marine bacteria in terms of chemotaxis and substrate utilization is still restricted to a few molecules and species.

## 2.4 Motility and Chemotaxis as Microbial Adaptations to Microscale Heterogeneity in the Ocean

The previous section described the vast array of nutrient gradients that prokaryotic cells may encounter in the water column. Homing in on these gradients using motility and chemotaxis can therefore be highly beneficial for bacteria and archaea. Cells are propelled by phosphorylation-triggered rotating flagella (Wadhams and Armitage 2004) and can reorient themselves toward food patches using sensitive sensing mechanisms (Lux and Shi 2004). While hotspots can be abundant, during a bloom of phytoplankton, for example, many regions of the ocean are characterized by low background nutrients, which result in lower biomass and sparser number of available hotspots. This implies that motility and chemotaxis in the marine environment need to be adapted to explore wide areas and efficiently sense a small increase in chemical concentration above background levels. Indeed, many marine bacteria possess motility adaptations including fast swimming and specific reorientation strategies that differentiate them from classic model systems such as *Escherichia coli* and are more suited to the harsh nutrient conditions of the ocean.

### 2.4.1 The Molecular Machinery of Chemotaxis

The molecular machinery underpinning chemotactic behaviors has been extensively studied in *E. coli*. Upon detection of the chemical gradient of a chemoeffector (a chemical that attracts or repels cells), the bacteria's sensory system (Fig. 2.6) triggers a change in the swimming pattern to bias movement toward the higher concentration of a chemoattractant or toward lower concentration of a chemorepellent. The sensory system of *E. coli* is sensitive enough to detect changes in receptor occupancy of a few molecules against background concentrations and can detect variation over five orders of magnitude (Kim et al. 2001; Sourjik and Berg 2002a).

When *E. coli*'s chemotactic sensory machinery encounters a steep gradient of a chemoattractant, the cell's run-time increases from 1 s to over 10 s. Chemoeffectors binding to the cell's receptors induce an excitatory pathway that results in the modulation of the flagellum's motor (Segall et al. 1982; Sourjik and Berg 2002b). Upon encounter of an attractant, the flagella move in a counterclockwise motion; conversely, flagella move clockwise upon encounter of a repellent or a lower concentration of an attractant (Berg and Tedesco 1975) inducing a change of orientation. The time interval between the onset of the stimulus and the clockwise-to-counter-clockwise transition is a linear function of the change in receptor occupancy (Berg and Tedesco 1975). Despite variation in the number and location of flagella among bacterial strains, all chemosensory pathways of chemotaxis rely on modulation of the rotation of the flagellar motor (Wadhams and Armitage 2004).

Gradient sensing is accomplished by comparing the cell's receptor occupancy through time. *E. coli* makes short-term comparisons up to 4 s in the past where the most recent 1 s is given a positive weighting and the previous 3 s a negative weighting (Segall et al. 1986). The cell responds according to the overall weighted sum (Segall et al. 1986), which leads to a chemical "memory" (Berg and Tedesco 1975). At the molecular level, the signaling pathway involved in chemotaxis relies on a histidine–aspartate phosphorelay pathway and is probably one of the best-described processes in biology. The pathway is composed of transmembrane chemoreceptors (methyl-accepting chemotaxis proteins, MCPs) that detect binding chemoeffectors (Fig. 2.6a). Chemotactic bacteria possess on average 14 different MCPs (Lacal et al. 2010); however, this number can vary greatly at the strain level from as few as one to as many as 90 (Alexandre et al. 2004; Salah Ud-Din and Roujeinikova 2017). With the help of the adaptor protein CheW, the MCPs are connected to the histidine protein kinase chemotaxis protein CheA, which can sense chemical inputs through the MCPs. Two diffusible response regulators, CheY and CheB, then compete for binding to CheA (Fig. 2.6a). The phosphorylated motor-binding protein CheY-P controls flagellar motor rotation by binding to the switch protein FliM, which leads to a reversal in the direction of the motor rotation (Wadhams and Armitage 2004) whereas the methylesterase CheB controls adaptation of the MCPs (Anand et al. 1998; Hess et al. 1988). An additional molecule, CheZ, is required to increase the rate of dephosphorylation of CheY-P to induce a time-efficient signal termination (McEvoy et al. 1999). Consequently, an extracellular decrease of chemoattractant concentration leads to a decreased rate of binding to the MCPs, which induces the trans-autophosphorylation of CheA and an increased amount of CheY-P in the cytoplasm through its direct phosphorylation (Fig. 2.6a). CheY-P then binds to the flagellar motor and stimulates a switch in rotation to a

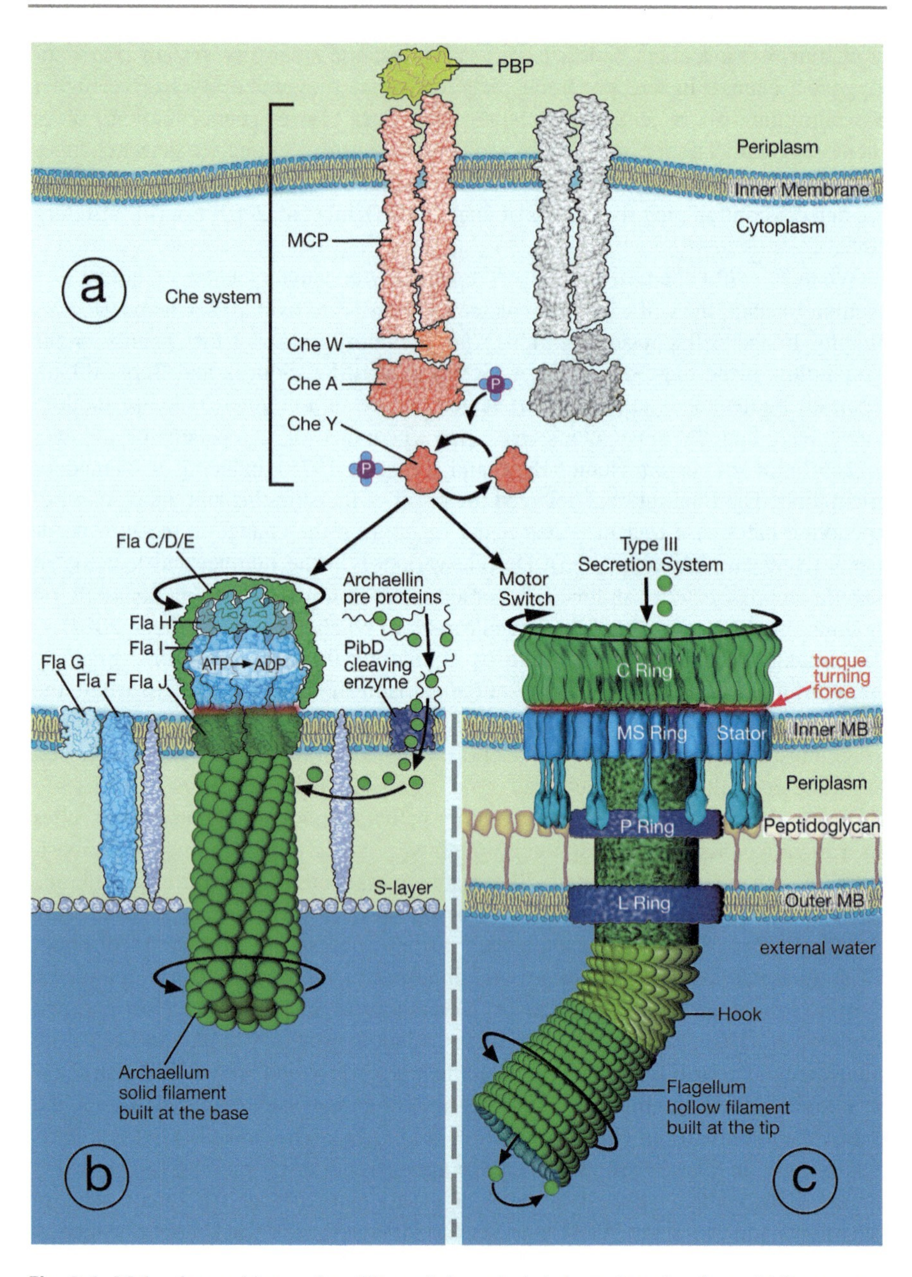


**Fig. 2.6** Molecular machinery of motility and chemotaxis in bacteria and archaea. (**a**) Representation of the chemotaxis sensing apparatus of both organisms showing the transduction of a signal (Periplasm Binding Protein PBP) from the receptors (Methyl-accepting Chemotaxis Protein MCPs) to CheW and to the phosphorylation of CheA. CheA phosphorylates CheY, and CheY-P then diffuses to the flagellum base. (**b**) In archaea, a rotation occurs when CheY binds to adaptor protein CheF (Schlesner et al. 2009) to navigate to the motor switch, constituted of the archaeal-specific proteins FlaC/D/E. FlaH, FlaI and FlaJ form a core motor platform. FlaF and FlaG provide a rigid structure between the S-layer and the rotating components of the motor (FlaJ) (Banerjee et al. 2015; Tsai et al. 2020). PibD, a prepilin peptidase, cleaves the N-terminus of the archaellins before

clockwise motion (Fig. 2.6a, b) resulting in the bacterium tumbling and hence changing swimming direction. The concentration of CheY-P is then decreased by the phosphatase CheZ. Simultaneously, the methylesterase activity of CheB is increased by phosphorylation from CheA-P, so that CheB-P induces demethylation of the MCPs thereby limiting the rate of CheA autophosphorylation. Consequently, the rate of switching in rotation then returns to the levels before stimulus and the cell is primed to react to any additional increase or decrease in chemoeffectors on the receptors. In the opposite case of an increase of chemoattractant concentration binding to the MCPs, the autophosphorylation of CheA is inhibited resulting in a reduction of the cytoplasmic CheY-P concentration and hence a decrease in the frequency of motor switching. Consequently, the cell swims longer in the same direction before tumbling. Additionally, the phosphorylation and activity of CheB are decreased so that the constitutive levels of the methyltransferase CheR then lead to a higher level of methylation of the MCPs. The MCPs are thus more capable of causing CheA autophosphorylation so that its rate returns to pre-stimulus levels and brings the bacterium back to a normal frequency of tumbling.

### 2.4.2 The Roles of Chemotaxis

Chemoattractants play at least two ecological roles. Most often they serve as high-quality bacterial substrates that bacteria can readily use to sustain growth (Cremer et al. 2019; Stocker et al. 2008). However, some microbial chemoattractants do not act as substrates but instead, serve only as signaling compounds, directing bacteria to ecologically advantageous microenvironments without being consumed (Seymour et al. 2017; Yang et al. 2015). For example, *Vibrio furnissii*, a chitin degrader, uses the water-soluble products of chitin hydrolysis as a cue to locate chitin (Bassler et al. 1991) and *V. coralliilyticus* uses dimethylsulfoniopropionate (DMSP) as a cue to locate its coral host (Garren et al. 2014) in both cases without the chemoattractant being metabolized. Chemoattractants as signaling molecules have also been reported in *B. subtilis* (Yang et al. 2015). In a comparison of the chemotactic response of *E. coli* and *B. subtilis* to a set of amino acids, the chemotactic response of the former was correlated with amino acid use, while no such correlation was found with the latter (Yang et al. 2015). This suggests that amino acids did not induce chemotaxis in *B. subtilis* because of their nutritional value but instead served as environmental cues.

**Fig. 2.6** (continued) assembly on the growing structure. (**c**) In bacteria, phosphorylated CheY binds to the switch complex and induces a change of rotation

### 2.4.3 Mechanics of Motility

At the molecular level, bacterial motility is achieved through the use of one or more helical flagella, which are used to propel cells and thus to explore and eventually exploit their environment (Stocker 2012). By utilizing a proton or sodium gradient (Berg 2008) molecular motors rotate each flagellum in a corkscrew motion and thereby propel the bacterium forward. These microscale movements are also not benefiting from inertia meaning that once the bacterium's flagella stop moving the cell will immediately stop its forward motion (within a distance of less than one hydrogen atom) (Purcell 1977).

Bacterial motility has been well studied in *E. coli* (Berg 1993, 2000, 2008) that possesses 4–8 proton-powered flagella (Berg 2008). Its motility pattern has been described as a "run and tumble" random walk (Berg 1993). Runs, periods of nearly straight-line swimming lasting 1–4 s, are generated by counterclockwise movement of the flagella that coalesces them into a bundle that propels the cell at 10–30 μm/s. Tumbles occur when at least one motor reverses direction and disrupts the flagella bundle, which triggers a very short (~0.1 s) random reorientation biased in the direction of the previous run (Berg 2008; Berg and Brown 1972).

Similarly to bacteria, archaea also have the ability to produce a propulsive force and direct their movement toward nutrient-rich hotspots in the ocean with the help of a flagellum-like filament, the archaellum (Fig. 2.6c) (Alam et al. 1984; Albers and Jarrell 2018; Jarrell and Albers 2012; Khan and Scholey 2018; Silverman and Simon 1974). Although this archaea-specific system has a similar function to the bacterial flagellum, its molecular organization is radically different (Jarrell and Albers 2012; Thomas et al. 2001). Indeed, the archaellum's structure consists of only 8–13 proteins, none of which share homologies with the 30 flagellar structural proteins called flagellins (Chaban et al. 2007; Lassak et al. 2012; Macnab 2003). The archaellum's assembly mechanism is analogous to that of bacterial type IV pili (Jarrell and Albers 2012; Jarrell et al. 1996). Additionally, whereas bacterial flagella are actuated by proton or sodium-driving forces, the archaellum's rotation is driven by ATP hydrolysis (Hirota and Imae 1983; Kinosita et al. 2016; Manson et al. 1977; Streif et al. 2008).

### 2.4.4 Abundance of Motile Prokaryotes

Although the most abundant marine bacterial taxa are nonmotile, such as *Pelagibacter ubique* (Morris et al. 2002) and *Prochlorococcus* (Liu et al. 1997; Moore et al. 1998), increasing evidence suggests that the fraction of marine microorganisms capable of motility can be important. Microscopy cell counts suggest that motile prokaryotes represent on average 10% of the total of bacterial and archaeal cells in coastal seawater samples but can increase to 80% over a short period of time (~12 h) upon enrichment with organic substrates (Mitchell et al. 1995a, b). Without enrichment, the motile fraction of bacteria ranges between 5 and 70% (Fenchel 2001; Grossart et al. 2001). This large fluctuation in the proportion of motile cells has been attributed to variation with depth, with seasonal and daily cycles, and with the amounts of dissolved and particulate organic matter in the water column (Buchan et al. 2014; Engel et al. 2011; Grossart et al. 2001). These surveys also indicate that motility appears to be widespread in eutrophic coastal regions and in productive surface waters, both of which are characterized by a high level of patchiness in the resource seascape. However, most approaches quantifying motility in bacterial communities are more than 20 years old and new methods enabling high-throughput quantification of motility in a variety of marine environments are needed.

### 2.4.5 Swimming Speed

Marine bacteria are typically much faster than the enteric bacterial models. This difference comes in part from the molecular motors they use, which are powered by sodium gradients across the cytoplasmic membrane instead of the proton gradients used by *E. coli* (Li et al. 2011; Magariyama et al. 1994). For example, the marine bacterium *Vibrio alginolyticus* swims faster with increasing sodium concentrations in surrounding bulk water (Muramoto et al. 1995; Son et al. 2013), and its flagellum rotate about 4–6 times faster than those of *E. coli* (Yorimitsu and Homma 2001). The swimming speed measured for marine isolates or marine bacteria in natural communities ranges from 45 to 230 μm/s (Grossart et al. 2001; Hütz et al. 2011; Johansen et al. 2002; Mitchell et al. 1995a, b; Muramoto et al. 1995; Seymour et al. 2010b; Shigematsu et al. 1995; Stocker et al. 2008; Xie et al. 2011) (Table 2.1). A survey revealed that most of the average swimming speeds of 84 marine isolates fell in the range 25–35 μm/s (Table 2.1) (Johansen et al. 2002). However, some species consistently swim faster than this average, such as *Pseudoalteromonas haloplanktis* (68–80 μm/s) (Seymour et al. 2010b; Stocker et al. 2008), *Thalassospira* (62 μm/s) (Hütz et al. 2011), and *Vibrio alginolyticus* (45–116 μm/s) (Muramoto et al. 1995; Xie et al. 2011) (Table 2.1). *Thiovolum majus* and *Ovobacter propellens,* residents of the sediment–water interface, are the fastest bacteria recorded so far, swimming at a striking 600 μm/s (Fenchel 1994) and 1000 μm/s (Fenchel and Thar 2004), respectively.

Compared to marine bacteria few studies exist on archaeal motility. The rod-shaped *Halobacterium salinarum* swims via the rotation of its archaella (Alam and Oesterhelt 1984; Alam et al. 1984). For these archaea, a simple back and forth movement was observed and the swimming speed of the cells was very low (2 μm $s^{-1}$). *Haloferax* sp. and *Haloarcula* sp. both exhibit "run and reverse" swimming patterns with low average speeds of ~2 μm $s^{-1}$ and ~ 2.3 μm $s^{-1}$, respectively (Thornton et al. 2020). The recent use of 3D-holographic microscopy and computer simulations revealed that halophilic archaea's swimming direction was stabilized by their archaellum, allowing for sustained directional swimming as well as energetic costs 100-fold lower than in common bacterial model systems (Thornton et al. 2020). However, not all archaea swim slowly. Two Euryarchaeota (*Methanocaldococcus jannaschii* and *M. villosus*) living in deep hydrothermal vents possess more than 50 polar archaella and are the fastest archaea observed so far,

**Table 2.1** Recorded swimming speeds of bacteria and archaea

| Bacterial/archaeal species | Environment | Speed (μm/s) | References |
|---|---|---|---|
| *Escherichia coli* | Enteric | 10–30 | Berg (2008), Berg and Brown (1972) |
| *Serratia marcescens* | Soil/enteric | 26 | Edwards et al. (2014) |
| *Pseudomonas aeruginosa* | Ubiquitous | 51–60 | Conrad et al. (2011), Hook et al. (2019) |
| *Bradyrhizobium diazoefficiens* | Soil | 27.5–29.8 | Quelas et al. (2016) |
| *Pseudomonas putida* | Soil | 44–75 | Harwood et al. (1989) |
| *Pseudomonas fluorescens* | Soil | 77–102 | Ping et al. (2013) |
| *Vibrio splendidus* | Marine | 20 | Johansen et al. (2002) |
| *Colwellia demingiae* | Marine | 17–27 | Johansen et al. (2002) |
| *Agrobacterium sanguineum* | Marine | 25 | Johansen et al. (2002) |
| *Vibrio cholerae* | Marine | 75 | Shigematsu et al. (1995) |
| *Pseudoalteromonas haloplanktis* | Marine | 68–80 | Seymour et al. (2010b) |
| *Thalassospira* sp. | Marine | 62 | Hütz et al. (2011) |
| *Vibrio alginolyticus* | Marine | 45–116 | Muramoto et al. (1995), Xie et al. (2011) |
| *Methanocaldococcus jannaschii* | Marine | ~400 | Herzog and Wirth (2012) |
| *Methanocaldococcus villosus* | Marine | ~500 | Herzog and Wirth (2012) |
| *Thiovulum majus* | Marine | 600 | Fenchel (1994) |
| *Ovobacter propellens* | Marine | 1000 | Fenchel and Thar (2004) |

reaching striking speeds of 400 and 500 μm $s^{-1}$, respectively (Herzog and Wirth 2012) (Table 2.1).

### 2.4.6 Why Do Marine Bacteria Swim Fast?

The ephemeral nature of many nutrient sources in the ocean implies that fast responses are beneficial to increase nutrient uptake. The primary parameter affecting the chemotactic response rate is the swimming speed: its importance in different ecological processes, including the colonization of particles and the uptake of dissolved nutrients, has been determined by numerical simulations (Kiørboe and Jackson 2001) and experiments (Stocker et al. 2008).

Fast swimming does not increase the flux of nutrients to bacteria. As the fluid flow generated by swimming decreases the thickness of the diffusion boundary layer surrounding a cell (see Sect. 2.2), it can induce an increased nutrient flux to the cell leading to a higher uptake rate of resources per unit time. However, this increased uptake rate is size-dependent, and sizeable advantages are only expected for very large cells (>10 μm), being negligible for most marine bacteria (Guasto et al. 2012). However, the effect of Brownian motion on bacteria might be an evolutionary reason for increased swimming speeds in bacteria found in the water column. The smaller a bacterium is the more prone it is to be redirected in a random direction due to Brownian motion while swimming, thereby disrupting directional swimming and chemotaxis (Berg 2008; Mitchell 1991). This effect can be quantified using the rotational diffusivity $D_R = kT/(8\nu\pi\mu R^3)$ of a spherical bacterium of radius $R$, where μ is the dynamic viscosity of water, $k$ is Boltzmann's constant and $T$ is the temperature in degrees Kelvin. For a bacterium of $R = 0.6$ μm the rotational diffusivity is 0.76 $rad^2/s$, which implies the generation of rotation of $(4D_R t)^{1/2}$ that is 100 degrees over 1 second, which will rapidly bring a cell off course. Even though cell shape (e.g., elongation and curvature) and the presence of a flagellum provide stability to marine bacteria against reorientation (Guadayol et al. 2017; Mitchell 1991; Schuech et al. 2019), Brownian motion effects impose a strong selective pressure for fast motility in small bacteria because faster cells will explore a longer distance before being spun off course.

### 2.4.7 Energetic Costs and Benefits of Motility

The high swimming speeds of marine bacteria come with an energetic cost. Early studies based on *E. coli* described bacterial motility as inexpensive (Purcell 1997), estimating that the costs of flagella synthesis and operation only amount to a modest ~0.1% of *E. coli*'s total energy expenditure (Macnab 1996). However, *E. coli*'s natural environment, the animal gut, harbors nutrient concentrations 2–4 orders of magnitude higher than those found in the ocean. In addition, marine bacteria, which swim about 3–5 times faster than *E.coli*, will incur an energy expenditure of about 10–25 times greater because of the propulsive power required in the viscosity-dominated regime in which bacteria live increases with the square of the swimming speed (Taylor and Stocker 2012). This energy demand imposes a strong selective pressure on bacterial motility in the ocean, a consideration supported by the large proportion of nonmotile marine bacteria that emphasizes the tradeoffs of motility and chemotaxis in the water column.

Mathematical simulations of bacterial competition in turbulent flow provide means to estimate the fitness advantage of chemotaxis (Taylor and Stocker 2012). Estimates of the potential gain of resources provided by chemotaxis and the energy expenditure due to swimming under realistic marine conditions suggest that if a bacterium is motile, its optimal swimming speed in the pelagic environment should be ~60 μm/s (Taylor and Stocker 2012; Watteaux et al. 2015). This theoretical value is close to the reported speeds of diverse marine bacterial taxa (Table 2.1) (Seymour et al. 2010b; Stocker et al. 2008). However, important parameters such as the costs of flagella and motor synthesis, production and activation of the signal transduction machinery, and ecological costs such as the effect on encounter rates with predators and viruses, have not yet been integrated in order to better estimate the cost of motility and chemotaxis.

One way that motile marine bacteria can save energy is by not swimming at a constant speed. Taylor and Stocker (2012) hypothesized that swimming speed may be adaptive and that cells might be able to regulate their speed upon encounter of chemical signals. This behavior, known as "chemokinesis," is supported by several observations. For example, *Pseudoalteromonas haloplanktis* displays a 20% increase in swimming speed when located within a patch of algal exudates (Seymour et al. 2009a). Similarly, *Vibrio coralliilyticus* swims 50% faster when exposed to the mucus of its coral host (Garren et al. 2014). Similar variations in swimming speeds over timescales of tens of seconds have been reported for natural assemblages of microorganisms under laboratory conditions (Grossart et al. 2001).

### 2.4.8 Swimming Patterns

The fast-swimming speeds of marine bacteria contribute to their high chemotactic performance but other factors also play a role. The chemotactic efficiency, $V_C/V_S$, representing the ratio between the chemotactic speed $V_C$ and the swimming speed $V_S$, is independent of the swimming speed as $V_C$ increases linearly with $V_S$ and therefore leads to a constant ratio. A perfectly directional response up the gradient excluding any random reorientation would result in a chemotactic efficiency of 1 whereas in the opposite case, a repulsion perfectly anti-directional to the gradient would result in $V_C/V_S = -1$. Between these extremes, a chemotactic efficiency of 0 characterizes purely random motion. While the chemotactic efficiency of *E. coli* typically ranges between 0.05 and 0.15 with exceptional peaks of 0.35 (Ahmed and Stocker 2008), marine bacteria achieve a chemotactic efficiency of up to 0.5 (Seymour et al. 2010a) in line with the idea that a rapid and directional response can provide an ecological advantage in a nutrient environment characterized by ephemeral hotspots.

Chemotaxis of *E. coli* has been further compared to the marine bacterium *P. haloplanktis* when subjected to 10-min nutrient pulses in microfluidics setups revealing that the chemotactic response of *P. haloplanktis* is almost 10 times faster than that of *E.coli* (Stocker et al. 2008). In addition, *P. haloplanktis* accumulated more strongly in the high concentration regions resulting ultimately in a 64–87% increase in potential nutrient uptake of the entire population and a ten-fold increase for the fastest 20% of bacteria (Stocker et al. 2008). Modeling revealed that this chemotactic performance could not be solely attributed to higher swimming speeds (68 μm/s for *P. haloplanktis* vs. 31 μm/s for *E. coli*) but resulted also from their highly directional swimming patterns (i.e., higher $V_C/V_S$). These results highlight the importance of considering the overall swimming behavior of a bacterium, rather than only its swimming speed, to fully understand the extent of its motility capacities.

This picture of motility and chemotactic performance is further completed by understanding how bacteria change direction along their course. *E. coli* uses a run-and-tumble swimming pattern: during each tumble, a cell's direction of motion is reoriented by a nearly random angle with the distribution of angles having a mean of 68° (Berg 2008). Yet, this is only one of several swimming strategies exhibited by bacteria (Mitchell and Kogure 2006). *Vibrios*, among other marine bacteria, swim using a single polar flagellum. This leads to a bidirectional movement, which was historically described as "run and reverse" (Johnson et al. 1992; Mitchell et al. 1996): when the flagellum rotates in one direction, the cell swims forward and when the flagellum reverses direction, the cell swims backward. This swimming pattern is marked by 180° reorientations of swimming direction, which would lead to constant back and forth swimming along the same line if Brownian rotational diffusion did not introduce randomness in the swimming directionality along each run. Run and reverse motility was historically considered as the most prevalent swimming pattern among marine bacteria (Johnson et al. 1992) and reversals have been proposed—on the basis of a mathematical model—to be more efficient than tumbles in enabling bacteria to stay close to a nutrient point source under the shear associated, for example, with turbulence (Luchsinger et al. 1999).

Recently, a new swimming pattern, "run-reverse-and-flick" motility, has been identified and appears to be widespread among marine bacteria. High-resolution imaging of *V. alginolyticus* (Xie et al. 2011) showed that cells follow a strict sequence of run, reverse (180° change of direction), and "flick," where the latter action is the previously unreported form of reorientation. Cell tracking and fluorescence labelling of the flagellum revealed that the flick, characterized by a normal distribution of reorientation angles with a mean of 90°, results from a large, whip-like deformation of the single polar flagellum. Using this hybrid motility (Stocker 2011), *V. alginolyticus* achieves a comparable exploration of its environment to that of *E. coli* and does so without requiring multiple flagella, which in the nutrient-poor ocean would be expensive to build. High-speed video microscopy has revealed that the flick occurs approximately 10 ms after the onset of a forward run rather than at the end of a backward run (Son et al. 2013). This timescale is faster than the time between two frames of standard cameras (33 ms), providing a potential reason for why flicks had not been detected before and why many or possibly all strains previously characterized as swimming in a "run-and-reverse" mode may actually be swimming in a "run-reverse-and-flick" mode.

The brief forward motion before a flick provided an important clue as to the mechanism of the flick: a compressive force exerted by the forward propulsion causes a mechanical buckling instability in the flagellum's hook. To what extent an actual "run-and-reverse" pattern is exhibited by bacteria in the ocean remains an open question but—on the grounds that the flick provides a much more effective and rapid form of reorientation than Brownian reorientation coupled with the fact that rapid navigation is critical to exploit transient microscale hotspots in the ocean—we propose that a minority of marine bacteria swim in a run-and-reverse mode, whereas we expect run-reverse-and-flick to be pervasive, considering the large fraction of marine bacteria that have a single flagellum (Leifson et al. 1964). Finally, the flick could be instrumental in allowing fast bacteria to accumulate at the top of nutrient gradients. For instance, Son et al. (2016) investigated the relationship between swimming speed, flicking motility, and high-performance chemotaxis by tracking large numbers of individual *V. alginolyticus* cells in controlled microfluidic gradients and found that the strength of bacterial accumulation at the peak of a gradient was swimming-speed dependent.

An additional mode of flagella-mediated movement has been described in bacteria using a single polar flagellum whereby cells wrap their flagellum around their body and swim in a screw-like motion to navigate through microenvironments (Kühn et al. 2017). Although this form of motility is unlikely to occur in the water column, it could be advantageous in marine sediments (Kühn et al. 2017) and has also been identified in marine symbionts (Kinosita et al. 2018) suggesting that it might play a role in symbiosis by aiding in host colonization (Raina et al. 2019). This alternative screw-like motion also indicates that there might be many other models of bacterial motility that are yet to be described.

Motile archaea explore marine seascapes by alternating forward and reverse swimming motions as the archaellum switches from clockwise to counter-clockwise rotation (Alam and Oesterhelt 1984; Kinosita et al. 2016; Shahapure et al. 2014). In the absence of stimuli, the best-studied motile archaea *Halobacterium salinarum* and *Haloferax volcanii* perform a random walk (Hildebrand and Schimz 1986; Quax et al. 2018) with swimming patterns more similar to the run-reverse-and-flick motion of *V. alginolyticus* (Xie et al. 2011) than to the run-and-tumble swimming of *E. coli*.

The higher chemotactic efficiency of marine bacteria might also result from differential signal transduction but this possibility has been not been thoroughly explored yet. The signal processing of marine bacteria has been suggested to be much faster than that of *E. coli*, to allow the detection of chemical gradients at higher swimming speeds, and this might translate into a higher turning frequency (Barbara and Mitchell 2003; Mitchell et al. 1996; Seymour et al. 2009a). In addition, it has been revealed that marine bacteria operate close to the theoretical limits of chemotactic precision allowing them to aggregate in microscale regions of high DOM concentration before these diffuse to background levels (Brumley et al. 2019).

The ocean imposes unique environmental constraints on chemotaxis including low nutrient concentrations, ephemeral gradients, and pervasive flow. It is thus not surprising that marine bacteria exhibit strong phenotypic differences compared to enteric bacteria such as *E. coli*, in the form of higher swimming speeds, different shapes, unique motility patterns, and higher levels of chemotactic performance.

## 2.5 Recent Insight from Omics Data

During the past 15 years, marine microbiology has been transformed by the advent of genomic approaches, which have provided unprecedented insights into the taxonomic and functional diversity of marine microbial communities (DeLong and Karl 2005; Sunagawa et al. 2015). Notably, these studies have also confirmed that genes involved in motility and chemotaxis are common, and their abundance is dynamic in marine bacterial communities.

### 2.5.1 Genomes of Marine Bacteria

While the most abundant clade of marine bacteria, *Pelagibacter ubique* (Giovannoni 2017; Morris et al. 2002), is nonmotile (Giovannoni et al. 2005) the genomes of many other marine bacteria, isolated from a wide variety of marine environments, frequently harbor genes involved in chemotaxis and motility (Gifford et al. 2013; Glöckner et al. 2003; López-Pérez et al. 2012; Ruby et al. 2005; Sunagawa et al. 2015; Thomas et al. 2008; Weiner et al. 2008). In particular, chemotaxis genes occur in multiple copies in many marine bacteria (Hamer et al. 2010) but are also found in archaea although less frequently (Salah Ud-Din and Roujeinikova 2017). According to a survey of sequenced genomes, aquatic bacteria typically contain a higher degree of duplication of genes associated with chemotaxis than bacteria that inhabit more environmentally stable environments (Alexandre et al. 2004).

Not surprisingly, chemotaxis genes are also abundant in the genomes of marine bacteria associated with animal hosts and organic surfaces attesting to the importance of active, directed motility in reaching these microenvironments (Gosink et al. 2002; Raina et al. 2019; Ruby et al. 2005; Thomas et al. 2008). These genes play a role in the colonization processes of both symbionts and pathogens. For example, chemotaxis and motility are essential for the attachment of the bacterial symbiont *Marinobacter adhaerens* to its diatom host *Thalassiosira weissflogii* (Sonnenschein et al. 2012). Similarly, the deletion of flagellar genes decreases the pathogenicity of *Edwardsiella tarda* to zebrafish, directly linking motility with the capacity of this pathogen to infect its host (Xu et al. 2014).

In some cases, the apparent absence of identifiable swimming and chemotaxis genes in the genomes of marine bacteria is equally illuminating. For instance, the model bacterium *Ruegeria pomeroyi* DSS-3 belongs to the *Roseobacter* clade, a group that commonly occurs in association with phytoplankton cells (Landa et al. 2017; Riemann et al. 2000) and often exhibits strong chemotactic performances (Miller and Belas 2004, 2006; Miller et al. 2004; Seymour et al. 2009b, 2010a). Analysis of *R. pomeroyi*'s genome has revealed the presence of genes involved in motility as well as in a suite of functions that are typically used for the organism's association with plankton and particles (Moran et al. 2004). While these characteristics all point to an organism that is likely to use chemotaxis to exploit microscale gradients, *R. pomeroyi*'s genome contains no homologs of known proteins involved in chemotaxis (Moran et al. 2004). Similarly, analysis of the genome of the marine non-flagellated motile cyanobacterium *Synechococcus* WH8102, which is chemotactic towards nitrogenous compounds (Willey and Waterbury 1989) has revealed that two unique large cell surface proteins are required for its motility: SwmA and SwmB (McCarren and Brahamsha 2007; McCarren et al. 2005). These findings suggest that marine bacteria may harbor as yet unrecognized motility and chemotaxis systems.

### 2.5.2 Metagenomics

Metagenomic surveys of marine microbial assemblages have revealed that the occurrence of chemotaxis and motility genes is strongly affected by environmental conditions (Dinsdale et al. 2008; Vega Thurber et al. 2009). Depth-related shifts in the occurrence of motility and chemotaxis genes were observed in a metagenomic analysis of the water column in the North Pacific Ocean with a higher representation of these genes in the photic zone (DeLong et al. 2006). This is consistent with the greater abundance of microenvironments enriched with phytoplankton produced organic matter within the upper, sunlit layers of the ocean. However, the more recent and much larger *Tara Oceans* campaign, which analyzed metagenomics data from 68 sites in epipelagic and mesopelagic waters across the globe revealed a significant enrichment of chemotaxis and motility genes directly below the photic zone (twilight zone) (Sunagawa et al. 2015). This enrichment of chemotaxis and motility genes is potentially of great utility to bacteria in the deep ocean to find and attach to sinking marine particles and aggregates but also to decrease their chance of encountering grazing predators (Matz and Jürgens 2005). The latter argument stands in contrast with the general understanding that swimming tends to increase encounters including with predators (Kiørboe 2008) highlighting instead a mechanism by which motility may help a bacterium escape from predators.

Metagenomics has also provided access to the genomes of uncultured microbes (Rusch et al. 2007; Venter et al. 2004). For example, members of the globally abundant Marine Group II archaea (order *Candidatus* Poseidoniales) harbor genes involved in motility, adhesion, and oligosaccharide degradation (Rinke et al. 2019; Tully 2019). These genomic capabilities suggest that members of the MGII archaea have a motile heterotrophic lifestyle exploiting oligosaccharide hotspots (e.g., phycospheres and particles) in the photic zone.

Metagenomics has shown that environmental variability can lead to shifts in the occurrence of motility and chemotaxis. Large increases in the abundance of motility and chemotaxis genes have been reported in coral-associated bacteria following temperature increases (Vega Thurber et al. 2009). These observations suggest that the prevalence of motility and chemotaxis varies strongly according to the physical and chemical features of specific marine habitats. Similarly, chemotaxis and motility gene abundance and regulation in the coral-associated microbiome is highly dependent on fine-scale chemical gradients emanating from the surfaces of corals ultimately impacting the microbial community structure of corals (Tout et al. 2014).

Perhaps one of the most intriguing observations relating to chemotaxis arising from metagenomic studies comes from a comparison of microbial and viral metagenomes across different environments (Dinsdale et al. 2008). High levels of proteins associated with motility and chemotaxis were observed in several viral metagenomes, which the authors suggest were not randomly acquired by the viral community. The role, if any, of these proteins in the phage is not clear but these observations indicate the potential for the horizontal transfer of genes involved in chemotaxis between different marine bacteria through phage infection.

### 2.5.3 Metatranscriptomics

Marine metatranscriptomic studies have shown that changing physicochemical conditions can shift the relative expression of motility and chemotaxis genes and have provided new insights into the processes determining when and where bacterial chemotaxis is most prevalent in the ocean. A temporal metatranscriptomic study of a coastal microbial assemblage revealed that transcripts for motility and chemotaxis followed both seasonal and daily patterns with higher levels of expression during the night (Gilbert et al. 2010). Daily variations in the transcription of genes for motility and chemotaxis may be associated with shifts in phytoplankton exudation rates or particulate organic carbon (POC) production, which would be consistent with previous direct measurements showing that increased bacterial motility levels in the early evening are correlated with POC production (Grossart et al. 2001). In a similar manner, significant upregulation of transcripts related to motility and chemotaxis have been recorded after the addition of dissolved organic substrates to a marine bacterial assemblage (McCarren et al. 2010) and after enrichment of a water sample from the North Pacific Subtropical Gyre with nutrient-rich deep-sea water (Shi et al. 2012).

Observations of increased expression of motility and chemotaxis genes in nutrient-amended samples are consistent with previous direct observations of increased bacterial motility following enrichment (Mitchell et al. 1995a, b). In both metatranscriptomic studies (McCarren et al. 2010; Shi et al. 2012), the increase in expression of motility and chemotaxis genes upon amendment occurred in parallel with an overall shift in community composition with a substantial increase in an *Alteromonas*-like population. This provides support for the hypothesis that increases in motility following enrichment are driven by shifts in community composition rather than directly by upregulation of expression in individuals. In contrast, expression of motility transcripts is decreased following bulk addition of DMSP (Vila-Costa et al. 2010). This finding is in line with observations that bulk additions of DMSP decrease bacterial chemotaxis to DMSP (Miller et al. 2004). These responses potentially occur because bulk DMSP additions eclipse the microscale DMSP cues surrounding individual phytoplankton cells decreasing the viability of chemotaxis as a strategy to find and exploit DMSP-rich hotspots (Seymour et al. 2010a).

Temperature is another environmental variable that can influence expression of chemotaxis and motility genes. For example, the marine bacterium *Photobacterium damselae* subsp. *damselae*, a facultative pathogen causing disease in fish and marine mammals, upregulates the expression of chemotaxis and flagellar genes at higher temperatures (Matanza and Osorio 2018). This is correlated with higher expression of virulence genes (Matanza and Osorio 2018), which highlights the importance of motility and chemotaxis in bacterial pathogenicity. In addition, increasing water temperatures considerably augment the performance of the coral pathogen *Vibrio coralliilyticus* in tracking the chemical signals of its coral host, *Pocillopora damicornis* (Garren et al. 2016). Indeed, when water temperature exceeded 30 °C the pathogen increased its chemotactic performance by >60%, and its swimming speed by >57% (Garren et al. 2016) substantially enhancing its ability to find its host.

The dynamic patterns in the occurrence of motility and chemotaxis genes in ocean metagenomes and transcriptomes confirm that these phenotypes are ecologically important features of natural marine bacterial assemblages that are often tightly coupled to the physicochemical nature of the environment.

## 2.6 Influence of Microscale Gradients on Large-Scale Processes

The physical, chemical, and biological processes that we have described above all take place over small spatial scales (micrometer) and short time periods (seconds to minutes). However, they underpin the behavior, physiology, ecological relationships, and genomic characteristics of planktonic marine microorganisms. An important question to answer is, do processes occurring at the microscale in the heterogeneous seascape inhabited by marine bacteria have large-scale impacts? Or can we instead neglect this heterogeneity and consider that its impact will "average out" over larger scales?

### 2.6.1 Impacts on Oceanic Primary Production

The productivity of the marine food web is governed by phytoplankton primary production, which means that interactions that directly affect phytoplankton growth have fundamental importance for ocean-scale processes. In addition to being controlled by dissolved nutrient availability in the bulk seawater phytoplankton growth is also influenced by processes occurring in the microenvironment surrounding their cells (i.e., the phycosphere). Reciprocal interactions with specific bacterial partners played out within this microenvironment can profoundly influence the provision of limiting nutrients and other essential growth factors to phytoplankton cells. In its simplest form, this reciprocal exchange can involve the uptake of exuded photosynthates (e.g., sugars) by the bacteria and the return of inorganic nutrients back to the phytoplankton cell (Azam and Malfatti 2007). However, more complex and specific chemical exchanges have been uncovered involving bacterial synthesis of important minerals, B-vitamins, and growth promoting hormones (Amin et al. 2009, 2012; Croft et al. 2005) that affect the growth and survival of phytoplankton cells.

Some microscale interactions may also negatively affect primary production in the ocean. Bacteria may outcompete phytoplankton for nutrients (Currie and Kalff 1984) while specific bacteria can inhibit phytoplankton cell division (van Tol et al. 2017) or produce algicidal compounds that kill these primary producers (Barak-Gavish et al. 2018; Furusawa et al. 2011; Seyedsayamdost et al. 2011). When considering the cumulative impact of these positive and negative relationships, it is clear that the microscale interactions between phytoplankton and bacteria influence phytoplankton growth and are a determinant of primary production in the ocean, ultimately affecting the functioning and productivity of marine ecosystems.

### 2.6.2 Impacts on Symbiont Recruitment

The acquisition of microbial symbionts enables host organisms to expand their metabolic capabilities, inhabit otherwise hostile environments, and carve new ecological niches, which promotes species diversity and ecosystem services (Margulis 1981; Ochman and Moran 2001). Many important marine symbioses such as those of corals, tube worms, squid, mussels, protists, and phytoplankton rely on the acquisition of microbial partners from the environment (Raina et al. 2019). The importance of bacterial motility and chemotaxis in the establishment and maintenance of symbiotic interactions is well established in a small number of model systems but is likely to be important across a wide range of hosts (Raina et al. 2019). One of the most well-studied model systems in the marine environment is the symbiosis between the bioluminescent bacterium *Aliivibrio fischeri* and the Hawaiian bobtail squid (*Euprymna scolopes*) where the host uses the light produced by the symbionts as camouflage against predators during its nocturnal foraging (Nyholm et al. 2000). In the few hours following hatching, bacterial symbionts are selectively taken up from the environment (Nyholm and McFall-Ngai 2004) and actively migrate toward the pores of the light organ using chemotaxis (Mandel et al. 2012). Another example of the use of motility and chemotaxis to recruit symbiotic partners is the marine macroalga *Ulva mutabilis*, which attracts its growth-enhancing symbiont *Roseovarius* by releasing the chemoattractant DMSP (Kessler et al. 2018). In addition, chemotaxis- and motility-deficient mutants of *Marinobacter adhaerens* were unable to locate and attach to their phytoplankton partners, negatively impacting the growth of the algal cells (Sonnenschein et al. 2012) and implying that chemotaxis is key to the establishment of a symbiotic exchange between bacteria and phytoplankton cells. As evidence of the ecological importance of symbioses to the fitness and survival of key marine organisms continues to emerge, the chemotactic encounter of symbiotic partners is likely to be a pervasive mechanism.

### 2.6.3 Impacts on Rates of Chemical Transformations

The behavioral responses of marine microorganisms to microscale heterogeneity in the water column are predicted to strongly affect the rates of carbon cycling through the base of the marine food web. Results derived from both experimental observations and mathematical models suggest that chemotaxis and motility significantly increase bacterial uptake rates of dissolved organic carbon (DOC) (Blackburn et al. 1997, 1998; Fenchel 2002; Smriga et al. 2016; Stocker et al. 2008). It is important to note that even in the absence of chemotactic bacteria, DOC derived from hotspots would ultimately diffuse into the bulk seawater and become available to non-chemotactic bacteria. This suggests that, while the rates of DOC uptake may increase due to chemotaxis, the absolute amounts of carbon cycled may not change (Stocker 2012; Stocker and Seymour 2012). However, behavioral exploitation of microscale DOC hotspots might enhance total carbon flux if these elevated concentrations of organic compounds support an increase in bacterial growth efficiency (Azam and Malfatti 2007). This process would ultimately lead to a higher proportion of DOC being converted into biomass and would therefore channel more carbon into the marine food web.

### 2.6.4 Impacts on Exchanges Between Ocean and Atmosphere

A large variety of biogenic volatile organic compounds (BVOCs) are produced by marine microorganisms and emitted to the atmosphere (Lawson et al. 2020; Moore et al. 2020). One of the best studied volatile compounds is the sulfur-containing dimethyl sulfide (DMS) because its release into the atmosphere represents the largest flux of biogenic sulfur on Earth and its subsequent oxidation forms sulfate aerosols that act as cloud condensation nuclei (Sievert et al. 2007; Simó 2001). The precursor of this gas is DMSP, which is produced in high concentrations by many phytoplankton taxa (with intracellular concentration reaching 1–2 M) (Caruana and Malin 2014; Keller 1989). At the scale of bacteria, large concentrations of DMSP are introduced into the environment via point source events including exudation into the phycosphere, viral lysis, and grazing events (Seymour et al. 2010a). Given the diffusivity of this molecule and its high concentration in patches, it is perhaps not surprising that DMSP is a potent chemoattractant for many species of marine bacteria (Garren et al. 2014; Miller et al. 2004; Seymour et al. 2010a; Zimmer-Faust et al. 1996). In addition, DMSP is also an important growth substrate, supporting up to 13% of the bacterial carbon demand and nearly all their reduced sulfur needs in surface waters (Kiene et al. 2000). However, not all marine bacteria use DMSP in the same way: some demethylate this compound to assimilate its sulfur and carbon into their cell (Howard et al. 2006) whereas others cleave DMSP and thereby produce the volatile DMS (Curson et al. 2011). These two competing pathways are often both present in marine bacteria (Curson et al. 2011) and chemotaxis toward DMSP has been demonstrated among bacterial strains that employ both the cleavage and demethylation pathways (Miller et al. 2004; Seymour et al. 2010a).

Twenty years ago, the DMSP availability hypothesis proposed that the relative importance of the two DMSP degradation pathways—and thus the amount of DMS produced—is regulated by the DMSP concentration in the environment (Kiene et al. 2000). According to this hypothesis, the utilization of DMSP in concentrated patches leads to the production of more DMS compared to utilization in dilute background concentrations. This long-standing hypothesis was recently validated experimentally, confirming that external DMSP concentration dictates the relative expression of the two pathways with an increase in DMSP cleavage (and therefore DMS production) measured close to the surface of phytoplankton (Gao et al. 2020). Bacterial exploitation of microscale DMSP hotspots such as the phycosphere surrounding a DMSP-producing phytoplankton cell, which is governed by motility and chemotaxis, is thus likely to be an important determinant of the release of sulfur into the atmosphere as DMS, influencing the cycling of sulfur.

### 2.6.5 Impacts on Exchanges Between Ocean and Sediments

The flux of sinking particles from the sunlit upper ocean to the deep ocean forms the basis of the biological carbon pump, which leads to the sequestration of carbon into marine sediments for millennia (Ducklow et al. 2001). This vertical carbon flux is responsible for the export of more than 50 Gt of carbon per year. From the perspective of a planktonic bacterium, sinking organic particles represent a localized resource hotspot. As particles sink, they are colonized and degraded by marine bacteria, which recycle the carbon they contain. Due to bacterial degradation, only 25% of the organic particles sink deeper than the photic zone and only 1% reach the ocean floor (Azam and Long 2001; Cho and Azam 1988). These particles are thus hotspots of microbial activity that influence the global biogeochemical cycles of carbon and nitrogen.

The decomposition of sinking particles involves specific behavioral and metabolic responses by marine bacteria and archaea. As discussed above (Sect. 2.2), motility and chemotaxis enhance the rate of encounter with particles by a factor of 100–1000 (Kiørboe and Jackson 2001; Kiørboe et al. 2001; Lambert et al. 2019). Observations of community assembly on model particles revealed a strong correlation between trophic level and motility (Datta et al. 2016). Early colonizers (arriving less than 48 hours after the exposure of particles to seawater) were not only motile and chemotactic but were also primary degraders of polymers (Datta et al. 2016). Conversely, late colonizers relied on metabolites from primary degraders to sustain their growth and were non-motile (Datta et al. 2016). These microscale processes have large-scale implications for carbon cycling because they directly control the quantity of particulate carbon that reaches the seafloor (Buesseler et al. 2007).

The metabolic activity of microbes on particles creates strong and persistent micrometer- to millimeter-scale oxygen gradients (Paerl and Prufert 1987). Important nitrogen transformation processes generally occur near oxic interfaces. As a result, particles are likely to support microscale partitioning of bacteria involved in nitrification (aerobic), denitrification (anaerobic), and nitrogen fixation (anaerobic) (Alldredge and Cohen 1987; Glud et al. 2015; Paerl and Prufert 1987). High levels of both nitrification and denitrification have been measured in organic aggregates derived from cyanobacteria (Klawonn et al. 2015). Similarly, direct stimulation of $N_2$ fixation has been measured in the presence of particles (Pedersen et al. 2018; Rahav et al. 2016) indicating that particles also represent important microscale hotspots for nitrogen cycling in the water column.

In summary, microbial processes occurring at the microscale in response to chemical gradients directly influence phytoplankton primary productivity, the recruitment of symbionts, the rate of biogeochemical transformations, the production of climate-active molecules, the cycling of limiting elements, and the long-term storage of carbon in the ocean. When reconsidering the question, does microscale heterogeneity matter, we can therefore safely answer in the affirmative. Microscale processes must be considered if we wish to achieve an accurate mechanistic understanding and realistic models of large-scale oceanic processes (Azam 1998; Stocker 2012).

## 2.7 Summary and Future Directions

- From the viewpoint of bacteria, seawater contains many nutrient hotspots and microhabitats that are either ephemeral or persistent. These hotspots arising from different micro- and macroorganisms, sinking particles, and decaying organic matter represent resource islands that can be exploited by copiotrophic bacteria for their growth.
- The physics of fluid dynamics and its impact on microorganisms at the microscale diverges from that ruling larger geophysical phenomena. In a world mostly dominated by diffusion, the microscale remains relatively unaffected by turbulence allowing the steady emission of chemical gradients that are accessible for uptake by heterotrophic microorganisms. Turbulence enhances microscale heterogeneity by stirring nutrients in the water column and creating microscopic nutrient filaments.
- The large distances separating microorganisms in the water column, relative to their body size, renders nutrient uptake highly challenging if it only relies on random encounters. Heterotrophic bacteria have therefore evolved active behaviors such as high-speed motility and sensitive chemotaxis to increase the frequency at which they encounter resource hotspots. The performance of marine bacteria differs from that of the well-studied enteric model organisms and has been demonstrated to yield higher profitability, through higher swimming speeds, efficient swimming patterns, and directed chemotaxis.
- The ocean's microscale seascape gives rise to a diverse range of interactions within multiple microhabitats such as the phycosphere or sinking marine particles. Gradients also mediate interactions between microorganisms and larger eukaryotes, such as corals and fishes, directly impacting the ecology and dynamics of the oceans.
- Although microscale behaviors and interactions may happen within a fraction of a drop of seawater, they have global-scale consequences. The impacts of these interactions do not average out over larger scales but instead microbial cycling of chemicals often occurs exclusively within localized microenvironments.

As we become more aware of the microscale complexity of bacterial behaviors and interactions ruling the foundations of marine microbial ecology and their global impact on biogeochemical cycles, it appears that the scale of classic sampling techniques used in oceanography (e.g., Niskin bottles) is fundamentally disconnected from the microscale interactions at work. New technologies and mathematical models (Słomka et al. 2020; Słomka and Stocker 2020) have been developed to decipher the microbial ecology of our oceans at more realistic scales. Single-cell genomics is beginning to reveal heterogeneity in gene expression (Blainey 2013; Gao et al. 2020; Kalisky and Quake 2011). Raman microscopy and mass spectrometry imaging now allow measurement of the chemical signatures of individual cells (Lee et al. 2019, 2020). Atomic force and electron microscopy are revealing the structural characteristics and the spatial configuration of cells (Mittelviefhaus et al. 2019; Turner et al. 2016). In addition, developments in microfluidics now allow researchers to unravel microbial behavior in response to chemical landscapes in the laboratory (Behrendt et al. 2020; Salek et al. 2019; Seymour et al. 2010a, b; Stocker et al. 2008) and directly in situ (Clerc et al. 2020; Lambert et al. 2017; Tout et al. 2015).

Despite these advances, key parameters needed to understand the survival of microbes in a sea of gradients are still poorly characterized. Work is required to determine (1) the fraction of motile bacteria in the ocean and how this fluctuates with daily cycles, seasons, nutrient abundance, and ocean depth; (2) the principal chemical currencies in the water column used for growth or as signaling molecules and their impact on microbial community assembly and composition; (3) the distribution of particle sizes and abundance through the water column and their impact on rates of bacterial encounter, degradation, and remineralization; and (4) the variables that drive relations between heterogeneity and diversity, motility, and chemotaxis. Obtaining realistic estimates of these parameters that truly represent the heterogeneity of the oceans will not be easy, but the payoff will be a better understanding of the exquisite adaptations of bacteria and archaea to the complexity of marine environments and their contributions to the element cycles and climate of our planet.

## References

Aaronson S (1978) Excretion of organic matter by phytoplankton in vitro. Limnol Oceanogr 23: 838–838

Ahmed T, Stocker R (2008) Experimental verification of the behavioral foundation of bacterial transport parameters using microfluidics. Biophys J 95:4481–4493

Alam M, Oesterhelt D (1984) Morphology, function and isolation of halobacterial flagella. J Mol Biol 176:459–475

Alam M, Claviez M, Oesterhelt D, Kessel M (1984) Flagella and motility behaviour of square bacteria. EMBO J 3:2899–2903

Albers S-V, Jarrell KF (2018) The archaellum: an update on the unique archaeal motility structure. Trends Microbiol 26:351–362

Alexandre G, Greer-Phillips S, Zhulin IB (2004) Ecological role of energy taxis in microorganisms. FEMS Microbiol Rev 28:113–126

Alldredge AL, Cohen Y (1987) Can microscale chemical patches persist in the sea? Microelectrode study of marine snow, fecal pellets. Science 235:689–691

Alldredge AL, Gotschalk CC (1990) The relative contribution of marine snow of different origins to biological processes in coastal waters. Cont Shelf Res 10:41–58

Alldredge AL, Silver MW (1988) Characteristics, dynamics and significance of marine snow. Prog Oceanogr 20:41–82

Alldredge AL, Passow U, Logan BE (1993) The abundance and significance of a class of large, transparent organic particles in the ocean. Deep Sea Res Pt Oceanogr Res Pap 40:1131–1140

Aluwihare LI, Repeta DJ, Chen RF (1997) A major biopolymeric component to dissolved organic carbon in surface sea water. Nature 387:166–169
Amin SA, Green DH, Hart MC, Küpper FC, Sunda WG, Carrano CJ (2009) Photolysis of iron–siderophore chelates promotes bacterial–algal mutualism. Proc Natl Acad Sci 106:17071–17076
Amin SA, Parker MS, Armbrust EV (2012) Interactions between diatoms and bacteria. Microbiol Mol Biol Rev 76:667–684
Amon RMW, Fitznar H-P, Benner R (2001) Linkages among the bioreactivity, chemical composition, and diagenetic state of marine dissolved organic matter. Limnol Oceanogr 46:287–297
Anand GS, Goudreau PN, Stock AM (1998) Activation of methylesterase CheB: evidence of a dual role for the regulatory domain. Biochemistry 37:14038–14047
Ankrah NYD, May AL, Middleton JL, Jones DR, Hadden MK, Gooding JR, LeCleir GR, Wilhelm SW, Campagna SR, Buchan A (2014) Phage infection of an environmentally relevant marine bacterium alters host metabolism and lysate composition. ISME J 8:1089–1100
Arnosti C (2004) Speed bumps and barricades in the carbon cycle: substrate structural effects on carbon cycling. Mar Chem 92:263–273
Azam F (1998) Microbial control of oceanic carbon flux: the plot thickens. Science 280:694–696
Azam F, Ammerman JW (1984) Cycling of organic matter by bacterioplankton in pelagic marine ecosystems: microenvironmental considerations. In: Fasham MJR (ed) Flows of energy and materials in marine ecosystems: theory and practice. Springer, Boston, pp 345–360
Azam F, Long RA (2001) Sea snow microcosms. Nature 414:495–498
Azam F, Malfatti F (2007) Microbial structuring of marine ecosystems. Nat Rev Microbiol 5:782–791
Banerjee A, Tsai C-L, Chaudhury P, Tripp P, Arvai AS, Ishida JP, Tainer JA, Albers S-V (2015) FlaF is a β-sandwich protein that anchors the archaellum in the archaeal cell envelope by binding the S-layer protein. Structure 23:863–872
Barak-Gavish N, Frada MJ, Ku C, Lee PA, DiTullio GR, Malitsky S, Aharoni A, Green SJ, Rotkopf R, Kartvelishvily E, et al. (2018) Bacterial virulence against an oceanic bloom-forming phytoplankter is mediated by algal DMSP. Sci Adv 4:eaau5716
Barbara GM, Mitchell JG (2003) Bacterial tracking of motile algae. FEMS Microbiol Ecol 44:79–87
Bar-On YM, Phillips R, Milo R (2018) The biomass distribution on earth. Proc Natl Acad Sci 115: 6506–6511
Bassler BL, Gibbons PJ, Yu C, Roseman S (1991) Chitin utilization by marine bacteria. Chemotaxis to chitin oligosaccharides by *Vibrio furnissii*. J Biol Chem 266:24268–24275
Behrendt L, Salek MM, Trampe EL, Fernandez VI, Lee KS, Kühl M, Stocker R (2020) PhenoChip: a single-cell phenomic platform for high-throughput photophysiological analyses of microalgae. Sci Adv 6:eabb2754
Bell W, Mitchell R (1972) Chemotactic and growth response of marine bacteria to algal extracellular products. Biol Bull 143:265–277
Bell WH, Lang JM, Mitchell R (1974) Selective stimulation of marine bacteria by algal extracellular products. Limnol Oceanogr 19:833–839
Berg HC (1993) Random walks in biology. Princeton University Press, Princeton
Berg HC (2000) Motile behavior of bacteria. Phys Today 53:24–29
Berg HC (2008) *E. coli* in motion. Springer, New York. https://doi.org/10.1007/b97370
Berg HC, Brown DA (1972) Chemotaxis in *Escherichia coli* analysed by three-dimensional tracking. Nature 239:500–504
Berg HC, Tedesco PM (1975) Transient response to chemotactic stimuli in *Escherichia coli*. Proc Natl Acad Sci 72:3235–3239
Biersmith A, Benner R (1998) Carbohydrates in phytoplankton and freshly produced dissolved organic matter. Mar Chem 63:131–144
Birtel J, Matthews B (2016) Grazers structure the bacterial and algal diversity of aquatic metacommunities. Ecology 97:3472–3484

Blackall LL, Wilson B, van Oppen MJH (2015) Coral-the world's most diverse symbiotic ecosystem. Mol Ecol 24:5330–5347
Blackburn N, Azam F, Hagström Å (1997) Spatially explicit simulations of a microbial food web. Limnol Oceanogr 42:613–622
Blackburn N, Fenchel T, Mitchell J (1998) Microscale nutrient patches in planktonic habitats shown by chemotactic bacteria. Science 282:2254–2256
Blainey PC (2013) The future is now: single-cell genomics of bacteria and archaea. FEMS Microbiol Rev 37:407–427
Bonnain C, Breitbart M, Buck KN (2016) The ferrojan horse hypothesis: iron-virus interactions in the ocean. Front Mar Sci 3:82
Broadbent AD, Jones GB (2004) DMS and DMSP in mucus ropes, coral mucus, surface films and sediment pore waters from coral reefs in the great barrier reef. Mar Freshw Res 55:849–855
Brumley DR, Carrara F, Hein AM, Yawata Y, Levin SA, Stocker R (2019) Bacteria push the limits of chemotactic precision to navigate dynamic chemical gradients. Proc Natl Acad Sci 116: 10792–10797
Buchan A, LeCleir GR, Gulvik CA, González JM (2014) Master recyclers: features and functions of bacteria associated with phytoplankton blooms. Nat Rev Microbiol 12:686–698
Buesseler KO, Lamborg CH, Boyd PW, Lam PJ, Trull TW, Bidigare RR, Bishop JKB, Casciotti KL, Dehairs F, Elskens M et al (2007) Revisiting carbon flux through the ocean's twilight zone. Science 316:567–570
Burdige DJ, Gardner KG (1998) Molecular weight distribution of dissolved organic carbon in marine sediment pore waters. Mar Chem 62:45–64
Burdige DJ, Komada T (2015) Chapter 12 – sediment pore waters. In: Hansell DA, Carlson CA (eds) Biogeochemistry of marine dissolved organic matter, 2nd edn. Academic Press, Boston, pp 535–577
Cai R, Zhou W, He C, Tang K, Guo W, Shi Q, Gonsior M, Jiao N (2019) Microbial processing of sediment-derived dissolved organic matter: implications for its subsequent biogeochemical cycling in overlying seawater. J Geophys Res Biogeosci 124:3479–3490
Carini P, Steindler L, Beszteri S, Giovannoni SJ (2013) Nutrient requirements for growth of the extreme oligotroph '*Candidatus Pelagibacter ubique*' HTCC1062 on a defined medium. ISME J 7:592–602
Caruana AMN, Malin G (2014) The variability in DMSP content and DMSP lyase activity in marine dinoflagellates. Prog Oceanogr 120:410–424
Chaban B, Ng SYM, Kanbe M, Saltzman I, Nimmo G, Aizawa S-I, Jarrell KF (2007) Systematic deletion analyses of the fla genes in the flagella operon identify several genes essential for proper assembly and function of flagella in the archaeon, *Methanococcus maripaludis*. Mol Microbiol 66:596–609
Cho BC, Azam F (1988) Major role of bacteria in biogeochemical fluxes in the ocean's interior. Nature 332:441–443
Cirri E, Pohnert G (2019) Algae−bacteria interactions that balance the planktonic microbiome. New Phytol 223:100–106
Clerc EE, Raina J-B, Lambert BS, Seymour J, Stocker R (2020) In situ chemotaxis assay to examine microbial behavior in aquatic ecosystems J Vis Exp:e61062
Cole JJ (1982) Interactions between bacteria and algae in aquatic ecosystems. Annu Rev Ecol Syst 13:291–314
Conrad JC, Gibiansky ML, Jin F, Gordon VD, Motto DA, Mathewson MA, Stopka WG, Zelasko DC, Shrout JD, Wong GCL (2011) Flagella and pili-mediated near-surface single-cell motility mechanisms in *P. aeruginosa*. Biophys J 100:1608–1616
Copping AE, Lorenzen CJ (1980) Carbon budget of a marine phytoplankton-herbivore system with carbon-14 as a tracer. Limnol Oceanogr 25:873–882
Cremer J, Honda T, Tang Y, Wong-Ng J, Vergassola M, Hwa T (2019) Chemotaxis as a navigation strategy to boost range expansion. Nature 575:658–663

Croft MT, Lawrence AD, Raux-Deery E, Warren MJ, Smith AG (2005) Algae acquire vitamin B12 through a symbiotic relationship with bacteria. Nature 438:90–93
Currie DJ, Kalff J (1984) A comparison of the abilities of freshwater algae and bacteria to acquire and retain phosphorus. Limnol Oceanogr 29:298–310
Curson ARJ, Todd JD, Sullivan MJ, Johnston AWB (2011) Catabolism of dimethylsulphoniopropionate: microorganisms, enzymes and genes. Nat Rev Microbiol 9: 849–859
Datta MS, Sliwerska E, Gore J, Polz MF, Cordero OX (2016) Microbial interactions lead to rapid micro-scale successions on model marine particles. Nat Commun 7:11965
DeLong EF, Karl DM (2005) Genomic perspectives in microbial oceanography. Nature 437:336–342
DeLong EF, Franks DG, Alldredge AL (1993) Phylogenetic diversity of aggregate-attached- vs. free-living marine bacterial assemblages. Limnol Oceanogr 38:924–934
DeLong EF, Preston CM, Mincer T, Rich V, Hallam SJ, Frigaard N-U, Martinez A, Sullivan MB, Edwards R, Brito BR et al (2006) Community genomics among stratified microbial assemblages in the ocean's interior. Science 311:496–503
Dinsdale EA, Edwards RA, Hall D, Angly F, Breitbart M, Brulc JM, Furlan M, Desnues C, Haynes M, Li L et al (2008) Functional metagenomic profiling of nine biomes. Nature 452: 629–632
Ducklow H, Steinberg D, Buesseler K (2001) Upper Ocean carbon export and the biological pump. Oceanography 14:50–58
Dührkop K, Shen H, Meusel M, Rousu J, Böcker S (2015) Searching molecular structure databases with tandem mass spectra using CSI:FingerID. Proc Natl Acad Sci 112:12580–12585
Durham BP, Sharma S, Luo H, Smith CB, Amin SA, Bender SJ, Dearth SP, Van Mooy BAS, Campagna SR, Kujawinski EB et al (2015) Cryptic carbon and sulfur cycling between surface ocean plankton. Proc Natl Acad Sci 112:453–457
Ebrahimi A, Schwartzman J, Cordero OX (2019) Cooperation and spatial self-organization determine rate and efficiency of particulate organic matter degradation in marine bacteria. Proc Natl Acad Sci 116:23309–23316
Edwards MR, Carlsen RW, Zhuang J, Sitti M (2014) Swimming characterization of *Serratia marcescens* for bio-hybrid micro-robotics. J Micro-Bio Robot 9:47–60
Elyakova LA, Pavlov GM, Isakov VV, Zaitseva II, Stepenchekova TA (1994) Molecular characteristics of subfractions of laminarin. Chem Nat Compd 30:273–274
Emerson S, Hedges J (2008) Chemical oceanography and the marine carbon cycle. Cambridge University Press, Cambridge
Engel A, Thoms S, Riebesell U, Rochelle-Newall E, Zondervan I (2004) Polysaccharide aggregation as a potential sink of marine dissolved organic carbon. Nature 428:929–932
Engel A, Händel N, Wohlers J, Lunau M, Grossart H-P, Sommer U, Riebesell U (2011) Effects of sea surface warming on the production and composition of dissolved organic matter during phytoplankton blooms: results from a mesocosm study. J Plankton Res 33:357–372
Fenchel T (1994) Motility and chemosensory behaviour of the Sulphur bacterium *Thiovulum majus*. Microbiology 140:3109–3116
Fenchel T (2001) Eppur si muove: many water column bacteria are motile. Aquat Microb Ecol 24: 197–201
Fenchel T (2002) Microbial behavior in a heterogeneous world. Science 296:1068–1071
Fenchel T, Thar R (2004) "Candidatus Ovobacter propellens": a large conspicuous prokaryote with an unusual motility behaviour. FEMS Microbiol Ecol 48:231–238
Ferguson RL, Sunda WG (1984) Utilization of amino acids by planktonic marine bacteria: importance of clean technique and low substrate additions1,2. Limnol Oceanogr 29:258–274
Flynn KJ, Clark DR, Xue Y (2008) Modeling the release of dissolved organic matter by phytoplankton. J Phycol 44:1171–1187
Fogg GE (1977) Excretion of organic matter by phytoplankton. Limnol Oceanogr 22:576–577

Follett CL, Repeta DJ, Rothman DH, Xu L, Santinelli C (2014) Hidden cycle of dissolved organic carbon in the deep ocean. Proc Natl Acad Sci 111:16706–16711
Fontanez KM, Eppley JM, Samo TJ, Karl DM, DeLong EF (2015) Microbial community structure and function on sinking particles in the North Pacific subtropical gyre. Front Microbiol 6:469
Fossing H, Gallardo VA, Jørgensen BB, Hüttel M, Nielsen LP, Schulz H, Canfield DE, Forster S, Glud RN, Gundersen JK et al (1995) Concentration and transport of nitrate by the mat-forming Sulphur bacterium *Thioploca*. Nature 374:713–715
Furusawa G, Yoshikawa T, Yasuda A, Sakata T (2011) Algicidal activity and gliding motility of *Saprospira* sp. SS98-5. Can J Microbiol 49:92–100
Ganesh S, Parris DJ, DeLong EF, Stewart FJ (2014) Metagenomic analysis of size-fractionated picoplankton in a marine oxygen minimum zone. ISME J 8:187–211
Gao C, Fernandez VI, Lee KS, Fenizia S, Pohnert G, Seymour JR, Raina J-B, Stocker R (2020) Single-cell bacterial transcription measurements reveal the importance of dimethylsulfoniopropionate (DMSP) hotspots in ocean sulfur cycling. Nat Commun 11:1942
Gärdes A, Iversen MH, Grossart H-P, Passow U, Ullrich MS (2011) Diatom-associated bacteria are required for aggregation of *Thalassiosira weissflogii*. ISME J 5:436–445
Garren M, Son K, Raina J-B, Rusconi R, Menolascina F, Shapiro OH, Tout J, Bourne DG, Seymour JR, Stocker R (2014) A bacterial pathogen uses dimethylsulfoniopropionate as a cue to target heat-stressed corals. ISME J 8:999–1007
Garren M, Son K, Tout J, Seymour JR, Stocker R (2016) Temperature-induced behavioral switches in a bacterial coral pathogen. ISME J 10:1363–1372
Gifford SM, Sharma S, Booth M, Moran MA (2013) Expression patterns reveal niche diversification in a marine microbial assemblage. ISME J 7:281–298
Gilbert JA, Field D, Swift P, Thomas S, Cummings D, Temperton B, Weynberg K, Huse S, Hughes M, Joint I et al (2010) The taxonomic and functional diversity of microbes at a temperate coastal site: a 'multi-omic' study of seasonal and diel temporal variation. PLoS One 5:e15545
Giovannoni SJ (2017) SAR11 bacteria: the most abundant plankton in the oceans. Annu Rev Mar Sci 9:231–255
Giovannoni SJ, Tripp HJ, Givan S, Podar M, Vergin KL, Baptista D, Bibbs L, Eads J, Richardson TH, Noordewier M et al (2005) Genome streamlining in a cosmopolitan oceanic bacterium. Science 309:1242–1245
Glöckner FO, Kube M, Bauer M, Teeling H, Lombardot T, Ludwig W, Gade D, Beck A, Borzym K, Heitmann K et al (2003) Complete genome sequence of the marine planctomycete *Pirellula* sp. strain 1. Proc Natl Acad Sci 100:8298–8303
Glud RN, Grossart H-P, Larsen M, Tang KW, Arendt KE, Rysgaard S, Thamdrup B, Nielsen TG (2015) Copepod carcasses as microbial hot spots for pelagic denitrification. Limnol Oceanogr 60:2026–2036
Goldman JC, McCarthy JJ, Peavey DG (1979) Growth rate influence on the chemical composition of phytoplankton in oceanic waters. Nature 279:210–215
Gosink KK, Kobayashi R, Kawagishi I, Häse CC (2002) Analyses of the roles of the three CheA homologs in chemotaxis of *vibrio cholerae*. J Bacteriol 184:1767–1771
Granum E (2002) Metabolism and function of b-1,3-glucan in marine diatoms. PhD Thesis, Department of Biotechnology, Faculty of Natural Sciences and Technology, Norwegian University of Science and Technology (NTNU), Trondheim, Norway
Granum E, Kirkvold S, Myklestad SM (2002) Cellular and extracellular production of carbohydrates and amino acids by the marine diatom *Skeletonema costatum*: diel variations and effects of N depletion. Mar Ecol Prog Ser 242:83–94
Grossart H-P, Simon M (1998) Bacterial colonization and microbial decomposition of limnetic organic aggregates (lake snow). Aq Microb Ecol 15:127–140
Grossart H, Riemann L, Azam F (2001) Bacterial motility in the sea and its ecological implications. Aq Microb Ecol 25:247–258

Guadayol Ò, Thornton KL, Humphries S (2017) Cell morphology governs directional control in swimming bacteria. Sci Rep 7:1–13
Guasto JS, Rusconi R, Stocker R (2012) Fluid mechanics of planktonic microorganisms. Annu Rev Fluid Mech 44:373–400
Gude S, Pinçe E, Taute KM, Seinen A-B, Shimizu TS, Tans SJ (2020) Bacterial coexistence driven by motility and spatial competition. Nature 578:588–592
Guidi L, Chaffron S, Bittner L, Eveillard D, Larhlimi A, Roux S, Darzi Y, Audic S, Berline L, Brum JR (2016) Plankton networks driving carbon export in the oligotrophic ocean. Nature 532:465–470
Haas AF, Wild C (2010) Composition analysis of organic matter released by cosmopolitan coral reef-associated green algae. Aq Biol 10:131–138
Hamer R, Chen P-Y, Armitage JP, Reinert G, Deane CM (2010) Deciphering chemotaxis pathways using cross species comparisons. BMC Syst Biol 4:3
Hansell DA (2013) Recalcitrant dissolved organic carbon fractions. Annu Rev Mar Sci 5:421–445
Hansell DA, Carlson CA (1998) Net community production of dissolved organic carbon. Glob Biogeochem Cycles 12:443–453
Hartmann AC, Petras D, Quinn RA, Protsyuk I, Archer FI, Ransome E, Williams GJ, Bailey BA, Vermeij MJA, Alexandrov T et al (2017) Meta-mass shift chemical profiling of metabolomes from coral reefs. Proc Natl Acad Sci 114:11685–11690
Harwood CS, Fosnaugh K, Dispensa M (1989) Flagellation of *pseudomonas putida* and analysis of its motile behavior. J Bacteriol 171:4063–4066
Hellebust JA (1965) Excretion of some organic compounds by marine phytoplankton. Limnol Oceanogr 10:192–206
Herzog B, Wirth R (2012) Swimming behavior of selected species of archaea. Appl Environ Microbiol 78:1670–1674
Hess JF, Oosawa K, Kaplan N, Simon MI (1988) Phosphorylation of three proteins in the signaling pathway of bacterial chemotaxis. Cell 53:79–87
Hildebrand E, Schimz A (1986) Integration of photosensory signals in *Halobacterium halobium*. J Bacteriol 167:305–311
Hirota N, Imae Y (1983) Na+−driven flagellar motors of an alkalophilic *bacillus* strain YN-1. J Biol Chem 258:10577–10581
Hodson RE, Azam F, Carlucci AF, Fuhrman JA, Karl DM, Holm-Hansen O (1981) Microbial uptake of dissolved organic matter in Mcmurdo sound, Antarctica. Mar Biol 61:89–94
Hollibaugh JT, Azam F (1983) Microbial degradation of dissolved proteins in seawater. Limnol Oceanogr 28:1104–1116
Hook AL, Flewellen JL, Dubern J-F, Carabelli AM, Zaid IM, Berry RM, Wildman RD, Russell N, Williams P, Alexander MR (2019) Simultaneous tracking of *Pseudomonas aeruginosa* motility in liquid and at the solid-liquid interface reveals differential roles for the flagellar stators. MSystems 4:5
Howard EC, Henriksen JR, Buchan A, Reisch CR, Bürgmann H, Welsh R, Ye W, González JM, Mace K, Joye SB et al (2006) Bacterial taxa that limit sulfur flux from the ocean. Science 314: 649–652
Hütz A, Schubert K, Overmann J (2011) *Thalassospira* sp. isolated from the oligotrophic eastern Mediterranean Sea exhibits chemotaxis toward inorganic phosphate during starvation. Appl Environ Microbiol 77:4412–4421
Isao K, Hara S, Terauchi K, Kogure K (1990) Role of sub-micrometre particles in the ocean. Nature 345:242–244
Jackson GA (1980) Phytoplankton growth and zooplankton grazing in oligotrophic oceans. Nature 284:439–441
Jackson GA (1987) Simulating chemosensory responses of marine microorganisms. Limnol Oceanogr 32:1253–1266
Jackson GA (1989) Simulation of bacterial attraction and adhesion to falling particles in an aquatic environment. Limnol Oceanogr 34:514–530

Jackson GA (1990) A model of the formation of marine algal flocs by physical coagulation processes. Deep Sea Res A 37:1197–1211
Jackson GA (2012) Seascapes: the world of aquatic organisms as determined by their particulate natures. J Exp Biol 215:1017–1030
Jacobsen TR, Azam F (1984) Role of bacteria in copepod fecal pellet decomposition: colonization, growth rates and mineralization. Bull Mar Sci 35:495–502
Jarrell KF, Albers S-V (2012) The archaellum: an old motility structure with a new name. Trends Microbiol 20:307–312
Jarrell KF, Bayley DP, Florian V, Klein A (1996) Isolation and characterization of insertional mutations in flagellin genes in the archaeon *Methanococcus voltae*. Mol Microbiol 20:657–666
Jenkinson IR, Sun XX, Seuront L (2015) Thalassorheology, organic matter and plankton: towards a more viscous approach in plankton ecology. J Plankton Res 37:1100–1109
Jiao N, Herndl GJ, Hansell DA, Benner R, Kattner G, Wilhelm SW, Kirchman DL, Weinbauer MG, Luo T, Chen F et al (2010) Microbial production of recalcitrant dissolved organic matter: long-term carbon storage in the global ocean. Nat Rev Microbiol 8:593–599
Johansen JE, Pinhassi J, Blackburn N, Zweifel UL, Hagström Å (2002) Variability in motility characteristics among marine bacteria. Aq Microb Ecol 28:229–237
Johnson AR, Wiens JA, Milne BT, Crist TO (1992) Animal movements and population dynamics in heterogeneous landscapes. Landsc Ecol 7:63–75
Jones AK, Cannon RC (1986) The release of micro-algal photosynthate and associated bacterial uptake and heterotrophic growth. Br Phycol J 21:341–358
Jørgensen BB, Revsbech NP (1983) Colorless sulfur bacteria, *Beggiatoa* spp. and *Thiovulum* spp., in $O_2$ and $H_2S$ microgradients. Appl Environ Microbiol 45:1261–1270
Jørgensen BB, Findlay AJ, Pellerin A (2019) The biogeochemical sulfur cycle of marine sediments. Front Microbiol 10:849
Jover LF, Effler TC, Buchan A, Wilhelm SW, Weitz JS (2014) The elemental composition of virus particles: implications for marine biogeochemical cycles. Nat Rev Microbiol 12:519–528
Kaiser K, Benner R (2012) Organic matter transformations in the upper mesopelagic zone of the North Pacific: chemical composition and linkages to microbial community structure. J Geophys Res Oceans 117:C1
Kalisky T, Quake SR (2011) Single-cell genomics. Nat Methods 8:311–314
Kallmeyer J, Pockalny R, Adhikari RR, Smith DC, D'Hondt S (2012) Global distribution of microbial abundance and biomass in subseafloor sediment. Proc Natl Acad Sci 109:16213–16216
Keller MD (1989) Dimethyl sulfide production and marine phytoplankton: the importance of species composition and cell size. Biol Oceanogr 6:375–382
Kessler RW, Weiss A, Kuegler S, Hermes C, Wichard T (2018) Macroalgal–bacterial interactions: role of dimethylsulfoniopropionate in microbial gardening by *Ulva* (*Chlorophyta*). Mol Ecol 27: 1808–1819
Khan S, Scholey JM (2018) Assembly, functions and evolution of archaella, flagella and cilia. Curr Biol 28:R278–R292
Kiene RP, Linn LJ, Bruton JA (2000) New and important roles for DMSP in marine microbial communities. J Sea Res 43:209–224
Kim C, Jackson M, Lux R, Khan S (2001) Determinants of chemotactic signal amplification in *Escherichia coli*. J Mol Biol 307:119–135
Kim S, Kramer RW, Hatcher PG (2003) Graphical method for analysis of ultrahigh-resolution broadband mass spectra of natural organic matter, the van Krevelen diagram. Anal Chem 75: 5336–5344
Kinosita Y, Uchida N, Nakane D, Nishizaka T (2016) Direct observation of rotation and steps of the archaellum in the swimming halophilic archaeon *Halobacterium salinarum*. Nat Microbiol 1:1–9

Kinosita Y, Kikuchi Y, Mikami N, Nakane D, Nishizaka T (2018) Unforeseen swimming and gliding mode of an insect gut symbiont, *Burkholderia* sp. RPE64, with wrapping of the flagella around its cell body. ISME J 12:838–848
Kiørboe T (2008) A mechanistic approach to plankton ecology. Princeton University Press, Princeton
Kiørboe T, Jackson GA (2001) Marine snow, organic solute plumes, and optimal chemosensory behavior of bacteria. Limnol Oceanogr 46:1309–1318
Kiørboe T, Ploug H, Thygesen U (2001) Fluid motion and solute distribution around sinking aggregates. I. Small-scale fluxes and heterogeneity of nutrients in the pelagic environment. Mar Ecol Prog Ser 211:1–13
Kiørboe T, Grossart H-P, Ploug H, Tang K (2002) Mechanisms and rates of bacterial colonization of sinking aggregates. Appl Environ Microbiol 68:3996–4006
Klawonn I, Bonaglia S, Brüchert V, Ploug H (2015) Aerobic and anaerobic nitrogen transformation processes in $N_2$ -fixing cyanobacterial aggregates. ISME J 9:1456–1466
Kühl M, Cohen Y, Dalsgaard T, Jørgensen BB, Revsbech NP (1995) Microenvironment and photosynthesis of zooxanthellae in scleractinian corals studied with microsensors for $O_2$, pH and light. Mar Ecol Prog Ser 117:159–172
Kühn MJ, Schmidt FK, Eckhardt B, Thormann KM (2017) Bacteria exploit a polymorphic instability of the flagellar filament to escape from traps. Proc Natl Acad Sci 114:6340–6345
Kujawinski EB, Longnecker K, Barott KL, Weber RJM, Kido Soule MC (2016) Microbial community structure affects marine dissolved organic matter composition. Front Mar Sci 3:45
Lacal J, García-Fontana C, Muñoz-Martínez F, Ramos J-L, Krell T (2010) Sensing of environmental signals: classification of chemoreceptors according to the size of their ligand binding regions. Environ Microbiol 12:2873–2884
Lambert BS, Raina J-B, Fernandez VI, Rinke C, Siboni N, Rubino F, Hugenholtz P, Tyson GW, Seymour JR, Stocker R (2017) A microfluidics-based in situ chemotaxis assay to study the behaviour of aquatic microbial communities. Nat Microbiol 2:1344–1349
Lambert BS, Fernandez VI, Stocker R (2019) Motility drives bacterial encounter with particles responsible for carbon export throughout the ocean. Limnol Oceanogr Lett 4:113–118
Lampert W (1978) Release of dissolved organic carbon by grazing zooplankton. Limnol Oceanogr 23:831–834
Lancelot C (1984) Extracellular release of small and large molecules by phytoplankton in the southern bight of the North Sea. Est Coast Shelf Sci 18:65–77
Landa M, Burns AS, Roth SJ, Moran MA (2017) Bacterial transcriptome remodeling during sequential co-culture with a marine dinoflagellate and diatom. ISME J 11:2677–2690
Lassak K, Neiner T, Ghosh A, Klingl A, Wirth R, Albers S-V (2012) Molecular analysis of the crenarchaeal flagellum. Mol Microbiol 83:110–124
Lauro FM, Bartlett DH (2008) Prokaryotic lifestyles in deep sea habitats. Extremophiles 12:15–25
Lauro FM, McDougald D, Thomas T, Williams TJ, Egan S, Rice S, DeMaere MZ, Ting L, Ertan H, Johnson J et al (2009) The genomic basis of trophic strategy in marine bacteria. Proc Natl Acad Sci 106:15527–15533
Lawson CA, Seymour JR, Possell M, Suggett DJ, Raina J-B (2020) The volatilomes of symbiodiniaceae-associated bacteria are influenced by chemicals derived from their algal partner. Front Mar Sci 7:106
Lee KS, Palatinszky M, Pereira FC, Nguyen J, Fernandez VI, Mueller AJ, Menolascina F, Daims H, Berry D, Wagner M (2019) An automated Raman-based platform for the sorting of live cells by functional properties. Nat Microbiol 4:1035–1048
Lee KS, Wagner M, Stocker R (2020) Raman-based sorting of microbial cells to link functions to their genes. Microb Cell 7:62
Lehman JT, Scavia D (1982a) Microscale nutrient patches produced by zooplankton. Proc Natl Acad Sci 79:5001–5005
Lehman JT, Scavia D (1982b) Microscale patchiness of nutrients in plankton communities. Science 216:729–730

Leifson E, Cosenza BJ, Murchelano R, Cleverdon RC (1964) Motile marine bacteria. Techniques, ecology, and general characteristics. J Bacteriol 87:652–666
Li N, Kojima S, Homma M (2011) Sodium-driven motor of the polar flagellum in marine bacteria *vibrio*. Genes Cells 16:985–999
Lidbury I, Murrell JC, Chen Y (2014) Trimethylamine N-oxide metabolism by abundant marine heterotrophic bacteria. Proc Natl Acad Sci 111:2710–2715
Liu H, Nolla H, Campbell L (1997) *Prochlorococcus* growth rate and contribution to primary production in the equatorial and subtropical North Pacific Ocean. Aq Microb Ecol 12:39–47
Liu Q, Lu X, Tolar BB, Mou X, Hollibaugh JT (2015) Concentrations, turnover rates and fluxes of polyamines in coastal waters of the South Atlantic bight. Biogeochemistry 123:117–133
Liu W, Cremer J, Li D, Hwa T, Liu C (2019) An evolutionarily stable strategy to colonize spatially extended habitats. Nature 575:664–668
Lønborg C, Middelboe M, Brussaard CPD (2013) Viral lysis of *Micromonas pusilla*: impacts on dissolved organic matter production and composition. Biogeochemistry 116:231–240
Long RA, Azam F (2001) Microscale patchiness of bacterioplankton assemblage richness in seawater. Aq Microb Ecol 26:103–113
López-Pérez M, Gonzaga A, Martin-Cuadrado A-B, Onyshchenko O, Ghavidel A, Ghai R, Rodriguez-Valera F (2012) Genomes of surface isolates of *Alteromonas macleodii*: the life of a widespread marine opportunistic copiotroph. Sci Rep 2:696
Luchsinger RH, Bergersen B, Mitchell JG (1999) Bacterial swimming strategies and turbulence. Biophys J 77:2377–2386
Lux R, Shi W (2004) Chemotaxis-guided movements in bacteria. Crit Rev Oral Biol Med Off Publ Am Assoc Oral Biol 15:207–220
Ma X, Coleman ML, Waldbauer JR (2018) Distinct molecular signatures in dissolved organic matter produced by viral lysis of marine cyanobacteria. Environ Microbiol 20:3001–3011
Macnab RM (1996) Flagella and motility. *Escherichia Coli* and *Salmonella*. In: Cellular and molecular biology. ASM Press, Washington DC, p 2822
Macnab RM (2003) How bacteria assemble flagella. Annu Rev Microbiol 57:77–100
Magariyama Y, Sugiyama S, Muramoto K, Maekawa Y, Kawagishi I, Imae Y, Kudo S (1994) Very fast flagellar rotation. Nature 371:752–752
Mandel MJ, Schaefer AL, Brennan CA, Heath-Heckman EAC, DeLoney-Marino CR, McFall-Ngai MJ, Ruby EG (2012) Squid-derived chitin oligosaccharides are a chemotactic signal during colonization by *Vibrio fischeri*. Appl Environ Microbiol 78:4620–4626
Manson MD, Tedesco P, Berg HC, Harold FM, van der Drift C (1977) A protonmotive force drives bacterial flagella. Proc Natl Acad Sci 74:3060–3064
Margulis L (1981) Symbiosis in cell evolution: life and its environment on the early earth. Freeman, San Francisco
Mari X, Passow U, Migon C, Burd AB, Legendre L (2017) Transparent exopolymer particles: effects on carbon cycling in the ocean. Prog Oceanogr 151:13–37
Matanza XM, Osorio CR (2018) Transcriptome changes in response to temperature in the fish pathogen *Photobacterium damselae* subsp. *damselae*: clues to understand the emergence of disease outbreaks at increased seawater temperatures. PLoS One 13:e0210118
Matz C, Jürgens K (2005) High motility reduces grazing mortality of planktonic bacteria. Appl Environ Microbiol 71:921–929
McCarren J, Brahamsha B (2007) SwmB, a 1.12-megadalton protein that is required for nonflagellar swimming motility in *Synechococcus*. J Bacteriol 189:1158–1162
McCarren J, Heuser J, Roth R, Yamada N, Martone M, Brahamsha B (2005) Inactivation of S*wmA* results in the loss of an outer cell layer in a swimming *Synechococcus* strain. J Bacteriol 187: 224–230
McCarren J, Becker JW, Repeta DJ, Shi Y, Young CR, Malmstrom RR, Chisholm SW, DeLong EF (2010) Microbial community transcriptomes reveal microbes and metabolic pathways associated with dissolved organic matter turnover in the sea. Proc Natl Acad Sci 107:16420–16427

McCarthy JJ, Goldman JC (1979) Nitrogenous nutrition of marine phytoplankton in nutrient-depleted waters. Science 203:670–672

McEvoy MM, Bren A, Eisenbach M, Dahlquist FW (1999) Identification of the binding interfaces on CheY for two of its targets the phosphatase *CheZ* and the flagellar switch protein FliM11. J Mol Biol 289:1423–1433

Middelboe M (2000) Bacterial growth rate and marine virus-host dynamics. Microb Ecol 40:114–124

Middelboe M, Jørgensen NOG (2006) Viral lysis of bacteria: an important source of dissolved amino acids and cell wall compounds. J Mar Biol Assoc 86:605–612

Miller TR, Belas R (2004) Dimethylsulfoniopropionate metabolism by *Pfiesteria*-associated *Roseobacter* spp. Appl Environ Microbiol 70:3383–3391

Miller TR, Belas R (2006) Motility is involved in *Silicibacter* sp. TM1040 interaction with dinoflagellates. Environ Microbiol 8:1648–1659

Miller TR, Hnilicka K, Dziedzic A, Desplats P, Belas R (2004) Chemotaxis of *Silicibacter* sp. strain TM1040 toward dinoflagellate products. Appl Environ Microbiol 70:4692–4701

Minniti G, Hagen LH, Porcellato D, Jørgensen SM, Pope PB, Vaaje-Kolstad G (2017) The skin-mucus microbial community of farmed Atlantic salmon (*Salmo salar*). Front Microbiol 8:2043

Mitchell JG (1991) The influence of cell size on marine bacterial motility and energetics. Microb Ecol 22:227–238

Mitchell JG, Kogure K (2006) Bacterial motility: links to the environment and a driving force for microbial physics. FEMS Microbiol Ecol 55:3–16

Mitchell JG, Okubo A, Fuhrman JA (1985) Microzones surrounding phytoplankton form the basis for a stratified marine microbial ecosystem. Nature 316:58–59

Mitchell JG, Pearson L, Bonazinga A, Dillon S, Khouri H, Paxinos R (1995a) Long lag times and high velocities in the motility of natural assemblages of marine bacteria. Appl Environ Microbiol 61:877–882

Mitchell JG, Pearson L, Dillon S, Kantalis K (1995b) Natural assemblages of marine bacteria exhibiting high-speed motility and large accelerations. Appl Environ Microbiol 61:4436–4440

Mitchell JG, Pearson L, Dillon S (1996) Clustering of marine bacteria in seawater enrichments. Appl Environ Microbiol 62:3716–3721

Mittelviefhaus M, Müller DB, Zambelli T, Vorholt JA (2019) A modular atomic force microscopy approach reveals a large range of hydrophobic adhesion forces among bacterial members of the leaf microbiota. ISME J 13:1878–1882

Møller EF (2005) Sloppy feeding in marine copepods: prey-size-dependent production of dissolved organic carbon. J Plankton Res 27:27–35

Möller KO, John MS, Temming A, Floeter J, Sell AF, Herrmann J-P, Möllmann C (2012) Marine snow, zooplankton and thin layers: indications of a trophic link from small-scale sampling with the video plankton recorder. Mar Ecol Prog Ser 468:57–69

Moore LR, Rocap G, Chisholm SW (1998) Physiology and molecular phylogeny of coexisting *Prochlorococcus* ecotypes. Nature 393:464–467

Moore ER, Davie-Martin CL, Giovannoni SJ, Halsey KH (2020) *Pelagibacter* metabolism of diatom-derived volatile organic compounds imposes an energetic tax on photosynthetic carbon fixation. Environ Microbiol 22:1720–1733

Mopper K, Schultz CA, Chevolot L, Germain C, Revuelta R, Dawson R (1992) Determination of sugars in unconcentrated seawater and other natural waters by liquid chromatography and pulsed amperometric detection. Environ Sci Technol 26:133–138

Moran MA, Buchan A, González JM, Heidelberg JF, Whitman WB, Kiene RP, Henriksen JR, King GM, Belas R, Fuqua C et al (2004) Genome sequence of *Silicibacter pomeroyi* reveals adaptations to the marine environment. Nature 432:910–913

Moran MA, Kujawinski EB, Stubbins A, Fatland R, Aluwihare LI, Buchan A, Crump BC, Dorrestein PC, Dyhrman ST, Hess NJ et al (2016) Deciphering Ocean carbon in a changing world. Proc Natl Acad Sci 113:3143–3151

Moriarty DJW, Iverson RL, Pollard PC (1986) Exudation of organic carbon by the seagrass *Halodule wrightii* Aschers and its effect on bacterial growth in the sediment. J Exp Mar Biol Ecol 96:115–126
Morris RM, Rappé MS, Connon SA, Vergin KL, Siebold WA, Carlson CA, Giovannoni SJ (2002) SAR11 clade dominates ocean surface bacterioplankton communities. Nature 420:806–810
Mühlenbruch M, Grossart H-P, Eigemann F, Voss M (2018) Mini-review: phytoplankton-derived polysaccharides in the marine environment and their interactions with heterotrophic bacteria. Environ Microbiol 20:2671–2685
Muramoto K, Kawagishi I, Kudo S, Magariyama Y, Imae Y, Homma M (1995) High-speed rotation and speed stability of the sodium-driven flagellar motor in *vibrio alginolyticus*. J Mol Biol 251: 50–58
Ni B, Colin R, Link H, Endres RG, Sourjik V (2020) Growth-rate dependent resource investment in bacterial motile behavior quantitatively follows potential benefit of chemotaxis. Proc Natl Acad Sci 117:595–601
Nyholm SV, McFall-Ngai M (2004) The winnowing: establishing the squid– *vibrio* symbiosis. Nat Rev Microbiol 2:632–642
Nyholm SV, Stabb EV, Ruby EG, McFall-Ngai MJ (2000) Establishment of an animal–bacterial association: recruiting symbiotic vibrios from the environment. Proc Natl Acad Sci 97:10231–10235
Ochman H, Moran NA (2001) Genes lost and genes found: evolution of bacterial pathogenesis and symbiosis. Science 292:1096–1099
Ochsenkühn MA, Schmitt-Kopplin P, Harir M, Amin SA (2018) Coral metabolite gradients affect microbial community structures and act as a disease cue. Comm Biol 1:1–10
Ogawa H, Amagai Y, Koike I, Kaiser K, Benner R (2001) Production of refractory dissolved organic matter by bacteria. Science 292:917–920
Osterholz H, Niggemann J, Giebel H-A, Simon M, Dittmar T (2015) Inefficient microbial production of refractory dissolved organic matter in the ocean. Nat Commun 6:7422
Overmann J, Lepleux C (2016) Marine bacteria and archaea: diversity, adaptations, and culturability. In: Stal LJ, Cretoiu MS (eds) The marine microbiome: an untapped source of biodiversity and biotechnological potential. Springer International Publishing, Cham, pp 21–55
Paerl HW, Prufert LE (1987) Oxygen-poor microzones as potential sites of microbial $N_2$ fixation in nitrogen-depleted aerobic marine waters. Appl Environ Microbiol 53:1078–1087
Panagiotopoulos C, Pujo-Pay M, Benavides M, Van Wambeke F, Sempere R (2019) The composition and distribution of semi-labile dissolved organic matter across the Southwest Pacific. Biogeosciences 16:105–116
Passow U (2002) Production of transparent exopolymer particles (TEP) by phyto- and bacterioplankton. Mar Ecol Prog Ser 236:1–12
Paul JH, DeFlaun MF, Jeffrey WH (1986) Elevated levels of microbial activity in the coral surface microlayer. Mar Ecol Prog Ser 33:29–40
Pedersen JN, Bombar D, Paerl RW, Riemann L (2018) Diazotrophs and $N_2$-fixation associated with particles in coastal estuarine waters. Front Microbiol 9:2759
Peduzzi P, Herndl GJ (1992) Zooplankton activity fueling the microbial loop: differential growth response of bacteria from oligotrophic and eutrophic waters. Limnol Oceanogr 37:1087–1092
Petroff A, Libchaber A (2014) Hydrodynamics and collective behavior of the tethered bacterium *Thiovulum majus*. Proc Natl Acad Sci 111:E537–E545
Petroff AP, Wu X-L, Libchaber A (2015) Fast-moving bacteria self-organize into active two-dimensional crystals of rotating cells. Phys Rev Lett 114:158102
Philippot L, Raaijmakers JM, Lemanceau P, van der Putten WH (2013) Going back to the roots: the microbial ecology of the rhizosphere. Nat Rev Microbiol 11:789–799
Ping L, Birkenbeil J, Monajembashi S (2013) Swimming behavior of the monotrichous bacterium *Pseudomonas fluorescens* SBW25. FEMS Microbiol Ecol 86:36–44

Ploug H, Grossart H-P (2000) Bacterial growth and grazing on diatom aggregates: respiratory carbon turnover as a function of aggregate size and sinking velocity. Limnol Oceanogr 45: 1467–1475

Pogoreutz C, Voolstra CR, Rädecker N, Weis V, Cardenas A, Raina J-B (2021) The coral holobiont highlights the dependence of cnidarian animal hosts on their associated microbes. In: Cellular dialogues in the holobiont. CRC Press, Boca Raton, FL, pp 91–118

Pollock FJ, McMinds R, Smith S, Bourne DG, Willis BL, Medina M, Thurber RV, Zaneveld JR (2018) Coral-associated bacteria demonstrate phylosymbiosis and cophylogeny. Nat Commun 9:4921

Poretsky RS, Sun S, Mou X, Moran MA (2010) Transporter genes expressed by coastal bacterioplankton in response to dissolved organic carbon. Environ Microbiol 12:616–627

Purcell EM (1977) Life at low Reynolds number. Am J Phys 45:3–11

Purcell EM (1997) The efficiency of propulsion by a rotating flagellum. Proc Natl Acad Sci 94: 11307–11311

Quax TEF, Altegoer F, Rossi F, Li Z, Rodriguez-Franco M, Kraus F, Bange G, Albers S-V (2018) Structure and function of the archaeal response regulator CheY. Proc Natl Acad Sci 115:E1259–E1268

Quelas JI, Althabegoiti MJ, Jimenez-Sanchez C, Melgarejo AA, Marconi VI, Mongiardini EJ, Trejo SA, Mengucci F, Ortega-Calvo J-J, Lodeiro AR (2016) Swimming performance of *Bradyrhizobium diazoefficiens* is an emergent property of its two flagellar systems. Sci Rep 6: 23841

Rahav E, Giannetto MJ, Bar-Zeev E (2016) Contribution of mono and polysaccharides to heterotrophic $N_2$ fixation at the eastern Mediterranean coastline. Sci Rep 6:27858

Raina J-B, Fernandez V, Lambert B, Stocker R, Seymour JR (2019) The role of microbial motility and chemotaxis in symbiosis. Nat Rev Microbiol 17:284–294

Ramsing NB, Kühl M, Jørgensen BB (1993) Distribution of sulfate-reducing bacteria, $O_2$, and $H_2S$ in photosynthetic biofilms determined by oligonucleotide probes and microelectrodes. Appl Environ Microbiol 59:3840–3849

Riemann L, Middelboe M (2002) Viral lysis of marine bacterioplankton: implications for organic matter cycling and bacterial clonal composition. Ophelia 56:57–68

Riemann L, Steward GF, Azam F (2000) Dynamics of bacterial community composition and activity during a mesocosm diatom bloom. Appl Environ Microbiol 66:578–587

Rinke C, Rubino F, Messer LF, Youssef N, Parks DH, Chuvochina M, Brown M, Jeffries T, Tyson GW, Seymour JR et al (2019) A phylogenomic and ecological analysis of the globally abundant marine group II archaea (*ca*. Poseidoniales Ord. Nov.). ISME J 13:663–675

Rohwer F, Breitbart M, Jara J, Azam F, Knowlton N (2001) Diversity of bacteria associated with the Caribbean coral *Montastraea franksi*. Coral Reefs 20:85–91

Rohwer F, Seguritan V, Azam F, Knowlton N (2002) Diversity and distribution of coral-associated bacteria. Mar Ecol Prog Ser 243:1–10

Rossel PE, Bienhold C, Boetius A, Dittmar T (2016) Dissolved organic matter in pore water of Arctic Ocean sediments: environmental influence on molecular composition. Org Geochem 97: 41–52

Ruby EG, Urbanowski M, Campbell J, Dunn A, Faini M, Gunsalus R, Lostroh P, Lupp C, McCann J, Millikan D et al (2005) Complete genome sequence of *Vibrio fischeri*: a symbiotic bacterium with pathogenic congeners. Proc Natl Acad Sci 102:3004–3009

Rusch DB, Halpern AL, Sutton G, Heidelberg KB, Williamson S, Yooseph S, Wu D, Eisen JA, Hoffman JM, Remington K et al (2007) The sorcerer II Global Ocean sampling expedition: Northwest Atlantic through eastern tropical Pacific. PLoS Biol 5:e77

Saba GK, Steinberg DK, Bronk DA (2011) The relative importance of sloppy feeding, excretion, and fecal pellet leaching in the release of dissolved carbon and nitrogen by *Acartia tonsa* copepods. J Exp Mar Biol Ecol 404:47–56

Salah Ud-Din AIM, Roujeinikova A (2017) Methyl-accepting chemotaxis proteins: a core sensing element in prokaryotes and archaea. Cell Mol Life Sci 74:3293–3303

Salek MM, Carrara F, Fernandez V, Guasto JS, Stocker R (2019) Bacterial chemotaxis in a microfluidic T-maze reveals strong phenotypic heterogeneity in chemotactic sensitivity. Nat Commun 10:1877
Sar N, Rosenberg E (1987) Fish skin bacteria: colonial and cellular hydrophobicity. Microb Ecol 13:193–202
Schlesner M, Miller A, Streif S, Staudinger WF, Müller J, Scheffer B, Siedler F, Oesterhelt D (2009) Identification of archaea-specific chemotaxis proteins which interact with the flagellar apparatus. BMC Microbiol 9:56
Schuech R, Hoehfurtner T, Smith DJ, Humphries S (2019) Motile curved bacteria are Pareto-optimal. Proc Natl Acad Sci 116:14440–14447
Schulz HN, Jørgensen BB (2001) Big bacteria. Annu Rev Microbiol 55:105–137
Segall JE, Manson MD, Berg HC (1982) Signal processing times in bacterial chemotaxis. Nature 296:855–857
Segall JE, Block SM, Berg HC (1986) Temporal comparisons in bacterial chemotaxis. Proc Natl Acad Sci 83:8987–8991
Seyedsayamdost MR, Case RJ, Kolter R, Clardy J (2011) The Jekyll-and-Hyde chemistry of *Phaeobacter gallaeciensis*. Nat Chem 3:331–335
Seymour JR, Mitchell JG, Pearson L, Waters RL (2000) Heterogeneity in bacterioplankton abundance from 4.5 millimetre resolution sampling. Aq Microb Ecol 22:143–153
Seymour JR, Marcos, Stocker R (2009a) Resource patch formation and exploitation throughout the marine microbial food web. Am Nat 173:E15–E29
Seymour JR, Ahmed T, Stocker R (2009b) Bacterial chemotaxis towards the extracellular products of the toxic phytoplankton *Heterosigma akashiwo*. J Plankton Res 31:1557–1561
Seymour JR, Simó R, Ahmed T, Stocker R (2010a) Chemoattraction to dimethylsulfoniopropionate throughout the marine microbial food web. Science 329:342–345
Seymour JR, Ahmed T, Durham WM, Stocker R (2010b) Chemotactic response of marine bacteria to the extracellular products of *Synechococcus* and *Prochlorococcus*. Aq Microb Ecol 59:161–168
Seymour JR, Amin SA, Raina J-B, Stocker R (2017) Zooming in on the phycosphere: the ecological interface for phytoplankton–bacteria relationships. Nat Microbiol 2:1–12
Shahapure R, Driessen RPC, Haurat MF, Albers S-V, Dame RT (2014) The archaellum: a rotating type IV pilus. Mol Microbiol 91:716–723
Shi Y, McCarren J, DeLong EF (2012) Transcriptional responses of surface water marine microbial assemblages to deep-sea water amendment. Environ Microbiol 14:191–206
Shigematsu M, Meno Y, Misumi H, Amako K (1995) The measurement of swimming velocity of *vibrio cholerae* and *Pseudomonas aeruginosa* using the video tracking method. Microbiol Immunol 39:741–744
Shoemaker KM, Duhamel S, Moisander PH (2019) Copepods promote bacterial community changes in surrounding seawater through farming and nutrient enrichment. Environ Microbiol 21:3737–3750
Shotts EB, Albert TF, Wooley RE, Brown J (1990) Microflora associated with the skin of the bowhead whale (*Balaena mysticetus*). J Wildl Dis 26:351–359
Sievert SM, Kiene RP, Schulz-Vogt HN (2007) The sulfur cycle. Oceanography 20:117–123
Silverman M, Simon M (1974) Flagellar rotation and the mechanism of bacterial motility. Nature 249:73–74
Simó R (2001) Production of atmospheric sulfur by oceanic plankton: biogeochemical, ecological and evolutionary links. Trends Ecol Evol 16:287–294
Simon M, Grossart H-P, Schweitzer B, Ploug H (2002) Microbial ecology of organic aggregates in aquatic ecosystems. Aq Microb Ecol 28:175–211
Słomka J, Stocker R (2020) Bursts characterize coagulation of rods in a quiescent fluid. Phys Rev Lett 124:258001
Słomka J, Alcolombri U, Secchi E, Stocker R, Fernandez VI (2020) Encounter rates between bacteria and small sinking particles. New J Phys 22:043016

Smith DC, Simon M, Alldredge AL, Azam F (1992) Intense hydrolytic enzyme activity on marine aggregates and implications for rapid particle dissolution. Nature 359:139–142

Smriga S, Fernandez VI, Mitchell JG, Stocker R (2016) Chemotaxis toward phytoplankton drives organic matter partitioning among marine bacteria. Proc Natl Acad Sci 113:1576–1581

Son K, Guasto JS, Stocker R (2013) Bacteria can exploit a flagellar buckling instability to change direction. Nat Phys 9:494–498

Son K, Menolascina F, Stocker R (2016) Speed-dependent chemotactic precision in marine bacteria. Proc Natl Acad Sci 113:8624–8629

Sonnenschein EC, Syit DA, Grossart H-P, Ullrich MS (2012) Chemotaxis of *Marinobacter adhaerens* and its impact on attachment to the diatom *Thalassiosira weissflogii*. Appl Environ Microbiol 78:6900–6907

Sourjik V, Berg HC (2002a) Receptor sensitivity in bacterial chemotaxis. Proc Natl Acad Sci 99: 123–127

Sourjik V, Berg HC (2002b) Binding of the *Escherichia coli* response regulator CheY to its target measured in vivo by fluorescence resonance energy transfer. Proc Natl Acad Sci 99:12669–12674

Stocker R (2011) Reverse and flick: hybrid locomotion in bacteria. Proc Natl Acad Sci 108:2635–2636

Stocker R (2012) Marine microbes see a sea of gradients. Science 338:628–633

Stocker R (2015) The 100 μm length scale in the microbial ocean. Aq Microb Ecol 76:189–194

Stocker R, Seymour JR (2012) Ecology and physics of bacterial chemotaxis in the ocean. Microbiol Mol Biol Rev 76:792–812

Stocker R, Seymour JR, Samadani A, Hunt DE, Polz MF (2008) Rapid chemotactic response enables marine bacteria to exploit ephemeral microscale nutrient patches. Proc Natl Acad Sci 105:4209–4214

Streif S, Staudinger WF, Marwan W, Oesterhelt D (2008) Flagellar rotation in the archaeon *Halobacterium salinarum* depends on ATP. J Mol Biol 384:1–8

Sunagawa S, Coelho LP, Chaffron S, Kultima JR, Labadie K, Salazar G, Djahanschiri B, Zeller G, Mende DR, Alberti A et al (2015) Structure and function of the global ocean microbiome. Science 348:1261359

Suttle CA (1994) The significance of viruses to mortality in aquatic microbial communities. Microb Ecol 28:237–243

Suttle CA (2007) Marine viruses — major players in the global ecosystem. Nat Rev Microbiol 5: 801–812

Suttle CA, Chan AM (1994) Dynamics and distribution of cyanophages and their effect on marine *Synechococcus* spp. Appl Environ Microbiol 60:3167–3174

Tang KW (2005) Copepods as microbial hotspots in the ocean: effects of host feeding activities on attached bacteria. Aq Microb Ecol 38:31–40

Taylor JD, Cunliffe M (2017) Coastal bacterioplankton community response to diatom-derived polysaccharide microgels. Environ Microbiol Rep 9:151–157

Taylor JR, Stocker R (2012) Trade-offs of chemotactic foraging in turbulent water. Science 338: 675–679

Taylor MW, Radax R, Steger D, Wagner M (2007) Sponge-associated microorganisms: evolution, ecology, and biotechnological potential. Microbiol Mol Biol Rev 71:295–347

Teeling H, Fuchs BM, Bennke CM, Krüger K, Chafee M, Kappelmann L, Reintjes G, Waldmann J, Quast C, Glöckner FO et al (2016) Recurring patterns in bacterioplankton dynamics during coastal spring algae blooms. elife 5:e11888

Thar R, Kühl M (2002) Conspicuous veils formed by vibrioid bacteria on sulfidic marine sediment. Appl Environ Microbiol 68:6310–6320

Thar R, Kühl M (2003) Bacteria are not too small for spatial sensing of chemical gradients: an experimental evidence. Proc Natl Acad Sci 100:5748–5753

Thomas NA, Bardy SL, Jarrell KF (2001) The archaeal flagellum: a different kind of prokaryotic motility structure. FEMS Microbiol Rev 25:147–174

Thomas T, Evans FF, Schleheck D, Mai-Prochnow A, Burke C, Penesyan A, Dalisay DS, Stelzer-Braid S, Saunders N, Johnson J et al (2008) Analysis of the *Pseudoalteromonas tunicata* genome reveals properties of a surface-associated life style in the marine environment. PLoS One 3:e3252

Thornton DCO (2014) Dissolved organic matter (DOM) release by phytoplankton in the contemporary and future ocean. Eur J Phycol 49:20–46

Thornton KL, Butler JK, Davis SJ, Baxter BK, Wilson LG (2020) Haloarchaea swim slowly for optimal chemotactic efficiency in low nutrient environments. Nat Commun 11:4453

Tout J, Jeffries TC, Webster NS, Stocker R, Ralph PJ, Seymour JR (2014) Variability in microbial community composition and function between different niches within a coral reef. Microb Ecol 67:540–552

Tout J, Jeffries TC, Petrou K, Tyson GW, Webster NS, Garren M, Stocker R, Ralph PJ, Seymour JR (2015) Chemotaxis by natural populations of coral reef bacteria. ISME J 9:1764–1777

Tripp HJ, Kitner JB, Schwalbach MS, Dacey JWH, Wilhelm LJ, Giovannoni SJ (2008) SAR11 marine bacteria require exogenous reduced Sulphur for growth. Nature 452:741–744

Trolldenier G (1987) Curl, E.a. and B. truelove: the rhizosphere. (advanced series in agricultural sciences, Vol. 15) springer-Verlag, Berlin-Heidelberg-New York-Tokyo, 1986. Z Pflanzenernähr Bodenkd 150:124–125

Tsai C-L, Tripp P, Sivabalasarma S, Zhang C, Rodriguez-Franco M, Wipfler RL, Chaudhury P, Banerjee A, Beeby M, Whitaker RJ et al (2020) The structure of the periplasmic FlaG–FlaF complex and its essential role for archaellar swimming motility. Nat Microbiol 5:216–225

Tully BJ (2019) Metabolic diversity within the globally abundant marine group II Euryarchaea offers insight into ecological patterns. Nat Commun 10:271

Turner RD, Hobbs JK, Foster SJ (2016) Atomic force microscopy analysis of bacterial cell wall peptidoglycan architecture. In: Hong H-J (ed) Bacterial Cell Wall homeostasis: methods and protocols. Springer, New York, pp 3–9

van Tol HM, Amin SA, Armbrust EV (2017) Ubiquitous marine bacterium inhibits diatom cell division. ISME J 11:31–42

Vega Thurber R, Willner-Hall D, Rodriguez-Mueller B, Desnues C, Edwards RA, Angly F, Dinsdale E, Kelly L, Rohwer F (2009) Metagenomic analysis of stressed coral holobionts. Environ Microbiol 11:2148–2163

Venter JC, Remington K, Heidelberg JF, Halpern AL, Rusch D, Eisen JA, Wu D, Paulsen I, Nealson KE, Nelson W et al (2004) Environmental genome shotgun sequencing of the Sargasso Sea. Science 304:66–74

Vila-Costa M, Rinta-Kanto JM, Sun S, Sharma S, Poretsky R, Moran MA (2010) Transcriptomic analysis of a marine bacterial community enriched with dimethylsulfoniopropionate. ISME J 4: 1410–1420

Visser AW, Jackson GA (2004) Characteristics of the chemical plume behind a sinking particle in a turbulent water column. Mar Ecol Prog Ser 283:55–71

Vorobev A, Sharma S, Yu M, Lee J, Washington BJ, Whitman WB, Ballantyne F, Medeiros PM, Moran MA (2018) Identifying labile DOM components in a coastal ocean through depleted bacterial transcripts and chemical signals. Environ Microbiol 20:3012–3030

Wadhams GH, Armitage JP (2004) Making sense of it all: bacterial chemotaxis. Nat Rev Mol Cell Biol 5:1024–1037

Watrous J, Roach P, Alexandrov T, Heath BS, Yang JY, Kersten RD, van der Voort M, Pogliano K, Gross H, Raaijmakers JM et al (2012) Mass spectral molecular networking of living microbial colonies. Proc Natl Acad Sci 109:E1743–E1752

Watteaux R, Stocker R, Taylor JR (2015) Sensitivity of the rate of nutrient uptake by chemotactic bacteria to physical and biological parameters in a turbulent environment. J Theor Biol 387: 120–135

Webster NS, Taylor MW (2012) Marine sponges and their microbial symbionts: love and other relationships. Environ Microbiol 14:335–346

Webster NS, Thomas T (2016) The sponge hologenome. MBio 7:2

Weinbauer M, Peduzzi P (1995) Effect of virus-rich high molecular weight concentrates of seawater on the dynamics of dissolved amino acids and carbohydrates. Mar Ecol Prog Ser 127:245–253
Weiner RM, Ii LET, Henrissat B, Hauser L, Land M, Coutinho PM, Rancurel C, Saunders EH, Longmire AG, Zhang H et al (2008) Complete genome sequence of the complex carbohydrate-degrading marine bacterium, *Saccharophagus degradans* strain 2-40T. PLoS Genet 4:e1000087
Wild C, Huettel M, Klueter A, Kremb SG, Rasheed MYM, Jørgensen BB (2004) Coral mucus functions as an energy carrier and particle trap in the reef ecosystem. Nature 428:66–70
Willey JM, Waterbury JB (1989) Chemotaxis toward nitrogenous compounds by swimming strains of marine *Synechococcus* spp. Appl Environ Microbiol 55:1888–1894
Williams PM, Druffel ERM (1987) Radiocarbon in dissolved organic matter in the central North Pacific Ocean. Nature 330:246–248
Xie L, Altindal T, Chattopadhyay S, Wu X-L (2011) Bacterial flagellum as a propeller and as a rudder for efficient chemotaxis. Proc Natl Acad Sci 108:2246–2251
Xu T, Su Y, Xu Y, He Y, Wang B, Dong X, Li Y, Zhang X-H (2014) Mutations of flagellar genes *fliC12*, *fliA* and *flhDC* of *Edwardsiella tarda* attenuated bacterial motility, biofilm formation and virulence to fish. J Appl Microbiol 116:236–244
Ya V, Jw D, Pa J, Bb K-B (1998) A predictive model of bacterial foraging by means of freely released extracellular enzymes. Microb Ecol 36:75–92
Yang Y, Pollard AM, Höfler C, Poschet G, Wirtz M, Hell R, Sourjik V (2015) Relation between chemotaxis and consumption of amino acids in bacteria. Mol Microbiol 96:1272–1282
Yorimitsu T, Homma M (2001) $Na^+$-driven flagellar motor of *Vibrio*. Biochim Biophys Acta Bioenerg 1505:82–93
Zehr JP, Weitz JS, Joint I (2017) How microbes survive in the open ocean. Science 357:646–647
Zhang C, Dang H, Azam F, Benner R, Legendre L, Passow U, Polimene L, Robinson C, Suttle CA, Jiao N (2018) Evolving paradigms in biological carbon cycling in the ocean. Natl Sci Rev 5: 481–499
Ziegler GR, Benado AL, Rizvi SSH (1987) Determination of mass diffusivity of simple sugars in water by the rotating disk method. J Food Sci 52:501–502
Zimmer-Faust RK, de Souza MP, Yoch DC (1996) Bacterial chemotaxis and its potential role in marine dimethylsulfide production and biogeochemical sulfur cycling. Limnol Oceanogr 41: 1330–1334

# Marine Cyanobacteria

3

Frédéric Partensky, Wolfgang R. Hess, and Laurence Garczarek

**Abstract**

Although numerous bacteria dwelling in the sunlit layer of oceans can exploit solar energy, cyanobacteria are the only ones to perform oxygenic photosynthesis and to produce organic carbon, a critical process that sustains the whole marine trophic web. Here, we review the advances on the two most abundant cyanobacterial genera of the ocean, *Prochlorococcus* and *Synechococcus,* and on diazotrophic cyanobacteria that by their ability to fix atmospheric dinitrogen constitute a major source of new nitrogen for the microbial community. Diazotrophic cyanobacteria are polyphyletic and display a large range of morphologies and lifestyles. These include both multicellular cyanobacteria such as the colonial *Trichodesmium* or the heterocyst-forming *Calothrix*, *Richelia*, and *Nodularia*, and unicellular cyanobacteria belonging to three major taxa: the symbiotic species *Candidatus* Atelocyanobacterium thalassa (UCYN-A) and the free-living genera *Crocosphaera watsonii* (UCYN-B) and *C. subtropica* (formerly *Cyanothece* sp.; UCYN-C). After several billions of years of evolution that led them to colonize most marine niches reached by solar light, cyanobacteria appear as truly fascinating organisms. They are the matter of a thriving field of research investigating their molecular ecology, contribution to biogeochemical cycles, and use for data mining and potential

F. Partensky (✉) · L. Garczarek
CNRS & Sorbonne Université, Station Biologique de Roscoff, UMR 7144 Adaptation and Diversity in the Marine Environment, Ecology of Marine Plankton Team (ECOMAP), Roscoff, France
e-mail: frederic.partensky@sb-roscoff.fr; laurence.garczarek@sb-roscoff.fr

W. R. Hess
Faculty of Biology, Institute of Biology 3, Genetics and Experimental Bioinformatics, University of Freiburg, Freiburg, Germany
e-mail: wolfgang.hess@biologie.uni-freiburg.de



L. J. Stal, M. S. Cretoiu (eds.), *The Marine Microbiome*, The Microbiomes of Humans, Animals, Plants, and the Environment 3,
https://doi.org/10.1007/978-3-030-90383-1_3

biotechnological applications such as, e.g., production of hydrogen, alkanes, or fluorophores.

**Keywords**

Diazotrophy · Photosynthesis · Phytoplankton · *Prochlorococcus* · *Synechococcus* · *Trichodesmium*

## 3.1 Introduction

A large proportion of bacteria thriving in the upper lit layer of oceans is able to use solar light as an energy source including aerobic anoxygenic photosynthetic bacteria (Auladell et al. 2019; Béjà et al. 2002; Boeuf et al. 2013; Koblížek 2015; Lehours et al. 2018; Yurkov and Hughes 2017) and proteorhodopsin-containing bacteria (Béjà et al. 2001; Boeuf et al. 2016; Giovannoni 2017; Pinhassi et al. 2016). Yet, cyanobacteria are the only bacteria capable of performing "oxygenic photosynthesis," a capacity they share with the eukaryotic microalgae and plants. Oxygenic photosynthesis requires two photosystems (PSI and PSII) connected via an electron transfer chain (Falkowski and Raven 2007). Photons are collected by antenna complexes coupled to photosystems and transferred to special chlorophylls (Chls) located in the core of both photosystems. Most often, like in red algae the light-harvesting antennae of cyanobacteria are large membrane-extrinsic water-soluble complexes called phycobilisomes that consist of phycobiliproteins binding numerous chromophores (phycobilins) and the terminal acceptors in photosystems are Chl *a* molecules (Bar-Eyal et al. 2018; Sidler 1994). Notable exceptions to this rule are the atypical marine cyanobacteria *Acaryochloris*, *Prochloron*, and *Prochlorococcus*, which possess integral membrane light-harvesting complexes and often use unusual Chls as terminal acceptors such as Chl *d* or divinyl-Chl *a* (for reviews, see Green 2019; Partensky and Garczarek 2003). Photon energy is used to break water molecules, a reaction that releases oxygen as a by-product, and produce adenosine triphosphate (ATP) as well as reducing power in the form of reduced ferredoxin and nicotinamide adenine dinucleotide phosphate (NADPH). NADPH and ATP are ultimately used to synthesize organic carbon from atmospheric $CO_2$ via the ribulose-1,5-bisphosphate carboxylase/oxygenase (RuBisCo) and the Calvin–Benson–Bassham cycle (Falkowski and Raven 2007). *Cyanobacteria* have recently been suggested to share a common ancestor with two non-oxyphototrophic bacterial lineages called *Melainabacteria* and *Sericytochromatia* (Soo et al. 2014; Soo et al. 2017; Shih et al. 2017). Based on this phylogenetic relatedness, Soo et al. (2017) proposed that the three lineages should be gathered into a single phylum called "*Cyanobacteria*" and that members of the monophyletic branch capable of oxygenic photosynthesis should be renamed "*Oxyphotobacteria*." However, this proposal is the matter of a vivid debate within the scientific community and a number of leading scientists have strongly rejected the idea to call "*Cyanobacteria*" organisms that are *not* capable of oxygenic photosynthesis (Garcia-Pichel et al. 2020). Instead they support the idea that *Cyanobacteria* should remain a separate phylum and reaffirmed that the correct definition of *Cyanobacteria* is the following: "organisms in the domain *Bacteria* able to carry out oxygenic photosynthesis with water as an electron donor and to reduce $CO_2$ as a source of carbon, or those secondarily evolved from such organisms."

In this chapter, we will make an overview of recent advances on the biology, ecology, evolution, and exploitability of marine cyanobacteria sensu (Garcia-Pichel et al. 2020). Besides their ability to use solar energy to synthesize organic carbon, which places them at the basis of the marine food web, they also perform several other important functions within the marine ecosystem. For instance, many marine cyanobacteria are diazotrophs, i.e., capable of fixing atmospheric dinitrogen ($N_2$) and transforming it into ammonium that is rapidly transferred to other members of the microbial community (Berthelot et al. 2016). The latter can be either free-living planktonic microbes, benthic organisms co-occurring with cyanobacteria in microbial mats, or eukaryotic partners in symbioses. Such mutually beneficial associations are frequently encountered in nutrient-depleted areas of the ocean (Foster and Zehr 2019). Some cyanobacteria, such as the marine genus *Trichodesmium* or the brackish water genus *Nodularia*, can develop large blooms that can be seen from space (McKinna 2015; Rousset et al. 2018) and may be toxic to other members of the planktonic community (Sacilotto Detoni et al. 2016; Sivonen et al. 1989).

Marine cyanobacteria constitute a unique resource for biotechnological applications. Notable examples are *Crocosphaera subtropica* (formerly *Cyanothece* sp.; cf. Sect. 3.5.3) strains that have a natural potential for high yield hydrogen production (Bandyopadhyay et al. 2010; Melnicki et al. 2012), the discovery of a biosynthetic pathway for short-to-medium chain alkanes in *Prochlorococcus* (Lea-Smith et al. 2015; Schirmer et al. 2010), ethanol production in *Synechococcus* sp. PCC 7002 by simultaneous inactivation of glycogen synthesis pathways and introduction of ethanologenic cassettes (Wang et al. 2020), or several strains producing beneficial metabolites (Jones et al. 2011; Mevers et al. 2014; Salvador-Reyes and Luesch 2015; Shao et al. 2015).

## 3.2 Marine Cyanobacteria and the Next Generation Sequencing Revolution

The availability of complete genome sequences is of utmost importance to understand the molecular bases of the biochemical and ecological potential of marine cyanobacteria. Thanks to their small genomes and large abundance in oceanic ecosystems the first marine representatives of this phylum to be sequenced were three strains of *Prochlorococcus* (Dufresne et al. 2003; Rocap et al. 2003) and one of *Synechococcus* (Palenik et al. 2003). Since these pioneering studies there has been a steady increase in the number of available genome sequences of marine cyanobacteria. Isolates of all major species of marine and brackish cyanobacteria known to date have been sequenced (Fig. 3.1) including (1) the planktonic free-living *Crocosphaera watsonii* (also called "unicellular cyanobacteria group B" or

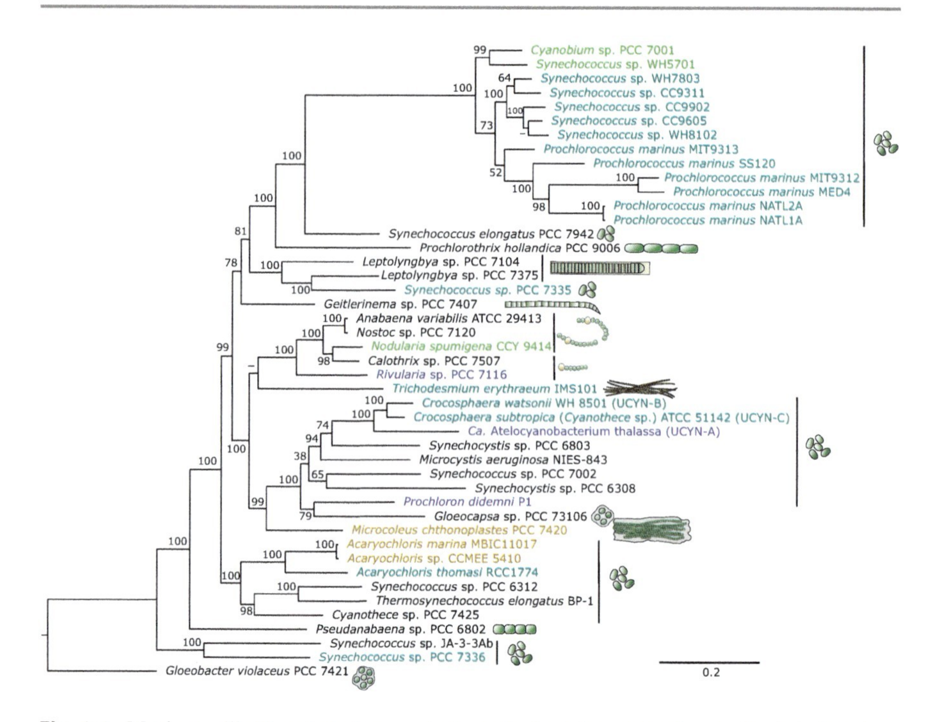


**Fig. 3.1** Maximum-likelihood phylogenomic tree of selected sequenced cyanobacteria based on 29 concatenated core markers. Species names are color-coded as follows: strictly marine planktonic strains (i.e., with obligate requirement for high salinities) are shown in lagoon blue, halotolerant planktonic strains (i.e., capable to grow over a large range of salinities) in green, marine benthic strains in beige, and symbiotic strains in purple. All other strains (freshwater, soils, etc.) are shown in black. *Cyanothece* sp. ATCC 51142, a representative of the unicellular cyanobacteria Group C (UCYN-C), has recently been renamed *Crocosphaera subtropica* by Mareš et al. (2019). Only nodes supported with a bootstrap value higher than 60% are indicated. Morphologies are indicated for each strain or group of strains. Adapted with permission from Supplementary Fig. S1 in Partensky et al. (2018)

"UCYN-B"; (Montoya et al. 2004), *Cyanothece*-like members of the UCYN-C group renamed *Crocosphaera subtropica* (Mareš et al. 2019), *Cyanobium* sp., *Nodularia spumigena*, and *Trichodesmium erythraeum*, (2) a few benthic species such as *Acaryochloris marina*, *Microcoleus chthonoplastes*, and *Rivularia* sp., but also (3) a number of symbionts including *Candidatus* Atelocyanobacterium thalassa (UCYN-A; (Thompson et al. 2012), as well as *Calothrix*, *Prochloron*, and *Richelia* spp. (the latter genus is missing from Fig. 3.1, but see Hilton et al. 2013).

Sequencing projects so far have exhibited a large bias toward *Prochlorococcus* and *Synechococcus*, which have the advantage to exhibit few DNA repeats and transposase genes considerably facilitating their assembly and closure compared to many other cyanobacteria (Biller et al. 2014a; Doré et al. 2020; Dufresne et al. 2008; Garczarek et al. 2021; Kettler et al. 2007; Lee et al. 2019; Palenik et al. 2006). Yet, the genomic diversity of a few other marine genera such as *Crocosphaera*, which contain hundreds of transposase genes, has also been well studied with ten different strains sequenced to date (Bench et al. 2013; Bombar et al. 2014). It must be stressed that the true genomic diversity of marine picocyanobacteria exceeds vastly the information obtained from laboratory strains, which is nowadays largely complemented by genomic analyses of single amplified genomes (SAGs) obtained from flow cytometry-sorted cells (see among others Berube et al. 2018; Kashtan et al. 2014; Malmstrom et al. 2013; Pachiadaki et al. 2019) and metagenome-assembled genomes (MAGs; see among others Delmont and Eren 2018; Engelberts et al. 2020; Rusch et al. 2010; Shi et al. 2011; Tully et al. 2018). For instance, sequencing of *Prochlorococcus* SAGs by Kashtan et al. (2014) revealed the occurrence of hundreds of coexisting *Prochlorococcus* sub-populations with distinct genomic backbones (cf. Sect. 3.4.5.1 for more details).

The tremendous power of next generation sequencing technologies coupled with rapid advancements in bioinformatics and data processing is currently revolutionizing our view of the genetic and functional diversity of marine cyanobacteria notably by allowing scientists to integrate comparative genomics, transcriptomics, and/or meta-omics information and to decipher their association with environmental factors using network approaches (see among others Doré et al. 2020; Garcia et al. 2020; Guidi et al. 2016; Guyet et al. 2020).

## 3.3 Cyanobacterial Origin and Evolution

### 3.3.1 The Advent of Cyanobacteria and Oxygenic Photosynthesis

Cyanobacteria are generally considered to be the most ancient microorganisms capable of oxygenic photosynthesis and held responsible for the "Great Oxidation Event" (GOE), i.e. the first sharp rise of atmospheric $O_2$ concentration in Earth history that occurred 2.3 $\pm$ 0.1 billion years (Gy) ago (Lyons et al. 2014). Yet, there is still a vivid controversy about when oxygenic photosynthesis arose with estimates based on geological or geochemical evidence spanning from 3.7 (Rosing and Frei 2004) to 2.3 Gy ago (Kirschvink and Kopp 2008). For instance, a study of the distribution of chromium isotopes and redox-sensitive metals in paleosols indicated that there were already notable levels of atmospheric oxygen (i.e., $3 \times 10^{-4}$ times present levels) about 3.0 Gy ago suggesting that ancestral cyanobacteria might have evolved by this time (Crowe et al. 2013). A similar controversy also exists about timing the advent of the *Cyanobacteria* phylum based on molecular analyses. For instance, by using cross-calibrated Bayesian relaxed molecular clock analyses of slowly evolving core proteins, Shih et al. (2017) suggested a fairly recent advent of this group about 2.0 Gy ago, i.e. *after* the GOE, a hypothesis that they explained by the occurrence of nowadays extinct stem lineages that would have evolved oxygenic photosynthesis after their divergence from *Melainabacteria* and would have sourced the $O_2$ fluxes that led to the GOE. In sharp contrast, by combining fossil record data and relaxed molecular clock phylogeny based on sequences of the UV-A sunscreen scytonemin, Garcia-Pichel et al. (2019) estimated the advent of the *Cyanobacteria* phylum at 3.6 ± 0.2 Gy, i.e. well *before* the GOE.

### 3.3.2 Evolutionary History of Marine Cyanobacteria

Besides their age other intriguing questions about early cyanobacteria deal with their morphology and habitat: did they first occur on land or in the ocean? The oldest microfossil interpreted with certainty as a cyanobacterium, *Eoentophysalis belcherensis*, dates back ~1.9 Gy and was discovered in silicified stromatolites originally located in intertidal mudflats (Demoulin et al. 2019; Hofmann 1976). Its modern counterpart, the coccoid cyanobacterium *Entophysalis major*, is one of the main mat-forming microorganisms (Golubic and Abed 2010). Yet, in many phylogenetic and phylogenomic studies on cyanobacterial evolution trees are rooted using the rock-dwelling *Gloeobacter violaceus*, which has the unique "primitive" characteristics to lack thylakoids (Fig. 3.1; Blank and Sánchez-Baracaldo 2010; Larsson et al. 2011; Mareš et al. 2013; Shih et al. 2013). Even if early cyanobacteria likely had a coccoid morphotype, filamentous forms must have evolved soon after possibly during or just after the GOE since they were prominent components of microbial mats during most of the Proterozoic Eon (2.5–0.54 Gy; Knoll and Semikhatov 1998). The fact that marine planktonic cyanobacteria lineages are not monophyletic within the cyanobacterial radiation but dispersed among terrestrial, freshwater, and marine benthic species in phylogenetic trees (Fig. 3.1; see also e.g. Shih et al. 2013) strongly suggests that ancestral cyanobacteria appeared first either in lakes or intertidal habitats and that the marine pelagic environment was colonized much later through several independent colonization events (Blank and Sánchez-Baracaldo 2010; Larsson et al. 2011; Sánchez-Baracaldo and Cardona 2020). The first marine planktonic lineages of cyanobacteria were likely diazotrophs and evolved only during the Neoproterozoic (0.54–1.0 Gy; Sánchez-Baracaldo 2015) possibly explaining why the open ocean was anoxic during most of the Proterozoic Eon (Reinhard et al. 2013).

### 3.3.3 Adaptation to Salinity

Comparative genomics studies have brought important insights into how cyanobacteria have adapted to the marine habitat. In particular, these studies showed that adaptation of cyanobacteria to salinity has been made possible by the development of specific machineries allowing cells to actively export inorganic ions (the "salt-out-strategy") and to increase the cellular osmolarity by accumulating small organic molecules called compatible solutes (Pade and Hagemann 2014). The nature of these compatible solutes strongly varies depending on habitats and/or lineages, with (1) glucosylglycerol, trehalose, and/or sucrose found in marine and brackish species, (2) glycine betaine in halophiles and stromatolite-forming cyanobacteria, (3) sucrose and glucosylglycerate in *Prochlorococcus*, and (4) trehalose in

*Crocosphaera watsonii* (Klähn et al. 2010; Scanlan et al. 2009; Teikari et al. 2018b). While the marine *Synechococcus* model strain WH7803 is able to accumulate glucosylglycerol and possibly glucosylglycerate and to take up glucosylglycerol, sucrose, and glycine betaine (Scanlan et al. 2009), these compounds appear to be involved in osmoregulation at different times of the cell cycle and to respond to distinct environmental conditions (Guyet et al. 2020). For *Trichodesmium*, no genes coding for known compatible solute biosynthesis had been identified during a first screening of its genome (Pade and Hagemann 2014). Yet, using NMR and liquid chromatography coupled to mass spectroscopy the quaternary ammonium compound N,N,N-trimethyl homoserine (or homoserine betaine), a previously unknown compatible solute, was identified as the main compatible solute in cultures and natural populations of *Trichodesmium* (Pade et al. 2016). The structure of this compound made it likely that it was synthesized by stepwise salt-regulated methylation of homoserine. Indeed, a transcriptomic analysis identified a single methyltransferase gene that was upregulated with increasing salinity. Biochemical assays of the recombinant protein, using L-homoserine as the precursor and S-adenosylmethionine as the methyl group donor, showed that this enzyme produced homoserine betaine (Pade et al. 2016).

### 3.3.4 Adaptation to Nitrogen Depletion

Another environmental factor that strongly influenced the evolution of marine cyanobacteria is the availability of nutrients. Nitrogen (N) limits marine productivity in the upper layer of ca. 75% of the global ocean while limitation by phosphorus (P) or iron (Fe) prevails in other areas (Bristow et al. 2017; Moore et al. 2001; Moore et al. 2013). Cyanobacteria have developed three main strategies to deal with inorganic N deprivation: (1) to fix atmospheric $N_2$, (2) to decrease their cell and genome size in order to limit their cellular requirement for N, and (3) to use organic N sources.

The first strategy has led to many different physiological and morphological adaptations aimed at circumventing the main drawback of $N_2$ fixation, i.e. the oxygen sensitivity of nitrogenase. The inhibition of nitrogenase by oxygen might have hindered the rise of atmospheric oxygen levels during the Proterozoic Eon (Allen et al. 2019). The endosymbiotic filamentous genera *Richelia* and *Calothrix* which fix $N_2$ during the day (Fig. 3.2) both form heterocysts, i.e. cells devoid of PSII and specialized in $N_2$ fixation. These heterocysts occur at the basis of the trichomes of *Richelia* and *Calothrix*. They are morphologically different from typical $CO_2$-fixing cells and can be easily recognized by their larger size, a thicker cell wall constituting a gas diffusion barrier, and by their reduced chlorophyll content compared to typical vegetative cells (see among others Hilton et al. 2013). The non-heterocystous *Trichodesmium* genus also primarily fixes $N_2$ during the day (Fig. 3.2), but this process is both spatially and temporally sequestered from oxygenic photosynthesis. Indeed, nitrogenase is localized in "non-granulated" cell types of the trichomes called "diazocytes" the frequency of which is lower at dawn

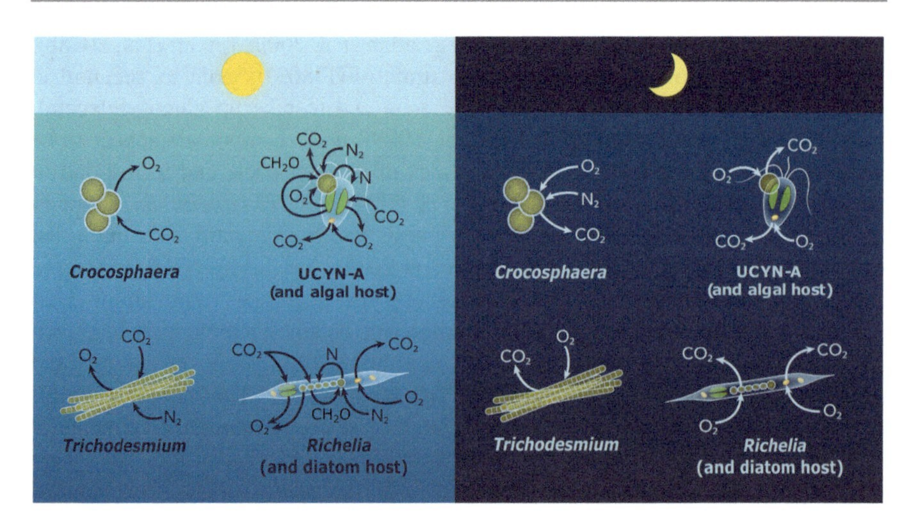


**Fig. 3.2** Major types of open ocean $N_2$-fixing cyanobacteria showing distinct physiological differences in the daily cycles of C and N metabolism, $O_2$, and nutrients. Cyanobacteria differ in whether they fix $N_2$ in the light or in the dark and have different adaptations for obtaining non-N nutrients that have implications for ecological distributions and magnitude of $N_2$ fixation. Adapted with permission from Fig. 3 in Zehr & Capone (2020)

and increased toward noon and is negatively affected by the presence of inorganic N (Bergman et al. 2013). The unicellular *Crocosphaera watsonii* displays a much more stringent temporal separation than *Trichodesmium* between photosynthesis and $N_2$ fixation, these processes being restricted to day and night, respectively (Fig. 3.2; Compaoré and Stal 2010; Shi et al. 2010). At last, the symbiotic *Candidatus* Atelocyanobacterium thalassa (UCYN-A) literally lost its PSII, making it the sole cyanobacterium *sensu* Garcia-Pichel et al. (2020) known so far to be incapable of oxygenic photosynthesis while it is an efficient $N_2$-fixer during the day (Fig. 3.2; Tripp et al. 2010; Zehr et al. 2008). All non-heterocystous $N_2$-fixing cyanobacteria synthesize specific hopanoid lipids that may limit $O_2$ diffusion across their membranes thus protecting nitrogenase from oxidation in $O_2$-saturated surface oceanic waters (Cornejo-Castillo and Zehr 2019).

The second strategy consists in a drastic decrease in cell size, a process that confers cells an advantage for nutrient uptake by increasing their surface to volume ratio. It is observed in the marine picocyanobacteria *Synechococcus* (average cell diameter ~ 1 μm) and is pushed to its limits in most *Prochlorococcus* lineages that have an average cell diameter of only about 0.7 μm (Chisholm et al. 1988; Waterbury et al. 1979). This decrease in cell size went hand in hand with a drastic genome streamlining, since *Prochlorococcus* genomes decreased in size by about 30% compared to the size of their ancestor via an extensive streamlining process (cf. Sect. 3.4.6; Doré et al. 2020; Dufresne et al. 2005; Kettler et al. 2007; Partensky and Garczarek 2010; Scanlan et al. 2009; Ting et al. 2007).

The third strategy, often combined with one of the two others, is to assimilate organic forms of N such as urea, amino acids, or amino sugars. Indeed, although marine cyanobacteria were long considered to have a strict photoautotrophic lifestyle there is more and more evidence that they are in fact mixotrophs because they are capable of assimilating organic forms of nutrients and carbon via a light-stimulated process called "photoheterotrophy" (Duhamel et al. 2018; Gómez-Pereira et al. 2013; Muñoz-Marín et al. 2013; Muñoz-Marín et al. 2020; Yelton et al. 2016; Zubkov 2009).

### 3.3.5 Adaptation to Spectral Niches

A third factor that primitive cyanobacteria had to deal with over the course of their adaptation to the marine habitat is the wide range of light qualities encountered in the water column, the vibrational modes of the $H_2O$ molecule delineating five underwater spectral niches, extending from the red niche found in turbid coastal waters to the violet niche in the nutrient-poorest open ocean waters (Holtrop et al. 2021; Kirk 1994; Stomp et al. 2007). Most cyanobacteria possess large extrinsic light-harvesting complexes called phycobilisomes constituted of an allophycocyanin core surrounded by six to eight rods with variable phycobiliprotein composition, namely phycocyanin (PC) alone or in combination with phycoerythrocyanin (PEC), phycoerythrin-I (PEI), and/or phycoerythrin-II (PEII; Bar-Eyal et al. 2018; Sidler 1994; Six et al. 2007). These phycobiliproteins exhibit distinct absorption properties depending upon the nature and relative proportions of covalently bound phycobilins, which can be phycourobilin (PUB; $A_{max}$: 495 nm), phycoerythrobilin (PEB; $A_{max}$: 550 nm), phycoviolobilin (PVB; $A_{max}$: 590 nm), and/or phycocyanobilin (PCB; $A_{max}$: 620 nm). While all phycobilisomes contain PCB, PVB is specific to PEC a pigment common in cyanobacteria thriving in soils, freshwater, hot springs, and marine benthic habitats. PEB is the major phycobilin in cyanobacteria living in habitats where green light is predominant such as coastal or mesotrophic marine waters. Finally, PUB is the dominant cyanobacterial chromophore in blue pelagic waters (Grébert et al. 2018; Holtrop et al. 2021; Lantoine and Neveux 1997). Both PCB and PEB derive from biliverdin IXa and exist as free pigments whereas PVB and PUB necessarily result from the binding and isomerization of PCB and PEB, respectively (Blot et al. 2009; Carrigee et al. 2021; Sanfilippo et al. 2019; Shukla et al. 2012; Zhao et al. 2000). Although PVB and PUB occur in different organisms the lyase-isomerases PecE-F and RpcG that catalyze the binding and isomerization reactions at the equivalent binding site (α-84) of PEC and PC, respectively, are chemically similar and phylogenetically closely related. This is a nice example of adaptation to blue light likely resulting from lateral gene transfer followed by a change in gene function (Blot et al. 2009). All oceanic free-living cyanobacteria possess PE-rich phycobilisomes characterized by high (*Crocosphaera*) or intermediate PUB:PEB ratios (*Trichodesmium*) while algal symbionts such as *Richelia* have a low PUB:PEB ratio possibly because this better complements the absorption properties of their host pigments (Neveux et al. 1999; Neveux et al. 2006; Ong and Glazer 1991; Six et al. 2007). *Synechococcus* is likely the most diversified group of marine cyanobacteria with regard to its pigment content, with some PEII-containing strains exhibiting constitutively low, medium, or high ratios of PUB:PEB while others called Type IV chromatic acclimating organisms are capable of varying this ratio depending on the ambient light color (Everroad et al. 2006; Grébert et al. 2021; Humily et al. 2013; Palenik 2001; Sanfilippo et al. 2019; Shukla et al. 2012; Six et al. 2007). This ability likely confers them a strong fitness advantage because they were shown to represent more than half of all *Synechococcus* cells of the world ocean (Grébert et al. 2018). In this context, *Prochlorococcus* constitutes an exception among marine planktonic cyanobacteria in that it has no phycobilisomes but instead possesses membrane-intrinsic antenna complexes (Pcb) binding divinyl derivatives of Chl *a* and *b*, a unique pigment complement allowing this microorganism to collect with even more efficacy than PUB the blue light prevailing at the bottom of the photic layer in open ocean waters (Garczarek et al. 2000; Goericke and Repeta 1992; Morel et al. 1993; Ting et al. 2002). *Prochlorococcus* has nevertheless retained a minimal set of phycobiliprotein genes from which small amounts of a chromophorylated phycoerythrin (PEIII) is produced (Hess et al. 1996; Hess et al. 1999; Hess et al. 2001). PEIII plays a minor role in light harvesting in *P. marinus* SS120 and might rather act in light sensing in these cells even though this hypothesis has not been experimentally validated yet (Steglich et al. 2005). A few other marine cyanobacteria have also lost their phycobilisomes either entirely like *Prochloron didemni*, a symbiont of ascidians (Lewin 1984), despite the presence in its genome of a *cpcBA* operon (Donia et al. 2011), or partially like some strains of the benthic cyanobacterium *Acaryochloris marina* that contain phycobiliprotein aggregates (Miyashita et al. 2003). When present these aggregates are localized in the inter-thylakoidal space and can efficiently transfer photon energy to PSII (Hu et al. 1999; Marquardt et al. 1997; Petrasek et al. 2005). Structurally, these aggregates are composed of phycocyanin and a vestigial allophycocyanin (APC) constituted of the sole β-APC subunit (Partensky et al. 2018). Like for *Prochlorococcus* the main light-harvesting complexes—and even the sole in *Prochloron* and in *A. marina* strains CCME5410 and HCIR111A that lack phycobiliprotein aggregates (Chan et al. 2007; Partensky et al. 2018)—are Pcb proteins binding either Chl *a* and *d* at a molar ratio of ca. 0.06 in most *Acaryochloris* species (Miyashita et al. 1997) or Chl *a* and *b* at a molar ratio of ca. 6 in *Prochloron didemni* and *A. thomasi* (Partensky et al. 2018; Withers et al. 1978). The occurrence of such large amounts of the near-infrared light absorbing Chl *d* ($A_{max}$: 705 nm) in most marine *Acaryochloris* species known so far is likely an adaptation to their shaded benthic niches. For instance, *A. marina* cells have been observed underneath *Prochloron*-containing didemnid ascidians a habitat enriched in far-red light because most visible light was filtered out (Kühl et al. 2005).

After several billions of years of evolution, which led them to colonize any single niche of the marine habitat reached by solar light, cyanobacteria appear as truly fascinating organisms and are the matter of an ebullient research area. The following paragraphs will detail recent advances on the main marine cyanobacterial genera.

## 3.4 *Prochlorococcus* and *Synechococcus*

### 3.4.1 Interest as Model Organisms in Marine Biology and Ecology

Despite their fairly recent discovery in the late twentieth century, the non-diazotrophic marine unicellular cyanobacteria *Prochlorococcus* (Chisholm et al. 1988; Chisholm et al. 1992) and *Synechococcus* (Waterbury et al. 1979) are nowadays the best known marine cyanobacteria at all scales of organization from the gene to the global ocean (Biller et al. 2015; Coleman and Chisholm 2007; Doré et al. 2020; Farrant et al. 2016; Flombaum et al. 2020; Garcia et al. 2020; Kent et al. 2019; Scanlan et al. 2009; Sohm et al. 2015). Indeed, these marine unicellular cyanobacteria have a number of advantages that make them particularly relevant model organisms for ecological, physiological, and evolutionary studies, including: (1) their abundance and ubiquity and thence strong contribution to global marine chlorophyll biomass and primary productivity (Buitenhuis et al. 2012; Flombaum et al. 2013; Flombaum et al. 2020; Partensky et al. 1999b; Partensky et al. 1999a), (2) their culturability (Hunter-Cevera et al. 2016; Moore et al. 2007; Morris et al. 2008; Waterbury and Willey 1988) that allowed refined comparative physiology analyses of representative isolates (Bagby and Chisholm 2015; Berube et al. 2015; Breton et al. 2019; Guyet et al. 2020; Humily et al. 2013; Krumhardt et al. 2013; Mella-Flores et al. 2012; Moore and Chisholm 1999; Paz-Yepes et al. 2013; Pittera et al. 2017; Pittera et al. 2018), as well as (3) their small genome sizes, which favored the sequencing of a large number of genomes, SAGs, and MAGs (Berube et al. 2018; Biller et al. 2014a; Doré et al. 2020; Engelberts et al. 2020; Garczarek et al. 2021; Kashtan et al. 2014; Malmstrom et al. 2013; Pachiadaki et al. 2019).

### 3.4.2 Global Abundance and Distribution

With cell densities of up to $3 \times 10^5$ cells $mL^{-1}$ in the upper layer of warm nutrient-poor central gyres and a distribution area extending between 40°S and 45°N *Prochlorococcus* is undoubtedly the most abundant photosynthetic organism on Earth with its annual mean global abundance estimated as $2.9 \pm 0.1 \times 10^{27}$ cells (Buitenhuis et al. 2012; Flombaum et al. 2013; Partensky et al. 1999b). *Prochlorococcus* always co-occurs with *Synechococcus* but the latter is even more widespread since its distribution extends to sub-polar areas and to brackish waters such as the Baltic Sea or estuaries (Cottrell and Kirchman 2009; Haverkamp et al. 2008; Hunter-Cevera et al. 2020; Larsson et al. 2014; Partensky et al. 1999a; Paulsen et al. 2016; Xia et al. 2015; Xia et al. 2017a; Xia et al. 2017b). *Synechococcus* abundance is typically two orders of magnitude lower than *Prochlorococcus* in warm central oceanic gyres but it often outcompetes *Prochlorococcus* in nutrient-rich regions and can even reach cell densities above $10^6$ cells $mL^{-1}$ in the Costa Rica Dome likely due to the high concentrations of cobalt and iron in this area (Ahlgren et al. 2014; Gutiérrez-Rodríguez et al. 2014; Saito et al. 2005). Although *Synechococcus* is globally less abundant than *Prochlorococcus* with an estimated mean global abundance of $7.0 \pm 0.3 \times 10^{26}$ cells its contribution to the oceanic net production is estimated to be twice higher (16.7% vs. 8.5%, respectively) due to the larger cell sizes and higher $CO_2$ fixation rates of *Synechococcus* (Flombaum et al. 2013; Li 1994). *Synechococcus* and its phages are key players of the plankton networks that drive carbon export and thence the oceanic biological pump (Guidi et al. 2016).

### 3.4.3 Phylogeny

Marine *Synechococcus* and *Prochlorococcus* are phylogenetically closely related and together with *Cyanobium* form a monophyletic branch called "Cluster 5" (Herdman et al. 2001), which long diverged from all other cyanobacteria including several freshwater *Synechococcus* species. The *Synechococcus* "genus" is polyphyletic (Fig. 3.1; Doré et al. 2020; Salazar et al. 2020; Sánchez-Baracaldo and Cardona 2020; Sánchez-Baracaldo 2015; Scanlan et al. 2009; Shih et al. 2013). All members of Cluster 5 share the presence of α-carboxysomes encapsulating a Form-IA RuBisCo (Badger and Price 2003). These proteins are phylogenetically more closely related to those of chemoautotrophic proteobacteria such as thiobacilli, than to the β-carboxysomes and Form-IB RuBisCO found in all other cyanobacteria. Yet, a comparative physiology study of strains possessing one or the other RuBisCo type could not find any obvious functional differences between these two types of carboxysomes (Whitehead et al. 2014).

The marine *Synechococcus/Cyanobium* radiation is divided into three deep-branching groups called sub-clusters (SC) 5.1 to 5.3 (Dufresne et al. 2008). All SC 5.1 lineages except one (clade VIII) are strictly marine and this SC largely predominates in open ocean waters (Fig. 3.3b and d; see also Ahlgren et al. 2020; Farrant et al. 2016; Sohm et al. 2015; Zwirglmaier et al. 2008). SC 5.2 encompasses halotolerant species thriving in estuaries and near coastal areas (Chen et al. 2006; Hunter-Cevera et al. 2016; Xia et al. 2015; Xia et al. 2017a) but also strictly freshwater species such as *Cyanobium gracile* PCC 6307 or *Vulcanococcus limneticus* (Di Cesare et al. 2018; Shih et al. 2013). Finally, SC 5.3 comprises both strictly marine members locally constituting an important component of *Synechococcus* communities in P-limited areas such as the Mediterranean Sea or the Gulf of Mexico (Fig. 3.3d; see also Farrant et al. 2016; Huang et al. 2012; Sohm et al. 2015) and freshwater members that can be abundant in lakes (Cabello-Yeves et al. 2017; Cabello-Yeves et al. 2018). A major diversification event seemingly occurred in SC 5.1 soon after its divergence from SC 5.2 and 5.3 (Urbach et al. 1998) so that, depending on the genetic marker, this SC nowadays accounts between 10 and 20 distinct clades (Ahlgren and Rocap 2012; Gutiérrez-Rodríguez et al. 2014; Huang et al. 2012; Penno et al. 2006; Scanlan et al. 2009; Xia et al. 2019), which can be further split into a number of sub-clades (Mazard et al. 2012).

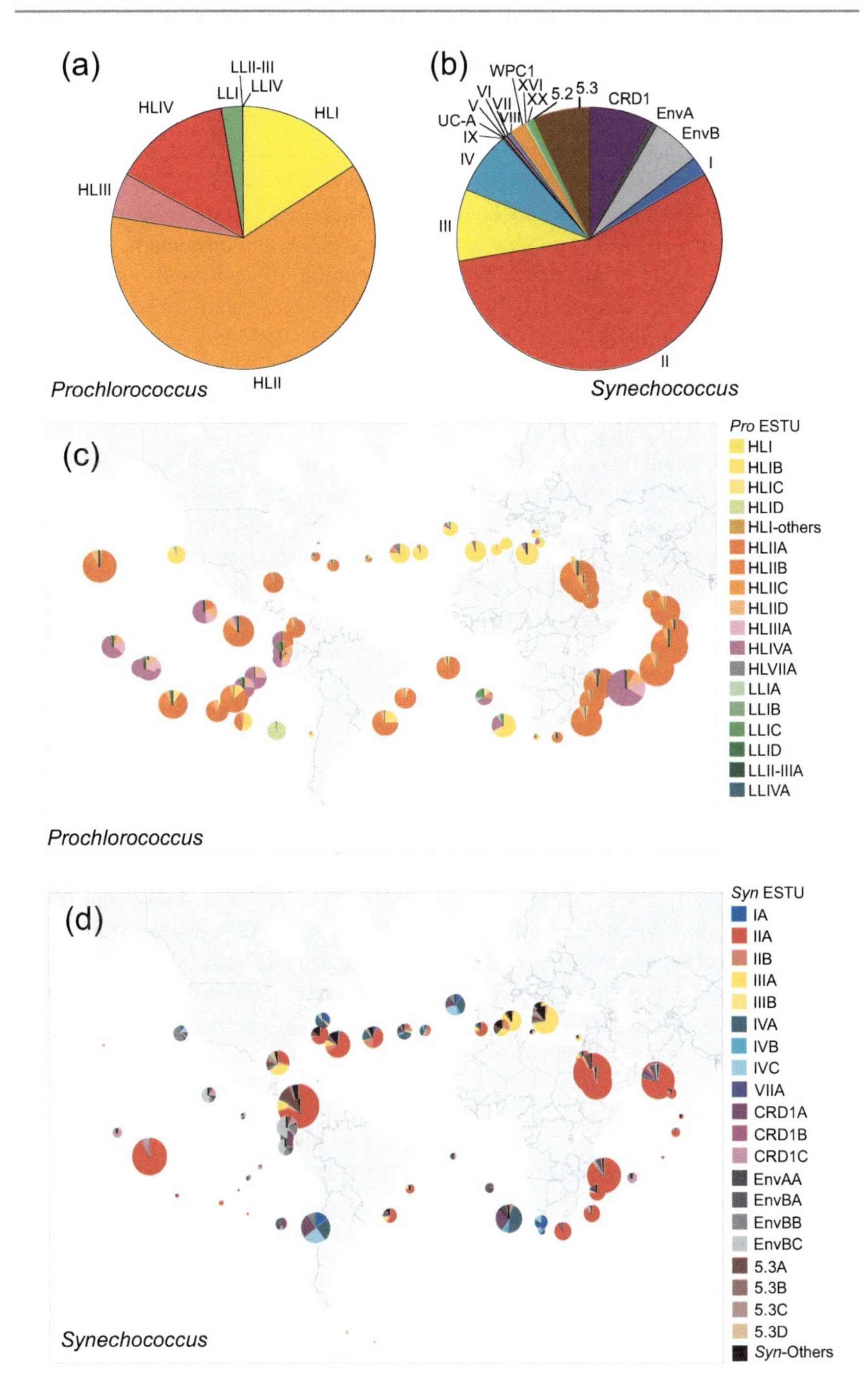


**Fig. 3.3** Relative abundance of *Prochlorococcus* and *Synechococcus* clades in surface (**a** and **b**) and global distribution along the *Tara* Oceans expedition transect of ecologically significant taxonomic units (ESTUs, defined as within-clade 94% OTUs sharing a similar distribution pattern; **c** and **d**). The size of circles is proportional to the number of *Prochlorococcus* or *Synechococcus*

### 3.4.4 The Wide Genomic Diversity of Marine Picocyanobacteria and Its Taxonomic Implications

Genomes of marine picocyanobacteria sequenced thus far exhibit a tremendous diversity of both nucleotide sequences and gene content (Doré et al. 2020; Dufresne et al. 2008; Kettler et al. 2007; Lee et al. 2019). Average amino acid identity (AAI) between pairs of genomes is ranging from 53.2 to 98.9% and intra-clade AAI is on average 91.0% (Doré et al. 2020). This large genomic diversity led one research team to split the *Prochlorococcus* branch into 5 distinct "genera" (Tschoeke et al. 2020) and the *Synechococcus* SC 5.1, 5.2, and 5.3 into 2, 3, and 2 genera, respectively, based on a "genus" delimitation of $\geq$70% AAI with further refinement using phylogenetic analyses. Yet, these taxonomic delineations were made without any attempt to match SC, clades, and/or sub-clades previously defined by other teams (see below for details). Using 53 (including 32 previously unpublished) marine or brackish *Synechococcus/Cyanobium* genomes Doré et al. (2020) drew a plot of AAI vs. 16S rRNA identity for the different pairs of genomes (Fig. 3.4) and showed that while there was a continuum of 16S rRNA identities ranging from 95.5 to 100% two major discontinuities could be defined based on the AAI: the first one at ca. 80% AAI discriminated (with a few exceptions) pairs of strains belonging to the same clade from pairs of strains belonging to different clades; the second one at 65% AAI set apart *Synechococcus* strains of the same SC from strains of different SC. However, there was *no* discontinuity at 70% AAI suggesting that this cut-off is irrelevant to define genera in this particular group (Fig. 3.4). The same holds true for the 95% AAI cut-off often used to delineate bacterial species (see, e.g., Konstantinidis and Tiedje 2005) that, when blandly applied to marine picocyanobacteria, leads to a myriad of species with no apparent ecological, biochemical, or physiological relevance (e.g., at least 137 species within the set of *Prochlorococcus* genomes analyzed by Tschoeke et al. 2020). Clearly, the taxonomy of *Cyanobacteria* Cluster 5 sensu (Herdman et al. 2001) needs to be revised but in order to define meaningful and consensual families, genera, and species within this group it is important not only to take into account *natural* genomic delineations as defined using key parameters such as the AAI but also to use solid phylogenies and to have a good knowledge about the physiology and ecological niches occupied by the candidate taxa.

### 3.4.5 Role of Environmental Factors in Genetic and Functional Diversification

#### 3.4.5.1 *Prochlorococcus*

The evolutionary history of marine picocyanobacteria has been strongly influenced by environmental factors (Biller et al. 2015; Martiny et al. 2015; Scanlan et al. 2009).

**Fig. 3.3** (continued) reads at the corresponding station. Reproduced with permission from Farrant et al. 2016 (**a** and **b**) or drawn using data from this study (**c** and **d**)

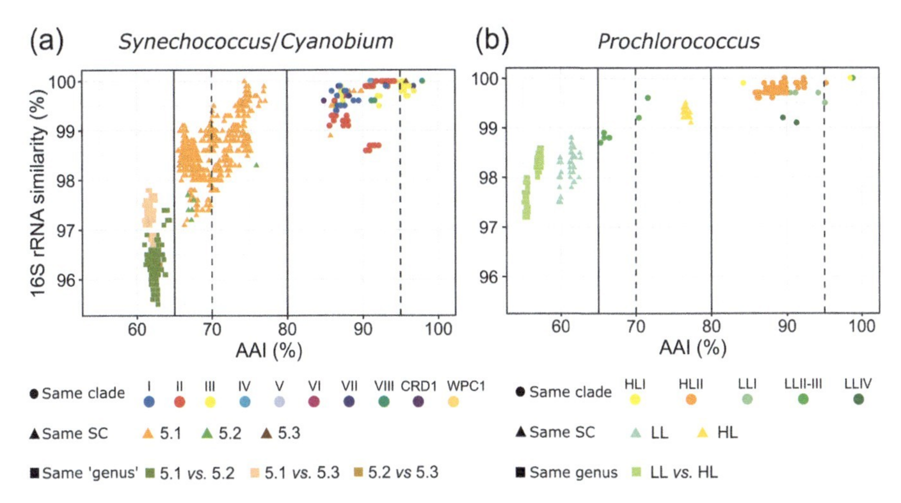


**Fig. 3.4** Relationships between 16S rRNA identity, AAI, and taxonomic information for *Synechococcus/Cyanobium* (**a**) and *Prochlorococcus* (**b**) genomes. Dots correspond to comparisons between pairs of genomes belonging to the same clade, triangles between pairs of genomes belonging to the same SC (or phototype, HL or LL, for *Prochlorococcus*) but different clades and squares between pairs of genomes belonging to different SC or phototype. Continuous vertical lines correspond to the "natural" AAI discontinuities mentioned by Doré et al. (2020), while the vertical dashed lines correspond to the limits for genus (70% AAI) and species (95% AAI) used by Tschoeke et al. (2020) for *Prochlorococcus* and Salazar et al. (2020) for the *Synechococcus/ Cyanobium* group. Adapted from Fig. 3.3b in Doré et al. (2020)

For *Prochlorococcus*, Moore and co-workers demonstrated the key role played by light in the vertical niche partitioning of high-light (HL) adapted ecotypes in the upper lit layer and low-light (LL) adapted ecotypes further down the euphotic layer (Moore et al. 1998). Yet, subsequent studies have considerably refined this picture by showing that there is a multiplicity of HL and LL ecotypes. Most LL clades (LLII-VII) are confined to the bottom of the euphotic layer with LLV-VI being specific of oxygen minimum zones (Lavin et al. 2010) while the LLI clade is more photo-tolerant (Huang et al. 2012; Johnson et al. 2006; Malmstrom et al. 2010; Partensky and Garczarek 2010). Indeed, in warm stratified oligotrophic waters LLI cells are most abundant at the base of the upper mixed layer and in temperate mixed waters they can even reach the surface, suggesting that they are capable to stand a temporary exposure to high irradiance (Johnson et al. 2006; Malmstrom et al. 2010; Thompson et al. 2018).

The genetic diversification of *Prochlorococcus* HL clades has been influenced by different environmental factors than for LL lineages. Indeed, Fe availability has likely conditioned the differentiation between the HLI-II clades, which thrive in Fe-replete waters, and HLIII-IV clades that co-occur in warm Fe-depleted areas notably in high-nutrient low-Chl (HNLC) areas of the Pacific Ocean (Fig. 3.3c; Farrant et al. 2016; Rusch et al. 2010; West et al. 2011). HNLC waters have been extensively explored during the *Tara* Oceans expedition, explaining the large proportion of HLIII-IV clades observed in this global metagenome dataset even though HLII remains globally the most abundant clade (Fig. 3.3a). Temperature has seemingly favored the separation between the HLI ecotype, which predominates in temperate waters at high latitudes and the HLII ecotype that preferentially thrives in warm (sub)tropical oligotrophic waters though the minor ecotype in each thermal niche is never outcompeted to extinction (Chandler et al. 2016; Johnson et al. 2006; Zinser et al. 2007). Shifts between HLI- and HLII-dominated *Prochlorococcus* communities sometimes occur over short geographical distances, e.g., between the Mediterranean Sea and the Red Sea (Farrant et al. 2016) or along the temperature gradient between the core and the outside of "Agulhas rings," huge anticyclonic eddies that are formed in the southern Indian Ocean and then drift across the South Atlantic Ocean (Fig. 3.3c; Villar et al. 2015). Lastly, P and N availability have also played a key role in the more recent evolutionary history of the *Prochlorococcus* genus. A number of studies have indeed shown that *Prochlorococcus* HL populations thriving in P-depleted areas possess many more genes involved in P metabolism or its regulation than populations (or strains isolated) from P-replete areas (Martiny et al. 2006; Martiny et al. 2009a). A similar phenomenon was also reported for genes involved in nitrite and/or nitrate assimilation the frequency of cells capable of assimilating these oxidized forms of nitrogen being positively correlated with decreased N availability (Aldunate et al. 2020; Astorga-Eló et al. 2015; Berube et al. 2015; Berube et al. 2016; Berube et al. 2019; Martiny et al. 2009b; Villar et al. 2015). Despite this apparent relationship between *Prochlorococcus* clades and community structure, several studies have shown the occurrence of a large within-clade genetic microdiversity. For instance, Kashtan et al. (2014) showed that *Prochlorococcus* HLII populations collected at three seasons at the BATS station off Bermuda Islands were composed of hundreds of closely related sub-populations possessing distinct genomic "backbones" each consisting of a different set of core genes associated with a defined set of accessory genes. These discrepancies were sufficient to allow differentiated responses of the HLII sub-populations to seasonal changes in the environment. Occurrence of within-clade sub-populations exhibiting distinct ecological niches has also been observed in other *Prochlorococcus* clades. Using the high-resolution marker *petB* Farrant et al. (2016) have defined "ecologically significant taxonomic units" (ESTUs) as within-clade operational taxonomic units (OTUs) at 94% nucleotide identity that occupied distinct niches along the *Tara* Oceans expedition transect. For instance, a minor *Prochlorococcus* LLI sub-population (ESTU LLIB) appeared to be adapted to Fe-limited surface waters like the major HLIIIA-IVA ESTUs, whereas a minor HLI sub-population (ESTU HLIC) thrived not only in cold temperate waters, as do typical HLI cells, but also in warm subtropical waters, thus extending the global niche occupied by these clades (Fig. 3.3c). Similarly, Larkin et al. (2016) observed that four different HLI sub-ecotypes had distinct seasonal and spatial patterns of diversity across latitudes. All these observations constitute strong pieces of evidence that genomic diversification is still ongoing within *Prochlorococcus* clades.

#### 3.4.5.2 *Synechococcus*

The temporal succession of physicochemical parameters that have driven the evolution of the *Synechococcus/Cyanobium* radiation is more difficult to establish because the rapid diversification of lineages that occurred within SC 5.1 (Urbach et al. 1998) has resulted in a fuzzier relationship between phylogeny and adaptation to specific environmental factors than for *Prochlorococcus*. Yet, temperature has played a major selective role since clades I and IV mainly occur in cold nutrient-rich waters, while clades II and III preferentially thrive in warm oligotrophic waters (Zwirglmaier et al. 2008). Like for *Prochlorococcus* HLI and HLII, differences in realized thermal niches between *Synechococcus* clades can be explained by the distinct thermophysiologies of these phylotypes as demonstrated by comparing temperature growth ranges and optima of representative isolates in culture (Breton et al. 2019; Mackey et al. 2013; Pittera et al. 2014; Varkey et al. 2016). Yet, it must be stressed that clade I and IV strains fall in distantly related branches in phylogenies made using 16S rRNA or other core gene markers (see, e.g., Doré et al. 2020; Dufresne et al. 2008; Scanlan et al. 2009; Tai and Palenik 2009). Furthermore, a study of the global distribution of *Synechococcus* at high taxonomic resolution (Fig. 3.3d) showed that a minor ESTU within clade II (ESTU IIB) was able to colonize cold niches (Farrant et al. 2016). Like for *Prochlorococcus* several recent studies have indicated that Fe availability could also be an important driving factor of the composition of *Synechococcus* communities. Indeed, CRD1 and EnvB clades —the latter, defined with the *petB* marker, corresponds to the CRD2 clade as defined with ITS (Ahlgren et al. 2020)— co-dominated in Fe-depleted areas (Ahlgren et al. 2014; Farrant et al. 2016; Sohm et al. 2015). Finally, P availability also seems to have influenced *Synechococcus* genetic diversification since both ESTUs IIIA and 5.3A appear to be adapted to P-limited areas such as the Mediterranean Sea and Gulf of Mexico (Fig. 3.3d) as confirmed by comparative genomics (Doré et al. 2020). The co-occurrence in the same niche of *Synechococcus* genotypes belonging to phylogenetically distant clades indicates that adaptation to specific environmental parameters (temperature, Fe, and/or P availability) likely happened several times independently during the evolution of the *Synechococcus* radiation through convergent evolution.

### 3.4.6 *Prochlorococcus* Genome Streamlining

The evolutionary history of the *Prochlorococcus* genus is also characterized by a major genome streamlining event that has affected most lineages (Biller et al. 2015; Doré et al. 2020; Dufresne et al. 2003; Kettler et al. 2007; Partensky and Garczarek 2010). While no comparable decrease in genome size has occurred during the evolution of marine *Synechococcus* it is worth mentioning that members of one sub-clade of clade II, representative of the most abundant *Synechococcus* population of the ocean, possess the smallest genomes known to date in this "genus" ($2.14 \pm 0.05$ Mbp) suggesting that some genome reduction occurred in this sub-clade relative to the rest of the *Synechococcus* population (Lee et al. 2019). In

*Prochlorococcus*, the streamlining process has led to an overall decrease of up to about one third of the genome size—corresponding to about 930 genes—compared to its most recent common ancestor with *Synechococcus* with a concomitant drop of the GC content down to ca. 30% in HL clades (Dufresne et al. 2005; Kettler et al. 2007; Partensky and Garczarek 2010). It is important to note that while many gene gains and losses have taken place in all *Prochlorococcus* lineages this gene flow was balanced in the basal LLIV lineage whose members have genome sizes of around 2.56 Mbp—i.e. in the range of *Synechococcus* SC 5.1 genomes: 2.11 to 3.31 Mbp—but strongly unbalanced toward losses in all other lineages, which have genome sizes between 1.48 and 1.92 Mbp (Biller et al. 2014a; Doré et al. 2020; Kettler et al. 2007). Yet, only about one hundred protein-coding genes are shared between all complete *Prochlorococcus* LLIV and *Synechococcus* genomes but lacking from all streamlined *Prochlorococcus* genomes (according to "phyletic pattern" searches in the Cyanorak v2.1 database; www.sb-roscoff.fr/cyanorak). This indicates that the different *Prochlorococcus* lineages have retained or lost different sets of genes.

For a free-living phototroph *Prochlorococcus* genomes encode only a small set of regulatory proteins, encompassing on average five different sigma factors, five to six two-component systems and eight types of other transcriptional regulators (Lambrecht et al. 2020; Scanlan et al. 2009). Although this appears to be a minimal regulatory system, it seems to be sufficient to address all relevant stress situations that *Prochlorococcus* cells have to face in the huge but relatively stable warm open ocean environment in which they thrive. Furthermore, riboswitches and RNA-based regulation also likely play an important role in the regulation of gene expression in *Prochlorococcus* (for a review, see Lambrecht et al. 2020).

### 3.4.7 Core, Accessory, and Pangenomes

The core genome of *Prochlorococcus* and *Synechococcus/Cyanobium* encompasses 1346 and 1713 genes, respectively—these numbers correspond to the "large core" genome, i.e. genes shared by more than 90% of members of the considered genus or group; Doré et al. (2020)—, while their accessory genome has a highly variable size (Dufresne et al. 2008; Kettler et al. 2007; Scanlan et al. 2009). A large fraction of accessory and unique genes, most of which of unknown function, is localized in hypervariable regions called genomic islands, which are likely involved in adaptation to the local environment and/or the resistance against grazers or phages (Avrani et al. 2011; Coleman et al. 2006; Dufresne et al. 2008; Palenik et al. 2006). The comparison of 81 non-redundant picocyanobacterial genomes has revealed that closely related strains share many more island genes than distantly related ones and that only a few gene exchanges had occurred between distantly related clades (Doré et al. 2020). By building a network of islands shared between different strains these authors also showed that *Prochlorococcus* HL strains share a large number of genomic regions that were already present in the common ancestor of all HL strains, then vertically transferred to all descendants, much like the phycobilisome region is shared by all *Synechococcus* strains (Dufresne et al. 2008; Six et al. 2007) while other islands are shared only by a subset of strains. *Prochlorococcus* isolated from P-depleted areas possess a number of specific genes involved in organic P acquisition, which are located in genomic islands (Martiny et al. 2006; Martiny et al. 2009a). In *Synechococcus* several strains possess island genes coding for "giant" proteins (>1 MDa) that owe their huge size to a high number of repeats (Dufresne et al. 2008). This includes the cell wall protein SwmB that, together with SwmA, is required for a unique form of swimming motility (McCarren and Brahamsha 2005; McCarren and Brahamsha 2007). Although this ability was known to be restricted to clade III members Doré et al. (2020) have suggested that not all clade III strains are motile since two out of eight strains of this clade did not possess *swmA* and *swmB* genes. Instead, they contained a 3-gene cluster composed of a *nfeD* homolog and two flotillin-like genes (*floT1-2*) that might be involved in the production of lipid rafts.

While the core genome of picocyanobacteria is quite small their pangenome is huge since each new sequenced genome contains on average 277 new genes (Baumdicker et al. 2010). Using a quantitative evolutionary model for the distributed genome, Baumdicker et al. (2012) predicted that the pangenome is finite and would contain 57,792 genes but a study based on more genomes raised this number to 84,872 genes (Biller et al. 2015). Yet, this vast "collective genome" might still miss some important gene functions since *Prochlorococcus* cells depend upon free-living "helper" bacteria from their immediate environment such as *Alteromonas*, e.g., to mitigate oxidative stress using exogenous catalases (Morris et al. 2008). This exchange might be reciprocal because *Prochlorococcus* cells can themselves release lipid vesicles containing proteins and nucleic acids that might be used for transferring material to their "helper" bacteria but also possibly be used as lures for cyanophages (Biller et al. 2014b).

### 3.4.8 Potential Biotechnological Value

Another advantage of the wealth of *Prochlorococcus*, *Synechococcus*, and *Cyanobium* genomes currently available (Biller et al. 2014a; Garczarek et al. 2021) is to inform us about the ability of these picocyanobacteria to synthesize products of potential biotechnological interest. Due to their small genome size, picocyanobacteria possess relatively few genes involved in the biosynthesis of secondary metabolites compared to their larger sized marine, freshwater, or extremophile counterparts (Mandal and Rath 2015). Yet, picocyanobacteria not only possess and express alkane biosynthesis genes (Klähn et al. 2014) but also produce and accumulate hydrocarbons, which may constitute between 0.02 and 0.37% of their dry cell weight (Lea-Smith et al. 2015; Schirmer et al. 2010). Thus, at the global ocean scale picocyanobacteria would produce about 540 million tons of hydrocarbons annually. A number of genes are involved in cell defense mechanisms including antibiotic resistance genes (Hatosy and Martiny 2015) and genes involved in the biosynthesis of microcin C, a "Trojan horse" antibiotic that plays a role in allelopathy between *Synechococcus* strains (Paz-Yepes et al. 2013). Some strains of *Prochlorococcus* and *Synechococcus* are prolific producers of prochlorosins, a particular class of lanthipeptides, ribosomally derived and posttranslationally modified peptide secondary metabolites. Lanthipeptides display diverse bioactivities (e.g., antifungal, antimicrobial, and antiviral) and therefore possess great potential for bioengineering and synthetic biology (Hetrick et al. 2018). Prochlorosins were discovered in *Prochlorococcus* sp. MIT9313, which possesses 29 *procA* genes encoding the peptide skeletons, which then are transformed by a single promiscuous enzyme into a library of highly diverse prochlorosins (Li et al. 2010). All 29 homologs were transcribed (Voigt et al. 2014). Interestingly, several *procA* homologs contain a second transcriptional start site within the coding region, pointing at potential additional diversity due to a transcriptional mechanism (Voigt et al. 2014). The capacity for the production of prochlorosins is frequent in the LLIV clade of *Prochlorococcus* but also found in representatives of *Synechococcus* belonging to SC 5.1 clades I and IX (Cubillos-Ruiz et al. 2017). Genome sequencing of marine cyanobacteria from environmental samples led to the discovery of 1.6 million open reading frames encoding these lanthipeptides (Cubillos-Ruiz et al. 2017). Despite their impressive distribution and variability the natural functions of prochlorosins have remained enigmatic thus far.

Other high value products are phycobiliproteins that have a great potential for nutraceutical, pharmaceutical, cosmetic, feed, and food industries (Pagels et al. 2019). Due to their high quantum yield and large Stokes shift phycobiliproteins are also widely used as fluorophores in fluorescence biotechnologies notably in antibody conjugates for surface labeling in flow cytometry and enzyme-linked immunosorbent assay (ELISA; Giepmans et al. 2006). In this context, although to our knowledge it has not been exploited for such purpose yet, it is noteworthy that PEII, a marine *Synechococcus*-specific phycobiliprotein with six chromophores per α,β monomer, is the most fluorescent of all phycobiliproteins and thus appears as a promising compound for fluorescence biotechnologies especially PEII forms that bind large amounts of PUB whose optical properties match well with blue lasers (Ong et al. 1984). The fact that most enzymes involved in PEI and/or PEII biosynthesis have been characterized in the last two decades (Carrigee et al. 2020, 2021; Grébert et al. 2021; Mahmoud et al. 2017; Sanfilippo et al. 2019) should now make it possible to produce these fluorophores using heterologous co-expression systems (Biswas et al. 2010). Also noteworthy, a small genomic island involved in Type IV chromatic acclimation, which exists in two configurations (CA4-A and -B; Humily et al. 2013), was characterized in marine *Synechococcus* (Grébert et al. 2021; Sanfilippo et al. 2016; Shukla et al. 2012) and could potentially be used in heterologous systems to control the expression of recombinant genes in a reversible manner by changes in light color (blue to green light shifts or reciprocally) rather than by the usual irreversible chemical induction.

Finally, metabolic modeling is an approach with great potential for better understanding cyanobacterial physiology and for optimizing the productivity of cultures in the perspective of biotechnological applications (Toyoshima et al. 2020; Zavřel et al. 2019). Although it has not been widely applied yet to marine cyanobacteria it is worth noting that a genome-scale metabolic model using a multi-omic machine learning pipeline has been proposed recently for *Synechococcus* sp. PCC 7002, a strain thought to have a large potential for renewable biofuels production (Vijayakumar et al. 2020).

## 3.5 Nitrogen-Fixing Cyanobacteria

### 3.5.1 Ecological Role and Importance of Diazotrophy in Marine Ecosystems

Nitrogen (N) is the predominant limiting nutrient for primary productivity in the upper sunlit layer throughout much of low-latitude oceans (Bristow et al. 2017; Canfield et al. 2010; Falkowski 1997; Moore et al. 2013). By assimilating dinitrogen ($N_2$), the simplest and most abundant N form in the atmosphere and in seawater, microbial diazotrophs constitute an important source of bioavailable N to oceanic surface waters and even the most important external N source before atmospheric and riverine inputs (Deutsch et al. 2007; Großkopf et al. 2012). Diazotrophy is a crucial marine biogeochemical process that compensates bioavailable N losses due to denitrification and anaerobic ammonium oxidation and sustains new primary production (Karl et al. 1997; Karl et al. 2002; Zehr and Kudela 2011). Most observations of marine diazotroph abundances have been made in the N-depleted waters of the tropical Atlantic and western tropical Pacific oceans, with sparse observations in the Indian and eastern Pacific oceans (Tang and Cassar 2019). However, two studies have revealed the occurrence of $N_2$ fixation in the cold nutrient-rich waters of the Arctic Ocean (Harding et al. 2018; Shiozaki et al. 2018). Some oceanic zones constitute hot spots of $N_2$ fixation, such as the western tropical South Pacific Ocean, with average fixation rates of 570 μmol $N.m^{-2}.d^{-1}$ over the whole area (Bonnet et al. 2017).

Even though cyanobacteria are not the sole $N_2$-fixers in marine ecosystems the contribution of heterotrophic diazotrophs such as some Gammaproteobacteria and Deltaproteobacteria is only starting to be elucidated (see among others Cornejo-Castillo and Zehr 2021) and it is generally considered that the marine $N_2$ fixation is mainly driven by diazotrophic cyanobacteria (Tang et al. 2019; Turk-Kubo et al. 2014; Zehr and Capone 2020; Zehr and Kudela 2011). A considerable fraction of the $N_2$ diazotrophic cyanobacteria fix is rapidly released in seawater as dissolved inorganic N and transferred to non-diazotrophic components of the microbial community (Berthelot et al. 2016; Knapp et al. 2016). The main genera of $N_2$-fixing marine cyanobacteria are *Trichodesmium, Nodularia, Richelia, Calothrix, Candidatus* Atelocyanobacterium (UCYN-A), and *Crocosphaera* (UCYN-B and -C; Fig. 3.2). Comparative genomic analyses showed that genes encoding the $N_2$ fixation machinery are generally clustered in a specific genomic region which, in some cases, may have been transferred laterally causing diazotrophy in taxa that are otherwise not known to fix nitrogen such as *Microcoleus chthonoplastes* (Bolhuis et al. 2010) or *Acaryochloris* sp. HICR111A (Pfreundt et al. 2012).

A data-driven model of the biogeography of the major diazotrophic cyanobacteria (Tang and Cassar 2019) predicted that (1) *Trichodesmium* is prevalent in the

30°N-30°S latitudinal range, dominating in the tropical Atlantic, western Indian, and western Pacific oceans, (2) UCYN-B exhibits a similar distribution but with smaller depth-integrated abundances, (3) UCYN-A distribution extends from tropical areas to temperate and polar regions, and (4) *Richelia* shows a more or less homogeneous distribution in the tropical ocean with hot spots in the Amazon River plume, the central North Pacific, and in the Mediterranean Sea. Abundance data for UCYN-C are still too scarce to get a global view of their distribution. Yet, local studies show that they can have a substantial impact even though they only represent a small share of the diazotrophic cells in the total microbial population. For instance, during the VAHINE mesocosm experiment performed in 2013 in the shallow water of the New Caledonia lagoon (Bonnet et al. 2016b) $N_2$ fixation rates reached >60 nmol N $L^{-1}$ $d^{-1}$, which are among the highest rates reported for marine waters (Bonnet et al. 2016a; Luo et al. 2012). $N_2$-fixing cyanobacteria of the UCYN-C type dominated the diazotroph community in mesocosms (Turk-Kubo et al. 2015). However, based on relative 16S rRNA gene copy numbers normalized by comparison with the flow cytometry counts of abundant marine picocyanobacteria a maximum of only 500 cells $mL^{-1}$ was calculated for these diazotrophs (Pfreundt et al. 2016). This matched reasonably well to the maximum of 100 UCYN-C *nifH* copies $mL^{-1}$ determined for the same population (Turk-Kubo et al. 2015) especially when taking into account that representative UCYN-C isolates, such as ATCC 51142 or TW3, usually contain two to three 16S rRNA gene copies per cell compared to the single-copy gene *nifH* (Taniuchi et al. 2008). In contrast, the total number of bacteria (heterotrophs and picocyanobacteria) was between 5 and $7 \times 10^6$ cells $mL^{-1}$, hence UCYN-C diazotrophs had a share of <0.01% in the total microbial population but impacted its biogeochemical properties in a profound way.

Like for *Prochlorococcus* and *Synechococcus* (cf. Sect. 3.4.5) accumulating evidence from genomic analyses now suggests that these groups also consist of genetically diverse physiologically and morphologically distinct sub-lineages that represent ecotypes adapted to specific conditions (Henke et al. 2018; Turk-Kubo et al. 2017; Zehr and Capone 2020).

### 3.5.2 Filamentous Marine Diazotrophs

The major genera of multicellular filamentous marine diazotrophs include: (1) the non-heterocystous colony-forming *Trichodesmium* (Capone 1997; Karl et al. 2002), (2) the heterocyst-forming *Richelia*, which lives in symbiosis with the diatoms *Rhizosolenia* (a diatom–diazotroph association also referred to as Het-1) or *Hemiaulus* (Het-2; Foster and Zehr 2006; Foster and Zehr 2019; Goebel et al. 2010; Janson et al. 1999; Pyle et al. 2020; Villareal 1991; Villareal 1992; Villareal 1994; Zeev et al. 2008), and (3) the heterocystous *Calothrix* (Het-3), which is commonly associated with the diatom *Chaetoceros* (Carpenter and Foster 2002; Foster and Zehr 2006; Goebel et al. 2010). These diatom–diazotroph associations are thought to be highly host specific, one symbiont type associating with a single host genus, though the driver of this specificity remains to be elucidated.

Other filamentous diazotrophs are found in coastal and reef environments such as the non-heterocystous genus *Lyngbya* (Omoregie et al. 2004; Woebken et al. 2015). Heterocystous cyanobacteria belonging to the genus *Nodularia* thrive in brackish water environments, such as the Baltic Sea (Ploug et al. 2011).

#### 3.5.2.1 *Trichodesmium*

Diazotrophic cyanobacteria of the genus *Trichodesmium* can form huge surface blooms of tens of thousands of $km^2$ in the tropical and subtropical ocean (Dupouy et al. 1988), constituting an important source of new N to these oligotrophic environments (Capone 1997; Davis and McGillicuddy Jr. 2006; Mahaffey et al. 2005). *Trichodesmium* is taxonomically close to filamentous cyanobacteria of the genus *Oscillatoria* (Larsson et al. 2011). It possesses the unique property among diazotrophic filamentous cyanobacteria to express the oxygen-sensitive nitrogenase and to fix $N_2$ during the day (Dugdale et al. 1961) concomitantly with photosynthetic oxygen evolution (Fig. 3.2). Nickel (Ni) availability seems to play a key role in controlling this process, since Ni-replete *Trichodesmium* cells can fix $N_2$ throughout most of the light:dark cycle (including the dark period) and fixation rates are several-fold higher in high-Ni than the low-Ni cultures (Rodriguez and Ho 2014). Whereas other diazotrophs such as *Calothrix*, *Nodularia*, or *Anabaena* develop heterocysts, i.e., differentiated cells specialized for nitrogen fixation (Muro-Pastor and Hess 2012), *Trichodesmium* uses a different not terminally differentiated cell type for this purpose called diazocyte (Berman-Frank et al. 2001; El-Shehawy et al. 2003; Sandh et al. 2009; Sandh et al. 2012). In addition, *Trichodesmium* can form colonies or multicellular aggregates of surprisingly diverse morphologies including threads (trichomes), radial puffs, vertically aligned fusiform tufts, and bowties (Hynes et al. 2012; Olson et al. 2015; Post et al. 2002; Webb et al. 2007). Although the physiological relevance of these morphologies is not well understood (Eichner et al. 2019), there is clear evidence for true multicellular behavior of *Trichodesmium* colonies, e.g., in the acquisition of mineral-rich dust particles (Kessler et al. 2020; Rubin et al. 2011).

Different factors have been discussed to influence termination of *Trichodesmium* blooms including bacteriophage-induced lysis (Hewson et al. 2004; Ohki 1999) or grazing by copepods (O'Neil and Roman 1994). An important factor seems to be the activation of an autocatalytic mechanism leading to programmed cell death resembling apoptosis in metazoans. While this process is well documented in marine phototrophs (Bidle and Falkowski 2004; Franklin et al. 2006) details of the involved mechanisms have remained enigmatic. In *Trichodesmium*, programmed cell death may be triggered by several environmental factors such as P or Fe starvation, high irradiance, or oxidative stress (Berman-Frank et al. 2004; Berman-Frank et al. 2007) and involves the activation of caspase-like activity (Berman-Frank et al. 2004). Analyses on *Trichodesmium* from the South Pacific Ocean demonstrated the stimulation of metacaspase expression and activity and the connection to programmed cell death induced mortality (Spungin et al. 2016; Spungin et al. 2019).

With a size of 7.75 Mbp and 4451 annotated genes, the genome of the reference strain *Trichodesmium erythraeum* IMS101 belongs to the larger cyanobacterial genomes known to date (Larsson et al. 2011; Shih et al. 2013). In sharp contrast with other free-living cyanobacteria genomes, which have an average coding capacity of ~85%, only 64% of the *T. erythraeum* genome encodes proteins (Larsson et al. 2011). This unusually high non-coding genome share is supported by metagenomic datasets (Walworth et al. 2015), hence it is typical for *Trichodesmium*. Transcription from such non-coding genome space can produce non-coding RNAs (sRNAs), which frequently have regulatory functions in cyanobacteria (Georg et al. 2014; Klähn et al. 2015; Kopf and Hess 2015). Indeed, the analysis of the primary transcriptome of *Trichodesmium erythraeum* IMS101, i.e. the sequencing of enriched nascent transcript starts, revealed that at least 40% of all promoters that are active under standard laboratory conditions produce non-protein-coding transcripts and that these accumulate in much larger amounts than mRNA (Pfreundt et al. 2014).

Among those non-coding transcripts is an actively splicing twintron (Pfreundt and Hess 2015) and a diversity generating retroelement (DGR). DGRs are widely distributed in the genomes of bacteria (Yan et al. 2019). They can generate mutations in their target genes and hence induce diversification of the encoded proteins, which likely is beneficial. In analogy to retrons, another class of retroelements in bacteria, it might be speculated that these have defense-related functions (Millman et al. 2020). However, experiments on the precise role of DGRs in natural populations are still lacking. DGRs consist of two components, a non-coding RNA called template repeat RNA that serves as a template for the second element, and an error-prone reverse transcriptase that converts this template into cDNA for recombination into the protein-coding region of the target gene(s) (Doulatov et al. 2004; Guo et al. 2008). Although there was previous evidence for the existence of this mechanism in *Trichodesmium* (Doulatov et al. 2004) the target genes remained unknown. The primary transcriptome enabled the exact definition of the DGR components in *Trichodesmium* and revealed 12 putative target genes (Pfreundt et al. 2014), which represent an unprecedented potential for in vivo protein diversification in bacteria. Although none of these genes possess a clear functional assignment some appear connected to putative signaling proteins (kinases) possibly constituting their receptor component. It can be speculated that at least some of the targeted proteins are involved in the defense against bacteriophages by systematic variation of a surface receptor. It is further possible that this system is involved in generating phenotypic variability by diversifying a few key genes. This could play a role, for instance, in the cell–cell recognition required for multicellular behavior and colony formation.

Transcriptomic analyses was also the key to establish the biosynthetic pathway for the previously unknown compatible solute N,N,N-trimethyl homoserine (homoserine betaine) via a stepwise methylation from L-homoserine and using S-adenosylmethionine as the methyl group donor (Pade et al. 2016; see also Sect. 3.3.3). To balance the typical open ocean salinity of ~35–37 grams per liter the cells accumulate substantial amounts of this compound and hence require also substantial amounts of fixed carbon. S-adenosylmethionine is involved as the methyl group donor in many different reactions including DNA methylation. In a seven-year evolution experiment higher DNA methylation levels in *Trichodesmium* correlated with phenotypic adaptation to enhanced $CO_2$ and increased $N_2$ fixation rates (Hutchins et al. 2015). In a follow-up analysis it was concluded that long-term m5C methylome modifications correlated with phenotypic adaptation to $CO_2$ (Walworth et al. 2020). It is truly fascinating that in *Trichodesmium* the presence of massive amounts of a trimethylated compatible solute, epigenetic modification, i.e., DNA methylation, and adaptation to enhanced $CO_2$ fixation rates, coincide with each other. Future research will show whether this is purely coincidental or linked in more intricate ways.

#### 3.5.2.2 *Nodularia*, a Bloom-Forming Cyanobacterium Specifically Adapted to Salinity Gradients

Almost every summer, massive blooms of toxic cyanobacteria occur in the central regions of the Baltic Sea. These cyanobacteria cope well with the salinity gradient and brackish conditions that characterize the Baltic Sea. The dominating genera within these blooms are the filamentous $N_2$-fixing *Aphanizomenon*, *Dolichospermum*, and *Nodularia*. *Dolichospermum* and *Nodularia* can produce toxins such as microcystins or the hepatotoxin nodularin (Fewer et al. 2013; Mazur-Marzec et al. 2012) and protease inhibitors of the pseudoaeruginosin family (Liu et al. 2015). Excess P combined with low N concentrations in surface brackish waters are thought to favor the growth and bloom formation of diazotrophic cyanobacteria in summer (Sellner 1997) particularly under stably stratified warm water conditions. Additionally, gas vesicles that provide buoyancy to *Nodularia* and related cyanobacteria lead to the formation of large surface scums in the absence of mixing.

*Nodularia* appears to have a selective advantage under the brackish conditions of the Baltic Sea (Möke et al. 2013). The question of what these advantages could be was addressed by genomic and transcriptomic analyses of representative strains. The genome sequence of *Nodularia* UHCC 0039 is 5.38 Mbp and contains 5108 protein-coding genes (Teikari et al. 2018b). From these, 52.8% (2699) were annotated to encode hypothetical proteins illustrating the uniqueness of the *Nodularia* group and that deeper analyses of their functions are required. Comparison to draft genome sequences of two other *Nodularia* isolates revealed a core genome consisting of 3627 genes. Among the shared and strongly conserved genes is a gene cluster for the biogenesis of chaperone usher fimbriae, pointing at an unknown role of this type of fimbriae, possibly in adhesion, colony formation, or during bloom development. *Nodularia* is known for its high tolerance to periods of increased P starvation (Degerholm et al. 2006). Therefore, it is of interest that *Nodularia* genomes possess *phn* gene clusters potentially enabling the uptake, degradation, and use of phosphonates as an alternative P source (Teikari et al. 2018a; Voß et al. 2013). Indeed, the analysis of axenic *Nodularia* strains revealed that they can utilize methylphosphonates as the sole P source and concomitantly release methane (Teikari et al. 2018a). Together with earlier observations on *Trichodesmium* (Beversdorf et al. 2010) and other cyanobacteria (Bižić et al. 2020) these results suggest that certain marine cyanobacteria can release methane through the degradation of methylphosphonates, hence contributing to the oceanic methane paradox that posits that methane concentrations in surface waters can be above the atmospheric equilibrium (Repeta et al. 2016).

Further insights were gained from transcriptome analyses after exposure of *Nodularia* cultures to high light and oxidative stress mimicking the extreme environmental conditions occurring in the surface layer of the Baltic Sea in summer (Kopf et al. 2015). The observed up-regulation of genes encoding enzymes for the biosynthesis of toxins implied that these compounds may play an important role in the acclimation of cells to conditions of bloom formation consistent with similar observations for the freshwater cyanobacterium *Microcystis* (Zilliges et al. 2011). The photosynthetic activity of *Nodularia spumigena* CCY9414 trichomes remained high also at the highest irradiances of 1200 μmol photons $m^{-2}$ $s^{-1}$ and there were signs of an increase in photorespiratory flux (Kopf et al. 2015). This observation is of interest for understanding the acclimation of cyanobacterial trichomes to the combination of high light, high $O_2$ partial pressure, and low nutrients, including low Fe and carbonates concentration in the surface layer. Indeed, in cyanobacteria photorespiration cooperates with Mehler-like reactions catalyzed by flavodiiron proteins to dissipate excess absorbed energy (Allahverdiyeva et al. 2011; Allahverdiyeva et al. 2013; Hackenberg et al. 2009). The observed activation of photorespiratory flux was also consistent with the observation of many upregulated genes encoding photorespiratory enzymes and flavodiiron proteins. Finally, the identification of many additional stress-induced genes encoding proteins of unknown functions (Kopf et al. 2015; Teikari et al. 2018a; Voß et al. 2013) further suggests the existence of stress-related mechanisms in surface cyanobacterial blooms occurring in the Baltic Sea, which remain to be identified.

#### 3.5.2.3 *Richelia* and *Calothrix*

Whereas *Richelia* is an obligatory symbiont, *Calothrix* is epiphytic on the surface of *Chaetoceros* cells (Carpenter and Foster 2002) and can even display a free-living lifestyle (Foster et al. 2010). These differences between obligatory or facultative interaction with their respective diatom host match distinct differences in the genetic capacity of these two cyanobacteria in a dramatic way. Indeed, the genome of the marine *Calothrix rhizosoleniae* strain SC01, isolated from outside the frustule of *Chaetoceros,* is at least 6.0 Mbp, a size similar to those of free-living heterocyst-forming cyanobacteria, whereas *Richelia intracellularis* HH01, isolated from inside the siliceous frustule of *Hemiaulus,* has a genome size of only 3.2 Mbp. Moreover, the *Richelia* genome lacks genes for ammonium transporters, nitrate/nitrite reductases, and glutamine:2-oxoglutarate aminotransferase, showing clear signs of adaptation as an obligate symbiont (Hilton et al. 2013). Modeling estimated that (1) the rate of $N_2$ fixation in these diatom–diazotroph associations is 5.5 times higher than without N transfer to the host, (2) 25% of fixed C from the host diatom is transferred to the symbionts to support the high rate of $N_2$ fixation, and (3) 82% of the N fixed ends up in the host (Inomura et al. 2020). The high degree of integration between symbiont and host was further demonstrated in the metatranscriptomic analysis of surface samples from the North Pacific Subtropical Gyre revealing the coordination of gene expression in the diatom-*Richelia* symbiosis over day–night transitions (Harke et al. 2019).

Despite some evidence of genome streamlining it is currently unknown whether *Richelia* species that are found in the diatom–diazotroph association with *Rhizosolenia* (Het-1) are obligate symbionts as well (Hilton et al. 2013; Villareal 1992).

### 3.5.3 Unicellular Marine Diazotrophs

Marine unicellular cyanobacterial diazotrophs are phylogenetically divided into three groups (Fig. 3.1): (1) the yet uncultured unicellular cyanobacteria group A (UCYN-A) provisionally called *Candidatus* Atelocyanobacterium thalassa (Tripp et al. 2010; Zehr et al. 2001; Zehr et al. 2008), which lives in association with unicellular haptophyte algae (Zehr and Capone 2020), (2) *Crocosphaera watsonii* (UCYN-B), and (3) the UCYN-C group for which several representative isolates exist in culture, including CCY0110, ATCC 51142, ATCC 51472, and TW3 (Reddy et al. 1993; Taniuchi et al. 2012). The taxonomy of the UCYN-C group was until recently particularly confusing since this group encompassed both strains named *Cyanothece* spp. and *Gloeocapsa* spp. (Taniuchi et al. 2012). Mareš et al. (2019) have revised UCYN-C and proposed that ATCC 51142 and 51472 be renamed *Crocosphaera subtropica* and CCY0110 be renamed *Crocosphaera chwakensis* (TW3 was not included in their analysis).

UCYN-A is a major player in marine biogeochemical cycles and particularly in the nitrogen cycle (Cabello et al. 2015; Goebel et al. 2010; Jardillier et al. 2010; Montoya et al. 2004; Zehr and Kudela 2011). Measurements in the tropical North Atlantic showed that UCYN-A contributed to the total $N_2$ fixation approximately as much as *Trichodesmium* (Martínez-Pérez et al. 2016). In pioneering work the streamlined genome of UCYN-A was sequenced after isolating cells by flow cytometry cell sorting (Tripp et al. 2010; Zehr et al. 2008). The lack of all genes encoding the photosystem II complex, the Calvin–Benson–Bassham cycle, as well as other pathways that are normally essential such as the tricarboxylic acid cycle led Tripp et al. (2010) and Zehr et al. (2008) to hypothesize that UCYN-A lives in symbiotic interaction, an assumption that was later confirmed (Hagino et al. 2013; Thompson et al. 2012; Thompson et al. 2014). Many insights into the relevance of symbiotic interactions were further gained from the analysis of the association of UCYN-A with their eukaryotic host (Krupke et al. 2014). Indeed, in this mutualistic relationship, UCYN-A provides fixed N to its host in exchange for fixed carbon. Yet, before any carbon is transported from the haptophyte to UCYN-A, transfer of N is required from UCYN-A to its host (Krupke et al. 2015). The UCYN-A group consists of at least four distinct clades (Cornejo-Castillo et al. 2019; Farnelid et al. 2016; Thompson et al. 2014) called UCYN-A1, UCYN-A2, UCYN-A3, and UCYN-A4 and the whole group is monophyletic within the clade that also includes UCYN-B and UCYN-C species (Bombar et al. 2014). With regard to the symbiotic hosts the small-sized UCYN-A1 sub-lineage is associated with an open ocean picoplanktonic prymnesiophyte while the larger sized UCYN-A2 lives in a symbiosis with the coastal nanoplanktonic coccolithophore *Braarudosphaera bigelowii* (Cabello et al. 2020; Hagino et al. 2013; Thompson et al. 2012; Thompson et al. 2014). Surprisingly, while UCYN-A2 shares 1159 of the 1200 UCYN-A1 protein-coding genes (96.6%) with high synteny, the average amino acid sequence identity between these orthologs is only 86% (Bombar et al. 2014). These authors suggested that the UCYN-A1 and UCYN-A2 variants may reflect adaptation of their respective prymnesiophyte hosts to two different niches, namely coastal and open ocean habitats. In fact, phylogenomic and Bayesian relaxed molecular clock analyses suggested that UCYN-A has diverged and co-evolved with its prymnesiophyte host for ~91 My since the late Cretaceous (Cornejo-Castillo et al. 2016).

The UCYN-B group gathers *Crocosphaera watsonii* and related cyanobacteria, which are mostly free-living diazotrophs, but are also capable of colonial aggregation (Foster et al. 2013) and to form symbioses with the diatom *Climacodium frauenfeldianum* (Carpenter and Janson 2000; Foster et al. 2011). Otherwise, it is a typical $N_2$-fixing cyanobacterium in the sense that it is photosynthetic during the day and fixes $N_2$ during the night (Fig. 3.2) and the peak of nitrogenase gene expression occurs just prior to the dark period (Wilson et al. 2017). Currently, there are genome sequences of ten *Crocosphaera watsonii* strains available with sizes ranging from 4.55 Mbp for WH0401 to 6.24 Mbp for WH8501 (Bench et al. 2013; Webb et al. 2009).

The ecology and genetic diversity of the UCYN-C group are less well known except that this group seems to be adapted to warmer waters compared to its UCYN-A and -B counterparts (Berthelot et al. 2017). However, one strain representative of this group, ATCC 51142 (recently renamed to *Crocosphaera subtropica;* Mareš et al. 2019), gained much attention as a platform strain in biotechnology and synthetic biology. Sequence analysis revealed a particular genome organization consisting of one major circular and one linear chromosome and four small plasmids (Welsh et al. 2008). The presence of two chromosomes from which one is linear is unique among cyanobacteria. Information on the 5.46 Mbp ATCC 51142 genome and its 5304 predicted protein-coding genes enabled a systems biology analysis of this UCYN-C group representative and popularized its use for applied research alike. Its transcriptome composition was addressed during diurnal cycles (Elvitigala et al. 2009; McDermott et al. 2011; Stöckel et al. 2008; Toepel et al. 2008) as was the dynamics of the cellular protein complement (Aryal et al. 2018; Stöckel et al. 2011). Other analyses targeting the proteome were performed under constant light conditions (Aryal et al. 2012) or under conditions supporting enhanced hydrogen production (Aryal et al. 2013). Studies addressing the metabolome were performed comparing mixotrophic and photoheterotrophic growth conditions (Feng et al. 2010). Variations were studied in the rhythms of respiration and nitrogen fixation (Bandyopadhyay et al. 2013) and a temperature-dependent ultradian metabolic rhythm was discovered (Červený et al. 2013).

Findings that ATCC 51142 can produce considerable amounts of hydrogen (Min and Sherman 2010) boosted a series of studies in this direction describing it as one of the most prolific natural producers of hydrogen (Bandyopadhyay et al. 2010) and advancing the design of custom-built photobioreactors (Melnicki et al. 2012). The availability of multiple omics datasets facilitated modeling of light-driven reductant partitioning and carbon fluxes (Vu et al. 2012), of batch culture growth (Sinetova et al. 2012), and of the relation between oxygenic photosynthesis and nitrogenase-mediated hydrogen production (Bernstein et al. 2015). More specialized work targeted α-glucan branching enzymes (Hayashi et al. 2015), the enzyme chlorophyllase as a biocatalyst (Chou et al. 2015), a novel α-dioxygenase suitable for the biotechnological production of odor-active methyl-branched aldehydes (Hammer et al. 2020), and a highly active extracellular carbonic anhydrase (Kupriyanova et al. 2019). ATCC 51142 has also served as a donor of genetic information. Already during the whole genome analysis, the strain was recognized to contain the largest intact contiguous cluster of nitrogen fixation-related genes (Welsh et al. 2008). This arrangement allowed engineering nitrogen fixation into a normally non-fixing cyanobacterium via transfer of this gene cluster (Liu et al. 2018; Mueller et al. 2016), providing promise for the engineering of more complex organisms including crop plants (Thiel 2019).

## 3.6 Concluding Remarks

Numerous studies on various aspects of marine cyanobacteria have been published since the previous version of this chapter (Hess et al. 2016), illustrating the strong and continued interest of the scientific community in these fascinating microorganisms. Like many fields of biology research on marine cyanobacteria has strongly benefited from the ever-increasing distribution and application of (meta)omics approaches. Thus, tremendous progress has notably been made in: (1) elucidating the molecular bases of adaptations and regulatory circuitries used by marine cyanobacteria to deal with salinity or temperature gradients, high light or UV stress, light color variations, as well as N, P, and/or Fe limitation, which are the main abiotic factors that these photosynthetic microorganisms have to face in the marine environment, (2) better understanding biotic interactions with their predators (e.g., phages, phagotrophs), symbiotic hosts, or co-occurring bacteria, and (3) deciphering the distribution patterns of the major phylotypes of cyanobacteria and how they relate to ecological niches.

Given their ubiquity and natural abundance marine cyanobacteria largely contribute to the biogeochemical cycles of two major elements, C and N, through their photosynthetic and, for many of them, diazotrophic abilities. It must be stressed that a number of studies modeling the effects of the ongoing human-induced global change on phytoplanktonic communities predicted that *Synechococcus* and free-living diazotrophs should have higher growth rates and relative abundances in the future warmer (predicted temperature increase range: +1 to +8 °C) and more acidic world Ocean (predicted pH decrease: −0.2 to −0.3 units) while adverse effects were predicted for eukaryotic microphytoplankton such as diatoms or coccolithophores (Dutkiewicz et al. 2015; Flombaum et al. 2013; Schmidt et al. 2020). Interestingly, while Fe starvation is known to limit the growth of diazotrophs and therefore their occurrence in Fe-depleted areas, it was recently shown that *Trichodesmium* cellular Fe content tends to decrease as temperature rises and that the optimum growth temperature of Fe-limited cells is ~5 °C higher than for Fe-replete cells (Jiang et al. 2018). Thus, the projections made by these authors predict that Fe use efficiencies of $N_2$ fixers could increase by ~76% by 2100 and potentially alleviate the prevailing Fe limitation. The situation is different for *Prochlorococcus* because a doubling of the present-day $CO_2$ concentration combined with higher temperature (+4 °C) had no effect on *Prochlorococcus* cell division and photosynthetic rates whereas under the same conditions *Synechococcus* grew more than twice faster and exhibited four times higher photosynthetic rates than the controls (Fu et al. 2007). This suggests that *Prochlorococcus* could have a competitive disadvantage compared to *Synechococcus* under the predicted year 2100 $CO_2$ and temperature regimes. Yet, this study was performed on single strain cultures and did not take into account a possible mutualistic effect between these two picocyanobacteria. Such an effect was recently evidenced by co-culturing experiments in which the growth rate of *Prochlorococcus* correlated positively with the *Synechococcus* cell density at high $CO_2$ concentrations (Knight and Morris 2020). Thus, as also previously demonstrated by the need of *Prochlorococcus* cells for catalases synthesized by "helper" heterotrophic bacteria to mitigate oxidative stress (Morris et al. 2008) it seems that *Prochlorococcus* strongly depends upon organisms co-occurring in its immediate vicinity, a probable result of the extensive genome streamlining that has affected most lineages.

Interactions with other members of the microbial community have been recognized as an important factor in the lifestyle of other marine cyanobacteria as well. *Trichodesmium* has a complex associated microbiome (Gradoville et al. 2017) with which it coordinates the regulation of gene expression (Frischkorn et al. 2018). However, when a *Trichodesmium* bloom collapses previously numerically underrepresented bacteria such as *Alteromonas* spp. may flourish (Hou et al. 2018). More generally, interactions occurring between cyanobacteria and other members of the marine planktonic community are only starting to be understood. They range from mutualistic interactions to saprophytism on the C and N fixed by cyanobacteria. In particular, the release of extracellular membrane vesicles into the marine environment, demonstrated for *Prochlorococcus* and *Synechococcus* (Biller et al. 2014b), is likely a general phenomenon in cyanobacteria and may play a role in carbon cycling, gene transfer, and viral defense, complementing more classical secretion modes. In this context, the analysis of marine *Synechococcus* exoproteome revealed the occurrence of transport systems for inorganic nutrients and of a large array of strain-specific exoproteins likely involved in mutualistic or hostile interactions (e.g., hemolysins, pilins, adhesins) as well as exoenzymes with a potential mixotrophic goal (e.g., exoproteases and chitinases; Christie-Oleza et al. 2015). Given the remarkable variety of lifestyles exhibited by marine cyanobacteria, it is likely that the majority of the interactions they have forged with other members of the microbial community still awaits to be discovered and this is clearly one of the next major challenges in this thriving research field.

**Acknowledgments** Authors of this review were all supported by the EU project MaCuMBA (grant agreement no. 311975). LG and FP have also received grants from the French "Agence Nationale de la Recherche" Programs CINNAMON (ANR-17-CE2–0014) and EFFICACY (ANR-19-CE02–0019). Authors also wish to thank the *Tara* Oceans expedition (http://oceans.taraexpeditions.org) for data sharing used in analyses used to draw Fig. 3.3 as well as Alexandra Calteau and Jonathan Zehr for sharing other figures used in this review.

## References

Ahlgren NA, Rocap G (2012) Diversity and distribution of marine *Synechococcus*: multiple gene phylogenies for consensus classification and development of qPCR assays for sensitive measurement of clades in the ocean. Front Microbiol 3:213. https://doi.org/10.3389/fmicb.2012.00213

Ahlgren NA, Noble A, Patton AP, Roache-Johnson K, Jackson L, Robinson D, McKay C, Moore LR, Saito MA, Rocap G (2014) The unique trace metal and mixed layer conditions of the Costa Rica upwelling dome support a distinct and dense community of *Synechococcus*. Limnol Oceanogr 59:2166–2184. https://doi.org/10.4319/lo.2014.59.6.2166

Ahlgren NA, Belisle BS, Lee MD (2020) Genomic mosaicism underlies the adaptation of marine *Synechococcus* ecotypes to distinct oceanic iron niches. Environ Microbiol 22:1801–1815. https://doi.org/10.1111/1462-2920.14893

Aldunate M, Henríquez-Castillo C, Ji Q, Lueders-Dumont J, Mulholland MR, Ward BB, von Dassow P, Ulloa O (2020) Nitrogen assimilation in picocyanobacteria inhabiting the oxygen-deficient waters of the eastern tropical north and South Pacific. Limnol Oceanogr 65:437–453. https://doi.org/10.1002/lno.11315

Allahverdiyeva Y, Ermakova M, Eisenhut M, Zhang P, Richaud P, Hagemann M, Cournac L, Aro EM (2011) Interplay between flavodiiron proteins and photorespiration in *Synechocystis* sp. PCC 6803. J Biol Chem 286:24007–24014. https://doi.org/10.1074/jbc.M111.223289

Allahverdiyeva Y, Mustila H, Ermakova M, Bersanini L, Richaud P, Ajlani G, Battchikova N, Cournac L, Aro E-M (2013) Flavodiiron·proteins Flv1 and Flv3 enable cyanobacterial growth and photosynthesis under fluctuating light. Proc Natl Acad Sci U S A 110:4111–4116. https://doi.org/10.1073/pnas.1221194110

Allen JF, Thake B, Martin WF (2019) Nitrogenase inhibition limited oxygenation of Earth's Proterozoic atmosphere. Trends Plant Sci 11:1022–1031. https://doi.org/10.1016/j.tplants.2019.07.007

Aryal UK, Stöckel J, Welsh EA, Gritsenko MA, Nicora CD, Koppenaal DW, Smith RD, Pakrasi HB, Jacobs JM (2012) Dynamic proteome analysis of *Cyanothece* sp. ATCC 51142 under constant light. J Proteome Res 11:609–619. https://doi.org/10.1021/pr200959x

Aryal UK, Callister SJ, Mishra S, Zhang X, Shutthanandan JI, Angel TE, Shukla AK, Monroe ME, Moore RJ, Koppenaal DW, Smith RD, Sherman L (2013) Proteome analyses of strains ATCC 51142 and PCC 7822 of the diazotrophic cyanobacterium *Cyanothece* sp. under culture conditions resulting in enhanced $H_2$ production. Appl Environ Microbiol 79:1070–1077. https://doi.org/10.1128/AEM.02864-12

Aryal UK, Ding Z, Hedrick V, Sobreira TJP, Kihara D, Sherman LA (2018) Analysis of protein complexes in the unicellular cyanobacterium *Cyanothece* ATCC 51142. J Proteome Res 17:3628–3643. https://doi.org/10.1021/acs.jproteome.8b00170

Astorga-Eló M, Ramírez-Flandes S, Delong EF, Ulloa O (2015) Genomic potential for nitrogen assimilation in uncultivated members of *Prochlorococcus* from an anoxic marine zone. ISME J 9:1264–1267. https://doi.org/10.1038/ismej.2015.21

Auladell A, Sánchez P, Sánchez O, Gasol JM, Ferrera I (2019) Long-term seasonal and interannual variability of marine aerobic anoxygenic photoheterotrophic bacteria. ISME J 13:1975–1987. https://doi.org/10.1038/s41396-019-0401-4

Avrani S, Wurtzel O, Sharon I, Sorek R, Lindell D (2011) Genomic island variability facilitates *Prochlorococcus*-virus coexistence. Nature 474:604–608. https://doi.org/10.1038/nature10172

Badger MR, Price GD (2003) $CO_2$ concentrating mechanisms in cyanobacteria: molecular components, their diversity and evolution. J Exp Bot 54:609–622. https://doi.org/10.1093/jxb/erg076

Bagby SC, Chisholm SW (2015) Response of *Prochlorococcus* to varying CO2:O2 ratios. ISME J 9:2232–2245. https://doi.org/10.1038/ismej.2015.36

Bandyopadhyay A, Stöckel J, Min H, Sherman LA, Pakrasi HB (2010) High rates of photobiological H2 production by a cyanobacterium under aerobic conditions. Nat Commun 1:139. https://doi.org/10.1038/ncomms1139

Bandyopadhyay A, Elvitigala T, Liberton M, Pakrasi HB (2013) Variations in the rhythms of respiration and nitrogen fixation in members of the unicellular diazotrophic cyanobacterial genus *Cyanothece*. Plant Physiol 161:1334–1346. https://doi.org/10.1104/pp.112.208231

Bar-Eyal L, Shperberg-Avni A, Paltiel Y, Keren N, Adir N (2018) Light harvesting in cyanobacteria: the phycobilisomes. In: Croce R, Grondelle R, Amerongen H, Stokkum I (eds) Light harvesting in photosynthesis. CRC Press, Boca Raton, FL, pp 77–93

Baumdicker F, Hess WR, Pfaffelhuber P (2010) The diversity of a distributed genome in bacterial populations. Ann Appl Probab 20:1567–1606. https://doi.org/10.1214/09-AAP657

Baumdicker F, Hess WR, Pfaffelhuber P (2012) The infinitely many genes model for the distributed genome of bacteria. Genome Biol Evol 4:443–456. https://doi.org/10.1093/gbe/evs016

Béjà O, Spudich EN, Spudich JL, Leclerc M, DeLong EF (2001) Proteorhodopsin phototrophy in the ocean. Nature 411:786–789. https://doi.org/10.1038/35081051

Béjà O, Suzuki MT, Heidelberg JF, Nelson WC, Preston CM, Hamada T, Eisen JA, Fraser CM, DeLong EF (2002) Unsuspected diversity among marine aerobic anoxygenic phototrophs. Nature 415:630–633. https://doi.org/10.1038/415630a

Bench SR, Heller P, Frank I, Arciniega M, Shilova IN, Zehr JP (2013) Whole genome comparison of six *Crocosphaera watsonii* strains with differing phenotypes. J Phycol 49:786–801. https://doi.org/10.1111/jpy.12090

Bergman B, Sandh G, Lin S, Larsson J, Carpenter EJ (2013) *Trichodesmium* - a widespread marine cyanobacterium with unusual nitrogen fixation properties. FEMS Microbiol Rev 37:286–302. https://doi.org/10.1111/j.1574-6976.2012.00352.x

Berman-Frank I, Lundgren P, Chen YB, Küpper H, Kolber Z, Bergman B, Falkowski P (2001) Segregation of nitrogen fixation and oxygenic photosynthesis in the marine cyanobacterium *Trichodesmium*. Science 294:1534–1537. https://doi.org/10.1126/science.1064082

Berman-Frank I, Bidle KD, Haramaty L, Falkowski PG (2004) The demise of the marine cyanobacterium, *Trichodesmium* spp., via an autocatalyzed cell death pathway. Limnol Oceanogr 49:997–1005. https://doi.org/10.4319/lo.2004.49.4.0997

Berman-Frank I, Rosenberg G, Levitan O, Haramaty L, Mari X (2007) Coupling between autocatalytic cell death and transparent exopolymeric particle production in the marine cyanobacterium *Trichodesmium*. Environ Microbiol 9:1415–1422. https://doi.org/10.1111/j.1462-2920.2007.01257.x

Bernstein HC, Charania MA, McClure RS, Sadler NC, Melnicki MR, Hill EA, Markillie LM, Nicora CD, Wright AT, Romine MF, Beliaev AS (2015) Multi-omic dynamics associate oxygenic photosynthesis with nitrogenase-mediated H2 production in *Cyanothece* sp. ATCC 51142. Sci Rep 5:16004. https://doi.org/10.1038/srep16004

Berthelot H, Bonnet S, Grosso O, Cornet V, Barani A (2016) Transfer of diazotroph-derived nitrogen towards non-diazotrophic planktonic communities: a comparative study between *Trichodesmium erythraeum*, *Crocosphaera watsonii* and *Cyanothece* sp. Biogeosciences 13:4005–4021. https://doi.org/10.5194/bg-13-4005-2016

Berthelot H, Benavides M, Moisander PH, Grosso O, Bonnet S (2017) High-nitrogen fixation rates in the particulate and dissolved pools in the Western tropical Pacific (Solomon and Bismarck seas). Geophys Res Lett 44:8414–8423. https://doi.org/10.1002/2017GL073856

Berube PM, Biller SJ, Kent AG, Berta-Thompson JW, Roggensack SE, Roache-Johnson KH, Ackerman M, Moore LR, Meisel JD, Sher D, Thompson LR, Campbell L, Martiny AC, Chisholm SW (2015) Physiology and evolution of nitrate acquisition in *Prochlorococcus*. ISME J 9:1195–1207. https://doi.org/10.1038/ismej.2014.211

Berube PM, Coe A, Roggensack SE, Chisholm SW (2016) Temporal dynamics of *Prochlorococcus* cells with the potential for nitrate assimilation in the subtropical Atlantic and Pacific oceans. Limnol Oceanogr 61:482–495. https://doi.org/10.1002/lno.10226

Berube PM, Biller SJ, Hackl T, Hogle SL, Satinsky BM, Becker JW, Braakman R, Collins SB, Kelly L, Berta-Thompson J, Coe A, Bergauer K, Bouman HA, Browning TJ, De Corte D, Hassler C, Hulata Y, Jacquot JE, Maas EW, Reinthaler T, Sintes E, Yokokawa T, Lindell D, Stepanauskas R, Chisholm SW (2018) Single cell genomes of *Prochlorococcus*, *Synechococcus*, and sympatric microbes from diverse marine environments. Sci Data 5: 180154. https://doi.org/10.1038/sdata.2018.154

Berube PM, Rasmussen A, Braakman R, Stepanauskas R, Chisholm SW (2019) Emergence of trait variability through the lens of nitrogen assimilation in *Prochlorococcus*. elife 8:e41043. https://doi.org/10.7554/eLife.41043

Beversdorf LJ, White AE, Björkman KM, Letelier RM, Karl DM (2010) Phosphonate metabolism of *Trichodesmium* IMS101 and the production of greenhouse gases. Limnol Oceanogr 55:1768–1778. https://doi.org/10.4319/lo.2010.55.4.1768

Bidle KD, Falkowski PG (2004) Cell death in planktonic, photosynthetic microorganisms. Nat Rev Microbiol 2:643–655. https://doi.org/10.1038/nrmicro956

Biller SJ, Berube PM, Berta-Thompson JW, Kelly L, Roggensack SE, Awad L, Roache-Johnson KH, Ding H, Giovannoni SJ, Rocap G, Moore LR, Chisholm SW (2014a) Genomes of diverse isolates of the marine cyanobacterium *Prochlorococcus*. Sci Data 1:140034. https://doi.org/10.1038/sdata.2014.34

Biller SJ, Schubotz F, Roggensack SE, Thompson AW, Summons RE, Chisholm SW (2014b) Bacterial vesicles in marine ecosystems. Science 343:183–186. https://doi.org/10.1126/science.1243457

Biller SJ, Berube PM, Lindell D, Chisholm SW (2015) *Prochlorococcus*: the structure and function of collective diversity. Nat Rev Microbiol 13:13–27. https://doi.org/10.1038/nrmicro3378

Biswas A, Vasquez YM, Dragomani TM, Kronfel ML, Williams SR, Alvey RM, Bryant DA, Schluchter WM (2010) Biosynthesis of cyanobacterial phycobiliproteins in *Escherichia coli*: Chromophorylation efficiency and specificity of all Bilin lyases from *Synechococcus* sp. strain PCC 7002. Appl Environ Microbiol 76:2729–2739. https://doi.org/10.1128/AEM.03100-09

Bižić M, Klintzsch T, Ionescu D, Hindiyeh MY, Günthel M, Muro-Pastor AM, Eckert W, Urich T, Keppler F, Grossart HP (2020) Aquatic and terrestrial cyanobacteria produce methane. Sci Adv 6:eaax5343. https://doi.org/10.1126/sciadv.aax5343

Blank CE, Sánchez-Baracaldo P (2010) Timing of morphological and ecological innovations in the cyanobacteria – a key to understanding the rise in atmospheric oxygen. Geobiology 8:1–23. https://doi.org/10.1111/j.1472-4669.2009.00220.x

Blot N, Wu X-JXJ, Thomas J-CJC, Zhang J, Garczarek L, Böhm S, Tu JMJ-M, Zhou M, Plöscher M, Eichacker L, Partensky F, Scheer H, Zhao KHK-H (2009) Phycourobilinin trichromatic phycocyanin from oceanic cyanobacteria is formed post-translationally by a phycoerythrobilin lyase-isomerase. J Biol Chem 284:9290–9298. https://doi.org/10.1074/jbc.M809784200

Boeuf D, Cottrell MT, Kirchman DL, Lebaron P, Jeanthon C (2013) Summer community structure of aerobic anoxygenic phototrophic bacteria in the western Arctic Ocean. FEMS Microbiol Ecol 85:417–432. https://doi.org/10.1111/1574-6941.12130

Boeuf D, Lami R, Cunnington E, Jeanthon C (2016) Summer abundance and distribution of proteorhodopsin genes in the western arctic ocean. Front Microbiol. https://doi.org/10.3389/fmicb.2016.01584

Bolhuis H, Severin I, Confurius-Guns V, Wollenzien UIA, Stal LJ (2010) Horizontal transfer of the nitrogen fixation gene cluster in the cyanobacterium *Microcoleus chthonoplastes*. ISME J 4: 121–130. https://doi.org/10.1038/ismej.2009.99

Bombar D, Heller P, Sánchez-Baracaldo P, Carter BJ, Zehr JP (2014) Comparative genomics reveals surprising divergence of two closely related strains of uncultivated UCYN-A cyanobacteria. ISME J 8:2530–2542. https://doi.org/10.1038/ismej.2014.167

Bonnet S, Berthelot H, Turk-Kubo K, Fawcett S, Rahav E, L'Helguen S, Berman-Frank I (2016a) Dynamics of N2 fixation and fate of diazotroph-derived nitrogen in a low nutrient low chlorophyll ecosystem: results from the VAHINE mesocosm experiment (New Caledonia). Biogeosciences 13:2653–2673. https://doi.org/10.5194/bg-13-2653-2016

Bonnet S, Moutin T, Rodier M, Grisoni JM, Louis F, Folcher E, Bourgeois B, Boré JM, Renaud M (2016b) Introduction to the project VAHINE: VAriability of vertical and tropHIc transfer of fixed $N_2$ in the south wEst Pacific. Biogeosciences 13:2803–2814. https://doi.org/10.5194/bg-13-2803-2016

Bonnet S, Caffin M, Berthelot H, Moutin T (2017) Hot spot of N2 fixation in the western tropical South Pacific pleads for a spatial decoupling between N2 fixation and denitrification. Proc Natl Acad Sci U S A 114:E2800–E2801. https://doi.org/10.1073/pnas.1619514114

Breton S, Jouhet J, Guyet U, Gros V, Pittera J, Demory D, Partensky F, Doré H, Ratin M, Maréchal E, Nguyen NA, Garczarek L, Six C (2019) Unveiling membrane thermoregulation strategies in marine picocyanobacteria. New Phytol. https://doi.org/10.1111/nph.16239

Bristow LA, Mohr W, Ahmerkamp S, Kuypers MMM (2017) Nutrients that limit growth in the ocean. Curr Biol 27:R431–R510. https://doi.org/10.1016/j.cub.2017.03.030

Buitenhuis ET, Li WKW, Vaulot D, Lomas MW, Landry MR, Partensky F, Karl DM, Ulloa O, Campbell L, Jacquet S, Lantoine F, Chavez F, Macias D, Gosselin M, McManus GB (2012) Picophytoplankton biomass distribution in the global ocean. Earth Syst Sci Data 4:37–46. https://doi.org/10.5194/essd-4-37-2012

Cabello AM, Cornejo-Castillo FM, Raho N, Blasco D, Vidal M, Audic S, de Vargas C, Latasa M, Acinas SG, Massana R (2015) Global distribution and vertical patterns of a prymnesiophyte-cyanobacteria obligate symbiosis. ISME J 12:693–706. https://doi.org/10.1038/ismej.2015.147

Cabello AM, Turk-Kubo KA, Hayashi K, Jacobs L, Kudela RM, Zehr JP (2020) Unexpected presence of the nitrogen-fixing symbiotic cyanobacterium UCYN-A in Monterey Bay, California. J Phycol 56:1521–1533. https://doi.org/10.1111/jpy.13045

Cabello-Yeves PJ, Haro-Moreno JM, Martin-Cuadrado AB, Ghai R, Picazo A, Camacho A, Rodriguez-Valera F (2017) Novel *Synechococcus* genomes reconstructed from freshwater reservoirs. Front Microbiol 8:1151. https://doi.org/10.3389/fmicb.2017.01151

Cabello-Yeves PJ, Picazo A, Camacho A, Callieri C, Rosselli R, Roda-Garcia JJ, Coutinho FH, Rodriguez-Valera F (2018) Ecological and genomic features of two widespread freshwater picocyanobacteria. Environ Microbiol 20:3757–3771. https://doi.org/10.1111/1462-2920.14377

Canfield DE, Glazer AN, Falkowski PG (2010) The evolution and future of Earth's nitrogen cycle. Science 330:192–196. https://doi.org/10.1126/science.1186120

Capone DG (1997) *Trichodesmium*, a globally significant marine cyanobacterium. Science 276: 1221–1229. https://doi.org/10.1126/science.276.5316.1221

Carpenter EJ, Foster RA (2002) Marine cyanobacterial symbioses. In: Rai AN, Bergman B, Rasmussen U (eds) Cyanobacteria in symbiosis. Springer, Dordrecht, pp 11–17. https://doi.org/10.1007/0-306-48005-0_2

Carpenter EJ, Janson S (2000) Intracellular cyanobacterial symbionts in the marine diatom *Climacodium frauenfeldianum* (Bacillariophyceae). J Phycol 36:540–544. https://doi.org/10.1046/j.1529-8817.2000.99163.x

Carrigee LA, Mahmoud RM, Sanfilippo JE, Frick JP, Strnat JA, Karty JA, Chen B, Kehoe DM, Schluchter WM (2020) CpeY is a phycoerythrobilin lyase for cysteine 82 of the phycoerythrin I α-subunit in marine *Synechococcus*. Biochim Biophys Acta Bioenerg 1861:148215. https://doi.org/10.1016/j.bbabio.2020.148215

Carrigee LA, Frick JP, Karty JA, Garczarek L, Partensky F, Schluchter WM (2021) MpeV is a lyase isomerase that ligates a doublylinked phycourobilin on the β-subunit of phycoerythrin I and II in marine *Synechococcus*. J Biol Chem 296:100031. https://doi.org/10.1074/jbc.ra120.015289
Červený J, Sinetova MA, Valledor L, Sherman LA, Nedbal L (2013) Ultradian metabolic rhythm in the diazotrophic cyanobacterium *Cyanothece* sp. ATCC 51142. Proc Natl Acad Sci U S A 110: 13210–13215. https://doi.org/10.1073/pnas.1301171110
Chan YW, Nenninger A, Clokie SJH, Mann NH, Scanlan DJ, Whitworth AL, Clokie MRJ (2007) Pigment composition and adaptation in free-living and symbiotic strains of *Acaryochloris marina*. FEMS Microbiol Ecol 61:65–73. https://doi.org/10.1111/j.1574-6941.2007.00320.x
Chandler JW, Lin Y, Gainer PJ, Post AF, Johnson ZI, Zinser ER (2016) Variable but persistent coexistence of *Prochlorococcus* ecotypes along temperature gradients in the ocean's surface mixed layer. Environ Microbiol Rep 8:272–284. https://doi.org/10.1111/1758-2229.12378
Chen F, Wang K, Kan J, Suzuki MT, Wommack KE (2006) Diverse and unique picocyanobacteria in Chesapeake Bay, revealed by 16S-23S rRNA internal transcribed spacer sequences. Appl Env Microbiol 72:2239–2243. https://doi.org/10.1128/AEM.72.3.2239-2243.2006
Chisholm SW, Olson RJ, Zettler ER, Goericke R, Waterbury JB (1988) A novel free-living prochlorophyte abundant in the oceanic euphotic zone. Nature 334:340–343. https://doi.org/10.1038/334340a0
Chisholm SW, Frankel SL, Goericke R, Olson RJ, Palenik B, Waterbury JB, West-Johnsrud L, Zettler ER (1992) *Prochlorococcus marinus* nov. gen. Nov. sp.: an oxyphototrophic marine prokaryote containing divinyl chlorophyll a and b. Arch Microbiol 157:297–300. https://doi.org/10.1007/BF00245165
Chou Y-L, Lee Y-L, Yen C-C, Chen L-FO, Lee L-C, Shaw J-F (2015) A novel recombinant chlorophyllase from cyanobacterium *Cyanothece* sp. ATCC 51142 for the production of bacteriochlorophyllide a. Biotechnol Appl Biochem 63:371–377. https://doi.org/10.1002/bab.1380
Christie-Oleza JA, Armengaud J, Guerin P, Scanlan DJ (2015) Functional distinctness in the exoproteomes of marine *Synechococcus*. Environ Microbiol 17:3781–3794. https://doi.org/10.1111/1462-2920.12822
Coleman ML, Chisholm SW (2007) Code and context: *Prochlorococcus* as a model for cross-scale biology. Trends Microbiol 15:398–407. https://doi.org/10.1016/j.tim.2007.07.001
Coleman ML, Sullivan MB, Martiny AC, Steglich C, Delong EF, Chisholm SW, Barry K, Chisholm SW (2006) Genomic islands and the ecology and evolution of *Prochlorococcus*. Science 311:1768–1770. https://doi.org/10.1126/science.1122050
Compaoré J, Stal LJ (2010) Oxygen and the light-dark cycle of nitrogenase activity in two unicellular cyanobacteria. Environ Microbiol 12:54–62. https://doi.org/10.1111/j.1462-2920.2009.02034.x
Cornejo-Castillo FM, Zehr JP (2019) Hopanoid lipids may facilitate aerobic nitrogen fixation in the ocean. Proc Natl Acad Sci U S A 116:18269–18271. https://doi.org/10.1073/pnas.1908165116
Cornejo-Castillo FM, Zehr JP (2021) Intriguing size distribution of the uncultured and globally widespread marine non-cyanobacterial diazotroph gamma-a. ISME J 15:124–128. https://doi.org/10.1038/s41396-020-00765-1
Cornejo-Castillo FM, Cabello AM, Salazar G, Sánchez-Baracaldo P, Lima-Mendez G, Hingamp P, Alberti A, Sunagawa S, Bork P, De Vargas C, Raes J, Bowler C, Wincker P, Zehr JP, Gasol JM, Massana R, Acinas SG (2016) Cyanobacterial symbionts diverged in the late cretaceous towards lineage-specific nitrogen fixation factories in single-celled phytoplankton. Nat Commun 7: 11071. https://doi.org/10.1038/ncomms11071
Cornejo-Castillo FM, Muñoz-Marín M d C, Turk-Kubo KA, Royo-Llonch M, Farnelid H, Acinas SG, Zehr JP (2019) UCYN-A3, a newly characterized open ocean sublineage of the symbiotic $N_2$-fixing cyanobacterium *Candidatus* Atelocyanobacterium thalassa. Environ Microbiol 21: 111–124. https://doi.org/10.1111/1462-2920.14429
Cottrell MT, Kirchman DL (2009) Photoheterotrophic microbes in the Arctic Ocean in summer and winter. Appl Environ Microbiol 75:4958–4966. https://doi.org/10.1128/AEM.00117-09

Crowe SA, Døssing LN, Beukes NJ, Bau M, Kruger SJ, Frei R, Canfield DE (2013) Atmospheric oxygenation three billion years ago. Nature 501:535–538. https://doi.org/10.1038/nature12426
Cubillos-Ruiz A, Berta-Thompson JW, Becker JW, Van Der Donk WA, Chisholm SW (2017) Evolutionary radiation of lanthipeptides in marine cyanobacteria. Proc Natl Acad Sci U S A 114: E5424–E5433. https://doi.org/10.1073/pnas.1700990114
Davis CS, McGillicuddy DJ Jr (2006) Transatlantic abundance of the $N_2$-fixing colonial cyanobacterium *Trichodesmium*. Science 312:1517–1520. https://doi.org/10.1126/science.1123570
Degerholm J, Gundersen K, Bergman B, Söderbäck E (2006) Phosphorus-limited growth dynamics in two Baltic Sea cyanobacteria, *Nodularia* sp. and *Aphanizomenon* sp. FEMS Microbiol Ecol 58:323–332. https://doi.org/10.1111/j.1574-6941.2006.00180.x
Delmont TO, Eren AM (2018) Linking pangenomes and metagenomes : the *Prochlorococcus* metapangenome. PeerJ 6:e4320. https://doi.org/10.7717/peerj.4320
Demoulin CF, Lara YJ, Cornet L, François C, Baurain D, Wilmotte A, Javaux EJ (2019) Cyanobacteria evolution: insight from the fossil record. Free Radic Biol Med 140:206–223. https://doi.org/10.1016/j.freeradbiomed.2019.05.007
Deutsch C, Sarmiento JL, Sigman DM, Gruber N, Dunne JP (2007) Spatial coupling of nitrogen inputs and losses in the ocean. Nature 445:163–167. https://doi.org/10.1038/nature05392
Di Cesare A, Cabello-Yeves PJ, Chrismas NAM, Sánchez-Baracaldo P, Salcher MM, Callieri C (2018) Genome analysis of the freshwater planktonic *Vulcanococcus limneticus* sp. nov. reveals horizontal transfer of nitrogenase operon and alternative pathways of nitrogen utilization. BMC Genomics 19:259. https://doi.org/10.1186/s12864-018-4648-3
Donia MS, Fricke WF, Partensky F, Cox J, Elshahawi SI, White JR, Phillippy AM, Schatz MC, Piel J, Haygood MG, Ravel J, Schmidt EW (2011) Complex microbiome underlying secondary and primary metabolism in the tunicate-*Prochloron* symbiosis. Proc Natl Acad Sci U S A 108: E1423–E1432. https://doi.org/10.1073/pnas.1111712108
Doré H, Farrant GK, Guyet U, Haguait J, Humily F, Ratin M, Pitt FD, Ostrowski M, Six C, Brillet-Guéguen L, Hoebeke M, Bisch A, Le Corguillé G, Corre E, Labadie K, Aury J-M, Wincker P, Choi DH, Noh JH, Eveillard D, Scanlan DJ, Partensky F, Garczarek L (2020) Evolutionary mechanisms of long-term genome diversification associated with niche partitioning in marine picocyanobacteria. Front Microbiol 11:567431. https://doi.org/10.3389/fmicb.2020.567431
Doulatov S, Hodes A, Dai L, Mandhana N, Liu M, Deora R, Simons RW, Zimmerly S, Miller JF (2004) Tropism switching in *Bordetella* bacteriophage defines a family of diversity-generating retroelements. Nature 431:476–481. https://doi.org/10.1038/nature02833
Dufresne A, Salanoubat M, Partensky F, Artiguenave F, Axmann IM, Barbe V, Duprat S, Galperin MY, Koonin EV, Le Gall F, Makarova KS, Ostrowski M, Oztas S, Robert C, Rogozin IB, Scanlan DJ, Tandeau de Marsac N, Weissenbach J, Wincker P, Wolf YI, Hess WR (2003) Genome sequence of the cyanobacterium *Prochlorococcus marinus* SS120, a nearly minimal oxyphototrophic genome. Proc Natl Acad Sci U S A 100:10020–10025. https://doi.org/10.1073/pnas.1733211100
Dufresne A, Garczarek L, Partensky F (2005) Accelerated evolution associated with genome reduction in a free-living prokaryote. Genome Biol 6:R14. https://doi.org/10.1186/gb-2005-6-2-r14
Dufresne A, Ostrowski M, Scanlan DJ, Garczarek L, Mazard S, Palenik BP, Paulsen IT, Tandeau de Marsac N, Wincker P, Dossat C, Ferriera S, Johnson J, Post AF, Hess WR, Partensky F, Tandeau de Marsac N, Wincker P, Dossat C, Ferriera S, Johnson J, Post AF, Hess WR, Partensky F (2008) Unraveling the genomic mosaic of a ubiquitous genus of marine cyanobacteria. Genome Biol 9:R90. https://doi.org/10.1186/gb-2008-9-5-r90
Dugdale RC, Menzel DW, Ryther JH (1961) Nitrogen fixation in the Sargasso Sea. Deep Sea Res 7: 297–300. https://doi.org/10.1016/0146-6313(61)90051-X
Duhamel S, Van Wambeke F, Lefevre D, Benavides M, Bonnet S (2018) Mixotrophic metabolism by natural communities of unicellular cyanobacteria in the western tropical South Pacific Ocean. Environ Microbiol 20:2743–2756. https://doi.org/10.1111/1462-2920.14111

Dupouy C, Petit M, Dandonneau Y (1988) Satellite detected cyanobacteria bloom in the south-western tropical Pacific. Int J Remote Sens 9:389–396. https://doi.org/10.1080/01431168808954862

Dutkiewicz S, Morris JJ, Follows MJ, Scott J, Levitan O, Dyhrman ST, Berman-Frank I (2015) Impact of ocean acidification on the structure of future phytoplankton communities. Nat Clim Chang 5:1002–1006. https://doi.org/10.1038/nclimate2722

Eichner M, Thoms S, Rost B, Mohr W, Ahmerkamp S, Ploug H, Kuypers MMM, de Beer D (2019) $N_2$ fixation in free-floating filaments of *Trichodesmium* is higher than in transiently suboxic colony microenvironments. New Phytol 222:852–863. https://doi.org/10.1111/nph.15621

El-Shehawy R, Lugomela C, Ernst A, Bergman B (2003) Diurnal expression of hetR and diazocyte development in the filamentous non-heterocystous cyanobacterium *Trichodesmium erythraeum*. Microbiology 149:1139–1146. https://doi.org/10.1099/mic.0.26170-0

Elvitigala T, Stöckel J, Ghosh BK, Pakrasi HB (2009) Effect of continuous light on diurnal rhythms in *Cyanothece* sp. ATCC 51142. BMC Genomics 10:226. https://doi.org/10.1186/1471-2164-10-226

Engelberts JP, Robbins SJ, de Goeij JM, Aranda M, Bell SC, Webster NS (2020) Characterization of a sponge microbiome using an integrative genome-centric approach. ISME J 14:1100–1110. https://doi.org/10.1038/s41396-020-0591-9

Everroad C, Six C, Partensky F, Thomas JC, Holtzendorff J, Wood AM (2006) Biochemical bases of type IV chromatic adaptation in marine *Synechococcus* spp. J Bacteriol 188:3345–3356. https://doi.org/10.1128/JB.188.9.3345-3356.2006

Falkowski PG (1997) Evolution of the nitrogen cycle and its influence on the biological sequestration of $CO_2$ in the ocean. Nature 387:272–275. https://doi.org/10.1038/387272a0

Falkowski PG, Raven JA (2007) Aquatic photosynthesis, 2nd edn. Princeton University Press, Princeton

Farnelid H, Turk-Kubo K, Del Carmen M-MM, Zehr JP (2016) New insights into the ecology of the globally significant uncultured nitrogen-fixing symbiont UCYN-A. Aquat Microb Ecol 77:125–138. https://doi.org/10.3354/ame01794

Farrant GK, Doré H, Cornejo-Castillo FM, Partensky F, Ratin M, Ostrowski M, Pitt FD, Wincker P, Scanlan DJ, Iudicone D, Acinas SG, Garczarek L (2016) Delineating ecologically significant taxonomic units from global patterns of marine picocyanobacteria. Proc Natl Acad Sci 113(24): E3365–E3374. https://doi.org/10.1073/pnas.1524865113

Feng X, Bandyopadhyay A, Berla B, Page L, Wu B, Pakrasi HB, Tang YJ (2010) Mixotrophic and photoheterotrophic metabolism in *Cyanothece* sp. ATCC 51142 under continuous light. Microbiology 156:2566–2574. https://doi.org/10.1099/mic.0.038232-0

Fewer DP, Jokela J, Paukku E, Österholm J, Wahlsten M, Permi P, Aitio O, Rouhiainen L, Gomez-Saez GV, Sivonen K (2013) New structural variants of aeruginosin produced by the toxic bloom-forming cyanobacterium *Nodularia spumigena*. PLoS One 8:e73618. https://doi.org/10.1371/journal.pone.0073618

Flombaum P, Gallegos JL, Gordillo RA, Rincon J, Zabala LL, Jiao N, Karl DM, Li WK, Lomas MW, Veneziano D, Vera CS, Vrugt JA, Martiny AC (2013) Present and future global distributions of the marine cyanobacteria *Prochlorococcus* and *Synechococcus*. Proc Natl Acad Sci U S A 110:9824–9829. https://doi.org/10.1073/pnas.1307701110

Flombaum P, Wang WL, Primeau FW, Martiny AC (2020) Global picophytoplankton niche partitioning predicts overall positive response to ocean warming. Nat Geosci 13:116–120. https://doi.org/10.1038/s41561-019-0524-2

Foster RA, Zehr JP (2006) Characterization of diatom-cyanobacteria symbioses on the basis of *nifH*, *hetR* and 16S rRNA sequences. Environ Microbiol 8:1913–1925. https://doi.org/10.1111/j.1462-2920.2006.01068.x

Foster RA, Zehr JP (2019) Diversity, genomics, and distribution of phytoplankton-cyanobacterium single-cell symbiotic associations. Annu Rev Microbiol 73:435–456. https://doi.org/10.1146/annurev-micro-090817-062650

Foster RA, Goebel NL, Zehr JP (2010) Isolation of *Calothrix Rhizosoleniae* (cyanobacteria) strain Sc01 from *Chaetoceros* (Bacillariophyta) spp. diatoms of the subtropical North Pacific Ocean. J Phycol 46:1028–1037. https://doi.org/10.1111/j.1529-8817.2010.00885.x

Foster RA, Kuypers MMM, Vagner T, Paerl RW, Musat N, Zehr JP (2011) Nitrogen fixation and transfer in open ocean diatom-cyanobacterial symbioses. ISME J 5:1484–1493. https://doi.org/10.1038/ismej.2011.26

Foster RA, Sztejrenszus S, Kuypers MMM (2013) Measuring carbon and $N_2$ fixation in field populations of colonial and free-living unicellular cyanobacteria using nanometer-scale secondary ion mass spectrometry. J Phycol 49:502–516. https://doi.org/10.1111/jpy.12057

Franklin DJ, Brussaard CPD, Berges JA (2006) What is the role and nature of programmed cell death in phytoplankton ecology? Eur J Phycol 41:1–14. https://doi.org/10.1080/09670260500505433

Frischkorn KR, Haley ST, Dyhrman ST (2018) Coordinated gene expression between *Trichodesmium* and its microbiome over day-night cycles in the North Pacific subtropical gyre. ISME J 12:997–1007. https://doi.org/10.1038/s41396-017-0041-5

Fu FX, Warner ME, Zhang Y, Feng Y, Hutchins DA (2007) Effects of increased temperature and $CO_2$ on photosynthesis, growth, and elemental ratios in marine *Synechococcus* and *Prochlorococcus* (cyanobacteria). J Phycol 43:485–496. https://doi.org/10.1111/j.1529-8817.2007.00355.x

Garcia CA, Hagstrom GI, Larkin AA, Ustick LJ, Levin SA, Lomas MW, Martiny AC (2020) Linking regional shifts in microbial genome adaptation with surface ocean biogeochemistry. Philos Trans R Soc B Biol Sci 375:20190244. https://doi.org/10.1098/rstb.2019.0254

Garcia-Pichel F, Lombard J, Soule T, Dunaj S, Wu SH, Wojciechowski MF (2019) Timing the evolutionary advent of cyanobacteria and the later great oxidation event using gene phylogenies of a sunscreen. MBio 10:e00561–e00519. https://doi.org/10.1128/mBio.00561-19

Garcia-Pichel F, Zehr JP, Bhattacharya D, Pakrasi HB (2020) What's in a name? The case of cyanobacteria. J Phycol 56:1–5. https://doi.org/10.1111/jpy.12934

Garczarek L, Hess WR, Holtzendorff J, van der Staay GWM, Partensky F (2000) Multiplication of antenna genes as a major adaptation to low light in a marine prokaryote. Proc Natl Acad Sci U S A 97:4098–4101. https://doi.org/10.1073/pnas.070040897

Garczarek L, Guyet U, Doré H, Farrant GK, Hoebeke M, Brillet-Guéguen L, Bisch A, Ferrieux M, Siltanen J, Corre E, Le Corguillé G, Ratin M, Pitt FD, Ostrowski M, Conan M, Siegel A, Labadie K, Aury J-M, Wincker P, Scanlan DJ, Partensky F (2021) Cyanorak v2.1: a scalable information system dedicated to the visualization and expert curation of marine and brackish picocyanobacteria genomes. Nucleic Acids Res 49:D667–D676. https://doi.org/10.1093/nar/gkaa958

Georg J, Dienst D, Schürgers N, Wallner T, Kopp D, Stazic D, Kuchmina E, Klähn S, Lokstein H, Hess WR, Wilde A (2014) The small regulatory RNA SyR1/PsrR1 controls photosynthetic functions in cyanobacteria. Plant Cell 26:3661–3679. https://doi.org/10.1105/tpc.114.129767

Giepmans BNG, Adams SR, Ellisman MH, Tsien RY (2006) The fluorescent toolbox for assessing protein location and function. Science 312:217–224. https://doi.org/10.1126/science.1124618

Giovannoni SJ (2017) SAR11 bacteria: the most abundant plankton in the oceans. Annu Rev Mar Sci 9:231–255. https://doi.org/10.1146/annurev-marine-010814-015934

Goebel NL, Turk KA, Achilles KM, Paerl R, Hewson I, Morrison AE, Montoya JP, Edwards CA, Zehr JP (2010) Abundance and distribution of major groups of diazotrophic cyanobacteria and their potential contribution to N fixation in the tropical Atlantic Ocean. Env Microbiol 12:3272–3289. https://doi.org/10.1111/j.1462-2920.2010.02303.x

Goericke R, Repeta DJ (1992) The pigments of *Prochlorococcus marinus*: the presence of divinyl-chlorophyll *a* and *b* in a marine procaryote. Limnol Oceanogr 37:425–433. https://doi.org/10.4319/lo.1992.37.2.0425

Golubic S, Abed RMM (2010) *Entophysalis* mats as environmental regulators. In: Seckbach J, Oren A (eds) Microbial mats. Springer, Dordrecht, pp 237–251. https://doi.org/10.1007/978-90-481-3799-2_12

Gómez-Pereira PR, Hartmann M, Grob C, Tarran GA, Martin AP, Fuchs BM, Scanlan DJ, Zubkov MV (2013) Comparable light stimulation of organic nutrient uptake by SAR11 and *Prochlorococcus* in the North Atlantic subtropical gyre. ISME J 7:603–614. https://doi.org/10.1038/ismej.2012.126

Gradoville MR, Crump BC, Letelier RM, Church MJ, White AE (2017) Microbiome of *Trichodesmium* colonies from the North Pacific subtropical gyre. Front Microbiol 8:1122. https://doi.org/10.3389/fmicb.2017.01122

Grébert T, Doré H, Partensky F, Farrant GK, Boss ES, Picheral M, Guidi L, Pesant S, Scanlan DJ, Wincker P, Acinas SG, Kehoe DM, Garczarek L (2018) Light color acclimation is a key process in the global ocean distribution of *Synechococcus* cyanobacteria. Proc Natl Acad Sci 115: E2010–E2019. https://doi.org/10.1073/pnas.1717069115

Grébert T, Nguyen AA, Pokhrel S, Joseph KL, Chen B, Ratin M, Dufour L, Haney AM, Trinidad J, Karty JA, Garczarek L, Schluchter WM, Kehoe DM, Partensky F (2021) Molecular basis of an alternative dual-enzyme system for light color acclimation of marine *Synechococcus* cyanobacteria. Proc Natl Acad Sci USA 118(9):e2019715118. https://doi.org/10.1073/pnas.2019715118

Green BR (2019) What happened to the phycobilisome? Biomol Ther 9:748. https://doi.org/10.3390/biom9110748

Großkopf T, Mohr W, Baustian T, Schunck H, Gill D, Kuypers MMM, Lavik G, Schmitz RA, Wallace DWR, LaRoche J (2012) Doubling of marine dinitrogen-fixation rates based on direct measurements. Nature 488:361–364. https://doi.org/10.1038/nature11338

Guidi L, Chaffron S, Bittner L, Eveillard D, Larhlimi A, Roux S, Darzi Y, Audic S, Berline L, Brum JR, Coelho LP, Espinoza JCI, Malviya S, Sunagawa S, Dimier C, Kandels-Lewis S, Picheral M, Poulain J, Searson S, Stemmann L, Not F, Hingamp P, Speich S, Follows M, Karp-Boss L, Boss E, Ogata H, Pesant S, Weissenbach J, Wincker P, Acinas SG, Bork P, De Vargas C, Iudicone D, Sullivan MB, Raes J, Karsenti E, Bowler C, Gorsky G (2016) Plankton networks driving carbon export in the global ocean. Nature 532:465–470. https://doi.org/10.1038/nature16942

Guo H, Tse LV, Barbalat R, Sivaamnuaiphorn S, Xu M, Doulatov S, Miller JF (2008) Diversity-generating retroelement homing regenerates target sequences for repeated rounds of codon rewriting and protein diversification. Mol Cell 31:813–823. https://doi.org/10.1016/j.molcel.2008.07.022

Gutiérrez-Rodríguez A, Slack G, Daniels EF, Selph KE, Palenik B, Landry MR (2014) Fine spatial structure of genetically distinct picocyanobacterial populations across environmental gradients in the Costa Rica dome. Limnol Oceanogr 59:705–723. https://doi.org/10.4319/lo.2014.59.3.0705

Guyet U, Nguyen NA, Doré H, Haguait J, Pittera J, Conan M, Ratin M, Corre E, Le Corguillé G, Brillet-Guéguen L, Hoebeke M, Six C, Steglich C, Siegel A, Eveillard D, Partensky F, Garczarek L (2020) Synergic effects of temperature and irradiance on the physiology of the marine *Synechococcus* strain WH7803. Front Microbiol 11:1707. https://doi.org/10.3389/fmicb.2020.01707

Hackenberg C, Engelhardt A, Matthijs HC, Wittink F, Bauwe H, Kaplan A, Hagemann M (2009) Photorespiratory 2-phosphoglycolate metabolism and photoreduction of $O_2$ cooperate in high-light acclimation of *Synechocystis* sp. strain PCC 6803. Planta 230:625–637. https://doi.org/10.1007/s00425-009-0972-9

Hagino K, Onuma R, Kawachi M, Horiguchi T (2013) Discovery of an endosymbiotic nitrogen-fixing cyanobacterium UCYN-A in *Braarudosphaera bigelowii* (Prymnesiophyceae). PLoS One 8:e81749. https://doi.org/10.1371/journal.pone.0081749

Hammer AK, Albrecht F, Hahne F, Jordan P, Fraatz MA, Ley J, Geissler T, Schrader J, Zorn H, Buchhaupt M (2020) Biotechnological production of odor-active methyl-branched aldehydes by a novel α-dioxygenase from *Crocosphaera subtropica*. J Agric Food Chem 68:10432–10440. https://doi.org/10.1021/acs.jafc.0c02035

Harding K, Turk-Kubo KA, Sipler RE, Mills MM, Bronk DA, Zehr JP (2018) Symbiotic unicellular cyanobacteria fix nitrogen in the Arctic Ocean. Proc Natl Acad Sci U S A 115:13371–13375. https://doi.org/10.1073/pnas.1813658115

Harke MJ, Frischkorn KR, Haley ST, Aylward FO, Zehr JP, Dyhrman ST (2019) Periodic and coordinated gene expression between a diazotroph and its diatom host. ISME J 13:118–131. https://doi.org/10.1038/s41396-018-0262-2

Hatosy SM, Martiny AC (2015) The ocean as a global reservoir of antibiotic resistance genes. Appl Environ Microbiol 81:7593–7599. https://doi.org/10.1128/AEM.00736-15

Haverkamp T, Acinas SG, Doeleman M, Stomp M, Huisman J, Stal LJ (2008) Diversity and phylogeny of Baltic Sea picocyanobacteria inferred from their ITS and phycobiliprotein operons. Environ Microbiol 10:174–188. https://doi.org/10.1111/j.1462-2920.2007.01442.x

Hayashi M, Suzuki R, Colleoni C, Ball SG, Fujita N, Suzuki E (2015) Crystallization and crystallographic analysis of branching enzymes from *Cyanothece* sp. ATCC 51142. Acta Crystallogr Sect Struct Biol Commun F71:1109–1113. https://doi.org/10.1107/S2053230X1501198X

Henke BA, Turk-Kubo KA, Bonnet S, Zehr JP (2018) Distributions and abundances of sublineages of the N2-fixing cyanobacterium *Candidatus* Atelocyanobacterium thalassa (UCYN-A) in the new Caledonian coral lagoon. Front Microbiol 9:554. https://doi.org/10.3389/fmicb.2018.00554

Herdman M, Castenholz RW, Waterbury JB, Rippka R (2001) Form-genus XIII. *Synechococcus*. In: Boone DR, Castenholz RW (eds) Bergey's manual of systematic bacteriology, 2nd edn. Springer-Verlag, New York, pp 508–512

Hess WR, Partensky F, Van Der Staay GWM, Garcia-Fernandez JM, Börner T, Vaulot D (1996) Coexistence of phycoerythrin and a chlorophyll *a/b* antenna in a marine prokaryote. Proc Natl Acad Sci U S A 93:11126–11130. https://doi.org/10.1073/pnas.93.20.11126

Hess WR, Steglich C, Lichtlé C, Partensky F (1999) Phycoerythrins of the oxyphotobacterium *Prochlorococcus marinus* are associated to the thylakoid membrane and are encoded by a single large gene cluster. Plant Mol Biol 40:507–521. https://doi.org/10.1023/A:1006252013008

Hess WR, Rocap G, Ting CS, Larimer F, Stilwagen S, Lamerdin J, Chisholm SW (2001) The photosynthetic apparatus of *Prochlorococcus*: insights through comparative genomics. Photosynth Res 70:53–71. https://doi.org/10.1023/A:1013835924610

Hess WR, Garczarek L, Pfreundt U, Partensky F (2016) Phototrophic microorganisms: the basis of the marine food web. In: Stal LJ, Cretoiu MS (eds) The marine microbiome: an untapped source of biodiversity and biotechnological potential, 1st edn. Springer International Publishing, Cham, pp 57–97. https://doi.org/10.1007/978-3-319-33000-6_3

Hetrick KJ, Walker MC, Van Der Donk WA (2018) Development and application of yeast and phage display of diverse lanthipeptides. ACS Cent Sci 4:458–467. https://doi.org/10.1021/acscentsci.7b00581

Hewson I, Govil SR, Capone DG, Carpenter EJ, Fuhrman JA (2004) Evidence of *Trichodesmium* viral lysis and potential significance for biogeochemical cycling in the oligotrophic ocean. Aquat Microb Ecol 36:1–8. https://doi.org/10.3354/ame036001

Hilton JA, Foster RA, Tripp HJ, Carter BJ, Zehr JP, Villareal TA (2013) Genomic deletions disrupt nitrogen metabolism pathways of a cyanobacterial diatom symbiont. Nat Commun 4:1767. https://doi.org/10.1038/ncomms2748

Hofmann HJ (1976) Precambrian microflora, belcher islands, Canada: significance and systematics. J Paleontol 50:1040–1073. https://www.jstor.org/stable/1303547

Holtrop T, Huisman J, Stomp M, Biersteker L, Aerts J, Grébert T, Partensky F, Garczarek L, van der Woerd HJ (2021) Vibrational modes of water predict spectral niches for photosynthesis in lakes and oceans. Nat Ecol Evol 5:55–66. https://doi.org/10.1038/s41559-020-01330-x

Hou S, López-Pérez M, Pfreundt U, Belkin N, Stüber K, Huettel B, Reinhardt R, Berman-Frank I, Rodriguez-Valera F, Hess WR (2018) Benefit from decline: the primary transcriptome of *Alteromonas macleodii* str. Te101 during *Trichodesmium* demise. ISME J 12:981–996. https://doi.org/10.1038/s41396-017-0034-4

Hu Q, Marquardt J, Iwasaki I, Miyashita H, Kurano N, Morschel E, Miyachi S (1999) Molecular structure, localization and function of biliproteins in the chlorophyll *a/d* containing oxygenic photosynthetic prokaryote *Acaryochloris marina*. Biochim Biophys Acta 1412:250–261. https://doi.org/10.1016/S0005-2728(99)00067-5

Huang S, Wilhelm SW, Harvey HR, Taylor K, Jiao N, Chen F (2012) Novel lineages of *Prochlorococcus* and *Synechococcus* in the global oceans. ISME J 6:285–297. https://doi.org/10.1038/ismej.2011.106

Humily F, Partensky F, Six C, Farrant GKK, Ratin M, Marie D, Garczarek L (2013) A gene island with two possible configurations is involved in chromatic acclimation in marine *Synechococcus*. PLoS One 8:e84459. https://doi.org/10.1371/journal.pone.0084459

Hunter-Cevera KR, Post AF, Peacock EE, Sosik HM (2016) Diversity of *Synechococcus* at the Martha's vineyard coastal observatory: insights from culture isolations, clone libraries, and flow cytometry. Microb Ecol 71:276–289. https://doi.org/10.1007/s00248-015-0644-1

Hunter-Cevera KR, Neubert MG, Olson RJ, Shalapyonok A, Solow AR, Sosik HM (2020) Seasons of *Syn*. Limnol Oceanogr 65:1085–1102. https://doi.org/10.1002/lno.11374

Hutchins DA, Walworth NG, Webb EA, Saito MA, Moran D, McIlvin MR, Gale J, Fu FX (2015) Irreversibly increased nitrogen fixation in *Trichodesmium* experimentally adapted to elevated carbon dioxide. Nat Commun 6:8155. https://doi.org/10.1038/ncomms9155

Hynes AM, Webb EA, Doney SC, Waterbury JB (2012) Comparison of cultured *Trichodesmium* (cyanophyceae) with species characterized from the field. J Phycol 48:196–210. https://doi.org/10.1111/j.1529-8817.2011.01096.x

Inomura K, Follett CL, Masuda T, Eichner M, Prášil O, Deutsch C (2020) Carbon transfer from the host diatom enables fast growth and high rate of $N_2$ fixation by symbiotic heterocystous cyanobacteria. Plan Theory 9:192. https://doi.org/10.3390/plants9020192

Janson S, Wouters J, Bergman B, Carpenter EJ (1999) Host specificity in the *Richelia*-diatom symbiosis revealed by *hetR* gene sequence analysis. Env Microbiol 1:431–438. https://doi.org/10.1046/j.1462-2920.1999.00053.x

Jardillier L, Zubkov MV, Pearman J, Scanlan DJ (2010) Significant $CO_2$ fixation by small prymnesiophytes in the subtropical and tropical Northeast Atlantic Ocean. ISME J 4:1180–1192. https://doi.org/10.1038/ismej.2010.36

Jiang HB, Fu FX, Rivero-Calle S, Levine NM, Sañudo-Wilhelmy SA, Qu PP, Wang XW, Pinedo-Gonzalez P, Zhu Z, Hutchins DA (2018) Ocean warming alleviates iron limitation of marine nitrogen fixation. Nat Clim Chang 8:709–712. https://doi.org/10.1038/s41558-018-0216-8

Johnson ZI, Zinser ER, Coe A, McNulty NP, Woodward EMS, Chisholm SW (2006) Niche partitioning among *Prochlorococcus* ecotypes along ocean-scale environmental gradients. Science 311:1737–1740. https://doi.org/10.1126/science.1118052

Jones AC, Monroe EA, Podell S, Hess WR, Klages S, Esquenazi E, Niessen S, Hoover H, Rothmann M, Lasken RS, Yates JR 3rd, Reinhardt R, Kube M, Burkart MD, Allen EE, Dorrestein PC, Gerwick WH, Gerwick L (2011) Genomic insights into the physiology and ecology of the marine filamentous cyanobacterium *Lyngbya majuscula*. Proc Natl Acad Sci U S A 108:8815–8820. https://doi.org/10.1073/pnas.1101137108

Karl D, Letelier R, Tupas L, Dore J, Christian J, Hebel D (1997) The role of nitrogen fixation in biogeochemical cycling in the subtropical North Pacific Ocean. Nature 388:533–538. https://doi.org/10.1038/41474

Karl D, Michaels A, Bergman B, Capone D, Carpenter E, Letelier R, Lipschultz F, Paerl H, Sigman D, Stal L (2002) Dinitrogen fixation in the world's oceans. Biogeochemistry 57–58: 47–98. https://doi.org/10.1023/A:1015798105851

Kashtan N, Roggensack SE, Rodrigue S, Thompson JW, Biller SJ, Coe A, Ding H, Marttinen P, Malmstrom RR, Stocker R, Follows MJ, Stepanauskas R, Chisholm SW (2014) Single-cell genomics reveals hundreds of coexisting subpopulations in wild *Prochlorococcus*. Science 344: 416–420. https://doi.org/10.1126/science.1248575

Kent AG, Baer SE, Mouginot C, Huang JS, Larkin AA, Lomas MW, Martiny AC (2019) Parallel phylogeography of *Prochlorococcus* and *Synechococcus*. ISME J 13:430–441. https://doi.org/10.1038/s41396-018-0287-6

Kessler N, Armoza-Zvuloni R, Wang S, Basu S, Weber PK, Stuart RK, Shaked Y (2020) Selective collection of iron-rich dust particles by natural *Trichodesmium* colonies. ISME J 14:91–103. https://doi.org/10.1038/s41396-019-0505-x

Kettler GC, Martiny AC, Huang K, Zucker J, Coleman ML, Rodrigue S, Chen F, Lapidus A, Ferriera S, Johnson J, Steglich C, Church GM, Richardson P, Chisholm SW (2007) Patterns and implications of gene gain and loss in the evolution of *Prochlorococcus*. PLoS Genet 3:e231. https://doi.org/10.1371/journal.pgen.0030231

Kirk JTO (1994) Light and photosynthesis in aquatic ecosystems. Cambridge University Press, Cambridge

Kirschvink JL, Kopp RE (2008) Palaeoproterozoic ice houses and the evolution of oxygen-mediating enzymes: the case for a late origin of photosystem II. Philos Trans R Soc B Biol Sci 363:2755–2765. https://doi.org/10.1098/rstb.2008.0024

Klähn S, Steglich C, Hess WR, Hagemann M (2010) Glucosylglycerate: a secondary compatible solute common to marine cyanobacteria from nitrogen-poor environments. Environ Microbiol 12:83–94. https://doi.org/10.1111/j.1462-2920.2009.02045.x

Klähn S, Baumgartner D, Pfreundt U, Voigt K, Schön V, Steglich C, Hess WR (2014) Alkane biosynthesis genes in cyanobacteria and their transcriptional organization. Front Bioeng Biotechnol 2:24. https://doi.org/10.3389/fbioe.2014.00024

Klähn S, Schaal C, Georg J, Baumgartner D, Knippen G, Hagemann M, Muro-Pastor AM, Hess WR (2015) The sRNA NsiR4 is involved in nitrogen assimilation control in cyanobacteria by targeting glutamine synthetase inactivating factor IF7. Proc Natl Acad Sci U S A 112:E6243–E6252. https://doi.org/10.1073/pnas.1508412112

Knapp AN, Fawcett SE, Martínez-Garcia A, Leblond N, Moutin T, Bonnet S (2016) Nitrogen isotopic evidence for a shift from nitrate-to diazotroph-fueled export production in VAHINE mesocosm experiments. Biogeosciences 16:4645–4657. https://doi.org/10.5194/bg-13-4645-2016

Knight MA, Morris JJ (2020) Co-culture with *Synechococcus* facilitates growth of *Prochlorococcus* under ocean acidification conditions. Environ Microbiol 22:876–4889. https://doi.org/10.1111/1462-2920.15277

Knoll AH, Semikhatov MA (1998) The genesis and time distribution of two distinctive Proterozoic stromatolite microstructures. PALAIOS 13:408–422. https://doi.org/10.1043/0883-1351

Koblížek M (2015) Ecology of aerobic anoxygenic phototrophs in aquatic environments. FEMS Microbiol Rev 39:854–870. https://doi.org/10.1093/femsre/fuv032

Konstantinidis KT, Tiedje JM (2005) Genomic insights that advance the species definition for prokaryotes. Proc Natl Acad Sci U S A 102:2567–2572. https://doi.org/10.1073/pnas.0409727102

Kopf M, Hess WR (2015) Regulatory RNAs in photosynthetic cyanobacteria. FEMS Microbiol Rev 39:301–315. https://doi.org/10.1093/femsre/fuv017

Kopf M, Möke F, Bauwe H, Hess WR, Hagemann M (2015) Expression profiling of the bloom-forming cyanobacterium *Nodularia* CCY9414 under light and oxidative stress conditions. ISME J 9:2139–2152. https://doi.org/10.1038/ismej.2015.16

Krumhardt KM, Callnan K, Roache-Johnson K, Swett T, Robinson D, Reistetter EN, Saunders JK, Rocap G, Moore LR (2013) Effects of phosphorus starvation versus limitation on the marine cyanobacterium *Prochlorococcus* MED4 I: uptake physiology. Env Microbiol 15:2114–2128. https://doi.org/10.1111/1462-2920.12079

Krupke A, Lavik G, Halm H, Fuchs BM, Amann RI, Kuypers MM (2014) Distribution of a consortium between unicellular algae and the $N_2$ fixing cyanobacterium UCYN-A in the North Atlantic Ocean. Env Microbiol 16:3153–3167. https://doi.org/10.1111/1462-2920.12431

Krupke A, Mohr W, LaRoche J, Fuchs BM, Amann RI, Kuypers MMM (2015) The effect of nutrients on carbon and nitrogen fixation by the UCYN-A-haptophyte symbiosis. ISME J 9: 1635–1647. https://doi.org/10.1038/ismej.2014.253

Kühl M, Chen M, Ralph PJ, Schreiber U, Larkum AW (2005) A niche for cyanobacteria containing chlorophyll *d*. Nature 433:820. https://doi.org/10.1038/433820a

Kupriyanova EV, Sinetova MA, Mironov KS, Novikova GV, Dykman LA, Rodionova MV, Gabrielyan DA, Los DA (2019) Highly active extracellular α-class carbonic anhydrase of *Cyanothece* sp. ATCC 51142. Biochimie 160:200–209. https://doi.org/10.1016/j.biochi.2019.03.009

Lambrecht SJ, Steglich C, Hess WR (2020) A minimum set of regulators to thrive in the ocean. FEMS Microbiol Rev 44:232–252. https://doi.org/10.1093/femsre/fuaa005

Lantoine F, Neveux J (1997) Spatial and seasonal variations in abundance and spectral characteristics of phycoerythrins in the tropical northeastern Atlantic Ocean. Deep Sea Res I 44:223–246. https://doi.org/10.1016/S0967-0637(96)00094-5

Larkin AA, Blinebry SK, Howes C, Lin Y, Loftus SE, Schmaus CA, Zinser ER, Johnson ZI (2016) Niche partitioning and biogeography of high light adapted *Prochlorococcus* across taxonomic ranks in the North Pacific. ISME J 10:1555–1567. https://doi.org/10.1038/ismej.2015.244

Larsson J, Nylander JA, Bergman B (2011) Genome fluctuations in cyanobacteria reflect evolutionary, developmental and adaptive traits. BMC Evol Biol 11:187. https://doi.org/10.1186/1471-2148-11-187

Larsson J, Celepli N, Ininbergs K, Dupont CL, Yooseph S, Bergman B, Ekman M (2014) Picocyanobacteria containing a novel pigment gene cluster dominate the brackish water Baltic Sea. ISME J 8:1892–1903. https://doi.org/10.1038/ismej.2014.35

Lavin P, Gonzalez B, Santibanez JF, Scanlan DJ, Ulloa O (2010) Novel lineages of *Prochlorococcus* thrive within the oxygen minimum zone of the eastern tropical South Pacific. Env Microbiol Rep 2:728–738. https://doi.org/10.1111/j.1758-2229.2010.00167.x

Lea-Smith DJ, Biller SJ, Davey MP, Cotton CAR, Perez Sepulveda BM, Turchyn AV, Scanlan DJ, Smith AG, Chisholm SW, Howe CJ (2015) Contribution of cyanobacterial alkane production to the ocean hydrocarbon cycle. Proc Natl Acad Sci 112:13591–13596. https://doi.org/10.1073/pnas.1507274112

Lee MD, Ahlgren NA, Kling JD, Walworth NG, Rocap G, Saito MA, Hutchins DA, Webb EA (2019) Marine *Synechococcus* isolates representing globally abundant genomic lineages demonstrate a unique evolutionary path of genome reduction without a decrease in GC content. Environ Microbiol 21:1677–1686. https://doi.org/10.1111/1462-2920.14552

Lehours A-C, Enault F, Boeuf D, Jeanthon C (2018) Biogeographic patterns of aerobic anoxygenic phototrophic bacteria reveal an ecological consistency of phylogenetic clades in different oceanic biomes. Sci Rep 8:4105. https://doi.org/10.1038/s41598-018-22413-7

Lewin RA (1984) *Prochloron* - a status report. Phycologia 23:203–208. https://doi.org/10.2216/i0031-8884-23-2-203.1

Li WKW (1994) Primary productivity of prochlorophytes, cyanobacteria, and eucaryotic ultraphytoplankton: measurements from flow cytometric sorting. Limnol Oceanogr 39:169–175. https://doi.org/10.4319/lo.1994.39.1.0169

Li B, Sher D, Kelly L, Shi Y, Huang K, Knerr PJ, Joewono I, Rusch D, Chisholm SW, Van Der Donk WA (2010) Catalytic promiscuity in the biosynthesis of cyclic peptide secondary metabolites in planktonic marine cyanobacteria. Proc Natl Acad Sci U S A 107:10430–10435. https://doi.org/10.1073/pnas.0913677107

Liu L, Budnjo A, Jokela J, Haug BE, Fewer DP, Wahlsten M, Rouhiainen L, Permi P, Fossen T, Sivonen K (2015) Pseudoaeruginosins, nonribosomal peptides in *Nodularia spumigena*. ACS Chem Biol 10:725–733. https://doi.org/10.1021/cb5004306

Liu D, Liberton M, Yu J, Pakrasi HB, Bhattacharyya-Pakrasi M (2018) Engineering nitrogen fixation activity in an oxygenic phototroph. MBio 9:e01029–e01018. https://doi.org/10.1128/mBio.01029-18

Luo Y-W, Doney SC, Anderson LA, Benavides M, Berman-Frank I, Bode A, Bonnet S, Boström KH, Böttjer D, Capone DG, Carpenter EJ, Chen YL, Church MJ, Dore JE, Falcón LI, Fernández A, Foster RA, Furuya K, Gómez F, Gundersen K, Hynes AM, Karl DM, Kitajima S, Langlois RJ, LaRoche J, Letelier RM, Marañón E, McGillicuddy DJ, Moisander PH, Moore CM, Mouriño-Carballido B, Mulholland MR, Needoba JA, Orcutt KM, Poulton AJ, Rahav E, Raimbault P, Rees AP, Riemann L, Shiozaki T, Subramaniam A, Tyrrell T, Turk-Kubo KA, Varela M, Villareal TA, Webb EA, White AE, Wu J, Zehr JP (2012) Database of diazotrophs in global ocean: abundance, biomass and nitrogen fixation rates. Earth Syst Sci Data 4:47–73. https://doi.org/10.5194/essd-4-47-2012

Lyons TW, Reinhard CT, Planavsky NJ (2014) The rise of oxygen in Earth's early ocean and atmosphere. Nature 506:307–315. https://doi.org/10.1038/nature13068

Mackey KRM, Paytan A, Caldeira K, Grossman AR, Moran D, Mcilvin M, Saito MA (2013) Effect of temperature on photosynthesis and growth in marine *Synechococcus* spp. Plant Physiol 163: 815–829. https://doi.org/10.1104/pp.113.221937

Mahaffey C, Michaels AF, Capone DG (2005) The conundrum of marine $N_2$ fixation. Am J Sci 305: 546–595. https://doi.org/10.2475/ajs.305.6-8.546

Mahmoud RM, Sanfilippo JE, Nguyen AA, Strnat JA, Partensky F, Garczarek L, Abo El Kassem N, Kehoe DM, Schluchter WM (2017) Adaptation to blue light in marine *Synechococcus* requires MpeU, an enzyme with similarity to phycoerythrobilin lyase isomerases. Front Microbiol 8:243. https://doi.org/10.3389/fmicb.2017.00243

Malmstrom RR, Coe A, Kettler GC, Martiny AC, Frias-Lopez J, Zinser ER, Chisholm SW (2010) Temporal dynamics of *Prochlorococcus* ecotypes in the Atlantic and Pacific oceans. ISME J 4: 1252–1264. https://doi.org/10.1038/ismej.2010.60

Malmstrom RR, Rodrigue S, Huang KH, Kelly L, Kern SE, Thompson A, Roggensack S, Berube PM, Henn MR, Chisholm SW (2013) Ecology of uncultured *Prochlorococcus* clades revealed through single-cell genomics and biogeographic analysis. ISME J 7:184–198. https://doi.org/10.1038/ismej.2012.89

Mandal S, Rath J (2015) Secondary metabolites of cyanobacteria and drug development. In: Extremophilic cyanobacteria for novel drug development. SpringerBriefs in pharmaceutical science & drug development. Springer, Cham, pp 23–43. https://doi.org/10.1007/978-3-319-12009-6_2

Mareš J, Hrouzek P, Kaňa R, Ventura S, Strunecký O, Komárek J (2013) The primitive thylakoid-less cyanobacterium *Gloeobacter* is a common rock-dwelling organism. PLoS One 8:e66323. https://doi.org/10.1371/journal.pone.0066323

Mareš J, Johansen JR, Hauer T, Zima J, Ventura S, Cuzman O, Tiribilli B, Kaštovsky J (2019) Taxonomic resolution of the genus *Cyanothece* (Chroococcales, cyanobacteria), with a treatment on Gloeothece and three new genera, *Crocosphaera*, *Rippkaea*, and *Zehria*. J Phycol 55: 578–610. https://doi.org/10.1111/jpy.12853

Marquardt J, Senger H, Miyashita H, Miyachi S, Morschel E (1997) Isolation and characterization of biliprotein aggregates from *Acaryochloris marina*, a *Prochloron*-like prokaryote containing mainly chlorophyll *d*. FEBS Lett 410:428–432. https://doi.org/10.1016/S0014-5793(97)00631-5

Martínez-Pérez C, Mohr W, Löscher CR, Dekaezemacker J, Littmann S, Yilmaz P, Lehnen N, Fuchs BM, Lavik G, Schmitz RA, LaRoche J, Kuypers MMM (2016) The small unicellular diazotrophic symbiont, UCYN-A, is a key player in the marine nitrogen cycle. Nat Microbiol 1: 16163. https://doi.org/10.1038/nmicrobiol.2016.163

Martiny AC, Coleman ML, Chisholm SW (2006) Phosphate acquisition genes in *Prochlorococcus* ecotypes: evidence for genome-wide adaptation. Proc Natl Acad Sci U S A 103:12552–12557. https://doi.org/10.1073/pnas.0601301103

Martiny AC, Huang Y, Li W (2009a) Occurrence of phosphate acquisition genes in *Prochlorococcus* cells from different ocean regions. Environ Microbiol 11:1340–1347. https://doi.org/10.1111/j.1462-2920.2009.01860.x

Martiny AC, Kathuria S, Berube PM (2009b) Widespread metabolic potential for nitrite and nitrate assimilation among *Prochlorococcus* ecotypes. Proc Natl Acad Sci U S A 106:10787–10792. https://doi.org/10.1073/pnas.0902532106

Martiny JBH, Jones SE, Lennon JT, Martiny AC (2015) Microbiomes in light of traits: a phylogenetic perspective. Science 350:aac9323. https://doi.org/10.1126/science.aac9323

Mazard S, Ostrowski M, Partensky F, Scanlan DJ (2012) Multi-locus sequence analysis, taxonomic resolution and biogeography of marine *Synechococcus*. Enviromental Microbiol 14:372–386. https://doi.org/10.1111/j.1462-2920.2011.02514.x

Mazur-Marzec H, Kaczkowska MJ, Blaszczyk A, Akcaalan R, Spoof L, Meriluoto J (2012) Diversity of peptides produced by *Nodularia spumigena* from various geographical regions. Mar Drugs 11:1–19. https://doi.org/10.3390/md11010001

McCarren J, Brahamsha B (2005) Transposon mutagenesis in a marine *Synechococcus* strain: isolation of swimming motility mutants. J Bacteriol 187:4457–4462. https://doi.org/10.1128/JB.187.13.4457-4462.2005

McCarren J, Brahamsha B (2007) SwmB, a 1.12-megadalton protein that is required for nonflagellar swimming motility in *Synechococcus*. J Bacteriol 189:1158–1162. https://doi.org/10.1128/JB.01500-06

McDermott JE, Oehmen CS, McCue LA, Hill E, Choi DM, Stöckel J, Liberton M, Pakrasi HB, Sherman LA (2011) A model of cyclic transcriptomic behavior in the cyanobacterium *Cyanothece* sp. ATCC 51142. Mol BioSyst 7:2407–2418. https://doi.org/10.1039/c1mb05006k

McKinna LIW (2015) Three decades of ocean-color remote-sensing *Trichodesmium* spp. in the World's oceans: a review. Prog Oceanogr 131:177–199. https://doi.org/10.1016/j.pocean.2014.12.013

Mella-Flores D, Six C, Ratin M, Partensky F, Boutte C, Le Corguillé G, Marie D, Blot N, Gourvil P, Kolowrat C, Garczarek L (2012) *Prochlorococcus* and *Synechococcus* have evolved different adaptive mechanisms to cope with light and UV stress. Front Microbiol 3:285. https://doi.org/10.3389/fmicb.2012.00285

Melnicki MR, Pinchuk GE, Hill EA, Kucek LA, Fredrickson JK, Konopka A, Beliaev AS (2012) Sustained $H_2$ production driven by photosynthetic water splitting in a unicellular cyanobacterium. MBio 3:e00197–e00112. https://doi.org/10.1128/mBio.00197-12

Mevers E, Matainaho T, Allara' M, Di Marzo V, Gerwick WH (2014) Mooreamide a: a cannabinomimetic lipid from the marine cyanobacterium *Moorea bouillonii*. Lipids 49:1127–1132. https://doi.org/10.1007/s11745-014-3949-9

Millman A, Bernheim A, Stokar-Avihail A, Fedorenko T, Voichek M, Leavitt A, Oppenheimer-Shaanan Y, Sorek R (2020) Bacterial retrons function in anti-phage defense. Cell 183:1551–1561. https://doi.org/10.1016/j.cell.2020.09.065

Min H, Sherman LA (2010) Hydrogen production by the unicellular, diazotrophic cyanobacterium *Cyanothece* sp. strain ATCC 51142 under conditions of continuous light. Appl Environ Microbiol 76:293–4301. https://doi.org/10.1128/AEM.00146-10

Miyashita H, Adachi K, Kurano N, Ikemoto H, Chihara M, Miyachi S (1997) Pigment composition of a novel oxygenic photosynthetic prokaryote containing chlorophyll *d* as the major chlorophyll. Plant Cell Physiol 38:274–281. https://doi.org/10.1093/oxfordjournals.pcp.a029163

Miyashita H, Ikemoto H, Kurano N, Miyachi S, Chihara M (2003) *Acaryochloris marina* gen. et sp. nov. (cyanobacteria), an oxygenic photosynthetic prokaryote containing chl *d* as a major pigment. J Phycol 39:1247–1253. https://doi.org/10.1111/j.0022-3646.2003.03-158.x

Möke F, Wasmund N, Bauwe H, Hagemann M (2013) Salt acclimation of *Nodularia spumigena* CCY9414 – a cyanobacterium adapted to brackish water. Aquat Microb Ecol 70:207–214. https://doi.org/10.3354/ame01656

Montoya JP, Holl CM, Zehr JP, Hansen A, Villareal TA, Capone DG (2004) High rates of $N_2$ fixation by unicellular diazotrophs in the oligotrophic Pacific Ocean. Nature 430:1027–1031. https://doi.org/10.1038/nature02824

Moore LR, Chisholm SW (1999) Photophysiology of the marine cyanobacterium *Prochlorococcus*: Ecotypic differences among cultured isolates. Limnol Oceanogr 44:628–638. https://doi.org/10.4319/lo.1999.44.3.0628

Moore LR, Rocap G, Chisholm SW (1998) Physiology and molecular phylogeny of coexisting *Prochlorococcus* ecotypes. Nature 393:464–467. https://doi.org/10.1038/30965

Moore JK, Doney SC, Glover DM, Fung IY (2001) Iron cycling and nutrient-limitation patterns in surface waters of the World Ocean. Deep-Sea Res II Top Stud Oceanogr 49:463–507. https://doi.org/10.1016/S0967-0645(01)00109-6

Moore LR, Coe A, Zinser ER, Saito MA, Sullivan MB, Lindell D, Frois-Moniz K, Waterbury J, Chisholm SW (2007) Culturing the marine cyanobacterium *Prochlorococcus*. Limnol Oceanogr Methods 5:353–362. https://doi.org/10.4319/lom.2007.5.353

Moore CM, Mills MM, Arrigo KR, Berman-Frank I, Bopp L, Boyd PW, Galbraith ED, Geider RJ, Guieu C, Jaccard SL, others (2013) Processes and patterns of oceanic nutrient limitation. Nat Geosci 6:701–710. https://doi.org/10.1038/ngeo1765

Morel A, Ahn Y-H, Partensky F, Vaulot D, Claustre H (1993) *Prochlorococcus* and *Synechococcus*: a comparative study of their optical properties in relation to their size and pigmentation. J Mar Res 51:617–649. https://doi.org/10.1357/0022240933223963

Morris JJ, Kirkegaard R, Szul MJ, Johnson ZI, Zinser ER (2008) Facilitation of robust growth of *Prochlorococcus* colonies and dilute liquid cultures by "helper" heterotrophic bacteria. Appl Environ Microbiol 74:4530–4534. https://doi.org/10.1128/AEM.02479-07

Mueller TJ, Welsh EA, Pakrasi HB, Maranas CD (2016) Identifying regulatory changes to facilitate nitrogen fixation in the non-diazotroph *Synechocystis* sp. PCC 6803. ACS Synth Biol 5:250–258. https://doi.org/10.1021/acssynbio.5b00202

Muñoz-Marín M del C, Luque I, Zubkov M V, Hill PG, Diez J, García-Fernández JM (2013) *Prochlorococcus* can use the Pro1404 transporter to take up glucose at nanomolar concentrations in the Atlantic Ocean. Proc Natl Acad Sci U S A 110:8597–8602. https://doi.org/10.1073/pnas.1221775110

Muñoz-Marín MC, Gómez-Baena G, López-Lozano A, Moreno-Cabezuelo JA, Díez J, García-Fernández JM (2020) Mixotrophy in marine picocyanobacteria: use of organic compounds by *Prochlorococcus* and *Synechococcus*. ISME J 14:1065–1073. https://doi.org/10.1038/s41396-020-0603-9

Muro-Pastor AM, Hess WR (2012) Heterocyst differentiation: from single mutants to global approaches. Trends Microbiol 20:548–557. https://doi.org/10.1016/j.tim.2012.07.005

Neveux J, Lantoine F, Vaulot D, Marie D, Blanchot J (1999) Phycoerythrins in the southern tropical and equatorial Pacific Ocean: evidence for new cyanobacterial types. J Geophys Res Ocean 10: 3311–3321. https://doi.org/10.1029/98jc02000

Neveux J, Tenório MMB, Dupouy C, Villareal TA (2006) Spectral diversity of phycoerythrins and diazotroph abundance in tropical waters. Limnol Oceanogr 51:1689–1698. https://doi.org/10.4319/lo.2006.51.4.1689

O'Neil JM, Roman MR (1994) Ingestion of the cyanobacterium *Trichodesmium* spp. by pelagic harpacticoid copepods *Macrosetella*, *Miracia* and *Oculosetella*. Hydrobiologia 292:235–240. https://doi.org/10.1007/BF00229946

Ohki K (1999) A possible role of temperate phage in the regulation of *Trichodesmium* biomass. Bull Inst Océanogr Monaco 19:287–291

Olson EM, McGillicuddy DJ Jr, Dyhrman ST, Waterbury JB, Davis CS, Solow AR (2015) The depth-distribution of nitrogen fixation by *Trichodesmium* spp. colonies in the tropical–subtropical North Atlantic. Deep Sea Res I 104:72–91. https://doi.org/10.1016/j.dsr.2015.06.012

Omoregie EO, Crumbliss LL, Bebout BM, Zehr JP (2004) Determination of nitrogen-fixing phylotypes in *Lyngbya* sp. and *Microcoleus chthonoplastes* cyanobacterial mats from Guerrero Negro, Baja California, Mexico. Appl Environ Microbiol 70:2119–2128. https://doi.org/10.1128/AEM.70.4.2119-2128.2004

Ong LJ, Glazer AN (1991) Phycoerythrins of marine unicellular cyanobacteria: I. Bilin types and locations and energy transfer pathways in *Synechococcus* spp phycoerythrins. J Biol Chem 266: 9515–9527. https://www.jbc.org/content/266/15/9515.long

Ong LJ, Glazer AN, Waterbury JB (1984) An unusual phycoerythrin from a marine cyanobacterium. Science 224:80–83. https://doi.org/10.1126/science.224.4644.80

Pachiadaki MG, Brown JM, Brown J, Bezuidt O, Berube PM, Biller SJ, Poulton NJ, Burkart MD, La Clair JJ, Chisholm SW, Stepanauskas R (2019) Charting the complexity of the marine microbiome through single-cell genomics. Cell 179:1623–1635. https://doi.org/10.1016/j.cell.2019.11.017

Pade N, Hagemann M (2014) Salt acclimation of cyanobacteria and their application in biotechnology. Life 5:25–49. https://doi.org/10.3390/life5010025

Pade N, Michalik D, Ruth W, Belkin N, Hess WR, Berman-Frank I, Hagemann M (2016) Trimethylated homoserine functions as the major compatible solute in the globally significant oceanic cyanobacterium *Trichodesmium*. Proc Natl Acad Sci U S A 113:13191–13196. https://doi.org/10.1073/pnas.1611666113

Pagels F, Guedes AC, Amaro HM, Kijjoa A, Vasconcelos V (2019) Phycobiliproteins from cyanobacteria: chemistry and biotechnological applications. Biotechnol Adv 37:422–443. https://doi.org/10.1016/j.biotechadv.2019.02.010

Palenik B (2001) Chromatic adaptation in marine *Synechococcus* strains. Appl Environ Microbiol 67:991–994. https://doi.org/10.1128/AEM.67.2.991-994.2001

Palenik B, Brahamsha B, Larimer FW, Land M, Hauser L, Chain P, Lamerdin J, Regala W, Allen EE, McCarren J, Paulsen I, Dufresne A, Partensky F, Webb EA, Waterbury J (2003) The genome of a motile marine *Synechococcus*. Nature 424:1037–1042. https://doi.org/10.1038/nature01943

Palenik B, Ren Q, Dupont CL, Myers GS, Heidelberg JF, Badger JH, Madupu R, Nelson WC, Brinkac LM, Dodson RJ, Durkin AS, Daugherty SC, Sullivan SA, Khouri H, Mohamoud Y, Halpin R, Paulsen IT (2006) Genome sequence of *Synechococcus* CC9311: insights into adaptation to a coastal environment. Proc Natl Acad Sci U S A 103:13555–13559. https://doi.org/10.1073/pnas.0602963103

Partensky F, Garczarek L (2003) The photosynthetic apparatus of chlorophyll *b*- and *d*-containing Oxychlorobacteria. In: AWD L, Douglas SE, Raven JA (eds) Photosynthesis in algae. Kluwer Academic Publisher, Dordrecht, pp 29–62

Partensky F, Garczarek L (2010) *Prochlorococcus*: advantages and limits of minimalism. Annu Rev Mar Sci 2:305–331. https://doi.org/10.1146/annurev-marine-120308-081034

Partensky F, Blanchot J, Vaulot D (1999a) Differential distribution and ecology of *Prochlorococcus* and *Synechococcus* in oceanic waters : a review. Bull Instit Océanogr Monaco 19:457–475

Partensky F, Hess WR, Vaulot D (1999b) *Prochlorococcus*, a marine photosynthetic prokaryote of global significance. Microbiol Mol Biol Rev 63:106–127. https://doi.org/10.1128/MMBR.63.1.106-127.1999

Partensky F, Six C, Ratin M, Garczarek L, Vaulot D, Grébert T, Marie D, Gourvil P, Probert I, Le Panse S, Gachenot M, Bouchier C, Rodríguez F, Garrido JL (2018) A novel species of the marine cyanobacterium *Acaryochloris* with a unique pigment content and lifestyle. Sci Rep 8: 9142. https://doi.org/10.1038/s41598-018-27542-7

Paulsen ML, Doré H, Garczarek L, Seuthe L, Müller O, Sandaa R-A, Bratbak G, Larsen A (2016) *Synechococcus* in the Atlantic gateway to the Arctic Ocean. Front Mar Sci 3:191. https://doi.org/10.3389/fmars.2016.00191

Paz-Yepes J, Brahamsha B, Palenik B (2013) Role of a microcin-C-like biosynthetic gene cluster in allelopathic interactions in marine *Synechococcus*. Proc Natl Acad Sci U S A 110:12030–12035. https://doi.org/10.1073/pnas.1306260110

Penno S, Lindell D, Post AF (2006) Diversity of *Synechococcus* and *Prochlorococcus* populations determined from DNA sequences of the N-regulatory gene *ntcA*. Environ Microbiol 8:1200–1211. https://doi.org/10.1111/j.1462-2920.2006.01010.x

Petrasek Z, Schmitt FJ, Theiss C, Huyer J, Chen M, Larkum A, Eichler HJ, Kemnitz K, Eckert HJ (2005) Excitation energy transfer from phycobiliprotein to chlorophyll *d* in intact cells of *Acaryochloris marina* studied by time- and wavelength-resolved fluorescence spectroscopy. Photochem Photobiol Sci 4:1016–1022. https://doi.org/10.1039/b512350j

Pfreundt U, Hess WR (2015) Sequential splicing of a group II twintron in the marine cyanobacterium *Trichodesmium*. Sci Rep 5:16829. https://doi.org/10.1038/srep16829

Pfreundt U, Stal LJ, Voß B, Hess WR (2012) Dinitrogen fixation in a unicellular chlorophyll *d*-containing cyanobacterium. ISME J 6:1367–1377. https://doi.org/10.1038/ismej.2011.199

Pfreundt U, Kopf M, Belkin N, Berman-Frank I, Hess WR (2014) The primary transcriptome of the marine diazotroph *Trichodesmium erythraeum* IMS101. Sci Rep 4:6187. https://doi.org/10.1038/srep06187

Pfreundt U, Van Wambeke F, Bonnet S, Hess WR (2016) Succession within the prokaryotic communities during the VAHINE mesocosms experiment in the New Caledonia lagoon. Biogeosciences 13:2319–2337. https://doi.org/10.5194/bg-13-2319-2016

Pinhassi J, DeLong EF, Béjà O, González JM, Pedrós-Alió C (2016) Marine bacterial and archaeal ion-pumping rhodopsins: genetic diversity, physiology, and ecology. Microbiol Mol Biol Rev 80:929–954. https://doi.org/10.1128/mmbr.00003-16

Pittera J, Humily F, Thorel M, Grulois D, Garczarek L, Six C (2014) Connecting thermal physiology and latitudinal niche partitioning in marine *Synechococcus*. ISME J 8:1221–1236. https://doi.org/10.1038/ismej.2013.228

Pittera J, Partensky F, Six C (2017) Adaptive thermostability of light-harvesting complexes in marine picocyanobacteria. ISME J 11:112–125. https://doi.org/10.1038/ismej.2016.102

Pittera J, Jouhet J, Breton S, Garczarek L, Partensky F, Maréchal É, Nguyen NA, Doré H, Ratin M, Pitt FD, Scanlan DJ, Six C (2018) Thermoacclimation and genome adaptation of the membrane lipidome in marine *Synechococcus*. Environ Microbiol 20:612–631. https://doi.org/10.1111/1462-2920.13985

Ploug H, Adam B, Musat N, Kalvelage T, Lavik G, Wolf-Gladrow D, Kuypers MMM (2011) Carbon, nitrogen and $O_2$ fluxes associated with the cyanobacterium *Nodularia spumigena* in the Baltic Sea. ISME J 5:1549–1558. https://doi.org/10.1038/ismej.2011.20

Post AF, Dedej Z, Gottlieb R, Li H, Thomas DN, El Absawi M, El Naggar A, El Gharabawi M, Sommer U (2002) Spatial and temporal distribution of *Trichodesmium* spp. in the stratified Gulf of Aqaba, Red Sea. Mar Ecol Prog Ser 239:241–250. https://doi.org/10.3354/meps239241

Pyle AE, Johnson AM, Villareal TA (2020) Isolation, growth, and nitrogen fixation rates of the *Hemiaulus-Richelia* (diatom-cyanobacterium) symbiosis in culture. PeerJ 8:e10115. https://doi.org/10.7717/peerj.10115

Reddy KJ, Haskell JB, Sherman DM, Sherman LA (1993) Unicellular, aerobic nitrogen-fixing cyanobacteria of the genus *Cyanothece*. J Bacteriol 175:1284–1292. https://doi.org/10.1128/jb.175.5.1284-1292.1993

Reinhard CT, Planavsky NJ, Robbins LJ, Partin CA, Gill BC, Lalonde SV, Bekker A, Konhauser KO, Lyons TW (2013) Proterozoic Ocean redox and biogeochemical stasis. Proc Natl Acad Sci U S A 110:5357–5362. https://doi.org/10.1073/pnas.1208622110

Repeta DJ, Ferrón S, Sosa OA, Johnson CG, Repeta LD, Acker M, Delong EF, Karl DM (2016) Marine methane paradox explained by bacterial degradation of dissolved organic matter. Nat Geosci 9:884–887. https://doi.org/10.1038/ngeo2837

Rocap G, Larimer FW, Lamerdin J, Malfatti S, Chain P, Ahlgren NA, Arellano A, Coleman M, Hauser L, Hess WR, Johnson ZI, Land M, Lindell D, Post AF, Regala W, Shah M, Shaw SL, Steglich C, Sullivan MB, Ting CS, Tolonen A, Webb EA, Zinser ER, Chisholm SW (2003) Genome divergence in two *Prochlorococcus* ecotypes reflects oceanic niche differentiation. Nature 424:1042–1047. https://doi.org/10.1038/nature01947

Rodriguez IB, Ho TY (2014) Diel nitrogen fixation pattern of *Trichodesmium*: the interactive control of light and Ni. Sci Rep 4:4445. https://doi.org/10.1038/srep04445

Rosing MT, Frei R (2004) U-rich Archaean Sea-floor sediments from Greenland – indications of >3700 ma oxygenic photosynthesis. Earth Planet Sci Lett 217:237–244. https://doi.org/10.1016/S0012-821X(03)00609-5

Rousset G, De Boissieu F, Menkes CE, Lefèvre J, Frouin R, Rodier M, Ridoux V, Laran S, Bonnet S, Dupouy C (2018) Remote sensing of *Trichodesmium* spp. mats in the western tropical South Pacific. Biogeosciences 15:5203–5219. https://doi.org/10.5194/bg-15-5203-2018

Rubin M, Berman-Frank I, Shaked Y (2011) Dust-and mineral-iron utilization by the marine dinitrogen-fixer *Trichodesmium*. Nat Geosci 4:529–534. https://doi.org/10.1038/ngeo1181

Rusch DB, Martiny AC, Dupont CL, Halpern AL, Venter JC (2010) Characterization of *Prochlorococcus* clades from iron-depleted oceanic regions. Proc Natl Acad Sci U S A 107: 16184–16189. https://doi.org/10.1073/pnas.1009513107

Sacilotto Detoni AM, Fonseca Costa LD, Pacheco LA, Yunes JS (2016) Toxic *Trichodesmium* bloom occurrence in the southwestern South Atlantic Ocean. Toxicon 110:51–55. https://doi.org/10.1016/j.toxicon.2015.12.003

Saito MA, Rocap G, Moffett JW (2005) Production of cobalt binding ligands in a *Synechococcus* feature at the Costa Rica upwelling dome. Limnol Oceanogr. https://doi.org/10.4319/lo.2005.50.1.0279

Salazar VW, Tschoeke DA, Swings J, Cosenza CA, Mattoso M, Thompson CC, Thompson FL (2020) A new genomic taxonomy system for the *Synechococcus* collective. Environ Microbiol 22:4557–4570. https://doi.org/10.1111/1462-2920.15173

Salvador-Reyes LA, Luesch H (2015) Biological targets and mechanisms of action of natural products from marine cyanobacteria. Nat Prod Rep 32:478–503. https://doi.org/10.1039/c4np00104d

Sánchez-Baracaldo P (2015) Origin of marine planktonic cyanobacteria. Sci Rep 5:17418. https://doi.org/10.1038/srep17418

Sánchez-Baracaldo P, Cardona T (2020) On the origin of oxygenic photosynthesis and cyanobacteria. New Phytol 225:1440–1446. https://doi.org/10.1111/nph.16249

Sandh G, El-Shehawy R, Diez B, Bergman B (2009) Temporal separation of cell division and diazotrophy in the marine diazotrophic cyanobacterium *Trichodesmium erythraeum* IMS101. FEMS Microbiol Lett 295:281–288. https://doi.org/10.1111/j.1574-6968.2009.01608.x

Sandh G, Xu L, Bergman B (2012) Diazocyte development in the marine diazotrophic cyanobacterium *Trichodesmium*. Microbiology 158:345–352. https://doi.org/10.1099/mic.0.051268-0

Sanfilippo JE, Nguyen AA, Karty JA, Shukla A, Schluchter WM, Garczarek L, Partensky F, Kehoe DM (2016) Self-regulating genomic island encoding tandem regulators confers chromatic acclimation to marine *Synechococcus*. Proc Natl Acad Sci U S A 113:6077–6082. https://doi.org/10.1073/pnas.1600625113

Sanfilippo JE, Nguyen AA, Garczarek L, Karty JA, Pokhrel S, Strnat JA, Partensky F, Schluchter WM, Kehoe DM (2019) Interplay between differentially expressed enzymes contributes to light color acclimation in marine *Synechococcus*. Proc Natl Acad Sci U S A 116:6457–6462. https://doi.org/10.1073/pnas.1810491116

Scanlan DJ, Ostrowski M, Mazard S, Dufresne A, Garczarek L, Hess WR, Post AF, Hagemann M, Paulsen I, Partensky F (2009) Ecological genomics of marine picocyanobacteria. Microbiol Mol Biol Rev 73:249–299. https://doi.org/10.1128/MMBR.00035-08

Schirmer A, Rude MA, Li X, Popova E, Del Cardayre SB (2010) Microbial biosynthesis of alkanes. Science 329:559–562. https://doi.org/10.1126/science.1187936

Schmidt K, Birchill AJ, Atkinson A, Brewin RJW, Clark JR, Hickman AE, Johns DG, Lohan MC, Milne A, Pardo S, Polimene L, Smyth TJ, Tarran GA, Widdicombe CE, Woodward EMS, Ussher SJ (2020) Increasing picocyanobacteria success in shelf waters contributes to long-term food web degradation. Glob Chang Biol 26:5574–5587. https://doi.org/10.1111/gcb.15161

Sellner KG (1997) Physiology, ecology, and toxic properties of marine cyanobacteria blooms. Limnol Oceanogr 42:1089–1104. https://doi.org/10.4319/lo.1997.42.5_part_2.1089

Shao C-L, Linington RG, Balunas MJ, Centeno A, Boudreau P, Zhang C, Engene N, Spadafora C, Mutka TS, Kyle DE, Gerwick L, Wang C-Y, Gerwick WH (2015) Bastimolide a, a potent

antimalarial polyhydroxy macrolide from the marine cyanobacterium *Okeania hirsuta*. J Org Chem 80:7849–7855. https://doi.org/10.1021/acs.joc.5b01264
Shi T, Ilikchyan I, Rabouille S, Zehr JP (2010) Genome-wide analysis of diel gene expression in the unicellular $N_2$-fixing cyanobacterium *Crocosphaera watsonii* WH 8501. ISME J 4:621–632. https://doi.org/10.1038/ismej.2009.148
Shi Y, Tyson GW, Eppley JM, DeLong EF (2011) Integrated metatranscriptomic and metagenomic analyses of stratified microbial assemblages in the open ocean. ISME J 5:999–1013. https://doi.org/10.1038/ismej.2010.189
Shih PM, Wu D, Latifi A, Axen SD, Fewer DP, Talla E, Calteau A, Cai F, Tandeau De Marsac N, Rippka R, Herdman M, Sivonen K, Coursin T, Laurent T, Goodwin L, Nolan M, Davenport KW, Han CS, Rubin EM, Eisen JA, Woyke T, Gugger M, Kerfeld CA (2013) Improving the coverage of the cyanobacterial phylum using diversity-driven genome sequencing. Proc Natl Acad Sci U S A. https://doi.org/10.1073/pnas.1217107110
Shih PM, Hemp J, Ward LM, Matzke NJ, Fischer WW (2017) Crown group Oxyphotobacteria postdate the rise of oxygen. Geobiology 15:19–29. https://doi.org/10.1111/gbi.12200
Shiozaki T, Fujiwara A, Ijichi M, Harada N, Nishino S, Nishi S, Nagata T, Hamasaki K (2018) Diazotroph community structure and the role of nitrogen fixation in the nitrogen cycle in the Chukchi Sea (western Arctic Ocean). Limnol Oceanogr 63:2191–2205. https://doi.org/10.1002/lno.10933
Shukla A, Biswas A, Blot N, Partensky F, Karty JA, Hammad LA, Garczarek L, Gutu A, Schluchter WM, Kehoe DM (2012) Phycoerythrin-specific bilin lyase-isomerase controls blue-green chromatic acclimation in marine *Synechococcus*. Proc Natl Acad Sci U S A 109:20136–20141. https://doi.org/10.1073/pnas.1211777109
Sidler WA (1994) Phycobilisome and phycobiliprotein structure. In: Bryant DA (ed) The molecular biology of cyanobacteria. Kluwer Academic Publishers, Dordrecht, pp 139–216. https://doi.org/10.1007/978-94-011-0227-8_7
Sinetova MA, Červený J, Zavřel T, Nedbal L (2012) On the dynamics and constraints of batch culture growth of the cyanobacterium *Cyanothece* sp. ATCC 51142. J Biotechnol 162:148–155. https://doi.org/10.1016/j.jbiotec.2012.04.009
Sivonen K, Kononen K, Carmichael WW, Dahlem AM, Rinehart KL, Kiviranta J, Niemela SI (1989) Occurrence of the hepatotoxic cyanobacterium *Nodularia spumigena* in the Baltic Sea and structure of the toxin. Appl Env Microbiol 55:1990–1995. https://doi.org/10.1128/aem.55.8.1990-1995.1989
Six C, Thomas JC, Garczarek L, Ostrowski M, Dufresne A, Blot N, Scanlan DJ, Partensky F (2007) Diversity and evolution of phycobilisomes in marine *Synechococcus* spp.: a comparative genomics study. Genome Biol 8:R259. https://doi.org/10.1186/gb-2007-8-12-r259
Sohm JA, Ahlgren NA, Thomson ZJ, Williams C, Moffett JW, Saito MA, Webb EA, Rocap G (2015) Co-occurring *Synechococcus* ecotypes occupy four major oceanic regimes defined by temperature, macronutrients and iron. ISME J 10:1–13. https://doi.org/10.1038/ismej.2015.115
Soo RM, Skennerton CT, Sekiguchi Y, Imelfort M, Paech SJ, Dennis PG, Steen JA, Parks DH, Tyson GW, Hugenholtz P (2014) An expanded genomic representation of the phylum cyanobacteria. Genome Biol Evol 6:1031–1045. https://doi.org/10.1093/gbe/evu073
Soo RM, Hemp J, Parks DH, Fischer WW, Hugenholtz P (2017) On the origins of oxygenic photosynthesis and aerobic respiration in cyanobacteria. Science 355:1436–1440. https://doi.org/10.1126/science.aal3794
Spungin D, Pfreundt U, Berthelot H, Bonnet S, AlRoumi D, Natale F, Hess WR, Bidle KD, Berman-Frank I (2016) Mechanisms of *Trichodesmium* bloom demise within the New Caledonia lagoon during the VAHINE mesocosm experiment. Biogeosciences 13:4187–4203. https://doi.org/10.5194/bg-13-4187-2016
Spungin D, Bidle KD, Berman-Frank I (2019) Metacaspase involvement in programmed cell death of the marine cyanobacterium *Trichodesmium*. Environ Microbiol. https://doi.org/10.1111/1462-2920.14512

Steglich C, Frankenberg-Dinkel N, Penno S, Hess WR (2005) A green light-absorbing phycoerythrin is present in the high-light-adapted marine cyanobacterium *Prochlorococcus* sp. MED4. Env Microbiol 7:1611–1618. https://doi.org/10.1111/j.1462-2920.2005.00855.x

Stöckel J, Welsh EA, Liberton M, Kunnvakkam R, Aurora R, Pakrasi HB (2008) Global transcriptomic analysis of *Cyanothece* 51142 reveals robust diurnal oscillation of central metabolic processes. Proc Natl Acad Sci U S A 105:6156–6161. https://doi.org/10.1073/pnas.0711068105

Stöckel J, Jacobs JM, Elvitigala TR, Liberton M, Welsh EA, Polpitiya AD, Gritsenko MA, Nicora CD, Koppenaal DW, Smith RD, Pakrasi HB (2011) Diurnal rhythms result in significant changes in the cellular protein complement in the cyanobacterium *Cyanothece* 51142. PLoS One 6:e16680. https://doi.org/10.1371/journal.pone.0016680

Stomp M, Huisman J, Stal LJ, Matthijs HCP (2007) Colorful niches of phototrophic microorganisms shaped by vibrations of the water molecule. ISME J 1:271–282. https://doi.org/10.1038/ismej.2007.59

Tai V, Palenik B (2009) Temporal variation of *Synechococcus* clades at a coastal Pacific Ocean monitoring site. ISME J 3:903–915. https://doi.org/10.1038/ismej.2009.35

Tang W, Cassar N (2019) Data-driven modeling of the distribution of diazotrophs in the global ocean. Geophys Res Lett 46:12258–12269. https://doi.org/10.1029/2019GL084376

Tang W, Wang S, Fonseca-Batista D, Dehairs F, Gifford S, Gonzalez AG, Gallinari M, Planquette H, Sarthou G, Cassar N (2019) Revisiting the distribution of oceanic $N_2$ fixation and estimating diazotrophic contribution to marine production. Nat Commun 10:831. https://doi.org/10.1038/s41467-019-08640-0

Taniuchi Y, Yoshikawa S, Maeda S-I, Omata T, Ohki K (2008) Diazotrophy under continuous light in a marine unicellular diazotrophic cyanobacterium, *Gloeothece* sp. 68DGA. Microbiology 154:1859–1865. https://doi.org/10.1099/mic.0.2008/018689-0

Taniuchi Y, Chen YL, Chen H-Y, Tsai M-L, Ohki K (2012) Isolation and characterization of the unicellular diazotrophic cyanobacterium group C TW3 from the tropical western Pacific Ocean. Environ Microbiol 14:641–654. https://doi.org/10.1111/j.1462-2920.2011.02606.x

Teikari JE, Fewer DP, Shrestha R, Hou S, Leikoski N, Mäkelä M, Simojoki A, Hess WR, Sivonen K (2018a) Strains of the toxic and bloom-forming *Nodularia spumigena* (cyanobacteria) can degrade methylphosphonate and release methane. ISME J 12:1619–1630. https://doi.org/10.1038/s41396-018-0056-6

Teikari JE, Hou S, Wahlsten M, Hess WR, Sivonen K (2018b) Comparative genomics of the Baltic Sea toxic cyanobacteria *Nodularia spumigena* UHCC 0039 and its response to varying salinity. Front Microbiol 9:356. https://doi.org/10.3389/fmicb.2018.00356

Thiel T (2019) Organization and regulation of cyanobacterial nif gene clusters: Implications for nitrogenase expression in plant cells. FEMS Microbiol Lett 366:fnz077. https://doi.org/10.1093/femsle/fnz077

Thompson AW, Foster RA, Krupke A, Carter BJ, Musat N, Vaulot D, Kuypers MMM, Zehr JP (2012) Unicellular cyanobacterium symbiotic with a single-celled eukaryotic alga. Science 337:1546–1550. https://doi.org/10.1126/science.1222700

Thompson A, Carter BJ, Turk-Kubo K, Malfatti F, Azam F, Zehr JP (2014) Genetic diversity of the unicellular nitrogen-fixing cyanobacteria UCYN-A and its prymnesiophyte host. Env Microbiol 16:3238–3249. https://doi.org/10.1111/1462-2920.12490

Thompson AW, van den Engh G, Ahlgren NA, Kouba K, Ward S, Wilson ST, Karl DM (2018) Dynamics of *Prochlorococcus* diversity and photoacclimation during short-term shifts in water column stratification at station ALOHA. Front Mar Sci 5:488. https://doi.org/10.3389/fmars.2018.00488

Ting CS, Rocap G, King J, Chisholm SW (2002) Cyanobacterial photosynthesis in the oceans: the origins and significance of divergent light-harvesting strategies. Trends Microbiol 10:134–142. https://doi.org/10.1016/S0966-842X(02)02319-3

Ting CS, Hsieh C, Sundararaman S, Mannella C, Marko M (2007) Cryo-electron tomography reveals the comparative three-dimensional architecture of *Prochlorococcus*, a globally important marine cyanobacterium. J Bacteriol 189:4485–4493. https://doi.org/10.1128/JB.01948-06

Toepel J, Welsh E, Summerfield TC, Pakrasi HB, Sherman LA (2008) Differential transcriptional analysis of the cyanobacterium *Cyanothece* sp. strain ATCC 51142 during light-dark and continuous-light growth. J Bacteriol 190:904–3913. https://doi.org/10.1128/JB.00206-08

Toyoshima M, Toya Y, Shimizu H (2020) Flux balance analysis of cyanobacteria reveals selective use of photosynthetic electron transport components under different spectral light conditions. Photosynth Res 143:31–43. https://doi.org/10.1007/s11120-019-00678-x

Tripp HJ, Bench SR, Turk KA, Foster RA, Desany BA, Niazi F, Affourtit JP, Zehr JP (2010) Metabolic streamlining in an open-ocean nitrogen-fixing cyanobacterium. Nature 464:90–94. https://doi.org/10.1038/nature08786

Tschoeke D, Salazar VW, Vidal L, Campeão M, Swings J, Thompson F, Thompson C (2020) Unlocking the genomic taxonomy of the *Prochlorococcus* collective. Microb Ecol 80:546–558. https://doi.org/10.1007/s00248-020-01526-5

Tully BJ, Graham ED, Heidelberg JF (2018) The reconstruction of 2,631 draft metagenome-assembled genomes from the global oceans. Sci Data 5:170203. https://doi.org/10.1038/sdata.2017.203

Turk-Kubo KA, Karamchandani M, Capone DG, Zehr JP (2014) The paradox of marine heterotrophic nitrogen fixation: abundances of heterotrophic diazotrophs do not account for nitrogen fixation rates in the eastern tropical South Pacific. Environ Microbiol 16:3095–3114. https://doi.org/10.1111/1462-2920.12346

Turk-Kubo KA, Frank IE, Hogan ME, Desnues A, Bonnet S, Zehr JP (2015) Diazotroph community succession during the VAHINE mesocosm experiment (New Caledonia lagoon). Biogeosciences 12:7435–7452. https://doi.org/10.5194/bg-12-7435-2015

Turk-Kubo KA, Farnelid HM, Shilova IN, Henke B, Zehr JP (2017) Distinct ecological niches of marine symbiotic $N_2$-fixing cyanobacterium *Candidatus* Atelocyanobacterium thalassa sublineages. J Phycol 53:451–461. https://doi.org/10.1111/jpy.12505

Urbach E, Scanlan DJ, Distel DL, Waterbury JB, Chisholm SW (1998) Rapid diversification of marine picophytoplankton with dissimilar light-harvesting structures inferred from sequences of *Prochlorococcus* and *Synechococcus* (cyanobacteria). J Mol Evol 46:188–201. https://doi.org/10.1007/PL00006294

Varkey D, Mazard S, Ostrowski M, Tetu SG, Haynes P, Paulsen IT (2016) Effects of low temperature on tropical and temperate isolates of marine *Synechococcus*. ISME J 10:1252–1263. https://doi.org/10.1038/ismej.2015.179

Vijayakumar S, Kaja-Mohideen Sheikh Mujibur Rahman P, Angione C (2020) A hybrid flux balance analysis and machine learning pipeline elucidates the metabolic response of cyanobacteria to different growth conditions. iScience 23:101818. https://doi.org/10.1016/j.isci.2020.101818

Villar E, Farrant GK, Follows M, Garczarek L, Speich S, Audic S, Bittner L, Blanke B, Brum JR, Brunet C, Casotti R, Chase A, Dolan JR, dí Ortenzio F, Gattuso JP, Grima N, Guidi L, Hill CN, Jahn O, Jamet JL, Le Goff H, Lepoivre C, Malviya S, Pelletier E, Romagnan JB, Roux S, Santini S, Scalco E, Schwenck SM, Tanaka A, Testor P, Vannier T, Vincent F, Zingone A, Dimier C, Picheral M, Searson S, Kandels-Lewis S, Tara Oceans Coordinators, Acinas SG, Bork P, Boss E, de Vargas C, Gorsky G, Ogata H, Pesant S, Sullivan MB, Sunagawa S, Wincker P, Karsenti E, Bowler C, Not F, Hingamp P, Iudicone D (2015) Environmental characteristics of Agulhas rings affect interocean plankton transport. Science 348:1261447. https://doi.org/10.1126/science.1261447

Villareal TA (1991) Nitrogen-fixation by the cyanobacterial symbiont of the diatom genus *Hemiaulus*. Mar Ecol Prog Ser 76:201–204. https://www.jstor.org/stable/i24825551

Villareal TA (1992) Marine nitrogen-fixing diatom-cyanobacteria symbioses. In: Carpenter EJ, Capone DG, Rueter J (eds) Marine pelagic cyanobacteria: *Trichodesmium* and other diazotrophs. Springer, Dordrecht, pp 163–175. https://doi.org/10.1007/978-94-015-7977-3_10

Villareal TA (1994) Widespread occurrence of the *Hemiaulus*-cyanobacterial symbiosis in the Southwest North Atlantic Ocean. Bull Mar Sci 54:1–7

Voigt K, Sharma CM, Mitschke J, Joke Lambrecht S, Voß B, Hess WR, Steglich C (2014) Comparative transcriptomics of two environmentally relevant cyanobacteria reveals unexpected transcriptome diversity. ISME J 8:2056–2068. https://doi.org/10.1038/ismej.2014.57

Voß B, Bolhuis H, Fewer DP, Kopf M, Möke F, Haas F, El-Shehawy R, Hayes P, Bergman B, Sivonen K, Dittmann E, Scanlan DJ, Hagemann M, Stal LJ, Hess WR (2013) Insights into the physiology and ecology of the brackish-water-adapted cyanobacterium *Nodularia spumigena* CCY9414 based on a genome-transcriptome analysis. PLoS One 8:e60224. https://doi.org/10.1371/journal.pone.0060224

Vu TT, Stolyar SM, Pinchuk GE, Hill EA, Kucek LA, Brown RN, Lipton MS, Osterman A, Fredrickson JK, Konopka AE, Beliaev AS, Reed JL (2012) Genome-scale modeling of light-driven reductant partitioning and carbon fluxes in diazotrophic unicellular cyanobacterium *Cyanothece* sp. ATCC 51142. PLoS Comput Biol 8:e1002460. https://doi.org/10.1371/journal.pcbi.1002460

Walworth N, Pfreundt U, Nelson WC, Mincer T, Heidelberg JF, Fu F, Waterbury JB, Glavina del Rio T, Goodwin L, Kyrpides NC, Land ML, Woyke T, Hutchins DA, Hess WR, Webb EA (2015) *Trichodesmium* genome maintains abundant, widespread noncoding DNA in situ, despite oligotrophic lifestyle. Proc Natl Acad Sci U S A 112:4251–4256. https://doi.org/10.1073/pnas.1422332112

Walworth NG, Lee MD, Dolzhenko E, Fu F-X, Smith AD, Webb EA, Hutchins DA (2020) Long-term m5C methylome dynamics parallel phenotypic adaptation in the cyanobacterium *Trichodesmium*. Mol biol Evol 38(3):927–939. https://doi.org/10.1093/molbev/msaa256

Wang M, Luan G, Lu X (2020) Engineering ethanol production in a marine cyanobacterium *Synechococcus* sp. PCC7002 through simultaneously removing glycogen synthesis genes and introducing ethanolgenic cassettes. J Biotechnol 317:1–4. https://doi.org/10.1016/j.jbiotec.2020.04.002

Waterbury J, Willey JM (1988) Isolation and growth of marine planktonic cyanobacteria. Methods Enzymol 167:105–112. https://doi.org/10.1016/0076-6879(88)67009-1

Waterbury JB, Watson SW, Guillard RRL, Brand LE (1979) Widespread occurrence of a unicellular, marine, planktonic, cyanobacterium. Nature 277:293–294. https://doi.org/10.1038/277293a0

Webb EA, Jakuba RW, Moffett JW, Dyhrman ST (2007) Molecular assessment of phosphorus and iron physiology in *Trichodesmium* populations from the western central and western South Atlantic. Limnol Oceanogr 52:2221–2232. https://doi.org/10.4319/lo.2007.52.5.2221

Webb EA, Ehrenreich IM, Brown SL, Valois FW, Waterbury JB (2009) Phenotypic and genotypic characterization of multiple strains of the diazotrophic cyanobacterium, *Crocosphaera watsonii*, isolated from the open ocean. Environ Microbiol 11:338–348. https://doi.org/10.1111/j.1462-2920.2008.01771.x

Welsh EA, Liberton M, Stöckel J, Loh T, Elvitigala T, Wang C, Wollam A, Fulton RS, Clifton SW, Jacobs JM, Aurora R, Ghosh BK, Sherman LA, Smith RD, Wilson RK, Pakrasi HB (2008) The genome of *Cyanothece* 51142, a unicellular diazotrophic cyanobacterium important in the marine nitrogen cycle. Proc Natl Acad Sci U S A 105:15094–15099. https://doi.org/10.1073/pnas.0805418105

West NJ, Lebaron P, Strutton PG, Suzuki MT (2011) A novel clade of *Prochlorococcus* found in high nutrient low chlorophyll waters in the south and equatorial Pacific Ocean. ISME J 5:933–944. https://doi.org/10.1038/ismej.2010.186

Whitehead L, Long BM, Price GD, Badger MR (2014) Comparing the in vivo function of α-carboxysomes and β-carboxysomes in two model cyanobacteria. Plant Physiol 165:398–411. https://doi.org/10.1104/pp.114.237941

Wilson ST, Aylward FO, Ribalet F, Barone B, Casey JR, Connell PE, Eppley JM, Ferron S, Fitzsimmons JN, Hayes CT, Romano AE, Turk-Kubo KA, Vislova A, Virginia Armbrust E, Caron DA, Church MJ, Zehr JP, Karl DM, De Long EF (2017) Coordinated regulation of

growth, activity and transcription in natural populations of the unicellular nitrogen-fixing cyanobacterium *Crocosphaera*. Nat Microbiol 2:17118. https://doi.org/10.1038/nmicrobiol.2017.118
Withers NW, Alberte RS, Lewin RA, Thornber JP, Britton G, Goodwin TW (1978) Photosynthetic unit size, carotenoids, and chlorophyll-protein composition of *Prochloron* sp., a prokaryotic green alga. Proc Natl Acad Sci U S A 75:2301–2305. https://doi.org/10.1073/pnas.75.5.2301
Woebken D, Burow LC, Behnam F, Mayali X, Schintlmeister A, Fleming ED, Prufert-Bebout L, Singer SW, Cortés AL, Hoehler TM, Pett-Ridge J, Spormann AM, Wagner M, Weber PK, Bebout BM (2015) Revisiting $N_2$ fixation in Guerrero Negro intertidal microbial mats with a functional single-cell approach. ISME J 9:485–496. https://doi.org/10.1038/ismej.2014.144
Xia X, Vidyarathna NK, Palenik B, Lee P, Liu H (2015) Comparison of the seasonal variations of *Synechococcus* assemblage structures in estuarine waters and coastal waters of Hong Kong. Appl Environ Microbiol 81:7644–7655. https://doi.org/10.1128/AEM.01895-15
Xia X, Guo W, Tan S, Liu H (2017a) *Synechococcus* assemblages across the salinity gradient in a salt wedge estuary. Front Microbiol 8:1254. https://doi.org/10.3389/fmicb.2017.01254
Xia X, Partensky F, Garczarek L, Suzuki K, Guo C, Cheung SY, Liu H, Yan Cheung S, Liu H (2017b) Phylogeography and pigment type diversity of *Synechococcus* cyanobacteria in surface waters of the northwestern Pacific Ocean. Environ Microbiol 19:142–158. https://doi.org/10.1111/1462-2920.13541
Xia X, Cheung S, Endo H, Suzuki K, Liu H (2019) Latitudinal and vertical variation of *Synechococcus* assemblage composition along 170° W transect from the South Pacific to the Arctic Ocean. Microb Ecol 77:333–342. https://doi.org/10.1007/s00248-018-1308-8
Yan F, Yu X, Duan Z, Lu J, Jia B, Qiao Y, Sun C, Wei C (2019) Discovery and characterization of the evolution, variation and functions of diversity-generating retroelements using thousands of genomes and metagenomes. BMC Genomics 20:595. https://doi.org/10.1186/s12864-019-5951-3
Yelton AP, Acinas SG, Sunagawa S, Bork P, Pedrós-Alió C, Chisholm SW (2016) Global genetic capacity for mixotrophy in marine picocyanobacteria. ISME J 10:2946–2957. https://doi.org/10.1038/ismej.2016.64
Yurkov V, Hughes E (2017) Aerobic anoxygenic phototrophs: four decades of mystery. In: Hallenbeck P (ed) Modern topics in the phototrophic prokaryotes. Springer, Cham, pp 193–214. https://doi.org/10.1007/978-3-319-46261-5_6
Zavřel T, Faizi M, Loureiro C, Poschmann G, Stühler K, Sinetova M, Zorina A, Steuer R, Červený J (2019) Quantitative insights into the cyanobacterial cell economy. elife 8:e42508. https://doi.org/10.7554/eLife.42508
Zeev EB, Yogev T, Man-Aharonovich D, Kress N, Herut B, Béjà O, Berman-Frank I (2008) Seasonal dynamics of the endosymbiotic, nitrogen-fixing cyanobacterium *Richelia intracellularis* in the eastern Mediterranean Sea. ISME J 2:911–923. https://doi.org/10.1038/ismej.2008.56
Zehr JP, Capone DG (2020) Changing perspectives in marine nitrogen fixation. Science 368: eaay9514. https://doi.org/10.1126/science.aay9514
Zehr JP, Kudela RM (2011) Nitrogen cycle of the open ocean: from genes to ecosystems. Annu Rev Mar Sci 3:197–225. https://doi.org/10.1146/annurev-marine-120709-142819
Zehr JP, Waterbury JB, Turner PJ, Montoya JP, Omoregie E, Steward GF, Hansen A, Karl DM (2001) Unicellular cyanobacteria fix N2 in the subtropical North Pacific Ocean. Nature 412: 635–638. https://doi.org/10.1038/35088063
Zehr JP, Bench SR, Carter BJ, Hewson I, Niazi F, Shi T, Tripp HJ, Affourtit JP (2008) Globally distributed uncultivated oceanic $N_2$-fixing cyanobacteria lack oxygenic photosystem II. Science 322:1110–1112. https://doi.org/10.1126/science.1165340
Zhao KH, Deng MG, Zheng M, Zhou M, Parbel A, Storf M, Meyer M, Strohmann B, Scheer H (2000) Novel activity of a phycobiliprotein lyase: both the attachment of phycocyanobilin and the isomerization to phycoviolobilin are catalyzed by the proteins PecE and PecF encoded by the

phycoerythrocyanin operon. FEBS Lett 469:9–13. https://doi.org/10.1016/S0014-5793(00)01245-X

Zilliges Y, Kehr J-C, Meissner S, Ishida K, Mikkat S, Hagemann M, Kaplan A, Börner T, Dittmann E (2011) The cyanobacterial hepatotoxin microcystin binds to proteins and increases the fitness of *Microcystis* under oxidative stress conditions. PLoS One 6:e17615. https://doi.org/10.1371/journal.pone.0017615

Zinser ER, Johnson ZI, Coe A, Karaca E, Veneziano D, Chisholm SW (2007) Influence of light and temperature on *Prochlorococcus* ecotype distributions in the Atlantic Ocean. Limnol Oceanogr 52:2205–2220. https://doi.org/10.4319/lo.2007.52.5.2205

Zubkov MV (2009) Photoheterotrophy in marine prokaryotes. J Plankton Res 31:933–938. https://doi.org/10.1093/plankt/fbp043

Zwirglmaier K, Jardillier L, Ostrowski M, Mazard S, Garczarek L, Vaulot D, Not F, Massana R, Ulloa O, Scanlan DJ (2008) Global phylogeography of marine *Synechococcus* and *Prochlorococcus* reveals a distinct partitioning of lineages among oceanic biomes. Environ Microbiol 10:147–161. https://doi.org/10.1111/j.1462-2920.2007.01440.x

# Marine Protists: A Hitchhiker's Guide to their Role in the Marine Microbiome

**4**

Charles Bachy, Elisabeth Hehenberger, Yu-Chen Ling, David M. Needham, Jan Strauss, Susanne Wilken, and Alexandra Z. Worden

**Abstract**

Diversity within marine microbiomes spans the three domains of life: microbial eukaryotes (i.e., protists), bacteria, and archaea. Although protists were the first microbes observed by microscopy, it took the advent of molecular techniques to begin to resolve their complex and reticulate evolutionary history. Symbioses between microbial entities have been key in this journey, and such interactions continue to shape the ecology of marine microbiomes. Nowadays, photosynthetic marine protists are appreciated for their activities as primary producers, rivalling land plant contributions in the global carbon cycle. Predatory protists are known for consuming prokaryotes and other protists, with some combining metabolisms into a mixotrophic lifestyle. Still, much must be learned about specific interactions and lifestyles, especially for uncultured groups recognized just by environmental sequences. With respect to the fate of protists in food webs, there are many paths to consider. Despite being in early stages of identifying

C. Bachy · E. Hehenberger · Y.-C. Ling · D. M. Needham · J. Strauss
Ocean EcoSystems Biology Unit, RD3, GEOMAR Helmholtz Centre for Ocean Research Kiel, Kiel, Germany
e-mail: ehehenberger@geomar.de; yling@geomar.de; dneedham@geomar.de; jstrauss@geomar.de

S. Wilken
Faculty of Science, Institute for Biodiversity and Ecosystem Dynamics, Amsterdam, Netherlands
e-mail: s.wilken@uva.nl

A. Z. Worden (✉)
Ocean EcoSystems Biology Unit, RD3, GEOMAR Helmholtz Centre for Ocean Research Kiel, Kiel, Germany

Kiel University, Kiel, Germany

Max Planck Institute for Evolutionary Biology, Plön, Germany
e-mail: azworden@geomar.de



L. J. Stal, M. S. Cretoiu (eds.), *The Marine Microbiome*, The Microbiomes of Humans, Animals, Plants, and the Environment 3,
https://doi.org/10.1007/978-3-030-90383-1_4

interactions, whether mutualistic or death-inducing infections by parasites and viruses, knowledge is advancing rapidly via methods for interrogation in nature without culturing. Here, we review marine protists, their evolutionary histories, diversity, ecological roles, and lifestyles in all layers of the ocean, with reference to how views have shifted over time through extensive investigation.

**Keywords**

Carbon cycle · Eukaryotic evolution · Marine food webs · Phytoplankton · Protistan evolution · Protistan interactions

*Many were increasingly of the opinion that they'd all made a big mistake in coming down from the trees in the first place. And some said that even the trees had been a bad move, and that no one should ever have left the oceans.*

From *A hitchhiker's guide to the galaxy—by Douglas Adams, 1979.*

## 4.1 Introduction: The Poetry and Beauty of Protists Through Time

Microbial communities comprise three domains of life: the Archaea and the Bacteria—domains containing cells that lack a nucleus—and single-celled Eukarya—organisms that do possess a nucleus—which are also known as protists. Protists are found throughout the ocean and in terrestrial environments, where they fulfill a vast array of ecological roles due to the wide variety of physiological capacities they collectively possess. The rise of eukaryotes is linked to the great oxygenation event (GOE), wherein cyanobacterial oxygenic photosynthesis drove a massive change in environmental conditions (Luo et al. 2016). The GOE spurred the advent of new metabolisms that functioned in an oxygenated environment, through merging of different cell types, resulting in eukaryotic cells which used oxygen for respiration (Planavsky et al. 2021). Symbioses played and still play major roles in diversification, including most known unicellular eukaryotic marine lineages. Protists not only have a nucleus, but other organelles as well, such as a mitochondrion where respiration occurs and a chloroplast (plastid) if photosynthetic. They can also have other "arrangements" involving more ephemeral capture and use of plastids. Interactions of protists with bacteria, archaea, viruses as well as with multicellular taxa in the ocean influence biogeochemical cycling, mortality, and even the metabolic potential of the protists themselves. Finally, protists can adopt a variety of cellular "organizations" ranging from free-living cells to colonial forms, with some being able to transform into truly multicellular forms and back to single cells. This breadth of interactions, behaviors, and relationships, connecting into deep time, drive the remarkable level of diversity that we see today. Known ecological roles of protists range from primary producers to predators, as well as from parasites and decomposers to organisms that blend multiple trophic modes. Although different "supergroups" can be parsed across eukaryotes based on their phylogenetic relationships, these divisions do not necessarily link to distinct trophic strategies, generating an even greater complexity with which microbial ecologists and microbiome scientists must grapple. In this chapter, key facets behind protistan diversity are introduced with a sprinkling of historical concepts as they have emerged and shifted through the centuries.

Through the 19th and 20th centuries scientists moved beyond skepticism about Antonie van Leeuwenhoek's (1632–1723) initial reports of having observed microbes or "animalcules"—as he called them—to deep appreciation of unicellular diversity. The studies by this Dutch microbiologist shifted from being infamous to famous as other scientists gained the possibility to observe microscopic cells, something upon which van Leeuwenhoek had a marked lead. Scholars and scientists then focused on how such microorganisms might influence the health of humans or on what these organisms could tell us about how cells function. Notable studies from individuals like Ernst Haeckel, who is credited with introducing the term "Protista", had a profound impact on both the scientific community and the public. Indeed, these scientific discussions and illustrations brought to light the infinite variety of microscopic organisms in which artists found objects of wonder, providing inspiration for Art Nouveau, a style of art and architecture that spread in Europe in the late nineteenth century. There are still traces today of how it took to the streets (Fig. 4.1), as well as how it permeated the salons of Europe as captured in the works of literary greats like Marcel Proust (Box 4.1). A little over 100 years from the publications of Haeckel and Proust, knowledge of protistan diversity and how protists arose and evolved is still advancing.

**Box 4.1**

"Ma tante n'habitait plus effectivement que deux chambres contiguës, restant l'après-midi dans l'une pendant qu'on aérait l'autre. C'étaient de ces chambres de province qui,—de même qu'en certains pays des parties entières de l'air ou de la mer sont illuminées ou parfumées par des myriades de protozoaires que nous ne voyons pas,—nous enchantent des mille odeurs qu'y dégagent les vertus, la sagesse, les habitudes, toute une vie secrète, invisible, surabondante et morale que l'atmosphère y tient en suspens; ..."

"My aunt's life was now practically confined to two adjoining rooms, in one of which she would rest in the afternoon while the other was being aired. They were rooms of that country order which (just as in certain climes whole tracts of air or ocean are illuminated or scented by myriads of protozoa which we cannot see) fascinate our sense of smell with the countless odors springing from their own special virtues, wisdom, habits, a whole secret system of life, invisible, superabundant and profoundly moral, which their atmosphere holds in solution;...".

From *Swann's Way—Combray—by Marcel Proust, first published in 1913.*

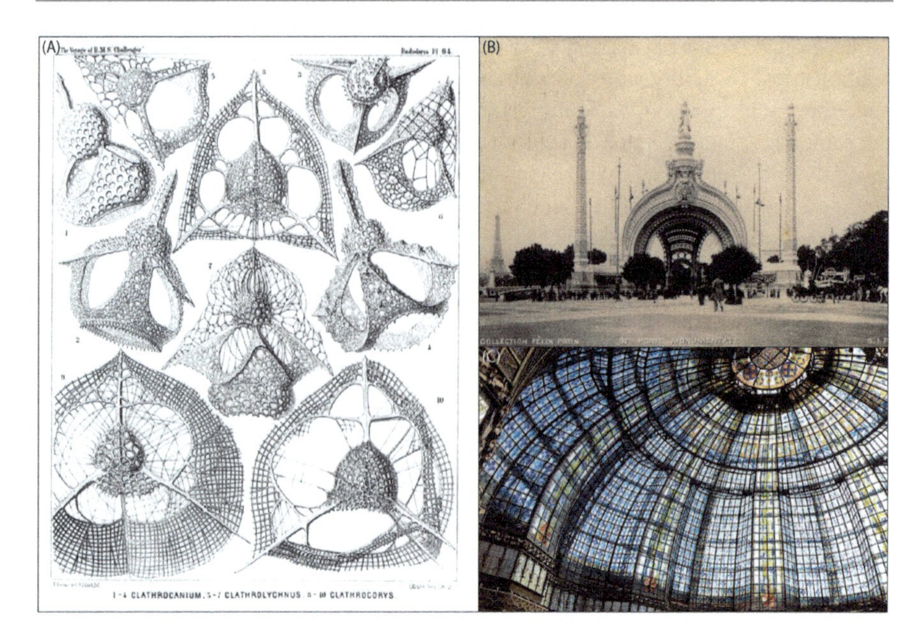


**Fig. 4.1** Art Nouveau inspired by protists—The first images of protists, even though "just" drawings, had a massive influence on the Art Nouveau (also called Modern style). (**a**) The plate depicts Radiolaria specimens as provided by Ernst Haeckel to the report on the scientific expedition made aboard the sailing ship *H.M.S.* Challenger (Haeckel 1887), which circumnavigated the globe between 1872 and 1876, traversing 70,000 nautical miles. This expedition is widely considered to have laid the foundations of oceanography as a discipline. (**b**) The radiolarian named *Clathrocanium reginae* inspired the architect René Binet to design the entrance for the 1900 Paris Exposition (photographer unknown), which was dismantled after the exhibition. There are still traces of the Art Nouveau architectural style in Europe, for example (**c**) the ceiling at the Brasserie Printemps in Paris, that appears to be inspired by the skeleton of the Radiolarian *Sethophormis eupilium* (*Litharachnium eupilium*) described by Haeckel

Marine protists are typically categorized by size class, trophic mode, or phylogenetic lineage, or using trait-based approaches which might integrate information from more than one of these categories. To what extent do marine microbiome scientists need to think about the evolutionary origins and relationships between protists, or between protists and other biological entities? Understanding of evolutionary relationships between eukaryotic groups has been advancing rapidly (Brown et al. 2018; Burki et al. 2020; Strassert et al. 2019). After more than a century of observations with the naked eye and successive generations of microscopes (Fig. 4.2), this improvement is largely due to application of the latest technologies in genomics, single cell/population genomics and transcriptomics to protists, as well as improvements in evolutionary models and methods for inferring phylogenetic relationships (Bhattacharya et al. 2014; Burki 2017; Cooney et al. 2020; Cuvelier et al. 2010; Gawryluk et al. 2019; Keeling et al. 2014; Krabberød et al. 2017).

Nevertheless, it might still seem attractive to keep studies of the dynamics in marine environments and future trajectories to a purely mechanistic science, in this

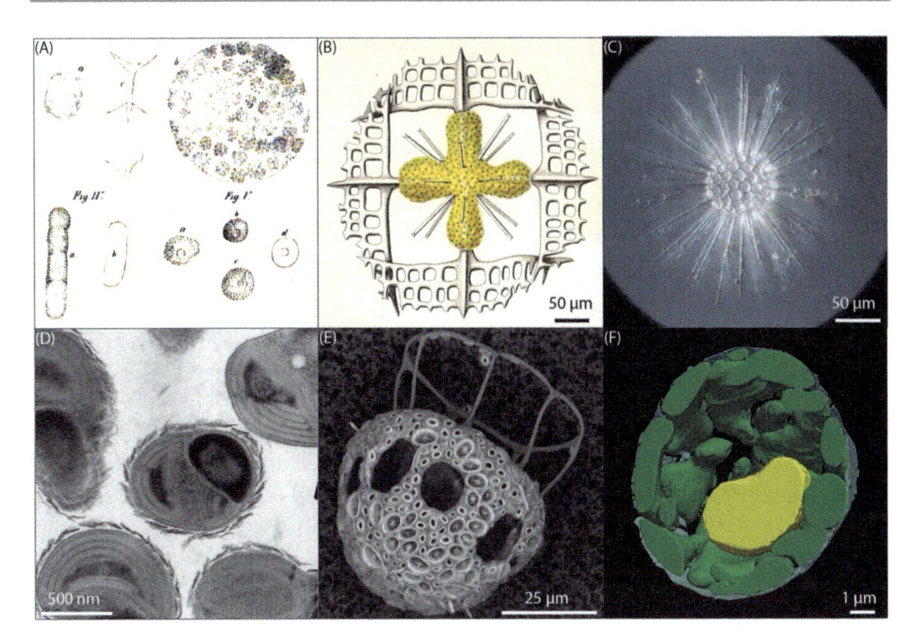


**Fig. 4.2** The beauty of protists—Antonie van Leeuwenhoek interpreted his first descriptions of motile microbial eukaryotes observed through his microscope as being little animals or "animalcules" (Van Leeuwenhoek 1677). Thereafter, a panoply of methods made it possible to show others the beauty and meticulous details of marine protists. (**a**) One of the first formal descriptions depicting planktonic specimens of collodarians (from Meyen (1834)). (**b**) Haeckel's drawing of *Lithoptera mulleri* and the symbiotic algae it contains (Haeckel and Ernst Heinrich Philipp 1862). (**c**) A *Cladococcus* skeleton from deep waters of Villefranche-sur-Mer imaged using an optical microscope (courtesy of John R. Dolan). (**d**) A Transmission Electron Microscopy image of the picophytoplanktonic prasinophyte alga *Bathycoccus calidus* (Worden Lab in collaboration with Danielle M. Jorgens). (**e**) Scanning Electron Microscopy image of a ciliate, the heterotrophic tintinnid *Dictyocysta*, covered with coccoliths from its haptophyte prey (image, Charles Bachy). (**f**) A three-dimensional representation of a haptophyte in the *Phaeocystis* genus and its subcellular structures (green, plastids and yellow, nucleus) found in endosymbiosis with a radiolarian similar to that in panel B (from Decelle et al. 2019, Copyright: J. Decelle & C. Uwizeye)

case by characterizing the cell biology and activities of microbial eukaryotes that exist today. However, information on the evolutionary origins of microbial eukaryotes provides important insights into how roles and capabilities change and how they might continue to change with accelerated perturbations to the ocean environment. Just as an example, evolutionary studies have revealed that the apicomplexan parasites that causes Malaria (*Plasmodium*) are descended from free-living algae (Dorrell et al. 2014; Gardner et al. 2002; McFadden et al. 1996). *Plasmodium* species still maintain a chloroplast vestige (termed apicoplast), although it has lost the genes related to photosynthetic function. These evolutionary changes to the primary nutrition of *Plasmodium* are fascinating and profoundly important in terms of the change in its ecological role even if the timeframe and triggers remain unclear. In fact, loss of photosynthesis has been documented in multiple marine protistan lineages, many having close relatives that are still photosynthetic, as well as other relatives that never were photosynthetic (Hadariová et al. 2018; Worden et al. 2015).

Recently, efforts have been made to summarize how genomic and transcriptomic datatypes could improve exploration of the physiology of marine protists (Caron et al. 2017) and be applied to study the ecological consequences of their activities and interactions with other biological entities (Worden et al. 2015). Indeed, single cell methods are being used to query possible trophic modes and cell attributes (Labarre et al. 2021; Seeleuthner et al. 2018; Sieracki et al. 2019; Wideman et al. 2019, 2020) and to identify host–virus pairs alongside virally-encoded molecular pathways that shape host biology (Castillo et al. 2019; Needham et al. 2019a, b; Yoon et al. 2011). Based on science to date we know that the ocean harbors a wealth of complex interactions that determine the composition and functioning of the marine microbiome, and most of which await elucidation. The importance of understanding ocean ecosystems was outlined by Dawn Wright in her examination of the challenges that lie ahead in ocean science (Box 4.2). These challenges extend to ocean microbial interactions and the need to study them in a contextualized manner reaching far beyond discovery science, because ocean productivity and biogeochemical cycling are driven by the microbiome and the evolution of the organisms within!

**Box 4.2**

"Today we map the ocean not only to increase fundamental scientific understanding of the ocean system but also to protect life and property, promote economic vitality, and inform ecosystem-based management and policy."

From *Swells, Soundings, and Sustainability, but... "Here Be Monsters"—by Dawn Wright, published in Oceanography (Wright* 2017*)*.

## 4.2 Evolutionary Relationships among Protists

The term "Protista" has been widely embraced since its introduction by Haeckel, who felt a term was needed to describe and group together the living beings that were neither animals nor plants (the latter then also including fungi). However, alongside this regrouping and other early attempts to classify organisms is the notion that protists represent "lower" and/or more ancient life forms, a view that ignores their immense diversity and complexity.

### 4.2.1 A Historical Perspective on Protistan Diversity

For many years the only described species of eukaryotes belonged to well-studied lineages of macroscopic animals, plants, and fungi, which represent only a minority of eukaryotic diversity. Study of the microbial world including the plethora of unicellular eukaryotes was hampered by the tiny size of its members and the availability of technology with appropriate resolution powers. After Leeuwenhoek's discoveries, protists were typically classified into distinct lineages based on their morphology and nutritional mode. In many studies, their naming and placement depended on the researcher's primary discipline, with zoologists and botanists using different taxonomic contexts to group photosynthetic taxa (named algae or phytoplankton) apart from heterotrophic taxa (named protozoa, Box 4.3).

**Box 4.3**

"Es ist schon von verschiedenen Seiten darauf aufmerksam gemacht worden, dass es sowohl für die Zoologie als für die Botanik ein großer Gewinn sein würde, wenn man die vielen zweifelhaften Lebewesen, die weder echte Tiere noch echte Pflanzen sind, in einem besonderen Mittelreiche oder Urwesenreiche vereinigen würde; ...".

"It has been pointed out by various parties, that both the fields of zoology and botany would greatly benefit, if the many dubious life forms, that are neither true animals nor true plants, were united in a special middle kingdom or kingdom of primeval life forms...".

From *Generelle Morphologie der Organismen: allgemeine Grundzüge der organischen Formen-Wissenschaft, mechanisch begründet durch die von C. Darwin reformierte Descendenz-Theorie. 1. Band (Berlin: Druck und Verlag von Georg Reimer)—by Ernst H. P. Haeckel, 1866.*

Major strides were made with comparative morphology studies and ultimately the application of transmission electron microscopy to photosynthetic protists, which was pioneered by Irene Manton in the 1950s. Manton was soon joined by Dorothy Pitelka, who studied heterotrophic protists, and collectively their efforts gave access to the subcellular scale, allowing morphological descriptions to be greatly improved. However, the enormous morphological diversity and often inconsistent descriptors thwarted attempts to elucidate the relationships between those groupings leading to protists being brought together as one kingdom of "lower" organisms at the base of animals, plants, and fungi (Whittaker 1969) (Fig. 4.3). Ultimately, protist lineages were regrouped and refined, although with disagreement, based on ultrastructural characteristics (Corliss 1984).

### 4.2.2 Developments in the Understanding of Evolution of Protists

The scientific view of protists being "lower" organisms was challenged by results from molecular phylogenetic analyses, the first of which used the gene coding for the small subunit of the ribosome, i.e., 18S rRNA, to resolve relationships, as reviewed in Taylor (2003). Using 18S rRNA genes, several research teams simultaneously

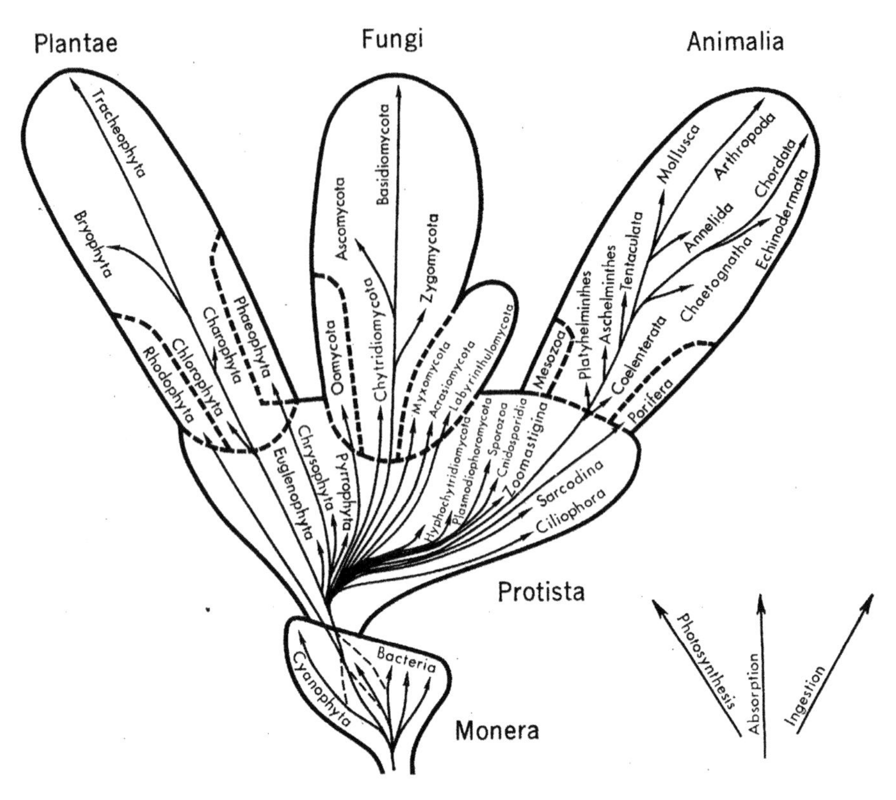


**Fig. 4.3** The classical five-kingdom system of the twentieth century—Now outdated and replaced, this view of the relationships between different life forms, as depicted by Robert Whittaker (Whittaker 1969), was a standard feature of biology textbooks in the last decades of the twentieth century. The advent of molecular phylogenetics and sequencing led to multiple revisions, and only with phylogenetics and studies of uncultivated taxa has a more stable framework been resolved, although there is still much to be learnt—especially as we try to understand the relationships between eukaryotes and how they arose from an evolutionary perspective

reported that there was considerable marine eukaryotic diversity even after pre-filtering water samples through 5, 3, or even 2 μm pore size filters to capture just the cells smaller than these pore sizes (Díez et al. 2001; López-García et al. 2001; Moon-van der Staay et al. 2001). While marine microbiologists had long been aware of small eukaryotic cells, particularly phytoplankton (Knight-Jones and Walne 1951; Murphy and Haugen 1985; Takahashi and Bienfang 1983), the extent of their diversity was shocking to the scientific world, as was the identification of sequences coming from small presumably heterotrophic protists. Collectively, environmental molecular 18S rRNA sequencing of marine and other habitats also revealed that the vast majority of eukaryotic phylogenetic diversity lies within the protists, see Sogin (2015).

It took the advent of "phylogenomics," phylogenetic reconstructions based on multiple homologous genes from each organism in the analysis (10s to 100s, or sometimes more), concatenated into datasets of ever-increasing size, to establish a clearer picture of the tree of eukaryotes (Baldauf 2003; Chan and Ragan 2013; Delsuc et al. 2005; Philippe et al. 2011). This picture was very different from prior conceptions wherein there was a concept of "crown eukaryotes" (bringing together many multicellular and macroscopic groups) versus "simple eukaryotes," which were protistan. These views were dismantled with implementation of improved methods for estimating evolutionary distances. Increasingly protists were shown to be interspersed with various macroscopic lineages throughout the eukaryotic tree, or in basal branches, or large groups with no known macroscopic lineages, e.g., Baldauf (2003). Indeed, phylogenomic analyses distributed eukaryotic diversity into a small number of enormous groups that were informally termed "supergroups." Thus, in contrast to the tumultuous history of eukaryotic tree structure through the centuries, its overall structure now changes more modestly, due to the establishment of the supergroup concept (Adl et al. 2005; Keeling et al. 2005; Simpson and Roger 2004). Nevertheless, the resolution of relationships continues to improve due to analytical advancements and genome and transcriptome sequencing of new groups of protists (Burki et al. 2020; Janouškovec et al. 2017; Lax et al. 2018).

### 4.2.3 Major Groups of Eukaryotes as of "Currently"

A well-resolved framework amenable to the phylogenetic study of eukaryotic relationships now exists, see Burki et al. (2020) and Strassert et al. (2019). This is central to the study of the numerous aspects of eukaryote evolution such as the invention of multicellularity (Brunet and King 2017; Ros-Rocher et al. 2021) and sex (Goodenough and Heitman 2014), but it is also an essential tool to interpret the diversity of environmental sequence data in ecological communities (Worden et al. 2015). A factor that hinders understanding of eukaryotic evolution is the propensity of protists to engage in endosymbioses events (Worden et al. 2015), just as eukaryogenesis itself is hypothesized to have occurred through endosymbiosis, or potentially other mechanisms of fusion, between different prokaryotic cell types (Imachi et al. 2020; López-García et al. 2017; Spang et al. 2015; Zaremba-Niedzwiedzka et al. 2017). Eukaryogenesis and discussions about the Last Eukaryotic Common Ancestor (LECA) will not be covered in this chapter; several excellent publications are available on these topics (López-García et al. 2017; O'Malley et al. 2019).

### 4.2.4 The Contribution of Plastid Acquisition and Evolution to the Generation of Eukaryotic Diversity

A key event in the natural history of eukaryotes was the uptake and ultimate integration of a cyanobacterium by a eukaryotic cell about a billion years ago. The endosymbiosis between a cyanobacterium and its eukaryotic host gave rise to the so-called primary plastid (or chloroplast) that is found in land plants, green and red

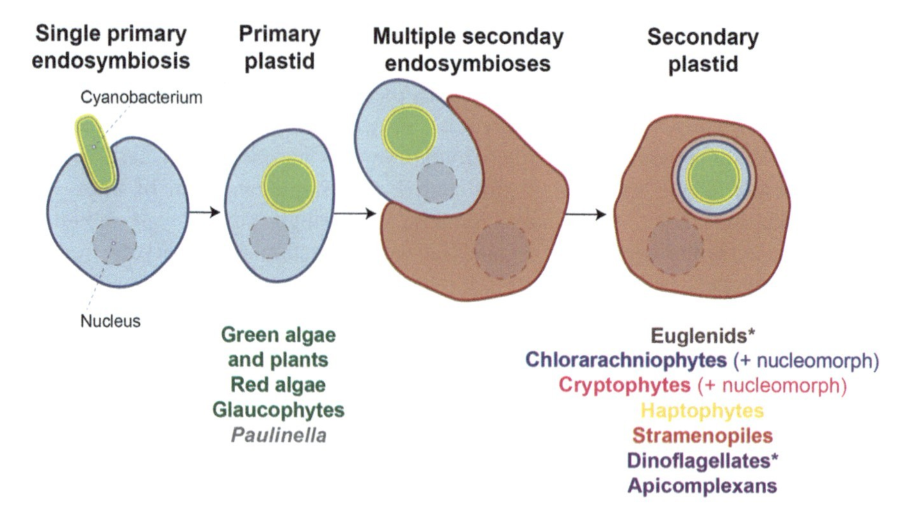


**Fig. 4.4** Evolutionary history of algal endosymbiosis and putative plastid losses—Mitochondria and plastids both are hypothesized to have arisen from the endosymbiotic uptake of different bacteria (an alphaproteobacterium and a cyanobacterium, respectively). These and subsequent events are still difficult to resolve, for example integration of the alphaproteobacterium that gave rise to the mitochondrion has recently been proposed to have occurred by a mechanism other than endosymbiosis. While plastid origins are still considered as being through endosymbiosis, tracing the evolution of plastids has been complicated by additional endosymbiosis events, and loss of photosynthesis in some lineages. The original or "primary" plastid that descended directly from the cyanobacterial endosymbiont is found in the archaeplastids (glaucophytes, red algae, green algae, and plants). But green and red algae have themselves been taken up by other eukaryotic lineages, resulting in "secondary" plastids characterized by the additional membranes and more complex protein-targeting systems present in euglenids, chlorarachniophytes, cryptophytes, haptophytes, stramenopiles, dinoflagellates, and apicomplexans. Some members of these groups are predatory mixotrophs, and others are purely heterotrophic (predatory, saprotrophic, or even parasitic) because photosynthesis or plastids have been lost, or had never been acquired. Although green algae are common in marine environments (e.g., picoprasinophytes such as *Bathycoccus*, *Ostreococcus*, and *Micromonas*), lineages resulting from secondary endosymbiotic partnerships include other important marine primary producers (e.g., diatoms, pelagophytes, haptophytes, and dinoflagellates) and represent incredible metabolic versatility. Secondary plastids are usually surrounded by four membranes but the asterisk (*) denotes those with three membranes, specifically the euglenids and dinoflagellates (Keeling 2013). It is interesting to speculate that the redundancy and reshuffling of characteristics resulting from mergers of distinct eukaryotic lineages favors new combinations of traits with strong ecological potential

algae, and glaucophytes (Fig. 4.4). Thus, this event was the source of photosynthesis as we know it in plants and algae, see, e.g., Sagan (1967); Sibbald and Archibald (2020) and the starting point for the evolution of diverse algal groups spread all over the eukaryotic tree of life, most of which thrived in the ocean. Much of that diversity arose because primary plastids have subsequently been "moved into" various other lineages via a eukaryote–eukaryote endosymbiosis, i.e., when an alga with a primary plastid was engulfed by another non-photosynthetic eukaryote, which then retained that alga (after quite some reductions in gene content) in what is termed a "secondary" plastid (Sibbald and Archibald 2020). Some lineages took this strategy a step further and engulfed secondary plastid-bearing algae in a tertiary endosymbiosis. The existence of even higher-order endosymbioses and plastids has been proposed in attempts to explain the distribution of plastids as seen in the current tree of eukaryotes (Archibald 2006; Ševčíková et al. 2015; Stiller et al. 2014).

Did multiple and higher-order endosymbiotic events occur? The answer to this question lies in the incongruence between the phylogenies of plastids themselves and the phylogenies of the algae hosting said plastids. While plastid phylogenies indicate that all plastids are seemingly closely related, the nuclear genomes of the algal hosts are not. Plastid-hosting lineages are found across the eukaryotic tree nested within lineages lacking plastids and heterotrophic lineages, resulting in a seemingly random distribution across the eukaryotic tree (Fig. 4.5). It is thought that there must have been at least two additional endosymbiotic events after the primary endosymbiosis, if not more, because both red and green algae have been taken up by other eukaryotes (Keeling 2013; Lane and Archibald 2008). Although the spread of the green primary plastid via secondary endosymbiosis is generally well understood, there are still some mysterious features, likely because we are still discovering photosynthetic lineages (Choi et al. 2017; Kim et al. 2011a), and because other taxa that are keys to the puzzle may now be extinct.

The evolutionary history of red algal plastids is highly contentious and has been discussed in dedicated publications, e.g., Sibbald and Archibald (2020). While the exact evolutionary path that plastids of red algal origin forged through different parts of the eukaryotic tree may not be known, detection of their presence (albeit not their absence) is usually more straightforward. Hence, the discovery of plastids in unexpected areas of the eukaryotic tree has provided essential pieces that bring the field closer to solving the puzzle of plastid evolution. "Red" plastids originating from secondary or even higher-order symbioses are found in cryptophytes (supergroup Cryptista), haptophytes (supergroup Haptista), and several lineages of alveolates and stramenopiles (Fig. 4.5). Application of multiple molecular biology methods allowed the discovery of novel branches of uncultured marine protists bearing plastids of probable "red" origin, such as the Rappemonads (Kawachi et al. 2021; Kim et al. 2011a) and other deep-branching plastid lineages (DPL1 and 2 (Choi et al. 2017)). However, these environmental clades await formal morphological descriptions. Most of the described lineages with red plastids are free-living photosynthetic algae, but not all plastid-bearing protists have retained the ability to perform photosynthesis and some are not even free-living. For example, it was discovered that a group of obligate intracellular parasites, the apicomplexans (e.g., *Plasmodium* and *Toxoplasma*), which are part of the major group Alveolates, have a plastid (McFadden et al. 1996). Subsequently, a free-living marine coral-sediment associated genus of photosynthetic protists, *Chromera*, was discovered, and found to be a close relative of apicomplexans and also helped to establish the red algal origin of the *Plasmodium* plastid (Janouskovec et al. 2010; Moore et al. 2008). Comparisons between *Plasmodium* and the chromerids (Janouškovec et al. 2015) have since led to the proposal that disruptions to the unique chloroplast transcript

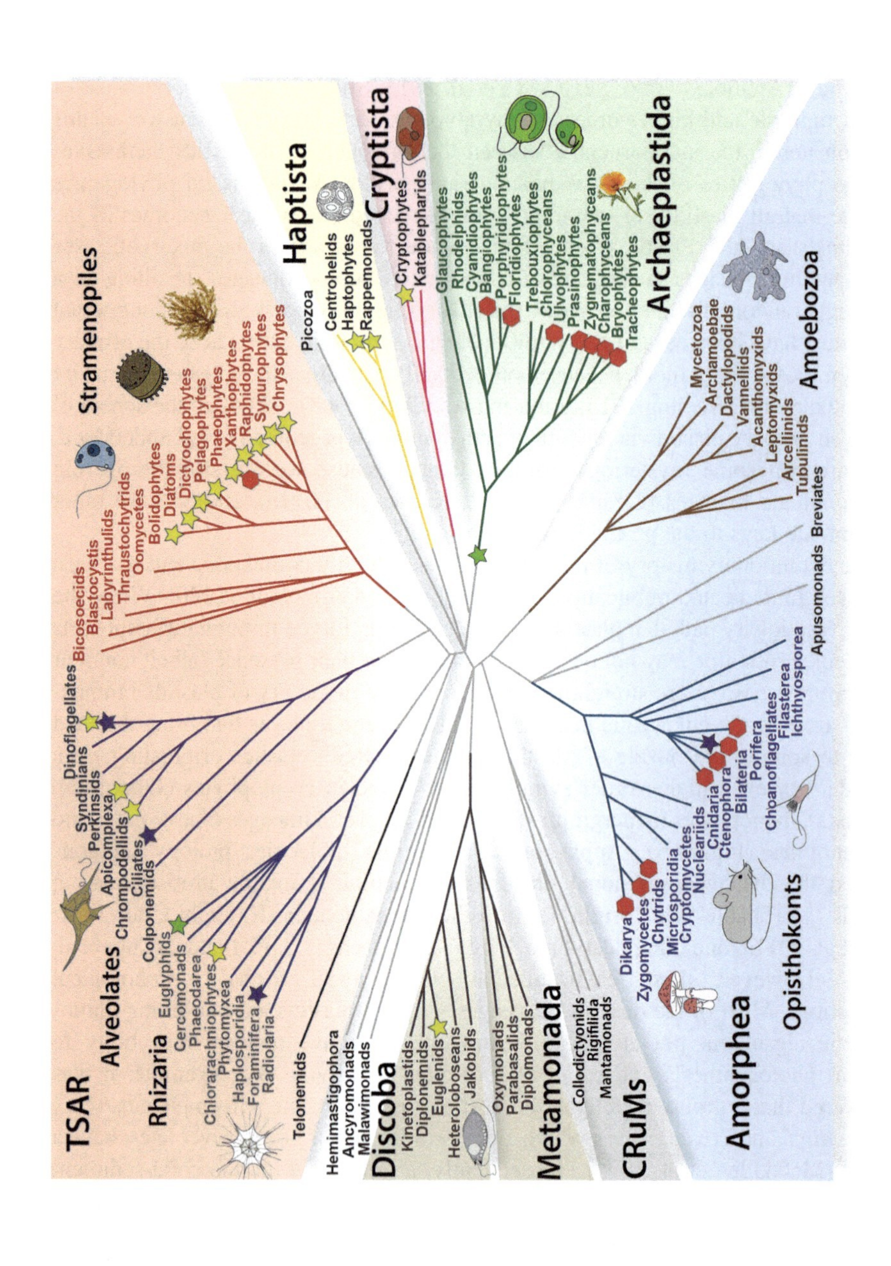
TSAR
Stramenopiles
Bicosoecids
Blastocystis
Labyrinthulids
Thraustochytrids
Oomycetes
Bolidophytes
Diatoms
Dictyochophytes
Pelagophytes
Phaeophytes
Xanthophytes
Raphidophytes
Synurophytes
Chrysophytes
Picozoa
Haptista
Centrohelids
Haptophytes
Rappemonads
Cryptista
Cryptophytes
Katablepharids
Glaucophytes
Rhodelphids
Cyanidiophytes
Bangiophytes
Porphyridiophytes
Floridiophytes
Trebouxiophytes
Chlorophyceans
Ulvophytes
Prasinophytes
Zygnematophyceans
Charophyceans
Bryophytes
Tracheophytes
Archaeplastida
Alveolates
Dinoflagellates
Syndinians
Perkinsids
Apicomplexa
Chrompodellids
Ciliates
Colponemids
Rhizaria
Euglyphids
Cercomonads
Phaeodarea
Chlorarachniophytes
Phytomyxea
Haplosporidia
Foraminifera
Radiolaria
Telonemids
Hemimastigophora
Ancyromonads
Malawimonads
Discoba
Kinetoplastids
Diplonemids
Euglenids
Heteroloboseans
Jakobids
Oxymonads
Parabasalids
Diplomonads
Metamonada
Collodictyonids
Rigifilida
Mantamonads
CRuMs
Mycetozoa
Archamoebae
Dactylopodids
Vannellids
Acanthomyxids
Leptomyxids
Arcellinids
Tubulinids
Amoebozoa
Breviates
Apusomonads
Dikarya
Zygomycetes
Chytrids
Microsporidia
Cryptomycetes
Nucleariids
Cnidaria
Ctenophora
Bilateria
Porifera
Choanoflagellates
Filasterea
Ichthyosporea
Opisthokonts
Amorphea

**Fig. 4.5** The eukaryotic tree of life—Eukaryote diversity has been subdivided into major lineages, often referred to as "supergroups." The supergroup concept emerged in large part due to support gained for relationships through the advent of phylogenomic methods. The Archaeplastida, containing land plants and green algae, red algae and glaucophytes, is distinguished by their primary plastids (green stars). A second, independent origin of a primary plastid has been reported within the Rhizaria, in the amoeba *Paulinella* (Nowack et al. 2008). The primary plastids within the Archaeplastida were spread across the tree of eukaryotes via secondary endosymbiotic events, i.e., the uptake of (red or green) primary algae by other eukaryotic lineages (see Fig. 4.4), and can be found in ecologically important marine lineages such as diatoms and dinoflagellates. Plastid remnants are also present in some unexpected groups such as the obligatory parasitic apicomplexans. Secondary plastids have spread further by tertiary endosymbiotic events and potentially even higher-order events (secondary and higher plastids are indicated by yellow stars). Purple stars mark lineages performing kleptoplasty, a specific form of mixotrophy. Such lineages sequester plastids from their prey and maintain them for variable amounts of time. Lineages containing at least one taxon that is multicellular are indicated by red hexagons. Although multicellularity has arisen more than once, and thus appears in a variety of lineages, the vast majority of eukaryotic life is unicellular. Figure adapted from Worden et al. (2015) with updates from Burki et al. (2020); Derelle et al. (2016); Keeling and Burki (2019); Parfrey and Lahr (2013); Sibbald and Archibald (2020)

processing machinery seen in chromerids resulted in the loss of photosynthesis in the ancestors of parasitic apicomplexans (Dorrell et al. 2014).

The major eukaryotic groups that are recognized today can be briefly characterized as follows. Note that these groupings are constantly being revised as new information and methods become available.

TSAR represents a combination of SAR, which stands for the clade uniting Stramenopiles, Alveolates, and Rhizaria, with the Telonemids. SAR emerged early after the implementation of phylogenomic methods and has been estimated to encompass half of all eukaryote diversity. The Telonemids, which contain only two described species and are considered to be a widespread (but not abundant) lineage of heterotrophic flagellates, were proposed to be a sister lineage of SAR, creating the even larger group TSAR (Strassert et al. 2019).

Stramenopiles comprise well-known microbial algae (e.g., diatoms and chrysophytes) but also macroscopic multicellular seaweeds (e.g., kelps) as well as an enormous diversity of free-living heterotrophic and mixotrophic protists. They also include several important pathogens of animals and plants, such as the infamous oomycete *Phytophthora*, causative agent of the potato blight that led to the Great Famine in the mid-nineteenth century Ireland. Stramenopiles also comprise many sequences from enigmatic groups called MASTs, which stands for MArine STramenopiles. MASTs were initially identified in environmental clone libraries and the organisms themselves remain largely uncultured (Massana et al. 2002). Currently there are 18 recognized MAST clades and only a few have been further elucidated in terms of functional roles, although most are considered to be heterotrophic and some have been demonstrated to consume bacterial cells using culture independent methods (Labarre et al. 2021; Lin et al. 2012; Massana et al. 2014). A distinctive feature of stramenopiles are two flagella (although they are not always present!) of unequal length (hence the alternative name “heterokont”) and hair-like structures termed mastigonemes on the longer flagellum (hence stramenopiles, or “straw-like pili”) (Cavalier-Smith 2018).

Alveolates include three hugely diverse and well-studied protist groups (ciliates, dinoflagellates, and apicomplexans) plus several smaller groups of parasites and flagellates. Ciliates are a major group of microbial predators and grazers in all known environments. Dinoflagellates are also extremely abundant in nature with a variety of lifestyles and include probably the most abundant marine eukaryotes, the parasitic MALVs (for Marine Alveolates). All apicomplexans described thus far are obligately associated with animals, most commonly as intracellular parasites (e.g., *Plasmodium*).

Rhizarians comprise a wide diversity of predominantly amoeboid protists with thin pseudopodia used more for feeding than for locomotion. This group also includes parasites of crop plants and invertebrates (e.g., *Plasmodiophora*) and even algae (chlorarachniophytes). In marine ecosystems the most prominent members of rhizarians are planktonic organisms such as the foraminifera and radiolarians that have been extensively described from the fossil record and are observed in open ocean and intertropical waters (Burki et al. 2016; Cavalier-Smith et al. 2015).

Haptista contains two main lineages: the haptophytes, which includes the still enigmatic uncultured rappemonads (Kim et al. 2011a), and the centrohelids (Burki et al. 2016; Cavalier-Smith et al. 2015). Haptophytes are mostly marine species that perform photosynthesis and can bloom to high density. Much attention has been given to the calcifying coccolithophorids (e.g., *Emiliania huxleyi*) because of their unique role in the biogeochemical cycles (connected to their calcium carbonate coccoliths) and their resulting sensitivity to climate change (Read et al. 2013; Taylor et al. 2017). Marine non-calcifying haptophytes are also important and diverse pico- and nano-phytoplankton (Cuvelier et al. 2010; Jardillier et al. 2010; Liu et al. 2009), and some appear to consume other cells (Hartmann et al. 2013). The other main Haptista lineage, the centrohelids, contains free-living heterotrophic protists characterized by distinctive radiating pseudopodia most often found in freshwater environments (Cavalier-Smith and von der Heyden 2007).

Cryptista contains the cryptomonads, algae best known for their red algal-derived plastids that retain a relict endosymbiont nucleus (the so-called nucleomorph) and which are central to the study of plastid origin and spread across eukaryotes. The group also contains the enigmatic heterotrophic katablepharids and the lone genus *Palpitomonas*. Cryptophytes are also noted for their importance in Antarctic waters (Mendes et al. 2018).

Archaeplastids are defined by the presence of the so-called primary plastids directly derived from the primary endosymbiosis with a cyanobacterium. They include green algae (from which land plants evolved), red algae and glaucophytes. A new group, *Rhodelphis*, was discovered that branches as sister group to red algae in phylogenomic analyses (Gawryluk et al. 2019). *Rhodelphis* cells are heterotrophic flagellates, but sequence data suggest that they have a non-photosynthetic primary plastid.

Amoebozoans are the second primarily amoeboid group besides the Rhizaria and include groups with pseudopodia for feeding and locomotion (Cavalier-Smith et al. 2016). The group also includes slime molds and flagellates as well as some important pathogens (e.g., *Entamoeba* causing amoebic dysentery). Several marine genera isolated from sediments have also been described (Kudryavtsev et al. 2018).

Opisthokonts include several protistan lineages, including a variety of heterotrophic flagellates, amoeboid protists, and parasites, and fungi as well as all multicellular animals. The term opisthokont refers to the presence of a single posterior flagellum. Opisthokonts, amoebozoans, and a few other small lineages are grouped as Amorphea. In marine and freshwater ecosystems, choanoflagellates have long been recognized as predators of bacteria and structural similarity with sponges was noted before molecular analysis confirmed that they are related to sponges, corals, and the rest of the animals (Brunet and King 2017; Leadbeater 2015).

CRuMS is an amalgamation of several former "orphan" taxa (see below): the Collodictyonids, Rigifilida, and the marine genus *Mantamonas* (Glücksman et al. 2011). Thus far these groups comprise solely free-living protists that have morphologically little in common, but it was recently found that they branch together in molecular phylogenies (Brown et al. 2018).

Discobids and Metamonads are two groups previously classified as the Excavate supergroup based on their distinctive morphology. Molecular phylogenies have mostly failed to support this grouping and therefore they are currently treated as two separate but possibly related groups. Discobids include photosynthetic euglenids (e.g., *Euglena*), parasites (e.g., *Trypanosoma*), and many free-living heterotrophic flagellates, such as the diplonemids, a great diversity of which are present in marine environments (Flegontova et al. 2016). Metamonads contain anaerobic protists including several pathogens (e.g., *Giardia* and *Trichomonas*) and symbionts in animal guts (e.g., *Trichonympha*).

Hemimastigophora or "hemimastigotes" are free-living soil protozoa with two rows of flagella. They were first noted in the nineteenth century but not cultured and genetic data were therefore lacking. Phylogenomic analyses based on transcriptomes sequenced from hand-picked cells of two genera indicated that hemimastigotes are one of the deepest branches within the eukaryotes. Because these analyses failed to place them as sister to any one of the "established" supergroups (or any "orphan"), it has been proposed that they should be considered as a new supergroup (Lax et al. 2018).

Orphan Taxa refers to several seemingly species-poor taxa for which phylogenomic analyses have thus far failed to provide convincing evolutionary relationships with other lineages. These so-called orphan taxa are all free-living heterotrophic protists and include e.g., the tiny and widespread marine group of Picozoa (Moreira and López-García 2014; Not et al. 2007b; Seenivasan 2013; Seenivasan et al. 2013).

## 4.3 Traits Distinguishing Protists from Other Marine Microbiome Members: Size and Cell Structure

The adequate representation of diversity in studies with the aim to address ecological questions has been a long-lasting challenge. Questions such as how elemental cycles function, how ecosystem stability is achieved, or what are the biogeography distributions of (micro)organisms all require some kind of simplification of the immense diversity of taxa, and the delineation of a tractable set of ecological roles fulfilled by diverse protists is no exception. Trait-based approaches have gained ground in ecology and oceanography (Kiørboe et al. 2018; Martini et al. 2020). These offer a representation of functional diversity that is independent from characterization of a large number of individual species and their assignment to pre-defined groups. Insights into community ecology thereby can be gained by understanding the traits that best characterize an ecological niche, how those traits affect fitness, and how they relate to one another in potential trade-offs. The resultant insights into community ecology include the potential for predicting response scenarios under future environmental conditions that could result in novel combinations of traits that are the most favorable under those conditions.

Detailed overviews of traits that are considered relevant based on current knowledge have been assembled for phytoplankton (Litchman and Klausmeier 2008) and zooplankton (Litchman et al. 2013) as well as for potential trade-offs found in mixotrophic organisms capable of both photosynthesis and phagocytosis (see below) (Andersen et al. 2015). Relevant traits are often classified into morphological, physiological, behavioral, and life-history traits, many of which are not yet understood for the uncultured majority of protistan diversity. Moreover, they are difficult to predict from genomic information alone (Keeling and Campo 2017). Thus, a major research pursuit is to identify functional traits in protists. An important feature of this type of research is to study mechanisms at the relevant scale. For example, results from the analysis of flow fields generated by beating flagella of swimming protists suggested a trade-off between maximizing resource acquisition (via swimming) and minimizing predation risk from flow-sensing predators (Nielsen and Kiørboe 2021). Below, we highlight just a few of the traits that perhaps have a larger influence on the biology and function of marine protists relative to their influence on the biology of bacteria and archaea.

### 4.3.1 Cell Size of Marine Protists

Unlike the size of marine archaea and most marine bacteria, the cell diameter of protists ranges from less than one to several hundred micrometers (Caron et al. 2017; Finkel et al. 2009). Thus, it has become standard practice to group microbial eukaryotes according to a series of size fractions. These fractions are based on cell diameter with the prefix pico- indicating 0.2–2.0 μm cell diameter, nano- indicating 2.0–20 μm, and micro- indicating 20–200 μm (Sieburth et al. 1978). Of course, the grouping "picoplankton" includes eukaryotes as well as bacteria and archaea and therefore further precision is needed. Thus, picoeukaryotes is the term used for the smallest eukaryotic cells and by current practice their size range is often defined as either 0.2–2.0 μm or 0.2–3.0 μm depending on the filter pore size used. Historically, other terms have also been employed for example, "ultraplankton" (variously, 0.2–3 to 10 μm, (Murphy and Haugen 1985; Pitta and Karakassis 2005; Reynolds 1973; Takahashi and Bienfang 1983), but usually <5 μm) and these terms have their own value because all of these fractions are to some extent arbitrary.

While it is somewhat arbitrary to lump organisms solely based on size, it can be helpful for considering the various life strategies and the competition processes that go on among different microbial populations, something eloquently addressed in 1992 by Sallie Chisholm (see Box 4.4). Moreover, due to their tiny size picoplanktonic organisms have low Reynolds numbers (Re) meaning that their movement is dominated by viscous forces rather than by inertial forces (Aris 1990). Because of their low Re, picoplankton does not sink through the water column as individual cells. They sink when aggregated into larger material (e.g., through predation, fecal pellet, and marine snow) or when other mechanisms occur, such as downwelling or other mesoscale oceanographic features (Omand et al. 2015). The viscosity of seawater also has evolutionary consequences for the energetics and mechanisms of directional movement by motile microbes such as the structure and placement of the flagellar apparatus (Barry et al. 2015; Brumley et al.

2015; Febvre-Chevalier and Febvre 1994). A corresponding cell size also puts different species of photosynthetic picoeukaryotes, such as the prasinophyte *Ostreococcus*, and photosynthetic bacteria, such as the cyanobacterium *Synechococcus*, under somewhat similar constraints in terms of the ratio of cell surface area to volume. This ratio impacts the efficiency of nutrient acquisition (since nutrient transporters are located on the surface of cells in cytoplasmic membranes) as well as cellular packaging of photosynthetic pigments (Raven 1998). These factors are critical to the success of these organisms in oligotrophic environments and overall, such size-based considerations are important for integrating organism dynamics into food webs and global biogeochemical cycles.

**Box 4.4**

"In reviewing this subject, it became clear to me that plankton ecologists fall out into two groups: Those who delight in finding the patterns in nature that can be explained by size, and those who delight in finding exceptions to the established size-dependent rules. I came to appreciate the degree to which the satisfaction of both groups is equally justified. The mechanisms underlying the size-dependent patterns have undoubtedly steered the general course of phytoplankton evolution, but the organisms that do not abide by the rules reveal the wonderful diversity of ways in which cells have managed to disobey the "laws" scripted for them. The simplicity of the general relationships serves as a stable backdrop against which the exceptions can shine. By understanding the forces that have driven the design of these exceptions, we can begin to understand the ecology that has shaped past and present planktonic ecosystems."

From *Phytoplankton Size—by Sallie W. Chisholm, published in 1992.*

### 4.3.2 Cellular Structure and Mosaic Genomes

Historically, the cell structure of Eukarya has been viewed as being different from Bacteria and Archaea (Doolittle 1998a). The cytological classification system formalized by Stanier and van Niel (Stanier and Van Niel 1962) laid out the criteria for distinguishing bacteria (including archaea) from protists. In this view, eukaryotic cells have complex structural features such as the membrane-enveloped true nucleus, a complex endomembrane system, and a cytoskeleton, while Bacteria and Archaea do not possess such features. These differences are not quite as clear-cut anymore, because some features that were considered to be characteristic for eukaryotes have also been discovered in Bacteria and Archaea, blurring the boundaries between the three domains of life (Grant et al. 2018; Oikonomou et al. 2016; Vellai and Vida 1999). Still, there are typical features of eukaryotes, such as the mitochondrion, although deviations have been identified (Karnkowska et al. 2016), as well as the capacity for endocytosis and exocytosis (Vellai and Vida 1999).

While partial compartmentalization has also been observed in Bacterial and Archaeal lineages, it is different from the membrane-delimited compartments that characterize eukaryotes (Diekmann and Pereira-Leal 2013). Eukaryotes originate from what is thought to have been a highly compartmentalized cell, LECA, a hypothetical lineage already containing a complex endomembrane system, thought to have given rise to all modern eukaryotes (Field and Dacks 2009). Protists and other eukaryotes have more or less retained this compartmentalized cell plan with multiple organelles, whereas Bacteria and Archaea tend to have only one compartment. Furthermore, endosymbionts or their remnants are ubiquitous in eukaryotes, such as mitochondria and plastids. In contrast, endosymbiosis appears to be rare in Bacteria and Archaea. Thus, the compartmentalized cell plan creates fundamental differences from the bulk of the known Bacteria and Archaea (Diekmann and Pereira-Leal 2013).

Cellular complexity has consequences for the protistan cell. Compounds acquired exogenously must be trafficked through multiple membranes, affecting both nutrient acquisition and energy allocation. Cell complexity also determines how ecological interactions can manifest, as well as possibilities for evolutionary adaptation. For example, the transient contribution of Auxiliary Metabolic Genes (AMGs) brought to their hosts by viruses is much discussed (Breitbart et al. 2018; Zimmerman et al. 2020). The fact that photosynthetic protists have a membrane around the plastid means that proteins must have a specific transit peptide to cross that membrane, and hence it would be difficult for a virally-encoded photosynthesis-related gene to function in the host cell, because it would not localize to the correct compartment (the plastid). In contrast, once a virus enters a cyanobacterium there is no such additional boundary for the AMGs it brings, and indeed photosynthesis-related genes are commonly seen in cyanophages and are highly active in augmenting host photosynthesis during infection (see below).

An expansion of genetic information, relative to prokaryotes, and its residence in the nucleus is a feature of eukaryotes, and indeed central to the evolution of eukaryotic cells. In 1986, Lynn Margulis and her son Dorion Sagan conveyed the intricate entwining of life forms and the importance of the nucleus to the lay audience (Box 4.5). Although controversial in the theory of eukaryotic evolution (Lynch and Marinov 2017), the expansion of genetic information in eukaryotic cells alongside increased cytoplasmic complexity and compartmentalization is proposed to have been facilitated by the increased energy supply that occurred through symbiotic integration of endocytosed bacteria (Lane 2011; Lane and Martin 2010; Vellai and Vida 1999). This proposed relief from energy constraints was purportedly paralleled by a more K-selected lifestyle of the Eukarya, compared to Bacteria and Archaea (Carlile 1982), and physiological optimization resulting in lower death rates (Kerszberg 2000).

It should be noted that following endosymbiosis there is usually a dramatic reduction of the gene content of the genome of the endosymbiont. In fact, the remnant genomes of endosymbionts can contain $<5\%$ of the genes found in their free-living relatives. Gene transfer from organelle "ancestors" to the "host" nucleus, a process called endosymbiotic gene transfer (EGT), also leads to genetic variation in eukaryotes (Gould et al. 2008; Kleine et al. 2009; Timmis et al. 2004). Overall, EGT from bacteria to eukaryotes has caused episodic transfer of bacterial genes to eukaryotic genomes (Ku et al. 2015) and also takes place between eukaryotes (Archibald 2015; Gould et al. 2008).

Endosymbiosis aside, it has been unclear to what extent protists and other eukaryotes transfer genetic material laterally between them. In the evolution of Bacteria and Archaea lateral or horizontal gene transfer (LGT or HGT) is known to play a major role (Soucy et al. 2015). Viruses can also act as vectors for gene transfer and likely facilitate LGT between eukaryotes (Gilbert and Cordaux 2017). Moreover, protists may acquire foreign DNA by ingesting and digesting prey (Doolittle 1998b). Collectively, EGT and potentially these other mechanisms of LGT are important modes of gene acquisition in eukaryotes and underpin the apparent mosaicism seen in the genomes of protists that reflects acquisition and retention of genetic material from different biological entities.

**Box 4.5**

"Life on earth is such a good story you cannot afford to miss the beginning...Beneath our superficial differences we are all of us walking communities of bacteria. The world shimmers, a pointillist landscape made of tiny living beings. Giant redwoods and whales, mosquitoes and mushrooms are intricate symbiotic networks, modular manifestations of the nucleated cell."

From *Four Billion Years of Evolution From Our Microbial Ancestors—by Lynn Margulis & Dorion Sagan, published in 1986.*

## 4.4 Metabolic Exchanges Between Microbiome Members

When considering microbiomes, the aim often is to understand the present function of that microbiome and consequences for the local habitat (e.g., most research on the human gut microbiome). However, it is important to recall that interactions between microorganisms in the early ocean are what eventually led to eukaryogenesis—the rise of protists—and similarly the plethora of symbioses and organismal functions seen in modern time have been shaped by cell-to-cell interactions in ancestral microbiomes. Interactions between microorganisms may simply involve exchange of metabolites or signaling compounds in the water column, but also occur through the range of physical interactions known as symbioses. The nuances of the term symbiosis have shifted over the years, and today the meaning tends toward being quite broad as outlined below.

### 4.4.1 Symbioses: Manifestation Is a Status Not an Identity

Symbioses can be mutually beneficial (mutualism), beneficial for one partner without incurring a cost for the other (commensalism), or beneficial for only one partner at a cost to the other (parasitism/pathogenicity). These distinctions are generally not clear-cut, in part because the costs and benefits can shift with changing environmental conditions. Examples are photo-endosymbiotic associations that are usually considered mutualistic since a heterotrophic host gains access to photosynthetically-fixed carbon from its (photosynthetic) endosymbiont, which receives nutrients in return. However, these associations may also represent an exploitation of the endosymbionts by their host (Decelle 2013) or under certain conditions can even be costly for the host, such as under low light conditions where photosynthesis does not operate well, but the endosymbiont still receives nutrients from the host (Lowe et al. 2016).

Parasites—One marine example of parasites is chytrid fungi that parasitize common marine diatoms (Garvetto et al. 2019; Hanic et al. 2009; Scholz et al. 2017). Chytrids are found throughout the ocean especially during and following diatom blooms and may play a role in the collapse of these phytoplankton blooms. Additionally, the chytrids and their host-specificity appear to influence diatom bloom dynamics and diversity (Chambouvet et al. 2019; Gsell et al. 2013).

Pathogens—Disease or death causing microorganisms are the subject of many biomedical studies but receive less attention in studies of the marine environment. One example is bacteria belonging to the newly discovered candidate phylum Dependentiae (also known as TM6) which can cause rapid infection and death of a heterotrophic flagellate host, the stramenopile *Spumella elongata* (Deeg et al. 2019), a stramenopile lineage with marine relatives. While this particular example is a freshwater pathogen-host system, similar interactions may also occur in the ocean. In regard to protists rather than bacteria that may be considered pathogens of protists, examples include the Oomyceta genera *Lagenisma* and *Olpidiopsis*, which infect the diatoms *Coscinodiscus* and *Rhizosolenia*, respectively (Buaya et al. 2017; Scholz et al. 2014).

Mutualists/Commensualists—An example of mutualism in the marine plankton is between diatoms and nitrogen-fixing bacteria (diazotrophs) also known as DDAs (diatom-diazotroph associations). This type of symbiosis is well documented between several diatom genera, especially *Rhizosolenia* and *Chaetoceros*, and the cyanobacteria *Richelia intracellularis* and *Calothrix rhizosoleniae* (Foster et al. 2011; Foster and Zehr 2006; Jahson et al. 1995). DDAs are found throughout the tropical and subtropical ocean (Monteiro et al. 2010). In addition, there are examples where the real impact of a symbiosis for each partner is still unclear with respect to the possible spectrum of commensalism to mutualism, principally because of the lack of knowledge of the cell biology underpinning the relationship.

### 4.4.2 Phycosphere and Metabolic Exchanges

The first attempts to maintain diverse photosynthetic protists in culture already revealed dependence on particular supplements. For example, for many years organic supplements in media for eukaryotic phytoplankton included B-vitamins (Droop 1957). Later it was realized that many of these phytoplankton did not actually need the vitamin, but did need one of the precursor moieties, for example hydroxylmeythl-pyrimidine (HMP) in the case of many haptophyte algae, or for some prasinophytes the thiazole moiety (Gutowska et al. 2017; McRose et al. 2014; Paerl et al. 2018). In nature these compounds would have been supplied by other microbes, while the phytoplankton themselves produce other compounds utilized by microbiome members. At this stage a wide range of metabolic interactions are known (Johnson et al. 2020), including the exchange of growth factors, essential nutrients, and carbon sources, which can either become available to other microbes as "public goods" upon release by the producer, or be exchanged in more targeted "trading relationships." In case of the photosynthetic protists unable to grow on a purely mineral medium, requirements can sometimes be met by the supply of filtered natural seawater indicating that the required compounds are present in sufficient concentrations and thus could be regarded as "public good." However, if the metabolite is depleted in the environment, the concentration gradient around the producer becomes steep and metabolite exchange requires a close cell proximity (van Tatenhove-Pel et al. 2021). Proximity can be attained by motile cells in brief encounters, but can also be semi-permanent in the case of stable co-associations such as symbioses.

Beyond the role of symbioses in protistan ecology, and the top-down controls on protists exerted by viruses and predation (see below), there are other mechanisms of interaction that occur in the aquatic environment. For example, many photosynthetic protists are large enough to have a viscous boundary layer termed the "phycosphere," in which bacteria can encounter higher concentrations of phytoplankton exudates and utilize them for growth (Raina et al. 2019; Seymour et al. 2017). Furthermore, a currency exchange can occur whereby compounds produced by particular bacteria can be used by the phytoplankton cells, while the bacteria themselves acquire other compounds from the phytoplanker, e.g., Amin et al. (2012, 2015).

### 4.4.3 The Holobiont Concept

The holobiont concept offers a contextual shift in biology that can help to describe and to understand biological interactions in marine ecosystems. This concept applies to organisms ranging from protists (Dittami et al. 2021) to animals, including humans (Pride et al. 2010; Simon et al. 2019; van de Guchte et al. 2018). The term "holobiont" was conceived in the 1940s (Meyer-Abich 1943) and later independently by Margulis (Margulis 1990) who coined the term in the context of symbiosis-driven evolutionary innovations. Today it usually refers to a close association between different individuals that form an anatomical, physiological, immunological, or evolutionary unit (Simon et al. 2019). The discovery of the extent of marine molecular microbial diversity and microbiome inter-connectivity suggests that the holobiont concept may apply more frequently in marine ecosystems than currently recognized (Lima-Mendez et al. 2015). Studies on corals (Apprill 2017; Thompson et al. 2014) and sponges (Pita et al. 2018) alongside fascinating foundational studies on squid (McFall-Ngai 2014; Nyholm and McFall-Ngai 2004; Tischler et al. 2019) have demonstrated that microbes that associate with this suite of multicellular marine animals are an integral part of the living system: the holobiont.

## 4.5 Shifting from a Functional Dichotomy to Recognizing the True Complexity of Marine Protists

Although the famous drawings by Haeckel made the wonders of protistan diversity accessible to a wide audience, a contemporary of Haeckel, Viktor Hensen was potentially the first to raise awareness of the functional importance of protists in the ocean. Hensen coined the term "plankton" for organisms drifting in the water and unable to swim against currents. He described plankton as "without doubt of great importance for the entire metabolism of the sea" (Smetacek 1999) and he hoped to link fisheries yields to the plankton productivity that supported them. At the time his idea was ridiculed, but is now seen as the foundation of quantitative ecology and biological oceanography. Hensen's work led to phytoplankton (and especially diatoms) being considered the "pasturage of the sea"—making them the first protists to be recognized as having global importance, as they support secondary production by metazoan zooplankton, such as copepods, which in turn are fed upon by fish. The importance of photosynthetic protists has now been proven and it is estimated that ~45% of global photosynthetic $CO_2$ fixation is performed by planktonic marine eukaryotes and cyanobacteria (Field et al. 1998).

It took much longer for the heterotrophic protists constituting the smallest size fractions of zooplankton to gain a comparable recognition if they even have by now. The realization that the larger size classes of plankton collected in nets, such as metazoan zooplankton, contributed only a minor fraction of overall respiration in seawater, steered attention toward microbes as metabolic hubs in the ocean (Pomeroy 1974). Additionally, it was recognized that phytoplankton excrete part of their photosynthetically-fixed carbon as dissolved organic matter (DOM) and that both DOM and particulate organic matter (POM) from these taxa could also become available when they are lysed by viruses, or during "sloppy feeding" by zooplankton. In turn, this free DOM and POM provid an energy source for heterotrophic microorganisms. Iterations between and through a web of carbon exchanges were recognized as involving consumption of bacterial and archaeal cells by heterotrophic nanoflagellates and consumption of these and other cells by larger protists, such as ciliates. Collectively, recognition of this complexity resulted in conceptualization of the microbial web (Pomeroy 1974; Sherr and Sherr 1988) and the microbial loop (Azam et al. 1983), both of which called into question the traditional view that aquatic microbial food webs were primarily a three-step food chain from primary producers to zooplankton and fish (Box 4.6). Further development of the microbial loop concept captured alternative flows of carbon to both lower and higher trophic levels that were less efficient than direct transfer from primary producers. Today, diverse protists are recognized as efficient bacterivores often controlling bacterial standing stocks and influencing community composition (Pernthaler 2005). Many of these protists can also feed efficiently on other eukaryotes collectively dominating herbivory in the ocean (Sherr and Sherr 2007). It has been estimated that about two-thirds of global planktonic primary production is removed through predation by protistan microzooplankton (Steinberg and Landry 2017). Thinking back to Hensen, if photosynthetic microbes are the "pasturage of the sea" then predatory protists would by analogy represent the "cows of the sea."

An important paradigm shift in plankton ecology addressed misconceptions inherent to the assignment as either "phytoplankter" or "zooplankter" (Flynn et al. 2012) which rooted back to the plant–animal dichotomy introduced by Carl von Linné in his "system of nature" first published in 1735 (von Linné 1735). The theory of endosymbiosis suggests that life forms could exist that maintain or combine both types of nutritional strategies (photosynthesis and phagocytosis). However, the maintenance of a mixed nutritional strategy was not considered because it was thought that as soon as the plastid was acquired, photosynthetic nutrition alone would be sufficient. This view is illustrated in the symbiont theory presented by Konstantin Mereschkowsky who concluded that plants evolved from animals through invasion by cyanobacteria, and that once photosynthetic, even a lion would become a peaceful creature, thriving by photosynthesis, with no interest in prey (Mereschkowsky 1905). Alas, Mereschkowsky and others at the time had no knowledge that many photosynthetic protists are also efficient predators!

As early as the 1950s, Ernst Georg Pringsheim used the term "mixotrophy" to describe a variety of lifestyles among flagellates that used preformed organic substances next to photosynthesis (Pringsheim 1958). Pringsheim's interest in these different nutritional requirements was rooted in an ambition to grow diverse photosynthetic flagellates in culture, which he found often required supplementation with organic compounds. Although photosynthetic protists with the capacity to ingest microbial prey had already been discovered (Biecheler 1936), their importance as bacterivores in aquatic microbial food webs was only reported in the late 1980s. Sparked by a seminal paper reporting high rates of bacterivory by photosynthetic protists (Bird and Kalff 1986), many reports of feeding by mixotrophic flagellates on bacteria or other groups of phytoplankton followed for the marine environment (Stoecker et al. 2017). The term "mixoplankton" was then introduced in an attempt to better represent the diverse planktonic protists capable of both photosynthesis and phagocytosis, clearly delineating between predatory mixotrophy and mixotrophy as a term referring to use of dissolved organic compounds by phytoplankton (Flynn et al. 2019). The combination of both predation and photosynthesis in the same cell has many consequences for the functional role of these protists. For instance, mixotrophs that acquire nutrients bound in their prey and use them to support a mainly photosynthetic lifestyle invalidate the assumption that primary productivity relies largely on the direct availability of dissolved inorganic nutrients. Moreover, the direct link between consumption of prey and photosynthetic carbon fixation in the same cell results in less release and recycling of nutrients and more efficient trophic transfer, potentially supporting higher biomass of top consumers (Mitra et al. 2014; Ward and Follows 2016). Both the access of mixotrophs to alternative resources (Rothhaupt 1996a) and their interaction with their prey by predation and competition (Wilken et al. 2014) can result in stronger suppression of prey abundances and allow them to outcompete specialists in resource-poor environments (Rothhaupt 1996b).

**Box 4.6**

"Unseen Strands in the Food Web: The new paradigm of the ocean's food web that is developing, as a result of recent studies of protistan activities and alternative pathways of organic matter, may contain many unseen strands. We are not certain how the long-recognized food web of diatoms and copepods fits into the expanded web which is gradually appearing. Quantitatively, large diatoms seem to be minor contributors to production, and net plankton seems to be a minor component of respiration; but if this is not the major link of photosynthesis to nekton, what is? Are the communities of upwellings really more efficient producers of nekton, and is the food web really different in them? These questions are important not only to the basic ecologist but to the fisheries scientist."

From *The Ocean's Food Web, A Changing Paradigm—by Larry Pomeroy,* 1974.

Predatory mixotrophs differ in their relative reliance on photosynthesis versus (phago)-heterotrophy for their nutrition and their inherent potential to perform photosynthesis (Mitra et al. 2016; Stoecker 1998). Constitutive predatory mixotrophs engage in phagocytosis and have an inherent photosynthetic potential through possession of their own stably integrated vertically inherited plastids, while non-constitutive mixotrophs derive their photosynthetic potential from their prey via either kleptoplasty or photo-endosymbiosis (Mitra et al. 2016). These different forms of mixotrophy imply different functionalities, for instance, in the degree to which their capacity to photosynthesize depends on prey availability.

With every paradigm shift, past or recent, the interaction network among protists as well as how they interact with Bacteria, Archaea, protists, and even multicellular eukaryotes has become more complex. Elemental flow no longer follows neatly distinguishable trophic levels but can be merged or arranged in loops and of course through the various manifestations of symbiotic interactions (Fig. 4.6). There is still much to learn about the functioning of microbial networks and particularly about the quantities and routes of carbon flow (Worden et al. 2015). One consideration is the balance between photosynthetic fixation of $CO_2$ into biomass of living cells and its release and remineralization through the action of heterotrophic bacteria. Small

**Fig. 4.6** The functional roles of protists in the ocean. Protists influence biogeochemical cycles in multiple ways, many of which involve direct interactions with other biological entities. This schematic depicts some of the major protistan roles discussed herein, with images of example organisms in white circles, as they link to the broader marine microbiome and the network of direct and indirect interactions it entails. Note that some of these categories overlap—for example, here we show a coral/dinoflagellate symbiosis, under symbiosis, but other categories can be considered symbioses, such as parasitism (see within chapter text). The image illustrating kleptoplasty is a confocal fluorescence micrograph of *Mesodinium rubrum* with cryptophyte organelles (courtesy of Matthew Johnson). Other images, as well as the overall figure, are adapted from Worden et al. (2015)

changes in this balance determine how much biomass will be moved into deep waters and potentially buried in sediments over long time scales, or released back into the atmosphere from the ocean surface waters. Although predatory protists are increasingly receiving attention, we still have much more knowledge, more cultured taxa, and even more genomic information from photosynthetic protists (Bhattacharya et al. 2014; Burki 2017; Cooney et al. 2020; Gawryluk et al. 2019; Keeling et al. 2014; Krabberød et al. 2017).

### 4.5.1 Pursuing Lines of Protistan Heterotrophy in the Sea

The role of heterotrophic protists as major bacterivores and herbivores in the ocean is established, but heterotrophic protists can have non-predatory modes for gaining nutrition. This includes protistan parasites that infect multicellular eukaryotes or other protists, and saprotrophs that utilize detrital organic matter often via the excretion of exoenzymes. The latter break down macromolecules into small enough units for uptake via osmotic transport, specialized transporters, or via endocytosis. The detailed characterization of these nutritional strategies traditionally relies on controlled laboratory studies of cultured protists. However, even the first molecular surveys of marine protistan communities revealed a large diversity of novel groups of small (presumably) heterotrophic protists that passed through filters of only a few micrometer pore size (Díez et al. 2001; López-García et al. 2001; Moon-van der Staay et al. 2001). Two groups that were and continue to be commonly retrieved at high sequence abundances were affiliated with the alveolates and stramenopiles, respectively, and were named accordingly MALV (Marine ALVeolates) and MAST (MArine STramenopiles). MALV include known species of Syndiniales (Guillou et al. 2008) that are obligate parasites, and using the guilt-by-association principle all MALV are commonly assumed to share this lifestyle although the possibility of more varied trophic modes has been suggested based on their paraphyly (Strassert et al. 2018). The MAST are also paraphyletic (Not et al. 2007a) and contain several independent lineages that appear to represent basal heterotrophic stramenopiles (Massana et al. 2014), some of which have been shown to be bacterivores (Lin et al. 2012; Massana et al. 2006).

While predation on other microbes and parasitism are probably the most common forms of nutrition in heterotrophic protists, novel nutritional modes are still being found. For instance, the analysis of the genome sequence of individual cells directly isolated from their natural environment suggested that some MAST lineages are photoheterotrophs. This idea arose because genes encoding microbial rhodopsins were observed, and these were hypothesized to use sunlight for photoheterotrophy, although their true cell biological role has yet to be shown (Labarre et al. 2021). Another study used single cell sequencing approaches on a few cells of the uncultured Picozoa and concluded that they may "feed" on phages (Brown et al. 2020). This evolutionarily distinct lineage was briefly in culture during which studies indicated that it lives on particulate organic matter (POM) (Seenivasan et al. 2013).

Predatory protists are perhaps the best characterized heterotrophic protists in pelagic marine environments. Their role in the carbon cycle is through the ingestion of prey via phagocytosis. Phagotrophic nutrition requires the internalization of prey through the invagination of the cell membrane and formation of a food vacuole followed by the modification of vacuolar conditions for the digestion of prey, absorption of nutrients from the vacuole into the cytoplasm, and finally egestion of any remaining material (Flannagan et al. 2012). The process of phagocytosis has been well characterized in metazoans because of its important role in immune responses. The variations occurring across the diverse protists that rely on phagocytotic processes for their nutrition are less well resolved despite the first observations of the process in the mid-twentieth century that came from the ciliate *Paramecium* (Mast 1947).

There are many different strategies and morphological adaptations within protistan lineages for finding, capturing, and ingesting their prey. Among the nano- and pico-sized flagellates that represent the majority of bacterivores in the ocean (Jürgens and Massana 2008), three main strategies are known for capturing prey in the dilute marine environment: filter feeding, direct interception feeding, and diffusion feeding. The first two rely on creating a feeding current toward the cell using undulating flagella. In the filter-feeding choanoflagellates, close relatives of metazoans, the feeding current passes through a collar of finely spaced microvilli from where the retained prey is transported to the cell surface and phagocytosed (Pettitt et al. 2002). The fine spacing of the microvilli allows capture of small food particles and also causes a strong flow resistance. How choanoflagellates create a flow strong enough to accomplish their high filtration rate remains unresolved and the presence of a flagellar vane has been postulated as a potential explanation (Nielsen et al. 2017). Direct interception feeding is common among cultured flagellates within the stramenopiles (Boenigk and Arndt 2002) and is speculated to be the feeding mode of many of the uncultured MAST groups (Labarre et al. 2021). Interception feeding requires direct contact with the prey at the protist's cell surface upon which ingestion is initiated. Since each food item is handled individually this allows selectivity based on physical or chemical properties of the prey item and preferential feeding based on size, prey species, or prey quality has been reported in many species of interception feeders (Gonzáez et al. 1993; Monger and Landry 1991). Finally, diffusion feeding predators remain motionless, waiting for prey to collide with pseudopodia extended from their cell body. This feeding strategy underlies the sun-like appearance of heliozoans and is also found among planktonic foraminifera, many of which are mixotrophs that host photosynthetic endosymbionts, and are able to capture prey ranging in size from bacteria to copepods.

Two alveolate groups have perhaps the widest array of different feeding strategies, the ciliates and dinoflagellates (Hansen and Calado 1999; Leander 2020). Both groups contain bacterivores that can dominate consumers of both heterotrophic and photosynthetic microorganisms in some marine habitats. Additionally, they contain specialized predators that are able to hunt and consume prey larger than themselves via phagocytosis. This usually requires immobilization of the prey, for which extrusive organelles are used, and appears to have evolved independently several times. The ciliate *Didinium* feeds on other ciliates at which it first shoots "toxicysts" to inject a toxin into its prey (Wessenberg and Antipa 1970). In a number of specialized dinoflagellates, extremely complex cell organelles termed nematocysts are used similarly. Nematocysts bear overall similarity to the harpoon-like cells of cnidarians with which they share their name. Nematocysts in dinoflagellates and cnidarians do not share their evolutionary origins, rather they are examples of convergent evolution over an enormous phylogenetic distance. Even colonial or multicellular prey organisms too large to be phagocytosed can be consumed by protists, for example chain-forming diatoms are fed upon by the dinoflagellate *Protoperidinium*, which extrudes a pseudopod-like structure, termed the pallium, to accomplish this (Gaines and Taylor 1984). The pallium stretches along the surface of the prey to enclose it, followed by the digestion and uptake of the cellular content, leaving only empty diatom frustules behind. Finally, large prey can also be consumed via injection of a tube-like structure to suck out prey cell contents as done by some dinoflagellates, using what is termed the peduncle.

Next to ciliates and dinoflagellates there are many less explored groups of small heterotrophic flagellates that feed upon other eukaryotes. These perform the impressive feat of "swallowing" cells that are almost their own size. In contrast to the heterotrophic flagellates identified by environmental surveys (e.g., MASTs), several predatory eukaryotrophic flagellates (i.e., feeding on eukaryotic prey) have been discovered and isolated using elaborate culturing efforts (Tikhonenkov et al. 2021). These efforts include identifying the eukaryotic prey for the eukaryotrophic predator of interest, providing this prey as well as its own source of nutrition (which in some cases can be bacteria). Predatory eukaryotrophic flagellates are rapid feeders and reproduce quickly in culture, hence they may play ecologically important roles in controlling other small flagellates (Tikhonenkov 2020). The position of eukaryotrophic flagellates in the "upper" trophic levels of the microbial food web may help to explain their relative scarcity in molecular surveys. In addition, such species often occupy regions of the eukaryotic tree that are not well resolved or are found at the "base" of large groups, which makes identification and placement of 18S rRNA gene amplicon sequences from them difficult, so that they are sometimes simply passed over in molecular diversity studies. However, from an evolutionary standpoint these small flagellates are frequently positioned as sisters to major eukaryotic groups, aiding understanding of the evolution of those groups (Tikhonenkov et al. 2020b). The complex life cycles and eukaryotrophic nutrition of novel unicellular animal-relatives can further help to understand possible features of the ancestor of animals and how the multicellularity of animals evolved (Hehenberger et al. 2017; Tikhonenkov et al. 2020a). An exciting example in the context of the origin of photosynthetic eukaryotes was the discovery of the aquatic eukaryotroph flagellate genus *Rhodelphis* which appears to be closely related to the typically photoautotrophic red algae (supergroup Archaeplastida). The discovery of a non-photosynthetic primary plastid in *Rhodelphis* combined with its predatory lifestyle suggests that the ancestor of red algae and *Rhodelphis* may have been a predatory mixotroph, raising questions about the long-standing idea that phagotrophy was lost early in the evolution of the Archaeplastida, before the divergence of the three major lineages (Fig. 4.4) (Gawryluk et al. 2019).

The large range of feeding mechanisms found among protists allows them to target different prey groups and size classes. Through this diversity in feeding strategies, predatory protists can collectively act as the main consumers of both bacteria and phytoplankton in the ocean (Calbet and Landry 2004), as well as feeding on other protists and archaea. However, because the size ranges of the prey organisms in these groups can overlap, hence many protists are in fact omnivores rather than exclusively bacterivores or herbivores. Conversely, not a single predatory protist would be able to feed on all of the diverse primary producers because they span several orders of magnitude in size. While bacterivorous protists are often treated as one functional group, the fact that they use different feeding strategies has implications for feeding preferences and strong prey size selectivity has been observed in many heterotrophic flagellates. This in turn can shape bacterial community size structure in which cells of intermediate size (~1 μm) are preferentially consumed by flagellate grazing, while both smaller and larger colony-forming cells persist (del Giorgio et al. 1996; Jürgens and Matz 2002). Prey motility is also thought to influence selectivity or at least feeding success, as motile prey presumably has both higher encounter rates with other cells, such as predators, and has the potential to escape ingestion (Harvey et al. 2013; Matz et al. 2002). Other prey characteristics influencing feeding selectivity by protists are not well understood. These include prey cell surface properties or compounds detected prior to ingestion, likely through the receptors that initiate phagocytosis by the predator (Roberts et al. 2011). Selective feeding also has technical implications for the methods to detect and quantify grazing rates. For example, some protists do not ingest the fluorescently labeled and heat-killed bacteria often used as tracers to detect and quantify consumption rates in field experiments, even if they do feed on the same bacterial strain when it is offered as live prey (Bock et al. 2021).

Parasitic protists—Parasitic protists are found across the eukaryotic tree of life and they can appear to infect a similar diversity of unicellular and multicellular hosts. Not surprisingly parasitic protists were first described in metazoan hosts. The first systematic description of parasitic life forms was by Francesco Redi who noted "animals living in animals" in 1648, and pointed to gregarines, which are now classified as apicomplexans (supergroup Alveolata). The Apicomplexa are a large phylogenetic group with many marine members and appear to consist predominantly of parasites, some of which reside in the intestines or coelom of invertebrates (Leander 2008).

The Alveolata contain other parasites, such as the perkinsids, a sister group to dinoflagellates, which includes parasites of other protists as well as bivalves. The basal dinoflagellate group syndiniales, also referred to as Marine ALVeolates (MALVs), harbors examples of obligate parasites. However, because most MALVs have not been cultured, the idea that they are parasitic comes from field studies (Chambouvet et al. 2008) and by analogy to their few described members (John et al. 2019). Syndiniales are particularly abundant in 18S rRNA PCR-based studies, including amplicon studies from marine ecosystems when sequencing the nano- and picoplankton size fraction (Guillou et al. 2008; Massana et al. 2004; Moon-van der Staay et al. 2001). Although syndiniales contain parasites of plankton ranging from protists to copepods, the host ranges of specific lineages generally remain unknown. Both syndinian and perkinsid parasites can infect dinoflagellates that cause harmful algal blooms and these parasites have also been implicated in rapid dinoflagellate species succession during blooms and even in being responsible for bloom termination (Chambouvet et al. 2008; Jephcott et al. 2016).

Other protists infect phytoplankton hosts and potentially play a role in controlling bloom formation or initiating bloom termination. Some members of the oomycetes, a stramenopile group once thought to be fungi, parasitize marine diatoms (Hanic et al. 2009). The chytrid fungi are important parasites of freshwater phytoplankton (Ibelings et al. 2004) and some chytrids infect marine diatoms (Gutiérrez et al. 2016) as do the novel chytrid-like-clade-1 (NCLC1) (Chambouvet et al. 2019). An awareness raised by these discoveries was that likely $<10\%$ of fungi species on the planet have been described (Blackwell 2011; Hawksworth and Lücking 2017; Jones 2011). Moreover, the ecological impact of those fungi infecting diatoms remains underexplored, although infections have been reported in upwelling regions where diatom blooms occur, suggesting a potentially important alteration of classical expectations of carbon cycling and food web dynamics in these regions. Finally, in the Arctic chytrid abundances reportedly correlate with sea-ice associated diatoms and predominantly occur in areas with ice melt, leading to questions about how parasite infection networks might change with future ice retreat (Kilias et al. 2020).

Osmotrophic and saprotrophic protists—Nearly all protists show some form of osmotrophy, which describes nutrition through direct uptake of dissolved organic substrates from the environment (Richards and Talbot 2018). In contrast to what the term might imply, this is not a passive process but rather involves specialized transport systems to facilitate active uptake of substrates. Lysotrophy (also called chemoheterotrophy) is a common form of osmotrophy that involves secretion of enzymes into the extracellular environment to break down larger substrates such as polymers (e.g., cellulose, lignin, lipids, and proteins) into their building blocks (e.g., sugars, fatty acids, amino acids) so that they can be taken up (Richards et al. 2006). When this mode of nutrition is used to utilize detrital organic matter, it is also referred to as saprotrophy. Extracellular digestion via lysotrophy can also be used to feed on living organisms in which case it would be considered parasitism.

The oomycetes, hyphochytriomycetes, and labyrinthulomycetes exhibit variations in their manner of osmotrophy (Amend et al. 2019; Cavalier-Smith 2018; Raghukumar 2002). Once thought to be fungi these lineages belong to the stramenopiles, unlike fungi which belong to the opisthokonts. Labyrinthulomycetes, which include both labyrinthulids and thraustochytrids, have been observed residing alongside fungi on marine snow (Bochdansky et al. 2017) and in marine sediments (Rodríguez-Martínez et al. 2020), raising the possibility that they utilize different enzymes than fungi, and thus different forms of organic matter which would result in effective niche partitioning. Environmental sequences of basal oomycetes are frequently detected in marine environments but generally cannot yet be associated with known species (Thines 2018). Additionally, ichthyosporeans, which are relatives of metazoans, have evolved osmotrophic nutrition independently of fungi and are found in the digestive tracts of some marine invertebrates (de Mendoza et al. 2015). While saprotrophic protists are present in marine habitats ranging from the surface layers to the sediments, there is still much to learn about their specific enzymes, activities, and substrate preferences.

Although it is known that different forms of osmotrophic nutrition are represented among protistan marine fungi, the ecological impacts of marine saprotrophic fungi are less well understood than of those that are parasitic (Grossart et al. 2019; Richards et al. 2012). Fungi isolated from seaweed have been demonstrated to degrade plant and algal biomass (Patyshakuliyeva et al. 2019); marine fungi also utilize phytoplankton derived polysaccharides (Chrismas and Cunliffe 2020; Cunliffe et al. 2017) and can be associated with phytoplankton blooms (Priest et al. 2021). Some marine fungi reside on aggregates of organic and detrital material referred to as marine snow (Bochdansky et al. 2017). Moreover, Arctic fungi have been reported to carry genes for degrading refractory compounds such as lignin and naphthalene alongside genes for nitrate assimilation. Active fungi have been detected in marine sediments, based on presence of ribosomal RNA (Rodríguez-Martínez et al. 2020), and can grow on zooplankton fecal pellets, a common carbon source in sediments (Hassett et al. 2019), suggesting they contribute to the degradation of organic matter in these deep-sea habitats. However, active diatoms (based on ribosomal RNA) have also been reported in sediments below 1000 m of overlying waters. Hence, it is important to tease apart sequence data coming from the resident community versus that coming from recently deposited surface water cells (Rodríguez-Martínez et al. 2020). Apart from their ecological roles, there is considerable interest in marine fungi for possible medical and industrial use of their enzymes. Marine fungi generate compounds that have been reported to have antibiotic and anticancer properties (Deshmukh et al. 2017). Additionally, their extracts can break down the cell wall of skin bacterial pathogens (Agrawal et al. 2020) and can degrade crude oil (Maamar et al. 2020).

Multiple lineages branching near the base of the fungal portion of the tree have been discovered in the last decade, including taxa grouped into the endoparasitic Opisthosporidia (Karpov et al. 2014), which contains the NCLC1 mentioned above that infect marine diatoms (Chambouvet et al. 2019). The Opisthosporidia as a whole relate to fungi in a manner similar to how choanoflagellates relate to animals (Brunet and King 2017). Another Opisthosporidia lineage detected in marine waters is the Cryptomycota, which includes the parasitic genus *Rozella* (Livermore and Mattes 2013; Richards et al. 2015). These and other discoveries have led to considerable restructuring of the fungal portion of the eukaryotic tree and the number of recognized fungal phyla has tripled over the last 20 years (James et al. 2020).

### 4.5.2 Non-constitutive Mixotrophy (Via Photosynthetic Endosymbionts and Kleptoplasty)

The British naturalist and poet Henry Baker was lucky that a friend sent him a specimen that seems to be the first recognized dinoflagellate (Box 4.7). The animalcule described in his friend's letter is now known as *Noctiluca*, a globally distributed marine dinoflagellate. It attracted the attention of Baker and others because of its bright bioluminescent blooms, earning it the common name "sea sparkle." Unusually big for a dinoflagellate (up to 2 mm), these "bladder"-like cells can host large populations of free-swimming endosymbionts that are green algae. This form of *Noctiluca* is called "green" *Noctiluca* in contrast to the "red" form that does not harbor endosymbionts (both types can form massive blooms). The photosynthetic endosymbionts contribute to growth, yet the green *Noctiluca* is still a voracious predator of other microbial eukaryotes including some species of their own dinoflagellate sisters (do Rosário Gomes et al. 2018). The green form of *Noctiluca* thus combines phototrophic and heterotrophic strategies to grow, representing a mixotrophic lifestyle. Because *Noctiluca* does not inherently possess the ability to fix carbon but needs to acquire that ability by engulfing and hosting photosynthetic symbionts, it has been termed a "non-constitutive" mixotroph. This strategy has resulted in blooms of the green mixotrophic *Noctiluca* over enormous expanses and appears to profit from changing oceanic conditions resulting from warming and anthropogenic inputs (do Rosário Gomes et al. 2014).

**Box 4.7**

"In the Glass of Sea Water I send with this are some of the Animalcules which cause the Sparkling Light in Sea Water; they may be seen by holding the Phial up against the Light, resembling very small Bladders or Air Bubbles..."

From *a letter from Mr. Joseph Sparshall to Henry Baker in "Employment for the microscope" (Henry Baker, 1753, Dodsley, London).*

A conceptual question arises in trying to place dinoflagellates in terms of functional classification. About half of the known dinoflagellates are photosynthetic and the other half lead a heterotrophic life. Those that are photosynthetic have plastids of a hodgepodge of various algal origins and "levels" of endosymbioses (Waller and Kořený 2017). As a whole, dinoflagellates display probably the most peculiar and complex plastid evolution among all plastid-bearing groups. Several lineages have either completely lost their ancestral plastid of red algal origin or have just lost the ability to perform photosynthesis but still harbor a more or less stably integrated plastid from a different alga. They are seemingly not too selective about which algal group they host as endosymbionts. Plastids from almost every other photosynthetic lineage have been identified in independent dinoflagellate lineages. This impressive flexibility of dinoflagellates with respect to losing or gaining a plastid from a variety of sources has likely played a key role in their niche expansion across aquatic environments. It also means that dinoflagellate lineages are distributed throughout the different functional modes we describe herein.

The relationship between non-constitutive mixotrophs and their endosymbionts can be complex, involving adaptations on several levels from both partners. A particularly captivating example is the interaction observed between the foraminiferan genera *Orbulina* and *Globigerinoides* and their dinoflagellate endosymbionts. Some foraminifera and several groups of radiolarians (both supergroup Rhizaria) can form easily-discernible associations with photosynthetic protists that were observed during the nineteenth century Challenger- Expedition (Tizard et al. 1885). Both foraminifera and radiolarians build intricate mineral skeletons composed of long spines and possess highly dynamic cytoplasmic strands that they can extend outside of their shells. In symbiotic foraminifera, the dinoflagellate endosymbionts dwell within vacuoles that are in turn attached to their network of cytoplasmic strands. The interaction between these two partners is characterized by a compelling diurnal pattern: at dawn the dinoflagellate symbionts move along the spines of the host to reside outside its shell during the day, at dusk they return into the inner cytoplasm inside the shell (Roger Anderson and Be 1976). Foraminifera and radiolarians are increasingly recognized as important players in open ocean communities due to their predatory activity. The mixotrophic (symbiotic) lineages also contribute to primary productivity and their large cells house up to several thousand photosynthesizing symbionts (Decelle et al. 2015).

One of the more peculiar ways to gain access to photosynthate occurs in non-constitutive mixotrophs that perform kleptoplasty or the “stealing of plastids.” Specifically, kleptoplasty involves the sequestration and retention of the plastid of an algal prey (plus sometimes other useful bits of the prey cell) while the rest is digested. Unlike the non-constitutive mixotrophy based on endosymbiosis, the road for the “stolen” alga, providing the photosynthetic ability, ends with uptake by the “host” cell. This behavior is found in a range of protists (and even animals). However, it is often challenging to distinguish lineages with kleptoplasts from those containing partially digested algae in their food vacuoles, unless observations are made for an extended period in culture.

The best described protistan examples of kleptoplasty are found in ciliates and dinoflagellates. *Mesodinium rubrum* is a globally distributed marine ciliate and represents one of the most common and abundant protists engaging in kleptoplasty. It can dominate ciliate biomass in the plankton and, during blooms, can dominate primary production, see Stoecker et al. (2017). This ciliate has evolved an elaborate scheme to make the best use of its stolen plastids, sequestering not just the photosynthetic organelles, but also the mitochondria and nuclei from its cryptophyte prey. Microscopy observations have revealed that the nuclei and plastids (together with the mitochondria) are packaged into two separate complexes surrounded by membranes upon ingestion. The ciliates and the stolen plastids divide as long as the prey nuclei are present. This indicates that the stolen nuclei still function in maintaining the plastids from the prey. Because the stolen prey nuclei begin to disappear before plastid numbers begin to decline, *Mesodinium* must recurrently steal cryptophyte nuclei (Johnson et al. 2007).

Unfortunately, the intricate strategy of *Mesodinium* to gain the benefits of photosynthesis has attracted a follower, the dinoflagellate *Dinophysis*, which appears to exploit the work done by the ciliate. *Dinophysis* has kleptoplasts of cryptophyte origin, but culturing efforts using a variety of prey, including cryptophytes, have not succeeded. Only when offered *Mesodinium* as a prey item did the dinoflagellate begins to grow and it was then shown that *Dinophysis* uses its peduncle to extract the cell contents of the ciliate, including the organelles it has stolen previously from its cryptophyte prey (Park et al. 2006), of which it only retains the plastids. It also needs to constantly reacquire these plastids, for which it has evolved the capacity to detect its prey via chemoreception. It appears to approach *Mesodinium* at low speed, which nevertheless can evade capture through escape jumps, to which *Dinophysis* responds by releasing mucus and/or using capture filaments to slow down and eventually immobilize the ciliate. The release of toxins is suspected to play a role as well (Jiang et al. 2018; Mafra et al. 2016).

Unlike *Dinophysis*, an abundant Antarctic dinoflagellate, the Ross Sea Dinoflagellate (RSD), can maintain its kleptoplasts for at least 30 months when starving (Sellers et al. 2014). The observed retention time of the kleptoplasts, which are stolen from the haptophyte *Phaeocystis antarctica*, is longer than in any other kleptoplastic systems currently known and suggests a tight integration of the kleptoplasts within the "host," putatively on the way to becoming fully integrated and stable plastids. Transcriptomic analyses of this singular relationship have revealed that RSD seems to maintain and employ kleptoplasts for photosynthetic functions as well as harbor and use the original secondary plastid found in "standard" photosynthetic dinoflagellates for plastidial metabolic pathways (Hehenberger et al. 2019). Kleptoplasty is not unique to protists since it has also been observed in sacoglossan sea slugs, which retain the plastid of their algal food (Händeler et al. 2009), and in two species of marine flatworms (Van Steenkiste et al. 2019).

### 4.5.3 Constitutive Mixotrophy

Because constitutive mixotrophs are defined as possessing an inherent capability to photosynthesize, they align with the description as being microalgae that feed on other microbes. This type of mixotrophy is found in the stramenopile groups chrysophytes and dictyochophytes, as well as dinoflagellates, cryptophytes, and haptophytes (Choi et al. 2020). Because the absence of a phagocytotic potential is difficult to prove (Wilken et al. 2019), the number of microalgae that were traditionally considered as purely photosynthetic but later found to also ingest prey has increased. An example are reports of feeding by coccolithophores (Avrahami and Frada 2020), although other mixotrophic members of the haptophytes had been known for many years (Frias-Lopez et al. 2009; Hansen and Hjorth 2002).

Much of the knowledge on the physiology of specific groups of constitutive mixotrophs comes from controlled laboratory experiments with cultured representatives. These data have been used to construct conceptual models of mixotrophy based on the relative importance of photosynthesis and phagotrophy and the environmental trigger that induces feeding (Jones 2000; Stoecker 1998). While a balanced contribution of both photosynthesis and phagotrophy to the overall nutrition seems to be rare, most constitutive mixotrophs are currently lumped into the category of being primarily photosynthetic. The haptophyte genus *Chrysochromulina*, the chrysophyte *Dinobryon*, many dinoflagellates, and probably most cryptophytes are considered capable of purely, or at least dominantly, photosynthetic growth. Current experimental work indicates that these taxa ingest prey when light for photosynthesis is inadequate, or when dissolved inorganic nutrients and other growth factors are limiting (Hansen 2011; Hansen and Hjorth 2002). However, some constitutive mixotrophs show a stronger reliance on heterotrophy, as is the case for several chrysophytes. Unfortunately, evolutionary relatedness does not align well with differences in physiological strategies of mixotrophs, and closely related strains can show divergent ecophysiologies differing in both resource requirements and responses to environmental conditions (Moeller et al. 2019; Wilken et al. 2020). This makes inferences of functional roles in nature difficult. Further, most species available in culture represent coastal rather than oceanic taxa, and it is unclear how the differences in nutrient availability in the latter might influence mixotrophic adaptations.

Constitutive mixotrophs, also known as "phagotrophic phytoflagellates", have been studied in natural communities through amendments with surrogate prey that have been fluorescently or radioactively labeled. This allows quantification of ingestion rates by heterotrophic versus pigmented flagellates, and the resulting studies have confirmed the important contribution of predatory mixotrophs to overall bacterivory in many marine habitats, especially the open ocean. However, these approaches often do not allow the taxonomic groups responsible for this predation to be distinguished, especially if morphological differences cannot be observed by fluorescence microscopy. Constitutive mixotrophs are often the main bacterivores in oligotrophic ecosystems as shown in the Atlantic subtropical gyres (Hartmann et al. 2012) and the Mediterranean Sea (Unrein et al. 2007), where fluorescence in-situ hybridization (FISH) approaches have suggested the quantitative importance of haptophytes to consumption, alongside dinoflagellates and chrysophytes (Hartmann et al. 2013; Unrein et al. 2014). Another important group in the open ocean are the dictyochophytes (see below). While the relative importance of constitutive mixotrophs in oligotrophic waters seems intuitive due to the benefit of feeding as a route of nutrient acquisition, there are also many examples of constitutive mixotrophs at high abundance in more eutrophic and coastal waters. In fact, many harmful algal bloom (HAB) species especially those belonging to dinoflagellates, haptophytes, and raphidophytes are mixotrophs (Flynn et al. 2018). Although many HAB species are intensively monitored, reports of their feeding behavior mainly come from experiments with cultured isolates and the role of mixotrophy in bloom formation is not well understood (Burkholder et al. 2008). Below, two important marine groups are discussed that are particularly complicated to categorize functionally, as described for dinoflagellates above.

Dictyochophytes are planktonic stramenopiles about which there is still much to be learned. The dictyochophytes display a variety of lifestyles ranging from planktonic photoautotrophs to mixotrophs (as shown for *Florenciella*) and bacterivores. With fossil records dating back to the Cretaceous period 145 to 66 million years ago (Preisig 1994), large scale molecular surveys have now revealed that dictyochophytes are abundant and diverse in the ocean, with most clades lacking any cultured representatives (Carradec et al. 2018; Choi et al. 2020; de Vargas et al. 2015). Because their pigments overlap with those of diatoms, pigment-based analyses thus far have incorporated dictyochophyte contributions as being from diatoms. Studies in the North Pacific subtropical gyre have shown that mixotrophic dictyochophytes graze on picocyanobacteria (Frias-Lopez et al. 2009). Furthermore, ecophysiological characterization of a dictyochophyte isolated from the same region, *Florenciella*, demonstrated increased prey ingestion rates under nutrient limitation (Li et al. 2021). Mixotrophic nutrition might thus explain the success of diverse dictyochophytes in the oligotrophic surface layer of strongly stratified subtropical oceans, as detected in a survey based on amplicon sequencing of the plastid 16S rRNA gene and single cell sorting of field samples using a flow cytometer (Choi et al. 2020).

Dictyochophytes can have a siliceous skeleton during one phase of their life cycle, leading to the entire lineage often being referred to as silicoflagellates, and spines on these skeletons are thought to reduce sinking rates (Han et al. 2019; Preisig 1994). Yet, so far, it is estimated that the silica skeletons of dictyochophytes make up a minor fraction (ca. 1–2%) of the siliceous component of marine sediments, indicating that they may be less abundant than diatoms, less prone to sinking, or are more actively degraded and utilized in the water column by other organisms. Most cultured representatives appear to propel themselves forward using their flagellum and some dictyochophytes (e.g., *Pseudochattonella*) can produce potent ichthyotoxins that detrimentally impact economically important fish species such as Atlantic salmon (Eckford-Soper and Daugbjerg 2016). A nuclear genome sequence for dictyochophytes is still lacking making it difficult to elucidate more of their cell biology, but complete plastid genomes of four cultured species and one uncultured dictyochophyte have been sequenced and analyzed (Choi et al. 2020; Han et al. 2019).

Chrysophytes are also difficult to assign to any one functional category. This diverse group of stramenopiles has more than 1000 described species including some marine representatives (Kristiansen and Škaloud 2017). Molecular surveys of plankton diversity have revealed the presence of novel clades of marine chrysophytes in particular in picoplanktonic cells with no cultured representatives (Choi et al. 2020; del Campo and Massana 2011; Seeleuthner et al. 2018). Chrysophytes contain both purely photosynthetic, mixotrophic and heterotrophic species, and among the photosynthetic chrysophytes that have been cultured, it seems that some become heterotrophic in the absence of light (Wilken et al. 2020). They typically live as solitary cells that are free-swimming but there are also filamentous and colonial forms that can grow as branched or unbranched chains. The cell surface of some chrysophytes is covered by silica scales and chrysophytes also produce siliceous resting cysts that accumulate in sediments. These cysts are abundant in deposits from the Paleocene (66–56 million years ago) while the oldest are from the Cretaceous (~145–66 million years ago).

### 4.5.4 Diversity and Importance of Photosynthetic Protists

Throughout their evolutionary history photosynthetic protists have been an important part of life and modifications of terrestrial ecosystems that facilitated the rise of animals. Through their photosynthetic activity following the rise of cyanobacteria, gaseous oxygen released has gradually changed the Earth's atmosphere and redox status to create the world as we know it (Lyons et al. 2014). As outlined above, today, marine phytoplankton contribute ~ half of annual global carbon fixation into organic carbon compounds, providing the basis for the marine food web and maintaining the oxygenated atmosphere and current $CO_2$ drawdown (Field et al. 1998). Although once thought to use only inorganic compounds, it is now widely accepted that most phytoplankton also use organic compounds. Based on the above sections, we see that the trophic modes of some photosynthetic lineages are complex because they can consume other cells (predatory mixotrophs). Photosynthetic taxa can also live in symbiosis with animals. A prominent example is *Symbiodinium*, a genus of dinoflagellates which is found in association with corals worldwide and with other marine animals, such as sea anemones (Baker 2003; Dixon et al. 2013; Liu et al. 2018; Pontasch et al. 2014), and with the calcifying ciliate *Tiarina* in open ocean waters (Mordret et al. 2016).

Photosynthetic protists are diverse in their size range, spanning three orders of magnitude from picoplankton to mesoplankton, and have representatives in almost all branches of the eukaryotic tree (Fig. 4.5). Historically, the diversity of photosynthetic protists was determined using microscope-based methods and morphological features. Applying such methods about 5000 species have been described (Sournia et al. 1991). Yet, this number largely underestimates the true biodiversity of photosynthetic protists because molecular surveys have revealed undescribed diversity including lineages for which we currently have only environmental sequence data and no cultured representatives, e.g., Massana and Pedrós-Alió (2008). It should be noted that molecular markers like 18S rRNA gene sequences can still underestimate diversity, with amplicon sequencing of some variable regions doing so even more (Monier et al. 2016), and also more generally for organisms with large population sizes and fast turnover rates, such as the prasinophyte algae (Leray and Knowlton 2016; Piganeau et al. 2011). Methods for studying photosynthetic protist communities have employed microscope-based morphological analysis, measurements of photosynthetic pigment signatures, flow cytometric cell counting and cell sorting, and molecular surveys, including species- or group-specific quantitative PCR and FISH, all of which added to our understanding of general ecological patterns (Karlusich et al. 2020).

Green algae were among the earliest eukaryotic algae in the ocean and are the product of the primary endosymbiosis event (Fig. 4.4). Prasinophytes are broadly distributed in the modern ocean and have been observed in the geological record

(Brocke et al. 2006), although their record is considered weaker than for some other algal groups with more robust cell structures. These unicellular green algae are also proposed to bear resemblance to the ancestral alga that gave rise to land plants (Lewis and McCourt 2004; Worden et al. 2015). The model green alga *Chlamydomonas reinhardtii* belongs to the Chlorophyceae or chlorophytes, alongside several other groups. The phylogenetic relationships and taxonomic levels of these classes and orders are under constant revision in part because the prasinophytes are clearly paraphyletic. It will likely take comprehensive phylogenomic analyses with even sampling of the different lineages to resolve this branch of the Archaeplastida tree.

Despite their ecological and evolutionary importance phylogenetic relationships within the prasinophytes are poorly resolved and information on physiological, morphological, and cellular characteristics are lacking for most so-called species (Duanmu et al. 2014; Marin and Melkonian 2010). The exception is the Mamiellophyceae class, for which a considerable body of literature exists for three genera that belong to the picoplankton size class, and therefore often termed "picoprasinophytes". Isolates of *Bathycocccus*, *Micromonas*, and *Ostreococcus* also have particularly small genomes (13–22 Mb) for eukaryotic cells and *Bathycocccus* and *Ostreococcus* especially seem to have engaged in an intriguing evolutionary process to reduce genome size, while *Micromonas* seems to have simply not expanded protein families as extensively as larger and multicellular archaeplastids (Moreau et al. 2012; Worden et al. 2009). The diminutive cell size of the Mamiellophyceae taxa renders a low ratio of cell surface area to volume, which provides a competitive uptake advantage relative to other eukaryotic phytoplankton in open ocean areas where nutrients can be scarce. Use of qPCR was instrumental in demonstrating that there are different *Ostreococcus* clades that rarely co-occur in nature, although initially they had been proposed to co-reside by partitioning the water column vertically, based on growth versus irradiance experiments on isolates in the laboratory. It now appears that one of the clades is better adapted to nutrient-rich "mesotrophic" conditions while the other is found in more nutrient poor "oligotrophic" environments (Demir-Hilton et al. 2011). Moreover, both clades are found in surface and deeper waters in their respective environments. A similar trend has been observed for two *Bathycoccus* clades using qPCR, with the difference that co-occurrence of both types is more common and they genetically less diverged than the *Ostreococcus* clades (Limardo et al. 2017; Simmons et al. 2016). Apart from the Mamiellophyceae, there are multiple other clades of prasinophytes, some with few cultured representatives (Tragin et al. 2016). Cell sizes within these clades appear to range from 2 to 20 μm. Prasinophytes as a whole abound in a wide range of marine habitats, serving not only as important primary producers but also as food for the predatory protists and in turn contribute to the food web of marine fauna (Bock et al. 2021; Tragin and Vaulot 2018; Worden et al. 2004). In addition, several Mamiellophyceae have recently been observed at high relative abundances in the North Atlantic spring bloom, which traditionally had been considered as diatom dominated (Bolaños et al. 2020).

While comprehensive quantitative maps of prasinophyte distributions are still lacking, changes have already been detected in connection to climate change. For example, in the Canadian Arctic, *Micromonas* has been increasing while larger algae such as diatoms are declining (Li et al. 2009; Worden et al. 2015). In addition, infection of prasinophytes by viruses (see below) is among the earliest known examples of viruses with marine algal hosts (Mayer and Taylor 1979). Finally, there is direct evidence for sexual reproduction in both *Nephroselmis olivacea* and *C. reinhardtii* (Goodenough et al. 2007; Suda et al. 1989), however, otherwise any evidence has largely been indirect. For example, the Mamiellophyceae genomes contain sex-related and meiotic genes (Worden et al. 2009) and comparative genome analyses indicate that sexual reproduction occurs in nature (Grimsley et al. 2010) but with a high prevalence of asexual division. For example, in *Ostreococcus* a minimum of 1 meiosis has been estimated for every 100,000 mitoses (Blanc-mathieu et al. 2017).

Green algae are sometimes found in close association with other protists. Some prasinophytes are observed in photosymbiosis with ciliates (Stoecker et al. 1988), but so far most symbiotic green algae are related to the "core chlorophytes" lineages, which are more common to freshwater environments. For example, the symbiont of the "green" *Noctiluca* is related to the class Pedinophyceae (Sweeney 1976; Wang et al. 2016) and members of the same class have been observed in association with radiolarians (Cachon and Caram 1979), although the latter has not been confirmed with molecular methods. Finally, although symbiotic green algae are found in a number of benthic foraminifers (Hallock 1999), in the iconic relationship with the flatworm *Symsagittifera roscoffensis* (Parke and Manton 1967), and in terrestrial lichens that cover a significant surface of land, they appear to participate in symbioses relatively infrequently in the marine water column.

Diatoms are one of the best studied photosynthetic protistan groups and belong to the stramenopiles. Diatoms are widespread in the plankton and benthos of marine and freshwater habitats occurring as solitary cells or chains of cells that are linked by hollow silica tubes (setae), mucilage, or chitin filaments. Fossil records and molecular phylogenetic analyses have been used to establish that the centric diatoms are the most ancient among diatoms, appearing ~150 million years ago (Cermeño 2016). Diatoms endured the Cretaceous-Tertiary mass extinction event 66 million years ago and thereafter their diversity increased (Benoiston et al. 2017). Their extraordinarily modern-day diversity comprises an estimated 100,000 species (Malviya et al. 2016) and, due to the many chain-forming species and species with large size (>100 μm), some were well represented in the early sampling campaigns of oceanographers like Viktor Hensen. Thus, the ecological relevance of diatoms was recognized early on and today their importance for fisheries and marine food chains is well established.

Diatoms have a diplontic life cycle that is often characterized by long periods (up to years) during which diploid cells divide mitotically alternated with brief periods (days) of sexual reproduction. They have an intricate silica cell wall, which is called a frustule and consists of two halves (called thecae) that overlap like a Petri dish (Hildebrand and Lerch 2015; Karlusich et al. 2020). It is now known that diatoms take up dissolved silicic acid and concentrate it in the cytoplasm within silica deposition vesicles (SDVs) near the plasma membrane (Heintze et al. 2020). Frustules may have a defensive role (Pančić et al. 2019) and are thought to impose a limitation that prevents diatoms from phagocytosing other cells.

Planktonic diatoms are important bloom-formers in nutrient-rich regions, such as coastal regions. They are well adapted to growth in mixed turbulent water where cells are shortly exposed to light and pulsed availability of nutrients because they can use their large central vacuole for nutrient storage. Coastal planktonic diatoms contribute importantly to total long-term organic carbon sequestration because considerable parts of coastal blooms sink rapidly (Armbrust 2009). They can also have "destructive" food web roles for example some species produce domoic acid, which is a neurotoxin that accumulates in higher trophic levels (Brunson et al. 2018). Finally, some genera including *Fragilariopsis* and *Pseudo-nitzschia* contain both benthic and planktonic species. The existence of versatility between benthic and planktonic lifestyles suggests that traits acquired while living in the benthos can also be beneficial during a planktonic lifestyle. Diatoms that have both benthic and planktonic lifestyles have been termed tychoplankton (Cahoon 2016).

Pelagophytes described to date are all marine. Two genera that have been successfully isolated are *Pelagomonas* and *Pelagococcus* which occur in the open ocean as well as transition zones beyond the truly coastal environment (Choi et al. 2020; Dupont et al. 2015; Worden et al. 2012). Their abundances suggest considerable contributions, and in the open ocean these are particularly important in the deep chlorophyll maximum (DCM) (Choi et al. 2020). The size of pelagophytes can range from about 3 to 5 μm, such as *Pelagomonas* and *Aureococcus*, to macroscopic sheets and flowing colonies up to 5 cm long (Schaffelke et al. 2004). Distinct morphological features within the pelagophytes are not known as there are multiple environmental clades that lack cultured representatives. An extracellular perforated theca has been proposed as a common feature of pelagophytes (Wetherbee et al. 2020). From an evolutionary perspective, the genome sequences of two pelagophytes (Grigoriev et al. 2021) provided new insights into how they differentiate from diatoms and to understanding of brown tide species, since one of the sequenced pelagophytes was *Aureococcus anophagefferens* a HAB that causes economic damage (Gobler et al. 2011). In the context of symbiosis, the dinoflagellate *Amphisolenia bidentata* hosts cyanobacteria and an undescribed pelagophyte species closely related to *Pelagomonas calceolata*, in an uncommon triumvirate association (Daugbjerg et al. 2013).

Haptophytes are also referred to as prymnesiophytes and have garnered much attention due to the coccolithophores, which form intricate calcified scales (coccoliths) that cover the cell. Coccoliths are built within Golgi vesicles prior to exocytosis on the cell surface and are major component of global biogenic calcium carbonate production (Billard and Inouye 2004). The unique light diffraction of coccoliths allows recognition of the presence of coccolithophores in satellite data. For example, annual blooms of the coccolithophore *E. huxleyi* have been observed in temperate North Atlantic waters near the UK. These blooms are thought to be supported by a high affinity for inorganic nutrients and mechanisms to maintain growth under high light that have been reported in *E. huxleyi* (Paasche 2001; Read et al. 2013). However, there are many other haptophyte groups, including many uncultured clades, that do not have coccoliths and these comprise a major fraction of global primary producer communities in several biogeochemical ocean provinces (Cuvelier et al. 2010). A characteristic feature of haptophytes as a whole is the haptonema (from Greek *hapis* touch and *nema* thread), which is similar to a flagellum but shows a different ultrastructural arrangement of microtubules and is used for swimming, surface attachment, or, in some lineages, for capturing prey (Kawachi et al. 1991).

The origin and evolutionary affiliation of haptophytes are contentious. Based on their plastids surrounded by four membranes, containing the chlorophylls a and c, various carotenoids, and the presence of the carbohydrate storage product beta-1,3-linked glucan, they were initially grouped with the stramenopiles (Cavalier-Smith 1981). Newer analyses place the haptophytes with a newly discovered deep-branching lineage, the rappemonads (Kim et al. 2011b), and the centrohelids, in a group termed the haptista (Burki et al. 2016) and new hypotheses have been developed for origins of their plastids (Dorrell et al. 2017). The earliest records of coccolith fossils correspond to the origin of calcifying haptophytes at ca. 220 mya. Coccolithophores have been used to calibrate molecular clock analyses and bio-stratigraphic dating since they are abundant microfossils in sediments. Haptophytes are estimated to have diverged around the onset of the Cryogenian "snowball Earth" (1031–637 mya) and extant haptophyte lineages diverged about 543 mya, early in the Cambrian period (Liu et al. 2010). This period was characterized by rapid and widespread diversification of life, however extant coccolithophores likely diversified from just a few lineages that survived the major extinction event at the Cretaceous-Tertiary boundary. In contrast, non-calcifying haptophytes were less affected by this extinction event (Medlin et al. 2008), possibly due to their ability to switch from phototrophy to mixotrophy.

Most haptophytes are non-calcifying, and many of these are predatory mixotrophs unlike the coccolithophores which generally lack this capacity (Anderson et al. 2018; Frias-Lopez et al. 2009; Kamennaya et al. 2018). In general, the mechanism of phagocytosis in mixotrophic haptophytes is not well known because cultures are lacking for the important marine lineages (Cuvelier et al. 2010; Frias-Lopez et al. 2009). Field studies indicate that uncultured pico- and nano-planktonic haptophytes are exceptionally diverse and they contribute considerably to primary production in the open ocean (Liu et al. 2009). Among cultured non-calcifying haptophytes that are environmentally important is the colony-forming genus *Phaeocystis*, which is found from the poles to the tropics. It forms dense blooms that are considered detrimental to growth and reproduction of zooplankton and shellfish (Schoemann et al. 2005). *Phaeocystis* has also been reported to live in close association with radiolarian hosts, in which the alga has "super-developed" plastids (see Fig. 4.2f) as compared to its free-living counterpart. Finally, in terms of the life cycle, most haptophytes are characterized by haploid and diploid stages that may occupy distinct ecological niches (Nöel et al. 2004). Both life cycle stages can grow independently by asexual division and can have distinct scale morphologies during each stage.

## 4.6 Distribution and Vertical Dimension of Protistan Diversity and Ecology: From the Sea Surface to Sediments

Protists are involved in major biogeochemical reactions and acclimate to environmental changes. As discussed throughout the above sections, many protists contribute to photosynthesis, generating organic matter in the photic zone (the sunlit portion of the water column) that fuels the marine food webs throughout the water column (Azam 1998; Ducklow et al. 2001; Gooday et al. 2020; Worden et al. 2015). Those with mineral structures often sink rapidly, bringing labile organic carbon to the deep ocean and sometimes accumulating in large deposits that are observed in the geological record. Thus, by sinking, other physical transport mechanisms, or by trophic interactions with other microbes and viruses that cause aggregation, or with multicellular zooplankton, protists contribute importantly to the biological carbon pump. The biological carbon pump refers to carbon dioxide that is removed from the atmosphere, fixed into organic material, and exported to the deep ocean where it is buried for millennia. Based on a back of the envelope calculation done in 2012, it has been estimated that if phytoplankton would stop its activities, the concentration of carbon dioxide in the atmosphere would rise by another 200 ppm and further accelerate global warming (Falkowski 2012). Photic zone processes, whether it be the fueling of higher trophic levels or contributions to the sequestration and burial of carbon in the deep sea, are intimately connected with ocean physics. Ocean physics from the small to large scale influences cell movement and metabolism, aggregation, dispersal, and many more aspects of ecology and biogeochemistry.

Vertical gradients in the ocean reflect the "dominance" of various microbial metabolisms, which shift dramatically from the surface to the dark ocean and into the ocean floor. Collectively, the resident microbes including protists influence the exchange and cycling of elements that occur in these connected but distinct environments. The photic zone is in fact a small portion of the ocean compared to these other deeper zones, with the dark ocean occupying 94.7% of the total ocean volume (Whitman et al. 1998). Microbial cell concentrations in the dark ocean are much lower than in the surface water, as are those of protists, albeit based on relatively little data. However, just by sheer volume the dark ocean contains about 1.8 times as many bacteria and archaea as does the photic zone (Orcutt et al. 2011; Sogin et al. 2006; Whitman et al. 1998).

### 4.6.1 Protists in the Photic Zone

In the open ocean, the photic zone ranges from the surface seawater to ~200 m deep (Fig. 4.7). A vertical gradient is often seen, with the DCM receiving less light than the surface water, but greater nutrient availability, in stratified water columns. As the name implies, the DCM contains the maximum amount of chlorophyll and typically displays the highest absolute abundances of protists (Rocke et al. 2015). In ecosystems that have pronounced seasonal stratification changes, protists tend to be distributed throughout the photic zone when "winter" mixing occurs and peak

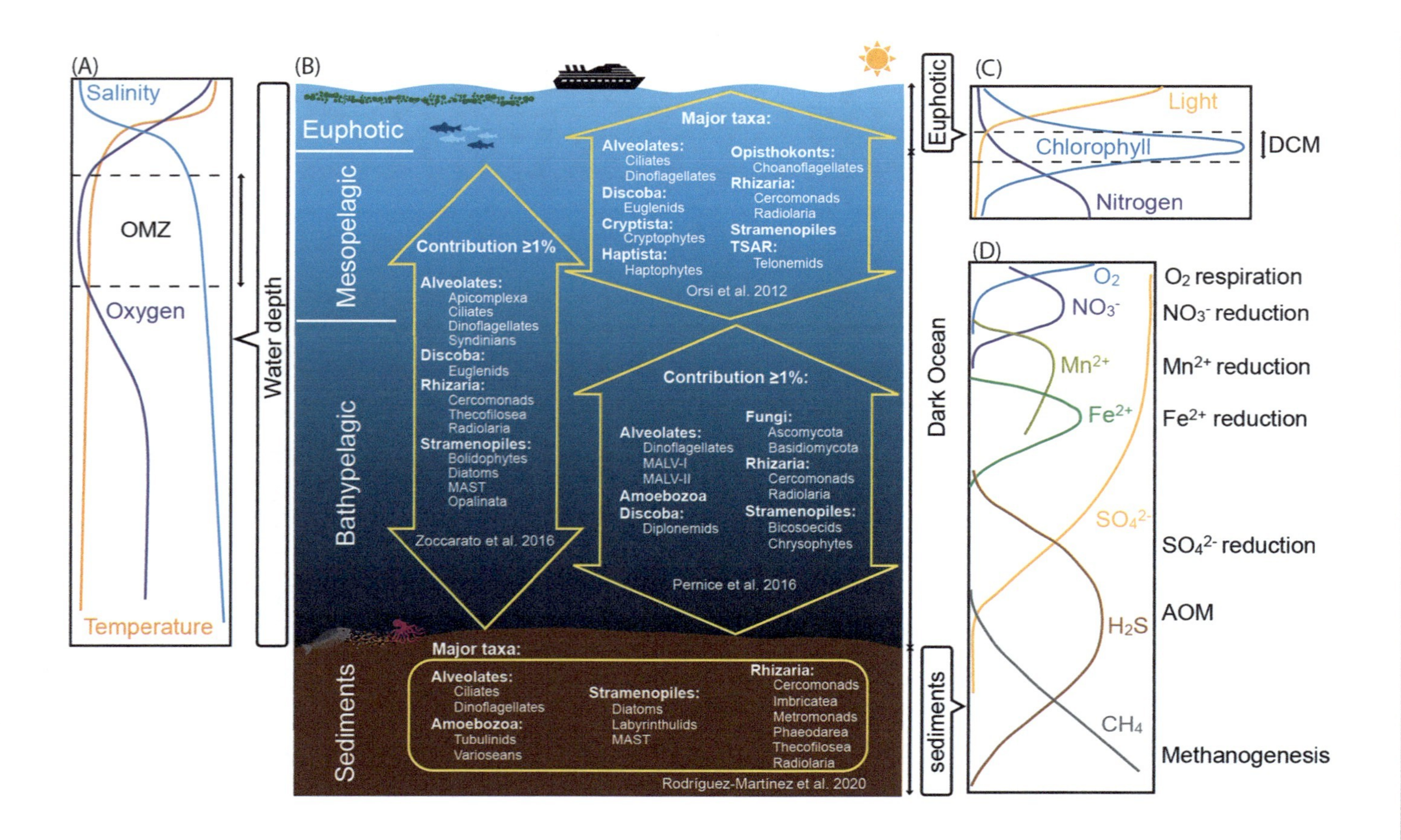

(A)
Salinity
OMZ
Oxygen
Temperature
Water depth
(B)
Euphotic
Mesopelagic
Bathypelagic
Sediments
Major taxa:
Alveolates:
Ciliates
Dinoflagellates
Discoba:
Euglenids
Cryptista:
Cryptophytes
Haptista:
Haptophytes
Opisthokonts:
Choanoflagellates
Rhizaria:
Cercomonads
Radiolaria
Stramenopiles
TSAR:
Telonemids
Orsi et al. 2012
Contribution ≥1%
Alveolates:
Apicomplexa
Ciliates
Dinoflagellates
Syndinians
Discoba:
Euglenids
Rhizaria:
Cercomonads
Thecofilosea
Radiolaria
Stramenopiles:
Bolidophytes
Diatoms
MAST
Opalinata
Zoccarato et al. 2016
Contribution ≥1%:
Alveolates:
Dinoflagellates
MALV-I
MALV-II
Amoebozoa
Discoba:
Diplonemids
Fungi:
Ascomycota
Basidiomycota
Rhizaria:
Cercomonads
Radiolaria
Stramenopiles:
Bicosoecids
Chrysophytes
Pernice et al. 2016
Major taxa:
Alveolates:
Ciliates
Dinoflagellates
Amoebozoa:
Tubulinids
Varioseans
Stramenopiles:
Diatoms
Labyrinthulids
MAST
Rhizaria:
Cercomonads
Imbricatea
Metromonads
Phaeodarea
Thecofilosea
Radiolaria
Rodríguez-Martínez et al. 2020
Euphotic
Dark Ocean
sediments
(C)
Light
Chlorophyll
Nitrogen
DCM
(D)
$O_2$
$NO_3^-$
$Mn^{2+}$
$Fe^{2+}$
$SO_4^{2-}$
$H_2S$
$CH_4$
$O_2$ respiration
$NO_3^-$ reduction
$Mn^{2+}$ reduction
$Fe^{2+}$ reduction
$SO_4^{2-}$ reduction
AOM
Methanogenesis

**Fig. 4.7** Current understanding of protistan distributions from surface marine waters to the seafloor. (**a**) Archetypical profile of temperature, oxygen, and salinity in oceanic environments. (**b**) The diversity described for protists reported thus far in each indicated depth horizon and in sediments. Boxes reflect the results of individual studies that employed amplicon sequencing (not visual observation). In that with results from (Orsi et al. 2012), protistan taxa shared between the photic and mesopelagic zones are shown. The boxes representing results of Zoccarato et al. (2016) and Pernice et al. (2016) include protistan taxa that contributed $\geq 1\%$ of the relative amplicon abundance in 18S rRNA gene sequencing surveys. The most abundant taxa within each major group (in bold) in marine sediments are displayed in the box based on results from Rodríguez-Martínez et al. (2020). (**c**) More resolved profiles of light, chlorophyll, and nitrogen of a representative photic zone section. The maximum concentration of chlorophyll is typically located within the photic zone, however note that biomass can be subducted to below the photic zone (Johnson and Omand 2021; Omand et al. 2015)! (**d**) Representative redox transitions in marine sediments on the scale of cm to km below the seafloor (Inagaki et al. 2016; Jørgensen and Kasten 2006). *OMZ* oxygen minimum zone, *DCM* deep chlorophyll maximum, *AOM* anaerobic oxidation of methane. Concentrations and distances not to scale

during the spring blooms triggered by the infusion of nutrients that comes along with winter mixing. In coastal areas and transition zones (moving from the coast to offshore) protistan abundances are often higher than in the open ocean due to overall higher nutrient availability.

The central focus of research on phytoplankton has long been photosynthesis and its metabolic product, $O_2$. Since the great oxygenation event, many complex life forms have evolved that are dependent on availability of oxygen (Sánchez-Baracaldo and Cardona 2020), although permanently anoxic environments also exist in which $O_2$ is toxic for the inhabitants. Primary producers are a source of food for zooplankton in the photic zone, including for heterotrophic protists that belong to the zooplankton and reportedly consume between 8 and 100% of the phytoplankton standing stock biomass (microscope counts) (Sherr and Sherr 2002). Vertical migrators also exist that come from darker waters below to graze in the photic zone. Limited data exists where specific or even bulk heterotrophic marine protists have been enumerated. This is because many are not amenable to fixation and time-consuming counting by microscopy is still the best method for enumerating heterotrophic taxa. To identify the taxon beyond being a heterotrophic predator, target sequences and FISH probes targeting those sequences are required.

Photosynthetic taxa do not extend below the photic zone (except as they exit the system as sinking blooms, aggregations, detritus, or through mesoscale processes). Laterally though, there is great variation with coastal and continental shelf regions often dominated by diatoms, dinoflagellates, and calcifying haptophytes (coccolithophores) that are able to form large blooms, while the open ocean is often dominated by cyanobacteria alongside picoplanktonic eukaryotic phytoplankton such as small prasinophytes, chrysochromulina-like haptophytes, as well as small stramenopiles like pelagophytes and chrysophytes. This emphasizes the important role of cell size in determining the global ecological patterns of photosynthetic protists (Peter and Sommer 2013), as discussed above. Protists have other adaptations to the low nutrient concentrations that frequently occur in the surface of stratified photic zones, such as high-affinity transporters, capacity to use organic nutrients and ingestion of particles (e.g., Arenovski et al. 1995; Finkel et al. 2009; Palenik and Morel 1990; Wilken et al. 2019; Zubkov and Tarran 2008). Each lineage and their individual members employ a plethora of diverse ecological strategies.

### 4.6.2 Protists in the Dark Ocean: Oxygen Minimum Zones and Sediments

Based on microscope and flow cytometry cell counts the absolute cell abundance of protists in the dark ocean appears to decline proportionally with that of bacteria and archaea. This indicates that predation by protists is taking place (Pernice et al. 2015) although little is known about protistan functional roles in the dark ocean. Without the availability of sunlight, microbial metabolisms in the dark ocean and sediments are based on redox reactions and rely on organic matter sinking from the surface (Orcutt et al. 2011).

The regions with the lowest oxygen saturation are called oxygen minimum zones (OMZs, Fig. 4.7). While there is no general agreement concerning threshold concentrations of oxygen, the major OMZs are defined by $O_2 < 20$ M and can reach concentrations as low as 1 M $O_2$ in the core (Paulmier and Ruiz-Pino 2009). With respect to protists, abundances in OMZs can fluctuate with seasonal shifts, as observed for example in the Canadian Pacific, relative abundances of ciliates and euglenozoans (i.e., diplonemids and symbiontids) increased in anoxic zones of the water column during summer (Orsi et al. 2012). In OMZs heterotrophic metabolism relies on other electron acceptors than oxygen such as nitrate and nitrite, while chemotrophic metabolisms can utilize different types of nitrogen (and other) compounds such as nitrite and ammonia as energy sources (Lam and Kuypers 2011). This metabolic diversity so far appears largely to be confined to bacterial and archaeal communities, although dissimilatory nitrate reduction has been reported in a fungus isolated from an OMZ (Kamp et al. 2015). Regardless, protists appear to control the abundance of nitrate-reducing and ammonia-oxidizing bacteria by preying on them (Orsi et al. 2012), and protists can consume up to 28% of the bacterial biomass based on data from microscope cell counts (Medina et al. 2017).

Going a little bit deeper, if most studies regarding marine protist diversity have focused on planktonic environments, the subseafloor sediments are estimated to contain $2.9 \times 10^{29}$ cells (Kallmeyer et al. 2012). This assessment is about five orders of magnitude higher than the numbers estimated for the entire ocean waters, i.e., $1.37 \times 10^{24}$ cells (Whitman et al. 1998). However, it should be noted that few studies have quantified protists in any of these environments, rather most of what we know is about their diversity through early environmental clone libraries and through amplicon sequencing. At this stage, a comprehensive systematic characterization of ocean habitats using amplicon sequence variants (ASVs) (Amir et al. 2017; Callahan et al. 2016; Eren et al. 2015) and metagenomics would be most beneficial for improving ecological resolution of different taxon distributions (Needham et al. 2017), even if broad-scale differences can be similar to earlier approaches of grouping sequences (Glassman and Martiny 2018).

To date, most deep-sea studies have found little overlap in community composition of protists in planktonic versus subseafloor environments; and that there are few relevant reference sequences from cultured organisms for categorizing these taxa (Forster et al. 2016). Among the classified protists, Rhizaria appear to be the dominant eukaryotes in surface sediments at depths ranging from 79 to 2939 m (Wu and Huang 2019). In subseafloor ecosystems, microorganisms have long been thought to compete for limited energy sources (Hoehler et al. 1998; Bradley et al. 2019, 2020), and differing dominant "metabolic guilds" are thought to have shaped the different redox zones over depth gradients (Jørgensen and Kasten 2006). These redox zones start with oxygen respiration at the subseafloor surface followed by nitrate-, iron-, and sulfate-reduction, anaerobic oxidation of methane, and methane generation (methanogenesis, Fig. 4.7), reflecting the typical profile of marine sediments. While some of this metabolic diversity again appears to be confined to bacteria and archaea, ciliates have symbiotic relationships with bacteria and archaea that are capable of aerobic methane oxidation, sulfate reduction, and methanogenesis in sulfidic marine sediments (Edgcomb et al. 2011). Additionally, in anoxic sediments, a new type of symbiosis between south-seeking magnetotactic protists and magnetite-containing *Deltaproteobacteria* has been observed (Monteil et al. 2019). Although the benefit for protists is not fully understood, it is thought that the motility of this symbiotic consortium along the geomagnetic field allows the protist to move toward locations that are optimal to them and allows the sulfate reducers to grow using the protist's metabolic products (Monteil et al. 2019). Finally, in marine sediments protists defy the dogma of redox zonation and often aggregate bacteria or archaea that would generally be partitioned over the three different redox zones.

### 4.6.3 Diversity of Marine Protists in the Vertical Dimension

We know by now that protist abundance and community composition are controlled by different environmental factors that are still poorly understood, especially in the dark ocean and marine sediments (Fig. 4.7). Based on the relative amplicon abundances, ciliates, dinoflagellates, and stramenopiles are considered important in deep and sediment ecosystems (Orsi et al. 2012; Pernice et al. 2016; Rodríguez-Martínez et al. 2020). In sediments underlying different water column depths the radiolarian Acantharea show higher relative amplicon abundances in cores collected from greater depths (881 and 957 m) than lesser depths (200 and 650 m) (Rodríguez-Martínez et al. 2020). They are also evenly detected in both the photic and mesopelagic zone which may be due to life cycle stages where flagellated swarmers are released from sinking cysts in deep waters (Decelle et al. 2013). Within stramenopiles, amplicon relative abundances indicate that in the dark ocean MAST, chrysophytes, bicosoecids, and diatoms are present, although it is difficult to discern which members reflect recent export versus in-situ growth (Pernice et al. 2016). Marine sediments also contain these stramenopile taxa (Rodríguez-Martínez et al. 2020). Cercozoans (members of the Rhizaria) display much higher relative amplicon sequences in the ocean water column (Orsi et al. 2012; Zoccarato et al. 2016) than in sediments (Rodríguez-Martínez et al. 2020). A molecular survey directly comparing diversity of benthic and pelagic environments suggests there is a higher diversity of protists in the former than the latter (Forster et al. 2016).

Differences in numerical abundances for inferring importance can be misleading. For example, in NE Atlantic deep-sea sediments at 2170 m depth, foraminifera and bacteria both account for 50% of algae degradation, but the biomass of the former is negligible compared to that of the bacteria (Moodley et al. 2002). Finally, rare populations are thought to maintain the diversity in different environments (Lennon and Jones 2011) and can quickly respond to environmental changes such as redox fluctuations (DeAngelis et al. 2010), tidal cycling (Ling et al. 2018), and even the deep-sea oil well blowout (Kleindienst et al. 2016). Rare protists have been reported to shift in relative amplicon abundances in response to fluctuations in OMZs (Orsi et al. 2012), chaotic flows (Villa Martín et al. 2020), and seasonal variations in surface seawater (Genitsaris et al. 2015).

Further investigation of protistan diversity, cell biology, and ecology, alongside their metabolic interactions with other microorganisms will advance the knowledge required to evaluate the efficiency and magnitude of global biogeochemical cycles. In just one study that addressed multiple marine zones including the mesopelagic and bathypelagic, it was determined that protists graze significantly upon bacterial and archaeal prey across the DCM (80–130 m), the upper mesopelagic zone (220 m), the deep Antarctic Intermediate Water (750 m), and the bathypelagic North Atlantic Deep Water (2500 m) (Rocke et al. 2015). Compared to the 8–100% removal rates reported in the photic zone (Sherr and Sherr 2002), these removal values ranged from 3.8–31.1% of the bacterial and archaeal standing stock biomass in the deep ocean (Rocke et al. 2015), or about 20% of the in-situ abundance of bacteria and archaea in the deep ocean oxycline (Edgcomb 2016).

## 4.7 Forces of Mortality

Biomass in the oceans is ~1.2% of that in terrestrial systems, yet productivity is roughly equal (Bar-On et al. 2018). This surprising fact is primarily due to ocean biomass being mainly microbial and having a fast turnover. High turnover rates in marine ecosystems come from strong top-down biomass "removal" imposed by grazing and viral infections, leading to turnover times of the microbial components on the order of days. Understanding the ecological and evolutionary implications of how these removal processes act on protists is important for understanding their influences on the marine microbiome and biogeochemical cycles.

### 4.7.1 Timeline of Virus Discovery

While viruses escaped the observations of Haeckel and Leeuwenhoek, they found their way into art and economic movements in the seventeenth century Europe. Indeed, when Leeuwenhoek was 5 years old the popularity of tulips in Holland had reached a zenith as extremely prized possessions ("tulip mania"). The most coveted tulips were those whose deep color was broken with white and the source of these stripes was chlorosis caused by a plant virus, one of the first documented viruses. As knowledge of viruses has increased it has become clear that their influence lies far beyond simply infecting and killing their hosts. Indeed, most viruses harbor genes that can augment host metabolism and influence the ecosystem in surprising ways.

The first formal description of a virus ("*Contagium vivum fluidum*" from latin: "contagious living fluid") was of the Tobacco Mosaic Virus (TMV) in 1892. TMV was defined as tiny (originally defined as being able to pass through a porcelain Chamberland filter of ~0.2 μm) obligate pathogen dependent on intracellular multiplication in hosts. After TMV, viruses of bacteria (bacteriophages) were discovered in ~1910 and subsequently revolutionized molecular biology. As early as 1958 there were observations that suggested viruses may infect the aquatic photosynthetic green alga *Chlorella* (Brown and Malcolm Brown 1972; Zavarzina and Protsenko 1958).

However, it was not until 1972 that the first virus of a protist was enriched and shown to be the causative agent of mortality in the amoebozoan *Entamoeba histolytica* (Diamond et al. 1972; Diamond and Mattern 1976; Wang and Wang 1991). The first virus isolated on a photosynthetic protist came from the multicellular green alga *Chara* (Gibbs et al. 1975) and shortly after the first virus of a marine phytoplankter was discovered, which infected the prasinophyte *Micromonas pusilla* (Mayer and Taylor 1979).

Despite these discoveries, the realization of the importance of viruses in ocean ecosystems did not become clear until two publications showed extraordinarily high abundances of free "living" particles of $>10^7$–$10^8$ per mL (Bergh et al. 1989) and estimated that up to 70% of marine microorganisms were infected at any given time (Proctor and Fuhrman 1990). Further experimental evidence of their importance came from experiments where the addition of concentrated viruses to natural seawater samples resulted in a considerable decrease in primary productivity by phytoplankton (Suttle et al. 1990). Over the subsequent decades, in-situ experiments, culturing studies, and natural observations have shown that viruses are an important source of mortality in the ocean and can impact community composition as well as facilitating genetic exchanges. Thus, these entities, which are only active once they have infected a host, contribute importantly to the ecology and evolution of the marine microbiome as well as to the cycling of nutrients and energy in the ocean (Breitbart et al. 2018; Zimmerman et al. 2020).

### 4.7.2 Current Perspectives on Viruses of Marine Protists

The current diversity of viruses of marine protists is substantial with most of the cultured viruses being those that infect photoautotrophs (Hyman and Abedon 2012). The diversity of known viruses of protists includes viruses made up of ssDNA, dsDNA, ssRNA, and dsRNA and ranges in size from a few genes (thousands of bases) to the thousands of genes (1–2 million bp), with the latter being akin to the size of small bacterial genomes. The characteristics of each of these virus types vary in terms of infectivity and burst sizes but rigorous comparative studies are still lacking.

Redirection of the carbon flow by viral infection of protists has been linked to phytoplankton bloom demise in the ocean (Brussaard 2004; Suttle et al. 1990; Weynberg 2018). For example, viruses associated with *E. huxleyi* blooms studied in the field and in mesocosm experiments appear to end these blooms (Brussaard et al. 1996; Schroeder et al. 2003; Vardi et al. 2012). Other phytoplankton species for which viruses are implicated in bloom decline include the pelagophyte *Aureococcus anophagefferens* (Gastrich et al. 2004; Gobler et al. 2007) and the haptophyte *Phaeocystis* (Brussaard et al. 2005; Ruardij et al. 2005). Still, there is much to learn about the degree to which viruses are the causative agent of phytoplankton bloom declines, especially since viruses commonly co-exist with their hosts without causing such dramatic events. Much less is known about viruses that infect heterotrophic protists although complex defenses against viral infection, e.g., virophages

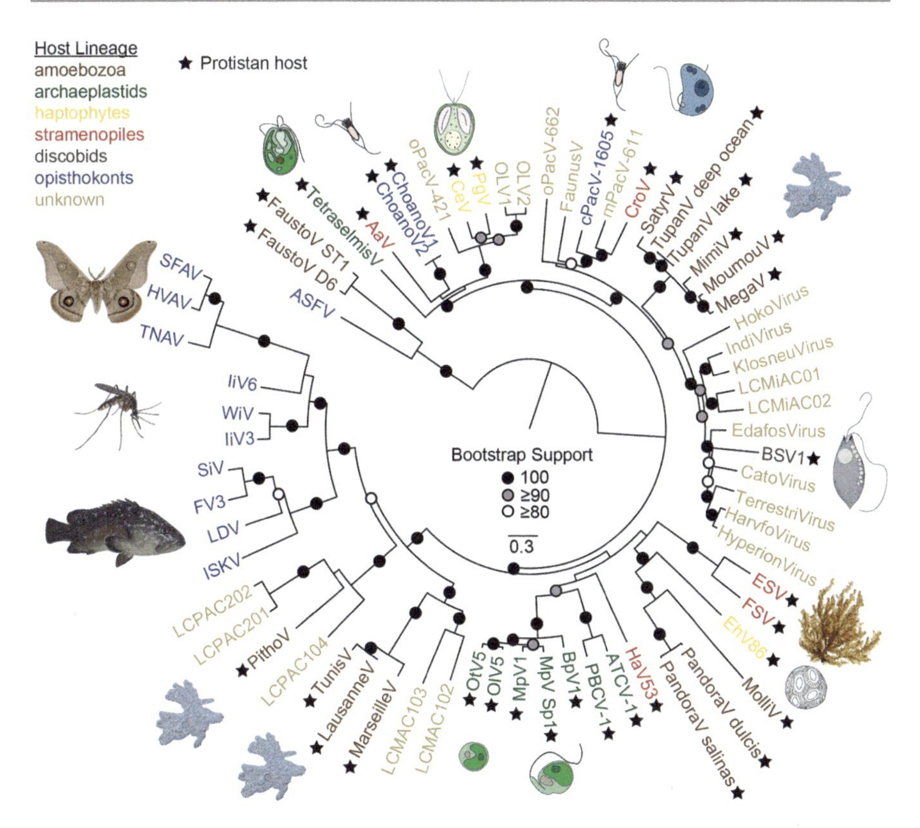


**Fig. 4.8** Maximum Likelihood phylogenomic tree of Nucleocytoplasmic Large DNA viruses (NCLDV) based on 10 putatively vertically inherited proteins. To build this type of tree, individual protein sequences need to be curated carefully to retain only those that are of high quality (for example, not truncated) and are not paralogs. This can be an issue because the growing number of NCLDV genomes are derived from metagenomics assemblies (MAGs). After curation, sequences are aligned, trimmed to remove ambiguously aligned or non-homologous positions, and an evolutionary model is selected for each individual gene, or on the entire group after concatenation (as done here). Subsequently, a phylogenetic tree is constructed (in this case with IQ-tree, under the model LG + C20 + F + G-PMSF) with calculations of bootstrap support (here, non-parametric bootstraps with 500 replicates). Here, statistical support is indicated when it exceeds 80% bootstrap support. The scale bar indicates the number of substitutions per site. NCLDV infect a variety of eukaryotic hosts, mainly protists, as indicated here for those recovered from hosts using cultivation and cultivation-independent studies, modified from Needham et al. (2019a). The color of each virus label indicates the host lineage with which the virus has been found to associate, as well as those with unknown hosts. Black stars indicate protistan hosts. The drawings of critters are not to scale. It is likely that the skew towards higher relative contributions of photosynthetic protists as hosts of NCLDV results from there being more photosynthetic protists in culture relative to heterotrophic and/or mixotrophic protists

have been observed (Mougari et al. 2019; Yau et al. 2011). Overall, viruses of protists that belong to different trophic modes or roles should be considered in conceptualizations and models of ecosystem dynamics (Taylor et al. 2014, 2018).

### 4.7.3 Diversity of Viruses Infecting Marine Protists

Probably the best studied group of viruses of marine protists are the dsDNA viruses from the nucleocytoplasmic large DNA virus (NCLDV) family. These include a wide diversity of viruses that infect different taxa within the haptophytes, amoebozoans, stramenopiles, and archaeplastids (Fig. 4.8). While bacteriophages and other viruses are generally small, the Mimiviridae is a family of NCLDV that is remarkable because of their large physical- and genome size, some of which are similar to small bacteria in these characteristics. In fact, their original discovery in association with the discovery of Amoebozoa was delayed for almost 10 years because they were erroneously considered to be bacteria (La Scola et al. 2003; Raoult et al. 2004). Mimiviruses (for "mimicking" bacteria) now have been isolated from a variety of environments and those from the ocean have been found in association with or co-cultured with a similar host range to all other NCLDVs (Mihara et al. 2018). Based on community surveys this viral group has been proposed to be as diverse or more diverse than bacteria and archaea together (Mihara et al. 2018). Notably, genes once thought to be hallmarks of cellular life are encoded by viruses: genes involved in transcription and translation are common across the Mimiviridae family including tRNAs, amino-acyl tRNA synthetases, transcription factors (Abrahão et al. 2018; Schulz et al. 2017), in addition to the replication machinery encoded by most viruses (such as DNA polymerases).

NCLDVs with sequenced genomes show an incredible array of genome potential with the ability to supplement functions their hosts already encode or bring new functions to the infected host. These can include genes to help with nutrient acquisition such as nitrogen- (in the form of ammonium) (Monier et al. 2017; Needham et al. 2019a) and phosphate transporters (Bachy et al. 2018) as well as supplementation of carbohydrate and fermentation metabolism (Schvarcz and Steward 2018). Diverse Mimiviruses also encode chitinases that may either be involved in degradation of host chitinaceous structures or the prey of heterotrophic protists that are infected (Needham et al. 2019b; Van Etten et al. 2017).

One striking difference between viruses of marine protists versus cyanobacteria is that the former do not carry genes involved in photosynthesis, genes that are common in cyanophages (Lindell et al. 2005; Sharon et al. 2009). This may be due to difficulties in viral acquisition of a photosystem protein that possesses the "correct" transit peptide, a peptide required to bring proteins across the chloroplast membrane of photosynthetic protists in order to function in photosynthesis (Bhattacharyya and Chakraborty 2018; Patron and Waller 2007). However, viruses of protists do encode other proteins that use light, not to be mistaken for photosystems involved in oxygenic photosynthesis. Viral rhodopsins were first reported in a virus of *Phaeocystis globosa* and in metagenome contigs putatively from eukaryotic viruses (Yutin and Koonin 2012). They have now been reported in uncultured viruses from choanoflagellates that were single cell sorted from the wild (Needham et al. 2019a, b). The presence of these microbial rhodopsins is independent of whether or not the host encodes such proteins. Hence, they can bring a new metabolic function to the host (Needham et al. 2019b). Furthermore, some

Mimiviruses encode the proteins required for biosynthesis of the photoreceptor pigments that rhodopsins need to function, i.e., retinal (Needham et al. 2019a, b). This is notable because likely host organisms do not appear to synthesize beta-carotene which is cleaved to retinal, and thus the viruses must encode multiple genes to do so which would then also be expressed by the infected host.

The dsRNA and ssRNA viruses are other types which infect aquatic protists including fungi, stramenopiles, and ciliates (Short 2012). In contrast to the NCLDV, these viruses are small, typically encoding only a few genes. The hallmark gene of RNA viruses is the RNA-dependent RNA polymerase (Gustavsen et al. 2014; Kaneko et al. 2021). While much of virus research in the ocean has focused on dsDNA viruses, RNA viromes are of great interest and there are some indications that these viruses may be much more abundant than previously thought (Kaneko et al. 2021; Steward et al. 2013; Urayama et al. 2018; Wolf et al. 2020; Zeigler Allen et al. 2017).

Finally, viruses influence host evolution by integrating into genomes, a phenomenon that has been described for the human genome as well (Griffiths 2001), and in plant genomes (Maumus et al. 2014). Likewise, the ability of viruses to integrate into protist genomes has been observed in cultured host–virus pairs (Delaroque and Boland 2008; Filée 2014; Gallot-Lavallée and Blanc 2017; Meints et al. 2008; Sharma et al. 2014). The consequences of this integration are unclear, but the viral material appears to reflect remnants of ancient infections and is not expressed at the same frequency as other genes. This suggests that these ancient remnant genes may have become nonfunctional and may not influence the host greatly, or are only expressed under certain conditions. Nevertheless, these observations complicate studies aiming at the description of extant viral diversity through metagenomic sequencing and must be considered when investigating protistan genomes.

### 4.7.4 Death of a Protist Via Predation

Forces of mortality are plentiful in the oceans. Protists can die from being eaten either by metazoans or by other protists, from being parasitized, through viral lysis, or through aggregation and sinking. The quantitative impact of mortality on protist populations has mainly been investigated for primary producers, which often appear to be eaten as fast as they reproduce, keeping populations stable despite rapid growth rates. A synthesis study suggested about two-thirds of the global phytoplankton productivity is removed through consumption by protistan zooplankton, while a large part of the remainder is eaten by larger metazoan zooplankton (Schmoker et al. 2013; Steinberg and Landry 2017). In this scenario, grazing already accounts for the loss of the majority of primary production. However, studies that also take viral lysis into account often find its importance quantitatively similar to grazing (Brussaard 2004). The relative importance of the forces causing protistan mortality also likely varies, for example, viruses may terminate phytoplankton blooms or have impacts that vary in different marine habitats or ocean regions (Mojica et al. 2016).

Despite the quantitative importance of grazing mortality for biogeochemical fluxes in the ocean, there is surprisingly little information about the organisms that consume protists. While there is some information on consumers of photosynthetic protists, this information is lacking for the predators that consume heterotrophic protists. In general, the dilution assays used to quantify community level grazing impact on phytoplankton are rarely combined with techniques that identify the actively grazing organisms. However, when combined with a detailed characterization of the microzooplankton community composition (Neuer and Cowles 1994) predators can be identified, leading to data suggesting that dinoflagellates are important herbivores in some environments (Sherr and Sherr 2007). Nevertheless, despite estimates of grazing mortality over large geographical ranges, the specific interactions responsible for these often remain elusive. It should be noted that if mixotrophic grazers control their grazing rates in response to nutrient or light availability (Anderson et al. 2017; Li et al. 2021), then, these environmental conditions will introduce considerations distinct from those for heterotrophic grazers (Edwards 2019). In cases where predator protistan–prey interactions have been tracked, the focus has been on predation on the smallest primary producers. For example, dinoflagellates, ciliates, telonemids, and the stramenopile MAST clades 1 and 3 have been shown to consume the picoeukaryote *Micromonas* in the Pacific Ocean (Orsi et al. 2018).

Knowledge of specific predator–prey interactions is also important in cases where overall community grazing mortality is low. Predators that selectively feed on rare community members or avoid dominant species can influence community composition and bloom dynamics. For example, despite high microzooplankton abundances and grazing on larger phytoplankton species, it has been suggested that low grazing pressure on the coccolithophore *E. huxleyi* is an important factor for bloom formation by this organism (Olson and Strom 2002). Results from laboratory experiments were taken to suggest that the haploid life-cycle stage of *E. huxleyi* is an inducible anti-predator response that might underpin low grazing pressure on this organism (Kolb and Strom 2013). The mechanistic basis for this hypothesis remains unknown as do the field implications. Both constitutive and inducible defenses against predators are common among photosynthetic protists (Pančić and Kiørboe 2018; Van Donk et al. 2011). For instance, the solid silica frustule of diatoms or the long spines found in some species may deter predators. Induced colony formation to increase particle size, or production of toxins may also both contribute to deterrence or negative effect on grazers.

As mentioned above almost nothing is known about predation on heterotrophic protists and their specific loss rates are rarely quantified. In contrast to the widespread assumption that protistan predators prefer prey of about a tenth of their own size, the diversity of feeding strategies found among different protists includes many predator-prey size relationships with large deviations from this assumption. Many heterotrophic protists are fed upon by other similarly sized protists as detected using double FISH staining for visualization of both predators and the protistan prey in their food vacuoles (Piwosz et al. 2021). Such trophic interactions among equally sized protists imply that increasing trophic levels in microbial food webs are not necessarily paralleled by increases in size. Hence, carbon may flow through the microbial food web via several trophic levels before reaching larger metazoan zooplankton which are fed upon by fish (Piwosz et al. 2021). Because organic carbon is considered lost at each trophic transfer (although utilized by other microbiome members in one way or another), a higher number of trophic levels will decrease food web efficiency. In contrast, a larger proportion of mixotrophic protists in the food web has been predicted to have the opposite effect—exhibiting higher efficiency (Stoecker et al. 2017; Ward and Follows 2016). Quantitation of the "exact" routes of carbon flow is needed to infer the food web efficiency that supports life at the highest trophic levels in the ocean, including fisheries of commercial interest.

## 4.8 Looking Forward

Studies of marine microbiomes are governed by prevailing conceptual and methodological challenges. Each of the topics below focuses on protists and could fill additional chapters, but will be discussed here in brief and without attention to prioritization order.

### 4.8.1 Classics: The Delineation of Protistan Species

Although the species is a basic unit and currency in ecological and evolutionary research in any environment, there is no consensus on how a microbial species is best defined. Traditionally, protists have been defined by their morphological features, but we now know that morphological features can change under varying environmental conditions (Pizay et al. 2009), and comparisons of morphological and molecular data as well as mating experiments have provided evidence for cryptic diversity (Amato et al. 2019; Sarno et al. 2005). Not least, small cells such as picoeukaryotic cells often lack distinctive features. The classical biological concept by Ernst Mayr (Mayr 1996) that defines a species as a member of an interbreeding population that is reproductively isolated from other such groups and capable of producing fertile descendants cannot be readily applied to most protists, due to a lack of knowledge on their sexual reproduction (Silva 2008). Thus, alternative species concepts have been proposed (De Queiroz 2007; Samadi and Barberousse 2006) but operationally objective criteria to define a microbial species are still lacking. Advances in resolving cryptic diversity have utilized metabarcoding/amplicon datasets for reconstructing phylogenetic networks (De Luca et al. 2021) and progress in this field will be crucial for understanding the population biology of protists that underlies their adaptive responses to contemporary and future environmental conditions.

### 4.8.2 Classics: Everything Is Everywhere, but, the Environment Selects Versus Endemism

Another challenging and highly debated concept is that of "*everything is everywhere*, but, *the environment selects*," which roots back to the nineteenth century (O'Malley 2007). It postulates that the abundance of microbial species is so large that their dispersal is never restricted by geographical barriers (Finlay 2002). This concept still dominates the ecological and evolutionary understanding of microbial distribution based on culture studies (de Wit and Bouvier 2006). However, environmental molecular surveys have provided evidence that barriers likely do exist for dispersal and while some species might be globally distributed, others are not (Casteleyn et al. 2010) or may exhibit niche partitioning so that they can co-reside (Foulon et al. 2008). For example, phytoplankton like the diatom *Fragilariopsis cylindrus* (Lundholm and Hasle 2010; Mock et al. 2017) and the prasinophyte *Micromonas polaris* (Simmons et al. 2015) are found in both Arctic and Antarctic environments, while the cryptophyte *Geminigera cryophilia* has only been observed in Antarctic environments (Taylor and Lee 1971), a finding which has held up in amplicon sequencing studies.

### 4.8.3 Classics: Diversity and Stability of Plankton Communities

The dazzling diversity of protists found in the ocean has long fascinated ecologists and led G. Evelyn Hutchinson to formulate the paradox of the plankton. He asked "why do so many planktonic species co-exist in a supposedly homogeneous habitat?" (Hutchinson 1961). Based on the competitive exclusion principle, phytoplankton species competing in a well-mixed environment for only few inorganic resources should outcompete each other resulting in a winner. Of course, we now know that most phytoplankton species utilize dissolved organic resources, not just inorganic compounds, and that many of them are capable of feeding on other microbes. Additional explanatory factors for the high diversity of photosynthetic protists include: (1) the temporal and spatial variability of surface ocean waters, which are now recognized as not being homogeneously mixed (Azam 1998); (2) the dynamics of species interactions that can result in oscillations and chaos, and thereby never settle toward an equilibrium (Record et al. 2013); (3) toxins produced by phytoplankton resulting in maintenance of diversity (Roy and Chattopadhyay 2007). Collectively, these findings might thus have (partly) resolved the paradox. Nevertheless, the mechanisms generating and maintaining the diversity of marine protists, and the relationship between diversity and ecosystem stability (another long-standing debate in ecology) require greater attention. With respect to marine microbial science and phytoplankton diversity, genomics and metagenomics have made clear that functional redundancy, which lumps organisms by what is perceived as their "main" biogeochemical function, is an inappropriate term that likely over predicts system resiliency and stability. Functional redundancy ignores evolutionary trajectories of organisms, the diversity of proteins they contain (many of which have uncharacterized functions that are likely important to how they make a living in the ocean), and different overall gene content in the genomes of different phytoplankton species. These factors underpin the biology of each species and determine how individual microbial species acclimate and adapt to changing ocean conditions as well as how they thrive or not in the novel community assemblies that arise from change. Hence, predicting the degree of diversity an ecosystem can sustain, or maybe more importantly the degree of diversity an ecosystem can afford to lose without ecosystem functions collapsing, is an urgent frontier for microbiome science given that we have entered the sixth period of mass extinction (Cavicchioli et al. 2019; Ceballos et al. 2015).

### 4.8.4 The Uncultured Majority: Quantifying Activities and Trophic Transfer

Beyond known microbiome interactions involving eukaryotes, there may be a plethora of interaction types that have yet to be discovered. In the course of studies on choanoflagellates aimed at improving understanding of the origins of animal multicellularity, it has become clear that bacteria can influence the behavior of these single-celled heterotrophic protists in culture. In particular, specific lineages of bacteria produce a compound that stimulates sexual reproduction in choanoflagellates (Woznica et al. 2017) as well as the transition to a truly multicellular state (Woznica et al. 2016). Although the degree and mechanisms by which such interactions occur in the dilute marine environment is still an open question, these types of discoveries highlight the fact that the many protists that remain uncultured to date may present many novel types of interactions. Unfortunately, this dearth of knowledge impedes identification of trophic linkages or the strength of interactions between different microbiome members, and how they might shift in future oceans.

What approaches can be used to tackle the uncultured majority? Advances in genomic and targeted metagenomic analyses of cultured, e.g., Bowler et al. (2010); Keeling et al. (2014); Mock et al. (2017); Moreau et al. (2012); Read et al. (2013); Worden et al. (2009) and wild algae (Cuvelier et al. 2010; Simmons et al. 2016; Teeling et al. 2012; Vannier et al. 2016; Worden et al. 2012), as well as marker-gene studies (Choi et al. 2017; de Vargas et al. 2015; Ibarbalz et al. 2019; Kim et al. 2011a; Not et al. 2007a; Pernice et al. 2016) have illustrated the tremendous diversity of uncultured taxa and have provided first insights into aspects of their biology and evolution. However, the natural distributions and activities of these algae are generally not known at the level of genetic differentiation that connects to their physiology and ecology. Importantly, much less is known from a genomic perspective about marine predatory protists, apart from e.g., the choanoflagellates (King et al. 2008; Richter et al. 2018) and there are few methodologies that quantify their activities in the field. Hence, the quantities of protistan carbon that moves into e.g., the microbial loop versus more directly into higher trophic levels are more or less unknown, although incredibly important for modeling efforts and for understanding how food networks might change as the ocean changes. Estimating the carbon flow along specific trophic interactions used to rely on tedious microscopy counts following FISH staining of groups of interest (Massana et al. 2009; Piwosz et al. 2021) but can now be supplemented by stable isotope techniques to track substrate uptake or prey ingestion into diverse consumers (Frias-Lopez et al. 2009; Orsi et al. 2018).

### 4.8.5 Bringing Cell Biology to Bear on the Protistan Role in the Marine Microbiome

As outlined throughout this chapter, protistan dynamics are essential parts of the carbon cycle and food networks, processes that determine the ultimate fate of the $CO_2$ fixed by algae through oxygenic photosynthesis (Behrenfeld et al. 2006; Field et al. 1998; Lomas et al. 2013; Steinberg et al. 2001; Worden et al. 2015). Despite the fundamental importance of their biological activities and the massive advances in the knowledge about their evolutionary relationships, mechanistic understanding of the factors that determine the growth, physiology, and fate of protists is often still lacking. Even actual host–virus pairs are typically unknown. Part of the challenge is the sheer diversity of marine protists and the fact that many remain uncultured. However, understanding the biology of protists is critically important for identifying how marine microbiomes will transition as the ocean changes. In addition to a lack of knowledge on specific field interactions with viruses and with predators, key parameters of population biology and sexual reproduction are mostly unknown. In particular, the ecological importance and evolutionary trade-offs of basic features such as the type of mixotrophy or even a motile versus non-motile lifestyle are unknown. Single-cell and targeted metagenomics have moved the field forward, as have transcriptomics and metaproteomics. However, only when these data types are engaged in ways that illuminate cellular responses can they be fully implemented in understanding physiology and ecology in a mechanistic manner. Without comprehension of organismal biology, many studies still rely on correlation analyses and yet most environmental parameters are not in steady state. Just imagine trying to address cancer biology without any knowledge of human physiology, cell boundaries, genome composition, genetics, or epidemiology. Marine science is well positioned to move beyond that place! A next step in this direction is to employ genetics to begin to recover functions for genes in marine protists. Genetic systems are now available for several environmentally relevant marine protists (Faktorová et al. 2020) that can be used to ascertain these functions and to perform studies examining mutant response relative to wild type to understand these functions in a cellular context.

### 4.8.6 Connecting Microbiome Members and Interactions to Ocean Physics and Chemistry

Understanding the interactions among marine protists as well as their interactions with bacteria, archaea, and viruses is challenging. Add to that their responses to the physical and chemical conditions of the environment and how that might modify activities and interactions and we start to tackle microbiome science. One important approach for addressing the complexity of these communities and their dynamics is through long-term time-series observations (Fuhrman et al. 2015; Giovannoni and Vergin 2012). Several marine time-series studies make measurements that are aimed at elucidating the diversity and dynamics of plankton alongside physical and chemical conditions. Unfortunately, protists are sometimes not included in time-series studies, while bacteria, archaea, and less commonly, viruses are more often considered. Nevertheless, time-series analyses have been used to examine protists at a variety of temporal scales. Here, we provide some examples for studies at different time scales, including diurnal (Hu et al. 2018; Madin et al. 2001; Needham et al. 2018), daily (Berdjeb et al. 2018; Fitzsimmons et al. 2015; Lie et al. 2013; Martin-Platero et al. 2018; Needham and Fuhrman 2016), monthly and inter-annual (Choi et al. 2020; Guadayol et al. 2009; Kim et al. 2013; Limardo et al. 2017; Massana et al. 2015; Pasulka et al. 2013; Steele et al. 2011; Wiltshire et al. 2008). Such time-series studies have often revealed strong seasonality in surface waters, which tend to be the focus. These studies also revealed strong correlations between different taxa or with physical and chemical conditions (Anderson and Harvey 2020; Genitsaris et al. 2015; Kim et al. 2013; Simon et al. 2015; Treusch et al. 2011), that can then be used to develop hypotheses on controls. Seasonal variations are also observed in bacterial, archaeal, and viral communities (Cram et al. 2014; Fuhrman et al. 2015; Giovannoni and Vergin 2012; Steinberg et al. 2001; Treusch et al. 2009). While the seasonality of these communities is often quite stable, they can be interrupted by periods of much higher variability, especially during bloom conditions (Lambert et al. 2019; Needham et al. 2018; Needham and Fuhrman 2016). Moreover, some locations such as the subtropical North Pacific Gyre (Ollison et al. 2021) reportedly have much weaker seasonality in bacterial, archaeal (Bryant et al. 2016), and viral communities (Luo et al. 2020). Hence, comprehensive, well-resolved, and long-term surveys from a variety of well-chosen global locations are needed to grasp the controls of short-term plankton dynamics and their influence on the ecosystem, as well as how these changes manifest under longer periods of natural variation (such as El Niño) and climate change (Fuhrman et al. 2015). Going forward, an understanding of ecosystem dynamics and of community composition will require integration of different scientific disciplines—with the diversity and activity of all three domains (Eukarya, Bacteria, and Archaea) and the viruses being addressed simultaneously (Needham et al. 2017). Small- and mesoscale physical phenomena as well as major currents, subduction, and upwelling must also be included because they all play a role in determining community composition (Bolaños et al. 2020) and influence carbon export (Guidi et al. 2016; Omand et al. 2015).

### 4.8.7 Climate Change and Conservation

Annual seasonal warming is a long-standing and important factor in determining phytoplankton community transitions and primary production (Lomas et al. 2013; Steinberg et al. 2001). What the marine microbiome is now confronted with is anthropogenically accelerated warming and acidification, which are predicted to change primary production over vast swaths of the ocean (Behrenfeld et al. 2006; Doney et al. 2009, 2012; Flombaum et al. 2013; Raven et al. 2005). Changes in primary production and species composition of resident communities have already been reported in the Arctic where the impact of warming is pronounced (Alexeev et al. 2012; Arrigo et al. 2008; Box et al. 2019; Graversen et al. 2008; Wassmann 2015). For example, picoeukaryotes like the tiny motile green alga *Micromonas* have increased in abundance in the Canadian Arctic while larger algae like diatoms have decreased (Li et al. 2009). These kinds of changes must have an impact on food webs but nothing is known about the consumers in the Arctic that might eat picoeukaryotes. Additionally, although conditions appear to be favorable now for *Micromonas* in the Canadian Arctic, a climate change laboratory study demonstrated that acidification via increased $pCO_2$ will result in a decreased motility of green algae, due to the loss of the flagellum (Wang et al. 2020). Hence, it is unclear how trajectories of Arctic phytoplankton will continue to change. The consequences of climate change for the myriad protists of the sea and the broader microbiome are grand challenges indeed.

How do we conserve communities that we still struggle to describe? Marine Protected Areas (MPAs) currently focus on fish and other economically important species (Edgar et al. 2014) or on hotspots of marine biodiversity such as coral reefs (McClanahan et al. 2006). It has been recommended that a substantial increase in ocean protection and strategic conservation planning that prioritizes highly protected MPAs would have multiple benefits on ecosystem services, protecting biodiversity, improving the yield of fisheries and securing marine carbon stocks (Sala et al. 2021). Modeling efforts could be used to start to tackle how protistan community composition, or invasion by one protistan group over another, might impact conservation efforts. Another approach is to recognize that there are likely more favorable versus less favorable biogeochemical states of marine habitats (Azam and Worden 2004). An example for a less favorable state brought about by human activity is coastal eutrophication, which can lead to mass occurrences of toxin producing photo- and mixotrophic protists in harmful algal blooms, which in turn impact other components of the marine food web (Glibert 2017). Identifying the "more" and the "less" desirable states requires assessment of the human populations living at the land–sea interface with respect to their needs and impacts, those of other living entities, and emergent feedback loops. As Cinda Scott explains, the best way to accomplish this recognition of the "whole living community" is to embrace the need for equity and for diverse voices to be heard (Box 4.8), something which should be prioritized in future marine microbiome research and more broadly across and between different shareholder groups and scientific disciplines.

**Box 4.8**

"At its very core, conservation has to be rooted in equity for ultimate success. Let's start reimagining how we can not only use Marine Protected Areas as a tool for conservation, but as a means to uplift, respect, and enhance others."

From *Cinda P. Scott, oral contribution, 2020.*

**Note from the Authors—** This chapter is not a comprehensive review but rather it reflects topics that the authors felt were interesting and important to consider with an emphasis on some of our favorites. Additionally, as authors, we struggled with inclusion of work by some scientists who have espoused despicable views with respect to human beings and origins or belief systems. We are aware of the concerning views and writings of these individuals, and in no way endorse them, rather we seek to present scientific ideas they formulated that facilitated the development of new concepts by the broader scientific community. In doing so, we recognize that rather than always thinking that we "stand on the shoulders of giants," it is important to remember that we are faulted human beings standing on the shoulders of other human beings and the faults of the individuals and systems in which they participated.

**Acknowledgements** We thank our various mentors and lab colleagues over the years who have inspired and challenged us to take on new ways of thinking. We thank the funding sources that have supported our research and time in writing this book chapter, especially the Dutch Research Council (NWO) Vidi grant VI.Vidi.193.101 (to SW), Gordon and Betty Moore Foundation 3788 and NSF DEB-1639033 (to AZW) and availability of data generated under the Simons Foundation International's BIOS-SCOPE program. AZW thanks the Hanse Wissenschaftskolleg (Delmenhorst Germany) and the Radcliffe Institute for Advanced Studies at Harvard (Cambridge, MA, USA) for fellowships that have facilitated thought on these topics. We also thank Lucas J. Stal, Mariana Silvia Cretoiu, and other chapter authors for their understanding of the challenges posed in completing this work with young children at home during the COVID epidemic, as was the case for multiple authors.

## References

Abrahão J, Silva L, Silva LS et al (2018) Tailed giant Tupanvirus possesses the most complete translational apparatus of the known virosphere. Nat Commun 9:749–749

Adl SM, Simpson AGB, Farmer MA et al (2005) The new higher level classification of eukaryotes with emphasis on the taxonomy of protists. J Eukaryot Microbiol 52:399–451

Agrawal S, Barrow CJ, Deshmukh SK (2020) Structural deformation in pathogenic bacteria cells caused by marine fungal metabolites: an in vitro investigation. Microb Pathog 146:104248

Alexeev VA, Esau I, Polyakov IV et al (2012) Vertical structure of recent arctic warming from observed data and reanalysis products. Clim Chang 111:215–239

Amato A, Kooistra WHCF, Montresor M (2019) Cryptic diversity: a long-lasting issue for diatomologists. Protist 170:1–7

Amend A, Burgaud G, Cunliffe M et al (2019) Fungi in the marine environment: open questions and unsolved problems. MBio 10:e01189–e01118

Amin SA, Parker MS, Armbrust EV (2012) Interactions between diatoms and bacteria. Microbiol Mol Biol Rev 76:667–684
Amin SA, Hmelo LR, van Tol HM et al (2015) Interaction and signalling between a cosmopolitan phytoplankton and associated bacteria. Nature 522:98–101
Amir A, McDonald D, Navas-Molina JA et al (2017) Deblur rapidly resolves single-nucleotide community sequence patterns. mSystems 2:e00191–e00116
Andersen KH, Aksnes DL, Berge T et al (2015) Modelling emergent trophic strategies in plankton. J Plankton Res 37:862–868
Anderson SR, Harvey EL (2020) Temporal variability and ecological interactions of parasitic marine Syndiniales in coastal protist communities. mSphere 5:e00209–e00220
Anderson R, Jürgens K, Hansen PJ (2017) Mixotrophic phytoflagellate bacterivory field measurements strongly biased by standard approaches: a case study. Front Microbiol 8:1398
Anderson R, Charvet S, Hansen PJ (2018) Mixotrophy in chlorophytes and haptophytes: effect of irradiance, macronutrient, micronutrient and vitamin limitation. Front Microbiol 9:1704
Apprill A (2017) Marine animal microbiomes: toward understanding host–microbiome interactions in a changing ocean. Front Mar Sci 4:222
Archibald JM (2006) Endosymbiosis: double-take on plastid origins. Curr Biol 16:R690–R692
Archibald JM (2015) Endosymbiosis and eukaryotic cell evolution. Curr Biol 25:R911–R921
Arenovski AL, Lim EL, Caron DA (1995) Mixotrophic nanoplankton in oligotrophic surface waters of the Sargasso Sea may employ phagotrophy to obtain major nutrients. J Plankton Res 17:801–820
Aris R (1990) Vectors, tensors and the basic equations of fluid mechanics. Courier Corporation, Chelmsford
Armbrust EV (2009) The life of diatoms in the world's oceans. Nature 459:185–192
Arrigo KR, van Dijken G, Pabi S (2008) Impact of a shrinking Arctic ice cover on marine primary production. Geophys Res Lett 35:L19603
Avrahami Y, Frada MJ (2020) Detection of phagotrophy in the marine phytoplankton group of the coccolithophores (Calcihaptophycidae, Haptophyta) during nutrient-replete and phosphate-limited growth. J Phycol 56:1103–1108
Azam F (1998) Microbial control of oceanic carbon flux: the plot thickens. Science 280:694–696
Azam F, Worden AZ (2004) Oceanography. Microbes, molecules, and marine ecosystems. Science 303:1622–1624
Azam F, Fenchel T, Field JG et al (1983) The ecological role of water-column microbes in the sea. Mar Ecol Prog Ser 10:257–263
Bachy C, Charlesworth CJ, Chan AM et al (2018) Transcriptional responses of the marine green alga *Micromonas pusilla* and an infecting prasinovirus under different phosphate conditions. Environ Microbiol 20:2898–2912
Baker AC (2003) Flexibility and specificity in coral-algal symbiosis: diversity, ecology, and biogeography of *Symbiodinium*. Ann Rev Ecol Evol Syst 34:661–689
Baldauf SL (2003) The deep roots of eukaryotes. Science 300:1703–1706
Bar-On YM, Phillips R, Milo R (2018) The biomass distribution on earth. Proc Natl Acad Sci 115: 6506–6511
Barry MT, Rusconi R, Guasto JS, Stocker R (2015) Shear-induced orientational dynamics and spatial heterogeneity in suspensions of motile phytoplankton. J R Soc Interface 12:20150791
Behrenfeld MJ, O'Malley RT, Siegel DA et al (2006) Climate-driven trends in contemporary ocean productivity. Nature 444:752–755
Benoiston A-S, Ibarbalz FM, Bittner L et al (2017) The evolution of diatoms and their biogeochemical functions. Phil Trans R Soc Lond B 372:20160397
Berdjeb L, Parada AE, Needham DM, Fuhrman JA (2018) Short-term dynamics and interactions of marine protist communities during the spring–summer transition. ISME J 12:1907–1917
Bergh O, Børsheim KY, Bratbak G, Heldal M (1989) High abundance of viruses found in aquatic environments. Nature 340:467–468

Bhattacharya D, Roy RS, Price DC, Schliep A (2014) Single-cell genomics of marine plankton: studying the single life of eukaryotic microbes. Biochemist 36:16–22
Bhattacharyya D, Chakraborty S (2018) Chloroplast: the Trojan horse in plant-virus interaction. Mol Plant Pathol 19:504–518
Biecheler B (1936) Observation de la capture et de la digestion des proies chez un Péridinien vert. J Crustacean Biol 122:1173–1175
Billard C, Inouye I (2004) What is new in coccolithophore biology? In: Thierstein HR, Young JR (eds) Coccolithophores: from molecular processes to global impact. Springer, Berlin, Heidelberg, pp 1–29
Bird DF, Kalff J (1986) Bacterial grazing by planktonic lake algae. Science 231:493–495
Blackwell M (2011) The fungi: 1, 2, 3 ... 5.1 million species? Am J Bot 98:426–438
Blanc-mathieu R, Krasovec M, Hebrard M et al (2017) Population genomics of picophytoplankton unveils novel chromosome hypervariability. Sci Adv 3:e1700239
Bochdansky AB, Clouse MA, Herndl GJ (2017) Eukaryotic microbes, principally fungi and labyrinthulomycetes, dominate biomass on bathypelagic marine snow. ISME J 11:362–373
Bock NA, Charvet S, Burns J et al (2021) Experimental identification and in silico prediction of bacterivory in green algae. ISME J 15:1987–2000
Boenigk J, Arndt H (2002) Bacterivory by heterotrophic flagellates: community structure and feeding strategies. Ant Leeuwenhoek 81:465–480
Bolaños LM, Karp-Boss L, Choi CJ et al (2020) Small phytoplankton dominate western North Atlantic biomass. ISME J 14:1663–1674
Bowler C, Vardi A, Allen AE (2010) Oceanographic and biogeochemical insights from diatom genomes. Annu Rev Mar Sci 2:333–365
Box JE, Colgan WT, Christensen TR et al (2019) Key indicators of Arctic climate change: 1971–2017. Environ Res Lett 14:045010
Bradley JA, Amend JP, LaRowe DE (2019) Survival of the fewest: microbial dormancy and maintenance in marine sediments through deep time. Geobiology 17(1):43–59
Bradley JA et al (2020) Widespread energy limitation to life in global subseafloor sediments. Sci Adv 6(32):eaba0697
Breitbart M, Bonnain C, Malki K, Sawaya NA (2018) Phage puppet masters of the marine microbial realm. Nat Microbiol 3:754–766
Brocke R, Fatka O, Wilde V (2006) Acritarchs and prasinophytes of the Silurian-Devonian GSSP (Klonk, Barrandian area, Czech Republic). Bull Geosci 81:27–41
Brown RM, Malcolm Brown R (1972) Algal viruses. Adv Virus Res 17:243–277
Brown MW, Heiss AA, Kamikawa R et al (2018) Phylogenomics places orphan protistan lineages in a novel eukaryotic super-group. Genome Biol Evol 10:427–433
Brown JM, Labonté JM, Brown J et al (2020) Single cell genomics reveals viruses consumed by marine protists. Front Microbiol 11:524828
Brumley DR, Rusconi R, Son K, Stocker R (2015) Flagella, flexibility and flow: physical processes in microbial ecology. Eur Phys J Spec Top 224:3119–3140
Brunet T, King N (2017) The origin of animal multicellularity and cell differentiation. Dev Cell 43: 124–140
Brunson JK, McKinnie SMK, Chekan JR et al (2018) Biosynthesis of the neurotoxin domoic acid in a bloom-forming diatom. Science 361:1356–1358
Brussaard CPD (2004) Viral control of phytoplankton populations--a review. J Eukaryot Microbiol 51:125–138
Brussaard CPD, Kempers RS, Kop AJ et al (1996) Virus-like particles in a summer bloom of Emiliania huxleyi in the North Sea. Aquat Microb Ecol 10:105–113
Brussaard CPD, Kuipers B, Veldhuis MJW (2005) A mesocosm study of *Phaeocystis globosa* population dynamics: I. regulatory role of viruses in bloom control. Harmful Algae 4:859–874
Bryant JA, Aylward FO, Eppley JM et al (2016) Wind and sunlight shape microbial diversity in surface waters of the North Pacific subtropical gyre. ISME J 10:1308–1322

Buaya AT, Ploch S, Hanic L et al (2017) Phylogeny of *Miracula helgolandica* gen. et sp. nov. and *Olpidiopsis drebesii* sp. nov., two basal oomycete parasitoids of marine diatoms, with notes on the taxonomy of *Ectrogella*-like species. Mycol Prog 16:1041–1050
Burkholder JM, Glibert PM, Skelton HM (2008) Mixotrophy, a major mode of nutrition for harmful algal species in eutrophic waters. Harmful Algae 8:77–93
Burki F (2017) The convoluted evolution of eukaryotes with complex plastids. In: Hirakawa Y (ed) Advances in botanical research. Academic Press, London, pp 1–30
Burki F, Kaplan M, Tikhonenkov DV et al (2016) Untangling the early diversification of eukaryotes: a phylogenomic study of the evolutionary origins of Centrohelida, Haptophyta and Cryptista. Proc Biol Sci 283:20152802
Burki F, Roger AJ, Brown MW, Simpson AGB (2020) The new tree of eukaryotes. Trends Ecol Evol 35:43–55
Cachon M, Caram B (1979) A symbiotic green alga, *Pedinomonas symbiotica* sp. nov. (Prasinophyceae), in the radiolarian *Thalassolampe margarodes*. Phycologia 18:177–184
Cahoon L (2016) Tychoplankton. In: Kennish MJ (ed) Encyclopedia of estuaries. Springer Netherlands, Dordrecht, pp 721–721
Calbet A, Landry MR (2004) Phytoplankton growth, microzooplankton grazing, and carbon cycling in marine systems. Limnol Oceanogr 49:51–57
Callahan BJ, Mcmurdie PJ, Rosen MJ et al (2016) DADA2: high resolution sample inference from amplicon data. Nat Methods 13:581–583
Carlile M (1982) Prokaryotes and eukaryotes: strategies and successes. Trends Biochem Sci 7:128–130
Caron DA, Alexander H, Allen AE et al (2017) Probing the evolution, ecology and physiology of marine protists using transcriptomics. Nat Rev Microbiol 15:6–20
Carradec Q, Pelletier E, Da Silva C et al (2018) A global ocean atlas of eukaryotic genes. Nat Commun 9:373
Casteleyn G, Leliaert F, Backeljau T et al (2010) Limits to gene flow in a cosmopolitan marine planktonic diatom. Proc Natl Acad Sci 107:12952–12957
Castillo YM, Mangot J-F, Benites LF et al (2019) Assessing the viral content of uncultured picoeukaryotes in the global-ocean by single cell genomics. Mol Ecol 28:4272–4289
Cavalier-Smith T (1981) Eukaryote kingdoms: seven or nine? Biosystems 14:461–481
Cavalier-Smith T (2018) Kingdom Chromista and its eight phyla: a new synthesis emphasising periplastid protein targeting, cytoskeletal and periplastid evolution, and ancient divergences. Protoplasma 255:297–357
Cavalier-Smith T, von der Heyden S (2007) Molecular phylogeny, scale evolution and taxonomy of centrohelid heliozoa. Mol Phylogenet Evol 44:1186–1203
Cavalier-Smith T, Chao EE, Lewis R (2015) Multiple origins of Heliozoa from flagellate ancestors: new cryptist subphylum Corbihelia, superclass Corbistoma, and monophyly of Haptista, Cryptista, Hacrobia and Chromista. Mol Phylogenet Evol 93:331–362
Cavalier-Smith T, Chao EE, Lewis R (2016) 187-gene phylogeny of protozoan phylum Amoebozoa reveals a new class (Cutosea) of deep-branching, ultrastructurally unique, enveloped marine Lobosa and clarifies amoeba evolution. Mol Phylogenet Evol 99:275–296
Cavicchioli R, Ripple WJ, Timmis KN et al (2019) Scientists' warning to humanity: microorganisms and climate change. Nat Rev Microbiol 17:569–586
Ceballos G, Ehrlich PR, Barnosky AD et al (2015) Accelerated modern human-induced species losses: entering the sixth mass extinction. Sci Adv 1:e1400253
Cermeño P (2016) The geological story of marine diatoms and the last generation of fossil fuels. Perspect Phycol 3:53–60
Chambouvet A, Morin P, Marie D, Guillou L (2008) Control of toxic marine dinoflagellate blooms by serial parasitic killers. Science 322:1254–1257
Chambouvet A, Monier A, Maguire F et al (2019) Intracellular infection of diverse diatoms by an evolutionary distinct relative of the fungi. Curr Biol 29:4093–4101
Chan CX, Ragan MA (2013) Next-generation phylogenomics. Biol Direct 8:3

Choi CJ, Bachy C, Jaeger GS et al (2017) Newly discovered deep-branching marine plastid lineages are numerically rare but globally distributed. Curr Biol 27:R15–R16

Choi CJ, Jimenez V, Needham DM et al (2020) Seasonal and geographical transitions in eukaryotic phytoplankton community structure in the Atlantic and Pacific oceans. Front Microbiol 11: 542372

Chrismas N, Cunliffe M (2020) Depth-dependent mycoplankton glycoside hydrolase gene activity in the open ocean-evidence from the Tara oceans eukaryote metatranscriptomes. ISME J 14: 2361–2365

Cooney EC, Okamoto N, Cho A et al (2020) Single-cell transcriptomics of Abedinium reveals a new early-branching dinoflagellate lineage. Genome Biol Evol 12:2417–2428

Corliss JO (1984) The kingdom Protista and its 45 phyla. Biosystems 17:87–126

Cram JA, Chow C-ET, Sachdeva R et al (2014) Seasonal and interannual variability of the marine bacterioplankton community throughout the water column over ten years. ISME J 9:563–563

Cunliffe M, Hollingsworth A, Bain C et al (2017) Algal polysaccharide utilisation by saprotrophic planktonic marine fungi. Fungal Ecol 30:135–138

Cuvelier ML, Allen AE, Monier A et al (2010) Targeted metagenomics and ecology of globally important uncultured eukaryotic phytoplankton. Proc Natl Acad Sci 107:14679–14684

Daugbjerg N, Jensen MH, Hansen PJ (2013) Using nuclear-encoded LSU and SSU rDNA sequences to identify the eukaryotic endosymbiont in *Amphisolenia bidentata* (Dinophyceae). Protist 164:411–422

De Luca D, Piredda R, Sarno D, Kooistra WHCF (2021) Resolving cryptic species complexes in marine protists: phylogenetic haplotype networks meet global DNA metabarcoding datasets. ISME J 15:1931–1942

de Mendoza A, Suga H, Permanyer J et al (2015) Complex transcriptional regulation and independent evolution of fungal-like traits in a relative of animals. elife 4:e08904

De Queiroz K (2007) Species concepts and species delimitation. Syst Biol 56:879–886

de Vargas C, Audic S, Henry N et al (2015) Eukaryotic plankton diversity in the sunlit ocean. Science 348:1261605

de Wit R, Bouvier T (2006) "Everything is everywhere, but, the environment selects"; what did Baas Becking and Beijerinck really say? Environ Microbiol 8:755–758

DeAngelis KM, Silver WL, Thompson AW, Firestone MK (2010) Microbial communities acclimate to recurring changes in soil redox potential status. Environ Microbiol 12:3137–3149

Decelle J (2013) New perspectives on the functioning and evolution of photosymbiosis in plankton: mutualism or parasitism? Commun Integr Biol 6:e24560

Decelle J, Martin P, Paborstava K et al (2013) Diversity, ecology and biogeochemistry of cyst-forming acantharia (radiolaria) in the oceans. PLoS One 8:e53598

Decelle J, Colin S, Foster RA (2015) Photosymbiosis in marine planktonic protists. In: Ohtsuka S, Suzaki T, Horiguchi T, Suzuki N, Not F (eds) Marine Protists: diversity and dynamics. Springer, Tokyo, pp 465–500

Decelle J, Stryhanyuk H, Gallet B et al (2019) Algal remodeling in a ubiquitous planktonic photosymbiosis. Curr Biol 29:968–978.e4

Deeg CM, Zimmer MM, George EE et al (2019) *Chromulinavorax destructans*, a pathogen of microzooplankton that provides a window into the enigmatic candidate phylum Dependentiae. PLoS Pathog 15:e1007801

del Campo J, Massana R (2011) Emerging diversity within chrysophytes, choanoflagellates and bicosoecids based on molecular surveys. Protist 162:435–448

del Giorgio PA, Gasol JM, Vaqué D et al (1996) Bacterioplankton community structure: Protists control net production and the proportion of active bacteria in a coastal marine community. Limnol Oceanogr 41:1169–1179

Delaroque N, Boland W (2008) The genome of the brown alga *Ectocarpus siliculosus* contains a series of viral DNA pieces, suggesting an ancient association with large dsDNA viruses. BMC Evol Biol 8:110

Delsuc F, Brinkmann H, Philippe H (2005) Phylogenomics and the reconstruction of the tree of life. Nat Rev Genet 6:361–375
Demir-Hilton E, Sudek S, Cuvelier ML et al (2011) Global distribution patterns of distinct clades of the photosynthetic picoeukaryote *Ostreococcus*. ISME J 5:1095–1107
Derelle R, López-García P, Timpano H, Moreira D (2016) A phylogenomic framework to study the diversity and evolution of stramenopiles (= heterokonts). Mol Biol Evol 33:2890–2898
Deshmukh SK, Prakash V, Ranjan N (2017) Marine fungi: A source of potential anticancer compounds. Front Microbiol 8:2536
Diamond LS, Mattern CF (1976) Protozoal viruses. Adv Virus Res 20:87–112
Diamond LS, Mattern CF, Bartgis IL (1972) Viruses of *Entamoeba histolytica*. I Identification of transmissible virus-like agents. J Virol 9:326–341
Diekmann Y, Pereira-Leal JB (2013) Evolution of intracellular compartmentalization. Biochem J 449:319–331
Díez B, Pedrós-Alió C, Massana R (2001) Study of genetic diversity of eukaryotic picoplankton in different oceanic regions by small-subunit rRNA gene cloning and sequencing. Appl Environ Microbiol 67:2932–2941
Dittami SM, Arboleda E, Auguet J-C et al (2021) A community perspective on the concept of marine holobionts: state-of-the-art, challenges, and future directions. PeerJ 9:e10911
Dixon AK, Needham DM, Al-Horani FA, Chadwick NE (2013) Microhabitat use and photoacclimation in the clownfish sea anemone *Entacmaea quadricolor*. J Mar Biol Assoc 94:473–480
do Rosário Gomes H, Goes JI, Matondkar SGP et al (2014) Massive outbreaks of *Noctiluca scintillans* blooms in the Arabian Sea due to spread of hypoxia. Nat Commun 5:4862
do Rosário Gomes H, McKee K, Mile A et al (2018) Influence of light availability and prey type on the growth and photo-physiological rates of the mixotroph *Noctiluca scintillans*. Front Mar Sci 5:374
Doney SC, Fabry VJ, Feely RA, Kleypas JA (2009) Ocean acidification: the other $CO_2$ problem. Annu Rev Mar Sci 1:169–192
Doney SC, Ruckelshaus M, Duffy JE et al (2012) Climate change impacts on marine ecosystems. Annu Rev Mar Sci 4:11–37
Doolittle WF (1998a) A paradigm gets shifty. Nature 392:15–16
Doolittle WF (1998b) You are what you eat: a gene transfer ratchet could account for bacterial genes in eukaryotic nuclear genomes. Trends Genet 14:307–311
Dorrell RG, Drew J, Nisbet RER, Howe CJ (2014) Evolution of chloroplast transcript processing in plasmodium and its chromerid algal relatives. PLoS Genet 10:e1004008
Dorrell RG, Gile G, McCallum G et al (2017) Chimeric origins of ochrophytes and haptophytes revealed through an ancient plastid proteome. elife 6:e23717
Droop MR (1957) Auxotrophy and organic compounds in the nutrition of marine phytoplankton. J Gen Microbiol 16:286–293
Duanmu D, Bachy C, Sudek S et al (2014) Marine algae and land plants share conserved phytochrome signaling systems. Proc Natl Acad Sci 111:15827–15832
Ducklow HW, Steinberg DK, Buesseler KO (2001) Upper Ocean carbon export and the biological pump. Oceanography 14:50–58
Dupont CL, McCrow JP, Valas R et al (2015) Genomes and gene expression across light and productivity gradients in eastern subtropical Pacific microbial communities. ISME J 9:1076–1092
Eckford-Soper L, Daugbjerg N (2016) The ichthyotoxic genus *Pseudochattonella* (Dictyochophyceae): distribution, toxicity, enumeration, ecological impact, succession and life history – a review. Harmful Algae 58:51–58
Edgar GJ, Stuart-Smith RD, Willis TJ et al (2014) Global conservation outcomes depend on marine protected areas with five key features. Nature 506:216–220
Edgcomb VP (2016) Marine protist associations and environmental impacts across trophic levels in the twilight zone and below. Curr Op Microbiol 31:169–175

Edgcomb VP, Leadbetter ER, Bourland W et al (2011) Structured multiple endosymbiosis of bacteria and archaea in a ciliate from marine sulfidic sediments: a survival mechanism in low oxygen, sulfidic sediments? Front Microbiol 2:55
Edwards KF (2019) Mixotrophy in nanoflagellates across environmental gradients in the ocean. Proc Natl Acad Sci 116:6211–6220
Eren AM, Morrison HG, Lescault PJ et al (2015) Minimum entropy decomposition: unsupervised oligotyping for sensitive partitioning of high-throughput marker gene sequences. ISME J 9:968–979
Faktorová D, Nisbet RER, Fernández Robledo JA et al (2020) Genetic tool development in marine protists: emerging model organisms for experimental cell biology. Nat Methods 17:481–494
Falkowski P (2012) Ocean science: the power of plankton. Nature 483:17–20
Febvre-Chevalier C, Febvre J (1994) Buoyancy and swimming in marine planktonic protists. In: Maddock L, Bone Q, Rayner JMV (eds) Mechanics and physiology of animal swimming. Cambridge University Press, Cambridge, pp 13–26
Field MC, Dacks JB (2009) First and last ancestors: reconstructing evolution of the endomembrane system with ESCRTs, vesicle coat proteins, and nuclear pore complexes. Curr Op Cell Biol 21: 4–13
Field CB, Behrenfeld MJ, Randerson JT et al (1998) Primary production of the biosphere: integrating terrestrial and oceanic components. Science 281:237–240
Filée J (2014) Multiple occurrences of giant virus core genes acquired by eukaryotic genomes: the visible part of the iceberg? Virology 466–467:53–59
Finkel ZV, Beardall J, Flynn KJ et al (2009) Phytoplankton in a changing world: cell size and elemental stoichiometry. J Plankton Res 32:119–137
Finlay BJ (2002) Global dispersal of free-living microbial eukaryote species. Science 296:1061–1063
Fitzsimmons JN, Hayes CT, Al-Subiai SN et al (2015) Daily to decadal variability of size-fractionated iron and iron-binding ligands at the Hawaii Ocean Time-Series Station ALOHA. Geochim Cosmochim Acta 171:303–324
Flannagan RS, Jaumouillé V, Grinstein S (2012) The cell biology of phagocytosis. Annu Rev Pathol 7:61–98
Flegontova O, Flegontov P, Malviya S et al (2016) Extreme diversity of diplonemid eukaryotes in the ocean. Curr Biol 26:3060–3065
Flombaum P, Gallegos JL, Gordillo RA et al (2013) Present and future global distributions of the marine cyanobacteria *Prochlorococcus* and *Synechococcus*. Proc Natl Acad Sci 110:9824–9829
Flynn KJ, Stoecker DK, Mitra A et al (2012) Misuse of the phytoplankton–zooplankton dichotomy: the need to assign organisms as mixotrophs within plankton functional types. J Plankton Res 35: 3–11
Flynn KJ, Mitra A, Glibert PM, Burkholder JM (2018) Mixotrophy in harmful algal blooms: by whom, on whom, when, why, and what next. In: Glibert PM, Berdalet E, Burford MA et al (eds) Global ecology and oceanography of harmful algal blooms. Springer International Publishing, Cham, pp 113–132
Flynn KJ, Mitra A, Anestis K et al (2019) Mixotrophic protists and a new paradigm for marine ecology: where does plankton research go now? J Plankton Res 41:375–391
Forster D, Dunthorn M, Mahé F et al (2016) Benthic protists: the under-charted majority. FEMS Microbiol Ecol 92:fiw120
Foster RA, Zehr JP (2006) Characterization of diatom–cyanobacteria symbioses on the basis of nifH, hetR and 16S rRNA sequences. Environ Microbiol 8:1913–1925
Foster RA, Kuypers MMM, Vagner T et al (2011) Nitrogen fixation and transfer in open ocean diatom–cyanobacterial symbioses. ISME J 5:1484–1484
Foulon E, Not F, Jalabert F et al (2008) Ecological niche partitioning in the picoplanktonic green alga *Micromonas pusilla*: evidence from environmental surveys using phylogenetic probes. Environ Microbiol 10:2433–2443

Frias-Lopez J, Thompson A, Waldbauer J, Chisholm SW (2009) Use of stable isotope-labelled cells to identify active grazers of picocyanobacteria in ocean surface waters. Environ Microbiol 11: 512–525

Fuhrman JA, Cram JA, Needham DM (2015) Marine microbial community dynamics and their ecological interpretation. Nat Rev Microbiol 13:133–146

Gaines G, Taylor FJR (1984) Extracellular digestion in marine dinoflagellates. J Plankton Res 6: 1057–1061

Gallot-Lavallée L, Blanc G (2017) A glimpse of Nucleo-cytoplasmic large DNA virus biodiversity through the eukaryotic genomics window. Viruses 9:17

Gardner MJ, Hall N, Fung E et al (2002) Genome sequence of the human malaria parasite plasmodium falciparum. Nature 419:498–511

Garvetto A, Badis Y, Perrineau M-M et al (2019) Chytrid infecting the bloom-forming marine diatom *Skeletonema* sp.: morphology, phylogeny and distribution of a novel species within the Rhizophydiales. Fungal Biol 123:471–480

Gastrich MD, Leigh-Bell JA, Gobler CJ et al (2004) Viruses as potential regulators of regional brown tide blooms caused by the alga, *Aureococcus anophagefferens*. Estuaries 27:112–119

Gawryluk RMR, Tikhonenkov DV, Hehenberger E et al (2019) Non-photosynthetic predators are sister to red algae. Nature 572:240–243

Genitsaris S, Monchy S, Viscogliosi E et al (2015) Seasonal variations of marine protist community structure based on taxon-specific traits using the eastern English Channel as a model coastal system. FEMS Microbiol Ecol 91:fiv034

Gibbs A, Skotnicki AH, Gardiner JE et al (1975) A tobamovirus of a green alga. Virology 64:571–574

Gilbert C, Cordaux R (2017) Viruses as vectors of horizontal transfer of genetic material in eukaryotes. Curr Op Virol 25:16–22

Giovannoni SJ, Vergin KL (2012) Seasonality in ocean microbial communities. Science 335:671–676

Glassman SI, Martiny JBH (2018) Broadscale ecological patterns are robust to use of exact sequence variants versus operational taxonomic units. mSphere 3:e00148–e00118

Glibert PM (2017) Eutrophication, harmful algae and biodiversity - challenging paradigms in a world of complex nutrient changes. Mar Pollut Bull 124:591–606

Glücksman E, Snell EA, Berney C et al (2011) The novel marine gliding zooflagellate genus *Mantamonas* (Mantamonadida Ord. N.: Apusozoa). Protist 162:207–221

Gobler CJ, Anderson OR, Gastrich MD, Wilhelm SW (2007) Ecological aspects of viral infection and lysis in the harmful brown tide alga *Aureococcus anophagefferens*. Aquat Microb Ecol 47: 25–36

Gobler CJ, Berry DL, Dyhrman ST et al (2011) Niche of harmful alga *Aureococcus anophagefferens* revealed through ecogenomics. Proc Natl Acad Sci 108:4352–4357

Gonzáez JM, Sherr EB, Sherr BF (1993) Differential feeding by marine flagellates on growing versus starving, and on motile versus nonmotile, bacterial prey. Mar Ecol Prog Ser 102:257–267

Gooday AJ, Schoenle A, Dolan JR, Arndt H (2020) Protist diversity and function in the dark ocean – challenging the paradigms of deep-sea ecology with special emphasis on foraminiferans and naked protists. Eur J Protistol 75:125721

Goodenough U, Heitman J (2014) Origins of eukaryotic sexual reproduction. Cold Spring Harb Perspect Biol 6:a016154

Goodenough U, Lin H, Lee J-H (2007) Sex determination in *Chlamydomonas*. Semin Cell Dev Biol 18:350–361

Gould SB, Waller RF, McFadden GI (2008) Plastid evolution. Annu Rev Plant Biol 59:491–517

Grant CR, Wan J, Komeili A (2018) Organelle formation in bacteria and archaea. Annu Rev Cell Dev Biol 34:217–238

Graversen RG, Mauritsen T, Tjernström M et al (2008) Vertical structure of recent Arctic warming. Nature 451:53–56

Griffiths DJ (2001) Endogenous retroviruses in the human genome sequence. Genome Biol 2: reviews1017.1
Grigoriev IV, Hayes RD, Calhoun S et al (2021) PhycoCosm, a comparative algal genomics resource. Nucleic Acids Res 49:D1004–D1011
Grimsley N, Péquin B, Bachy C et al (2010) Cryptic sex in the smallest eukaryotic marine green alga. Mol Biol Evol 27:47–54
Grossart H-P, Van den Wyngaert S, Kagami M et al (2019) Fungi in aquatic ecosystems. Nat Rev Microbiol 17:339–354
Gsell AS, de Senerpont Domis LN, Verhoeven KJF et al (2013) Chytrid epidemics may increase genetic diversity of a diatom spring-bloom. ISME J 7:2057–2059
Guadayol Ò, Peters F, Marrasé C et al (2009) Episodic meteorological and nutrient-load events as drivers of coastal planktonic ecosystem dynamics: a time-series analysis. Mar Ecol Prog Ser 381:139–155
Guidi L, Chaffron S, Bittner L et al (2016) Plankton networks driving carbon export in the oligotrophic ocean. Nature 532:465–465
Guillou L, Viprey M, Chambouvet A et al (2008) Widespread occurrence and genetic diversity of marine parasitoids belonging to Syndiniales (Alveolata). Environ Microbiol 10:3349–3365
Gustavsen JA, Winget DM, Tian X, Suttle CA (2014) High temporal and spatial diversity in marine RNA viruses implies that they have an important role in mortality and structuring plankton communities. Front Microbiol 5:703
Gutiérrez MH, Jara AM, Pantoja S (2016) Fungal parasites infect marine diatoms in the upwelling ecosystem of the Humboldt current system off Central Chile. Environ Microbiol 18:1646–1653
Gutowska MA, Shome B, Sudek S et al (2017) Globally important haptophyte algae use exogenous pyrimidine compounds more efficiently than thiamin. MBio 8:e01459–e01417
Hadariová L, Vesteg M, Hampl V, Krajčovič J (2018) Reductive evolution of chloroplasts in non-photosynthetic plants, algae and protists. Curr Genet 64:365–387
Haeckel E (1887) Report on the Radiolaria collected by HMS challenger during the years 1873–1876. Report of the voyage of HMS challenger. Zoology 18:i–clxxxviii
Haeckel EHPA, Ernst Heinrich Philipp (1862) Die Radiolarien (*Rhizopoda radiaria*): eine Monographie
Hallock P (1999) Symbiont-bearing foraminifera. In: Modern foraminifera. Springer, Berlin, pp 123–139
Han KY, Maciszewski K, Graf L et al (2019) Dictyochophyceae plastid genomes reveal unusual variability in their organization. J Phycol 55:1166–1180
Händeler K, Grzymbowski YP, Krug PJ, Wägele H (2009) Functional chloroplasts in metazoan cells - a unique evolutionary strategy in animal life. Front Zool 6:28
Hanic LA, Sekimoto S, Bates SS (2009) Oomycete and chytrid infections of the marine diatom *Pseudo-nitzschia pungens* (Bacillariophyceae) from Prince Edward Island, Canada. Botany 87: 1096–1105
Hansen PJ (2011) The role of photosynthesis and food uptake for the growth of marine mixotrophic dinoflagellates. J Eukaryot Microbiol 58:203–214
Hansen PJ, Calado AJ (1999) Phagotrophic mechanisms and prey selection in free-living dinoflagellates. J Eukaryot Microbiol 46:382–389
Hansen PJ, Hjorth M (2002) Growth and grazing responses of *Chrysochromulina ericina* (Prymnesiophyceae): the role of irradiance, prey concentration and pH. Mar Biol 141:975–983
Hartmann M, Grob C, Tarran GA et al (2012) Mixotrophic basis of Atlantic oligotrophic ecosystems. Proc Natl Acad Sci 109:5756–5760
Hartmann M, Zubkov MV, Scanlan DJ, Lepère C (2013) In situ interactions between photosynthetic picoeukaryotes and bacterioplankton in the Atlantic Ocean: evidence for mixotrophy. Environ Microbiol Rep 5:835–840
Harvey EL, Jeong HJ, Menden-Deuer S (2013) Avoidance and attraction: chemical cues influence predator-prey interactions of planktonic protists. Limnol Oceanogr 58:1176–1184

Hassett BT, Borrego EJ, Vonnahme TR et al (2019) Arctic marine fungi: biomass, functional genes, and putative ecological roles. ISME J 13:1484–1496
Hawksworth DL, Lücking R (2017) Fungal diversity revisited: 2.2 to 3.8 million species. In: Heitman J, Howlett BJ, Crous PW, Stukenbrock EH, James TY, NAR G (eds) The fungal Kingdom. American Society for Microbiology, Washington, DC. https://doi.org/10.1128/9781555819583.ch4
Hehenberger E, Tikhonenkov DV, Kolisko M et al (2017) Novel predators reshape Holozoan phylogeny and reveal the presence of a two-component signaling system in the ancestor of animals. Curr Biol 27:2043–2050
Hehenberger E, Gast RJ, Keeling PJ (2019) A kleptoplastidic dinoflagellate and the tipping point between transient and fully integrated plastid endosymbiosis. Proc Natl Acad Sci 116:17934–17942
Heintze C, Formanek P, Pohl D et al (2020) An intimate view into the silica deposition vesicles of diatoms. BMC Materials 2:11
Hildebrand M, Lerch SJL (2015) Diatom silica biomineralization: parallel development of approaches and understanding. Semin Cell Dev Biol 46:27–35
Hoehler TM, Alperin MJ, Albert DB, Martens CS (1998) Thermodynamic control on hydrogen concentrations in anoxic sediments. Geochim Cosmochim Acta 62:1745–1756
Hu SK, Connell PE, Mesrop LY, Caron DA (2018) A hard day's night: diel shifts in microbial eukaryotic activity in the North Pacific subtropical gyre. Front Mar Sci 5:351
Hutchinson GE (1961) The paradox of the plankton. Am Nat 95:137–145
Hyman P, Abedon ST (2012) Smaller fleas: viruses of microorganisms. Scientifica 2012:734023
Ibarbalz FM, Henry N, Brandão MC et al (2019) Global trends in marine plankton diversity across kingdoms of life. Cell 179:1084–1097
Ibelings BW, De Bruin A, Kagami M et al (2004) Host parasite interactions between freshwater phytoplankton and chytrid fungi (Chytridiomycota). J Phycol 40:437–453
Imachi H, Nobu MK, Nakahara N et al (2020) Isolation of an archaeon at the prokaryote–eukaryote interface. Nature 577:519–525
Inagaki F, Hinrichs K-U, Kubo Y, The IODP Expedition 337 Scientists (2016) IODP expedition 337: deep coalbed biosphere off Shimokita – microbial processes and hydrocarbon system associated with deeply buried coalbed in the ocean. Sci Drill 21:17–28
Jahson S, Rai AN, Bergman B (1995) Intracellular cyanobiont *Richelia intracellularis*: ultrastructure and immuno-localisation of phycoerythrin, nitrogenase, rubisco and glutamine synthetase. Mar Biol 124:1–8
James TY, Stajich JE, Hittinger CT, Rokas A (2020) Toward a fully resolved fungal tree of life. Annu Rev Microbiol 74:291–313
Janouskovec J, Horák A, Oborník M et al (2010) A common red algal origin of the apicomplexan, dinoflagellate, and heterokont plastids. Proc Natl Acad Sci 107:10949–10954
Janouškovec J, Tikhonenkov DV, Burki F et al (2015) Factors mediating plastid dependency and the origins of parasitism in apicomplexans and their close relatives. Proc Natl Acad Sci 112: 10200–10207
Janouškovec J, Tikhonenkov DV, Burki F et al (2017) A new lineage of eukaryotes illuminates early mitochondrial genome reduction. Curr Biol 27:3717–3724
Jardillier L, Zubkov MV, Pearman J, Scanlan DJ (2010) Significant $CO_2$ fixation by small prymnesiophytes in the subtropical and tropical Northeast Atlantic Ocean. ISME J 4:1180–1192
Jephcott TG, Alves-de-Souza C, Gleason FH et al (2016) Ecological impacts of parasitic chytrids, syndiniales and perkinsids on populations of marine photosynthetic dinoflagellates. Fungal Ecol 19:47–58
Jiang H, Kulis DM, Brosnahan ML, Anderson DM (2018) Behavioral and mechanistic characteristics of the predator-prey interaction between the dinoflagellate *Dinophysis acuminata* and the ciliate *Mesodinium rubrum*. Harmful Algae 77:43–54
John U, Lu Y, Wohlrab S et al (2019) An aerobic eukaryotic parasite with functional mitochondria that likely lacks a mitochondrial genome. Sci Adv 5:eaav1110

Johnson AR, Omand MM (2021) Evolution of a subducted carbon-rich filament on the edge of the North Atlantic gyre. J Geophys Res C: Oceans 49:438–460
Johnson MD, Oldach D, Delwiche CF, Stoecker DK (2007) Retention of transcriptionally active cryptophyte nuclei by the ciliate *Myrionecta rubra*. Nature 445:426–428
Johnson WM, Alexander H, Bier RL et al (2020) Auxotrophic interactions: a stabilizing attribute of aquatic microbial communities? FEMS Microbiol Ecol 96:fiaa115
Jones RI (2000) Mixotrophy in planktonic protists: an overview: Mixotrophy in planktonic protists. Freshw Biol 45:219–226
Jones EBG (2011) Are there more marine fungi to be described? Bot Mar 54:343–354
Jørgensen BB, Kasten S (2006) Sulfur cycling and methane oxidation. In: Schulz HD, Zabel M (eds) Marine Geochemistry. Springer, Berlin, Heidelberg, pp 271–309
Jürgens K, Massana R (2008) Protistan grazing on marine bacterioplankton. In: Kirchman DL (ed) Microbial ecology of the oceans. John Wiley & Sons, Inc., Hoboken, NJ, pp 383–441
Jürgens K, Matz C (2002) Predation as a shaping force for the phenotypic and genotypic composition of planktonic bacteria. Ant Leeuwenhoek 81:413–434
Kallmeyer J, Pockalny R, Adhikari RR et al (2012) Global distribution of microbial abundance and biomass in subseafloor sediment. Proc Natl Acad Sci 109:16213–16216
Kamennaya NA, Kennaway G, Fuchs BM, Zubkov MV (2018) "Pomacytosis"—semi-extracellular phagocytosis of cyanobacteria by the smallest marine algae. PLoS Biol 16:e2003502
Kamp A, Høgslund S, Risgaard-Petersen N, Stief P (2015) Nitrate storage and dissimilatory nitrate reduction by eukaryotic microbes. Front Microbiol 6:1492
Kaneko H, Blanc-Mathieu R, Endo H et al (2021) Eukaryotic virus composition can predict the efficiency of carbon export in the global ocean. iScience 24:102002
Karlusich JJP, Ibarbalz FM, Bowler C (2020) Exploration of marine phytoplankton: from their historical appreciation to the omics era. J Plankton Res 42:595–612
Karnkowska A, Vacek V, Zubáčová Z et al (2016) A eukaryote without a mitochondrial organelle. Curr Biol 26:1274–1284
Karpov SA, Mamkaeva MA, Aleoshin VV et al (2014) Morphology, phylogeny, and ecology of the aphelids (Aphelidea, Opisthokonta) and proposal for the new superphylum Opisthosporidia. Front Microbiol 5:112
Kawachi M, Inouye I, Maeda O, Chihara M (1991) The haptonema as a food-capturing device: observations on *Chrysochromulina hirta* (Prymnesiophyceae). Phycologia 30:563–573
Kawachi M, Nakayama T, Kayama M et al (2021) Rappemonads are haptophyte phytoplankton. Curr Biol 31:2395–2403
Keeling PJ (2013) The number, speed, and impact of plastid endosymbioses in eukaryotic evolution. Annu Rev Plant Biol 64:583–607
Keeling PJ, Burki F (2019) Progress towards the tree of eukaryotes. Curr Biol 29:R808–R817
Keeling PJ, Campo JD (2017) Marine protists are not just big bacteria. Curr Biol 27:R541–R549
Keeling PJ, Burger G, Durnford DG et al (2005) The tree of eukaryotes. Trends Ecol Evol 20:670–676
Keeling PJ, Burki F, Wilcox HM et al (2014) The marine microbial eukaryote transcriptome sequencing project (MMETSP): illuminating the functional diversity of eukaryotic life in the oceans through transcriptome sequencing. PLoS Biol 12:e1001889
Kerszberg M (2000) The survival of slow reproducers. J Theor Biol 206:81–89
Kilias ES, Junges L, Šupraha L et al (2020) Chytrid fungi distribution and co-occurrence with diatoms correlate with sea ice melt in the Arctic Ocean. Commun Biol 3:183
Kim E, Harrison JW, Sudek S et al (2011a) Newly identified and diverse plastid-bearing branch on the eukaryotic tree of life. Proc Natl Acad Sci 108:1496–1500
Kim E, Harrison JW, Sudek S et al (2011b) Newly identified and diverse plastid-bearing branch on the eukaryotic tree of life. Proc Natl Acad Sci 108:1496–1500
Kim DY, Countway PD, Jones AC et al (2013) Monthly to interannual variability of microbial eukaryote assemblages at four depths in the eastern North Pacific. ISME J 8:515–530

King N, Westbrook MJ, Young SL et al (2008) The genome of the choanoflagellate *Monosiga brevicollis* and the origin of metazoans. Nature 451:783–788
Kiørboe T, Visser A, Andersen KH (2018) A trait-based approach to ocean ecology. ICES J Mar Sci 75:1849–1863
Kleindienst S, Grim S, Sogin M et al (2016) Diverse, rare microbial taxa responded to the Deepwater horizon deep-sea hydrocarbon plume. ISME J 10:400–415
Kleine T, Maier UG, Leister D (2009) DNA transfer from organelles to the nucleus: the idiosyncratic genetics of endosymbiosis. Annu Rev Plant Biol 60:115–138
Knight-Jones EW, Walne PR (1951) *Chromulina pusilla* butcher; a dominant member of the ultraplankton. Nature 167:445–446
Kolb A, Strom S (2013) An inducible antipredatory defense in haploid cells of the marine microalga *Emiliania huxleyi* (Prymnesiophyceae). Limnol Oceanogr 58:932–944
Krabberød AK, Orr RJS, Bråte J et al (2017) Single cell transcriptomics, mega-phylogeny, and the genetic basis of morphological innovations in Rhizaria. Mol Biol Evol 34:1557–1573
Kristiansen J, Škaloud P (2017) Chrysophyta. In: Archibald JM, Simpson AGB, Slamovits CH (eds) Handbook of the protists. Springer International Publishing, Cham, pp 331–366
Ku C, Nelson-Sathi S, Roettger M et al (2015) Endosymbiotic origin and differential loss of eukaryotic genes. Nature 524:427–432
Kudryavtsev A, Pawlowski J, Smirnov A (2018) More amoebae from the deep-sea: two new marine species of *Vexillifera* (Amoebozoa, Dactylopodida) with notes on taxonomy of the genus. Eur J Protistol 66:9–25
La Scola B, Audic S, Robert C et al (2003) A giant virus in amoebae. Science 299:2033
Labarre A, López-Escardó D, Latorre F et al (2021) Comparative genomics reveals new functional insights in uncultured MAST species. ISME J 15:1767–1781
Lam P, Kuypers MMM (2011) Microbial nitrogen cycling processes in oxygen minimum zones. Annu Rev Mar Sci 3:317–345
Lambert S, Tragin M, Lozano J-C et al (2019) Rhythmicity of coastal marine picoeukaryotes, bacteria and archaea despite irregular environmental perturbations. ISME J 13:388–401
Lane N (2011) Energetics and genetics across the prokaryote-eukaryote divide. Biol Direct 6:35
Lane CE, Archibald JM (2008) The eukaryotic tree of life: endosymbiosis takes its TOL. Trends Ecol Evol 23:268–275
Lane N, Martin W (2010) The energetics of genome complexity. Nature 467:929–934
Lax G, Eglit Y, Eme L et al (2018) Hemimastigophora is a novel supra-kingdom-level lineage of eukaryotes. Nature 564:410–414
Leadbeater BSC (2015) The choanoflagellates. Cambridge University Press, Cambridge
Leander BS (2008) Marine gregarines: evolutionary prelude to the apicomplexan radiation? Trends Parasitol 24:60–67
Leander BS (2020) Predatory protists. Curr Biol 30:R510–R516
Lennon JT, Jones SE (2011) Microbial seed banks: the ecological and evolutionary implications of dormancy. Nat Rev Microbiol 9:119–130
Leray M, Knowlton N (2016) Censusing marine eukaryotic diversity in the twenty-first century. Phil Trans R Soc Lond B 371:20150331
Lewis LA, McCourt RM (2004) Green algae and the origin of land plants. Am J Bot 91:1535–1556
Li WKW, McLaughlin F, Lovejoy C, Carmack EC (2009) Smallest algae thrive as the Arctic Ocean freshens. Science 326:539–539
Li Q, Edwards KF, Schvarcz CR et al (2021) Plasticity in the grazing ecophysiology of *Florenciella* (Dichtyochophyceae), a mixotrophic nanoflagellate that consumes *Prochlorococcus* and other bacteria. Limnol Oceanogr 66:47–60
Lie AAY, Kim DY, Schnetzer A, Caron DA (2013) Small-scale temporal and spatial variations in protistan community composition at the San Pedro Ocean time-series station off the coast of southern California. Aquat Microb Ecol 70:93–110
Lima-Mendez G, Faust K, Henry N et al (2015) Ocean plankton. Determinants of community structure in the global plankton interactome. Science 348:1262073–1262073

Limardo AJ, Sudek S, Choi CJ et al (2017) Quantitative biogeography of picoprasinophytes establishes ecotype distributions and significant contributions to marine phytoplankton. Environ Microbiol 19:3219–3234
Lin Y-C, Campbell T, Chung C-C et al (2012) Distribution patterns and phylogeny of marine stramenopiles in the North Pacific Ocean. Appl Environ Microbiol 78:3387–3399
Lindell D, Jaffe JD, Johnson ZI et al (2005) Photosynthesis genes in marine viruses yield proteins during host infection. Nature 438:86–89
Ling Y-C, Gan HM, Bush M et al (2018) Time-resolved microbial guild responses to tidal cycling in a coastal acid-sulfate system. Environ Chem 15:2–17
Litchman E, Klausmeier CA (2008) Trait-based community ecology of phytoplankton. Annu Rev Ecol Evol Syst 39:615–639
Litchman E, Ohman MD, Kiørboe T (2013) Trait-based approaches to zooplankton communities. J Plankton Res 35:473–484
Liu H, Probert I, Uitz J et al (2009) Extreme diversity in noncalcifying haptophytes explains a major pigment paradox in open oceans. Proc Natl Acad Sci U S A 106:12803–12808
Liu H, Aris-Brosou S, Probert I, de Vargas C (2010) A time line of the environmental genetics of the haptophytes. Mol Biol Evol 27:161–176
Liu H, Stephens TG, González-Pech RA, et al (2018) *Symbiodinium* genomes reveal adaptive evolution of functions related to coral-dinoflagellate symbiosis. Commun Biol 1:95
Livermore JA, Mattes TE (2013) Phylogenetic detection of novel Cryptomycota in an Iowa (United States) aquifer and from previously collected marine and freshwater targeted high-throughput sequencing sets. Environ Microbiol 15:2333–2341
Lomas MW, Bates NR, Johnson RJ et al (2013) Two decades and counting: 24-years of sustained open ocean biogeochemical measurements in the Sargasso Sea. Deep Sea Res Part 2 Top Stud Oceanogr 93:16–32
López-García P, Rodríguez-Valera F, Pedrós-Alió C, Moreira D (2001) Unexpected diversity of small eukaryotes in deep-sea Antarctic plankton. Nature 409:603–607
López-García P, Eme L, Moreira D (2017) Symbiosis in eukaryotic evolution. J Theor Biol 434:20–33
Lowe CD, Minter EJ, Cameron DD, Brockhurst MA (2016) Shining a light on exploitative host control in a photosynthetic endosymbiosis. Curr Biol 26:207–211
Lundholm N, Hasle GR (2010) *Fragilariopsis* (Bacillariophyceae) of the Northern Hemisphere – morphology, taxonomy, phylogeny and distribution, with a description of *F. pacifica* sp. nov. Phycologia 49:438–460
Luo G, Ono S, Beukes NJ et al (2016) Rapid oxygenation of Earth's atmosphere 2.33 billion years ago. Sci Adv 2:e1600134
Luo E, Eppley JM, Romano AE et al (2020) Double-stranded DNA virioplankton dynamics and reproductive strategies in the oligotrophic open ocean water column. ISME J 14:1304–1315
Lynch M, Marinov GK (2017) Membranes, energetics, and evolution across the prokaryote-eukaryote divide. elife 6:e20437
Lyons TW, Reinhard CT, Planavsky NJ (2014) The rise of oxygen in Earth's early ocean and atmosphere. Nature 506:307–315
Maamar A, Lucchesi M-E, Debaets S et al (2020) Highlighting the crude oil bioremediation potential of marine fungi isolated from the port of Oran (Algeria). Diversity 12:196
Madin LP, Horgan EF, Steinberg DK (2001) Zooplankton at the Bermuda Atlantic time-series study (BATS) station: diel, seasonal and interannual variation in biomass, 1994–1998. Deep Sea Res Part 2 Top Stud Oceanogr 48:2063–2082
Mafra LL Jr, Nagai S, Uchida H et al (2016) Harmful effects of Dinophysis to the ciliate *Mesodinium rubrum*: implications for prey capture. Harmful Algae 59:82–90
Malviya S, Scalco E, Audic S et al (2016) Insights into global diatom distribution and diversity in the world's ocean. Proc Natl Acad Sci 113:E1516–E1525
Margulis L (1990) Words as battle cries—symbiogenesis and the new field of endocytobiology. Bioscience 40:673–677

Marin B, Melkonian M (2010) Molecular phylogeny and classification of the Mamiellophyceae class. Nov. (Chlorophyta) based on sequence comparisons of the nuclear- and plastid-encoded rRNA operons. Protist 161:304–336
Martini S, Larras F, Boyé A et al (2020) Functional trait-based approaches as a common framework for aquatic ecologists. Limnol Oceanogr 66:965–994
Martin-Platero AM, Cleary B, Kauffman K et al (2018) High resolution time series reveals cohesive but short-lived communities in coastal plankton. Nat Commun 9:266
Massana R, Pedrós-Alió C (2008) Unveiling new microbial eukaryotes in the surface ocean. Curr Op Microbiol 11:213–218
Massana R, Guillou L, Díez B, Pedrós-Alió C (2002) Unveiling the organisms behind novel eukaryotic ribosomal DNA sequences from the ocean. Appl Environ Microbiol 68:4554–4558
Massana R, Balagué V, Guillou L, Pedrós-Alió C (2004) Picoeukaryotic diversity in an oligotrophic coastal site studied by molecular and culturing approaches. FEMS Microbiol Ecol 50:231–243
Massana R, Terrado R, Forn I et al (2006) Distribution and abundance of uncultured heterotrophic flagellates in the world oceans. Environ Microbiol 8:1515–1522
Massana R, Unrein F, Rodríguez-Martínez R et al (2009) Grazing rates and functional diversity of uncultured heterotrophic flagellates. ISME J 3:588–596
Massana R, del Campo J, Sieracki ME et al (2014) Exploring the uncultured microeukaryote majority in the oceans: reevaluation of ribogroups within stramenopiles. ISME J 8:854–866
Massana R, Gobet A, Audic S et al (2015) Marine protist diversity in European coastal waters and sediments as revealed by high-throughput sequencing. Environ Microbiol 17:4035–4049
Mast SO (1947) The food-vacuole in paramecium. Biol Bull 92:31–72
Matz C, Boenigk J, Arndt H, Jürgens K (2002) Role of bacterial phenotypic traits in selective feeding of the heterotrophic nanoflagellate *Spumella* sp. Aquat Microb Ecol 27:137–148
Maumus F, Epert A, Nogué F, Blanc G (2014) Plant genomes enclose footprints of past infections by giant virus relatives. Nat Commun 5:4268
Mayer JA, Taylor FJR (1979) A virus which lyses the marine nanoflagellate *Micromonas pusilla*. Nature 281:299–301
Mayr E (1996) What is a species, and what is not? Phil Sci 63:262–277
McClanahan TR, Marnane MJ, Cinner JE, Kiene WE (2006) A comparison of marine protected areas and alternative approaches to coral-reef management. Curr Biol 16:1408–1413
McFadden GI, Reith ME, Munholland J, Lang-Unnasch N (1996) Plastid in human parasites. Nature 381:482
McFall-Ngai MJ (2014) The importance of microbes in animal development: lessons from the squid-vibrio symbiosis. Annu Rev Microbiol 68:177–194
McRose D, Guo J, Monier A et al (2014) Alternatives to vitamin B 1 uptake revealed with discovery of riboswitches in multiple marine eukaryotic lineages. ISME J 8:2517–2529
Medina LE, Taylor CD, Pachiadaki MG et al (2017) A review of protist grazing below the photic zone emphasizing studies of oxygen-depleted water columns and recent applications of in situ approaches. Front Mar Sci 4:105
Medlin LK, Sáez AG, Young JR (2008) A molecular clock for coccolithophores and implications for selectivity of phytoplankton extinctions across the K/T boundary. Mar Micropaleontol 67: 69–86
Meints RH, Ivey RG, Lee AM, Choi T-J (2008) Identification of two virus integration sites in the brown alga Feldmannia chromosome. J Virol 82:1407–1413
Mendes CRB, Tavano VM, Dotto TS et al (2018) New insights on the dominance of cryptophytes in Antarctic coastal waters: A case study in Gerlache Strait. Deep Sea Res Part 2 Top Stud Oceanogr 149:161–170
Mereschkowsky C (1905) Über Natur und Ursprung der Chromatophoren im Pflanzenreiche. Biologisches Centralblatt 25:293–604
Meyen FJF (1834) Reise um die Erde: ausgeführt auf dem königlich preussischen Seehandlungs-Schiffe Prinzess Louise, commandirt von Captain W. Wendt, in den Jahren 1830, 1831 und 1832. Sander'sche buchhandlung

Meyer-Abich A (1943) I. Das typologische Grundgesetz und seine Folgerungen für Phylogenie und Entwicklungsphysiologie. Acta Biotheor 7:1–80
Mihara T, Koyano H, Hingamp P et al (2018) Taxon richness of "Megaviridae" exceeds those of bacteria and archaea in the ocean. Microbes Environ 33:162–171
Mitra A, Flynn KJ, Burkholder JM et al (2014) The role of mixotrophic protists in the biological carbon pump. Biogeosciences 11:995–1005
Mitra A, Flynn KJ, Tillmann U et al (2016) Defining planktonic protist functional groups on mechanisms for energy and nutrient acquisition: incorporation of diverse mixotrophic strategies. Protist 167:106–120
Mock T, Otillar RP, Strauss J et al (2017) Evolutionary genomics of the cold-adapted diatom *Fragilariopsis cylindrus*. Nature 541:536–540
Moeller HV, Neubert MG, Johnson MD (2019) Intraguild predation enables coexistence of competing phytoplankton in a well-mixed water column. Ecology 100:e02874
Mojica KDA, Huisman J, Wilhelm SW, Brussaard CPD (2016) Latitudinal variation in virus-induced mortality of phytoplankton across the North Atlantic Ocean. ISME J 10:500–513
Monger BC, Landry MR (1991) Prey-size dependency of grazing by free-living marine flagellates. Mar Ecol Prog Ser 74:239–248
Monier A, Worden AZ, Richards TA (2016) Phylogenetic diversity and biogeography of the Mamiellophyceae lineage of eukaryotic phytoplankton across the oceans. Environ Microbiol Rep 8:461–469
Monier A, Chambouvet A, Milner DS et al (2017) Host-derived viral transporter protein for nitrogen uptake in infected marine phytoplankton. Proc Natl Acad Sci 114:E7489–E7498
Monteil CL, Vallenet D, Menguy N et al (2019) Ectosymbiotic bacteria at the origin of magnetoreception in a marine protist. Nat Microbiol 4:1088–1095
Monteiro FM, Follows MJ, Dutkiewicz S (2010) Distribution of diverse nitrogen fixers in the global ocean. Global Biogeochem Cycles 24:GB3017
Moodley L, Middelburg JJ, Boschker HTS et al (2002) Bacteria and foraminifera: key players in a short-term deep-sea benthic response to phytodetritus. Mar Ecol Prog Ser 236:23–29
Moon-van der Staay SY, De Wachter R, Vaulot D (2001) Oceanic 18S rDNA sequences from picoplankton reveal unsuspected eukaryotic diversity. Nature 409:607–610
Moore RB, Oborník M, Janouskovec J et al (2008) A photosynthetic alveolate closely related to apicomplexan parasites. Nature 451:959–963
Mordret S, Romac S, Henry N et al (2016) The symbiotic life of *Symbiodinium* in the open ocean within a new species of calcifying ciliate (*Tiarina* sp.). ISME J 10:1424–1436
Moreau H, Verhelst B, Couloux A et al (2012) Gene functionalities and genome structure in Bathycoccus prasinos reflect cellular specializations at the base of the green lineage. Genome Biol 13:R74
Moreira D, López-García P (2014) The rise and fall of Picobiliphytes: how assumed autotrophs turned out to be heterotrophs. BioEssays 36:468–474
Mougari S, Sahmi-Bounsiar D, Levasseur A et al (2019) Virophages of giant viruses: an update at eleven. Viruses 11:733
Murphy LS, Haugen EM (1985) The distribution and abundance of phototrophic ultraplankton in the North Atlantic. Limnol Oceanogr 30:47–58
Needham DM, Fuhrman JA (2016) Pronounced daily succession of phytoplankton, archaea and bacteria following a spring bloom. Nature Microbiol 1:16005–16005
Needham DM, Sachdeva R, Fuhrman JA (2017) Ecological dynamics and co-occurrence among marine phytoplankton, bacteria and myoviruses shows microdiversity matters. ISME J 11: 1614–1629
Needham DM, Fichot EB, Wang E et al (2018) Dynamics and interactions of highly resolved marine plankton via automated high frequency sampling. ISME J 12:2417–2432
Needham DM, Poirier C, Hehenberger E et al (2019a) Targeted metagenomic recovery of four divergent viruses reveals shared and distinctive characteristics of giant viruses of marine eukaryotes. Phil Trans R Soc Lond B 374:20190086

Needham DM, Yoshizawa S, Hosaka T et al (2019b) A distinct lineage of giant viruses brings a rhodopsin photosystem to unicellular marine predators. Proc Natl Acad Sci 116:20574–20583
Neuer S, Cowles TJ (1994) Protist herbivory in the Oregon upwelling system. Mar Ecol Prog Ser 113:147–162
Nielsen LT, Kiørboe T (2021) Foraging trade-offs, flagellar arrangements, and flow architecture of planktonic protists. Proc Natl Acad Sci 118:e2009930118
Nielsen LT, Asadzadeh SS, Dölger J et al (2017) Hydrodynamics of microbial filter feeding. Proc Natl Acad Sci 114:9373–9378
Nöel M-H, Kawachi M, Inouye I (2004) Induced dimorphic life cycle of a coccolithophorid, *Calyptrosphaera sphaeroidea* (Prymnesiophyceae, Haptophyta). J Phycol 40:112–129
Not F, Gausling R, Azam F et al (2007a) Vertical distribution of picoeukaryotic diversity in the Sargasso Sea. Environ Microbiol 9:1233–1252
Not F, Valentin K, Romari K et al (2007b) Picobiliphytes: a marine picoplanktonic algal group with unknown affinities to other eukaryotes. Science 315:253–255
Nowack ECM, Melkonian M, Glöckner G (2008) Chromatophore genome sequence of *Paulinella* sheds light on acquisition of photosynthesis by eukaryotes. Curr Biol 18:410–418
Nyholm SV, McFall-Ngai M (2004) The winnowing: establishing the squid–vibrio symbiosis. Nat Rev Microbiol 2:632–642
O'Malley MA (2007) The nineteenth century roots of "everything is everywhere.". Nat Rev Microbiol 5:647–651
O'Malley MA, Leger MM, Wideman JG, Ruiz-Trillo I (2019) Concepts of the last eukaryotic common ancestor. Nat Ecol Evol 3:338–344
Oikonomou CM, Chang Y-W, Jensen GJ (2016) A new view into prokaryotic cell biology from electron cryotomography. Nat Rev Microbiol 14:205–220
Ollison GA, Hu SK, Mesrop LY et al (2021) Come rain or shine: depth not season shapes the active protistan community at station ALOHA in the North Pacific subtropical gyre. Deep-Sea Res I Oceanogr Res Pap 170:103494
Olson MB, Strom SL (2002) Phytoplankton growth, microzooplankton herbivory and community structure in the southeast Bering Sea: insight into the formation and temporal persistence of an *Emiliania huxleyi* bloom. Deep Sea Res Part 2 Top Stud Oceanogr 49:5969–5990
Omand MM, D'Asaro EA, Lee CM et al (2015) Eddy-driven subduction exports particulate organic carbon from the spring bloom. Science 348:222–225
Orcutt BN, Sylvan JB, Knab NJ, Edwards KJ (2011) Microbial ecology of the dark ocean above, at, and below the seafloor. Microbiol Mol Biol Rev 75:361–422
Orsi W, Song YC, Hallam S, Edgcomb V (2012) Effect of oxygen minimum zone formation on communities of marine protists. ISME J 6:1586–1601
Orsi WD, Wilken S, Del Campo J et al (2018) Identifying protist consumers of photosynthetic picoeukaryotes in the surface ocean using stable isotope probing. Environ Microbiol 20:815–827
Paasche E (2001) A review of the coccolithophorid *Emiliania huxleyi* (Prymnesiophyceae), with particular reference to growth, coccolith formation, and calcification-photosynthesis interactions. Phycologia 40:503–529
Paerl RW, Sundh J, Tan D et al (2018) Prevalent reliance of bacterioplankton on exogenous vitamin B1 and precursor availability. Proc Natl Acad Sci 115:E10447–E10456
Palenik B, Morel FMM (1990) Amino acid utilization by marine phytoplankton: A novel mechanism. Limnol Oceanogr 35:260–269
Pančić M, Kiørboe T (2018) Phytoplankton defence mechanisms: traits and trade-offs. Biol Rev Camb Philos Soc 93:1269–1303
Pančić M, Torres RR, Almeda R, Kiørboe T (2019) Silicified cell walls as a defensive trait in diatoms. Proc Biol Sci 286:20190184
Parfrey LW, Lahr DJG (2013) Multicellularity arose several times in the evolution of eukaryotes (response to DOI https://doi.org/10.1002/bies.201100187). BioEssays 35:339–347

Park MG, Kim S, Kim HS et al (2006) First successful culture of the marine dinoflagellate *Dinophysis acuminata*. Aquat Microb Ecol 45:101–106
Parke M, Manton I (1967) The specific identity of the algal symbiont in *Convoluta roscoffensis*. J Mar Biol Assoc 47:445–464
Pasulka AL, Landry MR, Taniguchi DAA et al (2013) Temporal dynamics of phytoplankton and heterotrophic protists at station ALOHA. Deep Sea Res Part 2 Top Stud Oceanogr 93:44–57
Patron NJ, Waller RF (2007) Transit peptide diversity and divergence: A global analysis of plastid targeting signals. BioEssays 29:1048–1058
Patyshakuliyeva A, Falkoski DL, Wiebenga A et al (2019) Macroalgae derived fungi have high abilities to degrade algal polymers. Microorganisms 8:52
Paulmier A, Ruiz-Pino D (2009) Oxygen minimum zones (OMZs) in the modern ocean. Prog Oceanogr 80:113–128
Pernice MC, Forn I, Gomes A et al (2015) Global abundance of planktonic heterotrophic protists in the deep ocean. ISME J 9:782–792
Pernice MC, Giner CR, Logares R et al (2016) Large variability of bathypelagic microbial eukaryotic communities across the world's oceans. ISME J 10:945–958
Pernthaler J (2005) Predation on prokaryotes in the water column and its ecological implications. Nat Rev Microbiol 3:537–546
Peter KH, Sommer U (2013) Phytoplankton cell size reduction in response to warming mediated by nutrient limitation. PLoS One 8:e71528
Pettitt ME, Orme BAA, Blake JR, Leadbeater BSC (2002) The hydrodynamics of filter feeding in choanoflagellates. Eur J Protistol 38:313–332
Philippe H, Brinkmann H, Lavrov DV et al (2011) Resolving difficult phylogenetic questions: why more sequences are not enough. PLoS Biol 9:e1000602
Piganeau G, Eyre-Walker A, Jancek S et al (2011) How and why DNA barcodes underestimate the diversity of microbial eukaryotes. PLoS One 6:e16342
Pita L, Rix L, Slaby BM et al (2018) The sponge holobiont in a changing ocean: from microbes to ecosystems. Microbiome 6:46
Pitta P, Karakassis I (2005) Size distribution in ultraphytoplankton: a comparative analysis of counting methods. Environ Monit Assess 102:85–101
Piwosz K, Mukherjee I, Salcher MM et al (2021) CARD-FISH in the sequencing era: opening a new universe of protistan ecology. Front Microbiol 12:397
Pizay M-D, Lemée R, Simon N et al (2009) Night and day morphologies in a planktonic dinoflagellate. Protist 160:565–575
Planavsky NJ, Crowe SA, Fakhraee M et al (2021) Evolution of the structure and impact of Earth's biosphere. Nature Rev Earth Environment 2:123–139
Pomeroy LR (1974) The ocean's food web, a changing paradigm. Bioscience 24:499–504
Pontasch S, Scott A, Hill R et al (2014) Symbiodinium diversity in the sea anemone *Entacmaea quadricolor* on the east Australian coast. Coral Reefs 33:537–542
Preisig HR (1994) Siliceous structures and silicification in flagellated protists. In: Wetherbee R, Pickett-Heaps JD, Andersen RA (eds) The Protistan cell surface. Springer Vienna, Vienna, pp 29–42
Pride DT, Sun CL, Salzman J et al (2010) Analysis of streptococcal CRISPRs from human saliva reveals substantial sequence diversity within and between subjects over time. Genome Res 21: 126–136
Priest T, Fuchs B, Amann R, Reich M (2021) Diversity and biomass dynamics of unicellular marine fungi during a spring phytoplankton bloom. Environ Microbiol 23:448–463
Pringsheim EG (1958) Über Mixotrophie bei Flagellaten. Planta 52:405–430
Proctor LM, Fuhrman JA (1990) Viral mortality of marine bacteria and cyanobacteria. Nature 343: 60–62
Raghukumar S (2002) Ecology of the marine protists, the Labyrinthulomycetes (Thraustochytrids and Labyrinthulids). Eur J Protistol 38:127–145

Raina J-B, Fernandez V, Lambert B et al (2019) The role of microbial motility and chemotaxis in symbiosis. Nat Rev Microbiol 17:284–294
Raoult D, Audic S, Robert C et al (2004) The 1.2-megabase genome sequence of Mimivirus. Science 306:1344–1350
Raven JA (1998) The twelfth Tansley lecture. Small is beautiful: the picophytoplankton. Funct Ecol 12:503–513
Raven J, Caldeira K, Elderfield H, et al. (2005) Ocean acidification due to increasing atmospheric carbon dioxide The Royal Society Policy Document 12/05
Read BA, Kegel J, Klute MJ et al (2013) Pan genome of the phytoplankton *Emiliania* underpins its global distribution. Nature 499:209–213
Record NR, Pershing AJ, Maps F (2013) The paradox of the "paradox of the plankton.". ICES J Mar Sci 71:236–240
Reynolds N (1973) The estimation of the abundance of ultraplankton. Br Phycol J 8:135–146
Richards TA, Talbot NJ (2018) Osmotrophy. Curr Biol 28:R1179–R1180
Richards TA, Dacks JB, Jenkinson JM et al (2006) Evolution of filamentous plant pathogens: gene exchange across eukaryotic kingdoms. Curr Biol 16:1857–1864
Richards TA, Jones MDM, Leonard G, Bass D (2012) Marine fungi: their ecology and molecular diversity. Annu Rev Mar Sci 4:495–522
Richards TA, Leonard G, Mahé F et al (2015) Molecular diversity and distribution of marine fungi across 130 European environmental samples. Proc Biol Sci 282:20152243
Richter DJ, Fozouni P, Eisen M, King N (2018) Gene family innovation, conservation and loss on the animal stem lineage. elife 7:e34226–e34226
Roberts EC, Legrand C, Steinke M, Wootton EC (2011) Mechanisms underlying chemical interactions between predatory planktonic protists and their prey. J Plankton Res 33:833–841
Rocke E, Pachiadaki MG, Cobban A et al (2015) Protist community grazing on prokaryotic prey in deep ocean water masses. PLoS One 10:e0124505
Rodríguez-Martínez R, Leonard G, Milner DS et al (2020) Controlled sampling of ribosomally active protistan diversity in sediment-surface layers identifies putative players in the marine carbon sink. ISME J 14:984–998
Roger Anderson O, Be AWH (1976) The ultrastructure of a planktonic foraminifer, *Globigerinoides sacculifer* (Brady), and its symbiotic dinoflagellates. J Foraminiferal Res 6: 1–21
Ros-Rocher N, Pérez-Posada A, Leger MM, Ruiz-Trillo I (2021) The origin of animals: an ancestral reconstruction of the unicellular-to-multicellular transition. Open Biol 11:200359
Rothhaupt KO (1996a) Utilization of substitutable carbon and phosphorus sources by the mixotrophic chrysophyte *Ochromonas* sp. Ecology 77:706–715
Rothhaupt KO (1996b) Laboratorary experiments with a mixotrophic chrysophyte and obligately phagotrophic and photographic competitors. Ecology 77:716–724
Roy S, Chattopadhyay J (2007) Towards a resolution of "the paradox of the plankton": A brief overview of the proposed mechanisms. Ecol Complex 4:26–33
Ruardij P, Veldhuis MJW, Brussaard CPD (2005) Modeling the bloom dynamics of the polymorphic phytoplankter *Phaeocystis globosa*: impact of grazers and viruses. Harmful Algae 4:941–963
Sagan L (1967) On the origin of mitosing cells. J Theor Biol 14:225–274
Sala E, Mayorga J, Bradley D et al (2021) Protecting the global ocean for biodiversity, food and climate. Nature 592:297–402
Samadi S, Barberousse A (2006) The tree, the network, and the species. Biol J Linn Soc Lond 89: 509–521
Sánchez-Baracaldo P, Cardona T (2020) On the origin of oxygenic photosynthesis and cyanobacteria. New Phytol 225:1440–1446
Sarno D, Kooistra WHCF, Medlin LK et al (2005) Diversity in the genus *Skeletonema* (Bacillariophyceae). An assessment of the taxonomy of Costatum-like species with the description of four new species. J Phycol 41:151–176

Schaffelke B, Heimann K, Marshall PA, Ayling AM (2004) Blooms of *Chrysocystis fragilis* on the great barrier reef. Coral Reefs 23:514–514
Schmoker C, Hernández-León S, Calbet A (2013) Microzooplankton grazing in the oceans: impacts, data variability, knowledge gaps and future directions. J Plankton Res 35:691–706
Schoemann V, Becquevort S, Stefels J, et al. (2005) *Phaeocystis* blooms in the global ocean and their controlling mechanisms: a review. J Sea Res 53:43–66
Scholz B, Küpper FC, Vyverman W, Karsten U (2014) Eukaryotic pathogens (Chytridiomycota and Oomycota) infecting marine microphytobenthic diatoms – a methodological comparison. J Phycol 50:1009–1019
Scholz B, Vyverman W, Küpper FC et al (2017) Effects of environmental parameters on chytrid infection prevalence of four marine diatoms: a laboratory case study. Bot Mar 60:419–431
Schroeder DC, Oke J, Hall M et al (2003) Virus succession observed during an *Emiliania huxleyi* bloom. Appl Environ Microbiol 69:2484–2490
Schulz F, Yutin N, Ivanova NN et al (2017) Giant viruses with an expanded complement of translation system components. Science 356:82–85
Schvarcz CR, Steward GF (2018) A giant virus infecting green algae encodes key fermentation genes. Virology 518:423–433
Seeleuthner Y, Mondy S, Lombard V et al (2018) Single-cell genomics of multiple uncultured stramenopiles reveals underestimated functional diversity across oceans. Nat Commun 9:310
Seenivasan R (2013) Identification of a new eukaryotic phylum "Picozoa" in the oceanic environment. Doctoral Thesis, Universität zu Köln
Seenivasan R, Sausen N, Medlin LK, Melkonian M (2013) *Picomonas judraskeda* gen. et sp. nov.: the first identified member of the Picozoa phylum nov., a widespread group of picoeukaryotes, formerly known as "picobiliphytes." PLoS ONE 8:e59565
Sellers CG, Gast RJ, Sanders RW (2014) Selective feeding and foreign plastid retention in an Antarctic dinoflagellate. J Phycol 50:1081–1088
Ševčíková T, Horák A, Klimeš V et al (2015) Updating algal evolutionary relationships through plastid genome sequencing: did alveolate plastids emerge through endosymbiosis of an ochrophyte? Sci Rep 5:10134
Seymour JR, Amin SA, Raina J-B, Stocker R (2017) Zooming in on the phycosphere: the ecological interface for phytoplankton-bacteria relationships. Nat Microbiol 2:17065
Sharma V, Colson P, Giorgi R et al (2014) DNA-dependent RNA polymerase detects hidden giant viruses in published databanks. Genome Biol Evol 6:1603–1610
Sharon I, Alperovitch A, Rohwer F et al (2009) Photosystem I gene cassettes are present in marine virus genomes. Nature 461:258–262
Sherr E, Sherr B (1988) Role of microbes in pelagic food webs: A revised concept. Limnol Oceanogr 33:1225–1227
Sherr EB, Sherr BF (2002) Significance of predation by protists in aquatic microbial food webs. Ant Leeuwenhoek 81:293–308
Sherr EB, Sherr BF (2007) Heterotrophic dinoflagellates: a significant component of microzooplankton biomass and major grazers of diatoms in the sea. Mar Ecol Prog Ser 352: 187–197
Short SM (2012) The ecology of viruses that infect eukaryotic algae. Environ Microbiol 14:2253–2271
Sibbald SJ, Archibald JM (2020) Genomic insights into plastid evolution. Genome Biol Evol 12: 978–990
Sieburth JM, Smetacek V, Lenz J (1978) Pelagic ecosystem structure: heterotrophic compartments of the plankton and their relationship to plankton size fractions. Limnol Oceanogr 23:1256–1263
Sieracki ME, Poulton NJ, Jaillon O et al (2019) Single cell genomics yields a wide diversity of small planktonic protists across major ocean ecosystems. Sci Rep 9:6025
Silva PC (2008) Historical review of attempts to decrease subjectivity in species identification, with particular regard to algae. Protist 159:153–161

Simmons MP, Bachy C, Sudek S et al (2015) Intron invasions trace algal speciation and reveal nearly identical Arctic and Antarctic *Micromonas* populations. Mol Biol Evol 32:2219–2235
Simmons MP, Sudek S, Monier A et al (2016) Abundance and biogeography of picoprasinophyte ecotypes and other phytoplankton in the eastern North Pacific Ocean. Appl Environ Microbiol 82:1693–1705
Simon M, López-García P, Deschamps P et al (2015) Marked seasonality and high spatial variability of protist communities in shallow freshwater systems. ISME J 9:1941–1953
Simon J-C, Marchesi JR, Mougel C, Selosse M-A (2019) Host-microbiota interactions: from holobiont theory to analysis. Microbiome 7:5
Simpson AGB, Roger AJ (2004) The real "kingdoms" of eukaryotes. Curr Biol 14:R693–R696
Smetacek V (1999) Revolution in the ocean. Nature 401:647–647
Sogin ML (2015) Evolution of eukaryotic microorganisms and their small subunit ribosomal RNAs1. Integr Comp Biol 29:487–499
Sogin ML, Morrison HG, Huber JA et al (2006) Microbial diversity in the deep sea and the underexplored "rare biosphere.". Proc Natl Acad Sci 103:12115–12120
Soucy SM, Huang J, Gogarten JP (2015) Horizontal gene transfer: building the web of life. Nat Rev Genet 16:472–482
Sournia A, Chrdtiennot-Dinet M-J, Ricard M (1991) Marine phytoplankton: how many species in the world ocean? J Plankton Res 13:1093–1099
Spang A, Saw JH, Jørgensen SL et al (2015) Complex archaea that bridge the gap between prokaryotes and eukaryotes. Nature 521:173–179
Stanier RY, Van Niel CB (1962) The concept of a bacterium. Arch Mikrobiol 42:17–35
Steele JA, Countway PD, Xia LC et al (2011) Marine bacterial, archaeal and protistan association networks reveal ecological linkages. ISME J 5:1414–1425
Steinberg DK, Landry MR (2017) Zooplankton and the ocean carbon cycle. Annu Rev Mar Sci 9: 413–444
Steinberg DK, Carlson CA, Bates NR et al (2001) Overview of the US JGOFS Bermuda Atlantic Time-series Study (BATS): a decade-scale look at ocean biology and biogeochemistry. Deep Sea Res Part 2 Top Stud Oceanogr 48:1405–1447
Steward GF, Culley AI, Mueller JA et al (2013) Are we missing half of the viruses in the ocean? ISME J 7:672–679
Stiller JW, Schreiber J, Yue J et al (2014) The evolution of photosynthesis in chromist algae through serial endosymbioses. Nat Commun 5:5764
Stoecker DK (1998) Conceptual models of mixotrophy in planktonic protists and some ecological and evolutionary implications. Eur J Protistol 34:281–290
Stoecker DK, Silver MW, Michaels AE, Davis LH (1988) Enslavement of algal chloroplasts by four Strombidium spp. (Ciliophora, oligo trichida). Mar Microb Food Webs 3:79–100
Stoecker DK, Hansen PJ, Caron DA, Mitra A (2017) Mixotrophy in the marine plankton. Annu Rev Mar Sci 9:311–335
Strassert JFH, Karnkowska A, Hehenberger E et al (2018) Single cell genomics of uncultured marine alveolates shows paraphyly of basal dinoflagellates. ISME J 12:304–308
Strassert JFH, Jamy M, Mylnikov AP et al (2019) New phylogenomic analysis of the enigmatic phylum Telonemia further resolves the eukaryote tree of life. Mol Biol Evol 36:757–765
Suda S, Watanabe MM, Inouye I (1989) Evidence for sexual reproduction in the primitive green alga *Nephroselmis olivacea* (Prasinophyceae). J Phycol 25:596–600
Suttle CA, Chan AM, Cottrell MT (1990) Infection of phytoplankton by viruses and reduction of primary productivity. Nature 347:467–469
Sweeney BM (1976) *Pedinomonas noctilucae* (Prasinophyceae), the flagellate symbiotic in *Noctiluca* flagellate symbiotic in *Noctiluca* (Dinophyceae) in Southeast Asia. J Phycol 12: 460–464
Takahashi M, Bienfang PK (1983) Size structure of phytoplankton biomass and photosynthesis in subtropical Hawaiian waters. Mar Biol 76:203–211

Taylor FJRM (2003) The collapse of the two-kingdom system, the rise of protistology and the founding of the International Society for Evolutionary Protistology (ISEP). Int J Syst Evol Microbiol 53:1707–1714
Taylor DL, Lee CC (1971) A new cryptomonad from Antarctica: *Cryptomonas cryophila* sp. nov. Arch Mikrobiol 75:269–280
Taylor BP, Cortez MH, Weitz JS (2014) The virus of my virus is my friend: ecological effects of virophage with alternative modes of coinfection. J Theor Biol 354:124–136
Taylor AR, Brownlee C, Wheeler G (2017) Coccolithophore cell biology: chalking up progress. Annu Rev Mar Sci 9:283–310
Taylor BP, Weitz JS, Brussaard CPD, Fischer MG (2018) Quantitative infection dynamics of cafeteria Roenbergensis virus. Viruses 10:468
Teeling H, Fuchs BM, Becher D et al (2012) Substrate-controlled succession of marine bacterioplankton populations induced by a phytoplankton bloom. Science 336:608–611
Thines M (2018) Oomycetes. Curr Biol 28:R812–R813
Thompson JR, Rivera HE, Closek CJ, Medina M (2014) Microbes in the coral holobiont: partners through evolution, development, and ecological interactions. Front Cell Infect Microbiol 4:176
Tikhonenkov DV (2020) Predatory flagellates – the new recently discovered deep branches of the eukaryotic tree and their evolutionary and ecological significance. Protistology 14:15–22
Tikhonenkov DV, Hehenberger E, Esaulov AS et al (2020a) Insights into the origin of metazoan multicellularity from predatory unicellular relatives of animals. BMC Biol 18:39
Tikhonenkov DV, Strassert JFH, Janouškovec J et al (2020b) Predatory colponemids are the sister group to all other alveolates. Mol Phylogenet Evol 149:106839
Tikhonenkov DV, Gawryluk RMR, Mylnikov AP, Keeling PJ (2021) First finding of free-living representatives of Prokinetoplastina and their nuclear and mitochondrial genomes. Sci Rep 11: 2946
Timmis JN, Ayliffe MA, Huang CY, Martin W (2004) Endosymbiotic gene transfer: organelle genomes forge eukaryotic chromosomes. Nat Rev Genet 5:123–135
Tischler AH, Hodge-Hanson KM, Visick KL (2019) *Vibrio fischeri*–squid symbiosis. Encyclopedia Life Sci. https://doi.org/10.1002/9780470015902.a0028395
Tizard TH, Moseley HN, Buchanan JY, Murray J (1885) Narrative of the cruise of HMS challenger with a general account of the scientific results of the expedition. Report on the scientific results of the voyage of HMS challenger during the years 1873-76. Narrative 1:511–1110
Tragin M, Vaulot D (2018) Green microalgae in marine coastal waters: the ocean sampling day (OSD) dataset. Sci Rep 8:14020
Tragin M, Lopes dos Santos A, Christen R, Vaulot D (2016) Diversity and ecology of green microalgae in marine systems: an overview based on 18S rRNA gene sequences. Perspect Phycol 3:141–154
Treusch AH, Vergin KL, Finlay LA et al (2009) Seasonality and vertical structure of microbial communities in an ocean gyre. ISME J 3:1148–1163
Treusch AH, Demir-Hilton E, Vergin KL et al (2011) Phytoplankton distribution patterns in the northwestern Sargasso Sea revealed by small subunit rRNA genes from plastids. ISME J 6:481–492
Unrein F, Massana R, Alonso-Sáez L, Gasol JM (2007) Significant year-round effect of small mixotrophic flagellates on bacterioplankton in an oligotrophic coastal system. Limnol Oceanogr 52:456–469
Unrein F, Gasol JM, Not F et al (2014) Mixotrophic haptophytes are key bacterial grazers in oligotrophic coastal waters. ISME J 8:164–176
Urayama S-I, Takaki Y, Nishi S et al (2018) Unveiling the RNA virosphere associated with marine microorganisms. Mol Ecol Resources 18:1444–1455
van de Guchte M, Blottière HM, Doré J (2018) Humans as holobionts: implications for prevention and therapy. Microbiome 6:81
Van Donk E, Ianora A, Vos M (2011) Induced defences in marine and freshwater phytoplankton: a review. Hydrobiologia 668:3–19

Van Etten JL, Agarkova I, Dunigan DD et al (2017) Chloroviruses have a sweet tooth. Viruses 9:88
Van Leeuwenhoek A (1677) Observations, communicated to the publisher by Mr. Antony van Leewenhoeck, in a dutch letter of the 9th Octob. 1676. Here English'd: concerning little animals by him observed in rain-well-sea-and snow water; as also in water wherein pepper had lain infused. Philos Trans R Soc Lond 12:821–831
Van Steenkiste NWL, Stephenson I, Herranz M et al (2019) A new case of kleptoplasty in animals: marine flatworms steal functional plastids from diatoms. Sci Adv 5:eaaw4337
van Tatenhove-Pel RJ, Rijavec T, Lapanje A et al (2021) Microbial competition reduces metabolic interaction distances to the low µm-range. ISME J 15:688–701
Vannier T, Leconte J, Seeleuthner Y et al (2016) Survey of the green picoalga *Bathycoccus* genomes in the global ocean. Sci Rep 6:37900
Vardi A, Haramaty L, Van Mooy BAS et al (2012) Host-virus dynamics and subcellular controls of cell fate in a natural coccolithophore population. Proc Natl Acad Sci 109:19327–19332
Vellai T, Vida G (1999) The origin of eukaryotes: the difference between prokaryotic and eukaryotic cells. Proc Biol Sci 266:1571–1577
Villa Martín P, Buček A, Bourguignon T, Pigolotti S (2020) Ocean currents promote rare species diversity in protists. Sci Adv 6:eaaz9037
von Linné C (1735) Systema naturae; sive, Regna tria naturae: systematice proposita per classes, ordines, genera & species. Haak
Waller RF, Kořený L (2017) Plastid complexity in dinoflagellates: a picture of gains, losses, replacements and revisions. In: Hirakawa Y (ed) Advances in botanical research. Academic Press, Oxford, pp 105–143
Wang AL, Wang CC (1991) Viruses of the protozoa. Annu Rev Microbiol 45:251–263
Wang L, Lin X, Goes JI, Lin S (2016) Phylogenetic analyses of three genes of *Pedinomonas noctilucae*, the green endosymbiont of the marine dinoflagellate *Noctiluca scintillans*, reveal its affiliation to the order Marsupiomonadales (Chlorophyta, Pedinophyceae) under the reinstated name *Protoeuglena noctilucae*. Protist 167:205–216
Wang Y, Fan X, Gao G et al (2020) Decreased motility of flagellated microalgae long-term acclimated to CO2-induced acidified waters. Nature Clim Change 10:561–567
Ward BA, Follows MJ (2016) Marine mixotrophy increases trophic transfer efficiency, mean organism size, and vertical carbon flux. Proc Natl Acad Sci 113:2958–2963
Wassmann P (2015) Overarching perspectives of contemporary and future ecosystems in the Arctic Ocean. Prog Oceanogr 139:1
Wessenberg H, Antipa G (1970) Capture and ingestion of *Paramecium* by *Didinium nasutum*. J Protozool 17:250–270
Wetherbee R, Bringloe TT, Costa JF et al (2020) New pelagophytes show a novel mode of algal colony development and reveal a perforated theca that may define the class. J Phycol 57:396–411
Weynberg KD (2018) Viruses in marine ecosystems: from open waters to coral reefs. Adv Virus Res 101:1–38
Whitman WB, Coleman DC, Wiebe WJ (1998) Prokaryotes: the unseen majority. Proc Natl Acad Sci 95:6578–6583
Whittaker RH (1969) New concepts of kingdoms or organisms. Evolutionary relations are better represented by new classifications than by the traditional two kingdoms. Science 163:150–160
Wideman JG, Lax G, Leonard G et al (2019) A single-cell genome reveals diplonemid-like ancestry of kinetoplastid mitochondrial gene structure. Phil Trans R Soc Lond B 374:20190100
Wideman JG, Monier A, Rodríguez-Martínez R et al (2020) Unexpected mitochondrial genome diversity revealed by targeted single-cell genomics of heterotrophic flagellated protists. Nat Microbiol 5:154–165
Wilken S, Verspagen JMH, Naus-Wiezer S et al (2014) Comparison of predator-prey interactions with and without intraguild predation by manipulation of the nitrogen source. Oikos 123:423–432

Wilken S, Yung CCM, Hamilton M et al (2019) The need to account for cell biology in characterizing predatory mixotrophs in aquatic environments. Phil Trans R Soc Lond B 374: 20190090
Wilken S, Choi CJ, Worden AZ (2020) Contrasting mixotrophic lifestyles reveal different ecological niches in two closely related marine protists. J Phycol 56:52–67
Wiltshire KH, Malzahn AM, Wirtz K et al (2008) Resilience of North Sea phytoplankton spring bloom dynamics: an analysis of long-term data at Helgoland roads. Limnol Oceanogr 53:1294–1302
Wolf YI, Silas S, Wang Y et al (2020) Doubling of the known set of RNA viruses by metagenomic analysis of an aquatic virome. Nat Microbiol 5:1262–1270
Worden AZ, Nolan JK, Palenik B (2004) Assessing the dynamics and ecology of marine picophytoplankton: the importance of the eukaryotic component. Limnol Oceanogr 49:168–179
Worden AZ, Lee J-H, Mock T et al (2009) Green evolution and dynamic adaptations revealed by genomes of the marine picoeukaryotes *Micromonas*. Science 324:268–272
Worden AZ, Janouskovec J, McRose D et al (2012) Global distribution of a wild alga revealed by targeted metagenomics. Curr Biol 22:R675–R677
Worden AZ, Follows MJ, Giovannoni SJ et al (2015) Rethinking the marine carbon cycle: factoring in the multifarious lifestyles of microbes. Science 347:1257594–1257594
Woznica A, Cantley AM, Beemelmanns C et al (2016) Bacterial lipids activate, synergize, and inhibit a developmental switch in choanoflagellates. Proc Natl Acad Sci 113:7894–7899
Woznica A, Gerdt JP, Hulett RE et al (2017) Mating in the closest living relatives of animals is induced by a bacterial chondroitinase. Cell 170:1175–1183
Wright D (2017) Swells, soundings, and sustainability, but..."here be monsters.". Oceanography 30:209–221
Wu W, Huang B (2019) Protist diversity and community assembly in surface sediments of the South China Sea. Microbiol Open 8:e891
Yau S, Lauro FM, DeMaere MZ et al (2011) Virophage control of antarctic algal host–virus dynamics. Proc Natl Acad Sci 108:6163–6168
Yoon HS, Price DC, Stepanauskas R et al (2011) Single-cell genomics reveals organismal interactions in uncultivated marine protists. Science 332:714–717
Yutin N, Koonin EV (2012) Proteorhodopsin genes in giant viruses. Biol Direct 7:34–34
Zaremba-Niedzwiedzka K, Caceres EF, Saw JH et al (2017) Asgard archaea illuminate the origin of eukaryotic cellular complexity. Nature 541:353–358
Zavarzina NB, Protsenko AE (1958) On lysis of cultures *Chlorella pyrenoidosa* Pringsheim. Dokl Akad Nauk SSSR
Zeigler Allen L, McCrow JP, Ininbergs K et al (2017) The Baltic Sea virome: diversity and transcriptional activity of DNA and RNA viruses. mSystems 2:e00125–e00116
Zimmerman AE, Howard-Varona C, Needham DM et al (2020) Metabolic and biogeochemical consequences of viral infection in aquatic ecosystems. Nat Rev Microbiol 18:21–34
Zoccarato L, Pallavicini A, Cerino F et al (2016) Water mass dynamics shape Ross Sea protist communities in mesopelagic and bathypelagic layers. Prog Oceanogr 149:16–26
Zubkov MV, Tarran GA (2008) High bacterivory by the smallest phytoplankton in the North Atlantic Ocean. Nature 455:224–226

# Marine Fungi

5

Gaëtan Burgaud, Virginia Edgcomb, Brandon T. Hassett, Abhishek Kumar, Wei Li, Paraskevi Mara, Xuefeng Peng, Aurélie Philippe, Pradeep Phule, Soizic Prado, Maxence Quéméner, and Catherine Roullier

G. Burgaud (✉) · A. Philippe · M. Quéméner
University Brest, Laboratoire Universitaire de Biodiversité et Ecologie Microbienne, Plouzané, France
e-mail: gaetan.burgaud@univ-brest.fr; aurelie.philippe2@etudiant.univ-brest.fr; Maxence.Quemener@univ-brest.fr

V. Edgcomb · P. Mara
Department of Geology and Geophysics, Woods Hole Oceanographic Institution, Woods Hole, MA, USA
e-mail: vedgcomb@whoi.edu; pmara@whoi.edu

B. T. Hassett
Department of Arctic and Marine Biology, UiT – The Arctic University of Norway, Tromsø, Norway
e-mail: brandon.hassett@uit.no

A. Kumar
Institute of Bioinformatics, Bangalore, India
e-mail: abhishek@ibioinformatics.org

W. Li
College of Marine Life Sciences, Ocean University of China, Qingdao, China
e-mail: liwei01@ouc.edu.cn

X. Peng
School of Earth, Ocean and Environment, University of South Carolina, Columbia, SC, USA
e-mail: xpeng@ucsb.edu

P. Phule
Institute of Bioinformatics, Bangalore, India

Manipal Academy of Higher Education (MAHE), Manipal, Karnataka, India
e-mail: pradeep@ibioinformatics.org



L. J. Stal, M. S. Cretoiu (eds.), *The Marine Microbiome*, The Microbiomes of Humans, Animals, Plants, and the Environment 3,
https://doi.org/10.1007/978-3-030-90383-1_5

**Abstract**

Marine fungi are found in almost every marine habitat explored. From surface waters to kilometers below the seafloor fungi appear ubiquitous and contribute to global biogeochemical processes as saprotrophic degraders or parasites at numerous trophic levels. The purpose of this chapter is to review the increasing amount of knowledge on the diversity and adaptive capabilities of marine fungal communities along with their metabolic functions which can be hijacked and used for biotechnological applications. Specifically, the aim is to provide an overview of a number of innovative approaches to optimize the search for novel enzymes and bioactive compounds.

**Keywords**

Adaptation · Bioremediation · Biotechnology · Diversity · Marine fungi · Secondary metabolites

## 5.1 Introduction

The vast expanse of the ocean biome still holds many mysteries even though major advances have been achieved in understanding microbial diversity, biogeography, and ecology. There are still important knowledge gaps to fill regarding some microbial groups and especially marine fungi. While fungi are critically important on land for ecosystem functioning and carbon cycling their ecological roles in marine food webs and carbon cycling are still largely overlooked. And yet fungal communities have been found in almost every marine habitat explored from the surface of the ocean to kilometers below ocean sediments where they function mostly as saprotrophic degraders or parasites at numerous trophic levels. The aim of this chapter is to provide a state-of-the-art summary of existing knowledge on the diversity and adaptive capabilities of marine fungal communities along with their functions which can be hijacked and used for biotechnological applications.

S. Prado
Unité Molécules de Communication et Adaptation des Micro-organismes (UMR 7245), Sorbonne Universités, Muséum National d'Histoire Naturelle, Centre National De La Recherche Scientifique (CNRS), Paris, France
e-mail: soizic.prado@mnhn.fr

C. Roullier
Mer Molécules Santé—EA 2160, UFR des Sciences Pharmaceutiques et Biologiques, Université de Nantes, Nantes, France
e-mail: catherine.roullier@univ-nantes.fr

## 5.2 From Culture-Based to Next-Generation Sequencing Methods to Access Marine Fungal Life

The fungi are a fascinating group that circumscribes morphologically diverse microorganisms with non-trivial roles in virtually any ecosystem. A clear answer to the question "what is a fungus" is not easily given because the quest to identify synapomorphy/synapomorphies to unify the fungi has failed (Naranjo-Ortiz and Gabaldón 2019) and led to Bruns' Law: "there is no fungal synapomorphy, get used to it" (Richards et al. 2017). Moreover, as molecular phylogenetics guides taxonomy the limits of the fungal kingdom are expanding as new phyla are being introduced (Galindo et al. 2020). Knowing this, it appears that a clear answer to the question "what is a marine fungus" is not straightforward. The definition of marine fungi has been subject of discussion for a long time (see Rédou et al. 2016a). However, a consensual definition has now been proposed and seems to be adopted by the scientific community (Pang et al. 2016). In this study, a consortium of marine mycologists proposed a broad definition of a marine fungus "as any fungus that is recovered repeatedly from marine habitats because (1) it is able to grow and/or sporulate (on substrata) in marine environments, (2) it forms symbiotic relationships with other organisms, or (3) it adapts and evolves at the genetic level or is metabolically active in a marine environment." This definition appears consistent with the one proposed by Rédou et al. (2016a) indicating that "Marine fungi display long-term presence and metabolic activities in a marine habitat as revealed by their adaptations (ecophysiological profile), active metabolism (rRNA), gene expression (mRNA), catalytic functions (proteome), or specific metabolites (metabolome) resulting from their biotic and abiotic interactions." In this chapter these definitions of marine fungi are adopted.

There are many ways of targeting fungi in a marine sample from using microscopy to omic-based approaches. Direct observations of fungal structures have been largely used in early studies (e.g., Barghoorn and Linder 1944; Johnson and Sparrow 1961; Kohlmeyer and Kohlmeyer 1979) and are still used in numerous studies to qualitatively and quantitatively highlight the fungal diversity through cell visualization and quantification (e.g., Bochdansky et al. 2017; Burgaud et al. 2010; Gutiérrez et al. 2010; Hassett et al. 2019; Priest et al. 2020) but also cell morphology (e.g., Mitchison-Field et al. 2019). However, culture-dependent and molecular methods have clearly stolen the show.

Culturing has the advantage that it allows to generate genetic libraries of marine fungal isolates. These libraries can then be analyzed by using a variety of genetic markers or even be fully sequenced for further genomic or comparative genomic approaches coupled with (comparative) transcriptomics. This is often necessary in order to make a definitive species assignment. Oftentimes, short high-throughput sequencing reads that target conserved genetic loci (e.g., the 18S small ribosomal subunit) do not have sufficient taxonomic resolution to reliably identify species or genera, especially if clade segregation was based on full-length concatenated sequences. To this end, culturing approaches can be more advantageous. For example, Kumar et al. (2018) analyzed two marine fungi from the North Sea, namely

*Calcarisporium* sp. and *Pestalotiopsis* sp. and unraveled numerous biosynthetic gene clusters of which some are involved in the production of secondary metabolites with antimicrobial activities. These data are unobtainable by targeted loci metabarcoding sequencing analyses. A transcriptomic-based study has also demonstrated that fungi can adapt to changes in temperature, salinity, and acidity by up- or down-regulation of stress-related genes. This was shown for the marine fungus *Aspergillus terreus* strain NTOU4989 which was isolated from a hydrothermal vent system under a combined set of growth conditions of different temperature, pH, and salinity (Pang et al. 2020). Culturing has the important advantage of providing an opportunity to thoroughly characterize isolates through ecophysiological screening including tolerance to temperature, salinity, pH, and hydrostatic pressure (e.g., Burgaud et al. 2009; Burgaud et al. 2010; Rédou et al. 2015). Moreover, cultures allow to study their metabolic potential as well as their ability to grow on different carbon sources (e.g., Quémener et al. 2020). Numerous culturing techniques can be used to understand the culturable diversity of fungi of a specific marine habitat such as direct plating (e.g., Li et al. 2014; Vrijmoed 2000; Zuccaro et al. 2004), particle filtration, and dilution to extinction plating coupled with mid/high-throughput culturing (e.g., Quémener et al. 2020), or in situ culturing (see Overy et al. 2019). Some other culturing-based approaches appear promising such as those using microcapillaries (see L'Haridon et al. 2016), isolation chip (Nichols et al. 2010), and fluorescent activated cell sorter combined with microfluidics (Lambert et al. 2017). Although promising, such techniques have not been applied to understand the diversity and richness of marine fungi. However, culturing is facing some issues. The most important of these issues is the culturability of presumed fastidious organisms, which explains why it is thought that potentially as few as 1% of marine fungi have been identified so far (Gladfelter et al. 2019).

These fastidious organisms, the anticipated presence of obligate symbionts, and the laborious and resource-intensive efforts associated with culturing fungi from samples collected from often remote locations pose a major constraint on its broad implementation. To this end, molecular methods have provided a less-biased and sometimes more efficient way of assessing the full fungal community. Earlier applications of molecular methods utilized to study marine fungi have been based on molecular fingerprinting and cloned sequences (e.g., Fell and Newell 1998). Now the wider implementation of Next-Generation Sequencing (NGS) platforms has become the gold standard although cloning-based approaches are still useful and utilized (Hassett et al. 2017). This era of omics has ushered in the ability to generate independent, yet comparative broad assessments of marine fungal richness and diversity by targeting nucleotide sequences primarily through rRNA gene metabarcoding of taxonomically informative loci (18S, 28S, ITS). As the field continues to advance more nuanced approaches aimed at inventorying the metabolically active fraction through mRNA analyses (e.g., Orsi et al. 2013a) and ecologically targeted approaches with DNA stable isotope probing methods are being utilized (Cunliffe et al. 2017). As the focus begins to shift from understanding diversity toward functionality microarray analyses (e.g., Hassett et al. 2019) and shotgun sequencing (Chrismas and Cunliffe 2020) are being utilized in order to thoroughly assess the functional gene repertoire and gene expression patterns of fungi in the marine environment. The efficacy of these high-throughput approaches to resolve longstanding questions in marine mycology is rapidly increasing but still depends on reference databases. As of 2019, only ~50% of all marine fungi were represented by a nucleotide sequence and less than 20% of marine fungal genera were represented in the NCBI RefSeq database (Hassett et al. 2019). Balancing the limitations of culturing and molecular approaches often requires a dual approach to thoroughly understand marine fungal communities in terms of their diversity, activity, structuration, and ecological functions.

## 5.3 Habitat Specific Community Composition or over-Dispersion?

The picture of marine fungal diversity remains largely pixelated even though in the past decade new studies have highlighted the importance of this group of microorganisms. Answering the question whether marine fungi are over-dispersed or endemic would help to better understand the broad-scale habitat and geographic differences among marine fungal communities.

### 5.3.1 Plant-Based Habitats

Saprophytic fungi occurring on plant substrates are the most well-studied group of marine fungi. Since the monumental study of wood-inhabiting fungi by Barghoorn and Linder (1944) many new species of marine fungi have been described from wood in diverse habitats including wood buried in sandy beaches, decaying wood in mangroves, and drift or trapped wood on rocky shores. Wood-inhabiting marine fungi form fruiting bodies on wood and cause soft-rot decay mainly by producing cellulases and laccases (Bucher et al. 2004). Marine Dothideomycetes and Sordariomycetes belonging to the Ascomycota are dominant together with a few Basidiomycota. These identifications were based on the observation of fruiting bodies and also on culture-independent techniques using 454 pyrosequencing of the ribosomal ITS and of the 18S rRNA gene (Arfi et al. 2012a; Jones et al. 2015). Marine lignicolous basidiomycetes are mainly intertidal species and belong to the Agaricomycetes with reduced and enclosed fruiting bodies, loss of ballistospory, and evolution of spore appendages (Hibbett and Binder 2001). Marine lignicolous ascomycetes are phylogenetically diverse but mainly belonging to the Pleosporales in the Dothideomycetes and Microascales (Halosphaeriaceae) in the Sordariomycetes class forming exposed or immersed perithecia (Jones et al. 2015). As revealed by Rämä et al. (2014) Hypocreales and Helotiales also represent important taxonomic groups in which fungal communities from Arctic intertidal and seafloor logs were obtained using culture-based techniques. Apothecium-type of ascomata is uncommon with only 10 described species (Baral and Rämä 2015) possibly due to the inability to withstand wave action (Suetrong and Jones 2006).

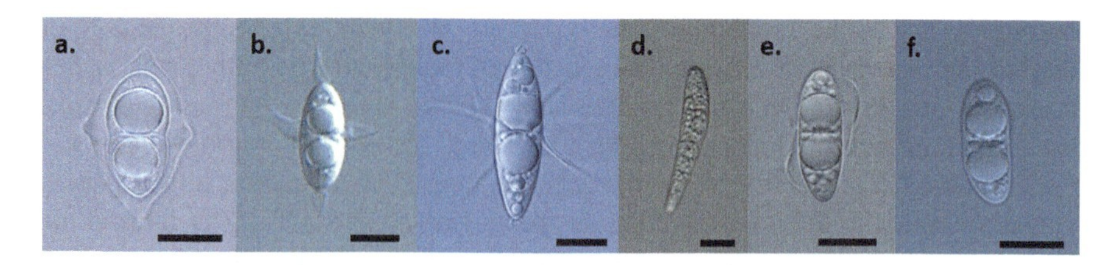


**Fig. 5.1** Plant-inhabiting marine fungi from diverse habitats. (**a**) Ascospore of *Ebullia octonae* with a sheath. (**b**) Ascospore of *Halosphaeriopsis appendiculata* with polar and equatorial spoon-shaped appendages. (**c**) Ascospore of *Halosphaeriopsis mediosetigera*. (**d**) Clavate-shaped ascospore of *Buergenerula spartinae*. (**e**) Ascospore of *Natantispora retorquens* with bipolar unfurling appendages. (**f**) Ascospore of *Lignincola laevis* (Scale bar 10 μm)

These ascomycetes have evolved diverse morphologies to adapt to a marine lifestyle, i.e. deliquescing asci and ascospore appendages of different morphology and ontogeny (Fig. 5.1a–c) (Pang 2002). Two of the largest lineages of marine lignicolous ascomycetes, the Halosphaeriaceae and the Lulworthiales, were derived from terrestrial ancestors (Spatafora et al. 1998). Phylogenetic studies of the ribosomal RNA genes revealed further independent lineages into the marine environment in the Ascomycota: Dyfrolomycetales (Hyde et al. 2013), Tirisporellales, Torpedosporales (Jones et al. 2015), and Savoryellales (Boonyuen et al. 2011). This confirmed the high diversity of Ascomycota on wood substrates.

Marine fungi also grow on the decaying intertidal part of saltmarsh plants such as *Spartina* spp., *Juncus roemerianus*, and *Phragmites australis*, and the palm *Nypa fruticans*. Many species of marine fungi are host-/substrate-specific (Calado and Barata 2012; Pilantanapak et al. 2005). In particular, fungi associated with *Spartina* spp. have been well studied in US and Portuguese salt marshes where they are involved in nutrient cycling (Newell and Wasowski 1995). Diverse laccase genes were detected from the fungal community associated with *Spartina alterniflora,* which may suggest their involvement in lignin mineralization (Lyons et al. 2003). A total of 132 species of marine fungi have been documented to live saprophytically on *Spartina* spp. with the dominant groups belonging to the classes Dothideomycetes and Sordariomycetes (Calado and Barata 2012). *Phaeosphaeria halima*, *Phaeosphaeria spartinicola*, *Mycosphaerella* sp., *Byssothecium obiones*, and *Buergenerula spartinae* (Fig. 5.1d) are common taxa living on *Spartina* spp. in US saltmarshes (Buchan et al. 2002; Newell et al. 1996; Walker and Campbell 2010). Many of these species have fully functional asci for forcible expulsion of spores (Newell 2001). These species are also common in Portuguese saltmarshes along with *Natantispora retorquens* (Fig. 5.1e) (Barata 2002; Calado et al. 2015). Based on an automated ribosomal intergenic spacer analysis (ARISA) the fungal community composition of *J. roemerianus* appears to be different from the one of *Spartina alterniflora* in the US saltmarsh. This suggests a host/substrate specificity of these plants (Torzilli et al. 2006). A total of 136 taxa have been recorded on *J. roemerianus* (Calado and Barata 2012). Many of those are not marine but grow on the terrestrial exposed parts of the plant. Common taxa on *J. roemerianus* include

*Loratospora aestuarii*, *Papulosa amerospora*, *Aropsiclus junci*, *Anthostomella poecila*, *Physalospora citogerminans*, *Scirrhia annulata*, *Massarina ricifera*, and *Tremateia halophile* (Newell and Porter 2000). Over 300 fungal species most of which asexual have been documented on the intertidal plant *P. australis* (Calado and Barata 2012). Common fungal species include *Cladosporium* spp., *Colletotrichum* sp., *Didymella glacialis*, *Halosarpheia phragmiticola*, *Lignincola laevis* (Fig. 5.1f), *Phaeosphaeria* sp., *Phoma* sp., *Phomatospora berkeleyi*, *Phomopsis* sp., *Septoriella* spp., and *Trichoderma* sp. (Luo and Pang 2014).

Mangrove plants support diverse fungal communities. Currently, about 500 fungi are known from mangrove habitats associated with 69 mangrove plants, sediments, and seawater with data from 80 countries (Jones et al. 2019; Schmit and Shearer 2004). Eighteen species identified as Ascomycota (e.g., *Antennospora quadricornuta*, *Halorosellinia oceanica*, *Sammeyersia grandispora*), asexual morphs (i.e., *Bactrodesmium linderi*, *Hydraena pygmaea*, *Periconia prolifica*), and Basidiomycota (i.e., *Calathella mangrovei*, *Halocyphina villosa*) have been listed as the core mangrove fungi (Jones 2011; Jones et al. 2019; Pang et al. 2011). From the palm tree *Nypa fruticans* in Brunei Hyde (1992) discovered 43 species whereas another study reported 135 species with 90 Ascomycota, 3 Basidiomycota, and 42 asexual fungi (Loilong et al. 2012). A higher fungal diversity occurred on the leaf base compared to the other tissues of this intertidal plant including inflorescence, leaf, leaf midrib, rachis, and aerial parts (Hyde and Alias 2000). Host specificity is pronounced with an estimated 40 endemic species (Hyde and Alias 2000). For example, *Aniptodera intermedia* and *Linocarpon appendiculatum* were found only on *N. fruticans* although this palm tree grows alongside mangrove tree species (Besitulo et al. 2010; Loilong et al. 2012). Using 454 pyrosequencing of the ITS of rRNA gene regions, Arfi et al. (2012a) found that the Agaricomycetes was the dominant fungal class in mangrove soil and that fungi may play a role in the decomposition of organic matter in the anoxic organic-rich sediments. However, the pathogenic genera such as *Diaporthe*, *Mycosphaerella*, *Phaeoramularia*, and *Ramulispora* were well represented in the leaves and branches of the two mangrove trees *Avicennia marina* and *Rhizophora stylosa* (Arfi et al. 2012b). The shift in community composition between sediments and healthy leaves highlights a switch in the ecological function of fungi from the submerged parts to aerial parts of mangrove trees.

Sea grasses have also been examined for endophytic fungi albeit colonization frequency and diversity were relatively low compared to terrestrial plants (Alva et al. 2002; Devarajan et al. 2002). Ascomycetes mostly belonging to the orders *Eurotiales*, *Hypocreales*, and *Capnodiales* appear dominant on seagrasses (Sakayaroj et al. 2012; Venkatachalam et al. 2015). Common genera include *Aspergillus*, *Cladosporium*, *Paecilomyces*, and *Penicillium*, which are typical asexual fungi of seawater and sediment (Sakayaroj et al. 2012; Venkatachalam et al. 2015). The few basidiomycetes found on *Enhalus acoroides* may represent mycorrhizal relationships (Sakayaroj et al. 2010) suggesting that fungal communities occurring on plant-based habitats are diverse and play different ecological roles. This trend was confirmed in a study focusing on the seagrass *Posidonia oceanica* sampled in the coastal waters of the Elba Island in Italy (Poli et al. 2020). Ascomycetes represented 97% of the culturable diversity and, except for the genera *Penicillium* and *Aspergillus* for which representatives were isolated from all sites and plant parts, the diversity was homogeneously distributed in the classes Dothideomycetes (mainly *Pleosporales* and *Capnodiales*), Eurotiomycetes (mainly *Eurotiales*), and Sordariomycetes (mainly *Hypocreales* and *Microascales*). This was consistent with the results of an ITS-based metabarcoding study where the fungal alpha-diversity on different tissues (leaves, root, and rhizome) was significantly lower than that of the surrounding sediments (Ettinger and Eisen 2019; Wainwright et al. 2019a). Fungal communities on plants were dominated by the classes Dothideomycetes, Eurotiomycetes, Agaricomycetes, and Saccharomycetes for the seagrass *Enhalus acoroides* collected from Singapore and Peninsular Malaysia while the mycobiome of the seagrass *Zostera marina* appeared more complex in terms of diversity with the occurrence of the classes Sordariomycetes, Dothideomycetes, Saccharomycetes along with the basidiomycetous classes Agaricomycetes, Cystobasidiomycetes, and Malasseziomycetes as well as the *Chytridiomycota* and *Aphelidomycota*.

### 5.3.2 Coastal Waters

Compared to plant-based habitats the marine water column has much lower quantities of organic carbon substrate available for heterotrophic communities such as fungi. While the open ocean on one end of the spectrum is characterized by the lowest number and amounts of substrates and consequently fungal abundance and diversity, coastal waters on the other end of the spectrum harbor diverse fungal communities. Both culturing and metabarcoding methods revealed that fungi from the *Dikarya* phyla *Ascomycota* and *Basidiomycota* dominate most marine environments (Duan et al. 2018; Gao et al. 2010; Li et al. 2016a, 2016b, 2018; Taylor and Cunliffe 2016; Wang et al. 2018, 2019). For example, in Chinese coastal regions *Ascomycota* and *Basidiomycota* are dominant in seawater and sediment (Li et al. 2016a, 2016b, 2018; Wang et al. 2018, 2019) with large abundances of Dothideomycetes, Leotiomycetes, Eurotiomycetes, Agaricomycetes, Malasseziomycetes, and Tremellomycetes.

Most studies mentioned above were based on metabarcoding methods targeting the internal transcribed spacer (ITS) region. In studies where the small and large subunits of the ribosomal RNA genes were targeted early-diverging fungi from the phylum *Chytridiomycota* were recognized as representing a large part of the overall community (Picard 2017; Richards et al. 2015; Wang et al. 2017). Chytrid fungi in the marine environment are best known as diatom parasites (Gutiérrez et al. 2016; Hassett and Gradinger 2016) and probably play an important role in the transfer of organic matter and nutrients in marine food webs. The low representation of *Chytridiomycota* in studies targeting the ITS region is likely a result of the paucity of the reference sequences belonging to marine basal fungal clades, i.e., *Chytridiomycota* and *Cryptomycota* in the UNITE database (Kõljalg et al. 2013) as many marine chytrid observations are based on morphological characteristics (e.g., Sparrow 1973).

In coastal ecosystems the main factors controlling fungal diversity include salinity, dissolved oxygen, and nutrient conditions. Salinity is recognized as one of the most important elements determining fungal diversity and community (Hassett et al. 2019; Jones 2000, 2011; Rojas-Jimenez et al. 2019; Taylor and Cunliffe 2016). Transitions in fungal communities along a salinity gradient were observed in Rhode Island (Mohamed and Martiny 2011) and the Delaware Bay (Burgaud et al. 2013), respectively. In the coastal waters off Plymouth, UK, mycoplankton alpha-diversity was negatively correlated with salinity with the highest values during periods of decreased salinity (Taylor and Cunliffe 2016). In the East China, Sea fungal communities in different water masses were mainly influenced by dissolved oxygen and water depth (Li et al. 2018). Nutrient conditions including ammonia, phosphate, and silicate were significantly correlated with fungal OTU richness or sequence read abundance in the coastal ecosystems of Plymouth (Taylor and Cunliffe, 2016) and North Carolina (Duan et al. 2018), respectively. Also, in the coastal waters of Hawaiian and North Carolina, positive correlations between fungal diversity and phytoplankton biomass were found (Gao et al. 2010; Duan et al. 2018). A similar pattern was observed in the upwelling ecosystem off the coast of Chile (Gutiérrez et al. 2011). A study based on 18S rRNA metabarcoding coupled to CARD-FISH highlighted the fungal diversity and biomass during a phytoplanktonic bloom in the North Sea and underscored the occurrence of *Cryptomycota* reaching cell concentrations similar to those in freshwater habitats (Priest et al. 2020).

The coastal mycobiome is strongly influenced by riverine input and ocean currents. Some terrestrial fungi that are well known from habitats such as soil and plants were often abundant in coastal areas suggesting the dispersal of terrestrial fungi to the marine environment (Amend et al. 2019; Richards et al. 2012). This dispersal occurs mainly via the riverine inputs, which enhance mycoplankton richness and shape community composition in coastal waters (Taylor and Cunliffe 2016; Wang et al. 2019). In the East China Sea, the coastal water mass harbored a high abundance of the typical terrestrial and freshwater fungal genus *Byssochlamys* suggesting the riverine inputs of fungi by the Yangtze River (Li et al. 2018). Other studies also detected the riverine inputs of fungi to coastal ecosystems (Wang et al. 2018, 2019). Planktonic fungi can travel long distances by hitchhiking on ocean currents. For example, the Kuroshio current contributes to the passive dispersal of fungi especially species affiliated to the genus *Aspergillus* along the shelf of the East China Sea (Li et al. 2018), which would lead to a great influence on biogeographic distribution pattern of marine fungi on a regional or even a global scale.

### 5.3.3 Algae

Macroalgae are colonized by various microorganisms collectively referred to as the microbiota. These microorganisms are interacting with their host throughout its life cycle impacting the physiological state of the host (Egan et al. 2013; Singh and

Reddy 2015; Wahl et al. 2012). The key role of these associated microorganisms in the algal development led to the concept of the holobiont in which the algae and its microbiota are defined as an entity (Egan et al. 2013). While the bacteria have been extensively studied macroalgae harbor also a large diversity of fungi which needs to be included in the study of algal microbiota. The first report of an obligate mycophycobiosis between the Fucales *Ascophyllum nodosum*, *Pelvetia canaliculata*, and the fungal endosymbiont *Stigmidium ascophylli* (formerly *Mycosphaerella ascophylli*) dates back more than a century (Cotton 1907; Stanley 1991) and it has been suggested that the symbiont may protect the algae host from desiccation while obtaining nutrients in exchange (Decker and Garbary 2005; Garbary and London 1995; Garbary and Macdonald 1995). Similarly, the fungal symbiont *Turgidosculum ulvae* colonizes the inner tissue of the green alga *Blidingia minima* and induces dark spots that are never consumed by host predatory gastropods (Kohlmeyer and Volkmann-Kohlmeyer 2003).

Many filamentous fungi colonize the algal inner tissues without causing any apparent damage or disease (Debbab et al. 2012; Porras-Alfaro and Bayman 2011) and such asymptomatic colonization by marine fungi remains mostly uninvestigated (Fries 1979; Harvey and Goff 2010; Loque et al. 2010; Jones and Pang 2012; Zuccaro et al. 2003). Thanks to DNA sequencing the fungal diversity in marine substrata has been unraveled and now it constitutes the second biggest known marine reservoir of fungi after sponges (Rateb and Ebel 2011). This diversity encompasses mutualistic symbionts, opportunistic pathogens, parasites, and saprophytes (Jones and Pang 2012; Rédou et al. 2016a; Richards et al. 2012; Zuccaro and Mitchell 2005). Studies based on a culturing approach showed that *Ascomycota* is the dominant endophyte of algae (Flewelling et al. 2013a, b; Godinho et al. 2013; Zuccaro et al. 2003, 2008). The study of the culturable fungal endophytic community associated with the inner tissues of the brown algae *L. digitata*, *S. latissima*, *A. nodosum*, and *P. caniculata* confirmed the dominance of the *Ascomycota* and pointed out that the species belonged especially to the Sordariomycetes and the Dothideomycetes (Vallet et al. 2018). The proportion of taxa recovered within the main orders and classes were more or less similar to those described for marine fungi associated with plants or algae when compared to the public SSU rRNA reference databases (Panzer et al. 2015). The taxonomic diversity and abundance of isolates differed between the algal organs suggesting a potential tissue and host preference. This pattern of fungal colonization may be explained by the differences in chemical composition and defense in algal species and organs (Cosse et al. 2009; Megan et al. 2001; Thomas et al. 2014). However, aside from the cosmopolitan *Paradendryphiella arenaria*, which occurred in all four plant species investigated, few marine fungi sensu stricto or unknown species have been previously isolated. Instead, most of the recovered strains were closely related to terrestrial phytopathogens (37%), endophytes (21%), or a miscellaneous group of lignivore, soil-borne, and air-borne saprophytes (28%). Several genera, i.e., *Acremonium*, *Coniothyrium*, *Botryotinia*, *Phaeosphaeria*, and *Cordyceps* were not previously isolated from marine algal hosts (Flewelling et al. 2013b; Godinho et al. 2013). In particular, sequences matching the phytopathogens *Phoma exigua* and

*Botryotinia fuckeliana* were retrieved leading to the hypothesis that these strains might be opportunistic pathogens plausibly able to colonize an otherwise compromised alga.

Microbiota associated to the brown alga *Saccharina latissima* studied by high-throughput Illumina-based DNA sequencing highlighted that the fungal community was dominated by *Ascomycota* (54.6%) and *Basidiomycota* (45.3%) in particular *Mycosphaerellaceae*, *Psathyrellaceae*, and *Bulleribasidiaceae* (Tourneroche et al. 2020). This result is consistent with previous metabarcoding studies on fungal communities associated with macroalgae which reported a predominance of *Ascomycota*, followed by *Basidiomycota* (Agusman and Dan-qing 2017; Wainwright et al. 2019b). However, important differences between the seaweed associated microbiota and the surrounding seawater microbiota were recorded suggesting that the seaweed has the capacity to recruit its microbiota. This phenomenon described as a "microbial gardening" hypothesis has been demonstrated in the green alga *Ulva* which enriches its environment by attracting and stimulating the growth of microorganisms necessary for its morphogenesis (Wichard et al. 2015). The fungal communities associated with *S. latissima* algal tissues were highly diverse and it turned out that the fungal diversity was spatially organized within the sampled algal tissues. Hitherto, no core fungal microbiota has been identified from numerous algal samples through snapshots of microbial diversity, suggesting that a functional core microbiota should be given more consideration to better understand the ecology of the host.

### 5.3.4 Deep-Sea and Deep Subsurface

Fungi are thought to play various roles in the deep sea including the decomposition of organic matter (Ivarsson et al. 2015a), in mineral weathering (Bengtson et al. 2014), and in manganese (Ivarsson et al. 2015b) and arsenic (Dekov et al. 2013) cycling. Deep-sea fungi also play a role as putative symbionts with chemoautotrophic bacteria and archaea in oceanic crust habitats (Bengtson et al. 2014). In terms of diversity and more precisely morphology a study of fungal marker genes in 11 deep-sea samples from around the world (1500–4000 m depth) suggested that certain yeast forms dominated the low-diversity fungal communities (Bass et al. 2007). However, this trend does not appear to be generalizable mostly because of the wide array of habitats in the deep sea.

#### 5.3.4.1 Deep-Sea Habitats

Fungi can be transported into the deep sea along with sinking organic material (algae, plant material, wood, and particulate organic carbon) as active cells, spores, or mycelial filaments (Lorenz and Molitoris 1997). Initial investigations of microbial eukaryotes in the deep sea utilized "universal" eukaryotic PCR primers that likely underestimated the diversity and abundance of fungi obscuring their relevance. Interest in the possible importance of fungi for deep-sea ecosystems was galvanized when living fungi were found in deep-sea sediments of the deepest ocean realm the

Mariana Trench (Takami et al. 1997, 1999). Since then, only a low number of studies of bathypelagic and abyssopelagic zones and unique deep-sea habitats such as hydrothermal systems have been conducted that specifically investigated fungal diversity.

Our knowledge of fungal diversity is in fact based on studies of marker genes (predominantly the internal transcribed spacer (ITS) of rRNA genes from environmental samples) and documentation of the fruiting bodies and culture characteristics of isolates from deep-sea sediment samples (Bass et al. 2007; Nagano et al. 2010; Raghukumar et al. 2004; Singh et al. 2010, 2012; Xu et al. 2014, 2016; Zhang et al. 2016). Marker gene sequencing has the advantage of detecting rare species and taxa that are currently unculturable or present only as vegetative mycelia as well as to recover the signatures of many putatively novel taxa (Lai et al. 2007; Nagano et al. 2017; Wang et al. 2018; Xu et al. 2016, 2017, 2019; Zhang et al. 2016).

Several studies provided insights into the extent of undescribed fungal diversity in deep-sea hydrothermal vent habitats. These include a culture-based study of yeast diversity in Mid-Atlantic Ridge sediments that revealed that 33% of the isolates were new phylotypes (Gadanho and Sampaio 2005). Burgaud et al. (2009) examined the culturable diversity of filamentous fungi associated with shrimps, mussels, alvinellids, tubeworms, sediments, hydrothermal chimney rocks, and other types of samples from hydrothermal sites at the mid-Atlantic Ridge, South-west Pacific-Lau Basin, and the East Pacific Rise. The authors suggested that isolates from animals may be opportunistic pathogens or facultative parasites. Physiological studies on the strains in the collection of Burgaud et al. (2009) (including previously unknown phylotypes) included the assessment of their ability to grow under marine conditions. This demonstrated that not all fungal isolates should be considered as terrestrial "stowaways." Analyses of sediments collected from the East Pacific Rise and the Mid-Atlantic Ridge and Lucky Strike hydrothermal sites using marker genes and culturing efforts revealed previously unidentified species of *Ascomycota* and *Basidiomycota* as well as a novel ancient evolutionary lineage of *Chytridiomycota* (Le Calvez et al. 2009). Hydrothermal sediments along the Mid-Oceanic Ridge in the East Pacific and the South Indian Oceans were investigated using culture-based and culture-independent approaches and revealed that *Ascomycota* dominated over *Basidiomycota* (Tang et al. 2020). In addition to some putatively novel taxa (less than 97% similarity to sequences in GenBank) culture-based approaches recovered 97 isolates belonging to 7 genera and 10 species including *Penicillium, Rhodotorula, Meyerozyma, Ophiocordyceps, Vishniacozyma, Aspergillus*, and *Phoma*. Most (5/7) genera were also detected using marker gene approaches (5.8S ITS along with 18S rRNA genes) confirming the importance of the classes Eurotiomycetes, Dothideomycetes, and Sordariomycetes in the marine environment from coastal waters to the deep biosphere. Fungi have also been found living in association with deep-sea animals at hydrothermal sites along the Mid-Atlantic Ridge, South Pacific Basins, and East Pacific Rise (Burgaud et al. 2010) and their growth potential under elevated hydrostatic pressure has been evaluated (Burgaud et al. 2015). Whether these are commensal, pathogenic, or mutualistic associations and whether the nature of the association can change under different conditions remain to be determined. ITS marker gene sequences reveal many of the same dominant genera in samples from the northwest Pacific Magellan Seamount area (Luo et al. 2020). However, illustrating the unique fungal community composition that can be found at different locations at the Magellan Seamounts, *Basidiomycota* comprised the majority of OTUs (44%) with representation also from the phyla *Ascomycota* (25%) and minor representation from the phyla *Mortierellomycota*, *Chytridiomycota*, *Mucoromycota*, *Glomeromycota*, and *Monoblepharidomycota* (Luo et al. 2020). Five genera were common in most of the samples including *Aspergillus, Cladosporium, Fusarium, Chaetomium,* and *Penicillium* (representing once again the classes Eurotiomycetes, Sordariomycetes, and Dothideomycetes) all of which have been reported worldwide in marine settings. Studies of deep-sea hydrothermal sediments suggest there is much we still have to learn about fungal diversity in these habitats.

Examination of whale falls in the South Atlantic Ocean revealed a considerable diversity of previously unidentified fungi indicating that still much is unknown about this kingdom in the marine environment (Nagano et al. 2020). Molecular studies show that deep-branching basal (or "lower") fungi and more precisely *Cryptomycota* are common in deep-sea methane cold-seep sediments (Nagahama et al. 2011). The deep sea appears to be a hot spot for discovery of additional early-diverging lineages that will help to elucidate the evolution of fungi. *Ascomycota* dominated ribosomal RNA ITS marker gene libraries from sediments of subtropical southern and northern Yellow Sea and the Bohai Sea where 816 operational taxonomic units that included 130 known genera, 36 orders, 14 classes, and 5 phyla were identified (Li et al. 2016a, b). Yeasts affiliated to the *Basidiomycota* and *Ascomycota* (particularly the genera *Rhodotorula, Rhodosporidium, Candida, Debaryomyces*, and *Cryptococcus*) appeared to be typical of deep-sea environments.

Fungal diversity in deep-sea sediments appears to be highly variable and possibly correlated with environmental factors that include sediment sources, organic carbon and nitrogen levels, geographical distance from land, latitude, temperature, salinity, and depth (Li et al. 2016a, b), although some researchers did not find such a strong correlation between these factors and fungal community composition (Luo et al. 2020). An increasing number of studies of the deep-sea suggest that fungal populations and saprophytic fungi contribute to the turnover of organic carbon and likely are important players in many nutrient cycles. For example, deep-sea fungi identified as *Dikarya* may play an important role in the degradation of lignin in deep marine realms because this group of fungi is present and is known to possess the ability to degrade lignin (Nagahama et al. 2011). Culture-based studies of deep-sea isolates from marine hypoxic sediments have shown that some fungi participate in denitrification under anaerobic conditions (Jebaraj and Raghukumar 2010).

#### 5.3.4.2 Deep Subsurface Sediments and Oceanic Crust

Knowledge of deep-sea fungi has now been extended into the deep subsurface biosphere, which includes sedimented and lithified realms, and which represents one of the largest reservoirs of microbial communities. The "deep biosphere" is defined here as ocean floor sediments that can only be accessed with drilling and are too deep to be accessed with typical over-the-side gravity coring or submersible-assisted push coring. Microbial communities in deep biosphere habitats must cope with increasing temperature and hydrostatic/lithostatic pressures, low water activity (low water availability for biochemical reactions), and often nutrient-poor conditions (Gaboyer et al. 2019). Sufficient sediment pore spaces and veins or fractures in rocks are also required for subsurface microbial life to provide connectivity to fluids that can supply nutrients and electron donors and acceptors. Fungi and/or their biosignatures have been detected in deep (up to thousands of meters below the seafloor, mbsf) subsurface sediments of the Peru Margin, North Pond, Hydrate Ridge, Central Indian Basin, Benguela Upwelling System, South Pacific Gyre, Eastern Equatorial Pacific, Canterbury Basin, and off the Shimokita Peninsula in Japan (Ciobanu et al. 2014; Damare et al. 2006; Edgcomb et al. 2010; Inagaki et al. 2015; Liu et al. 2017; Orsi et al. 2013a, b; Rédou et al. 2014, 2015), the basaltic upper ocean crust (Ivarsson et al. 2012, 2016), and even in the gabbroic lower ocean crust (Quémener et al. 2020). Fungal diversity and culturability generally decrease with depth below seafloor (Rédou et al. 2014, 2015). Although many of these studies provide evidence that fungi are metabolically active in the deep subsurface their role in and impact on biogeochemical cycles and their interactions with other microbiota are uncertain.

Fungal activities reported (on the basis of mRNA) in deep subsurface sediments and crust include organic matter recycling (Orsi et al. 2013a; Orsi 2018; Pachiadaki et al. 2016; Quémener et al. 2020) and biosynthesis of antimicrobial compounds, presumably used in competition with other microorganisms for available carbon and energy (Navarri et al. 2016). Fungal communities are thought to have a symbiotic interdependence with chemoautotrophic bacteria in basalts (Bengtson et al. 2014) and to participate in degradation of refractory organic matter and in manganese and arsenic cycling (Dekov et al. 2013; Ivarsson et al. 2015b). Additionally, anaerobic fungal lineages may support the energy needs of subsurface methanogens and hydrogen-consuming bacteria through the production of hydrogen as an end product of their fermentative energy metabolism (Drake and Ivarsson 2017; Drake et al. 2017; Hackstein et al. 2019; Ivarsson et al. 2016). Many fungal taxa detected in the deep biosphere appear to be ubiquitous taxa to deep sea and deep subsurface biosphere habitats including the genera *Penicillium, Aspergillus, Cladosporium,* and *Debaryomyces* (Hirayama et al. 2015; Li et al. 2019; Manohar et al. 2014; Nagano et al. 2010, 2017; Pang et al. 2019; Polinski et al. 2019; Zhang et al. 2016). As an example, members of *Aspergillus* have a wide distribution in marine biomes from coastal waters (Li et al. 2019) to the deep-sea (Burgaud et al. 2009) and deep subsurface sediments (Damare et al. 2006; Xu et al. 2017). *Aspergillus* species are known that they adapt to high hydrostatic pressures (Damare et al. 2006) and deep-sea yeasts affiliated to *Candida/Debaryomyces* as well as other fungal genera can grow at pressures ~25-40 MPa (Burgaud et al. 2015; Raghukumar et al. 2010). Laboratory tests of growth of fungal isolates from the lower ocean crust using diverse substrates suggest their ability to utilize a wide array of carbon compounds (Quémener et al. 2020). In addition to utilizing necromass and fluid-derived carbon such metabolic flexibility may be an important adaptation to survive in the lower ocean crust where sources of carbon and energy delivered via fluid flows are likely ephemeral and variable. Expression of genes unique to fatty acid metabolism and peroxisomal biogenesis suggests that fungi maintain membrane integrity and may use fatty acids as an energy source (Quémener et al. 2020). This is consistent with laboratory studies that demonstrated fungal growth on fatty acids as a sole carbon source under aerobic conditions (Hynes et al. 2006). Recycling of proteins into central metabolism may be an additional adaptation to cope with limited carbon availability. Carbon-related stress responses can induce expression of hydrolases, cell wall degrading enzymes, and secondary metabolites (e.g., polyketides) as a means for fungal survival. Detection of expressed fungal genes for these processes as well as genes associated with cell division and senescence in deep crustal samples suggests a heretofore unknown ratio of senescent vs. active cells (Quémener et al. 2020). Future studies of the deep subsurface biosphere will hopefully include microscopy approaches that can tell us more about the physical associations of fungi with organic matter and about the abundance of active/vegetative cells in different habitats.

### 5.3.5 Polar Waters

The strong seasonal fluxes of temperature and light in high-latitude regions provide unparalleled opportunities to study the effects of environmental parameters on biological systems, which can aid in the elucidation of phenomena in regions with less pronounced natural gradients or fluxes. Within polar environments semi-porous sea ice is among the more unique physical features, which can harbor high densities of biota in close proximity (generally μm–mm scale) within brine channels. As ice crystals thermodynamically form in marine environments the water molecules assemble into a lattice structure that physically excludes ions, which consequently accumulate into reticulate networks that are conceptualized as brine channels. The salinity and volume of these brine channels are partially governed by temperature: as the ice gets colder the salinity of brine increases as channel volumes decrease (Cox and Weeks 1983) (Fig. 5.2). The hypersalinity of sea ice brine and the constrained space within brine channels can exert strong selection on organisms (Caron and Gast 2010) including those seeking refuge from predators (Bluhm et al. 2017) and those not adapted to exist in the presence of elevated ion concentrations (Firth et al. 2016).

The rapid return of light in the spring fuels the development of massive under-ice algal blooms with associated primary productivity rates upwards to 23.0 mg C $m^{-3}$ $h^{-1}$ (Gradinger 2009). In turn, this strong seasonal pulse of fresh primary production supports the life histories of diverse secondary producers including osmotrophic eukaryotes. With the return of summer comes seasonally elevated temperatures that melt snow, ice, and thaws surface terrestrial soils leading to location-variable seasonal freshening of polar seas (e.g., Bendtsen et al. 2014; Porter et al. 2019) and terrestrial export of organic material (Kaiser et al. 2017; Wadham et al. 2019). Although polar environments experience strong seasonality of light, temperature, and sea ice coverage the Arctic and Antarctic ecosystems are

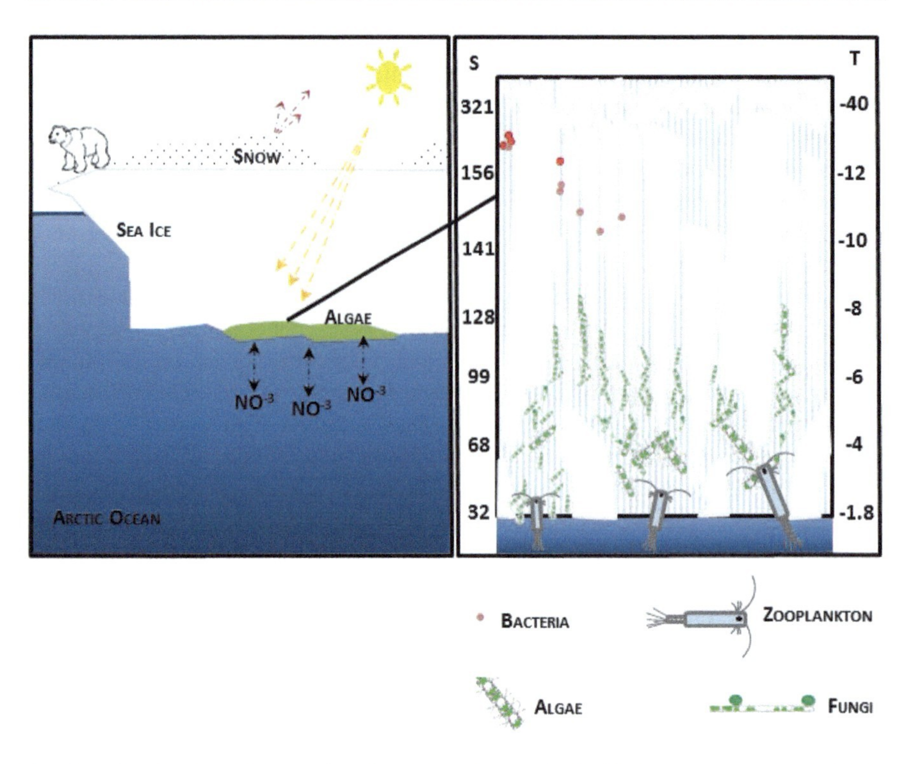


**Fig. 5.2** Simplified and basic sea ice schematic and conceptually illustrating organisms inhabiting brine channels. Temperature (T) and its corresponding salinity (S) according to the equations of Cox and Weeks (1983) are displayed on the vertical axes. The brine volume fraction increases with temperature which is not represented here

markedly different: (1) The Arctic Ocean is surrounded by land, whereas the Antarctic is surrounded by water, (2) The Arctic Ocean's close proximity to land leads to it receiving ~11% of the global river discharge (McClelland et al. 2012) and is consequently disproportionately affected by terrestrial processes. In contrast, the Southern Ocean is considered a more or less closed system (Fraser et al. 2018), and (3) The Arctic has in recent history contained larger quantities of multi-year ice, whereas the Antarctic is primarily a first-year ice-covered region. These differences probably result in diverging patterns of microbial richness and diversity (e.g., Bowman et al. 2012) and certainly confound the effective likening of associated biology within these different systems.

The biogeographical patterns of polar marine fungi as with nearly all marine fungal species are currently nebulous because under-sampling due to the limited number of studies prevents to discern true endemism. Although nebulous, there are emerging patterns of the relationship between specific environmental variables and marine fungal communities but these relationships appear to be scale-dependent. Specifically, at global scales pelagic marine fungi do not exhibit latitudinal gradients of species richness or diversity (Hassett et al. 2020; Tisthammer et al. 2016) indicating that polar seas should not be unique in fungal species richness, even though new species have been described (e.g., Fell et al. 1969; Pang et al. 2008) and unique clades of fungi molecularly detected (Comeau et al. 2016; Hassett et al. 2017). Species checklists from culturing efforts have been discussed elsewhere for the Arctic (Rämä et al. 2017) and Antarctic (Godinho et al. 2019; Ogaki et al. 2019). At regional scales throughout the world marine fungal community composition regularly co-occurs with salinity (Hassett et al. 2020; Jeffries et al. 2016; Rojas-Jimenez et al. 2019; Taylor and Cunliffe 2016). However, Arctic marine fungal communities are frequently heterogenous (Perini et al. 2019) with weak to no relationships detected between salinity and geography (Comeau et al. 2016; Hassett et al. 2017). In polar regions, ascribing an effect of salinity on the composition of marine fungal communities is challenging and should be done cautiously especially in the Arctic, which contains many distinct hydro-morphological domains (Bluhm et al. 2015). Specifically: (1) the co-occurrence of a unique microbial community with low salinity during the sea ice melt season is most likely a detection of distinct brine channel communities being released into the water column (e.g., Rapp et al. 2018) as opposed to the result of stable environmental filtering/selection processes, (2) the disproportionally large quantities of freshwater that are discharged into polar waters are also source of terrestrial organisms (e.g., Collins et al. 2010) thereby creating unique communities which co-occur at lower salinity, and (3) sea ice brine salinity is calculated as a function of temperature. Hence, without carefully controlled experiments it is difficult to separate the covarying effects of temperature from salinity.

Lower latitude regional drivers of fungal communities do not appear to be robust uniform predictors of polar marine fungal community composition. However, local environmental differences such as specific substrate availability (e.g., Lacerda et al. 2020) and photon flux can explain the distribution of polar marine fungal communities. Specifically, the effect of higher light transmission into the water column at polynyas and under low localized snowpack on sea ice can result in a Chytridiomycota-predominated system (Hassett and Gradinger 2016; Hassett et al. 2017; Terrado et al. 2011). Many of these Arctic marine chytrids are parasites of diatoms (Hassett and Gradinger 2016; Hassett et al. 2020; Horner and Schrader 1982). The effect of light on *Chytridiomycota* and their parasitic activity on algae has been documented in experimental studies (Muehlstein et al. 1987; Scholz et al. 2017; Tao et al. 2020). Emerging evidence demonstrates that *Chytridiomycota* possess gene clusters associated with virulence (van de Vossenberg et al. 2019). If marine *Chytridiomycota* also possess virulence-associated genes like many diatoms possess defense response pathways such as those involving oxylipins (Johnson et al. 2020), marine *Chytridiomycota* may be disproportionately governed by the genetics of host and parasite according to various disease paradigms (e.g., Scholthof 2007), as opposed to exclusively environmental factors such as salinity. Reversible encystment of chytrid zoospores to host surfaces (Doggett and Porter 1994) lends additional evidence to the importance of biological interactions in governing local abundances. In addition to the factors that affect biological interactions spatially constrained brine channels harbor elevated proportions of *Chytridiomycota* relative to open water (Comeau et al. 2016; Hassett et al. 2019). These brine channels also harbor elevated diatom densities (Gradinger 2009). Consequently, if *Chytridiomycota* abundances are governed by the molecular underpinnings of host–parasite interactions, then these elevated concentrations of both host diatoms and parasitic *Chytridiomycota* are likely amplified by principles of epidemiology such as lower dispersal time of infectious propagules in spatially constrained spaces. The atypical salinity patterns within sea ice could indeed serve as an amplification of *Chytridiomycota* abundances thereby creating an ideal niche for parasitic flagellated fungi. Saprotrophic chytrids have been previously reported in the Arctic Ocean (Sparrow 1973) and in association with fecal pellets (Hassett et al. 2019) although their diversity and ecological contributions remain to be elucidated.

In summary, the Arctic and Antarctic contain many cosmopolitan fungi, which support hypotheses of over-dispersal in marine realms. However, fungi in polar marine environments shift into habitat-specific communities seemingly driven by a preponderance of *Chytridiomycota*. The unique feature of constrained space within sea ice brine channels, hypersalinity of this brine, and high concentration of hosts has allowed this habitat-specific phenomenon to be putatively identified. This underscores the potential of the large gradients and fluxes in polar environments for answering outstanding questions in marine science.

In conclusion, fungi have been reported from every marine realm. From the Arctic to the Antarctic from surface waters to kilometers below the seafloor marine fungi are seemingly ubiquitous and certainly contributing to global biogeochemical processes. As the phylogenetic limits of the fungal kingdom both expand to include additional clades while existing clades are segregated into additional phyla it is expected that the reported marine fungal diversity increases with increasing sampling efforts especially among early-diverging taxa. This will lead to answers to outstanding questions of biogeographical patterns of species diversity, which will allow conclusions regarding over-dispersal.

## 5.4 Adaptation of Marine Fungi

Many surveys have highlighted a high diversity of ubiquitous marine fungal species with a dominance of the filamentous genera *Penicillium*, *Cladosporium*, *Aspergillus*, *Fusarium*, and *Trichoderma* and the yeast genera *Candida*, *Rhodotorula*, *Cryptococcus*, and *Hortaea*. Detection of such fungi related to known freshwater or terrestrial groups in the ocean is surprising and suggests effective adaptive capabilities of these fungi for biotic and abiotic stresses related to the marine environment. The term "adaptation" is defined as any adjustment of an organism that makes it better suited to live in a given environment (Rédou et al. 2016a) either in terms of adaptive evolution (at the genetic level) or adaptive response (at the expression level). Evidence of real activity in the marine environment has been obtained using microscopic features, e.g., the germination of *Acremonium fuci* conidia only in the presence of *Fucus serratus* algal tissues (Zuccaro et al. 2004), or ecophysiological features, e.g., a shift from terrestrially-adapted to marine-adapted fungal lifestyles along a sediment core (Rédou et al. 2015), or an ability to grow under elevated hydrostatic pressure (Burgaud et al. 2015; Damare et al. 2006). Modern high-throughput omics approaches based on rRNA and mRNA sequencing have also allowed to reveal the activity and functions of numerous ubiquitous fungal species in different marine habitats (Edgcomb et al. 2010; Orsi et al. 2013a; Orsi 2018; Pachiadaki et al. 2016; Quémener et al. 2020).

Comparative genomics/transcriptomics of marine vs. terrestrial isolates of the same species appears as powerful approaches to gain insights into the physiological capabilities, evolution, and adaptation of marine fungi. Assuming that habitats control genome evolution or genome adaptations it should be possible to observe the up- or down-regulation of specific genes or pathways when comparing marine to terrestrial representatives of the same species. Based on available data on halophilic fungi, namely *Eurotium rubrum*, *Hortaea werneckii*, and *Wallemia ichthyophaga* comparative genomics highlighted a high number of genes coding for proteins with a higher proportion of acidic amino acid residues, of genes related to stress response (A-/B-barrel proteins, catalases), of genes coding for hydrophobins and polyol synthesis, and of genes related to DNA processing and damage (Kis-Papo et al. 2014; Lenassi et al. 2013; Zajc et al. 2013). Complementary comparative transcriptomics revealed that most of these genes were highly expressed under high salinity conditions.

Evidence of adaptation to the marine environment can also be figured out based on marine fungal secondary metabolism. Genomic and transcriptomic analyses of *Scopulariopsis brevicaulis*, a well-known terrestrial fungus, isolated from a sponge revealed the gene/gene clusters responsible for the synthesis of anti-cancerous scopularides only found in marine conspecific *S. brevicaulis* (Kumar et al. 2015). Available genomes of marine fungi such as *Corollospora maritima* and *Lindra thalassiae* are available at the Joint Genome Institute. An overview of their secondary metabolite gene clusters revealed many non-ribosomal peptide synthetase (NRPS) genes, polyketide synthase (PKS) genes, and terpene-encoding genes in their genomes that may be involved in the production of unique secondary metabolites and could be explained as an adaptation to life in the marine environment. Metabolomics thus appears as another interesting approach to provide concrete evidence regarding the real in situ activity and interactions of ubiquitous marine fungi through their ability to produce a wide spectrum of secondary metabolites, which differ from those of their terrestrial counterparts (Bhakuni and Rawat 2005).

Despite their ecological importance marine fungi have been largely overlooked by marine microbiologists because they were considered by them to be inactive in the marine environment. Studies highlighting their presence, activity, and function in numerous marine habitats and their putative importance paved the road for an integrated analysis and to delve deeper into their adaptation to the marine environment and their capacities to cope with changes such as hydrostatic pressure, salinity, and temperature (see Rédou et al. 2016a).

The many ways of marine fungi to cope with biotic and abiotic stresses thanks to their adaptive capabilities make them an interesting subject for biotechnology.

Numerous capabilities can be hijacked from their first ecological functions and used for biotechnological applications with the idea to contribute with solutions to pressing societal/environmental challenges. The most important applications recognized to date are enzymes. For example, enzymes able to degrade complex polymers (e.g., hydrocarbons and plastics) can be applied in bioremediation. In addition, secondary metabolites of marine fungi are known for their antimicrobial activities.

## 5.5 Accessing the Bioremediation Potential of Marine Fungi

Marine fungi are interesting as producers of enzymes with an industrial value such as cellulases, amylases, xylanases, lipases, proteases, and laccases (Rao et al. 2017). The adaptive potential of marine fungi providing them with a high ability to withstand many types of stressful conditions allows for the synthesis of a wide array of enzymes with different temperature and pH optima (Bonugli-Santos et al. 2015). To date, marine fungal enzymes have been mostly obtained from marine fungi isolated from seawater, sediments, or mangroves and, to a lesser extent, from deep-sea habitats.

### 5.5.1 Degradation of Hydrocarbons

Fungi have a high tolerance to hydrocarbons (Al-Nasrawi 2012) and more than 100 genera (Prince 2005) play important roles in the biodegradation of hydrocarbons in soils and sediments. Filamentous fungi such as *Cladosporium* and *Aspergillus* are among those known to participate in aliphatic hydrocarbon degradation. The genera *Cunninghamella, Penicillium, Fusarium, Mucor*, and *Aspergillus* are among those known to take part in the degradation of aromatic hydrocarbons (Al-Nasrawi 2012; Passarini et al. 2011; Steliga et al. 2012). While most filamentous fungi investigated to date are unable to fully mineralize aromatic hydrocarbons, they facilitate the degradation of the more recalcitrant hydrocarbons in the environment by secreting extracellular enzymes that transform these compounds into intermediates of lower environmental toxicity and increased susceptibility to bacterial decomposition (Steliga et al. 2012). The poor bioavailability of many hydrocarbon components is considered to be a major rate limiting factor in the hydrocarbon remediation process (Das et al. 2014). Biosurfactants act as surface-active amphiphilic compounds with a hydrophobic and hydrophilic moiety that interact with phase boundaries in a heterogeneous system to solubilize organic compounds (Sen et al. 2017). The entire phenomenon enhances the bioavailability of contaminants for microbial degradation through better solubilization of hydrocarbons in water or water in hydrocarbons (Banat et al. 2014). Chemical surfactants exist such as carboxylates, sulfonates, and sulfates but biosurfactants have several advantages over these such as lower toxicity and higher biodegradability (Shekhar et al. 2015). While a majority of biosurfactants described is of bacterial origin with producers affiliated to *Pseudomonas,*

*Acinetobacter,* and *Bacillus* the importance of the production of biosurfactants by yeasts and filamentous fungi is increasingly recognized. Fungal producers are affiliated to the yeasts *Candida, Pseudozyma*, and *Rhodotorula* (Sajna et al. 2015; Sen et al. 2017) and to the filamentous fungi *Cunninghamella, Fusarium, Phoma, Cladophialophora, Exophiala, Aspergillus,* and *Penicillium* (Lima et al. 2016; Silva et al. 2014) all of them with numerous representatives in the marine environment.

### 5.5.2 Degradation of Plastics

Plastics have become the most common form of waste in the environment and represent a major and growing environmental threat at the global scale. An estimation indicated that of the 8300 million metric tons (Mt) of virgin plastics produced so far ~80% accumulated in landfills or in the environment (Geyer et al. 2017). Annual plastic waste input from land into the ocean varies from 4.8 to 12.7 Mt. representing 1.75 to 4.65% of the 275 Mt. of plastic waste generated annually (Jambeck et al. 2015). In terms of global composition of marine litter plastics account for ~62%. Plastics also contribute to 49% of the litter composition in the seafloor and 81% at the sea surface (Sánchez 2020).

Essential characteristics that are responsible for the resistance of plastics to biodegradation include a long chain polymer structure made of carbon, silicon, hydrogen, nitrogen, oxygen, and chloride, high molecular weight (MW), absence of a favorable functional group, hydrophobicity, and crystallinity (Urbanek et al. 2018). Nevertheless, plastics are rapidly covered by organic matter collectively referred to as the "ecocorona," which decreases the hydrophobicity of the surfaces and facilitates microbial colonization (Wright et al. 2020). Microplastics thus constitute a new ecological niche for microbial communities defined as the "plastisphere" (Amaral-Zettler et al. 2020; Zettler et al. 2013).

Not only bacteria but also fungi form biofilms on plastic surfaces mainly dominated by *Chytridiomycota*, *Cryptomycota*, and *Ascomycota* (Jacquin et al. 2019). In the framework of an exposure experiment (PET drinking bottles deployed at several stations in the North Sea) and using a metabarcoding approach, Oberbeckmann et al. (2016) revealed different microeukaryotic communities including fungi and more precisely the *Ascomycota*, *Basidiomycota*, and *Chytridiomycota*. These observations were supported by another metabarcoding approach that revealed similar phyla on PolyEthylene (PE) samples during an exposure experiment in the Belgian part of the North Sea (De Tender et al. 2017) leading to the identification of the species *Cladosporium cladosporioides*, *Fusarium redolens*, and *Mortierella alpina* as putative PE degraders. The abundant presence of *Chytridiomycota* could result in different kinds of parasitic or saprotrophic relationships with the primary producers that are also present in the plastic biofilms, mediating the carbon, nutrient, and energy transfer into the food webs (Kettner et al. 2017). Moreover, marine plastics are often covered with polysaccharide-rich diatom biofilms, which could explain the attachment and association of fungi with biofilms on plastics (Lacerda et al. 2020).

While some fungi may be hitchhikers on plastics benefiting from a substrate covered with organic matter, some fungi appear to have the ability to biodegrade plastics as, for example, *Zalerion maritimum*. This marine fungus is able to biodegrade PE, which is one of the most widely used polymers and thus one of the main polymers found in the environment (Paço et al. 2017).

The biodegradation of plastics takes place in 4 steps: (1) Microbial biofilm formation decreasing the buoyancy and hydrophobicity of the plastic (Rai et al. 2020). Fungi have the ability to produce hydrophobins, which represent an important tool for bioremediation purposes since they act as natural biosurfactants increasing attachment to plastic hydrophobic substrate and thus its bioavailability (Sánchez 2020). (2) The generation of biofilms is followed by biodeterioration that weakens the polymer integrity and produces cracks and pores. (3) After biodeterioration, the carbon skeleton of plastics is destabilized through enzymatic depolymerization with amidases, oxidases, laccases, and peroxidases (e.g., the conversion of polymers into monomers that are assimilated into microbial biomass). (4) Mineralization as a final but rate-determining step where the polymers are completely degraded resulting in the release of final products such as, for example, $CO_2$ and $H_2O$ (Rai et al. 2020).

Thus, fungi including marine species have (1) abilities to produce hydrophobin for surface coating to attach hyphae/cells/spores to hydrophobic substrates, (2) a machinery of unspecific enzymes able to catalyze diverse reaction mechanisms, and (3) cellular ability to penetrate three-dimensional substrates making them highly convenient for plastic degradation (Sánchez 2020).

## 5.6 Hints to Ecological Roles Inferred from Secondary Metabolites

### 5.6.1 Secondary Metabolites (or Specialized Metabolites): A Definition

Organisms grow and develop using abiotic and/or biotic resources they find in their environment. The so-called primary metabolism is dedicated to the production of what is inherent to their development such as cell membranes, proteins for basic cellular functions, or nucleic acids for DNA. These primary constitutive metabolites are necessary for growth as they are involved in energy storage, cell machinery, and structure of organisms. These metabolites are universal and common in organisms. In contrast, the so-called secondary metabolites are produced in response to a plethora of different stimuli and are not considered to be essential for the organism. However, secondary metabolites often respond to diverse needs such as defense against predators, communication with other organisms, or adaptation to environmental changes. Actually, secondary metabolites are produced by organisms for specific functions which can be sometimes essential for their survival at a specific moment (e.g., mate recognition or settlement cues). Even if the production of these metabolites is not universal and often limited to certain species under specific conditions, they are not less important. Secondary metabolites are produced to fulfill specific functions or to interact with specific biological targets. For all these reasons, the term "secondary", which has been used for a long time, tends to be replaced by "specialized," to avoid any misinterpretation on "secondary" as being less important than "primary" and to better reflect the distribution of these metabolites to restricted taxa.

Thanks to their intrinsic privileged interactions with living material, many secondary metabolites are used as valuable compounds for therapeutic targets thus providing interesting leads for the pharmaceutical industry. Among the drugs that have been approved since 1980, more than one third actually corresponds to compounds derived from natural products (mainly being secondary metabolites) and the drug discovery pipeline still continues to add new drugs of natural origin (Newman and Cragg 2020).

Microorganisms such as bacteria and fungi are invaluable for the discovery of drugs and/or lead compounds. Indeed, these microorganisms produce a huge variety of antimicrobial agents and the screening for secondary metabolites produced by microorganisms became highly generalized after the discovery of penicillin. Metabolites from fungi also lead to the discovery of potent compounds for other applications than antibiotics. This is, for instance, the case for lovastatin the lead compound for a series of drugs that lower cholesterol levels and cyclosporine currently used to suppress the immune response after transplantation operations (Numata et al. 1993). Secondary metabolites are central to the ecology of many fungi and allow their diverse interactions with other organisms. For instance, gliotoxin is a virulence factor and penicillin is an antibiotic whose ecological role lies in fungal defense against bacteria while α-amatoxin is considered to serve as a constitutive chemical defense.

Secondary metabolites are produced from elementary bricks provided by the primary metabolism such as amino acids, sugars, and fatty acids. Various biosynthetic pathways have been described leading to different classes of compounds such as polyketides, alkaloids, or peptides. Surprisingly, the number of building blocks is small compared to the huge number of compounds that can be built. These pathways are governed by chemical reactions, many of which are catalyzed by specific enzymes encoded in the genome of the organisms. Indeed, most fungal secondary metabolites are encoded by biosynthetic gene clusters (BGCs). Each cluster typically contains the majority of the genes participating in the production of a given secondary metabolite. These genes are located adjacent to each other in the genome and are therefore called "clustered." BGCs contain genes encoding proteins governing the biosynthesis of the backbone of the metabolite (e.g., polyketide synthases (PKSs), non-ribosomal peptide synthetases (NRPSs), prenyltransferases, and terpene cyclases) as well as genes encoding for tailoring enzymes that modify this backbone (epimerases, methyl-transferases, and hydroxylases), proteins involved in metabolite transport, transcription factors involved in regulation of the BGCs expression, and proteins that confer resistance to the activity of the secondary metabolite (Rédou et al. 2016a).

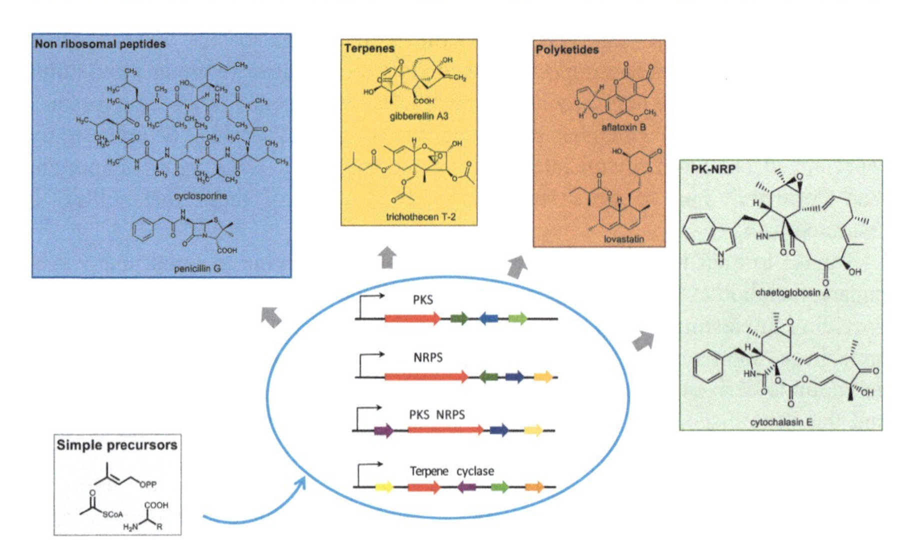


**Fig. 5.3** Schematic fungal BGCs and their secondary metabolites

The availability of a growing number of sequenced fungal genomes and their analyses by bioinformatics revealed that the number of genes encoding the highly conserved PKS and NRPS is much larger than anticipated, even in strains that have been extensively studied for the formation of natural products. Thus, fungi possess more potential secondary metabolic biosynthetic clusters than reported metabolites of which the identity, structure, and function often remain unknown. In fungi, the main biosynthetic pathways described include polyketides, terpenes, non-ribosomal peptides, alkaloids, and shikimate pathways (Fig. 5.3). With this machinery in hands, fungi are known to be prolific producers of secondary metabolites. The diversity of structures produced by fungi from elemental bricks arising from the primary metabolism appears huge, if not infinite.

### 5.6.2 Marine Fungal Chemodiversity

In the marine environment (micro)organisms have developed adaptations to cope with numerous stimuli through expression of specific biosynthetic pathways for defense or communication. It is well-known that there is an extremely harsh competition between (micro)organisms in the marine environment. On some tropical coral reefs, for example, there can be as many as 1000 species per $m^2$ meaning that the organisms need to find ways to co-exist. This is where chemical cues and chemical communication take place.

While in 1992, only 15 fungal metabolites had been described from the marine environment (Fenical and Jensen 1993); this number increased tremendously over the last decades (Blunt et al. 2016; Bugni and Ireland 2004; Daletos et al. 2018; Rateb and Ebel 2011). So far, more than 3500 secondary metabolites have been isolated from fungi collected in the marine environment. According to the open-source Natural Product Atlas (NPA, accessed in October 2020) (van Santen et al. 2019), marine fungal secondary metabolites now represent approximately 22% of all metabolites isolated from fungi.

Marine fungi have yielded many clinically-relevant natural products, including antibiotics (Blunt et al. 2018; Demain 2014; Rateb and Ebel 2011; Svahn et al. 2012). The antibiotics include gliotoxin from a deep-sea strain of *Aspergillus* sp. (Fan et al. 2016) and indanonaftol A from a marine *Aureobasidium* sp. (Biabani and Laatsch 1998). The biosynthesis of various secondary metabolites by fungi usually coincides with their life cycle (Calvo et al. 2002) and with fungal strategies to inhabit and survive in diverse ecological niches. Identification of fungal secondary metabolites can be challenging since it can require appropriate culture conditions that mimic the conditions of different extreme habitats such as those found in the deep sea and deep biosphere. As an example, few secondary fungal metabolites have been successfully isolated under high-pressure conditions (>90 atm) to date because of the technological constraints of recovering and culturing piezophile or piezotolerant fungal strains at elevated pressures (Kumar et al. 2015). The bioactive molecules that have been isolated from deep sea fungi include various secondary metabolites with biotechnological potential (Nicoletti and Trincone 2016). The majority of secondary metabolites isolated so far are derived primarily but not exclusively from strains belonging to two genera: *Penicillium* and *Aspergillus* (Petersen et al. 2020; Arifeen et al. 2020). These include biomolecules with anti-inflammatory, cytotoxic, antimicrobial, antifungal, and antiviral activities (Wang et al. 2015).

New polyketides (aspilactonol, aspiketolactonols, aspyronol, epiaspinonediol), alkaloids (meleagrin D and E, sorbicillamines, brevicompanines D–H, circumdatin F and G, and cyclopiamides), steroid and terpenoid derivatives (sterolic acid, breviones) were isolated from *Aspergillus* and/or *Penicillium* strains collected from deep sea hydrothermal vents (2255 m), and from deep sea sediments in the Pacific Ocean (~5000 m depth) and South China Sea (4593 m depth) (Chen et al. 2014; Fredimoses et al. 2015; Li et al. 2012; Xu et al. 2015; Zhang et al. 2018). Alkaloids from *Penicillium* sp. exhibit anti-inflammatory activity and weak or moderate cytotoxicity against cancer cell lines (Du et al. 2009). The terpenoids and steroid metabolites showed antiviral effects against HIV-1 and cytotoxicity against lung and breast cancer cells (Li et al. 2012). The cytotoxicity of two new C9 polyketides (spyronol and epiaspinonediol) and wentilactones (terpenoids) identified in *Aspergillus* sp. was tested against human leukemia cells and they turned out to be potential antitumor agents (Xu et al. 2015). Phenolic compounds from *Aspergillus* sp. (from 2326 and 3002 m depth; South China Sea) showed antifouling activities and possibly activate transcription factors related to detoxification mechanisms (Wu et al. 2016). Phenolic compounds from deep sea *Aspergillus* sp. were identified as potential antiviral agents against Herpes Simplex Virus 1 (HSV-1) (Wu et al. 2016). *Alternaria* sp. isolated from sediments at 3927 m depth (South China Sea) produced new perylenequinones that inhibit transcriptional and epigenetic regulators (e.g., BRD4) that participate in cancer development (Ding et al. 2017).

Fungal polyketides (engyodontiumones, clindanones A–B), thiazole analogs (acaromyester A), and polypeptides (simplicilliumtides A–I) from *Engyodontium* sp., *Cladosporium* sp., *Simplicillium* sp., and *Acaromyces* sp. isolated from deep-sea sediments (3415 and 3739 m depth) in the South China Sea and the Indian Ocean (3471 and 4571 m depth) likely exhibit moderate to pronounced antitumor effects against lymphomas and cancer cells (Gao et al. 2016; Yao et al. 2014; Zhang et al. 2016). Other secondary fungal metabolites include terpenes (spirograterpene A, conidiogenone C and I, aspewentin A and D–H, asperethers A–E, asperolides D and E, and guignarderemophilane F) and cyclic lactones (butanolide A). These were all isolated from *Penicillium* sp. from deep waters (2284 m depths) and deep sediments (1393 m depth) at Prydz Bay of Antarctica. These compounds present antiallergic effects similar to common antihistamines (e.g., loratadine) and are thought to inhibit the protein tyrosine phosphatase 1B (PTP1B) which production is implicated in type-2 diabetes (Zhou et al. 2017). Fungal strains that belong to the family Lindgomycetaceae, isolated from deep sediments in the Greenland Sea, produce lindgomycin that shows antibiotic activity similar to chloramphenicol (Wu et al. 2015). Canescenin A and B (polypeptides) and nitrogen-bearing heterocyclic compounds (e.g., piperazine derivatives) isolated from deep-sea *Penicillium* sp., *Aspergillus* sp., and *Dichotomomyces* sp. have also been scanned for antibacterial activity (Dasanayaka et al. 2020; Fan et al. 2016; Wang et al. 2016). Dichotocejpins isolated from *Dichotomomyces* sp. seems to exhibit antifungal activity because of their inhibitory activity against α-glucosidase (Fan et al. 2016). Terpenes and polyketides from the piezotolerant deep-sea fungi *Ascotricha* sp. and *Phialocephala* sp. have cytotoxic and antifungal activities (Ganesh Kumar et al. 2019; Zhang et al. 2018). In silico studies demonstrated that ascotrichin isolated from *Ascotricha* sp. can fit into serotonin receptors presenting a potential target for the development of drugs related to central nervous system disorders (Ganesh Kumar et al. 2019).

### 5.6.3 Marine Fungal SMs and Specificity to the Marine Environment

While some of the secondary metabolites have also been found in terrestrial fungi, others may be specific to the marine environment. By clustering all fungal metabolites described so far in the literature based on structure similarity it became clear that compounds isolated from the marine environment (highlighted in blue on Fig. 5.4) represent diverse skeletons. Some clusters of compounds were already described as produced by strains from the terrestrial environment while others such as spiromastixones were only found in marine organisms such as the deep-sea-derived fungus *Spiromastix* sp. (Niu et al. 2014). These molecules were able to decrease oxidized low-density lipoprotein-induced lipid over-accumulation and

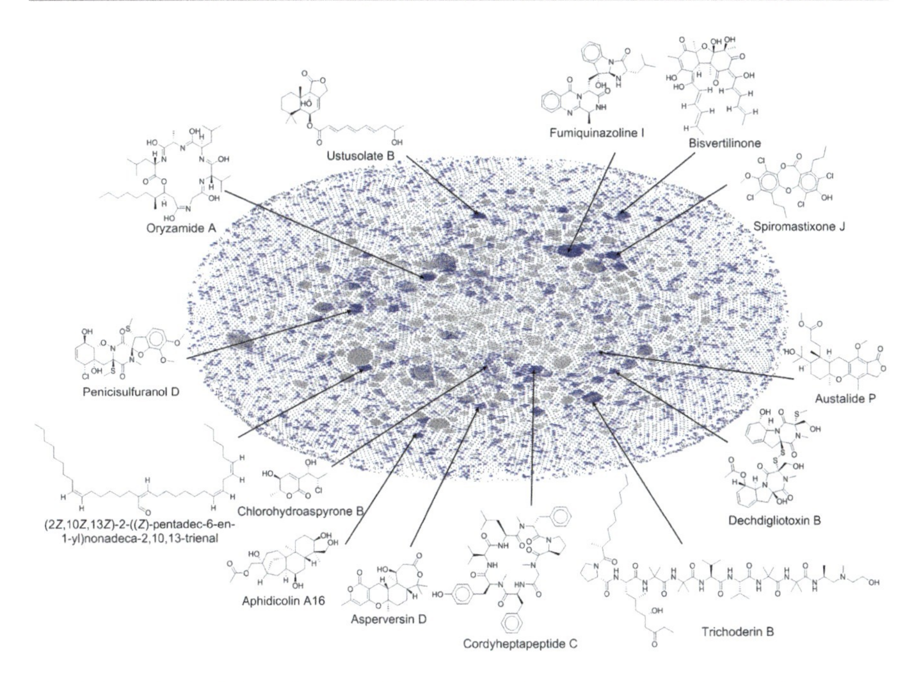


**Fig. 5.4** Similarity chart of the 17,441 fungal metabolites described in the Natural Product Atlas with metabolites first isolated from strains collected in the marine environment highlighted in blue. Each structure is represented by a dot and structures are clustered based on their similarity using the "Frag Fp" descriptor (on "data warrior" software, Sander et al. 2015). Some representative examples of natural skeletons isolated from the marine environment (blue clusters) are given

decrease intracellular cholesterol concentration, which is of potential interest in the treatment of atherosclerosis (Wu et al. 2015).

Specificity to the marine environment still remains difficult to assess on a chemical basis. Many examples have shown that after being isolated from a marine fungus the same molecule was later identified from a terrestrial species (e.g., communesins, Hayashi et al. 2004; Numata et al. 1993). Detecting an original compound produced by a fungus collected in the marine environment does not preclude it might be also found from a terrestrial strain. However, investigation of the chemodiversity produced by marine fungi in comparison to all fungal metabolites described so far in the NPA database shows some interesting points. First, while the overall chemical space occupied by fungal metabolites whether marine or not appears quite similar (Fig. 5.5), some specific physicochemical properties from fungal metabolites isolated in the marine environment could be highlighted. For example, drug likeness and halogen count in the molecules isolated from this environment appeared to be much higher (Fig. 5.6), which is consistent with the marine environment being rich in halogens, which leads to marine organisms to include these atoms in their metabolism. In fact, some studies already revealed the potential of marine fungal strains to produce halogenated metabolites (Roullier et al. 2016). While cheminformatic analyses of fungal metabolites have

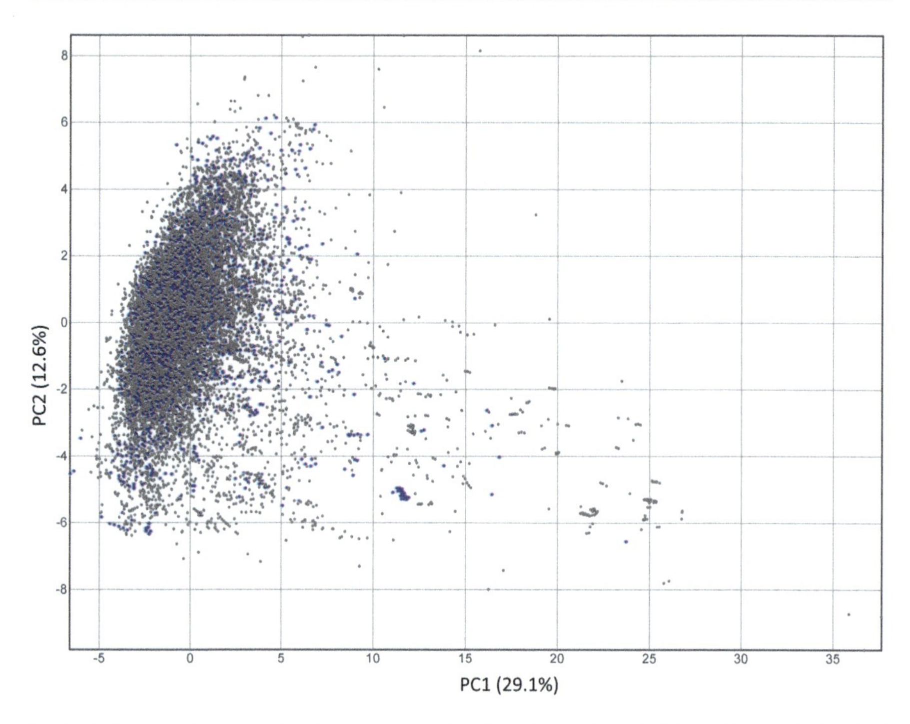


**Fig. 5.5** Representation of chemical space overlapping of fungal metabolites with metabolites first isolated from strains collected in the marine environment highlighted in blue. This graph was generated using "data warrior" software (Sander et al. 2015) after calculating different chemical properties and performing a principal component analysis

already shown that they are as structurally diverse as other natural products and cover relevant chemical space with drug-like physicochemical properties (González-Medina et al. 2016), the representation here shows that the marine environment may be even more interesting in this perspective. Marine fungal products then represent promising candidates for therapeutics. So far, plinabulin is the only marine fungal synthetic analog that has entered clinical trials now reaching phase III for the treatment of non-small cell lung cancer (Gomes et al. 2015).

### 5.6.4 New Methods to Access the Marine Fungal Metabolome

Numerous approaches are available to investigate the fungal metabolome either based on culture-independent methods (e.g., genome mining, genetic engineering, heterologous expression) or culture-based ones (e.g., OSMAC, cocultivation, time-scale studies) (See Rédou et al. 2016a). Some new methods have been developed and seem ready to be used to access the marine fungal metabolome such as, for example, microscale platforms (Barkal et al. 2016), the development of CRISPR-Cas9 systems for filamentous fungi (Nødvig et al. 2015; Zhang et al. 2017), or molecular

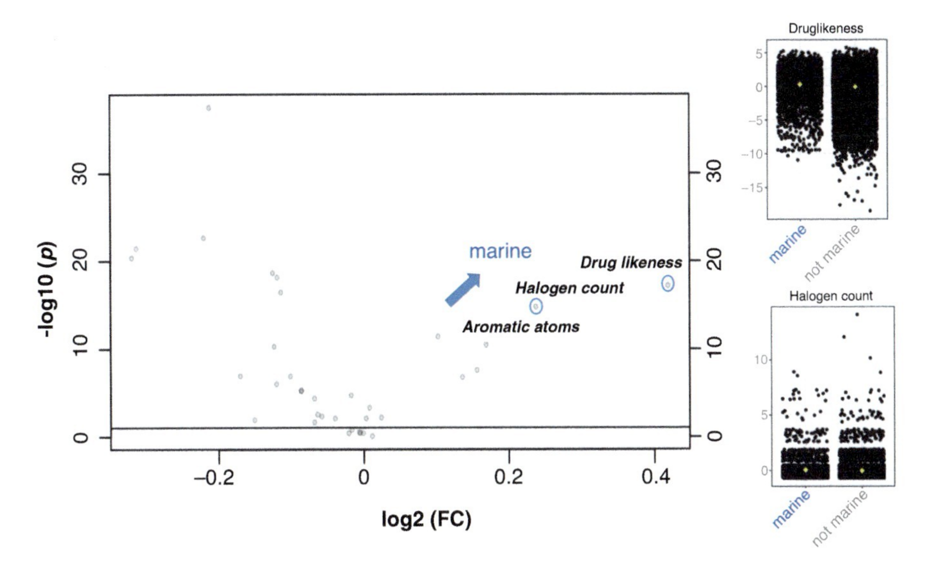


**Fig. 5.6** Volcano plot presenting the most significant factors differentiating marine fungal metabolites (based on the same physicochemical properties calculated for the PCA)

networking (Oppong-Danquah et al. 2018). It is not the aim to describe all these approaches here but rather to list the putative next strategies to use in order to foster new research and generate novel knowledge on marine fungal metabolomics (see also Hautbergue et al. 2018; Hoang et al. 2018; Overy et al. 2019).

### 5.6.5 Marine Fungal Chemical Ecology: Ecological Role of Marine Fungal Metabolites

Efforts over the past decade to expand culture collections of terrestrial, freshwater, marine, and deep biosphere microbial fungi coupled with expanding databases with fungal genomic data are providing insights into the extent of fungal (including novel) diversity as well as the diverse metabolic capabilities of fungi.

Numerous studies have been done on the ecological role of chemical mediation by marine fungi. For instance, it has been shown that endophytic fungi associated with brown algae are able to protect the algal model *Ectocarpus siliculosus* against pathogenic infection through the production of fungal metabolites belonging to the pyrenocine series. A protection effect of the pyrenocines has been observed against oomycetes (i.e., *Olpidiopsis pyropiae* and *P. porphyrae*) infecting the economically important red seaweed *Pyropia* (Vallet et al. 2018). These findings suggest that brown algae-derived endophytes may determine the infection outcome of algal pathogens by chemically protecting their host through the production of chemicals. Brown algae-derived endophytic fungi interact with their host but are also likely to interact with bacteria within the algal holobiont to maintain the host–microbiota equilibrium and thus contributing to maintain health of their hosts (Deveau et al. 2018; Hassani et al. 2018). The mechanisms underlying bacterial–fungal homeostasis remain unclear although they appear crucial for the macroalgal physiology. Evidence of a specific cell–cell signaling named "quorum sensing" among bacterial and fungal endophytes interactions of brown algae was provided by Tourneroche et al. (2019). Quorum sensing (QS), which allows bacteria to coordinate gene expression based on the density of specific signaling molecules, is of particular interest because it is essential for virulence, colonization, biofilm formation, and toxin production (Atkinson and Williams 2009). While the inhibitory activity against quorum sensing had been previously detected in marine fungi (Martín-Rodríguez et al. 2014), Atkinson and Williams (2009) demonstrate the importance of chemical communication, especially quorum sensing and quorum quenching (inhibition of QS), among bacteria and fungi sharing the same marine holobiont (*S. latissima*). The key role of QS in marine fungi functioning has been emphasized by a study carried out on the marine fungus *Paradendryphiella salina* isolated from several healthy brown macrophyte species (Vallet et al. 2020). Indeed, *P. salina* produces novel α-hydroxy γ-butenolides that interfere with the bacterial quorum sensing system inhibiting QS phenotypes of the pathogenic bacterial *Pseudomonas aeruginosa*. Ultra-performance liquid chromatography–high-resolution mass spectrometry (UHPLC-HRMS)-based comparative metabolomics revealed the presence of the main α-hydroxy γ-butenolides among all the *P. salina* strains isolated from different hosts as well as a high metabolic variability related to the alga-host species suggesting that the harboring algal species might have a substantial influence on fungal metabolism. Collectively, these findings strengthen the hypothesis of a key role of microbial chemical signaling which may occur within the algal holobiont.

In challenging habitats where there may be limited carbon availability the stress responses of fungi likely include processes that are known to induce expression of hydrolases, cell wall degrading enzymes, and secondary metabolites as means of survival (García-Lepe et al. 1997; Kim et al. 2011). This idea that interactions are mainly driven by chemical mediation is supported by metatranscriptome data from deep biosphere samples that show expression of genes associated with hydrolytic and polyketide biosynthetic genes (Quémener et al. 2020). Polyketides can have antimicrobial and antifungal properties and are frequently associated with fungal survival under nutrient limitation where competition for scarce resources is intense (Sheridan et al. 2015). This was recently using culture-based approaches highlighting deep subsurface fungi as able to produce antimicrobial compounds, presumably used in competition with other microorganisms for available carbon and energy (Navarri et al. 2016).

## 5.7 From (Meta)Genomes to Bioactive Molecules

Rapid progresses in DNA-sequencing methods in the last decade from Sanger sequencing to various types of short read sequencing as well as advances in modern biology and biotechnology approaches have reached a point, where exploring genetic information for any species of interest has become a quest requiring only a few months of time. This era of genome sequencing is often referred to as the era of "Desktop genomics." Short read sequencing methods are collectively called Next-Generation Sequencing (NGS) methods and the primary NGS platforms include: the Roche GS-FLX 454 pyrosequencer, MiSeq, HiSeq, and Genome Analyzer II platforms (Illumina), SOLiD system (Life Technologies/Applied Biosystems), Ion Torrent and Ion Proton (Life Technologies), and the PacBio RS II system (Pacific Biosciences) (Culligan et al. 2013; Metzker 2010).

Fungal genome sequencing and analyses began with yeast models (Goffeau 2000; Goffeau et al. 1996) and were followed by the filamentous fungus *Neurospora crassa* (Galagan et al. 2003). Within 14–15 years ~50 fungal genomes were obtained using Sanger sequencing. In 2010 the *Sordaria macrospora* genome was sequenced using more than one NGS method (Nowrousian et al. 2010). This accelerated other fungal genome sequencing projects using NGS methods. To date, over 1000 fungal genomes have been sequenced thanks to the 1000 fungal genomes project aiming to cover all groups of fungi (Grigoriev et al. 2014). However, there are only a few marine fungal genomes sequenced to date, which are primarily derived from EU-funded projects, like "Marine Fungi" (FP7, 2011–2014) and "MaCuMBA: Marine Microorganisms: Cultivation Methods for Improving their Biotechnological Applications" (FP7, 2012–2016).

Fungi generally produce four different types of secondary metabolites (SMs), namely polyketides (PKS), non-ribosomal peptides (NRPS), terpene or hybrid molecules such as PKS-NRPS or PKS-terpenes (Keller et al. 2005). The majority of fungal secondary metabolites derives from either non-ribosomal peptide synthetases (NRPSs) or polyketide synthases (PKSs) while only a few are hybrid NRPS-PKS (Brakhage 2013). Generally, SM-encoding genes are organized in discrete clusters (Brakhage 2013; Keller et al. 2005) called biosynthetic gene clusters (BGCs). The genomes of filamentous fungi possess up to 70 BGCs encompassing from 30–80 kb regions on fungal chromosomes or scaffolds (Bok et al. 2015). Production of SMs is regulated by cluster-specific transcription factors and global regulators within BGCs (Brakhage 2013; Keller et al. 2005). SMs are highly relevant for drug discovery particularly for new antibiotics and pharmaceuticals (Newman and Cragg 2020). There is huge potential in fungal bioactive compounds which are expected to serve many roles for future drugs (Montaser and Luesch 2011).

Here, we propose a general pipeline for genome sequencing and annotation of marine fungi using short and long read sequencing (Fig. 5.7).

1. The first step is collecting samples from the targeted marine environment.
2. The second step is bringing samples to the laboratory to establish enrichment cultures under laboratory conditions, to isolate individual fungal strains, and to extract genomic DNA. If the aim is to process a metagenomic analysis, then DNA extraction directly from marine samples can be performed with described methods and some modifications (see Kumar et al. 2015).
3. The third step is genome sequencing using next-generation methods like the Illumina platforms, which are also cost-effective choices. However, a hybrid

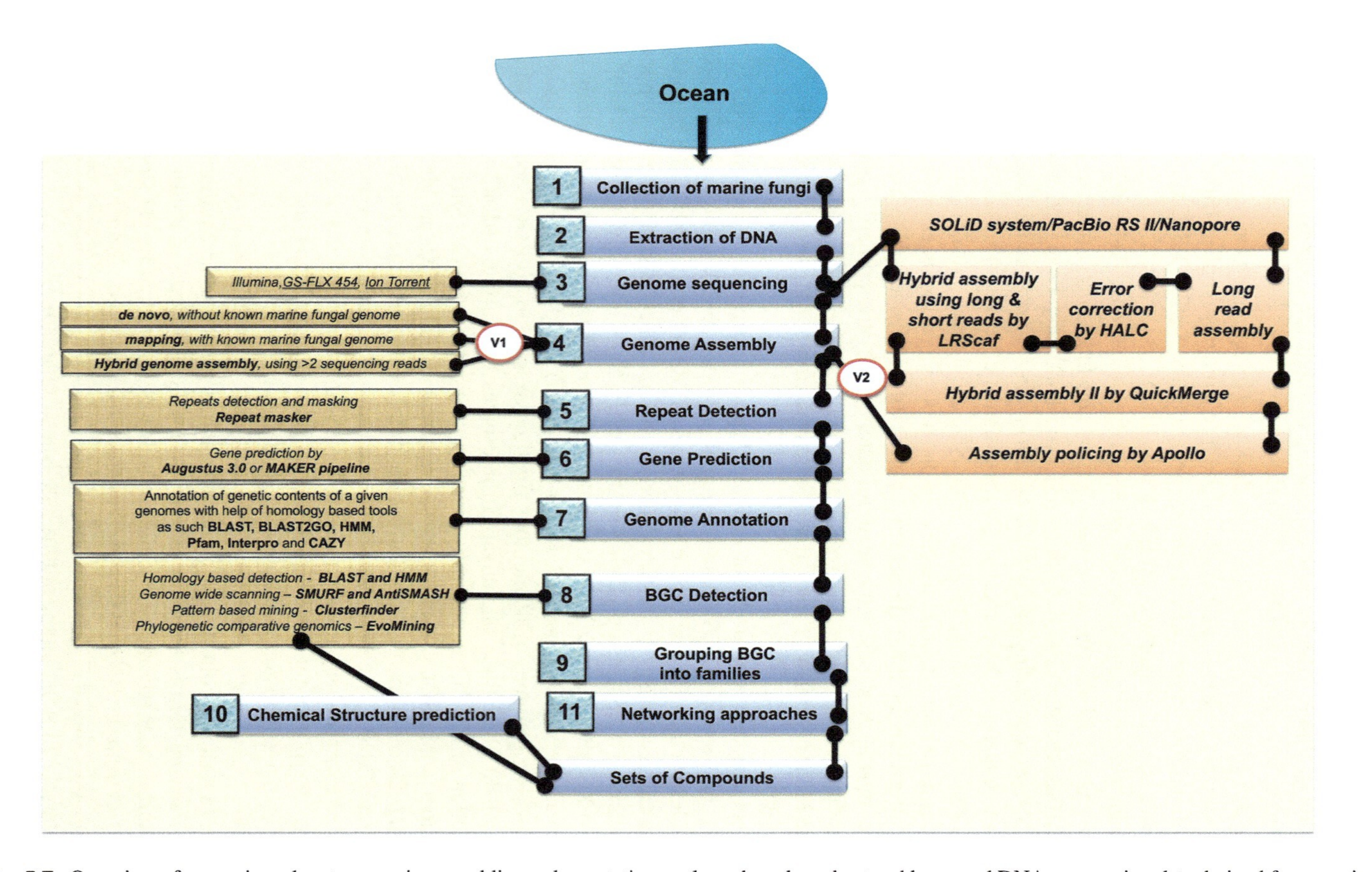


**Fig. 5.7** Overview of genomic and metagenomic assemblies and annotation analyses based on short and long read DNA-sequencing data derived from marine fungi originating from the marine environment

assembly of data from two or more methods is recommended. For instance, the genome of *Scopulariopsis brevicaulis* LF580 has been sequenced using a combination of three methods (Kumar et al. 2015).

4. The fourth step is genome assembly, which assembles reads into genomic fragments called contigs or larger scaffolds. Genome assembly can be achieved in two ways. De novo genome assembly is used when a closely related fungal genome is not known and reads are assembled by genome assemblers without prior knowledge of chromosomes. Alternatively, in case the genome of a closely related fungus is available smaller reads can be mapped to the chromosomes of that close relative's fungal genome and this method is called genome mapping. When more than one type of genome sequencing reads is obtained using different sequencing platforms a hybrid genome assembly can be performed and this can increase the quality and coverage of the overall assembly by allowing longer reads to bridge gaps or scaffold shorter reads. There are several genome assemblers available for genome assembly and can be chosen depending on the requirements and types of genomic reads. A good example is documented in the genome assembly of *Scopulariopsis brevicaulis* LF580 (Kumar et al. 2015) with (a) an initial genome assembly performed using Roche 454 reads and the Newbler assembler (Margulies et al. 2005) coupled with (b) an assembly based on a de Bruijn graph through a de novo assembler in the CLC Genomics workbench (Knudsen and Knudsen 2013) which used different types of reads from Illumina and Ion Torrent and (c) a hybrid genome assembly that was achieved using three different types of reads. The genome assembly normally produces genomes assembled into contigs with contigs joined to form scaffolds, which leads this method to be called "genome scaffolding."

   The Roche 454 sequencing platform has been fading out of use while third generation sequencing platforms that produce longer reads like PacBio/Nanopore became more important. These sequencing methods have slowly become the standard for generating hybrid assemblies. Long read-sequencers minimize the issue of read length from the second-generation sequencing technologies (e.g., Illumina) however they often contain up to 15% sequencing errors. To decrease sequencing errors tools like HALC (Bao and Lan 2017) can be used, however Canu (Koren et al. 2017) has its own built-in tool to correct sequencing errors. The output of error correction tools can be used for long read sequence assemblers such as Canu which assembles sequencing data into long-range scaffolds. Similarly, HALC output can be submitted for hybrid assembly using tools such as LRScaf (Qin et al. 2019) which combines draft genome assembly of short reads with long reads sequencing data. This hybrid assembly approach produces a more accurate assembly than short read assemblers alone. Furthermore, to refine genome assembly a LRScaf hybrid assembly can be combined with a Canu assembly using QuickMerge (Chakraborty et al. 2016) tools to obtain more continuity in final genome assembly output.

5. Step five is the detection of repeat elements and masking of repetitive elements, which is an essential step as these elements hamper gene predictions in the next step. This detection and masking are performed using RepeatMasker and

RepeatProteinMasker software programs (http://www.repeatmasker.org) using the fungal transposon species library as an input. This library contains all known fungal transposon elements.

6. Step six is gene prediction which can be performed using either the Augustus suite (Stanke and Morgenstern 2005) or within the MAKER 2.0 pipeline (Cantarel et al. 2008).
7. Step seven is gene annotation which can be performed with the help of homology detection tools such as BLAST homology searches (Altschul et al. 1997), the OMICSBOX tool (Gotz et al. 2008), and the Kyoto Encyclopedia of Genes and Genomes (KEGG) database (Kanehisa et al. 2010). Predicted proteins of a given genome can be scanned for protein domains using Pfam (Finn et al. 2014) and InterPro (Hunter et al. 2012) protein domain collections using HMMER 3.0 (Finn et al. 2011). Additionally, gene annotations can be performed when the objective of the analyses is known. In case it is necessary to annotate carbohydrate active enzymes of a marine fungal genome scanning the entire proteome against either the CAZy database (Levasseur et al. 2013; Lombard et al. 2014) or the dbCAN database (Yin et al. 2012) can be used. This type of analysis is often known as analysis of genes of interests (Kumar et al. 2014). For example, when plant pathogen fungal genomes are sequenced the sets of pathogen specific genes are looked upon for annotation.
8. Step eight is the detection of BGCs and is performed following four different steps. Initially, putative biosynthetic genes are identified using either BLAST (Altschul et al. 1997) or HMMER 3.0 (Finn et al. 2011). This scanning is performed using a collection of known biosynthetic genes. Secondly, two state-of-the-art tools, namely Secondary Metabolite Unique Regions Finder (SMURF) (Khaldi et al. 2010) and antibiotics & Secondary Metabolite Analysis SHell (antiSMASH) (Blin et al. 2013) are recommended for whole genome scanning of putative BGCs. Furthermore, the functional domains of PKSs and NRPSs can be identified as previously described (Hansen et al. 2014), using a combination of tools, namely antiSMASH (Blin et al. 2013), NCBI Conserved Domain Database (Marchler-Bauer et al. 2011), InterPro (Hunter et al. 2012), and the PKS/NRPS Analysis Web-site (Bachmann and Ravel 2009). Pattern-based approaches can be employed by using *clusterfinder* modules within antiSMASH (Blin et al. 2013). Phylogenetic and comparative genome methods can also be used for BGC detection and analysis with help of genome browsers and specialized tools like EvoMining (Cruz-Morales et al. 2015).
9. Steps nine, ten, and eleven aim at to first group similar BGCs into families based on sequence and domain similarities. Then, mass-spectroscopy-based networking approaches can help in the detection of molecules predicted by genomes. Predictions of small molecule structure using knowledge of BGCs are laborious tasks although antiSMASH (Blin et al. 2013) makes an attempt to predict structures. This is challenging because our knowledge of fungal BGCs is limited and rigorous classification and indexing systems for BGCs do not yet exist. The scientific community has determined a standard for carefully cataloging BGCs, potentially important enzymes, and structural records present in the literature and

this will be incorporated into genome standards such as the Minimum Information about a Biosynthetic Gene cluster (MIBiG) (Medema et al. 2015). This will help in the near future for accurately predicting molecules from genomes. Further details for genome mining of BGCs can be found in the recent reviews (Boddy 2014; Medema and Fischbach 2015; Weber 2014).

Initial studies of marine fungal BGCs have been performed within the framework of EU-funded projects, namely "Marine Fungi" (FP7, 2011–2014) and "MaCuMBA: Marine Microorganisms: Cultivation Methods for Improving their Biotechnological Applications" (FP7, 2012–2016) during which five marine fungal genomes were sequenced and their BGCs were mined for further genetic engineering and production of selected bioactive compounds (Kumar et al. 2015; Kumar et al. 2018; Rédou et al. 2016b, c). These five marine fungi (namely *Scopulariopsis brevicaulis, Pestalotiopsis* sp., *Calcarisporium* sp., *Cadophora* sp., and *Cryptococcus* sp.) were sequenced using different sequencing techniques and expression profiles of SM-genes were analyzed (Kumar et al. 2015, 2018; Rédou et al. 2016b, c). These studies revealed that marine-derived fungi have high contents of SM-encoded genes and not all SM clusters are active under one particular condition as became evident from RNA-Seq based expression profiling of SM clusters (Kumar et al. 2015, 2018; Rédou et al. 2016b, c). These genomic analyses yielded evidence for several promising bioactive compounds with potential roles as antibiotics and anti-cancer metabolites.

Marine-derived fungi are considered as an excellent source of bioactive compounds and their genomes harbor many biosynthetic gene clusters (BGCs) from which these natural compounds are produced (Kumar et al. 2015, 2018; Rédou et al. 2016b, c). The genome sequence of the marine fungal strain of *Scopulariopsis brevicaulis* LF580 revealed 18 BGC clusters including one hybrid NRPS-PKS cluster (Fig. 5.8) which codes for anti-cancerous scopularides (Kumar et al. 2015). This cluster and remaining clusters were characterized using UV-mediated mutagenesis for the production of SM compounds through RNA-sequencing, proteomics, and comparative genomic approaches (Kramer et al. 2015; Lukassen et al. 2015). The other marine fungi *Pestalotiopsis* sp., *Calcarisporium* sp., *Cadophora* sp., and *Cryptococcus* sp. contain a total of 67, 60, 15, and 4 BGCs, respectively, emphasizing the high variability of such SM-encoding genes in marine fungal genomes.

Computational genome mining approaches alone cannot solve all problems associated with identifying fungal BGCs. Several BGC are silent under standard laboratory conditions leaving a plethora of bioactive compounds uncharacterized (Brakhage 2013). Hence, screening each fungus under a wide range of conditions appears essential to activate silent clusters.

Taken together, genome mining and detection of BGC involved in the production of bioactive compounds hold great potential. This is because initial genome mining reveals several different types of BGCs in marine fungal genomes. Heterologous expression systems have already been developed for the entire set of intact fungal BGCs using fungal artificial chromosomes (Bok et al. 2015). It is a matter of time

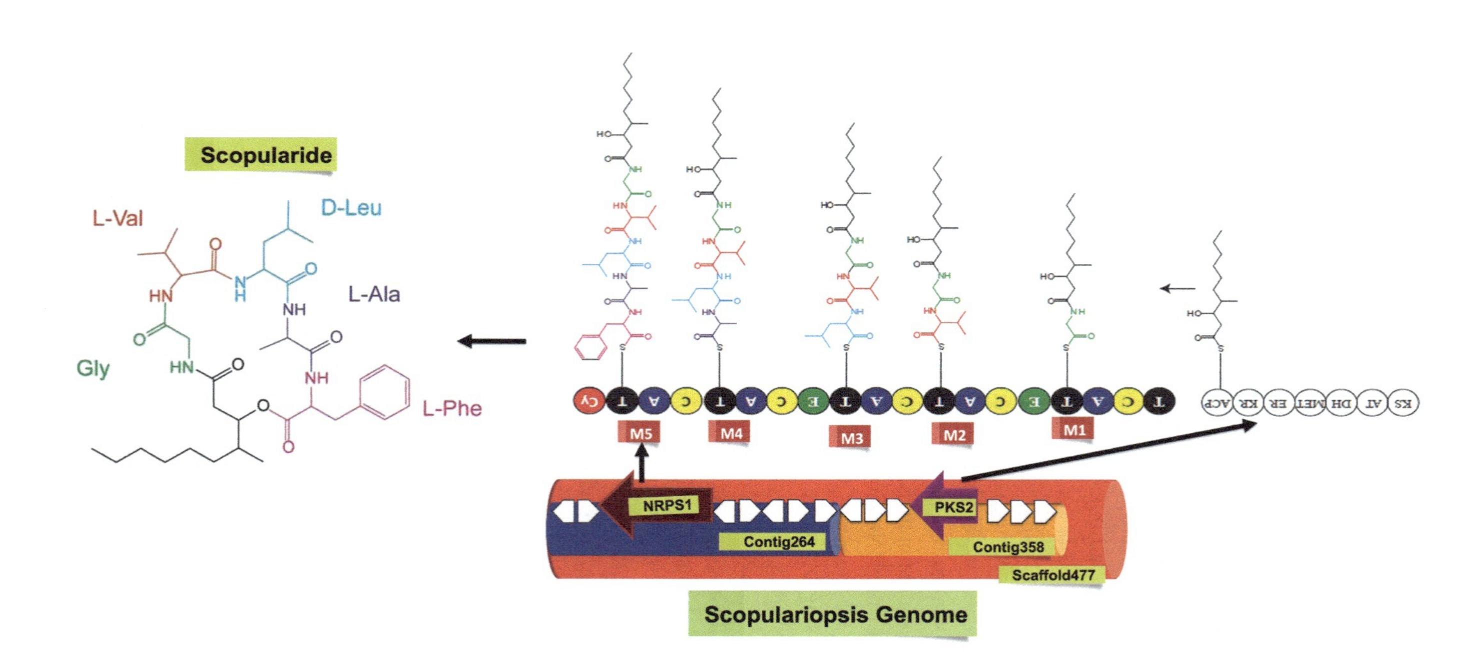


**Fig. 5.8** An example of genome mining strategies for the marine fungal strain *Scopulariopsis brevicaulis* LF580 illustrates production of anti-cancerous scopularides by the hybrid BGC of NRPS1 (five modular M1-M5) and PKS2. Modified from Kumar et al. (2015)

before BGCs will be characterized in heterologous systems. This holds great promise for the future of drug discovery for which genome mining of BGCs plays a fast and cost-effective role.

**Acknowledgements** AK is a recipient of Ramalingaswami Re-Retry Faculty Fellowship (Grant; BT/RLF/Re-entry/38/2017) from Department of Biotechnology (DBT), Government of India (GOI). BTH is supported by the Seasonal Ice Zone Ecology Project (SIZE). VE acknowledges funding from the National Science Foundation grants OCE-1658031 and OCE-1829903. GB and AP acknowledge funding from the French National Research Agency grant ANR-19-CE04–0001-01 (MycoPLAST).

## References

Agusman A, Dan-qing F (2017) Fungal community structure of macroalga *Ulva intestinalis* revealed by MiSeq sequencing. Squalen Bull Mar Fish Postharv Biotechnol 12:99–106

Al-Nasrawi H (2012) Biodegradation of crude oil by fungi isolated from the Gulf of Mexico. J Bioremed Biodegr 3:1–6

Altschul SF, Madden TL, Schaffer AA, Zhang J, Zhang A, Miller W, Lipman DJ (1997) Gapped BLAST and PSI-BLAST: a new generation of protein database search programs. Nucleic Acids Res 25:3389–3402

Alva PME, Pointing SB, Pena-Muralla R, Hyde KD (2002) Do sea grasses harbour endophytes? In: Hyde K (ed) Fungi in marine environments. Fungal Diversity Press, Hong Kong, pp 167–178

Amaral-Zettler LA, Zettler ER, Mincer TJ (2020) Ecology of the plastisphere. Nat Rev Microbiol 18:139–151

Amend A, Burgaud G, Cunliffe M, Edgcomb VP, Ettinger CL, Gutiérrez MH, Heitman J, Hom EFY, Ianiri G, Jones AC, Kagami M, Picard KT, Quandt CA, Raghukumar S, Riquelme M, Stajich J, Vargas-Muniz J, Walker AK, Yarden O, Gladfelter AS (2019) Fungi in the marine environment: open questions and unsolved problems. mBio 10:e01189–e01118

Arfi Y, Marchand C, Wartel M, Record E (2012a) Fungal diversity in anoxic-sulfidic sediments in a mangrove soil. Fungal Ecol 5:282–285

Arfi Y, Buée M, Marchand C, Levasseur A, Record E (2012b) Multiple marker pyrosequencing reveals highly diverse and host-specific fungal communities on the mangrove trees *Avicennia marina* and *Rhizophora stylosa*. FEMS Microbiol Ecol 79:433–444

Arifeen MZ, Ma YN, Xue YR, Liu CH (2020) Deep-sea fungi could be the new arsenal for bioactive compounds. Mar Drug 18:9

Atkinson S, Williams P (2009) Quorum sensing and social networking in the microbial world. J R Soc Interface 6:959–978

Bachmann BO, Ravel J (2009) Methods for *in silico* prediction of microbial polyketide and nonribosomal peptide biosynthetic pathways from DNA sequence data. Method Enzymol 458:181–217

Banat IM, Satpute SK, Cameotra SS, Patil R, Nyayanit NV (2014) Cost effective technologies and renewable substrates for biosurfactants' production. Front Microbiol 5:697

Bao E, Lan L (2017) HALC: high throughput algorithm for long read error correction. BMC Bioinformatics 18:204

Baral H, Rämä T (2015) Morphological update on *Calycina* marina (*Pezizellaceae*, *Helotiales*, *Leotiomycetes*), a new combination for *Laetinaevia marina*. Bot Mar 58:523–534

Barata M (2002) Fungi on the halophyte *Spartina maritima* in salt marshes. In: Hyde K (ed) Fungi in marine environments. Fungal Diversity Press, Hong Kong, pp 179–193

Barghoorn ES, Linder DH (1944) Marine fungi: their taxonomy and biology. Farlowia 1:395–467

Barkal LJ, Theberge AB, Guo CJ, Spraker J, Rappert L, Berthier J, Brakke KA, Wang CCC, Beebe DJ, Keller NP, Berthier E (2016) Microbial metabolomics in open microscale platforms. Nat Commun 7:10610

Bass D, Howe A, Brown N, Barton H, Demidova M, Michelle H, Li L, Sanders H, Watkinson SC, Willcock S, Richards TA (2007) Yeast forms dominate fungal diversity in the deep oceans. Proc Royal Soc B 274:3069–3077

Bendtsen J, Mortensen J, Rysgaard S (2014) Seasonal surface layer dynamics and sensitivity to runoff in a high Arctic fjord (Young sound/Tyrolerfjord, 74°N). J Geophys Res Oceans 119: 6461–6478

Bengtson S, Ivarsson M, Astolfo A, Belivanova V, Broman C, Marone F, Stampanoni M (2014) Deep-biosphere consortium of fungi and prokaryotes in Eocene subseafloor basalts. Geobiology 12:489–496

Besitulo A, Moslem MA, Hyde KD (2010) Occurrence and distribution of fungi in a mangrove forest on Siargao Island, Philippines. Bot Mar 53:535–543

Bhakuni DS, Rawat DS (2005) Bioactive metabolites of marine algae, fungi and bacteria. In: Bioactive marine natural products. Anamaya Publishers and Springer, New Delhi, pp 1–25

Biabani MAF, Laatsch H (1998) Advances in chemical studies on low-molecular weight metabolites of marine fungi. J Prakt Chem 340:589–607

Blin K, Medema MH, Kazempour D, Fischbach MA, Breitling R, Takano E, Weber T (2013) antiSMASH 2.0 -a versatile platform for genome mining of secondary metabolite producers. Nucleic Acids Res 41:204–212

Bluhm BA, Kosobokova KN, Carmack EC (2015) A tale of two basins: an integrated physical and biological perspective of the deep Arctic Ocean. Prog Oceanogr 139:89–121

Bluhm BA, Swadling KM, Gradinger R (2017) Sea ice as a habitat for macrograzers. In: Thomas DN (ed) Sea Ice, 3rd edn. Wiley-Blackwell, Oxford, pp 394–414

Blunt JW, Copp BR, Keyzers RA, Munro MHG, Prinsep MR (2016) Marine natural products. Nat Prod Rep 33:382–431

Blunt JW, Carroll AR, Copp BR, Davis RA, Keyzers RA, Prinsep MR (2018) Marine natural products. Nat Prod Rep 35:8–53

Bochdansky AB, Clouse MA, Herndl GJ (2017) Eukaryotic microbes, principally fungi and labyrinthulomycetes, dominate biomass on bathypelagic marine snow. ISME J 11:362–373

Boddy CN (2014) Bioinformatics tools for genome mining of polyketide and non-ribosomal peptides. J Ind Microbiol Biotechnol 41:443–450

Bok JW, Ye R, Clevenger KD, Mead D, Wagner M, Krerowicz A, Albright JC, Goering AW, Thomas PM, Kelleher NL, Keller NP, Wu CC (2015) Fungal artificial chromosomes for mining of the fungal secondary metabolome. BMC Genomics 16:343

Bonugli-Santos RC, dos Santos Vasconcelos MR, Passarini MR, Vieira GA, Lopes VC, Mainardi PH, dos Santos JA, de Azevedo DL, Otero IVR, da Silva Yoshida AM, Feitosa VA, Pessoa A, Sette LD (2015) Marine-derived fungi: diversity of enzymes and biotechnological applications. Front Microbiol 6:269

Boonyuen N, Chuaseeharonnachai C, Suetrong S, Sri-Indrasutdhi V, Sivichai S, Jones EB, Pang KL (2011) Savoryellales (Hypocreomycetidae, Sordariomycetes): a novel lineage of aquatic ascomycetes inferred from multiple-gene phylogenies of the genera *Ascotaiwania*, *Ascothailandia*, and *Savoryella*. Mycologia 103:1351–1371

Bowman JS, Rasmussen S, Blom N, Deming JW, Rysgaard S, Sicheritz-Ponten T (2012) Microbial community structure of Arctic multiyear sea ice and surface seawater by 454 sequencing of the 16S RNA gene. ISME J 6:11–20

Brakhage AA (2013) Regulation of fungal secondary metabolism. Nat Rev Microbiol 11:21–32

Buchan A, Lyons SY, Moreta JIL, Moran MA (2002) Analysis of internal transcribed spacer (ITS) regions of rRNA genes in fungal communities in a southeastern U.S. salt marsh. Microb Ecol 43:329–340

Bucher VVC, Pointing SB, Hyde KD, Reddy CA (2004) Production of wood decay enzymes, loss of mass, and lignin solubilization in wood by diverse tropical freshwater fungi. Microb Ecol 48: 331–337

Bugni TS, Ireland CM (2004) Marine-derived fungi: a chemically and biologically diverse group of microorganisms. Nat Prod Rep 21:143–163

Burgaud G, Calvez T, Arzur D, Vandenkoornhuyse P, Barbier G (2009) Diversity of culturable marine filamentous fungi from deep-sea hydrothermal vents. Environ Microbiol 11:588–600

Burgaud G, Arzur D, Durand L, Cambon-Bonavita MA, Barbier G (2010) Marine culturable yeasts in deep-sea hydrothermal vents: species richness and association with fauna. FEMS Microb Ecol 73:121–133

Burgaud G, Woehlke S, Redou V, Orsi W, Beaudoin D, Barbier G, Biddle JF, Edgcomb VP (2013) Deciphering the presence and activity of fungal communities in marine sediments using a model estuarine system. Aquat Microb Ecol 70:45–62

Burgaud G, Hué NTM, Arzur D, Coton M, Perrier-Cornet JM, Jebbar M, Barbier G (2015) Effects of hydrostatic pressure on yeasts isolated from deep-sea hydrothermal vents. Res Microbiol 166: 700–709

Calado ML, Barata M (2012) Salt marsh fungi. In: EBG J, Pang K-L (eds) Marine fungi and fungal-like organisms. De Gruyter, Berlin, pp 345–381

Calado MD, Carvalho L, Pang KL, Barata M (2015) Diversity and ecological characterization of sporulating higher filamentous marine fungi associated with *Spartina maritima* (Curtis) Fernald in two Portuguese salt marshes. Microb Ecol 70:612–633

Calvo AM, Wilson RA, Bok JW, Keller NP (2002) Relationship between secondary metabolism and fungal development. Microbiol Mol Biol Rev 66:447–459

Cantarel BL, Korf I, Robb SM, Parra G, Ross E, Moore B, Holt C, Alvarado AS, Yandell M (2008) MAKER: an easy-to-use annotation pipeline designed for emerging model organism genomes. Genome Res 18:188–196

Caron DA, Gast RJ (2010) Heterotrophic protists associated with sea ice. In: Thomas DN, Dieckmann GS (eds) Sea Ice, 2nd edn. Wiley-Blackwell, Oxford, pp 327–356

Chakraborty M, Baldwin-Brown JG, Long AD, Emerson JJ (2016) Contiguous and accurate de novo assembly of metazoan genomes with modest long read coverage. Nucleic Acids Res 44: e147

Chen XW, Li CW, Cui CB, Hua W, Zhu TJ, Gu QQ (2014) Nine new and five known polyketides derived from a deep sea-sourced *aspergillus* sp. 16-02-1. Mar Drugs 12:3116–3137

Chrismas N, Cunliffe M (2020) Depth-dependent mycoplankton glycoside hydrolase gene activity in the open ocean - evidence from the Tara oceans eukaryote metatranscriptomes. ISME J 14: 2361–2365

Ciobanu MC, Burgaud G, Dufresne A, Breuker A, Rédou V, Maamar SB, Vandenkoornhuyse P, Alain K (2014) Microorganisms persist at record depths in the subseafloor of the Canterbury basin. ISME J 8:1370–1380

Collins RE, Rocap G, Deming JW (2010) Persistence of bacterial and archaeal communities in sea ice through an Arctic winter. Environ Microbiol 12:1828–1841

Comeau AM, Vincent WF, Bernier L, Lovejoy C (2016) Novel chytrid lineages dominate fungal sequences in diverse marine and freshwater habitats. Sci Rep 6:30120

Cosse A, Potin P, Leblanc C (2009) Patterns of gene expression induced by Oligoguluronates reveal conserved and environment-specific molecular defense responses in the brown alga *Laminaria digitata*. New Phytol 182:239–250

Cotton AD (1907) Notes on marine pyrenomycetes. Trans Brit Mycol Soc 3:92–99

Cox GFN, Weeks WF (1983) Equations for determining the gas and brine volumes in sea-ice samples. J Glaciol 29:306–316

Cruz-Morales P, Martínez-Guerrero CE, Morales-Escalante MA, Yáñez-Guerra LA, Kopp JF, Feldmann J, Ramos-Aboites HE, Barona-Gomez F (2015) Recapitulation of the evolution of biosynthetic gene clusters reveals hidden chemical diversity on bacterial genomes. bioRxiv:020503

Culligan EP, Sleator RD, Marchesi JR, Hill C (2013) Metagenomics and novel gene discovery: promise and potential for novel therapeutics. Virulence 5:1–14

Cunliffe M, Hollingsworth A, Bain C, Sharma V, Taylor JD (2017) Algal polysaccharide utilisation by saprotrophic planktonic marine fungi. Fungal Ecol 30:135–138

Daletos G, Ebrahim W, Ancheeva E, El-Neketi M, Song W, Lin W, Proksch P (2018) Natural products from deep-sea-derived fungi - a new source of novel bioactive compounds? Curr Med Chem 25:186–207

Damare S, Raghukumar C, Raghukumar S (2006) Fungi in deep-sea sediments of the central Indian Basin. Deep Sea Res Part I Oceanogr Res Pap 53:14–27

Das P, Yang XP, Ma LZ (2014) Analysis of biosurfactants from industrially viable pseudomonas strain isolated from crude oil suggests how rhamnolipids congeners affect emulsification property and antimicrobial activity. Front Microbiol 5:696

Dasanayaka SAHK, Nong XH, Liang X, Liang JQ, Amin M, Qi SH (2020) New dibenzodioxocinone and pyran-3, 5-dione derivatives from the deep-sea-derived fungus *Penicillium canescens* SCSIO z053. J Asian Nat Prod Res 22:338–345

De Tender C, Schlundt C, Devriese LI, Mincer TJ, Zettler ER, Amaral-Zettler LA (2017) A review of microscopy and comparative molecular-based methods to characterize "Plastisphere" communities. Anal Methods 9:2132–2143

Debbab A, Aly A, Proksch P (2012) Endophytes and associated marine derived fungi—ecological and chemical perspectives. Fungal Diver 57:45–83

Decker RJ, Garbary DJ (2005) Ascophyllum and its symbionts. VIII. Interactions among *Ascophyllum nodosum* (Phaeophyceae), *Mycophycias ascophylli* (ascomycetes) and *Elachista fucicola* (Phaeophyceae). Algae 20:363–368

Dekov V, Bindi L, Burgaud G, Petersen S, Asael D, Rédou V, Fouquet Y, Pracejus B (2013) Inorganic and biogenic as-sulfide precipitation at seafloor hydrothermal fields. Mar Geol 342: 28–38

Demain AL (2014) Importance of microbial natural products and the need to revitalize their discovery. J Ind Microbiol Biot 41:185–201

Devarajan P, Suryanarayanan T, Geetha V (2002) Endophytic fungi associated with the tropical seagrass *Halophila ovalis* (Hydrocharitaceae). Ind J Mar Sci 31:73–74

Deveau A, Bonito G, Uehling J, Paoletti M, Becker M, Bindschedler S, Hacquard S, Hervé V, Labbé J, Lastovetsky OA, Mieszkin S, Millet LJ, Vajna B, Junier P, Bonfante P, Krom BP, Olsson S, Dirk van Elsas J, Wick LY (2018) Bacterial-fungal interactions: ecology, mechanisms and challenges. FEMS Microbiol Rev 42:335–352

Ding H, Zhang D, Zhou B, Ma Z (2017) Inhibitors of BRD4 protein from a marine-derived fungus *Alternaria* sp. NH-F6. Mar Drugs 15:76

Doggett MS, Porter D (1994) Further evidence for host-specific variants in *Zygorhizidium planktonicum*. Mycologia 87:161–171

Drake H, Ivarsson M (2017) The role of anaerobic fungi in fundamental biogeochemical cycles in the deep biosphere. Fungal Biol Rev 32:20–25

Drake H, Ivarsson M, Bengtson S, Heim C, Siljeström S, Whitehouse MJ, Broman C, Belivanova V, Astrom ME (2017) Anaerobic consortia of fungi and sulfate reducing bacteria in deep granite fractures. Nat Commun 8:1–9

Du L, Li D, Zhu T, Cai S, Wang F, Xiao X, Gu Q (2009) New alkaloids and diterpenes from a deep ocean sediment derived fungus *Penicillium* sp. Tetrahedron 65:1033–1039

Duan YB, Xie ND, Song ZQ, Ward CS, Yung CM, Hunt DE, Johnson ZI, Wang GY (2018) High-resolution time-series reveals seasonal patterns of planktonic fungi at a temperate coastal ocean site (Beaufort, North Carolina, USA). Appl Environ Microbiol 84:1–14

Edgcomb VP, Beaudoin D, Gast R, Biddle J, Teske A (2010) Marine subsurface eukaryotes: the fungal majority. Environ Microbiol 13:172–183

Egan S, Harder T, Burke C, Steinberg P, Kjelleberg S, Thomas T (2013) The seaweed holobiont: understanding seaweed–bacteria interactions. FEMS Microbiol Rev 37:462–476

Ettinger CL, Eisen JA (2019) Characterization of the mycobiome of the seagrass, *Zostera marina*, reveals putative associations with marine chytrids. Front Microbiol 10:2476

Fan Z, Sun ZH, Liu Z, Chen YC, Liu HX, Li HH, Zhang WM (2016) Dichotocejpins A–C: new diketopiperazines from a deep-sea-derived fungus *Dichotomomyces cejpii* FS110. Mar Drugs 14:164

Fell JW, Newell SY (1998) Biochemical and molecular methods for the study of marine fungi. In: Cooksey K (ed) Molecular approaches to the study of the ocean. Chapman & Hall, London, pp 259–283

Fell JW, Statzel AC, Hunter IL, Phaff HJ (1969) *Leucosporidium* gen. Nov. the heterobasidiomycetous stage of several yeasts of the genus *Candida*. Antonie Van Leeuwenhoek 35:433–462

Fenical W, Jensen PR (1993) Marine microorganisms: a new biomedical resource. In: Attaway DH, Zaborsky OR (eds) Marine biotechnology, Pharmaceutical and bioactive natural products, vol 1. Plenum, New York, pp 419–457

Finn RD, Clements J, Eddy SR (2011) HMMER web server: interactive sequence similarity searching. Nucleic Acids Res 39:W29–W37

Finn RD, Bateman A, Clements J, Coggill P, Eberhardt RY, Eddy SR, Heger A, Hetherington K, Holm L, Mistry J, Sonnhammer ELL, Tate J, Punta M (2014) Pfam: the protein families database. Nucleic Acids Res 42:222–230

Firth E, Carpenter SD, Sørensen HL, Collins RE, Deming JW (2016) Bacterial use of choline to tolerate salinity shifts in sea-ice brines. Elementa Science of the Anthropocene 4:000120

Flewelling AJ, Ellsworth KT, Sanford J, Forward E, Johnson JA, Gray CA (2013a) Macroalgal endophytes from the Atlantic Coast of Canada: a potential source of antibiotic natural products? Microorganisms 1:175–187

Flewelling AJ, Johnson JA, Gray CA (2013b) Isolation and bioassay screening of fungal endophytes from North Atlantic marine macroalgae. Bot Mar 56:287–297

Fraser CI, Morrison AK, Hogg AM, Macaya EC, van Sebille E, Ryan PG, Padovan A, Jack C, Valdivia N, Waters JM (2018) Antarctica's ecological isolation will be broken by storm-driven dispersal and warming. Nat Clim Chang 8:704–708

Fredimoses M, Zhou X, Ai W, Tian X, Yang B, Lin X, Xian JY, Liu Y (2015) Westerdijkin a, a new hydroxyphenylacetic acid derivative from deep sea fungus *aspergillus westerdijkiae* SCSIO 05233. Nat Prod Res 29:158–162

Fries N (1979) Physiological characteristics of *Mycosphaerella ascophylli*, a fungal endophyte of the marine brown alga *Ascophyllum nodosum*. Physiol Plant 45:117–121

Gaboyer F, Burgaud G, Edgcomb V (2019) The deep subseafloor and biosignatures. In: Cavalazzi B, Westall F (eds) Biosignatures for astrobiology. Springer, Cham, pp 87–109

Gadanho M, Sampaio JP (2005) Occurrence and diversity of yeasts in the mid-Atlantic ridge hydrothermal fields near the Azores archipelago. Microb Ecol 50:408–417

Galagan JE, Calvo SE, Borkovich KA, Selker EU, Read ND et al (2003) The genome sequence of the filamentous fungus *Neurospora crassa*. Nature 422:859–868

Galindo LJ, López-Garcia P, Torroella G, Karpov S, Moreira D (2020) Phylogenomics of a new fungal phylum reveals multiple waves of reductive evolution across Holomycota. BioRxiv. https://doi.org/10.1101/2020.11.19.389700

Ganesh Kumar A, Balamurugan K, Vijaya Raghavan R, Dharani G, Kirubagaran R (2019) Studies on the antifungal and serotonin receptor agonist activities of the secondary metabolites from piezotolerant deep-sea fungus *Ascotricha* sp. Mycology 10:92–108

Gao Z, Johnson ZI, Wang GY (2010) Molecular characterization of the spatial diversity and novel lineages of mycoplankton in Hawaiian coastal waters. ISME J 4:111–120

Gao XW, Liu HX, Sun ZH, Chen YC, Tan YZ, Zhang WM (2016) Secondary metabolites from the deep-sea derived fungus *Acaromyces ingoldii* FS121. Molecules 21:371

Garbary DJ, London JF (1995) The *Ascophyllum Polysiphonial Mycosphaerella* symbiosis V. fungal infection protects *a. nosodum* from desiccation. Bot Mar 38:529–534

Garbary DJ, MacDonald KA (1995) The *Ascophyllum/Polysiphonia/Mycosphaerella* symbiosis. IV. Mutualism in the *Ascophyllum/Mycosphaerella* interaction. Bot Mar 38:221–225

García-Lepe R, Nuero OM, Reyes F, Santamaría F (1997) Lipases in autolysed cultures of filamentous fungi. Lett Appl Microbiol 25:127–130

Geyer R, Jambeck JR, Law KL (2017) Production, use, and fate of all plastics ever made. Sci Adv 3: e1700782

Gladfelter AS, James TY, Amend AS (2019) Marine fungi. Curr Biol 29:191–195

Godinho VM, Furbino LE, Santiago IF, Pellizzari FM, Yokoya NS, Pupo D, Rosa LH (2013) Diversity and bioprospecting of fungal communities associated with endemic and cold-adapted macroalgae in Antarctica. ISME J 7:1434–1451

Godinho VM, de Paula MTR, Silva DAS, Paresque K, Martins AP, Colepicolo P, Rosa CA, Rosa LH (2019) Diversity and distribution of hidden cultivable fungi associated with marine animals of Antarctica. Fungal Biol 123:507–516

Goffeau A (2000) Four years of post-genomic life with 6,000 yeast genes. FEBS Lett 480:37–41

Goffeau A, Barrell BG, Bussey H, Davis RW, Dujon B, Feldmann H, Galibert F, Hoheisel JD, Jacq C, Johnston M, Louis EJ, Mewes HW, Murakami Y, Philippsen P, Tettelin H, Oliver SG (1996) Life with 6000 genes. Science 274:546–657

Gomes NGM, Lefranc F, Kijjoa A, Kiss R (2015) Can some marine-derived fungal metabolites become actual anticancer agents? Mar Drugs 13:3950–3991

González-Medina M, Prieto-Martínez FD, Naveja JJ, Méndez-Lucio O, El-Elimat T, Pearce CJ, Oberlies NH, Figueroa M, Medina-Franco JL (2016) Chemoinformatic expedition of the chemical space of fungal products. Future Med Chem 8:1399–1412

Gotz S, Garcia-Gomez JM, Terol J, Williams TD, Nagaraj SH, Nueda MJ, Robles M, Talon M, Dopazo J, Conesa A (2008) High-throughput functional annotation and data mining with the Blast2GO suite. Nucleic Acids Res 36:3420–3435

Gradinger R (2009) Sea-ice algae: major contributors to primary production and algal biomass in the Chukchi and Beaufort seas during May/June 2002. Deep Sea Res Part II: Top Stud Oceanogr 56:1201–1212

Grigoriev IV, Nikitin R, Haridas S, Kuo A, Ohm R, Otillar R, Riley R, Salamov A, Zhao X, Korzeniewski F, Smirnova T, Nordberg H, Dubchak I, Shabalov I (2014) MycoCosm portal: gearing up for 1000 fungal genomes. Nucleic Acids Res 42:D699–D704

Gutiérrez MH, Pantoja S, Quiñones RA, González RR (2010) First record of filamentous fungi in the coastal upwelling ecosystem off Central Chile. Gayana 74:66–73

Gutiérrez MH, Pantoja S, Tejos E, Quiñones RA (2011) The role of fungi in processing marine organic matter in the upwelling ecosystem off Chile. Mar Biol 158:205–219

Gutiérrez MH, Jara AM, Pantoja S (2016) Fungal parasites infect marine diatoms in the upwelling ecosystem of the Humboldt current system off Central Chile. Environ Microbiol 18:1646–1653

Hackstein JHP, Baker SE, van Hellemond JJ, Tielens AGM (2019) Hydrogenosomes of anaerobic fungi: an alternative way to adapt to anaerobic environments. In: Tachezy J (ed) Hydrogenosomes and mitosomes: mitochondria of anaerobic eukaryotes. Springer, Cham, pp 159–175

Hansen FT, Gardiner DM, Lysøe E, Fuertes PR, Tudzynski B, Wiemann P, Sondergaard TE, Giese H, Bordersen DE, Sorensen JL (2014) An update to polyketide synthase and nonribosomal synthetase genes and nomenclature in fusarium. Fungal Genet Biol 75:20–29

Harvey JBJ, Goff LJ (2010) Genetic covariation of the marine fungal symbiont *Haloguignardia irritans* (Ascomycota, Pezizomycotina) with its algal hosts *Cystoseira* and *Halidrys* (Phaeophyceae, Fucales) along the west coast of North America. Fungal Biol 114:82–95

Hassani MA, Durán P, Hacquard S (2018) Microbial interactions within the plant holobiont. Microbiome 6:58

Hassett BT, Gradinger R (2016) Chytrids dominate arctic marine fungal communities. Environ Microbiol 18:2001–2009

Hassett BT, Ducluzeau AL, Collins RE, Gradinger R (2017) Spatial distribution of aquatic marine fungi across the western arctic and sub-arctic. Environ Microbiol 19:475–484

Hassett BT, Borrego EJ, Vonnahme TR, Rama T, Kolomiets MV, Gradinger R (2019) Arctic marine fungi: biomass, functional genes, and putative ecological roles. ISME J 13:1484–1496

Hassett BT, Vonnahme TR, Peng X, Jones EBG, Heuze C (2020) Global diversity and geography of planktonic marine fungi. Bot Mar 63:121–139

Hautbergue T, Jamin EL, Debrauwer L, Puel O, Oswald IP (2018) From genomics to metabolomics, moving toward an integrated strategy for the discovery of fungal secondary metabolites. Nat Prod Rep 35:147–173

Hayashi H, Matsumoto H, Akiyama K (2004) New insecticidal compounds, communesins C, D and E, from *Penicillium expansum* link MK-57. Biosci Biotechnol Biochem 68:753–756

Hibbett DS, Binder M (2001) Evolution of marine mushrooms. Biol Bull 201:319–322

Hirayama H, Abe M, Miyazaki J, Sakai S, Takai K (2015) Data report: cultivation of microorganisms from basaltic rock and sediment cores from the north pond on the western flank of the mid-Atlantic ridge. IODP expedition 336. Sci Technol (JAMSTEC) 2:15

Hoang TPT, Roullier C, Boumard MC, Robiou du Pont T, Nazih H, Gallard JF, Pouchus YF, Beniddir MA, Grovel O (2018) Metabolomics-driven discovery of meroterpenoids from a mussel-derived *Penicillium ubiquetum*. J Nat Prod 81:2501–2511

Horner R, Schrader GC (1982) Relative contributions of ice algae, phytoplankton, and benthic microalgae to primary production in nearshore regions of the Beaufort Sea. Arctic 35:485–503

Hunter S, Jones P, Mitchell A, Apweiler R, Attwood TK, Bateman A, Bernard T, Binns D, Bork P, Burge S, de Castro E, Coggill P, Corbett M, Das U, Dougherty L, Duquenne L, Finn RD, Fraser M, Gough J, Haft D, Hulo N, Kahn D, Kelly E, Letunic I, Lonsdale D, Lopez R, Madra M, Maslen J, McAnulla C, McDowall J, McMenamin C, Mi H, Mutowo-Muellnet P, Mulder N, Natale D, Orengo C, Pesseat S, Punta M, Quinn AF, Rivoire C, Sangrador-Vegas A, Selengut JD, Sigrist CJA, Scheremetjew M, Tate J, Thimmajanarthanan M, Thomas PD, Wu CH, Yeats C, Yong SY (2012) InterPro in 2011: new developments in the family and domain prediction database. Nucleic Acids Res 40:306–312

Hyde KD (1992) Fungi from decaying intertidal fronds of *Nypa fruticans*, including three new genera and four new species. Bot J Linn Soc 110:95–110

Hyde KD, Alias SA (2000) Biodiversity and distribution of fungi associated with decomposing Nypa palm. Biodivers Conserv 9:393–402

Hyde KDJE, Liu JK, Ariyawansha H, Boehm E, Boonmee S, Braun U, Chomnunti P, Crous PW, Dai D, Diederich P, Dissanayake A, Doilom M, Doveri F, Hongsanan S, Jayawardena R, Lawrey JD, Li YM, Liu YX, Lücking R, Monkai J, Muggia L, Nelsen MP, Pang KL, Phookamsak R, Senanayake I, Shearer CA, Suetrong S, Tanaka K, Thambugala KM, Wijayawardene N, Wikee S, Wu HX, Zhang Y, Aguirre-Hudson B, Alias SA, Aptroot A, Bahkali AH, Bezerra JL, Bhat JD, Camporesi E, Chukeatirote E, Gueidan C, Hawksworth DL, Hirayama K, De Hoog S, Kang JC, Knudsen K, Li WJ, Li XH, Liu ZY, Mapook A, McKenzie EHC, Miller AN, Mortimer PE, Phillips AJL, Raja HA, Scheuer C, Schumm F, Taylor J, Tian Q, Tibpromma S, Wanasinghe DN, Wang Y, Xu JC, Yacharoen S, Yan JY, Zhang M (2013) Families of dothideomycetes. Fungal Divers 63:1–313

Hynes MJ, Murray SL, Duncan A, Khew GS, Davis MA (2006) Regulatory genes controlling fatty acid catabolism and peroxisomal functions in the filamentous fungus *aspergillus nidulans*. Eukaryot Cell 5:794–805

Inagaki F, Hinrichs KU, Kubo Y, Boles MW, Heuer VB, Long WL, Hoshino T, Ijiri A, Imachi H, Ito M, Kaneko M, Lever MA, Lin YS, Methé BA, Morita A, Morono Y, Tanikawa W, Bihan M, Bowden SA, Elvert M, Glombitza C, Gross D, Harrington GJ, Hori T, Li K, Limmer D, Liu CH, Murayama M, Ohkouchi N, Ono S, Park YS, Phillips SC, Prieto-Mollar X, Pukey M, Riedinger N, Sanada Y, Sauvage J, Snyder G, Susilawati R, Takan OY, Tasumi E, Terada T, Tomaru H, Trembath-Reichert E, Wang DT, Yamada Y (2015) Exploring deep marine microbial life in coal-bearing sediment down to ~2.5 km below the ocean floor. Science 349:420–424

Ivarsson M, Bengtson S, Belivanova V, Stampanoni M, Marone F, Tehler A (2012) Fossilized fungi in subseafloor Eocene basalts. Geology 40:163–166

Ivarsson M, Bengtson S, Skogby H, Lazor P, Broman C, Belivanova V, Marone F (2015a) A fungal-prokaryotic consortium at the basalt-zeolite interface in subseafloor igneous crust. PLoS One 10:e0140106

Ivarsson M, Peckmann J, Tehler A, Broman C, Bach W, Behrens K, Reitner J, Bottcher ME, Ivarsson LN (2015b) Zygomycetes in vesicular basanites from Vesteris Seamount, Greenland Basin–a new type of cryptoendolithic fungi. PLoS One 10:e0133368
Ivarsson M, Bengtson S, Neubeck A (2016) The igneous oceanic crust – Earth's largest fungal habitat? Fungal Ecol 20:249–255
Jacquin J, Cheng J, Odobel C, Pandin C, Conan P, Pujo-Pay M, Barbe V, Meistertzheim L, Ghiglione JF (2019) Microbial ecotoxicology of marine plastic debris: a review on colonization and biodegradation by the "plastisphere". Front Microbiol 10:865
Jambeck JR, Geyer R, Wilcox C, Siegler TR, Perryman M, Andrady A, Law KL (2015) Plastic waste inputs from land into the ocean. Science 347:768–771
Jebaraj CS, Raghukumar C (2010) Nitrate reduction by fungi in marine oxygen-depleted laboratory microcosms. Bot Mar 53:469–474
Jeffries TC, Curlevski NJ, Brown MV, Harrison DP, Doblin MA, Petrou K, Ralph PJ, Seymour JR (2016) Partitioning of fungal assemblages across different marine habitats. Environ Microbiol Rep 8:235–238
Johnson TW Jr, Sparrow FK Jr (1961) Fungi in oceans and estuaries. Weinheim, J. Cramer, 668 pp
Johnson MD, Edwards BR, Beaudoin DJ, Van Mooy BAS, Vardi A (2020) Nitric oxide mediates oxylipin production and grazing defense in diatoms. Environ Microbiol 22:629–645
Jones EBG (2000) Marine fungi: some factors influencing biodiversity. Fungal Divers 4:53–73
Jones EBG (2011) Fifty years of marine mycology. Fungal Divers 50:73–112
Jones EBG, Pang KL (2012) Marine fungi and fungal-like organisms (marine and freshwater botany). Walter de Gruyter, Berlin, p 528
Jones EBG, Suetrong S, Sakayaroj J, Bahkali AH, Abdel-Wahab MA, Boekhout T, Pang KL (2015) Classification of marine Ascomycota, Basidiomycota, Blastocladiomycota and Chytridiomycota. Fungal Divers 3:1–72
Jones EBG, Pang KL, Abdel-Wahab MA, Scholz B, Hyde KD, Boekhout T, Ebel R, Rateb ME, Henderson L, Sakayaroj J, Suetrong S, Dayarathne MC, Kumar V, Raghukumar S, Sridhar KR, Bahkali AHA, Gleason FH, Norphanphoun C (2019) An online resource for marine fungi. Fungal Divers 96:347–433
Kaiser K, Canedo-Oropeza M, McMahon R, Amon RMW (2017) Origins and transformations of dissolved organic matter in large Arctic rivers. Sci Rep 7:13064
Kanehisa M, Goto S, Furumichi M, Tanabe M, Hirakawa M (2010) KEGG for representation and analysis of molecular networks involving diseases and drugs. Nucleic Acids Res 38:D355–D360
Keller NP, Turner G, Bennett JW (2005) Fungal secondary metabolism - from biochemistry to genomics. Nat Rev Microbiol 3:937–947
Kettner MT, Rojas-Jimenez K, Oberbeckmann S, Labrenz M, Grossart HP (2017) Microplastics alter composition of fungal communities in aquatic ecosystems. Environ Microbiol 19:4447–4459
Khaldi N, Seifuddin FT, Turner G, Haft D, Nierman WC, Wolf KH, Fedorova ND (2010) SMURF: genomic mapping of fungal secondary metabolite clusters. Fungal Genet Biol 47:736–741
Kim Y, Islam N, Moss BJ, Nandakumar MP, Marten MR (2011) Autophagy induced by rapamycin and carbon-starvation have distinct proteome profiles in *aspergillus nidulans*. Biotechnol Bioeng 108:2705–2715
Kis-Papo T, Weig AR, Riley R, Peršoh D, Salamov A, Sun H, Lipzen A, Wasser SP, Rambold G, Grigoriev IV, Nevo E (2014) Genomic adaptations of the halophilic Dead Sea filamentous fungus *Eurotium rubrum*. Nat Commun 5:1–8
Knudsen T, Knudsen B (2013) CLC genomics Benchwork 6. CLC Genomic Work
Kohlmeyer J, Kohlmeyer E (1979) Marine mycology: the higher Fung. Acad Press, New York
Kohlmeyer J, Volkmann-Kohlmeyer B (2003) Marine ascomycetes from algae and animal hosts. Bot Mar 46:285–306
Kõljalg U, Nilsson RH, Abarenkov K, Tedersoo L, Taylor AFS, Bahram M, Bates ST, Bruns TD, Bengtsson-Palme J, Callaghan TM, Douglas B, Drenkhan T, Eberhardt U, Duenas M, Grebnc T,

Griffith GW, Hartmann M, Kirk PM, Kohout P, Larsson E, Lindahl BD, Lücking R, Martin MP, Matheny PB, Nguyen NH, Niskanen T, Oja J, Peay KG, Pintner U, Peterson M, Poldmaa K, Saag L, Saar I, Schubler A, Scott JA, Senes C, Smith ME, Suija A, Taylor DL, Telleria MT, Wiss M, Larsson KH (2013) Towards a unified paradigm for sequence based identification of fungi. Mol Ecol 22:5271–5277

Koren S, Walenz BP, Berlin K, Miller JR, Bergman NH, Phillippy AM (2017) Canu: scalable and accurate long-read assembly via adaptive k-mer weighting and repeat separation. Genome Res 27:722–736

Kramer A, Beck HC, Kumar A, Kristensen LP, Imhoff JF, Labes A (2015) Proteomic analysis of anti-cancerous scopularide production by a marine *microascus brevicaulis* strain and its UV-mutant. PLoS One 10:e0140047

Kumar A, Congiu L, Lindstrom L, Piiroinen S, Vidotto M, Grapputo A (2014) Sequencing, de novo assembly and annotation of the Colorado potato beetle, *Leptinotarsa decemlineata*, transcriptome. PLoS One 9:e86012

Kumar A, Henrissat B, Arvas M, Syed MF, Thieme N, Bnz JP, Sørensen JL, Record E, Pöggeler S, Kempken F (2015) De novo assembly and genome analyses of the marine-derived *Scopulariopsis brevicaulis* strain LF580 unravels life-style traits and anticancerous scopularide biosynthetic gene cluster. PLoS One 10:e0140398

Kumar A, Sørensen JL, Hansen FT, Arvas M, Syed MF, Hassan L, Bnz JP, Record E, Henrissat B, Pöggeler S, Kempken F (2018) Genome sequencing and analyses of two marine fungi from the North Sea unraveled a plethora of novel biosynthetic gene clusters. Sci Rep 8:1–16

L'Haridon S, Markx GH, Ingham CJ, Paterson L, Duthoit F, Le Blay G (2016) New approaches for bringing the uncultured into culture. In: Stal LJ, Cretoiu MS (eds) The marine microbiome, an untapped source of biodiversity and biotechnological potential. Springer, Basel, pp 401–434

Lacerda ALF, Proietti MC, Secchi ER, Taylor JD (2020) Diverse groups of fungi are associated with plastics in the surface waters of the Western South Atlantic and the Antarctic peninsula. Mol Ecol 29:1903–1918

Lai X, Cao L, Tan H, Fang S, Huang Y, Zhou S (2007) Fungal communities from methane hydrate-bearing deep-sea marine sediments in South China Sea. ISME J 1:756–762

Lambert BS, Raina JB, Fernandez VI, Rinke C, Siboni N, Rubino F, Stocker R (2017) A microfluidics-based in situ chemotaxis assay to study the behaviour of aquatic microbial communities. Nat Microbiol 2:1344–1349

Le Calvez T, Burgaud G, Mahe S, Barbier G, Vandenkoornhuyse P (2009) Fungal diversity in deep-sea hydrothermal ecosystems. Appl Environ Microbiol 75:6415–6421

Lenassi M, Gostinčar C, Jackman S, Turk M, Sadowski I, Nislow C, Jones S, Birol I, Cimerman NG, Plemenitaš A (2013) Whole genome duplication and enrichment of metal cation transporters revealed by de novo genome sequencing of extremely halotolerant black yeast *Hortaea werneckii*. PLoS One 8:e71328

Levasseur A, Drula E, Lombard V, Coutinho PM, Henrissat B (2013) Expansion of the enzymatic repertoire of the CAZy database to integrate auxiliary redox enzymes. Biotechnol Biofuels 6:41

Li Y, Ye D, Shao Z, Cui C, Che Y (2012) A sterol and spiroditerpenoids from a *Penicillium* sp. isolated from a deep sea sediment sample. Mar Drugs 10:497–508

Li L, Singh P, Liu Y, Pan S, Wang G (2014) Diversity and biochemical features of culturable fungi from the coastal waters of southern China. AMB Express 4:60

Li W, Wang MM, Bian XM, Guo JJ, Cai L (2016a) A high-level fungal diversity in the intertidal sediment of Chinese seas presents the spatial variation of community composition. Front Microbiol 7:2098

Li W, Wang MM, Wang XG, Cheng XL, Guo JJ, Bian XM, Cai L (2016b) Fungal communities in sediments of subtropical Chinese seas as estimated by DNA metabarcoding. Sci Rep 6:26528

Li W, Wang MM, Pan HQ, Burgaud G, Liang SK, Guo JJ, Luo T, Li Z, Zhang S, Cai L (2018) Highlighting patterns of fungal diversity and composition shaped by ocean currents using the East China Sea as a model. Mol Ecol 27:564–576

Li W, Wang MM, Burgaud G, Yu HM, Cai L (2019) Fungal community composition and potential depth-related driving factors impacting distribution pattern and trophic modes from epi- to abyssopelagic zones of the Western Pacific Ocean. Microbial Ecol 78:820–831
Lima JMS, Pereira JO, Batista IH, Neto PDQC, dos Santos JC, de Araújo SP, de Azevedo JL (2016) Potential biosurfactant producing endophytic and epiphytic fungi, isolated from macrophytes in the Negro River in Manaus, Amazonas, Brazil. Afr J Biotechnol 15:1217–1223
Liu CH, Huang X, Xie TN, Duan N, Xue YR, Zhao TX, Lever MA, Hinrichs KU, Inagaki F (2017) Exploration of cultivable fungal communities in deep coal-bearing sediments from ~1.3 to 2.5 km below the ocean floor. Environ Microbiol 19:803–818
Loilong A, Sakayaroj J, Rungjindamai N, Choeyklin R, Jones EBG (2012) Biodiversity of fungi on the palm *Nypa fruticans*. In: Marine fungi and fungal-like organisms. Walter de Gruyter GmbH & Co. KG, Berlin, p 273
Lombard V, Golaconda Ramulu H, Drula E, Coutinho PM, Henrissat B (2014) The carbohydrate-active enzymes database (CAZy) in 2013. Nucleic Acids Res 42:490–495
Loque CP, Medeiros AO, Pellizzari FM, Oliveira EC, Rosa CA, Rosa LH (2010) Fungal community associated with marine macroalgae from Antarctica. Polar Biol 33:641–648
Lorenz R, Molitoris HP (1997) Cultivation of fungi under simulated deep-sea conditions. Mycol Res 101:1355–1365
Lukassen MB, Saei W, Sondergaard TE, Tamminen A, Kumar A, Kempken F, Wiebe MG, Sørensen JL (2015) Identification of the Scopularide biosynthetic gene cluster in *Scopulariopsis brevicaulis*. Mar Drugs 13:4331–4343
Luo ZH, Pang KL (2014) Fungi on substrates in marine environment. In: Misra JK, Tewari JP, Deshmukh SK, Vágvölgyi C (eds) Progress in mycological research, Fungi in/on various substrates, vol III. CRC Press, Baca Raton, pp 97–114
Luo Y, Wei X, Yang S, Gao YH, Luo ZH (2020) Fungal diversity in deep-sea sediments from the Magellan seamounts as revealed by a metabarcoding approach targeting the ITS2 regions. Mycology 11:214–229
Lyons JI, Newell SY, Buchan A, Moran MA (2003) Diversity of ascomycete laccase gene sequences in a southeastern US salt marsh. Microb Ecol 45:270–281
Manohar CS, Menezes LD, Ramasamy KP, Meena RM (2014) Phylogenetic analyses and nitrate-reducing activity of fungal cultures isolated from the permanent, oceanic oxygen minimum zone of the Arabian Sea. Can J Microbiol 61:217–226
Marchler-Bauer A, Lu S, Anderson JB, Chitsaz F, Derbyshire MK, DeWese-Scott C, Fong JH, Geer LY, Geer RC, Gonzales NR, Gwadz M, Hurwitz DI, Jackson JD, Ke Z, Lanczycki CJ, Lu F, Marchloer GH, Mullokandov M, Omelchenko MV, Robertson CL, Song JS, Thanki N, Yamashita RA, Zhang D, Zhang N, Zheng C, Bryant SH (2011) CDD: a conserved domain database for the functional annotation of proteins. Nucleic Acids Res 39:225–229
Margulies M, Egholm M, Altman WE, Attiya S, Bader JS et al (2005) Genome sequencing in microfabricated high-density picolitre reactors. Nature 437:376–380
Martín-Rodríguez AJ, Reyes F, Martín J, Pérez-Yépez J, León-Barrios M, Couttolenc A, Espinoza C, Trigos A, Martin VS, Norte M, Fernandez JJ (2014) Inhibition of bacterial quorum sensing by extracts from aquatic fungi: first report from marine endophytes. Mar Drugs 12: 5503–5526
McClelland JW, Holmes RM, Dunton KH, Macdonald RW (2012) The Arctic Ocean estuary. Estuar Coasts 35:353–368
Medema MH, Fischbach MA (2015) Computational approaches to natural product discovery. Nat Chem Biol 11:639–648
Medema MH, Kottmann R, Yilmaz P, Cummings M, Biggins JB et al (2015) Minimum information about a biosynthetic gene cluster. Nat Chem Biol 11:625–631
Megan ND, Kathryn LVA, David OD (2001) Spatial patterns in macroalgal chemical defenses. In: JB MC, Baker BJ (eds) Marine chemical ecology. CRC Press, Boca Raton, FL, pp 301–324
Metzker ML (2010) Sequencing technologies - the next generation. Nat Rev Genet 11:31–46

Mitchison-Field LM, Vargas-Muñiz JM, Stormo BM, Vogt EJ, Van Dierdonck S, Pelletier JF, Ehrlich C, Lew DJ, Field CM, Gladfelter AS (2019) Unconventional cell division cycles from marine-derived yeasts. Curr Biol 29:3439–3456

Mohamed DJ, Martiny JB (2011) Patterns of fungal diversity and composition along a salinity gradient. ISME J 5:379–388

Montaser R, Luesch H (2011) Marine natural products: a new wave of drugs? Future Med Chem 3: 1475–1489

Muehlstein LK, Amon JP, Leffler DL (1987) Phototaxis in the marine fungus *Rhizophycium littoreum*. Appl Environ Microbiol 53:1819–1821

Nagahama T, Takahashi E, Nagano Y, Abdel-Wahab M, Miyazaki M (2011) Molecular evidence that deep-branching fungi are major fungal components in deep-sea methane cold-seep sediments. Environ Microbiol 13:2359–2370

Nagano Y, Nagahama T, Hatada Y, Nunoura T, Takami H, Miyazaki J, Takai K, Horikoshi K (2010) Fungal diversity in deep-sea sediments – the presence of novel fungal groups. Fungal Ecol 3:316–325

Nagano Y, Miura T, Nishi S, Lima AO, Nakayama C, Pellizari VH, Fujikura K (2017) Fungal diversity in deep-sea sediments associated with asphalt seeps at the Sao Paulo plateau. Deep Sea Res Part II Top Studies Oceanogr 146:59–67

Nagano Y, Miura T, Tsubouchi T, Lima AO, Kawato M, Fujiwara Y, Fujikura K (2020) Cryptic fungal diversity revealed in deep-sea sediments associated with whale-fall chemosynthetic ecosystems. Mycology 11:263–278

Naranjo-Ortiz MA, Gabaldón T (2019) Fungal evolution: diversity, taxonomy and phylogeny of the fungi. Bio Rev 94:2101–2137

Navarri M, Jégou C, Meslet-Cladière L, Brillet B, Barbier G, Burgaud G, Fleury Y (2016) Deep subseafloor fungi as an untapped reservoir of amphipathic antimicrobial compounds. Mar Drugs 14:50

Newell SY (2001) Spore-expulsion rates and extents of blade occupation by ascomycetes of the smooth-cordgrass standing-decay system. Bot Mar 44:277–285

Newell S, Porter D (2000) Microbial secondary production from salt marsh-grass shoots, and its known and potential fates. In: Weinstein M, Kreeger D (eds) Concepts and controversies in tidal marsh ecology. Springer, Netherlands, pp 159–185

Newell SY, Wasowski J (1995) Sexual productivity and spring intramarsh distribution of a key saltmarsh microbial secondary producer. Estuaries 18:241–249

Newell SY, Porter D, Lingle WL (1996) Lignocellulolysis by ascomycetes (fungi) of a saltmarsh grass (smooth cordgrass). Microsc Res Tech 33:32–46

Newman DJ, Cragg GM (2020) Natural products as sources of new drugs over the nearly four decades from 01/1981 to 09/2019. J Nat Prod 83:770–803

Nicoletti R, Trincone A (2016) Bioactive compounds produced by strains of Penicillium and Talaromyces of marine origin. Mar Drugs 14:37

Niu S, Liu D, Hu X, Proksch P, Shao Z, Lin W (2014) Spiromastixones A–O, antibacterial chlorodepsidones from a deep-sea-derived *Spiromastix* sp. fungus. J Nat Prod 77:1021–1030

Nødvig CS, Nielsen JB, Kogle ME, Mortensen UH (2015) A CRISPR-Cas9 system for genetic engineering of filamentous fungi. PLoS One 10:e0133085

Nowrousian M, Stajich JE, Chu M, Engh I, Espagne E, Halliday K, Kamerewerd J, Kempken F, Knab B, Kuo HC (2010) *De novo* assembly of a 40 Mb eukaryotic genome from short sequence reads: *Sordaria macrospora*, a model organism for fungal morphogenesis. PLoS Genet 6: e100891

Numata A, Takahashi C, Ito Y, Takada T, Kawai K, Usami Y, Matsumura E, Imachi M, Ito T, Hasegawa T (1993) Communesins, cytotoxic metabolites of a fungus isolated from a marine alga. Tet Lett 34:2355–2358

Oberbeckmann S, Osborn AM, Duhaime MB (2016) Microbes on a bottle: substrate, season and geography influence community composition of microbes colonizing marine plastic debris. PLoS One 11:e0159289

Ogaki MB, de Paula MT, Ruas D, Pellizzari FM, García-Laviña CX, Rosa LH (2019) Marine fungi associated with Antarctic macroalgae. In: Castro-Sowinski S (ed) The ecological role of micro-organisms in the Antarctic environment. Springer polar science. Springer, Cham, pp 239–255

Oppong-Danquah E, Parrot D, Blümel M, Labes A, Tasdemir D (2018) Molecular networking-based metabolome and bioactivity analyses of marine-adapted fungi co-cultivated with phytopathogens. Front Microbiol 9:2072

Orsi WD (2018) Ecology and evolution of seafloor and subseafloor microbial communities. Nat Rev Microbiol 16:671–683

Orsi WD, Edgcomb VP, Christman GD, Biddle JF (2013a) Gene expression in the deep biosphere. Nature 499:205–208

Orsi W, Biddle JF, Edgcomb V (2013b) Deep sequencing of subseafloor eukaryotic rRNA reveals active fungi across marine subsurface provinces. PLoS One 8:e56335

Overy DP, Rämä T, Oosterhuis R, Walker AK, Pang KL (2019) The neglected marine fungi, *sensu stricto*, and their isolation for natural products' discovery. Mar Drugs 17:42

Pachiadaki MG, Rédou V, Beaudoin DJ, Burgaud G, Edgcomb VP (2016) Fungal and prokaryotic activities in the marine subsurface biosphere at Peru margin and Canterbury Basin inferred from RNA-based analyses and microscopy. Front Microbiol 7:846

Paço A, Duarte K, da Costa JP, Santos PSM, Pereira R, Pereira ME, Freitas AC, Duarte AC, Rocha-Santos TAP (2017) Biodegradation of polyethylene microplastics by the marine fungus *Zalerion maritimum*. Sci Total Environ 586:10–15

Pang KL (2002) Systematics of the Halosphaeriales: which morphological characters are important? In: Hyde K (ed) Fungi in marine environments. Fungal Diversity Press, Hong Kong, pp 35–57

Pang KL, Chiang MWL, Vrijmoed LLP (2008) *Havispora longyearbyenensis* gen. et sp. nov.: an arctic marine fungus from Svalbard, Norway. Mycologia 100:291–295

Pang KL, Jheng JS, Jones EBG (2011) Marine mangrove fungi of Taiwan. National Taiwan Ocean University Press, Keelung, pp 1–131

Pang KL, Overy DP, Jones EG, da Luz CM, Burgaud G, Walker AK et al (2016) 'Marine fungi' and 'marine-derived fungi' in natural product chemistry research: toward a new consensual definition. Fungal Biol Rev 30:163–175

Pang K, Guo SY, Chen IA, Burgaud G, Luo ZH, Dahms HU, Hwang JS, Lin YL, Huang JS, Ho TW, Tsang LM, Chiang MWL, Cha HJ (2019) Insights into fungal diversity of a shallow-water hydrothermal vent field at Kueishan Island, Taiwan by culture-based and metabarcoding analyses. PLoS One 14:e0226616

Pang KL, Chiang MWL, Guo SY, Shih CY, Dahms HU, Hwang JS, Cha HJ (2020) Growth study under combined effects of temperature, pH and salinity and transcriptome analysis revealed adaptations of *aspergillus terreus* NTOU4989 to the extreme conditions at Kueishan Island hydrothermal vent Field, Taiwan. PLoS One 15:e0233621

Panzer K, Yilmaz P, Weiss M, Reich L, Richter M, Wiese J, Schmaljohann R, Labes A, Imhoff JF, Glöckner FO, Reich M (2015) Identification of habitat-specific biomes of aquatic fungal communities using a comprehensive nearly full-length 18S rRNA dataset enriched with contextual data. PLoS One 10:e0134377

Passarini MRZ, Rodrigues MVN, da Silva M, Sette LD (2011) Marine-derived filamentous fungi and their potential application for polycyclic aromatic hydrocarbon bioremediation. Mar Pollut Bull 62:364–370

Perini L, Gostinčar C, Gunde-Cimerman N (2019) Fungal and bacterial diversity of Svalbard subglacial ice. Sci Rep 9:20230

Petersen LE, Kellermann MY, Schupp PJ (2020) Secondary metabolites of marine microbes: from natural products chemistry to chemical ecology. In: Jungblut S, Liebich V, Bode-Dalby M (eds) YOUMARES 9 – the oceans: our research, our future: proceedings of the 2018 conference for YOUng MArine RESearcher in Oldenburg, Germany. Springer International Publishing, Cham, pp 159–180

Picard KT (2017) Coastal marine habitats harbor novel early-diverging fungal diversity. Fungal Ecol 25:1–13

Pilantanapak A, Jones EBG, Eaton Rod A (2005) Marine fungi on *Nypa fruticans* in Thailand. Bot Mar 48:365–373

Poli A, Bovio E, Ranieri L, Varese GC, Prigione V (2020) Fungal diversity in the Neptune Forest: comparison of the mycobiota of *Posidonia oceanica*, *Flabellia petiolata*, and *Padina pavonica*. Front Microbiol 11:933

Polinski JM, Bucci JP, Gasser M, Bodnar AG (2019) Metabarcoding assessment of prokaryotic and eukaryotic taxa in sediments from Stellwagen Bank national marine sanctuary. Sci Rep 9:1–8

Porras-Alfaro A, Bayman P (2011) Hidden fungi, emergent properties: endophytes and microbiomes. Annu Rev Phytopathol 49:291–315

Porter DF, Springer SR, Padman L, Fricker HA, Tinto KJ, Riser SC, Bell RE, Ice Team ROSETTA (2019) Evolution of the seasonal surface mixed layer of the Ross Sea, Antarctica, observed with autonomous profiling floats. J Geophys Res Oceans 124:4934–4953

Priest T, Fuchs B, Amann R, Reich M (2020) Diversity and biomass dynamics of unicellular marine fungi during a spring phytoplankton bloom. Environ Microbiol 23:448–463

Prince RC (2005) The microbiology of marine oil spill bioremediation. In: Oliver B, Magot M (eds) Petroleum microbiology. ASM Press, Washington, DC, pp 317–335

Qin M, Wu S, Li A, Zhao F, Feng H et al (2019) LRScaf: improving draft genomes using long noisy reads. BMC Genomics 20:955

Quémener M, Mara P, Schubotz F, Beaudoin D, Li W, Pachiadaki M, Sehein TR, Sylvan JB, Li J, Barbier G, Edgcomb VP, Burgaud G (2020) Meta-omics highlights the diversity, activity and adaptations of fungi in deep ocean crust. Environ Microbiol 22:3950–3967

Raghukumar C, Raghukumar S, Sheelu G, Gupta SM, Nath BN, Rao BR (2004) Buried in time: culturable fungi in a deep-sea sediment core from the Chagos trench. Indian Ocean Deep-Sea Res Part 1 51:1759–1768

Raghukumar C, Damare S, Singh P (2010) A review on deep-sea fungi: occurrence, diversity and adaptations. Bot Mar 53:479–492

Rai PK, Lee J, Brown RJ, Kim KH (2020) Micro-and nanoplastic pollution: behavior, microbial ecology, and remediation technologies. J Clean Prod 291:125240

Rämä T, Nordén J, Davey ML, Mathiassen GH, Spatafora JW, Kauserud H (2014) Fungi ahoy! Diversity on marine wooden substrata in the high north. Fungal Ecol 8:46–58

Rämä T, Hassett BT, Bubnova E (2017) Arctic marine fungi: from filaments and flagella to operational taxonomic units and beyond. Bot Mar 60:433–452

Rao TE, Imchen M, Kumavath R (2017) Marine enzymes: production and applications for human health. Adv Food Nut Res 80:149–163

Rapp JZ, Fernández-Méndez M, Bienhold C, Boetius A (2018) Effects of ice-algal aggregate export on the connectivity of bacterial communities in the Central Arctic Ocean. Front Microbiol 9: 1035

Rateb ME, Ebel R (2011) Secondary metabolites of fungi from marine habitats. Nat Prod Rep 28: 290–344

Rédou V, Ciobanu MC, Pachiadaki MG, Edgcomb V, Alain K, Barbier G, Burgaud G (2014) In-depth analyses of deep subsurface sediments using 454-pyrosequencing reveals a reservoir of buried fungal communities. FEMS Microbiol Ecol 90:908–921

Rédou V, Navarri M, Meslet-Cladière L, Barbier G, Burgaud G (2015) Species richness and adaptation of marine fungi from deep-subseafloor sediments. Appl Environ Microbiol 81: 3571–3583

Rédou V, Vallet M, Meslet-Cladière L, Kumar A, Pang KL, Pouchus YF et al (2016a) Marine fungi. In: Stal LJ, Cretoiu MS (eds) The marine microbiome: an untapped source of biodiversity and biotechnological potential. Springer International Publishing, Cham, pp 99–153

Rédou V, Kumar A, Hainaut M, Henrissat B, Record E et al. (2016b) Draft genome sequence of the deep-sea ascomycetous filamentous fungus Cadophora malorum Mo12 from the mid-Atlantic ridge reveals its biotechnological potential. Genome Announc 4(4), e00467–16.

Rédou V, Kumar A, Hainaut M, Henrissat B, Record E et al. (2016c). Draft genome sequence of the deep-sea basidiomycetous yeast Cryptococcus sp. strain Mo29 reveals its biotechnological potential. Genome Announc 4(4), e00461–16.
Richards TA, Jones MDM, Leonard G, Bass D (2012) Marine fungi: their ecology and molecular diversity. Annu Rev Mar Sci 4:495–522
Richards TA, Leonard G, Mahé F, del Campo J, Romac S, Jones MDM et al (2015) Molecular diversity and distribution of marine fungi across 130 European environmental samples. Proc R Soc B 282:2015–2243
Richards TA, Leonard G, Wideman JG (2017) What defines the "kingdom" fungi? The Fungal Kingdom, pp. 57–77
Rojas-Jimenez K, Rieck A, Wurzbacher C, Jürgens K, Labrenz M, Grossart HP (2019) A salinity threshold separating fungal communities in the Baltic Sea. Front Microbiol 10:680
Roullier C, Guitton Y, Valery M, Amand S, Prado S, Robiou du Pont T, Grovel O, Pouchus YF (2016) Automated detection of natural halogenated compounds from LC-MS profiles–application to the isolation of bioactive chlorinated compounds from marine-derived fungi. Anal Chem 88:9143–9150
Sajna KV, Sukumaran RK, Gottumukkala LD, Pandey A (2015) Crude oil biodegradation aided by biosurfactants from *Pseudozyma* sp. NII 08165 or its culture broth. Bioresour Technol 191:133–139
Sakayaroj JPS, Supaphon O, Jones EBG, Phongpaichit S (2010) Phylogenetic diversity of endophyte assemblages associated with the tropical seagrass *Enhalus acoroides* in Thailand. Fungal Divers 42:27–45
Sakayaroj J, Preedanon S, Phongpaichit S, Buatong J, Chaowalit P, Rukachaisirikul V (2012) Diversity of endophytic and marine-derived fungi associated with marine plants and animals. In Marine Fungi (pp. 291–328). Chapter 16. De Gruyter.
Sánchez C (2020) Fungal potential for the degradation of petroleum-based polymers: an overview of macro-and microplastics biodegradation. Biotechnol Adv 40:107501
Sander T, Freyss J, von Korff M, Rufener C (2015) DataWarrior: an open-source program for chemistry aware data visualization and analysis. J Chem Inf Model 55:460–473
Schmit JP, Shearer CA (2004) Geographical and host distribution of lignicolous mangrove microfungi. Bot Mar 47:496–500
Scholthof KBG (2007) The disease triangle: pathogens, the environment and society. Nat Rev Microbiol 5:152–156
Scholz B, Küpper FC, Vyverman W, Ólafsson HG, Karsten U (2017) Chytridiomycosis of marine diatoms-the role of stress physiology and resistance in parasite-host recognition and accumulation of defense molecules. Mar Drugs 15:26
Sen S, Borah SN, Bora A, Deka S (2017) Production, characterization, and antifungal activity of a biosurfactant produced by *Rhodotorula babjevae* YS3. Microb Cell Factories 16:95
Shekhar S, Sundaramanickam A, Balasubramanian T (2015) Biosurfactant producing microbes and their potential applications: a review. Crit Rev Env Sci Tech 45:1522–1554
Sheridan KJ, Dolan SK, Doyle S (2015) Endogenous cross-talk of fungal metabolites. Front Microbiol 5:732
Silva RDCF, Almeida DG, Rufino RD, Luna JM, Santos VA, Sarubbo LA (2014) Applications of biosurfactants in the petroleum industry and the remediation of oil spills. Int J Mol Sci 15: 12523–12542
Singh RP, Reddy CR (2015) Unraveling the functions of the macroalgal microbiome. Front Microbiol 6:1488
Singh P, Raghukumar C, Verma P, Shouche Y (2010) Phylogenetic diversity of culturable fungi from the deep-sea sediments of the central Indian Basin and their growth characteristics. Fungal Divers 40:89–102
Singh RP, Raghukumar C, Verma P, Shouche Y (2012) Assessment of fungal diversity in deep-sea sediments by multiple primer approach. World J Microbiol Biotechnol 28:659–667
Sparrow FK (1973) Three monocentric chytrids. Mycologia 65:1331–1336

Spatafora JW, Volkmann-Kohlmeyer B, Kohlmeyer J (1998) Independent terrestrial origins of the Halosphaeriales (marine Ascomycota). Am J Bot 85:1569–1580
Stanke M, Morgenstern B (2005) AUGUSTUS: a web server for gene prediction in eukaryotes that allows user-defined constraints. Nucleic Acids Res 33:465–467
Stanley SJ (1991) The autecology and ultrastructure interaction between Mycosphaerella ascophylli Cotton, Lautitia danica (Berlese) Schatz, Mycaureola dilsea Maire et Chemin and their respective marine algal hosts. PhD Thesis, University of Portsmouth, UK
Steliga T, Jakubowicz P, Kapusta P (2012) Changes in toxicity during in situ bioremediation of weathered drill wastes contaminated with petroleum hydrocarbons. Bioresour Technol 125:1–10
Suetrong S, Jones EBG (2006) Marine discomycetes: a review. Indian J Mar Sci 35:291–296
Svahn KS, Göransson U, El-Seedi H, Bohlin L, Larsson DJ, Olsen B, Chryssanthou E (2012) Antimicrobial activity of filamentous fungi isolated from highly antibiotic-contaminated river sediment. Infect Ecol Epidemiol 2:11591
Takami H, Inoue A, Fuji F, Horikoshi K (1997) Microbial flora in the deepest sea mud of the Mariana trench. FEMS Microbiol Lett 152:279–285
Takami H, Kobata K, Nagahama T, Kobayashi H, Inoue A, Horikoshi K (1999) Biodiversity in deep-sea sites located near the south part of Japan. Extremophiles 3:97–102
Tang X, Yu L, Xu W, Zhang X, Xu X, Wang Q, Wei S, Qiu Y (2020) Fungal diversity of deep-sea sediments in mid-oceanic ridge area of the East Pacific and the south Indian oceans. Bot Mar 63: 183–196
Tao Y, Wolinska J, Hölker F, Agha R (2020) Light intensity and spectral distribution affect chytrid infection of cyanobacteria via modulation of host fitness. Parasitology 147:1206–1215
Taylor JD, Cunliffe M (2016) Multi-year assessment of coastal planktonic fungi reveals environmental drivers of diversity and abundance. ISME J 10:2118–2128
Terrado R, Medrinal E, Dasilva C, Thaler M, Vincent WF, Lovejob C (2011) Protist community composition during spring in an Arctic flaw lead polynya. Polar Biol 34:1901–1914
Thomas F, Cosse A, Le Panse S, Kloareg B, Potin P, Leblanc C (2014) Kelps feature systemic defense responses: insights into the evolution of innate immunity in multicellular eukaryotes. New Phytol 204:567–576
Tisthammer KH, Cobian GM, Amend AS (2016) Global biogeography of marine fungi is shaped by the environment. Fungal Ecol 19:39–46
Torzilli AP, Sikaroodi M, Chalkley D, Gillevet PM (2006) A comparison of fungal communities from four salt marsh plants using automated ribosomal intergenic spacer analysis (ARISA). Mycologia 98:690–698
Tourneroche A, Lami R, Hubas C, Blanchet E, Vallet M, Escoubeyrou K, Paris A, Prado S (2019) Bacterial-fungal interactions in the kelp Endomicrobiota drive autoinducer-2 quorum sensing. Front Microbiol 10:1693
Tourneroche A, Lami R, Burgaud G, Coulon-Dommart I, Li X, Gachon C, Gèze M, Boeuf D, Prado S (2020) The bacterial and fungal microbiota of *Saccharina latissima* (Laminariales, Phaeophyceae). Front Mar Sci 7:587566
Urbanek AK, Rymowicz W, Mirończuk AM (2018) Degradation of plastics and plastic-degrading bacteria in cold marine habitats. Appl Microbiol Biotechnol 102:7669–7678
Vallet M, Strittmatter M, Murúa P, Lacoste S, Dupont J, Hubas C, Genta-Jouve G, Gachon CMM, Kim GH, Prado S (2018) Chemically-mediated interactions between macroalgae, their fungal endophytes, and protistan pathogens. Front Microbiol 9:3161
Vallet M, Chong YM, Tourneroche A, Genta-Jouve G, Hubas C, Lami R, Gachon C, Klochkova T, Chan KG, Prado S (2020) Novel α-hydroxy γ-butenolides of kelp endophytes disrupt bacterial cell-to-cell signaling. Front Mar Sci 7:601
Van de Vossenberg BTLH, Warris S, Nguyen HDT, van Gent-Pelzer MPE, Joly DL, van de Geest HC, Bonants PJM, Smith DS, Lévesque A, van der Lee TAJ (2019) Comparative genomics of chytrid fungi reveal insights into the obligate biotrophic and pathogenic lifestyle of *Synchytrium endobioticum*. Sci Rep 9:8672

van Santen JA, Jacob G, Singh AL, Aniebok V, Balunas MJ, Bunsko D, Neto FC, Castaño-Espriu L, Chang C, Clark TN, Cleary Little JL, Delgadillo DA, Dorrestein PC, Duncan KR, Egan JM, Galey MM, Haeckl FPJ, Hua A, Hughes AH, Iskakova D, Khadilkar A, Lee JH, Lee S, LeGrow N, Liu DY, Macho JM, McCaughey CS, Medema MH, Neupane RP, O'Donnell TJ, Paula JS, Sanchez JM, Shaikh AF, Soldatou S, Terlouw BR, Tran TA, Valentine M, van der Hooft JJJ, Vo DA, Wang M, Wilson D, Zink KE, Linington RG (2019) The natural products atlas: an open access knowledge base for microbial natural products discovery. ACS Cent Sci 5: 1824–1833

Venkatachalam A, Thirunavukkarasu N, Suryanarayanan TS (2015) Distribution and diversity of endophytes in seagrasses. Fungal Ecol 13:60–65

Vrijmoed LLP (2000) Isolation and culture of higher filamentous fungi. In: Hyde KD, Pointing SB (eds) Marine mycology—a practical approach. Fungal Diversity Press, Hong Kong, pp 1–20

Wadham JL, Hawkings JR, Tarasov L, Gregoire LJ, Spencer RGM, Gutjahr M, Ridgwell A, Kohfeld KE (2019) Ice sheets matter for the global carbon cycle. Nature Comm 10:3567

Wahl M, Goecke F, Labes A, Dobretsov S, Weinberger F (2012) The second skin: ecological role of epibiotic biofilms on marine organisms. Front Microbiol 3:292

Wainwright BJ, Zahn GL, Zushi J, Lee NLY, Ooi JLS, Lee JN, Huang D (2019a) Seagrass-associated fungal communities show distance decay of similarity that has implications for seagrass management and restoration. Ecol Evol 9:11288–11297

Wainwright BJ, Bauman AG, Zahn GL, Todd PA, Huang D (2019b) Characterization of fungal biodiversity and communities associated with the reef macroalga *Sargassum ilicifolium* reveals fungal community differentiation according to geographic locality and algal structure. Mar Biodivers 49:2601–2608

Walker AK, Campbell J (2010) Marine fungal diversity: a comparison of natural and created salt marshes of the north-Central Gulf of Mexico. Mycologia 102:513–521

Wang YT, Xue YR, Liu CH (2015) A brief review of bioactive metabolites derived from deep-sea fungi. Mar Drugs 13:4594–4616

Wang J, He W, Huang X, Tian X, Liao S, Yang B, Wang F, Zhou X, Liu Y (2016) Antifungal new oxepine-containing alkaloids and xanthones from the deep-sea-derived fungus *aspergillus versicolor* SCSIO 05879. J Agric Food Chem 64:2910–2916

Wang YP, Guo XH, Zheng PF, Zou SB, Li GH, Gong J (2017) Distinct seasonality of chytrid-dominated benthic fungal communities in the neritic oceans (Bohai Sea and North Yellow Sea). Fungal Ecol 30:55–66

Wang YQ, Sen B, He YD, Xie ND, Wang GY (2018) Spatiotemporal distribution and assemblages of planktonic fungi in the coastal waters of the Bohai Sea. Front Microbiol 9:584

Wang YQ, Sen K, He YD, Xie YX, Wang GY (2019) Impact of environmental gradients on the abundance and diversity of planktonic fungi across coastal habitats of contrasting trophic status. Sci Total Environ 683:822–833

Weber T (2014) In silico tools for the analysis of antibiotic biosynthetic pathways. Int J Med Microbiol 304:230–235

Wichard T, Charrier B, Mineur F, Bothwell JH, Clerck OD, Coates JC (2015) The green seaweed Ulva: a model system to study morphogenesis. Front Plant Sci 6:72

Wright RJ, Erni-Cassola G, Zadjelovic V, Latva M, Christie-Oleza JA (2020) Marine plastic debris: a new surface for microbial colonization. Environ Sci Technol 54:11657–11672

Wu C, Chen R, Liu M, Liu D, Li X, Wang S, Niu S, Guo P, Lin W (2015) Spiromastixones inhibit foam cell formation via regulation of cholesterol efflux and uptake in RAW264.7 macrophages. Mar Drugs 13:6352–6365

Wu Z, Wang Y, Liu D, Proksch P, Yu S, Lin W (2016) Antioxidative phenolic compounds from a marine-derived fungus *aspergillus versicolor*. Tetrahedron 72:50–57

Xu W, Pang KL, Luo ZH (2014) High fungal diversity and abundance recovered in the deep-sea sediments of the Pacific Ocean. Microb Ecol 68:688–698

Xu R, Xu GM, Li XM, Li CS, Wang BG (2015) Characterization of a newly isolated marine fungus *aspergillus dimorphicus* for optimized production of the anti-tumor agent wentilactones. Mar Drugs 13:7040–7054

Xu W, Luo ZH, Guo S, Pang KL (2016) Fungal community analysis in the deep-sea sediments of the Pacific Ocean assessed by comparison of ITS, 18S and 28S ribosomal DNA regions. Deep Sea Res Part I Oceanogr Res Pap 109:51–60

Xu W, Guo S, Pang KL, Luo ZH (2017) Fungi associated with chimney and sulfide samples from a south mid-Atlantic ridge hydrothermal site: distribution, diversity and abundance. Deep Sea Res Part I Oceanogr Res Pap 123:48–55

Xu W, Gao Y, Gong L, Li M, Pang KL, Luo ZH (2019) Fungal diversity in the deep-sea hadal sediments of the yap trench by cultivation and high throughput sequencing methods based on ITS rRNA gene. Deep Sea Res Part I Oceanogr Res Pap 145:125–136

Yao Q, Wang J, Zhang X, Nong X, Xu X, Qi S (2014) Cytotoxic polyketides from the deep-sea-derived fungus *Engyodontium album* DFFSCS021. Mar Drugs 12:5902–5915

Yin Y, Mao X, Yang J, Chen X, Mao F et al (2012) dbCAN: a web resource for automated carbohydrate-active enzyme annotation. Nucleic Acids Res 40:445–451

Zajc J, Liu Y, Dai W, Yang Z, Hu J, Gostinčar C, Gunde-Cimerman N (2013) Genome and transcriptome sequencing of the halophilic fungus *Wallemia ichthyophaga*: haloadaptations present and absent. BMC Genomics 14:1–21

Zettler ER, Mincer TJ, Amaral-Zettler LA (2013) Life in the "plastisphere": microbial communities on plastic marine debris. Environ Sci Technol 47:7137–7146

Zhang X, Feng X, Wang F (2016) Diversity and metabolic potentials of subsurface crustal microorganisms from the western flank of the mid-Atlantic ridge. Front Microbiol 7:363

Zhang MM, Wong FT, Wang Y, Luo S, Lim YH, Heng E, Yeo WL, Cobb RE, Enghiad B, Ang EL, Zhao H (2017) CRISPR-Cas9 strategy for activation of silent *Streptomyces* biosynthetic gene clusters. Nat Chem Biol 13:607–609

Zhang X, Li SJ, Li JJ, Liang ZZ, Zhao CQ (2018) Novel natural products from extremophilic fungi. Mar Drugs 16:194

Zhou Y, Li YH, Yu HB, Liu XY, Lu XL, Jiao BH (2017) Furanone derivative and sesquiterpene from Antarctic marine-derived fungus *Penicillium* sp. S-1-18. J Asian Nat Prod Res 20:1108–1115

Zuccaro A, Mitchell JI (2005) Fungal communities of seaweeds. Mycol Ser 23:533

Zuccaro A, Schulz B, Mitchell JI (2003) Molecular detection of ascomycetes associated with *Fucus serratus*. Mycol Res 107:1451–1466

Zuccaro A, Summerbell RC, Gams W, Schroers HJ, Mitchell JI (2004) A new *Acremonium* species associated with *Fucus* spp., and its affinity with a phylogenetically distinct marine *Emericellopsis* clade. Stud Mycol 50:283–297

Zuccaro A, Schoch CL, Spatafora JW, Kohlmeyer J, Draeger S, Mitchell JI (2008) Detection and identification of fungi intimately associated with the brown seaweed *Fucus serratus*. Appl Environ Microbiol 74:931–941

# Marine Viruses: Agents of Chaos, Promoters of Order

6

Marcos Mateus

**Abstract**

Marine viruses are considered the most enigmatic form of life in the oceans. They are, simultaneously, agents of chaos and promoters of order. Whether or not they can be considered living entities, they are active agents of infection and drivers of host diversity. Consequently, they modulate the dynamic changes of populations of marine bacteria, archaea, and eukaryotes. Marine viruses are highly abundant, diverse, and active components of marine environments and play a crucial role in the ecology and biogeochemistry of marine ecosystems. Their ecological importance across different marine biomes is attested by the profusion of studies published over the previous decades. Regardless of the substantial scientific output so far, the study on the nature and role of marine viruses is far from exhausted. Novel insights have, concurrently, tested hypotheses, consolidated previously knowledge collected, and pushed forward new lines of research. This chapter describes some of these insights published over the past 5 years, by addressing topics that range across various fields of marine virology.

**Keywords**

Bacteriophages · Marine virome · Viral metagenomics · Viral shunt · Virioplankton · Viromics

M. Mateus (✉)
Marine, Environment and Technology Centre/Laboratory for Robotics and Engineering Systems (MARETEC/LARSyS), Instituto Superior Técnico, Universidade de Lisboa, Lisbon, Portugal
e-mail: marcos.mateus@tecnico.ulisboa.pt



L. J. Stal, M. S. Cretoiu (eds.), *The Marine Microbiome*, The Microbiomes of Humans, Animals, Plants, and the Environment 3,
https://doi.org/10.1007/978-3-030-90383-1_6

## 6.1 Introduction

It only takes a quick browse on the bibliography to notice an inaugural recurrent statement: viruses are the most abundant biological entities in the oceans. This is indeed an astonishing fact. No less surprising is that, despite their impressive numbers, viruses account for less than 1% of the total biomass in marine systems. To mention that viruses exist wherever life is found has become a prosaic observation in recent years. They are by far the most numerous, genetically diverse, and pervasive biological entities in all ecosystems of the planet (Fuhrman 1999; Lipkin 2010; Suttle 2005). Estimates for the world's ocean point to impressive figures: about $10^{23}$ phytoplankton and bacterial cells get infected by viruses every second (Knowles et al. 2016) in a total estimate of $10^{28}$ infections per day (Suttle 2007). Due to their high abundance and rate of infection, viral-induced cell lysis releases large amounts of dissolved organic matter, which promotes marine microbial activity, and modulates the lifespan, gene flow, and metabolic outputs of microbes. As important controllers of host diversity and nutrient recycling, viruses are central components of the marine biome, essential in the dynamics of marine food webs, and key players in global biogeochemical cycles. The study of marine viruses has thrived over the past decades and, judging alone by recent years, the incessant, fast-paced, scientific output on the subject shows that it is far from exhausted. Instead, new knowledge on the dynamic population changes and role of viruses on the marine biota points to several unknowns that beget more in-depth studies. The stimulating essence of this field of research was surely one of the reasons that lead Brussaard et al. (2016) to state that this "is an exciting time of discovery" in the previous edition of this volume. Five years later, it still is! In a sense, just the surface layer of the deep waters of knowledge on marine viruses is scratched.

This chapter addresses recent advances in this vibrant field of research. A note must be made, however: this chapter is not a mere continuation or update on the chapter by Brussaard et al. (2016) because much of its content is still relevant and up-to-date. Instead, it tackles marine viruses from a different perspective, adding new insights to the aforementioned work. The compilation of themes here included explores the abundant body of knowledge produced recently, alongside some future challenges. The chapter starts by visiting some of the main tenets on the role of viruses in marine systems and how they have been established by novel insights. Then it goes through an overview on the treasure of knowledge condensed in review papers published over the past 5 years. This section is followed by two other dealing with the profusion of viruses in marine ecosystems, an important theme in many recent works, and a brief description of recent developments in methodological approaches in metagenomics and the processing of omics-data. Finally, the inclusion of viruses in marine ecosystem modelling is briefly addressed. Hopefully, this chapter will underpin the idea of marine viruses as agents of chaos and promoters of order in the world's ocean.

## 6.2 Consolidating the Role of Marine Viruses

The role of viruses in marine systems has received increasing recognition over the last four decades, catapulting these biological entities from apparent obscurity to the spotlight of marine ecology and biogeochemistry. Marine virology became a hot research topic mostly owing to four major lines of evidence (Cai et al. 2016): marine viruses (i) are the most abundant organisms in the ocean, (ii) they keep microbial numbers below the carrying capacity of the system through viral lysis, (iii) they reprogram the metabolic output of their hosts, and (iv) they increase the release of dissolved organic matter into the water.

### 6.2.1 Revisiting the Evidence

The sheer number of viruses in the ocean is impressive: an average concentration of ten million particles per millilitre (Breitbart 2012). Nonetheless, the ongoing development of new methodologies and equipment has revealed other important facts: viruses comprise ~94% of the nucleic acid-containing particles in the ocean and stand as a large reservoir of genetic diversity (Angly et al. 2006; Jacquet et al. 2010). Moreover, the compilation of viral metagenomic data over the past decade has revealed a huge viral diversity, making the marine virome (the nucleic acid complement of all viruses in a certain environment) the largest pool of unexplored genetic diversity on the globe (Middelboe and Brussaard 2017).

Viruses play a decisive role in determining the microbial composition of marine systems by controlling the population size and by horizontally transferring genes between hosts (Rohwer and Thurber 2009; Thingstad 2000). As an outcome, viruses keep microbial populations below the carrying capacity of the system. Viral action induces a shift in dominant bacterial groups by infecting and bursting more susceptible community members and, simultaneously, fomenting the abundance of other groups that take advantage of the substrate made available by the "viral shunt". Hence, the vital action of viral lysis as a driver of bacterial diversity is consistent with the "killing the winner" hypothesis (Heinrichs et al. 2020), a model for the food web that assumes trade-offs between competition and defence strategies when the winner (dominant clade) for nutrient acquisition is lysed (killed) by viruses.

Early studies on marine viral activity focused mainly on the outcome of lyses-induced mortality on phytoplankton and bacterioplankton. Important as they were, they soon pointed to the need for a closer analysis of the many processes involved in the infection-burst cycle. Advances in understanding the biogeochemical impact of viruses soon followed, narrowing on the metabolic reprogramming of host cells during lytic viral infection and how this process modifies the flow of energy and nutrients. New findings suggest that the host metabolism is transformed upon infection with viruses expressing host-derived genes haltering host metabolism (Hurwitz et al. 2016). An example of such a new finding reports that virus-encoded metabolic genes seem to decrease the energetic and biosynthetic restrictions to viral production (Zimmerman et al. 2020). Such advances consolidate previous studies revealing that viruses reprogram the metabolic output of their hosts by the expression of virally encoded auxiliary metabolic genes with key roles in facilitating biochemical and metabolic processes (Breitbart 2012). Another example comes from cyanophages that are able to transcribe and express photosynthesis genes during lytic infection, thereby sustaining photosynthesis in the host during the phage lytic cycle, which might potentially lead to an increase in viral propagation (Thompson et al. 2011).

Following cell lysis, new viral particles are released to the environment along with host cell materials that end up being remineralized. The lytic process leading to the release of materials (dissolved organic matter mostly) termed as "viral shunt" (Wilhelm and Suttle 1999) impedes the flow of carbon and nutrients to higher trophic levels, keeping them readily available to heterotrophic bacteria. As such, virus-mediated mortality promotes net respiration and dissolved organic matter cycling (Jacquet et al. 2010) and shunts nutrients between particulate and dissolved phases (Proctor and Fuhrman 1990; Rohwer and Thurber 2009). This process highlights the potential of viruses to modify the efficiency of the carbon pump (Suttle 2007), which is the export of carbon fixed by photosynthesis to the deep ocean via sinking. Weitz et al. (2015) hypothesized that, besides preventing the transfer to higher trophic levels, the viral shunt is also able to decrease the carbon export efficiency (the proportion of primary production that sinks to the deep ocean) by holding nutrients and carbon in the euphotic zone.

While the release of DOM following cell lysis after infection is now a settled matter, a deeper look has revealed that this process is more complex than previously assumed. Experimenting with viral lysis of picocyanobacteria, the autotrophs responsible for half of the primary production in the ocean, Zhao et al. (2019) concluded that the released viral-induced dissolved organic matter (vDOM) composes a complex matrix of labile organic matter that is used by bacterioplankton. The lysed materials induced an enrichment in bacterial community biodiversity indexes which enabled key bacterial species to form a complex relationship with vDOM substrate. Hence, this would denote a potential correspondence between bacterial populations and the chemical diversity of the DOM components.

Besides the already established "viral shunt" a "viral shuttle" was also revealed (Sullivan et al. 2017). This process describes the association of viruses with sinking material, i.e., virus-infected cells form larger particles that not just sink faster, but also lead to preferential grazing by heterotrophic protists stimulating the growth of grazers. This implies a synergetic effect between the "viral shunt" and "viral shuttle" on the biological carbon pump, a hypothesis recently proposed by Kaneko et al. (2021).

### 6.2.2 The Nutrient Connexion

For many years, the importance of viruses in ocean biogeochemistry was mostly associated with their role in host lysis (Breitbart et al. 2007; Fuhrman 1999; Suttle 2005). The impact of viruses on nutrient cycles was studied for a long time, but not the close connection between viruses and nutrients. Eventually, it became clear that marine virus particles contain a considerable quantity of macronutrients (carbon, nitrogen, and phosphorus) and the virus-host stoichiometric mismatch had the potential to drive the differential release of nutrients upon cell lysis (Gazitúa et al. 2021; Jover et al. 2014). Experiments using phytoplankton host-virus culture systems showed that the viral production that impacts nutrient availability is, in turn, influenced by the availability of nitrogen and phosphorus (Maat and Brussaard 2016; Mojica and Brussaard 2014). The stress of phosphorus limitation in virally infected phytoplankton host, for instance, may lead to a prolonged latency period that delays the release of progeny viruses from the host cell possibly due to host energy deficiency. N-stress decreased the number of newly formed viruses released per lysed host cell (viral burst size). Viral activity is a determinant of nutrient availability in the water and, at the same time, strongly impaired by the nutrient limitation of their host cells.

If viruses' particles contribute to the macronutrient pools, their involvement in the micro-nutrient cycles is also expected. For many years the studies addressing the influence of viruses on micronutrient cycles were limited and even more scarce were studies addressing the chemical contribution of the viral particles themselves. This seeming gap in knowledge, however, has been explored over the past few years revealing a much more complex virus mediation in nutrient availability. If viruses contain or interact with trace metals such as iron, as hypothesized by Bonnain et al. (2016), then their abundance in the ocean should translate into a major influence on biogeochemical cycling of this micronutrient. The effect of the viral shunt on Fe concentrations and bioavailability, for instance, has received attention. Fe is not only released by lysis, but its solubility is also increased by the immediate release of Fe-binding dissolved organic ligands. Previous studies seem to support such a claim. More than 99.9% of dissolved iron in the ocean is bound to organic ligands and a correlation between Fe-binding ligand concentration and biological activity is found. The release of dissolved organic ligands in the colloidal fraction after viral lysis may well stand as an important contribution to this pool of ligands. So far, the connexion of viruses to Fe seems far from simple, since iron also mediates the phage–host relationship. According to the "Ferrojan Horse Hypothesis" proposed by Bonnain et al. (2016), phages use iron as a Trojan horse to exploit the host iron-uptake mechanism: the phage tail fibres contain iron which is recognized by the host siderophore-bound iron receptor, allowing the phage to attach to the bacterial cell and proceed to puncture the cell membrane and infect the host with its nucleic acid.

There is now substantial evidence that supports the claim that Fe is central in viral activity because there is an apparent limiting role of Fe, similarly to what had already been confirmed for N and P. Low diversity and abundance of diatom viruses were observed in a chronically Fe-limited region of the subarctic northeast Pacific and a considerable decrease in viral replication was reported in a coastal upwelling region of the California Current under transient iron limitation (Kranzler et al. 2021). Similar results were obtained in Fe-limited cultures of *Chaetoceros tenuissimus*, a bloom forming centric diatom, which exhibited delayed virus-mediated mortality and decreased viral replication. Such outcomes lead the authors to hypothesize that

Fe-limitation enables diatoms to escape viral lysis and subsequent remineralization in the photic zone, thus enhancing carbon export efficiency and silica burial in Fe-limited areas or during specific periods when conditions are less favourable. Previous observations had already tackled the effect of Fe-limitation on phytoplankton host–virus dynamics in Fe-limited monocultures of the abundant algae *Micromonas pusilla* and *Phaeocystis globose* infected with their respective viruses, MpV and PgV (Slagter et al. 2016). In both host–virus systems, Fe-stress induced a considerable decrease in the virus burst size and, in the case of MpV, virus progeny showed highly decreased infectivity. Again, this outcome indicates that Fe-limitation has the potential to decrease viral infection and, consequently, phytoplankton loss, thereby affecting through the viral shunt the Fe-cycling along with the cycling of macronutrients.

## 6.3 Marine Viruses Reviewed

Several works have been published over the past 5 years to report the progress and define the state-of-the-art in many fields of marine virology. Much of the knowledge produced has been compiled in review papers and therefore these reviews should not be neglected in an overview of recent advances. However, such publications can only be briefly described for obvious reasons related to size and scope. As such, only a summary is presented here (Table 6.1) and, for simplicity, discussed in broad categories of subjects, even though some reviews are transversal to many. More details on each topic can be found in the countless references provided in each review.

### 6.3.1 The Ecology of Marine Viruses

Zimmerman et al. (2020) offer one of the most thorough reviews on the role of viruses across the full spectra of natural systems (not exclusive to marine environments). Their work reviews recent developments that brought new insights to the biogeochemical impact of viruses, with special emphasis on the metabolic reprogramming of host cells during lytic viral infection and how they lead to shifts in the flow of energy and nutrients in aquatic ecosystems. The identity and activity of viruses have also been the theme of recent reviews. Sun et al. (2020), for instance, provide an overview of the continued work to characterize giant aquatic viruses, while Sadeghi et al. (2021) condensed the knowledge to date about RNA viruses, a major fraction of marine virus assemblages, infecting known host species of marine unicellular eukaryotes.

Other works converge on the dynamics of virus–host interactions. The work of Weynberg et al. (2017) discuss the interaction between marine prasinoviruses and their host, the prasinophyte *Ostreococcus*. Drawing from their extensive revision, Weynberg et al. (2017) conclude that this host–virus system should be a model for the study of cellular and molecular processes in the marine environment. McMinn

**Table 6.1** Overview of some reviews addressing marine virus published since 2016 (sorted by year and author name)

| References | Title and author keywords |
|---|---|
| Sadeghi et al. (2021) | RNA viruses in aquatic unicellular eukaryotes<br>*Protist, +ssRNA viruses, Marnaviridae, RdRp, ecology, sequence diversity, viral species, metagenomics, virus isolation, protist viruses* |
| McMinn et al. (2020) | Minireview: The role of viruses in marine photosynthetic biofilms<br>*Algae, bacteria, microphytobenthos, sea ice, virus* |
| Sun et al. (2020) | Host range and coding potential of eukaryotic giant viruses<br>*Nucleo-cytoplasmic large DNA viruses (NCLDVs), algae, protists, co-phylogeny, host switch, auxiliary genes, virus-encoded metabolism, gene repertoire, genome evolution, lateral gene transfers* |
| Zimmerman et al. (2020) | Metabolic and biogeochemical consequences of viral infection in aquatic ecosystems<br>*Biogeochemistry, marine microbiology, microbial bio-oceanography, microbial ecology, virus, water microbiology, host interactions* |
| Rahlff (2019) | The virioneuston: A review on viral–bacterial associations at air–water interfaces<br>*Aerosols, air, bacterioneuston, bubbles, particles, phages, sea interaction, surface microlayer, surfactants, transparent exopolymer particles (TEP)* |
| Yau and Seth-Pasricha (2019) | Viruses of polar aquatic environments<br>*Antarctica, arctic, DNA viruses, freshwater, polar regions, RNA viruses, saline, viruses* |
| Breitbart et al. (2018) | Phage puppet masters of the marine microbial realm<br>*Bacteriophages, microbial ecology, microbiology, systems biology* |
| Coy et al. (2018). | Viruses of eukaryotic algae: Diversity, methods for detection, and future directions<br>*Eukaryotic algal virus, algal-NCLDV, picornavirales, phytoplankton* |
| Evans et al. (2018) | Algal viruses: The (atomic) shape of things to come<br>*Algal virus, atomic force microscopy, electron microscopy, imaging, infection dynamics, virus structure* |
| Horas et al. (2018) | Why are algal viruses not always successful?<br>*Stressors, algal viruses, intrinsic and extrinsic factors, viral life cycle traits, temperature, sunlight, effects, latency period, burst size, host resistance* |
| Engel et al. (2017) | The ocean's vital skin: Toward an integrated understanding of the sea surface microlayer<br>*Sea surface microlayer, air-sea exchange, neuston, aerosols, surface films, gas exchange* |
| Hayes et al. (2017) | Metagenomic approaches to assess bacteriophages in various environmental niches<br>*Virome, phage, marine, microbiota* |
| Mateus (2017) | Bridging the gap between knowing and modeling viruses in marine systems—An upcoming frontier<br>*Plankton modelling, mechanistic models, viruses, marine food webs, microbial loop* |
| Weynberg et al. (2017) | Marine prasinoviruses and their tiny plankton hosts: A review<br>*Virus–host interactions, marine virus ecology, virus-driven evolution* |

(continued)

**Table 6.1** (continued)

| References | Title and author keywords |
|---|---|
| Hurwitz et al. (2016) | Computational prospecting the great viral unknown*bacteriophage, bioinformatics, metagenomics, phage, virome, virus* |
| Record et al. (2016) | Quantifying tradeoffs for marine viruses<br>*Marine virus, host, trade-off, trait, model* |

et al. (2020) analysed available bibliography on virus–bacteria and virus–algae interactions, but exclusively focussing on photosynthetic biofilms of microphytobenthos and sea ice algae. Of relevance in their review is the hypotheses raised to explain the low infection rates in face of the high density of cells in biofilms. McMinn et al. (2020) conceived the development of resistance, the influence of environmental characteristics, or the insufficiency of studies as reasons that prevented drawing any conclusions. The apparent sub-optimal performance of viruses had already been addressed before in a review written by Horas et al. (2018), in which the factors implicated in lowering viral success are enumerated and labelled as intrinsic factors (related to the life cycle traits of the virus) and extrinsic factors (external to the virus and related to their environment).

The review by Breitbart et al. (2018) summarizes the output of numerous studies on the ecology of marine viruses, with a special focus on the interactions between phages and their bacterial hosts and on the contribution of phage particles to the dissolved organic matter pool. In addition, Breitbart et al. (2018) emphasize the role of phages as "puppet masters" of their bacterial hosts, i.e., they control the metabolism of infected bacteria by expressing auxiliary metabolic genes and redirecting host gene expression patterns. This review finalizes with a hypothesis describing successional patterns of bacteria and phages, termed the "royal family model", asserting that only a small number of phages appear to continually dominate a given marine ecosystem, despite the high richness and seasonal differences.

The function of viruses in specific oceanic regions and ecosystems has also received attention in published reviews. Yau and Seth-Pasricha (2019), for instance, looked at the diversity of viruses in aquatic polar regions mostly as an outcome of the advances in genomics-enabled technologies over the previous decade. The ocean–atmosphere interface also received attention in two different reviews addressing viral abundance and activity at the surface microlayers (SML). Engel et al. (2017) identified gaps in the current knowledge of SML but only addressed viruses tangentially. Subsequently, Rahlff (2019) abridged the relevant literature on the dispersal and infection mechanisms of virioneuston.

### 6.3.2 Methodological Approaches

The intense advance in knowledge on the functioning and role of marine viruses observed over the past decade is, in part, a consequence of methodological improvements and new instrumentation as reported in some of the reviews.

Important progress in the field of viral metagenomics has fuelled a constant flow of information on viral communities as reviewed by Hayes et al. (2017). Coy et al. (2018) also addressed the development of molecular and bioinformatic approaches in their review, browsing through the history of eukaryotic algal virus research, highlighting seeming future opportunities created by such methodologies. While the identification of viral metagenomes (viromes) keeps providing new insight, sometimes the efforts of deep sequencing viral communities fail to find a match to known proteins. This shortcoming requires new tools to process the huge amount of information generated, thus bringing novel insights into the diversity and functional roles of viruses. Hurwitz et al. (2016) describe such tools in an overview of current computational methodologies for virome analysis. In a similar approach as that of Coy et al. (2018), Evans et al. (2018) describe the evolution of the virus structure with an emphasis on electron- and atomic force microscopy as instruments for the direct observation of algae–virus interactions. These authors also advance some predictions on the direction of future algal virus studies, made possible by modern high-speed atomic force microscopy methods and instrumentation.

### 6.3.3 Numerical Modelling

The inclusion of viruses in marine ecological models was also addressed in review papers. Record et al. (2016), for instance, dealt with the question of how a trait-based approach can contribute to the understanding of marine virus ecology, by synthesizing current knowledge on virus traits with a particular focus on the quantification of the associated trade-offs. Relying on a nutrient-susceptible-infected-virus model as a framework, Record et al. (2016) have used virulence, host range, and cost of resistance to illustrate how quantification of trade-offs can help to explain observed patterns, generate hypotheses, and improve theoretical understanding of virus ecology. Facing the challenge that the high biodiversity of viruses and the complex virus–host interactions pose to link modelling work with experimental work, Record et al. (2016) also discuss trade-offs as a connection between theory supporting model building and its application in experimental design. On a different approach, Mateus (2017) has focused specifically on mechanistic approaches, tackling the limitations and opportunities of including viruses in marine ecosystem models. Available models are reviewed in Mateus (2017) based on their complexity, i.e., on the way viruses or viral activity are included in models, from simple density-dependent mortality rates up to a more detailed modelling of virus-host dynamics and subsequent differentiated mortality products.

## 6.4 The Omnipresence of Virus in the Sea

The pervasive nature of viruses is probably the second most common remark found in the literature and a predictable corollary of the first, i.e., that they are the most abundant biological entities in the ocean. The presence of virus across all regions of the marine environment has long been established (Angly et al. 2006; Brum et al. 2015; Finke et al. 2017) but not necessarily its magnitude and relevance; these are part of the unfolding knowledge brought by incessant research. An astonishing outcome of the enduring study of marine viruses is that they are no longer considered abundant in the oceans but omnipresent, not passive components of the ecosystems but active forces, critical for the functioning of ecosystems. Early studies have already established the abundance of virus particles in seawater, but subsequent findings shed new light on the persistence and ubiquity of these biological entities (Gregory et al. 2019) and revealed that, on average, viruses outnumber microbial cells approximately ten-fold at the surface and 16-fold in deeper waters. Therefore, it is no overstatement to replace the famous remark "we live in a dancing matrix of viruses" (Thomas and Parker 1974) with "drowning in viruses" (Handley and Virgin 2019) as a more marine-oriented and pertinent remark.

### 6.4.1 Different Environments, Same Incidence

Being obligate parasites, the persistence of viruses is tightly coupled with their hosts. Hence, host traits such as abundance, size, and physiological state determine not just their abundance, but also the biogeography of their viruses (Chow and Suttle 2015; Gazitúa et al. 2021; Gran-Stadniczeñko et al. 2019). The widespread geographic distribution has been observed for the most abundant family of marine RNA viruses, the Marnaviridae (Vlok et al. 2019) but other groups also follow the same pattern. Studies made along a latitude gradient in the North Atlantic Ocean (from a subtropical region to a temperate region) once again revealed the abundance and diversity of viruses, as well as their role in the regulation of heterotrophic microbial communities, pointing to virus-mediated mortality as the primary loss process for bacteria with an average of 55% of the total mortality (Mojica and Brussaard 2020). On a more regional scale, viral activity has been studied in estuarine transitional zones that are known to be highly dynamic and more diverse habitats, but of which the spatiotemporal distribution of viruses is still largely unknown (Cai et al. 2016; Jasna et al. 2017). So far, an unequivocal conclusion seems to be that these estuarine regions harbour their unique viral assemblages (Labbé et al. 2018).

The research on viral activity on the poles has intensified over the past decade, mostly fuelled by advances in genomics. Making up 14% of the Earth's biosphere, these polar regions are particularly interesting because grazing by macrofauna is limited and viruses are important agents of mortality. They also modulate community dynamics at seasonal and spatial scales (Yau and Seth-Pasricha 2019). A high diversity of viromes of the order Caudovirales has been reported for both the Arctic and Antarctic in a pattern consistent with previously described viromes from the Pacific Ocean and a range of different biomes (Yang et al. 2019). Similar observations have been made for the Arctic Ocean (Gregory et al. 2019; Sandaa et al. 2018; Vaqué et al. 2021) while a higher presence of nucleocytoplasmic large DNA viruses (mainly chloroviruses and mimiviruses) seems to occur in the Southern Ocean surrounding the Antarctic (Flaviani et al. 2018).

The discoveries of the presence of viruses across the marine system are not exclusive to small particles with small genomes containing a few protein-encoding genes. Viruses with double-stranded DNA (dsDNA) genomes larger than 300 kb (and up to 1.2 Mb), which encode hundreds of proteins, are also discovered and characterized with increasing frequency (Sun et al. 2020). Giruses are an example (Claverie 2006; van Etten 2011) like those that infect the coccolithophore alga *Emiliania huxleyi* (termed EhV viruses) and which terminate the massive blooms of these algae in the ocean. Such large viruses can be found in wide coastal and mid-oceanic areas at high latitudes in both hemispheres. Nucleocytoplasmic large DNA viruses (NCLDVs) that infect diverse eukaryotes are also ubiquitous in marine environments but still little is known about their biogeography and ecology (Endo et al. 2020).

### 6.4.2 From Surface to Bottom, and deeper

Viruses also proliferate at the air–water interface of marine ecosystems as part of the neuston (the microorganisms dwelling on the biofilm-like surface habitats). Unlike bacterial communities in this sea-surface microlayer, little is known about the viruses found there, the virioneuston (Engel et al. 2017). In a review on this subject written by Rahlff (2019) viruses are labelled as the most enigmatic biological entities in the SML, because of their decisive ecological impact on the microbial loop with a potential influence on the major air-water exchange processes. Some of these viruses that infect phytoplankton cells can leave the SML to the atmosphere and be transported for hundreds of kilometres, while retaining their infectious capacity, ending up infecting distant phytoplankton populations (Sharoni et al. 2015). In many regions of the ocean it is possible to observe photosynthetic biofilms of algae extending from the air–water interface down to the bottom sediments, linking these two boundaries of the water column. These biological structures, with a geographical distribution extending from coastal areas to the open ocean, and from the equator to the poles, are also infested and, to a certain degree, shaped by the activity of viruses (McMinn et al. 2020).

While the abundance, biodiversity, and distribution of viral communities in the surface layers of the ocean have been well studied, their presence in deeper regions and sub-seafloor are poorly examined until now. However, new studies have unveiled additional details about the "enigmatic virosphere" (Yau and Seth-Pasricha 2019) and the hadal biosphere and sub-seafloor biosphere provide good examples. Mostly overlooked for decades, these ecosystems were the subject of viral studies lately. Manea et al. (2019), for instance, have addressed the activity and interactions of benthic viruses and microbes in three hadal trenches (Japan, Izu-Ogasawara, and Mariana trenches) and nearby abyssal sites. Their results reveal that hadal trenches support higher abundances and biomass of microbes when compared with the surrounding abyssal sites, and that the high microbial biomass of hadal trenches promotes high rates of viral infection and cell lysis. Consequently, the viral component of the hadal trenches is highly dynamic and enhances microbial biomass production by releasing large amounts of highly labile and promptly available organic material.

Also zooming on marine trenches (Mariana, Yap, and Kermadec Trenches) Jian et al. (2021) tackled viruses from a community diversity perspective. Based on the analysis of a viral genome dataset composed of 19 microbial metagenomes derived from seawater and sediment samples from the trenches, the results of Jian et al. (2021) highlight the genomic novelty and environment-driven diversity of viral communities at such sites. Again, viruses appeared to have a decisive role influencing the entire hadal ecosystem, by reprogramming the metabolism of their hosts and therefore controlling the key microbial communities. These and similar studies seem to demonstrate that viral lysis is the main mortality factor for microorganisms in deep-sea sediments, and a factor that induces changes in the composition of the microbial community and the release of cellular components to overlying water columns (Heinrichs et al. 2020, 2021).

If knowledge on viral communities is still incipient for the deeper reaches of the ocean, even less is known about them at the sub-seafloor biosphere. Novel findings in the Baltic Sea sub-seafloor also point to an important role of viruses in biogeochemical processes by controlling the microbial dynamics in these regions (Cai et al. 2019). A highly diverse community of viruses is reported for such environments with densities of up to $1.8 \times 10^{10}$ viruses $cm^{-3}$. In addition, these findings confirm the high potential of viral production down to 37 m below the seafloor in 6000-years-old sediments, advancing the suggestion that viruses of phototrophic hosts may persist in marine sediments for thousands of years. Such a claim seems to imply that besides omnipresent, viruses may as well be perpetual.

## 6.5 Recent Developments in Viral Research

The progress of marine virology owes much to the emergence of metagenomics and its widespread application to the study of the viral metagenome (virome). Not only metagenomics opened the way to assess viral diversity across a wide range of marine environments, but it also proved to be a fertile field of research to discover novel viruses. The ongoing progress in the study of viruses is in no way exclusive to the fields in which metagenomics plays a crucial role, but also extends to other disciplines. Visualization techniques are such an example. Recent developments have focused on the direct observation of the morphological dynamics of infection, using electron microscopy and atomic force microscopy, but also highlight the potential in modern high-speed atomic force microscopy methods and instrumentation (Evans et al. 2018). Nonetheless, given their central role in the impressive advancement of the study of the marine virome, only metagenomics and innovative ways to process and analyze omics-data are briefly addressed in this section.

### 6.5.1 The Endless Harvest in the Field of Metagenomics

The term metagenomics dates back to 1998 (Handelsman et al. 1998) and refers to the direct sequencing and analysis of all genetic material contained in an environmental sample (Thomas et al. 2012). Metagenomic analysis relies on two main methods: marker gene amplification metagenomics, typically using the 16S ribosomal RNA gene (Handelsman 2009), and the entire sequencing of the nucleic acid present in a sample termed shotgun metagenomics (Jovel et al. 2016). With these two approaches, metagenomics becomes the most effective and all-inclusive method to disclose and analyze the genome of uncultured microbial populations (Duhaime and Sullivan 2012). Viral metagenomics has expanded considerably as demonstrated by the review written by Hayes et al. (2017). Since 2002, when the first application of viral metagenomics to uncultured marine samples was reported by Breitbart et al. (2002), virome studies have been applied to countless marine ecosystems. To showcase the capacity of metagenomics, Coutinho et al. (2017) reported a data set of 27,346 marine virome contigs that includes 44 complete genomes, but they still emphasized that even with the improvements in metagenomics much of the virome remains uncharacterized.

The present power of metagenomics is best illustrated by its ability to process small volumes of water and determine the microbial diversity contained therein. Less than a decade ago, such analyses using the early sequencing technologies required sample volumes of tens of litres to as much as a thousand litres of water, to allow the processing of a substantial quantity of DNA, usually in the order of micrograms (Hoeijmakers et al. 2011; Lima-Mendez et al. 2015). Newer technologies, however, require smaller quantities (nanograms) of DNA. A good demonstration of this new technology is found in Flaviani et al. (2017) who assumes that only 250 mL of seawater is enough to identify the dominant microbial taxa (from viruses to protists) in any marine environment. As evidence for this supposition, Flaviani et al. (2017) reported a total of 834 bacterial/archaeal, 346 eukaryotic, and 254 unique virus phylotypes in such a small volume of seawater. In addition, using a metagenomic-barcoding comparative analysis, they also concluded that viruses are the probable origin of microbial environmental DNA (meDNA).

Metagenomics is a verified method to explore microbial and viral diversity, and will surely be decisive to unfold the many unknowns around host–virus relationships. As stated by Yau and Seth-Pasricha (2019) in their review on viruses in polar environments, one of the final goals of metagenomics is to answer the unresolved question in many marine biomes: "which virus infects what host?" The combination of metagenomic and metatranscriptomics analyses may help to achieve this goal as illustrated by the study of Zeigler Allen et al. (2017) on the Baltic Sea virome. Far from simply producing massive volumes of sequencing data, metagenomic studies are continually yielding new insights, whether they come from the discovery of novel genes and enzymes (Culligan et al. 2014) or from revealing the dynamics of the viral community composition and function and the impact that viruses may have on their hosts.

### 6.5.2 Novel Applications, Innovative Methodologies, New Protocols

Metagenomics is virtually transversal to all fields of research dealing with viral activity in the oceans and new examples of its application are a constant in the literature on marine viruses. Hurwitz et al. (2016) mentioned the use of metagenomics to understand the interaction between uncultured bacteria, their phages, and the environment. Particular emphasis was put on how viromes provide new insight into auxiliary metabolic genes (phage-encoded host genes) capable of reprogramming host metabolism during infection. Another example is the study of the viruses of Bacteroidetes, one of the most abundant heterotrophic bacterial taxa and major recycler of phytoplankton-derived organic matter (Tominaga et al. 2020). Likewise, giant viruses infecting the microalga *Emiliania huxleyi* have also been studied with the specific aim of unravelling their life cycle and hijacking strategies (Ku et al. 2020). The viral transcriptional trajectory obtained from clustering single cells based on viral expression profiles revealed exclusive viral genetic programs, composed by genes with distinct promoter elements controlling sequential expression.

The progress in new protocols to profile marine viruses means targeting specific clades, usually those identified as having a relevant role on marine systems. Megaviridae, for instance, a family of giant ubiquitous viruses infecting unicellular eukaryotes, have an important impact on marine microbial community composition and dynamics. However, the scarce detection of Megaviridae sequences in metagenomes, allied to the limited reference sequences used to design specific primers for this viral group, prevented a detailed characterization of their diversity and biogeography. This led to the development of a tool to examine the richness of natural Megaviridae communities at the population level in natural environments consisting of a set of 82 degenerated primers (MEGAPRIMER) targeting DNA polymerase genes (polBs) of Megaviridae (Li et al. 2018). Subsequently, Prodinger et al. (2020) have improved the methodology to profile Mimiviridae, a group of viruses with large genomes and virions, using varying PCR conditions and purification protocols to streamline the previously proposed meta-barcoding MEGAPRIMER procedure. This novel methodology yielded similar results to the original protocol, but decreased the required amount of meDNA by 90%.

Another new protocol to estimate host-virus relationships in diatoms of the genus *Chaetoceros* was advanced, leading to the detection of previously unrecognized viruses that specifically infect *Chaetoceros tenuissimus* and *C. setoensis* cells (Tomaru et al. 2015; Tomaru and Kimura 2020). On a more generalist approach, Beckett and Weitz (2018) have faced the challenge in determining in situ rates of viral-induced lysis and contributed with a discussion on the implications of using the modified dilution method and alternative dilution-based approaches in such estimations. Coutinho et al. (2017) proposed a new method for host prediction, confirming that viruses infect dominant members of the marine microbiome. Finally, and on a rather different approach, Tsiola et al. (2020) suggest using viral metagenome sequencing methodologies in samples impacted by human-related activities, in order to assess the relevance of a potential contribution of viromics to define the ecological quality status in coastal waters.

The development of sampling techniques for the quantification of viruses has also been extended to media other than water. In sediments, for instance, the interference that the sediment matrix imposed on flow cytometric quantification of viruses has been a long-recognized limitation. To circumvent this shortcoming Heinrichs et al. (2021) developed a new protocol using a Nycodenz density gradient in the process of separating viruses from sediment particles, leading to a faster and more accurate enumeration of marine sediment viruses, but also enabling virus sorting, targeted viromics, and single-virus sequencing.

### 6.5.3 Tackling Omics-Data

Viral diversity is overwhelming but regardless of the build-up of viral metagenomic data over the past decade, the marine virome is still a large pool of unexplored genetic diversity. The enormous amount of virome information in omics-data presents a problem that can only be tackled with more comprehensive reference databases and computational tools, in order to fully explore the viral taxonomy and gene function (Kieft et al. 2020). Only the use of such post-processing tools will enable to probe viral "dark matter", the fraction of sequence data that cannot be attributed to any specific viral taxonomy or function, accounting for something between 70% and 90% of the sequences in viral libraries used for metagenomic analysis (Handley and Virgin 2019). Such vast unknown sequence data is mostly attributed to the lack of universal gene markers and database representatives, but also to insufficiently advanced identification tools. Hence, to cope with such limitations and harvest the full potential of the data that remains hidden in metagenomic/ metatranscriptomic sequence datasets many computational tools have been developed in recent years.

VirSorter2 (Guo et al. 2021), ViralRecall (Aylward and Moniruzzaman 2021), and VirION2 (Zablocki et al. 2021) are just a few examples of the recently developed methodologies added to the virosphere exploring tool-kit. VirSorter2 is a virus identification tool (for both DNA and RNA viruses) aiming to improve the accuracy and range of virus sequence detection, by leveraging genome-informed database advances across an assemblage of customized automatic classifiers. Targeting large marine viruses, ViralRecall is a bioinformatic tool for the identification of NCLDVs signatures in omics-data. The method relies on a library of giant virus orthologous groups (GVOGs) to identify sequences with signatures of NCLDVs. This approach avoids previous limitations associated with the particularly high diversity of genes encoded in the genomes and the high level of sequence divergence found between NCLDV families. VirION2 is not just a computational tool, but merges a short- and long-read metagenomics wet-lab with informatics to enhance the utility of long-read viral metagenomics and achieve accurate gene prediction. Their advocates reported an optimized error-correction strategy using long- and short-read data (99.97%

accuracy), thus avoiding the limitations of high DNA requirements and sequencing errors.

## 6.6 Emergent Themes

The profusion of studies on marine viruses published over the past few years attests to their importance on marine ecosystems. While many questions have been answered, many more have been raised. Past mysteries that once surrounded viral action in the ocean have been solved, but new unknowns keep pushing forward the research on these elusive particles. Among the vast number of research topics on marine viruses, several emerging themes will surely open numerous investigation venues. Being impractical to address an exhaustive list of such themes, some deserve special attention and are briefly discussed here.

### 6.6.1 Resistance to Infection

Viruses rely on host cell machinery for their dissemination. As such, they depend on a reliable host population. However, hosts are known to develop resistance to their viruses and this adaptation is no exception in the marine environment. While most studies focused on bacterial resistance to viral infection (Avrani et al. 2011, 2012; Avrani and Lindell 2015), there are indications that the same mechanism may occur in eukaryotic primary producers, as demonstrated for two groups of phytoplankton, *Ostreococcus tauri* (Heath et al. 2017; Thomas et al. 2011) and the coccolithophorid *Emiliania huxleyi* (Ruiz et al. 2017).

Defence responses shared by viral host cells include multiple mechanisms for destroying the viral DNA upon infection and resistance to phage adsorption gained by mutational changes in the cell surface receptors. To address these dynamics, Thomas et al. (2011) set to address the following question: how do host cells survive when they coexist with their viruses? The experiments of Thomas et al. (2011) with clonal lines of three picoeukaryotic green algae (*Bathycoccus* sp., *Micromonas* sp., and *Ostreococcus tauri*) showed acquired resistance to specific viruses following a round of infection. Two mechanisms of resistance operate in *O. tauri*: the algae acquired tolerance to their virus and released them consistently and viruses attached to the host cells did not develop new particles. The host blocks the virus from the point of adsorption by acting on the key at the interface of virus and host, the glycans (the carbohydrate modifications on proteins or lipids that act as ligands), in response to initial infection events.

For Yau et al. (2018), the strong selective pressure of abundant prasinoviruses in eutrophic coastal waters may confer *O. tauri* the ability to resist an extensive set of independently isolated viruses, suggesting a fast adaptative mechanism of resistance to the virus. Such rationale seems consistent with the observation that more closely related viruses can lyse a similar range of host strains (Bellec et al. 2014; Clerissi et al. 2012). If so, resistance to a particular virus strain may well confer resistance to genetically related strains. Mordecai et al. (2017), for instance, reported a different life cycle strategy of the dsDNA EhV (*E. huxleyi* viruses) where the presence of viral RNA in the virus-resistant haploid cell of *E. huxleyi* pointed not just to a novel infection mechanism, but also to the co-existence of viruses and host. Resistance to viral infection was also reported for the small marine phytoplankton prasinophytes in the form of an immunity chromosome (Weynberg et al. 2017).

Despite the advantage, developing and maintaining resistance to viruses is often associated with a fitness cost, reported to vary in form and magnitude. Such cost can be expressed by a decrease in growth rates (Frickel et al. 2016; Lennon et al. 2007), increased susceptibility to other viruses (Clerissi et al. 2012; Marston et al. 2012), impaired competitive ability (Bohannan et al. 2002), and loss or lessened function of receptor proteins that impair substrate uptake or enzyme secretion (Seed et al. 2012). Besides, the maintenance of mechanisms for virus inactivation may require additional energy. Early studies with the marine picoplankton *O. tauri* suggested that resistance was associated with decreased host fitness in terms of growth rate (Thomas et al. 2011). However, no detectable cost of resistance (measured by growth rate or competitive ability) was detected for *O. tauri* suggesting that, regardless of the selection environment, the shift from susceptibility to resistance is more common than a shift from resistance to susceptibility (Heath et al. 2017). The apparent lack of a direct cost of resistance may be the result of an interaction between environmental conditions and virus–host co-evolution dynamics. Finally, acquiring resistance does not necessarily mean that it becomes a permanent trait. Experiments using *Escherichia coli* have shown that resistance to infection can eventually be decreased or even be lost, once viruses disappeared and the selection pressure is removed (Meyer et al. 2010).

### 6.6.2 Ocean Acidification

Over the course of the past century, anthropogenic activities have increased atmospheric $pCO_2$ at an estimated rate of ~1.8 ppm per year with the ocean surface layers absorbing ~25% of these $CO_2$ emissions (Le Quéré et al. 2016). By the end of the century, the average pH of the ocean surface waters is predicted to fall from ~8.1 to ~7.8 (Feely et al. 2009). The resulting increase in $CO_2$ and $HCO_3^-$ concentrations and decreased pH has been labelled as ocean acidification (Caldeira and Wickett 2003). Shifts in the water pH are particularly relevant for marine plankton, especially for species that possess calcium carbonate structures (e.g., coccolithophores) because less alkaline conditions lead to higher dissolution rates of carbonate.

Increase in $CO_2$ concentration is associated with ocean acidification and may induce unequal effects among phytoplankton groups/species, as reported for the increase in photosynthetic efficiency of the marine coccolithophore *E. huxleyi*, when compared to the diatom *Skeletonema costatum* or the flagellate *Phaeocystis globosa* (Rost et al. 2003). An obvious outcome of such an effect is the shift in phytoplankton taxonomic composition, dominance, and species succession in the ocean (Bach and Taucher 2019; Schulz et al. 2017). Changes in the pH also seem to influence the virus–host dynamics. For instance, Highfield et al. (2017) looked at the effects of elevated $pCO_2$ on *E. huxleyi* genetic diversity and EhVs in large volume enclosures and found that EhV diversity was much lower in the high-$pCO_2$ treatment enclosure, where the growth of *E. huxleyi* was not inhibited.

The effect of ocean acidification on marine viruses and their plankton hosts is controversial and even more in combination with other anthropogenic stressors like eutrophication. In this regard, Malits et al. (2021) observed that under an ocean acidification scenario, phage production and ensuing organic carbon release rates were considerably decreased in the nutrient-replete winter condition. Malits et al. (2021) concluded that eutrophication was a stimulus to viral production regardless of the season or initial conditions. These authors also proposed an antagonistic interplay between these two stressors with high nutrient loads lowering the negative effect of ocean acidification on viral lysis.

So far, available reports seem to provide contradictory results. No obvious effects of higher $pCO_2$ have been detected on lytic viral production in acidification mesocosm experiments in distinct environments: the Mediterranean waters (Tsiola et al. 2017), the Arctic (Vaqué et al. 2021), and the Baltic Sea (Crawfurd et al. 2017). Consequently, results on the effect of ocean acidification on virioplankton activity are still inconclusive, impairing any extrapolation to assess its subsequent impact on the carbon and nutrient fluxes on the microbial food web. Future work will surely shed some light on these processes because virus infection is generally implicated as a major factor in terminating phytoplankton blooms and, as pointed out by Highfield et al. (2017), no study of the effect of ocean acidification in phytoplankton can be complete if it does not include an assessment of viruses.

### 6.6.3 Response to Climate Change

The relationship between virus and climate change is not a new topic but one that has received special attention in recent years. Understanding the role of algal viruses under changing environmental conditions such as temperature may be particularly relevant in vulnerable regions like the poles, where climate changes effects are already visible (Maat et al. 2017; Piedade et al. 2018). Virus dynamics are affected by shifting environmental factors because viruses exert control on their host. Temperature, for instance, impacts virus infectivity and proliferation by altering the metabolic activity of the host and by changing lytic effects (Demory et al. 2017; Mojica and Brussaard 2014; Toseland et al. 2013). Kimura and Tomaru (2017) reported that viral proliferation in *C. tenuissimus* cells changes according to water temperature. Williamson and Paul (2006) even suggested that temperature rise promotes the switch from a lysogenic to a lytic cycle in marine phage–host systems, a process that impacts the entire viral production.

Other studies on the effect of temperature increase in viral infectivity report a shortening of the latent periods (up to 50%) and an increase in the burst size (up to 40%) in the Arctic picoeukaryote community (Maat et al. 2017). The magnitude of response to temperature was, nonetheless, high for the different viruses and host strains assessed suggesting virus-specific effects. Moreover, environmental change may probably drive virus selection and host population dynamics by inducing distinct responses in different viruses infecting the same host strain, as previously suggested by Nagasaki and Yamaguchi (1998).

Until now, thermal stability in marine virus-host interactions has been reported mostly for phage–bacterium systems (Borriss et al. 2003; D'Amico et al. 2006). More recently, however, a study in the English Channel waters revealed that temperature-regulated growth rates of *Micromonas* strains, a ubiquitous genus belonging to the picophytoplankton fraction, were responsible for shortened latent periods and increased viral burst sizes upon infection (Demory et al. 2017). Similar conclusions were drawn by Maat et al. (2017) for *Micromonas polaris* in Arctic waters, pointing to the fact that the response to temperature can affect the Arctic picoeukaryote community composition both in the short term (seasonal cycles) and long term (global warming). After investigating the impact of temperature change on virus infectivity and production, Maat et al. (2017) hypothesized that the increase of temperature associated with climate change will lead to an increase in viral infectivity and stimulate virus production due to shorter latent periods and higher burst sizes.

A curious link between warming and viral activity has been investigated in the absorption of marine viruses to glacier-delivered fine sediments (after ice melting) followed by its removal from the water column by sinking (Maat et al. 2019). Substantial losses of algal and bacterial viruses were observed with a magnitude directly related to higher sediment load and viral abundance. The glacier-derived sediments lowered host mortality by decreasing virus infection and delaying host growth, thus exerting a strong impact on microbial interactions. Finally, the study by Maat et al. (2017) also showed that the adsorption of viruses to sediment is reversible, meaning that desorbed viruses still retained the ability to infect their respective hosts. These authors ultimately concluded that the sediment delayed host lysis and virus production.

Eventually, the importance of climate change does not just rests on direct and indirect consequences of the activity of marine viruses, but also on the cascading effects on the biogeochemical cycles, food webs, and the metabolic balance of the ocean (Danovaro et al. 2011). Knowing the role of viruses in the ocean, a central question is whether they will exacerbate or attenuate the magnitude of climate changes on marine ecosystems. Moreover, future ocean acidification and warming may, individually and cumulatively, affect the biogeochemistry of the ocean through shifts in phytoplankton species composition and sinking rates. As an example, warming and increased $pCO_2$ contribute together to the greatly decreased proportion of diatoms to dinoflagellates (Feng et al. 2021), a shift that may have a considerable impact on the behaviour of viruses and their role in the carbon and nutrient fluxes.

### 6.6.4 Viral Action during Harmful Algal Blooms

The role of viruses on the dynamics of harmful algal blooms (HABs) is a topic that will surely receive special attention during the coming years. The recent increase in the occurrence, geographic expansion, and persistence of HABs has prompted researchers to investigate their triggers and to identify the factors that control their dynamics, in an attempt to effectively manage and control such blooms (Anderson et al. 2012). Viruses are among the many biological agents considered to be a biosecure way of influencing HAB occurrence and dynamics (Du et al. 2020). However, while the shifts in algal communities during HABs have been widely investigated, the characteristics of viral communities during such episodes are poorly understood. The important role that viruses have in phytoplankton blooms has been explored in many works on diatoms and cyanobacteria, but only a few studies have addressed successional trajectories of virioplankton communities in dinoflagellate blooms, which are common in HABs. While the role of viruses is not clear, metagenomic studies of the dinoflagellate *Gymnodinium catenatum* blooms support the participation of viruses in the evolution of the bloom, delaying its spreading and potentially inducing its termination (Garretto et al. 2019).

## 6.7 Viruses and Marine Models

Viruses have a decisive role in marine ecosystems but still their inclusion in marine models remains the exception rather than the rule (Lindemann et al. 2017; Mateus 2017). The need to account for viruses in the marine ecosystem and geochemical models is indisputable given their central role in the composition, diversity, and functioning of primary producers and mineralizers (Jacquet et al. 2010; Sandaa et al. 2017), up to their contribution in the modulation of nutrient cycling (Gazitúa et al. 2021). Some authors propose modelling studies including viruses as a way to comprehend (and predict) the extent to which the Arctic phytoplankton community can be influenced by changes in infection dynamics associated with temperature changes (Maat et al. 2017; Piedade et al. 2018). Danovaro et al. (2011) go even further on such claims, attesting that the inclusion of the viral component on ocean climate models will improve the accuracy of the predictions of the climate change impacts. Despite this apparent inevitability, equipping plankton ecosystem models with a realistic description of the viral activity is far from being an achievable goal anytime soon.

### 6.7.1 Different Modelling Approaches

The link between the host and viral performance has been characterized empirically and numerous theories have been advanced for understanding intracellular dynamics. Such theories, however, are too detailed to be included in models without adding too much (and unnecessary) complexity. Nonetheless, the inclusion of viruses in models not only has increased, but also diversified in the complexity with which viral activity is described (Jacquet et al. 2010; Mateus 2017). Several models have been developed with the specific aim of exploring the viruses-host relationship. Of particular interest has been the role of viruses on the microbial food web and their potential impact on the biogeochemical cycles in the ocean. Mateus (2017) provided a thorough description of the many available models to date, along with an in-depth look at their assumptions and components. However, just like marine virus research in general, numerical modelling is also subject to fast developments. Consequently, new models or model components (algorithms) have been proposed ever since.

Early approaches treated the "problem" of marine viruses by addressing viral activity as a black-box model where elements go in (algae, bacteria) and flow out (lysed products made of among other organic matter and nutrients). The mechanisms inside the box were not necessarily unknown, but either lacked a more detailed description or were too complex to be fully described by the mathematical formulae contained in those models. As such, only the global effects of the box were analyzed by studying the inputs and outputs. This approach is still valid in current modelling approaches, sometimes with additional components in the model that increase its complexity, but still lacking a full mechanistic approach. For instance, the infection phenomena can lack detail and only its results analyzed, if the infection output of viruses is to be determined. Such "grey-box" models have been explored (Larsen et al. 2015; Sandaa et al. 2017) and an example was proposed by Thingstad and Våge (2019) to modulate the balance between defensive and competitive abilities in the host community. Their model also addressed the variation in the equilibrium of the food web structure and virus-to-host ratios, as well as the stable coexistence of viruses and predators. This scheme can be used essentially for large-scale studies when the impact of viruses on the food webs is complex, but still can be integrated into a simple representation of their influence on the microbial community composition and on the element cycles in the water. On the other end of the complexity, spectrum models have explored the effect of traits, the dynamics of the virus–host system, and how they interact with environmental factors (Beckett and Weitz 2018; Talmy et al. 2019). Flynn et al. (2021) advanced these models by a detailed mechanistic formulation for the interactions between a single lytic-virus – phytoplankton-host couple.

### 6.7.2 Challenges Ahead

Modelling marine viruses is still an upcoming frontier (Mateus 2017) but a promising endeavour nonetheless. Models of marine viruses will eventually catch up with the incessant developments in other disciplines. But until then, there are many challenges to be faced even if we ignore the ever-increasing demand of computational power required to run complex marine models. Some are the same posed for ecological models such as the plasticity in organism traits, relationships between physiological traits, and environmental forcing, or diversity among organisms. On a more specific approach, models for viruses will have to address host–phage interactions with detail if viral plasticity responses across systems are desired (Choua and Bonachela 2019). Viral mutation rates will possibly need some attention in models as many viruses evolve rapidly (Peck and Lauring 2018), adapting to environmental pressure and potentially changing the microbial community around them. Besides, the role of competitive and defensive traits in host diversity will also deserve some attention in models to achieve a realistic description of virus abundance and activity, and more accurate estimates on the impact of shunting material out of the predatory pathway (Thingstad and Våge 2019).

The diversified nature of data available to improve predictive models will surely be one of the most demanding problems or, in the words of Hellweger (2020), the grand challenge. The variety is substantially, extending from molecular observational data (e.g., gene, transcript) to the usual type of data used in ecosystem models (e.g., nutrient concentrations, biomass). To cope with the considerable differences among data sets, models necessarily need to become more mechanistically detailed and complex, a development that, according to Hellweger (2020), implies technical and cultural challenges for the ecological modelling community. In the meantime, the improvement of numerical models will surely continue providing a singular contribution to the ongoing study of viral dynamics in the ocean, but it will also offer new insights into some of the emergent themes discussed here.

## 6.8 Concluding Remarks

From the invasive and destructive process of infection, replication, and lysis, to the developing action in microbial populations and controlling role in elemental and energy cycles in the ocean, viruses are, paradoxically, agents of chaos and promoter of order in the marine ecosystem. Viruses organize microbial communities by disrupting populations of bacteria and algae. In doing so, the tiniest of all biological entities displays the astonishing capacity to transform entire ecosystems. This is surely one of the reasons why marine viruses received so much attention. The spectacular scientific advances of the past decades made this an exciting, stimulating, and prolific period in the study of marine viruses. The content of this chapter exposes such progress tangentially. At best, it emphasized the unique role of these biological entities in the marine biome.

New frontiers are drawn every day as the mysteries of the great viral unknown are revealed. This means that the massive output of knowledge seen over the last decades will surely be matched by equivalent scientific production in years to come. Metagenomic tools will undoubtedly keep on being improved, new viromes discovered in many more marine ecosystems, the virus–host dynamics understood with unprecedented detail, new roles in the elemental cycles of the ocean revealed, and viruses will eventually reclaim their due place in computational models.

**Acknowledgements** This work is dedicated to the memory of Professor Martin Sprung, for his enthusiastic commitment to unveil the mysteries of marine life to his students.

## References

Anderson DM, Cembella AD, Hallegraeff GM (2012) Progress in understanding harmful algal blooms: paradigm shifts and new technologies for research, monitoring, and management. Annu Rev Mar Sci 4:143–176. https://doi.org/10.1146/annurev-marine-120308-081121
Angly FE, Felts B, Breitbart M et al (2006) The marine viromes of four oceanic regions. PLoS Biol 4:2121–2131. https://doi.org/10.1371/journal.pbio.0040368
Avrani S, Lindell D (2015) Convergent evolution toward an improved growth rate and a reduced resistance range in Prochlorococcus strains resistant to phage. Proc Natl Acad Sci 112:E2191–E2200. https://doi.org/10.1073/pnas.1420347112
Avrani S, Schwartz DA, Lindell D (2012) Virus-host swinging party in the oceans. Mob Genet Elements 2:88–95. https://doi.org/10.4161/mge.20031
Avrani S, Wurtzel O, Sharon I et al (2011) Genomic island variability facilitates Prochlorococcusvirus coexistence. Nature 474:604–608. https://doi.org/10.1038/nature10172
Aylward FO, Moniruzzaman M (2021) ViralRecall-A flexible command-line tool for the detection of giant virus signatures in 'omic data. Viruses 13:150. https://doi.org/10.3390/v13020150
Bach LT, Taucher J (2019) $CO_2$ effects on diatoms: a synthesis of more than a decade of ocean acidification experiments with natural communities. Ocean Sci 15:1159–1175. https://doi.org/10.5194/os-15-1159-2019
Beckett SJ, Weitz JS (2018) The effect of strain level diversity on robust inference of virus-induced mortality of phytoplankton. Front Microbiol 9:1850. https://doi.org/10.3389/fmicb.2018.01850
Bellec L, Clerissi C, Edern R et al (2014) Cophylogenetic interactions between marine viruses and eukaryotic picophytoplankton. BMC Evol Biol 14:59. https://doi.org/10.1186/1471-2148-14-59
Bohannan BJM, Kerr B, Jessup CM et al (2002) Trade-offs and coexistence in microbial microcosms. Antonie van Leeuwenhoek. Int J Gen Mol Microbiol 81:107–115. https://doi.org/10.1023/A:1020585711378
Bonnain C, Breitbart M, Buck KN (2016) The Ferrojan horse hypothesis: iron-virus interactions in the ocean. Front Mar Sci 3:82. https://doi.org/10.3389/fmars.2016.00082
Borriss M, Helmke E, Hanschke R, Schweder T (2003) Isolation and characterization of marine psychrophilic phage-host systems from Arctic Sea ice. Extremophiles 7:377–384. https://doi.org/10.1007/s00792-003-0334-7
Breitbart M (2012) Marine viruses: truth or dare. Annu Rev Mar Sci 4:425–448. https://doi.org/10.1146/annurev-marine-120709-142805
Breitbart M, Bonnain C, Malki K, Sawaya NA (2018) Phage puppet masters of the marine microbial realm. Nat Microbiol 3:754–766. https://doi.org/10.1038/s41564-018-0166-y
Breitbart M, Salamon P, Andresen B et al (2002) Genomic analysis of uncultured marine viral communities. Proc Natl Acad Sci 99:14250–14255. https://doi.org/10.1073/pnas.202488399
Breitbart M, Thompson LR, Suttle CA, Sullivan MB (2007) Exploring the vast diversity of marine viruses. Oceanography 20:135–139. https://doi.org/10.5670/oceanog.2007.58
Brum JR, Cesar Ignacio-Espinoza J, Roux S et al (2015) Patterns and ecological drivers of ocean viral communities. Science 348:1261498. https://doi.org/10.1126/science.1261498
Brussaard CPD, Baudoux AC, Rodríguez-Valera F (2016) Marine viruses. In: Stal LJ, Cretoiu MS (eds) The marine microbiome: An untapped source of biodiversity and biotechnological potential. Springer International Publishing, pp 155–183
Cai L, Jørgensen BB, Suttle CA et al (2019) Active and diverse viruses persist in the deep sub-seafloor sediments over thousands of years. ISME J 13:1857–1864. https://doi.org/10.1038/s41396-019-0397-9
Cai L, Zhang R, He Y et al (2016) Metagenomic analysis of virioplankton of the subtropical Jiulong River estuary, China. Viruses 8:35. https://doi.org/10.3390/v8020035
Caldeira K, Wickett ME (2003) Anthropogenic carbon and ocean pH. Nature 425:365. https://doi.org/10.1038/425365a

Choua M, Bonachela JA (2019) Ecological and evolutionary consequences of viral plasticity. Am Nat 193:346–358. https://doi.org/10.1086/701668
Chow CET, Suttle CA (2015) Biogeography of viruses in the sea. Annu Rev Virol 2:41–66. https://doi.org/10.1146/annurev-virology-031413-085540
Claverie JM (2006) Viruses take center stage in cellular evolution. Genome Biol 7:110. https://doi.org/10.1186/gb-2006-7-6-110
Clerissi C, Desdevises Y, Grimsley N (2012) Prasinoviruses of the marine green alga *Ostreococcus tauri* are mainly species specific. J Virol 86:4611–4619. https://doi.org/10.1128/jvi.07221-11
Coutinho FH, Silveira CB, Gregoracci GB et al (2017) Marine viruses discovered via metagenomics shed light on viral strategies throughout the oceans. Nat Commun 8:15955. https://doi.org/10.1038/ncomms15955
Coy SR, Gann ER, Pound HL et al (2018) Viruses of eukaryotic algae: diversity, methods for detection, and future directions. Viruses 10:487. https://doi.org/10.3390/v10090487
Crawfurd K, Alvarez-Fernandez S, Mojica K et al (2017) Alterations in microbial community composition with increasing $fCO_2$: a mesocosm study in the eastern Baltic Sea. Biogeosciences 14:3831–3849. https://doi.org/10.5194/bg-14-3831-2017
Culligan EP, Sleator RD, Marchesi JR, Hill C (2014) Metagenomics and novel gene discovery: promise and potential for novel therapeutics. Virulence 5:399–412. https://doi.org/10.4161/viru.27208
D'Amico S, Collins T, Marx JC et al (2006) Psychrophilic microorganisms: challenges for life. EMBO Rep 7:385–389. https://doi.org/10.1038/sj.embor.7400662
Danovaro R, Corinaldesi C, Dell'Anno A et al (2011) Marine viruses and global climate change. FEMS Microbiol Rev 35:993–1034. https://doi.org/10.1111/j.1574-6976.2010.00258.x
Demory D, Arsenieff L, Simon N et al (2017) Temperature is a key factor in *micromonas*–virus interactions. ISME J 11:601–612. https://doi.org/10.1038/ismej.2016.160
Du X-P, Cai Z-H, Zuo P et al (2020) Temporal variability of virioplankton during a *Gymnodinium catenatum* algal bloom. Microorganisms 8:107. https://doi.org/10.3390/microorganisms8010107
Duhaime MB, Sullivan MB (2012) Ocean viruses: rigorously evaluating the metagenomic sample-to-sequence pipeline. Virology 434:181–186. https://doi.org/10.1016/j.virol.2012.09.036
Endo H, Blanc-Mathieu R, Li Y et al (2020) Biogeography of marine giant viruses reveals their interplay with eukaryotes and ecological functions. Nat Ecol Evol 4:1639–1649. https://doi.org/10.1038/s41559-020-01288-w
Engel A, Bange HW, Cunliffe M et al (2017) The ocean's vital skin: toward an integrated understanding of the sea surface microlayer. Front Mar Sci 4:165. https://doi.org/10.3389/fmars.2017.00165
Evans C, Payton O, Picco L, Allen M (2018) Algal viruses: the (atomic) shape of things to come. Viruses 10:490. https://doi.org/10.3390/v10090490
Feely RA, Doney SC, Cooley SR (2009) Ocean acidification: present conditions and future changes in a high-$CO_2$ world. Oceanography 22:36–47. https://doi.org/10.5670/oceanog.2009.95
Feng Y, Chai F, Wells ML et al (2021) The combined effects of increased $pCO_2$ and warming on a coastal phytoplankton assemblage: from species composition to sinking rate. Front Mar Sci 8:622319. https://doi.org/10.3389/fmars.2021.622319
Finke J, Hunt B, Winter C et al (2017) Nutrients and other environmental factors influence virus abundances across oxic and hypoxic marine environments. Viruses 9:152. https://doi.org/10.3390/v9060152
Flaviani F, Schroeder D, Balestreri C et al (2017) A pelagic microbiome (viruses to protists) from a small cup of seawater. Viruses 9:47. https://doi.org/10.3390/v9030047
Flaviani F, Schroeder DC, Lebret K et al (2018) Distinct oceanic microbiomes from viruses to protists located near the Antarctic circumpolar current. Front Microbiol 9:1474. https://doi.org/10.3389/fmicb.2018.01474
Flynn KJ, Kimmance SA, Clark DR et al (2021) Modelling the effects of traits and abiotic factors on viral lysis in phytoplankton. Front Mar Sci 8:460. https://doi.org/10.3389/fmars.2021.667184

Frickel J, Sieber M, Becks L (2016) Eco-evolutionary dynamics in a coevolving host-virus system. Ecol Lett 19:450–459. https://doi.org/10.1111/ele.12580
Fuhrman JA (1999) Marine viruses and their biogeochemical and ecological effects. Nature 399: 541–548. https://doi.org/10.1038/21119
Garretto A, Hatzopoulos T, Putonti C (2019) virMine: automated detection of viral sequences from complex metagenomic samples. PeerJ 7:e6695. https://doi.org/10.7717/peerj.6695
Gazitúa MC, Vik DR, Roux S et al (2021) Potential virus-mediated nitrogen cycling in oxygen-depleted oceanic waters. ISME J 15:981–998. https://doi.org/10.1038/s41396-020-00825-6
Gran-Stadniczeñko S, Krabberød AK, Sandaa RA et al (2019) Seasonal dynamics of algae-infecting viruses and their inferred interactions with protists. Viruses 11:1043. https://doi.org/10.3390/v11111043
Gregory AC, Zayed AA, Conceição-Neto N et al (2019) Marine DNA viral macro- and microdiversity from pole to pole. Cell 177:1109–1123. https://doi.org/10.1016/j.cell.2019.03.040
Guo J, Bolduc B, Zayed AA et al (2021) VirSorter2: a multi-classifier, expert-guided approach to detect diverse DNA and RNA viruses. Microbiome 9:37. https://doi.org/10.1186/s40168-020-00990-y
Handelsman J (2009) Metagenetics: spending our inheritance on the future. Microb Biotechnol 2: 138–139. https://doi.org/10.1111/j.1751-7915.2009.00090_8.x
Handelsman J, Rondon MR, Brady SF et al (1998) Molecular biological access to the chemistry of unknown soil microbes: a new frontier for natural products. Chem Biol 5:R245–R249. https://doi.org/10.1016/S1074-5521(98)90108-9
Handley SA, Virgin HW (2019) Drowning in viruses. Cell 177:1084–1085. https://doi.org/10.1016/j.cell.2019.04.045
Hayes S, Mahony J, Nauta A, van Sinderen D (2017) Metagenomic approaches to assess bacteriophages in various environmental niches. Viruses 9:127. https://doi.org/10.3390/v9060127
Heath S, Knox K, Vale P, Collins S (2017) Virus resistance is not costly in a marine alga evolving under multiple environmental stressors. Viruses 9:39. https://doi.org/10.3390/v9030039
Heinrichs ME, De Corte D, Engelen B, Pan D (2021) An advanced protocol for the quantification of marine sediment viruses via flow cytometry. Viruses 13:102. https://doi.org/10.3390/v13010102
Heinrichs ME, Tebbe DA, Wemheuer B et al (2020) Impact of viral lysis on the composition of bacterial communities and dissolved organic matter in deep-sea sediments. Viruses 12:922. https://doi.org/10.3390/v12090922
Hellweger FL (2020) Combining molecular observations and microbial ecosystem modeling: a practical guide. Annu Rev Mar Sci 12:267–289. https://doi.org/10.1146/annurev-marine-010419-010829
Highfield A, Joint I, Gilbert JA et al (2017) Change in *Emiliania huxleyi* virus assemblage diversity but not in host genetic composition during an ocean acidification mesocosm experiment. Viruses 9:41. https://doi.org/10.3390/v9030041
Hoeijmakers WAM, Bártfai R, Françoijs KJ, Stunnenberg HG (2011) Linear amplification for deep sequencing. Nat Protoc 6:1026–1036. https://doi.org/10.1038/nprot.2011.345
Horas EL, Theodosiou L, Becks L (2018) Why are algal viruses not always successful? Viruses 10: 474. https://doi.org/10.3390/v10090474
Hurwitz BL, U'Ren JM, Youens-Clark K (2016) Computational prospecting the great viral unknown. FEMS Microbiol Lett 363:fnw077. https://doi.org/10.1093/femsle/fnw077
Jacquet S, Miki T, Noble R et al (2010) Viruses in aquatic ecosystems: important advancements of the last 20 years and prospects for the future in the field of microbial oceanography and limnology. Adv Oceanogr Limnol 1:97–141. https://doi.org/10.1080/19475721003743843
Jasna V, Parvathi A, Pradeep Ram AS et al (2017) Viral-induced mortality of prokaryotes in a tropical monsoonal estuary. Front Microbiol 8:895. https://doi.org/10.3389/fmicb.2017.00895

Jian H, Yi Y, Wang J et al (2021) Diversity and distribution of viruses inhabiting the deepest ocean on earth. ISME J. https://doi.org/10.1038/s41396-021-00994-y
Jovel J, Patterson J, Wang W et al (2016) Characterization of the gut microbiome using 16S or shotgun metagenomics. Front Microbiol 7:459. https://doi.org/10.3389/fmicb.2016.00459
Jover LF, Effler TC, Buchan A et al (2014) The elemental composition of virus particles: implications for marine biogeochemical cycles. Nat Rev Microbiol 12:519–528. https://doi.org/10.1038/nrmicro3289
Kaneko H, Blanc-Mathieu R, Endo H et al (2021) Eukaryotic virus composition can predict the efficiency of carbon export in the global ocean. iScience 24:102002. https://doi.org/10.1016/j.isci.2020.102002
Kieft K, Zhou Z, Anantharaman K (2020) VIBRANT: automated recovery, annotation and curation of microbial viruses, and evaluation of viral community function from genomic sequences. Microbiome 8:90. https://doi.org/10.1186/s40168-020-00867-0
Kimura K, Tomaru Y (2017) Effects of temperature and salinity on diatom cell lysis by DNA and RNA viruses. Aquat Microb Ecol 79:79–83. https://doi.org/10.3354/ame01818
Knowles B, Silveira CB, Bailey BA et al (2016) Lytic to temperate switching of viral communities. Nature 531:466–470. https://doi.org/10.1038/nature17193
Kranzler CF, Brzezinski MA, Cohen NR et al (2021) Impaired viral infection and reduced mortality of diatoms in iron-limited oceanic regions. Nat Geosci 14:231–237. https://doi.org/10.1038/s41561-021-00711-6
Ku C, Ku C, Sheyn U et al (2020) A single-cell view on alga-virus interactions reveals sequential transcriptional programs and infection states. Sci Adv 6:eaba4137. https://doi.org/10.1126/sciadv.aba4137
Labbé M, Raymond F, Lévesque A et al (2018) Communities of phytoplankton viruses across the transition zone of the St. Lawrence estuary. Viruses 10:672. https://doi.org/10.3390/v10120672
Larsen A, Egge JK, Nejstgaard JC et al (2015) Contrasting response to nutrient manipulation in Arctic mesocosms are reproduced by a minimum microbial food web model. Limnol Oceanogr 60:360–374. https://doi.org/10.1002/lno.10025
Le Quéré C, Andrew RM, Canadell JG et al (2016) Global carbon budget 2016. Earth Syst Sci Data 8:605–649. https://doi.org/10.5194/essd-8-605-2016
Lennon JT, Khatana SAM, Marston MF, Martiny JBH (2007) Is there a cost of virus resistance in marine cyanobacteria? ISME J 1:300–312. https://doi.org/10.1038/ismej.2007.37
Li Y, Hingamp P, Watai H et al (2018) Degenerate PCR primers to reveal the diversity of giant viruses in coastal waters. Viruses 10:496. https://doi.org/10.3390/v10090496
Lima-Mendez G, Faust K, Henry N et al (2015) Determinants of community structure in the global plankton interactome. Science 348:1262073. https://doi.org/10.1126/science.1262073
Lindemann C, Aksnes DL, Flynn KJ, Menden-Deuer S (2017) Editorial: modeling the plankton-enhancing the integration of biological knowledge and mechanistic understanding. Front Mar Sci 4:358. https://doi.org/10.3389/fmars.2017.00358
Lipkin WI (2010) Microbe hunting. Microbiol Mol Biol Rev 74:363–377. https://doi.org/10.1128/mmbr.00007-10
Maat D, Biggs T, Evans C et al (2017) Characterization and temperature dependence of Arctic *Micromonas polaris* viruses. Viruses 9:134. https://doi.org/10.3390/v9060134
Maat DS, Brussaard CPD (2016) Both phosphorus- and nitrogen limitation constrain viral proliferation in marine phytoplankton. Aquat Microb Ecol 77:87–97. https://doi.org/10.3354/ame01791
Maat DS, Prins MA, Brussaard CPD (2019) Sediments from arctic tide-water glaciers remove coastal marine viruses and delay host infection. Viruses 11:123. https://doi.org/10.3390/v11020123
Malits A, Boras JA, Balagué V et al (2021) Viral-mediated microbe mortality modulated by ocean acidification and eutrophication: consequences for the carbon fluxes through the microbial food web. Front Microbiol 12:635821. https://doi.org/10.3389/fmicb.2021.635821

Manea E, Dell'anno A, Rastelli E et al (2019) Viral infections boost prokaryotic biomass production and organic C cycling in hadal trench sediments. Front Microbiol 10:1952. https://doi.org/10.3389/fmicb.2019.01952

Marston MF, Pierciey FJ, Shepard A et al (2012) Rapid diversification of coevolving marine *Synechococcus* and a virus. Proc Natl Acad Sci 109:4544–4549. https://doi.org/10.1073/pnas.1120310109

Mateus MD (2017) Bridging the gap between knowing and modeling viruses in marine systems-an upcoming frontier. Front Mar Sci 3:284. https://doi.org/10.3389/FMARS.2016.00284

McMinn A, Liang Y, Wang M (2020) Minireview: the role of viruses in marine photosynthetic biofilms. Mar Life Sci Technol 2:203–208. https://doi.org/10.1007/s42995-020-00042-2

Meyer JR, Agrawal AA, Quick RT et al (2010) Parallel changes in host resistance to viral infection during 45,000 generations of relaxed selection. Evolution 64:3024–3034. https://doi.org/10.1111/j.1558-5646.2010.01049.x

Middelboe M, Brussaard CPD (2017) Marine viruses: key players in marine ecosystems. Viruses 9:302. https://doi.org/10.3390/v9100302

Mojica KDA, Brussaard CPD (2014) Factors affecting virus dynamics and microbial host-virus interactions in marine environments. FEMS Microbiol Ecol 89:495–515. https://doi.org/10.1111/1574-6941.12343

Mojica KDA, Brussaard CPD (2020) Significance of viral activity for regulating heterotrophic prokaryote community dynamics along a meridional gradient of stratification in the Northeast Atlantic Ocean. Viruses 12:1293. https://doi.org/10.3390/v12111293

Mordecai GJ, Verret F, Highfield A, Schroeder DC (2017) Schrödinger's Cheshire cat: are haploid *Emiliania huxleyi* cells resistant to viral infection or not? Viruses 9:51. https://doi.org/10.3390/v9030051

Nagasaki K, Yamaguchi M (1998) Effect of temperature on the algicidal activity and the stability of HaV (*Heterosigma akashiwo* virus). Aquat Microb Ecol 15:211–216. https://doi.org/10.3354/ame015211

Peck KM, Lauring AS (2018) Complexities of viral mutation rates. J Virol 92:e01031–e01017. https://doi.org/10.1128/JVI.01031-17

Piedade GJ, Wesdorp EM, Montenegro-Borbolla E et al (2018) Influence of irradiance and temperature on the virus mpov-45t infecting the arctic picophytoplankter *Micromonas polaris*. Viruses 10:676. https://doi.org/10.3390/v10120676

Proctor LM, Fuhrman JA (1990) Viral mortality of marine bacteria and cyanobacteria. Nature 343:60–62. https://doi.org/10.1038/343060a0

Prodinger F, Endo H, Gotoh Y et al (2020) An optimized metabarcoding method for Mimiviridae. Microorganisms 8:506. https://doi.org/10.3390/microorganisms8040506

Rahlff J (2019) The virioneuston: a review on viral–bacterial associations at air–water interfaces. Viruses 11:191. https://doi.org/10.3390/v11020191

Record NR, Talmy D, Våge S (2016) Quantifying tradeoffs for marine viruses. Front Mar Sci 3:251. https://doi.org/10.3389/fmars.2016.00251

Rohwer F, Thurber RV (2009) Viruses manipulate the marine environment. Nature 459:207–212. https://doi.org/10.1038/nature08060

Rost B, Riebesell U, Burkhardt S, Sültemeyer D (2003) Carbon acquisition of bloom-forming marine phytoplankton. Limnol Oceanogr 48:55–67. https://doi.org/10.4319/lo.2003.48.1.0055

Ruiz E, Oosterhof M, Sandaa RA et al (2017) Emerging interaction patterns in the *Emiliania huxleyi*-EhV system. Viruses 9:61. https://doi.org/10.3390/v9030061

Sadeghi M, Tomaru Y, Ahola T (2021) RNA viruses in aquatic unicellular eukaryotes. Viruses 13:362. https://doi.org/10.3390/v13030362

Sandaa R-A, Pree B, Larsen A et al (2017) The response of heterotrophic prokaryote and viral communities to labile organic carbon inputs is controlled by the predator food chain structure. Viruses 9:238. https://doi.org/10.3390/v9090238

Sandaa RA, Storesund JE, Olesin E et al (2018) Seasonality drives microbial community structure, shaping both eukaryotic and prokaryotic host–viral relationships in an arctic marine ecosystem. Viruses 10:715. https://doi.org/10.3390/v10120715
Schulz KG, Bach LT, Bellerby RGJ et al (2017) Phytoplankton blooms at increasing levels of atmospheric carbon dioxide: experimental evidence for negative effects on prymnesiophytes and positive on small picoeukaryotes. Front Mar Sci 4:64. https://doi.org/10.3389/fmars.2017.00064
Seed KD, Faruque SM, Mekalanos JJ et al (2012) Phase variable O antigen biosynthetic genes control expression of the major protective antigen and bacteriophage receptor in *vibrio cholerae* O1. PLoS Pathog 8:1002917. https://doi.org/10.1371/journal.ppat.1002917
Sharoni S, Trainic M, Schatz D et al (2015) Infection of phytoplankton by aerosolized marine viruses. Proc Natl Acad Sci 112:6643–6647. https://doi.org/10.1073/pnas.1423667112
Slagter HA, Gerringa LJA, Brussaard CPD (2016) Phytoplankton virus production negatively affected by iron limitation. Front Mar Sci 3:156. https://doi.org/10.3389/fmars.2016.00156
Sullivan MB, Weitz JS, Wilhelm S (2017) Viral ecology comes of age. Environ Microbiol Rep 9: 33–35. https://doi.org/10.1111/1758-2229.12504
Sun T-W, Yang C-L, Kao T-T et al (2020) Host range and coding potential of eukaryotic giant viruses. Viruses 12:1337. https://doi.org/10.3390/v12111337
Suttle CA (2005) Viruses in the sea. Nature 437:356–361. https://doi.org/10.1038/nature04160
Suttle CA (2007) Marine viruses - major players in the global ecosystem. Nat Rev Microbiol 5:801–812. https://doi.org/10.1038/nrmicro1750
Talmy D, Beckett SJ, Zhang AB et al (2019) Contrasting controls on microzooplankton grazing and viral infection of microbial prey. Front Mar Sci 6:182. https://doi.org/10.3389/fmars.2019.00182
Thingstad TF (2000) Elements of a theory for the mechanisms controlling abundance, diversity, and biogeochemical role of lytic bacterial viruses in aquatic systems. Limnol Oceanogr 45:1320–1328. https://doi.org/10.4319/lo.2000.45.6.1320
Thingstad TF, Våge S (2019) Host–virus–predator coexistence in a grey-box model with dynamic optimization of host fitness. ISME J 13:3102–3111. https://doi.org/10.1038/s41396-019-0496-7
Thomas L, Parker T (1974) The lives of a cell. Viking Books, New York, NY
Thomas R, Grimsley N, Escande M-L et al (2011) Acquisition and maintenance of resistance to viruses in eukaryotic phytoplankton populations. Environ Microbiol 13:1412–1420. https://doi.org/10.1111/j.1462-2920.2011.02441.x
Thomas T, Gilbert J, Meyer F (2012) Metagenomics - a guide from sampling to data analysis. Microb Inform Exp 2:3. https://doi.org/10.1186/2042-5783-2-3
Thompson LR, Zeng Q, Kelly L et al (2011) Phage auxiliary metabolic genes and the redirection of cyanobacterial host carbon metabolism. Proc Natl Acad Sci 108:E757–E764. https://doi.org/10.1073/pnas.1102164108
Tomaru Y, Kimura K (2020) Novel protocol for estimating viruses specifically infecting the marine planktonic diatoms. Diversity 12:225. https://doi.org/10.3390/d12060225
Tomaru Y, Toyoda K, Kimura K (2015) Marine diatom viruses and their hosts: resistance mechanisms and population dynamics. Perspect Phycol 2:69–81. https://doi.org/10.1127/pip/2015/0023
Tominaga K, Morimoto D, Nishimura Y et al (2020) In silico prediction of virus-host interactions for marine Bacteroidetes with the use of metagenome-assembled genomes. Front Microbiol 11: 738. https://doi.org/10.3389/fmicb.2020.00738
Toseland A, Daines SJ, Clark JR et al (2013) The impact of temperature on marine phytoplankton resource allocation and metabolism. Nat Clim Chang 3:979–984. https://doi.org/10.1038/nclimate1989
Tsiola A, Michoud G, Fodelianakis S et al (2020) Viral metagenomic content reflects seawater ecological quality in the coastal zone. Viruses 12:806. https://doi.org/10.3390/v12080806
Tsiola A, Pitta P, Giannakourou A et al (2017) Ocean acidification and viral replication cycles: frequency of lytically infected and lysogenic cells during a mesocosm experiment in the NW

Mediterranean Sea. Estuar Coast Shelf Sci 186:139–151. https://doi.org/10.1016/j.ecss.2016.05.003
van Etten JL (2011) Another really, really big virus. Viruses 3:32–46. https://doi.org/10.3390/v3010032
Vaqué D, Boras JA, Arrieta JM et al (2021) Enhanced viral activity in the surface microlayer of the arctic and antarctic oceans. Microorganisms 9:317. https://doi.org/10.3390/microorganisms9020317
Vlok M, Lang AS, Suttle CA (2019) Marine RNA virus quasispecies are distributed throughout the oceans. mSphere 4:e00157–e00119. https://doi.org/10.1128/mSphereDirect.00157-19
Weitz JS, Stock CA, Wilhelm SW et al (2015) A multitrophic model to quantify the effects of marine viruses on microbial food webs and ecosystem processes. ISME J 9:1352–1364. https://doi.org/10.1038/ismej.2014.220
Weynberg K, Allen M, Wilson W (2017) Marine prasinoviruses and their tiny plankton hosts: a review. Viruses 9:43. https://doi.org/10.3390/v9030043
Wilhelm SW, Suttle CA (1999) Viruses and nutrient cycles in the sea. Bioscience 49:781–788. https://doi.org/10.2307/1313569
Williamson SJ, Paul JH (2006) Environmental factors that influence the transition from lysogenic to lytic existence in the φHSIC/Listonella pelagia marine phage-host system. Microb Ecol 52:217–225. https://doi.org/10.1007/s00248-006-9113-1
Yang Q, Gao C, Jiang Y et al (2019) Metagenomic characterization of the viral community of the south scotia ridge. Viruses 11:95. https://doi.org/10.3390/v11020095
Yau S, Caravello G, Fonvieille N et al (2018) Rapidity of genomic adaptations to prasinovirus infection in a marine microalga. Viruses 10:441. https://doi.org/10.3390/v10080441
Yau S, Seth-Pasricha M (2019) Viruses of polar aquatic environments. Viruses 11:189. https://doi.org/10.3390/v11020189
Zablocki O, Michelsen M, Burris M et al (2021) VirION2: a short and long-read sequencing and informatics workflow to study the genomic diversity of viruses in nature. PeerJ 9:e11088. https://doi.org/10.7717/peerj.11088
Zeigler Allen L, McCrow JP, Ininbergs K, et al. (2017) The Baltic Sea virome: diversity and transcriptional activity of DNA and RNA viruses. mSystems 2:e00125-16 doi:https://doi.org/10.1128/mSystems.00125-16
Zhao Z, Gonsior M, Schmitt-Kopplin P et al (2019) Microbial transformation of virus-induced dissolved organic matter from picocyanobacteria: coupling of bacterial diversity and DOM chemodiversity. ISME J 13:2551–2565. https://doi.org/10.1038/s41396-019-0449-1
Zimmerman AE, Howard-Varona C, Needham DM et al (2020) Metabolic and biogeochemical consequences of viral infection in aquatic ecosystems. Nat Rev Microbiol 18:21–34. https://doi.org/10.1038/s41579-019-0270-x

# Evolutionary Genomics of Marine Bacteria and Archaea

7

Carolina A. Martinez-Gutierrez and Frank O. Aylward

**Abstract**

The ocean harbors an enormous diversity of bacteria and archaea whose metabolic activities have global biogeochemical impacts. Evolutionary genomic studies play an important role in marine microbiology by providing insights into the selective pressures that influence the diversity and functioning of marine microbial lineages, shedding light on their niche specialization, and helping to define their broader ecological roles. The scope of evolutionary genomic studies in the ocean has dramatically expanded in recent years owing to advances in sequencing technology, metagenomics, and single-cell sequencing. Historically the collection of sequenced genomes has been composed of a small number of cultured microbes, but recent advances have opened a window into the genomics of a broad diversity of lineages throughout the biosphere. This broad genomic representation of lineages across the tree of life has enabled comparative genomic analyses that help to identify the ecological and evolutionary forces that drive the diversification of microorganisms in the ocean. In this chapter, we review some of the salient themes that have emerged in the last few decades of these studies. We discuss the main findings of these evolutionary genomic studies and their implications for our understanding of the diversity and functioning of microbial life in the ocean.

**Keywords**

Genome evolution · Microbial diversity · Microbial oceanography · Ocean microbiome · Marine viruses

C. A. Martinez-Gutierrez (✉) · F. O. Aylward (✉)
Department of Biological Sciences, Virginia Tech, Blacksburg, VA, USA
e-mail: cmartinez@vt.edu; faylward@vt.edu



L. J. Stal, M. S. Cretoiu (eds.), *The Marine Microbiome*, The Microbiomes of Humans, Animals, Plants, and the Environment 3,
https://doi.org/10.1007/978-3-030-90383-1_7

## 7.1 Introduction

The ocean plays a central role in Earth's biogeochemistry (Falkowski 1998; Falkowski et al. 2008). Due to their overwhelming abundance and high activity, microorganisms mediate the majority of nutrient transformations in the marine environment and are therefore key engines of Earth's biogeochemical cycles (Arrigo 2005; Brown et al. 2014; Bunse and Pinhassi 2017). A major fraction of the oxygen produced globally is the result of the activity of marine phototrophs (Kasting 2002), and higher trophic levels derive their energy and nutrients thanks to the activities of marine microorganisms (Azam 2004; Brown et al. 2014; Falkowski et al. 2008; Pomeroy et al. 2007).

Despite their important roles, marine microorganisms were not always recognized for their influence on global processes. Because they are not readily visible, progress in understanding ocean microbes has often depended on the development of methods that allowed for their examination directly in the environment (Sherr and Sherr 2008). Before sequence-based surveys of microbial diversity became common practice, most of the research that aimed at understanding microbial evolution and diversity was based on the study of a few cultured microorganisms and relied on methods derived from clinical microbiology (Salazar and Sunagawa 2017; Sherr and Sherr 2008). Early studies of microbial population genetics were based on culture-dependent methodologies such as electrophoretic profiles of well-studied cultured microbes and rarely included marine microorganisms (Witzel 1990). Throughout the mid- to late twentieth Century, it was recognized that the number of microbial colonies that grew on solid agar media when plating environmental samples were significantly lower than the number of cells observed through direct microscopic counts (Staley and Konopka 1985). Now known as the "Great Plate Count Anomaly," this discrepancy is often used to emphasize the importance of culture-independent methods for studying microbial life in the biosphere, and especially the ocean (Salazar and Sunagawa 2017).

In recent years the study of microbial communities has been aided by advances in DNA sequencing. This has subsequently led to breakthroughs in marker gene surveys of microbial diversity, community metagenomics, and single-cell genomics (Lasken and McLean 2014; Moran 2008; Tringe and Hugenholtz 2008). It is now possible to routinely recover nearly complete microbial genomes directly from environmental samples (Sharon and Banfield 2013). These approaches have generated a plethora of new data that are publicly available, which allow for the analysis of the phylogenetic diversity, genome evolution, and physiology of widespread marine microbial groups (Sherr and Sherr 2008). Progress in the evolutionary genomics of marine bacteria and archaea has often been hampered by the lack of sequenced genomes because it relies on genomic comparisons within and between different lineages (Heidelberg et al. 2010). Hence, these newly available genomic data have spurred numerous insights into the how and what of marine microbial diversity.

Rather than attempting to provide a review of the enormous number of genomic studies of marine bacteria and archaea that have been published to date, we have instead focused this chapter on five major themes that highlight important insights in

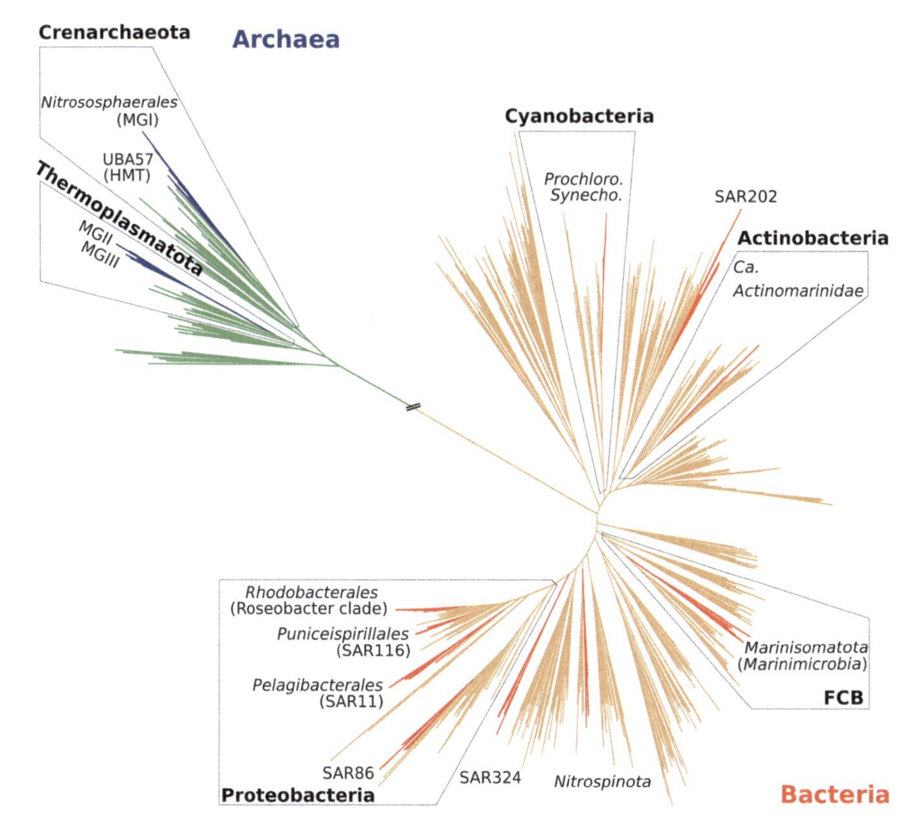


**Fig. 7.1** Phylogenetic relationships of the major planktonic archaeal and bacterial clades. The phylogenetic tree was reconstructed based on the representative genomes from the Genome Taxonomy Database (GTDB; v89) (Parks et al. 2018) using a Maximum Likelihood approach and a concatenated alignment of 30 ribosomal proteins and RNA polymerase subunits. Taxonomy shown is based on the GTDB, and other common names are provided in parentheses. Abbreviations: *MG* Marine Group, *HMT* heterotrophic marine Thaumarchaeota, *FCB* superphylum named after the phyla Fibrobacteres, Chlorobi, and Bacteroidetes, *Prochloro. Prochlorococcus, Synecho. Synechococcus*. Green and orange branches represent Archaeal and Bacterial clades, respectively, and dark branches are the major marine groups. MGII is taxonomically classified as *Poseidoniales*

marine evolutionary genomics. These are: (1) the origin of genomic diversity in marine bacteria and archaea, (2) genome streamlining, (3) ecological factors that determine genome composition, (4) genomics of the dark ocean, and (5) virus–host interactions as drivers of genome evolution. Much of the insight into marine microbial genomics to date has been garnered through the analysis of certain cosmopolitan marine clades that are particularly dominant in the ocean and distributed across the tree of life (Fig. 7.1). We provide particular emphasis on these clades, although recent studies have begun to shed light on numerous other lineages that will likely be an important emphasis of future work in this field (Yilmaz et al. 2015).

## 7.2 The Origins of Genomic Diversity in Marine Microbial Populations

Microorganisms inhabiting the ocean harbor an enormous functional and phylogenetic diversity (DeLong 1997; Eguíluz et al. 2019). The microevolutionary processes that lead to the origin and maintenance of this diversity are driven by the interplay of four primary evolutionary forces: mutation, recombination or horizontal gene transfer (HGT), genetic drift, and selection (Feil 2004; Futuyma 1986; Lynch 2007). HGT is particularly frequent in bacteria and archaea and it can occur even between distantly related groups, thereby playing a central role in the acquisition of novel traits and the overall genome evolution of microbial lineages (Bansal et al. 2011; Gogarten et al. 2005). High levels of HGT can blur population boundaries and thereby complicate application of the species concept to bacteria and archaea (Rosselló-Mora and Amann 2001). Therefore, ecological populations or "ecotypes" are usually used to distinguish between groups of individuals that are clustered based on genotypic and phenotypic criteria as well as ecological similarities (Cohan 2006; Cohan and Perry 2007; Cordero and Polz 2014). For simplicity, in this chapter we will refer to these clusters as populations.

New genes can be introduced into a population through the duplication of existing genes or the introduction of exogenous genes via HGT. In the case of gene duplication, new paralogs can subsequently diverge from the original sequence and evolve new functions (i.e., neofunctionalization), thereby generating genomic novelty (Kunin and Ouzounis 2003; McDaniel et al. 2010; Snel et al. 2002). Although genes acquired through duplication may display an important role in the survival of microorganisms under changing environmental conditions (Sanchez-Perez et al. 2008), paralogs are often not retained simply because newly acquired gene copies are quickly eliminated before they evolve a new function (Kirchberger et al. 2020; Mira et al. 2001). This is especially true for marine lineages that exhibit genome streamlining and thereby have few paralogs in their genome (Giovannoni et al. 2014). Evidence, therefore, suggests that HGT is generally a more prominent force introducing novelty into bacterial and archaeal genomes (Koonin et al. 2001; Treangen and Rocha 2011).

HGT is a particularly important force that drives microbial diversification in marine systems (Sobecky and Hazen 2009). In addition to the well-studied routes of HGT such as those mediated by plasmids, temperate viruses, and other mobile genetic elements, Gene Transfer Agents (GTAs) and extracellular vesicles are ubiquitous in marine systems and may constitute an important and previously unrecognized route of gene exchange (Biller et al. 2014; McDaniel et al. 2010). GTAs are virus-derived gene clusters that package diverse DNA fragments and can potentially transfer them between disparate lineages (Lang et al. 2017). They have been found in diverse bacterial and archaeal lineages, but the ability to transfer DNA has only been confirmed for a few cases (Lang et al. 2017; Shakya et al. 2017). Extracellular vesicles are another potential means of HGT; these vesicles are produced by a wide variety of marine bacteria and archaea (Deatherage and Cookson 2012) and contain diverse DNA fragments, although the mechanism of DNA incorporation remains unclear (Biller et al. 2014, 2017). One study found that most vesicles did not contain microscopically detectable DNA when using fluorescent stains, suggesting that only a few vesicles may contain large quantities of DNA (Biller et al. 2017). Nonetheless, extracellular vesicles have been shown to facilitate HGT on some occasions (Klieve et al. 2005; Renelli et al. 2004; Yaron et al. 2000). Together with GTAs, extracellular vesicles comprise a potentially important route of gene exchange in marine systems.

HGT plays a prominent role in influencing the composition of bacterial and archaeal genomes over long time scales, and it has likely promoted the diversification of many marine lineages by allowing them to occupy their current niches. Prevalent HGT was found in pelagic ammonia-oxidizing archaea (Marine Group I: MGI) and Euryarchaeota (Marine Groups II and III: MGII and MGIII, respectively), where a fosmid-based analysis has shown that up to 25% of the genes analyzed may have been acquired through HGT (Brochier-Armanet et al. 2011). Most of the putatively transferred genes are predicted to be involved in processes such as energy metabolism and transport of metabolites across membranes (Brochier-Armanet et al. 2011; Deschamps et al. 2014). Another study showed that the transfer of genes from bacteria to ammonia-oxidizing archaea (AOA) may have played a role in the transition of this group from terrestrial environments to the ocean (Ren et al. 2019). The genes analyzed here may have facilitated the adaptation to different environmental conditions along the water column. In another example, a genomic analysis of the nitrite oxidizer *Nitrospina gracilis* revealed that a large fraction of its encoded proteins shows homology with *Deltaproteobacteria, Gammaproteobacteria,* and *Nitrospira,* suggesting frequent HGT between these groups. Genes involved in anaerobic nitrite oxidation in *N. gracilis* were probably obtained from anaerobic ammonia oxidizers (Lücker et al. 2013). This suggests that the capability to oxidize nitrite was acquired under anaerobic conditions and subsequently distributed to other areas of the ocean. Lastly, comparative genomic analyses of Ca. Marinimicrobia, a candidate phylum comprising lineages specialized in epipelagic or mesopelagic environments, has revealed a central role of HGT in the diversification of this marine candidate phylum. Distinct marinimicrobial clades that inhabited the same environment had convergently acquired similar genomic repertoires, indicating that parallel HGT events had played a role in their niche partitioning (Getz et al. 2018).

Although bacteria and archaea can acquire genes from different sources, their genomes tend to remain small and compact over time (Bobay and Ochman 2017). Mutations in the genomes of bacteria and archaea are biased toward deletions (Mira et al. 2001). Thus, newly acquired genes resulting from duplications or HGT are typically fixed and retained in populations only if they provide a fitness benefit, whereas genes that do not represent a significant advantage will be expected to be lost or pseudogenized within a few generations. The fraction of successfully incorporated genes due to HGT or duplications is therefore low (Batut et al. 2014; Kuo and Ochman 2009). The overall effect of this deletional bias is related to the strength of selection relative to genetic drift, which is ultimately determined by the effective population size ($N_e$) (Box 7.1) (Bobay and Ochman 2017). Given their

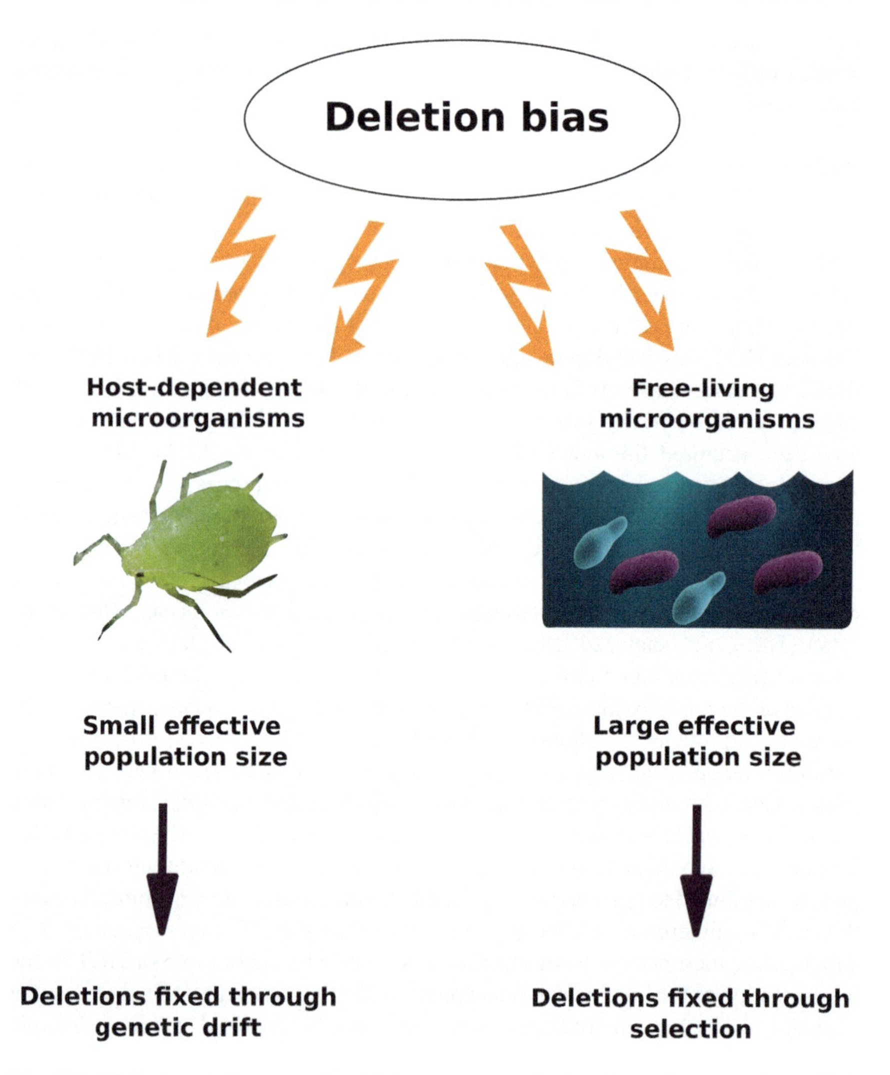


**Fig. 7.2** Distinct evolutionary paths driving genome reduction in marine vs endosymbiotic bacteria. Adapted from Giovannoni et al. (2014)

broad distribution, most marine bacteria and archaea have extremely large $N_e$ when compared with multicellular life, and thus selection can be considered a strong evolutionary force acting on their populations (Brockhurst et al. 2019). This is in stark contrast to organisms with a small $N_e$, such as obligate parasites and endosymbionts, where selection is weak relative to genetic drift. In these cases, deleterious mutations may reach fixation through stochastic processes (Fig. 7.2) (Mira et al. 2001).

**Box 7.1 Effective population size and its role on microbial evolution**

Scientists apply two different metrics when studying microbial population dynamics in nature or in culture: census population size ($N_c$) and effective population size ($N_e$). $N_c$ refers to the number of organisms within a given population. $N_e$ is a complex concept that reflects neutral diversity within a population, that is, the size of a population evolving in the absence of selection that would generate as much neutral diversity as is actually observed in the population (Fraser et al. 2009; Luo et al. 2014a). Empirical evidence suggests that there is a mismatch of several orders of magnitude between both metrics for most of the archaeal and bacterial clades studied so far. Such discrepancy is associated with processes like bottlenecks, in which neutral diversity is low due to a drastic decrease of the population in a short period of time. Other processes like selective sweeps may decrease neutral diversity because the population adapts to a new environment and rapidly fixes genes correlated with such conditions (Fraser et al. 2009; Luo et al. 2014a). $N_e$ is an important evolutionary concept because it determines the way evolution in a population occurs. Populations with small $N_e$ will tend to evolve under stronger genetic drift relative to selection, whereas in large $N_e$ microorganisms evolve predominantly through selection. Although some marine microbial populations have been described as enormous, all populations in nature are finite and evolve through the interplay of selection and drift (Bobay and Ochman 2017). Estimates of $N_e$ are difficult in microorganisms because its calculation requires mutation rate measurements and the detection of intraspecific nucleotide diversity at synonymous sites. Both values are unknown for most microbial species, particularly those that have not been cultured (Fraser et al. 2009; Luo et al. 2014a). Several studies suggest that abundant marine oligotrophs like SAR11 and *Prochlorococcus* have large $N_e$ (Giovannoni 2017; Kashtan et al. 2014; Wilhelm et al. 2007) and a study based on metagenomic data showed high intrapopulation sequence diversity in SAR11 (López-Pérez et al. 2020). These findings are consistent with prolonged habitat specialization and the absence of drastic population reductions in the recent past (Batut et al. 2014). Abundant bacteria appear to experience high levels of purifying selection concomitant with their large $N_e$, indicating a potential to purge deleterious mutations and spread favorable variants (Fraser et al. 2009; Luo et al. 2014a; Martinez-Gutierrez and Aylward 2019). However, $N_e$ estimations are still needed for a broad range of bacteria and archaea in order to obtain a comprehensive picture of the way marine microbes evolve.

## 7.3 Streamlining: Genome Simplification in the Open Ocean

Many bacteria and archaea that live in the surface waters of the ocean are adapted to the characteristic low-nutrient state of these epipelagic environments (Azam and Malfatti 2007; Giovannoni and Stingl 2005). The genomes of many epipelagic microorganisms share some peculiar features including small size (typically <2 Mbp), short intergenic regions, low %GC content, and few paralogous genes. The evolution of these features is thought to be driven by a process of genome streamlining, whereby natural selection in nutrient-depleted environments favors cellular economization leading to small genomes with little extraneous coding material and simple gene regulation (Giovannoni et al. 2014). Although streamlining was first noted in well-studied cultured lineages like *Prochlorococcus marinus* and *Pelagibacter ubique,* these genomic features were soon found in other abundant marine groups such as *Oceanospirillales, Euryarchaeota, Thaumarchaeota,* Ca. Marinimicrobia, *Puniceispirillales,* the Roseobacter clade, and marine members of the *Actinobacteria*, *Sphingomonadaceae*, and *Dadabacteria* (Dupont et al. 2012; Getz et al. 2018; Ghai et al. 2013; Graham and Tully 2020; Lauro et al. 2009; Luo et al. 2014b; Martin-Cuadrado et al. 2015; Orellana et al. 2019; Santoro et al. 2015; Swan et al. 2013). These comparative genomics studies have demonstrated that streamlined genomic features have evolved independently in a wide variety of marine lineages, providing a remarkable example of inter-domain convergent evolution. Some clades, such as the *Pelagibacterales* appear to contain exclusively streamlined genomes, while others, such as the Roseobacter and Ca. Marinimicrobia groups consist of a mixture of both, streamlined and non-streamlined representatives (Getz et al. 2018; Giovannoni 2017; Luo et al. 2014b).

There has been some debate over the drivers of genome streamlining in marine bacteria and archaea, especially regarding the question of whether all genomic features resulting from this process can be viewed as purely adaptive (reviewed in (Batut et al. 2014)). This debate has been motivated in part by the observation that many parasitic and endosymbiotic bacteria exhibit genomic features in some ways akin to streamlined genomes, including small size and low %GC content. The high level of genetic drift experienced by endosymbionts and parasites is a consequence of their small effective population size (Box 7.1), which is driven in part by the extreme population bottlenecks experienced during transmission (Moran 1996). This leads to the fixation of deleterious mutations, including the loss of genes that provide an adaptive benefit, and an overall pattern of genomic erosion (Fig. 7.2). The loss of functional genes observed in host-dependent microorganisms often does not occur in parallel with other features indicative of streamlining; the reduced genomes of host-dependent microorganisms are therefore not the product of adaptive processes (Wolf and Koonin 2013). Marine bacteria and archaea span vast ocean basins and have enormous effective population sizes, and thus genome simplification in this group of microorganisms is unlikely due to high genetic drift (Wolf and Koonin 2013). Although some work has shown that weakly deleterious mutations and low rates of recombination may lower the effective population size of bacteria compared to their actual population size (Price and Arkin 2015) it is unlikely that this decreases $N_e$ to a size where genetic drift is the dominant force in genome evolution.

Indeed, several studies have indicated that streamlined features of marine genomes can at least in part be attributed to adaptation. One study concluded that low %GC content in SAR11 is maintained by strong purifying selection (Luo et al. 2015), while another study showed that streamlined Ca. Marinimicrobia experience higher levels of purifying selection compared to their non-streamlined relatives (Martinez-Gutierrez and Aylward 2019). A similar result was found between high-light adapted *Prochlorococcus* and its close relative *Sychechococcus* (Hu and Blanchard 2009). This indicates that the inherent deletional bias of bacterial genomes causes genes that are not essential to undergo relaxed selection and subsequent erosion and deletion from the population, while genes that provide a selective benefit are retained (Bobay and Ochman 2017; Mira et al. 2001). Hence, genome streamlining is expected to be prevalent in stable environments such as oceanic gyres where environmental conditions and nutrient concentrations do not change substantially throughout the year. The loss of costly regulatory machinery used to respond to changing conditions would not be selected against in such environments (Giovannoni et al. 2014).

Although genetic drift in marine bacterial and archaeal populations is unlikely to give rise to genome streamlining, there are two other non-adaptive processes that could potentially give rise to some of these genomic features: population bottlenecks that occurred in the distant past and evolutionary constraints that may cause traits to co-vary. Regarding population bottlenecks, some studies have suggested that population reduction events in the distant past may have contributed to the current features of streamlined genomes irrespective of current selective pressures (Luo et al. 2017; Luo et al. 2014b). For example, one study evaluating radical substitutions found evidence for population bottlenecks in the early evolution of SAR11 and *Prochlorococcus* (Luo et al. 2017). Adaptive radiations are often accompanied by relaxed selection and it is therefore plausible that many marine lineages experienced higher levels of genetic drift and, hence, more non-adaptive deletions during the early stages of their radiation into the ocean. The inference of selective pressures in ancient adaptive radiations is tantalizingly difficult to ascertain, however, and some studies have come to the opposing conclusion that early genome reduction in picocyanobacteria was adaptive (Sun and Blanchard 2014). Regardless, the hypothesis of high genetic drift during early marine radiations need not contradict the adaptive nature of many streamlined genomic features: it is also possible that many such features evolved during a period of relaxed selection but were subsequently selected for as effective population sizes increased and purifying selection strengthened.

Another complication with ascertaining the adaptive benefit of streamlined genomic features is that many of them co-vary and may therefore be the product of evolutionary constraints rather than bona fide adaptations. One example of linked genomic traits is %GC content and the nitrogen content of encoded amino acids (often referred to as the N-ARSC: Nitrogen Atoms per Residue Side Chain). Codons with low %GC content preferentially code for amino acids with low nitrogen content and selection for either %GC content or protein nitrogen content will therefore alter the other (Bragg and Hyder 2004). Several studies have shown that both N-ARSC and %GC content are significantly lower in microbial communities that reside in N-depleted waters. One study demonstrated that these features were significantly lower in pelagic compared to coastal marine microbial communities (Grzymski and Dussaq 2012) and another study showed that N-ARSC and %GC were significantly lower in surface water communities compared to those sampled below the nutricline in the North Pacific Subtropical Gyre (Mende et al. 2017). G-C base pairs contain one more nitrogen atom than A-T pairs, and a transition to low %GC content genomes would therefore also lead to nitrogen savings. It is therefore difficult to disentangle the selective pressures acting on these traits. Given that proteins make up a larger portion of the nitrogen content of cells it is plausible that the primary selective force acting on N-depleted microbial populations is a lowering of the N-ARSC, thereby leading to the decrease of %GC content as an evolutionary by-product. In support of this view, it was shown that *Prochlorococcus marinus* shifts its transcriptome during N depletion to produce proteins with a lower N content (Read et al. 2017). Since *Prochlorococcus* is already highly streamlined, this shift in transcription can be considered as evidence for an additional selective pressure to lower the proteome N content during nitrogen limitation.

Lastly, it has been postulated that interactions between sympatric microbes may be a driver of reductive genome evolution. The Black Queen Hypothesis (BQH) (Morris et al. 2012) posits that gene losses in free-living marine microorganisms may lead to a dependence on co-existing microbes to substitute for lost metabolic capabilities. The BQH proposes that under nutrient-limited conditions, as is usually the case in the open ocean, streamlined microorganisms may lose functions when these are provided by other microorganisms in the community ("common goods") thereby decreasing the cost of gene maintenance, transcription, and translation. The BQH is supported by the absence of the gene that encodes catalase-peroxidase (*katG)* in *Prochlorococcus*, which is considered to be essential for this organism because it removes hydrogen peroxide (HOOH), a by-product of oxygenic photosynthesis that causes oxidative stress. Other microorganisms may act as a sink of HOOH, relieving phototrophs like *Prochlorococcus* from the necessity to maintain the genetic potential to protect themselves from oxidative stress (Morris et al. 2012). According to the BQH, selection for the loss of redundant genetic material could be an important driver of genome minimization in oligotrophic environments (Mas et al. 2016; Morris et al. 2012).

## 7.4 Ecological Factors Influencing Genome Composition

The ecological niche of a microbe plays a central role in determining its genomics, both in terms of gene content and overall genome structure. Given the ubiquity of HGT and the fast rates at which microbial populations can evolve, closely related marine microorganisms that inhabit different niches often harbor substantially different gene complements that provide niche-specific adaptations. Conversely, distantly-related microbes inhabiting the same environment can sometimes evolve convergently and thereby acquire similar genomic traits. By examining both trends of niche partitioning and parallel evolution it is often possible to gain valuable insight into the ecological factors and selective pressures that a microbial population may experience.

Perhaps the most dramatic example of convergent evolution of gene content involves proteorhodopsin (PR), a light-driven proton pump found in diverse microorganisms in the ocean. First identified in the SAR86 lineage of *Gammaproteobacteria* (Beja 2000), PRs were subsequently found in diverse epipelagic microbial lineages (Frigaard et al. 2006; Giovannoni et al. 2005; Torre et al. 2003; Yutin and Koonin 2012). Experimental work has provided evidence that PR aids in the survival of microorganisms that encode it and may increase their growth rates at low nutrient concentrations (Gómez-Consarnau et al. 2007, 2010). This likely provides an adaptive benefit when thriving in the prevailing oligotrophic conditions of many marine habitats (DeLong and Béjà 2010). PR genes appear to be transferred readily across large phylogenetic distances (Frigaard et al. 2006) likely because of the simple modular structure of these gene cassettes, which typically include the PR gene itself and sometimes genes involved in carotenoid biosynthesis (Pinhassi et al. 2016). It has been estimated that between 17 and 80% of marine bacteria and archaea in the sunlit ocean encode a PR in their genome (DeLong and Béjà 2010; Moran and Miller 2007) emphasizing how divergent lineages have repeatedly converged on the same trait, presumably because it both provides a sufficient adaptive benefit and it is readily transferred into new genomes without disrupting existing metabolic or regulatory networks.

Another example of genomic changes that occur concomitantly with niche specialization is highlighted in marine picocyanobacteria. The genera *Synechococcus* and *Prochlorococcus*, both abundant picocyanobacteria in the ocean, have distinct niche specializations despite their close phylogenetic relationship (Sánchez-Baracaldo et al. 2019). *Synechococcus* dominates in coastal and temperate waters whereas *Prochlorococcus* prevails in warm-oligotrophic waters (Zwirglmaier et al. 2008). Such niche differentiation results in greater phenotypic flexibility and regulatory capacity in *Synechococcus,* while *Prochlorococcus* exhibits larger population sizes and strong specialization to nutrient-depleted environments (Moore et al. 1995; Partensky and Garczarek 2010; Rocap et al. 2003). *Prochlorococcus* has lost the large extrinsic antenna complexes (phycobilisomes) that are present in *Synechococcus*; instead, *Prochlorococcus* possesses simpler compact antennae within thylakoid membranes that are in direct contact with the photosystems (Berube et al. 2018; Partensky and Garczarek 2010; Sánchez-Baracaldo et al. 2019). This difference may be responsible for the discrepancy in light absorption properties of both clades, since *Prochlorococcus* cells are specialized for the capture of blue wavelengths that prevail in oligotrophic waters (Ting et al. 2002). In addition to physiological changes, genomic disparities have accompanied the specialization of different habitats in these picocyanobacteria. High-light adapted *Prochlorococcus* ecotypes have undergone a reduction in genome size and %GC content as part of streamlining and adaptation to the oligotrophic environment, while these changes have not been observed in *Synechococcus* (Partensky and Garczarek 2010; Rocap et al. 2003). *Prochlorococcus* likely experiences higher purifying selection than *Synechococcus* (Hu and Blanchard 2009; Wolf and Koonin 2013), which may explain why these genomic trends of genome reduction and lower %GC are more prominent in the former. Different ecotypes partitioned based on light availability are also present within *Prochlorococcus*; the high-light adapted ecotype (HL) is more abundant in surface waters and tends to have small genomes and low %GC content whereas low-light adapted *Prochlorococcus* are dominant in deeper waters and tend to have larger genomes with higher %GC content (Kettler et al. 2007; Rocap et al. 2003). Both ecotypes differ in light absorption properties, and within each ecotype there are groups that are different in terms of growth temperature and abundance (Rocap et al. 2003). Analyses have shown that the HL-adapted ecotype has likely evolved from deeper waters (Martiny et al. 2009; Partensky and Garczarek 2010; Paul et al. 2010).

Besides *Prochlorococcus*, many other marine microbial lineages exhibit clear changes in genomic properties that are linked with the depth in which they live. Several studies examining shifts in microbial community composition as well as genomic features across a depth gradient have been conducted at Station ALOHA in the North Pacific Subtropical Gyre (DeLong 2006; Konstantinidis et al. 2009; Mende et al. 2017). Mende et al. (2017) studied the evolutionary and ecological processes affecting the genomes of bacteria and archaea at Station ALOHA. This study revealed a marked vertical transition in which organisms of the same clade showed a decrease in %GC content above the deep chlorophyll maximum (DCM), as well as reduction in genome size and intergenic spacer length. These genomic traits are therefore correlated with the physicochemical features that vary with depth. Another study at Station ALOHA that compared the microbial communities residing in the surface to those present at 4000 m suggested that %GC content, as well as codon usage changes, may be associated with differences in the strength of purifying selection and ultimately disparities in the effective population sizes (Konstantinidis et al. 2009). This suggests that surface and deep-water environments are subjected to substantially different evolutionary regimes, both in terms of selective pressures themselves as well as their overall strength. A similar trend was found in a study that compared the codon usage of coastal and open ocean microorganisms (Grzymski and Dussaq 2012), suggesting that nitrogen limitation may represent a strong selective pressure driving adaptations toward lower %GC content and the minimization of protein synthesis costs in many marine systems.

Genomic changes that occur coincident with ecological transitions across depth have been studied in detail in Ca. Marinimicrobia, in which streamlining occurred multiple times independently in distantly related epipelagic clades. This emphasizes that similar environmental features can drive convergent evolution of genomic features such as low %GC content, short intergenic regions, and low encoded nitrogen content (Getz et al. 2018) (Fig. 7.3). Moreover, epipelagic Ca. Marinimicrobia experienced higher levels of purifying selection than their mesopelagic relatives, which is consistent with the strong purifying selection often associated with streamlining (Martinez-Gutierrez and Aylward 2019). In addition to

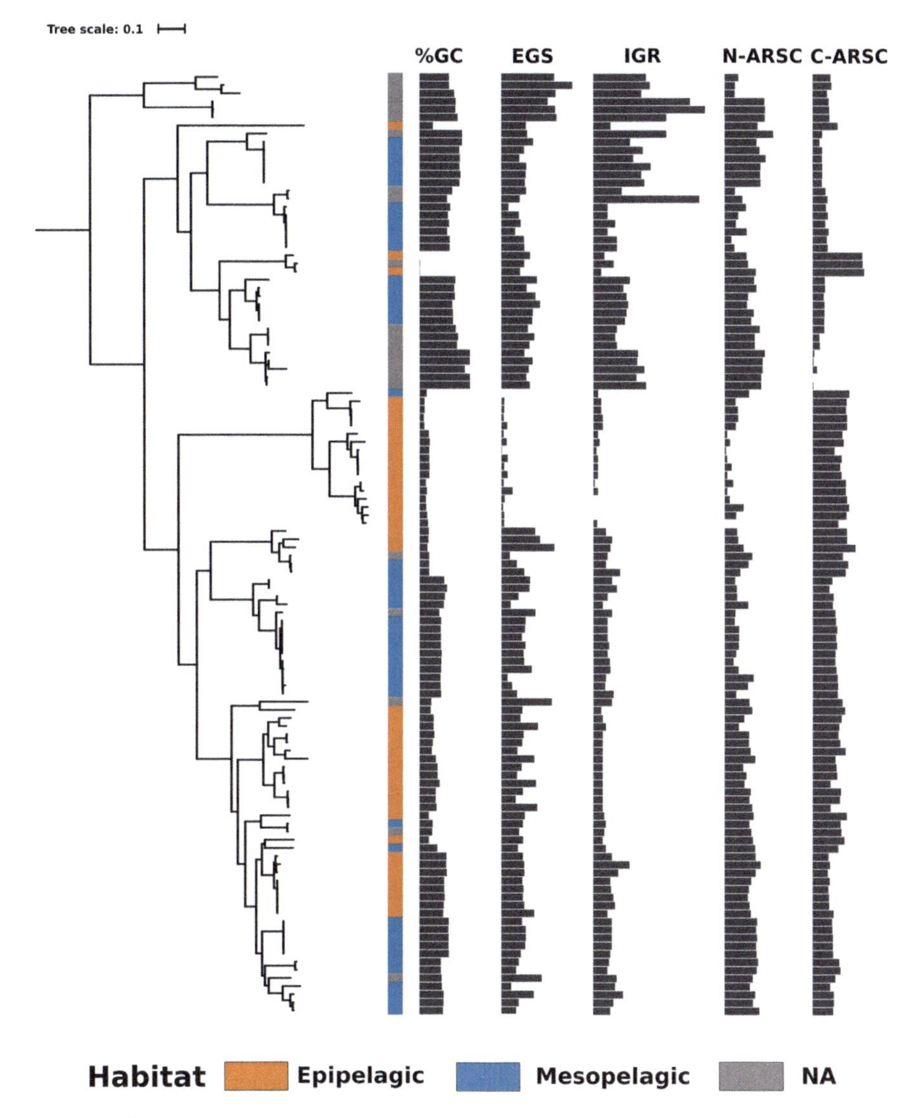


**Fig. 7.3** Habitat transitions in Ca. Marinimicrobia clades. Maximum likelihood phylogenetic tree reconstructed using a concatenated alignment of 30 ribosomal proteins and RNA polymerase subunits. The colored strip shows the habitat for each tip on the tree. Abbreviations: %GC, %GC content (range, 27.16–50.55%); EGS, estimated genome size (range, 1.06–3.53 Mbp); IGR, median intergenic region length (range, 7–78 bp); N-ARSC, Nitrogen atoms per residue side chain (range, 0.3–0.35); C-ARSC, Carbon atoms per residue side chain (range, 2.97–3.20)

differences in bulk genomic features, epipelagic and mesopelagic Ca. Marinimicrobia also differed in their encoded functional repertoires and cellular bioenergetics. Whereas epipelagic Ca. Marinimicrobia often encoded proteorhodopsins and a $Na^{+}$-translocating NADH dehydrogenase (NQR), their mesopelagic counterparts acquired nitrate respiratory machinery, a $H^+$-translocating NADH dehydrogenase complex (NDH), and different cytochromes possibly involved in the use of alternative electron acceptors (Getz et al. 2018). The repeated acquisition of similar genes by distantly related Ca. Marinimicrobia emphasizes the important role of HGT in facilitating parallel niche diversification across large phylogenetic distances.

Further studies have provided insight into ecologically driven genomic changes involving other habitats. Using a comparative genomic approach, one study reconstructed the genomic consequences of habitat transitions in the *Methylophilaceae* group in the *Betaproteobacteria* (Salcher et al. 2019). This study was able to identify two key transitions in this group; one between freshwater sediment to a pelagic freshwater habitat and another from a pelagic freshwater to a marine habitat. The authors found evidence for a sudden and pronounced loss of genes coincident with the transition from sediment to freshwater pelagic habitat. The subsequent transition from freshwater to marine pelagic environment was not accompanied by further genome reduction, but rather by the acquisition of genes that facilitate adaptation to higher salinities. Another study used a similar approach to reconstruct the genomic changes in *Flavobacteria* that took place coincident with ecological transitions from marine to terrestrial systems (Zhang et al. 2019). This study found similar patterns of gene loss and gain, notably the loss of genes involved in the tolerance of salinity in the ocean. Similar to the Ca. Marinimicrobia example discussed above, this study also found that the presence of either the NQR or NDH dehydrogenase was strongly correlated with the environment in which the *Flavobacteria* were found. These studies demonstrate that divergent microbial lineages will often evolve convergent genomic features depending on their habitat, underscoring how environment-specific selective pressures are a major force that can determine how marine microbial populations evolve.

Although transitions between environments are important factors that determine microbial diversification, it is also possible that ecological transitions drive niche specialization within a single environment. Due to the high dispersibility and large population size of marine bacteria and archaea, sympatric speciation may occur as the result of initial ecological adaptations followed by a decrease of interpopulation gene flow caused by the separation of microhabitats (Shapiro et al. 2012). It is thought that population-specific mutations and the introduction of new genes through HGT trigger the differentiation of populations until they become distinct genotypic clusters. This has been experimentally explored among coastal *Vibrionaceae* strains; here numerous sympatric but ecologically distinct populations of *Vibrio splendidus* have been found, possibly as a result of adaptations to planktonic vs particle-associated lifestyles (Hunt et al. 2008). Given the large heterogeneity that can be found within small volumes of seawater, it is likely that this route of niche specialization and concomitant sympatric differentiation may be a major force that drives diversification in marine microbes that reside in the same environment.

## 7.5 Genome Evolution in the Dark Ocean

The dark ocean represents the vast majority of the ocean by volume and comprises a wide variety of marine habitats, including sediments, oceanic crust, cold seeps, and hydrothermal vents (Arístegui et al. 2009; Orcutt et al. 2011). The water column below the photic zone is considered to be an extreme environment due to low temperature, high hydrostatic pressure, and the absence of solar radiation (Orcutt et al. 2011). Moreover, the environmental conditions in the dark ocean can vary dramatically in chemical composition and consequently, metabolically diverse bacterial and archaeal lineages reside in this environment. Given the metabolic diversity of microbes that reside in the dark ocean, breakthroughs in genomics have been highly valuable in examining the predicted physiology and biogeochemical roles of bacteria and archaea in this environment. For purposes of this section, we will focus on recent research on several planktonic lineages inhabiting the water column of the dark ocean. Several reviews discuss other deep-sea habitats and provide more detail into some of the groups discussed here (Corinaldesi 2015; Dick 2019; Lauro and Bartlett 2008; Orcutt et al. 2011; Orsi 2018; Santoro et al. 2019).

Several recent studies focusing on heterotrophic lineages of bacteria and archaea in the dark ocean have begun to unveil their genomic and physiological diversity. One study found that the broadly distributed SAR202 clade in the *Chloroflexi* phylum represents up to 30% of the microbial community in the dark ocean and is broadly distributed in mesopelagic and bathypelagic waters (Mehrshad et al. 2018). This study identified two major clades of SAR202 that were partitioned by depth and had genome sizes ranging from 1.2 to 3.4 Mbp and %GC content ranging from 41 to 55%. These authors revealed the capacity to metabolize organosulfur compounds and oxidize sulfite in many of the genomes of SAR202, suggesting that this lineage may play an important role in sulfur cycling in the dark ocean. Other studies examining single-cell genomes belonging to this clade identified numerous catabolic enzymes including monooxygenases, dioxygenases, and racemases that were predicted to target recalcitrant carbon compounds. Paralogous expansion of these catabolic enzymes was linked to SAR202 subclades that reside in different locales, suggesting that duplication of these genes has played an important role in the niche expansion in this group (Landry et al. 2017; Saw et al. 2020). Saw et al. (2020) also recovered proteorhodopsin-encoding genes in SAR202 genomes from surface waters, indicating that this group is not exclusively found in the dark ocean.

The Marine Group II Euryarchaeota (MGII), recently named Ca. Poseidoniia, are also abundant constituents of planktonic communities in both the sunlit and dark ocean, where they are thought to be major contributors to organic matter remineralization (Rinke et al. 2019; Zhang et al. 2015). MGII was first discovered in coastal marine environments (DeLong 1992; Fuhrman et al. 1992) but this group was subsequently found in pelagic waters where it is consistently associated with the deep chlorophyll maximum and below (Santoro et al. 2019). MGII is a group of aerobic heterotrophs with the genomic potential to degrade a broad range of substrates as evidenced by genes encoding di- and oligo-peptidases as well as enzymes for the degradation of amino acids and fatty acids (Tully 2019). Members belonging to MGII show a remarkable spatial partitioning in terms of their genomic properties; representatives from surface waters show shorter genomes and harbor photolyase and proteorhodopsin-encoding genes whereas those thriving in the deep ocean encode the potential for flagellar-based adhesion and respiration of nitrate, possibly as an adaptation to oxygen limitation (Rinke et al. 2019; Tully 2019). The acquisition of diverse genes from different sources has facilitated the niche adaptation of different MGII lineages, underscoring the importance of HGT for the diversification of this group.

Several studies have noted that genomes of bacteria and archaea from the mesopelagic are generally larger than those from the epipelagic, but there are exceptions to this trend. One notable example is a lineage of heterotrophic marine *Thaumarchaeota* (HMT) that branches within a sister lineage to the ammonia oxidizing archaea (AOA) (Aylward and Santoro 2020; Reji and Francis 2020). This group, sometimes also referred to as the psl12-like group, is widespread in the dark ocean and has been found in waters ranging from a depth of 150 m near Monterey Bay (Reji and Francis 2020) to bathypelagic and abyssopelagic waters of the South Pacific and South Atlantic (Aylward and Santoro 2020). This HMT group has some genomic characteristics of streamlining including notably small genomes of near 1 Mbp and a low %GC content of 33%. Unlike their AOA relatives, the HMT lack ammonia monooxygenase and hydroxypropionate/hydroxybutyrate carbon fixation pathway. Instead, they encode a type iii-a RuBisCO and diverse pyrroloquinoline quinone (PQQ)-dependent dehydrogenases that are highly expressed and likely play a key role in their heterotrophic lifestyle. PQQ-dehydrogenases comprise as much as 3% of the HMT genomes and their prevalence is likely the product of a combination of paralogous expansion and widespread HGT (Aylward and Santoro 2020). More features of this lineage are likely to become uncovered in the future given its widespread occurrence throughout the dark ocean.

Due to the low energy flux that characterizes the dark ocean, chemolithoautotrophic microorganisms that derive energy from reduced inorganic compounds are among the most abundant at these depths (Bach et al. 2006; Orcutt et al. 2011). Perhaps the most well-studied example is the Marine Group I (MGI) ammonia-oxidizing archaea, which can comprise up to 40% of the cells in some marine habitats (Karner et al. 2001). Genome sizes of complete MGI genomes in the NCBI database range from 1.23 to 2.17 Mbp with the smaller genomes exhibiting signatures of genome streamlining (Santoro et al. 2015). Several studies based on single-cell genomes found that MGI populations are strongly depth-stratified (Luo et al. 2014; Swan et al. 2014) confirming previous observations based on marker genes (Francis et al. 2005; Hallam et al. 2006; Nicol et al. 2014). A large-scale comparative analysis of the radiation of ammonia oxidizing archaea throughout terrestrial and marine habitats revealed that the genomic repertoires of this group varied with environmental variables such as hydrostatic pressure and the availability of ammonia and phosphorus, suggesting that the genomic content is largely structured by partitioning into spatially defined niches (Qin et al. 2020). Another study found that the acquisition of a V-type ATPase through HGT was a salient feature in deep water adapted MGI genomes including several that were recovered from hadopelagic waters (Wang et al. 2019). Although the physiological implications of the V-Type ATPase in deep water AOA remain unclear, it is noteworthy that this was also found in some acidophilic AOA that may use it for pumping protons at low pH (Wang et al. 2019).

Another group that contributes to carbon fixation and nitrogen cycling in the deep ocean is the nitrite oxidizing bacteria (NOB), a polyphyletic group involved in the oxidation of nitrite to nitrate (Bock and Wagner 2006). Although most of the NOB representatives only have the metabolic capability to perform the last step of nitrification, *Nitrospira* strains able to oxidize ammonia to nitrate have been isolated (Comammox) (Daims et al. 2015; van Kessel et al. 2015). The most abundant marine members of the NOB belong to the phylum *Nitrospinae* and it has been estimated that they fix 15 to 45% of the inorganic carbon in the mesopelagic waters of the western North Atlantic, and are therefore a major contributor to nitrification in deep waters (Pachiadaki et al. 2017). Analysis of the *Nitrospina gracilis* genome revealed signatures of extensive HGT from distantly related groups; it is thought that the subunit A of the nitrite oxidoreductase (NxrA) found in *N. gracilis* was obtained from anaerobic ammonium oxidizers (Lücker et al. 2013). This as well as the presence of rTCA cycle for carbon fixation suggests that the *Nitrospinae* may have originated from an anaerobic or microaerophilic environment and subsequently colonized aerobic habitats (Lücker et al. 2013). This is consistent with another study that suggests that oxygen levels played an important role in the diversification of *Nitrospinae* into different marine habitats (Sun et al. 2019).

Lastly, carbon fixation in the dark ocean is also associated with sulfur oxidation as shown by metagenomic reconstructions of the ubiquitous SAR324 clade in the *Deltaproteobacteria* (Swan et al. 2011). In addition to autotrophy, metabolic predictions of this group revealed the presence of genes that encode enzymes involved in the oxidation of carbon monoxide and hydrocarbons, as well as methylotrophy (Brown et al. 2014; Sheik et al. 2014; Swan et al. 2011). Furthermore, genes involved in motility may also play a role in the attachment to particulate organic matter. These observations suggest that the high abundance and ability to inhabit diverse environments observed in SAR324 may be associated with their broad metabolic capabilities (Giovannoni and Vergin 2012; Swan et al. 2011). Regarding genomic composition, SAR324 genomes tend to be small (1.4–2.9 Mbp) and harbor a large amount of repetitive sequences and signatures of prophage integration, indicating that interactions with viruses may play an important role in SAR324 genome evolution (Cao et al. 2016).

## 7.6 Virus–Host Interactions Influencing Genome Evolution in Bacteria and Archaea

Viruses are omnipresent biological entities that play a key role in genome evolution and determine the abundance of microbial lineages in the ocean (Breitbart et al. 2018; Wommack et al. 2000). It is estimated that every second $10^{23}$ viral infections occur in the ocean, killing approximately 20% of the marine microbial biomass per day. Marine viruses, therefore, exert strong selective pressure on their hosts (Suttle 2007). As a consequence, viruses alter the flow of energy and nutrients, and determine the selective pressures experienced by their hosts (Brum et al. 2015; Fuhrman 1999; Lindell et al. 2007; Suttle 2007; Zimmerman et al. 2020).

Host evolution is driven in part by constant arms races with their associated viruses that lead to counter-adaptations and the increase of diversity within populations (Lindell et al. 2007). This phenomenon is known as the Red Queen hypothesis (Stern and Sorek 2011). Many studies have reported long-term stability of viral genotypes in the ocean, often across broad geographic distances (Aylward et al. 2017; Breitbart et al. 2004; Breitbart and Rohwer 2005; Hevroni et al. 2020; Marston and Martiny 2016; Short and Suttle 2005); this indicates that viruses coexist with host populations for long periods in a pervasive arms race dynamics. A long-term analysis performed on marine microbial and viral communities indicated that, although the viral composition is stable throughout the year, there is a rise and fall of variants within populations (Ignacio-Espinoza et al. 2020). This is probably the result of the fluctuating selection from virus–host defenses and counter-defenses as proposed by the Red Queen hypothesis. Additionally, experimental evidence indicates that co-occurrence of viruses and hosts speeds up the evolution of both groups (Thingstad et al. 2014). For example, phages promote microdiversity in cultures of the marine bacteria *Flavobacterium* and *Prochlorococcus* by driving strain succession towards resistance (Avrani et al. 2011; Middelboe et al. 2009). Marine populations under phage pressure are composed of subpopulations with different viral susceptibility. It is thought that the diversity generated by phage-driven selective pressure may facilitate adaptations in fluctuating environments (Williams 2013). For example, this selective pressure may drive the alteration of polysaccharides, glycoproteins, or outer membrane proteins because these structures can serve as recognition sites for phages. In *Alteromonas* populations these genes are located in flexible genomic islands notable for their microdiversity (López-Pérez and Rodriguez-Valera 2016; Rodriguez-Valera et al. 2009).

Viruses mobilize $10^{25}$–$10^{28}$ bp of DNA per year in the world's oceans (Sandaa 2008). They mediate the transfer of genetic material across microbial populations and thereby provide the host with potentially new adaptive traits that may aid subsequent diversification (Brum et al. 2015; Rodriguez-Valera et al. 2009; Sandaa 2008; Touchon et al. 2017). Viral genomes often contain auxiliary metabolic genes (AMGs) derived from their host (Breitbart et al. 2007; Hurwitz et al. 2013; Roux et al. 2016). Some of the most well-studied examples are T4-like phages that infect *Prochlorococcus* and *Synechococcus* (Lindell et al. 2007; Sullivan et al. 2010). Contigs of T4-like phages recovered from marine metagenomes often contain photosystem I genes (PSI), which are thought to form a monomeric PSI complex that can tunnel reducing power from the electron transport chain of the host to this PSI-related function complex during the infection process, possibly contributing to phage fitness (Sullivan et al. 2010). Fridman et al. (2017) successfully cultivated a *Prochlorococcus*-infecting myovirus that encoded genes involved in both photosystems I and II, emphasizing the complex manipulation of host physiology by viruses during infection. The evolution of host photosystem genes may also be driven in part by cyanophages because viral genes are able to recombine and transfer back into the host's gene pool (Lindell et al. 2004; Sullivan et al. 2006, 2010). Aside from photosystem genes, a wide variety of other phage-encoded AMGs have been identified, including genes involved in carbon, nitrogen, and sulfur metabolism (Ahlgren et al. 2019; Anantharaman et al. 2014; Hurwitz et al. 2013). Also, ammonia monooxygenase genes (*amoC*) have been identified in globally distributed viruses that infect marine AOA (Ahlgren et al. 2019). Ahlgren et al. (2019) found viral *amoC* expression, providing evidence that viruses play a role in nitrogen cycling.

Although phages are considered to be the major cause of death of bacteria and archaea in the ocean, not all phages immediately lyse their host upon infection. Lysogenic phages can integrate into the host chromosome and be maintained until specific environmental conditions trigger induction (Breitbart 2012). Culture-based and bioinformatic approaches have estimated that about half of all marine bacteria harbor prophages, which can mediate defense against future infections by providing prophage-induced viral immunity (Bondy-Denomy et al. 2016; Breitbart 2012; Zhao et al. 2019). Prophages may also provide novel functions to the host and can become fixed in a population if they provide fitness benefits to the recipient cell (Bondy-Denomy and Davidson 2014; Paul 2008; Rodriguez-Valera et al. 2009). For example, a region similar to the *Escherichia coli* defective prophage CP4–57 involved in biofilm development has been found in most *Alteromonas* genomes described so far (López-Pérez et al. 2014), suggesting that the prevalence and maintenance of this prophage in *Alteromonas* species may provide an advantage for survival. Prophages and prophage-like sequences have also been found in the streamlined organisms SAR11 and *Prochlorococcus*, respectively (Malmstrom et al. 2013; Morris et al. 2020). This suggests that viral integration can also influence the evolution of even the smallest streamlined genomes.

## 7.7 Outlook

Comparative genomics has provided an important window into the ecology, evolution, and physiology of bacterial and archaeal lineages in the ocean. Here we have discussed some common themes in the evolutionary genomics of a selected number of marine bacterial and archaeal clades. Many more clades will continue to be discovered (Yilmaz et al. 2015) and each of them will have unique features and evolutionary histories that must be taken into account to develop a comprehensive understanding of their evolution, ecology, and activities in the ocean. Metagenomics, single-cell sequencing, and culture-based methods will continue to increase the genomic representation of clades of marine bacteria and archaea in the future. These data will provide an important basis for addressing key questions in important frontiers of microbial oceanography. These include a better understanding of the drivers and consequences of microdiversity in microbial populations, the phylogenetic relationships between major marine clades in the Tree of Life, reconstruction of key evolutionary innovations that have allowed for successful lineages to occupy their current ecological niches, and the extent to which marine viruses determine the ecologies and evolutionary trajectories of their hosts, among many others. Given the wide-ranging nature of these questions, continued research into the evolutionary genomics of marine microbes will be a critical component of future research.

**Acknowledgments** Given the broad scope of topics discussed in this chapter, it is inevitable that many important studies could not be discussed, and we apologize to all colleagues whose important work could not be included in this chapter. This work was supported by a Simons Foundation Early Career Award in Marine Microbial Ecology and Evolution to FOA.

## References

Ahlgren NA, Fuchsman CA, Rocap G, Fuhrman JA (2019) Discovery of several novel, widespread, and ecologically distinct marine Thaumarchaeota viruses that encode amoC nitrification genes. ISME J 13:618–631

Anantharaman K, Duhaime MB, Breier JA, Wendt KA, Toner BM, Dick GJ (2014) Sulfur oxidation genes in diverse deep-sea viruses. Science 344:757–760. https://doi.org/10.1126/science.1252229

Arístegui J, Gasol JM, Duarte CM, Herndl GJ (2009) Microbial oceanography of the dark ocean's pelagic realm. Limnol Oceanogr 54:1501–1529. https://doi.org/10.4319/lo.2009.54.5.1501

Arrigo KR (2005) Marine microorganisms and global nutrient cycles. Nature 437:349–355

Avrani S, Wurtzel O, Sharon I, Sorek R, Lindell D (2011) Genomic island variability facilitates *Prochlorococcus*-virus coexistence. Nature 474:604–608

Aylward FO, Boeuf D, Mende DR, Wood-Charlson EM, Vislova A, Eppley JM, Romano AE, DeLong EF (2017) Diel cycling and long-term persistence of viruses in the ocean's euphotic zone. Proc Natl Acad Sci U S A 114:11446–11451. https://doi.org/10.1073/pnas.1714821114

Aylward FO, Santoro AE (2020) Heterotrophic Thaumarchaea with small genomes are widespread in the dark ocean. mSystems 5:1–20. https://doi.org/10.1128/mSystems.00415-20

Azam F (2004) Oceaonography: microbes, molecules, and marine ecosystems. Science 303:1622–1624. https://doi.org/10.1126/science.1093892

Azam F, Malfatti F (2007) Microbial structuring of marine ecosystems. Nat Rev Microbiol 5:782–791

Bach W, Edwards KJ, Hayes JM, Sievert S, Huber JA, Sogin ML (2006) Energy in the dark: fuel for life in the deep ocean and beyond. Eos 87:73–79. https://doi.org/10.1029/2006eo070002

Bansal MS, Banay G, Peter Gogarten J, Shamir R (2011) Detecting highways of horizontal gene transfer. J Comput Biol 18:1087–1114. https://doi.org/10.1089/cmb.2011.0066

Batut B, Knibbe C, Marais G, Daubin V (2014) Reductive genome evolution at both ends of the bacterial population size spectrum. Nat Rev Microbiol 12:841–850

Beja O (2000) Bacterial rhodopsin: evidence for a new type of phototrophy in the sea. Science 289:1902–1906. https://doi.org/10.1126/science.289.5486.1902

Berube P, Biller S, Hackl T et al (2018) Single cell genomes of *Prochlorococcus*, *Synechococcus*, and sympatric microbes from diverse marine environments. Sci Data 5:1–11. https://doi.org/10.1038/sdata.2018.154

Biller SJ, McDaniel LD, Breitbart M, Rogers E, Paul JH, Chisholm SW (2017) Membrane vesicles in sea water: heterogeneous DNA content and implications for viral abundance estimates. ISME J 11:394–404

Biller SJ, Schubotz F, Roggensack SE, Thompson AW, Summons RE, Chisholm SW (2014) Bacterial vesicles in marine ecosystems. Science 343:183–186

Bobay LM, Ochman H (2017) The evolution of bacterial genome architecture. Front Genet 8:1–6. https://doi.org/10.3389/fgene.2017.00072

Bock E, Wagner M (2006) Oxidation of inorganic nitrogen compounds as an energy source. In: Dworkin M, Falkow S, Rosenberg E, Schleifer KH, Stackebrandt E (eds) The prokaryotes: a handbook on the biology of bacteria. Springer, pp 457–495. https://doi.org/10.1007/0-387-30742-7_16

Bondy-Denomy J, Davidson AR (2014) When a virus is not a parasite: the beneficial effects of prophages on bacterial fitness. J Microbiol 52:235–242. https://doi.org/10.1007/s12275-014-4083-3

Bondy-Denomy J, Qian J, Westra ER, Buckling A, Guttman DS, Davidson AR, Maxwell KL (2016) Prophages mediate defense against phage infection through diverse mechanisms. ISME J 10:2854–2866. https://doi.org/10.1038/ismej.2016.79

Bragg JG, Hyder CL (2004) Nitrogen versus carbon use in prokaryotic genomes and proteomes. Proc R Soc Lond B Biol Sci 271:374–377

Breitbart M (2012) Marine viruses: truth or dare. Annu Rev Mar Sci 4:425–448. https://doi.org/10.1146/annurev-marine-120709-142805

Breitbart M, Bonnain C, Malki K, Sawaya NA (2018) Phage puppet masters of the marine microbial realm. Nat Microbiol 3:754–766

Breitbart M, Miyake JH, Rohwer F (2004) Global distribution of nearly identical phage-encoded DNA sequences. FEMS Microbiol Lett 236:249–256

Breitbart M, Rohwer F (2005) Here a virus, there a virus, everywhere the same virus? Trends Microbiol 13:278–284

Breitbart M, Thompson L, Suttle C, Sullivan M (2007) Exploring the vast diversity of marine viruses. Oceanography 20:135–139. https://doi.org/10.5670/oceanog.2007.58

Brochier-Armanet C, Deschamps P, López-García P, Zivanovic Y, Rodríguez-Valera F, Moreira D (2011) Complete-fosmid and fosmid-end sequences reveal frequent horizontal gene transfers in marine uncultured planktonic archaea. ISME J 5:1291–1302

Brockhurst MA, Harrison E, Hall JPJ, Richards T, McNally A, MacLean C (2019) The ecology and evolution of pangenomes. Curr Biol 29:1094–1103

Brown MV, Ostrowski M, Grzymski JJ, Lauro FM (2014) A trait based perspective on the biogeography of common and abundant marine bacterioplankton clades. Mar Genomics 15: 17–28

Brum JR, Ignacio-Espinoza JC, Roux S, Doulcier G et al (2015) Ocean plankton. Patterns and ecological drivers of ocean viral communities. Science 348:1261498

Bunse C, Pinhassi J (2017) Marine bacterioplankton seasonal succession dynamics. Trends Microbiol 25:494–505

Cao H, Dong C, Bougouffa S, Li J, Zhang W, Shao Z, Bajic VB, Qian PY (2016) Delta-proteobacterial SAR324 group in hydrothermal plumes on the south mid-Atlantic ridge. Sci Rep 6:1–9

Cohan FM (2006) Towards a conceptual and operational union of bacterial systematics, ecology, and evolution. Phil Trans R Soc Lond B Biol Sci 361:1985–1996

Cohan FM, Perry EB (2007) A systematics for discovering the fundamental units of bacterial diversity. Curr Biol 17:373–386

Cordero OX, Polz MF (2014) Explaining microbial genomic diversity in light of evolutionary ecology. Nat Rev Microbiol 12:263–273

Corinaldesi C (2015) New perspectives in benthic deep-sea microbial ecology. Front Mar Sci 2:1–12. https://doi.org/10.3389/fmars.2015.00017

Daims H, Lebedeva EV, Pjevac P, Han P, Herbold C, Albertsen M, Jehmlich N, Palatinszky M, Vierheilig J, Bulaev A, Kirkegaard RH, von Bergen M, Rattei T, Bendinger B, Nielsen PH, Wagner M (2015) Complete nitrification by *Nitrospira* bacteria. Nature 528:504–509

Deatherage BL, Cookson BT (2012) Membrane vesicle release in bacteria, eukaryotes, and archaea: a conserved yet underappreciated aspect of microbial life. Infect Immun 80:1948–1957

DeLong EF (1992) Archaea in coastal marine environments. Proc Natl Acad Sci U S A 89:5685–5689

DeLong EF (1997) Marine microbial diversity: the tip of the iceberg. Trends Biotechnol 15:203–207

DeLong EF (2006) Community genomics among stratified microbial assemblages in the ocean's interior. Science 311:496–503. https://doi.org/10.1126/science.1120250

DeLong EF, Béjà O (2010) The light-driven proton pump proteorhodopsin enhances bacterial survival during tough times. PLoS Biol 8:e1000359

Deschamps P, Zivanovic Y, Moreira D, Rodriguez-Valera F, López-García P (2014) Pangenome evidence for extensive interdomain horizontal transfer affecting lineage core and shell genes in uncultured planktonic Thaumarchaeota and Euryarchaeota. Genome Biol Evol 6:1549–1563

Dick GJ (2019) The microbiomes of deep-sea hydrothermal vents: distributed globally, shaped locally. Nat Rev Microbiol 17:271–283

Dupont CL, Rusch DB, Yooseph S et al (2012) Genomic insights to SAR86, an abundant and uncultivated marine bacterial lineage. ISME J 6:1186–1199

Eguíluz VM, Salazar G, Fernández-Gracia J, Pearman JK, Gasol JM, Acinas SG, Sunagawa S, Irigoien X, Duarte CM (2019) Scaling of species distribution explains the vast potential marine prokaryote diversity. Sci Rep 9:1–8

Falkowski PG (1998) Biogeochemical controls and feedbacks on ocean primary production. Science 281:200–206

Falkowski PG, Fenchel T, DeLong EF (2008) The microbial engines that drive Earth's biogeochemical cycles. Science 320:1034–1039

Feil EJ (2004) Small change: keeping pace with microevolution. Nat Rev Microbiol 2:483–495

Francis CA, Roberts KJ, Beman JM, Santoro AE, Oakley BB (2005) Ubiquity and diversity of ammonia-oxidizing archaea in water columns and sediments of the ocean. Proc Natl Acad Sci U S A 102:14683–14688

Fraser C, Alm EJ, Polz MF, Spratt BG, Hanage WP (2009) The bacterial species challenge: making sense of genetic and ecological diversity. Science 323:741–746

Fridman S, Flores-Uribe J, Larom S et al (2017) A myovirus encoding both photosystem I and II proteins enhances cyclic electron flow in infected *Prochlorococcus* cells. Nat Microbiol 2: 1350–1357

Frigaard NU, Martinez A, Mincer TJ, DeLong EF (2006) Proteorhodopsin lateral gene transfer between marine planktonic bacteria and archaea. Nature 439:847–850

Fuhrman JA (1999) Marine viruses and their biogeochemical and ecological effects. Nature 399: 541–548

Fuhrman JA, McCallum K, Davis AA (1992) Novel major archaebacterial group from marine plankton. Nature 356:148–149

Futuyma DJ (1986) Evolutionary biology, 2nd edn. Sinauer Associates Inc.

Getz EW, Tithi SS, Zhang L, Aylward FO (2018) Parallel evolution of genome streamlining and cellular bioenergetics across the marine radiation of a bacterial phylum. mBio 9:1–14. https://doi.org/10.1128/mBio.01089-18

Ghai R, Mizuno CM, Picazo A, Camacho A, Rodriguez-Valera F (2013) Metagenomics uncovers a new group of low GC and ultra-small marine Actinobacteria. Sci Rep 3:1–8. https://doi.org/10.1038/srep02471

Giovannoni SJ (2017) SAR11 bacteria: the most abundant plankton in the oceans. Annu Rev Mar Sci 9:231–255

Giovannoni SJ, Bibbs L, Cho JC, Stapels MD, Desiderio R, Vergin KL, Rappé MS, Laney S, Wilhelm LJ, Tripp HJ, Mathur EJ, Barofsky DF (2005) Proteorhodopsin in the ubiquitous marine bacterium SAR11. Nature 438:82–85

Giovannoni SJ, Cameron Thrash J, Temperton B (2014) Implications of streamlining theory for microbial ecology. ISME J 8:1553–1565

Giovannoni SJ, Stingl U (2005) Molecular diversity and ecology of microbial plankton. Nature 437: 343–348. https://doi.org/10.1038/nature04158

Giovannoni SJ, Vergin KL (2012) Seasonality in ocean microbial communities. Science 335:671–676

Gogarten JP, Peter Gogarten J, Townsend JP (2005) Horizontal gene transfer, genome innovation and evolution. Nat Rev Microbiol 3:679–687. https://doi.org/10.1038/nrmicro1204

Gómez-Consarnau L, Akram N, Lindell K, Pedersen A, Neutze R, Milton DL, González JM, Pinhassi J (2010) Proteorhodopsin phototrophy promotes survival of marine bacteria during starvation. PLoS Biol 8:e1000358. https://doi.org/10.1371/journal.pbio.1000358

Gómez-Consarnau L, González JM, Coll-Lladó M, Gourdon P, Pascher T, Neutze R, Pedrós-Alió C, Pinhassi J (2007) Light stimulates growth of proteorhodopsin-containing marine Flavobacteria. Nature 445:210–213. https://doi.org/10.1038/nature05381

Graham ED, Tully BJ (2020) Marine Dadabacteria exhibit genome streamlining and phototrophy-driven niche partitioning. ISME J 15:1248–1256. https://doi.org/10.1038/s41396-020-00834-5

Grzymski JJ, Dussaq AM (2012) The significance of nitrogen cost minimization in proteomes of marine microorganisms. ISME J 6:71–80

Hallam SJ, Mincer TJ, Schleper C, Preston CM, Roberts K, Richardson PM, DeLong EF (2006) Pathways of carbon assimilation and ammonia oxidation suggested by environmental genomic analyses of marine Crenarchaeota. PLoS Biol 4:e95

Heidelberg KB, Gilbert JA, Joint I (2010) Marine genomics: at the interface of marine microbial ecology and biodiscovery. Microb Biotechnol 3:531–543

Hevroni G, Flores-Uribe J, Béjà O, Philosof A (2020) Seasonal and diel patterns of abundance and activity of viruses in the Red Sea. Proc Natl Acad Sci U S A 117:29738–29747

Hu J, Blanchard JL (2009) Environmental sequence data from the Sargasso Sea reveal that the characteristics of genome reduction in *Prochlorococcus* are not a harbinger for an escalation in genetic drift. Mol Biol Evol 26:1191–1191. https://doi.org/10.1093/molbev/msn299

Hunt DE, David LA, Gevers D, Preheim SP, Alm EJ, Polz MF (2008) Resource partitioning and sympatric differentiation among closely related bacterioplankton. Science 320:1081–1085

Hurwitz BL, Hallam SJ, Sullivan MB (2013) Metabolic reprogramming by viruses in the sunlit and dark ocean. Genome Biol 14:1–14

Ignacio-Espinoza JC, Ahlgren NA, Fuhrman JA (2020) Long-term stability and red queen-like strain dynamics in marine viruses. Nat Microbiol 5:265–271

Karner MB, DeLong EF, Karl DM (2001) Archaeal dominance in the mesopelagic zone of the Pacific Ocean. Nature 409:507–510

Kashtan N, Roggensack SE, Rodrigue S et al (2014) Single-cell genomics reveals hundreds of coexisting subpopulations in wild *Prochlorococcus*. Science 344:416–420

Kasting JF (2002) Life and the evolution of Earth's atmosphere. Science 296:1066–1068. https://doi.org/10.1126/science.1071184

Kettler GC, Martiny AC, Huang K et al (2007) Patterns and implications of gene gain and loss in the evolution of *Prochlorococcus*. PLoS Genet 3:e231

Kirchberger PC, Schmidt M, Ochman H (2020) The ingenuity of bacterial genomes. Ann Rev Microbiol 74:815–834. https://doi.org/10.1146/annurev-micro-020518-115822

Klieve AV, Yokoyama MT, Forster RJ, Ouwerkerk D, Bain PA, Mawhinney EL (2005) Naturally occurring DNA transfer system associated with membrane vesicles in cellulolytic *Ruminococcus* spp. of ruminal origin. Appl Environ Microbiol 71:4248–4253

Konstantinidis KT, Braff J, Karl DM, DeLong EF (2009) Comparative metagenomic analysis of a microbial community residing at a depth of 4,000 meters at station ALOHA in the North Pacific subtropical gyre. Appl Environ Microbiol 75:5345–5355. https://doi.org/10.1128/aem.00473-09

Koonin EV, Makarova KS, Aravind L (2001) Horizontal gene transfer in prokaryotes: quantification and classification. Ann Rev Microbiol 55:709–742

Kunin V, Ouzounis CA (2003) The balance of driving forces during genome evolution in prokaryotes. Genome Res 13:1589–1594

Kuo CH, Ochman H (2009) The fate of new bacterial genes. FEMS Microbiol Rev 33:38–43. https://doi.org/10.1111/j.1574-6976.2008.00140.x

Landry Z, Swan BK, Herndl GJ, Stepanauskas R, Giovannoni SJ (2017) SAR202 genomes from the dark ocean predict pathways for the oxidation of recalcitrant dissolved organic matter. mBio 8: 1–19. https://doi.org/10.1128/mBio.00413-17
Lang AS, Westbye AB, Beatty JT (2017) The distribution, evolution, and roles of gene transfer agents in prokaryotic genetic exchange. Ann Rev Virol 4:87–104
Lasken RS, McLean JS (2014) Recent advances in genomic DNA sequencing of microbial species from single cells. Nat Rev Genet 15:577–584
Lauro FM, Bartlett DH (2008) Prokaryotic lifestyles in deep sea habitats. Extremophiles 12:15–25
Lauro FM, McDougald D, Thomas T et al (2009) The genomic basis of trophic strategy in marine bacteria. Proc Natl Acad Sci U S A 106:15527–15533
Lindell D, Jaffe JD, Coleman ML et al (2007) Genome-wide expression dynamics of a marine virus and host reveal features of co-evolution. Nature 449:83–86
Lindell D, Sullivan MB, Johnson ZI, Tolonen AC, Rohwer F, Chisholm SW (2004) Transfer of photosynthesis genes to and from *Prochlorococcus* viruses. Proc Natl Acad Sci U S A 101: 11013–11018
López-Pérez M, Gonzaga A, Ivanova EP, Rodriguez-Valera F (2014) Genomes of *Alteromonas australica*, a world apart. BMC Genomics 15:1–13
López-Pérez M, Haro-Moreno JM, Coutinho FH, Martinez-Garcia M, Rodriguez-Valera F (2020) The evolutionary success of the marine bacterium SAR11 analyzed through a metagenomic perspective. mSystems 5:1–13. https://doi.org/10.1128/mSystems.00605-20
López-Pérez M, Rodriguez-Valera F (2016) Pangenome evolution in the marine bacterium *Alteromonas*. Genome Biol Evol 8:1556–1570
Lücker S, Nowka B, Rattei T, Spieck E, Daims H (2013) The genome of *Nitrospina gracilis* illuminates the metabolism and evolution of the major marine nitrite oxidizer. Front Microbiol 4:1–17. https://doi.org/10.3389/fmicb.2013.00027
Luo H, Huang Y, Stepanauskas R, Tang J (2017) Excess of non-conservative amino acid changes in marine bacterioplankton lineages with reduced genomes. Nat Microbiol 2:1–9
Luo H, Swan BK, Stepanauskas R, Hughes AL, Moran MA (2014a) Comparing effective population sizes of dominant marine alphaproteobacteria lineages. Environ Microbiol Rep 6:167–172
Luo H, Swan BK, Stepanauskas R, Hughes AL, Moran MA (2014b) Evolutionary analysis of a streamlined lineage of surface ocean Roseobacters. ISME J 8:1428–1439
Luo H, Thompson LR, Stingl U, Hughes AL (2015) Selection maintains low genomic GC content in marine SAR11 lineages. Mol Biol Evol 32:2738–2748
Luo H, Tolar BB, Swan BK, Zhang CL, Stepanauskas R, Ann Moran M, Hollibaugh JT (2014) Single-cell genomics shedding light on marine Thaumarchaeota diversification. ISME J 8:732–736
Lynch M (2007) The origins of genome architecture, 1st edn. Sinauer Associates Inc.
Malmstrom RR, Rodrigue S, Huang KH, Kelly L, Kern SE, Thompson A, Roggensack S, Berube PM, Henn MR, Chisholm SW (2013) Ecology of uncultured *Prochlorococcus* clades revealed through single-cell genomics and biogeographic analysis. ISME J 7:184–198. https://doi.org/10.1038/ismej.2012.89
Marston MF, Martiny JBH (2016) Genomic diversification of marine cyanophages into stable ecotypes. Environ Microbiol 18:4240–4253
Martin-Cuadrado AB, Garcia-Heredia I, Moltó AG, López-Úbeda R, Kimes N, López-García P, Moreira D, Rodriguez-Valera F (2015) A new class of marine Euryarchaeota group II from the Mediterranean deep chlorophyll maximum. ISME J 9:1619–1634
Martinez-Gutierrez CA, Aylward FO (2019) Strong purifying selection is associated with genome streamlining in epipelagic *Marinimicrobia*. Genome Biol Evol 11:2887–2894
Martiny AC, Tai APK, Veneziano D, Primeau F, Chisholm SW (2009) Taxonomic resolution, ecotypes and the biogeography of *Prochlorococcus*. Environ Microbiol 11:823–832
Mas A, Jamshidi S, Lagadeuc Y, Eveillard D, Vandenkoornhuyse P (2016) Beyond the black queen hypothesis. ISME J 10:2085–2091

McDaniel LD, Young E, Delaney J, Ruhnau F, Ritchie KB, Paul JH (2010) High frequency of horizontal gene transfer in the oceans. Science 330:50
Mehrshad M, Rodriguez-Valera F, Amoozegar MA, López-García P, Ghai R (2018) The enigmatic SAR202 cluster up close: shedding light on a globally distributed dark ocean lineage involved in sulfur cycling. ISME J 12:655–668
Mende DR, Bryant JA, Aylward FO, Eppley JM, Nielsen T, Karl DM, DeLong EF (2017) Environmental drivers of a microbial genomic transition zone in the ocean's interior. Nat Microbiol 2:1367–1373. https://doi.org/10.1038/s41564-017-0008-3
Middelboe M, Holmfeldt K, Riemann L, Nybroe O, Haaber J (2009) Bacteriophages drive strain diversification in a marine Flavobacterium: implications for phage resistance and physiological properties. Environ Microbiol 11:1971–1982. https://doi.org/10.1111/j.1462-2920.2009.01920.x
Mira A, Ochman H, Moran NA (2001) Deletional bias and the evolution of bacterial genomes. Trends Genet 17:589–596
Moore LR, Goericke R, Chisholm SW (1995) Comparative physiology of *Synechococcus* and *Prochlorococcus*: influence of light and temperature on growth, pigments, fluorescence and absorptive properties. Mar Ecol Prog Ser 116:259–275. https://doi.org/10.3354/meps116259
Moran MA (2008) Genomics and metagenomics of marine prokaryotes. In: Kirchman DL (ed) Microbial ecology of the oceans, 2nd edn. Wiley, pp 91–129
Moran MA, Miller WL (2007) Resourceful heterotrophs make the most of light in the coastal ocean. Nat Rev Microbiol 5:792–800. https://doi.org/10.1038/nrmicro1746
Moran NA (1996) Accelerated evolution and Muller's rachet in endosymbiotic bacteria. Proc Natl Acad Sci U S A 93:2873–2878. https://doi.org/10.1073/pnas.93.7.2873
Morris JJ, Lenski RE, Zinser ER (2012) The black queen hypothesis: evolution of dependencies through adaptive gene loss. mBio 3:1–7. https://doi.org/10.1128/mBio.00036-12
Morris RM, Cain KR, Hvorecny KL, Kollman JM (2020) Lysogenic host-virus interactions in SAR11 marine bacteria. Nat Microbiol 5:1011–1015. https://doi.org/10.1038/s41564-020-0725-x
Nicol GW, Leininger S, Schleper C (2014) Distribution and activity of ammonia-oxidizing archaea in natural environments. In: Ward B, Arp D, Klotz M (eds) Nitrification. ASM Press, pp. 157–178 doi:https://doi.org/10.1128/9781555817145.ch7
Orcutt BN, Sylvan JB, Knab NJ, Edwards KJ (2011) Microbial ecology of the dark ocean above, at, and below the seafloor. Microbiol Mol Biol Rev 75:361–422. https://doi.org/10.1128/mmbr.00039-10
Orellana LH, Ben Francis T, Krüger K, Teeling H, Müller MC, Fuchs BM, Konstantinidis KT, Amann RI (2019) Niche differentiation among annually recurrent coastal marine group II Euryarchaeota. ISME J 13:3024–3036
Orsi WD (2018) Ecology and evolution of seafloor and subseafloor microbial communities. Nat Rev Microbiol 16:671–683
Pachiadaki MG, Sintes E, Bergauer K et al (2017) Major role of nitrite-oxidizing bacteria in dark ocean carbon fixation. Science 358:1046–1051
Parks DH, Chuvochina M, Waite DW, Rinke C, Skarshewski A, Chaumeil PA, Hugenholtz P (2018) A standardized bacterial taxonomy based on genome phylogeny substantially revises the tree of life. Nat Biotechnol 36:996–1004
Partensky F, Garczarek L (2010) Prochlorococcus: advantages and limits of minimalism. Annu Rev Mar Sci 2:305–331
Paul JH (2008) Prophages in marine bacteria: dangerous molecular time bombs or the key to survival in the seas? ISME J 2:579–589. https://doi.org/10.1038/ismej.2008.35
Paul S, Dutta A, Bag SK, Das S, Dutta C (2010) Distinct, ecotype-specific genome and proteome signatures in the marine cyanobacteria *Prochlorococcus*. BMC Genomics 11:2–15
Pinhassi J, DeLong EF, Béjà O, González JM, Pedrós-Alió C (2016) Marine bacterial and archaeal ion-pumping rhodopsins: genetic diversity, physiology, and ecology. Microbiol Mol Biol Rev 80:929–954. https://doi.org/10.1128/mmbr.00003-16

Pomeroy L, Williams PL, Azam F, Hobbie J (2007) The microbial loop. Oceanography 20:28–33. https://doi.org/10.5670/oceanog.2007.45
Price MN, Arkin AP (2015) Weakly deleterious mutations and low rates of recombination limit the impact of natural selection on bacterial genomes. mBio 6:e01302–e01315
Qin W, Zheng Y, Zhao F et al (2020) Alternative strategies of nutrient acquisition and energy conservation map to the biogeography of marine ammonia-oxidizing archaea. ISME J 14:2595–2609. https://doi.org/10.1038/s41396-020-0710-7
Read RW, Berube PM, Biller SJ, Neveux I, Cubillos-Ruiz A, Chisholm SW, Grzymski JJ (2017) Nitrogen cost minimization is promoted by structural changes in the transcriptome of N-deprived *Prochlorococcus* cells. ISME J 11:2267–2278
Reji L, Francis CA (2020) Metagenome-assembled genomes reveal unique metabolic adaptations of a basal marine Thaumarchaeota lineage. ISME J 14:2105–2115
Renelli M, Matias V, Lo RY, Beveridge TJ (2004) DNA-containing membrane vesicles of Pseudomonas aeruginosa PAO1 and their genetic transformation potential. Microbiology 150: 2161–2169
Ren M, Feng X, Huang Y et al (2019) Phylogenomics suggests oxygen availability as a driving force in Thaumarchaeota evolution. ISME J 13:2150–2161
Rinke C, Rubino F, Messer LF et al (2019) A phylogenomic and ecological analysis of the globally abundant marine group II archaea (*ca.* Poseidoniales Ord. Nov.). ISME J 13:663–675
Rocap G, Larimer FW, Lamerdin J et al (2003) Genome divergence in two *Prochlorococcus* ecotypes reflects oceanic niche differentiation. Nature 424:1042–1047
Rodriguez-Valera F, Martin-Cuadrado AB, Rodriguez-Brito B, Pasić L, Thingstad TF, Rohwer F, Mira A (2009) Explaining microbial population genomics through phage predation. Nat Rev Microbiol 7:828–836
Rosselló-Mora R, Amann R (2001) The species concept for prokaryotes. FEMS Microbiol Rev 25: 39–67
Roux S, Brum JR, Dutilh BE et al (2016) Ecogenomics and potential biogeochemical impacts of globally abundant ocean viruses. Nature 537:689–693
Salazar G, Sunagawa S (2017) Marine microbial diversity. Curr Biol 27:489–494
Salcher MM, Schaefle D, Kaspar M, Neuenschwander SM, Ghai R (2019) Evolution in action: habitat transition from sediment to the pelagial leads to genome streamlining in Methylophilaceae. ISME J 13:2764–2777
Sánchez-Baracaldo P, Bianchini G, Di Cesare A, Callieri C, Chrismas NAM (2019) Insights into the evolution of picocyanobacteria and phycoerythrin genes (mpeBA and cpeBA). Front Microbiol 10:1–17. https://doi.org/10.3389/fmicb.2019.00045
Sanchez-Perez G, Mira A, Nyiro G, Pasić L, Rodriguez-Valera F (2008) Adapting to environmental changes using specialized paralogs. Trends Genet 24:154–158
Sandaa RA (2008) Burden or benefit? Virus-host interactions in the marine environment. Res Microbiol 159:374–381. https://doi.org/10.1016/j.resmic.2008.04.013
Santoro AE, Dupont CL, Richter RA, Craig MT, Carini P, McIlvin MR, Yang Y, Orsi WD, Moran DM, Saito MA (2015) Genomic and proteomic characterization of "*Candidatus* Nitrosopelagicus brevis": an ammonia-oxidizing archaeon from the open ocean. Proc Natl Acad Sci U S A 112:1173–1178
Santoro AE, Richter RA, Dupont CL (2019) Planktonic marine archaea. Annu Rev Mar Sci 11:131–158
Saw JHW, Nunoura T, Hirai M, Takaki Y, Parsons R, Michelsen M, Longnecker K, Kujawinski EB, Stepanauskas R, Landry Z, Carlson CA, Giovannoni SJ (2020) Pangenomics analysis reveals diversification of enzyme families and niche specialization in globally abundant SAR202 bacteria. mBio 11:1–18. https://doi.org/10.1128/mBio.02975-19
Shakya M, Soucy SM, Zhaxybayeva O (2017) Insights into origin and evolution of α-proteobacterial gene transfer agents. Virus Evol 3:1–13

Shapiro BJ, Friedman J, Cordero OX, Preheim SP, Timberlake SC, Szabó G, Polz MF, Alm EJ (2012) Population genomics of early events in the ecological differentiation of bacteria. Science 336:48–51

Sharon I, Banfield JF (2013) Genomes from metagenomics. Science 342:1057–1058. https://doi.org/10.1126/science.1247023

Sheik CS, Jain S, Dick GJ (2014) Metabolic flexibility of enigmatic SAR324 revealed through metagenomics and metatranscriptomics. Environ Microbiol 16:304–317

Sherr E, Sherr B (2008) Understanding roles of microbes in marine pelagic food webs: a brief history. In: Kirchman DL (ed) Microbial ecology of the oceans, 2nd edn. Wiley, pp 27–44. https://doi.org/10.1002/9780470281840.ch2

Short CM, Suttle CA (2005) Nearly identical bacteriophage structural gene sequences are widely distributed in both marine and freshwater environments. Appl Environ Microbiol 71:480–486

Snel B, Bork P, Huynen MA (2002) Genomes in flux: the evolution of archaeal and proteobacterial gene content. Genome Res 12:17–25

Sobecky PA, Hazen TH (2009) Horizontal gene transfer and mobile genetic elements in marine systems. Methods Mol Biol 532:435–453

Staley JT, Konopka A (1985) Measurement of in situ activities of nonphotosynthetic microorganisms in aquatic and terrestrial habitats. Ann Rev Microbiol 39:321–346

Stern A, Sorek R (2011) The phage-host arms race: shaping the evolution of microbes. BioEssays 33:43–51

Sullivan MB, Huang KH, Ignacio-Espinoza JC et al (2010) Genomic analysis of oceanic cyanobacterial myoviruses compared with T4-like myoviruses from diverse hosts and environments. Environ Microbiol 12:3035–3056

Sullivan MB, Lindell D, Lee JA, Thompson LR, Bielawski JP, Chisholm SW (2006) Prevalence and evolution of core photosystem II genes in marine cyanobacterial viruses and their hosts. PLoS Biol 4:1344–1357. https://doi.org/10.1371/journal.pbio.0040234

Sun X, Kop LFM, Lau MCY, Frank J, Jayakumar A, Lücker S, Ward BB (2019) Uncultured *Nitrospina*-like species are major nitrite oxidizing bacteria in oxygen minimum zones. ISME J 13:2391–2402

Sun Z, Blanchard JL (2014) Strong genome-wide selection early in the evolution of *Prochlorococcus* resulted in a reduced genome through the loss of a large number of small effect genes. PLoS One 9:1–15. https://doi.org/10.1371/journal.pone.0088837

Suttle CA (2007) Marine viruses — major players in the global ecosystem. Nat Rev Microbiol 5: 801–812. https://doi.org/10.1038/nrmicro1750

Swan BK, Chaffin MD, Martinez-Garcia M et al (2014) Genomic and metabolic diversity of marine group I Thaumarchaeota in the mesopelagic of two subtropical gyres. PLoS One 9:1–9. https://doi.org/10.1371/journal.pone.0095380

Swan BK, Martinez-Garcia M, Preston CM et al (2011) Potential for chemolithoautotrophy among ubiquitous bacteria lineages in the dark ocean. Science 333:1296–1300

Swan BK, Tupper B, Sczyrba A et al (2013) Prevalent genome streamlining and latitudinal divergence of planktonic bacteria in the surface ocean. Proc Natl Acad Sci U S A 110:11463–11468

Thingstad TF, Våge S, Storesund JE, Sandaa RA, Giske J (2014) A theoretical analysis of how strain-specific viruses can control microbial species diversity. Proc Natl Acad Sci U S A 111: 7813–7818

Ting CS, Rocap G, King J, Chisholm SW (2002) Cyanobacterial photosynthesis in the oceans: the origins and significance of divergent light-harvesting strategies. Trends Microbiol 10:134–142

Torre JR, Christianson LM, Beja O, Suzuki MT, Karl DM, Heidelberg J, DeLong EF (2003) Proteorhodopsin genes are distributed among divergent marine bacterial taxa. Proc Natl Acad Sci U S A 100:12830–12835. https://doi.org/10.1073/pnas.2133554100

Touchon M, Moura de Sousa JA, Rocha EP (2017) Embracing the enemy: the diversification of microbial gene repertoires by phage-mediated horizontal gene transfer. Curr Opin Microbiol 38: 66–73

Treangen TJ, Rocha EPC (2011) Horizontal transfer, not duplication, drives the expansion of protein families in prokaryotes. PLoS Genet 7:1–12. https://doi.org/10.1371/journal.pgen.1001284
Tringe SG, Hugenholtz P (2008) A renaissance for the pioneering 16S rRNA gene. Curr Opin Microbiol 11:442–446
Tully BJ (2019) Metabolic diversity within the globally abundant marine group II Euryarchaea offers insight into ecological patterns. Nat Commun 10:1–12
van Kessel MAHJ, Speth DR, Albertsen M, Nielsen PH, Op den Camp HJM, Kartal B, Jetten MSM, Lücker S (2015) Complete nitrification by a single microorganism. Nature 528:555–559
Wang B, Qin W, Ren Y et al (2019) Expansion of Thaumarchaeota habitat range is correlated with horizontal transfer of ATPase operons. ISME J 13:3067–3079
Wilhelm LJ, Tripp HJ, Givan SA, Smith DP, Giovannoni SJ (2007) Natural variation in SAR11 marine bacterioplankton genomes inferred from metagenomic data. Biol Direct 2:1–19
Williams HTP (2013) Phage-induced diversification improves host evolvability. BMC Evol Biol 13:1–17
Witzel KP (1990) Approaches to bacterial population dynamics. In: Overbeck J, Chróst RJ (eds) Aquatic microbial ecology: biochemical and molecular approaches. Springer, pp 96–128. https://doi.org/10.1007/978-1-4612-3382-4_5
Wolf YI, Koonin EV (2013) Genome reduction as the dominant mode of evolution. BioEssays 35: 829–837
Wommack KE, Eric Wommack K, Colwell RR (2000) Virioplankton: viruses in aquatic ecosystems. Microbiol Mol Biol Rev 64:69–114. https://doi.org/10.1128/mmbr.64.1.69-114.2000
Yaron S, Kolling GL, Simon L, Matthews KR (2000) Vesicle-mediated transfer of virulence genes from *Escherichia coli* O157:H7 to other enteric bacteria. Appl Environ Microbiol 66:4414–4420
Yilmaz P, Yarza P, Rapp JZ, Glöckner FO (2015) Expanding the world of marine bacterial and archaeal clades. Front Microbiol 6:1–29
Yutin N, Koonin EV (2012) Proteorhodopsin genes in giant viruses. Biol Direct 7:1–6. https://doi.org/10.1186/1745-6150-7-34
Zhang CL, Xie W, Martin-Cuadrado AB, Rodriguez-Valera F (2015) Marine group II archaea, potentially important players in the global ocean carbon cycle. Front Microbiol 6:1–9
Zhang H, Yoshizawa S, Sun Y, Huang Y, Chu X, González JM, Pinhassi J, Luo H (2019) Repeated evolutionary transitions of flavobacteria from marine to non-marine habitats. Environ Microbiol 21:648–666
Zhao Y, Qin F, Zhang R, Giovannoni SJ, Zhang Z, Sun J, Du S, Rensing C (2019) Pelagiphages in the Podoviridae family integrate into host genomes. Environ Microbiol 21:1989–2001
Zimmerman AE, Howard-Varona C, Needham DM, John SG, Worden AZ, Sullivan MB, Waldbauer JR, Coleman ML (2020) Metabolic and biogeochemical consequences of viral infection in aquatic ecosystems. Nat Rev Microbiol 18:21–34
Zwirglmaier K, Jardillier L, Ostrowski M, Mazard S, Garczarek L, Vaulot D, Not F, Massana R, Ulloa O, Scanlan DJ (2008) Global phylogeography of marine *Synechococcus* and *Prochlorococcus* reveals a distinct partitioning of lineages among oceanic biomes. Environ Microbiol 10:147–161

# Part II

# Marine Habitats

# Towards a Global Perspective of the Marine Microbiome

# 8

Silvia G. Acinas, Marta Sebastián, and Isabel Ferrera

**Abstract**

Marine microbes play fundamental roles in nutrient cycling and climate regulation at a planetary scale. The field of marine microbial ecology has experienced major breakthroughs following the application of high-throughput sequencing and culture-independent methodologies that have pushed the exploration of the marine microbiome to an unprecedented scale. This chapter overviews how the advances in gene- and genome-centric approaches as well as in culturing and single cell physiological methodologies in conjunction with global oceanographic circumnavigation expeditions and long-term time series are fueling our understanding of the biogeography, temporal dynamics, functional diversity, and evolutionary processes of microbial populations. We discuss how the joint effort of all those integrative approaches will help to boost our knowledge of the marine microbiome to reach a predictive understanding of how it is going to evolve in future scenarios.

**Keywords**

Biogeography · Cultures · Evolution · Marine microbiome · MAGs · SAGs

S. G. Acinas (✉) · M. Sebastián
Department of Marine Biology and Oceanography, Institut de Ciències del Mar (ICM-CSIC), Barcelona, Spain
e-mail: sacinas@icm.csic.es

I. Ferrera
Centro Oceanográfico de Málaga, Instituto Español de Oceanografía, Málaga, Spain



L. J. Stal, M. S. Cretoiu (eds.), *The Marine Microbiome*, The Microbiomes of Humans, Animals, Plants, and the Environment 3,
https://doi.org/10.1007/978-3-030-90383-1_8

# 8.1 Marine Microbial Ecology: Opening the Black Box

## 8.1.1 Major Breakthroughs before the -Omics Revolution

As Sydney Brenner, Nobel Prize in Physiology or Medicine 2002 said, "*Progress in science depends on new techniques, new discoveries and new ideas, probably in that order.*" The field of marine microbial ecology has been no exception to this quote as it has seen major breakthroughs in the last 60 years following the application of new technologies. For example, the use of epifluorescence microscopy for the estimation of bacterial abundance unveiled that traditional plate counting methods were underestimating the real values by several orders of magnitude (Jannasch and Jones 1959) because most bacterial species do not form colonies on solid media (what was later coined as the "Great Plate Count Anomaly", Staley and Konopka 1985). Likewise, epifluorescence microscopy was also crucial to uncover the presence of abundant and ubiquitous unicellular cyanobacteria in the surface ocean (Johnson and Sieburth 1979). Initial studies using radioisotopes to estimate respiration and primary production also revealed that much of the respiration, dissolved organic matter (DOM) processing, and primary production were carried out by picoplanktonic organisms (Li et al. 1983; Williams 1970, 1981). The application of techniques commonly used in the biomedical field like flow cytometry also constituted a tipping point in our understanding of marine microbial ecology. For example, it allowed the discovery of *Prochlorococcus*, the most abundant primary producer in the ocean (Chisholm et al. 1988), and enabled the delineation of two populations of heterotrophic bacteria based on their nucleic acid content (Li et al. 1995; Robertson and Button 1989). This separation in the DNA content was initially attributed to a difference in the physiological state with the high nucleic acid (HNA) cells representing the active fraction of the community (Lebaron et al. 2001; Servais et al. 2003). However, this view was later challenged when it was shown that low nucleic acid (LNA) cells were also important contributors to bacterial metabolism in the sea (Longnecker et al. 2005; Sherr et al. 2006), and that the nucleic acid content may partially reflect the genome size of the cells (Vila-Costa et al. 2012). The addition of flow cytometry to the microbial ecologist's set of tools fueled a series of studies combining different stains probing cellular activity or growth (see Del Giorgio and Gasol 2008 for a review on this topic). These studies were crucial to open the black box of microbial communities unveiling the large heterogeneity in the physiological status of cells in the environment, which has an impact on their contribution to ecosystem function as we will expand on below.

The application of molecular tools like the 16S rRNA gene PCR amplification and sequencing from environmental DNA enabled access to the diversity of the "uncultured majority" (Amann et al. 1995; Rappé and Giovannoni 2003), which profoundly revolutionized the field of marine microbial ecology. These techniques opened a new venue to understand and classify marine microbial diversity and resulted in, among others, the discovery of the most abundant bacteria in the ocean, the SAR11 clade (Giovannoni et al. 1990), elusive for culturing using traditional methods. The rRNA approach also fueled the implementation of fluorescent in situ hybridization (FISH) methods to visualize and enumerate specific phylogenetic groups in natural samples. The combination of FISH with radioisotopic methods such as microautoradiography (MAR) was pivotal in unraveling niche partitioning in regard to the use of organic substrates between the different components of bacterial communities (Alonso-Sáez et al. 2012a; Cottrell and Kirchman 2000). Likewise, the estimation of the net growth rates of microorganisms belonging to various phylogenetic groups by using FISH has provided evidence for the important role that numerically unremarkable microbes can play in ecosystem functioning (Kirchman 2016). The advent of high-throughput sequencing approaches represented another revolution in microbial ecology that shed light on the composition of marine microbial communities, the relative abundance of their components, and the discovery of the "rare biosphere" (Sogin et al. 2006). The latter comprises low abundant taxa that act as reservoir for most phylogenetic and functional diversity (Pedrós-Alió 2012).

### 8.1.2 It Is Not Always Black and White: The Discovery of Photoheterotrophs

One of the most important findings in microbial oceanography in the last decades was the discovery of photoheterotrophs as an important component of the bacterioplankton. This discovery challenged the classic portrayal of bacterioplankton composed of photoautotrophic microorganisms as primary producers and of chemoheterotrophic microorganisms as consumers. Early genomic analyses of natural occurring marine bacterioplankton reported that an uncultured bacterium harbored a gene coding for proteorhodopsin (PR), a light-dependent proton pump, unveiling a new type of phototrophy in the ocean (Béjà et al. 2000). That same year, using infrared fluorometry Kolber et al. (2000) detected high signals of bacteriochlorophyll *a* (BChla) in the surface oligotrophic ocean suggesting that aerobic anoxygenic phototrophic (AAP) bacteria are a substantial component of the marine microbiome. These two reports represented the beginning of a change of paradigm in the field of marine microbial ecology that demanded to consider the direct effects of light on heterotrophic processes and, consequently, to rethink the models of organic carbon fluxes in the ocean.

Metagenomics and other molecular approaches have revealed an unsuspected large diversity among PR and AAP-containing bacteria and have shown that the genes responsible for photoheterotrophy are widespread among the most abundant microbial taxa in the surface ocean (DeLong and Béjà 2010; Koblížek 2015). Proton-pumping rhodopsins are found in marine Proteobacteria (including the SAR11 clade), Bacteroidetes, and Euryarchaeota (Pinhassi et al. 2016). These rhodopsins consist of only one opsin protein with a covalently bound pigment (retinal) and this simple structure has favored their lateral gene transfer and spreading among distant taxa, even across the domains Bacteria and Archaea (Frigaard et al. 2006). Contrarily, the machinery for light harvesting and energy synthesis in AAP bacteria consists of several pigments and proteins. Thus, even though the genes involved in aerobic anoxygenic photosynthesis have also been laterally transferred in marine bacteria, these events of gene transfer has been constrained to the Proteobacteria, mostly to the Alpha- and Gammaproteobacteria. Besides the size of coding operons, rhodopsins and bacteriochlorophyll systems display differences in the cost of biosynthesis and benefits of phototrophy (Kirchman and Hanson 2013). While the broad occurrence of marine photoheterotrophs has been well described, less is known about their physiology and ecology. Some axenic cultures of AAP and PR-containing bacteria have shown that they can use light to grow more efficiently (i.e., Gómez-Consarnau et al. 2007; Hauruseu and Koblížek 2012) and this has also been confirmed for natural populations of AAPs (Ferrera et al. 2017). In contrast, other PR-containing isolates do not grow better with light (González et al. 2008), and it has been reported that proteorhodopsins can also promote survival during starvation (Gómez-Consarnau et al. 2010). In addition to proton-pumping rhodopsins, other types of rhodopsins have been discovered in marine bacteria and archaea including chloride- and sodium-pumping rhodopsins (see Pinhassi et al. 2016), xanthorhodopsins (proton pumps associated with a particular carotenoid molecule named salinixanthin, Balashov et al. 2005), or sensory rhodopsins such as the newly discovered heliorhodopsins (Pushkarev et al. 2018). Further, rhodopsins, including heliorhodopsins, have been reported in marine viruses and single-celled eukaryotes (Needham et al. 2019). In short, following the discovery of photoheterotrophs it seems now clear that, in the marine environment, photoheterotrophy is not just an exception but probably the rule. In fact, there is growing evidence that bacteriochlorophyll- and rhodopsin-based dual phototrophy may have evolved in nature and is awaiting discovery (Zeng et al. 2020).

### 8.1.3 Are all Microorganisms Equally Active in the Ocean?

When single cell approaches started to be used in the late 70s (e.g., Hoppe 1976) it became evident that the activity of marine microbes is not homogeneous but on the contrary is tremendously heterogeneous. Within a given microbial community cells can be dead, injured, dormant (i.e., in a non-growth state), active but constrained by the availability of a certain resource, slow- or fast growing. Any of these physiological states will have strong implications for the role that such cells play in the environment. Yet, one of the major challenges that microbiologists face is how to define the physiological state of a microorganism.

A variety of methods has been developed in order to differentiate active- from non-active cells, probing cellular division, membrane integrity, respiratory activity, substrate uptake, or protein synthesis (see Del Giorgio and Gasol 2008; Sebastián and Gasol 2019; Singer et al. 2017 for reviews on this topic). These methods target different processes and differ in resolution, and thus the delineation of active cells is not always consistent (see Del Giorgio and Gasol 2008 for a comparison of methods). Therefore, the categorization within the various physiological states is purely operational. For example, using the tetrazolium salt 5-cyano-2,3-ditolyl tetrazolium chloride (CTC) as an indicator of activity Del Giorgio and Scarborough

(1995) found that the percentage of active cells represented only 5% of the cells in the oligotrophic ocean, which is consistent with the microautoradiography (MAR) values that were obtained using $^{3}$H-thymidine as substrate (Longnecker et al. 2010). However, when these authors used $^{33}$P-phosphate instead of thymidine the percentage of active cells increased up to 50%. This indicates that the categorization of active cells using substrate uptake methods like MAR strongly relies on the substrate of choice. Despite the variability found in the proportion of active cells, these single cell techniques have been highly informative, particularly when coupled with other approaches that allow taxonomic characterization such as fluorescence in situ hybridization (MAR-FISH). These techniques have shown that some members of the community are more active than others and that there is heterogeneity in the levels of activity even within a given population (e.g., Alonso-Sáez et al. 2007, 2012a; Cottrell and Kirchman 2000).

The delineation of dormant cells is also important because it has been hypothesized that these cells constitute a seed bank of taxonomic and functional diversity that guarantees the persistence and long-term maintenance of the diversity and function of the community (Lennon and Jones 2011). Yet, we still lack a method to assess whether or not a cell is dormant. Some attempts have been made to characterize the active and inactive members of bacterial and archaeal communities through the joint analysis of ribosomal RNA and DNA of individual taxa, using RNA:DNA ratios below 1 as a threshold to delineate "inactivity" or dormancy (Bowsher et al. 2019; Campbell et al. 2011; Jones and Lennon 2010; Kearns et al. 2016). However, the delineation of the inactive members of a community is sometimes problematic (Steven et al. 2017) because some taxa accumulate ribosomes during dormancy in order to be able to respond swiftly to favorable conditions (see Blazewicz et al. 2013 for a review on this topic). Despite these caveats, the sequencing of ribosomal RNA has provided valuable information about marine taxa that have or lack the potential for protein synthesis and how this changes over space or time (Campbell and Kirchman 2013; Campbell et al. 2009, 2011; Ghiglione et al. 2009; Hugoni et al. 2013; Hunt et al. 2013; Zhang et al. 2014).

Labeling with the thymidine substitute 5-bromo-2′-deoxyuridine (BrdU) has also been used to define the active members of the community (Galand et al. 2013; Pernthaler et al. 2002). This and other approaches that link activity and identity, such as stable isotope probing, have been utilized to identify the major players in a certain biogeochemical process or follow the uptake of specific compounds (Bryson et al. 2017; Mou et al. 2008; Nelson and Carlson 2012; Orsi et al. 2016; Taubert et al. 2017). Other methods that are used in marine microbial studies but are still in their infancy include Bioorthogonal non-canonical amino acid tagging (BONCAT) (Couradeau et al. 2019; Hatzenpichler et al. 2016; Leizeaga et al. 2017; Samo et al. 2014) and Raman micro-spectroscopy (Berry et al. 2015; Huang et al. 2007; Lorenz et al. 2017). These methods could have a tremendous potential to address the physiological status of individual cells as will be expanded on below.

Altogether these studies have shed light on microbial processes in the ocean and identified the major players driving them. Yet, accurate knowledge is lacking of how many cells are active, how dynamic is the transition between activity and inactivity of individual cells, and what are the factors driving these transitions. Furthermore, most studies are temporally and geographically restricted and are particularly focused on the sunlit ocean. When inexpensive techniques like BONCAT will be broadly applied by the research community these gaps of knowledge may be filled. This is crucial to understand microbial function and to predict how it is going to evolve in future scenarios.

## 8.2 The Marine Microbiome over Space and Time

### 8.2.1 The Beginning of the Global Exploration of the Marine Microbiome

Microbial biogeography relies on the description of how microbial communities are distributed in space, in the vertical as well as in the horizontal dimension. While multiple studies that were restricted to particular biogeographical areas provided hints on the spatial distribution of marine microorganisms (e.g., Agogué et al. 2011; DeLong et al. 2006; Pham et al. 2008; Pommier et al. 2010) it was not until the global oceanographic circumnavigations took place (Fig. 8.1) when the exploration of the worldwide distribution of marine microorganisms became feasible. The pioneering large-scale survey was the Global Ocean Sampling Expedition (GOS) that was launched in 2003 in the Sargasso Sea. This survey continued as a several years expedition across the globe although most of the data were generated between 2004 and 2006 from a transect running from the North Atlantic to the South Pacific through the Panama Canal. The GOS expedition indeed represents the first approximation to the microbial diversity of the global surface ocean (Rusch et al. 2007; Venter et al. 2004). The GOS unveiled ~1300 different 16S rRNA gene sequences in surface seawater samples from the Sargasso Sea (Venter et al. 2004) and thousands of gene families in the surface ocean (Rusch et al. 2007). Contemporary to the GOS, the

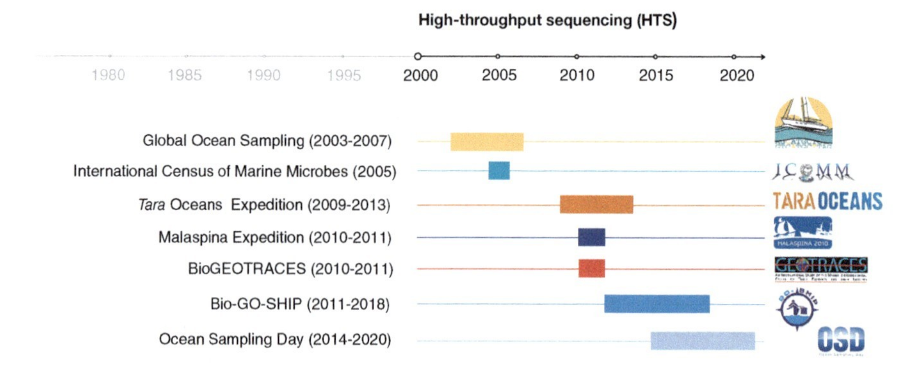


**Fig. 8.1** The most important global marine circumnavigations for the exploration of marine microbiomes and when they took place. Figure courtesy of Dr. Marta Royo-Llonch from SHOOK Studio

International Census of Marine Microbes (ICoMM) was launched in 2005 aiming at using standardized sampling and data analysis procedures to study microbial diversity in a multitude of marine habitats including pelagic and benthic systems using 454-pyrosequencing of the 16S rRNA gene (Amaral-Zettler et al. 2010). At that time, the ICoMM represented the most comprehensive diversity picture of global ocean bacterial and archaeal communities and reported remarkable horizontal and vertical large-scale segregation showing contrasting patterns for the different explored habitats (pelagic, benthic, and anoxic ecosystems, or vents, Zinger et al. 2011). This study showed that alike yet remote habitats can harbor similar communities in the pelagic realm supporting the Baas-Becking and Beijerinck hypothesis that *"everything is everywhere, but the environment selects"* (Baas Becking 1934; Beijerinck 1913). Hence, marine planktonic bacteria display an unlimited potential for dispersal while abiotic environmental filtering is responsible for their different distributions in the global ocean. Furthermore, it was the ICoMM initiative that allowed the discovery of the abovementioned "rare biosphere" (Pedrós-Alió 2012; Sogin et al. 2006).

After the GOS and the ICoMM initiatives the main surveys with focus on the worldwide exploration of the marine microbiome have been the *Tara* Oceans Expedition (2009–2013) (Karsenti et al. 2011), the Malaspina 2010 Expedition (Duarte 2015), the Ocean Sampling Day, which is a simultaneous sampling campaign of the world's coasts on the summer solstice that has been carried out since 2014 (Kopf et al. 2015; Tragin and Vaulot 2018), the BioGEOTRACES (Biller et al. 2018), the Bio-GO-SHIP programs (Larkin et al. 2021) (Fig. 8.1), and other initiatives that are in the planning phase. Between 2009 and 2013, the *Tara* Oceans Expedition sampled the global ocean from surface waters to the mesopelagic layer. It used standardized sampling procedures to obtain seven size fractions of planktonic diversity, from viruses to small metazoans, and comprised a large sequencing effort (>30 Gb/sample) (Pesant et al. 2015). One of the main legacies of this expedition was the generation of the Ocean Microbial Reference Gene Catalog (OMR-GC) containing a total of >40 million-non-redundant genes of the global marine microbiome of which over 80% of the sequences were new (Sunagawa et al. 2015). This catalog was updated with the OMR-GCv2 integrating the metagenomes of the Arctic region and the metatranscriptomes of the global ocean (Salazar et al. 2019). The multidisciplinary and extensive effort of the *Tara* Oceans consortium has resulted in many important outcomes (see Sunagawa et al. 2020 for a review) such as the genetic repertoire of bacteria and archaea (Sunagawa et al. 2015) including more than 500 bacterial and archaeal metagenome assembled genomes (MAGs) from the polar Arctic Ocean (Royo-Llonch et al. 2021) and eukaryotes (Carradec et al. 2018) in the sunlit and mesopelagic global ocean, the diversity of eukaryotic plankton (De Vargas et al. 2015) and viruses (Brum et al. 2015; Gregory et al. 2019), and the planktonic interactions occurring in the photic ocean (Lima-Mendez et al. 2015). Other relevant achievements from the *Tara* Oceans Expedition were studies on the potential contribution of unexpected components such as bacteria, archaea, and viruses to carbon export in the nutrient-depleted oligotrophic ocean (Guidi et al. 2016), the potential impact of ocean warming on community composition and gene expression (Salazar et al. 2019), or the discovery of latitudinal gradients of diversity for most planktonic groups (Ibarbalz et al. 2019). Furthermore, data from the *Tara* Oceans Expedition confirmed that the geographic distance only plays a subordinate role in determining the taxonomic and functional microbial community composition in the photic open ocean whereas environmental selection seems to be an important driver of microbial biogeography (Sunagawa et al. 2015).

Between 2010 and 2011, the Malaspina 2010 Circumnavigation Expedition (Duarte 2015) sampled the marine microbiome in the tropical and subtropical oceans from the surface down to bathypelagic waters (~4000 m depth). This expedition showed that shifts towards communities enriched in rare taxa in the sunlit ocean reflect environmental transitions (Ruiz-González et al. 2019) and it explored the role of dispersion on planktonic and micro-nektonic organisms (Villarino et al. 2018). The Malaspina Expedition also contributed with an assessment of the diversity and biogeography of deep-sea pelagic bacteria and archaea (Salazar et al. 2016) as well as provided an account of the diversity of heterotrophic protists in the deep ocean, particularly unveiling the special relevance of fungal taxa (Pernice et al. 2015). This expedition also unraveled that the particle-association lifestyle is a phylogenetically conserved trait in bathypelagic microorganisms (Salazar et al. 2015), and provided a metabolic characterization of the deep ocean microbiome based on Metagenome Assembled Genomes (Acinas et al. 2021). Nonetheless, the Malaspina -omics datasets have not yet been fully exploited and new inputs are expected in the coming years, such as those derived from ongoing detailed analyses of vertical profiles and of specific water masses or insights in the microbiome associated with the deep scattering layer in the ocean.

The datasets of the BioGEOTRACES and Bio-GO-SHIP programs have recently been made available (Biller et al. 2018; Larkin et al. 2021) and these will surely push further our knowledge of the marine microbiome. The BioGEOTRACES initiative spans 610 metagenomes collected from diverse regions of the Pacific and Atlantic oceans (Biller et al. 2018). It also adds a temporal dimension to the marine microbiome by providing metagenomes collected every month during two years at the stations HOT (North Pacific) and BATS (North Atlantic), which are part of long-term time series programs (Karl and Church 2014; Steinberg et al. 2001) and for which a suite of physicochemical and biological data are available. The Bio-GO-SHIP program aims at providing high-resolution spatiotemporal sampling of the marine microbiome in order to link microbial traits with ecosystem function and biochemical fluxes. This program has released 720 globally distributed surface ocean metagenomes from samples collected every 4–6 h representing a median distance between sampling stations of only 26 km (Larkin et al. 2021) and thus providing a much higher spatial resolution than other global expeditions such as *Tara* Oceans (~700 km) or bioGEOTRACES (~200 km).

Overall, these global expeditions together with other local and regional studies have unveiled some general microbial patterns across horizontal and vertical scales in the ocean. These general trends include among others: (i) the key concept that microbial communities are formed by a few abundant taxa and a long tail of low abundant taxa (the rare biosphere) (Sogin et al. 2006), (ii) that temperature is one of the main drivers to predict the taxonomic and functional gene composition of microbial communities in epipelagic waters of the open ocean (Sunagawa et al. 2015), (iii) that phosphorus exerts a strong selective pressure in the surface ocean (Coleman and Chisholm 2010; Grote et al. 2012), (iv) that the latitudinal gradient of diversity observed in macroorganisms, meaning that species richness increases in latitudes closer to the equator and mid-latitudes, also applies to most marine planktonic microorganisms (Amend et al. 2013; Ibarbalz et al. 2019; Sul et al. 2013), (v) that the microbial community composition of contrasting polar zones, even though being distant from each other, is more similar than that of microbial communities of temperate or tropical latitudes (Cao et al. 2020; Ghiglione et al. 2012; Royo-Llonch et al. 2021; Sul et al. 2013), (vi) a vertical segregation between photic and aphotic microbial communities (Amaral-Zettler et al. 2010; DeLong et al. 2006; Sunagawa et al. 2015), and (vii) the existence of a vertical connectivity between surface and deep ocean communities (Cram et al. 2015; Mestre et al. 2018; Parada and Fuhrman 2017; Ruiz-González et al. 2020).

Despite the relevant information generated by all these studies, the myriad of microbial processes ocurring in the ocean are far from understood. For instance, there is an increasing recognition of the role of sub-mesoscale hydrographic features in key processes such as the carbon pump (Boyd et al. 2019; Resplandy et al. 2019) or the dispersion of marine microbes, pollutants, and microplastics, but studies on the changes of the marine microbiome at fine spatial scales are still missing. Current initiatives looking at this heterogeneity like the EXPORTS program (https://oceanexports.org/about.html) and further implementation of genomic sensors (Scholin et al. 2017) will help to address this issue. Similarly, the aim of the recently launched AtlantECO EU project (https://www.atlanteco.eu) is a better understanding and integration of the marine microbiome in the context of ocean circulation and the presence of pollutants, e.g. plastics. AtlantECO also pursues to assess the role of the marine microbiome in driving the dynamics of the Atlantic ecosystem at basin and regional scales. The joint effort of all those initiatives will help to boost our knowledge of the marine microbiome.

### 8.2.2 Seasonality and Temporal Dynamics of Marine Microbial Communities

Besides studies aiming at understanding how microbial communities vary across space, increasing efforts are being invested towards exploring how they change over time. The establishment of microbial observatories across the globe has allowed the monitoring of microbial communities over time from short- to long-term scales (see reviews by Bunse and Pinhassi 2017; Buttigieg et al. 2018). Defining seasonality is essential to understand how microbes react to changes in environmental conditions or perturbations. Time series of marine microbiomes also allow addressing fundamental ecological questions such as which patterns of biodiversity are present in an ecosystem, how are these patterns governed, how stable and predictable are microbial communities, how do species interact or what is the ecological niche of a given taxon. Seasonality had been observed in phytoplankton blooms but it was only after the molecular revolution that it could be investigated in bacterioplankton communities. The pioneer application of fingerprinting methods and clone libraries to samples collected from marine microbial observatories over 1–2 year periods elucidated community shifts over seasons and demonstrated the existence of temporal niches for specific organisms (Alonso-Sáez et al. 2007; Brown et al. 2005; Ghiglione et al. 2005). Nevertheless, time series of many consecutive years are required to test the robustness of seasonal patterns. Such long-term time series with large sampling efforts have been undertaken in oceanic and coastal monitoring stations; the San Pedro Ocean Time Series (SPOT) and the Hawaii Ocean Time Series (HOT) in the Pacific Ocean, the Bermuda Atlantic Time Series (BATS) in the Atlantic Ocean, the Western Channel Observatory in the English Channel, or the Service d'Observation du Laboratoire Arago (SOLA Station; Banyuls-sur-Mer, France) and the Blanes Bay Microbial Observatory (BBMO) in the Mediterranean Sea are some examples of such long-term programs (for a detailed list, see Bunse and Pinhassi 2017; Buttigieg et al. 2018).

Over a decade ago the application of high-resolution molecular fingerprinting methods over monthly samples from SPOT revealed remarkably repeatable and predictable seasonal patterns in the distribution and abundance of microbial taxa (Fuhrman et al. 2006). High-throughput sequencing of bacterioplankton communities confirmed this observation at higher resolution in different locations (Cram et al. 2015; Eiler et al. 2011; Fuhrman et al. 2015; Gilbert et al. 2012; Lambert et al. 2019) and unveiled that also the rare members of the bacterioplankton showed seasonality (Alonso-Sáez et al. 2015). Moreover, long-term time series captured the occasional blooming of some rare members of the community which became dominant when the conditions were favorable (Gilbert et al. 2012). Likewise, microbial eukaryotes showed recurrent seasonal patterns (Giner et al. 2019; Lambert et al. 2019). In addition, changes in community composition were accompanied by repeatable shifts in alphadiversity (Gilbert et al. 2012; Giner et al. 2019). Besides, the seasonal patterns of relevant phylogenetic groups (Díez-Vives et al. 2019; Salter et al. 2015; Vergin et al. 2013) or of certain functional groups (AAPs, Auladell et al. 2019) have been studied, revealing remarkably similar patterns to those of the whole communities (Fig. 8.2).

Current methodologies allow to investigate the dynamics of individual taxa, unveiling that closely related populations can represent distinct ecotypes that temporally occupy different niches (Auladell et al. 2019, 2021; Chafee et al. 2018). Further, analysis of finely resolved taxonomic units in combination with high frequency sampling over multiple years has shown that regardless of interannual variation in phytoplankton blooms recurrent modules of co-varying microbes exist (Chafee et al. 2018). Beyond long-term studies, high frequency sampling over a phytoplankton bloom has shown that biological interactions among bacteria, archaea, and eukaryotic microorganisms may play critical roles in controlling plankton diversity and dynamics (Needham and Fuhrman 2016), contradicting the traditional view that blooms are mainly controlled by physical- and chemical processes. Daily sampling over periods of several months has also revealed that

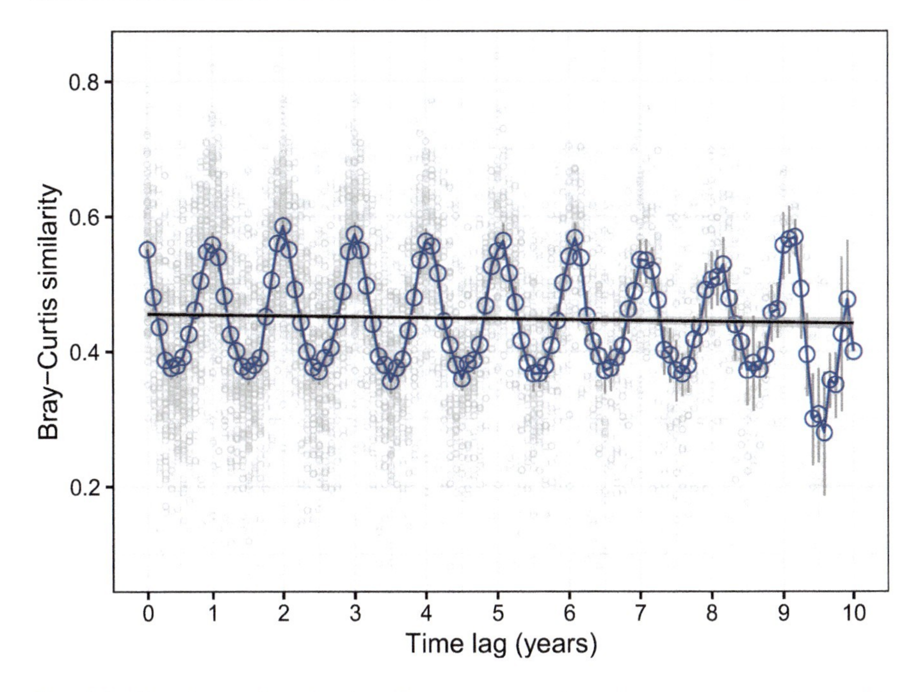


**Fig. 8.2** Time-decay plot showing the recurrence of aerobic anoxygenic phototrophic communities in the Blanes Bay Microbial Observatory (NW Mediterranean). Bray–Curtis similarity between samples is plotted against the time lag between each comparison. Blue dots represent mean values for each time lag and gray vertical bars the standard error (background gray dots show each comparison). A linear regression with 95% confidence intervals is shown (modified from Auladell et al. 2019)

microbial plankton is organized in clearly defined but ephemeral communities whose turnover is rapid, mirroring environmental variability (Martin-Platero et al. 2018). On shorter time scales, metatranscriptomics has shown evidence of diel transcriptional oscillations of both phototrophic and chemotrophic microorganisms (Ottesen et al. 2013, 2014) as well as of viruses (Aylward et al. 2017; Kolody et al. 2019).

Altogether, long time series have provided evidence for seasonal and interannual recurrence of some microbial taxa, highly resolved time series have demonstrated that communities fluctuate on a daily and monthly scale along with changes in environmental conditions, and sampling on an hourly scale has unveiled diel periodicity of gene transcription. These studies demonstrate that the temporal scale of sampling is directly linked to the scale of temporal variability that we are able to capture. Highly resolved sampling over long periods will be eased by the development of automated samplers. This, in combination with the decreasing analytical and computing costs, will provide further insights into the stability and reproducibility of the short-term changes over longer periods of time. Moreover, increasing efforts on obtaining data from high and low latitudes will help defining a more global picture of the temporal dynamics of marine microorganisms. All these data will be useful to feed model-based analyses aimed at better predicting the temporal dynamics of marine microbial communities in future scenarios.

## 8.3 Approaches to Link Taxonomy and Function of Marine Bacteria and Archaea

### 8.3.1 The Genome-Centric Approaches: Single Amplified Genomes (SAGs) and Metagenome Assembled Genomes (MAGs)

Whereas direct analysis of metagenomes can provide a community overview, other strategies have been developed to either access individual environmental genomes without the need for cultivation (Single Amplified Genomes, SAGs) or group the community's metagenomic information into meaningful genomic units reflective of a population of close taxa as in Metagenome Assembled Genomes (MAGs). Moreover, properly assigning function to taxonomy, which has been an essential goal in the microbial ecology of uncultured microorganisms, requires the genetic information to be considered in a genomic context.

#### 8.3.1.1 Single-Amplified Genomes (SAGs)

SAGs are generated from the direct amplification of DNA from previously sorted individual cells, its sequencing and assembly (Fig. 8.3). SAGs may represent

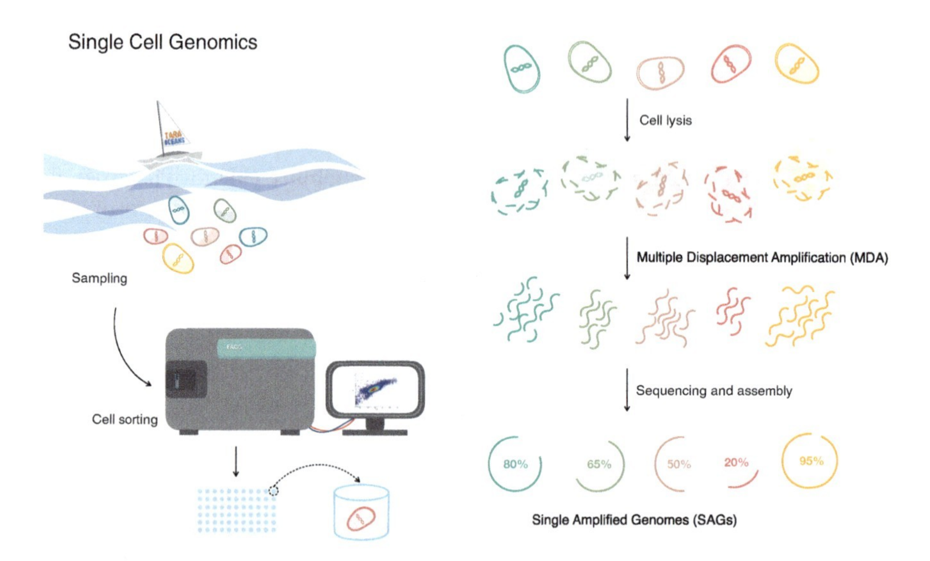


**Fig. 8.3** Simplified workflow for SAGs generation from seawater samples. After sample collection cells are sorted in a flow cytometer. Individual cells are then lyzed and nucleic acids undergo Multiple Displacement Amplification (MDA). After that, genomes are sequenced and assembled resulting in SAGs of variable quality, from low to high coverage. Figure courtesy of Dr. Marta Royo-Llonch from SHOOK Studio (https://www.instagram.com/shookstudio/?hl=en)

environmental genomes sequenced from the most fundamental units of life (Blainey 2013; Stepanauskas 2012; Sieracki 2007; Woyke et al. 2009). Individual cells are selected from a sample by fluorescence-activated cell sorting (FACS) or microfluidics. Subsequently, the cells undergo lysis and whole genome amplification through various methodologies: degenerate oligonucleotide-primed PCR, multiple displacement amplification (MDA), or WGA-X, an MDA method that utilizes a thermostable mutant of the phi29 polymerase (Stepanauskas et al. 2017). This is followed by sequencing and genome assembly. SAGs retrieves all the DNA molecules of a cell, unveiling microbial intimate interactions in their natural environment otherwise overlooked like infections, symbioses, and predation (Castillo et al. 2019; Labonté et al. 2015; Martinez-Garcia et al. 2014; Roux et al. 2014; Yoon et al. 2011). Because SAGs circumvents the taxonomic binning used in metagenome assembly, it improves the understanding of microevolutionary processes in the environment (Kashtan et al. 2014), providing as well unique reference genome datasets from uncultured microbes. Indeed, the analysis of 2715 partial SAGs from the tropical and subtropical euphotic ocean enabled the functional and taxonomic annotation of about 80% of metagenomic reads from diverse oceanographic cruises and marine stations (Pachiadaki et al. 2019).

Some of the valuable lessons learnt from SAGs studies include: (i) genome streamlining is a prevalent feature in the oligotrophic surface ocean (Swan et al. 2013), (ii) the extensive microdiversity and co-existence of hundreds of genomes within *Prochlorococcus* populations (Kashtan et al. 2014), (iii) the reconstruction of novel uncultured bacterial species (Royo-Llonch et al. 2020), (iv) the ubiquity of light harvesting and secondary metabolite biosynthetic pathways across microbial lineages (Pachiadaki et al. 2019), (v) the impact of chemolithoautotrophic microorganisms such as the SAR324 clade and the Gammaproteobacteria ARCTIC96BD-19 (Swan et al. 2011), (vi) the overlooked role of the nitrite-oxidizing bacteria (Pachiadaki et al. 2017), or (vii) the adaptation to anoxic niches such as Oxygen Minimum Zones (OMZ) of unique SAR11 lineages with capacity for nitrate respiration (Tsementzi et al. 2016).

#### 8.3.1.2 Metagenome Assembled Genomes (MAGs)

Metagenomic reads can be assembled into contigs and later binned into the so-called Metagenome Assembled Genomes (Fig. 8.4). MAGs are composite genomes of closely related populations from natural communities. The first attempts to reconstruct genomes from metagenomic DNA sequences of environmental communities started in the early 2000s (Martín et al. 2006; Tyson et al. 2004) but more reliable methods and larger-scale results emerged during the last decade (Albertsen et al. 2013; Alneberg et al. 2014; Parks et al. 2017; Sharon and Banfield 2013; Wrighton et al. 2012). Nowadays, thousands of genomes have been reconstructed from marine metagenomes both from discrete sampling events and from global circumnavigations (Acinas et al. 2021; Delmont and Eren 2018; Delmont et al. 2018; Royo-Llonch et al. 2021; Tully et al. 2017). Just recently, a large-scale study used 10,450 metagenomes sampled from a variety of habitats, including marine environments, and recovered 52,515 medium- and high-quality MAGs,

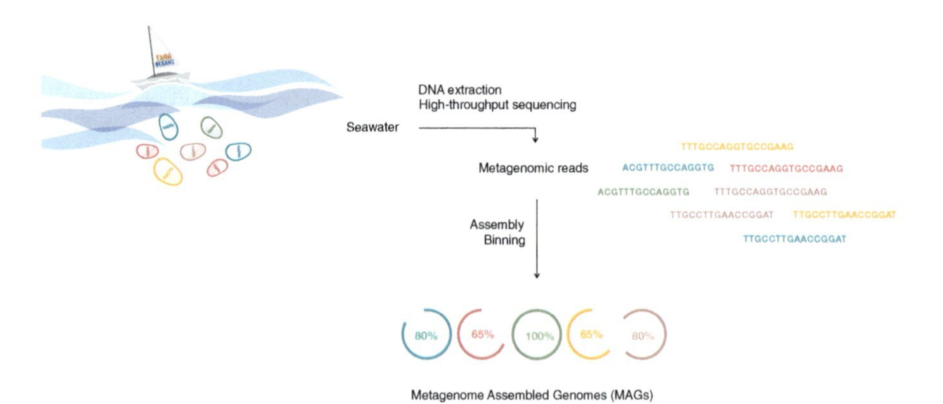


**Fig. 8.4** Simplified workflow for MAGs generation from seawater samples. Community DNA is extracted from seawater samples and undergoes high-throughput sequencing, which yields large amounts of metagenomic reads. Reads are then assembled into contigs which are later binned to form Metagenome Assembled Genomes (MAGs) of varying coverage levels. Figure courtesy of Dr. Marta Royo-Llonch from SHOOK Studio (https://www.instagram.com/shookstudio/?hl=en)

which constitute the Genomes from Earth's Microbiomes (GEM) catalog, of which 8578 represent marine microbial MAGs (Nayfach et al. 2021).

There has been an increase in the development of tools for estimating genome completion and contamination (Eren et al. 2015; Parks et al. 2015) improving the approaches to retrieve completed (circularized curated, no gaps) Metagenome Assembled Genomes (CMAGs) (Chen et al. 2020) and guidelines on genome quality standards and complementary analyses for the correct deposition in public databases (Bowers et al. 2017; Konstantinidis et al. 2017). However, a community consensus on the pipeline for MAG reconstruction is missing since different strategies in each step of the process (e.g., assemblies, binning, or annotation) are still used.

The reconstruction of bacterial and archaeal MAGs has provided insights into the ecology and evolution of marine microbial taxa unveiling, among other things: (i) the prevalence of diazotrophs in the surface ocean belonging to the Proteobacteria and Planctomycetes (Delmont et al. 2018), (ii) the potential for primary productivity in a globally distributed AAP bacterium (Graham et al. 2018), (iii) the metabolic diversity within Marine Group II Euryarchaea (Tully 2019), (iv) the biogeography and evolutionary processes within the SAR11 clade (Delmont et al. 2019), (v) the widespread potential for mixotrophy found in the genomes of uncultured bacteria and archaea in the bathypelagic ocean (Acinas et al. 2021) or (vi) transcriptional patters of unique bacteria and archaea polar Arctic MAGs (Royo-Llonch et al. 2021).

Both SAGs and MAGs have been successfully combined to derive conclusions from organisms and their ecosystems and to improve the binning quality of single cell assemblies or metagenomes. Finally, the emergence of new pipelines for analysing bacterial and archaeal individual genomes, metagenome assembled genomes, and single-amplified genomes such as MetaSanity (Neely et al. 2020) will facilitate genome quality evaluation, phylogenetic and functional annotation through a variety of integrated programs. It is clear that these two genome-centric approaches will enhance the understanding of functional and evolutionary processes of the prevalently uncultured marine microbes and will as well serve as a starting point for future experimental validation processes of their potential metabolisms, for in situ taxonomic quantification using CARD-FISH, or for providing essential information in order to design strategies for isolation of key microorganisms into culture.

### 8.3.2 The Relevance of Culturing Marine Bacteria in the -Omics Era

Despite the fact that gene- and genome-centric approaches have allowed the description of marine microbial diversity from diverse habitats at an unprecedented scale, as overviewed in this chapter, isolates are still a necessary and complementary resource of knowledge. Isolating bacteria and archaea in the laboratory is a fundamental requirement to investigate their physiology under different scenarios, to test ecological hypotheses raised from metagenomics and genome-centric studies, to have access to their complete genomes, to assess the function of novel genes (Muller et al. 2013), to interpret multi-species interactions in co-culture experiments (Stomp et al. 2004, 2008), or to investigate evolutionary principles and population dynamics in long-term monitoring efforts (Good et al. 2017; Rosenzweig et al. 1994). At the same time, genomes from isolates are important to update and improve the existing databases that are needed for the correct annotation of sequencing data (Giovannoni and Stingl 2007; Gutleben et al. 2017). Moreover, isolation is still the only current option for the official procedures for classification and characterization of novel prokaryotic species (Parker et al. 2019) although new efforts to create guidelines for nomenclature of uncultured microorganisms are being developed (Murray et al. 2020). Finally, the short generation times and the nearly 4 billion years of evolution of marine microorganisms have resulted in an enormous biodiversity and a plethora of metabolic pathways and thus, having access to pure cultures, represents an excellent opportunity for biotechnology research (Luna 2015), including bioremediation of polluted ecosystems.

There is little overlap between taxa retrieved by molecular techniques and those retrieved by isolation (Crespo et al. 2016; Lekunberri et al. 2014). This is mainly due to the fact that molecular techniques usually recover the abundant bacteria present in a given environment, while cultures often retrieve those taxa that belong to the rare biosphere (Pedrós-Alió 2012; Sogin et al. 2006). Isolation is thus still essential to decipher the full spectra of diversity of the marine ecosystem (Sanz-Sáez et al. 2020). New culture-dependent techniques have been developed to expand the range of bacteria that can be cultured like microfluidics (Boitard et al. 2015; Ma et al. 2014), culturing chips (Gao et al. 2013; Hesselman et al. 2012; Ingham et al. 2007), manipulation of single cells (Ben-Dov et al. 2009; Park et al. 2011), high-throughput culturing techniques termed "culturomics" (Giovannoni and Stingl 2007; Lagier et al. 2012), or culturing following large-scale dilution to extinction (Henson et al.

2020). A common feature of all these new techniques is that they are based on the same principles of recreating the nutrient conditions of natural environments and overcoming the tendency of rapidly growing cells to outcompete species that reproduce more slowly. The use of these methodologies allowed the isolation of previously uncultured groups such as the alphaproteobacterium *Candidatus* Pelagibacter ubique within the SAR11 clade (Morris et al. 2002; Rappé et al. 2002), or the chemolithotrophic ammonia-oxidizing archaeon *Nitrosopumilus maritimus* (Könneke et al. 2005). The existence of these and other cultures has been essential to test hypotheses derived from genomic data and confirm the role of some proteins in the cell's physiology. For example, experiments with isolates have unveiled light-stimulated growth in some bacteria harboring proteorhodopsin (Gómez-Consarnau et al. 2007), whereas in other bacteria this protein is involved in cell survival (Gómez-Consarnau et al. 2010) or in improving cell fitness (González et al. 2008; Steindler et al. 2011). Likewise, experiments with *Candidatus* Pelagibacter ubique have made possible to understand the growth requirements of bacteria with a limited genetic repertoire (Carini et al. 2013; Tripp 2013; Tripp et al. 2008). Altogether, these culturing depending approaches are fundamental to microbial ecologists to fully understand the ecology, function, and biotechnological potential of microorganisms in marine ecosystems.

### 8.3.3 Shedding Light on the Active Microbiome

Genome-centric approaches such as SAGs and MAGs have increased our understanding of the metabolic capabilities of uncultured bacteria and archaea and culturing efforts are increasing the array of model organisms to be used for carrying out physiological studies. However, the ultimate goal of microbial ecologists is to understand the activity and function of the different microbes in situ and how they are affected by changes in environmental conditions. Only then we will be able to have a predictive understanding of microbial processes in the ocean.

Different approaches have been used to unravel the role of key taxa driving microbial processes (Berry et al. 2015; Hall et al. 2011; Musat et al. 2012; Singer et al. 2017). Microautoradiography (MAR) coupled with fluorescence in situ hybridization (FISH) has been undoubtedly the most widely used technique to assess the groups of bacteria or archaea that are taking part in the uptake of a given substrate (Alonso-Sáez et al. 2012a; Cottrell and Kirchman 2000; Sintes and Herndl 2006), but it suffers from the poor taxonomic resolution of FISH. Stable isotope probing (SIP), which involves the incubation with a stable isotope-labeled substrate and the downstream analysis of heavy-isotope enriched cellular components such as DNA, RNA, or proteins has also been widely used (see Dumont and Murrell 2005; Musat et al. 2012 for reviews on this technique). Nano-scale secondary ion mass spectrometry (nanoSIMS) is a SIP-based technique that enables the quantification of stable isotopes with high spatial resolution. NanoSIMS has been crucial for unveiling the interactions between individual microbial cells and biochemical processes (see Mayali 2020 and references therein) and particularly relevant for the study of metabolic fluxes between symbionts and their hosts (Foster et al. 2011; Thompson et al. 2012). However, NanoSIMS is usually coupled with FISH and it is therefore also limited by its phylogenetic resolution. An alternative approach is applying nanoSIMS to an isotope-labeled whole community RNA hybridized to a phylogenetic microarray. This method is called Chip-SIP and it increases the taxonomic resolution to the level of individual taxa (Mayali et al. 2012; Bryson et al. 2017).

Given the short half live of mRNA, metatranscriptomics provides information about the gene expression profiling in near-real time conditions (Moran et al. 2013). This approach has been pivotal to elucidate daily transcriptional oscillations in heterotrophic bacterioplankton (Ottesen et al. 2014) and in viral genes in host cell assemblages (Aylward et al. 2017; Kolody et al. 2019). Besides, degradation pathways may be identified based on shifts in the proportion of certain transcripts upon substrate additions (Li et al. 2014; McCarren et al. 2010; Mou et al. 2011; Vila-Costa et al. 2010), and the fluctuations of transcripts in the mRNA pool are highly informative for how cells sense shifts in environmental conditions and the machinery involved in the response to these shifts (Moran et al. 2013). Nevertheless, high sequencing depth is required to detect transcripts of functional genes that are not involved in core metabolic pathways or transcripts of rare microbes. Metaproteomics is an emergent field in ocean studies (see Saito et al. 2019 for a review), and it has been useful to shed light on transport functions and microbial nutrient utilization (Morris et al. 2010; Sowell et al. 2009), nutrient stresses (Saito et al. 2014) as well as on changes on substrate utilization by microbial communities in the water column (Bergauer et al. 2018). However, metaproteomics is limited to well described proteins produced by the abundant fraction of microbial communities (Saito et al. 2019). Targeted meta-omics with SIP (Chen and Murrell 2010; Coyotzi et al. 2016; Grob et al. 2015) or targeted Single Cell Genomics with fluorescently labeled substrates (Doud et al. 2020; Martinez-Garcia et al. 2012) is a powerful alternative to circumvent the limitation of metatranscriptomics and metaproteomics to only the abundant fraction of the community. They allow the detection of rare taxa participating in a given biochemical process and facilitate the identification of the different enzymes involved. Similarly, other approaches like Bioorthogonal non-canonical amino acid tagging (BONCAT) or Raman micro-spectroscopy (see Hatzenpichler et al. 2020 and references therein) present great prospect for targeted meta-omics. BONCAT is a sensitive technique that uses a synthetic amino acid that upon incorporation can be fluorescently detected via copper-catalyzed alkyne–azide click chemistry (Dieterich et al. 2006). It has been applied to environmental samples to identify protein-synthesizing cells and the taxonomic identification of these cells has been performed by FISH (Hatzenpichler et al. 2014; Sebastián et al. 2019) or by 16S rRNA tagging after fluorescence-activated cell sorting (Couradeau et al. 2019; Hatzenpichler et al. 2016; Reichart et al. 2020). BONCAT has also the potential to study short-term proteomic responses of bacteria (Bagert et al. 2016) and opens new avenues of research to study the proteins involved in relevant biochemical processes. Likewise, Raman activated cell sorting after incubation with heavy water or other isotope-labeled substrates (Berry et al. 2015; Lee et al. 2019; Wang et al. 2013;

Zhang et al. 2015) can be used for subsequent single cell genomics or mini-metagenomics of selected populations (Yu et al. 2017).

Although single cell RNA-Seq has not made its way yet into environmental studies it has the potential to unravel transcriptional heterogeneity of individual bacterial cells (Imdahl et al. 2020) or to discern different stages of viral–host interactions among a single population and monitor how the host metabolism is rewired during viral infection (Ku et al. 2020). All these emerging techniques will surely contribute to a better understanding of microbial processes in the ocean and the key organisms involved, which is central to assess how changes in diversity in the future will impact global biogeochemical cycles.

## 8.4 What Have we Learnt from the Exploration of the Marine Microbiome?

### 8.4.1 The Unknown Marine Microbial Diversity

Current predictions indicate that the ocean may be home to $\sim 10^{10}$ microbial species of which only $\sim 10^4$ have been cultured (Locey and Lennon 2016). Moreover, the few cultured microbial species often represent rare members of microbial communities while the most abundant taxa remain largely elusive and begun only to be elucidated after the development of culturing-independent methodologies (Rappé and Giovannoni 2003). Certainly, although the "Great Plate Count Anomaly" was known at the time (Staley and Konopka 1985), the pioneer studies that investigated bacterioplankton diversity in the 90s led among others to the discovery of SAR11 and were groundbreaking because they brought to light that the most abundant organisms in the ocean were in fact unknown (Fuhrman et al. 1993; Giovannoni et al. 1990). Ever since, sequencing DNA from the ocean has continuously unveiled hitherto unknown microorganisms, which has expanded the tree of life to a great extent. Only a negligible fraction of the diversity detected in molecular surveys has been eventually cultured such as the SAR11 (*Candidatus* Pelagibacter ubique) (Rappé et al. 2002), while most of the observed diversity remains uncultured and therefore largely unknown (the so-called "microbial dark matter"). Indeed, most microbial isolates belong to only a few phyla (Hug et al. 2016) and most phyla do not have a cultured representative. Yet, metagenomics and single cell genomics have provided genetic information of many uncultured lineages including those represented by many of the first clones reported in the 90s. This has helped to get insights on the potential functional and ecological roles of these uncultured lineages (Parks et al. 2017; Rinke et al. 2013). For example, the SAR86, an abundant marine clade belonging to the Gammaproteobacteria and detected for the first time in clone libraries constructed almost 30 years ago (Britschgi and Giovannoni 1991) and for which no isolate exists, shares traits with SAR11 such as the presence of proteorhodopsin (Béjà et al. 2000) and metabolic streamlining, but also displays distinct carbon compound specialization that might possibly avoid competition with SAR11 (Dupont et al. 2012). Moreover, analyses of the SAR86 pangenome indicate that the clade is composed of different ecotypes with unique geographic distributions (Hoarfrost et al. 2020). The SAR406 clade, first reported by Gordon and Giovannoni (1996) and now referred to as the "candidate" phylum *Marinimicrobia*, represents a deeply branching lineage of bacteria that is abundant in the aphotic zone. *Marinimicrobia* possesses the potential to degrade complex carbohydrate compounds as well as performing nitrate reduction, dissimilatory nitrite reduction to ammonia, and sulfur reduction (Thrash et al. 2017; Wright et al. 2014). The SAR202, also common in the aphotic zone, appears to metabolize multiple organosulfur compounds, oxidize sulfite, and oxidize recalcitrant organic compounds and thus are predicted to play major roles in the sulfur and carbon cycles in the aphotic water column (Landry et al. 2017; Mehrshad et al. 2018; Saw et al. 2020).

In addition, SAGs and MAGs have unveiled the existence of many new "candidate" phyla (Rinke et al. 2013; Parks et al. 2017). These "candidate" phyla are often detected in marine environments other than seawater, for example, in hydrothermal vents, blue holes, marine sediments, or associated with marine animals (Dudek et al. 2017; He et al. 2020; Parks et al. 2017; Rinke et al. 2013). However, analyses of the *Tara* Oceans metagenomes have revealed the presence of diverse members of "candidate" archaeal and bacterial lineages in the pelagic realm (Delmont et al. 2018; Lannes et al. 2019; Parks et al. 2017; Royo-Llonch et al. 2021; Tully et al. 2018). Moreover, many unknown clades within the "known" phyla are constantly being discovered in the marine water column (Parks et al. 2017; Yilmaz et al. 2016) and the publication of the genomic catalog of Earth's microbiomes has disclosed a breadth of phylogenetic diversity from multiple biomes including the marine aquatic biome (Nayfach et al. 2021), highlighting the ocean as a reservoir of hidden diversity.

### 8.4.2 Insights into New Metabolic Capacities of Uncultured Microorganisms

Although the wealth of genomic data of the "uncultured majority" in the ocean is exponentially increasing, there is still a long way to go to be able to interpret these data (Ferrera et al. 2015). Approximately 50% of the predicted genes detected in the ocean have an unknown function (Sunagawa et al. 2015) and the other half has an assigned putative function based on sequence homology to a gene from a distantly related cultured isolate for which the function has been experimentally demonstrated. Indeed, experiments with isolates have demonstrated that homologous genes yield different phenotypes in distinct microbes as is the case with proteorhodopsins mentioned earlier in this chapter. Despite this, homology searches in global metagenomic datasets have been pivotal to explore the potential relevance of different processes once the genes involved in a particular process are identified.

Like the discovery of proteorhodopsin changed the concept of phototrophy in the ocean, the discovery of ammonia oxidation genes in a genomic fragment belonging to Thaumarchaeota in the marine metagenome from the Sargasso Sea (Venter et al.

2004) changed our understanding of the global nitrogen cycle. Until then, nitrification was thought to be performed by a few low abundant bacterial genera. The abundant Thaumarchaeota were recognized as the organisms that exerted the primary control on ammonia oxidation in oligotrophic waters (Martens-Habbena et al. 2009; Wuchter et al. 2006). Genomic data of the "uncultured majority" was also fundamental to unveil other aspects of the nitrogen cycle such as that cyanate and urea may serve as potential substrates for nitrification (Alonso-Sáez et al. 2012b; Pachiadaki et al. 2017; Shi et al. 2011; Yakimov et al. 2011), which was subsequently tested by using cultures or single cell culture-independent approaches (Kitzinger et al. 2019, 2020; Palatinszky et al. 2015).

Likewise, our perspective of the life of microorganisms in the aphotic ocean has changed with the increase in genomic data from the deep-sea realm. In addition to heterotrophic and chemolithotrophic microorganisms, the deep ocean contains also mixotrophs that may play an important role in the carbon cycle (Acinas et al. 2021; Anantharaman et al. 2013; Pachiadaki et al. 2017; Sheik et al. 2014; Swan et al. 2011; Tang et al. 2016). These mixotrophic microorganisms may obtain energy from the oxidation of a plethora of compounds including ammonia, nitrite, CO, sulfur, hydrogen, and recalcitrant DOM (Acinas et al. 2021; Anantharaman et al. 2013; Landry et al. 2017; Martin-Cuadrado et al. 2009; Mehrshad et al. 2018; Pachiadaki et al. 2017; Sheik et al. 2014; Swan et al. 2011). Besides Thaumarchaeota (Reinthaler et al. 2010), other key players participating in carbon fixation in the dark ocean have been identified such as the ubiquitous SAR324 (Swan et al. 2011), the Thiomicrospirales (SUP05 cluster, Mattes et al. 2013), and *Nitrospina* (Pachiadaki et al. 2017). Moreover, a novel pathway for carbon fixation, the reductive glycine pathway, has just been described (Sánchez-Andrea et al. 2020), and the wealth of genomic data available allows the assessment of the distribution and potential relevance of this pathway in the ocean (Fig. 8.5).

After the discovery that the cosmopolitan SAR11 clade produces methane in phosphorus deficient waters as a by-product of the decomposition of methylphosphonate (Carini et al. 2014), the role of the vast oligotrophic gyres as a source of methane to the atmosphere became clear. The genes coding for the enzymes of this pathway of methanogenesis is widespread in marine bacterial genomes (Villarreal-Chiu et al. 2012) and in marine metagenomes (Sosa et al. 2019). Similarly, a relevant strategy to cope with phosphorus stress was unveiled after the identification of a phospholipase C responsible for lipid remodeling in a soil bacterium (Zavaleta-Pastor et al. 2010). This led to the discovery that lipid substitution is a widespread strategy to decrease the phosphorus demand of heterotrophic bacteria in the vast phosphorus depleted waters of the ocean (Carini et al. 2015; Sebastián et al. 2016) similarly to what had been previously observed in phytoplankton (Van Mooy et al. 2009). The discovery of the strategy of lipid substitution to overcome phosphorus depletion shows the need to implement the flexible stoichiometry of planktonic cells in biogeochemical budgets in which a fixed stoichiometry is assumed to be the rule.

Strategies to deal with the vast number of genes with unknown function are constantly being developed. Some approaches involve the functional screening of

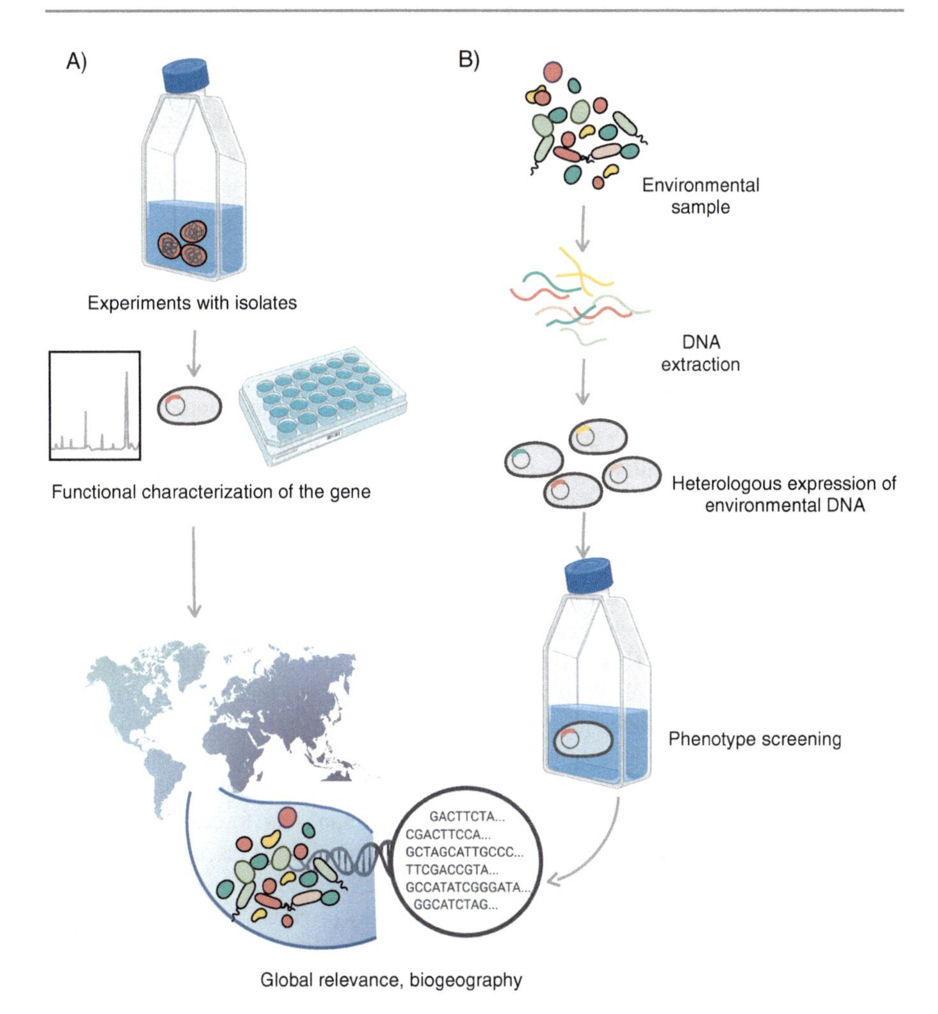


**Fig. 8.5** Approaches to unveil new functions in uncultured microorganisms. (**a**) Novel genes are identified and functionally characterized in experiments with isolates from diverse environments (e.g., ocean, soil, gut). Search for these genes in global environmental -omics datasets provides information about the relevance and biogeography of the novel genes, their taxonomic distribution, and the conditions under which these genes are expressed. (**b**) Genes with unknown function can also be identified through functional metagenomics in which environmental DNA fragments are cloned into a plasmid and expressed in a surrogate host. Then, the phenotype can be screened (e.g., enzymatic activity assays, development of color in rhodopsin containing cells). After the function is identified its biogeography and global relevance can be assessed in the global -omics databases. Some items have been created with BioRender.com

metagenomic libraries (e.g., Colin et al. 2015; Pushkarev et al. 2018) while others rely on computational workflows that narrow down the potential role of genes based on the clustering of their coding sequence spaces and their contextualization with genomic and environmental information (Vanni et al. 2020). A large fraction of the unknown genes is phylogenetically conserved and may be relevant for niche adaptation (Vanni et al. 2020). Thus, identifying these unknown genes is crucial to gain a deeper insight of microbial processes in the ocean. Given that sequencing effort has been directedly linked to the increase in the rate of gene discovery (Duarte et al. 2020), it is expected that many new genes will be uncovered in the near future. This calls for ways to unveil the function of these "unknown" genes.

### 8.4.3 Delineation of Ecological Meaningful Units of Uncultured Microorganisms

Bacterial and archaeal populations, and by extension species, are fundamental units of ecology and evolution (VanInsberghe et al. 2020). Identifying genetically and ecologically congruent units remains a great challenge in microbial ecology because bacteria and archaea reproduce asexually. Their genomes are subjected to homologous recombination events (Fraser et al. 2007; Papke et al. 2007) and lateral gene transfer between similar or distant relatives frequently occurs (Doolittle and Papke 2006; Fernández-Gómez et al. 2012). Furthermore, even within delineated species, there is no homogeneity in the genetic content or in the total nucleotide composition of bacteria and archaea. This variability is known as microdiversity (Acinas et al. 2004; Fuhrman and Campbell 1998) and it can be seen as "bushy tips" of distinct sequence clusters in phylogenetic reconstructions (Cohan 2001; Giovannoni 2004), by clusters of gene orthologous groups from genomes (Thompson et al. 2019), or by the identification of gene-flow discontinuities derived from recent genetic exchanges (Arevalo et al. 2019). Microdiversity may persist thanks to forces like periodic selection (Cohan 2001) and homologous recombination (Fraser et al. 2007; Konstantinidis and DeLong 2008; Shapiro et al. 2012; Whitaker et al. 2005) in which gene gain and losses are often involved and may promote divergence between populations favoring the delineation of ecotypes (Cordero et al. 2012). The "ecotype concept" describes a collection of strains that show some ecological distinctiveness within its species. Ecotypes preserve nearly the full phenotypic and ecological potential of the species with slight changes in their genetic repertoire that enables them to exploit a slightly different ecological niche. However, microbial speciation occurs in a continuous spectrum in which microbial populations are in constant evolutionary tradeoffs between gene flow and natural selection (Shapiro and Polz 2015; Shapiro et al. 2012) and therefore the delineation of ecological meaningful microbial populations or units remains challenging.

Despite the "species concept" still remains highly controversial, a widely accepted view of microbial species is the "pangenome concept" that classifies the genetic repertoire of a species into the core genome and the flexible genome. The former includes the shared genes between all individuals categorized as the same species and the latter includes the gene pool that is partially shared or strain-specific (Mira et al. 2010; Tettelin et al. 2005). Thus, the genomes of multiple representatives of a species are needed to accurately define the genetic potential and size of the pangenome. Even though the core genome contains the essence of the species and is

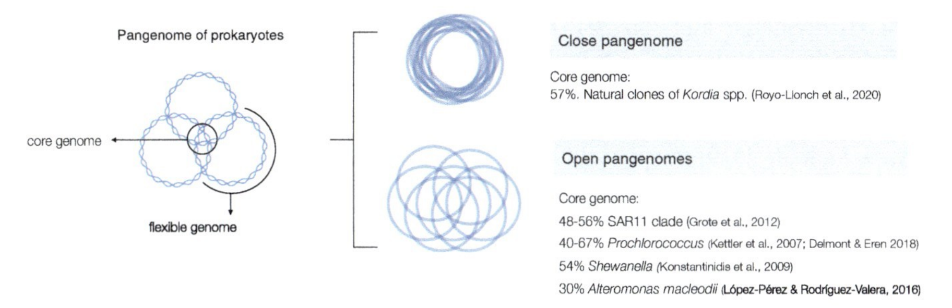


**Fig. 8.6** Visual representation of the pangenome concept for genomes of bacteria and archaea. The pangenome classifies de genetic repertoire of a species into the core genome, that includes all genes shared between all individuals categorized as the same species, and the flexible genome, which includes the gene pool that is partially shared or strain-specific. Depending on the genome size and genetic diversity of each species, the pangenome can be open or close (some examples from the literature are shown). Figure courtesy of Dr. Marta Royo-Llonch from SHOOK Studio (https://www.instagram.com/shookstudio/?hl=en)

indispensable, it is the flexible genome that can confer selective advantages like niche adaptation, new host colonization, or antibiotic resistance, and contributes to the species diversity (Tettelin et al. 2005, 2008). The size of the pangenome is dynamic and pangenomes can be more open or close depending on the multitude of niches in which the species is able to live (Medini et al. 2005) (Fig. 8.6).

In marine ecosystems, the pangenome exploration was first performed with genomes from bacterial isolates including *Prochlorococcus marinus* (Kettler et al. 2007), *Synechococcus* (Dufresne et al. 2008), *Shewanella* (Konstantinidis et al. 2009), *Alteromonas macleodii* (López-Pérez and Rodriguez-Valera 2016), and *Vibrio alginolyticus* (Chibani et al. 2020) but generally using a limited number of genomes. A large-scale analysis of pangenomes using ~7000 high-quality cultured genomes from 155 phylogenetically diverse species belonging to ten phyla revealed the important role of environmental preferences and phylogeny in explaining the majority of variation of pangenome features across different species (Maistrenko et al. 2020). The advent of Single Cell Genomics, however, has enabled the study of the pangenome concept in uncultured microbial taxa (Fig. 8.6). Some pioneer studies have explored the pangenomes of uncultured *Prochlorococcus* (Kashtan et al. 2014; Thompson et al. 2019), SAR11 lineages (Grote et al. 2012; Haro-Moreno et al. 2020; Thrash et al. 2014; Tsementzi et al. 2016), or uncultured Bacteroidetes relatives of *Kordia* sp. (Royo-Llonch et al. 2020).

Further, the combination of pangenomes and fragment recruitment analyses on marine metagenomes has led to the "metapangenomes concept" that uses the delineation of single amino acid variants to explore the biogeography of distinct populations. This has been successfully applied on single cell genomes of *Prochlorococcus* (Delmont and Eren 2018) and the SAR11 clade (Delmont et al. 2019) (Fig. 8.6). Similarly, the "genomospecies concept" represents a species that can be differentiated from others based on the average nucleotide identity (ANI) value by comparative genomics, phylogenomic and fragment recruitment analyses (Haro-Moreno et al. 2020). Alternatively, the reverse ecology approach (Arevalo et al. 2019; Shapiro and Polz 2015) used on isolate's genomes or single cell genomes relies on the identification of microbial populations as gene-flow units using network analyses. This approach enables the delineation of ecologically relevant populations and the identification of genes that have been under recent positive selection (Arevalo et al. 2019; Shapiro et al. 2012). All these alternatives will become more powerful when applied at large scale using uncultured genomes from different oceanic regions spanning a wider array of phylogenetic taxa, which will altogether enhance our knowledge on the evolution processes driving microbial speciation.

## 8.5 Future Perspectives

Despite last years' integrative research has boosted the knowledge of the marine microbiome to an unprecedented scale, as overviewed in this chapter, it is still insufficient for understanding the functioning of the microbiome and how it will respond to climate change-driven alterations. Here, we discuss some of the issues that in our opinion will become increasingly important in marine microbiome research.

- *Large-scale sequencing and inputs from new technologies.* Analyses of marine microbial communities from global circumnavigations have pointed out that only the "tip of the iceberg" has been unraveled and that even larger sequencing efforts are necessary to cover the immense genetic diversity of marine microbiomes. The generation of high-quality genomic references (using long read sequencing such as Nanopore) should be a priority to enhance the genome completeness and quality of SAGs and MAGs.
- *Large-scale single cell genomics studies.* Extending the analyses of SAGs at large scale from a wide array of phylogenetic taxa combined with cutting-edge analyses of population genomics will enhance our knowledge of the evolutionary processes driving microbial speciation.
- *Increasing efforts in high-throughput culturomics.* Large marine microbial culture collections will improve the microbial reference gene catalogs and will also be a fundamental tool to investigate the physiology of marine bacteria and archaea and to test physiological traits inferred from -omics data.
- *The scale matters.* Choosing the right spatial and temporal scales is essential to fully comprehend microbial processes in the ocean and to understand the triggers driving changes in community composition and ecosystem function. Despite the increasing effort to move from local to global scales, analyses at the fine scale are still lacking, such as how sub-mesoscale oceanographic features drive changes in the composition and function of the marine microbiome. Other unique features like the deep scattering layer, which is a hotspot for microbial activity, deserves further investigation. Likewise, at the temporal scale, highly resolved sampling over long periods must be carried out to be able to investigate the stability and repeatability of short-term changes.

- *Expand environmental genomic datasets, particularly on the temporal dimension.* Global expeditions have an incalculable value. Yet, they provide a snapshot of the marine microbiome at the moment of sampling. Repeated global expeditions at different times of the year in combination with time series from different latitudes will be necessary to have a complete view of the marine microbiome.
- *Beyond DNA sequencing.* DNA-based approaches have been highly informative on the diversity, potential metabolic capabilities, and evolution of the uncultured majority, but more effort should be invested towards understanding how this genetic information translates into function. The extended use of other -omics approaches, such as metatranscriptomics, metaproteomics, and metabolomics, will help to identify which metabolic processes are actually occurring in the environment and their magnitude. Likewise, the application of next generation physiological approaches such as BONCAT and Raman will be pivotal to experimentally validate information inferred from -omics data. A multifaceted approach in microbiome research is needed to grasp the relevance that the observed diversity has for ecosystem functioning.

**Acknowledgments** The authors have received funding from projects MAGGY (CTM2017-87736-R) from the Spanish Ministry of Economy and Competitiveness and AtlantECO from the European Union's Horizon 2020 Research and Innovation program under grant agreement N° 862923 to SGA, and projects ECLIPSE (PID2019-110128RB-I00) to IF and MIAU (RTI2018-101025-B-I00) to MS, both from the Spanish Ministry of Science and Innovation. SGA and MS have had the institutional support of the "Severo Ochoa Centre of Excellence" accreditation (CEX2019-000928-S) and IF received the support of the Fundación BBVA through the "Becas Leonardo a Investigadores y Creadores Culturales" 2019 Program. This publication reflects the views of only the authors, and neither the European Union nor the Fundación BBVA can be held responsible for any use which may be made of the information contained therein. We would like to thank Dr. Marta Royo-Llonch and Dr. Isabel Sanz-Sáez for useful discussions on genome-centric and culturing studies, to Adrià Auladell for sharing his data and for discussions on time series, and to SHOOK Studio for the illustration of some of the figures of the present chapter.

## References

Acinas SG, Klepac-Ceraj V, Hunt DE et al (2004) Fine-scale phylogenetic architecture of a complex bacterial community. Nature 430:551–554

Acinas SG, Sánchez P, Salazar G et al (2021) Deep ocean metagenomes provide insight into the metabolic architecture of bathypelagic microbial communities. Commun Biol 4:604

Agogué H, Lamy D, Neal PR et al (2011) Water mass-specificity of bacterial communities in the North Atlantic revealed by massively parallel sequencing. Mol Ecol 20:258–274

Albertsen M, Hugenholtz P, Skarshewski A et al (2013) Genome sequences of rare, uncultured bacteria obtained by differential coverage binning of multiple metagenomes. Nat Biotechnol 31: 533–538

Alneberg J, Bjarnason BS, De Bruijn I et al (2014) Binning metagenomic contigs by coverage and composition. Nat Methods 11:1144–1146

Alonso-Sáez L, Balagué V, Sà EL et al (2007) Seasonality in bacterial diversity in north-west Mediterranean coastal waters: assessment through clone libraries, fingerprinting and FISH. FEMS Microbiol Ecol 60:98–112

Alonso-Sáez L, Díaz-Pérez L, Morán XAG (2015) The hidden seasonality of the rare biosphere in coastal marine bacterioplankton. Environ Microbiol 10:3766–3780

Alonso-Sáez L, Sánchez O, Gasol JM (2012a) Bacterial uptake of low molecular weight organics in the subtropical Atlantic: are major phylogenetic groups functionally different? Limnol Oceanogr 57:798–808

Alonso-Sáez L, Waller AS, Mende DR et al (2012b) Role for urea in nitrification by polar marine archaea. Proc Natl Acad Sci 109:17989–17994

Amann RI, Ludwig W, Schleifer KH (1995) Phylogenetic identification and in situ detection of individual microbial cells without cultivation. Microbiol Rev 59:143–169

Amaral-Zettler L, Artigas LF, Baross J et al (2010) A global census of marine microbes. Life in the world's oceans: diversity, distribution and abundance. Wiley-Blackwell, Oxford, UK, pp 221–245

Amend AS, Oliver TA, Amaral-Zettler LA et al (2013) Macroecological patterns of marine bacteria on a global scale. J Biogeogr 40:800–811

Anantharaman K, Breier JA, Sheik CS, Dick GJ (2013) Evidence for hydrogen oxidation and metabolic plasticity in widespread deep-sea sulfur-oxidizing bacteria. Proc Natl Acad Sci 110: 330–335

Arevalo P, VanInsberghe D, Elsherbini J et al (2019) A reverse ecology approach based on a biological definition of microbial populations. Cell 178:820–834

Auladell A, Barberán A, Logares R et al (2021) Seasonal niche differentiation among closely related marine bacteria. ISME J. https://doi.org/10.1038/s41396-021-01053-2

Auladell A, Sánchez P, Sánchez O et al (2019) Long-term seasonal and interannual variability of marine aerobic anoxygenic photoheterotrophic bacteria. ISME J 13:1975–1987

Aylward FO, Boeuf D, Mende DR et al (2017) Diel cycling and long-term persistence of viruses in the ocean's euphotic zone. Proc Natl Acad Sci 114:11446–11451

Baas Becking LGM (1934) Geobiologie of inleiding tot de milieukunde. Van Stockum WP and Zoon NV, Den Haag, The Netherlands

Bagert JD, van Kessel JC, Sweredoski MJ et al (2016) Time-resolved proteomic analysis of quorum sensing in *Vibrio harveyi*. Chem Sci 7:1797–1806

Balashov SP, Imasheva ES, Boichenko VA et al (2005) Xanthorhodopsin: a proton pump with a light-harvesting carotenoid antenna. Science 309:2061–2064

Beijerinck MW (1913) De infusies en de ontdekking der bakterien. Johannes Müller

Béjà O, Aravind L, Koonin EV et al (2000) Bacterial rhodopsin: evidence for a new type of phototrophy in the sea. Science 289:1902–1906

Ben-Dov E, Kramarsky-Winter E, Kushmaro A (2009) An *in situ* method for cultivating microorganisms using a double encapsulation technique. FEMS Microbiol Ecol 68:363–371

Bergauer K, Fernandez-Guerra A, Garcia JAL et al (2018) Organic matter processing by microbial communities throughout the Atlantic water column as revealed by metaproteomics. Proc Natl Acad Sci 115:E400–E408

Berry D, Mader E, Lee TK et al (2015) Tracking heavy water (D2O) incorporation for identifying and sorting active microbial cells. Proc Natl Acad Sci 112:194–203

Biller SJ, Berube PM, Dooley K et al (2018) Marine microbial metagenomes sampled across space and time. Sci Data 5:180176

Blainey PC (2013) The future is now: single-cell genomics of bacteria and archaea. FEMS Microbiol Rev 37:407–427

Blazewicz SJ, Barnard RL, Daly RA, Firestone MK (2013) Evaluating rRNA as an indicator of microbial activity in environmental communities: limitations and uses. ISME J 7:2061–2068

Boitard L, Cottinet D, Bremond N et al (2015) Growing microbes in millifluidic droplets. Eng Life Sci 15:318–326

Bowers RM, Kyrpides NC, Stepanauskas R et al (2017) Minimum information about a single amplified genome (MISAG) and a metagenome-assembled genome (MIMAG) of bacteria and archaea. Nat Biotechnol 35:725–731

Bowsher AW, Kearns PJ, Shade A (2019) 16S rRNA/rRNA gene ratios and cell activity staining reveal consistent patterns of microbial activity in plant-associated soil. mSystems 4:e00003–e00019

Boyd PW, Claustre H, Levy M et al (2019) Multi-faceted particle pumps drive carbon sequestration in the ocean. Nature 568:327–335

Britschgi TB, Giovannoni SJ (1991) Phylogenetic analysis of a natural marine bacterioplankton population by rRNA gene cloning and sequencing. Appl Environ Microbiol 57:1707–1713

Brown MV, Schwalbach MS, Hewson I, Fuhrman JA (2005) Coupling 16S-ITS rDNA clone libraries and automated ribosomal intergenic spacer analysis to show marine microbial diversity: development and application to a time series. Environ Microbiol 7:1466–1479

Brum JR, Ignacio-Espinoza JC, Roux S et al (2015) Patterns and ecological drivers of ocean viral communities. Science 348:1261498

Bryson S, Li Z, Chavez F et al (2017) Phylogenetically conserved resource partitioning in the coastal microbial loop. ISME J 11:2781–2792

Bunse C, Pinhassi J (2017) Marine bacterioplankton seasonal succession dynamics. Trends Microbiol 25:494–505

Buttigieg PL, Fadeev E, Bienhold C et al (2018) Marine microbes in 4D–using time series observation to assess the dynamics of the ocean microbiome and its links to ocean health. Curr Opin Microbiol 43:169–185

Campbell BJ, Kirchman DL (2013) Bacterial diversity, community structure and potential growth rates along an estuarine salinity gradient. ISME J 7:210–220

Campbell BJ, Yu L, Heidelberg JF, Kirchman DL (2011) Activity of abundant and rare bacteria in a coastal ocean. Proc Natl Acad Sci 108:12776–12781

Campbell BJ, Yu L, Straza TR, Kirchman DL (2009) Temporal changes in bacterial rRNA and rRNA genes in Delaware (USA) coastal waters. Aquat Microb Ecol 57:123–135

Cao S, Zhang W, Ding W et al (2020) Structure and function of the Arctic and Antarctic marine microbiota as revealed by metagenomics. Microbiome 8:47

Carini P, Steindler L, Beszteri S, Giovannoni SJ (2013) Nutrient requirements for growth of the extreme oligotroph '*Candidatus* Pelagibacter ubique' HTCC1062 on a defined medium. ISME J 7:592–602

Carini P, Van Mooy BAS, Thrash JC et al (2015) SAR11 lipid renovation in response to phosphate starvation. Proc Natl Acad Sci 112:7767–7772

Carini P, White AE, Campbell EO, Giovannoni SJ (2014) Methane production by phosphate-starved SAR11 chemoheterotrophic marine bacteria. Nat Commun 5:4346

Carradec Q, Pelletier E, Da Silva C et al (2018) A global ocean atlas of eukaryotic genes. Nat Commun 9:373

Castillo YM, Mangot J, Benites LF et al (2019) Assessing the viral content of uncultured picoeukaryotes in the global-ocean by single cell genomics. Mol Ecol 28:4272–4289

Chafee M, Fernàndez-Guerra A, Buttigieg PL et al (2018) Recurrent patterns of microdiversity in a temperate coastal marine environment. ISME J 12:237–252

Chen LX, Anantharaman K, Shaiber A et al (2020) Accurate and complete genomes from metagenomes. Genome Res 30:315–333

Chen Y, Murrell JC (2010) When metagenomics meets stable-isotope probing: progress and perspectives. Trends Microbiol 18:157–163

Chibani CM, Roth O, Liesegang H, Wendling CC (2020) Genomic variation among closely related *Vibrio alginolyticus* strains is located on mobile genetic elements. BMC Genomics 21:1–14

Chisholm SW, Olson RJ, Zettler ER et al (1988) A novel free-living prochlorophyte abundant in the oceanic euphotic zone. Nature 334:340–343

Cohan FM (2001) Bacterial species and speciation. Syst Biol 50:513–524

Coleman ML, Chisholm SW (2010) Ecosystem-specific selection pressures revealed through comparative population genomics. Proc Natl Acad Sci 107:18634–18639

Colin PY, Kintses B, Gielen F et al (2015) Ultrahigh-throughput discovery of promiscuous enzymes by picodroplet functional metagenomics. Nat Commun 6:10008

Cordero OX, Ventouras LA, DeLong EF, Polz MF (2012) Public good dynamics drive evolution of iron acquisition strategies in natural bacterioplankton populations. Proc Natl Acad Sci 109: 20059–20064
Cottrell MT, Kirchman DL (2000) Natural assemblages of marine proteobacteria and members of the *Cytophaga-Flavobacter* cluster consuming low- and high-molecular-weight dissolved organic matter. Appl Environ Microbiol 66:1692–1697
Couradeau E, Sasse J, Goudeau D et al (2019) Probing the active fraction of soil microbiomes using BONCAT-FACS. Nat Commun 10:2770
Coyotzi S, Pratscher J, Murrell JC, Neufeld JD (2016) Targeted metagenomics of active microbial populations with stable-isotope probing. Curr Opin Biotechnol 41:1–8
Cram JA, Chow C-ET, Sachdeva R et al (2015) Seasonal and interannual variability of the marine bacterioplankton community throughout the water column over ten years. ISME J 9:563–580
Crespo BG, Wallhead PJ, Logares R, Pedrós-Alió C (2016) Probing the rare biosphere of the North-West Mediterranean Sea: an experiment with high sequencing effort. PLoS One 11:e0159195
De Vargas C, Audic S, Henry N et al (2015) Eukaryotic plankton diversity in the sunlit ocean. Science 348:1261605
Del Giorgio PA, Gasol JM (2008) Physiological structure and single-cell activity in marine bacterioplankton. In: Kirchman DL (ed) Microbial ecology of the oceans, 2nd edn. John Wiley & Sons, Inc., Hoboken, NJ, USA, pp 243–298
Del Giorgio PA, Scarborough G (1995) Increase in the proportion of metabolically active bacteria along gradients of enrichment in freshwater and marine plankton: implications for estimates of bacterial growth and production rates. J Plankton Res 17:1905–1924
Delmont TO, Eren AM (2018) Linking pangenomes and metagenomes: the *Prochlorococcus* metapangenome. PeerJ 6:e4320
Delmont TO, Kiefl E, Kilinc O et al (2019) Single-amino acid variants reveal evolutionary processes that shape the biogeography of a global SAR11 subclade. elife 8:e46497
Delmont TO, Quince C, Shaiber A et al (2018) Nitrogen-fixing populations of Planctomycetes and Proteobacteria are abundant in surface ocean metagenomes. Nat Microbiol 3:804–813
DeLong EF, Béjà O (2010) The light-driven proton pump proteorhodopsin enhances bacterial survival during tough times. PLoS Biol 8:e1000359
DeLong EF, Preston CM, Mincer TJ et al (2006) Community genomics among stratified microbial assemblages in the ocean's interior. Science 311:496–503
Dieterich DC, Link AJ, Graumann J et al (2006) Selective identification of newly synthesized proteins in mammalian cells using bioorthogonal noncanonical amino acid tagging (BONCAT). Proc Natl Acad Sci 103:9482–9487
Díez-Vives C, Nielsen S, Sánchez P et al (2019) Delineation of ecologically distinct units of marine Bacteroidetes in the Northwestern Mediterranean Sea. Mol Ecol 28:2846–2859
Doolittle WF, Papke RT (2006) Genomics and the bacterial species problem. Genome Biol 7:116
Doud DFR, Bowers RM, Schulz F et al (2020) Function-driven single-cell genomics uncovers cellulose-degrading bacteria from the rare biosphere. ISME J 14:659–675
Duarte CM (2015) Seafaring in the 21st century: the Malaspina 2010 Circumnavigation Expedition. Limnol Oceanogr Bull 24:11–14
Duarte CM, Ngugi DK, Alam I et al (2020) Sequencing effort dictates gene discovery in marine microbial metagenomes. Environ Microbiol 22:4589–4603
Dudek NK, Sun CL, Burstein D et al (2017) Novel microbial diversity and functional potential in the marine mammal oral microbiome. Curr Biol 27:3752–3762
Dufresne A, Ostrowski M, Scanlan DJ et al (2008) Unravelling the genomic mosaic of a ubiquitous genus of marine cyanobacteria. Genome Biol 9:1–16
Dumont MG, Murrell JC (2005) Stable isotope probing–linking microbial identity to function. Nat Rev Microbiol 3:499–504
Dupont CL, Rusch DB, Yooseph S et al (2012) Genomic insights to SAR86, an abundant and uncultivated marine bacterial lineage. ISME J 6:1186–1199

Eiler A, Hayakawa DH, Rappé MS (2011) Non-random assembly of bacterioplankton communities in the subtropical North Pacific Ocean. Front Microbiol 2:140
Eren AM, Esen ÖC, Quince C et al (2015) Anvi'o: an advanced analysis and visualization platform for 'omics data. PeerJ 3:e1319
Fernández-Gómez B, Fernàndez-Guerra A, Casamayor EO et al (2012) Patterns and architecture of genomic islands in marine bacteria. BMC Genomics 13:347
Ferrera I, Sánchez O, Kolářová E, Koblížek M, Gasol JM (2017) Light enhances the growth rates of natural populations of aerobic anoxygenic phototrophic bacteria. ISME J 11:2391–2393
Ferrera I, Sebastian M, Acinas SG, Gasol JM (2015) Prokaryotic functional gene diversity in the sunlit ocean: stumbling in the dark. Curr Opin Microbiol 25:33–39
Foster RA, Kuypers MMM, Vagner T et al (2011) Nitrogen fixation and transfer in open ocean diatom-cyanobacterial symbioses. ISME J 5:1484–1493
Fraser C, Hanage WP, Spratt BG (2007) Recombination and the nature of bacterial speciation. Science 315:476–480
Frigaard NU, Martinez A, Mincer TJ, DeLong EF (2006) Proteorhodopsin lateral gene transfer between marine planktonic bacteria and archaea. Nature 439:847–850
Fuhrman JA, Campbell L (1998) Microbial microdiversity. Nature 393:410–411
Fuhrman JA, Cram JA, Needham DDM (2015) Marine microbial community dynamics and their ecological interpretation. Nat Rev Microbiol 13:133–146
Fuhrman JA, Hewson I, Schwalbach MS et al (2006) Annually reoccurring bacterial communities are predictable from ocean conditions. Proc Natl Acad Sci 103:13104–13109
Fuhrman JA, McCallum K, Davis AA (1993) Phylogenetic diversity of subsurface marine microbial communities from the Atlantic and Pacific oceans. Appl Environ Microbiol 59:1294–1302
Galand PE, Alonso-Sáez L, Bertilsson S, Lovejoy C, Casamayor EO (2013) Contrasting activity patterns determined by BrdU incorporation in bacterial ribotypes from the Arctic Ocean in winter. Front Microbiol 4:118
Gao W, Navarroli D, Naimark J et al (2013) Microbe observation and cultivation array (MOCA) for cultivating and analyzing environmental microbiota. Microbiome 1:4
Ghiglione JF, Conan P, Pujo-Pay M (2009) Diversity of total and active free-living vs. particle-attached bacteria in the euphotic zone of the NW Mediterranean Sea. FEMS Microbiol Lett 299: 9–21
Ghiglione JF, Galand PE, Pommier T et al (2012) Pole-to-pole biogeography of surface and deep marine bacterial communities. Proc Natl Acad Sci 109:17633–17638
Ghiglione JF, Larcher M, Lebaron P (2005) Spatial and temporal scales of variation in bacterioplankton community structure in the NW Mediterranean Sea. Aquat Microb Ecol 40: 229–240
Gilbert JA, Steele JA, Caporaso JG et al (2012) Defining seasonal marine microbial community dynamics. ISME J 6:298–308
Giner CR, Balagué V, Krabberød AK et al (2019) Quantifying long-term recurrence in planktonic microbial eukaryotes. Mol Ecol 28:923–935
Giovannoni S (2004) Oceans of bacteria. Nature 430:515–516
Giovannoni S, Stingl U (2007) The importance of culturing bacterioplankton in the 'omics' age. Nat Rev Microbiol 5:820–826
Giovannoni SJ, Britschgi TB, Moyer CL, Field KG (1990) Genetic diversity in Sargasso Sea bacterioplankton. Nature 345:60–63
Gómez-Consarnau L, Akram N, Lindell K et al (2010) Proteorhodopsin phototrophy promotes survival of marine bacteria during starvation. PLoS Biol 8:e1000358
Gómez-Consarnau L, González JM, Coll-Lladó M et al (2007) Light stimulates growth of proteorhodopsin-containing marine Flavobacteria. Nature 445:210–213
González JM, Fernández-Gómez B, Fernàndez-Guerra A et al (2008) Genome analysis of the proteorhodopsin-containing marine bacterium *Polaribacter* sp. MED152 (Flavobacteria). Proc Natl Acad Sci 105:8724–8729

Good BH, McDonald MJ, Barrick JE et al (2017) The dynamics of molecular evolution over 60,000 generations. Nature 551:45–50

Gordon DA, Giovannoni SJ (1996) Detection of stratified microbial populations related to *Chlorobium* and *Fibrobacter* species in the Atlantic and Pacific oceans. Appl Environ Microbiol 62:1171–1177

Graham ED, Heidelberg JF, Tully BJ (2018) Potential for primary productivity in a globally-distributed bacterial phototroph. ISME J 12:1861–1866

Gregory AC, Zayed AA, Conceição-Neto N et al (2019) Marine DNA viral macro-and microdiversity from pole to pole. Cell 177:1109–1123

Grob C, Taubert M, Howat AM et al (2015) Combining metagenomics with metaproteomics and stable isotope probing reveals metabolic pathways used by a naturally occurring marine methylotroph. Environ Microbiol 17:4007–4018

Grote J, Thrash JC, Huggett MJ et al (2012) Streamlining and core genome conservation among highly divergent members of the SAR11 clade. MBio 3:e00252–e00212

Guidi L, Chaffron S, Bittner L et al (2016) Plankton networks driving carbon export in the oligotrophic ocean. Nature 532:465–470

Gutleben J, Chaib M, Mares D et al (2017) The multi-omics promise in context: from sequence to microbial isolate. Crit Rev Microbiol 44:212–219

Hall EK, Singer GA, Pölzl M et al (2011) Looking inside the box: using Raman microspectroscopy to deconstruct microbial biomass stoichiometry one cell at a time. ISME J 5:196–208

Haro-Moreno JM, Rodriguez-Valera F, Rosselli R et al (2020) Ecogenomics of the SAR11 clade. Environ Microbiol 22:1748–1763

Hatzenpichler R, Connon SA, Goudeau D et al (2016) Visualizing in situ translational activity for identifying and sorting slow-growing archaeal–bacterial consortia. Proc Natl Acad Sci 113: E4069–E4078

Hatzenpichler R, Krukenberg V, Spietz RL, Jay ZJ (2020) Next-generation physiology approaches to study microbiome function at single cell level. Nat Rev Microbiol 18:241–256

Hatzenpichler R, Scheller S, Tavormina PL et al (2014) *In situ* visualization of newly synthesized proteins in environmental microbes using amino acid tagging and click chemistry. Environ Microbiol 16:2568–2590

Hauruseu D, Koblížek M (2012) Influence of light on carbon utilization in aerobic anoxygenic phototrophs. Appl Environ Microbiol 78:7414–7419

He P, Xie L, Zhang X et al (2020) Microbial diversity and metabolic potential in the stratified Sansha Yongle blue hole in the South China Sea. Sci Rep 10:5949

Henson MW, Lanclos VC, Pitre DM et al (2020) Expanding the diversity of bacterioplankton isolates and modeling isolation efficacy with large-scale dilution-to-extinction cultivation. Appl Environ Microbiol 86:1–23

Hesselman MC, Odoni DI, Ryback BM et al (2012) A multi-platform flow device for microbial (co-) cultivation and microscopic analysis. PLoS One 7:e36982

Hoarfrost A, Nayfach S, Ladau J et al (2020) Global ecotypes in the ubiquitous marine clade SAR86. ISME J 14:178–188

Hoppe HG (1976) Determination and properties of actively metabolizing heterotrophic bacteria in the sea, investigated by means of micro-autoradiography. Mar Biol 36:291–302

Huang WE, Stoecker K, Griffiths R et al (2007) Raman-FISH: combining stable-isotope Raman spectroscopy and fluorescence in situ hybridization for the single cell analysis of identity and function. Environ Microbiol 9:1878–1889

Hug LA, Baker BJ, Anantharaman K et al (2016) A new view of the tree of life. Nat Microbiol 1: 16048

Hugoni M, Taib N, Debroas D et al (2013) Structure of the rare archaeal biosphere and seasonal dynamics of active ecotypes in surface coastal waters. Proc Natl Acad Sci 110:6004–6009

Hunt DE, Lin Y, Church MJ et al (2013) Relationship between abundance and specific activity of bacterioplankton in open ocean surface waters. Appl Environ Microbiol 79:177–184

Ibarbalz FM, Henry N, Brandão MC et al (2019) Global trends in marine plankton diversity across kingdoms of life. Cell 179:1084–1097
Imdahl F, Vafadarnejad E, Homberger C et al (2020) Single-cell RNA-sequencing reports growth-condition-specific global transcriptomes of individual bacteria. Nat Microbiol 5:1202–1206
Ingham CJ, Sprenkels A, Bomer J et al (2007) The micro-petri dish, a million-well growth chip for the culture and high-throughput screening of microorganisms. Proc Natl Acad Sci 104:18217–18222
Jannasch HW, Jones GE (1959) Bacterial populations in sea water as determined by different methods of enumeration. Limnol Oceanogr 4:128–139
Johnson PW, Sieburth JMN (1979) Chroococcoid cyanobacteria in the sea: a ubiquitous and diverse phototrophic biomass. Limnol Oceanogr 24:928–935
Jones SE, Lennon JT (2010) Dormancy contributes to the maintenance of microbial diversity. Proc Natl Acad Sci 107:5881–5886
Karl DM, Church MJ (2014) Microbial oceanography and the Hawaii Ocean time-series programme. Nat Rev Microbiol 12:699–713
Karsenti E, Acinas SG, Bork P et al (2011) A holistic approach to marine eco-systems biology. PLoS Biol 9:e1001177
Kashtan N, Roggensack SE, Rodrigue S et al (2014) Single-cell genomics reveals hundreds of coexisting subpopulations in wild *Prochlorococcus*. Science 344:416–420
Kearns PJ, Angell JH, Howard EM et al (2016) Nutrient enrichment induces dormancy and decreases diversity of active bacteria in salt marsh sediments. Nat Commun 7:12881
Kettler GC, Martiny AC, Huang K et al (2007) Patterns and implications of gene gain and loss in the evolution of *Prochlorococcus*. PLoS Genet 3:2515–2528
Kirchman DL (2016) Growth rates of microbes in the oceans. Annu Rev Mar Sci 8:285–309
Kirchman DL, Hanson TE (2013) Bioenergetics of photoheterotrophic bacteria in the oceans. Environ Microbiol Rep 5:188–199
Kitzinger K, Marchant HK, Bristow LA et al (2020) Single cell analyses reveal contrasting life strategies of the two main nitrifiers in the ocean. Nat Commun 11:767
Kitzinger K, Padilla CC, Marchant HK et al (2019) Cyanate and urea are substrates for nitrification by Thaumarchaeota in the marine environment. Nat Microbiol 4:234–243
Koblížek M (2015) Ecology of aerobic anoxygenic phototrophs in aquatic environments. FEMS Microbiol Rev 39:854–870
Kolber ZS, Van Dover CL, Niederman RA, Falkowski PG (2000) Bacterial photosynthesis in surface waters of the open ocean. Nature 407:177–179
Kolody BC, McCrow JP, Allen LZ et al (2019) Diel transcriptional response of a California current plankton microbiome to light, low iron, and enduring viral infection. ISME J 13:2817–2833
Könneke M, Bernhard AE, de la Torre JR et al (2005) Isolation of an autotrophic ammonia-oxidizing marine archaeon. Nature 437:543–546
Konstantinidis KT, DeLong EF (2008) Genomic patterns of recombination clonal divergence and environment in marine microbial populations. ISME J 2:1052–1065
Konstantinidis KT, Rosselló-Móra R, Amann R (2017) Uncultivated microbes in need of their own taxonomy. ISME J 11:2399–2406
Konstantinidis KT, Serres MH, Romine MF et al (2009) Comparative systems biology across an evolutionary gradient within the *Shewanella* genus. Proc Natl Acad Sci 106:15909–15914
Kopf A, Bicak M, Kottmann R et al (2015) The ocean sampling day consortium. Gigascience 4:27
Ku C, Sheyn U, Sebé-Pedrós A et al (2020) A single-cell view on alga-virus interactions reveals sequential transcriptional programs and infection states. Sci Adv 6:eaba4137
Labonté JM, Swan BK, Poulos B et al (2015) Single-cell genomics-based analysis of virus-host interactions in marine surface bacterioplankton. ISME J 9:2386–2399
Lagier JC, Armougom F, Million M et al (2012) Microbial culturomics: paradigm shift in the human gut microbiome study. Clin Microbiol Infect 18:1185–1193
Lambert S, Tragin M, Lozano JC et al (2019) Rhythmicity of coastal marine picoeukaryotes, bacteria and archaea despite irregular environmental perturbations. ISME J 13:388–401

Landry Z, Swa BK, Herndl GJ et al (2017) SAR202 genomes from the dark ocean predict pathways for the oxidation of recalcitrant dissolved organic matter. mBio 8:e00413–e00417
Lannes R, Olsson-Francis K, Lopez P, Bapteste E (2019) Carbon fixation by marine ultrasmall prokaryotes. Genome Biol Evol 11:1166–1177
Larkin AA, Garcia CA, Garcia N et al (2021) High spatial resolution global ocean metagenomes from Bio-GO-SHIP repeat hydrography transects. Sci Data 8:107
Lebaron P, Servais P, Agogué H, Courties C, Joux F (2001) Does the high nucleic acid content of individual bacterial cells allow us to discriminate between active cells and inactive cells in aquatic systems? Appl Environ Microbiol 67:1775–1782
Lee KS, Palatinszky M, Pereira FC et al (2019) An automated Raman-based platform for the sorting of live cells by functional properties. Nat Microbiol 4:1035–1048
Leizeaga A, Estrany M, Forn I, Sebastián M (2017) Using click-chemistry for visualizing *in situ* changes of translational activity in planktonic marine bacteria. Front Microbiol 8:2360
Lekunberri I, Gasol JM, Acinas SG et al (2014) The phylogenetic and ecological context of cultured and whole genome-sequenced planktonic bacteria from the coastal NW Mediterranean Sea. Syst Appl Microbiol 37:216–228
Lennon JT, Jones SE (2011) Microbial seed banks: the ecological and evolutionary implications of dormancy. Nat Rev Microbiol 9:119–130
Li WKW, Rao DVS, Harrison WG et al (1983) Autotrophic picoplankton in the tropical ocean. Science 219:292–295
Li M, Jain S, Baker BJ et al (2014) Novel hydrocarbon monooxygenase genes in the metatranscriptome of a natural deep-sea hydrocarbon plume. Environ Microbiol 16:60–71
Li WKW, Jellett JF, Dickie PM (1995) DNA distributions in planktonic bacteria stained with TOTO or TO-PRO. Limnol Oceanogr 40:1485–1495
Lima-Mendez G, Faust K, Henry N et al (2015) Determinants of community structure in the global plankton interactome. Science 348:1262073
Locey KJ, Lennon JT (2016) Scaling laws predict global microbial diversity. Proc Natl Acad Sci 113:5970–5975
Longnecker K, Lomas MW, Van Mooy BA (2010) Abundance and diversity of heterotrophic bacterial cells assimilating phosphate in the subtropical North Atlantic Ocean. Environ Microbiol 12:2773–2782
Longnecker K, Sherr BF, Sherr EB (2005) Activity and phylogenetic diversity of bacterial cells with high and low nucleic acid content and electron transport system activity in an upwelling ecosystem. Appl Environ Microbiol 71:7737–7749
López-Pérez M, Rodriguez-Valera F (2016) Pangenome evolution in the marine bacterium *Alteromonas*. Genome Biol Evol 8:1556–1570
Lorenz B, Wichmann C, Stöckel S, Rösch P, Popp J (2017) Cultivation-free Raman spectroscopic investigations of bacteria. Trends Microbiol 25:413–424
Luna GM (2015) Biotechnological potential of marine microbes. In: Kim S (ed) Springer handbook of marine biotechnology. Springer, Berlin, Heidelberg, pp 651–661
Ma L, Kim J, Hatzenpichler R et al (2014) Gene-targeted microfluidic cultivation validated by isolation of a gut bacterium listed in human microbiome Project's Most wanted taxa. Proc Natl Acad Sci 111:9768–9773
Maistrenko OM, Mende DR, Luetge M et al (2020) Disentangling the impact of environmental and phylogenetic constraints on prokaryotic within-species diversity. ISME J 14:1247–1259
Martens-Habbena W, Berube PM, Urakawa H et al (2009) Ammonia oxidation kinetics determine niche separation of nitrifying archaea and bacteria. Nature 461:976–979
Martín HG, Ivanova N, Kunin V et al (2006) Metagenomic analysis of two enhanced biological phosphorus removal (EBPR) sludge communities. Nat Biotechnol 24:1263–1269
Martin-Cuadrado AB, Ghai R, Gonzaga A, Rodriguez-Valera F (2009) CO dehydrogenase genes found in metagenomic fosmid clones from the deep Mediterranean Sea. Appl Environ Microbiol 75:7436–7444

Martin-Platero AM, Cleary B, Kauffman K et al (2018) High resolution time series reveals cohesive but short-lived communities in coastal plankton. Nat Commun 9:266

Martinez-Garcia M, Brazel DM, Swan BK et al (2012) Capturing single cell genomes of active polysaccharide degraders: an unexpected contribution of *Verrucomicrobia*. PLoS One 7:e35314

Martinez-Garcia M, Santos F, Moreno-Paz M et al (2014) Unveiling viral–host interactions within the 'microbial dark matter'. Nat Commun 5:4542

Mattes TE, Nunn BL, Marshall KT et al (2013) Sulfur oxidizers dominate carbon fixation at a biogeochemical hot spot in the dark ocean. ISME J 7:2349–2360

Mayali X (2020) NanoSIMS: microscale quantification of biogeochemical activity with large-scale impacts. Annu Rev Mar Sci 12:449–467

Mayali X, Weber PK, Brodie EL et al (2012) High-throughput isotopic analysis of RNA microarrays to quantify microbial resource use. ISME J 6:1210–1221

McCarren J, Becker JW, Repeta DJ et al (2010) Microbial community transcriptomes reveal microbes and metabolic pathways associated with dissolved organic matter turnover in the sea. Proc Natl Acad Sci 107:16420–16427

Medini D, Donati C, Tettelin H et al (2005) The microbial pan-genome. Curr Opin Genet Dev 15: 589–594

Mehrshad M, Rodriguez-Valera F, Amoozegar MA et al (2018) The enigmatic SAR202 cluster up close: shedding light on a globally distributed dark ocean lineage involved in sulfur cycling. ISME J 12:655–668

Mestre M, Ruiz-González C, Logares R et al (2018) Sinking particles promote vertical connectivity in the ocean microbiome. Proc Natl Acad Sci 115:E6799–E6807

Mira A, Martín-Cuadrado AB, D'Auria G, Rodríguez-Valera F (2010) The bacterial pan-genome: a new paradigm in microbiology. Int Microbiol 13:45–57

Moran MA, Satinsky B, Gifford SM et al (2013) Sizing up metatranscriptomics. ISME J 7:237–243

Morris RM, Nunn BL, Frazar C et al (2010) Comparative metaproteomics reveals ocean-scale shifts in microbial nutrient utilization and energy transduction. ISME J 4:673–685

Morris RM, Rappé MS, Connon SA et al (2002) SAR11 clade dominates ocean surface bacterioplankton communities. Nature 420:806–810

Mou X, Sun S, Edwards RA, Hodson RE, Moran MA (2008) Bacterial carbon processing by generalist species in the coastal ocean. Nature 451:708–711

Mou X, Vila-Costa M, Sun et al (2011) Metatranscriptomic signature of exogenous polyamine utilization by coastal bacterioplankton. Environ Microbiol Rep 3:798–806

Muller EEL, Glaab E, May P et al (2013) Condensing the omics fog of microbial communities. Trends Microbiol 21:325–333

Murray AE, Freudenstein J, Gribaldo S et al (2020) Roadmap for naming uncultivated archaea and bacteria. Nat Microbiol 5:987–994

Musat N, Foster R, Vagner T et al (2012) Detecting metabolic activities in single cells, with emphasis on nanoSIMS. FEMS Microbiol Rev 36:486–511

Nayfach S, Roux S, Seshadri R et al (2021) A genomic catalog of Earth's microbiomes. Nat Biotechnol 39:499–509

Needham DM, Fuhrman JA (2016) Pronounced daily succession of phytoplankton, archaea and bacteria following a spring bloom. Nat Microbiol 1:16005

Needham DM, Yoshizawa S, Hosaka T et al (2019) A distinct lineage of giant viruses brings a rhodopsin photosystem to unicellular marine predators. Proc Natl Acad Sci 116:20574–20583

Neely CJ, Graham ED, Tully BJ (2020) MetaSanity: an integrated microbial genome evaluation and annotation pipeline. Bioinformatics 36:4341–4344

Nelson CE, Carlson CA (2012) Tracking differential incorporation of dissolved organic carbon types among diverse lineages of Sargasso Sea bacterioplankton. Environ Microbiol 14:1500–1516

Orsi WD, Smith JM, Liu S et al (2016) Diverse, uncultivated bacteria and archaea underlying the cycling of dissolved protein in the ocean. ISME J 10:2158–2173

Ottesen EA, Young CR, Eppley JM et al (2013) Pattern and synchrony of gene expression among sympatric marine microbial populations. Proc Natl Acad Sci 110:E488–E497
Ottesen EA, Young CR, Gifford SM et al (2014) Multispecies diel transcriptional oscillations in open ocean heterotrophic bacterial assemblages. Science 345:207–212
Pachiadaki MG, Brown JM, Brown J et al (2019) Charting the complexity of the marine microbiome through single-cell genomics. Cell 179:1623–1635
Pachiadaki MG, Sintes E, Bergauer K et al (2017) Major role of nitrite-oxidizing bacteria in dark ocean carbon fixation. Science 358:1046–1051
Palatinszky M, Herbold C, Jehmlich N et al (2015) Cyanate as an energy source for nitrifiers. Nature 524:105–108
Papke RT, Zhaxybayeva O, Feil EJ et al (2007) Searching for species in haloarchaea. Proc Natl Acad Sci 104:14092–14097
Parada AE, Fuhrman JA (2017) Marine archaeal dynamics and interactions with the microbial community over 5 years from surface to seafloor. ISME J 11:2510–2525
Park J, Kerner A, Burns MA, Lin XN (2011) Microdroplet-enabled highly parallel co-cultivation of microbial communities. PLoS One 6:e17019
Parker CT, Tindall BJ, Garrity GM (2019) International code of nomenclature of prokaryotes. Int J Syst Evol Microbiol 69:S1–S111
Parks DH, Imelfort M, Skennerton CT et al (2015) CheckM: assessing the quality of microbial genomes recovered from isolates, single cells, and metagenomes. Genome Res 25:1043–1055
Parks DH, Rinke C, Chuvochina M et al (2017) Recovery of nearly 8,000 metagenome-assembled genomes substantially expands the tree of life. Nat Microbiol 2:1533–1542
Pedrós-Alió C (2012) The rare bacterial biosphere. Annu Rev Mar Sci 4:449–466
Pernice MC, Forn I, Gomes A et al (2015) Global abundance of planktonic heterotrophic protists in the deep ocean. ISME J 9:782–792
Pernthaler A, Pernthaler J, Schattenhofer M, Amann R (2002) Identification of DNA-synthesizing bacterial cells in coastal North Sea plankton. Appl Environ Microbiol 68:5728–5736
Pesant S, Not F, Picheral M et al (2015) Open science resources for the discovery and analysis of *Tara* oceans data. Sci Data 2:150023
Pham VD, Konstantinidis KT, Palden T, DeLong EF (2008) Phylogenetic analyses of ribosomal DNA-containing bacterioplankton genome fragments from a 4000 m vertical profile in the North Pacific subtropical gyre. Environ Microbiol 10:2313–2330
Pinhassi J, DeLong EF, Béjà O, González JM, Pedrós-Alió C (2016) Marine bacterial and archaeal ion-pumping rhodopsins: genetic diversity, physiology, and ecology. Microbiol Mol Biol Rev 80:929–954
Pommier T, Neal P, Gasol J et al (2010) Spatial patterns of bacterial richness and evenness in the NW Mediterranean Sea explored by pyrosequencing of the 16S rRNA. Aquat Microb Ecol 61: 221–233
Pushkarev A, Inoue K, Larom S et al (2018) A distinct abundant group of microbial rhodopsins discovered using functional metagenomics. Nature 558:595–599
Rappé MS, Connon SA, Vergin KL, Giovannoni SJ (2002) Cultivation of the ubiquitous SAR11 marine bacterioplankton clade. Nature 418:630–633
Rappé MS, Giovannoni SJ (2003) The uncultured microbial majority. Annu Rev Microbiol 57:369–394
Reichart NJ, Jay ZJ, Krukenberg V et al (2020) Activity-based cell sorting reveals responses of uncultured archaea and bacteria to substrate amendment. ISME J 14:2851–2861
Reinthaler T, van Aken HM, Herndl GJ (2010) Major contribution of autotrophy to microbial carbon cycling in the deep North Atlantic's interior. Deep Sea Res Part II Top Stud Oceanogr 57:1572–1580
Resplandy L, Lévy M, McGillicuddy DJ (2019) Effects of eddy-driven subduction on ocean biological carbon pump. Global Biogeochem Cycles 33:1071–1084
Rinke C, Schwientek P, Sczyrba A et al (2013) Insights into the phylogeny and coding potential of microbial dark matter. Nature 499:431–437

Robertson BR, Button DK (1989) Characterizing aquatic bacteria according to population, cell size, and apparent DNA content by flow cytometry. Cytometry 10:70–76
Rosenzweig RF, Sharp RR, Treves DS, Adams J (1994) Microbial evolution in a simple unstructured environment: genetic differentiation in *Escherichia coli*. Genetics 137:903–917
Roux S, Hawley AK, Torres Beltran M et al (2014) Ecology and evolution of viruses infecting uncultivated SUP05 bacteria as revealed by single-cell- and meta-genomics. elife 3:e03125
Royo-Llonch M, Sánchez P, González JM et al (2020) Ecological and functional capabilities of an uncultured *Kordia* sp. Syst Appl Microbiol 43:126045
Royo-Llonch M, Sánchez P, Ruiz-González C et al (2021) Compendium of 530 metagenome-assembled bacterial and archaeal genomes from the polar Arctic Ocean. Nat Microbiol. https://doi.org/10.1038/s41564-021-00979-9
Ruiz-González C, Logares R, Sebastián M et al (2019) Higher contribution of globally rare bacterial taxa reflects environmental transitions across the surface ocean. Mol Ecol 28:1930–1945
Ruiz-González C, Mestre M, Estrada M et al (2020) Major imprint of surface plankton on deep ocean prokaryotic structure and activity. Mol Ecol 29:1820–1838
Rusch DB, Halpern AL, Sutton G et al (2007) The Sorcerer II Global Ocean Sampling Expedition: Northwest Atlantic through eastern tropical Pacific. PLoS Biol 5:e77
Saito MA, Bertrand EM, Duffy ME et al (2019) Progress and challenges in ocean metaproteomics and proposed best practices for data sharing. J Proteome Res 18:1461–1476
Saito MA, McIlvin MR, Moran DM et al (2014) Multiple nutrient stresses at intersecting Pacific Ocean biomes detected by protein biomarkers. Science 345:1173–1177
Salazar G, Cornejo-Castillo FM, Benítez-Barrios V et al (2016) Global diversity and biogeography of deep-sea pelagic prokaryotes. ISME J 10:596–608
Salazar G, Cornejo-Castillo FM, Borrull E et al (2015) Particle-association lifestyle is a phylogenetically conserved trait in bathypelagic prokaryotes. Mol Ecol 24:5692–5706
Salazar G, Paoli L, Alberti A et al (2019) Gene expression changes and community turnover differentially shape the global ocean metatranscriptome. Cell 179:1068–1083
Salter I, Galand PE, Fagervold SK et al (2015) Seasonal dynamics of active SAR11 ecotypes in the oligotrophic Northwest Mediterranean Sea. ISME J 9:347–360
Samo TJ, Smriga S, Malfatti F, Sherwood BP, Azam F (2014) Broad distribution and high proportion of protein synthesis active marine bacteria revealed by click chemistry at the single cell level. Front Mar Sci 1:48
Sánchez-Andrea I, Guedes IA, Hornung B et al (2020) The reductive glycine pathway allows autotrophic growth of *Desulfovibrio desulfuricans*. Nat Commun 11:5090
Sanz-Sáez I, Salazar G, Sánchez P et al (2020) Diversity and distribution of marine heterotrophic bacteria from a large culture collection. BMC Microbiol 20:1–16
Saw JHW, Nunoura T, Hirai M et al (2020) Pangenomics analysis reveals diversification of enzyme families and niche specialization in globally abundant SAR202 bacteria. MBio 11:e02975–e02919
Scholin C, Birch J, Jensen S et al (2017) The quest to develop ecogenomic sensors: a 25-year history of the environmental sample processor (ESP) as a case study. Oceanography 30:100–113
Sebastián M, Estrany M, Ruiz-González C et al (2019) High growth potential of long-term starved deep ocean opportunistic heterotrophic bacteria. Front Microbiol 10:760
Sebastián M, Gasol JM (2019) Visualization is crucial for understanding microbial processes in the ocean. Philos Trans R Soc B Biol Sci 374:20190083
Sebastián M, Smith AF, González JM et al (2016) Lipid remodelling is a widespread strategy in marine heterotrophic bacteria upon phosphorus deficiency. ISME J 10:968–978
Servais P, Casamayor EO, Courties C et al (2003) Activity and diversity of bacterial cells with high and low nucleic acid content. Aq Microb Ecol 33:41–51
Shapiro BJ, Friedman J, Cordero OX et al (2012) Population genomics of early events in the ecological differentiation of bacteria. Science 336:48–51
Shapiro BJ, Polz MF (2015) Microbial Speciation. Cold Spring Harb Perspect Biol 7:a018143

Sharon I, Banfield JF (2013) Genomes from metagenomics. Science 342:1057–1058
Sheik CS, Jain S, Dick GJ (2014) Metabolic flexibility of enigmatic SAR324 revealed through metagenomics and metatranscriptomics. Environ Microbiol 16:304–317
Sherr EB, Sherr BF, Longnecker K (2006) Distribution of bacterial abundance and cell-specific nucleic acid content in the Northeast Pacific Ocean. Deep Sea Res Part I Oceanogr Res Pap 53: 713–725
Shi Y, Tyson GW, Eppley JM, DeLong EF (2011) Integrated metatranscriptomic and metagenomic analyses of stratified microbial assemblages in the open ocean. ISME J 5:999–1013
Singer E, Wagner M, Woyke T (2017) Capturing the genetic makeup of the active microbiome in situ. ISME J 11:1949–1963
Sintes E, Herndl GJ (2006) Quantifying substrate uptake by individual cells of marine bacterioplankton by catalyzed reporter deposition fluorescence in situ hybridization combined with microautoradiography. Appl Environ Microbiol 72:7022–7028
Sogin ML, Morrison HG, Huber JA et al (2006) Microbial diversity in the deep sea and the underexplored "rare biosphere". Proc Natl Acad Sci 103:12115–12120
Sosa OA, Repeta DJ, DeLong EF et al (2019) Phosphate-limited ocean regions select for bacterial populations enriched in the carbon–phosphorus lyase pathway for phosphonate degradation. Environ Microbiol 21:2402–2414
Sowell SM, Wilhelm LJ, Norbeck AD et al (2009) Transport functions dominate the SAR11 metaproteome at low-nutrient extremes in the Sargasso Sea. ISME J 3:93–105
Staley JT, Konopka A (1985) Measurement of in situ activities of nonphotosynthetic microorganisms in aquatic and terrestrial habitats. Ann Rev Microbiol 39:321–346
Steinberg DK, Carlson CA, Bates NR et al (2001) Overview of the US JGOFS Bermuda Atlantic time-series study (BATS): a decade-scale look at ocean biology and biogeochemistry. Deep Sea Res Part II Top Stud Oceanogr 48:1405–1447
Steindler L, Schwalbach MS, Smith DP et al (2011) Energy starved *Candidatus* Pelagibacter ubique substitutes light-mediated ATP production for endogenous carbon respiration. PLoS One 6: e19725
Stepanauskas R (2012) Single cell genomics: an individual look at microbes. Curr Opin Microbiol 15:613–620
Stepanauskas R, Fergusson EA, Brown J et al (2017) Improved genome recovery and integrated cell-size analyses of individual uncultured microbial cells and viral particles. Nat Commun 8:84
Stepanauskas R, Sieracki ME (2007) Matching phylogeny and metabolism in the uncultured marine bacteria, one cell at a time. Proc Natl Acad Sci 104:9052–9057
Steven B, Hesse C, Soghigian J, Dunbar J (2017) Simulated rRNA/DNA ratios show potential to misclassify active populations as dormant. Appl Environ Microbiol 83:AEM.00696–AEM.00617
Stomp M, Huisman J, de Jongh F et al (2004) Adaptive divergence in pigment composition promotes phytoplankton biodiversity. Nature 432:104–107
Stomp M, van Dijk MA, van Overzee HMJ et al (2008) The timescale of phenotypic plasticity and its impact on competition in fluctuating environments. Am Nat 172:E169–E185
Sul WJ, Oliver TA, Ducklow HW et al (2013) Marine bacteria exhibit a bipolar distribution. Proc Natl Acad Sci 110:2342–2347
Sunagawa S, Acinas SG, Bork P et al (2020) *Tara* Oceans: towards global ocean ecosystems biology. Nat Rev Microbiol 18:428–445
Sunagawa S, Coelho LP, Chaffron S et al (2015) Structure and function of the global ocean microbiome. Science 348:1261359
Swan BK, Martinez-Garcia M, Preston CM et al (2011) Potential for chemolithoautotrophy among ubiquitous bacteria lineages in the dark ocean. Science 333:1296–1300
Swan BK, Tupper B, Sczyrba A et al (2013) Prevalent genome streamlining and latitudinal divergence of planktonic bacteria in the surface ocean. Proc Natl Acad Sci 110:11463–11468
Tang K, Yang Y, Lin D et al (2016) Genomic, physiologic, and proteomic insights into metabolic versatility in *Roseobacter* clade bacteria isolated from deep-sea water. Sci Rep 6:35528

Taubert M, Grob C, Howat AM et al (2017) Methylamine as a nitrogen source for microorganisms from a coastal marine environment. Environ Microbiol 19:2246–2257

Tettelin H, Masignani V, Cieslewicz MJ et al (2005) Genome analysis of multiple pathogenic isolates of *Streptococcus agalactiae*: implications for the microbial "pan-genome". Proc Natl Acad Sci 102:13950–13955

Tettelin H, Riley D, Cattuto C, Medini D (2008) Comparative genomics: the bacterial pan-genome. Curr Opin Microbiol 11:472–477

Thompson AW, Foster RA, Krupke A et al (2012) Unicellular cyanobacterium symbiotic with a single-celled eukaryotic alga. Science 337:1546–1550

Thompson LR, Haroon MF, Shibl AA et al (2019) Red Sea SAR11 and *Prochlorococcus* single-cell genomes reflect globally distributed pangenomes. Appl Environ Microbiol 85:1–18

Thrash JC, Seitz KW, Baker BJ et al (2017) Metabolic roles of uncultivated bacterioplankton lineages in the northern Gulf of Mexico "dead zone". mBio 8:e01017–e01017

Thrash JC, Temperton B, Swan BK et al (2014) Single-cell enabled comparative genomics of a deep ocean SAR11 bathytype. ISME J 8:1440–1451

Tragin M, Vaulot D (2018) Green microalgae in marine coastal waters: the ocean sampling day (OSD) dataset. Sci Rep 8:14020

Tripp HJ (2013) The unique metabolism of SAR11 aquatic bacteria. J Microbiol 51:147–153

Tripp HJ, Kitner JB, Schwalbach MS et al (2008) SAR11 marine bacteria require exogenous reduced Sulphur for growth. Nature 452:741–744

Tsementzi D, Wu J, Deutsch S et al (2016) SAR11 bacteria linked to ocean anoxia and nitrogen loss. Nature 536:179–183

Tully BJ (2019) Metabolic diversity within the globally abundant marine group II Euryarchaea offers insight into ecological patterns. Nat Commun 10:271

Tully BJ, Graham ED, Heidelberg JF (2018) The reconstruction of 2,631 draft metagenome-assembled genomes from the global oceans. Sci Data 5:170203

Tully BJ, Sachdeva R, Graham ED, Heidelberg JF (2017) 290 metagenome-assembled genomes from the Mediterranean Sea: a resource for marine microbiology. PeerJ 5:e3558

Tyson GW, Chapman J, Hugenholtz P et al (2004) Community structure and metabolism through reconstruction of microbial genomes from the environment. Nature 428:37–43

Van Mooy BAS, Fredricks HF, Pedler BE et al (2009) Phytoplankton in the ocean use non-phosphorus lipids in response to phosphorus scarcity. Nature 458:69–72

VanInsberghe D, Arevalo P, Chien D, Polz MF (2020) How can microbial population genomics inform community ecology? Philos Trans R Soc B Biol Sci 375:20190253

Vanni C, Schechter MS, Acinas SG et al. (2020) Light into the darkness: unifying the known and unknown coding sequence space in microbiome analyses. bioRxiv 2020.06.30.180448

Venter JC, Remington K, Heidelberg JF et al (2004) Environmental genome shotgun sequencing of the Sargasso Sea. Science 304:66–74

Vergin KL, Beszteri B, Monier A et al (2013) High-resolution SAR11 ecotype dynamics at the Bermuda Atlantic time-series study site by phylogenetic placement of pyrosequences. ISME J 7: 1322–1332

Vila-Costa M, Rinta-Kanto JM, Sun S et al (2010) Transcriptomic analysis of a marine bacterial community enriched with dimethylsulfoniopropionate. ISME J 4:1410–1420

Vila-Costa M, Gasol JM, Sharma S, Moran MA (2012) Community analysis of high- and low-nucleic acid-containing bacteria in NW Mediterranean coastal waters using 16S rDNA pyrosequencing. Environ Microbiol 14:1390–1402

Villarino E, Watson JR, Jönsson B et al (2018) Large-Scale Ocean connectivity and planktonic body size. Nat Commun 9:142

Villarreal-Chiu JF, Quinn JP, McGrath JW (2012) The genes and enzymes of phosphonate metabolism by bacteria, and their distribution in the marine environment. Front Microbiol 3:19

Wang Y, Ji Y, Wharfe ES et al (2013) Raman activated cell ejection for isolation of single cells. Anal Chem 85:10697–10701

Whitaker RJ, Grogan DW, Taylor JW (2005) Recombination shapes the natural population structure of the hyperthermophilic archaeon *Sulfolobus islandicus*. Mol Biol Evol 22:2354–2361
Williams PJL (1970) Heterotrophic utilization of dissolved organic compounds in the sea I. Size distribution of population and relationship between respiration and incorporation of growth substrates. J Mar Biol Assoc United Kingdom 50:859–870
Williams PJL (1981) Microbial contribution to overall marine plankton metabolism: direct measurements of respiration. Oceanol Acta 4:359–364
Woyke T, Xie G, Copeland A et al (2009) Assembling the marine metagenome, one cell at a time. PLoS One 4:e5299
Wright JJ, Mewis K, Hanson NW et al (2014) Genomic properties of marine group a bacteria indicate a role in the marine sulfur cycle. ISME J 8:455–468
Wrighton KC, Thomas BC, Sharon I et al (2012) Fermentation, hydrogen, and sulfur metabolism in multiple uncultivated bacterial phyla. Science 337:1661–1665
Wuchter C, Abbas B, Coolen MJL et al (2006) Archaeal nitrification in the ocean. Proc Natl Acad Sci 103:12317–12322
Yakimov MM, La Cono V, Smedile F et al (2011) Contribution of crenarchaeal autotrophic ammonia oxidizers to the dark primary production in Tyrrhenian deep waters (Central Mediterranean Sea). ISME J 5:945–961
Yilmaz P, Yarza P, Rapp JZ, Glöckner FO (2016) Expanding the world of marine bacterial and archaeal clades. Front Microbiol 6:1524
Yoon HS, Price DC, Stepanauskas R et al (2011) Single-cell genomics reveals organismal interactions in uncultivated marine protists. Science 332:714–717
Yu FB, Blainey PC, Schulz F et al (2017) Microfluidic-based mini-metagenomics enables discovery of novel microbial lineages from complex environmental samples. elife 6:e26580
Zavaleta-Pastor M, Sohlenkamp C, Gao JL et al (2010) Sinorhizobium meliloti phospholipase C required for lipid remodeling during phosphorus limitation. Proc Natl Acad Sci 107:302–307
Zeng Y, Chen X, Madsen AM et al (2020) Potential rhodopsin-and bacteriochlorophyll-based dual phototrophy in a high Arctic glacier. MBio 11:e02641–e02620
Zhang Q, Zhang P, Gou H et al (2015) Towards high-throughput microfluidic Raman-activated cell sorting. Analyst 140:6163–6174
Zhang Y, Zhao Z, Dai M, Jiao N, Herndl GJ (2014) Drivers shaping the diversity and biogeography of total and active bacterial communities in the South China Sea. Mol Ecol 23:2260–2274
Zinger L, Amaral-Zettler LA, Fuhrman JA et al (2011) Global patterns of bacterial beta-diversity in seafloor and seawater ecosystems. PLoS One 6:e24570

# The Pelagic Light-Dependent Microbiome 9

Julie LaRoche and Brent M. Robicheau

**Abstract**

The pelagic sunlit zone or euphotic zone is defined as the lit surface of the ocean down to a depth where enough light penetrates to allow photosynthesis. Typically, the depth at which photosynthesis still exceeds respiration coincides with the depth at which the light intensity is attenuated to 1% of the sea surface irradiance. The composition of the light-dependent pelagic microbiome is both a function of light availability and spectral quality, as well as the strength of the water column stratification relative to turbulent mixing. Solar radiation provides the energy source for the pelagic light-dependent microbiome. However, additional environmental factors affecting the surface layer of the ocean make this environment spatially and temporally variable. Latitude has a combined effect on photoperiod, seasonal light levels, and temperature regimes. Overall, the microbiome inhabiting the euphotic zone in the ocean is adapted to steep gradients in light, temperature, salinity, and nutrient concentrations that can occur on local spatial scales. The three main categories of microorganisms that conserve energy from sunlight are the cyanobacteria and eukaryotic microalgae who carry out oxygenic photosynthesis, the aerobic anoxygenic photosynthetic bacteria (AAnPB) that perform photoheterotrophy, and proteorhodopsin-containing bacteria which synthesize ATP via a light-driven proton pump. Organisms within all three groups are taxonomically diverse and are adapted to cope with rapidly changing environmental conditions. Although the energy flow through the euphotic zone is controlled by primary producers that carry out carbon fixation via oxygenic photosynthesis, the microbiome in this layer of the ocean is additionally composed of heterotrophic, mixotrophic, and symbiotic

J. LaRoche (✉) · B. M. Robicheau
Department of Biology, Dalhousie University, NS, Halifax, Canada
e-mail: julie.laroche@dal.ca



L. J. Stal, M. S. Cretoiu (eds.), *The Marine Microbiome*, The Microbiomes of Humans, Animals, Plants, and the Environment 3,
https://doi.org/10.1007/978-3-030-90383-1_9

microorganisms who interact with primary producers either in a predatory role or as nutrient recyclers.

**Keywords**

AAnPB · Light-dependent microbiome · Pelagic environment · Photoheterotrophy · Phototrophic bacteria · Proteorhodopsin

## 9.1 Introduction

The pelagic zone of the ocean comprises the entire water column from the surface to the abyss. Within this larger zone, the region at the sea surface that receives sunlight is interchangeably called the epipelagic or euphotic (photic) zone. Within the epipelagic zone, several diverse groups of microbes have developed the ability to conserve energy from absorbed solar radiation. The three known microbial groups able to utilize solar radiation are the aerobic anoxygenic photosynthetic bacteria (AAnPB) that perform photoheterotrophy (Yurkov and Beatty 1998), the proteorhodopsin-containing microorganisms that synthesize ATP via a light-driven proton pump (Béja et al. 2001; Gómez-Consarnau et al. 2007), and the oxygen-evolving photosynthetic organisms (Falkowski et al. 2004). Light-driven oxygenic photosynthesis by cyanobacteria and the eukaryotic microalgae overwhelmingly accounts for the bulk of oceanic primary production with important consequences for the composition of the marine ecosystem of the epipelagic zone but also throughout the entire oceanic realm (Field et al. 1998). With a standing stock of 1 Gt C, oceanic primary producers support consumer biomass that is five times larger. That achievement is in contrast with terrestrial ecosystems, where the biomass is locked in trees and is only possible in the ocean because of the rapid growth rate and turnover of the microbially-dominated primary production relative to a much slower turnover of the consumer biomass (Bar-On et al. 2018). The epipelagic microbiome is also populated by heterotrophic bacteria, archaea, and eukaryotes (protists) that are metabolically diverse and adapted to thrive by consuming the primary production synthesized in the epipelagic zone (Field et al. 1998).

Sunlight radiation is rapidly absorbed in the epipelagic zone resulting in a change in the spectral quality and a rapid attenuation with depth (Holtrop et al. 2020). The depth of light penetration and the spectral changes play an important role in determining the composition of the microbiome in the epipelagic zone. The light environment within the epipelagic zone is dependent on several additional factors such as turbidity, the concentration of colored dissolved organic matter (cDOM), and the concentration of phytoplankton (Kirk 1994). The geographic location within the global ocean will greatly affect the light intensity reaching the surface of the ocean as well as the daily photoperiod and seasonal cycle. Finally, phototrophic primary production, which is central to the microbiome occupying the epipelagic zone, is controlled by the light regimes within this zone as well as by a complex interplay between the opposing forces of stratification and turbulence within the upper water column (Franks 2015). The epipelagic zone is also unique because it covers wide gradients of temperature, salinity, and nutrient conditions both spatially and temporally (Longhurst 1995), making it difficult to assign a specific microbiome composition or predict the dominant species for the microbiome. However, general adaptations and acclimation mechanisms to the variable light and nutrient conditions experienced in the euphotic zone are key to survival in this environment.

Although the microbiome of the epipelagic zone is highly diverse in various regions of the ocean (Righetti et al. 2019), the dominant photoautotrophs present within a given oceanic region will determine the primary productivity. This serves as the key currency fueling higher trophic levels and determines the sequestration of organic carbon to the deeper layers of the ocean through the biological carbon pump (BCP) and the microbial carbon pump (MCP) (Jiao et al. 2010; Polimene et al. 2017). In this chapter, the quantitative and spectral aspects of light absorption in the water column will be reviewed, emphasizing how the light properties affect the selection of phototrophic microbes. The availability of inorganic macro- and micronutrients, as well as global patterns of nutrient limitation, play an equally important role in controlling photoautotrophic growth (Moore et al. 2013), which ultimately affects the diversity and function of the regional microbiome. The general overview of light as a variable in the photic zone is then followed by a discussion of how phytoplankton and their associated network of microorganisms are affected by the light conditions and the stability of the stratification within the surface water column. The balance between nutrient and light availability is an important driver of microbial adaptation in the euphotic zone. Finally, specific examples from selected regions of the global ocean demonstrate how spatial and temporal gradients affect the composition of epipelagic microbiomes.

## 9.2 Sunlight as the Dominant Source of Energy in the Epipelagic Zone

Sunlight drives many processes in the euphotic zone of the pelagic realm. Foremost, it warms up the surface of the ocean and supplies photosynthetically active radiation (PAR). Ranging in wavelength between 400 to 700 nm, PAR drives photosynthetic activity in photoautotrophs, including marine phytoplankton. At the surface of the earth, absorption of PAR by pure water is highest at the long (red) wavelengths and decreases towards the short (violet/blue) wavelengths reaching a minimum in the ultraviolet (UV) range at 344 nm (Holtrop et al. 2020). In any water body, including the marine environment, light decays exponentially with depth (Kirk 1994). The depth of the euphotic zone (Ez), which is relevant to photoautotrophic primary production, commonly corresponds to the 1% light level ($Ez_{1.0}$) and furthermore represents the light level at which photosynthetic organic matter production is no longer significant (Falkowski 1994). Other metrics to determine the depth of the euphotic zone have relied on defining a primary production zone (PPZ) based on the depth where chlorophyll fluorescence corresponds to 10% of the fluorescence measured at the surface (Owens et al. 2015). The PPZ depth generally coincides with the 0.1% light level ($Ez_{0.1}$) and with the compensation depth at which gross primary production is balanced by respiration (Marra et al. 2014). In general, the $Ez_{0.1}$ is up to tens of meters deeper than the $Ez_{1.0}$ (Buesseler et al. 2020). Here, an inclusive view of the euphotic zone down to the depth of 0.1% of PAR ($Ez_{0.1}$) is taken for the purpose of defining the light-dependent microbiome, given that the $Ez_{0.1}$ would also encompass photoautotrophs that are located deeper than the deep chlorophyll maximum (DCM). It is also noteworthy that blue light penetrates deeper than other wavelengths in the water column reaching the 1% level at a depth that is generally the 0.1% light level for the longer wavelengths within the PAR (Marra et al. 2014). The depth of the euphotic zone is spatially and temporally variable depending on latitude, season, and short-lived meteorological events (Buesseler et al. 2020). As noted by Buesseler et al. (2020) the absolute photon flux received at the sea surface and propagated at depth rather than the PAR ultimately controls photosynthesis and photoautotrophic primary production and defines the depth of the euphotic zone.

The spectral quality of incoming sunlight is altered as it penetrates the surface layer of the ocean, with the physical properties of pure water, the non-algal particles (e.g., minerals and detritus), cDOM and non-photosynthetic matter, as well as pigment-containing microbes all contributing to the observed underwater irradiance spectra (Letelier et al. 2017). Depth profiles of light spectra have shown that the exponential attenuation of light with depth, coupled with the physical properties of cDOM in water, produced distinct spectral niches that are exploited by phototrophic microbes such as photosynthetic bacteria and phytoplankton (Holtrop et al. 2020; Stomp et al. 2007). Holtrop et al. (2020) refined the concept of spectral niches by providing a theoretical framework based on a radiative transfer model that takes into consideration both absorption and scattering. This new model refined the predictions that five light niches in the violet, blue, green, orange, and red parts of the spectrum are captured by the known extent pigments found in photosynthetic microbes.

## 9.3 UV Radiation (UVR) in the Euphotic Zone and its Effect on the Microbiome

### 9.3.1 UVR in the Atmosphere

Total solar irradiance is more or less constant at the surface of the earth, with variations of ~0.1%. In contrast, variations in UVR within the range of 100–400 nm are associated with events of sunspots and faculae and can be ~1% and ~ 10% at the wavelengths 300 nm and 200 nm, respectively (Dudok de Wit et al. 2017). UVR is categorized into UV-C (100–280 nm), UV-B (280–315 nm), and UV-A (315–400 nm) (Bais et al. 2015). Several factors such as zenith angle, atmospheric ozone, clouds, aerosol, surface albedo, and height above sea level contribute to the variability in UVR before reaching the surface of the ocean. Depending on the atmospheric conditions, UV-C radiation does not reach the surface of the earth at present because it is absorbed in the atmosphere by molecular oxygen and stratospheric ozone, while UV-B is only partially absorbed. In clear skies, the ratio of UV-B and UV-A against PAR is highest with a small solar zenith angle (smallest solar zenith angle is at mid-day) while UV-B/PAR and UV-A/PAR decrease by 97% and 20%, respectively, as the solar zenith angle increases. As a result, UV-B radiation displays considerably more diurnal and latitudinal variation compared to UV-A (Bais et al. 2019).

### 9.3.2 Factors Affecting UVR Absorption in Seawater

Radiation is attenuated by both dissolved and particulate organic matter (POM) and also by inorganic matter such as silt and sand (more important in coastal areas). The penetration of UVR in the euphotic zone depends on the particle load and the concentration and composition of cDOM (Zepp et al. 2007). Humic acid and other colored or chromophoric DOM (cDOM) preferentially absorb energetic short wavelengths. Thus, UVR penetrates to a shallow depth in rivers and lakes but deeper in the ocean. Because cDOM absorbs shorter wavelengths more strongly than longer wavelengths, UV-B will be more attenuated than UV-A and PAR (Tedetti and Sempéré 2006). Hence, cDOM plays a role in seawater similar to that of ozone in the atmosphere. UVR will penetrate to a greater depth in clear oceanic waters where cDOM is low and will penetrate to a shallower depth in coastal areas where cDOM is high. POM such as bacteria, phytoplankton, and organic debris also affect and decrease the transparency of seawater. Decaying plankton is the source of DOM in the open ocean; however, UVR absorption also has an effect on the photodegradation of DOM, making it more available to the heterotrophic bacteria (Zepp et al. 2007). At high latitudes, springtime ozone depletion can lead to a seasonal increase in UVR. This phenomenon is well documented for the Southern Ocean but may also be important in the Arctic Ocean (Tedetti and Sempéré 2006). Tidal rhythm and change in transparency will also affect UV light. Furthermore, the ranges of UV-B and UV-A penetrations in the euphotic zone vary widely geographically. For example, the range of UV-B penetration is 6–15 m while that of UV-A is 8–46 m in oligotrophic areas, such as the Mediterranean Sea, contrasting with coastal waters where penetration of UV-B and UV-A range between 0.1–5 m and 0.33–11.5 m, respectively (Smyth 2011).

### 9.3.3 Global Distribution of UVR in the Ocean

Although incoming UVR is currently highest between 20°N and 30°S UVR is expected to increase in the Arctic because of melting sea ice and the ensuing decrease in albedo. For example, predictions are that at the present rate of change in albedo UV-B radiation could increase by ten-fold by the end of the twenty-first century (Bais et al. 2015). In the oligotrophic open ocean waters, UV-A and UV-B can penetrate to depths of 38 and 17 m, respectively (Tedetti and Sempéré 2006). The center of the South Pacific Gyre (SPG) shows some of the lowest chlorophyll concentrations, with a deep chlorophyll maximum situated at the 185–190 m depth (Tedetti et al. 2007). Tedetti et al. (2007) reported 10% irradiance depths (Z10%) of 28 m and 110 m for UV-B and UV-A, respectively, at an ultra-oligotrophic station in the SPG. Similarly, the Red Sea and the Indian S. Subtropical Gyre Province (ISSG) are among the clearest waters in the world, and there UVR is likely to influence the composition of the microbiome and the development of adaptation strategies to combat the detrimental effects of UVR (Overmans and Agustí 2019; Iuculano et al. 2019). In contrast, upwelling regions in the Pacific have the highest levels of cDOM (N. Pacific Equatorial Countercurrent Province). Although cDOM plays an important role in determining the attenuation of UVR in the marine environment, other factors such as vertical mixing, length of exposure to UVR, and the presence of DNA repair mechanisms or UV-absorbing molecules altogether will affect the impact on marine microbial communities. Overall, the regions of the ocean receiving most of the UV radiation are the oligotrophic gyres because there are many factors that limit productivity in those regions. The main factor is the limited nutrient flux reaching the ocean surface waters because of the strong stratification of the water column of the oligotrophic gyres (Wang et al. 2021). However, the incoming UV radiation in those regions is constant and predictable and therefore, it is expected that the microbes inhabiting the surface of the gyres may have developed adaptations that protect them from UV-B. In other words, the system of incoming radiation and damage-repair cycle may be at steady-state. Therefore, the biology of the oligotrophic central gyres may not be the most affected by UV damage. In contrast, Antarctic waters are subjected to a variable level of ozone (i.e., ozone measured historically in this region) created by the ozone hole, and the damage to the microbial community may be more difficult to avoid in a fluctuating environment (Buma et al. 2001; Rijkenberg et al. 2005).

### 9.3.4 Detrimental Effects on the Microbiome and Adaptations to UVR

Harmful UVR radiation penetrating the euphotic zone has detrimental effects on the microbiome. Several known biological effects of radiation are found in the UVR region, particularly regarding the potential damage afflicted to proteins and DNA. Both UV-A and UV-B can cause damage to DNA (reviewed by Sinha and Häder 2002). However, UV-A damage is caused indirectly by radicals, while the damage from UV-B is direct through absorption (Castenholz and Garcia-Pichel 2013). UV-B radiation with its shorter and more energetic wavelength range is directly absorbed by cellular biomolecules that have absorption maxima in the 280–315 nm range, including mainly DNA and proteins that are damaged when exposed to these short-wavelength radiations (Buma et al. 1996). Also, UV-B does not require the presence of oxygen to damage the biomolecules. For example, D1, one of two major proteins of the photosystem II reaction center, is degraded by UV-B (Bouchard et al. 2005; Campbell et al. 1998). There are also indications that UV-B is detrimental to other enzymes involved in redox reactions and those containing iron atoms (Zhu et al.

2020). In photosynthetic organisms, UV-A affects biomolecules by creating reactive oxygen species (ROS) that are damaging also for other targets in the cell. For instance, in cyanobacteria, UVR was detrimental to processes such as tetrapyrrole synthesis, nutrient uptake, motility, energy transfer in light harvesting, nitrogen- and carbon fixation by nitrogenase and rubisco, respectively (Gao and Garcia-Pichel 2011). Adaptations that cope with UVR effects are DNA repair, de novo synthesis of proteins less sensitive to UVR, the detoxification of ROS, the synthesis of antioxidants such as carotenoid, vitamin C, tocopherol, and reduced glutathione, as well as enzymes such as superoxide dismutase and catalase (Gao and Garcia-Pichel 2011). Other mechanisms developed by microbes that help to protect cells include, for instance, escape behavior such as vertical migration and formation of colonies (Bebout and Garcia-Pichel 1995). Another common strategy is the production of UV-absorbing compounds to act as sunscreens and release the absorbed high energy as heat (Gao and Garcia-Pichel 2011). Sunscreens, however, are not the best strategy for microorganisms, especially not for bacteria and archaea because their cells are too small to pack these compounds intracellularly. In contrast, the presence of mycosporine-like amino acid (MAA) metabolites (one of these microbial sunscreens) is more widely observed in larger unicellular microorganisms that synthesize these compounds as an alternative approach to repair (Gao and Garcia-Pichel 2011). Some larger phytoplankton like *Trichodesmium* possesses MAA that absorb at UV-B and UV-A wavelengths (330 nm, 360 nm) as an adaptation for living in environments that receive high UVR doses (Dupouy et al. 2018). This is especially important for the occasionally blooming *Trichodesmium* species, which has gas vesicles to keep it at the air-sea interface where it can directly capture Fe-rich Saharan dust grains that are important to decrease UVR exposure in addition to supplying cellular iron (Langlois et al. 2012). Further, it was demonstrated that *Trichodesmium* is polyploid (Sargent et al. 2016), providing yet another strategy for this species to counteract DNA damage by having backup gene copies. As the examples with *Trichodesmium* demonstrate, microbes may have a wide array of adaptive solutions to counteract the detrimental effects of UVR at the ocean surface that can include colony and aggregate formations, genome polyploidy, and the harvesting of dust particles. One should note, though, that these strategies are not limited to *Trichodesmium*. Many other large phytoplankton are polyploid and, therefore likely also benefit from having large numbers of chromosomes that can decrease the effects of DNA damage that could be detrimental to essential genes (Watanabe 2020).

DNA's peak absorbance at 260 nm makes it a prime target for damage by UV-B radiation, and consequently, DNA repair mechanisms have developed as an adaptation to UVR. Several types of DNA repairs have been reviewed by Núñez-Pons et al. (2018). Most of the commonly identified DNA repair mechanisms are ancient in origin and depend on photoreactivation to remove DNA lesions through the action of a photolyase, an enzyme found in bacteria and archaea and in some eukaryotes (Lucas-Lledo and Lynch 2009). Photolyases have been identified in Antarctic marine bacteria and diatoms (Coesel et al. 2009; Núñez-Pons et al. 2018). Other types of defenses include de novo synthesis of damaged proteins, such as in the case of photosystem II repair. For instance, the de novo synthesis of the D1 protein is an important response to UVR damage for Antarctic phytoplankton communities (Bouchard et al. 2005). Additionally, heat shock proteins are also involved in the stress response to UVR. Finally, several reports have shown that hard cell coverings such as those found in diatoms and coccolithophorids protect against UVR (Aguirre et al. 2018; De Tommasi et al. 2018). Xu et al. (2016) observed that calcified strains of *Emiliania huxleyi* were more resistant to UVR than naked strains from the same species; hence hard cell coverings may further protect against the negative effects of UVR.

### 9.3.5 The Overall Effects of UV-B on Net Community Production (NCP) in the Upper Global Ocean

Climate change will likely increase the exposure time of phytoplankton and other plankton to UVR because of the shoaling of the upper mixed layer (Gao et al. 2012; Nonoyama et al. 2019). The surface layer of the subtropical and tropical ocean is subjected to warming resulting in stratification and increased penetration of UV-B. Some studies reported that net primary production (NPP) in the subtropical and tropical ocean is negatively affected by the interaction of UV-B and temperature, which act in opposite directions (Garcia-Corral et al. 2017a). Results of global ocean surveys indicate that ocean warming may lead to a shift in the metabolic balance between phototrophic and heterotrophic microorganisms with a shift towards smaller phytoplankton (picophytoplankton) (Bolaños et al. 2020) exacerbated by the increased penetration of UV-B due to stronger stratification of the euphotic zone. Experimental work has also shown that UV-B affects net community production differently between phototrophs and heterotrophs with a larger effect on the former (Garcia-Corral et al. 2017b).

## 9.4 Macro and Micronutrient Limitation in the Euphotic Zone

### 9.4.1 Nutrient Limitation

Microbes, like all other living cells, require the essential elements carbon, nitrogen, and phosphorus for macromolecule synthesis and growth. It has long been recognized that together with sunlight the supply of macro- and micronutrients is essential for phytoplankton growth and thus for sustaining the marine microbiome of the euphotic zone (Geider and La Roche 2002; Redfield 1958). Nitrate, nitrite, and ammonia are the forms of dissolved fixed inorganic nitrogen (DIN) preferred for growth by phytoplankton and many heterotrophic bacteria and archaea. Large regions of the oceanic euphotic zones are permanently stratified and depleted in one or more essential nutrients such as DIN and phosphate, leading to nutrient limitation of phytoplankton primary production. The concept of nutrient limitation and limiting nutrient refers to the nutrient that is present in the smallest amount in an environment (Moore et al. 2013). Nutrient co-limitation is invoked under many different circumstances, for example, when two essential elements have been drawn down simultaneously. This is often the case for DIN and phosphate because those two macronutrients are drawn down in a ratio close to the Redfield ratio on the surface of the ocean (Geider and La Roche 2002; Redfield 1958). Although the debate is still open on whether nitrogen or phosphorus is the ultimate limiting nutrient in the ocean (Tyrrell 1999) it is clear that both phosphate and fixed nitrogen are present in very low to limiting concentrations in large regions of the oceans and are at times considered to co-limit phytoplankton production (Mills et al. 2008; Moore et al. 2008). Several micronutrients have vertical profiles corresponding to biological nutrients in addition to the classical vertical profiles for nitrate, phosphate, and silicate (Morel et al. 2020). The trace metals Co, Cd, and Zn all show depletion in the surface lit pelagic zone indicating their incorporation into biological molecules (also see Sunda (2012) for a review of trace metals with respect to phytoplankton). Unlike nitrate, however, micronutrients display patterns of surface depletion that are more geographically confined, indicating regional or seasonal impact (Morel et al. 2020). The elemental stoichiometry of microbes and phytoplankton in particular varies according to their taxonomic affiliation and reflects their niche preferences (Liefer et al. 2019; Quigg et al. 2011). Several micronutrients can also substitute for each other either within the same protein or macromolecule or by completely exchanging them. Alternatively, co-limitation has also been described as a co-dependence of nutrients, as might be the case between a macro- and a micronutrient (Morel et al. 2020; Saito et al. 2008). Numerous examples are available demonstrating the dependence of the assimilation of a macronutrient on a particular enzyme that is itself dependent on a trace element for activity (Morel et al. 2020; Sunda and Huntsman 1995). An example of these substitutions is the replacement of Zinc by Cobalt or Cadmium in the carbonic anhydrase enzyme involved in the acquisition of $CO_2$ in phototrophs.

### 9.4.2 Nitrogen Limitation

In large regions of the surface ocean, primary production is limited by the availability of DIN such as nitrate, nitrite, and ammonium. DIN is mostly undetectable in the central gyres of large ocean basins, followed by phosphorus which is often a secondary limiting nutrient as was verified by shipboard-based nutrient bioassay experiments (Moore et al. 2013). It has been postulated that in an N-depleted environment, one adaptive mechanism to decrease the nitrogen demand would be to minimize gene length to save on cellular N (Dlugosch et al. 2021). Such a mechanism may also affect the transcription of genes in *Prochlorococcus* (Read et al. 2017). A more extensive analysis of marine microbial genomes by Shenhav and Zeevi (2020) has revealed that nitrate played a key role as an environmental factor in determining the purifying selection on genomes resulting in the conservation of N-resources according to the principle of resource driven-selection. This, of course, would be highly relevant to the euphotic zone of the pelagic environment, where nitrogen limitation covers large areas of the surface ocean (Shenhav and Zeevi 2020).

### 9.4.3 Phosphorus Limitation

In the euphotic zone, nitrogen and phosphorus are usually drawn down by the microbial community in a stochiometric N:P ratio that is close to 16:1 (Moreno and Martiny 2018). Phosphorus limitation of primary production has been observed in the North Atlantic (Moore et al. 2013). Indications that phosphorus in the euphotic zone is in low supply generally and particularly in the North Atlantic is also supported by the substitution of sulfolipids and other non-phosphorus lipids in picocyanobacterial and other phytoplankton (Van Mooy et al. 2009). Similarly, the suggestion that phosphorus is a sparse resource in the lit ocean surface is supported by the widely distributed presence of a high-affinity transporter for phosphate (Fiore et al. 2020) and the use of glycerophosphate and phosphonate by *Trichodesmium* in regions of low phosphorus (Dyhrman et al. 2006).

### 9.4.4 Silica Limitation

Silica can be a limiting factor for diatoms. In polar regions where the spring bloom is dominated by diatoms, silica may become limiting after a long bloom period. For instance, silica limitation in the Arctic Ocean was identified as a factor responsible for the demise of the diatom bloom (Krause et al. 2019). However, re-analysis of global datasets showed that Pacific diatom blooms were less likely to become Si limited than those in the Sargasso Sea in the Atlantic Ocean. This is due to the fact that silica cycling in the Pacific and Atlantic oceans is different and shows different seasonal patterns, although both occurred in oligotrophic areas (Kemp and Villareal 2018).

### 9.4.5 Iron Limitation

Iron is an important cofactor in many essential enzymes and proteins involved in energy acquisition both in photosynthetic and heterotrophic microbes. Despite the high abundance of iron, its solubility at the pH of seawater is low, rendering this essential trace element biologically difficult to obtain in the euphotic zone (Jickells 2005). The large regions of the ocean considered iron limited have been coined high nitrate low chlorophyll (HNLC) because in these regions, nitrate is not drawn down by phytoplankton growth that remains low (Strzepek et al. 2019). HNLC regions where chronic iron limitation exists are found in the Subarctic Pacific, the Eastern Pacific equatorial upwelling, and the Southern Ocean (Boyd et al. 2007). The seasonal iron limitation at the tail end of the spring phytoplankton bloom has been confirmed as well in the Eastern North Atlantic Ocean (Moore et al. 2013). In photoautotrophs, the large cellular Fe quota is tied to the photosynthetic apparatus. Consequently, eukaryotic phytoplankton have adapted to tolerate chronically low iron concentrations in HNLC areas and other regions of the ocean. Such adaptations include replacement of ferredoxin for flavodoxin (Cohen et al. 2021; LaRoche et al. 1996), change in the stoichiometry of components in the photosynthetic apparatus (Strzepek and Harrison 2004), storage of iron in ferritin (Lampe et al. 2018), and the presence of iron stress-induced proteins (ISIP) as a novel iron acquisition system found in marine phytoplankton (Allen et al. 2008; McQuaid et al. 2018) and particularly diatoms (Behnke and LaRoche 2020; Caputi et al. 2019). Within eukaryotic phytoplankton species, the diatom *Pseudo-nitzschia* has been revealed as an indicator species for microbial communities inhabiting surface waters subjected to chronically low iron availability due to its ability to maintain cellular iron quotas that are 10 times higher than other centric diatoms in HNLC regions through the luxury accumulation of iron and enhanced storage capacity through ferritin (Twining et al. 2020). Similarly, *Prochlorococcus* HLIII and IV ecotypes and *Synechococcus* clade III are dominant in HNLC regions (Caputi et al. 2019; Marchetti 2019). Diazotrophs in general and cyanobacterial diazotrophs, in particular, have high iron demands because of the requirements for the nitrogenase enzyme that is rich in iron. For example, *Trichodesmium,* in particular, is often seen in areas of high Saharan dust deposition (Ratten et al. 2015) and is able to directly concentrate dust particles to the center of colonies to directly dissolve Fe (Kessler et al. 2020; Langlois et al. 2012).

Physiological and transcriptional studies with field samples have shown that heterotrophic bacteria are also affected by Fe-limitation, with coastal bacteria decreasing their growth rate more than oceanic bacteria. The acclimation to iron limitation is achieved via changes in carbon metabolism. In heterotrophic microbes, most redox enzymes and proteins involved in energy conservation also contain iron cofactors. However, the iron requirements of heterotrophic bacteria and archaea are less well known. The bacterium with the lowest known iron cellular quota is *Alteromonas* (Blain and Tagliabue 2016; Obernosterer et al. 2015). Although *Alteromonas* is a copiotroph with a wide range of adaptation to nutrient limitation in general, this bacterium expresses different genes when subjected to either iron or carbon limitation. For instance, under iron limitation, *Alteromonas* will express specific Ton-B dependent transporters (TBDT) and siderophore genes (Manck et al. 2020). The glyoxylate shunt has also repeatedly been identified as an adaptive response to oxidative stress, including iron limitation. For instance, Pelagibacteriaceae (SAR11 clade) are among the bacterial groups that are abundant in the iron-limited HNLC regions. In response to iron limitation, the glyoxylate shunt is triggered in *Cand.* Pelagibacter ubique (Beier et al. 2015; Koedooder et al. 2018). The glyoxylate pathway is a two-step pathway present in aerobic bacteria and is an alternative route within the TCA cycle. The glyoxylate cycle, in contrast to the TCA cycle, leads to the recycling of the carbon molecules for biosynthesis and avoids the loss of carbon atoms when iron is scarce. Although the genome of *Pelagibacter* is small and streamlined, it contains all genes necessary for the glyoxylate cycle. It has been postulated that the glyoxylate shunt and the presence of proteorhodopsin work in concert as adaptations to iron limitation (Beier et al. 2015; Obernosterer et al. 2015). In heterotrophic bacteria, the components of the main biochemical pathways affected by iron limitation are the complexes I-IV of the electron transport chain, where iron atoms are cofactors in the various redox enzymes involved in ATP synthesis. During conditions of iron limitation, the glyoxylate pathway functions in preventing the release of $CO_2$ and instead diverts the carbon molecules to fuel anabolic pathways such as fatty acid and oxaloacetate synthesis (Koedooder et al. 2018). There is also a reorganization of the metabolism to optimize oxygen respiration (Koedooder et al. 2018, 2020). *Pelagibacter* does not appear to have direct adaptation to Fe-limitation as other heterotrophic bacteria have developed, nor does it have the ability to synthesize and use high-affinity siderophore as uptake systems for $Fe^{2+}$, as is the case in some other bacterial species. Instead, studies in the Southern Ocean have shown that *Pelagibacter* has a full complex of the glyoxylate shunt pathway and can use small two-carbon molecules that induce the glyoxylate shunt pathway (Debeljak et al. 2019). In contrast, Flavobacteriaceae display competitive Fe uptake systems containing bacterioferritin. Overall, *Pelagibacter* has an advantage over other groups when both Fe and carbon are limited (Debeljak et al. 2019), while they can also trigger the glyoxylate shunt pathway.

Saharan dust deposition is an important source of iron in the North Atlantic, and it has been linked to blooms of *Trichodesmium* and of other marine bacteria. For example, marine *Vibrio*, a ubiquitous genus in the ocean with a well-developed system for iron uptake, is usually present in marine microbial communities, even though in small numbers. Although *Vibrio* is typically rare in the marine environment, dust events, especially in the tropical Atlantic, can cause blooms of this genus. Blooms of *Vibrio* were stimulated by the natural occurrence of Saharan dust plumes and also by experimental manipulative experiments with dust addition (Westrich et al. 2016, 2018). The biomass of *Vibrio* is much larger than the biomass per cell of *Pelagibacter,* and this suggests that it can contribute to a much larger extent to the carbon cycle. Furthermore, *Vibrio* cells with their larger size are subject to preferential grazing and viral lysis when in bloom—two pathways that release DOM into the environment. Proteorhodopsin (PR) is an important adaptation for survival in the lit pelagic zone where dissolved organic carbon and Fe are limiting. This is because iron limitation in bacteria interferes with the respiratory electron transport chain, and therefore the light-sensitive proton pump may play an important role in compensating for the lower respiratory electron transport. However, studies with Vibrionaceae showed that PR is not directly associated with Fe-limitation but is a more general starvation-stress response that may play an important role in survival during the stationary phase of growth (Koedooder et al. 2020). Overall, it is to be expected that nutrient limitation of various kinds will lead to the selection of specific types of microbes, and this is especially true for the pelagic euphotic microbiome.

## 9.5 Subsurface Chlorophyll Maximum Layer (SCML) and Subsurface Biomass Maximum Layer (SBML)

The subsurface chlorophyll maximum layer (SCML) or deep chlorophyll maximum (DCM) is the result of several environmental conditions that favor the accumulation of phytoplankton at depth. The presence of the DCM is widely observed in the ocean. However, it is important to distinguish DCM from subsurface biomass maximum layer (SBML) because the two are not always correspondingly observed at the same depth, and they are not formed by exactly the same processes. A comprehensive critical review on the subject of DCM formation and persistence of this oceanographic feature (Cullen 2015) has identified the following three key processes that lead to the accumulation of chlorophyll concentration at depth. First, the maximum growth rate of phytoplankton is highest near the nutricline because the shallower depths at the ocean surface within a stable water column are usually depleted or low in essential nutrients (Cullen 2015). Second, although there is enough light for photosynthesis to occur throughout the euphotic zone (particularly above $Ez_{0.1}$), photon capture is optimized by the process of photo-acclimation, resulting in an increase in light-harvesting antenna and their associated pigments in low light conditions experienced at the bottom of the euphotic zone (Falkowski and LaRoche 1991). Third, additional physiological adaptation to modify buoyancy or vertical migration allows certain groups to migrate to the bottom of the euphotic zone (Wirtz and Smith 2020). The DCM is widely distributed with a scale ranging between 1–10 m in thickness and can range horizontally as sub-mesoscale features to ocean basins or gyres in oligotrophic areas. The DCM is maintained by the balance between sunlight penetration that is sufficient to maintain photosynthesis and the supply of nutrients from the deep water below the nutricline. However, the characteristics of the DCM vary depending on the trophic level. In eutrophic coastal waters, phytoplankton accumulates at the nutricline primarily via sinking, and there the SBML and the DCM are often coincident. In oligotrophic systems, the DCM and the SBML are usually decoupled because the DCM is often the product of an adaptation or acclimation to low light manifested by an enhanced cellular chlorophyll concentration. In addition, the DCM in the oligotrophic gyres is a semi-permanent feature that forms just above the pycnocline, and that persists throughout the year.

The DCM is increasing in importance in the Arctic Ocean during the phytoplankton growing season. For instance, increasing sea ice retreat in the Arctic Ocean and a longer open-water season are resulting in nutrient depletion in the surface layers of the euphotic zone and the seasonal formation of a DCM at a depth where the nutrient flux from deep water sustains the primary productivity of phytoplankton (Zhuang et al. 2020). In the Arctic Ocean, the SCML and SBML are usually coincident, as is often the case also for coastal areas where a seasonal DCM can form. Similarly, the formation of the DCM in the Southern Ocean is an annually recurrent feature, however, its formation and maintenance are driven by factors other than the co-limitation by light and nutrients, as is the case elsewhere. In particular, factors such as sea ice retreat, subduction events, eddy formation and propagation have been suggested as controlling factors for the formation of the DCM in the Southern Ocean (Baldry et al. 2020). The large diatoms form sinking aggregates during the development of severe iron limitation that occurs in the seasonal mixed layer but re-establish their buoyancy at the base of the euphotic zone just above the nutrient-rich deeper waters (Baldry et al. 2020).

## 9.6 Mixotrophy in the Euphotic Zone

### 9.6.1 Defining Mixotrophy

In the euphotic zone, mixotrophy is a strategy used by microbes to expand their trophic range by exploiting light energy to alleviate carbon limitation, thereby gaining a competitive edge over the specialized heterotrophic microbes that cannot utilize light. Conversely, mixotrophy in the form of predation, phagotrophy, or osmotrophy can be employed by microbes to alleviate light or nutrient requirements encountered by obligate photoautotrophs in low light or in nutrient-limited waters, respectively (Edwards 2019).

Mixotrophs can be classified into constitutive mixotrophs (mixotrophic nanoflagellates such as *Prymnesium parvum*), and at the other end of the scale, the photosynthetic zooplankton or heterotrophs that acquire chloroplast through kleptoplastidy or retaining intact phytoplankton for several days before digesting their prey (Flynn and Mitra 2009). Mixotrophy is also distributed among photosynthetic eukaryotes that can employ a mode of phagotrophic predation to supplement photosynthesis. Mixotrophs are widely distributed from high to low latitude and from coastal to open oceans, and mixotrophy can be thought of as a strategy of a trophic generalist (Edwards 2019). Mixotrophy has also been associated with oligotrophic environments as a way to supplement nutrients. However, Edwards (2019) has argued that the increased light availability in the highly stratified oligotrophic water column may be more important in the adaptation to mixotrophy than the low nutrients. In the context of the euphotic zone, the main types of mixotrophic organisms are those able to utilize light as well as organic nutrients.

### 9.6.2 Mixotrophy in Bacteria and Archaea

Mixotrophy is widespread in bacteria and archaea and primarily involves photoheterotrophy (Olson et al. 2018). In photosynthetic cyanobacteria such as *Prochlorococcus* and *Synechococcus,* mixotrophy can supplement both the nutrient requirements and energy acquisition (Muñoz-Marín et al. 2020). Examination of the Tara Oceans metagenome collection indicated that universal mixotrophy in picocyanobacteria was more pronounced in deep water (Yelton et al. 2016), possibly as a means to alleviate light limitation. Bacterial vesicles may play a role in the type of mixotrophy associated with picocyanobacteria, in addition to the roles that they play in horizontal gene transfer (HGT) and quorum sensing (Biller et al. 2014). In

*Prochlorococcus,* vesicles contain lipids, proteins, and nucleic acids. The possibility of co-evolution between *Prochlorococcus* and the SAR11 clade has been raised with an exchange of glycolate or pyruvate from *Prochlorococcus* to SAR11 in exchange for malate from SAR11 to *Prochlorococcus* (Yelton et al. 2016). Mixotrophy in cyanobacteria implies the use of organic carbon sources (Muñoz-Marín et al. 2020). For example, glucose transporters are ubiquitous in all *Prochlorococcus* strains examined to date (Roth-Rosenberg et al. 2020).

In addition to photosynthetic cyanobacteria, proteorhodopsin-containing microorganisms and aerobic anoxygenic phototrophic bacteria (AAnP) can harvest light. Proteorhodopsin is widely distributed in marine bacteria and archaea and, although the estimates vary, up to 70% of bacterioplankton inhabiting the euphotic zone may contain proteorhodopsin likely acquired by widespread horizontal gene transfer (HGT) of these genes between taxonomically diverse microbes (Pinhassi et al. 2016). The proteorhodopsin system, widely distributed among archaea and bacteria, possesses a similar basic structure in all these organisms examined to date. It is composed of a single protein (opsin) that binds the pigment retinal and works by absorbing light and generating a proton motive force (PMF) that can then be used for motility, ATP synthesis, and nutrient transport. The proteorhodopsin system absorbs either in the blue or green wavelengths for deep water or shallow water, respectively (reviewed in Pinhassi et al. 2016).

Although less taxonomically diverse and less numerically abundant, the AAnP are nevertheless an important component of the bacterioplankton regionally, at times reaching an abundance of up to 30% of the bacteria in surface waters (Kirchman and Hanson 2013). The AAnP have a complex bacteriochlorophyll-containing light-harvesting system with a separate reaction center and a short electron transport chain. However, they do not evolve oxygen and require organic carbon for growth (Kirchman and Hanson 2013). While the machinery used by the AAnP bacteria is more costly to synthesize, the energetic gain is one order of magnitude higher for the AAnP than for the PR system, although for both, it is much smaller than the energetic gain obtained from oxygen-evolving photosynthetic apparatus of cyanobacteria (Kirchman and Hanson 2013). Theoretical calculations are supported by laboratory experiments that demonstrated that AAnP bacteria derive a significant advantage from light as opposed to the proteorhodopsin-containing microorganisms that do not appear to gain an advantage for growth (Gómez-Consarnau et al. 2019; Kirchman and Hanson 2013) although the PR system may improve survival under stress conditions (Gómez-Consarnau et al. 2019).

### 9.6.3 Mixotrophy in Eukaryotic Microbes

Arguably, the widespread ability for mixotrophy and its diversified expression in microbial eukaryotes is a testimony to its usefulness as an ecological strategy (Flynn and Mitra 2009). There is a wide spectrum of metabolic diversity in the mixotrophs. In photosynthetic microbial eukaryotes, mixotrophy can be activated to alleviate light limitation or as an acquisition strategy for dissolved organic nutrients, including carbon, macro-and micronutrients, and vitamins. The acquisition of energy- or nutrient-rich compounds can be achieved by assimilating prey such as bacteria (including cyanobacteria) and picoeukaryotes (Stoecker et al. 2017). Mixotrophic phytoflagellates are a group of mixotrophs that harbors a vertically transmitted chloroplast, but they are also motile and can ingest prey. The microalgae within this group are phylogenetically diverse, including taxa belonging to chrysophytes, haptophytes, cryptophytes, silicoflagellates, and chlorophytes. Some haptophytes also produce toxins that kill their prey, allowing them to target preys that are much larger than themselves (Tillmann 2003).

Mixotrophy is also widely distributed and metabolically diverse among dinoflagellates. Although dinoflagellates can be facultative mixotrophs, those that harbor a vertically inherited chloroplast and dominate surface waters during stratified conditions in marine waters are often considered to be photoautotrophs because they require light to grow (reviewed in Stoecker et al. 2017). Within these photoautotrophic dinoflagellates, carbon acquired through phagotrophy depends on the species, irradiance, availability of prey, and nutrient concentrations. Dinoflagellates that lack their own vertically transmitted chloroplasts but inhabit nutrient-poor oceanic surface waters may harbor cyanobacterial ecto- or endosymbionts while others are able to sequester functional chloroplasts from their prey items or retain intact ingested cryptophytes and haptophytes cells (Stoecker et al. 2017). Adaptation to mixotrophy can also include the presence of proteorhodopsin as, for example, seen in *Oxyrrhis marina* (Guo et al. 2014), an adaptation most likely acquired through HGT (Stoecker et al. 2017).

In *Mesodinium*, a well-studied clade of obligate mixotrophic ciliate, the red-pigmented species *M. rubrum*, can form large red-tides in coastal areas and accounts for a large fraction of the primary production and total chlorophyll. Their red-pigmented cryptophyte prey is part of the *Geminigera/Plagioselmis/Teleaulax* clade (Nishitani and Yamaguchi 2018). Finally, in their review, Stoecker et al. (2017) also included the poorly understood Rhizarian symbioses with haptophytes as examples of mixotrophy.

## 9.7 The Fate of the Ocean Pelagic Lit-Zone Microbiome

The fate of the microbiome in the euphotic zone of the pelagic environment depends largely on geographical location and, at higher latitude, the timing within a seasonal cycle. Several abiotic processes can contribute to the fate of microbes in the euphotic zone. However, the standing stock of a microbial community within the euphotic zone depends on interacting biological and abiotic processes (Carlson et al. 2007). Plankton can be grazed and lysed by predators or viruses, or they aggregate and sink, contributing to the biological carbon pump (BCP) (Polimene et al. 2017). The microbial communities can be transported out of their local environment through advection by ocean currents (Richter et al. 2020), eddies (Browning et al. 2021; Doblin et al. 2016), or subduction via the mixed layer pump (Gardner et al. 1995; Lacour et al. 2019). Traditionally, the biological carbon pump (BCP) has been described as the major process leading to the removal of primary production from the euphotic zone (Agusti et al. 2015). The rapid sinking of phytoplankton to 4000 m has been confirmed by the presence of healthy diatoms and dinoflagellates in deep water samples recovered from circumnavigation expeditions such as the Malaspina 2010 (Agusti et al. 2015). However, smaller diatoms also contribute to the BCP (Tréguer et al. 2018). The ubiquitous genus *Minidiscus* exhibits a high sinking rate and can contribute to global carbon export to deep water. Aggregates of *Synechococcus* have also been recovered from deep water samples (Guidi et al. 2016), indicating that in the open ocean, small picoplankton are an important contribution to the BCP. Furthermore, as determined from the Tara Oceans expedition data, the biological carbon pump's effectiveness in the North Atlantic was driven mainly by the presence of *Synechococcus* forming an assemblage with *Cobetia*, *Pseudoalteromonas*, *Idiomarina*, *Vibrio*, and *Arcobacter* (Guidi et al. 2016). Although *Synechococcus* was an important indicator species to assess carbon export in the deep ocean, *Prochlorococcus* and SAR11 did not seem to play a role in this process.

The microbial carbon pump (MCP) has been proposed as a complementary mechanism that results in carbon removal from the rapidly cycling carbon pool by the progressive microbial conversion of labile dissolved organic carbon (DOC) into a slowly degrading refractory DOC (Jiao et al. 2010; Legendre et al. 2015). In the euphotic zone, the labile components of the DOM are utilized by heterotrophic microbes leaving behind the less biodegradable components of the DOM, which are usually enriched in refractory carbon compounds (Jiao et al. 2010). The formation of refractory DOC is also promoted by the high light levels and UV radiation that are found in the euphotic zone (Jiao et al. 2010). The preference of heterotrophic microbes for certain compounds within the DOM pools results in a DOM that is increasingly less biodegradable by microbes, eventually becoming refractory with a turnover time that reaches 100 s of years (Legendre et al. 2015). Although these two mechanisms (BCP and MCP) operate concurrently in the marine environment, the balance between them varies depending on the trophic state of the specific environment. The MCP:BCP ratio is influenced by the concentration of nutrients, and it has been proposed that in the future stronger stratification will increase the contribution of the MCP relative to the BCP (Polimene et al. 2017). The ocean is one of the largest reservoirs of DOM, and most of it is derived from phytoplankton production in the lit-zone of the pelagic environment (Wagner et al. 2020). The BCP and the MCP are linked, and their relative contribution will depend to a large extent on the composition of the microbial and phytoplankton communities growing in the euphotic zone.

Biogenic particles and DOM can also be removed from the euphotic zone by the mixed layer pump, which invokes the deep mixing of the water column resulting from rapid heat exchange between the lower atmosphere and the sea surface as, for example, can be experienced between day and night (Gardner et al. 1995). The mixed layer pump is most effective at sequestering organic carbon into the deep sea during the unstable period before seasonal stratification stabilizes the water column, and it operates mainly at high latitude (Dall'Olmo et al. 2016; Lacour et al. 2019).

## 9.8 The Marine Microbiome of the Euphotic Zone

In the previous sections, the general environmental conditions that determine the community composition of microbes were explored, and the specific adaptations that increase microbial fitness in the lit pelagic zone were reviewed. Here, the expected community composition of the microbiomes relevant to major habitats in the euphotic zone is summarized, ranging from the permanently stratified euphotic zone of oligotrophic gyres to the seasonally productive polar regions. Furthermore, a non-exhaustive list of important taxa inhabiting the euphotic zone in selected regions of the world's ocean is provided. Results from the Tara Oceans expedition have shown that there is a latitudinal diversity gradient observed in the euphotic zone (Ibarbalz et al. 2019) and that global warming may increase plankton diversity at high latitude. Temperature ranging in the surface layer of the ocean between −2 °C and 31 °C has been identified as a major driver of microbial diversity in the euphotic zone (Sunagawa et al. 2015, 2020). However, locally, microbial communities are mainly determined by depth (Sunagawa et al. 2020), reflecting specific adaptations to light intensity preferences and strategies for nutrient acquisition as discussed in previous sections of this chapter. In general, the observations from the Tara Oceans expedition indicate a decrease in the diversity of bacterioplankton as well as a decline in cyanobacteria at higher latitudes with an increase in the relative abundance of photosynthetic eukaryotes, mainly of diatoms, as the habitat changes from the tropical areas to polar regions (Ibarbalz et al. 2019). However, the effects of temperature and latitude are difficult to tease apart given that both the light regime and seawater temperature vary as a function of latitude (Fuhrman et al. 2008). It is important to keep in mind also that, although the surface oceanic waters are connected by strong currents globally, plankton dispersal is limited by local selection and adaptation of the microbiome (Ward et al. 2021).

Although marine microbial communities are geographically diverse, some cosmopolitan species of bacteria, archaea, and microbial eukaryotes exhibit biogeographic distributions that reflect their evolutionary success (Lima-Mendez et al. 2015; Tong et al. 2019; Westrich et al. 2016). The bacterial SAR11 clade is ubiquitous in the global ocean, albeit represented by a large pangenome that accommodates numerous ecotypes adapted to specific niches within the euphotic zone and deeper into the aphotic zone (Giovannoni 2017; López-Pérez et al. 2020). Similarly, *Synechococcus* is also broadly distributed in the euphotic zone of the global ocean due to multiple ecological strategies that range from acclimation to spectral irradiance (Grébert et al. 2018, 2021) to their mixotrophic capacity (Yelton et al. 2016). Among bacterial and archaeal taxa, *Vibrio*, *Alteromonas*, SAR86, and the MGII clade of the Euryarchaeota are also considered ubiquitous (Boeuf et al. 2016; López-Pérez et al. 2020; Man et al. 2003; Santoro et al. 2019; Westrich et al. 2018). However, closer genome inspection of the ecotypes occupying different niches will likely reveal genetic diversity at the functional level reflecting specific adaptation within these clades to the various local environmental conditions (Grébert et al. 2018; Man et al. 2003). Among microbial eukaryotes, a similar pattern of cosmopolitan species has been observed for the coccolithophorid *Emiliania huxleyi*

(Tong et al. 2019), diplonemid protists (Flegontova et al. 2016), several Ciliophora clades (Canals et al. 2020), and several diatom species like, for example, the genus *Minidiscus* (Leblanc et al. 2018; Sunagawa et al. 2020). The symbiosis between the cyanobacterial diazotroph *Cand.* Atelocyanobacterium thalassa and its haptophyte host is also widely distributed throughout the global ocean and are an important contributor to global marine nitrogen fixation (Krupke et al. 2013; Martínez-Pérez et al. 2016; Mills et al. 2020; Moreira-Coello et al. 2017; Ratten et al. 2015; Zehr et al. 2016).

### 9.8.1 Central Oligotrophic Gyres

It is now well established that oxygenic photosynthetic bacteria belonging to *Prochlorococcus* are the dominant microbe inhabiting the euphotic zone of tropical and subtropical gyres of large oceanic basins (Berube et al. 2019; Pierella Karlusich et al. 2020). Metagenomic studies and global surveys have now shown that genome remodeling of *Prochlorococcus* through viral infections and vesicle shedding has led to many ecotypes adapted to various environmental conditions (Biller et al. 2014). As the smallest and most abundant cyanobacterial genus known to date, *Prochlorococcus* is widely distributed in surface warm waters, reaching cell density of $10^8$ cells $L^{-1}$ and displays strong niche partitioning of ecotypes with depth and latitude driven by gradients of light and temperature (Johnson et al. 2006; Malmstrom et al. 2010). The warm waters of the central oligotrophic gyres are also populated by mixotrophic nanoflagellate grazers like *Florenciellla* (Dichtyochophyceae) that prey on *Prochlorococcus* (Li et al. 2021). The SAR11 clade (or *Pelagibacterales* order) is the heterotrophic counterpart to *Prochlorococcus* in terms of size, cell density, and in diversity of ecotypes. In fact, within the euphotic zone of oligotrophic waters, *Prochlorococcus* and the SAR11 clades co-occur (Becker et al. 2019), although the distribution of SAR11 is not restricted to the euphotic zone. SAR11 ecotypes inhabiting the euphotic zone contain light-driven proteorhodopsin, but the ecological advantage of maintaining the light-driven proton pump is not clear (Giovannoni 2017). In addition to the dominant *Prochlorococcus* and SAR11 clades, tropical waters are home to a wide range of clades belonging to proteobacteria, bacteroidetes, and cyanobacteria as determined from single-cell genome sequences (SAG) (Pachiadaki et al. 2019). SAGs originating from tropical waters also uncovered candidate phyla such as *Cand.* Luxescamonaceae an alphaproteobacterium that can potentially carry out anoxygenic photosynthesis. In their study, Pachiadaki et al. (2019) found that 58% of the SAGs contained rhodopsin genes used for photoheterotrophy, thereby supporting the widespread presence of mixotrophy. Bacteria and archaea that are generally observed in the marine tropical microbiome include members of *Roseobacter*, *Verrucomicrobia*, cyanobacteria, the archaeal Nitrosopumilales *Cand.* Nitrosopelagicus, and marine Group II, as well as some heterotrophic marine bacteria such as *Alteromonas* and *Erythrobacter* (López-Pérez et al. 2020). The cyanobacterial diazotrophs *Trichodesmium* and *Crocosphaera* are also sometimes abundant in these regions because the low or depleted fixed nitrogen concentrations in the surface waters create a natural niche for diazotrophy (Moisander et al. 2010; Ratten et al. 2015; Tang and Cassar 2019).

Diatoms, dinoflagellates, haptophytes, and chlorophytes comprise the bulk of the eukaryotic marine phytoplankton diversity, with more than 95% of the described species, while euglenophytes, cryptophytes, and chlorarachniophytes are much less diverse (Pierella Karlusich et al. 2020). Estimates from the Tara Oceans expedition revealed that there are approximately 110,000 eukaryotic OTUs (operational taxonomic units) belonging to unicellular plankton (de Vargas et al. 2015). The OTUs confirmed that dinoflagellates and diatoms were the dominant nanoplankton in the ocean (Pierella Karlusich et al. 2020). In general, picoeukaryotes are numerically less abundant than their bacterial and archaeal counterparts reaching cell densities of $10^6$ cells $L^{-1}$ in oligotrophic waters. However, the contribution of picoeukaryotes to biomass and primary productivity is considerable given their much larger cell size compared to most bacteria and archaea (Worden and Not 2008). The abundance of photosynthetic picoeukaryotes covaries with *Synechococcus* in surface waters, but their abundance increases with depth, especially abundant in the DCM (Worden and Not 2008). For example, pelagophytes (*Pelagomonas calceolata*) are often abundant in the DCM of oligotrophic warm waters (Choi et al. 2020). Other groups that are often present in the subtropical and tropical waters include prasinophytes and haptophytes (Choi et al. 2020). In oceanic waters, the dominant chlorophyte is the picoplanktonic green alga prasinophyte clade VII (Dos Santos et al. 2017).

### 9.8.2 Higher Latitudes

Seasonal cycles of light and temperature at higher latitudes lead to a successional pattern in microbial community composition (Bolaños et al. 2021; Choi et al. 2020; Fuhrman et al. 2006, 2008). In coastal areas, photosynthetic picoeukaryotes are typically ~$5 \times 10^6$ cells $L^{-1}$ (Worden and Not 2008) except that they can also form blooms reaching $2 \times 10^7$ cells $L^{-1}$. *Micromonas* is an important picoeukaryote in temperate and high latitudes and dominates in the English Channel and coastal and polar fronts (Bolaños et al. 2021). In upwelling regions, prymnesiophytes and chrysophytes are most abundant (Worden and Not 2008). Northern regions of the Northwest Atlantic are linked to the Arctic Ocean through the Labrador Sea, where waters of Atlantic and Arctic origin meet (Zorz et al. 2019). The microbial community composition in these regions is seasonally influenced by either Arctic water or the Gulf Stream (Bolaños et al. 2020; Zorz et al. 2019), leading to strong succession in the microbial community that is nevertheless predictable with a spring bloom dominated by various diatoms species as well as by *Micromonas, Bathycoccus, Ostreococcus*, and the colony-forming *Phaeocystis* (Bolaños et al. 2020). Cold water diatoms are the predominant phytoplankton in these waters, and *Chaetoceros, Fragilariopsis, Coscinodiscus*, and Thalassiosirales are the common taxa (Fragoso et al. 2018). Once thought to be relegated to warm oligotrophic waters, symbiotic cyanobacterial and heterotrophic diazotrophs belonging to the Cluster III phylotypes were detected from metabarcoding of *nifH,* a marker gene for diazotrophy thereby expanding the range of marine diazotrophs to high latitudes, including the Arctic Ocean (Shiozaki et al. 2018). Subsequently, the $N_2$ fixation rates measured in the Arctic Ocean were attributed to the symbiotic *Cand.* Atelocyanobacterium thalassa (Harding et al. 2018). Polar SAR11 ecotypes have been identified and partially characterized relative to the global distribution of SAR11 clades (Kraemer et al. 2020), and SAR11 ecotypes endemic to the Arctic environment were characterized while others were found in both the Southern Ocean and the Arctic Ocean showing a bi-polar distribution (Kraemer et al. 2020). The Arctic SAR11 lineages were present in the euphotic zone and in the DCM. The specific ecological role of SAR11 in the Arctic is unclear and needs to be investigated. Archaea (Thaumarchaeota) are found at the base of the euphotic zone in the Arctic (Pedneault et al. 2014) and in the North Pacific Ocean (Church et al. 2010).

## 9.9 The Microbiome of the Euphotic Zone in the Future Ocean

Although sunlight radiation and photoperiod will remain constant as a function of latitude in the future, ocean warming will have far-reaching effects on major ocean currents, on the strength of water column stratification and, consequently, on the mixed layer depth and nutrient availability. Global warming is predicted to have a major impact on ocean circulation, resulting in changes in microbial community composition and productivity patterns (Cavicchioli et al. 2019; Flombaum et al. 2020; Oziel et al. 2020). The potential consequences of these predicted changes on the microbiome of the euphotic zone are multi-fold. In particular, the distribution ranges of phytoplankton species will be affected, and primary production patterns may also change. For example, picophytoplankton like *Prochlorococcus*, *Synechococcus,* and picoeukaryotic phytoplankton species may increase in biomass at low latitudes and expand their niche to higher latitudes (Flombaum et al. 2020). In this regard, the polar regions are changing at an especially rapid rate, with a decrease in glacier and seasonal sea ice covering having already caused major ecosystem changes (Freyria et al. 2021). The Atlantification of the Arctic Ocean, caused in part from increased transport of North Atlantic waters through the European Arctic Corridor, is resulting in a poleward expansion of the distribution range of temperate phytoplankton species such as the coccolithophorid *Emiliania huxleyi* (Oziel et al. 2020). Overall, establishing the effects of global change on marine microbial communities of the euphotic zone will require time-series observations to track the seasonal and annual variability in microbial community composition and productivity level at strategic locations throughout the world ocean, building on the network of established time-series sitess and initiating new ones at locations sensitive to change (Buttigieg et al. 2018; Wietz et al. 2021).

## References

Aguirre LE, Ouyang L, Elfwing A, Hedblom M, Wulff A, Inganäs O (2018) Diatom frustules protect DNA from ultraviolet light. Sci Rep 8:1–6

Agusti S, González-Gordillo JI, Vaqué D, Estrada M, Cerezo MI, Salazar G et al (2015) Ubiquitous healthy diatoms in the deep sea confirm deep carbon injection by the biological pump. Nat Commun 6:1–8

Allen AE, LaRoche J, Maheswari U, Lommer M, Schauer N, Lopez PJ et al (2008) Whole-cell response of the pennate diatom *Phaeodactylum tricornutum* to iron starvation. Proc Natl Acad Sci U S A 105:10438–10443

Bais AF, Bernhard G, McKenzie RL, Aucamp PJ, Young PJ, Ilyas M et al (2019) Ozone-climate interactions and effects on solar ultraviolet radiation. Photochem Photobiol Sci 18:602–640

Bais AF, McKenzie RL, Bernhard G, Aucamp PJ, Ilyas M, Madronich S et al (2015) Ozone depletion and climate change: impacts on UV radiation. Photochem Photobiol Sci 14:19–52

Baldry K, Strutton PG, Hill NA, Boyd PW (2020) Subsurface chlorophyll-a maxima in the Southern Ocean. Front Mar Sci 7:671

Bar-On YM, Phillips R, Milo R (2018) The biomass distribution on earth. Proc Natl Acad Sci U S A 115:6506–6511

Bebout BM, Garcia-Pichel F (1995) UV B-induced vertical migrations of cyanobacteria in a microbial mat. Appl Environ Microbiol 61:4215–4222

Becker JW, Hogle SL, Rosendo K, Chisholm SW (2019) Co-culture and biogeography of *Prochlorococcus* and SAR11. ISME J 13:1506–1519

Behnke J, LaRoche J (2020) Iron uptake proteins in algae and the role of iron starvation-induced proteins (ISIPs). Eur J Phycol 55:339–360

Beier S, Gálvez MJ, Molina V, Sarthou G, Quéroué F, Blain S et al (2015) The transcriptional regulation of the glyoxylate cycle in SAR11 in response to iron fertilization in the Southern Ocean. Environ Microbiol Rep 7:427–434

Béja O, Spudich EN, Spudich JL, Leclerc M, DeLong EF (2001) Proteorhodopsin phototrophy in the ocean. Nature 411:786–789

Berube PM, Rasmussen A, Braakman R, Stepanauskas R, Chisholm SW (2019) Emergence of trait variability through the lens of nitrogen assimilation in *Prochlorococcus*. elife 8:e41043

Biller SJ, Schubotz F, Roggensack SE, Thompson AW, Summons RE, Chisholm SW (2014) Bacterial vesicles in marine ecosystems. Science 343:183–186

Blain S, Tagliabue A (2016) In: Tréguer P (ed) Iron cycle in oceans. ISTE Ltd, London, UK

Boeuf D, Lami R, Cunnington E, Jeanthon C (2016) Summer abundance and distribution of proteorhodopsin genes in the western arctic ocean. Front Microbiol 7:1–12

Bolaños LM, Choi CJ, Worden AZ, Baetge N, Carlson CA, Giovannoni S (2021) Seasonality of the microbial community composition in the North Atlantic. Front Mar Sci 8:1–16

Bolaños LM, Karp-Boss L, Choi CJ, Worden AZ, Graff JR, Haëntjens N et al (2020) Small phytoplankton dominate western North Atlantic biomass. ISME J 14:1663–1674

Bouchard JN, Roy S, Ferreyra G, Campbell DA, Curtosi A (2005) Ultraviolet-B effects on photosystem II efficiency of natural phytoplankton communities from Antarctica. Polar Biol 28:607–618

Boyd PW, Jickells T, Law CS, Blain S, Boyle EA, Buesseler KO et al (2007) Mesoscale iron enrichment experiments 1993-2005: synthesis and future directions. Science 315:612–617

Browning TJ, Al-Hashem AA, Hopwood MJ, Engel A, Belkin IM, Wakefield ED et al (2021) Iron regulation of North Atlantic eddy phytoplankton productivity. Geophys Res Lett 48: e2020GL091403

Buesseler KO, Boyd PW, Black EE, Siegel DA (2020) Metrics that matter for assessing the ocean biological carbon pump. Proc Natl Acad Sci U S A 117:9679–9687

Buma AGJ, van Hannen EJ, Veldhuis MJW, Gieskes WWC (1996) UV-B induces DNA damage and DNA synthesis delay in the marine diatom *Cyclotella* sp. Sci Mar 60:101–106

Buma AGJ, De Boer MK, Boelen P (2001) Depth distributions of DNA damage in Antarctic marine phyto- and bacterioplankton exposed to summertime UV radiation. J Phycol 37:200–208
Buttigieg PL, Fadeev E, Bienhold C, Hehemann L, Pierre O, Boetius A (2018) Marine microbes in 4D–using time series observation to assess the dynamics of the ocean microbiome and its links to ocean health. Curr Opin Microbiol 43:169–185
Campbell D, Eriksson MJ, Öquist G, Gustafsson P, Clarke AK (1998) The cyanobacterium *Synechococcus* resists UV-B by exchanging photosystem II reaction-center D1 proteins. Proc Natl Acad Sci U S A 95:364–369
Canals O, Obiol A, Muhovic I, Vaqué D, Massana R (2020) Ciliate diversity and distribution across horizontal and vertical scales in the open ocean. Mol Ecol 29:2824–2839
Caputi L, Carradec Q, Eveillard D, Kirilovsky A, Pelletier E, Pierella Karlusich JJ et al (2019) Community-level responses to iron availability in open ocean plankton ecosystems. Global Biogeochem Cycles 33:391–419
Carlson CA, Del Giorgio PA, Herndl GJ (2007) Microbes and the dissipation of energy and respiration: from cells to ecosystems. Oceanography 20:89–100
Castenholz RW, Garcia-Pichel F (2013) Cyanobacterial responses to UV radiation. In: Whitton BA (ed) Ecology of cyanobacteria II: their diversity in space and time. Springer, Netherlands, pp 481–499
Cavicchioli R, Ripple WJ, Timmis KN, Azam F, Bakken LR, Baylis M, Behrenfeld MJ et al (2019) Scientists' warning to humanity: microorganisms and climate change. Nat Rev Microbiol 17: 569–586
Choi CJ, Jimenez V, Needham DM, Poirier C, Bachy C, Alexander H et al (2020) Seasonal and geographical transitions in eukaryotic phytoplankton community structure in the Atlantic and Pacific oceans. Front Microbiol 11:542372
Church MJ, Wai B, Karl DM, DeLong EF (2010) Abundances of crenarchaeal *amoA* genes and transcripts in the Pacific Ocean. Environ Microbiol 12:679–688
Coesel S, Mangogna M, Ishikawa T, Heijde M, Rogato A, Finazzi G et al (2009) Diatom PtCPF1 is a new cryptochrome/photolyase family member with DNA repair and transcription regulation activity. EMBO Rep 10:655–661
Cohen NR, McIlvin MR, Moran DM, Held NA, Saunders JK, Hawco NJ et al (2021) Dinoflagellates alter their carbon and nutrient metabolic strategies across environmental gradients in the Central Pacific Ocean. Nat Microbiol 6:173–186
Cullen JJ (2015) Subsurface chlorophyll maximum layers: enduring enigma or mystery solved? Annu Rev Mar Sci 7:207–239
Dall'Olmo G, Dingle J, Polimene L, Brewin RJW, Claustre H (2016) Substantial energy input to the mesopelagic ecosystem from the seasonal mixed-layer pump. Nat Geosci 9:820–823
De Tommasi E, Congestri R, Dardano P, De Luca AC, Managò S, Rea I et al (2018) UV-shielding and wavelength conversion by centric diatom nanopatterned frustules. Sci Rep 8:1–14
De Vargas C, Audic S, Henry N, Decelle J, Mahé F, Logares R et al (2015) Eukaryotic plankton diversity in the sunlit ocean. Science 348:1–12
Debeljak P, Toulza E, Beier S, Blain S, Obernosterer I (2019) Microbial iron metabolism as revealed by gene expression profiles in contrasted Southern Ocean regimes. Environ Microbiol 21:2360–2374
Dlugosch L, Poehlein A, Wemheuer B, Pfeiffer B, Giebel H-A, Daniel R, Simon M (2021) Nitrogen availability drives gene length of dominant prokaryotes and diversity of genes acquiring Nitrogen-species in oceanic systems. bioRxiv 2021.01.10.426031:1–26
Doblin MA, Petrou K, Sinutok S, Seymour JR, Messer LF, Brown MV et al (2016) Nutrient uplift in a cyclonic eddy increases diversity, primary productivity and iron demand of microbial communities relative to a western boundary current. PeerJ 4:e1973
Dos Santos AL, Gourvil P, Tragin M, Noël MH, Decelle J, Romac S et al (2017) Diversity and oceanic distribution of prasinophytes clade VII, the dominant group of green algae in oceanic waters. ISME J 11:512–528

Dudok de Wit T, Kopp G, Fröhlich C, Schöll M (2017) Methodology to create a new total solar irradiance record: making a composite out of multiple data records. Geophys Res Lett 44:1196–1203
Dupouy C, Frouin R, Tedetti M, Maillard M, Rodier M, Lombard F et al (2018) Diazotrophic *Trichodesmium* impact on UV-vis radiance and pigment composition in the western tropical South Pacific. Biogeosciences 15:5249–5269
Dyhrman ST, Chappell PD, Haley ST, Moffett JW, Orchard ED, Waterbury JB et al (2006) Phosphonate utilization by the globally important marine diazotroph *Trichodesmium*. Nature 439:68–71
Edwards KF (2019) Mixotrophy in nanoflagellates across environmental gradients in the ocean. Proc Natl Acad Sci U S A 116:6211–6220
Falkowski PG (1994) The role of phytoplankton photosynthesis in global biogeochemical cycles. Photosynth Res 39:235–258
Falkowski PG, Katz ME, Knoll AH, Quigg A, Raven JA, Schofield O et al (2004) The evolution of modern eukaryotic phytoplankton. Science 305:354–360
Falkowski PG, Laroche J (1991) Acclimation to spectral irradiance in algae. J Phycol 27:8–14
Field CB, Behrenfeld MJ, Randerson JT, Falkowski P (1998) Primary production of the biosphere: integrating terrestrial and oceanic components. Science 281:237–240
Fiore C, Alexander H, Kido Soule M, Kujawinski E (2020) A phosphate starvation response gene (psr1-like) is present and expressed in *Micromonas pusilla* and other marine algae. Aquat Microb Ecol 86:29–46
Flegontova O, Flegontov P, Malviya S, Audic S, Wincker P, de Vargas C et al (2016) Extreme diversity of diplonemid eukaryotes in the ocean. Curr Biol 26:3060–3065
Flombaum P, Wang WL, Primeau FW, Martiny AC (2020) Global picophytoplankton niche partitioning predicts overall positive response to ocean warming. Nat Geosci 13:116–120
Flynn KJ, Mitra A (2009) Building the "perfect beast": modelling mixotrophic plankton. J Plankton Res 31:965–992
Fragoso GM, Poulton AJ, Yashayaev IM, Head EJH, Johnsen G, Purdie DA (2018) Diatom biogeography from the Labrador Sea revealed through a trait-based approach. Front Mar Sci 5:297
Franks PJS (2015) Marine science. ICES J Mar Sci 72:1897–1907
Freyria NJ, Joli N, Lovejoy C (2021) A decadal perspective on north water microbial eukaryotes as Arctic Ocean sentinels. Sci Rep 11:1–14
Fuhrman JA, Hewson I, Schwalbach MS, Steele JA, Brown MV, Naeem S (2006) Annually reoccurring bacterial communities are predictable from ocean conditions. Proc Natl Acad Sci U S A 103:13104–13109
Fuhrman JA, Steele JA, Hewson I, Schwalbach MS, Brown MV, Green JL et al (2008) A latitudinal diversity gradient in planktonic marine bacteria. Proc Natl Acad Sci U S A 105:7774–7778
Gao K, Helbling E, Häder D, Hutchins D (2012) Responses of marine primary producers to interactions between ocean acidification, solar radiation, and warming. Mar Ecol Prog Ser 470:167–189
Gao Q, Garcia-Pichel F (2011) Microbial ultraviolet sunscreens. Nat Rev Microbiol 9:791–802
Garcia-Corral LS, Holding JM, Carrillo-de-Albornoz P, Steckbauer A, Pérez-Lorenzo M, Navarro N et al (2017a) Effects of UVB radiation on net community production in the upper global ocean. Glob Ecol Biogeogr 26:54–64
Garcia-Corral LS, Holding JM, Carrillo-de-Albornoz P, Steckbauer A, Pérez-Lorenzo M, Navarro N et al (2017b) Temperature dependence of plankton community metabolism in the subtropical and tropical oceans. Global Biogeochem Cycles 31:1141–1154
Gardner WD, Chung SP, Richardson MJ, Walsh ID (1995) The oceanic mixed-layer pump. Deep Sea Res Part II: Topical Studies in Oceanography 42:757–775
Geider RJ, La Roche J (2002) Redfield revisited: variability of C:N:P in marine microalgae and its biochemical basis. Eur J Phycol 37:1–17

Giovannoni SJ (2017) SAR11 bacteria: the most abundant plankton in the oceans. Annu Rev Mar Sci 9:231–255
Gómez-Consarnau L, González JM, Coll-Lladó M, Gourdon P, Pascher T, Neutze R et al (2007) Light stimulates growth of proteorhodopsin-containing marine Flavobacteria. Nature 445:210–213
Gómez-Consarnau L, Raven JA, Levine NM, Cutter LS, Wang D, Seegers B et al (2019) Microbial rhodopsins are major contributors to the solar energy captured in the sea. Sci Adv 5:1–8
Grébert T, Doré H, Partensky F, Farrant GK, Boss ES, Picheral M et al (2018) Light color acclimation is a key process in the global ocean distribution of *Synechococcus* cyanobacteria. Proc Natl Acad Sci U S A 115:E2010–E2019
Grébert T, Nguyen AA, Pokhrel S, Joseph KL, Ratin M, Dufour L et al (2021) Molecular basis of an alternative dual-enzyme system for light color acclimation of marine *Synechococcus* cyanobacteria. Proc Natl Acad Sci U S A 118:e2019715118
Guidi L, Chaffron S, Bittner L, Eveillard D, Larhlimi A, Roux S et al (2016) Plankton networks driving carbon export in the oligotrophic ocean. Nature 532:465–470
Guo Z, Zhang H, Lin S (2014) Light-promoted rhodopsin expression and starvation survival in the marine dinoflagellate *Oxyrrhis marina*. PLoS One 9:1–23
Harding K, Turk-Kubo KA, Sipler RE, Mills MM, Bronk DA, Zehr JP (2018) Symbiotic unicellular cyanobacteria fix nitrogen in the Arctic Ocean. Proc Natl Acad Sci U S A 115:13371–13375
Holtrop T, Huisman J, Stomp M, Biersteker L, Aerts J, Grébert T et al (2020) Vibrational modes of water predict spectral niches for photosynthesis in lakes and oceans. Nat Ecol Evol 5:55–66
Ibarbalz FM, Henry N, Brandão MC, Martini S, Busseni G, Byrne H et al (2019) Global trends in marine plankton diversity across kingdoms of life. Cell 179:1084–1097
Iuculano F, Álverez-Salgado XA, Otero J, Catalá TS, Sobrino C, Duarte CM et al (2019) Patterns and drivers of UV absorbing chromophoric dissolved organic matter in the euphotic layer of the open ocean. Front Mar Sci 6:1–19
Jiao N, Herndl GJ, Hansell DA, Benner R, Kattner G, Wilhelm SW et al (2010) Microbial production of recalcitrant dissolved organic matter: long-term carbon storage in the global ocean. Nat Rev Microbiol 8:593–559
Jickells TD (2005) Global iron connections between desert dust, ocean biogeochemistry, and climate. Science 308:67–71
Johnson ZI, Zinser ER, Coe A, Mcnulty NP, Woodward EMS, Chisholm SW (2006) Niche partitioning among *Prochlorococcus* ecotypes along ocean-scale environmental gradients. Science 311:1737–1741
Kemp AES, Villareal TA (2018) The case of the diatoms and the muddled mandalas: time to recognize diatom adaptations to stratified waters. Prog Oceanogr 167:138–149
Kessler N, Armoza-Zvuloni R, Wang S, Basu S, Weber PK, Stuart RK et al (2020) Selective collection of iron-rich dust particles by natural *Trichodesmium* colonies. ISME J 14:91–103
Kirchman DL, Hanson TE (2013) Bioenergetics of photoheterotrophic bacteria in the oceans. Environ Microbiol Rep 5:188–199
Kirk JTO (1994) Light and photosynthesis in aquatic ecosystems. Cambridge University Press, Cambridge
Koedooder C, Guéneuguès A, Van Geersdaële R, Vergé V, Bouget FY, Labreuche Y et al (2018) The role of the glyoxylate shunt in the acclimation to iron limitation in marine heterotrophic bacteria. Front Mar Sci 5:1–12
Koedooder C, Van Geersdaële R, Guéneuguès A, Bouget F-Y, Obernosterer I, Blain S (2020) The interplay between iron limitation, light, and carbon in the proteorhodopsin containing Photobacterium angustum S14. FEMS Microbiol Ecol 103:1–12
Kraemer S, Ramachandran A, Colatriano D, Lovejoy C, Walsh DA (2020) Diversity and biogeography of SAR11 bacteria from the Arctic Ocean. ISME J 14:79–90
Krause JW, Schulz IK, Rowe KA, Dobbins W, Winding MHS, Sejr MK et al (2019) Silicic acid limitation drives bloom termination and potential carbon sequestration in an Arctic bloom. Sci Rep 9:1–11

Krupke A, Musat N, LaRoche J, Mohr W, Fuchs BM, Amann RI et al (2013) In situ identification and $N_2$ and C fixation rates of uncultivated cyanobacteria populations. Syst Appl Microbiol 36: 259–271

LaRoche J, Boyd PW, McKay RML, Geider RJ (1996) Flavodoxin as an in situ marker for iron stress in phytoplankton. Nature 382:802–805

Lacour L, Briggs N, Claustre H, Ardyna M, Dall'Olmo G (2019) The intraseasonal dynamics of the mixed layer pump in the subpolar North Atlantic Ocean: a biogeochemical-Argo float approach. Global Biogeochem Cycles 33:266–281

Lampe RH, Mann EL, Cohen NR, Till CP, Thamatrakoln K, Brzezinski MA et al (2018) Different iron storage strategies among bloom-forming diatoms. Proc Natl Acad Sci 115:E12275–E12284

Langlois R, Mills MM, Ridame C, Croot P, LaRoche J (2012) Diazotrophic bacteria respond to Saharan dust additions. Mar Ecol Prog Ser 470:1–14

Leblanc K, Quéguiner B, Diaz F, Cornet V, Michel-Rodriguez M, Durrieu De Madron X et al (2018) Nanoplanktonic diatoms are globally overlooked but play a role in spring blooms and carbon export. Nat Commun 9:1–12

Legendre L, Rivkin RB, Weinbauer MG, Guidi L, Uitz J (2015) The microbial carbon pump concept: potential biogeochemical significance in the globally changing ocean. Prog Oceanogr 134:432–450

Letelier RM, White AE, Bidigare RR, Barone B, Church MJ, Karl DM (2017) Light absorption by phytoplankton in the North Pacific subtropical gyre. Limnol Oceanogr 62:1526–1540

Li Q, Edwards KF, Schvarcz CR, Selph KE, Steward GF (2021) Plasticity in the grazing ecophysiology of *Florenciella* (Dichtyochophyceae), a mixotrophic nanoflagellate that consumes *Prochlorococcus* and other bacteria. Limnol Oceanogr 66:47–60

Liefer JD, Garg A, Fyfe MH, Irwin AJ, Benner I, Brown CM et al (2019) The macromolecular basis of phytoplankton C:N:P under nitrogen starvation. Front Microbiol 10:1–16

Lima-Mendez G, Faust K, Henry N, Decelle J, Colin S, Carcillo F et al (2015) Determinants of community structure in the global plankton interactome. Science 348:1–10

Longhurst A (1995) Seasonal cycles of pelagic production and consumption. Prog Oceanogr 36:77–167

López-Pérez M, Haro-Moreno JM, Coutinho FH, Martinez-Garcia M, Rodriguez-Valera F (2020) The evolutionary success of the marine bacterium SAR11 analyzed through a metagenomic perspective. mSystems 5:1–13

Lucas-Lledo JI, Lynch M (2009) Evolution of mutation rates: phylogenomic analysis of the photolyase/cryptochrome family. Mol Biol Evol 26:1143–1153

Malmstrom RR, Coe A, Kettler GC, Martiny AC, Frias-lopez J, Zinser ER et al (2010) Temporal dynamics of *Prochlorococcus* ecotypes in the Atlantic and Pacific oceans. ISME J 4:1252–1264

Man D, Wang W, Sabehi G, Aravind L, Post AF, Massana R et al (2003) Diversification and spectral tuning in marine proteorhodopsins. EMBO J 22:1725–1731

Manck LE, Espinoza JL, Dupont CL, Barbeau KA (2020) Transcriptomic study of substrate-specific transport mechanisms for iron and carbon in the marine copiotroph *Alteromonas macleodii*. mSystems 5:1–16

Marchetti A (2019) A global perspective on iron and plankton through the Tara oceans lens. Global Biogeochem Cycles 33:239–242

Marra JF, Lance VP, Vaillancourt RD, Hargreaves BR (2014) Resolving the ocean's euphotic zone. Deep Res Part I Oceanogr Res Pap 83:45–50

Martínez-Pérez C, Mohr W, Löscher CR, Dekaezemacker J, Littmann S, Yilmaz P et al (2016) The small unicellular diazotrophic symbiont, UCYN-A, is a key player in the marine nitrogen cycle. Nat Microbiol 1:16163

McQuaid JB, Kustka AB, Oborník M, Horák A, McCrow JP, Karas BJ et al (2018) Carbonate-sensitive phytotransferrin controls high-affinity iron uptake in diatoms. Nature 555:534–537

Mills MM, Moore CM, Langlois R, Milne A, Achterberg EP, Nachtigall K et al (2008) Nitrogen and phosphorus co-limitation of bacterial productivity and growth in the oligotrophic sub-tropical North Atlantic. Limnol Oceanogr 53:824–834

Mills MM, Turk-Kubo KA, van Dijken GL, Henke BA, Harding K, Wilson ST et al (2020) Unusual marine cyanobacteria/haptophyte symbiosis relies on N2 fixation even in N-rich environments. ISME J 14:2395–2406

Moisander PH, Beinart RA, Hewson I, White AE, Johnson KS, Carlson CA et al (2010) Unicellular cyanobacterial distributions broaden the oceanic $N_2$ fixation domain. Science 327:1512–1514

Moore CM, Mills MM, Arrigo KR, Berman-Frank I, Bopp L, Boyd PW et al (2013) Processes and patterns of oceanic nutrient limitation. Nat Geosci 6:701–710

Moore CM, Mills MM, Langlois R, Milne A, Achterberg EP, LaRoche J et al (2008) Relative influence of nitrogen and phosphorous availability on phytoplankton physiology and productivity in the oligotrophic sub-tropical North Atlantic Ocean. Limnol Oceanogr 53:291–305

Moreira-Coello V, Mouriño-Carballido B, Marañón E et al (2017) Biological $N_2$ fixation in the upwelling region off NW Iberia: magnitude, relevance, and players. Front Mar Sci 4:303

Morel FMM, Lam PJ, Saito MA (2020) Trace metal substitution in marine phytoplankton. Annu Rev Earth Planet Sci 48:491–517

Moreno AR, Martiny AC (2018) Ecological stoichiometry of ocean plankton. Annu Rev Mar Sci 10:43–69

Muñoz-Marín MC, Gómez-Baena G, López-Lozano A, Moreno-Cabezuelo JA, Díez J, García-Fernández JM (2020) Mixotrophy in marine picocyanobacteria: use of organic compounds by *Prochlorococcus* and *Synechococcus*. ISME J 14:1065–1073

Nishitani G, Yamaguchi M (2018) Seasonal succession of ciliate *Mesodinium* spp. with red, green, or mixed plastids and their association with cryptophyte prey. Sci Rep 8:1–9

Nonoyama K, Nawaly G, Tsuji M et al (2019) Metabolic innovations underpinning the origin and diversification of the diatom chloroplast. Biomol Ther 9:322

Núñez-Pons L, Avila C, Romano G, Verde C, Giordano D (2018) UV-protective compounds in marine organisms from the southern ocean. Mar Drugs 16:336

Obernosterer I, Fourquez M, Blain S (2015) Fe and C co-limitation of heterotrophic bacteria in the naturally fertilized region off the Kerguelen Islands. Biogeosciences 12:1983–1992

Olson DK, Yoshizawa S, Boeuf D, Iwasaki W, Delong EF (2018) Proteorhodopsin variability and distribution in the North Pacific subtropical gyre. ISME J 12:1047–1060

Overmans S, Agustí S (2019) Latitudinal gradient of UV attenuation along the highly transparent Red Sea Basin. Photochem Photobiol 95:1267–1279

Owens SA, Pike S, Buesseler KO (2015) Thorium-234 as a tracer of particle dynamics and upper ocean export in the Atlantic Ocean. Deep Res Part II Top Stud Oceanogr 116:42–59

Oziel L, Baudena A, Ardyna M, Massicotte P, Randelhoff A, Sallée JB, Ingvaldsen RB, Devred E, Babin M (2020) Faster Atlantic currents drive poleward expansion of temperate phytoplankton in the Arctic Ocean. Nat Commun 11:1–8

Pachiadaki MG, Brown JM, Brown J, Bezuidt O, Berube PM, Biller SJ et al (2019) Charting the complexity of the marine microbiome through single-cell genomics. Cell 179:1623–1635

Pedneault E, Galand PE, Potvin M, Tremblay JÉ, Lovejoy C (2014) Archaeal *amoA* and *ureC* genes and their transcriptional activity in the Arctic Ocean. Sci Rep 4:4661

Pierella Karlusich JJ, Ibarbalz FM, Bowler C (2020) Phytoplankton in the Tara Ocean. Annu Rev Mar Sci 12:233–265

Pinhassi J, DeLong EF, Béjà O, González JM, Pedrós-Alió C (2016) Marine bacterial and archaeal ion-pumping rhodopsins: genetic diversity, physiology, and ecology. Microbiol Mol Biol Rev 80:929–954

Polimene L, Sailley S, Clark D, Mitra A, Allen JI (2017) Biological or microbial carbon pump? The role of phytoplankton stoichiometry in ocean carbon sequestration. J Plankton Res 39:180–186

Quigg A, Irwin AJ, Finkel ZV (2011) Evolutionary inheritance of elemental stoichiometry in phytoplankton. Proc R Soc B Biol Sci 278:526–534

Ratten J-M, LaRoche J, Desai DK, Shelley RU, Landing WM, Boyle E et al (2015) Sources of iron and phosphate affect the distribution of diazotrophs in the North Atlantic. Deep Res Part II Top Stud Oceanogr 116:332–341

Read RW, Berube PM, Biller SJ, Neveux I, Cubillos-Ruiz A, Chisholm SW et al (2017) Nitrogen cost minimization is promoted by structural changes in the transcriptome of N-deprived Prochlorococcus cells. ISME J 11:2267–2278

Redfield A (1958) The biological control of chemical factors in the environment. Amercian Sci 46: 205–221
Richter D, Watteaux R, Vannier T, Leconte J, Frémont P, Reygondeau G, Maillet N, et al. (2020) Genomic evidence for global ocean plankton biogeography shaped by large-scale current systems. bioRxiv 867739:1–38
Righetti D, Vogt M, Gruber N, Psomas A, Zimmermann NE (2019) Global pattern of phytoplankton diversity driven by temperature and environmental variability. Sci Adv 5:1–11
Rijkenberg MJA, Fischer AC, Kroon JJ, Gerringa LJA, Timmermans KR, Wolterbeek HT et al (2005) The influence of UV irradiation on the photoreduction of iron in the Southern Ocean. Mar Chem 93:119–129
Roth-Rosenberg D, Aharonovich D, Luzzatto-Knaan T, Vogts A, Zoccarato L, Eigemann F et al (2020) Prochlorococcus cells rely on microbial interactions rather than on chlorotic resting stages to survive long-term nutrient starvation. MBio 11:1–13
Saito MA, Goepfert TJ, Ritt JT (2008) Some thoughts on the concept of colimitation: three definitions and the importance of bioavailability. Limnol Oceanogr 53:276–290
Santoro AE, Richter RA, Dupont CL (2019) Planktonic marine archaea. Annu Rev Mar Sci 11:131–158
Sargent EC, Hitchcock A, Johansson SA, Langlois R, Moore CM, LaRoche J et al (2016) Evidence for polyploidy in the globally important diazotroph *Trichodesmium*. FEMS Microbiol Lett 363: fnw244
Shenhav L, Zeevi D (2020) Resource conservation manifests in the genetic code. Science 370:683–687
Shiozaki T, Fujiwara A, Ijichi M, Harada N, Nishino S, Nishi S et al (2018) Diazotroph community structure and the role of nitrogen fixation in the nitrogen cycle in the Chukchi Sea (western Arctic Ocean). Limnol Oceanogr 63:2191–2205
Sinha RP, Häder DP (2002) UV-induced DNA damage and repair: a review. Photochem Photobiol Sci 1:225–236
Smyth TJ (2011) Penetration of UV irradiance into the global ocean. J Geophys Res Ocean 116:1–12
Stoecker DK, Hansen PJ, Caron DA, Mitra A (2017) Mixotrophy in the marine plankton. Annu Rev Mar Sci 9:311–335
Stomp M, Huisman J, Stal LJ, Matthijs HCP (2007) Colorful niches of phototrophic microorganisms shaped by vibrations of the water molecule. ISME J 1:271–282
Strzepek RF, Boyd PW, Sunda WG (2019) Photosynthetic adaptation to low iron, light, and temperature in Southern Ocean phytoplankton. Proc Natl Acad Sci U S A 116:4388–4393
Strzepek RF, Harrison PJ (2004) Photosynthetic architecture differs in coastal and oceanic diatoms. Nature 431:689–692
Sunagawa S, Acinas SG, Bork P, Bowler C, Acinas SG, Babin M et al (2020) Tara oceans: towards global ocean ecosystems biology. Nat Rev Microbiol 18:428–445
Sunagawa S, Coelho LP, Chaffron S, Kultima JR, Labadie K, Salazar G et al (2015) Structure and function of the global ocean microbiome. Science 348:1261359
Sunda WG (2012) Feedback interactions between trace metal nutrients and phytoplankton in the ocean. Front Microbiol 3:204
Sunda WG, Huntsman SA (1995) Cobalt and zinc interreplacement in marine phytoplankton: biological and geochemical implications. Limnol Oceanogr 40:1404–1417
Tang W, Cassar N (2019) Data-driven modeling of the distribution of diazotrophs in the global ocean. Geophys Res Lett 46:12258–12269
Tedetti M, Sempéré R (2006) Penetration of ultraviolet radiation in the marine environment. A review. Photochem Photobiol 82:389–397
Tedetti M, Sempéré R, Vasilkov A, Charrière B, Nérini D, Miller WL et al (2007) High penetration of ultraviolet radiation in the south East Pacific waters. Geophys Res Lett 34:1–5
Tillmann U (2003) Kill and eat your predator: a winning strategy of the planktonic flagellate *Prymnesium parvum*. Aquat Microb Ecol 32:73–84
Tong S, Hutchins DA, Gao K (2019) Physiological and biochemical responses of *Emiliania huxleyi* to ocean acidification and warming are modulated by UV radiation. Biogeosciences 16:561–572

Tréguer P, Bowler C, Moriceau B, Dutkiewicz S, Gehlen M, Aumont O et al (2018) Influence of diatom diversity on the ocean biological carbon pump. Nat Geosci 11:27–37
Twining BS, Antipova O, Chappell PD, Cohen NR, Jacquot JE, Mann EL et al (2020) Taxonomic and nutrient controls on phytoplankton iron quotas in the ocean. Limnol Oceanogr Lett 6:96–106
Tyrrell T (1999) The relative influences of nitrogen and phosphorus on oceanic primary production. Nature 400:525–531
Van Mooy BAS, Fredricks HF, Pedler BE, Dyhrman ST, Karl DM, Koblížek M et al (2009) Phytoplankton in the ocean use non-phosphorus lipids in response to phosphorus scarcity. Nature 458:69–72
Wagner S, Schubotz F, Kaiser K, Hallmann C, Waska H, Rossel PE et al (2020) Soothsaying DOM: a current perspective on the future of oceanic dissolved organic carbon. Front Mar Sci 7:1–17
Wang Y, Lee Z, Wei J, Shang S, Wang M, Lai W (2021) Extending satellite ocean color remote sensing to the near-blue ultraviolet bands. Remote Sens Environ 253:112228
Ward BA, Cael BB, Collins S, Young CR (2021) Selective constraints on global plankton dispersal. Proc Natl Acad Sci U S A 118:e2007388118
Watanabe S (2020) Cyanobacterial multi-copy chromosomes and their replication. Biosci Biotechnol Biochem 84:1309–1321
Westrich JR, Ebling AM, Landing WM, Joyner JL, Kemp KM, Griffin DW et al (2016) Saharan dust nutrients promote vibrio bloom formation in marine surface waters. Proc Natl Acad Sci U S A 113:5964–5969
Westrich JR, Griffin DW, Westphal DL, Lipp EK (2018) Vibrio population dynamics in mid-Atlantic surface waters during Saharan dust events. Front Mar Sci 5:1–9
Wietz M, Bienhold C, Metfies K, Torres-Valdés S, Von Appen W-J, Salter I, Boetius A, Wegener A (2021) The polar night shift: annual dynamics and drivers of microbial community structure in the Arctic Ocean physical oceanography of the Polar Seas. bioRxiv 2021.04.08.436999
Wirtz K, Smith SL (2020) Vertical migration by bulk phytoplankton sustains biodiversity and nutrient input to the surface ocean. Sci Rep 10:1–13
Worden AZ, Not F (2008) Ecology and diversity of picoeukaryotes. In: Kirchman DL (ed) Microbial ecology of the oceans, 2nd edn. Wiley-Blackwell, New Jersey, pp 159–205
Xu J, Bach LT, Schulz KG, Zhao W, Gao K, Riebesell U (2016) The role of coccoliths in protecting *Emiliania huxleyi* against stressful light and UV radiation. Biogeosciences 13:4637–4643
Yelton AP, Acinas SG, Sunagawa S, Bork P, Pedrós-Alió C, Chisholm SW (2016) Global genetic capacity for mixotrophy in marine picocyanobacteria. ISME J 10:2946–2957
Yurkov VV, Beatty JT (1998) Aerobic anoxygenic phototrophic bacteria. Microbiol Mol Biol Rev 62:695–724
Zehr JP, Shilova IN, Farnelid HM, Muñoz-Marín M del C, Turk-Kubo KA, Fowler D, et al. (2016) Unusual marine unicellular symbiosis with the nitrogen-fixing cyanobacterium UCYN-A. Nat Microbiol 2:16214
Zepp RG, Erickson DJ, Paul ND, Sulzberger B (2007) Interactive effects of solar UV radiation and climate change on biogeochemical cycling. Photochem Photobiol Sci 6:286–300
Zhu Z, Fu F, Qu P, Mak EWK, Jiang H, Zhang R et al (2020) Interactions between ultraviolet radiation exposure and phosphorus limitation in the marine nitrogen-fixing cyanobacteria *Trichodesmium* and *Crocosphaera*. Limnol Oceanogr 65:363–376
Zhuang Y, Jin H, Chen J, Ren J, Zhang Y, Lan M et al (2020) Phytoplankton community structure at subsurface chlorophyll maxima on the Western Arctic shelf: patterns, causes, and ecological importance. J Geophys Res Biogeosciences 125:1–15
Zorz J, Willis C, Comeau AM, Langille MGI, Johnson CL, Li WKW et al (2019) Drivers of regional bacterial community structure and diversity in the Northwest Atlantic Ocean. Front Microbiol 10:1–24

# Microbial Inhabitants of the Dark Ocean

# 10

Federico Baltar and Gerhard J. Herndl

**Abstract**

The dark ocean refers to the oceanic water column deeper than 200 m, which includes the mesopelagic (200–1000 m), bathypelagic (1000–4000 m), abyssopelagic (4000–6000 m), and hadalpelagic (>6000 m) realms. This vast system is in terms of volume the largest habitat of the biosphere, harboring the largest reservoir of microbes in aquatic systems, and most of the heterotrophic microbial biomass and production of the global ocean. Despite its relevance, knowledge on the phylogenetic and functional diversity of the dark ocean's microbes is rather limited compared to the euphotic waters. In this chapter, recent advances in our understanding of the dark ocean's microbiome are summarized. The spatial and temporal heterogeneity of microbial communities and surface-deep ocean connectivity are described as well as their functional diversity. The chapter is concluding with a section summarizing the limited information available on the phylogenetic and functional microbial diversity of the abyssal and hadal realm. With this chapter, we hope to call the attention of the need to deepen our understanding on the dark ocean's microbiome.

F. Baltar (✉)
Department of Functional and Evolutionary Ecology, University of Vienna, Vienna, Austria
e-mail: federico.baltar@univie.ac.at

G. J. Herndl
Department of Functional and Evolutionary Ecology, University of Vienna, Vienna, Austria

NIOZ, Department of Marine Microbiology and Biogeochemistry, Royal Netherlands Institute for Sea Research, Utrecht University, Texel, The Netherlands
e-mail: gerhard.herndl@univie.ac.at



L. J. Stal, M. S. Cretoiu (eds.), *The Marine Microbiome*, The Microbiomes of Humans, Animals, Plants, and the Environment 3,
https://doi.org/10.1007/978-3-030-90383-1_10

**Keywords**

Abyssipelagic · Bathypelagic · Deep ocean · Functional diversity · Mesopelagic · Microbes

## 10.1 The Dark Ocean: The Largest Habitat in the Biosphere

The dark ocean (i.e., >200 m depth) encompasses the mesopelagic (200–1000 m), bathypelagic (1000–4000 m), abyssopelagic (4000–6000 m), and hadalpelagic (>6000 m) zone and is in terms of volume the largest habitat in the biosphere (~ $1.3 \times 10^{18}$ $m^3$). Despite being the least explored aquatic habitat on the planet it is the major reservoir of organic carbon (mainly in the form of dissolved organic carbon) in the biosphere (Benner 2002; Hansell and Carlson 1998; Libes 1992) and it harbors more than 98% of the global dissolved inorganic carbon (DIC) pool (Gruber et al. 2004). The deep ocean is also the largest reservoir of microbes in aquatic systems (Whitman et al. 1998) supporting about 75% and 50% of the microbial biomass and production, respectively, of the global ocean (Arístegui et al. 2009).

There is growing evidence that the dark ocean plays a central role in ocean's biogeochemistry and holds a unique reservoir of high phylogenetic and functional microbial diversity (Arístegui et al. 2009; Herndl and Reinthaler 2013; Nagata et al. 2010) albeit much less information is available on this diversity than for surface waters. This is basically due to the difficulty related to the sampling and accessibility of the deep ocean. Also, until recently the dark ocean has been viewed as a homogenous system based on the rather stable physicochemical parameters (i.e., temperature, salinity, organic and inorganic nutrient concentrations). Consequently, it has been suggested that the dark ocean harbors also a rather stable microbial assemblage with low metabolic activity. Only over the last two decades, the notion emerged of the dark ocean as a site harboring a diverse and active community of Bacteria and Archaea (Baltar et al. 2010a; Herndl et al. 2005; Kirchman et al. 2007). It has therefore become evident that the knowledge on the dark ocean's microbiome needs to be increased in order to better understand the phylogenetic composition and the metabolic functions of the microbial communities and their role in the biogeochemical cycles in the largest habitat of the biosphere—the dark ocean.

## 10.2 The Dark Ocean's Microbiome

### 10.2.1 Bacteria Versus Archaea

The main difference in terms of community composition between the sunlit surface waters and the deep ocean is the high abundance of Archaea relative to Bacteria in the dark ocean (10–50% of their total cell abundance) (Herndl et al. 2005; Karner et al. 2001), particularly of the marine Crenarchaeota Group I/Thaumarchaeota (referred to as Crenarchaeota in the rest of the chapter) and Euryarchaeota Group

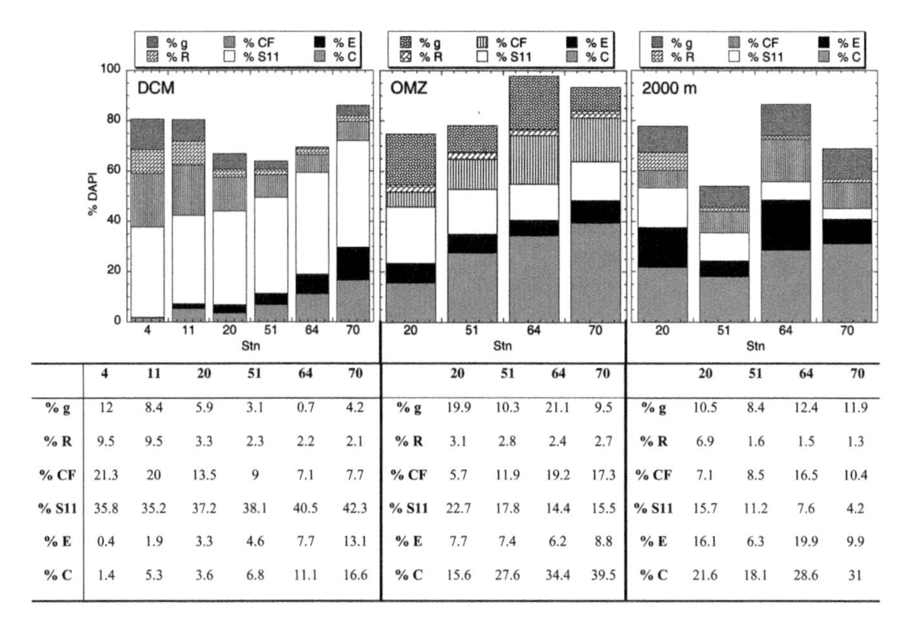


| | 4 | 11 | 20 | 51 | 64 | 70 |
|---|---|---|---|---|---|---|
| % g | 12 | 8.4 | 5.9 | 3.1 | 0.7 | 4.2 |
| % R | 9.5 | 9.5 | 3.3 | 2.3 | 2.2 | 2.1 |
| % CF | 21.3 | 20 | 13.5 | 9 | 7.1 | 7.7 |
| % S11 | 35.8 | 35.2 | 37.2 | 38.1 | 40.5 | 42.3 |
| % E | 0.4 | 1.9 | 3.3 | 4.6 | 7.7 | 13.1 |
| % C | 1.4 | 5.3 | 3.6 | 6.8 | 11.1 | 16.6 |

| | 20 | 51 | 64 | 70 |
|---|---|---|---|---|
| % g | 19.9 | 10.3 | 21.1 | 9.5 |
| % R | 3.1 | 2.8 | 2.4 | 2.7 |
| % CF | 5.7 | 11.9 | 19.2 | 17.3 |
| % S11 | 22.7 | 17.8 | 14.4 | 15.5 |
| % E | 7.7 | 7.4 | 6.2 | 8.8 |
| % C | 15.6 | 27.6 | 34.4 | 39.5 |

| | 20 | 51 | 64 | 70 |
|---|---|---|---|---|
| % g | 10.5 | 8.4 | 12.4 | 11.9 |
| % R | 6.9 | 1.6 | 1.5 | 1.3 |
| % CF | 7.1 | 8.5 | 16.5 | 10.4 |
| % S11 | 15.7 | 11.2 | 7.6 | 4.2 |
| % E | 16.1 | 6.3 | 19.9 | 9.9 |
| % C | 21.6 | 18.1 | 28.6 | 31 |

**Fig. 10.1** Relative abundances of bacterial and archaeal groups detected by horseradish peroxidase-oligonucleotide probes and CARD-FISH scaled to DAPI counts at the deep chlorophyll maximum (DCM), the oxygen minimum zone (OMZ), and at 2000 m depth. C: Marine Crenarchaeota Group I; E: marine Euryarchaeota Group II; S11: SAR11; CF: Bacteroidetes; R: Roseobacter; g: Gammaproteobacteria. From Baltar et al. (2007)

II (Baltar et al. 2007; Church et al. 2003; Varela et al. 2008). Therefore, the relative abundance of bacterial taxa such as *Roseobacter* and SAR 11 decreases with depth (Baltar et al. 2007) (Fig. 10.1). SAR 11 is a cosmopolitan group making up >50% of the bacterial and archaeal abundance in the surface waters and <25% of the mesopelagic assemblages (Baltar et al. 2007; DeLong et al. 2006; Morris et al. 2002) (Fig. 10.1).

SAR11 is composed of different ecotypes including a specific ecotype that is relatively more abundant and specially adapted to deep waters (Thrash et al. 2014) (Fig. 10.2). The existence of different ecotypes (oligotypes) in surface versus deep ocean waters is not restricted to SAR 11 but has been shown for other abundant taxa such as *Alteromonas macleodii* (Ivars-Martínez et al. 2008) indicating adaptations to contrasting environments.

Besides this vertical decrease in the ratio of Bacteria to Archaea in the water column a lateral succession between Bacteria and Archaea has been found in epipelagic waters where the Bacteria: Archaea ratio decreases from more eutrophic coastal to offshore oligotrophic waters (Baltar et al. 2007) (Fig. 10.1). This coastal-ocean pattern in the ratio between Bacteria and Archaea is also detectable in the mesopelagic layers of productive regions but it is not evident in the bathypelagic zone (Baltar et al. 2007) (Fig. 10.1). These shifts in the relative abundance between Archaea (Crenarchaeota and Euryarchaeota) and Bacteria (Baltar et al. 2007; Herndl

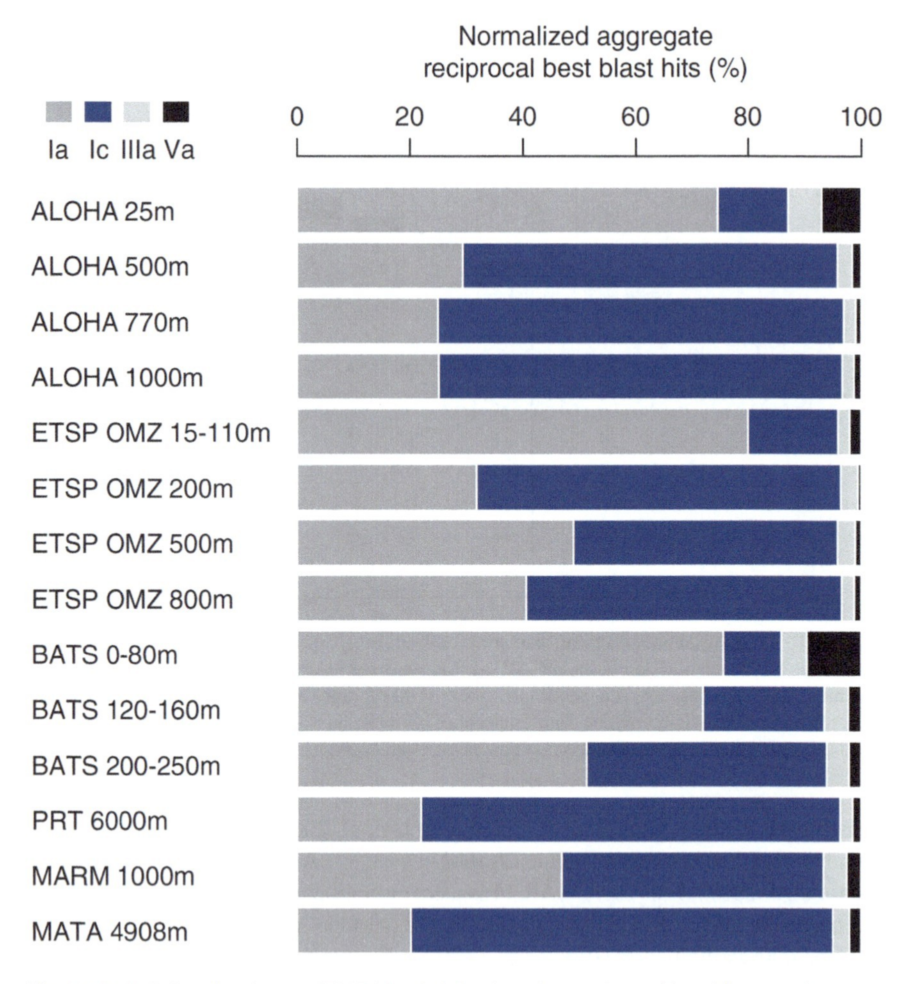


**Fig. 10.2** Relative abundance of SAR11 subclades based on reciprocal best blast recruitment of metagenomic sequences. From Thrash et al. (2014)

et al. 2005; Teira et al. 2006) indicate different growth requirements. Herndl et al. (2005) reported that up to 20% of archaeal cells fix dissolved inorganic carbon (DIC) in bathypelagic waters (1000–3000 m) of the North Atlantic whereas Bacteria contributed <2% to the DIC-fixing cells. A subsequent study suggested that at least some of the Crenarchaeota in the bathypelagic realm might also be heterotrophic in subtropical deep waters (Agogué et al. 2008). It was estimated that ~83% of archaeal carbon was derived from DIC fixation as revealed by determining the natural radiocarbon of Crenarchaeota-specific lipids from the mesopelagic North Pacific (Ingalls et al. 2006). Another study based on the radiocarbon signature of microbial DNA collected also from mesopelagic Pacific waters indicated that i) both DIC and fresh DOC (most likely from sinking POC) were used substantially, ii) ambient dissolved organic carbon (DOC) was not the main source of carbon, and iii) the DIC fixation of the total community correlated with the archaeal *amoA* gene and Crenarchaeota 16S rRNA gene abundance (Hansman et al. 2009). Thus, it is obvious that archaeal DIC fixation has been an understudied and underestimated process in the dark ocean. Rate and budget calculations suggest that DIC fixation is in the same order of magnitude as bacterial and archaeal heterotrophic carbon biomass production (Reinthaler et al. 2010). The contribution of DIC fixation to the microbial carbon demand in the dark ocean is equivalent to that from the sinking particulate organic carbon (POC) flux (Baltar et al. 2010c). Still, at the same time when this was discovered another enigma emerged: what would be the source of energy fueling that dark DIC fixation? Based on the studies performed with a Crenarchaeota isolate (*Nitrosopumilus maritimus*) it was found that oxidation of ammonia (nitrification) is the energy source (Beman et al. 2008; Könneke et al. 2005; Wuchter et al. 2006). However, the ammonia supply rate in the dark ocean is not sufficient to account for the ubiquitous DIC fixation rates measured (Herndl and Reinthaler 2013) and thus other energy sources must be present. Reduced sulfur compounds are also potential energy sources for DIC fixation albeit associated with bacterial DIC fixation (Swan et al. 2011). However, the oxygenated oceanic water column makes it likely that chemolithoautotrophy in the dark ocean is associated with suboxic microniches such as in deep-water marine snow (Swan et al. 2011). Nitrite oxidation by *Nitrospina* and *Nitrospira* was revealed as another key DIC fixation mechanism in the dark ocean (Pachiadaki et al. 2017; Zhang et al. 2020).

### 10.2.2 Diversity and Community Composition of the Dark ocean's Microbiome

The understanding of the phylogeny and functional role of the dark ocean's microbiome progressed tremendously from the development of high throughput sequencing techniques. The first application of high throughput sequencing revealed an unprecedented diversity of deep ocean microbes with a large number of rare taxa coined the "rare biosphere" (Sogin et al. 2006). A striking finding was that despite the known decrease with a depth of bacterial and archaeal abundance a several-fold higher diversity was detected in the deep ocean than in surface waters (Quince et al. 2008; Sogin et al. 2006). Although Alphaproteobacteria and Gammaproteobacteria dominate most of the deep waters there are specific taxa increasing with depth (e.g., Acidobacteria, Actinobacteria, Bacteroidetes, Firmicutes, Gemmatimonadetes, Lentisphaera, Nitrospirae, Planctomycetes and Verrucomicrobia, and Deltaproteobacteria) (Lauro and Bartlett 2008). As mentioned by Nagata et al. (2010), the majority of these bacterial groups that are apparently specially adapted to dark ocean conditions are also dominating in soils (Table 10.1).

The major deep ocean taxa found in different ocean basins are remarkably similar. In the deep waters of the subtropical Pacific (DeLong et al. 2006) and the Ionian Sea (Martín-Cuadrado et al. 2007) most of the Bacteria were affiliated to Alphaproteobacteria, Gammaproteobacteria, Deltaproteobacteria, Actinobacteria,

**Table 10.1** Examples of bacterial groups increasing in abundance with water column depth. General characteristics include functional and physiological features, taxonomic signatures, and typical habitats (in parentheses) that characterize previously known species and genera belonging to the corresponding major phylogenetic group. Adapted from Nagata et al. (2010)

| Microbial group | Examples | General characteristics | References |
|---|---|---|---|
| *Acidobacteria* phylum | *Holophaga, Geothrix, Acidobacterium* | (Soil, sediment) | (Jones et al. 2009) |
| *Bacteroidetes* phylum | *Bacteroidetes, Flavobacteria, Sphingobacteria* | Biopolymer degraders | (Kirchman 2002) |
| *Actinobacteria* phylum | *Frankia, Mycobacterium, Streptomyces* | High GC% gram positives (soil, sediment) | (Jensen and Lauro 2008) |
| *Deltaproteobacteria* subphylum | *Myxoccocus, Bdellovibrio* | Diverse lifestyles, sulfate reducers (soil, sediment) | (Rodionov et al. 2004) |
| *Planctomycetes* phylum | *Planctomyces, Pirellula.* | Biopolymer degraders, anaerobic ammonia oxidizers (sewage, sediment) | (Woebken et al. 2007) |
| *Firmicutes* phylum | *Bacillus, clostridium, Desulfotomaculum* | Low GC% gram positives (soil, sediment) | (Onyenwoke et al. 2004) |
| *Nitrospirae* phylum | *Nitrospira, Leptospirillum, Thermodesulfovibrio* | Nitrite and iron oxidizers, sulfate reducers (aquatic, terrestrial, sewage) | (Maixner et al. 2008) |
| *Gemmatimonadetes* phylum | *Gemmatimonas aurantiaca* | Polyphosphate accumulation (sludge, soil, sediment) | (Zhang et al. 2003) |
| *Lentisphaerae* phylum | *Lentisphaera araneosa, Victivallis vadensis* | Transparent exopolymer particles production (marine, anaerobic digester) | (Cho et al. 2004) |
| *Verrucomicrobia* phylum | *Acidimethylosilex, Alterococcus, Chthoniobacter* | Degraders of plant saccharides, low pH methanotrophs (soils, animals, mud volcano) | (Wagner and Horn 2006) |

Chloroflexi, and Planctomycetes. Among Archaea, the marine Crenarchaeota are commonly increasing in relative abundance from the euphotic to the deep mesopelagic layers (Santoro et al. 2019; Teira et al. 2006).

There is a pronounced transition between the euphotic and the dark ocean microbiome which is also reflected in changes in its alpha-diversity. Taxonomic diversity and richness exhibit a sharp shift below the euphotic zone with higher abundance of unique operational taxonomic units (OTUs) in surface waters followed by a peak at 125 m and 200 m and a drop below 200 m (Mende et al. 2017) (Fig. 10.3). These changes in bacterial and archaeal diversity are paralleled by shifts in physicochemical parameters (Fig. 10.3).

Expectedly, the vertical zonation of bacterial and archaeal groups between the euphotic and the dark ocean is accompanied by a shift in the repertoire of functional genes (DeLong et al. 2006). Besides the strong and expected shift from a community

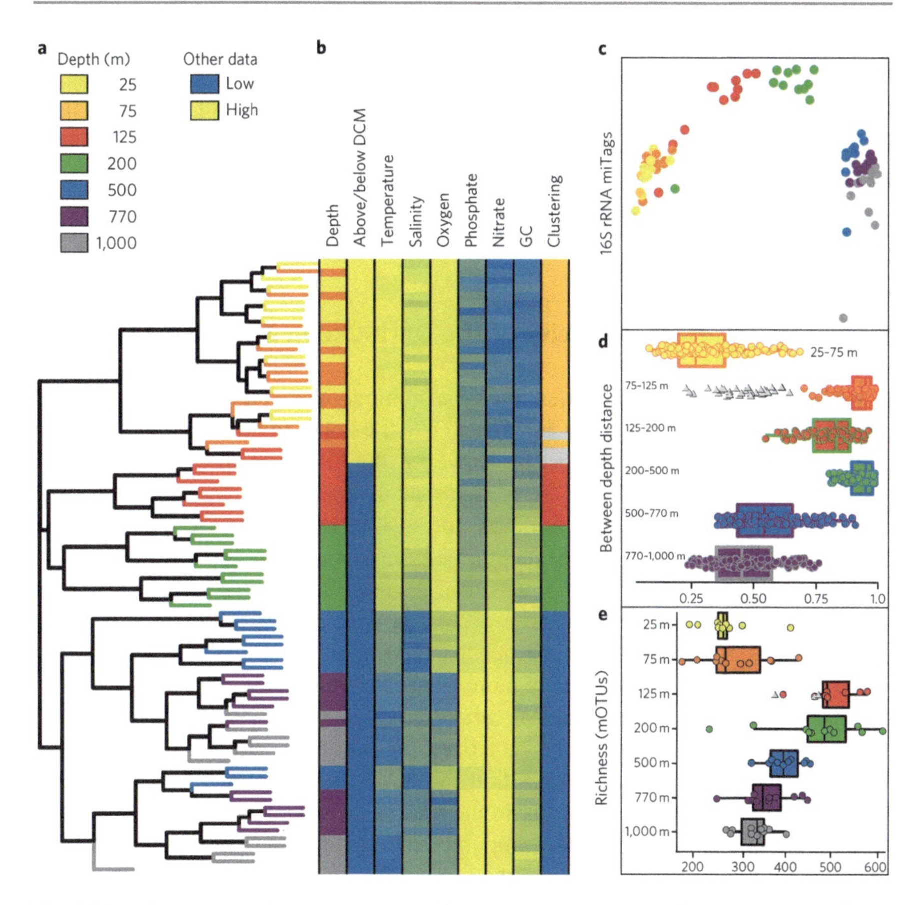


**Fig. 10.3** Quantitative relationships of microbiome genes and taxa and as a function of depth, time, and environmental variables at Station ALOHA. (**a**) Dendrogram displaying the Bray–Curtis distances between sample mOTU abundance profiles across the time series. (**b**) Environmental data represented by a heatmap, ranging from blue (low) to yellow (high). (**c**) Non-metric multidimensional scaling (NMDS) plot of small subunit ribosomal RNA miTag OTU abundance profiles. (**d**) Bray–Curtis distances of mOTU abundances between samples at adjacent depths. Gray triangles at 125 m represent comparisons between 75 m samples and 125 m samples located above the DCM. Whiskers (error bars) show the lowest datum still within the 1.5 interquartile range (IQR) of the lower quartile, and the highest datum still within 1.5 IQR of the upper quartile. (**e**) mOTU richness colored according to depth as in **a**, except for the gray triangles at 125 m, which represent samples located above the DCM. Whiskers are defined as in (**d**) All figures are based on 79 samples. From Mende et al. (2017)

dominated by photoautotrophs (in the euphotic zone) to heterotrophs (in the dark ocean), there are differences in the relative abundance of microbial genes with depth. Strong shifts with depth in the water column have been detected in carbon and energy metabolism as well as in attachment, motility, gene mobility, and host–viral interactions (DeLong et al. 2006). This strong transition between euphotic and aphotic layers is also associated with a community-wide shift of bacterial and archaeal genomes and proteomes in which microbes tend to exhibit larger genome size, higher genomic GC content, and proteins with higher N but lower C content at greater depths (Fig. 10.4) (Mende et al. 2017). These authors found that differences in nutrient availability are associated with the community-wide genome changes concluding that nutrient limitation is a key driver in the evolution and characteristics of the core bacterial and archaeal genomes and proteomes.

### 10.2.3 Spatial Heterogeneity of the Dark Ocean Microbiome

It was assumed for a long time that microbial communities in the deep ocean would be rather stable in terms of space and time due to the stable environmental conditions presumed to occur in the dark ocean. However, the first assessments of the spatial and temporal variability of deep ocean microbial diversity and community composition revealed that the deep ocean was not such a homogenous environment as anticipated. Hewson et al. (2006) reported a heterogeneous bacterial community in the bathypelagic (3000 m) North Pacific and Atlantic Oceans also revealing that the similarity in community composition decreased with increasing distance between sampling sites. Consistently, the mesopelagic Northeast Atlantic was shown to be in terms of bacterial activity relatively more heterogeneous than the epipelagic probably due to mesoscale hydrographical variations (Baltar et al. 2012). These findings suggest that the dark ocean is characterized by substantial heterogeneity. Spatial heterogeneity in community composition and the dispersal of Bacteria and Archaea play key roles in carbon processing in the marine environment (Miki et al. 2008). Thus, constraining the spatial and temporal heterogeneity and connectivity of deep ocean microorganisms become important research foci.

In-depth phylogenetic analyses along the core of the main water masses of the Atlantic revealed that different water masses exhibit differences in microbial community composition (Agogué et al. 2011; Frank et al. 2016). Yet, connectivity of deep ocean pelagic microbial communities appears to be restricted due to the limited mixing between water masses (Agogué et al. 2011; Hamdan et al. 2013) or modulated by advection (Wilkins et al. 2013), which strongly limits the dispersion of the deep ocean microbiome. Large-scale sampling expeditions such as “TARA Oceans” or “Malaspina” substantially increased knowledge on the diversity, ecology, and biogeochemical role of the dark ocean’s microbiome. Prior to these huge sampling efforts, most of the available studies on the phylogenetic and functional diversity of the deep ocean were restricted to a single location or oceanic basin. These sampling campaigns allowed, by using the same sampling and processing/analysis protocols, for the first ocean basin-scale comparison of surface and deep ocean microbiomes. This provided an unprecedented synoptic view of the global phylogenetic and functional diversity of the microorganisms inhabiting the dark ocean (Acinas et al. 2019; Bork et al. 2015; Ruiz-González et al. 2020; Salazar et al. 2016, 2019; Sunagawa et al. 2015, 2020). The combination of circumnavigational sampling and the application of high-throughput sequencing allowed for the

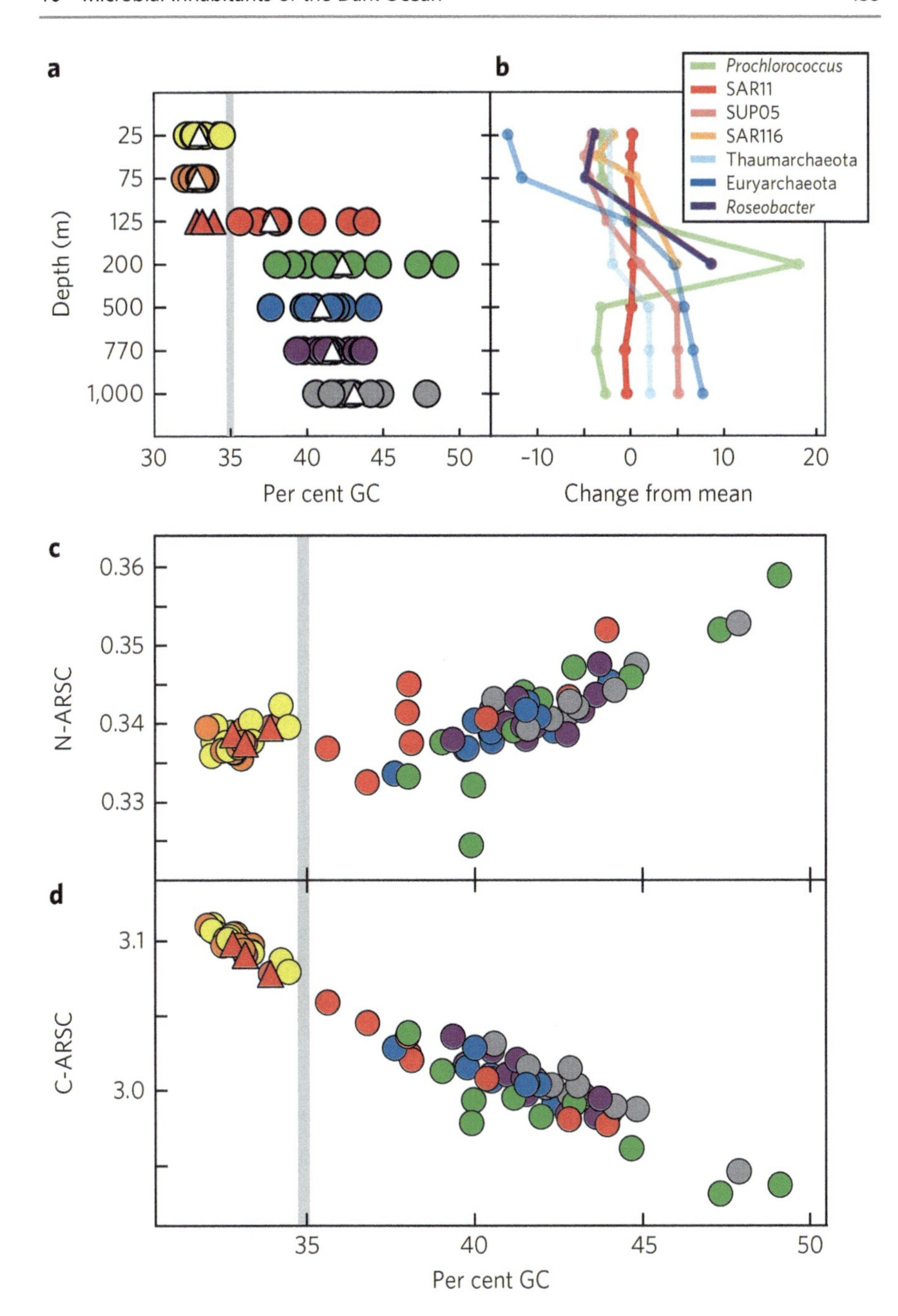


**Fig. 10.4** Microbiome GC content, N-ARSC and C-ARSC (i.e., average number of nitrogen and carbon atoms, respectively, per amino acid residue side chain) versus depth at Station ALOHA. (**a**) Weighted average GC content of all assembled genes in bulk microbial communities. Red triangles indicate 125 m samples collected during periods when the DCM was located below a depth of 125 m. The vertical gray line highlights the partitioning of GC values in samples located above (left) and below (right) the DCM. (**b**) Average difference between the GC content of mOTU genes that map to select taxa at a given depth, and the overall GC mean across all samples. (**c**, **d**) Weighted

identification of the most widespread OTUs in the global deep ocean and its phylogenetic and functional diversity (Salazar et al. 2016) (Fig. 10.5).

Salazar et al. (2016) found that approximately half of the OTUs detected are affiliated to previously unknown taxa. They found that the global dark ocean harbors a modest number of microbial phylotypes (~3600 OTUs) because on average 42% of the OTUs detected in one sample were also detected in another randomly selected sample. In a previous study, Salazar et al. also distinguished between particle-attached and free-living communities (Salazar et al. 2015). They found major differences in the phylogenetic composition of free-living and particle-associated dark ocean bacteria, which suggests niche partitioning. These global-ocean results derived from bathypelagic communities are also consistent with studies performed at specific sites in deep and surface waters (Crespo et al. 2013; Duret et al. 2019; Eloe et al. 2011; Ghiglione et al. 2007; Mestre et al. 2020; Steiner et al. 2020). Most of the bathypelagic Bacteria and Archaea (60%) were detected either particle-attached or free-living but rarely in both (micro-)environments. This pattern appears to be phylogenetically conserved indicating that the ambient deep waters and the particles represent two different ecological niches and that from an evolutionary perspective transitions from one to the other environments are rare (Salazar et al. 2015). Both fractions, free-living (<0.8 μm size fraction) and particle-attached (0.8–20 μm) bacterial and archaeal communities exhibit consistent differences in alpha- and beta-diversity at a global scale.

The preference for a particle-associated or free-living life mode of specific taxa can also be defined via the "particle-association niche index" (PAN index; Stegen et al. 2012, 2013). Using the PAN index, Salazar et al. (2015) grouped the main taxa into preferentially particle-associated or free-living members. Archaea (both Crenarchaeota and Euryarchaeota) exhibited a strong preference for a free-living lifestyle (Salazar et al. 2015). Bacterial taxa with a preference for a free-living life mode included taxa abundant and specific to the lower meso- and bathypelagic waters such as members of the SAR86, SAR324, SAR406, and SAR202 clades (Agogué et al. 2011; Baltar et al. 2016a; Ghiglione et al. 2012; Landry et al. 2017) suggesting that the bulk of the bacterial and archaeal community in the bathypelagic realm consists of free-living microorganisms. Yet, particle-attached microorganisms are usually more active than free-living ones (Ghiglione et al. 2007; Kirchman and Ducklow 1993) and can compensate their lower abundance with higher activity. The particle-attached Bacteria included members of the Bacteroidetes, Firmicutes, Planctomycetes, Deltaproteobacteria clade OM27, and Desulfuromonadales (Salazar et al. 2015).

---

**Fig. 10.4** (continued) average of N-ARSC (**c**) and C-ARSC (**d**) values of all Station ALOHA genes as a function of their corresponding GC content across all samples. Samples to the left of the vertical gray line were collected above and to the right below the DCM, respectively. Sample points are colored by the depth of origin as in Fig. 10.3. All figures are based on 79 samples. From Mende et al. (2017)

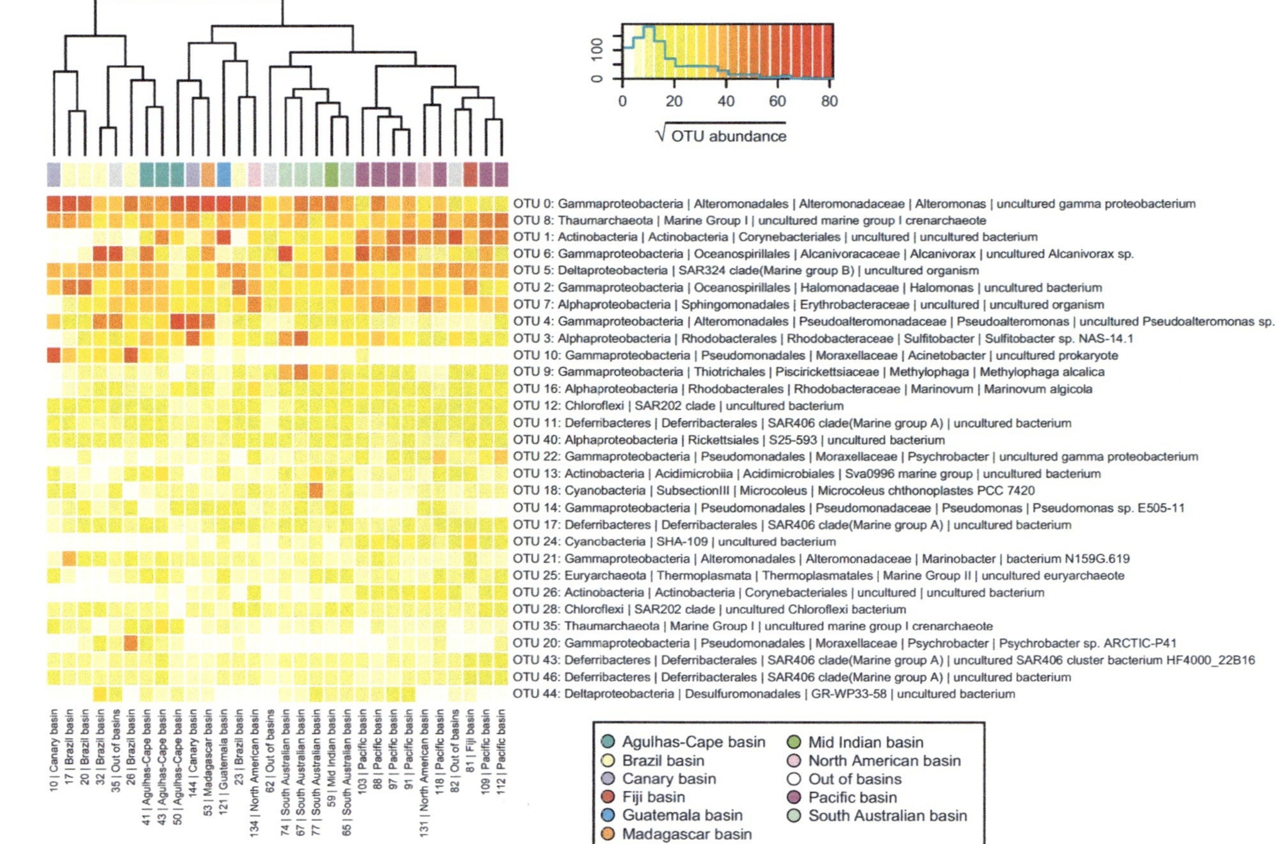
0
100
0
20
40
60
80
√OTU abundance
OTU 0: Gammaproteobacteria | Alteromonadales | Alteromonadaceae | Alteromonas | uncultured gamma proteobacterium
OTU 8: Thaumarchaeota | Marine Group I | uncultured marine group I crenarchaeote
OTU 1: Actinobacteria | Actinobacteria | Corynebacteriales | uncultured | uncultured bacterium
OTU 6: Gammaproteobacteria | Oceanospirillales | Alcanivoracaceae | Alcanivorax | uncultured Alcanivorax sp.
OTU 5: Deltaproteobacteria | SAR324 clade(Marine group B) | uncultured organism
OTU 2: Gammaproteobacteria | Oceanospirillales | Halomonadaceae | Halomonas | uncultured bacterium
OTU 7: Alphaproteobacteria | Sphingomonadales | Erythrobacteraceae | uncultured | uncultured organism
OTU 4: Gammaproteobacteria | Alteromonadales | Pseudoalteromonadaceae | Pseudoalteromonas | uncultured Pseudoalteromonas sp.
OTU 3: Alphaproteobacteria | Rhodobacterales | Rhodobacteraceae | Sulfitobacter | Sulfitobacter sp. NAS-14.1
OTU 10: Gammaproteobacteria | Pseudomonadales | Moraxellaceae | Acinetobacter | uncultured prokaryote
OTU 9: Gammaproteobacteria | Thiotrichales | Piscirickettsiaceae | Methylophaga | Methylophaga alcalica
OTU 16: Alphaproteobacteria | Rhodobacterales | Rhodobacteraceae | Marinovum | Marinovum algicola
OTU 12: Chloroflexi | SAR202 clade | uncultured bacterium
OTU 11: Deferribacteres | Deferribacterales | SAR406 clade(Marine group A) | uncultured bacterium
OTU 40: Alphaproteobacteria | Rickettsiales | S25-593 | uncultured bacterium
OTU 22: Gammaproteobacteria | Pseudomonadales | Moraxellaceae | Psychrobacter | uncultured gamma proteobacterium
OTU 13: Actinobacteria | Acidimicrobiia | Acidimicrobiales | Sva0996 marine group | uncultured bacterium
OTU 18: Cyanobacteria | SubsectionIII | Microcoleus | Microcoleus chthonoplastes PCC 7420
OTU 14: Gammaproteobacteria | Pseudomonadales | Pseudomonadaceae | Pseudomonas | Pseudomonas sp. E505-11
OTU 17: Deferribacteres | Deferribacterales | SAR406 clade(Marine group A) | uncultured bacterium
OTU 24: Cyanobacteria | SHA-109 | uncultured bacterium
OTU 21: Gammaproteobacteria | Alteromonadales | Alteromonadaceae | Marinobacter | bacterium N159G.619
OTU 25: Euryarchaeota | Thermoplasmata | Thermoplasmatales | Marine Group II | uncultured euryarchaeote
OTU 26: Actinobacteria | Actinobacteria | Corynebacteriales | uncultured | uncultured bacterium
OTU 28: Chloroflexi | SAR202 clade | uncultured Chloroflexi bacterium
OTU 35: Thaumarchaeota | Marine Group I | uncultured marine group I crenarchaeote
OTU 20: Gammaproteobacteria | Pseudomonadales | Moraxellaceae | Psychrobacter | Psychrobacter sp. ARCTIC-P41
OTU 43: Deferribacteres | Deferribacterales | SAR406 clade(Marine group A) | uncultured SAR406 cluster bacterium HF4000_22B16
OTU 46: Deferribacteres | Deferribacterales | SAR406 clade(Marine group A) | uncultured bacterium
OTU 44: Deltaproteobacteria | Desulfuromonadales | GR-WP33-58 | uncultured bacterium
10 | Canary basin
17 | Brazil basin
20 | Brazil basin
32 | Brazil basin
35 | Out of basins
26 | Brazil basin
41 | Agulhas-Cape basin
43 | Agulhas-Cape basin
50 | Agulhas-Cape basin
144 | Canary basin
53 | Madagascar basin
121 | Guatemala basin
23 | Brazil basin
134 | North American basin
62 | Out of basins
74 | South Australian basin
67 | South Australian basin
77 | South Australian basin
59 | Mid Indian basin
65 | South Australian basin
103 | Pacific basin
88 | Pacific basin
97 | Pacific basin
91 | Pacific basin
131 | North American basin
118 | Pacific basin
82 | Out of basins
81 | Fiji basin
109 | Pacific basin
112 | Pacific basin
Agulhas-Cape basin
Brazil basin
Canary basin
Fiji basin
Guatemala basin
Madagascar basin
Mid Indian basin
North American basin
Out of basins
Pacific basin
South Australian basin

**Fig. 10.5** Heatmap representing the square root of abundances (number of reads) of the 30 most abundant OTUs (rows) along the 30 stations (columns). Subsampled abundances to the minimum sequencing depth (10,617 reads per sample) have been used for comparison and data from the two size fractions within a station was summed after subsampling. The deep-oceanic basins to which each station belongs are indicated at the top (see color legend). Taxonomical annotation for each OTU is based on the SILVA taxonomic assignment of each OTU representative sequence. OTUs are ordered top to bottom based on their global abundance in the whole data set. From Salazar et al. (2016)

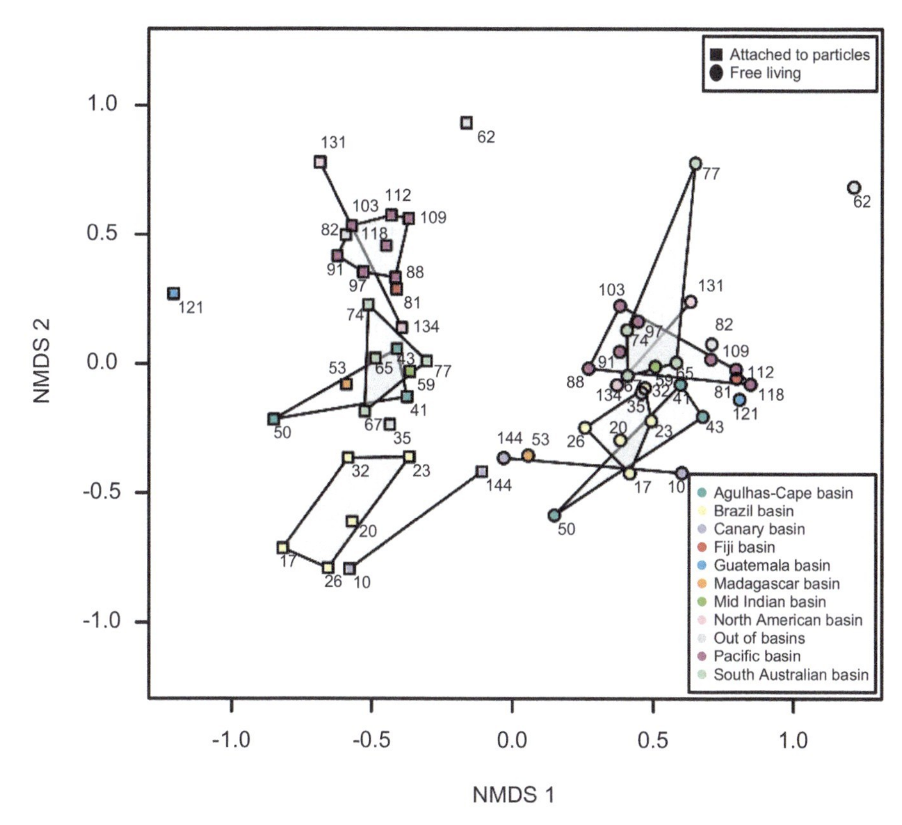


**Fig. 10.6** Non-metric multi-dimensional scaling (NMDS) analysis of beta-diversity (Bray–Curtis distances) for the 60 samples in the data set based on iTags (i.e., amplicon sequencing of the V4 region of the 16S rRNA gene with the Illumina MiSeq platform). Size-fraction is coded with point style (squares, attached and circles, free-living) and deep-oceanic basins following color codes (see legends). Numbers close to each sample represent the station. From Salazar et al. (2016)

For both particle-attached and free-living communities water masses act as key factors controlling their geographical distribution (Fig. 10.6).

However, both the free-living and the particle-attached microorganisms are controlled by different factors: changes in temperature and depth seem to determine the composition of free-living communities whereas those that are particle-attached are more influenced by global circulation and age of water masses (Salazar et al. 2016). These authors also found evidence of dispersal limitation causing basin-specificity in particle-attached communities but not in the free-living fraction. These findings might indicate that a fraction of the particles inhabited by microorganisms in the dark ocean are not sinking from the sunlit surface waters but are neutrally buoyant or slow-sinking particles that are produced autochthonously at depth (Baltar et al. 2009; Baltar et al. 2010c; Herndl and Reinthaler 2013).

### 10.2.4 Surface: Deep Ocean Connectivity of the Microbiome

Sinking particles are the main providers of organic carbon to the dark ocean microbial community (Arístegui et al. 2009; Boyd et al. 2019; Herndl and Reinthaler 2013). Thus, a connection between epipelagic conditions and the composition of the deep ocean microbial community has been reported (Cram et al. 2015a, b; Parada and Fuhrman 2017). This connection between surface and deep-water communities is driven by sinking particles carrying surface microbial cells down the water column (Mestre et al. 2018). Because the microbial community composition of sinking particles is ultimately depending on the environmental conditions and the microorganisms present in the euphotic zone a major imprint of the latter on the microbial community of deep-water particles is evident (Poff et al. 2021; Ruiz-González et al. 2020). This is consistent with previous findings indicating that the nutritional quality and quantity of sinking particles are determined by phytoplankton community composition and grazing (Bach et al. 2019; Boyd and Newton 1995; Guidi et al. 2016). Thus, euphotic planktonic communities can impact dark ocean microbial assemblages by controlling the properties of the sinking material. Consistently, the observed transmission from epi- to mesopelagic waters of temporal patterns in the composition of free-living microbial communities is associated to nutrients derived from sinking particles (Cram et al. 2015a; Parada and Fuhrman 2017). Also, since the microbial community associated to phytoplankton changes during their succession (Baltar et al. 2016b; Bunse et al. 2016; Grossart et al. 2005; Needham and Fuhrman 2016) the composition of the phytoplankton community dictates which microorganisms are exported with the sinking particles. This connection between surface phytoplankton and deep-water microbial assemblages might not always and/or everywhere occur. An analysis by Salazar et al. (2015) using the Malaspina expedition data did not find differences in dark ocean microbial communities among the biogeographic provinces as defined by Longhurst (1998) which exhibit different phytoplankton assemblages.

When analyzing the microbial communities of particles up to 200 μm in size it was found that most of the taxa detected in the bathypelagic were also found in the epipelagic realm (Mestre et al. 2018). These authors also found that the spatial differences between sampling sites were similar for particle-attached communities at 3 m and 4000 m depth. To confirm the mechanism behind this surface to deep ocean connection Ruiz-González et al. (2020) compared the changes in community composition from the same locations as Mestre et al. (2018). They found that the spatial distribution of bathypelagic Bacteria and Archaea was mostly linked to the changes in the surface primary productivity and the abundance of surface dinoflagellate and ciliate abundance, but not so much by the bathypelagic environmental conditions (Ruiz-González et al. 2020).

### 10.2.5 Temporal Heterogeneity of the Dark Ocean's Microbiome

Although synoptic studies (done at a single site or on a global scale) provide critical information on the spatial variability in the community composition of the dark ocean's microbiome it is also important to consider the temporal dynamics of the ocean and perform time-series studies to better constrain the full dynamics of the dark ocean's microbiome. Particularly so since available evidence points to a key link between the euphotic and the dark ocean's microbiome via sinking particles, which will fluctuate with time and seasons (Poff et al. 2021).

The first study focusing on the seasonal dynamics of the dark ocean microbiome was performed by Cram et al. (2015a). They found that seasonal patterns in the diversity of the free-living microbial community were pronounced in the surface waters, weak at mid-depths (deep chlorophyll maximum, at 150 m and at 500 m), and again pronounced at the deepest layer (890 m) sampled. Also, samples collected in the same season over a period of 10 years were similar but those obtained from contrasting seasons were dissimilar at both, 5 m and 890 m depth (Fig. 10.7) (Cram et al. 2015a). The highest levels of richness and the inverse Simpson Index were detected in the winter at 5 m and in spring at 890 m depth. These authors also reported that once seasonal/interannual variability was factored out environmental factors explained the variability of the microbial assemblage at 5 m and 890 m but not at the other depths. Cram et al. (2015a) suggest that the microbial community in the dark ocean is connected to the surface water community via fast-sinking particles and migratory organisms. Based on these findings they concluded that the "seasonal" free-living dark ocean taxa are probably those whose ecology is strongly influenced by these particles.

In a subsequent study, using the same dataset, Cram et al. (2015b) used microbial association networks to investigate the links between community dynamics and environmental parameters within and between depth layers. They found concurrent and lagged shifts in community composition between depths (Fig. 10.8). Thus, they concluded that free-living communities in the dark ocean are linked to environmental conditions and/or communities in overlying waters (Cram et al. 2015b). This connection between communities from different depth layers is probably due to sinking particles and/or migrating organisms transporting nutrients across otherwise stratified waters (Maas et al. 2020).

Additionally, recent evidence from a time-series study comparing seasonal dynamics of surface (2 m) and mesopelagic (500 m) waters in the South Pacific Ocean suggests that the connection between surface and dark ocean prokaryotic communities is not simply unidirectional (vertical sinking) (Wenley et al. 2021). These authors identified two main mechanisms connecting surface and mesopelagic microbial communities, which change seasonally. In spring/summer (the productive season) the deep-water community is seeded via sinking particles and in winter via deep mixing bringing typically deep-water taxa into the surface waters (Wenley et al. 2021).

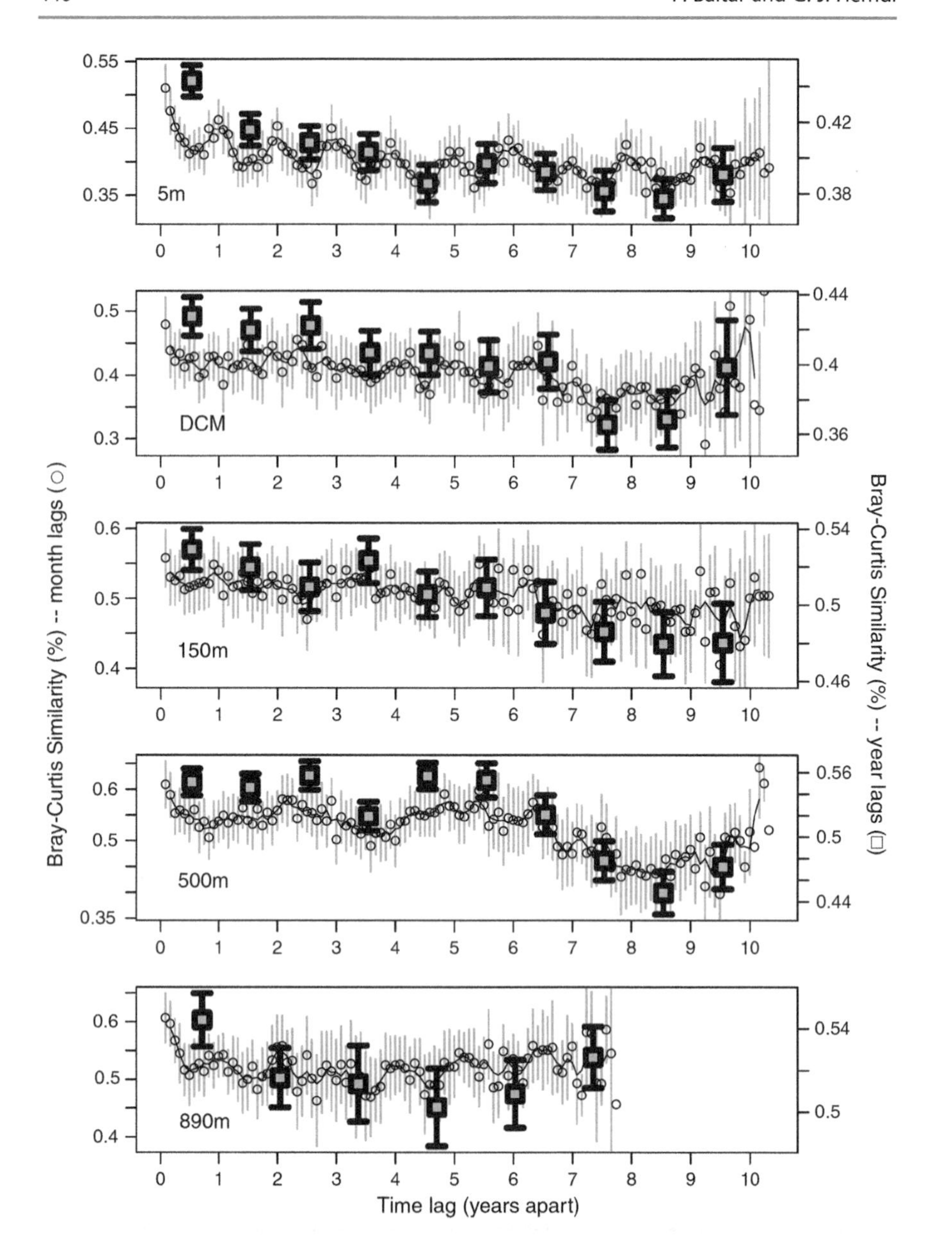


**Fig. 10.7** Mean Bray-Curtis similarities of all pairs of samples (*y*-axis) separated by different intervals of time (*x*-axis). Circles represent the mean similarity of all pairs of samples taken a given number of months apart (intermonthly; left *y*-axis). Thus, the first circle is the mean Bray-Curtis similarity of all pairs of samples taken 15–45 days (B1 month) apart, the second circle is the mean similarity of all pairs of samples taken 46–76 days (B2 months), the twelfth circle (aligned with the 1-year lag tick mark) represents samples taken 12 months apart and so on. Squares represent the mean similarity of samples taken a given number of years apart (interannual; right *y*-axis). Accordingly, the first square represents the mean Bray-Curtis similarity of all pairs of samples taken less than 1 year apart, the second is all pairs of samples taken between 1 and 2 years apart and so on. Error bars, for both types of data points, represent 95% confidence intervals of the mean

## 10.3 Functional Diversity of the Dark Ocean's Microbiome

The functional diversity and the associated ecological and biogeochemical role of dark ocean microorganisms have received considerable attention in recent years. Generally, it has been found that the phylogenetic composition of the microbial community and its functional diversity is depth stratified in the oceanic water column (Bergauer et al. 2017; DeLong et al. 2006; Ruiz-González et al. 2020; Zhao et al. 2020). One of the main peculiarities of the functional traits of the dark ocean's microbiome is the role of chemolithoautotrophic metabolisms in the oxygenated deep-waters of the global ocean mostly associated to ammonia oxidation by Crenarchaeota (Wuchter et al. 2006), *Nitrospina* and *Nitrospira* nitrite oxidation (Bayer et al. 2021; Pachiadaki et al. 2017; Zhang et al. 2020) and other Gammaproteobacteria with the potential to utilize reduced sulfur compounds (Swan et al. 2011). Nitrifying Crenarchaeota exhibit differences in the ammonia monooxygenase (*amoA*) gene depending apparently on the prevailing ammonia concentrations in the different water layers of the deep ocean (Sintes et al. 2013). A low-affinity ammonia monooxygenase enzyme is expressed in lower euphotic and mesopelagic waters where generally higher ammonia concentrations are detected versus a high-affinity ammonia monooxygenase in bathypelagic waters (Sintes et al. 2013, 2016). Another functional peculiarity of the dark ocean's microbiome is the capability of some of its members such as SAR202 to utilize recalcitrant dissolved organic matter, which makes up the majority of the bulk organic carbon in the dark ocean (Landry et al. 2017). Since fixing inorganic carbon requires energy (Hügler and Sievert 2011), and energy sources are generally limited in the dark ocean, mixotrophy (i.e., the capability to fix inorganic carbon and to perform heterotrophy (Baltar and Herndl 2019; Dick et al. 2008; Sorokin 2003)) has been suggested as a particularly favorable and efficient strategy for the dark ocean Bacteria and Archaea (Acinas et al. 2019; Baltar and Herndl 2019; Baltar et al. 2016a).

A global ocean analysis of 58 bathypelagic metagenomes from the Malaspina expedition uncovered the functional potential of dark ocean microbes producing a vast gene collection covering 1.12 million genes (i.e., nonredundant unique gene clusters) of which 71% were novel (Acinas et al. 2019). Thus, these genes had not been found in global epipelagic surveys (Sunagawa et al. 2015). A strong dichotomy was found between the functional characteristics of particle-attached and free-living microbes (Fig. 10.9). These findings indicate that dark ocean's particle-attached and free-living communities represent contrasting lifestyles and provide another line of evidence suggesting the heterogeneous nature of the dark ocean (Zhao et al. 2020).

**Fig. 10.7** (continued) similarity. Points with nonoverlapping error bars suggest statistically significant differences in mean similarity between samples taken different distances apart in time. For instance, samples taken 6 months apart in the surface (the sixth circular data point) are statistically less similar than samples taken 1 month apart (first circular point), while samples taken 12 months apart (twelfth circular point) are not less similar than samples taken 1 month apart. From Cram et al. (2015a)

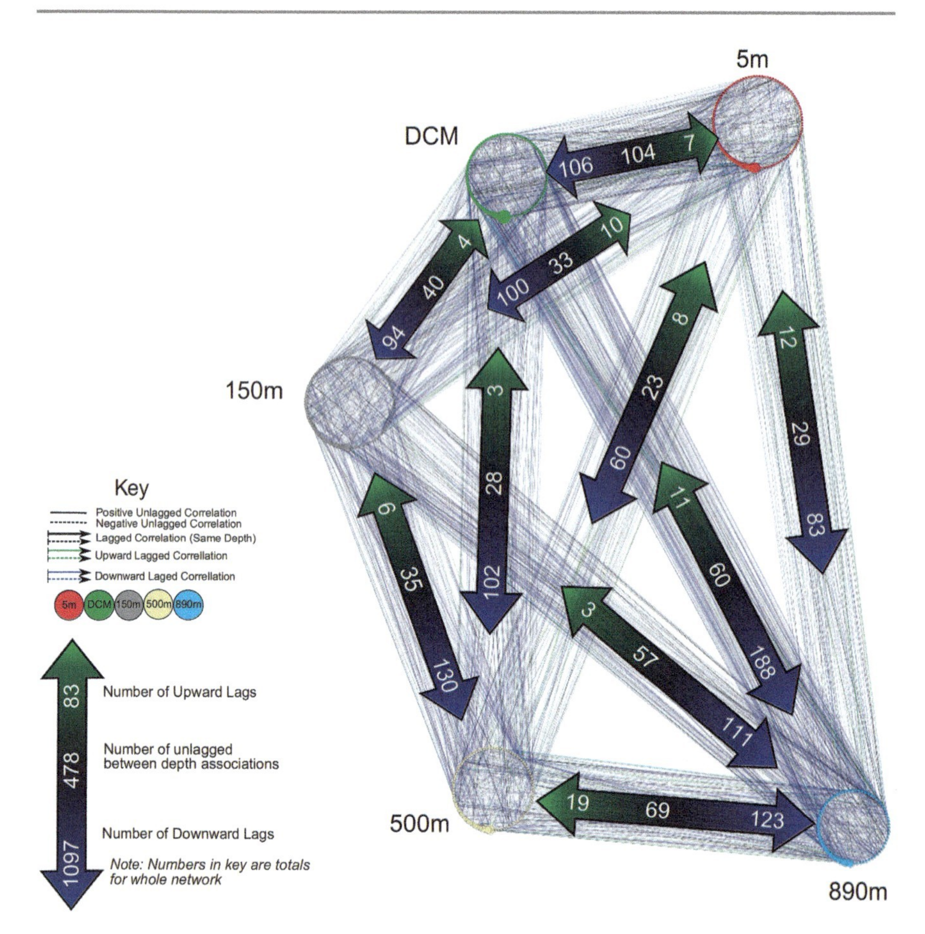


**Fig. 10.8** Association network showing significant, time-lagged and non-time-lagged correlations between bacterial nodes both within and between depths. Nodes (circles) represent bacterial OTUs at each depth. Edges (lines) represent correlations, sometimes time-lagged, between the bacterial OTUs. Shown are bacteria that occur at least 25 times and edges that have lagged Spearman correlations such that $P < 0.01$, $Q < 0.05$. Edges are color-coded to indicate whether they represent unlagged correlations, "downward" time-lagged correlations in which changes in the shallower node precede changes in the deeper node, or "upward" lagged correlations in which changes in the deeper node precede changes in the shallower node. Large arrows show the total numbers of edges of each type connecting depths. From Cram et al. (2015b)

Dissolved inorganic carbon fixation pathways (particularly the Calvin-Benson-Bassham cycle and the 3-hydroxypropionate bicycle) are ubiquitously distributed in the global dark ocean in both, particle-attached and free-living communities (Fig. 10.10) (Acinas et al. 2019).

Nitrification genes are linked mainly to the free-living microbial community (Acinas et al. 2019). In contrast, genes associated to $H_2$ and CO oxidation ($H_2$ oxidation genes in 24% of the samples and CO oxidation genes in all samples) were mainly associated to particle-attached microorganisms. Thus, $H_2$ and CO are major

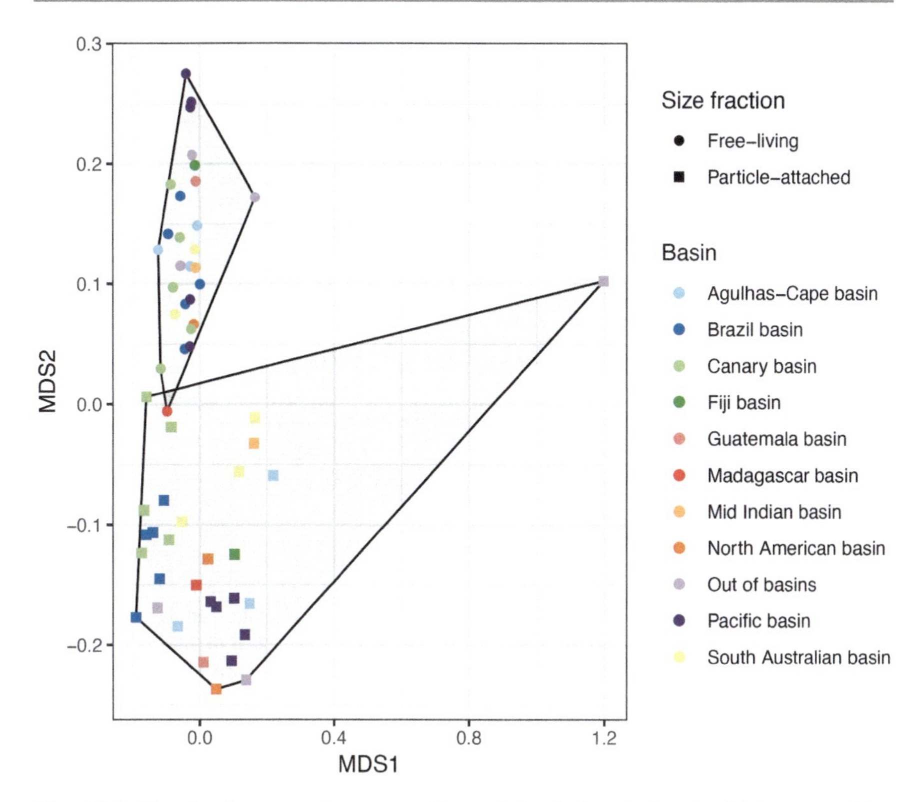


**Fig. 10.9** Functional community composition of the bathypelagic microbial communities. Non-metric multidimensional scaling (NMDS) of the microbial communities based on the functional compositional similarity (Bray-Curtis distances) among the 58 samples in the dataset based on clusters of KEGG orthologous groups (KOs). Size-fraction is coded by a symbol (squares, particle-attached and circles, free-living microorganisms) and the main deep-ocean basins by color codes (see legends). From Acinas et al. (2019)

energy sources in the dark ocean. Moreover, genes involved in anaerobic metabolisms (e.g., dissimilatory nitrate reduction, sulfate reduction, methanogenesis, and denitrification) were also detected in 60–100% of the samples (Acinas et al. 2019). The abundant and widespread presence of genes indicative of anaerobic metabolism in the oxic water column is clear evidence of the widespread existence of hypoxic and anoxic microniches in the dark ocean. Interestingly, 24% of the bathypelagic metagenomic assembled genomes (MAGs) produced in their study included mixotrophy potential, and those mixotrophic MAGs were detected in >2/3 of the samples (Acinas et al. 2019). These authors concluded that the metabolic versatility mixotrophy offers (and the capability to switch between autotrophic and heterotrophic metabolism) are a widespread and common trait of dark ocean microbes.

Informative metagenomic analyses reveal only the potential to perform specific metabolic reactions and thus, do not necessarily translate into RNA or protein

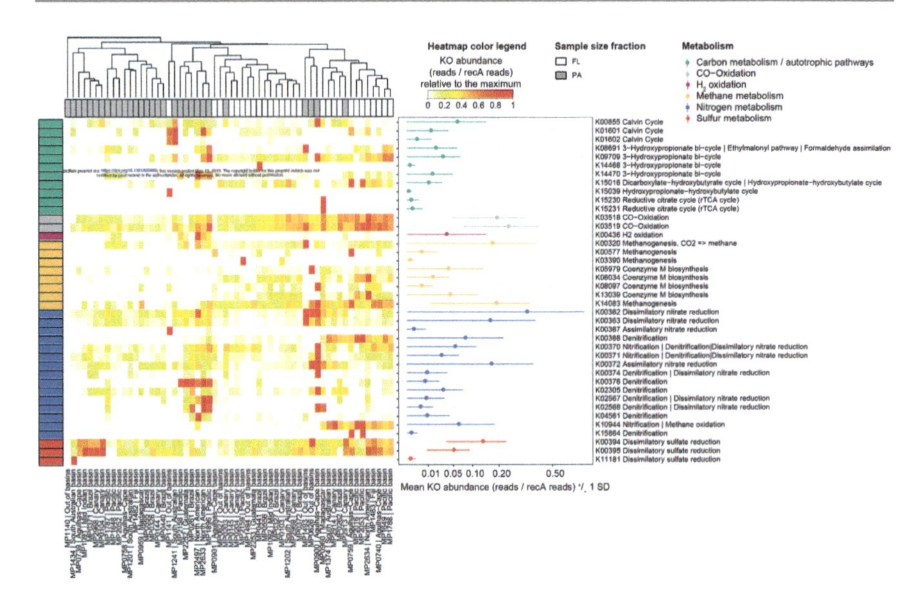


**Fig. 10.10** Heat map of selected markers genes for different metabolic pathways across the 58 metagenomes. A total of 40 marker genes (KOs, *y*-axis) were indicative of different metabolic processes detected in the Malaspina samples (*x*-axis). KO abundance was normalized by *recA* single copy gene as a proxy for copy number per cell. The general metabolism assignation is color coded (see legend in the upper right) and the KEGG module(s) assignation used in the KO label is also indicated. The relative abundance across samples for each KO is shown in the heatmap. The mean ($\pm$ 1 s.d.) untransformed abundance of each KO across all samples (reads/*recA* reads) is presented in the right panel. From Acinas et al. (2019)

expression. In the only available global deep ocean metatranscriptomics/metagenomics analysis an unexpected disparity was found between metagenomic and metatranscriptomic richness patterns (Fig. 10.11) (Salazar et al. 2019). It was suggested that the non-transcribed fraction of a given metagenome is higher in mesopelagic relative to euphotic waters (Salazar et al. 2019). This pattern could be either associated to a higher prevalence of dormant or dead cells in the mesopelagic than in the epipelagic layers or due to the prevalence of genome streamlining in epipelagic waters (Swan et al. 2013) where the abundance of genes per genome is generally lower (Mende et al. 2017).

The largest differences in genes and transcripts between epi- and mesopelagic waters were, as expected, photosynthesis marker genes (*psaA* and *psbA*), and genes encoding the subunits of RuBisCO (*rbcL* and *rbcS*) (Fig. 10.12) (Salazar et al. 2019). These authors also found low RuBisCO gene expression levels in mesopelagic waters affiliated to chemolithoautotrophs. Also, genes and transcripts indicative for denitrification (*napA*, *nirS*, *norB*, and *nosZ*) were more abundant in mesopelagic compared to epipelagic waters (Salazar et al. 2019). In addition, *nifH* gene expression was detected in mesopelagic Arctic waters (Salazar et al. 2019).

Although metatranscriptomic analyses are closer to the actually realized function of microbes than metagenomics, the real actors of the biogeochemical

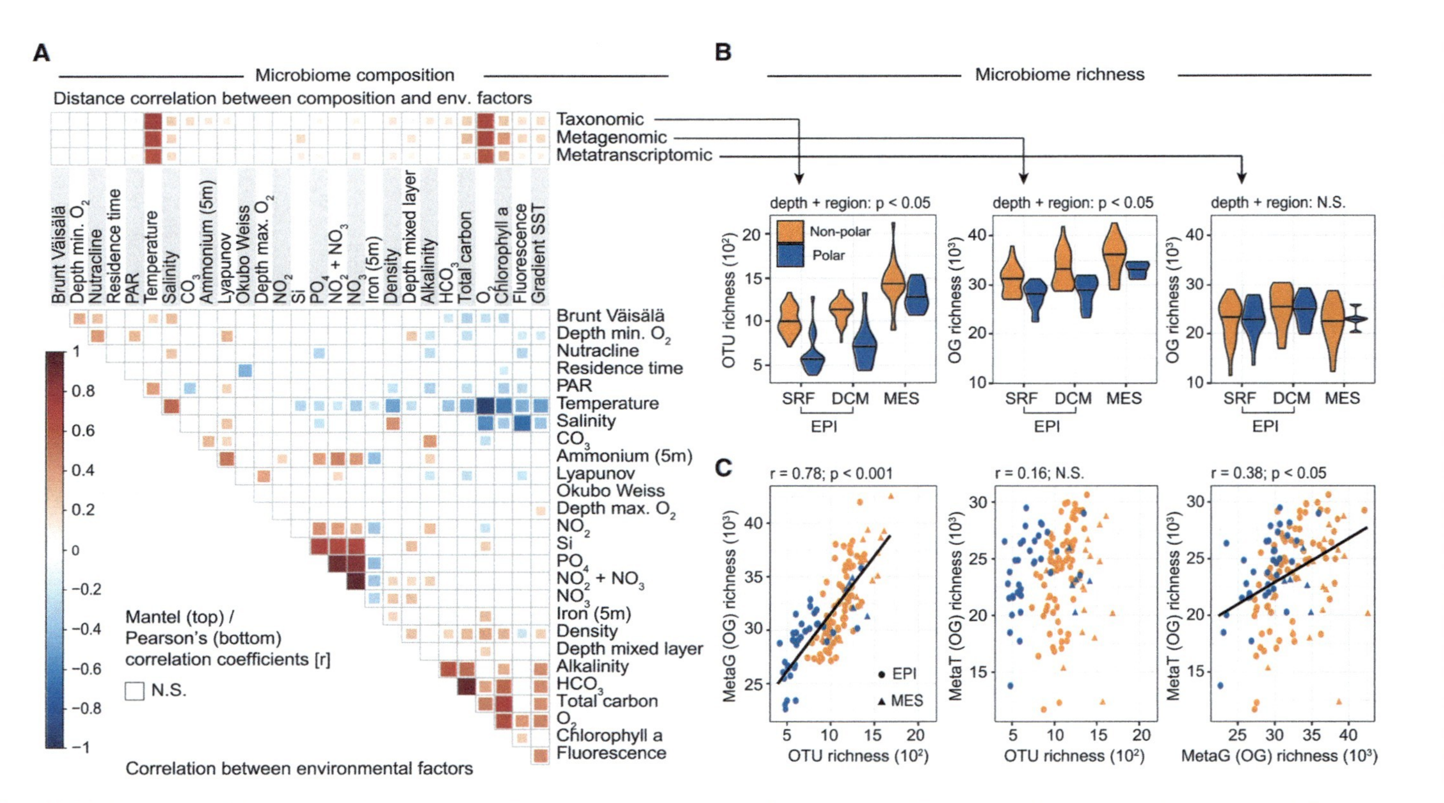


**Fig. 10.11** Patterns and drivers of global ocean microbiome composition across depth layers and between polar and nonpolar regions. (**a**) Taxonomic, metagenomic, and metatranscriptomic composition of epipelagic samples (based on mitags and the normalized abundances of eggNOG-derived orthologous groups (OGs) from metagenomic and metatranscriptomic data, respectively) were related to each of 27 environmental factors using partial (geographic distance corrected) Mantel tests with 10,000 permutations and Bonferroni correction. Pairwise comparisons of environmental factors are shown below, with a color

transformations are the proteins. The few available metaproteomic studies have already shed light on the functional role of dark ocean microbes. The only drawback though is that the level of resolution of metaproteomic analysis is still not as deep as that of metatranscriptomics. Shifts in the community composition with depth were compared to changes in transporter proteins using metaproteomic and metagenomic analyses of microbial communities collected from 100- to 5000 m depth in the Atlantic Ocean (Bergauer et al. 2017). That study revealed the vertical dynamics of transport and uptake of solutes by dark ocean microorganisms indicating a high expression level of transport proteins in the bathypelagic realm despite the pronounced decrease of organic matter with depth. However, the relative quantity of individual transporter systems differed with depth (Bergauer et al. 2017). For instance, TRAP transporters were preferentially detected in bathypelagic samples (Fig. 10.13).

Bergauer et al. (2017) concluded that while the phylogenetic composition of the communities changes with depth, the transporters (composition and substrate specificities) are ubiquitous, however, their abundance changes with depth.

A multi-omics (integrating metagenomics, metatranscriptomics, and metaproteomics) study focused on a global survey of two key enzyme groups initiating organic matter assimilation (i.e., peptidases and carbohydrate-active enzymes (CAZymes)) (Zhao et al. 2020). This study revealed a genomic and functional adaptation of deep ocean Bacteria and Archaea in response to the low concentration and diverse nature of the deep-sea dissolved organic matter pool. The main adaptation to the deep ocean conditions is an increasing proportion of genes encoding secreted enzymes with depth (Fig. 10.14), which is consistent with the previously reported high proportion of dissolved versus total extracellular enzymatic activities in the deep sea measured with substrate analogs (Baltar et al. 2010b, 2013).

**Fig. 10.11** (continued) gradient denoting Spearman's correlation coefficients. Temperature is the best explanatory variable for all of the profiles in the epipelagic ocean (taxonomic profile: Pearson's $r = 0.75$; metagenomic profile: Pearson's $r = 0.69$; metatranscriptomic profile: Pearson's $r = 0.64$; all $p < 0.05$), followed by oxygen concentration, which is highly correlated to temperature (Pearson's $r =$ _0.72). A more detailed description of the variables is available in https://doi.org/10.5281/zenodo.3473199. (**b**) Compositional richness of polar and nonpolar microbiomes across three depth layers. Taxonomic and functional metagenomic richness (numbers of OTUs and OGs, respectively) increases with depth, although the richness is consistently lower in polar samples than in nonpolar samples (two-way ANOVA: $p < 0.05$ for depth layers and polar/nonpolar, for both taxonomic and metagenomic functional richness). By contrast, there was no significant difference in functional metatranscriptomic richness (number of OGs), either across depths or between polar and nonpolar samples (two-way ANOVA: $p > 0.05$ for depth layers and polar/nonpolar). Violin plots represent the (mirrored) density distribution of the data with the median shown as a horizontal line. (**c**) Correlations among species richness (number of OTUs), functional metagenomic (metaG) richness, and metatranscriptomic (metaT) richness (number of OGs). Data were rarefied before richness computation (STAR methods). Pearson's correlation was used for all comparisons (OTU-metaG; $r = 0.78$, $p < 0.001$; OTU- metaT: $r = 0.16$, $p = 0.06$; metaG-metaT: $r = 0.39$, $p < 0.05$). The solid line corresponds to the best linear fit. N.S., not significant ($p > 0.05$). From Salazar et al. (2019)

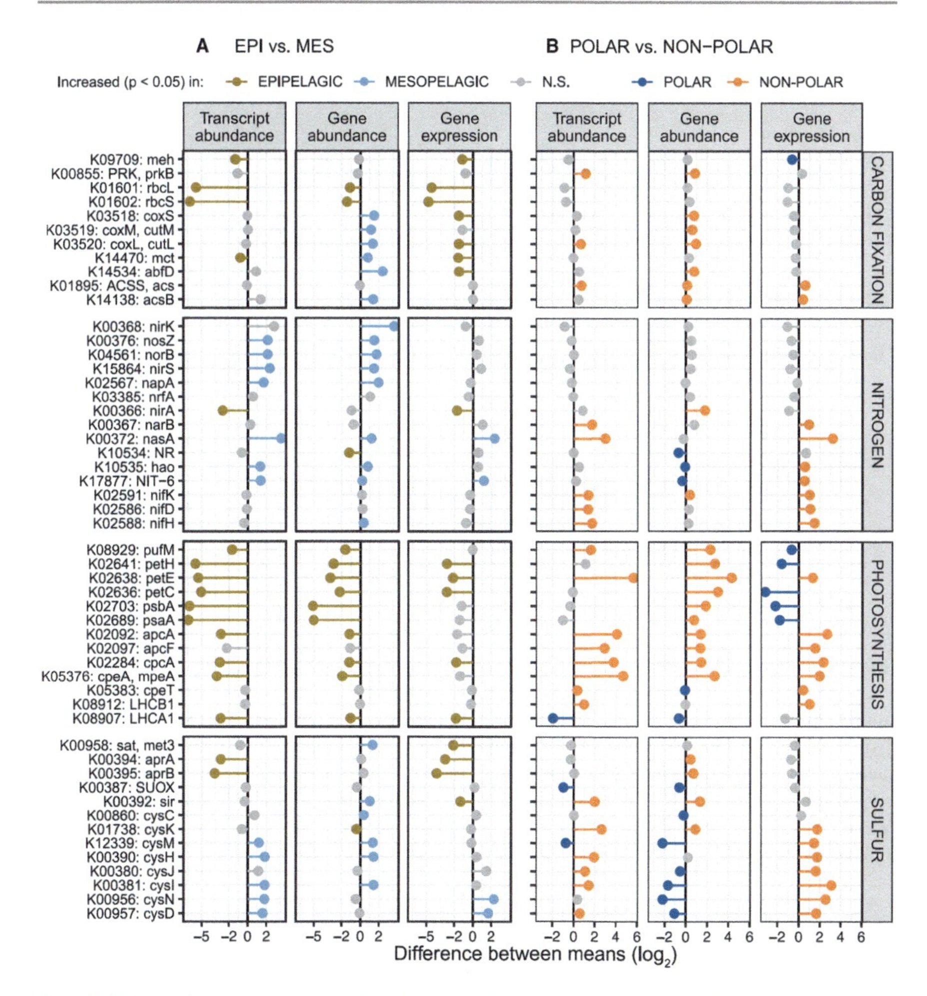


**Fig. 10.12** Differences in gene abundance and expression determine differential transcript abundances of metabolic marker genes across depth layers and between polar and nonpolar regions (A and B). Differences in the abundance of genes and transcripts and the gene expression level of metabolic marker genes (KOs) were determined (**a**) between epipelagic and mesopelagic layers and (**b**) between polar and nonpolar regions. The data points show the differences in the mean transcript abundances, mean gene abundances, and mean gene expression (i.e., transcript abundance normalized by gene abundance) of KOs. Differences were computed using log2-transformed values (STAR methods) and tested for significance by Mann-Whitney tests. Differences were considered significant if $p$ values after Holm correction were smaller than 0.05. Only epipelagic samples were used for the data shown in (**b**). From Salazar et al. (2019)

Thus, according to the "foraging theory" increasing secretory enzymatic activities with depth strongly suggest a preferential particle-attached lifestyle of dark ocean microorganisms (Arístegui et al. 2009; Baltar et al. 2009, 2010b). Consistent with the metaproteomic study on membrane-associated transporters of Bergauer et al. (2017), Zhao et al. (2020) also found that the composition of genes encoding specific

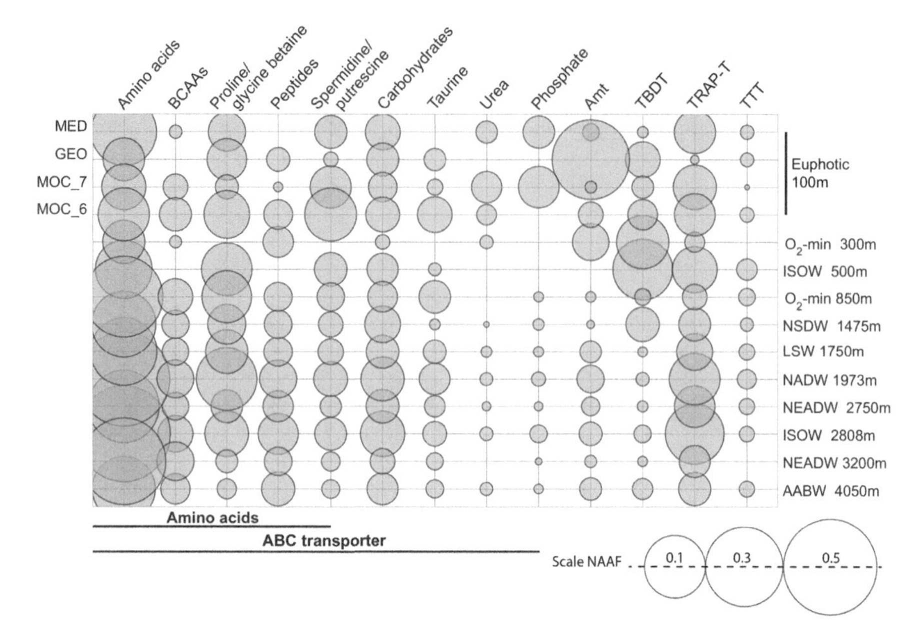


**Fig. 10.13** Vertical expression profiles of selected transporters were analyzed in a semiquantitative manner based on NAAF (normalized area abundance factors) values. Transporter proteins were grouped by the predicted substrate specificity of the substrate-binding proteins. *BCAAs* branched-chain amino acids; *TRAP-T* ATP-independent periplasmic transporters; *TTT* tripartite tricarboxylate transporters; *TBDT* TonB-dependent transporters. *ISOW* Iceland-Scotland Overflow Water, *$O_2$-min* oxygen minimum zone, *MSOW* Mediterranean Sea Outflow Water, *NADW* North Atlantic Deep Water, *NEADW* North East Atlantic Deep Water, *AABW* Antarctic Bottom Water. From Bergauer et al. (2017)

enzymatic functions (both total and secretory) was remarkably constant throughout the water column despite the depth-stratified phylogenetic composition of the microbial community (Fig. 10.15). This points to a high level of functional redundancy within microbial communities across the water column of the dark ocean.

## 10.4 Abyssal and Hadal Phylogenetic and Functional Diversity

Most of the limited research performed on the dark ocean's microbiome focused on the mesopelagic and bathypelagic waters. Very limited information is available, however, on the abyssal- and hadalpelagic zones. In particular, hadal zones (waters below 6000 m depth) comprising almost exclusively the trenches harbor the least explored biomes on Earth. In a detailed study of the phylogenetic composition from surface to the deepest trench on Earth (Challenger Deep) a shift in the community composition was found not only from the epi- to the meso- and bathypelagic realm but also from the abyssopelagic into the hadalpelagic layers (Fig. 10.16) (Nunoura et al. 2015). SAR11 dominated the communities in aphotic waters down to 2000 m

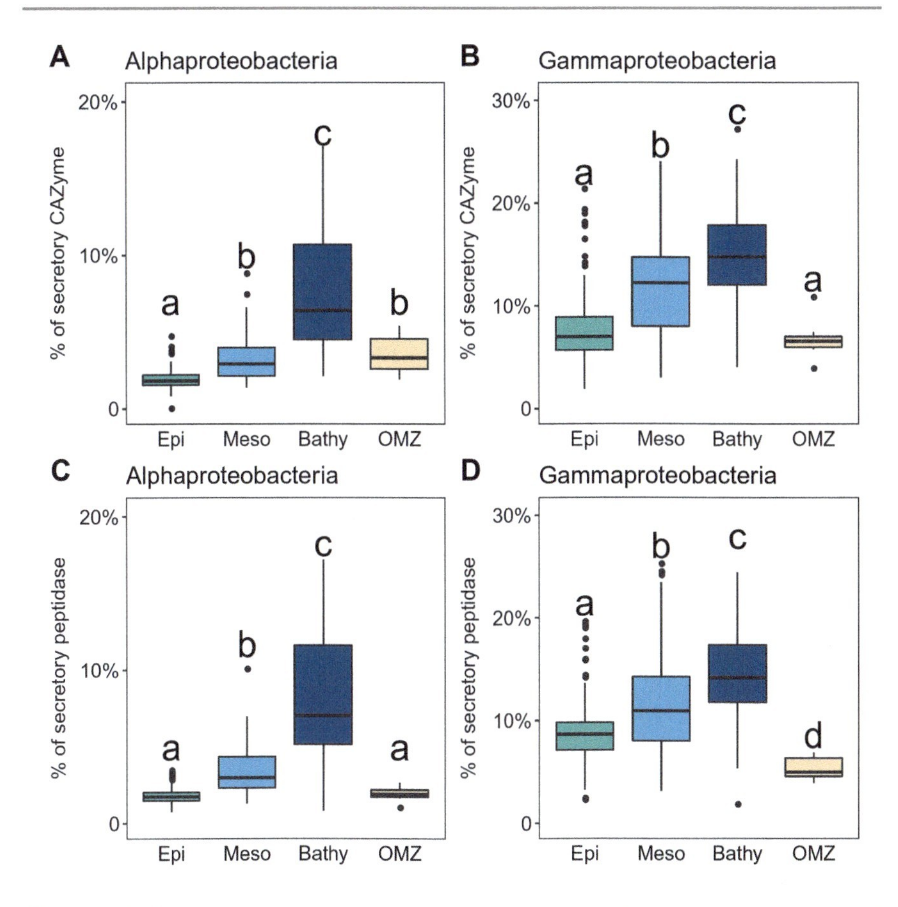


**Fig. 10.14** Percentage of genes encoding secretory CAZymes and peptidases of Alpha-proteobacteria and Gammaproteobacteria increasing with depth. (**a** and **c**), Alpha-proteobacteria; (**b** and **d**), Gammaproteobacteria. Box shows median and IQR; whiskers show 1.5 × IQR of the lower and upper quartiles or range; outliers extend to the data range. Statistics are based on Wilcoxon test, and letters are used to show statistical significance ($p < 0.05$); a shared letter means no significant difference. Epipelagic ($n = 216$); mesopelagic ($n = 68$); bathypelagic ($n = 54$); OMZ ($n = 7$). From Zhao et al. (2020)

depth and decreased drastically in abundance at 3000 m (Nunoura et al. 2015). Although Crenarchaeota was the major group below 150 m depth their abundance decreased with increasing depth (Nunoura et al. 2015). The abundance of SAR324 increased at 150 m depth and was the most abundant group down to 5000 m. In contrast, SAR406 and Bacteroidetes became dominant below 6000 m depth whereas they only comprised a minor fraction above 5000 m (Nunoura et al. 2015). *Halomonas* sp. and *Pseudomonas stutzeri* dominated at the bottom waters of the trench (9000 and 10,241 m depth) (Fig. 10.16).

Nunoura et al. (2015) also found a niche separation among nitrifying populations (both ammonia and nitrate oxidizers) with high abundances of *Nitrospina* and

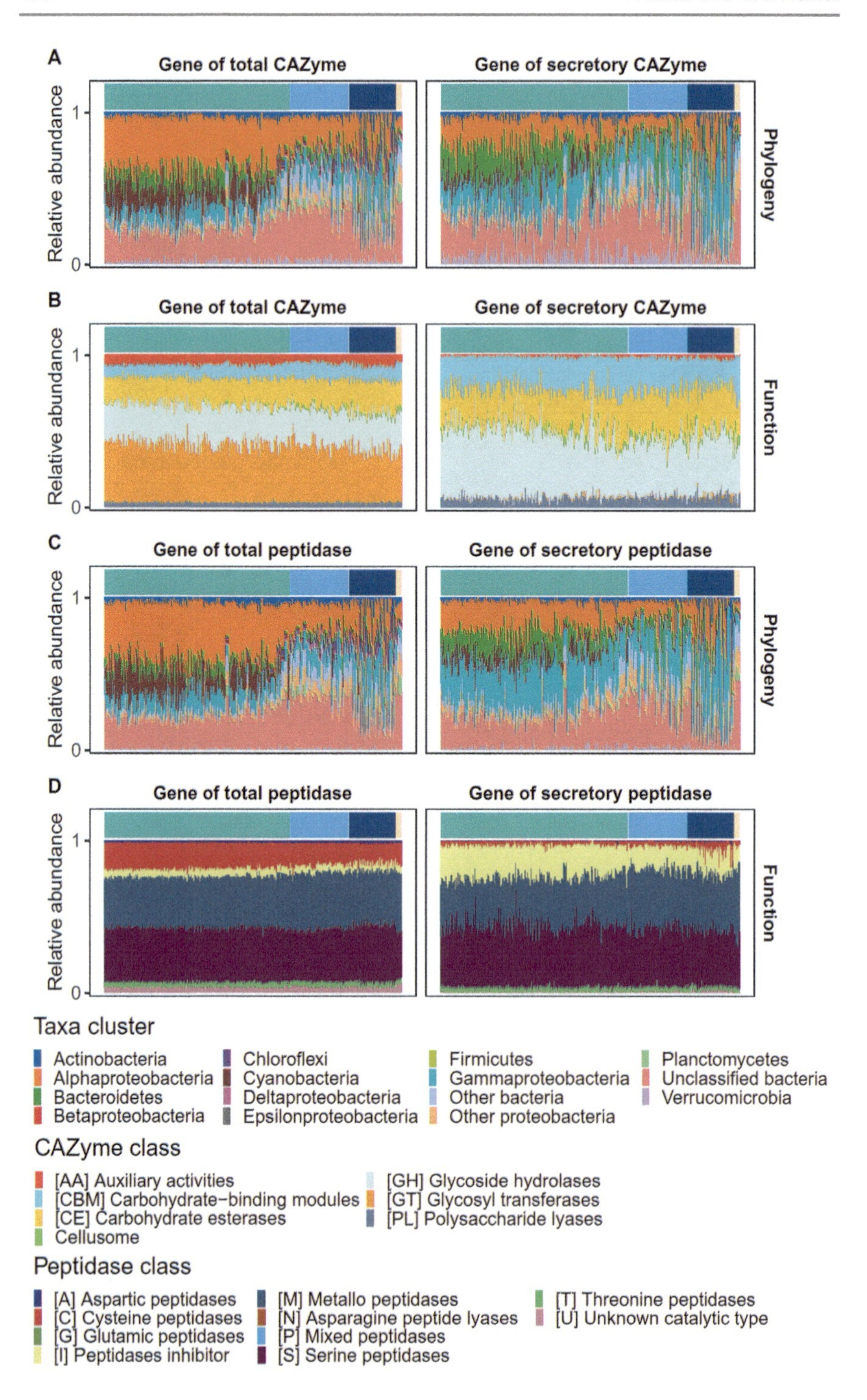


**Fig. 10.15** Phylogenetic affiliation and functional classification of genes encoding bacterial CAZymes and peptidases throughout the water column. Taxonomic variability at the phylum

Crenarchaeota between 150 and 2000 m depth shifting to a predominance of *Nitrospira* and *Nitrosomonas* in the trench waters (Fig. 10.17). Among the Crenarchaeota, the β-subgroup dominated just below the photic zone (150–300 m), the δ-subgroup was found <2000 m, the α-subgroup dominated in the trench waters >6000 m, and the γ-subgroup was found throughout the water column below 150 m depth. In summary, Nunoura et al. (2015) found that the hadal community composition differed from that in abyssal waters. In hadal waters, chemolithoautotrophs decreased in abundance with depth while heterotrophic microbes became dominant. Thus, the hadal microbiome is different from the one in the abyssopelagic mostly due to the partial isolation of the hadal waters from the overlying waters. The more heterotrophic microbial community in hadal waters seems to be supported by the endogenous recycling of organic matter that is associated with the trench geomorphology.

## 10.5 Summary

The major taxonomic groups of Bacteria and Archaea are present throughout the oceanic water column from the euphotic to the hadal waters as revealed by 16S rRNA sequencing. On a finer taxonomic level, however, the microbial community is depth stratified in the ocean water column apparently adapted to the specific nutrient conditions in the different water layers (Agogué et al. 2011; DeLong et al. 2006). Also, there are ecotypes (also coined oligotypes) present in specific depth layers such as shown for the abundant SAR11 (Thrash et al. 2014) or *Alteromonas macleodii* (Ivars-Martínez et al. 2008). On a functional gene level adaptations to specific physicochemical conditions have been found such as for nitrifying Crenarchaeota exhibiting differences in the *amoA* gene where different ammonia monooxygenases are expressed depending on the ammonia concentrations in the water column (Sintes et al. 2013). Some specific Bacteria such as SAR202 are increasing even in absolute abundance from the lower euphotic layer to the bathypelagic waters (Varela et al. 2008). SAR202 is specifically adapted to utilize refractory dissolved organic matter in the deep ocean (Landry et al. 2017). On the proteome level, differences between the different depth layers in the water column are minor on the transporter level (Bergauer et al. 2017) albeit clearly detectable on the expression level of carbohydrate active enzymes and peptidase among different phylogenetic taxa (Zhao et al. 2020). Overall, there is a surprisingly high chemolithoautotrophic potential in the microbial community of the dark ocean. With increasing sampling efforts and using

**Fig. 10.15** (continued) level (class level for Proteobacteria) of genes encoding CAZymes (**a**) and peptidases (**c**); functional composition of genes encoding CAZymes (**b**) and peptidases (**d**). Color bars along *x*-axis indicate samples from different depths: green, epipelagic; light blue, mesopelagic; dark blue, bathypelagic; sandy yellow, oxygen minimum zone. From Zhao et al. (2020)

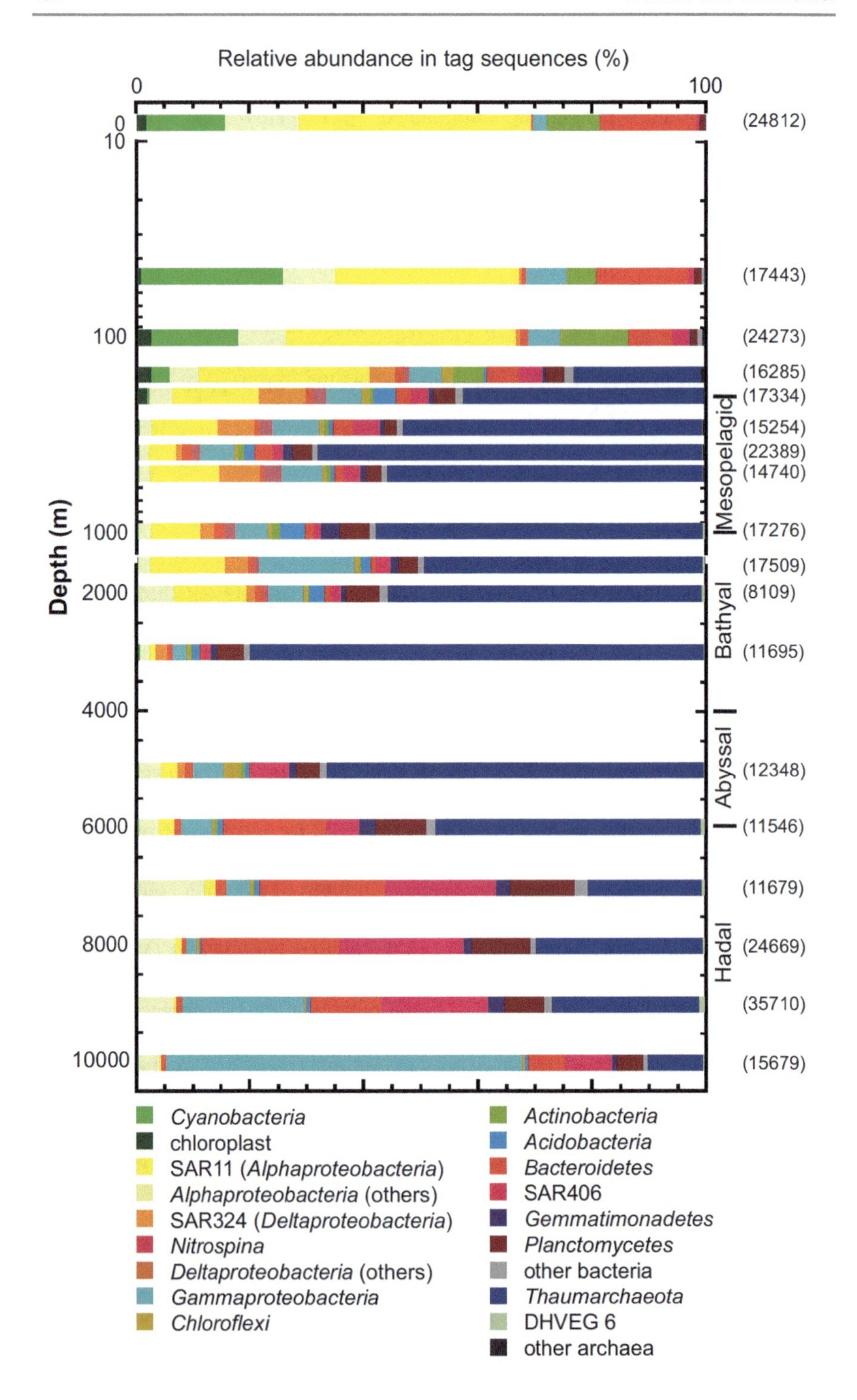


**Fig. 10.16** Bacterial and archaeal SSU rRNA gene community composition along the water column of the Challenger Deep. Numbers in parentheses indicate the number of tag sequences. From Nunoura et al. (2015)

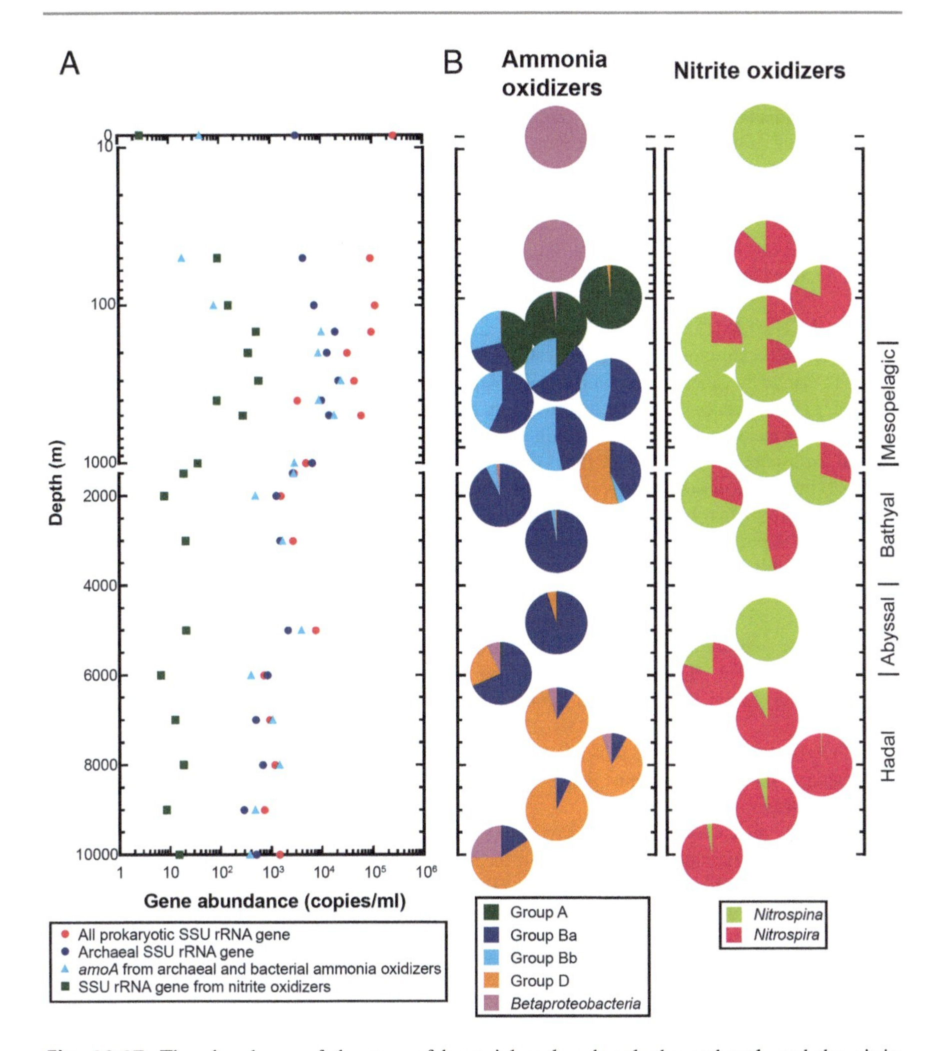


**Fig. 10.17** The abundance of the sum of bacterial and archaeal, the archaeal, and the nitrite oxidizer SSU rRNA and amoA genes (**a**), and the abundance of subgroups of amoA genes and SSU rRNA genes of nitrite oxidizing bacteria (**b**) along the water column in the Challenger Deep obtained by quantitative-PCR. From Nunoura et al. (2015)

more advanced technologies light will be shed on the ecological and biogeochemical role of the dark ocean's microbiome.

**Acknowledgments** GJH was supported by the Austrian Science Fund (FWF) project ARTEMIS (P28781-B21), the projects I486-B09 and P23234-B11, and by the European Research Council (ERC) under the European Community's Seventh Framework Programme (FP7/2007-2013)/ERC grant agreement 268595 (MEDEA project). FB was supported by the Austrian Science Fund (FWF) project OCEANIDES (P34304-B), by a University of Otago Research Grant and a Rutherford Discovery Fellowship (Royal Society of New Zealand).

## References

Acinas SG, Sánchez P, Salazar G, Cornejo-Castillo FM, Sebastián M, Logares R, Sunagawa S, Hingamp P, Ogata H, Lima-Mendez G (2019) Metabolic architecture of the deep ocean microbiome. bioRxiv: 635680

Agogué H, Brink M, Dinasquet J, Herndl GJ (2008) Major gradients in putatively nitrifying and non-nitrifying archaea in the deep North Atlantic. Nature 456:788–791

Agogué H, Lamy D, Neal PR, Sogin ML, Herndl GJ (2011) Water mass-specificity of bacterial communities in the North Atlantic revealed by massively parallel sequencing. Mol Ecol 20:258–274

Arístegui J, Gasol JM, Duarte CM, Herndl GJ (2009) Microbial oceanography of the dark ocean's pelagic realm. Limnol Oceanogr 54:1501–1529

Bach LT, Stange P, Taucher J, Achterberg EP, Algueró-Muñiz M, Horn H, Esposito M, Riebesell U (2019) The influence of plankton community structure on sinking velocity and remineralization rate of marine aggregates. Global Biogeochem Cycles 33:971–994

Baltar F, Arísstegui J, Gasol JM, Herndl GJ (2012) Microbial functioning and community structure variability in the mesopelagic and epipelagic waters of the subtropical Northeast Atlantic Ocean. Appl Environ Microbiol 78:3309–3316

Baltar F, Arístegui J, Gasol JM, Hernández-León S, Herndl GJ (2007) Strong coast–ocean and surface–depth gradients in prokaryotic assemblage structure and activity in a coastal transition zone region. Aq Microb Ecol 50:63–74

Baltar F, Arístegui J, Gasol JM, Lekunberri I, Herndl GJ (2010a) Mesoscale eddies: hotspots of prokaryotic activity and differential community structure in the ocean. ISME J 4:975–988

Baltar F, Arístegui J, Gasol JM, Sintes E, van Aken HM, Herndl GJ (2010b) High dissolved extracellular enzymatic activity in the deep Central Atlantic Ocean. Aq Microb Ecol 58:287–302

Baltar F, Arístegui J, Gasol JM, Sintes E, Herndl GJ (2009) Evidence of prokaryotic metabolism on suspended particulate organic matter in the dark waters of the subtropical North Atlantic. Limnol Oceanogr 54:182–193

Baltar F, Arístegui J, Gasol JM, Yokokawa T, Herndl GJ (2013) Bacterial versus archaeal origin of extracellular enzymatic activity in the Northeast Atlantic deep waters. Microb Ecol 65:277–288

Baltar F, Arístegui J, Sintes E, Gasol JM, Reinthaler T, Herndl GJ (2010c) Significance of non-sinking particulate organic carbon and dark $CO_2$ fixation to heterotrophic carbon demand in the mesopelagic Northeast Atlantic. Geophys Res Lett 37:L09602

Baltar F, Herndl GJ (2019) Ideas and perspectives: is dark carbon fixation relevant for oceanic primary production estimates? Biogeosciences 16:3793–3799

Baltar F, Lundin D, Palovaara J, Lekunberri I, Reinthaler T, Herndl GJ, Pinhassi J (2016a) Prokaryotic responses to ammonium and organic carbon reveal alternative $CO_2$ fixation pathways and importance of alkaline phosphatase in the mesopelagic North Atlantic. Front Microbiol 7:1670

Baltar F, Palovaara J, Unrein F, Catala P, Horňák K, Šimek K, Vaqué D, Massana R, Gasol JM, Pinhassi J (2016b) Marine bacterial community structure resilience to changes in protist predation under phytoplankton bloom conditions. ISME J 10:568–581

Bayer B, Saito MA, Mcilvin ME, Lücker S, Moran DM, Lankiewicz TS, Dupont CL, Santoro AE (2021) Metabolic versatility of the nitrite-oxidizing bacterium *Nitrospira marina* and its proteomic response to oxygen-limited conditions. ISME J 15:1025–1039

Beman JM, Popp BN, Francis CA (2008) Molecular and biogeochemical evidence for ammonia oxidation by marine Crenarchaeota in the Gulf of California. ISME J 2:429–441

Benner R (2002) Chemical composition and reactivity. In: Hansell DA, Carlson CA (eds) Biogeochemistry of marine dissolved organic matter. Elsevier Science, New York, pp 59–90

Bergauer K, Fernandez-Guerra A, Garcia JA, Sprenger RR, Stepanauskas R, Pachiadaki MG, Jensen ON, Herndl GJ (2017) Organic matter processing by microbial communities throughout the Atlantic water column as revealed by metaproteomics. Proc Natl Acad Sci 115:E400–E408

Bork P, Bowler C, de Vargas C, Gorsky G, Karsenti E, Wincker P (2015) Tara oceans studies plankton at planetary scale. Science 348:873–873

Boyd P, Newton P (1995) Evidence of the potential influence of planktonic community structure on the interannual variability of particulate organic carbon flux. Deep Sea Res I 42:619–639

Boyd PW, Claustre H, Levy M, Siegel DA, Weber T (2019) Multi-faceted particle pumps drive carbon sequestration in the ocean. Nature 568:327–335

Bunse C, Bertos-Fortis M, Sassenhagen I, Sildever S, Sjöqvist C, Godhe A, Gross S, Kremp A, Lips I, Lundholm N (2016) Spatio-temporal interdependence of bacteria and phytoplankton during a Baltic Sea spring bloom. Front Microbiol 7:517

Cho JC, Vergin KL, Morris RM, Giovannoni SJ (2004) Lentisphaera araneosa gen. nov., sp. nov, a transparent exopolymer producing marine bacterium, and the description of a novel bacterial phylum, Lentisphaerae. Environ Microbiol 6:611–621

Church MJ, DeLong EF, Ducklow HW, Karner MB, Preston CM, Karl DM (2003) Abundance and distribution of planktonic *archaea* and *bacteria* in the waters west of the Antarctic peninsula. Limnol Oceanogr 48:1893–1902

Cram JA, Chow C-ET, Sachdeva R, Needham DM, Parada AE, Steele JA, Fuhrman JA (2015a) Seasonal and interannual variability of the marine bacterioplankton community throughout the water column over ten years. ISME J 9:563–580

Cram JA, Xia LC, Needham DM, Sachdeva R, Sun F, Fuhrman JA (2015b) Cross-depth analysis of marine bacterial networks suggests downward propagation of temporal changes. ISME J 9: 2573–2586

Crespo BG, Pommier T, Fernández-Gómez B, Pedrós-Alió C (2013) Taxonomic composition of the particle-attached and free-living bacterial assemblages in the Northwest Mediterranean Sea analyzed by pyrosequencing of the 16S rRNA. Microbiol Open 2:541–552

DeLong EF, Preston CM, Mincer T, Rich V, Hallam SJ, Frigaard NU, Martinez A, Sullivan MB, Edwards R, Brito BR, Chisholm SW, Karl DM (2006) Community genomics among stratified microbial assemblages in the ocean's interior. Science 311:496–503

Dick GJ, Podell S, Johnson HA, Rivera-Espinoza Y, Bernier-Latmani R, McCarthy JK, Torpey JW, Clement BG, Gaasterland T, Tebo BM (2008) Genomic insights into Mn (II) oxidation by the marine alphaproteobacterium *Aurantimonas* sp. strain SI85-9A1. Appl Environ Microbiol 74: 2646–2658

Duret MT, Lampitt RS, Lam P (2019) Prokaryotic niche partitioning between suspended and sinking marine particles. Environ Microbiol Rep 11:386–400

Eloe EA, Shulse CN, Fadrosh DW, Williamson SJ, Allen EE, Bartlett DH (2011) Compositional differences in particle-associated and free-living microbial assemblages from an extreme deep-ocean environment. Environ Microbiol Rep 3:449–458

Frank AH, Garcia JA, Herndl GJ, Reinthaler T (2016) Connectivity between surface and deep waters determines prokaryotic diversity in the North Atlantic deep water. Environ Microbiol 18: 2052–2063

Ghiglione J-F, Galand PE, Pommier T, Pedrós-Alió C, Maas EW, Bakker K, Bertilson S, Kirchman DL, Lovejoy C, Yager PL (2012) Pole-to-pole biogeography of surface and deep marine bacterial communities. Proc Natl Acad Sci 109:17633–17638

Ghiglione JF, Mevel G, Pujo-Pay M, Mousseau L, Lebaron P, Goutx M (2007) Diel and seasonal variations in abundance, activity, and community structure of particle-attached and free-living bacteria in NW Mediterranean Sea. Microb Ecol 54:217–231

Grossart HP, Levold F, Allgaier M, Simon M, Brinkhoff T (2005) Marine diatom species harbour distinct bacterial communities. Environ Microbiol 7:860–873

Gruber N, Friedlingstein P, Field CB, Valentini R, Heimann M, Richey JE, Romero-Lankao P, Schulze D, Chenille C-TA (2004) The vulnerability of the carbon cycle in the 21st century: an assessment of carbon-climate-human interactions. In: Field CB, Raupach MR (eds) The global carbon cycle: integrating humans, climate, and the natural world. Island Press, Washington, DC, pp 45–76

Guidi L, Chaffron S, Bittner L, Eveillard D, Larhlimi A, Roux S, Darzi Y, Audic S, Berline L, Brum JR (2016) Plankton networks driving carbon export in the oligotrophic ocean. Nature 532:465–470

Hamdan LJ, Coffin RB, Sikaroodi M, Greinert J, Treude T, Gillevet PM (2013) Ocean currents shape the microbiome of Arctic marine sediments. ISME J 7:685–696

Hansell DA, Carlson CA (1998) Deep-ocean gradients of dissolved organic carbon. Nature 395: 263–266

Hansman RL, Griffin S, Watson JT, Druffel ERM, Ingalls AE (2009) The radiocarbon signature of microorganisms in the mesopelagic ocean. Proc Natl Acad Sci 106:6513–6518

Herndl GJ, Reinthaler T (2013) Microbial control of the dark end of the biological pump. Nat Geosci 6:718–724

Herndl GJ, Reinthaler T, Teira E, Hv A, Veth C, Pernthaler A, Pernthaler J (2005) Contribution of *archaea* to total prokaryotic production in the deep Atlantic Ocean. Appl Environ Microbiol 71: 2303–2309

Hewson I, Steele JA, Capone DG, Fuhrman JA (2006) Remarkable heterogeneity in meso- and bathypelagic bacterioplantkon assemblage composition. Limnol Oceanogr 51:1274–1283

Hügler M, Sievert SM (2011) Beyond the Calvin cycle: autotrophic carbon fixation in the ocean. Annu Rev Mar Sci 3:261–289

Ingalls AE, Shah SR, Hansman RL, Aluwihare LI, Santos GM, Druffel ERM, Pearson A (2006) Quantifying archaeal community autotrophy in the mesopelagic ocean using natural radiocarbon. Proc Natl Acad Sci 103:6442–6447

Ivars-Martínez E, Martín-Cuadrado AB, D'Auria G, Mira A, Ferriera S, Johnson J, Friedman R, Rodriguez-Valera F (2008) Comparative genomics of two ecotypes of the marine planktonic copiotroph *Alteromonas macleodii* suggests alternative lifestyles associated with different kinds of particulate organic matter. ISME J 2:1194–1212

Jensen PR, Lauro FM (2008) An assessment of actinobacterial diversity in the marine environment. Ant Leeuwenhoek 94:51–62

Jones RT, Robeson MS, Lauber CL, Hamady M, Knight R, Fierer N (2009) A comprehensive survey of soil acidobacterial diversity using pyrosequencing and clone library analyses. ISME J 3:442–453

Karner MB, DeLong EF, Karl DM (2001) Archaeal dominance in the mesopelagic zone of the Pacific Ocean. Nature 409:507–510

Kirchman DL (2002) The ecology of Cytophaga–Flavobacteria in aquatic environments. FEMS Microbiol Ecol 39:91–100

Kirchman DL, Ducklow HW (1993) Estimating conversion factors for thymidine and leucine methods for measuring bacterial production. In: Kemp PF, Sherr BF, Sherr EB, Cole JJ (eds) Handbook of methods in aquatic microbial ecology. Lewis Publishers, Boca Raton, pp 513–517

Kirchman DL, Elifantz H, Dittel AI, Malmstrom RR, Cottrell MT (2007) Standing stock and activity of archaea and bacteria in the western Arctic Ocean. Limnol Oceanogr 52:495–507

Könneke M, Bernhard AE, JRDL T, Walker CB, Waterbury JB, Stahl DA (2005) Isolation of an autotrophic ammonia-oxidizing marine archaeon. Nature 437:543–546

Landry Z, Swan BK, Herndl GJ, Stepanauskas R, Giovannoni SJ (2017) SAR202 genomes from the dark ocean predict pathways for the oxidation of recalcitrant dissolved organic matter. MBio 8: e00413–e00417

Lauro F, Bartlett DH (2008) Prokaryotic lifestyles in deep sea habitats. Extremophiles 12:15–25

Libes SM (1992) An introduction to marine biogeochemistry. Wiley, New York

Longhurst A (1998) Ecological geography of the sea. Academic Press, San Diego

Maas AE, Liu S, Bolaños LM, Widner B, Parsons RJ, Kujawinski EB, Blanco Bercial L, Carlson CA (2020) Migratory zooplankton excreta and its influence on prokaryotic communities. Front Mar Sci 7:1014

Maixner F, Wagner M, Lücker S, Pelletier E, Schmitz-Esser S, Hace K, Spieck E, Konrat R, Le Paslier D, Daims H (2008) Environmental genomics reveals a functional chlorite dismutase in the nitrite-oxidizing bacterium '*Candidatus* Nitrospira defluvii'. Environ Microbiol 10:3043–3056

Martín-Cuadrado AB, López-García R, Alba JC, Moreira D, Monticelli L, Strittmatter A, Gottschalk G, Rodriguez-Valera F (2007) Metagenomics of the deep Mediterranean, a warm bathypelagic habitat. PLoS One 2:e914

Mende DR, Bryant JA, Aylward FO, Eppley JM, Nielsen T, Karl DM, DeLong EF (2017) Environmental drivers of a microbial genomic transition zone in the ocean's interior. Nature Microbiol 2:1367–1373

Mestre M, Höfer J, Sala MM, Gasol JM (2020) Seasonal variation of bacterial diversity along the marine particulate matter continuum. Front Microbiol 11:1590

Mestre M, Ruiz-González C, Logares R, Duarte CM, Gasol JM, Sala MM (2018) Sinking particles promote vertical connectivity in the ocean microbiome. Proc Natl Acad Sci 115:E6799–E6807

Miki T, Yokokawa T, Nagata T, Yamamura N (2008) Immigration of prokaryotes to local environments enhances remineralization efficiency of sinking particles: a metacommunity model. Mar Ecol Progr Ser 366:1–14

Morris RM, Rappé MS, Connon SA, Vergin KL, Siebold WA, Carlson CA, Giovannoni SJ (2002) SAR11 clade dominates ocean surface bacterioplankton communities. Nature 420:806–810

Nagata T, Tamburini C, Arístegui J, Baltar F, Bochdansky AB, Fonda-Umani S, Fukuda H, Gogou A, Hansell DA, Hansman RL (2010) Emerging concepts on microbial processes in the bathypelagic ocean–ecology, biogeochemistry, and genomics. Deep Sea Res Pt II: Top Stud Oceanogr 57:1519–1536

Needham DM, Fuhrman JA (2016) Pronounced daily succession of phytoplankton, archaea and bacteria following a spring bloom. Nature Microbiol 1:1–7

Nunoura T, Takaki Y, Hirai M, Shimamura S, Makabe A, Koide O, Kikuchi T, Miyazaki J, Koba K, Yoshida N (2015) Hadal biosphere: insight into the microbial ecosystem in the deepest ocean on earth. Proc Natl Acad Sci 112:E1230–E1236

Onyenwoke RU, Brill JA, Farahi K, Wiegel J (2004) Sporulation genes in members of the low G+C gram-type-positive phylogenetic branch (Firmicutes). Arch Microbiol 182:182–192

Pachiadaki MG, Sintes E, Bergauer K, Brown JM, Record NR, Swan BK, Mathyer ME, Hallam SJ, Lopez-Garcia P, Takaki Y, Nunoura T, Woyke T, Herndl GJ, Stepanauskas R (2017) Major role of nitrite-oxidizing bacteria in dark ocean carbon fixation. Science 358:1046–1051

Parada AE, Fuhrman JA (2017) Marine archaeal dynamics and interactions with the microbial community over 5 years from surface to seafloor. ISME J 11:2510–2525

Poff KE, Leu AO, Eppley JM, Karl DM, DeLong EF (2021) Microbial dynamics of elevated carbon flux in the open ocean's abyss. Proc Natl Acad Sci 118:E2018269118

Quince C, Curtis TP, Sloan WT (2008) The rational exploration of microbial diversity. ISME J 2: 997–1006

Reinthaler T, van Aken HM, Herndl GJ (2010) Major contribution of autotrophy to microbial carbon cycling in the deep North Atlantic's interior. Deep Sea Res Pt II: Top Stud Oceanogr 57: 1572–1580

Rodionov DA, Dubchak I, Arkin A, Alm E, Gelfand MS (2004) Reconstruction of regulatory and metabolic pathways in metal-reducing δ-proteobacteria. Genome Biol 5:R90

Ruiz-González C, Mestre M, Estrada M, Sebastián M, Salazar G, Agustí S, Moreno-Ostos E, Reche I, Álvarez-Salgado XA, Morán XAG (2020) Major imprint of surface plankton on deep ocean prokaryotic structure and activity. Mol Ecol 29:1820–1838

Salazar G, Cornejo-Castillo FM, Benítez-Barrios V, Fraile-Nuez E, Álvarez-Salgado XA, Duarte CM, Gasol JM, Acinas SG (2016) Global diversity and biogeography of deep-sea pelagic prokaryotes. ISME J 10:596–608

Salazar G, Cornejo-Castillo FM, Borrull E, Díez-Vives C, Lara E, Vaqué D, Arrieta JM, Duarte CM, Gasol JM, Acinas SG (2015) Particle-association lifestyle is a phylogenetically conserved trait in bathypelagic prokaryotes. Mol Ecol 24:5692–5706

Salazar G, Paoli L, Alberti A, Huerta-Cepas J, Ruscheweyh H-J, Cuenca M, Field CM, Coelho LP, Cruaud C, Engelen S (2019) Gene expression changes and community turnover differentially shape the global ocean metatranscriptome. Cell 179:1068–1083

Santoro AE, Richter RA, Dupont CL (2019) Planktonic marine archaea. Annu Rev Mar Sci 11:131–158

Sintes E, Bergauer K, De Corte D, Yokokawa T, Herndl GJ (2013) Archaeal *amoA* gene diversity points to distinct biogeography of ammonia-oxidizing Crenarchaeota in the ocean. Environ Microbiol 15:1647–1658

Sintes E, De Corte D, Haberleitner E, Herndl GJ (2016) Geographic distribution of archaeal ammonia oxidizing ecotypes in the Atlantic Ocean. Front Microbiol 7:77

Sogin ML, Morrison HG, Huber JA, Welch DM, Huse SM, Neal PR, Arrieta JM, Herndl GJ (2006) Microbial diversity in the deep sea and the under-explored "rare biosphere". Proc Natl Acad Sci 103:12115–12120

Sorokin DY (2003) Oxidation of inorganic sulfur compounds by obligately organotrophic bacteria. Microbiology 72:641–653

Stegen JC, Lin X, Fredrickson JK, Chen X, Kennedy DW, Murray CJ, Rockhold ML, Konopka A (2013) Quantifying community assembly processes and identifying features that impose them. ISME J 7:2069–2079

Stegen JC, Lin X, Konopka AE, Fredrickson JK (2012) Stochastic and deterministic assembly processes in subsurface microbial communities. ISME J 6:1653–1664

Steiner PA, Betel J, Fadeev E, Obiol A, Sintes E, Rattei T, Herndl GJ (2020) Functional seasonality of free-living and particle-associated prokaryotic communities in the coastal Adriatic Sea. Front Microbiol 11:2875

Sunagawa S, Acinas SG, Bork P, Bowler C, Eveillard D, Gorsky G, Guidi L, Iudicone D, Karsenti E, Lombard F (2020) Tara oceans: towards global ocean ecosystems biology. Nature Rev Microbiol 18:428–445

Sunagawa S, Coelho LP, Chaffron S, Kultima JR, Labadie K, Salazar G, Djahanschiri B, Zeller G, Mende DR, Alberti A (2015) Structure and function of the global ocean microbiome. Science 348:1261359

Swan BK, Martinez-Garcia M, Preston CM, Sczyrba A, Woyke T, Lamy D, Reinthaler T, Poulton NJ, Masland EDP, Gomez ML, Sieracki ME, DeLong EF, Herndl GJ, Stepanauskas R (2011) Potential for chemolithoautotrophy among ubiquitous bacteria lineages in the dark ocean. Science 333:1296–1300

Swan BK, Tupper B, Sczyrba A, Lauro FM, Martinez-Garcia M, González JM, Luo H, Wright JJ, Landry ZC, Hanson NW (2013) Prevalent genome streamlining and latitudinal divergence of planktonic bacteria in the surface ocean. Proc Natl Acad Sci 110:11463–11468

Teira E, Lebaron P, Hv A, Herndl GJ (2006) Distribution and activity of bacteria and archaea in the deep water masses of the North Atlantic. Limnol Oceanogr 51:2131–2144

Thrash JC, Temperton B, Swan BK, Landry ZC, Woyke T, DeLong EF, Stepanauskas R, Giovannoni SJ (2014) Single-cell enabled comparative genomics of a deep ocean SAR11 bathytype. ISME J 8:1440–1451

Varela M, van Aken HM, Sintes E, Herndl GJ (2008) Latitudinal trends of Crenarchaeota and bacteria in the meso- and bathypelagic water masses of the eastern North Atlantic. Environ Microbiol 10:110–124

Wagner M, Horn M (2006) The Planctomycetes, Verrucomicrobia, Chlamydiae and sister phyla comprise a superphylum with biotechnological and medical relevance. Curr Opin Biotechnol 17:241–249

Wenley J, Currie K, Lockwood S, Thomson B, Baltar F, Morales SE (2021) Seasonal prokaryotic community linkages between surface and deep ocean water. Front Mar Sci 659641

Whitman WB, Coleman DC, Wiebe WJ (1998) Prokaryotes: the unseen majority. Proc Natl Acad Sci 95:6578–6583

Wilkins D, Van Sebille E, Rintoul SR, Lauro FM, Cavicchioli R (2013) Advection shapes Southern Ocean microbial assemblages independent of distance and environment effects. Nature Comm 4:1–7

Woebken D, Teeling H, Wecker P, Dumitriu A, Kostadinov I, DeLong EF, Amann R, Glöckner FO (2007) Fosmids of novel marine Planctomycetes from the Namibian and Oregon coast upwelling systems and their cross-comparison with planctomycete genomes. ISME J 1:419–435

Wuchter C, Herfort L, Coolen MJL, Abbas B, van Bleijswijk J, Timmers P, Strous M, Teira E, Herndl GJ, Middelburg JJ, Schouten S, Damsté JSS (2006) Archaeal nitrification in the ocean. Proc Natl Acad Sci 103:12317–12322

Zhang H, Sekiguchi Y, Hanada S, Hugenholtz P, Kim H, Kamagata Y, Nakamura K (2003) Gemmatimonas aurantiaca gen. nov., sp. nov., a gram-negative, aerobic, polyphosphate-accumulating micro-organism, the first cultured representative of the new bacterial phylum Gemmatimonadetes phyl. nov. J Med Microbiol 53:1155–1163

Zhang Y, Qin W, Hou L, Zakem EJ, Wan X, Zhao Z, Liu L, Hunt KA, Jiao N, Kao S-J (2020) Nitrifier adaptation to low energy flux controls inventory of reduced nitrogen in the dark ocean. Proc Natl Acad Sci 117:4823–4830

Zhao Z, Baltar F, Herndl GJ (2020) Linking extracellular enzymes to phylogeny indicates a predominantly particle-associated lifestyle of deep-sea prokaryotes. Sci Adv 6:eaaz4354

# The Subsurface and Oceanic Crust Prokaryotes

# 11

Mohamed Jebbar

**Abstract**

The deep subseafloor is a vast yet almost underexplored habitat for life on Earth featuring large-scale global flows of water, heat, and dissolved chemical compounds. Finding out where and how life exists in this large and fragmented ecosystem is crucial because the chemical reactions that occur in this environment have implications for the broader ocean system. In the past decades a number of major oceanographic expeditions have been devoted to characterizing the deep-sea crustal and sediment microbiomes, which has become possible because of the advances in deep-sea drilling and observation technologies. Culture- and more importantly non-culture approaches that are based on the assembly of genomes from massive metagenomic data, have shed light on several unknown phyla that have no cultured representatives with metabolic characteristics that show adaptations to the extreme physicochemical conditions of deep-sea subsurface environments.

**Keywords**

Archaea · Bacteria · Deep biosphere · Extreme conditions · Metabolic features · Oceanic subsurface

M. Jebbar (✉)
University Brest, CNRS, Ifremer, UMR 6197, Laboratoire de Microbiologie des Environnements Extrêmes LM2E, Plouzané, France
e-mail: mohamed.jebbar@univ-brest.fr



L. J. Stal, M. S. Cretoiu (eds.), *The Marine Microbiome*, The Microbiomes of Humans, Animals, Plants, and the Environment 3,
https://doi.org/10.1007/978-3-030-90383-1_11

## 11.1 Introduction

Because of the enormous overall volume of the geosphere and hydrosphere these parts of the planet Earth provide ample room for life but most of these habitats are located in compartments that do not receive sunlight and, hence, do not rely on photosynthesis (Fig. 11.1). Deep-sea ecosystems whose food web is based on microbial chemosynthesis are considered to be extreme environments in which certain adapted animal communities combine the production and consumption of chemoautotrophic microorganisms (Jannasch and Taylor 1984; Monaco and Prouzet 2015). These microorganisms are free-living or associated with invertebrates in case they live in a symbiotic relationship with them or as an epibiont living attached to these animals. These chemoautotrophic (generate energy by the oxidation of a reduced substrate and fix $CO_2$) microorganisms support an exceptionally high production and a large standing stock of biomass of these deep-sea environments. These ecosystems are mainly located in accretionary zones on ridges, on passive continental margins, and in subduction zones and are linked to sources of reduced compounds: sulfide, methane, or other simple hydrocarbons (Fig. 11.1).

The oceanic crust is the upper part of the oceanic lithosphere (Seibold and Berger 2017). The oceanic lithosphere lies on the upper mantle's plastic part, the asthenosphere, and is located (with rare exception) under the oceanic layer forming the ocean floor (Seibold and Berger 2017). The oceanic crust is the largest hydrogeological reservoir on Earth containing fluids in thermodynamic disequilibrium with the basaltic crust. Little is known about the microbial ecosystems in this vast domain or about the microorganisms that inhabit it and exploit the chemically favorable conditions for their metabolic activities. A major part of the oceanic crust is covered with sediments that were deposited by sinking material from the water column but there are also many areas where the rocky seafloor is directly exposed to the overlying water column (Fig. 11.1).

Marine sediments on the seafloor are formed from the accumulation of organic and inorganic particles that are sinking from the water column. The composition of these sediments depends on their origin. They may be composed of the remains of marine microorganisms such as diatoms and coccolithophorids and organic matter derived from them, remains of terrestrial plants, minerals of marine origin (such as from hydrothermal vents), or sand and mud produced from terrestrial erosion. The majority of marine sediments are a desert-like abyssal plain where the hydrostatic pressure and temperature are similar to the oceanic deep water (on average 0–4 °C and 38 MPa) (Fig. 11.1).

Subsurface deep marine sediments represent one of Earth's most extensive microbial habitats that cover over two thirds of the Earth's surface. Active microorganisms (bacteria and archaea) are widely distributed in these bottom deposits (Teske and Sørensen 2008). Subsurface sediments are sedimentary layers with microbial communities well distinct from those in the water column. Microorganisms living in subsurface sediments seem to have different microbial imprints compared to those of microorganisms in the water column, reflecting an

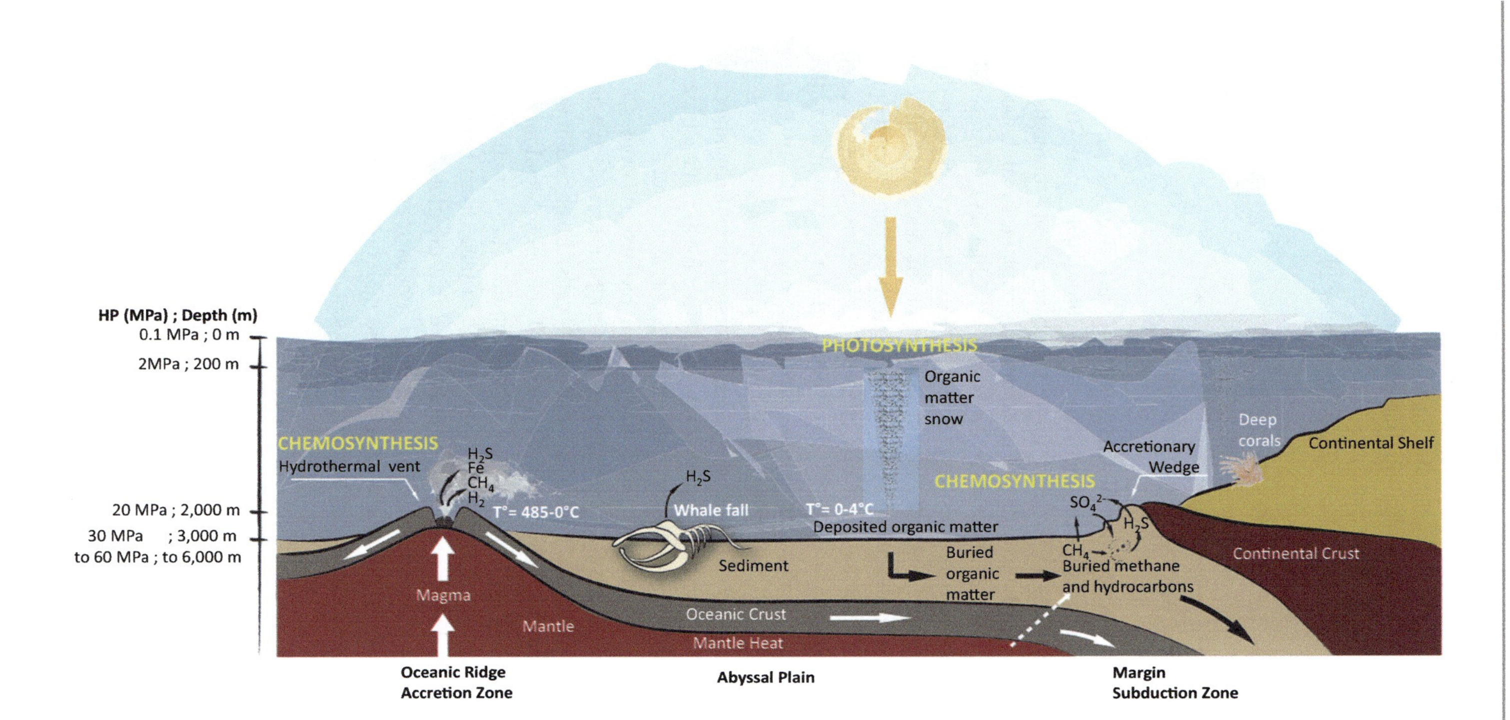


**Fig. 11.1** Cross-section of the ocean illustrating the different ecosystems of the deep sea. (Adapted from Sarrazin Jozee, Desbruyeres Daniel (2015). Hydrothermal Vents: Oases at Depth. In Marine Ecosystems: Diversity and Functions. 2015. André Monaco, Patrick Prouzet (Eds.) From the Seas and Oceans Set coordinated by André Mariotti and Jean-Charles Pomerol. John Wiley & Sons, Inc. ISTE. Print ISBN:9781848217829, Online ISBN:9781119116219, DOI: https://doi.org/10.1002/9781119116219. Chap.6, pp. 225–292 (Wiley). https://archimer.ifremer.fr/doc/00648/75995/)

adaptation to the distinct physicochemical conditions of subsurface sediments in contrast to the water column. (Teske and Sørensen 2008).

In this chapter the knowledge on bacterial and archaeal life in the deep subseafloor (oceanic crust and deep-sea sediments) environment is reviewed and further insight on the phylogeny and metabolic features of cultured and uncultured phyla is provided.

## 11.2 Deep Subseafloor Exploration

During the past fifty years scientific ocean drilling has revealed the widespread occurrence of microorganisms living beneath the seafloor (Heuer et al. 2019). Many of the bacteria and archaea on Earth are found in subsurface environments but their physiology has been poorly investigated or adequately explained mainly because the lack of laboratory cultures of these microorganisms (Hoehler and Jørgensen 2013).

Bacteria and archaea are responsible for many fundamental processes in deep-sea sediments, including the oxidation of organic matter, the production of methane and other hydrocarbons, and the removal of sulfate from the ocean. The geographical global distribution of sedimentary microbes in the deep sea is largely unexplored, let alone explained. Whitman et al. (1998) calculated the microbial abundance of subsurface sediments as $35.5–10^{29}$ cells, encompassing 55–86% of the Earth's microbial biomass and 27–33% of the total of the Earth's living biomass. In their estimations it was assumed that the average relationship of cell abundance versus depth at six sites in the Pacific Ocean was representative for the characterization of the sedimentary microbial abundance in the global ocean as a whole.

Based on quantifications of intact phospholipid biomarkers from six Pacific Ocean sites and one Black Sea site Lipp et al. (2008) estimated the microbial abundance in the subseafloor sediments as $5–10^{30}$ cells. Later, Kallmeyer et al. (2012) showed that the total abundance of microbial cells in subseafloor sediments varies between sites by about five orders of magnitude. This variation correlated strongly with the average sedimentation rate and the distance from land. Based on these correlations, Kallmeyer et al. (2012) estimated that the global abundance of microbial cells in subsurface sediments is $2.9–10^{29}$ cells (corresponding to 4.1 petagrams (Pg) C and ~0.6% of the total living biomass on Earth). This estimate of sedimentary microbial abundance below the sedimentary seafloor is roughly equal to previous estimates of total microbial abundance in seawater and total microbial abundance in soil. It is lower than previous estimates of sedimentary microbial abundance in the subseafloor sediment. As a result, Kallmeyer et al. (2012) estimated that the total number of microbes and the total living biomass on Earth are, respectively, 50–78% and 10–45% lower than previous estimates (Lipp et al. 2008; Whitman et al. 1998). The total amount of subsurface biomass is still debated (Hinrichs and Inagaki 2012; Kallmeyer et al. 2012; Parkes et al. 2014) and the factors that determine the ultimate limit of life and the geographical frontiers of the Earth's habitability remain to be resolved.

Microbial cells are present throughout the entire sedimentary sequence at all deep subseafloor explored sites in both organic matter rich and nutrient poor sediments (D'Hondt et al. 2015; Kallmeyer et al. 2012) but still not much is known about these vast ecosystems. Molecular signatures and observations of dividing cells in several million years old deep-sea sediments (below 4500 m water column and up to 1.6 km below the seafloor) extended the limits of life below the seafloor from 842 to 1626 meters below the seafloor (mbsf) (Roussel et al. 2008). The IODP (Integrated Ocean Drilling Program) Leg 317 cruise drilling to 1922 mbsf (bathymetry 344 m) in the deep marine sediments of the Canterbury Basin revealed the molecular signatures of bacteria, archaea, and microbial fungi (Ciobanu et al. 2014). This study extended the depth limit of life below the seafloor for bacteria (1922 mbsf) and microeukaryotes (1740 mbsf) (Ciobanu et al. 2014). Another study detected microbial communities in ~40° to 60 °C sediments associated with lignite coal beds 2.5 km below the ocean floor in the Pacific Ocean off Japan (Inagaki et al. 2015). This study showed that microbial methanogenesis is potentially operational in these deep buried marine sediments (Inagaki et al. 2015). A dynamic post-drilling response of crustal microbial ecosystems to changing physical and chemical conditions was recorded in experiments carried out in young (3.5 Ma) basaltic crust on the eastern flank of the Juan de Fuca Ridge (Orcutt et al. 2011). Twisted stalks showing a biogenic iron oxyhydroxide signature covered the surface of mineral substrates in the observatories. These biosignatures are indicative of colonization by iron-oxidizing bacteria during an initial phase of cold, oxic, and iron-rich conditions following the setting up of the observatory (Orcutt et al. 2011). Subsequently, after thermal and chemical recovery under warmer and reducing conditions the microbial community of the observatory changed. Hence, this represented the naturally occurring conditions in the regional crustal fluids (Orcutt et al. 2011). The recovered bacterial community was dominated by Firmicutes of which the metabolic potential was unknown but possibly involved N and/or S cycling. The archaeal community was characterized by low diversity. The experiment of Orcutt et al. (2011) documented the in-situ conditions within a natural hydrological system and demonstrated the power of observational experiments to investigate the subsurface basaltic biosphere, which is considered the most extensive but poorly understood biotope on Earth.

The subseafloor sedimentary sites of the south pacific gyre (SPG) are characterized by low biomass and metabolic activity of the microbial community (D'Hondt et al. 2009). Along the sediment cores of the SPG the average cell abundances were 3 to 4 orders of magnitude lower than at the same depth in any previously explored deep-sea subsurface community. However, the net respiration rates of the subsurface sediment community at each SPG site were only 1 to 3 orders of magnitude lower than those at previously surveyed sites. Due to the low respiration rate and the thin layer of sediment the interstitial waters of the sediment column in most of the SPG area are oxygenated. Hence, the SPG sedimentary microbial community is mostly aerobic whereas those of previously explored subsurface communities are not (D'Hondt et al. 2009).

Scientific attention has increasingly focusing on the ability of microorganisms to withstand long periods of low energy supply. The microorganisms that live in such highly stable oligotrophic conditions are estimated to catabolize $10^4$ to $10^6$ times more slowly than typical microorganisms that thrive under nutrient-rich conditions. These microorganisms turnover biomass over time scales of centuries to millennia rather than hours to days, and persist at energy fluxes 1000 times lower than typical culture-based estimates of maintenance requirements (Hoehler and Jørgensen 2013). Considerably less is known about microbial life under low energy conditions. Such microorganisms are able to maintain their essential cellular functions during extended periods of energy limitation. The conditions that prevent growth may have a profound influence on the evolution of these microbes similar as on microorganisms that grow and replicate under conditions of energy sufficiency (Lever et al. 2015).

In these oligotrophic environments the cells are mostly viable. The microbial life recovered from the 460,000-year-old deep subsurface sediments (Morono et al. 2011) has retained the metabolic potentials for carbon and nitrogen assimilation and growth. Up to 76% and 22% of the cells in the sediment incorporated, respectively, organic and inorganic carbon into their biomass. This agrees well with previous observations that the microbial community in organic-rich subsurface sediments on the continental margins is predominantly composed of heterotrophs (Biddle et al. 2006; Lipp et al. 2008). Further studies on subsurface sediments have shown that microorganisms persist in a dormant state. Endospores are as abundant as vegetative cells and microbial activity is low, resulting in microbial biomass turnover times of several hundred to several thousand years (Lomstein et al. 2012). From model calculations it was deduced that biomass production is supported by organic carbon deposited millions of years ago from photosynthetic organisms in the photic zone of the ocean and that the microbial necromass is recycled on time scales of hundreds of thousands of years (Lomstein et al. 2012).

## 11.3 Deep-Sea Biosphere Bacteria and Archaea

The current Living Tree Project (LTP) lists 39 phyla, 99 classes, 234 orders, 582 families, 3261 genera, 17,137 species, and 431 subspecies of bacteria and archaea. The nearly complete 16S rRNA gene sequences represent essentially all bacterial and archaeal type species (Ludwig et al. 2021).

These described and validated species represent only a small fraction of the bacteria and archaea. The genomic approach has increased the number of recognized phyla (at least 92 bacterial phyla, 27 archaeal phyla, and 5 eukaryotic supergroups) of which approximately two thirds do not have cultured representatives (Baker et al. 2020; Castelle and Banfield 2018; Hug et al. 2016). The tree of life has expanded dramatically as a result of a major effort using high-throughput sequencing to obtain whole or nearly whole genomes of previously cryptic or unknown microbial lineages (Hug et al. 2016). The newly developed representation of the rapidly evolving tree represents the current genomic sampling of life, illustrating the progress made in the last two decades since the release of the first genomes (Hug et al. 2016).

The recovery of the genomes of organisms belonging to phyla that lack isolated and cultured representatives (the so-called candidate phyla) delineated major new microbial lineages, namely the candidate bacterial radiation phyla (CPR) (Hug et al. 2016), archaeal superphylum DPANN (Diapherotrites, Parvarchaeota, Aenigmarchaeota, Nanoarchaeota, Nanohaloarchaeota, Micrarchaeota, Pacearcheota, Woesearchaeota, Altiarchaeota, Huberarchaeota) (Baker et al. 2020; Castelle et al. 2015; Hug et al. 2016; Rinke et al. 2013), and archaeal superphylum ASGARD (Lokiarchaeota, Thorarchaeota, Odinarchaeota, Heimdallarchaeota, Helarcheota) (Seitz et al. 2016, 2019; Spang et al. 2015; Zaremba-Niedzwiedzka et al. 2017). The analysis of the tree of life reveals the depth of evolutionary history within the bacterial domain. Especially the recognition of the CPR appears to subdivide the bacterial domain (Hug et al. 2016). The CPR and DPANN organisms are thought to be mainly symbionts or episymbionts of other microbial community members (Castelle and Banfield 2018). Some genes and functions encoded in the genomes of ASGARD archaea are considered to be typically eukaryotic. Their inclusion in phylogenetic analyses results in the positioning of eukaryotes as a branch within the archaea (Spang et al. 2015). These newly discovered genomes have modified the tree of life and altered the way we understand the evolution of the metabolic processes in biogeochemical cycles (Hug et al. 2016). Even more importantly, these phylogenetic analyses highlight the large fraction of diversity that is currently only accessible through culture-independent genomic approaches.

## 11.4 Deep Subseafloor Archaea

Archaea have attracted particular interest because their proportion and diversity in the deep biosphere is higher than the bacteria, which is the opposite in the shallow biosphere.

Besides numerous members of the superphyla ASGARD and DPANN listed above, the genomic signatures of members of the superphylum TACK (Thaumarchaeota, Aigarchaeota, Crenarchaeota, Korarchaeota, Verstraetearchaeota, Nezhaarchaeota, Bathyarchaeota, Geothermarchaeota, Geoarcheota, Marsarchaeota) as well as genomic signatures of organisms belonging to the phylum *Euryarchaeota* (the only phylum not belonging to a superphylum) were also found in the deep-sea subsurface biosphere (Baker et al. 2020, 2021).

The phylum Euryarchaeota (Oren 2019) is one of the two major groups of archaea with cultured representatives isolated from a variety of extreme and moderate environments in the shallow and deep biosphere. Many species of Euryarchaeota were isolated from deep-sea hydrothermal vents (Jebbar et al. 2015), deep-sea brine (L'Haridon et al. 2020), and deep-sea sediments (Zeng et al. 2013).

Metagenomic analysis of samples from subseafloor observatory CORKs (Circulation Obviation Retrofit Kits) installed along the flank of the Juan de Fuca Ridge in the Pacific Ocean that deliver pristine fluids from the underlying igneous basement aquifer (Lin et al. 2012), has revealed a distinctive microbial community, including a number of new Archaeoglobi (Jungbluth et al. 2016, 2017). A nearly complete

Archaeoglobi metagenomic assembled genome (MAG; *Candidatus* Polytropus marinifundus gen. nov., sp. nov.) has been recovered from the deep subsurface along the flank of the Juan de Fuca. *Candidatus* P. marinifundus is close to the root of the phylogenetic tree of the class *Archaeoglobi*. The genome of this organism contains genes that encode for β-oxidation, which potentially enables *Candidatus* P. marinifundus to metabolize alkanes (Boyd et al. 2019). It has, therefore, been proposed to change the name Archaeoglobi to Allopolytropaceae (Al.lo.po.ly.tro. pa.ce'ae. N.L. masc. n. Allopolytropus a (*Candidatus*) generic name; −aceae ending to denote a family; N.L. fem. pl. n. Allopolytropaceae the Allopolytropus family) (Oren and Garrity 2021). Indeed, a thermophilic, lithoautotrophic, sulfate-reducing archaeon, *Archaeoglobus sulfaticallidus*, was isolated from the black rust formed on the steel surface of a borehole observatory (CORK 1026B) recovered during IODP Expedition 301 on the eastern flank of the Juan de Fuca Ridge in the eastern Pacific Ocean (Steinsbu et al. 2010). Only a dozen species of Archaeoglobales have been cultured, originating mainly from deep-sea hydrothermal vents. The Archaeoglobales are mostly involved in sulfate reduction in hot environments and some are capable of chemolithotrophic growth and β-oxidation. The family *Archaeoglobaceae* belongs to the order Archaeoglobales, class Archaeoglobi ((Parte et al. 2020) http://www.bacterio.net/index.html) is composed of three genera. *Archaeoglobus* comprises five species with validly published names: *A. fulgidus* (Stetter 1988), *A. profundus* (Burggraf et al. 1990), *A. veneficus* (Huber et al. 1997), *A. infectus* (Mori et al. 2008), and *A. sulfaticallidus* (Steinsbu et al. 2010). *Geoglobus* comprises two species: *G. acetivorans* (Slobodkina et al. 2009) and *G. ahangari* (Kashefi 2002) and *Ferroglobus* has one validly named species *F. placidus* (Hafenbradl et al. 1996). *A. neptunius* was isolated from a fragment of the hydrothermal diffuser at the TAG vent field (26°13′69"N 44°82′61"W, 3625 m water depth) on the Mid-Atlantic Ridge (https://doi.org/10.17600/18000004) and the authors of this work propose a reclassification of the genus *Archaeoglobus* (Slobodkina et al. 2021).

Several genomes assembled from metagenomes of the subsurface biosphere belong to the phylum *Euryarchaeota*. This includes genomes of *Hadesarchaeota* (Baker et al. 2016; Biddle et al. 2006) of which no cultured representatives have yet been obtained. These organisms have streamlined genomes and their gene content suggests that they may couple carbon monoxide and $H_2$ oxidation to nitrite reduction (Baker et al. 2020).

Among the archaea the ASGARD superphylum has attracted much attention. Metagenomic characterization of the deep-sea archaeal group/marine benthic group-B (also known as Lokiarchaeota) and the ASGARD archaea superphylum led to the theory that the eukaryotes originated from an archaeon that was closely related to these lineages (Spang et al. 2015; Zaremba-Niedzwiedzka et al. 2017). The evolution of the eukaryotic cell is not precisely known. Various evolutionary models have been proposed among which the most widely accepted are the symbiogenic models that conceive the merging of an archaeal host and an alphaproteobacterial endosymbiont (Eme et al. 2017; Koonin 2015; López-García and Moreira 2015, 2020; Martin et al. 2015). The genomes of the ASGARD superphylum contain a unique repertoire of proteins found only in eukarya (eukaryotic signature proteins) including those with homology to eukaryotic actin, the endosomal sorting complexes required for transport (ESCRT-I and -III complexes), and the ubiquitin modifier system (Spang et al. 2015; Zaremba-Niedzwiedzka et al. 2017). Further metagenomic analysis has suggested that the ASGARD archaea have a broad range of physiological characteristics, notably hydrogen-dependent anaerobic autotrophy (Sousa et al. 2016), peptide- or short-chain hydrocarbon-dependent organotrophy (Dombrowski et al. 2018; Liu et al. 2018; Seitz et al. 2016, 2019; Spang et al. 2019) and rhodopsin-based phototrophy (Bulzu et al. 2019; Pushkarev et al. 2018). After 12 years of enrichment culture a co-culture of a *Lokiarchaeum* and a *Methanogenium* was finally obtained from deep-sea methane-seep sediment of the Nankai Trough (Kumano area, Japan) at 2533 m water depth (Imachi et al. 2020). This member of the Lokiarchaeota was grown in a syntrophic association with a sulfate-reducing bacterium (*Halodesulfovibrio*) and a methanogen (*Methanogenium*) (Imachi et al. 2020). The isolate, *Prometheoarchaeum syntrophicum* (strain MK-D1), is a co-cultured representative of the ASGARD superphylum and its physiological and genomic features show that it is anaerobic, degrades amino acids, and produces hydrogen and formate, which are used by the bacterial partner (Imachi et al. 2020). Microscopic analyses showed that the cells are small cocci (approximately 300–750 nm in diameter (average 550 nm)) that generally form aggregates with the cells embedded in a matrix of extracellular polymer substances (EPS). Such small cells seem to be one of the morphological characteristics of microorganisms living in the oceanic subsurface and deep biosphere environments (Braun et al. 2016; Hinrichs and Inagaki 2012).

In oceanic subsurface environments, molecular signatures and genome assemblages from metagenomes have demonstrated the importance of the TACK superphylum (Baker et al. 2020). This archaeal superphylum has many cultured representatives notably of the phyla *Crenarchaeota* (Garrity and Holt 2001) and *Thaumarchaeota* (Brochier-Armanet et al. 2008) that were isolated from various environments including deep sea and deep subseafloor. Among the TACK superphylum, the Bathyarchaeota are among the most abundant and active groups of microorganisms in marine sediment (Lloyd et al. 2013; Meng et al. 2014). Based on metagenome analysis it has been suggested that the Bathyarchaeota play a globally important role in the carbon cycling in the marine environment by the fermentation of complex organic substances (e.g., lignin) (Yu et al. 2018), by acetogenesis (He et al. 2016), and by methanogenesis (Evans et al. 2015). The discovery of *mcr* genes (Methanogenesis metabolism gene marker) in the genome of Bathyarchaeota was the first occurring outside the Euryarchaeota. Given that some Mcr proteins are involved in butane oxidation (Laso-Pérez et al. 2016) it is attractive to suggest that *Bathyarchaeota* are capable of oxidizing short-chain hydrocarbons in addition to methane. The efforts to enrich and isolate uncultured archaea (e.g., Bathyarchaeota) from marine sediments using a refined combination of microbial culture methods have enabled to maintain in co-culture A representative of a subgroup of the Bathyarchaeota with bacteria affiliated to the genera

*Pseudomonas* and *Glutamicibacter* using culture media supplemented with a variety of substrates (e.g., methane, sulfate, and lignin) (Hu et al. 2021).

## 11.5 Deep Subseafloor Bacteria

Deep-sea subsurface environments are characterized by different extreme conditions (such as pressure, temperature, pH, oxygen availability, availability of electron donors and acceptors, carbon-, nitrogen-, and sulfur sources, porosity, water activity). In order to increase the number and diversity of isolates of microorganisms from these environments the Deep-IsoBUG device has been developed (Parkes et al. 2009). When this device is coupled with the HYACINTH pressure-retaining drilling and core storage system and the PRESS core cutting and processing system, it enables deep sediments to be handled without depressurization (up to 25 MPa) and anaerobic enrichments and isolation of microorganisms to be conducted at a pressure of up to 100 MPa (Parkes et al. 2009). Subsurface gas hydrate sediments from the Indian continental shelf, the Cascadia margin, and the Gulf of Mexico have been used in combination with this new culturing system. It achieved the highest cell concentrations in enrichments at in-situ pressure (14 MPa) using a variety of media, although growth was possible up to at least 80 MPa (Parkes et al. 2009). Molecular analysis showed that only bacterial 16S rRNA gene sequences were present, including those of aerobic *Carnobacterium*, *Clostridium*, *Marinilactibacillus*, *Bacteroidetes*, and *Pseudomonas*, as well as the obligate anaerobic bacteria *Acetobacterium* and *Clostridium* (Parkes et al. 2009).

The dominant sediment bacterial types have not received as much attention as the archaea. The bacteria tend to comprise the well-known taxa, including the Proteobacteria, Bacteroidetes, Chlamydiae, Firmicutes, Chloroflexi, Gemmatimonadetes, Thermotogales, and the Planctomycetes, as well as several common candidate phyla, including OP1 (Acetothermia), OP3 (Caldiserica), OP8 (Aminicenantes), OP9 or JS1 (Atribacteria), and OP10 (Armatimonadetes) (Baker et al. 2021; Durbin and Teske 2011). For example, the "Atribacteria," is a globally distributed candidate phylum that appears restricted to anaerobic environments and is predominant in organic- and hydrocarbon-rich sediments (Lee et al. 2018). The "Atribacteria" phylogenomic analysis confirms the monophyly of this candidate phylum that lack cultured representatives. The metagenomic analysis of metabolic potential showed that the "Atribacteria" are likely to be anaerobic heterotrophs that lack respiratory capacity, with some lineages predicted to specialize in either primary fermentation of carbohydrates or secondary fermentation of organic acids such as propionate (Nobu et al. 2016). Additional sequencing of deep-sea sediments has continued to expand phylum-level bacterial diversity and allowed the description of 56 phyla among which 40 phyla without cultured representatives (Dombrowski et al. 2018). Lineages that have been cultured from sediments such as Gammaproteobacteria (Boden et al. 2017) and Deltaproteobacteria (Bale et al. 1997; Cao et al. 2016) are capable of specialized metabolic processes such as sulfur oxidation and sulfate reduction, respectively.

The upper surface of many deep-sea sediments is oxic, often rich in recalcitrant organic matter that allows the growth of aerobic heterotrophic microorganisms that perform aerobic respiration. Deeper into the sediment the oxygen concentration drops and followed by a zone where sulfate reduction occurs. Below that an intermediate methane sulfate transition zone is followed by the methane production zone. Methanogenesis is a characteristic metabolism of some archaea, whereas methanotrophy is found in archaea and in bacteria from deep-sea sediments (Bessette et al. 2017; Ruff et al. 2019; Evans et al. 2019). *Methyloprofundus sedimenti* WF1T, a gammaproteobacterial methane-oxidizing bacterium, was isolated from marine surface sediment (0-1 cm) in Monterey Canyon off the coast of California, USA (36.708 °N 122.105 °W; 1828 m below sea-level) in close proximity to a whale fall. Its 16S rRNA gene sequence shares 98% identity with uncultured free-living methanotrophs and the methanotrophic endosymbionts of deep-sea mussels (Tavormina et al. 2015). This isolate represents the first cultivar from the "deep sea-1" clade of marine methanotrophs, which includes members that participate in methane oxidation in sediments and the water column as well as in mussels where they live as endosymbionts. Cells of strain WF1T were elongated cocci, approximately 1.5 μm in diameter, and occur single, in pairs, and in clumps (Tavormina et al. 2015). *Methyloprofundus sedimenti* WF1T grows in liquid media at an optimal temperature of 23 °C and depends on the presence of methane or methanol (Tavormina et al. 2015). Atmospheric nitrogen could serve as the sole nitrogen source for WF1T, a capacity that had not been demonstrated previously in members of *Methylobacter* (Bowman et al. 1993; Tavormina et al. 2015).

## 11.6 Conclusions

The deep-sea subsurface is the largest ecosystem and the last frontier on Earth that is poorly accessible and underexplored because of the technical and technological challenges to be faced. Nevertheless, oceanographic campaigns and in particular the IODP drilling expeditions have explored many locations in the ocean and at varying depths in order to study the sedimentary and oceanic crustal samples. Such environments are characterized by a combination of one or more extreme conditions such as pH, temperature, hydrostatic pressure, availability and accessibility of energy resources, and water activity. Estimates of the number of bacterial and archaeal cells vary depending on the locations explored and the markers and techniques used but are invariably large and are at least equivalent to the number of microbial cells on the Earth's surface. Attempts to culture bacteria and archaea from the oceanic subsurface biosphere resulted in the isolation of a considerable number of species often belonging to phyla with other cultured representatives with high growth rates. Such isolates are sometimes not consistent with the low energy metabolism that characterizes the deep subseafloor and most of the microorganisms thriving in this environment. Hence, diversity estimates of bacteria and archaea that are based on traditional molecular approaches or on the isolation of culturable organisms have frequently neglected environmentally important phyla that resist culturing under laboratory conditions.

The omics revolution of the past decades represents a major breakthrough because it uses approaches that do not require culturing of the microorganisms in order to produce high quality genomic data that can be used for in-depth analyses of phylogeny and of metabolic features. Thus, the reconstruction of single genomes from complex deep subseafloor environments has revealed a vast unexplored diversity of bacteria and archaea comprising new lineages which have considerably modified the topology of the tree of life and advanced the understanding of the origin and evolution of life. As an illustration, new phyla of archaea within the ASGARD superphylum share a common ancestor with eukaryotes and have advanced the understanding of the origin of eukaryotes and of the increasing complexity of cellular organization. Despite the major advances achieved through metagenomics there are still many undescribed lineages that are broadly distributed in the deep-sea subsurface. These novel and undescribed lineages need to be explored with the aim of unveiling their metabolic pathways and to improve knowledge of their environmental biogeochemical roles, in particular their role in the linking of the carbon and nutrient cycles in the deep subseafloor.

## References


Baker BJ, Appler KE, Gong X (2021) New microbial biodiversity in marine sediments. Annu Rev Mar Sci 13:161–175. https://doi.org/10.1146/annurev-marine-032020-014552

Baker BJ, De Anda V, Seitz KW, Dombrowski N, Santoro AE, Lloyd KG (2020) Diversity, ecology and evolution of archaea. Nat Microbiol 5:887–900. https://doi.org/10.1038/s41564-020-0715-z

Baker BJ, Saw JH, Lind AE, Lazar CS, Hinrichs K-U, Teske AP, Ettema TJG (2016) Genomic inference of the metabolism of cosmopolitan subsurface archaea, Hadesarchaea. Nat Microbiol 1:16002. https://doi.org/10.1038/nmicrobiol.2016.2

Bale SJ, Goodman K, Rochelle PA, Marchesi JR, Fry JC, Weightman AJ, Parkes RJ (1997) Desulfovibrio profundus sp. nov., a novel barophilic sulfate-reducing bacterium from deep sediment layers in the Japan Sea. Int J Syst Bacteriol 47:515–521. https://doi.org/10.1099/00207713-47-2-515

Bessette S, Moalic Y, Gautey S, Lesongeur F, Godfroy A, Toffin L (2017) Relative abundance and diversity of bacterial methanotrophs at the oxic–anoxic interface of the Congo deep-sea fan. Front Microbiol 8:715. https://doi.org/10.3389/fmicb.2017.00715

Biddle JF, Lipp JS, Lever MA, Lloyd KG, Sorensen KB, Anderson R, Fredricks HF, Elvert M, Kelly TJ, Schrag DP, Sogin ML, Brenchley JE, Teske A, House CH, Hinrichs K-U (2006) Heterotrophic archaea dominate sedimentary subsurface ecosystems off Peru. Proc Natl Acad Sci 103:3846–3851. https://doi.org/10.1073/pnas.0600035103

Boden R, Scott KM, Williams J, Russel S, Antonen K, Rae AW, Hutt LP (2017) An evaluation of *Thiomicrospira*, *Hydrogenovibrio* and *Thioalkalimicrobium*: reclassification of four species of *Thiomicrospira* to each *Thiomicrorhabdus* gen. Nov. and *Hydrogenovibrio*, and reclassification of all four species of *Thioalkalimicrobium* to *Thiomicrospira*. Int J Syst Evol Microbiol 67: 1140–1151. https://doi.org/10.1099/ijsem.0.001855

Bowman JP, Sly LI, Nichols PD, Hayward AC (1993) Revised taxonomy of the methanotrophs: description of *Methylobacter* gen. Nov., emendation of *Methylococcus*, validation of

*Methylosinus* and *Methylocystis* species, and a proposal that the family Methylococcaceae includes only the group I methanotrophs. Int J Syst Bacteriol 43:735–753
Boyd JA, Jungbluth SP, Leu AO, Evans PN, Woodcroft BJ, Chadwick GL, Orphan VJ, Amend JP, Rappé MS, Tyson GW (2019) Divergent methyl-coenzyme M reductase genes in a deep-subseafloor Archaeoglobi. ISME J 13:1269–1279. https://doi.org/10.1038/s41396-018-0343-2
Braun S, Morono Y, Littmann S, Kuypers M, Aslan H, Dong M, Jørgensen BB, BAA L (2016) Size and carbon content of sub-seafloor microbial cells at Landsort deep, Baltic Sea. Front Microbiol 7:1375. https://doi.org/10.3389/fmicb.2016.01375
Brochier-Armanet C, Boussau B, Gribaldo S, Forterre P (2008) Mesophilic Crenarchaeota: proposal for a third archaeal phylum, the Thaumarchaeota. Nat Rev Microbiol 6:245–252. https://doi.org/10.1038/nrmicro1852
Bulzu P-A, Andrei A-Ş, Salcher MM, Mehrshad M, Inoue K, Kandori H, Beja O, Ghai R, Banciu HL (2019) Casting light on Asgardarchaeota metabolism in a sunlit microoxic niche. Nat Microbiol 4:1129–1137. https://doi.org/10.1038/s41564-019-0404-y
Burggraf S, Jannasch HW, Nicolaus B, Stetter KO (1990) Archaeoglobus profundus sp. nov., represents a new species within the sulfate-reducing Archaebacteria. Syst Appl Microbiol 13: 24–28. https://doi.org/10.1016/S0723-2020(11)80176-1
Cao J, Gayet N, Zeng X, Shao Z, Jebbar M, Alain K (2016) Pseudodesulfovibrio indicus gen. nov., sp. nov., a piezophilic sulfate-reducing bacterium from the Indian Ocean and reclassification of four species of the genus Desulfovibrio. Int J Syst Evol Microbiol 66:3904–3911. https://doi.org/10.1099/ijsem.0.001286
Castelle CJ, Banfield JF (2018) Major new microbial groups expand diversity and alter our understanding of the tree of life. Cell 172:1181–1197. https://doi.org/10.1016/j.cell.2018.02.016
Castelle CJ, Wrighton KC, Thomas BC, Hug LA, Brown CT, Wilkins MJ, Frischkorn KR, Tringe SG, Singh A, Markillie LM, Taylor RC, Williams KH, Banfield JF (2015) Genomic expansion of domain archaea highlights roles for organisms from new phyla in anaerobic carbon cycling. Curr Biol 25:690–701. https://doi.org/10.1016/j.cub.2015.01.014
Ciobanu M-C, Burgaud G, Dufresne A, Breuker A, Rédou V, Ben Maamar S, Gaboyer F, Vandenabeele-Trambouze O, Lipp JS, Schippers A, Vandenkoornhuyse P, Barbier G, Jebbar M, Godfroy A, Alain K (2014) Microorganisms persist at record depths in the subseafloor of the Canterbury Basin. ISME J 8:1370–1380. https://doi.org/10.1038/ismej.2013.250
D'Hondt S, Inagaki F, Zarikian CA, Abrams LJ, Dubois N, Engelhardt T, Evans H, Ferdelman T, Gribsholt B, Harris RN, Hoppie BW, Hyun J-H, Kallmeyer J, Kim J, Lynch JE, McKinley CC, Mitsunobu S, Morono Y, Murray RW, Pockalny R, Sauvage J, Shimono T, Shiraishi F, Smith DC, Smith-Duque CE, Spivack AJ, Steinsbu BO, Suzuki Y, Szpak M, Toffin L, Uramoto G, Yamaguchi YT, Zhang G, Zhang X-H, Ziebis W (2015) Presence of oxygen and aerobic communities from sea floor to basement in deep-sea sediments. Nat Geosci 8:299–304. https://doi.org/10.1038/ngeo2387
D'Hondt S, Spivack AJ, Pockalny R, Ferdelman TG, Fischer JP, Kallmeyer J, Abrams LJ, Smith DC, Graham D, Hasiuk F, Schrum H, Stancin AM (2009) Subseafloor sedimentary life in the South Pacific gyre. Proc Natl Acad Sci 106:11651–11656. https://doi.org/10.1073/pnas.0811793106
Dombrowski N, Teske AP, Baker BJ (2018) Expansive microbial metabolic versatility and biodiversity in dynamic Guaymas Basin hydrothermal sediments. Nat Commun 9:4999. https://doi.org/10.1038/s41467-018-07418-0
Durbin AM, Teske A (2011) Microbial diversity and stratification of South Pacific abyssal marine sediments: South Pacific abyssal sediment microbial communities. Environ Microbiol 13:3219–3234. https://doi.org/10.1111/j.1462-2920.2011.02544.x
Eme L, Spang A, Lombard J, Stairs CW, Ettema TJG (2017) Archaea and the origin of eukaryotes. Nat Rev Microbiol 15:711–723. https://doi.org/10.1038/nrmicro.2017.133

Evans PN, Boyd JA, Leu AO, Woodcroft BJ, Parks DH, Hugenholtz P, Tyson GW (2019) An evolving view of methane metabolism in the archaea. Nat Rev Microbiol 17:219–232. https://doi.org/10.1038/s41579-018-0136-7

Evans PN, Parks DH, Chadwick GL, Robbins SJ, Orphan VJ, Golding SD, Tyson GW (2015) Methane metabolism in the archaeal phylum Bathyarchaeota revealed by genome-centric metagenomics. Science 350:434–438. https://doi.org/10.1126/science.aac7745

Garrity GM, Holt JG (2001) Phylum AI. Crenarchaeota phy. Nov. In: Boone DR, Castenholz RW, Garrity GM (eds) Bergey's manual® of systematic bacteriology. Springer, New York, NY. https://doi.org/10.1007/978-0-387-21609-6_16

Hafenbradl D, Keller M, Dirmeier R, Rachel R, Roßnagel P, Burggraf S, Huber H, Stetter KO (1996) Ferroglobus placidus gen. nov., sp. nov., a novel hyperthermophilic archaeum that oxidizes $Fe^{2+}$ at neutral pH under anoxic conditions. Arch Microbiol 166:308–314. https://doi.org/10.1007/s002030050388

He Y, Li M, Perumal V, Feng X, Fang J, Xie J, Sievert SM, Wang F (2016) Genomic and enzymatic evidence for acetogenesis among multiple lineages of the archaeal phylum Bathyarchaeota widespread in marine sediments. Nat Microbiol 1:16035. https://doi.org/10.1038/nmicrobiol.2016.35

Heuer V, Lever M, Morono Y, Teske A (2019) The limits of life and the biosphere in earth's interior. Oceanography 32:208–211. https://doi.org/10.5670/oceanog.2019.147

Hinrichs K-U, Inagaki F (2012) Downsizing the deep biosphere. Science 338:204–205. https://doi.org/10.1126/science.1229296

Hoehler TM, Jørgensen BB (2013) Microbial life under extreme energy limitation. Nat Rev Microbiol 11:83–94. https://doi.org/10.1038/nrmicro2939

Hu H, Natarajan VP, Wang F (2021) Towards enriching and isolation of uncultivated archaea from marine sediments using a refined combination of conventional microbial cultivation methods. Mar Life Sci Technol 3:231–242. https://doi.org/10.1007/s42995-021-00092-0

Huber H, Jannasch H, Rachel R, Fuchs T, Stetter KO (1997) Archaeoglobus veneficus sp. nov., a novel facultative chemolithoautotrophic hyperthermophilic sulfite reducer, isolated from abyssal black smokers. Syst Appl Microbiol 20:374–380. https://doi.org/10.1016/S0723-2020(97)80005-7

Hug LA, Baker BJ, Anantharaman K, Brown CT, Probst AJ, Castelle CJ, Butterfield CN, Hernsdorf AW, Amano Y, Ise K, Suzuki Y, Dudek N, Relman DA, Finstad KM, Amundson R, Thomas BC, Banfield JF (2016) A new view of the tree of life. Nat Microbiol 1:16048. https://doi.org/10.1038/nmicrobiol.2016.48

Imachi H, Nobu MK, Nakahara N, Morono Y, Ogawara M, Takaki Y, Takano Y, Uematsu K, Ikuta T, Ito M, Matsui Y, Miyazaki M, Murata K, Saito Y, Sakai S, Song C, Tasumi E, Yamanaka Y, Yamaguchi T, Kamagata Y, Tamaki H, Takai K (2020) Isolation of an archaeon at the prokaryote–eukaryote interface. Nature 577:519–525. https://doi.org/10.1038/s41586-019-1916-6

Inagaki F, Hinrichs K-U, Kubo Y, Bowles MW, Heuer VB, Hong W-L, Hoshino T, Ijiri A, Imachi H, Ito M, Kaneko M, Lever MA, Lin Y-S, Methe BA, Morita S, Morono Y, Tanikawa W, Bihan M, Bowden SA, Elvert M, Glombitza C, Gross D, Harrington GJ, Hori T, Li K, Limmer D, Liu C-H, Murayama M, Ohkouchi N, Ono S, Park Y-S, Phillips SC, Prieto-Mollar X, Purkey M, Riedinger N, Sanada Y, Sauvage J, Snyder G, Susilawati R, Takano Y, Tasumi E, Terada T, Tomaru H, Trembath-Reichert E, Wang DT, Yamada Y (2015) Exploring deep microbial life in coal-bearing sediment down to 2.5 km below the ocean floor. Science 349:420–424. https://doi.org/10.1126/science.aaa6882

Jannasch HW, Taylor CD (1984) Deep-sea microbiology. Annu Rev Microbiol 38:487–514

Jebbar M, Franzetti B, Girard E, Oger P (2015) Microbial diversity and adaptation to high hydrostatic pressure in deep-sea hydrothermal vents prokaryotes. Extremophiles 19:721–740. https://doi.org/10.1007/s00792-015-0760-3

Jungbluth SP, Amend JP, Rappé MS (2017) Metagenome sequencing and 98 microbial genomes from Juan de Fuca ridge flank subsurface fluids. Sci Data 4:170037. https://doi.org/10.1038/sdata.2017.37

Jungbluth SP, Bowers RM, Lin H-T, Cowen JP, Rappé MS (2016) Novel microbial assemblages inhabiting crustal fluids within mid-ocean ridge flank subsurface basalt. ISME J 10:2033–2047. https://doi.org/10.1038/ismej.2015.248

Kallmeyer J, Pockalny R, Adhikari RR, Smith DC, D'Hondt S (2012) Global distribution of microbial abundance and biomass in subseafloor sediment. Proc Natl Acad Sci 109:16213–16216. https://doi.org/10.1073/pnas.1203849109

Kashefi K (2002) Geoglobus ahangari gen. nov., sp. nov., a novel hyperthermophilic archaeon capable of oxidizing organic acids and growing autotrophically on hydrogen with Fe(III) serving as the sole electron acceptor. Int J Syst Evol Microbiol 52:719–728. https://doi.org/10.1099/ijs.0.01953-0

Koonin EV (2015) Origin of eukaryotes from within archaea, archaeal eukaryome and bursts of gene gain: eukaryogenesis just made easier? Philos Trans R Soc B Biol Sci 370:20140333. https://doi.org/10.1098/rstb.2014.0333

Laso-Pérez R, Wegener G, Knittel K, Widdel F, Harding KJ, Krukenberg V, Meier DV, Richter M, Tegetmeyer HE, Riedel D, Richnow H-H, Adrian L, Reemtsma T, Lechtenfeld OJ, Musat F (2016) Thermophilic archaea activate butane via alkyl-coenzyme M formation. Nature 539: 396–401. https://doi.org/10.1038/nature20152

Lee YM, Hwang K, Lee JI, Kim M, Hwang CY, Noh H-J, Choi H, Lee HK, Chun J, Hong SG, Shin SC (2018) Genomic insight into the predominance of candidate phylum Atribacteria JS1 lineage in marine sediments. Front Microbiol 9:2909. https://doi.org/10.3389/fmicb.2018.02909

Lever MA, Rogers KL, Lloyd KG, Overmann J, Schink B, Thauer RK, Hoehler TM, Jørgensen BB (2015) Life under extreme energy limitation: a synthesis of laboratory- and field-based investigations. FEMS Microbiol Rev 39:688–728. https://doi.org/10.1093/femsre/fuv020

L'Haridon S, Haroun H, Corre E, Roussel E, Chalopin M, Pignet P, Balière C, la Cono V, Jebbar M, Yakimov M, Toffin L (2020) Methanohalophilus profundi sp. nov., a methylotrophic halophilic piezophilic methanogen isolated from a deep hypersaline anoxic basin. Syst Appl Microbiol 43: 126107. https://doi.org/10.1016/j.syapm.2020.126107

Lin H-T, Cowen JP, Olson EJ, Amend JP, Lilley MD (2012) Inorganic chemistry, gas compositions and dissolved organic carbon in fluids from sedimented young basaltic crust on the Juan de Fuca ridge flanks. Geochim Cosmochim Acta 85:213–227. https://doi.org/10.1016/j.gca.2012.02.017

Lipp JS, Morono Y, Inagaki F, Hinrichs K-U (2008) Significant contribution of archaea to extant biomass in marine subsurface sediments. Nature 454:991–994

Liu Y, Zhou Z, Pan J, Baker BJ, Gu J-D, Li M (2018) Comparative genomic inference suggests mixotrophic lifestyle for Thorarchaeota. ISME J 12:1021–1031. https://doi.org/10.1038/s41396-018-0060-x

Lloyd KG, Schreiber L, Petersen DG, Kjeldsen KU, Lever MA, Steen AD, Stepanauskas R, Richter M, Kleindienst S, Lenk S, Schramm A, Jørgensen BB (2013) Predominant archaea in marine sediments degrade detrital proteins. Nature 496:215–218. https://doi.org/10.1038/nature12033

Lomstein BAA, Langerhuus AT, D'Hondt S, Jørgensen BB, Spivack AJ (2012) Endospore abundance, microbial growth and necromass turnover in deep sub-seafloor sediment. Nature 484:101–104. https://doi.org/10.1038/nature10905

López-García P, Moreira D (2015) Open questions on the origin of eukaryotes. Trends Ecol Evol 30:697–708. https://doi.org/10.1016/j.tree.2015.09.005

López-García P, Moreira D (2020) The syntrophy hypothesis for the origin of eukaryotes revisited. Nat Microbiol 5:655–667. https://doi.org/10.1038/s41564-020-0710-4

Ludwig W, Viver T, Westram R, Francisco Gago J, Bustos-Caparros E, Knittel K, Amann R, Rossello-Mora R (2021) Release LTP_12_2020, featuring a new ARB alignment and improved 16S rRNA tree for prokaryotic type strains. Syst Appl Microbiol 44:126218. https://doi.org/10.1016/j.syapm.2021.126218

Martin WF, Garg S, Zimorski V (2015) Endosymbiotic theories for eukaryote origin. Philos Trans R Soc B Biol Sci 370:20140330. https://doi.org/10.1098/rstb.2014.0330

Meng J, Xu J, Qin D, He Y, Xiao X, Wang F (2014) Genetic and functional properties of uncultivated MCG archaea assessed by metagenome and gene expression analyses. ISME J 8: 650–659. https://doi.org/10.1038/ismej.2013.174

Monaco A, Prouzet P (2015) Hydrothermal vents: oases at depth. In: Monaco A, Prouzet P (eds) Marine ecosystems. Wiley, Hoboken, NJ, USA, pp 225–292

Mori K, Maruyama A, Urabe T, K-i S, Hanada S (2008) Archaeoglobus infectus sp. nov., a novel thermophilic, chemolithoheterotrophic archaeon isolated from a deep-sea rock collected at Suiyo seamount, Izu-Bonin arc, western Pacific Ocean. Int J Syst Evol Microbiol 58:810–816. https://doi.org/10.1099/ijs.0.65422-0

Morono Y, Terada T, Nishizawa M, Ito M, Hillion F, Takahata N, Sano Y, Inagaki F (2011) Carbon and nitrogen assimilation in deep subseafloor microbial cells. Proc Natl Acad Sci 108:18295–18300. https://doi.org/10.1073/pnas.1107763108

Nobu MK, Dodsworth JA, Murugapiran SK, Rinke C, Gies EA, Webster G, Schwientek P, Kille P, Parkes RJ, Sass H, Jørgensen BB, Weightman AJ, Liu W-T, Hallam SJ, Tsiamis G, Woyke T, Hedlund BP (2016) Phylogeny and physiology of candidate phylum 'Atribacteria' (OP9/JS1) inferred from cultivation-independent genomics. ISME J 10:273–286. https://doi.org/10.1038/ismej.2015.97

Orcutt BN, Bach W, Becker K, Fisher AT, Hentscher M, Toner BM, Wheat CG, Edwards KJ (2011) Colonization of subsurface microbial observatories deployed in young ocean crust. ISME J 5: 692–703. https://doi.org/10.1038/ismej.2010.157

Oren A, Garrity GM (2021) Candidatus list no. 2. Lists of names of prokaryotic Candidatus taxa. Int J Syst Evol Microbiol 71:4671. https://doi.org/10.1099/ijsem.0.004671

Oren A (2019) Euryarchaeota. In: ELS. Wiley, Chichester

Parkes RJ, Cragg B, Roussel E, Webster G, Weightman A, Sass H (2014) A review of prokaryotic populations and processes in sub-seafloor sediments, including biosphere:geosphere interactions. Mar Geol 352:409–425. https://doi.org/10.1016/j.margeo.2014.02.009

Parkes RJ, Sellek G, Webster G, Martin D, Anders E, Weightman AJ, Sass H (2009) Culturable prokaryotic diversity of deep, gas hydrate sediments: first use of a continuous high-pressure, anaerobic, enrichment and isolation system for subseafloor sediments (DeepIsoBUG). Environ Microbiol 11:3140–3153. https://doi.org/10.1111/j.1462-2920.2009.02018.x

Parte AC, Sardà Carbasse J, Meier-Kolthoff JP, Reimer LC, Göker M (2020) List of prokaryotic names with standing in nomenclature (LPSN) moves to the DSMZ. Int J Syst Evol Microbiol 70:5607–5612. https://doi.org/10.1099/ijsem.0.004332

Pushkarev A, Inoue K, Larom S, Flores-Uribe J, Singh M, Konno M, Tomida S, Ito S, Nakamura R, Tsunoda SP, Philosof A, Sharon I, Yutin N, Koonin EV, Kandori H, Béjà O (2018) A distinct abundant group of microbial rhodopsins discovered using functional metagenomics. Nature 558:595–599. https://doi.org/10.1038/s41586-018-0225-9

Rinke C, Schwientek P, Sczyrba A, Ivanova NN, Anderson IJ, Cheng J-F, Darling A, Malfatti S, Swan BK, Gies EA, Dodsworth JA, Hedlund BP, Tsiamis G, Sievert SM, Liu W-T, Eisen JA, Hallam SJ, Kyrpides NC, Stepanauskas R, Rubin EM, Hugenholtz P, Woyke T (2013) Insights into the phylogeny and coding potential of microbial dark matter. Nature 499:431–437. https://doi.org/10.1038/nature12352

Roussel EG, Bonavita M-AC, Querellou J, Cragg BA, Webster G, Prieur D, Parkes RJ (2008) Extending the sub-sea-floor biosphere. Science 320:1046–1046. https://doi.org/10.1126/science.1154545

Ruff SE, Felden J, Gruber-Vodicka HR, Marcon Y, Knittel K, Ramette A, Boetius A (2019) In situ development of a methanotrophic microbiome in deep-sea sediments. ISME J 13:197–213. https://doi.org/10.1038/s41396-018-0263-1

Seibold E, Berger W (2017) The sea floor: an introduction to marine geology, 4th edn. Springer International Publishing, Cham

Seitz KW, Dombrowski N, Eme L, Spang A, Lombard J, Sieber JR, Teske AP, Ettema TJG, Baker BJ (2019) Asgard archaea capable of anaerobic hydrocarbon cycling. Nat Commun 10:1822. https://doi.org/10.1038/s41467-019-09364-x

Seitz KW, Lazar CS, Hinrichs K-U, Teske AP, Baker BJ (2016) Genomic reconstruction of a novel, deeply branched sediment archaeal phylum with pathways for acetogenesis and sulfur reduction. ISME J 10:1696–1705. https://doi.org/10.1038/ismej.2015.233

Slobodkina GB, Kolganova TV, Querellou J, Bonch-Osmolovskaya EA, Slobodkin AI (2009) Geoglobus acetivorans sp. nov., an iron(III)-reducing archaeon from a deep-sea hydrothermal vent. Int J Syst Evol Microbiol 59:2880–2883. https://doi.org/10.1099/ijs.0.011080-0

Slobodkina G, Allioux M, Merkel A, Cambon-Bonavita M-A, Alain K, Jebbar M, Slobodkin A (2021) Physiological and genomic characterization of a hyperthermophilic Archaeon Archaeoglobus neptunius sp. nov. isolated from a deep-sea hydrothermal vent warrants the reclassification of the genus Archaeoglobus. Front Microbiol 12:679245. https://doi.org/10.3389/fmicb.2021.679245

Sousa FL, Neukirchen S, Allen JF, Lane N, Martin WF (2016) Lokiarchaeon is hydrogen dependent. Nat Microbiol 1:16034. https://doi.org/10.1038/nmicrobiol.2016.34

Spang A, Saw JH, Jørgensen SL, Zaremba-Niedzwiedzka K, Martijn J, Lind AE, van Eijk R, Schleper C, Guy L, Ettema TJG (2015) Complex archaea that bridge the gap between prokaryotes and eukaryotes. Nature 521:173–179. https://doi.org/10.1038/nature14447

Spang A, Stairs CW, Dombrowski N, Eme L, Lombard J, Caceres EF, Greening C, Baker BJ, Ettema TJG (2019) Proposal of the reverse flow model for the origin of the eukaryotic cell based on comparative analyses of Asgard archaeal metabolism. Nat Microbiol 4:1138–1148. https://doi.org/10.1038/s41564-019-0406-9

Steinsbu BO, Thorseth IH, Nakagawa S, Inagaki F, Lever MA, Engelen B, Ovreas L, Pedersen RB (2010) Archaeoglobus sulfaticallidus sp. nov., a thermophilic and facultatively lithoautotrophic sulfate-reducer isolated from black rust exposed to hot ridge flank crustal fluids. Int J Syst Evol Microbiol 60:2745–2752. https://doi.org/10.1099/ijs.0.016105-0

Stetter KO (1988) Archaeoglobus fulgidus gen. nov., sp. nov.: a new taxon of extremely thermophilic archaebacteria. Syst Appl Microbiol 10:172–173. https://doi.org/10.1016/S0723-2020(88)80032-8

Tavormina PL, Hatzenpichler R, McGlynn S, Chadwick G, Dawson KS, Connon SA, Orphan VJ (2015) Methyloprofundus sedimenti gen. nov., sp. nov., an obligate methanotroph from ocean sediment belonging to the 'deep sea-1' clade of marine methanotrophs. Int J Syst Evol Microbiol 65:251–259. https://doi.org/10.1099/ijs.0.062927-0

Teske A, Sørensen KB (2008) Uncultured archaea in deep marine subsurface sediments: have we caught them all? ISME J 2:3–18. https://doi.org/10.1038/ismej.2007.90

Whitman WB, Coleman DC, Wiebe WJ (1998) Prokaryotes: the unseen majority. Proc Natl Acad Sci 95:6578–6583

Yu T, Wu W, Liang W, Lever MA, Hinrichs K-U, Wang F (2018) Growth of sedimentary *Bathyarchaeota* on lignin as an energy source. Proc Natl Acad Sci 115:6022–6027. https://doi.org/10.1073/pnas.1718854115

Zaremba-Niedzwiedzka K, Caceres EF, Saw JH, Bäckström D, Juzokaite L, Vancaester E, Seitz KW, Anantharaman K, Starnawski P, Kjeldsen KU, Stott MB, Nunoura T, Banfield JF, Schramm A, Baker BJ, Spang A, Ettema TJG (2017) Asgard archaea illuminate the origin of eukaryotic cellular complexity. Nature 541:353–358. https://doi.org/10.1038/nature21031

Zeng X, Zhang X, Jiang L, Alain K, Jebbar M, Shao Z (2013) Palaeococcus pacificus sp. nov., an archaeon from deep-sea hydrothermal sediment. Int J Syst Evol Microbiol 63:2155–2159. https://doi.org/10.1099/ijs.0.044487-0

# The Microbiome of Coastal Sediments 12

Graham J. C. Underwood, Alex J. Dumbrell, Terry J. McGenity, Boyd A. McKew, and Corinne Whitby

**Abstract**

Coastal zones are among the most productive marine environments and many are highly impacted by anthropogenic activity. Coastal zones are key regions for the transformation of land-based inputs of nutrients and pollutants and provide many essential ecosystem services for human society. Periods of tidal exposure and submergence, coupled with seasonal variation in land-based inputs, result in intertidal habitats characterized by highly variable environmental conditions that pose crucial adaptive challenges for organisms. This review focuses on the microbiome of coastal sediments consisting of protists (especially diatoms), bacteria, archaea, and fungi. The diversity, distribution, production, adaptations, and interactions between these groups are reviewed. Coastal microbiomes are characterized by high rates of biogeochemical activity. Photoautotrophic diatoms exhibit complex patterns of behavior to cope with a highly variable light climate. Multiple species–species interactions between autotrophs and heterotrophs contribute to the cycling of carbon and nitrogen. In sediments, autotrophic and heterotrophic processes are closely coupled both spatially and temporally. Bacteria and archaea control the nitrogen- and carbon cycles while taxonomic diversity is influenced by gradients of organic matter, nitrogen compounds, sulfide, and oxygen. Fungi are important components of coastal salt marsh sediment microbiomes but their role in unvegetated sediments is less well understood. This review considers the high human impact on coastal sediments and the importance of nutrient gradients and pollution pressures (hydrocarbons) in affecting diversity and species distribution.

---

G. J. C. Underwood (✉) · A. J. Dumbrell · T. J. McGenity · B. A. McKew · C. Whitby
School of Life Sciences, University of Essex, Essex, UK
e-mail: gjcu@essex.ac.uk; adumb@essex.ac.uk; tjmcgen@essex.ac.uk; boyd.mckew@essex.ac.uk; cwhitby@essex.ac.uk



L. J. Stal, M. S. Cretoiu (eds.), *The Marine Microbiome*, The Microbiomes of Humans, Animals, Plants, and the Environment 3,
https://doi.org/10.1007/978-3-030-90383-1_12

**Keywords**

Behavior · Biogeochemistry · Estuarine gradients · Microbial diversity · Pollution · Species–species interactions

## 12.1 Introduction

Coastal zones are among the most productive marine environments. Located at the interface between marine, freshwater, and terrestrial environments and receiving inputs from both the open ocean and from the land, coastal zones consist of a matrix of diverse habitats positioned along various physical, chemical, and biological gradients. Organisms living within transitional coastal zones have to be adapted to major gradients of conditions that can be subject to seasonal variability. In addition to seasonal changes (particularly present at temperate- and polar latitudes), there is a strong influence of the lunar tidal cycle (bi-weekly) resulting in periods of varying length of aerial exposure and saline water submergence in intertidal environments. These exposure cycles can result in important changes in environmental conditions on an hourly basis. Thus, the coastal zone is characterized by highly variable environmental conditions that pose considerable adaptive challenges for organisms living within it. Despite these challenges, coastal habitats support characteristic microbiomes (defined as a characteristic microbial community with distinct physiological and chemical properties and activities resulting in the formation of specific ecological niches, Berg et al. 2020) that underpin the ecological functioning of these habitats. Coastal microbiomes play crucial roles in biogeochemical cycling, food webs, and habitat modification, resulting in the provision of important ecosystem services to human society.

The coastal zone encompasses a wide range of habitats. Rocky shores are extensive worldwide and are generally characterized by steep spatial gradients from land to sea. Rocky shores host abundant communities of macroscopic organisms but their microbiology is less well described (Maggi et al. 2017). The impervious nature of rocky shores means that the influence of microbes present as thin epilithic biofilms is strongly affected by cycles of desiccation, extreme salinity fluctuations, macroalgal spore settlement and germination, and grazing by macroinvertebrates. This chapter does not focus on these environments and readers are referred to Dal Bello et al. (2017) and Notman et al. (2016).

Estuaries and intertidal flats are among the most productive of the various coastal ecosystems and provide important ecosystem services such as the provisioning of food resources, water purification, carbon storage, and coastal storm surge tidal defense (Waltham et al. 2020). Estuaries are also some of the most human-modified environments because of historic and current concentration of human populations and industries along their coastlines (Henderson et al. 2020; Van Niekerk et al. 2013). This has resulted in a considerable degradation of estuarine ecosystems, which changed the ecological processes that govern their health and ecosystem services (Duarte et al. 2020; Van Niekerk et al. 2019; Waltham et al. 2020). Coastal habitats (salt marsh, mangrove, seagrass, muddy and sandy intertidal flats) are important zones of nutrient cycling (Nedwell et al. 2016) and valuable sites of organic carbon generation and accumulation ("blue carbon," the carbon stored in the sediment, living and non-living above- and below-ground biomass of salt marsh and seagrass habitats (Alongi 2020; Beaumont et al. 2014; Burden et al. 2019; Legge et al. 2020; Waltham et al. 2020). Because coastal environments are located in the transition zone between the land and the sea, they are particularly susceptible to pollution, including excess nutrients (particularly inorganic N and P), heavy metals, pesticides, pharmaceuticals, numerous industrial persistent organic pollutants, and plastics. One of the major types of organic pollutants in coastal ecosystems are petroleum hydrocarbons from crude oil and its many refined products.

Estuarine sedimentary systems are extensively distributed across the globe. All estuaries have their own characteristics influenced by the local geology and catchment features and the local tidal range, from microtidal ($<2$ m tidal range), mesotidal (2–4 m tidal range), and macrotidal ($>4$ m tidal range). Tidal range and wind and wave-climate are major factors influencing the geomorphology of a coastal-estuarine system. A typical meso- or macro-tidal estuary is usually characterized by a well-mixed salinity gradient from freshwater to fully marine with fine-grained sediment and mudflats in the sheltered regions of the estuary toward its head and mixed- and sandy sediments toward the mouth where tides and wind-driven waves and currents are stronger (Baas et al. 2019; Green and Coco 2014; Zhu et al. 2020). In agricultural and populous catchments, most nutrient loading is land derived. Hence, an estuarine gradient reflects covarying conditions of increasing salinity, decreasing nutrient loading, increasing sediment particle size, and varying levels of tidal exposure (Nedwell et al. 2016). Approximately perpendicular to this linear gradient is the gradient of tidal exposure with upper shores often colonized by vascular macrophytes (salt marshes in temperate regions and mangroves in the tropics, Alongi 2020) and by sand dune habitats on wind-dominated sandy shores (Galiforni-Silva et al. 2020). Mid-tide level shores tend to be dominated by micro- and macro-algal mats, and the lower shores more physically disturbed, but also colonized by biogenetic reefs of bivalves or polychaete worms. These environmental gradients, the large surface area provided by sediment particles, and accounting for areal and depth dimensions result in an extensive mosaic of habitats that support productive and diverse microbiomes (Heip et al. 1995; Luna et al. 2013; Underwood and Kromkamp 1999).

## 12.2 Coastal Autotrophic Microbiomes: Microphytobenthic Biofilms

On intertidal mud and sand flats and in shallow subtidal systems where sunlight reaches the sediment surface diverse and abundant microbial biofilms occur. Collectively these assemblages are termed microphytobenthos (MPB) or benthic microalgae (BMA) biofilms, terms which emphasize the important role played by the photoautotrophic components of these complex agglomerates of autotrophic and heterotrophic protists, bacteria, archaea, and fungi (An et al. 2020; Chen et al. 2017; Cibic et al. 2019; Pinckney 2018; Sahan et al. 2007; Underwood and Kromkamp 1999). The photoautotrophic diatoms (Stramenopiles, Bacillariophyceae) are major components of most MPB or BMA biofilms with net primary production of 29–314 g C $m^{-2}$ $y^{-1}$ (Pinckney 2018; Underwood and Kromkamp 1999). The primary production of MPB provides the main energy resource to biofilm consumers (protozoans and metazoans) and their predators (Christianen et al. 2017; Green et al. 2012; Herman et al. 2000; Hope et al. 2020), while heterotrophic bacteria and archaea are the primary remineralizers of MPB-derived organic material, including volatile compounds and detrital organic matter present in the sediment (Acuña Alvarez et al. 2009; Bohórquez et al. 2017; Gaubert-Boussarie et al. 2020; Luna et al. 2013; Nedwell et al. 2016).

Two types of microphytobenthic biofilms are recognized: transient microbial biofilms that form and reform over daily and weekly timescales and more permanent microbial mats. Microbial mats are characterized by higher biomass and are usually dominated by cyanobacteria. Microbial mats show long-term temporal persistence (months to years) such that a macroscopic structure is formed, and they are often closed systems with much internal recycling of nutrients (Long et al. 2013; Stal et al. 2019). Stromatolites are a particular type of microbial mat that possess a laminated calcified structure, which is considered to be the outcome of an intense coupling between microbial (cyanobacteria, heterotrophic bacteria, archaea, eukarya) and geochemical processes leading to a remnant geological formation. A specialized type of a coastal microbial mat is the supratidal microbialite. These microbialites have been found in the supratidal zone of rocky shores in South Africa, Australia, and the U.K., where there is a freshwater input (Rishworth et al. 2020). Microbial mats and stromatolite microbiomes have been reviewed by Stal (2016), Stal et al. (2019), and Rishworth et al. (2020) and are not considered further here.

### 12.2.1 Diversity of Microphytobenthos in Coastal Sediments

Transient marine benthic biofilms have a high potential species richness of photoautotrophs. Although both photosynthetic and heterotrophic microeukaryotes such as flagellates and ciliates are present (Chen et al. 2017; Gong et al. 2015; Massana et al. 2015), their ecology and importance in intertidal biofilm is in most cases unresolved. A few genera of cyanobacteria (e.g., *Lyngbya, Oscillatoria*) and motile euglenophytes (e.g., *Euglena deses, E. proxima*) are found and are often in high abundance in transient biofilms (Bellinger et al. 2005; Kingston 1999; Perkins et al. 2002; Underwood et al. 2005) (Table 12.1). However, the dominant group of MPB in terms of biomass and activity are benthic diatoms with well over 1500 benthic diatom (morpho)species described from different geographical regions (Witkowski et al. 2000). Within a particular environment however, especially on estuarine intertidal mudflats it is more usual to find only a few (20+) species that are numerically dominant within MPB assemblages (Forster et al. 2006; Park et al. 2014; Redzuan and Underwood 2020, 2021; Ribeiro et al. 2013, 2020; Sahan et al.

**Table 12.1** Representative taxa of microphytobenthos found in European coastal sediment microbiomes, characterized by life form and habitat. Note that sediment type represents a continuum of sediment grain size and properties, and individual taxa may occur across this gradient

| Epipelon (clays and muds) (silts and silty sand) | Epipsammon (sand) | Tychoplankton (resuspended) |
|---|---|---|
| Diatoms (Stramenopiles, Bacillariophyceae) | | |
| *Navicula phyllepta, N. gregaria, N. perminuta, N. flanatica, N. Spartinetensis, N. salinarum N. peregrina, N. digitoradiata,, N. arenaria* | *Planothidium delicatulum Biremis lucens, Achnanthes* sp., *Nitzschia frustulum* | *Rhaphoneis minutissima, Rhaphoneis amphiceros* |
| *Gyrosigma limosum, G. fasciola, G. accuminatum Gyrosigma balticum* | *Amphora ovalis, A. salina. A.* c.f. *tenuissima* | *Cymatosira belgica Staurosira construens* |
| *Pleurosigma angulatum, Scolioneis tumida Hantzschia virgata, Tropidoneis vitrea* | *Opephora guenter-grassi* | *Thalassiosira* sp., *Actinoptychus senarius, Odontella aurita* |
| *Nitzschia* c.f. *panduriformis, Nitzschia sigma, Tryblionella apiculata* | *Dimeregramma minor* | *Opephora* sp. |
| *Cylindrotheca gracilis. C. signata, C. closterium* | *Catenula adhaerens* | |
| Euglenids (Stramenopiles, Euglenophyceae) | | |
| *Euglena deses, E. proxima* | | |
| Cyanobacteria | | |
| *Microcoleus chthonoplastes, Lyngbya aestuarii, spirulina* sp. | *Merismopedia glauca* | |
| *Oscillatoria limosa, O. princeps* | | |

Taken from: (Sabbe 1993; Underwood 1994; Underwood et al. 1998; Hamels et al. 1998; Bellinger et al. 2005; Forster et al. 2006; Ribeiro et al. 2013; Redzuan and Underwood 2020, 2021)

2007; Thornton et al. 2002; Underwood 1994; Underwood and Barnett 2006). The majority of the literature on benthic diatom diversity relies on microscopy-based identification and a morphology-based taxonomy, an approach which is time-consuming and requires a high level of expertise (Ribeiro et al. 2020). Where detailed studies have been conducted it is clear that deterministic (niche-based) factors rather than neutral factors determine the community composition of the abundant species (Plante et al. 2016, 2021; Thornton et al. 2002). Taxonomic composition of the dominant components in the microbiome is strongly influenced by sediment particle size (the balance of sands, silts, and clays) selecting for a range of highly motile biraphid epipelic (mud inhabiting), less motile mono- or biraphid epipsammic (attached to sand grains), or araphid, diatom taxa

(Table 12.1) (Hamels et al. 1998; Sabbe 1993; Underwood and Barnett 2006). The distribution of sediment types corresponds to gradients of physical energy, salinity, and water flow, with sands present in the more exposed marine sediments and clays and silts settling in the more sheltered, low-energy, upper reaches of estuaries, often exposed to a greater range of salinity conditions over tidal and seasonal cycles (Baas et al. 2019; Green and Coco 2014). These gradients are major factors that determine both alpha- and beta diversity in MPB microbiomes (Gong et al. 2015; Park et al. 2014; Plante et al. 2016, 2021; Ribeiro et al. 2013; Witkowski et al. 2000).

Salinity within an estuarine gradient and position on the shore, which relates to the degree of tidal exposure and period of subtidal disturbance, are important controls on diatom species distribution (Forster et al. 2006; Oppenheim 1991; Peletier 1996; Ribeiro et al. 2003, 2013; Sahan et al. 2007; Thornton et al. 2002; Underwood 1994; Underwood et al. 1998). These physical factors vary with seasonal changes in irradiance, thermal stress, and winter mixing and storminess and alter the species composition of estuarine benthic diatom communities (Oppenheim 1991; Underwood 1994, 2005). For example, species such as *Fallacia pygmaea* and *Navicula salinarum* are found in cold and warm months, respectively (Admiraal et al. 1984). Inorganic nutrient concentrations are a strong driver of dominant taxa. They correlate with microphytobenthic biomass and species composition on spatial and temporal scales (Thornton et al. 2002; Underwood et al. 1998).

Many patterns of species distribution in estuaries are based on correlative field surveys. Because of covarying gradients, especially of exposure at low tide, sediment type, salinity, and nutrient concentrations, it is not clear how much variability in community composition occurs in the absence of changes in nutrients (or by changes in other variables such as sediment particle size distribution) and over what time scales such variability operates. Experimental manipulations have shown that nutrients are a selective force in determining species composition (Sullivan 1999; Underwood et al. 1998; Underwood and Provot 2000). High concentrations of ammonium and sulfide (often due to sewage inputs or organic enrichment) can be inhibitory or selective for particularly resistant taxa (Admiraal 1984). Significant decrease in Chl *a* and changes in species composition in the Ems Dollard estuary occurred between 1979 and 1993 after the lowering of organic waste input from local potato starch industries onto adjacent mudflats (Peletier 1996). Small spatial scale (10–100 m) patterns in biomass and species composition were documented in the Colne Estuary (Thornton et al. 2002; Underwood et al. 1998), which have been experimentally demonstrated to relate to species-specific preferences (Underwood and Provot 2000) and tolerance to sulfide and anoxia (McKew et al. 2013).

High throughput sequencing (HTS) methodologies have been applied to the coastal benthic eukaryotic microbiome in the last decade. The results of these analyses have in general supported the conclusions of microscope-based approaches about patterns of richness, distribution of species, and correlation with environmental variables. However, HTS report higher operational taxonomic unit (OTU) richness than that recognized by morphological taxonomic approaches. Chen et al. (2017) found 6627 benthic microeukaryotic OTUs (18S rRNA gene, 97% sequence similarity cutoff) at Xiamen Island, China; Plante et al. (2021) identified 4411 different diatom OTUs (18S rRNA gene, sequence similarity threshold of 98%) in South Carolina sediments, while An et al. (2020) reported 9582 diatom OTU (*rbcL* gene, 98% similarity cut off) on Korean intertidal mudflats. Chen et al. (2017) found strong deterministic control of the abundant and conditionally rare taxa (total phosphorus, total nitrogen, salinity, phosphate, and total oxidized nitrogen for the dominant taxa) in benthic microbial eukaryote communities but rare species showed no spatial, environmental, or distance-decay pattern. Comparisons of the diatoms present on mudflat-salt marsh transitions on a number of barrier island sites in South Carolina, USA, revealed that 95% of all OTUs were rare, less than 0.1% of total sequence count and that a few key taxa (e.g., *Navicula* and *Gyrosigma* species) were dominant (Plante et al. 2021). Spatial effects (dispersal limitation) and spatially-structured environmental factors affected these dominant taxa that caused the significant differences in beta diversity between island sites. The presence of planktonic or tychoplankton taxa, e.g., *Thalassiosira* was also recorded in benthic samples (Plante et al. 2021). Within single geographical sites factors such as physical disturbance and sediment type (sands, muds) determined the dominance patterns of a few key taxa but with greater neutral or stochastic process elements influencing the composition of the different patches (Plante et al. 2016).

High throughput sequencing provides a much greater resolution of the richness of the eukaryotic members of the sediment microbiome than revealed by morphological microscopic approaches. This higher richness is due in part to a higher intensity of sampling (greater sample sizes) and because of multiple copies of target genes within single cells, which varies between taxa (Gong and Marchetti 2019). But it could also be caused by cryptic diversity within some groups of diatoms. Vanelslander et al. (2009) found that the common estuarine diatom morphospecies *Navicula phyllepta* was in fact two different species separated spatially along the salinity gradient of the Western Scheldt estuary. Clonal cultures of *Nitzschia inconspicua* isolated from across a range of freshwater, brackish, and marine habitats in the River Ebro were paraphyletic with six different genotypes and a range of different reproductive strategies and salinity tolerances (Rovira et al. 2015). Growth of two clonal cultures of *Cylindrotheca closterium*, isolated from the oligosaline and mesosaline regions of the Colne estuary, showed different growth optima to salinity and nitrogen gradients (Underwood and Provot 2000). Morphological approaches with their much lower sample sizes (usually 200–400 individuals) cannot sample as many rare taxa as HTS and cannot resolve cryptic diversity. Moreover, rare valves require greater identification skills and may just be reported as unidentified by the untrained eye. These factors hamper the understanding of community assembly processes for microphytobenthos. Comparison of morphological species lists from spatially-distant sites, especially when studied by different research teams, is difficult because of the judgment required for many morphological species. HTS data lends itself to larger spatial analyses (Clark et al. 2017) and avoids individuals' decisions on species attribution. However, ecological interpretation of HTS datasets also relies on accurate reference databases and consideration of the issues of numbers of copies of phylogenetically informative genes when translating to relative abundance (Gong and Marchetti 2019). There is a need for better taxonomic alignment in the libraries used to assign HTS DNA sequence data to coastal marine morphological species in order to minimize the risk of misidentification and to ensure that the latest taxonomic phylogenies are reported as is being done for freshwater diatoms (Pérez-Burillo et al. 2020). This is particular necessary in order to understand the role of the many rare taxa that HTS detects in coastal microbiomes. Are these OTUs representing cryptic diversity within recognized species complexes or are these new, unknown, species or false identifications made by the original sequence depositors? These challenges need addressing to understand the rules for community assembly in these coastal microbiomes.

Intertidal diatom-rich biofilms exhibit a positive relationship between assemblage biodiversity (species richness and Shannon diversity of abundant taxa) and net primary production (Forster et al. 2006). Experimental studies with cultured benthic diatoms show niche complementarity and transgressive over-yielding (increased biovolumes) for mixtures of up to eight species combinations but antagonistic interactions were present between some taxa (Koedooder et al. 2019; Vanelslander et al. 2009). Facilitation, possibly through mixotrophic growth on organic substrates produced by other diatoms, or by associated heterotrophic bacteria, was shown for *Cylindrotheca closterium*, which grew strongly in the spent media from *Navicula* cultures (Vanelslander et al. 2009). The nature of the competitive interactions between species is not well described. *Nitzschia* c.f. *pellucida* releases cyanogen bromide immediately after the onset of light. Cyanogen bromide is toxic to other diatom species that thrive in the immediate vicinity and cause their death (Vanelslander et al. 2012). There is generally a negative relationship between biofilm biomass and diatom diversity in mudflat biofilms with lower shore beach sediments exposed to a greater level of disturbance having a higher diversity and a more even species distribution than upper and mid tide sites that support high biomass (Forster et al. 2006; Hill-Spanik et al. 2019; Underwood 1994). At times, conditions can be favorable for the rapid growth of just a few taxa or even a single species may "bloom" resulting in a biofilm with low species diversity and high biomass (Forster et al. 2006; Underwood 1994; Underwood et al. 1998). These studies were conducted in high nutrient status estuaries with traditional microscopic assessments of composition. In lower nutrient environments a positive relationship between biofilm biomass and Shannon diversity (determined by HTS) has been found (Plante et al. 2021). When comparing biomass-diversity relationships between sediment types, epipsammic habitats can have both lower biomass and lower alpha-diversity than mudflat sites (Plante et al. 2016). This illustrates how more detailed studies on the causes and patterns of benthic diatom assemblage composition and functioning are needed.

### 12.2.2 Adaptations of Photoautotrophs to Living in Intertidal Sediments

The consequence of periods of tidal exposure for photosynthetic microorganisms living on intertidal sediments is that they can experience high incident radiation (including UVB) at varying times over the day and over a year (Laviale et al. 2015; Mouget et al. 2008; Waring et al. 2007). Additionally, sediment disturbance and mobilization by waves and tidal flows mixes cells out of the photic zone of the sediment surface (De Jonge and van Beusekom 1995; Redzuan and Underwood 2020; Savelli et al. 2019). Autotrophic microphytobenthos has adapted to these environmental pressures of high irradiation and sediment disturbance in various ways. Motility is a key adaptation for MPB taxa living in mud and silty sediments allowing them to (re)position at locations with a favorable light climate within the sediment. Motile MPB is primarily composed of pennate diatoms (motile taxa found on mud are termed epipelon) and euglenophytes, as well as some taxa of filamentous cyanobacteria (Underwood and Kromkamp 1999). Upon tidal exposure the populations of phototrophic organisms undergo mass vertical migration, which brings cells to the surface (Consalvey et al. 2004; Jesus et al. 2009) in order to be able to photosynthesize. This vertical migration has been recognized for over a century (Perkins 1960) and is a macroscale feature visible to the naked eye as a color change of the sediment and is even detectable by remote sensing (Méléder et al. 2020; Savelli et al. 2020).

The motility of MPB exhibits a number of features of ecological relevance. There is an underlying endogenous rhythm of motility, which is maintained for a number of days in the absence of light or tidal stimuli. This rhythmicity is detectable in terms of changing cell density at the sediment surface, intensity of photosynthetic pigments, and rate of photosynthesis and carbohydrate production (Coelho et al. 2011; Haro et al. 2019; Perkins 1960; Round and Palmer 1966; Serôdio et al. 1997; Smith and Underwood 1998). These patterns of rhythmicity are evidence for a circadian rhythm and circadian rhythm regulator genes such as *kaiA, kaiB, kaiC*, and peroxiredoxin (*prx*) are expressed by cyanobacteria and diatoms in microbial mats (Hörnlein et al. 2018). Circadian rhythms of activity have also been found in the expression patterns of conserved gene transcripts for photosynthesis (PSI and PSII) and $CO_2$ fixation (RuBisCO) of cyanobacteria and diatoms in intertidal cyanobacterial mats (Hörnlein et al. 2018). Circadian rhythms can be entrained by external stimuli. MPB shows entrainment of rhythmicity by tidal exposure cycles as well as by day-night light cycles (Haro et al. 2019; Hörnlein et al. 2018; Perkins 1960; Serôdio et al. 1997). Haro et al. (2019) found that endogenous rhythms of migration and net community primary production were lost after exposure to continuous light for 3 days but were reset by re-imposition of an alternating light regime. Although easily demonstrated and reported quite some time ago (Perkins 1960; Round and Palmer 1966) the mechanism by which MPB maintains its endogenous vertical migration rhythms in synchrony with the progressive daily movement of the tidal exposure window (Happey-Wood and Jones 1988) is not yet resolved.

Taxon-specific differences are present within the mass movement of the whole community during tidal emersion-immersion cycles. Some taxa only appear at the sediment surface for short periods of the exposure period (e.g., Round and Palmer 1966; Round 1979; Underwood et al. 2005) or have a specific movement in response to light stimuli (*Gyrosigma balticum*, Jönnson et al. (1994)). Barranguet et al. (1998)

proposed that cells of different species micro-migrated into and out of the surface photic zone of the biofilm during illumination thus avoiding photo-inhibition while maintaining an overall high assemblage photosynthesis. Single cell imaging of intact biofilms provided evidence for this (Oxborough et al. 2000; Underwood et al. 2005), correlating time and light intensity with the species present at the sediment surface during tidal exposure.

Microphytobenthos shows strong behavioral responses to light intensity with populations migrating down into sediment to avoid high light (Perkins et al. 2002, 2010; Prins et al. 2020; Underwood 2002). MPB can also detect spectral composition with high and low intensities of blue and red light generating different patterns of surface-active biomass and photo-acclimation in diatom-rich mudflat biofilms (Prins et al. 2020). Benthic diatoms can also sense UVB radiation and will move away from the surface even when PAR light intensities are constant (Waring et al. 2007). Benthic diatoms respond to light through a combination of positive and negative phototaxis (directional movement) and photokinesis (changing speed) (Cohn et al. 2016). There are differences in response between diatom species. *Navicula perminuta* exhibits positive phototaxis at low levels of blue light and negative phototaxis at high intensities as well as a photokinetic response to red light. However, under the same conditions *Cylindrotheca closterium* only displayed photokinetic responses to red light and no blue light response (McLachlan et al. 2009). *Craticula cuspidata*, *Stauroneis phoenicenteron*, and *Pinnularia viridis* have different positive and negative photophobic (changing direction of movement) responses to red, green, and blue light (Cohn et al. 2016). Light sensing appears to take place at the apices of the cell valves and is therefore not necessarily directly associated with the chloroplast (Cohn et al. 2016; McLachlan et al. 2012). High intensity blue light at the apex of *N. perminuta* causes an increase in intracellular calcium concentration along the line of the raphe in the cell wall followed by a reversal of the direction of movement (McLachlan et al. 2012). Diatoms possess genes that code for phytochromes, cryptochromes, aureochromes, and other light receptor proteins for harvesting red/far red and blue light (Blommaert et al. 2020; König et al. 2017; Mann et al. 2020). Benthic diatoms in sediment experience a gradient of the light spectrum with red light most rapidly attenuated and blue light penetrating deepest (Lassen et al. 1992). Differential motility behavior modulated by spectral quality would allow cells to position themselves in a favorable light climate within the narrow photic zone of intertidal sediments.

Benthic diatoms are photo-physiologically flexible and are able to use rapid photochemical and non-photochemical quenching (NPQ), the xanthophyll cycle (XC), and longer-term acclimation of Chl *a* and other photopigments in order to maintain high rates of primary production in a rapidly varying light climate over tidal emersion and during the year (Barnett et al. 2015; Juneau et al. 2015; Prins et al. 2020; Underwood et al. 2005; Waring et al. 2010). The ability to dissipate light energy through NPQ and XC is particularly important for non-motile species that cannot migrate away from damaging light conditions. Experimental work has demonstrated differences in the ability of diatom species to induce high levels of NPQ and XC (Barnett et al. 2015; Blommaert et al. 2018). Epipsammic diatoms, which are found attached to sand grains, have high capacity for non-photochemical quenching (NPQ), while epipelic diatoms have lower potential NPQ, and non-motile tychoplankton, which live under low light conditions in frequently-mixed and resuspended sediments, also possess a low capacity for NPQ (Barnett et al. 2015; Blommaert et al. 2018).

### 12.2.3 Distribution of MPB Biomass in Coastal Sediments

The biomass of MPB present at any location and time is a consequence of the physical and environmental conditions of the preceding period. MPB shows rapid growth responses and can increase their biomass over a period of a few days when conditions are conducive to growth (Nedwell et al. 2016). Nutrients, light availability, and sediment type are major controls of biomass (Cibic et al. 2019; Underwood and Kromkamp 1999), while grazing and desiccation (McKew et al. 2011; Savelli et al. 2018), physical disturbance by waves, wind, and tides (de Jonge and van Beusekom 1995; Redzuan and Underwood 2020, 2021; Savelli et al. 2018), and macrofauna (birds) (Booty et al. 2020) may have local impacts.

Combined, these environmental factors produce general patterns of higher biomass on upper intertidal flats and in sheltered regions of estuaries (Daggers et al. 2020; Underwood and Kromkamp 1999). Seasonal patterns are variable between locations. In temperate zones biomass may peak at any time throughout the year, though summer peaks are common. However, summer declines have also been reported as the result of grazing or temperature stress (Daggers et al. 2020; Nedwell et al. 2016; Park et al. 2014; Savelli et al. 2018; Underwood and Paterson 2003). This seasonal variability reflects in part the spatial patchiness of MPB, which occurs on a range of scales from cm to km (Redzuan and Underwood 2021; Spilmont et al. 2011; Taylor et al. 2013; Weerman et al. 2012). Only a few long-term (>3 years) data sets of MPB biomass exist, and these indicate that sediment type, exposure or tidal position, windiness, and, to a lesser extent, air temperature are the main drivers of biomass. De Jonge et al. (2012) found similar inter-annual patterns of biomass at different stations in the Ems estuary (Netherlands) and a long-term positive relationship between biomass and annual air temperatures with higher Chl *a* content during the 1990s during the monitoring period from 1976 to 1999. Van der Wal et al. (2010) used remote sensing data to determine MPB biomass on mudflats in the southern North Sea over the period 2001 to 2009 and found broad synchrony in the patterns of occurrence and biomass between estuaries, although stronger relationships were present within regional data sets (e.g., within Dutch estuaries). Weather and summer temperatures strongly influenced MPB biomass in the Loire estuary (France) from 1993 to 1998 and from 2006 to 2010 (Benyoucef et al. 2014).

### 12.2.4 Interactions between Photoautotrophs and Chemoheterotrophs and the Turnover of Organic Carbon in Coastal Microbiomes

Photosynthetic activity by MPB produces oxygen and a variety of labile carbon compounds. MPB, particularly diatoms, produces extracellular polymeric substances (EPS) as well as low molecular weight labile carbon compounds (Bellinger et al. 2005, 2009; Underwood and Paterson 2003). EPS molecules are important in creating a biofilm matrix that increases sediment stability (Baas et al. 2019; Hope et al. 2020) and provide protection from desiccation and salinity stress (Steele et al. 2014). The production of these molecules is variable and is moderated by environmental factors (e.g., light and nutrients, Staats et al. 2000; Underwood and Paterson 2003) and rhythms of vertical migration (Hanlon et al. 2006; Perkins et al. 2001; Smith and Underwood 1998). These environmental factors drive a distinct seasonality in the balance of labile and recalcitrant exudates produced over a year (Moerdijk-Poortvliet et al. 2018a, b). There is evidence from freshwater studies that EPS production and biofilm formation by diatoms is enhanced by the presence of certain bacterial taxa (Bruckner et al. 2008, 2011; Grossart et al. 2005). These interactions appear to be species-specific. Bacterial-diatom interactions have negative effects on estuarine diatom biomass in cultures with a single diatom species but are neutral in co-cultures of different diatom species (Koedooder et al. 2019). Different benthic diatom taxa promoted the growth of assemblages of sediment bacteria. For example, *Seminavis robusta* cultures supported the Alphaproteobacteria *Thalassospira* sp., *Roseobacter* sp., and *Kordiimonadaceae* sp. and the Bacteroidetes *Mangrovimonas* sp. and *Owenweeksia* sp., while monocultures of the diatoms *Cylindrotheca closterium* and *Navicula phyllepta* had different bacterial assemblage profiles (Koedooder et al. 2019). There is evidence of antagonistic interactions between the bacterial assemblage associated with a certain diatom species with other benthic diatom taxa (Stock et al. 2019). This suggests that different diatom species may have their own associated bacterial microbiome.

Microphytobenthos-derived dissolved organic carbon compounds contribute 30 to 50% of the total organic matter in the sediments (Bellinger et al. 2009) and represent the key source of labile organic carbon (Nedwell et al. 2016). The importance of this carbon source varies in different intertidal habitats (sandy to muddy; temperate to tropical) (Cook et al. 2007; Oakes et al. 2010, 2012). MPB-fixed carbon has a characteristic $^{\delta13}C$ signal that can be tracked through food webs (Christianen et al. 2017). EPS $^{13}C$-carbon has been tracked into the phospholipid fatty acids (PLFA) and RNA of various bacterial groups (Bellinger et al. 2009; Gihring et al. 2009; Middelburg et al. 2000; Taylor et al. 2013). Major utilizers of diatom EPS in aerobic sediments are Alphaproteobacteria, Gammaproteobacteria, and Bacteriodetes, and in anaerobic conditions Deltaproteobacteria (Bohórquez et al. 2017; McKew et al. 2013; Miyatake et al. 2014; Taylor et al. 2013). A subset of Alphaproteobacteria and Gammaproteobacterial taxa was adapted to utilize diatom EPS before it became available to the rest of the bacterial assemblage (Taylor et al. 2013). Different bacterial groups (for example, Sphingobacteria and *Tenacibaculum* (Bacteroidetes), two classes of Verrucomicrobia (Verrucomicrobiae and Opitutae)) grow preferentially on labile and refractory diatom EPS (Bohórquez et al. 2017; Underwood et al. 2019). Turnover rates of these different DOC fractions vary under aerobic and anaerobic conditions. Anaerobic conditions lead to preferential breakdown of refractory compounds and enhance the growth of Firmicutes (Clostridia, Lachnospiraceae, Peptostreptococcaceae, Ruminococcaceae, and other unclassified Clostridiales) and sulfate-reducing Deltaproteobacteria (Desulfobacteraceae and Desulfobulbaceae) (McKew et al. 2013). There are close linkages between photoautotrophic and chemoheterotrophic microorganisms present in the coastal sediment microbiome and evidence of antagonistic, synergistic, and mutualistic interactions. Hörnlein et al. (2018) proposed the *Choirmaster-Choir* theory. This theory predicts that the rhythmic release of photosynthate and other metabolites is controlled by the circadian clock of the photoautotrophic members of the microbiome (cyanobacteria, diatoms) and dictates the genetic clocks of other microbes either directly or in association with external Zeitgebers such as light and temperature, which results in a synchronized activity during a 24-h cycle. This idea remains to be further explored.

## 12.3 Nitrogen Cycling in the Marine Coastal Microbiome

The dominant heterotrophic bacteria found in aerobic coastal sediments are Actinobacteria, Alphaproteobacteria, Gammaproteobacteria, Chloroflexi, Verrumicrobiae, and Bacteriodetes (Bohórquez et al. 2017; McKew et al. 2013; Yao et al. 2019; Yi et al. 2020). The relative abundance of the different groups is strongly influenced by sediment characteristics particularly by sediment grain size, organic content, pH, and nitrogen and phosphorous availability (Yao et al. 2019; Yi et al. 2020). In estuarine environments, salinity and freshwater inputs influence taxonomic composition, with Actinobacteria and Betaproteobacteria more abundant in lower salinity zones or during periods of higher rainfall. For example, Actinobacteria, Chloroflexi, and Verrucomicrobia showed significant differences between rainfall seasons in the Yangtze estuary (Yi et al. 2020). However, overall salinity-related changes in assemblages appear to be more pronounced in estuarine bacterioplankton assemblages (Gołębiewski et al. 2017; Osterholz et al. 2018) than in estuarine benthic assemblages where organic matter and sediment properties are most influential (McKew et al. 2013; Yao et al. 2019; Yi et al. 2020). Marine coastal sediment microbiomes exhibit profound depth profiles of the distribution of taxonomic groups of bacteria reflecting the gradient of electron acceptors (Böer et al. 2009; Webster et al. 2010; Wilms et al. 2006). In muds with high organic carbon content, high rates of bacterial activity, and limited diffusion of oxygen, the anaerobic zone can be as near as a few millimeters below the surface and sometimes even reaches the surface. In the anaerobic zone the processing of organic carbon is largely driven by sulfate reduction and nitrogen cycling with Delta- and

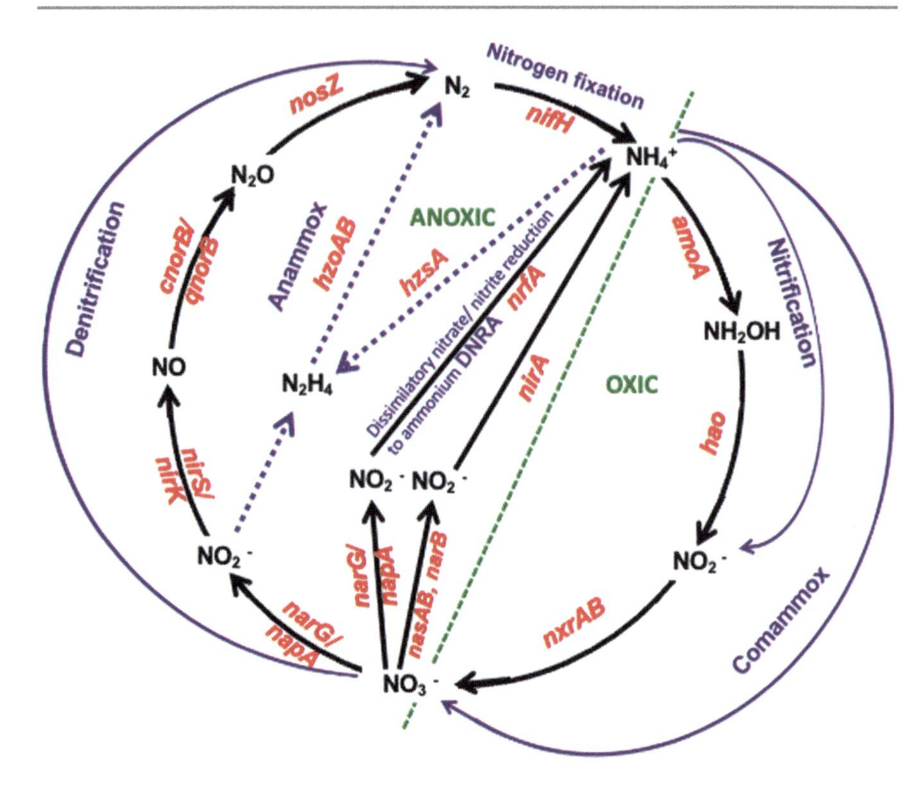


**Fig. 12.1** The nitrogen cycle in coastal sediments indicating transformations (purple), functional genes (red), and oxic/anoxic zones (green)

Epsilonproteobacteria, including sulfate-reducing bacteria, e.g. Desulfomonadales, as well as Archaea as prominent players (McKew et al. 2013; Nedwell et al. 2016; Webster et al. 2010).

The nitrogen (N) cycle (Fig. 12.1) is mediated by metabolically diverse groups of microorganisms. The location of the different processes in the sediment is determined by sediment redox state. Microbial-driven N transformations are especially crucial in coastal systems, which often receive high anthropogenic N inputs (e.g., via fluvial discharges) resulting in organic matter breakdown and oxygen depletion (Nedwell et al. 2016). Characterizing N cycle communities in the environment by traditional microbiological methods has been problematic due to the difficulties in obtaining pure cultures of the responsible microorganisms. However, molecular methods and HTS techniques have enabled researchers to uncover the functional nitrogen-cycle microbiome of these ecosystems.

### 12.3.1 Nitrogen Cycling in Aerobic Coastal Sediments: Nitrification and Aerobic Ammonia Oxidation and Comammox

Autotrophic ammonia oxidation is the rate-limiting step in nitrification and important in the N cycle (Fig. 12.1). During nitrification, aerobic ammonia-oxidizing bacteria (AOB) and archaea (AOA) oxidize ammonium by ammonia monooxygenase (encoded by *amoA*) (McTavish et al. 1993). With AOB, the second step is the dehydrogenation of hydroxylamine to nitrite by hydroxylamine oxidoreductase (encoded by *hao*) (Arp et al. 2002). However, AOA genome data does not appear to contain *hao* gene homologs, and an alternative mechanism has been proposed (Hallam et al. 2006). Nitrite is oxidized to nitrate by nitrite-oxidizing bacteria (NOB) (e.g., *Nitrospira*). Previously, based on 16S rRNA gene sequencing autotrophic ammonia oxidation was thought to be restricted to two monophyletic lineages of aerobic ammonia-oxidizing bacteria (AOB) (Head et al. 1993). The first lineage belongs to the Betaproteobacteria (Beta-AOB) (e.g., *Nitrosomonas, Nitrosospira*) and the second lineage belongs to the Gammaproteobacteria (Gamma-AOB) (e.g., *Nitrosococcus* sp.) (Head et al. 1993). However, metagenome libraries from seawater (Venter et al. 2004) and soil (Treusch et al. 2005) revealed putative genes involved in ammonia oxidation from uncultured Thaumarchaeota. AOA *amoA* gene sequences form five clusters four with cultured representatives (*Nitrosocaldus*, *Nitrososphaera*, *Nitrosopumilus*, *Nitrosotalea*) and the fifth is known as "*Nitrososphaera* sister" cluster (Pester et al. 2011).

Aerobic ammonia oxidizers are found in most environments (Table 12.1) (Francis et al. 2005; Jiang et al. 2009; Phillips et al. 1999; Stehr et al. 1995; Whitby et al. 1999, 2001). In some ecosystems AOA outnumber AOB often by a factor of 10 to 1000 (Beman et al. 2008; Wuchter et al. 2006). This is the case, for example, in North Sea coastal sediments (Lipsewers et al. 2014) and coastal waters (Smith et al. 2014a), suggesting a greater contribution of AOA to nitrification in these systems (Jiang et al. 2009). However, in some coastal and estuarine sediments AOB are more abundant than AOA (Caffrey et al. 2007). In a hypernutrified temperate estuary (Colne, U.K.) with gradients of salinity and ammonia concentration, benthic AOB (notably *Nitrosomonas* spp.) were significantly more abundant (by 100-fold) than AOA, suggesting that AOB were the main contributors to nitrification (Li et al. 2015a). Seasonal differences in nitrification in coastal sediments have been observed with the highest rates often in the summer (Li et al. 2015a). However, in North Sea coastal sediments, AOA 16S rRNA gene transcriptional activity was higher in the winter despite the lower abundance of these organisms (Lipsewers et al. 2014, 2017). In contrast, higher AOA abundances were found in the winter in the North Sea water column, which was attributed to ammonia availability and the lack of competition for ammonia with phytoplankton (Pitcher et al. 2011; Wuchter et al. 2006). Differences in spatial distribution between *Nitrosospira* and *Nitrosomonas* have also been found. For example, in freshwater lake sediments, *N. europaea* was present in littoral sediments whilst *N. eutropha* was found in profundal sediments, whilst members of *Nitrosospira* were ubiquitous (Whitby et al. 1999, 2001); and in the water column of the Mediterranean Sea, where different members of the beta-proteobacterial ammonia oxidizers were associated with particulate material and planktonic samples (Phillips et al. 1999).

The oxidation of ammonia via nitrite to nitrate was originally considered to be a two-step process catalyzed by two functionally distinct groups of chemolithoautotrophs (ammonia oxidizers and nitrite oxidizers). However, a nitrifying bacterium belonging to *Nitrospira* was discovered and sequencing of its genome revealed that it has all the genes necessary for the oxidation of ammonia and nitrite (Daims et al. 2016). The discovery of the complete oxidation of ammonia to nitrate in one organism (comammox) (Fig. 12.1) has changed the paradigm that this process requires two distinct functional groups of microbes and raises questions about the role of comammox *Nitrospira* in N-cycling.

Comammox organisms belong as far as known to the *Nitrospira* lineage II (Daims et al. 2016; Koch et al. 2019). Based on *amoA* gene sequences from metagenomes comammox bacteria comprise two clades, A and B (Daims et al. 2016; Palomo et al. 2018; Van Kessel et al. 2015). Putative comammox (clade A) *amoA* gene sequences were previously misidentified as "unusual" methanotroph *pmoA* genes relating to *Crenothrix* in the Gammaproteobacteria (Stoecker et al. 2006) or presumed to belong to a methanotroph from Alphaproteobacteria (clade B, *amoA*) (Radajewski et al. 2002). Clade A is further delineated in two groups: clades A.1 and A.2 (Xia et al. 2018). Since comammox bacteria do not form a monophyletic group within the *Nitrospira,* lineage II comammox and canonical nitrite-oxidizing *Nitrospira* cannot be distinguished by 16S rRNA-based methods (Pjevac et al. 2017). Comammox bacteria have been found in various habitats (Table 12.2) with high proportions in estuarine and coastal environments (Xia et al. 2018). In the open ocean, however, comammox *amoA* genes were either rarely detected (Daims et al. 2016) or absent (Xia et al. 2018).

### 12.3.2 Environmental Factors Influencing Nitrification and Ammonia Oxidation

Although AOA and AOB coexist there is evidence of niche differentiation linked to various environmental factors (e.g., temperature, ammonium concentration, oxygen, pH, salinity, light, macrofaunal activity) (Caffrey et al. 2007; Cao et al. 2011; Dang et al. 2010; Erguder et al. 2009; Scarlett et al. 2021; Stehr et al. 1995). Ammonium concentration and availability are major factors for niche partitioning of AOA versus AOB with lower concentrations generally favoring AOA (Clark et al. 2020; Martens-Habbena et al. 2009). In some coastal sediments low phosphate availability selects for AOA over AOB (Lipsewers et al. 2014). In subsurface sediments *Nitrosomonas* dominated and was linked to nitrite concentration (Cao et al. 2012). In estuarine sediments, decreased dissolved oxygen altered AOB *amoA* expression but not AOA (Abell et al. 2010). Phytoplankton may also outcompete nitrifiers for substrates in surface waters (Smith et al. 2014b). In estuarine sediments benthic microalgae have a high demand for ammonium (Thornton et al. 1999) and can outcompete AOB, reducing the rates of nitrification (Risgaard-Petersen 2003).

**Table 12.2** Representative N-cycle microorganisms found in various coastal marine and estuarine environments

| N-cycle microorganisms | Environment | Reference |
|---|---|---|
| **Aerobic ammonia oxidizing bacteria (AOB)** | | |
| *N. europaea*, *N. communis*, *N. oligotropha* (cluster 6a), *N. marina* (cluster 6b), *Nitrosospira* spp. (clusters 13–15) | Coastal environments (Jiaozhou Bay) | Dang et al. (2010) |
| *Nitrosomonas* spp. | Hypernutrified estuary (Bahía del Tóbari) | Beman and Francis (2006) |
| *Nitrosomonas* spp.<br>*Nitrosospira* spp. | Estuarine (brackish) (Westerschelde estuary)<br>Estuarine (marine) (Westerschelde estuary) | Sahan and Muyzer (2008) |
| Estuarine/marine *Nitrosospira*-like cluster and *Nitrosomonas*-like cluster | Estuarine sediments (Elkhorn Slough) | Wankel et al. (2010) |
| *Nitrosospira* spp., *Nitrosomonas* spp. (*N. marina*, *N. oligotropha*, *N. ureae*, *N. eutropha*) | Wetland sediments of subtropical coastal mangroves | Wang et al. (2013) |
| *Nitrosospira* sp. and *Nitrosolobus multiformis* | Coastal sediments (North Sea) | Lipsewers et al. (2014) |
| **Aerobic ammonia oxidizing archaea (AOA)** | | |
| *Nitrosopumilus maritimus*, *Nitrososphaera gargensis* | Wetland sediments of subtropical coastal mangroves | Wang et al. (2013) |
| *Nitrosopumilus maritimus* | Mangrove sediments, South China Sea sediments | Li et al. (2011b); Cao et al. (2012) |
| *Nitrosopumilus* subclusters 12 and 16 (stable marine cluster), *Nitrosopumilus* subcluster 4.1 (estuarine cluster) | Marine coastal sediments (North Sea) | Lipsewers et al. (2014) |
| *Nitrosopumilus maritimus*, *Nitrosphaera gargensis* | Subtropical macrotidal estuarine sediments | Abell et al. (2010) |
| **Comammox** | | |
| Clade A.1 (e.g., *Cand.* Nitrospira nitrificans, *Cand.* Nitrospira nitrosa, clade A.2, clade B | Various sediments (tidal flat, saltmarsh, coastal), coastal waters | Xia et al. (2018) |
| **Anammox** | | |
| *Scalindua* spp., *Kuenenia* spp. | Wetland sediments of subtropical coastal mangroves | Wang et al. (2013) |
| Uncultured Planctomycetes, *Cand.* Scalindua spp., *Cand.* Brocadia spp., *Cand.* Kuenenia spp. | Anoxic basin on Black Sea | Kuypers et al. (2003) |

(continued)

**Table 12.2** (continued)

| N-cycle microorganisms | Environment | Reference |
|---|---|---|
| **Denitrification** | | |
| *Anaeormyxobacter delahogens* 2CP-C, *Thermus thermophilus* strain HB8, *Geobacter metallireducans, Rhodoferax ferrireducans, Halomonas halodenitrificans, Rhodobacter sphaeroides, Cupriavidus necator, Hahella chejuensis, Shewanella* spp., *vibrio* spp., *Saccharophagus degradans, Rhodopseudomonas palustris, Magnetospirillum magneticum* AMB1, *helicobacter hepaticus* ATCC51449, *pseudomonas* spp. | Hypernutrified estuarine sediments | Smith et al. (2007) |
| *Alcaligenes* spp. *(A. faecalis, A. xylosoxidans) pseudomonas* spp. *(P. Stutzeri), Bradyrhizobium japonicum, Blastobacter denitrificans* | Sediments within the oxygen-deficient zone, Pacific coast | Liu et al. (2003) |
| **Dissimilatory nitrate reduction to ammonium (DNRA)** | | |
| *Shewanella frigidimarina, Chlorobium phaeobacteroides* | Hypernutrified estuarine sediments | Smith et al. (2007) |
| **Nitrogen fixation** | | |
| *Azotobacter* spp., *Azospirillum* spp., *campylobacter* spp., *Beggiatoa* spp., *Enterobacter* spp., *Klebsiella* spp., *vibrio* spp., *Desulfobacter* spp., *Desulfovibrio* spp., *clostridium* spp. cyanobacteria including unicellular and non-heterocystous species, (Chromatiaceae, Chlorobiaceae, Chloroflexaceae, Rhodospirillaceae), archaea (e.g. *Methanococcus* spp., *Methanosarcina* spp.) | Marine/ seagrass sediment, estuarine sediment, salt marsh sediment, *Spartina* roots, *Zostera* roots, beach sediment, intertidal sediments, seawater | Herbert (1999) |

Differential sensitivity to pollutants between AOB and AOA in coastal environments has also been found with higher Beta-AOB diversity in polluted sites whilst AOA were unaffected (Cao et al. 2011). Agriculturally-impacted estuarine sediments were dominated by AOA and *Nitrosomonas* spp. *amoA* sequences, whilst *Nitrosospira* spp. dominated less impacted sites (Wankel et al. 2010). Distinct clusters of *Nitrosomonas* and *Nitrosospira* lineages have been found in eutrophic coastal sediments subjected to inputs from nearby wastewater treatment plants and polluted rivers (Dang et al. 2010). Silver nanoparticles inhibited AOB-driven nitrification but not AOA in a temperate eutrophic estuary (Beddow et al. 2017). Addition of titanium nanoparticles resulted in increased ammonium fluxes from sediments into overlying water, which could be due to lower rates of ammonia oxidation and nitrification, as well as a decrease of net MPB primary productivity (Passarelli et al. 2020). Beta-AOB *N. europaea* and *N. communis* lineages also thrive in heavy metal-polluted environments and in environments with high ammonium concentrations (Dang et al. 2010; Stein et al. 2007). In estuaries, *Nitrosospira*-like lineages appear to be better adapted than *Nitrosomonas* (Cao et al. 2011) and Beta-AOB (particularly the *N. oligotropha* lineage) and could be used as bioindicators of pollution in coastal systems (Dang et al. 2010).

Comammox bacteria are functionally versatile and adaptative to many environments (Hu and He 2017). Comammox bacteria exhibit niche partitioning influenced by various environmental factors (Shi et al. 2020) and differences in abundance among clades have been found (Xia et al. 2018). Co-occurrences of comammox with canonical ammonia oxidizers indicate a potential functional differentiation between these groups (Bartelme et al. 2017; Palomo et al. 2018; Pjevac et al. 2017) and may depend on whether the main activity of comammox in an environment is ammonia oxidation or nitrite oxidation (Xia et al. 2018). Comammox bacteria may outnumber AOB (Xia et al. 2018) and can functionally outcompete other canonical nitrifiers in highly oligotrophic systems (Hu and He 2017). However, which factors drive niche specialization between comammox and canonical ammonia oxidizers currently remains unknown.

### 12.3.3 Nitrogen Cycling in Anaerobic Coastal Sediments: Anammox, Denitrification, and Dissimilatory Reduction of Nitrate to Ammonium

Anaerobic ammonia oxidation (anammox) involves the conversion of ammonium and nitrite to $N_2$ in the absence of oxygen (Fig. 12.1). Some anammox bacteria are facultative chemoorganotrophs that can also metabolize organic compounds, notably formate, acetate, and propionate (Kartal et al. 2007; Strous et al. 2006), allowing anammox bacteria to adopt a "disguised" denitrifying lifestyle (Kartal et al. 2007). Anammox bacteria form a monophyletic order of the Brocadiales within the Planctomycetes (Jetten et al. 2010), and consist of five candidate genera: *Candidatus* Kuenenia (Strous et al. 2006); *Candidatus* Brocadia (Oshiki et al. 2011; Strous et al. 1999); *Candidatus* Anammoxoglobus (Kartal et al. 2007); *Candidatus* Jettenia (Quan et al. 2008), and *Candidatus* Scalindua (Schmid et al. 2003).

Anammox bacteria are found in virtually any anoxic environment that contains fixed N (Table 12.2). Although anammox is responsible for a large proportion of $N_2$ production in marine sediments, in eutrophic coastal sediments, and saline tidal marsh sediments, anammox is not important relative to denitrification (Koop-Jakobsen and Giblin 2010; Thamdrup and Dalsgaard 2002). Although anammox bacteria have been found in coastal and estuarine sediments (Li et al. 2011a; Tal et al. 2005; Trimmer et al. 2003) and coastal mangrove wetlands (Cao et al. 2011; Li et al. 2011a), greater anammox bacterial diversity occurs in the Oxygen Minimum Zones (OMZs) of oceans (Woebken et al. 2009). Anammox bacteria are abundant and active in oxygenated upper sediments and bioturbated marine coastal sediments in the North Sea (Lipsewers et al. 2014). High anammox bacterial abundances have also been found in surface sediments of hypernutrified estuarine tidal flats (Zhang et al. 2013). In some environments anammox bacteria are scarce like suboxic and anoxic aquatic systems where low anammox bacterial diversity was found and comprised mostly of *Scalindua* (Penton et al. 2006).

Denitrification is fundamental in the N cycle releasing nitric oxide (NO), nitrous oxide ($N_2O$), and dinitrogen gas ($N_2$) to the atmosphere (Fig. 12.1). As denitrifying bacteria belong to different phylogenetic groups the 16S rRNA gene is not very useful for analyzing denitrifier communities. Instead, functional genes involved in denitrification have been targeted, e.g., *napA*, *narG* (nitrate reductase), *nirS*, *nirK* (nitrite reductases), and *nosZ* (nitrous oxide reductase) (Nogales et al. 2002) (Fig. 12.1). Denitrifiers are facultative organoheterotrophic anaerobes that constitute a phylogenetically diverse group spanning >50 different genera (Jones and Hallin 2010; Zumft 1997). Most denitrifiers belong to the alpha-, beta-, gamma-, and epsilon-Proteobacteria (Braker and Conrad 2011). The most frequently isolated denitrifying bacteria belong to the Pseudomonads (Herbert 1999). Denitrification has also been found among Firmicutes, Actinomycetes, Bacteroidetes, Aquificaceae, and Archaea (Braker and Conrad 2011). Denitrification is also widespread among Foraminifera, *Gromiida* (Piña-Ochoa et al. 2010; Risgaard-Petersen et al. 2006), and fungi (Braker and Conrad 2011).

Denitrification is widely distributed in the environment (Table 12.2). In the ocean however, denitrification is geographically restricted to a few oceanic regions (e.g., OMZs and hemipelagic sediments) (Jayakumar et al. 2009) and distinct *nirS* and *nirK* populations have been found within the oxygen-deficient zone in marine sediments (Liu et al. 2003). In eutrophic estuaries, denitrification can mediate the lowering of N load and contribute to eutrophication control (Nogales et al. 2002). Indeed, in coastal and estuarine sediments denitrification can remove >50% of inorganic N inputs from terrestrial systems (Nedwell et al. 2016; Rivera-Monroy et al. 2010; Seitzinger 1988). In coastal and estuarine sediments denitrification rates are generally higher than in shallower waters (Herbert 1999).

In addition to denitrification, microbial nitrate reduction may also take place via alternative pathways. Dissimilatory nitrate reduction to ammonium (DNRA) (Fig. 12.1) is particularly important in organic-rich sediments (King and Nedwell 1987; Laverman et al. 2006) and tends to retain bioavailable N in aquatic ecosystems. DNRA is common in bacteria (e.g., Proteobacteria, Firmicutes, Verrucomicrobia, Planctomycetes, Acidobacteria, Chloroflexi, *Beggiatoa*, *Thioploca*, and *Chlorobia*) (Papaspyrou et al. 2014; Preisler et al. 2007), and also occurs in eukaryotes (e.g., diatoms, fungi) (Pajares and Ramos 2019). Anammox bacteria may also perform DNRA in the presence of small organic compounds (Kartal et al. 2007) or ammonium might be released from fermentative reactions (Herbert 1999; Lam et al. 2009). DNRA is commonly found in environments low in oxygen, such as OMZs (Lam et al. 2009) and sediments with steep oxygen gradients (Kamp et al. 2011). DNRA has also been found in the Namibian inner-shelf bottom waters (Kartal et al. 2007) and deep-sea sediments (Pajares and Ramos 2019).

### 12.3.4 Environmental Factors Influencing the Anaerobic Nitrogen Cycling Biome

Anammox is controlled by several environmental factors including salinity (Sonthiphand et al. 2014), temperature (Qian et al. 2018), organic matter content (Trimmer and Engström 2011), and inorganic N availability (Trimmer et al. 2005). Interactions between AOA, AOB, and anammox bacteria have been shown where nitrifiers supply nitrite to anammox (Lam et al. 2007, 2009). In mangrove sediments, positive correlations occur with AOA diversity and abundance and anammox *hzo* gene abundances (Li et al. 2011a, b; Li and Gu 2013), suggesting that complex interactions exist between anammox bacteria and ammonia oxidizers. Sulfide may also inhibit anammox bacteria (Dalsgaard et al. 2003; Jensen et al. 2008).

Nitrogen removal via denitrification may cause a decrease in N availability, which in coastal environments can severely impact primary producers and levels of eutrophication (Herbert 1999; Seitzinger 1988). Numerous environmental factors (e.g., N availability and concentration, temperature, oxygen concentration, water depth, organic matter quality and quantity, bioturbation) affect denitrifier distribution and abundance (Braker et al. 2000; Dang et al. 2009; Liu et al. 2003; Prokopenko et al. 2011; Zhang et al. 2014). Denitrification rates also show distinct seasonal patterns driven largely by temperature, nitrate, and availability of organic carbon (Kaplan et al. 1977). Denitrification rates decrease in the spring (in estuarine sediments) (Jørgensen and Sorensen 1988) and in the summer (in subtropical macrotidal estuarine sediments) where *nirS*:*nirK* ratios are negatively correlated with temperature (Abell et al. 2010). Nitrate concentration and oxygen have an impact on denitrifying communities (Liu et al. 2003) and nitrate availability drives *nirS* communities whilst *nirK* communities respond to other parameters (Jones and Hallin 2010). To date, the ecological function of these denitrifying communities and the factors that determine the composition of *nirS*/*nirK* communities remains unknown (Jones and Hallin 2010). Sulfide also decreases denitrification rates (Porubsky et al. 2009). Yet, paradoxically in sulfidic sediments some microorganisms use sulfide as an electron donor for denitrification (Bowles et al. 2012). Bioturbated sediments from large burrowing macrofauna also increase coupled nitrification-denitrification (Laverock et al. 2011; Papaspyrou et al. 2014).

Seasonal and spatial differences in DNRA have been found with increased rates in the summer throughout sediment depths compared to other times when activity was restricted to deeper sediments (Jørgensen 1989). In intertidal and subtidal environments DNRA may change on a daily basis due to the growth and photosynthetic activity of benthic microalgae. Photosynthetically evolved oxygen diffuses into the surface of the sediment during daylight which inhibits DNRA (Herbert 1999). MPB photosynthesis can decrease the rate of denitrification of nitrate that diffuses into the sediment from the water column (Dw) but stimulates the rate of coupled nitrification-denitrification (Dong et al. 2000; Risgaard-Petersen 2003). In estuaries, high *nrfA* gene abundances (encoding cytochrome c nitrite reductase) have been found and change along gradients of salinity and nitrate (Papaspyrou et al. 2014).

### 12.3.5 Nitrogen Fixation in Coastal Sediments

Biological nitrogen fixation involves specialized groups of autotrophic and heterotrophic bacteria and archaea that possess molybdenum (Mo)–Fe protein (dinitrogenase) (encoded by *nifDK*) and Fe protein (dinitrogenase reductase) (encoded by *nifH*) (Fig. 12.1). Oxygen exposure deactivates nitrogenase and oxygenic phototrophs must separate dinitrogen fixation from oxygenic photosynthesis either spatially (e.g., in heterocysts) or temporally (Berman-Frank et al. 2003).

Nitrogen-fixing organisms (diazotrophs) are a diverse group of bacteria and archaea that include members of the Chromatiaceae, Chlorobiaceae, Chloroflexaceae, Rhodospirillaceae, and chemoautotrophic bacteria and archaea (Bergman et al. 1997; Capone 1988; Raymond et al. 2004). Marine diazotrophs mainly include non-heterocystous, heterocystous, symbiotic, and unicellular cyanobacteria (e.g., Ca. Atelocyanobacterium thalassa [UCYN-A]; *Crocosphaera watsonii* [UCYN-B] and *Cyanothece* [UCYN-C]) (Capone 1988; Martinez-Perez et al. 2016; Pajares and Ramos 2019). Other marine diazotrophs include heterotrophic bacteria (e.g., *Klebsiella*), anoxygenic phototrophic bacteria (e.g., *Chlorobium, Chromatium*), strict anaerobic chemotrophs (e.g., *Clostridium, Desulfovibrio*), methanogenic Euryarchaeota and Planctomycetes (Pajares and Ramos 2019). Nitrogen-fixing eukaryotes are not known and it seems that these organisms solved the problem by entering in symbiosis with nitrogen-fixing bacteria (Kuypers et al. 2018).

The main factors that affect marine diazotroph distribution are oxygen, light, temperature, inorganic N, phosphorus, iron, and organic matter (Pajares and Ramos 2019). In estuaries and coastal regions, UCYN-A are highly abundant (Moreira-Coello et al. 2019), along with heterotrophic bacteria (Pajares and Ramos 2019). Several factors influence nitrogen fixation activity in benthic sediments including carbon availability, temperature, light, pH, oxygen, inorganic N, salinity, and trace metal availability (Herbert 1999). Organic carbon availability is generally the main factor limiting the nitrogen fixation in unvegetated sediments (Herbert 1999). In unvegetated shallow coastal lagoons and intertidal sediments where light is not limiting, dense communities of benthic nitrogen-fixing cyanobacteria may occur (Herbert 1999; Stal 2016; Stal et al. 2019). In tropical coastal marine lagoons sediment nitrogen fixation contributes 11% of the annual N input (Hanson and Gundersen 1977) and high rates occur in temperate sediments, mudflats, and salt marshes, especially in organically rich sediments (Nedwell and Aziz 1980; Herbert 1999). Cyanobacterial mats (both temperate and tropical) exhibit high nitrogen fixation rates linked to dark-light cycles and are under the control of circadian clocks (Herbert 1999; Hörnlein et al. 2018; Stal 2016; Stal et al. 2019). High nitrogen fixation rates have been found in salt marsh sediments which has been attributed to organic compounds excreted from plant roots coupled to plant photosynthetic activity (Moriarty and O'Donohue 1993; Whiting et al. 1986) whilst rates in bare marine sediments were low (Herbert 1999).

## 12.4 Archaea in Marine Sediment Microbiomes

### 12.4.1 An Array of Coastal Archaea: Marine Group III (Putative Pontarchaea), Asgard Archaea, Marine Benthic Group D, and Woesarchaeota

Archaea are an important component in the surface sediments of intertidal communities with an abundance of one to two orders of magnitude lower than bacteria (Li et al. 2012; McKew et al. 2011; Wang et al. 2020). Deeper in the sediment, e.g., in the sulfate-methane transition zone, they can be in equal abundance (Li et al. 2012; Wang et al. 2020). Until the 1990s the domain Archaea was divided into Euryarchaeota and Crenarchaeota but this view is rapidly changing (see Baker et al. 2020). Methanogens were the only Archaea in coastal environments that were well known. Advances in sequencing technology unveiled the uncultured archaeal diversity in coastal settings. Evidence for the presence of Archaea in non-extreme environments such as the open ocean was thanks to the pioneering work of Norman Pace, Ed DeLong, Jed Fuhrman and colleagues (DeLong 1992; Fuhrman et al. 1992; Pace 1997). Archaea were described from Colne Point salt marsh in Essex, U.K. by Munson et al. (1997) who detected 16S rRNA gene sequences of methanogens, haloarchaea, and an archaeal lineage that was distinct from any known taxon. Subsequently, this unknown archaeal taxon was detected elsewhere, e.g., from the deep sea (Fuhrman and Davis 1997) and continental shelf samples (Vetriani et al. 1998). It became known as Marine Group III (MG-III) Euryarchaeota with a proposed phylum-level reassignment to Pontarchaea (Li et al. 2015b). Further surveys, using fosmid clones and metagenome assembled genomes (MAGs), revealed the distribution and putative functions of MG-III. For example, Haro-Moreno et al. (2017) showed that MG-III phylotypes living in the photic zone probably have a photoheterotrophic lifestyle, which they based on the presence of photolyase and rhodopsin genes as well as of genes for peptide and lipid uptake and degradation. It remains to be seen whether the coastal MG-III found by Munson et al. (1997) are similar to epipelagic or bathypelagic phylotypes (Haro-Moreno et al. 2017).

Kim et al. (2005) found MG-III Euryarchaeota in tidal flat sediments from Ganghwa Island, Korea, together with many sequences that were considered to be Crenarchaeota, which had not been detected in the Colne Estuary salt marshes by Munson et al. (1997). This phylogenetic lineage was referred to as Marine Benthic Group B (MBG-B) by Vetriani et al. (1998), a sister group of the Deep-Sea Archaeal Group (DSAG), which have been reclassified as members of the Asgard Archaea. Specifically, MBG-B are now known as Thorarchaeota (Seitz et al. 2016) and DSAG as Lokiarchaeota (Spang et al. 2015). Phylogenomic analysis places Eukaryotes within the archaea most closely related to the Asgard archaea, which possess a range of eukaryote features. The classification of the Asgard archaea has contributed to redefining the tree of life from three domains into one with two domains (Williams et al. 2020). Thorarchaeota (MBG-B) have been found in a number of different estuaries (Fig. 12.2; Zou et al. 2020a), including the Colne, Essex, U.K. (Webster

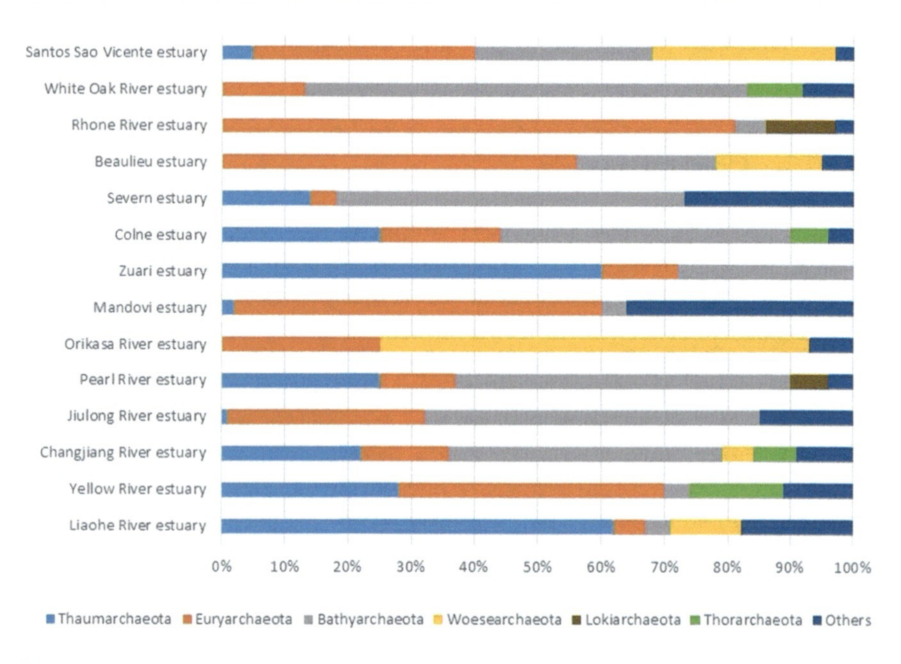


**Fig. 12.2** Relative abundance of archaeal taxa in estuarine sediments, based on 16S rRNA gene clone libraries. This figure is produced from data collated by Zou et al. (2020a). Refer to Zou et al. (2020a) for details of the source papers and the primers pairs used

et al. 2015). Thorarchaeota from White Oak River estuary, North Carolina, USA, have the genetic capacity for protein degradation with the formation of acetate as well as for the reduction of elemental sulfur and thiosulfate and therefore this group of organisms may play an important role in carbon and sulfur cycling in estuarine sediments (Seitz et al. 2016). Lokiarchaeota in Namibian shelf sediments anaerobically consume necromass and extracellular polymeric substances from diatoms and also fixed $CO_2$ via the $H_2$-dependent Wood–Ljungdahl pathway (Orsi et al. 2019). This archaeal mixotrophic activity was more rapid than that of bacteria emphasizing that archaea should not be overlooked in sediment biogeochemical processes (Orsi et al. 2019).

Another archaeal group that is commonly found in estuaries is the Marine Benthic Group D or MBG-D (also called DHVE1), which has been variously called Thermoprofundales or Izemarchaea (Baker et al. 2020; Zhou et al. 2019). MBG-D were the second most abundant archaea in the Pearl River Estuary, China, after the Bathyarchaeota (Wang et al. 2020; see also Zou et al. 2020b) and were also detected in sediments along the Colne Estuary, U.K. (Webster et al. 2015). MBG-D seem to be heterotrophic with the capacity for degrading proteins (Lloyd et al. 2013). In addition, the potential for mixotrophic growth was revealed upon the reconstruction of MBG-D genomes from a mangrove sediment (Zhou et al. 2019).

Woesearchaeota, formerly known as DHVEG-6 (Liu et al. 2018a, b), are globally distributed in many environments including estuaries where they are usually a minor component of the microbial community but occasionally reach a high abundance (Fig. 12.2; Zou et al. 2020a). In *Zostera marina* seagrass beds and nearby bare sediment in Rongcheng Bay, Yellow Sea, China, Woesearchaeota (42% of Archaea) were the most abundant archaea followed by Bathyarchaeota (21%) and Thaumarchaeota (17%) with specific subclades of Woesearchaeota and Bathyarchaeota enriched in the vegetated areas (Zheng et al. 2019).

### 12.4.2 Bathyarchaeota (Miscellaneous Crenarchaeota Group) and Thaumarchaeota Are Generally the Most Abundant Archaea in Marine Sediments

There are two other even more widely distributed and abundant examples of novel archaeal taxa in estuarine environments. The Marine Group I, which together with species from terrestrial environments belongs to the Thaumarchaeota, consists predominantly of ammonia-oxidizing Archaea, which are discussed in Sect. 12.3.1. The Marine Group I are abundant in estuarine sediments (Fig. 12.2) and dominated in a large-scale study of eastern Chinese marginal seas (Liu et al. 2020). The Miscellaneous Crenarchaeota Group (MCG), now known as Bathyarchaeota, is one of the most abundant phyla on Earth and generally the most abundant archaea in estuarine sediments (Fig. 12.2; Li et al. 2012). Bathyarchaeota have an anaerobic organoheterotrophic lifestyle (Seyler et al. 2014), probably degrading proteins (Lloyd et al. 2013), carbohydrates (Lazar et al. 2016), aromatic (Dong et al. 2019; Meng et al. 2014), and aliphatic (Dong et al. 2019) compounds, as well as a variety of other organic matter (Seyler et al. 2014). The aforementioned taxonomic groups and the Bathyarchaeota are phyla and hence comprise a variety of different microorganisms with an array of genetic and functional capacities, which will likely reflect their ecological distribution in estuarine sediments. Bathyarchaeota, for example, comprise 25 subclades (Zhou et al. 2018). Lazar et al. (2014) propose that the Bathy-6 in contrast to other lineages prefers suboxic sediment with minimal free sulfide. Bathy-6 also has the genetic capacity to take up and catabolize a wide range of carbohydrates and proteins (Lazar et al. 2016) and may be able to carry out dissimilatory nitrite reduction to ammonium (DNRA) (Lazar et al. 2016). By performing diverse enrichments from estuarine sediments Yu et al. (2018) showed that subclade Bathy-8 grew on lignin as an energy source. Then, by using lipid stable-isotope probing, these authors demonstrated that lignin-degrading cultures used bicarbonate as a carbon source. This organoautotrophic growth on an abundant biopolymer may partially explain its dominance particularly in estuarine sediments that receive input from plant debris (Yu et al. 2018).

Thus, a variety of different archaea contribute to the turnover of organic matter in coastal sediments while some of them may be autotrophic at the same time. There is a need to better understand the contribution of archaea to benthic cycling of carbon, sulfur, and nitrogen as well as their interactions with other organisms. Obtaining enriched or pure cultures, as was done for Lokiarchaeota (Imachi et al. 2020), will be necessary in order to understand the ecophysiology of archaea in coastal sediments.

### 12.4.3 Archaea Drive the Methane Cycle in Coastal Sediments

Strictly anaerobic methane-producing archaea perform the final step in the anaerobic degradation of organic matter. Much is known about methanogens, primarily because many strains from different classes have been isolated and studied in detail. All characterized methanogens belong to the phylum Euryarchaeota and include: Methanobacteriales, Methanococcales, Methanomicrobiales, Methanosarcinales, Methanopyrales, Methanocellales, and Methanomassiliicoccales (Lyu and Liu 2018) and the class Methanonatronarchaeia (Sorokin et al. 2017, 2018). The first-described methanogens use the major products of microbial fermentation either hydrogen plus $CO_2$ (hydrogenotrophic) or acetate (acetoclastic) (Thauer et al. 2008). Fermentation and methanogenesis occur when energetically favorable electron acceptors such as oxygen, nitrate, and sulfate have been depleted such as is the case in deeper coastal sediments (Wilms et al. 2006, 2007). However, methanogenesis also occurs near the surface of coastal sediments, where sulfate reducers outcompete methanogens for hydrogen and acetate. Here, methanogens coexist with sulfate-reducing bacteria by using non-competitive methylated substrates, such as methylamine (e.g., Oremland et al. 1982). Methylamines, methyl sulfides, and other methylated compounds are common in marine and hypersaline environments as breakdown products of osmolytes (McGenity and Sorokin 2018) and also as components of lipid polar head groups, e.g., choline (Jameson et al. 2018). Methylotrophic methanogenesis could be distinguished into two mechanisms: (1) hydrogen-independent carried out by several representatives of the Methanosarcinales, and (2) hydrogen-dependent carried out by several other groups (Feldewert et al. 2020). Hydrogen-dependent methylotrophic methanogens also appear to compete with $H_2$-utilizing sulfate-reducing bacteria as long as the partial pressure of hydrogen is low and there is a supply of suitable $C_1$-compounds owing to their superior affinity for hydrogen (Feldewert et al. 2020).

A summary of the numerous investigations on coastal/estuarine methanogenesis is beyond the scope of this chapter. For a discussion of methanogenesis in the Colne Estuary, see Nedwell et al. (2016). The Colne Estuary, U.K., is typical for many global estuaries as methane production occurs along the length of the estuary together with sulfate reduction but at a rate almost two orders of magnitude lower than sulfate reduction (Nedwell et al. 2004). In the Colne Estuary a change from acetoclastic and hydrogenotrophic taxa to methylotrophic (*Methanococcoides*) and versatile (*Methanosarcina*) taxa from the head to the mouth was observed (Webster et al. 2015). The salinity (and, hence, sulfate) gradient that characterizes estuaries, together with proximity to land and sea, are major reasons why these environments have a higher diversity of methanogens than other ecosystems (Wen et al. 2017). For example, *Methanoregula* is typically freshwater while *Methanococcoides* is typically marine but both are common in estuaries (Wen et al. 2017). An investigation on methanogens and methanogenesis in mangrove sediments showed that the dominant taxa were Methanomicrobiales and Methanosarcinales together with putative hydrogen-dependent methyl-reducing methanogens *Candidatus* Methanofastidiosa and Methanomassiliicoccales, the latter exhibiting the highest activity (Zhang et al.

2020). Thus, even for a well-known archaeal process much has to be learned about the diversity of the responsible microbes.

It came as a surprise that methylotrophic methanogenesis may be a property of non-euryarchaeal candidate phyla within the archaea such as Verstraetearchaeota (Vanwonterghem et al. 2016) and Bathyarchaeota (Evans et al. 2015). This was based on the possession of *mcr* genes coding for methyl coenzyme M reductase (Mcr) complex, which catalyzes the terminal step of methanogenesis. Subsequently, *mcr* in *Candidatus* Syntrophoarchaeum was proposed to code for an enzyme involved in short-chain alkane oxidation and its sequence was similar to the *mcr* sequences from Verstraetearchaeota and Bathyarchaeota (Evans et al. 2019). Thus, it is supposed that these archaea are not methanogens but oxidize short-chain alkanes.

Consumption of short-chain alkanes may not be a common process in estuaries because propane and butane are not present in large amounts. However, anaerobic oxidation of methane produced by methanogens occurs ubiquitously especially in the sulfate-methane transition zone (SMTZ) (Boetius et al. 2000; Hoehler et al. 1994) and is a near-quantitative sink for the methane produced (Egger et al. 2018). This process is carried out by polyphyletic groups of uncultured archaea, which are related to methanogenic Euryarchaeota and referred to as ANME (anaerobic methane oxidizers) (Evans et al. 2019; Knittel et al. 2018). The original mechanistic explanation for anaerobic oxidation of methane was that it occurs as a syntrophic process in which the ANME methanotrophs convert methane to hydrogen, which is consumed by associated bacteria most typically in marine sediments by sulfate-reducing bacteria (Boetius et al. 2000). However, the precise mechanism of anaerobic oxidation of methane is debated in terms of: 1) the main interacting bacterial species and their terminal electron acceptors (e.g., sulfate, Fe III, Mn IV, and nitrate), 2) the internal metabolic processes in the ANME methanotrophs, which genetically resemble methanogens, and 3) the energetics and mode of exchange e.g., metabolite transfer or direct interspecies electron exchange (McGlynn 2017).

In estuaries, which overall are a methane source, more methane reaches the atmosphere as salinity decreases due to a combination of greater methane production and less effective anaerobic removal where sulfate reduction is lower (Dean et al. 2018). However, there remains much to be learned about the sources and sinks of methane in coastal environments and particularly in estuaries. For example, in a brackish Baltic Sea estuary, anaerobic oxidation of methane was identified as an important process presumably coupled to iron III and manganese IV reduction (Myllykangas et al. 2020).

### 12.4.4 Haloarchaea Are Consistently Present and Locally Abundant in Coastal Sediments

Extremely halophilic Euryarchaeota belonging to the Halobacteria (more commonly referred to as haloarchaea) dominate in coastal environments where seawater evaporates to create hypersaline conditions such as in sabkhas, hypersaline lagoons, as well as artificial salt pans (McGenity and Oren 2012). However, Munson et al.

(1997) reported that haloarchaea were abundant in a temperate salt marsh. This observation led Purdy et al. (2004) to culture haloarchaea from creek or saltmarsh pan sediments, aerobically, over a range of salinities, with either glucose or glycerol as carbon and energy sources, and with antibiotics to inhibit growth of bacteria and eukarya. They isolated three taxa of haloarchaea one of which had strains that grew slowly at seawater salinity and optimally with 10% NaCl, a property that is unusual for haloarchaea (Purdy et al. 2004). Subsequently, haloarchaea have been found in coastal environments across the globe sometimes locally at high abundance. This is particularly true for members of *Haladaptatus*, which are likely to contribute to carbon cycling during periods of desiccation.

## 12.5 The Coastal Fungal Microbiome

Fungi are a ubiquitous component of all ecosystems. They support the decomposition of lignocellulosic compounds (Bani et al. 2019; Francioli et al. 2020), provide industry-relevant bioactive products (Overy et al. 2019), and in vegetated habitats (e.g., coastal marshes) produce mycorrhizal networks that facilitate nutrient uptake by plants (Smith and Read 2008). In addition, through the differential accumulation of fungal pathogens, they can promote plant biodiversity and productivity (Mommer et al. 2018).

Within coastal marshes, mycorrhizal associations sustain specific interactions between plants and fungi that are beneficial to survival and growth. For example, the arbuscular mycorrhizal (AM) fungi (Phylum: Glomeromycota), which are obligate plant-root endosymbionts of most terrestrial plant species, decrease salt stress, and increase water uptake in plants growing in coastal marshes (Evelin et al. 2009). They may also decrease the impacts of localized hypoxia experienced by plant roots during tidal inundation. AM fungi play similar roles in plant nutrient acquisition (P and N uptake in exchange for plant-derived C) in coastal marshes as they do in most terrestrial habitats (Fitter 2005). The rewards to a given plant species of this symbiosis depend both on the identification of the AM fungal species present and on the soil/sediment nutrient levels (Hoeksema et al. 2010). Thus, these interactions influence plant competition dynamics at the ecosystem scale (van der Heijden 2002). In salt marshes the interaction between nutrient levels and AM fungi can influence plant zonation via changing the competitive ability of different plant species (Daleo et al. 2008), although this is less extensively studied than for terrestrial habitats. For example, in the presence of AM fungi and at low nutrient concentrations *Spartina densiflora* has a competitive advantage over *S. alterniflora* but this is reversed by increased nutrient concentrations and/or suppression of AM fungi (Daleo et al. 2008). Other fungal phyla (e.g., Basidiomycota and Ascomycota) comprise species that provide the primary route for the decomposition of vascular plant litter and the remineralization of carbon in salt marshes, although bacteria may supersede fugal decomposers under more saline conditions (Cortes-Tolalpa et al. 2018). Some specificity in fungal decomposer communities appears to be present with different fungal species decomposing different plants and being present at different geographic locations (Calado et al. 2019; Cortes-Tolalpa et al. 2018; Lyons et al. 2010). Decomposition activity shows limited spatial variability on marshes (Buchan et al. 2003) and is largely uninfluenced by changes in salinity (Connolly et al. 2014), although which fungal decomposers are present may be affected by salinity (see below). However, decomposition rates vary seasonally (Buchan et al. 2003) and related microbial activity varies across diel cycles (Kuehn et al. 2004). Invasive (non-native) plant species affect bacteria-to-fungal ratios in marshes (Zhang et al. 2018), potentially altering decomposition rates via the introduction of novel substrates. This link between decomposition and substrate type means that there is also a certain degree of top-down control on fungal saprotrophs, where the modification of the physical structure of aboveground vegetation by animal grazers can alter both recalcitrant autochthonous input and the capacity to trap less-recalcitrant allochthonous inputs of plant material (Mueller et al. 2017).

Fungal biodiversity within coastal marshes is regulated by the identity and abundance of the plant species present, tidal inundation and salinity, alongside environmental gradients in physiochemistry that covary with salinity (Alzarhani et al. 2019; Mohamed and Martiny 2011). Across three marshes of different salinity (27–33, 15–25, and 0–10 ppt) fungal species richness was predominantly influenced by the presence of plant species and less influenced by salinity and other environmental gradients (Mohamed and Martiny 2011). This most likely reflects increased heterogeneity in the microhabitats that fungi occupy and differential patterns of co-occurrence (and or host specificity) across fungi and plant species. In contrast, the community composition of fungi from the same marshes was primarily determined by the underlying salinity gradient and not by the identity of plant species present, reflecting levels of halotolerance across fungal species limiting their occurrences to marshes within their salinity tolerances (Mohamed and Martiny 2011). However, larger-scale studies examining multiple marshes ($n = 3$) in high- and low-salinity environments (33–43 and 3.3–5.9 ppt) have shown contrasting results; with abiotic factors primarily determined the patterns of fungal richness while the interaction between abiotic and biotic factors determined community composition (Alzarhani et al. 2019). Moreover, the relationship between abiotic variables and fungal species richness was not generalizable across marshes and the relative influence of abiotic and biotic factors on community composition also varied. Subsequently, statistical models relating fungal biodiversity to the abiotic and/or biotic factors on a particular salt marsh performed poorly at predicting fungal biodiversity on other marshes despite the similarities between these environments (Alzarhani et al. 2019). These context-dependencies can be attributed, among other things, to differences in the functional groups of fungi present in salt marshes where certain abiotic or biotic variables were more strongly related to specific functional groups than over others (Alzarhani et al. 2019).

In coastal marine and aquatic estuarine environments adjacent to salt marshes much less is known about the diversity and functionality of fungi. Typically, estuaries support more diverse fungal communities than coastal marine environments, which in turn are more diverse than oceanic environments (Jeffries et al. 2016). This gradient of fungal diversity reflects the flow of terrestrial matter into estuarine systems and the role of the critical transition zone between freshwater and marine systems that determines benthic biodiversity (Levin et al. 2001). Fungal turnover is highest in estuarine environments when compared to other marine systems because terrestrial, freshwater, and non-halotolerant species give way to halotolerant fungi found in near-shore environments (Burgaud et al. 2013). In coastal environments that experience extreme salt stress (e.g., hypersaline lagoons) fewer fungal species are present (e.g., *Trimmatostroma* spp., *Emericella* spp., and *Phaeotheca* spp.) and in general there are only a few known halophilic fungal species (Gostinčar et al. 2010). In the absence of terrestrial plant species with which to interact, the major functional role of fungi in these environments is the decomposition of lignocellulosic compounds and recycling of vascular plant litter (Newell 1996), alongside those that are pathogens. This also includes lignin degraders (Bucher et al. 2004), which contribute to the primary decomposition of woody debris in estuarine environments (Poole and Price 1972; Tsui and Hyde 2004). The primarily saprophytic role fungi play in estuary environments has led to them being dominated by Basidiomycota and Ascomycota (Burgaud et al. 2013; Wang et al. 2019), but with a reduction in terrestrial subsidies in coastal waters, many species of Chytridiomycota become abundant (Jeffries et al. 2016; Sun et al. 2014). Generalizing these patterns of presence of fungal phyla is problematic as the dominance of species from different phyla changes with geographic locations. For example, in the Baltic Sea, low-salinity areas (< 8 ppt) contain fungal communities compositionally similar to those in local freshwaters, but higher-salinity areas (> 8 ppt) contain fungal communities similar to those in marine systems (Rojas-Jimenez et al. 2019). However, in these locations Basidiomycota and Ascomycota dominate in the marine environment and Chytridiomycota in freshwaters (Rojas-Jimenez et al. 2019). It is also worth noting that DNA sequences from Glomeromycota (AM fungi) have been detected during surveys in near-shore fungal communities, which if they originate from spores suggests a previously underexplored dispersal route (Lacerda et al. 2020). Given the important role of fungi in degrading complex organic material close species–species interactions and biogeochemical coupling between the fungal and algal, bacterial, and archaeal constituents are expected in coastal sediment biomes. These questions are currently unanswered.

## 12.6 Impacts of Oil Pollution on Coastal Microbiomes

An estimated 1.3 million tons of oil enters the marine environment each year (National Research Council 2003). This includes oil from natural seeps and spills associated with the extraction and transportation of petroleum (e.g., tanker, pipeline, and coastal facility spills). The largest offshore oil spill in history was The Deepwater Horizon spill resulting from the Macondo well blowout that resulted in the release of 134 million gallons of crude oil into the Gulf of Mexico. Despite the deposition of large quantities of oil in deep water systems, or being biodegraded, chemically dispersed, or burned in situ, large quantities of oil still reached coastal ecosystems and contaminated 2100 km of coastline. This pollution caused serious negative effects on marine life and coastal saltmarsh, seagrass, and reef systems (Beyer et al. 2016). Many coastal environments are particularly vulnerable to oil spills because many oil refineries are situated at the coast or at large estuaries. The majority of the 20 largest oil tanker spills to date also occurred close to the coast when tanker vessels ran aground (ITOPF 2019). Oil spills have major effects on coastal ecosystems such as mass mortality of invertebrates, birds and mammals. For example, the Prestige oil spill caused a decrease of 66% of total species richness on Spanish Galician beaches (de La Huz et al. 2005) and the Exxon Valdez spill on the Alaskan coast caused major sea otter (Monson et al. 2000) and seabird (Piatt and Ford 1996) mortalities. Oil can also have toxic effects on microorganisms because the accumulation of hydrocarbon molecules in the membrane can result in loss of membrane integrity and impaired cellular homeostasis (Sikkema et al. 1995). This can have major impacts on key microbial ecosystem services such as coastal nitrogen cycling (Horel et al. 2014; Zhao et al. 2020). For example, some ammonia-oxidizing bacteria and archaea are, respectively, 100 and 1000 times more sensitive to hydrocarbon toxicity than model heterotrophs (Urakawa et al. 2019). Whilst large oil spills are thankfully rare, coastal ecosystems are threatened continually with chronic oil and hydrocarbon pollution from rivers and land runoff (National Research Council 2003) and intensive industrial and recreational activities around coasts particularly near estuaries and harbors (Duran et al. 2015; McGenity 2014; Nogales and Bosch 2019).

Crude oil contains a complex mix of hydrocarbons that includes saturated aliphatic hydrocarbons such as cycloalkanes, linear *n*-alkanes (ranging from short chains to long chains with over 40 carbon atoms), and branched alkanes such as pristine and phytane (Weisman 1998). There are also many aromatic hydrocarbons such as the monoaromatic BTEX compounds (benzene, toluene, ethylbenzene, and xylene) and polyaromatic hydrocarbons that include a wide range of both parent and methylated 2- to 5-ring compounds such as naphthalenes, phenanthrenes, pyrenes, and perylenes. In addition to these two main classes of hydrocarbons there is a variety of large and highly recalcitrant asphaltene and resin compounds. Whilst certain processes remove some components of oil from the environment (e.g., evaporation of the lighter fractions, chemical- or photo-oxidation), unless oil is physically removed, the primary loss route will be via natural biodegradation by hydrocarbon-degrading microbes that utilize hydrocarbons as their carbon and energy source (Harayama et al. 1999; Head et al. 2006; McGenity et al. 2012). Because of the complexity of oil, its biodegradation requires a diverse consortium of species that can degrade different hydrocarbons. There is niche partitioning between the different species in the consortium in the utilization of different hydrocarbon substrates (Head et al. 2006; McGenity et al. 2012; McKew et al. 2007).

### 12.6.1 Diversity of Hydrocarbon-Degrading Microbes in Coastal Sediments

High concentrations of hydrocarbons can dramatically alter the composition of coastal microbial communities leading to large decreases in species richness and diversity coupled with the selection for specialist hydrocarbon-degrading bacteria (Head et al. 2006; McGenity et al. 2012). In estuarine and coastal sediments hydrocarbons are particularly used by Alpha- and Gammaproteobacteria (Chronoupolou et al. 2013; Coulon et al. 2012; Greer 2010). The selection for specific bacterial species may be influenced by numerous factors including the concentration or type of oil and/or its degree of weathering (Head et al. 2006) or environmental conditions such as temperature or the concentration of key nutrients such as nitrogen and phosphorus (Coulon et al. 2007). However, many common patterns are observed globally, such as an increase in the relative abundance of obligate hydrocarbonoclastic bacteria (OHCB; Yakimov et al. 2007) in oxygenated oil-contaminated marine sediments. The OHCB include key genera such as *Alcanivorax*, *Thalassolituus*, *Oleiphilus*, *Oleispira*, and *Oleibacter*, which typically degrade alkanes as well as *Cycloclasticus,* which degrade a wide range of PAHs. The name "OHCB" is a slightly misleading because these organisms are not truly "obligate." For example, in pure culture *Alcanivorax* degrades some other compounds (Radwan et al. 2019) and also some polyesters (Zadjelovic et al. 2020). However, there is still very limited evidence that the OHCB are competitive for non-hydrocarbon substrates in the environment and their lifestyle is often restricted to the use of hydrocarbons or their fatty acid or alcohol derivatives. The OHCB are often in low abundance in marine environments when hydrocarbons are absent but respond quickly and grow rapidly in response to oil pollution. Their streamlined genomes are specifically geared toward a hydrocarbon-degrading lifestyle (Kube et al. 2013; Schneiker et al. 2006; Yakimov et al. 2007) and their marine distribution is truly global (Yakimov et al. 2007). In muddy and sandy coastal sediments, OHCB such as *Alcanivorax, Oleibacter, Cycloclasticus*, and *Marinobacter hydrocarbonoclasticus* tend to increase in abundance and often dominate the bacterial community after the addition of crude oil (Chronoupolou et al. 2013; Coulon et al. 2012; Kostka et al. 2011; Thomas et al. 2020).

*Alcanivorax borkumensis* (Yakimov et al. 1998), a specialist *n*-alkane and branched alkane degrader, was first isolated from North Sea sediments and was the first OHCB to have its genome sequenced (Schneiker et al. 2006). Since its discovery 14 named species and a large diversity of unclassified *Alcanivorax* have been recorded in the NCBI database, including species isolated from deep sea sediments such as *A. dieselolei* (Liu and Shao 2005), *A. pacificus* (Lai et al. 2011), and *A. mobilis* (Yang et al. 2018), or from intertidal sediments such as *A. jadensis* (Fernandez-Martinez et al. 2003) and *A. gelatiniphagus* (Kyoung Kwon et al. 2015). *Alcanivorax* is often dominant in oil-contaminated intertidal sediments globally, including in the Gulf of Mexico, in the Atlantic Galician coast, and in Mediterranean beaches following the Deepwater Horizon (Newton et al. 2013; Rodriguez-R et al.

2015), Prestige (Acosta-González et al. 2015), and Agia Zoni (Thomas et al. 2020) oil spills, respectively.

Another key alkane degrader in intertidal sediments is *Thalassolituus*, the type strain of which was isolated from harbor seawater/sediment samples in Milazzo, Italy (Yakimov et al. 2004). *Thalassolituus* is a highly competitive *n*-alkane degrader in estuarine environments and mudflats (McKew et al. 2007; Sanni et al. 2015) but is also found globally including deep water environments such as the oil plume from the DWH oil spill in the Gulf of Mexico (Camilli et al. 2010). *Oleibacter* related to the type strain *O. marinus* 201 (Teramoto et al. 2011) or a variety of unclassified strains is important in coastal fine-grained (Chronoupolou et al. 2013; Coulon et al. 2012; Sanni et al. 2015) and sandy (Thomas et al. 2020) sediment communities. However, like *Alcanivorax* and *Thalassolituus, Oleibacter* species are not specific to oil-degrading communities within coastal sediments as they (along with many other oil-degrading bacteria) are also found in a variety of marine environments, including for example seawater at 10,400 m in the Challenger Deep at the southern end of the Mariana Trench (Liu et al. 2019). This suggests that it is the availability of hydrocarbon that selects for oil degraders rather than the specific environmental conditions themselves. Low temperature often results in the selection for *Oleispira*. Bacteria closely related to the psychrophilic alkane-degrading *Oleispira antarctica* (Gregson et al. 2020; Kube et al. 2013; Yakimov et al. 2003) are important in oil-contaminated temperate coastal microbial communities at winter temperatures (4 °C) (Coulon et al. 2007), as well as in cold environments such as deep arctic sediments (Dong et al. 2015) and sea-ice (Gerdes et al. 2005).

*Cycloclasticus* often plays the primary role in PAH degradation in coastal environments (Chronoupolou et al. 2013; Coulon et al. 2012; Duran and Cravo-Laureau 2016; Kasai et al. 2002; McKew et al. 2007; Sanni et al. 2015; Thomas et al. 2020). This genus comprises a wide variety of PAH-degrading species, including *C. pugetii* (Dyksterhouse et al. 1995), *C. oligotrophus* (Wang et al. 1996), *C. spirillensus* (Chung and King 2001), and *C. zancles* (Messina et al. 2016). There is a bivalve- and sponge symbiont lineage that can also degrade short-chain alkanes (Rubin-Blum et al. 2017).

In addition to these specialist hydrocarbon-degrading genera there are many other species from more nutritionally versatile genera that degrade hydrocarbons and many are regularly found in oil-polluted coastal sediments. There are too many to consider here but they include, for example, species from genera such as *Marinobacter* (particularly *Marinobacter hydrocarbonoclasticus* (Gauthier et al. 1992)), *Alteromonas, Erythrobacter, Idiomarina, Microbacterium, Psuedomonas, Pseudoalteromonas, Rhodococcus, Roseovarius, Shewanella, Sphingomonas, Vibrio*, and *Xanthomonas* (e.g. see Prince et al. (2018) for a review of all hydrocarbon degraders, Goñi-Urriza and Duran (2018) for the role of bacteria in hydrocarbon degradation in coastal microbial mats, Greer (2010) for a review of bacterial diversity in hydrocarbon-polluted estuaries and sediments, and Supplementary Table 12.1 in Thomas et al. (2020) for many such observations in sandy coastal sediments).

Whilst aerobic biodegradation of hydrocarbons dominates in sediments, those that become buried in anoxic sediments can remain there for decades (Reddy et al. 2002) as anaerobic biodegradation of hydrocarbons is slow in comparison with aerobic biodegradation. Aerobic hydrocarbon degraders rely on oxygen not only for respiration; oxygen is also required for the primary step of hydrocarbon degradation catalyzed by oxygenase enzyme systems (Wang and Shao 2013; Wang et al. 2018). Consequently, anaerobic bacteria must employ alternative pathways of biodegradation in the absence of oxygen. Due to the abundance of sulfate in coastal sediments many of the known anaerobic hydrocarbon oxidizers in coastal sediments are sulfate-reducing bacteria related to *Desulfosarcina*, *Desulfococcus*, *Desulfonema*, *Desulfobacula*, *Desulfotomaculum*, *Desulfotignum*, and *Geobacter* (McGenity 2014; Païssé et al. 2008; Rabus et al. 2016). Also, marine sediments are subjected to oscillations in oxygen concentration as the result of the tides, burrowing, and oxygenic phototrophic activity, which can encourage the growth of certain phylotypes of *Alcanivorax* (Terrisse et al. 2017).

The relative abundance of genera with potential hydrocarbon-degrading abilities can be used to estimate hydrocarbon exposure in an environment using the Ecological Index of Hydrocarbon Exposure (Lozada et al. 2014), which is based on microbial composition determined by 16S rRNA gene sequencing, as the numbers of hydrocarbon-degrading bacteria typically correlate with the concentration of hydrocarbons in sediments (Thomas et al. 2020).

### 12.6.2 Association of Hydrocarbon-Degrading Bacteria with Photoautotrophs

Many hydrocarbon-degrading bacteria in intertidal sediments have close association with photoautotrophs. Hydrocarbons can alter the composition of phototrophic communities considerably, for example, by inhibiting enzyme activities and photosynthesis (Megharaj et al. 2000). This toxicity may favor hydrocarbon-resistant species. For example, cyanobacteria belonging to *Phormidium, Planktotrix*, and *Oscillatoria* have shown varying degrees of tolerance to oil pollution (Van Bleijswijk and Muyzer 2004). In sediment mesocosms with oil-polluted fine-sediments that were dominated with the hydrocarbonoclastic bacteria *Alcanivorax*, *Cycloclasticus*, and *Oleibacter,* there was also an increased abundance of MPB, primarily due to a ten-fold increase in the abundance of cyanobacteria (Chronoupolou et al. 2013; Coulon et al. 2012). This increase was attributed to a lower grazing pressure and/or nitrogen depletion, which encouraged the growth of diazotrophic cyanobacteria. Some microalgae coexist with hydrocarbon-degrading bacteria (Amin et al. 2009; Chernikova et al. 2020; Gutierrez et al. 2013) and diatom-OHCB floating biofilms have been seen in mudflat sediments after an experimental oil spill (Coulon et al. 2012). There are numerous hypothesized ways that phototrophs enhance hydrocarbon degradation including direct degradation (although evidence for this is limited), supplying key resources (e.g., oxygen, N, Fe) to hydrocarbon-degrading bacteria, or assisting in immobilizing hydrocarbon-degrading bacteria within EPS. Phototrophs in turn can benefit from higher $CO_2$ concentrations from hydrocarbon-degrading bacterial respiration (e.g., see reviews by Abed 2019; Ardelean 2014; McGenity 2014; McGenity et al. 2012).

### 12.6.3 Mechanisms of Oil Biodegradation

Generally, the bioavailability and rate of degradation of hydrocarbons decreases with increasing carbon number. Saturated hydrocarbons are often degraded at higher rates than light aromatics. The high-molecular-weight aromatics and polar compounds are degraded at low rates (Leahy and Colwell 1990). Most hydrocarbon-degrading bacteria will typically degrade a small range of either aliphatic or aromatic compounds, although some bacteria may possess pathways for catabolism of both aliphatic and PAH compounds such as some *Pseudomonas* (Whyte et al. 1997), *Rhodococcus* (Andreoni et al. 2000), or *Colwellia* (Mason et al. 2014) strains. Uptake via transport systems lowers the substrate concentration around the cell driving diffusive flux of hydrophobic hydrocarbons toward the cell (Harms et al. 2010), whilst the production of extracellular or cell-bound surface-active compounds (e.g., glucolipid biosurfactant produced by *Alcanivorax* (Yakimov et al. 1998)) increases bioavailability by decreasing the interfacial tension between water and oil (Marchant and Banat 2012). Aerobic hydrocarbon-degrading bacteria are equipped with a wide array of genes that code for monooxygenase enzyme systems to activate hydrocarbons such as the two integral-membrane non--heme iron alkane monooxygenase systems AlkB1 and AlkB2 and three heme-containing P450 cytochromes employed by *Alcanivorax* (Gregson et al. 2019; Schneiker et al. 2006; Yakimov et al. 1998) that convert medium-chain *n*-alkanes or branched alkanes to a primary alcohol that can be further degraded by an array of alcohol- and aldehyde dehydrogenases. Long-chain alkanes can be biodegraded with AlmA flavin-binding monooxygenases (Wang and Shao 2014) or similar sub-terminal Baeyer-Villiger monooxygenases in *Thalassolituus oleivorans* (Gregson et al. 2018). Dioxygenase systems are typically employed by aerobic PAH-degraders such as the PhnA1, PhnA2, PhnA3, and PhnA4 proteins (alpha and beta subunits of an iron-sulfur protein, a ferredoxin and a ferredoxin reductase, respectively) that make up a PAH dioxygenase system in *Cycloclaticus* strain A5 (Kasai et al. 2003).

Fine-grained intertidal sediments such as mudflats are typically anoxic below 1–2 mm. This prevents the activation of hydrocarbons using oxygenase enzymes. The exact mechanisms of the anaerobic activation of hydrocarbons are less well understood but may include direct carboxylation, methylation followed by addition to fumarate, or even utilization of nitrite to activate alkanes (McGenity et al. 2012; Meckenstock and Mouttaki 2011; Rabus et al. 2016; Widdel and Musat 2010; Zedelius et al. 2011).

**Acknowledgements** The authors wish to acknowledge the many Ph.D., postdoctoral researchers, and technical staff who have contributed to the research of the Ecology and Environmental

Microbiology group at the University of Essex, and whose work is cited in this chapter. The following funding bodies and awards are gratefully acknowledged: U.K. Natural Environment Research Council (NERC) Coastal Biodiversity and Ecosystem Services programme (NE/J01561X/1); NERC EPStromNet (NE/V00834X/1); NERC SSB programme, module Blue Carbon, (NE/K001914/1); NERC Quantifying a marine ecosystem's response to a catastrophic oil spill. (NE/R016569/1); NERC PRINCE-A new dynamic for Phosphorus in RIverbed Nitrogen Cycling (NE/P011624/1); European Union's Horizon 2020 research and innovation programme, grant agreement No 702217; Defra funding "Ascertaining Predisposing Factors that Affect Oak Health in the U.K. and Advancements toward Management for Resilient Oak Populations"; Eastern ARC Academic Research Consortium; The Royal Society, Understanding microalgal biofilm contributions to sediment "blue carbon" in contrasting salt marsh habitats in the U.S. and Europe (IES\R1\201260).

## References

Abed RMM (2019) Phototroph-heterotroph oil-degrading partnerships. In: McGenity T (ed) Microbial communities utilizing hydrocarbons and lipids: members, metagenomics and ecophysiology. Handbook of hydrocarbon and lipid microbiology. Springer, Cham, pp 37–50

Abell GCJ, Revill AT, Smith C et al (2010) Archaeal ammonia oxidizers and nirS-type denitrifiers dominate sediment nitrifying and denitrifying populations in a subtropical macrotidal estuary. ISME J 4:286–300

Acosta-González A, Martirani-von Abercron SM, Rosselló-Móra R et al (2015) The effect of oil spills on the bacterial diversity and catabolic function in coastal sediments: a case study on the prestige oil spill. Environ Sci Pollut Res Int 22:15200–15214

Acuña Alvarez L, Exton DA, Suggett DJ et al (2009) Characterization of marine isoprene-degrading communities. Environ Microbiol 11:3280–3291

Admiraal W (1984) The ecology of estuarine sediment inhabiting diatoms. Progr Phyc Res 3:269–322

Admiraal W, Peletier H, Brouwer T (1984) The seasonal succession patterns of diatom species on an intertidal mudflat: an experimental analysis. Oikos 42:30–40

Alongi DM (2020) Carbon balance in salt marsh and mangrove ecosystems: a global synthesis. J Mar Sci Eng 8:767. https://doi.org/10.3390/jmse8100767

Alzarhani AK, Clark DR, Underwood GJC et al (2019) Are drivers of root-associated fungal community structure context specific? ISME J 13:1330–1344

Amin SA, Green HD, Hart MC et al (2009) Photolysis of iron-siderophore chelates promotes bacterial-algal mutualism. Proc Nat Acad Sci 106:17071–17076

An SM, Choi DH, Noh JH (2020) High-throughput sequencing analysis reveals dynamic seasonal succession of diatom assemblages in a temperate tidal flat. Estuar Coast Shelf Sci 237:106686

Andreoni V, Bernasconi S, Colombo M et al (2000) Detection of genes for alkane and naphthalene catabolism in *Rhodococcus* sp. strain 1BN. Environ Microbiol 2:572–577

Ardelean II (2014) The involvement of cyanobacteria in petroleum hydrocarbons degradation: fundamentals, applications and perspectives. In: Davison D (ed) Cyanobacteria: ecological importance, biotechnological uses and risk management. Nova Science Publishers, New York, pp 41–60

Arp DJ, Sayavedra-Soto LA, Hommes NG (2002) Molecular biology and biochemistry of ammonia oxidation by *Nitrosomonas europaea*. Arch Microbiol 178:250–255

Baas JH, Baker ML, Malarkey J et al (2019) Integrating field and laboratory approaches for ripple development in mixed sand–clay–EPS. Sedimentol 66:2749–2768

Baker BJ, De Anda V, Seitz KW et al (2020) Diversity, ecology and evolution of archaea. Nat Microbiol 5:887–900

Bani A, Borruso L, Nicholass KJM et al (2019) Site-specific microbial decomposer communities do not imply faster decomposition: results from a litter transplantation experiment. Microorganisms 7:349

Barnett A, Méléder V, Blommaert L et al (2015) Growth form defines physiological photoprotective capacity in intertidal benthic diatoms. ISME J 9:32–45

Barranguet C, Kromkamp J, Peene J (1998) Factors controlling primary production and photosynthetic characteristics of intertidal microphytobenthos. Mar Ecol Prog Ser 173:117–126

Bartelme RP, McLellan SL, Newton RJ (2017) Freshwater recirculating aquaculture system operations drive biofilter bacterial community shifts around a stable nitrifying consortium of ammonia-oxidizing archaea and comammox *Nitrospira*. Front Microbiol 8:101

Beaumont NJ, Jones L, Garbutt A et al (2014) The value of carbon sequestration and storage in coastal habitats. Estuar Coast Shelf Sci 137:32–40

Beddow J, Stolpe B, Cole PA et al (2017) Nanosilver inhibits nitrification and reduces ammonia-oxidizing bacterial but not archaeal amoA gene abundance in estuarine sediments. Environ Microbiol 19:500–510

Bellinger BJ, Abdullahi AS, Gretz MR et al (2005) Biofilm polymers: relationship between carbohydrate biopolymers from estuarine mudflats and unialgal cultures of benthic diatoms. Aquat Microb Ecol 38:169–180

Bellinger BJ, Underwood GJC, Ziegler SE et al (2009) Significance of diatom-derived polymers in carbon flow dynamics within estuarine biofilms determined through isotopic enrichment. Aquat Microb Ecol 55:169–187

Beman JM, Francis CA (2006) Diversity of ammonia-oxidizing archaea and bacteria in the sediments of a hypernutrified subtropical estuary: Bahia del Tobari, Mexico. Appl Environ Microbiol 72:7767–7777

Beman JM, Popp BN, Francis CA (2008) Molecular and biogeochemical evidence for ammonia oxidation by marine Crenarchaeota in the Gulf of California. ISME J 2:429–441

Benyoucef I, Blandin E, Lerouxel A et al (2014) Microphytobenthos interannual variations in a north-European estuary (Loire estuary, France) detected by visible-infrared multispectral remote sensing. Estuar Coast Shelf Sci 136:43–52

Berg G, Rybakova D, Fischer D et al (2020) Microbiome definition re-visited: old concepts and new challenges. Microbiome 8:103

Bergman B, Gallon JR, Rai AN et al (1997) $N_2$ fixation by non-heterocystous cyanobacteria. FEMS Microbiol Rev 19:139–185

Berman-Frank I, Lundgren P, Falkowski P (2003) Nitrogen fixation and photosynthetic oxygen evolution in cyanobacteria. Res Microbiol 154:157–164

Beyer J, Trannum HC, Bakke T et al (2016) Environmental effects of the Deepwater horizon oil spill: a review. Mar Pollut Bull 110:28–51

Blommaert L, Lavaud J, Vyverman W et al (2018) Behavioural versus physiological photoprotection in epipelic and epipsammic benthic diatoms. Eur J Phycol 53:146–155

Blommaert L, Vancaester E, Huysman MJJ et al (2020) Light regulation of LHCX genes in the benthic diatom *Seminavis robusta*. Front Mar Sci 7:192

Böer SI, Hedtkamp SIC, van Beusekom JEE et al (2009) Time- and sediment depth-related variations in bacterial diversity and community structure in subtidal sands. ISME J 3:780–791

Boetius A, Ravenschlag K, Schubert CJ et al (2000) A marine microbial consortium apparently mediating anaerobic oxidation of methane. Nature 407:623–626

Bohórquez J, McGenity TJ, Papaspyrou S et al (2017) Different types of diatom-derived extracellular polymeric substances drive changes in heterotrophic bacterial communities from intertidal sediments. Front Microbiol 8:245

Booty JM, Underwood GJC, Parris A et al (2020) Wading birds affect ecosystem functioning on an intertidal mudflat. Front Mar Sci 7:685

Bowles M, Nigro L, Teske A et al (2012) Denitrification and environmental factors influencing nitrate removal in Guaymas Basin hydrothermally altered sediments. Front Microbiol 3:377

Braker G, Conrad R (2011) Diversity, structure, and size of $N_2O$-producing microbial communities in soils—what matters for their functioning? Adv Appl Microbiol 75:33–70

Braker G, Zhou JZ, Wu LY et al (2000) Nitrite reductase genes (*nirK* and *nirS*) as functional markers to investigate diversity of denitrifying bacteria in Pacific northwest marine sediment communities. Appl Environ Microbiol 66:2096–2104

Bruckner CG, Bahulikar R, Rahalkar M et al (2008) Bacteria associated with benthic diatoms from Lake Constance: phylogeny and influences on diatom growth and secretion of extracellular polymeric substances. Appl Environ Microbiol 74:7740–7749

Bruckner CG, Rehm C, Grossart H-P et al (2011) Growth and release of extracellular organic compounds by benthic diatoms depend on interactions with bacteria. Environ Microbiol 13: 1052–1063

Buchan A, Newell SY, Butler M et al (2003) Dynamics of bacterial and fungal communities on decaying salt marsh grass. Appl Environ Microbiol 69:6676–6687

Bucher VVC, Hyde KD, Pointing SB et al (2004) Production of wood decay enzymes, mass loss and lignin solubilization in wood by marine ascomycetes and their anamorphs. Fungal Divers 15:1–14

Burden A, Garbutt A, Evans CD (2019) Effect of restoration on saltmarsh carbon accumulation in eastern England. Biol Lett 15:20180773

Burgaud G, Woehlke S, Rédou V et al (2013) Deciphering the presence and activity of fungal communities in marine sediments using a model estuarine system. Aquat Microb Ecol 70:45–62

Caffrey JM, Bano N, Kalanetra K et al (2007) Ammonia oxidation and ammonia-oxidizing bacteria and archaea from estuaries with differing histories of hypoxia. ISME J 1:660–662

Calado ML, Carvalho L, Barata M et al (2019) Potential roles of marine fungi in the decomposition process of standing stems and leaves of *Spartina maritima*. Mycologia 111:371–383

Camilli R, Reddy CM, Yoerger DR et al (2010) Tracking hydrocarbon plume transport and biodegradation at Deepwater horizon. Science 330:201–204

Cao H, Hong Y, Li M et al (2012) Community shift of ammonia-oxidizing bacteria along an anthropogenic pollution gradient from the Pearl River Delta to the South China Sea. Appl Microbiol Biotechnol 94:247–259

Cao H, Hong Y, Li M et al (2011) Diversity and abundance of ammonia-oxidizing prokaryotes in sediments from the coastal Pearl River estuary to the South China Sea. Ant Leeuwenhoek 100: 545–556

Capone DG (1988) Benthic nitrogen fixation. In: Blackburn TH, Sorensen J (eds) Nitrogen cycling in coastal marine sediments. John Wiley and Sons, New York, pp 85–123

Chen W, Pan Y, Yu L et al (2017) Patterns and processes in marine microeukaryotic community biogeography from Xiamen coastal waters and intertidal sediments, Southeast China. Front Microbiol 8:1912

Chernikova TN, Bargiela R, Toshchakov SV et al (2020) Hydrocarbon-degrading bacteria *Alcanivorax* and *Marinobacter* associated with microalgae *Pavlova lutheri* and *Nannochloropsis oculata*. Front Microbiol 11:2650

Christianen MJA, Middelburg JJ, Holthuijsen SJ et al (2017) Benthic primary producers are key to sustain the Wadden Sea food web: stable carbon isotope analysis at landscape scale. Ecology 98: 1498–1512

Chronoupolou P-M, Fahy A, Coulon F et al (2013) Impact of a simulated oil spill on benthic phototrophs and nitrogen-fixing bacteria. Environ Microbiol 15:241–252

Chung WK, King GM (2001) Isolation, characterization, and polyaromatic hydrocarbon degradation potential of aerobic bacteria from marine macrofaunal burrow sediments and description of *Lutibacterium anuloederans* gen. Nov., sp. nov., and *Cycloclasticus spirillensus* sp. nov. Appl Environ Microbiol 67:5585–5592

Cibic T, Fazi S, Nasi F et al (2019) Natural and anthropogenic disturbances shape benthic phototrophic and heterotrophic microbial communities in the Po River Delta system. Estuar Coast Shelf Sci 222:168–182

Clark DR, McKew B, Dong L et al (2020) Mineralization and nitrification: archaea dominate ammonia-oxidising communities in grassland soils. Soil Biol Biochem 143:107725
Clark DR, Mégane Mathieu M, Mourot L et al (2017) Biogeography at the limits of life: do extremophilic microbial communities show biogeographic regionalisation? Glob Ecol Biogeogr 26:1435–1446
Coelho H, Vieira S, Serôdio J (2011) Endogenous versus environmental control of vertical migration by intertidal benthic microalgae. Eur J Phycol 46:271–281
Cohn SA, Dunbar S, Ragland R et al (2016) Analysis of light quality and assemblage composition on diatom motility and accumulation rate. Diatom Res 31:173–184
Connolly CT, William V, Sobczak WV, SEG F (2014) Salinity effects on phragmites decomposition dynamics among the Hudson River's freshwater tidal wetlands. Wetlands 34:575–582
Consalvey M, Paterson DM, Underwood GJC (2004) The ups and downs of life in a benthic biofilm: migration of benthic diatoms. Diatom Res 19:181–202
Cook PLM, Veuger B, Böer S et al (2007) Effect of nutrient availability on carbon and nitrogen incorporation and flows through benthic algae and bacteria in near-shore sandy sediment. Aq Microb Ecol 49:165–180
Cortes-Tolalpa L, Norder J, van Elsas JD et al (2018) Halotolerant microbial consortia able to degrade highly recalcitrant plant biomass substrate. Appl Microbiol Biotechnol 102:2913–2927
Coulon F, Chronopoulou PM, Fahy A et al (2012) Central role of dynamic tidal biofilms dominated by aerobic hydrocarbonoclastic bacteria and diatoms in the biodegradation of hydrocarbons in coastal mudflats. Appl Environ Microbiol 78:3638–3648
Coulon F, McKew BA, Osborn AM et al (2007) Effects of temperature and biostimulation on oil-degrading microbial communities in temperate estuarine waters. Environ Microbiol 9:177–186
Daggers TD, Herman PMJ, van der Wal D (2020) Seasonal and spatial variability in patchiness of microphytobenthos on intertidal flats from Sentinel-2 satellite imagery. Front Mar Sci 7:392
Daims H, Lucker S, Wagner M (2016) A new perspective on microbes formerly known as nitrite-oxidizing bacteria. Trends Microbiol 24:699–712
Dal Bello M, Rindi L, Benedetti-Cecchi L (2017) Legacy effects and memory loss: how contingencies moderate the response of rocky intertidal biofilms to present and past extreme events. Glob Change Biol 23:3259–3268
Daleo P, Alberti J, Canepuccia A et al (2008) Mycorrhizal fungi determine salt-marsh plant zonation depending on nutrient supply. J Ecol 96:431–437
Dalsgaard T, Canfield DE, Petersen J et al (2003) $N_2$ production by the anammox reaction in the anoxic water column of Golfo Dulce, Costa Rica. Nature 422:606–608
Dang H, Li J, Chen R et al (2010) Diversity, abundance, and spatial distribution of sediment ammonia-oxidizing betaproteobacteria in response to environmental gradients and coastal eutrophication in Jiaozhou Bay, China. Appl Environ Microbiol 76:4691–4702
Dang H, Wang C, Li J et al (2009) Diversity and distribution of sediment *nirS*-encoding bacterial assemblages in response to environmental gradients in the eutrophied Jiaozhou Bay, China. Microb Ecol 58:161–169
De Jonge VN, de Boer WF, de Jong DJ et al (2012) Long-term mean annual microphytobenthos chlorophyll *a* variation correlates with air temperature. Mar Ecol Prog Ser 468:43–56
De Jonge VN, van Beusekom JEE (1995) Wind- and tide-induced resuspension of sediment microphytobenthos from the tidal flats in the ems estuary. Limnol Oceanogr 40:766–778
de la Huz R, Lastra M, Junoy J et al (2005) Biological impacts of oil pollution and cleaning in the intertidal zone of exposed sandy beaches: preliminary study of the "Prestige" oil spill. Estuar Coast Shelf Sci 65:19–29
Dean JF, Middelburg JJ, Röckmann T et al (2018) Methane feedbacks to the global climate system in a warmer world. Rev Geophys 56:207–250
DeLong EF (1992) Archaea in coastal marine environments. Proc Natl Acad Sci 89:5685–5689
Dong C, Bai X, Sheng H et al (2015) Distribution of PAHs and the PAH-degrading bacteria in the deep-sea sediments of the high-latitude Arctic Ocean. Biogeosciences 12:2163–2177

Dong LF, Thornton DCO, Nedwell DB et al (2000) Denitrification in the sediments of the river Colne estuary, England. Mar Ecol Prog Ser 203:109–122
Dong X, Greening C, Rattray JE et al (2019) Metabolic potential of uncultured bacteria and archaea associated with petroleum seepage in deep-sea sediments. Nat Commun 10:1816
Duarte CM, Agusti S, Barbier E et al (2020) Rebuilding marine life. Nature 580:39–51
Duran R, Cravo-Laureau C (2016) Role of environmental factors and microorganisms in determining the fate of polycyclic aromatic hydrocarbons in the marine environment. FEMS Microbiol Rev 40:814–830
Duran R, Cuny P, Bonin P et al (2015) Microbial ecology of hydrocarbon-polluted coastal sediments. Environ Sci Pollut Res 22:15195–15199
Dyksterhouse SE, Gray JP, Herwig RP et al (1995) Cycloclasticus pugetii gen. nov., sp. nov., an aromatic hydrocarbon-degrading bacterium from marine sediments. Int J Syst Bacteriol 45:116–123
Egger M, Riedinger N, Mogollo'n JM, et al. (2018) Global diffusive fluxes of methane in marine sediments. Nat Geosci 11:421–425
Erguder TH, Boon N, Wittebolle L et al (2009) Environmental factors shaping the ecological niches of ammonia-oxidizing archaea. FEMS Microbiol Rev 33:855–869
Evans PN, Boyd JA, Leu AO et al (2019) An evolving view of methane metabolism in the archaea. Nat Rev Microbiol 17:219–232
Evans PN, Parks DH, Chadwick GL et al (2015) Methane metabolism in the archaeal phylum Bathyarchaeota revealed by genome-centric metagenomics. Science 350:434–438
Evelin H, Kapoor R, Giri B (2009) Arbuscular mycorrhizal fungi in alleviation of salt stress: a review. Ann Bot 104:1263–1280
Feldewert C, Lang K, Brune A (2020) The hydrogen threshold of obligately methyl775 reducing methanogens. FEMS Microbiol Lett 17:fnaa137
Fernandez-Martinez J, Pujalte MJ, Garcia-Martinez J et al (2003) Description of *Alcanivorax venustensis* sp. nov. and reclassification of *Fundibacter jadensis* DSM 12178T (Bruns and Berthe-Corti 1999) as *Alcanivorax jadensis* comb. nov., members of the emended genus *Alcanivorax*. Int J Syst Evol Microbiol 53:331–338
Fitter AH (2005) Darkness visible: reflections on underground ecology. J Ecol 93:231–243
Forster RM, Creach V, Sabbe K et al (2006) Biodiversity-ecosystem function relationship in microphytobenthic diatoms of the Westerschelde estuary. Mar Ecol Prog Ser 311:192–201
Francioli D, van Rijssel SQ, van Ruijven J et al (2020) Plant functional group drives the community structure of saprophytic fungi in a grassland biodiversity experiment. Plant Soil. https://doi.org/10.1007/s11104-020-04454-y
Francis CA, Roberts KJ, Beman JM et al (2005) Ubiquity and diversity of ammonia-oxidizing archaea in water columns and sediments of the ocean. Proc Natl Acad Sci 102:14683–14688
Fuhrman JA, Davis AA (1997) Widespread archaea and novel bacteria from the deep sea as shown by 16S rRNA gene sequences. Mar Ecol Prog Ser 150:275–285
Fuhrman JA, McCallum K, Davis AA (1992) Novel major archaebacterial group from marine plankton. Nature 356:148–149
Galiforni-Silva F, Wijnberg KM, Hulscher SJMH (2020) On the relation between beach-dune dynamics and shoal attachment processes: a case study in Terschelling (NL). J Mar Sci Eng 8: 541
Gaubert-Boussarie J, Prado S, Hubas C (2020) An untargeted metabolomic approach for microphytobenthic biofilms in intertidal mudflats. Front Mar Sci 7:250
Gauthier MJ, Lafay B, Christen R et al (1992) Marinobacter hydrocarbonoclasticus gen. nov., sp. nov., a new, extremely halotolerant, hydrocarbon-degrading marine bacterium. Int J Syst Bacteriol 42:568–576
Gerdes B, Brinkmeyer R, Dieckmann G et al (2005) Influence of crude oil on changes of bacterial communities in Arctic Sea-ice. FEMS Microbiol Ecol 53:129–139
Gihring TM, Humphrys M, Mills HJ et al (2009) Identification of phytodetritus-degrading microbial communities in sublittoral Gulf of Mexico sands. Limnol Oceanogr 54:1073–1083

Gołębiewski M, Całkiewicz J, Creer S et al (2017) Tideless estuaries in brackish seas as possible freshwater-marine transition zones for bacteria: the case study of the Vistula river estuary. Environ Microbiol Rep 9:129–143

Gong J, Shi F, Ma B et al (2015) Depth shapes α- and β-diversities of microbial eukaryotes in surficial sediments of coastal ecosystems. Environ Microbiol 17:3722–3737

Gong W, Marchetti A (2019) Estimation of 18S gene copy number in marine eukaryotic plankton using a next-generation sequencing approach. Front Mar Sci 6:219

Goñi-Urriza M, Duran R (2018) Impact of petroleum contamination on microbial mats. In: McGenity T (ed) Microbial communities utilizing hydrocarbons and lipids: members, metagenomics and ecophysiology. Handbook of hydrocarbon and lipid microbiology. Springer, Cham, pp 19–36

Gostinčar C, Grube M, de Hoog GS et al (2010) Extremotolerance in fungi: evolution on the edge. FEMS Microbiol Ecol 71:2–11

Green BC, Smith DJ, Grey J et al (2012) High site fidelity and low site connectivity of temperate salt marsh fish populations: a stable isotope approach. Oecologia 168:245–255

Green M, Coco G (2014) Review of wave-driven sediment resuspension and transport in estuaries. Rev Geophys 52:77–117

Greer CW (2010) Bacterial diversity in hydrocarbon-polluted rivers, estuaries and sediments. In: Timmis KN, McGenity TJ, van der Meer JR, de Lorenzo V (eds) Handbook of hydrocarbon and lipid microbiology. Springer, Berlin, Heidelberg, pp 2329–2338

Gregson BH, Metodieva G, Metodiev MV et al (2018) Differential protein expression during growth on medium versus long-chain alkanes in the obligate marine hydrocarbon-degrading bacterium *Thalassolituus oleivorans* MIL-1. Front Microbiol 9:3130

Gregson BH, Metodieva G, Metodiev MV et al (2019) Differential protein expression during growth on linear versus branched alkanes in the obligate marine hydrocarbon-degrading bacterium *Alcanivorax borkumensis* SK2T. Environ Microbiol 21:2347–2359

Gregson BH, Metodieva G, Metodiev MV et al (2020) Protein expression in the obligate hydrocarbon-degrading psychrophile *Oleispira Antarctica* RB-8 during alkane degradation and cold tolerance. Environ Microbiol 22:1870–1883

Grossart HP, Levold F, Allgaier M et al (2005) Marine diatom species harbour distinct bacterial communities. Environ Microbiol 7:860–873

Gutierrez T, Green DH, Nichols PD et al (2013) Polycyclovorans algicola gen. Nov., sp. nov., an aromatic-hydrocarbon-degrading marine bacterium found associated with laboratory cultures of marine phytoplankton. Appl Environ Microbiol 79:205–214

Hallam SJ, Mincer TJ, Schleper C et al (2006) Pathways of carbon assimilation and ammonia oxidation suggested by environmental genomic analyses of marine Crenarchaeota. PLoS Biol 4: e95

Hamels I, Sabbe K, Muylaert K et al (1998) Organisation of microbenthic communities in intertidal estuarine flats, a case study from the Molenplaat (Westerschelde estuary, the Netherlands). Eur J Protistol 34:308–320

Hanlon ARM, Bellinger B, Haynes K et al (2006) Dynamics of extracellular polymeric substance (EPS) production and loss in an estuarine, diatom-dominated, microalgal biofilm over a tidal emersion-immersion period. Limnol Oceanogr 51:79–93

Hanson RB, Gundersen K (1977) Relationship between nitrogen fixation (acetylene reduction) and the C:N ratio in a polluted coral reef system, Kaneohe Bay, Hawaii. Est Coast Mar Sci 5:437–444

Happey-Wood CM, Jones P (1988) Rhythms of vertical migration and motility in intertidal diatoms with particular reference to *Pleurosigma angulatum.* Diat Res 3:83–93

Harayama S, Kishira H, Kasai Y et al (1999) Petroleum biodegradation in marine environments. J Mol Microbiol Biotechnol 1:63–70

Harms H, Smith KEC, Wick LY (2010) Microorganism–hydrophobic compound interactions. In: Timmis KN, McGenity TJ, van der Meer JR, de Lorenzo V (eds) Handbook of hydrocarbon and lipid microbiology. Springer, Berlin, Heidelberg, pp 1479–1490

Haro S, Bohórquez J, Lara M et al (2019) Diel patterns of microphytobenthic primary production in intertidal sediments: circadian photosynthetic rhythm and migration. Sci Rep 9:13376

Haro-Moreno JM, Rodriguez-Valera F, López-García P et al (2017) New insights into marine group III Euryarchaeota, from dark to light. ISME J 11:1102–1117

Head IM, Hiorns WD, Embley TM et al (1993) The phylogeny of autotrophic ammonia oxidizing bacteria as determined by analysis of 16S ribosomal RNA gene-sequences. J Gen Microbiol 139:1147–1153

Head IM, Jones DM, Röling WFM (2006) Marine microorganisms make a meal of oil. Nat Rev Microbiol 4:173–182

Heip CHR, Goosen NK, Herman PMJ et al (1995) Production and consumption of biological particles in temperate tidal estuaries. Oceanogr Mar Biol Ann Rev 33:1–149

Henderson CJ, Gilby BL, Schlacher TA et al (2020) Low redundancy and complementarity shape ecosystem functioning in a low-diversity ecosystem. J Anim Ecol 89:784–794

Herbert RA (1999) Nitrogen cycling in coastal marine ecosystems. FEMS Microbiol Rev 5:563–590

Herman PMJ, Middelburg JJ, Widdows J et al (2000) Stable isotopes as trophic tracers: combining field sampling and manipulative labelling of food resources for macrobenthos. Mar Ecol Prog Ser 204:79–92

Hill-Spanik KM, Smith AS, Plante CJ (2019) Recovery of benthic microalgal biomass and community structure following beach renourishment at Folly Beach, South Carolina. Estuaries Coast 42:157–172

Hoehler TM, Alperin MJ, Albert DB et al (1994) Field and laboratory studies of 534 methane oxidation in an anoxic marine sediment: evidence for a methanogen-sulfate 535 reducer consortium. Glob Biogeochem Cycles 8:451–463

Hoeksema JD, Chaudhary VB, Gehring CA et al (2010) A meta-analysis of context-dependency in plant response to inoculation with mycorrhizal fungi. Ecol Lett 13:394–407

Hope JA, Paterson DM, Thrush SF (2020) The role of microphytobenthos in soft-sediment ecological networks and their contribution to the delivery of multiple ecosystem services. J Ecol 108:815–830

Horel A, Bernard RJ, Mortazavi B (2014) Impact of crude oil exposure on nitrogen cycling in a previously impacted *Juncus roemerianus* salt marsh in the northern Gulf of Mexico. Environ Sci Pollut Res Int 21:6982–6993

Hörnlein C, Confurius-Guns V, Stal LJ et al (2018) Daily rhythmicity in coastal microbial mats. NPJ Biofilms Microbiomes 4:11

Hu H, He J (2017) Comammox—a newly discovered nitrification process in the terrestrial nitrogen cycle. J Soils Sediments 17:2709–2717

Imachi H, Nobu MK, Nakahara N et al (2020) Isolation of an archaeon at the prokaryote-eukaryote interface. Nature 577:519–525

ITOPF (2019) Oil tanker spill statistics 2019. Available http://www.itopf.org/

Jameson E, Stephenson J, Jones H et al (2018) Deltaproteobacteria (*Pelobacter*) and Methanococcoides are responsible for choline-dependent methanogenesis in a coastal saltmarsh sediment. ISME J 13:277–289

Jayakumar A, O'Mullan GD, Naqvi SWA et al (2009) Denitrifying bacterial community composition changes associated with stages of denitrification in oxygen minimum zones. Microb Ecol 58:350–362

Jeffries TC, Curlevski NJ, Brown MV et al (2016) Marine fungal biogeography. Environ Microbiol Rep 8:235–238

Jensen MM, Kuypers MMM, Lavik G et al (2008) Rates and regulation of anaerobic ammonium oxidation and denitrification in the Black Sea. Limnol Oceanogr 53:23–36

Jesus B, Brotas V, Ribeiro L et al (2009) Adaptations of microphytobenthos assemblages to sediment type and tidal position. Cont Shelf Res 29:1624–1634

Jetten MSM, Op den Camp HJM, Kuenen JG et al (2010) Description of the order Brocadiales. In: Krieg NR, Ludwig W, Whitman WB, Hedlund BP, Paster BJ, Staley JT, Ward N, Brown D,

Parte A (eds) Bergey's manual of systematic bacteriology, vol 4. Springer, Heidelberg, pp 596–603
Jiang H, Dong H, Yu B et al (2009) Diversity and abundance of ammonia-oxidizing archaea and bacteria in Qinghai Lake, northwestern China. Geomicrobiol J 26:199–211
Jones C, Hallin S (2010) Ecological and evolutionary factors underlying global and local assembly of denitrifier communities. ISME J 4:633–641
Jönnson B, Sundbäck K, Nilsson C (1994) An upright life-form of an epipelic motile diatoms: on the behaviour of *Gyrosigma balticum*. Eur J Phycol 29:11–15
Jørgensen KS (1989) Annual pattern of denitrification and nitrate ammonification in an estuarine sediment. Appl Environ Microbiol 55:1841–1847
Jørgensen KS, Sorensen J (1988) Two annual maxima of nitrate reduction and denitrification in estuarine sediment (Norsminde Fjord, Denmark). Mar Ecol Prog Ser 94:267–274
Juneau P, Barnett A, Méléder V et al (2015) Combined effect of high light and high salinity on the regulation of photosynthesis in three diatom species belonging to the main growth forms of intertidal flat inhabiting microphytobenthos. J Exp Mar Biol Ecol 463:95–104
Kamp A, Beer D, Nitsch JL et al (2011) Diatoms respire nitrate to survive dark and anoxic conditions. Proc Nat Acad Sci 108:5649–5654
Kaplan WA, Teal JM, Valiela I (1977) Denitrification in saltmarsh sediments: evidence for seasonal temperature selection among populations of denitrifiers. Microb Ecol 3:193–224
Kartal B, Kuypers MM, Lavik G et al (2007) Anammox bacteria disguised as denitrifiers: nitrate reduction to dinitrogen gas via nitrite and ammonium. Environ Microbiol 9:635–642
Kasai Y, Kishira H, Harayama S (2002) Bacteria belonging to the genus *Cycloclasticus* play a primary role in the degradation of aromatic hydrocarbons released in a marine environment. Appl Environ Microbiol 68:5625–5633
Kasai Y, Shindo K, Harayama S et al (2003) Molecular characterization and substrate preference of a polycyclic aromatic hydrocarbon dioxygenase from *Cycloclasticus* sp. strain A5. Appl Environ Microbiol 69:6688–6697
Kim BS, Oh HM, Kang H et al (2005) Archaeal diversity in tidal flat sediment as revealed by 16S rDNA analysis. J Microbiol 43:144–151
King D, Nedwell DB (1987) The adaptation of nitrate reducing bacterial communities in estuarine sediments in response to overlying nitrate load. FEMS Microb Ecol 45:15–21
Kingston MB (1999) Effect of light on vertical migration and photosynthesis of *Euglena proxima* (Euglenophyta). J Phycol 35:245–253
Knittel K, Wegener G, Boetius A (2018) Anaerobic methane oxidizers. In: McGenity TJ (ed) Microbial communities utilizing hydrocarbons and lipids: members, metagenomics and ecophysiology. Springer International Publishing, Cham, pp 113–132
Koch H, van Kessel MAHJ, Lücker S (2019) Complete nitrification: insights into the ecophysiology of comammox *Nitrospira*. Appl Microbiol Biotechnol 103:177–189
Koedooder C, Stock W, Willems A et al (2019) Diatom-bacteria interactions modulate the composition and productivity of benthic diatom biofilms. Front Microbiol 10:1255
König S, Eisenhut M, Bräutigam A et al (2017) The influence of a cryptochrome on the gene expression profile in the diatom *Phaeodactylum tricornutum* under blue light and in darkness. Plant Cell Physiol 58:1914–1923
Koop-Jakobsen K, Giblin AE (2010) The effect of increased nitrate loading on nitrate reduction via denitrification and DNRA in salt marsh sediments. Limnol Oceanog 55:789–802
Kostka JE, Prakash O, Overholt WA et al (2011) Hydrocarbon-degrading bacteria and the bacterial community response in Gulf of Mexico beach sands impacted by the Deepwater horizon oil spill. Appl Environ Microbiol 77:7962–7974
Kube M, Chernikova T, Al-Ramahi Y et al (2013) Genome sequence and functional genomic analysis of the oil-degrading bacterium *Oleispira Antarctica*. Nat Commun 4:2156
Kuehn KA, Steiner D, Gessner MO (2004) Dielmineralization patterns of standing-dead plant litter: implications for $CO_2$ flux from wetlands. Ecology 85:2504–2518

Kuypers MM, Sliekers AO, Lavik G et al (2003) Anaerobic ammonium oxidation by anammox bacteria in the Black Sea. Nature 422:608–611

Kuypers MMM, Marchant HK, Kartal B (2018) The microbial nitrogen cycling network. Nat Rev Microbiol 16:263–276

Kwon KK, Oh JH, Yang S-H et al (2015) Alcanivorax gelatiniphagus sp. nov., a marine bacterium isolated from tidal flat sediments enriched with crude oil. Int J Syst Evol Microbiol 65:2204–2208

Lacerda ALDF, Proietti MC, Secchi ER et al (2020) Diverse groups of fungi are associated with plastics in the surface waters of the Western South Atlantic and the Antarctic peninsula. Mol Ecol 29:1903–1918

Lai Q, Wang L, Liu Y et al (2011) Alcanivorax pacificus sp. nov., isolated from a deep-sea pyrene-degrading consortium. Int J Syst Evol Microbiol 61:1370–1374

Lam P, Jensen MM, Lavik G et al (2007) Linking crenarchaeal and bacterial nitrification to anammox in the Black Sea. Proc Nat Acad Sci 104:7104–7109

Lam P, Lavik G, Jensen MM et al (2009) Revising the nitrogen cycle in the Peruvian oxygen minimum zone. Proc Natl Acad Sci U S A 106:4752–4757

Lassen C, Ploug H, Jørgensen BB (1992) Microalgal photosynthesis and spectral scalar irradience in coastal marine sediments of Limfjorden, Denmark. Limnol Oceanogr 37:760–772

Laverman AM, Cappellen PV, Van Rotterdam-Los D et al (2006) Potential rates and pathways of microbial nitrate reduction in coastal sediments. FEMS Microbiol Ecol 58:179–192

Laverock B, Gilbert JA, Tait K et al (2011) Bioturbation: impact on the marine nitrogen cycle. Biochem Soc Trans 39:315–320

Laviale M, Barnett A, Ezequiel J et al (2015) Response of intertidal benthic microalgal biofilms to a coupled light–temperature stress: evidence for latitudinal adaptation along the Atlantic coast of southern Europe. Environ Microbiol 17:3662–3677

Lazar CS, Baker BJ, Seitz K et al (2016) Genomic evidence for distinct carbon substrate preferences and ecological niches of Bathyarchaeota in estuarine sediments. Environ Microbiol 18:1200–1211

Lazar CS, Biddle JF, Meador TB et al (2014) Environmental controls on intragroup diversity of the uncultured benthic archaea of the miscellaneous Crenarchaeotal group lineage naturally enriched in anoxic sediments of the white Oak River estuary (North Carolina, USA). Environ Microbiol 17:2228–2238

Leahy JG, Colwell RR (1990) Microbial degradation of hydrocarbons in the environment. Microbiol Rev 54:305–315

Legge O, Johnson M, Hicks N et al (2020) Carbon on the northwest European shelf: contemporary budget and future influences. Front Mar Sci 7:143

Levin LA, Boesch DF, Covich A et al (2001) The function of marine critical transition zones and the importance of sediment biodiversity. Ecosystems 4:430–451

Li J, Nedwell DB, Beddow J et al (2015a) amoA gene abundances and nitrification potential rates suggest that benthic ammonia-oxidizing bacteria (AOB) not archaea (AOA) dominate N cycling in the Colne estuary, UK. Appl Environ Microbio 81:159–165

Li M, Baker BJ, Anantharaman K et al (2015b) Genomic and transcriptomic evidence for scavenging of diverse organic compounds by widespread deep-sea archaea. Nat Commun 6:8933

Li M, Cao H, Hong Y et al (2011b) Spatial distribution and abundances of ammonia-oxidizing archaea (AOA) and ammonia-oxidizing bacteria (AOB) in mangrove sediments. Appl Microbiol Biotechnol 89:1243–1254

Li M, Gu J (2013) Community structure and transcript responses of anammox bacteria, AOA, and AOB in mangrove sediment microcosms amended with ammonium and nitrite. Appl Microbiol Biotechnol 97:9859–9874

Li M, Hong Y-G, Cao H-L et al (2011a) Mangrove trees affect the community structure and distribution of anammox bacteria at an anthropogenic-polluted mangrove in the Pearl River Delta reflected by 16S rRNA and hydrazine oxidoreductase (HZO) encoding gene analyses. Ecotoxicology 20:1780–1790

Li Q, Wang F, Chen Z et al (2012) Stratified active archaeal communities in the sediments of Jiulong River estuary China. Front Microbiol 3:311–314

Lipsewers YA, Bale NJ, Hopmans EC et al (2014) Seasonality and depth distribution of the abundance and activity of ammonia oxidizing microorganisms in marine coastal sediments (North Sea). Front Microbiol 5:472–483

Lipsewers YA, Vasquez Cardenas D, Seitaj D et al (2017) Impact of seasonal hypoxia on activity and community structure of chemolithoautotrophic bacteria in a coastal sediment. Appl Environ Microbiol 83:e03517–e03516

Liu C, Shao Z (2005) Alcanivorax dieselolei sp. nov., a novel alkane degrading bacterium isolated from sea water and deep-sea sediment. Int J Syst Evol Microbiol 55:1181–1186

Liu J, Zheng Y, Lin H et al (2019) Proliferation of hydrocarbon-degrading microbes at the bottom of the Mariana trench. Microbiome 7:47

Liu J, Zhu S, Liu X et al (2020) Spatiotemporal dynamics of the archaeal community in coastal sediments: assembly process and co-occurrence relationship. ISME J 14:1463–1478

Liu X, Li M, Castelle CJ et al (2018b) Insights into the ecology, evolution, and metabolism of the widespread Woesearchaeotal lineages. Microbiome 6:102

Liu X, Pan J, Liu Y (2018a) Diversity and distribution of archaea in global estuarine ecosystems. Sci Total Environ 637-638:349–358

Liu XD, Tiquia SM, Holguin G (2003) Molecular diversity of denitrifying genes in continental margin sediments within the oxygen deficient zone off the Pacific coast of Mexico. Appl Environ Microbiol 69:3549–3560

Lloyd K, Schreiber GL, Petersen DG et al (2013) Predominant archaea in marine sediments degrade detrital proteins. Nature 496:215–220

Long RA, Eveillard D, Franco SLM et al (2013) Antagonistic interactions between heterotrophic bacteria as a potential regulator of community structure of hypersaline microbial mats. FEMS Microbiol Ecol 83:74–81

Lozada M, Marcos MS, Commendatore MG et al (2014) The bacterial community structure of hydrocarbon-polluted marine environments as the basis for the definition of an ecological index of hydrocarbon exposure. Microbes Environ 29:269–276

Luna GM, Corinaldesi C, Rastelli E et al (2013) Patterns and drivers of bacterial α- and β-diversity across vertical profiles from surface to subsurface sediments. Environ Microbiol Rep 5:731–739

Lyons JI, Alber M, Hollibaugh JT (2010) Ascomycete fungal communities associated with early decaying leaves of Spartina spp. from Central California estuaries. Oecologia 162:435–442

Lyu Z, Liu Y (2018) Diversity and taxonomy of methanogens. In: AJM S, Sousa DZ (eds) Biogenesis of hydrocarbons. Handbook of hydrocarbon and lipid microbiology, 2nd edn. Springer, Cham, pp 19–77

Maggi E, Rindi L, Dal Bello M et al (2017) Spatio-temporal variability in Mediterranean rocky shore microphytobenthos. Mar Ecol Prog Ser 575:17–29

Mann M, Serif M, Wrobel T et al (2020) The aureochrome photoreceptor PtAUREO1a is a highly effective blue light switch in diatoms. iScience. https://doi.org/10.1016/j.isci.2020.101730

Marchant R, Banat IM (2012) Microbial biosurfactants: challenges and opportunities for future exploitation. Trends Biotechnol 30:558–565

Martens-Habbena W, Berube PM, Urakawa H et al (2009) Ammonia oxidation kinetics determine niche separation of nitrifying archaea and bacteria. Nature 461:976–979

Martinez-Perez C, Mohr W, Löscher CR et al (2016) The small unicellular diazotrophic symbiont, UCYN-A, is a key player in the marine nitrogen cycle. Nat Microbiol 1:16163

Mason OU, Han J, Woyke T et al (2014) Single-cell genomics reveals features of a *Colwellia* species that was dominant during the Deepwater horizon oil spill. Front Microbiol 5:332

Massana R, Gobet A, Audic S et al (2015) Protist diversity in European coastal areas. Environ Microbiol 17:4035–4049

McGenity TJ (2014) Hydrocarbon biodegradation in intertidal wetland sediments. Curr Opin Biotechnol 27:46–54

McGenity TJ, Folwell BD, McKew BA et al (2012) Marine crude-oil biodegradation: a central role for interspecies interactions. Aquat Biosyst 8:1–19

McGenity TJ, Oren A (2012) Hypersaline environments. In: Bell EM (ed) Life at extremes: environments, organisms and strategies for survival. CAB International, UK, pp 402–437

McGenity TJ, Sorokin D (2018) Methanogens and methanogenesis in hypersaline environments. In: AJM S, Sousa DZ (eds) Biogenesis of hydrocarbons. Handbook of hydrocarbon and lipid microbiology, 2nd edn. Springer, Cham, pp 283–309

McGlynn SE (2017) Energy metabolism during anaerobic methane oxidation in ANME archaea. Microbes Environ 32:5–13

McKew BA, Coulon F, Osborn AM et al (2007) Determining the identity and roles of oil-metabolizing marine bacteria from the Thames estuary, UK. Environ Microbiol 9:165–176

McKew BA, Dumbrell A, Taylor JD et al (2013) Differences between aerobic and anaerobic degradation of microphytobenthic biofilm-derived organic matter within intertidal sediments. FEMS Microbiol Ecol 84:495–509

McKew BA, Taylor JD, McGenity TJ et al (2011) Resistance and resilience of benthic biofilm communities from a temperate saltmarsh to desiccation and rewetting. ISME J 5:30–41

McLachlan DH, Brownlee C, Taylor AR et al (2009) Light inducted motile responses of the estuarine benthic diatoms *Navicula perminuta* and *Cylindrotheca closterium* (Bacillariophyceae). J Phycol 45:592–599

McLachlan DH, Underwood GJC, Taylor AR et al (2012) Calcium release from intracellular stores is necessary for the photophobic motility response in the benthic diatom *Navicula perminuta*. J Phycol 48:675–681

McTavish H, Fuchs JA, Hooper AB (1993) Sequence of the gene coding for ammonia monooxygenase in *Nitrosomonas europaea*. J Bacteriol 175:2436–2444

Meckenstock RU, Mouttaki H (2011) Anaerobic degradation of non-substituted aromatic hydrocarbons. Curr Opin Biotechnol 22:406–414

Megharaj M, Singleton I, McClure NC et al (2000) Influence of petroleum hydrocarbon contamination on microalgae and microbial activities in a long-term contaminated soil. Arch Environ Contam Toxicol 38:439–445

Méléder V, Savelli R, Barnett A et al (2020) Mapping the intertidal microphytobenthos gross primary production part I: coupling multispectral remote sensing and physical modeling. Front Mar Sci 7:520

Meng J, Xu J, Qin D et al (2014) Genetic and functional properties of uncultivated MCG archaea assessed by metagenome and gene expression analyses. ISME J 8:650–659

Messina E, Denaro R, Crisafi F et al (2016) Genome sequence of obligate marine polycyclic aromatic hydrocarbons-degrading bacterium *Cycloclasticus* sp. 78-ME, isolated from petroleum deposits of the sunken tanker Amoco Milford Haven, Mediterranean Sea. Mar Genomics 25:11–13

Middelburg JJ, Barranguet C, Boschker HTS et al (2000) The fate of intertidal microphytobenthos carbon: An *in situ* $^{13}$C-labelling study. Limnol Oceanogr 45:1224–1334

Miyatake T, Moerdijk-Poortvliet TCW, Stal LJ et al (2014) Tracing carbon flow from microphytobenthos to major bacterial groups in an intertidal marine sediment by using an *in situ* $^{13}$C pulse-chase method. Limnol Oceanogr 59:1275–1287

Moerdijk-Poortvliet TCW, Beauchard O, Stal LJ et al (2018b) Production and consumption of extracellular polymeric substances in an intertidal diatom mat. Mar Ecol Prog Ser 592:77–95

Moerdijk-Poortvliet TCW, van Breugel P, Sabbe K et al (2018a) Seasonal changes in the biochemical fate of carbon fixed by benthic diatoms in intertidal sediments. Limnol Oceanogr 63:550–569

Mohamed D, Martiny J (2011) Patterns of fungal diversity and composition along a salinity gradient. ISME J 5:379–388

Mommer L, Cotton TEA, Raaijmakers JM et al (2018) Lost in diversity: the interactions between soil-borne fungi, biodiversity and plant productivity. New Phytol 218:542–553

Monson DH, Doak DF, Ballachey BE et al (2000) Long-term impacts of the Exxon Valdez oil spill on sea otters, assessed through age-dependent mortality patterns. Proc Natl Acad Sci 97:6562–6567

Moreira-Coello V, Mouriño-Carballido B, Marañón E et al (2019) Temporal variability of diazotroph community composition in the upwelling region off NW Iberia. Sci Rep 9:3737

Moriarty DJW, O'Donohue MJ (1993) Nitrogen fixation in seagrass communities during summer in the Gulf of Carpentaria, Australia. Aust J Mar Freshw Res 44:117–125

Mouget J-L, Perkins R, Consalvey M et al (2008) Migration or photoacclimation to prevent high irradiance and UV-B damage in marine microphytobenthic communities. Aq Microb Ecol 52: 223–232

Mueller P, Granse D, Nolte S et al (2017) Top-down control of carbon sequestration: grazing affects microbial structure and function in salt marsh soils. Ecol Appl 27:1435–1450

Munson MA, Nedwell DB, Embley TM (1997) Phylogenetic diversity of archaea in sediment samples from a coastal salt marsh. Appl Environ Microbiol 63:4729–4733

Myllykangas JP, Rissanen AJ, Hietanen S et al (2020) Influence of electron acceptor availability and microbial community structure on sedimentary methane oxidation in a boreal estuary. Biogeochemistry 148:291–309

National Research Council (NRC) (2003) Oil in the sea III: inputs, fates, and effects. National Academy Press, Washington, DC

Nedwell D, Aziz S (1980) Heterotrophic nitrogen fixation in an intertidal salt marsh sediment. Est Coast Mar Sci 10:699–702

Nedwell DB, Embley TM, Purdy KJ (2004) Sulphate reduction, methanogenesis and phylogenetics of the sulphate reducing bacterial communities along an estuarine gradient. Aq Microb Ecol 37: 209–217

Nedwell DB, Underwood GJC, McGenity TJ et al (2016) The Colne estuary: a long-term microbial ecology observatory. Adv Ecol Res 55:227–281

Newell SY (1996) Established and potential impacts of eukaryotic mycelial decomposers in marine/ terrestrial ecotones. J Exp Mar Biol Ecol 200:187–206

Newton R, Huse SM, Morrison HG et al (2013) Shifts in the microbial community composition of Gulf Coast beaches following beach oiling. PLoS One 8:e74265

Nogales B, Bosch R (2019) Microbial communities in hydrocarbon-polluted harbors and marinas. In: McGenity T (ed) Microbial communities utilizing hydrocarbons and lipids: members, metagenomics and ecophysiology. Handbook of hydrocarbon and lipid microbiology. Springer, Cham, pp 63–80

Nogales B, Timmis KN, Nedwell DB et al (2002) Detection and diversity of expressed denitrification genes in estuarine sediments after reverse transcription-PCR amplification from mRNA. Appl Environ Microbiol 68:5017–5025

Notman GM, McGill RAR, Hawkins SJ et al (2016) Macroalgae contribute to the diet of *Patella vulgata* from contrasting conditions of latitude and wave exposure in the UK. Mar Ecol Prog Ser 549:113–123

Oakes JM, Eyre BD, Middelburg JJ (2012) Transformation and fate of microphytobenthos carbon in subtropical shallow subtidal sands: a $^{13}$C-labeling study. Limnol Oceanogr 57:1846–1856

Oakes JM, Eyre BD, Middelburg JJ et al (2010) Composition, production, and loss of carbohydrates in subtropical shallow subtidal sandy sediments: rapid processing and long-term retention revealed by $^{13}$C-labeling. Limnol Oceanogr 55:2126–2138

Oppenheim DR (1991) Seasonal changes in epipelic diatoms along an intertidal shore, Berrow flats, Somerset. J Mar Biol Assoc UK 71:579–596

Oremland R, Marsh L, Polcin S (1982) Methane production and simultaneous sulphate reduction in anoxic, salt marsh sediments. Nature 296:143–145

Orsi WD, Vuillemin A, Rodriguez P et al (2019) Metabolic activity analyses demonstrate that Lokiarchaeon exhibits homoacetogenesis in sulfidic marine sediments. Nat Microbiol 5:248–255

Oshiki M, Shimokawa M, Fujii N et al (2011) Physiological characteristics of the anaerobic ammonium-oxidizing bacterium '*Candidatus Brocadia sinica*'. Microbiology 157:1706–1713
Osterholz H, Kirchman DL, Niggemann J et al (2018) Diversity of bacterial communities and dissolved organic matter in a temperate estuary. FEMS Microbiol Ecol 94:fiy119
Overy DP, Rämä T, Oosterhuis R et al (2019) The neglected marine fungi, sensu stricto, and their isolation for natural products discovery. Mar Drugs 17:42
Oxborough K, Hanlon ARM, Underwood GJC et al (2000) In vivo estimation of the photosystem II photochemical efficiency of individual microphytobenthic cells using high-resolution imaging of chlorophyll *a* fluorescence. Limnol Oceanogr 45:1420–1425
Pace NR (1997) A molecular view of microbial diversity and the biosphere. Science 276:734–740
Païssé S, Coulon F, Goñi-Urriza M et al (2008) Structure of bacterial communities along a hydrocarbon contamination gradient in a coastal sediment. FEMS Microbiol Ecol 66:295–305
Pajares S, Ramos R (2019) Processes and microorganisms involved in the marine nitrogen cycle: knowledge and gaps. Front Mar Sci 6:739
Palomo A, Pedersen AG, Fowler SJ et al (2018) Comparative genomics sheds light on niche differentiation and the evolutionary history of comammox *Nitrospira*. ISME J 12:1779–1793
Papaspyrou S, Smith CJ, Dong LF et al (2014) Nitrate reduction functional genes and nitrate reduction potentials persist in deeper estuarine sediments why? PLoS One 9:e94111
Park J, Kwon B-O, Kim M et al (2014) Microphytobenthos of Korean tidal flats: a review and analysis on floral distribution and tidal dynamics. Ocean Coast Manag 102:471–482
Passarelli C, Cui X, Valsami-Jones E et al (2020) Environmental context determines the impact of titanium oxide and silver nanoparticles on the functioning of intertidal microalgal biofilms. Environ Sci Nano 7:3020–3035
Peletier H (1996) Long-term changes in intertidal estuarine diatom assemblages related to reduced input of organic waste. Mar Ecol Prog Ser 137:265–271
Penton CR, Devol AH, Tiedje JM (2006) Molecular evidence for the broad distribution of anaerobic ammonium-oxidizing bacteria in freshwater and marine sediments. Appl Environ Microbiol 72: 6829–6832
Pérez-Burillo J, Trobajo R, Vasselon V et al (2020) Evaluation and sensitivity analysis of diatom DNA metabarcoding for WFD bioassessment of Mediterranean rivers. Sci Total Environ 727: 138445
Perkins EJ (1960) The diurnal rhythm of the littoral diatoms of the river Eden estuary, fife. J Ecol 48:725–728
Perkins RG, Lavaud J, Serôdio J et al (2010) Vertical cell movement is a primary response of intertidal benthic biofilms to increasing light dose. Mar Ecol Prog Ser 416:93–103
Perkins RG, Oxborough K, Hanlon ARM et al (2002) Can chlorophyll fluorescence be used to estimate the rate of photosynthetic electron transport within microphytobenthic biofilms? Mar Ecol Prog Ser 228:47–56
Perkins RG, Underwood GJC, Brotas V et al (2001) Responses of microphytobenthos to light: primary production and carbohydrate allocation over an emersion period. Mar Ecol Prog Ser 223:101–112
Pester M, Rattei T, Flechl S et al (2011) amoA-based consensus phylogeny of ammonia-oxidizing archaea and deep sequencing of amoA genes from soils of four different geographic regions. Environ Microbiol 14:525–539
Phillips CJ, Smith Z, Embley TM et al (1999) Phylogenetic differences between particle-associated and planktonic ammonia-oxidizing bacteria of the beta-subdivision of the class Proteobacteria in the northwestern Mediterranean Sea. Appl Environ Microbiol 65:779–786
Piatt JF, Ford RG (1996) How many seabirds were killed by the Exxon Valdez oil spill? Am Fish Soc Symp 18:712–719
Piña-Ochoa E, Høgslund S, Geslin E et al (2010) Widespread occurrence of nitrate storage and denitrification among foraminifera and Gromiida. Proc Natl Acad Sci 107:1148–1153
Pinckney JL (2018) A mini-review of the contribution of benthic microalgae to the ecology of the continental shelf in the South Atlantic bight. Estuaries Coast 41:2070–2078

Pitcher A, Wuchter C, Siedenberg K et al (2011) Crenarchaeol tracks winter blooms of ammonia-oxidizing Thaumarchaeota in the coastal North Sea. Limnol Oceanogr 56:2308–2318
Pjevac P, Schauberger C, Poghosyan L et al (2017) AmoA-targeted polymerase chain reaction primers for the specific detection and quantification of comammox Nitrospira in the environment. Front Microbiol 8:11
Plante C, Fleer V, Jones ML (2016) Neutral processes and species sorting in benthic microalgal community assembly: effects of tidal resuspension. J Phycol 52:827–839
Plante CJ, Hill-Spanik K, Cook M et al (2021) Environmental and spatial influences on biogeography and community structure of saltmarsh benthic. Estuaries Coast 44:147–161
Poole NJ, Price PC (1972) Fungi colonizing wood submerged in the Medway estuary. T Brit Mycoll Soc 59:2
Porubsky WP, Weston NB, Joye SB (2009) Benthic metabolism and the fate of dissolved inorganic nitrogen in intertidal sediments. Estuar Coast Shelf Sci 83:392–402
Preisler A, de Beer D, Lichtschlag A et al (2007) Biological and chemical sulfide oxidation in a *Beggiatoa* inhabited marine sediment. ISME J 1:341–353
Prince RC, Amande TJ, McGenity TJ (2018) Prokaryotic hydrocarbon degraders. In: McGenity TJ (ed) Taxonomy, genomics and ecophysiology of hydrocarbon-degrading microbes. Springer Nature, Cham, pp 1–41
Prins A, Deleris P, Hubas C, Jesus B (2020) Effect of light intensity and light quality on diatom behavioral and physiological photoprotection. Front Mar Sci 7:203
Prokopenko MG, Sigman DM, Berelson WM et al (2011) Denitrification in anoxic sediments supported by biological nitrate transport. Geochim Cosmochim Acta 75:7180–7199
Purdy KJ, Cresswell-Maynard TD, Nedwell DB et al (2004) Isolation of haloarchaea that grow at low salinities. Environ Microbiol 6:591–595
Qian G, Wang J, Kan J et al (2018) Diversity and distribution of anammox bacteria in water column and sediments of the eastern Indian Ocean. Int Biodeterior Biodegr 133:52–62
Quan ZX, Rhee SK, Zuo JE et al (2008) Diversity of ammonium-oxidizing bacteria in a granular sludge anaerobic ammonium-oxidizing (anammox) reactor. Environ Microbiol 10:3130–3139
Rabus R, Boll M, Heider J et al (2016) Anaerobic microbial degradation of hydrocarbons: from enzymatic reactions to the environment. J Mol Microbiol Biotechnol 26:5–28
Radajewski S, Webster G, Reay DS et al (2002) Identification of active methylotroph populations in an acidic forest soil by stable isotope probing. Microbiology 148:2331–2342
Radwan SS, Khanafer MM, Al-Awadhi HA (2019) Ability of the so-called obligate hydrocarbonoclastic bacteria to utilize nonhydrocarbon substrates thus enhancing their activities despite their misleading name. BMC Microbiol 19:41
Raymond J, Siefert JL, Staples CR et al (2004) The natural history of nitrogen fixation. Mol Biol Evol 21:541–554
Reddy CM, Eglinton TI, Hounshell A et al (2002) The West Falmouth oil spill after thirty years: the persistence of petroleum hydrocarbons in marsh sediments. Environ Sci Technol 36:4754–4760
Redzuan NS, Underwood GJC (2021) The importance of weather and tides on the resuspension and deposition of microphytobenthos (MPB) on intertidal mudflats. Estuar Coastal Shelf Sci 251:107190
Redzuan NS, Underwood GJC (2020) Movement of microphytobenthos and sediment between mudflats and salt marsh during spring tides. Front Mar Sci 7:496
Ribeiro L, Brotas V, Hernández-Fariñas T et al (2020) Assessing alternative microscopy-based approaches to species abundance description of intertidal diatom communities. Front Mar Sci 7:36
Ribeiro L, Brotas V, Mascarell G et al (2003) Taxonomic survey of the microphytobenthic communities of two Tagus estuary mudflats. Acta Oecol 24:S117–S123
Ribeiro L, Brotas V, Rincé Y et al (2013) Structure and diversity of intertidal benthic diatom assemblages in contrasting shores: a case study from the Tagus estuary. J Phycol 49:258–270
Risgaard-Petersen N (2003) Coupled nitrification-denitrification in autotrophic and heterotrophic estuarine sediments: on the influence of benthic microalgae. Limnol Oceanogr 48:93–105

Risgaard-Petersen N, Langezaal A, Ingvardsen S et al (2006) Evidence for complete denitrification in a benthic foraminifer. Nature 443:93–96

Rishworth GM, Dodd C, Perissinotto R et al (2020) Modern supratidal microbialites fed by groundwater: functional drivers, value and trajectories. Earth Sci Rev 210:103364

Rivera-Monroy VH, Lenaker P, Twilley RR et al (2010) Denitrification in coastal Louisiana: a spatial assessment and research needs. J Sea Res 63:157–172

Rodriguez-R LM, Overholt WA, Hagan C et al (2015) Microbial community successional patterns in beach sands impacted by the Deepwater horizon oil spill. ISME J 9:1928–1940

Rojas-Jimenez K, Rieck A, Wurzbacher C et al (2019) A salinity threshold separating fungal communities in the Baltic Sea. Front Microbiol 10:680

Round FE (1979) Occurrence and rhythmic behaviour of *Tropidoneis lepidoptera* in the epipelon of Barnstable Harbor, Massachusetts, USA. Mar Biol 54:215–217

Round FE, Palmer JD (1966) Persistent, vertical-migration rhythms in benthic microflora. II. Field and laboratory studies on diatoms from the banks of the river Avon. J Mar Biol Assoc UK 46: 191–214

Rovira L, Trobajoa R, Satob S et al (2015) Genetic and physiological diversity in the diatom *Nitzschia inconspicua*. J Eukaryot Microbiol 62:815–832

Rubin-Blum M, Antony CP, Borowski C et al (2017) Short-chain alkanes fuel mussel and sponge *Cycloclasticus* symbionts from deep-sea gas and oil seeps. Nat Microbiol 2:17093

Sabbe K (1993) Short-term fluctuations in benthic diatom numbers on an intertidal sandflat in the Westerschelde estuary (Zeeland, the Netherlands). Hydrobiol 269(270):275–284

Sahan E, Muyzer G (2008) Diversity and spatio-temporal distribution of ammonia-oxidizing archaea and bacteria in sediments of the Westerschelde estuary. FEMS Microbiol Ecol 64: 175–186

Sahan E, Sabbe K, Creach V et al (2007) Community structure and seasonal dynamics of diatom biofilms and associated grazers in intertidal mudflats. Aq Microb Ecol 47:253–266

Sanni GO, Coulon F, McGenity TJ (2015) Dynamics and distribution of bacterial and archaeal communities in oil-contaminated temperate coastal mudflat mesocosms. Environ Sci Pollut Res Int 22:15230–15247

Savelli R, Dupuy C, Barille L et al (2018) On biotic and abiotic drivers of the microphytobenthos seasonal cycle in a temperate intertidal mudflat: a modelling study. Biogeosciences 15:7243–7271

Savelli R, Méléder V, Cugier P et al (2020) Mapping the intertidal microphytobenthos gross primary production, part II: merging remote sensing and physical-biological coupled modeling. Front Mar Sci 7:521

Savelli R, Bertin X, Orvain F et al (2019) Impact of chronic and massive resuspension mechanisms on the microphytobenthos dynamics in a temperate intertidal mudflat. J Geophys Res 124:3752–3777

Scarlett K, Denman S, Clark DR et al (2021) Relationships between nitrogen cycling microbial community abundance and composition reveal the indirect effect of soil pH on oak decline. ISME J 15:623–635

Schmid M, Walsh K, Webb R et al (2003) Candidatus "Scalindua brodae", sp. nov., Candidatus "Scalindua wagneri", sp. nov., two new species of anaerobic ammonium oxidizing bacteria. Syst Appl Microbiol 26:529–538

Schneiker S, Dos Santos VAPM, Bartels D et al (2006) Genome sequence of the ubiquitous hydrocarbon-degrading marine bacterium *Alcanivorax borkumensis*. Nat Biotechnol 24:997–1004

Seitz KW, Lazar CS, Hinrichs K-U et al (2016) Genomic reconstruction of a novel, deeply branched sediment archaeal phylum with pathways for acetogenesis and sulfur reduction. ISME J 10: 1696–1705

Seitzinger SP (1988) Denitrification in fresh-water and coastal marine ecosystems-ecological and geochemical significance. Limnol Oceanogr 33:702–724

Serôdio J, Marques da Silva J, Catarino F (1997) Nondestructive tracing of migratory rhythms of intertidal benthic microalgae using in vivo chlorophyll a fluorescence. J Phycol 33:542–553
Seyler LM, McGuinness LM, Kerkhof LJ (2014) Crenarchaeal heterotrophy in salt marsh sediments. ISME J 8:1534–1543
Shi Y, Jiang YY, Wang SY et al (2020) Biogeographic distribution of comammox bacteria in diverse terrestrial habitats. Sci Total Environ 717:137257
Sikkema J, de Bont JAM, Poolman B (1995) Mechanisms of membrane toxicity of hydrocarbons. Microbiol Rev 59:201–222
Smith CJ, Nedwell DB, Dong LF et al (2007) Diversity and abundance of nitrate reductase genes (*narG* and *napA*), nitrite reductase genes (*nirS* and *nrfA*), and their transcripts in estuarine sediments. Appl Environ Microbiol 73:3612–3622
Smith DJ, Underwood GJC (1998) Exopolymer production by intertidal epipelic diatoms. Limnol Oceanogr 43:1578–1591
Smith JM, Cascotti KL, Chavez FP et al (2014a) Differential contributions of archaeal ammonia oxidizer ecotypes to nitrification in coastal surface waters. ISME J 8:1704–1714
Smith JM, Chavez FP, Francis CA (2014b) Ammonium uptake by phytoplankton regulates nitrification in the sunlit ocean. PLoS One 9:e108173
Smith SE, Read DJ (2008) Mycorrhizal symbiosis, 3rd edn. Academic Press, San Diego, CA, USA
Sonthiphand P, Hall MW, Neufeld JD (2014) Biogeography of anaerobic ammonia-oxidizing (anammox) bacteria. Front Microbiol 5:399
Sorokin DY, Makarova K, Abbas B et al (2017) Discovery of extremely halophilic methyl-reducing euryarchaea provides insight into the evolutionary origin of methanogenesis. Nat Microbiol 2: 17081
Sorokin DY, Merkel AY, Abbas B et al (2018) Methanonatronarchaeum thermophilum gen. nov., sp. nov. and 'Candidatus Methanohalarchaeum thermophilum', extremely halo(natrono)philic methyl-reducing methanogens from hypersaline lakes comprising a new euryarchaeal class Methanonatronarchaeia classis nov. Int J Syst Evol Microbiol 68:2199–2208
Spang A, Saw JH, Jørgensen SL et al (2015) Complex archaea that bridge the gap between prokaryotes and eukaryotes. Nature 521:173–179
Spilmont N, Seuront L, Meziane T et al (2011) There's more to the picture than meets the eye: sampling microphytobenthos in a heterogeneous environment. Estuar Coast Shelf Sci 95:470–476
Staats N, Stal LJ, Mur LR (2000) Exopolysaccharide production by the epipelic diatom *Cylindrotheca closterium*: effects of nutrient conditions. J Exp Mar Biol Ecol 249:3–27
Stal LJ (2016) Coastal sediments: transition from land to sea. In: Stal LJ, Cretoiu MS (eds) The marine microbiome. Springer International Publishing, Switzerland, pp 283–304
Stal LJ, Bolhuis H, Cretoiu MS (2019) Phototrophic marine benthic microbiomes: the ecophysiology of these biological entities. Environ Microbiol 21:1529–1551
Steele DJ, Franklin DJ, Underwood GJC (2014) Protection of cells from salinity stress by extracellular polymeric substances in diatom biofilms. Biofouling 30:987–998
Stehr G, Biittcher B, Dittberner P et al (1995) The ammonia-oxidizing nitrifying population of the river Elbe estuary. FEMS Microbiol Ecol 17:177–186
Stein LY, Arp DJ, Berube PM et al (2007) Whole-genome analysis of the ammonia-oxidizing bacterium, *Nitrosomonas eutropha* C91: implications for niche adaptation. Environ Microbiol 9:2993–3007
Stock W, Blommaert L, De Troch M et al (2019) Host specificity in diatom–bacteria interactions alleviates antagonistic effects. FEMS Microbiol Ecol 95:fiz171
Stoecker K, Bendinger B, Schoning B et al (2006) Cohn's *Crenothrix* is a filamentous methane oxidizer with an unusual methane monooxy- genase. Proc Natl Acad Sci 103:2363–2367
Strous M, Fuerst JA, Kramer EH et al (1999) Missing lithotroph identified as new planctomycete. Nature 400:446–449
Strous M, Pelletier E, Mangenot S et al (2006) Deciphering the evolution and metabolism of an anammox bacterium from a community genome. Nature 440:790–794

Sullivan MJ (1999) Applied diatom studies in estuarine and shallow coastal environments. In: Stoermer EF, Smol JP (eds) The diatoms: applications for the environmental and earth sciences. Cambridge University Press, Cambridge, pp 334–351
Sun Z, Li G, Wang C et al (2014) Community dynamics of prokaryotic and eukaryotic microbes in an estuary reservoir. Sci Rep 4:6966
Tal Y, Watts JE, Schreier HJ (2005) Anaerobic ammonia-oxidizing bacteria and related activity in Baltimore inner harbor sediment. Appl Environ Microbiol 71:1816–1821
Taylor JD, McKew BA, Kuhl A et al (2013) Microphytobenthic extracellular polymeric substances (EPS) in intertidal sediments fuel both generalist and specialist EPS-degrading bacteria. Limnol Oceanogr 58:1463–1480
Teramoto M, Ohuchi M, Hatmanti A et al (2011) Oleibacter marinus gen. nov., sp. nov., a novel bacterium that degrades petroleum aliphatic hydrocarbons in the tropical marine environment. Int J Syst Evol Microbiol 61:375–380
Terrisse F, Cravo-Laureau C, Noël C et al (2017) Variation of oxygenation conditions on a hydrocarbonoclastic microbial community reveals *Alcanivorax* and *Cycloclasticus* ecotypes. Front Microbiol 8:1549
Thamdrup B, Dalsgaard T (2002) Production of $N_2$ through anaerobic ammonium oxidation coupled to nitrate reduction in marine sediments. Appl Environ Microbiol 68:1312–1318
Thauer RK, Kaster A-K, Seedorf H et al (2008) Methanogenic archaea: ecologically relevant differences in energy conservation. Nat Rev Microbiol 6:579–591
Thomas GE, Cameron TC, Campo P et al (2020) Bacterial community legacy effects following the Agia Zoni II oil-spill, Greece. Front Microbiol 11:1706
Thornton DCO, Dong LF, Underwood GJC et al (2002) Factors affecting microphytobenthic biomass, species composition and production in the Colne estuary (UK). Aq Microb Ecol 27: 285–300
Thornton DCO, Underwood GJC, Nedwell DB (1999) Effect of illumination and emersion period on the exchange of ammonium across the estuarine sediment-water interface. Mar Ecol Prog Ser 184:11–20
Treusch AH, Leininger S, Kletzin A et al (2005) Novel genes for nitrite reductase and Amo-related proteins indicate a role of uncultivated mesophilic crenarchaeota in nitrogen cycling. Environ Microbiol 7:1985–1995
Trimmer M, Engström P (2011) Distribution, activity, and ecology of anammox bacteria in aquatic environments. In: Ward BB, Arp DJ, Klotz MG (eds) Nitrification. ASM Press, Washington, DC, pp 201–235
Trimmer M, Nicholls J, Deflandre B (2003) Anaerobic ammonium oxidation measured in sediments along the Thames estuary, United Kingdom. Appl Environ Microbiol 69:6447–6454
Trimmer M, Nicholls JC, Morley N et al (2005) Biphasic behavior of anammox regulated by nitrite and nitrate in an estuarine sediment. Appl Environ Microbiol 71:1923–1930
Tsui CKM, Hyde KD (2004) Biodiversity of fungi on submerged wood in a stream and estuary in the tai Ho Bay, Hong Kong. Fungal Divers 15:205–220
Underwood GJC (1994) Seasonal and spatial variation in epipelic diatom assemblages in the Severn estuary. Diatom Res 9:451–472
Underwood GJC (2002) Adaptations of tropical marine microphytobenthic assemblages along a gradient of light and nutrient availability in Suva lagoon, Fiji. Eur J Phycol 37:449–462
Underwood GJC (2005) Microalgal (microphytobenthic) biofilms in shallow coastal waters: how important are species? Proc Calif Acad Sci 56:162–169
Underwood GJC, Barnett M (2006) What determines species composition in microphytobenthic biofilms? In: Kromkamp J (ed) Functioning of microphytobenthos in estuaries. Proceedings of the Microphytobenthos symposium, Amsterdam, the Netherlands, august 2003. Royal Netherlands Academy of Arts and Sciences, Amsterdam, pp 121–138
Underwood GJC, Kromkamp J (1999) Primary production by phytoplankton and microphytobenthos in estuaries. Adv Ecol Res 29:93–153

Underwood GJC, Michel C, Meisterhans G et al (2019) Organic matter from Arctic Sea ice loss alters bacterial community structure and function. Nat Clim Chang 9:170–176
Underwood GJC, Paterson DM (2003) The importance of extracellular carbohydrate production by marine epipelic diatoms. Adv Bot Res 40:184–240
Underwood GJC, Perkins RG, Consalvey M et al (2005) Patterns in microphytobenthic primary productivity: species-specific variation in migratory rhythms and photosynthetic efficiency in mixed-species biofilms. Limnol Oceanogr 50:755–767
Underwood GJC, Phillips J, Saunders K (1998) Distribution of estuarine benthic diatom species along salinity and nutrient gradients. Eur J Phycol 33:173–183
Underwood GJC, Provot L (2000) Determining the environmental preferences of four estuarine epipelic diatom taxa – growth across a range of salinity, nitrate and ammonium conditions. Eur J Phycol 35:173–182
Urakawa H, Rajan S, Feeney ME et al (2019) Ecological response of nitrification to oil spills and its impact on the nitrogen cycle. Environ Microbiol 21:18–33
Van Bleijswijk J, Muyzer G (2004) Genetic diversity of oxygenic phototrophs in microbial mats exposed to different levels of oil pollution. Ophelia 58:157–164
van der Heijden MGA (2002) Arbuscular mycorrhizal fungi as a determinant of plant diversity: in search for underlying mechanisms and general principles. In: van der Heijden MGA, Sanders IR (eds) Mycorrhizal ecology. Springer-Verlag, Berlin, Germany, pp 243–266
van der Wal D, Wielemaker-van den Dool A, Herman PMJ (2010) Spatial synchrony in intertidal benthic algal biomass in temperate coastal and estuarine ecosystems. Ecosystems 13:338–351
Van Kessel MAHJ, Speth DR, Albertsen M et al (2015) Complete nitrification by a single microorganism. Nature 528:555–559
Van Niekerk L, Adams JB, Bate GC et al (2013) Country-wide assessment of estuary health: an approach for integrating pressures and ecosystem response in a data limited environment. Estuar Coast Shelf Sci 130:239–251
Van Niekerk L, Adams JB, Lamberth SJ, et al. (eds) (2019) South African national biodiversity assessment 2018: technical report. Volume 3: Estuarine Realm. CSIR report number CSIR/SPLA/EM/EXP/2019/0062/A. South African National Biodiversity Institute, Pretoria. Report Number: http://hdl.handle.net/20.500.12143/6373
Vanelslander B, De Wever A, Van Oostende N et al (2009) Complementarity effects drive positive diversity effects on biomass production in experimental benthic diatom biofilms. J Ecol 97: 1075–1082
Vanelslander B, Paul C, Grueneberg J et al (2012) Daily bursts of biogenic cyanogen bromide (BrCN) control biofilm formation around a marine benthic diatom. Proc Nat Acad Sci 109: 2412–2417
Vanwonterghem I, Evans PN, Parks DH et al (2016) Methylotrophic methanogenesis discovered in the archaeal phylum Verstraetearchaeota. Nat Microbiol 1:16170
Venter JC, Remington K, Heidelberg JF et al (2004) Environmental genome shotgun sequencing of the Sargasso Sea. Science 304:66–74
Vetriani C, Reysenbach A-L, Dore J (1998) Recovery and phylogenetic analysis of archaeal rRNA sequences from continental shelf sediments. FEMS Microbiol Lett 161:83–88
Waltham NJ, Elliott M, Lee SY et al (2020) UN decade on ecosystem restoration 2021-2030 – what chance for success in restoring coastal ecosystems? Front Mar Sci 7:71. https://doi.org/10.3389/Fmars.2020.00071
Wang W, Shao Z (2013) Enzymes and genes involved in aerobic alkane degradation. Front Microbiol 4:116
Wang W, Shao Z (2014) The long-chain alkane metabolism network of *Alcanivorax dieselolei*. Nat Commun 5:5755
Wang W, Tao J, Liu H et al (2020) Contrasting bacterial and archaeal distributions reflecting different geochemical processes in a sediment core from the Pearl River estuary. AMB Expr 10: 16

Wang W, Wang L, Shao Z (2018) Polycyclic aromatic hydrocarbon (PAH) degradation pathways of the obligate marine PAH degrader *Cycloclasticus* sp. strain P1. Appl Environ Microbiol 84: e01261–e01218

Wang Y, Feng Y, Ma X et al (2013) Seasonal dynamics of ammonia/ammonium-oxidizing prokaryotes in oxic and anoxic wetland sediments of subtropical coastal mangrove. Appl Microbiol Biotechnol 97:7919–7934

Wang Y, Lau PC, Button DK (1996) A marine oligobacterium harboring genes known to be part of aromatic hydrocarbon degradation pathways of soil pseudomonads. Appl Environ Microbiol 62: 2169–2173

Wang YQ, Sen K, He YD et al (2019) Impact of environmental gradients on the abundance and diversity of planktonic fungi across coastal habitats of contrasting trophic status. Sci Total Environ 683:822–833

Wankel SD, Mosier AC, Hansel CM et al (2010) Spatial variability in nitrification rates and ammonia-oxidizing microbial communities in the agriculturally impacted Elkhorn Slough estuary, California. Appl Environ Microbiol 77:269–280

Waring J, Baker NR, Underwood GJC (2007) Responses of estuarine intertidal microphytobenthic algal assemblages to enhanced UV-B radiation. Glob Chang Biol 13:1398–1413

Waring J, Klenell M, Bechtold U et al (2010) Light-induced responses of oxygen photoreduction, reactive oxygen species production and scavenging in two diatom species. J Phycol 46:1206–1217

Webster G, O'Sullivan LA, Meng Y et al (2015) Archaeal community diversity and abundance changes along a natural salinity gradient in estuarine sediments. FEMS Microbiol Ecol 91:1–18

Webster G, Rinna J, Roussel EG et al (2010) Prokaryotic functional diversity in different biogeochemical depth zones in tidal sediments of the Severn estuary, UK, revealed by stable-isotope probing. FEMS Microbiol Ecol 72:179–197

Weerman EJ, Van Belzen J, Rietkerk M et al (2012) Changes in diatom patch-size distribution and degradation in a spatially self-organized intertidal mudflat ecosystem. Ecology 93:608–618

Weisman WE (1998) Analysis of petroleum hydrocarbons in environmental media. Amherst Scientific Publishers, Massachusetts

Wen X, Yang S, Horn F et al (2017) Global biogeographic analysis of methanogenic archaea identifies community-shaping environmental factors of natural environments. Front Microbiol 8:1339

Whitby C, Saunders JR, Pickup RW et al (2001) A comparison of ammonia-oxidiser populations in eutrophic and oligotrophic basins of a large freshwater lake. Ant Leeuwenhoek 79:179–201

Whitby C, Saunders JR, Pickup RW et al (1999) Phylogenetic differentiation of two closely related *Nitrosomonas* spp. that inhabit different sediment environments in an oligotrophic freshwater lake. Appl Environ Microbiol 65:4855–4862

Whiting GJ, Gandy EL, Yoch DC (1986) Tight coupling of root-associated nitrogen fixation and plant photosynthesis in the salt marsh grass *Spartina alterniflora* and carbon dioxide enhancement of nitrogenase activity. Appl Environ Microbiol 52:108–113

Whyte LG, Bourbonniere L, Greer CW (1997) Biodegradation of petroleum hydrocarbons by psychrotrophic *pseudomonas* strains possessing both alkane (alk) and naphthalene (nah) catabolic pathways. Appl Environ Microbiol 63:3719–3723

Widdel F, Musat F (2010) Energetic and other quantitative aspects of microbial hydrocarbon utilization. In: Timmis KN, McGenity TJ, van der Meer JR, de Lorenzo V (eds) Handbook of hydrocarbon and lipid microbiology. Springer, Berlin Heidelberg, pp 731–763

Williams TA, Cox CJ, Foster PG et al (2020) Phylogenomics provides robust support for a two-domains tree of life. Nature Ecol Evol 4:138–147

Wilms R, Sass H, Köpke B et al (2006) Specific bacterial, archaeal, and eukaryotic communities in tidal-flat sediments along a vertical profile of several meters. Appl Environ Microbiol 72:2756–2764

Wilms R, Sass H, Köpke B et al (2007) Methane and sulfate profiles within the sub-surface of a tidal flat are reflected by the distribution of sulfate-reducing bacteria and methanogenic archaea. FEMS Microbiol Ecol 59:611–621

Witkowski A, Lange-Bertalot H, Metzeltin D (2000) Diatom flora of marine coasts. Iconographia diatomologica, vol 7. Koeltz Scientific Books, Königstein, Germany

Woebken D, Lam P, Kuypers MM et al (2009) A microdiversity study of anammox bacteria reveals a novel *Candidatus Scalindua* phylotype in marine oxygen minimum zones. Environ Microbiol 10:3106–3119

Wuchter C, Abbas B, Coolen MJ et al (2006) Archaeal nitrification in the ocean. Proc Natl Acad Sci 103:12317–12322

Xia F, Wang JG, Zhu T et al (2018) Ubiquity and diversity of complete ammonia oxidizers (Comammox). Appl Environ Microbiol 84:e01390–e01318

Yakimov MM, Giuliano L, Denaro R et al (2004) Thalassolituus oleivorans gen. nov., sp. nov., a novel marine bacterium that obligately utilizes hydrocarbons. Int J Syst Evol Microbiol 54:141–148

Yakimov MM, Giuliano L, Gentile G et al (2003) Oleispira antarctica gen. nov., sp. nov., a novel hydrocarbonoclastic marine bacterium isolated from Antarctic coastal sea water. Int J Syst Evol Microbiol 53:779–785

Yakimov MM, Golyshin PN, Lang S et al (1998) Alcanivorax borkumensis gen. Nov., sp. nov., a new, hydrocarbon-degrading and surfactant producing marine bacterium. Int J Syst Bacteriol 48:339–348

Yakimov MM, Timmis KN, Golyshin PN (2007) Obligate oil-degrading marine bacteria. Curr Opin Biotechnol 3:257–266

Yang S, Li M, Lai Q et al (2018) Alcanivorax mobilis sp. nov., a new hydrocarbon-degrading bacterium isolated from deep-sea sediment. Int J Syst Evol Microbiol 68:1639–1643

Yao Z, Du S, Liang C et al (2019) Bacterial community assembly in a typical estuarine marsh with multiple environmental gradients. Appl Environ Microbiol 85:e02602–e02618

Yi J, Lo LSH, Cheng J (2020) Dynamics of microbial community structure and ecological functions in estuarine intertidal sediments. Front Mar Sci 7:585970

Yu T, Wu W, Liang W et al (2018) Growth of sedimentary Bathyarchaeota on lignin as an energy source. Proc Natl Acad Sci 115:6022–6027

Zadjelovic V, Chhun A, Quareshy M et al (2020) Beyond oil degradation: enzymatic potential of *Alcanivorax* to degrade natural and synthetic polyesters. Environ Microbiol 22:1356–1369

Zedelius J, Rabus R, Grundmann O et al (2011) Alkane degradation under anoxic conditions by a nitrate reducing bacterium with possible involvement of the electron acceptor in substrate activation. Environ Microbiol Rep 3:125–135

Zhang C-J, Pan J, Liu Y et al (2020) Genomic and transcriptomic insights into methanogenesis potential of novel methanogens from mangrove sediments. Microbiome 8:94

Zhang P, Neher DA, Li B et al (2018) The impacts of above- and belowground plant input on soil microbiota: invasive *Spartina alterniflora* versus native *Phragmites australis*. Ecosystems 21: 469–481

Zhang X, Agogué H, Dupuy C et al (2013) Relative abundance of ammonia oxidizers, denitrifiers, and Anammox bacteria in sediments of hypernutrified estuarine tidal flats and in relation to environmental conditions. Clean-Soil Air Water 42:815–823

Zhang Y, Xie X, Jiao N et al (2014) Diversity and distribution of *amoA*-type nitrifying and *nirS*-type denitrifying microbial communities in the Yangtze River estuary. Biogeosciences 11: 2131–2145

Zhao Y, Chen W, Wen D (2020) The effects of crude oil on microbial nitrogen cycling in coastal sediments. Environ Int 139:105724

Zheng P, Wang C, Zhang X et al (2019) Community structure and abundance of archaea in a *Zostera marina* meadow: a comparison between seagrass-colonized and bare sediment sites. Archaea 2019:5108012

Zhou Z, Liu Y, Lloyd KG et al (2019) Genomic and transcriptomic insights into the ecology and metabolism of benthic archaeal cosmopolitan, Thermoprofundales (MBG-D archaea). ISME J 13:885–901

Zhou Z, Pan J, Wang F et al (2018) Bathyarchaeota: globally distributed metabolic generalists in anoxic environments. FEMS Microbiol Rev 42:639–655

Zhu Z, van Belzen J, Zhu Q et al (2020) Vegetation recovery on neighboring tidal flats forms an Achilles' heel of saltmarsh resilience to sea level rise. Limnol Oceanogr 65:51–62

Zou D, Liu H, Li M (2020a) Community, distribution, and ecological roles of estuarine archaea. Front Microbiol 11:2060

Zou D, Pan J, Liu Z et al (2020b) The distribution of Bathyarchaeota in surface sediments of the Pearl River estuary along salinity gradient. Front Microbiol 11:285

Zumft WG (1997) Cell biology and molecular basis of cell biology and molecular basis of denitrification. Microbiol Mol Biol Rev 61:533–610

# Symbiosis in the Ocean Microbiome

# 13

Jonathan P. Zehr and David A. Caron

**Abstract**

Symbiotic interactions are widespread in Earth's ecosystems including the marine environment. "Living together" describes a spectrum of interactions ranging from predation and parasitism to the positive interactions of commensalism and mutualism most commonly associated with the term symbiosis. Many well-known symbioses in the marine environment involve associations between microbes and multicellular organisms such as corals but there are diverse microbe-microbe symbiotic interactions that have been described for decades if not centuries from microscopic observations. Microbe-microbe symbioses have been challenging to study in part because of their small size, our inability to establish and culture them in the laboratory, and the ineffectiveness or inappropriateness of the methods that have been used to study macroscopic species. However, technical advances in nucleic acid sequencing, bioinformatics, isotopic approaches, and imaging have begun to provide new insights into these diverse and abundant interactions. The application of culture-independent approaches has revealed that microbial interactions in the marine microbiome range from metabolite exchanges between free-living planktonic cells to epibiotic and intracellular endosymbiotic interactions that bridge the symbiosis – organelle transition. Here we provide a brief overview of symbiosis and then focus on two specific vignettes in the oceanic plankton—$N_2$-fixing and planktonic rhizarian symbioses—that illustrate

J. P. Zehr (✉)
University of California Santa Cruz, Santa Cruz, CA, USA
e-mail: zehrj@ucsc.edu

D. A. Caron
University of Southern California, Los Angeles, CA, USA
e-mail: dcaron@usc.edu

 
L. J. Stal, M. S. Cretoiu (eds.), *The Marine Microbiome*, The Microbiomes of Humans, Animals, Plants, and the Environment 3,
https://doi.org/10.1007/978-3-030-90383-1_13

how cutting-edge approaches and methodologies are providing new insights into the establishment and functioning of these associations.

**Keywords**

Microbial associations · Mutualism · Nitrogen fixation · Parasitism · Rhizaria · Symbiotic cyanobacteria

## 13.1 Introduction

Symbioses, intimate interactions between two or more organisms are ubiquitous in the ocean and have been important drivers of the evolution of life on Earth. Early observations of macroscopic microbial associations such as the cyanobacterial-fungal interaction in lichens led to the definition of the term symbiosis (DeBary 1879; Frank 1877; Sapp 2004). At that time, debate focused on whether the observed relationships were parasitic or a more beneficial and mutual interaction (Sapp 2004). The term symbiosis is now used to define a broad spectrum of obligate or facultative associations that have beneficial, detrimental, or neutral effects on their partners. Pure neutralism where neither organism has an effect on the other may not exist except in concept, but commensalism, where one organism benefits while the other is unaffected, is common. Mutualism, often confused with the broader term symbiosis, refers to partner organisms that mutually benefit from their association. Many symbioses are assumed to be mutualistic, although demonstrating the benefits to all partners within an association is often difficult and sometimes a matter of perspective. Parasitism, and by the broadest definition of symbiosis, predation, are common forms of symbioses in which the interaction is beneficial to one partner but clearly detrimental to the other(s).

In the marine environment, symbioses involving multicellular organisms from fish to corals are found throughout ocean habitats and across all domains of life (Apprill 2020). These include diverse associations such as bioluminescent bacteria associated with light organs in fish and squid (McFall-Ngai 2014), chemosynthetic bacteria associated with hydrothermal vent annelids and molluscs (Childress and Fisher 1992), sulfide-oxidizing bacteria with bivalves and seagrasses (Heide et al. 2012), and photosynthetic dinoflagellates in corals (Rosset et al. 2020). Microbial symbioses have increasingly become the focus of diverse research efforts as powerful new tools provide the ability to better characterize these relationships (Egan et al. 2020).

These well-known associations are partnerships between animals and their associated microbes but symbioses involving only single-celled species also abound (Wernegreen 2017), and may involve Bacteria including cyanobacteria, Archaea, and protists (predominantly microscopic, single-celled, eukaryotic organisms, Archibald et al. (2017)). These relationships are diverse and common in many habitats and are presumed to be analogs of early evolutionary events of organelle acquisition in eukaryotes. They range from loose associations to endosymbioses that in some cases have progressed towards organelle acquisition (Keeling et al. 2015). The marine microbiome (restricted here to microbe-microbe interactions) is a broad concept that spans the microbial species and/or genomes inhabiting the diverse habitats of the sea, from the dilute waters of the open ocean, to the complex sediment environment, and the many invertebrates and vertebrates that inhabit pelagic and benthic environments. The oceanic water column habitat, the subject of this overview, is dominated by microorganisms and includes many known and other yet-to-be-discovered symbiotic interactions.

Early microbiological investigations of microbes relied on microscopy and culturing in order to observe cells and obtain them for experimentation. Through these studies, it was realized that many microorganisms were more easily cultured as assemblages, or enrichments, presumably due to metabolic reliance on one another, but it was difficult to determine the underlying mechanisms of symbiosis or identify and work with the uncultured partners. Characterization and understanding of the metabolic interactions taking place within such "microbial consortia" using traditional methods has proven challenging, and the concept is still rudimentary with respect to the principles underlying the vast array of symbiotic interactions that occur among co-occurring microorganisms. Advances in culture-independent techniques, global environmental gene surveys, and advanced isotope and visualization technologies have facilitated the study of marine microbial interactions (Decelle et al. 2020, 2021; LeKieffre et al. 2018; Lima-Mendez et al. 2015; Meyer and Weis 2012). This chapter focuses on recent advances in understanding the symbiotic relationships within several important marine microbiomes (associations between protists and bacteria, specifically cyanobacteria) of the water column of the open ocean, perhaps the largest habitat. The diversity of these associations is briefly summarized and the reader is pointed to excellent reviews on these topics. From there, this chapter delves into examples of symbioses involving a few specific groups of planktonic microbes (mutualistic $N_2$ fixation, Sect. 13.3 and planktonic rhizarian protists, Sect. 13.4), and how cutting-edge technologies and methodological approaches are currently being employed to understand the metabolic and physiological interactions between the partners of these associations.

## 13.2 Physical Relationships and the Breadth of Microbial Symbioses

Virtually all microbes in the ocean interact to some degree, physically or energetically. Microbial interactions involve diverse physical and metabolic/chemical associations and range from those between unattached (but chemically-interacting) organisms, to attached (episymbiotic) relationships, and finally when one or more species is contained within the other (endosymbiosis) (Fig. 13.1). Beyond predator-prey or virus/parasite-host interactions, many of these relationships are beneficial to one of the associates but inconsequential to the other. One or both of the partners can require symbiosis in an obligate relationship. Obligate dependencies tend to require close physical association, such as surface attachment or intracellular localization.

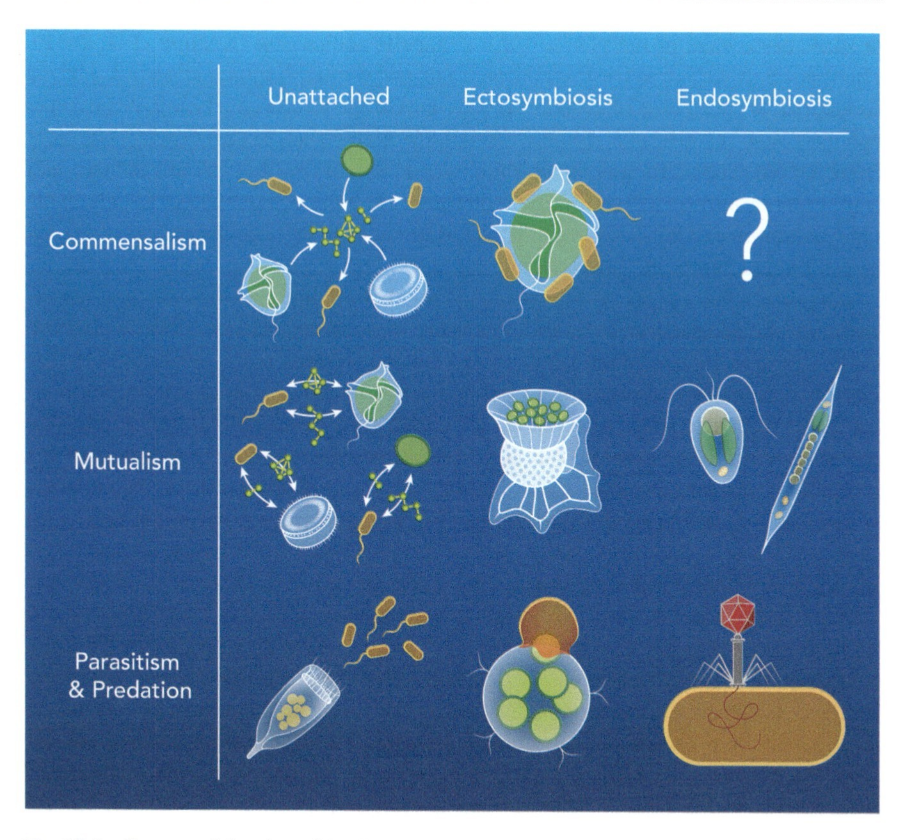


**Fig. 13.1** Conceptual drawing of the diversity of symbiotic interactions ranging from commensalism to predation/parasitism (*y*-axis) and for physical relationships ranging from unattached but interacting partners to endosymbiotic situations (*x*-axis). Unattached commensalisms and mutualisms (left column) are typically established and mediated via chemical communication between the partners (note molecules and directional arrows), or interactions between free-living predators and prey. Ectosymbioses (middle column) involve the exchange of metabolites or nutrients between intimately associated partners (commensalism, mutualism) or predation/parasitism in which an attached parasite/predator extracts cells or cell contents from its prey (in the example shown, a vampyrellid amoeba is extracting algal cells from an algal colony). Endosymbioses (right column) are intimate physical relationships in which the host and symbionts share benefits (mutualism) or those in which one of the partners benefits to the detriment of the other (parasitism/predation). Examples of endosymbiotic commensalism are rare, or at least difficult to definitively confirm as true commensalisms

Microbial symbiotic interactions are generally based on diverse nutritional benefits (Braga 2016) and/or protection. Common nutritional benefits are acquired photosynthesis by otherwise heterotrophic hosts (Decelle et al. 2015; Norris 1996; Not et al. 2016; Stoecker et al. 2017), chemolithotrophic metabolisms of bacteria or archaea (Wrede et al. 2012), vitamin supplementation (Cruz-López et al. 2018), element acquisition (Zehr and Capone 2020), and general nutrition via the translocation of amino acids, carbohydrates, fatty acids, or other organic compounds (Balzano et al. 2015; LeKieffre et al. 2018; Liu et al. 2019).

### 13.2.1 Unattached Microbial Interactions

Open ocean microorganisms live in a dilute environment with respect to each other and nutrient molecules (Zehr et al. 2017). Cells are free-floating, swimming, or attached to other microorganisms or particles. Diffusion rapidly provides for exchanges of carbon compounds and nutrients among cells that are many cell diameters apart. Thus, generic metabolic dependencies exist between, for example, phytoplankton and bacteria that might exchange an organic carbon source for inorganic nutrients (Fig. 13.1, left column). Phytoplankton are generally a source of substrate for bacteria, and bacteria often provide specific nutritional advantages as sources of nutrients, vitamins, or trace elements (Cruz-López et al. 2018; Seymour et al. 2017; Yarimizu et al. 2018). At the microscale, boundary layers surrounding cells provide the opportunity for tighter metabolic interactions and most interactions happen at scales $\leq$100 μm (Cordero and Datta 2016; Stocker 2012). Microbial interactions between chemoheterotrophic bacteria and phytoplankton have been known for decades, starting with the concept of the phycosphere, which was suggested to be analogous to the rhizosphere in terrestrial ecosystems (Bell and Mitchell 1972; Cole 1982; Johansson et al. 2019; Seymour et al. 2017). There are many examples of loose physical associations of algae and bacteria whose interactions have played important roles over the course of evolution (Ramanan et al. 2016).

Algae growing in culture or in natural phytoplankton blooms have specific associated bacteria, although the nature of most interactions is poorly known (Buchan et al. 2014; Not et al. 2016; Sapp et al. 2007; Schäfer et al. 2002). It has long been assumed that microbial interactions were based on the exchange of metabolites. Recent work has identified a variety of specific metabolic interactions among planktonic microbes. Video microscopy has shown bacteria using chemotaxis to find lysing cells, which is important in particle-rich environments such as coastal waters (Smriga et al. 2016). Metabolic profiling of a diatom showed the effect of co-cultured bacteria on diatom metabolite profiles, although changes in growth rates were not detected (Paul et al. 2013). It has been shown that common oceanic phytoplankton can provide sulfur compounds to abundant free-living heterotrophic bacteria such as SAR11 (Durham et al. 2015). However, the metabolic interactions between microorganisms can be very specific. Through work with co-occurring bacteria in a diatom culture, Amin et al. (2015) demonstrated complex metabolic interactions that underlie positive effects on the growth of the individual microorganisms. Based on transcriptome analysis, *Sulfitobacter* sp. SA11 and *Phaeodactylum multiseries* grown together appeared to exchange carbon and nitrogen compounds for nutrition and possibly also for cell signaling and gene regulation. When grown together, the diatom had increased levels of transcripts for tryptophan synthesis and transport whereas *Sulfitobacter* appeared to increase transcript levels for indole acetic acid (IAA) production from tryptophan. IAA is not known to be utilized in bacteria but plays an important role in stimulating growth and cell division in plants and algae. The interactions are much more complex, however, since this alone could not explain the enhancement of growth. Other metabolic dependencies, such as the uptake of nitrate by *Sulfitobacter* and supply of ammonium to *P. multiseries* were suggested.

More complex and possibly evolutionarily important interactions have also been documented. Research on the biochemical interactions between choanoflagellates and some bacteria has revealed that certain bacteria are capable of producing compounds that elicit or inhibit changes between the solitary and colonial forms of these minute protists (Cantley et al. 2016; Woznica et al. 2016). Three distinct compounds produced by the bacterium *Algoriphagus machipongonensis* activate, enhance, or inhibit colony formation in the choanoflagellate *Salpingoeca rosetta* (Woznica et al. 2016). This remarkable responsiveness of the choanoflagellate life stage implies that chemically-mediated microbial interactions observed in symbioses may have influenced the development of coloniality in protists and by extension the evolution of multicellular species such as animals (Alegado et al. 2012; King et al. 2008).

There may be many metabolic dependencies among microbes in the open ocean: between free-floating, unattached cells, cells within the boundary layer, or attached to other cells (e.g., phytoplankton), or in close proximity to other cells on particles. The interactions at these spatial scales are complex, differ between different types of molecules, and are related to cell- and molecule sizes. Much is yet to be learned about the diversity of these types of interactions, which are at one end of the spectrum of symbiotic associations.

### 13.2.2 Ectosymbioses

The surfaces of organisms, including the intestinal linings of many animals (technically, still external to the animal), serve as substrata and often a source of nutrition for many types of microorganisms (Fig. 13.1, middle column; Fig. 13.2). Commensalism is perhaps most common among these associations, although that is unproven. All macroscopic and many microscopic organisms carry a myriad of microorganisms that may benefit from improved nutrition, improved environmental conditions, or protection from predators due to their associations with hosts but in turn cause no apparent harm or benefit to their hosts. These microbial associates, typically referred to as microbiomes, are not covered in detail in this review.

Numerous examples of species-specific bacterial taxa that colonize a variety of protists have been documented, although such symbioses are better characterized from non-marine ecosystems. For example, methanogenic Archaea are common ectosymbionts of ciliated protistan species in the rumen of cattle, benefiting from the production of hydrogen and formate produced by their hosts (Ushida 2018; Vogels et al. 1980). Another classic example is the colonization of some heterotrophic protists (protozoa) by ectosymbiotic bacteria in the hindgut of the termite,

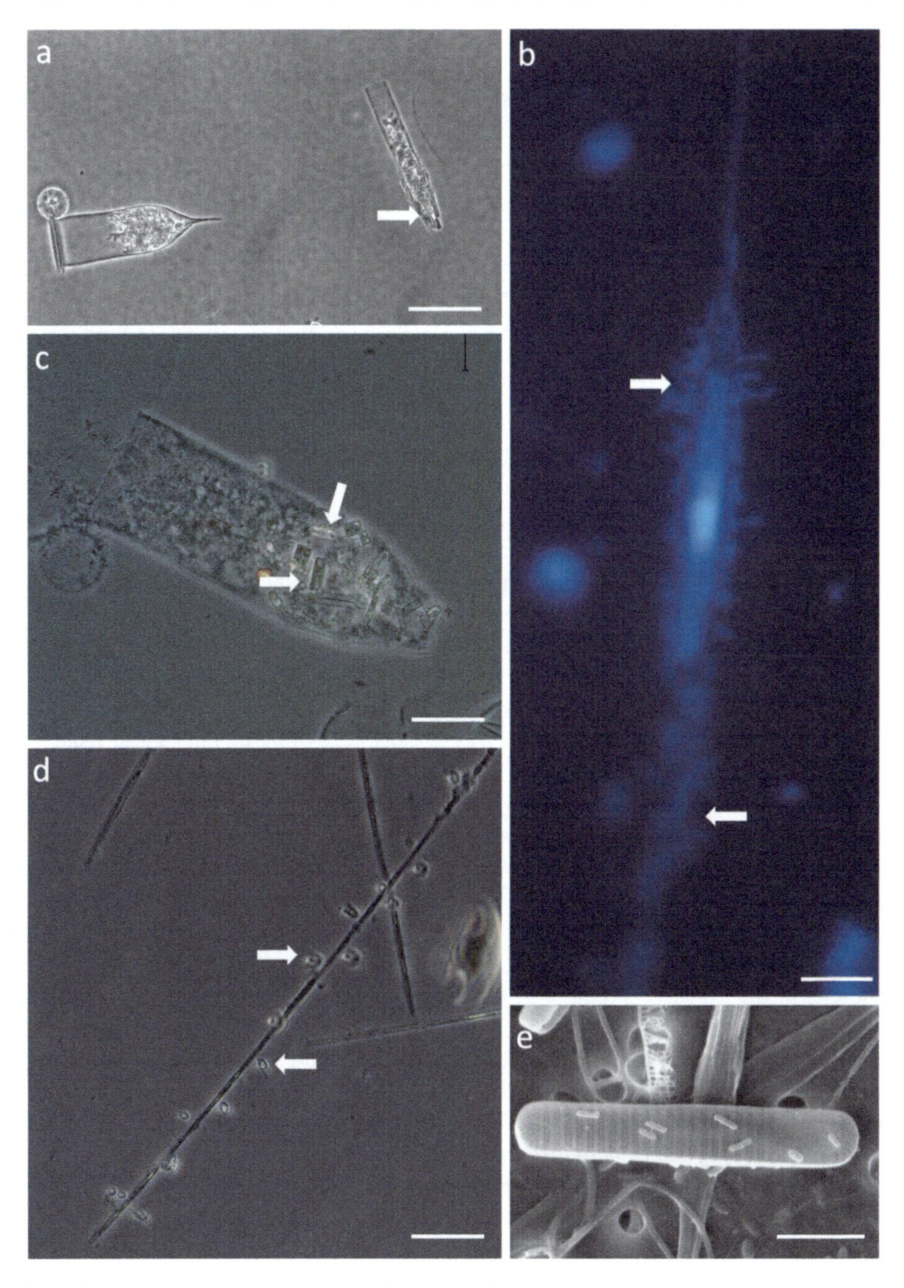


**Fig. 13.2** Examples of ectocommensal relationships among microorganisms in the plankton. Algae are often found attached to protistan microzooplankton, or vice versa, including this tintinnid ciliate with multiple small diatoms attached to the outside of its lorica (arrows in **a**, **c**). The diatoms may obtain organic matter or nutrients released as waste by the host. A different tintinnid species in the same sample shows no colonization (**a**, specimen on left), implying specificity in the ectocommensalism on the right. An epifluorescence micrograph of a pennate diatom reveals many DAPI-stained bacteria (arrows) attached to its frustule (**b**). Minute heterotrophic flagellated protists (arrows in **d**) attached to a chain of thin diatoms feed on unattached suspended free-living

*Cryptotermes cavifrons* (Tamm 1982). The bacteria presumably benefit nutritionally from their association with the protists. It is unclear if the protists benefit directly from the presence of these ectosymbionts, although the host does obtain motility from the flagellar activity of the bacteria and may benefit somehow from the acquired motility. Similarly, ectosymbiotic bacteria instill magnetotactic behavior in some marine protists of anoxic sediments, a relationship that is thought to be mutualistic (Monteil et al. 2019).

Ectosymbiotic associations in marine systems are often non-specific involving random assortments of hosts and symbionts. A variety of small heterotrophic protists, for example, are commonly observed as ectosymbionts on larger photosynthetic protists (Taylor 1982). Diatoms are commonly observed with bacteria or various minute ciliated and flagellated protistan ectosymbionts attached to their siliceous frustules (Fig. 13.2b,d,e), or vice versa (Fig. 13.2a,c). The heterotrophic protists are typically bacterivorous and presumably benefit by feeding at somewhat higher bacterial abundances that characterize the phycosphere of diatoms or other microalgae (Seymour et al. 2017). Physical protection from suspension-feeding zooplankton that might consume small free-living protists may also be one of the benefits for the hitchhikers. While hosts and symbionts may constitute somewhat random pairings in most ectosymbioses, some relationships are remarkably species-specific. The bacterial ectosymbionts of the ciliated protist, *Zoothamnium niveum*, for example, are composed of a single bacterium that uniformly covers the exterior of the ciliate rather than a random assortment of bacterial types suggesting that the association itself is not random but a species-specific symbiotic interaction (Bauer-Nebelsick et al. 1996).

Some (perhaps most?) ectosymbiotic associations are comprised of a mixture of mutualistic, commensal, and even parasitic interactions. *Trichodesmium*, a colony-forming, free-living diazotrophic cyanobacterium, has long been known to have associated microorganisms including diverse proteobacteria, cyanobacteria, protists, and metazoa (Hewson et al. 2009; Sheridan et al. 2002; Siddiqui et al. 1992). An analysis of microbiomes of *Trichodesmium* colonies sampled from a number of stations in the Atlantic Ocean showed that there were diverse epibionts including alpha- and gamma Proteobacteria, that these assemblages differed from the surrounding water or particles, and that the same taxa were often present on colonies from numerous sampling sites (Frischkorn et al. 2017). The metabolic capabilities of the associates overlapped with *Trichodesmium* itself but extended the metabolic repertoire in the *Trichodesmium* aggregates suggesting that complex complementary metabolisms may contribute to *Trichodesmium*'s ecological success in the environment (Frischkorn et al. 2017; Gradoville et al. 2017). Similar findings of a complex yet commonly-occurring microbial community of *Trichodesmium* colonies have been reported by Lee et al. (2018) who concluded that the cyanobacterium's

**Fig. 13.2** (continued) bacteria. Scanning electron micrograph of a pennate diatom shows several bacteria attached to its frustule (**e**). Marker bars are 50 (**a**), 10 (**b**, **d**, **e**) and 25 μm (**c**). Panel (**d**) courtesy of Richard Weinberg, University of Southern California

ectosymbionts may play a role in colony-level nitrogen cycling. *Trichodesmium* also provides an example of "symbioses within symbioses." Anderson observed a large amoeba associated with *Trichodesmium* colonies in the Sargasso Sea that harbored at least two different endosymbiotic bacteria (Anderson 1977). Such findings indicate that complex symbiotic associations are probably the rule rather than the exception.

Ectoparasites also exist although many of these are situations that may be transitory in nature, beginning with attachment to the exterior of the host and transitioning into an intracellular invasion. Colonization and ultimate invasion of diatoms by stramenopile (heterokont) flagellates in the genus *Pirsonia* (Kühn 1998; Kühn et al. 2004) and parasitic infections of diatoms and other phytoplankton by oomycete species are well-known examples of such cellular invasions (Garvetto et al. 2018; Hanic et al. 2009). A striking example involves the vampyrellid amoebae originally described from freshwater ecosystems but now identified from a variety of marine environments (Berney et al. 2013). These species attach to and excavate holes in the cell walls of some planktonic algae, and extricate chloroplasts from the algal cells or invade the cell to consume the contents from the inside (Hess 2017). These relationships muddy the lines between ecto- and endosymbiosis, parasitism, and predation, but they illustrate the potential for complexity and specificity of species interactions in the plankton.

Finally, the nature of some ectosymbioses (and other types of symbioses) may change with environmental situations. Ectocommensal oomycete protists may shift the relationship from a commensal interaction to a parasitic one if the host somehow becomes susceptible to attack. Many bacteria also appear to display this opportunistic ability, exhibiting algicidal activity under certain circumstances (Mayali and Azam 2004).

### 13.2.3 Endosymbioses

Perhaps the best known and most thoroughly studied symbioses are those that involve endosymbiosis. These are perhaps best known since they can be visualized. These interactions span mutualism to parasitism (Fig. 13.1, right column). Endosymbiosis greatly expands the capabilities of some macroorganisms with implications for adaptation and even evolution of new species (Kiers and West 2015) although how symbiosis has played a role is controversial (O'Malley 2015). Endosymbiosis with multicellular organisms takes on many forms and can involve complex cellular or organ development in the host. Single-celled symbiosis, the symbiosis between two unicellular (or sometimes colonial or filamentous) microorganisms provides simpler model systems for studying various stages of evolution of endosymbiosis and organelle evolution (Nowack and Melkonian 2010; Zehr 2015), although presenting methodological challenges because of their small size (Douglas and Raven 2003).

Intracellular symbiosis spans a spectrum of interactions from mutualistic relationships among partners to organelle acquisition where a symbiont has been subsumed into the cellular/genetic machinery of the host although few good model systems to study the transition from endosymbiosis to organelle exist (Douglas and Raven 2003). One well-known example is the amoeba *Paulinella* which harbors a cyanobacterial symbiont. The symbiont contains a highly reduced genome and extensive protein trafficking between host and symbiont has been demonstrated thereby exhibiting some characteristics of a plastid (Mackiewicz et al. 2012; Meheust et al. 2016; Nowack and Grossman 2012; Nowack and Melkonian 2010; Singer et al. 2017). The $N_2$-fixing cyanobacteria *Richelia*, UCYN-A, and the spheroid bodies of the freshwater diatom *Rhopalodia* (Trapp et al. 2012), are endosymbiotic and suggestive of evolution towards a $N_2$-fixing organelle, or "nitroplast" (Zehr 2015). Little is known about the biogeography, ecology, or biology of most of these symbioses.

There is a wide variety of presumedly mutualistic endosymbioses involving protistan hosts in the marine environment (Decelle et al. 2012; Gast and Caron 2001; Shaked and de Vargas 2006; Stoecker et al. 2017). The endosymbionts range from bacteria to eukaryotic algae and are found in a wide variety of heterotrophic protists. Photosynthetic endosymbionts of protists are particularly common and include dinoflagellates, haptophytes, prasinophytes, diatoms, and cyanobacteria. The potential benefit of "acquired photosynthetic ability" in otherwise heterotrophic protists is obvious to the host but less clear for the endosymbiont and has resulted in considerable research to understand how these associations are established and maintained and whether they constitute truly mutualistic relationships.

Parasitic and pathogenic interactions can be endosymbiotic and have evolutionary paths in common that are reflected in the pathogen and parasite symbiont genomes (Ochman and Moran 2001). While presumed to be common and important to the abundances of hosts in the ocean, the specific relationships and impacts of microbial pathogens on vulnerable hosts are still poorly known (Bratbak et al. 1996; Mayali and Azam 2004). Predatory microbes include diverse taxa that act in a variety of ways (Guerrero et al. 1986; Pasternak et al. 2013) including *Bdellovibrio*, which enter the periplasm of their hosts in order to replicate and lyse the prey (Sockett 2009). However, others exhibit a range of epibiotic or other life strategies that are evident in the proteomes (Pasternak et al. 2013). Predatory (Guerrero et al. 1986) or parasitic (Wang and Wu 2014) interactions have been suggested to be involved in the evolution of mitochondria.

Viruses are well-known obligate endosymbionts of bacteria, archaea, and eukaryotes and as such included in this category of pathogenic interactions. Much research has been performed during the past few decades to understand viral–host interactions in the ocean and those relationships have been the topic of many focused reviews (Breitbart 2012; Fuhrman and Suttle 1993; Tomaru et al. 2015; Weinbauer 2004).

Well-known protist-protist parasitic symbioses also exist in the plankton. Scholz et al. (2016) summarized the diverse phylogenetic groups of protists that function as parasites in many of these interactions. Knowledge of the extent of parasitic symbioses in the plankton has expanded rapidly in recent decades although the existence of these lifestyles among some protists has been known for over a century (noted in Coats (1999)). In particular, gene sequencing surveys of environmental

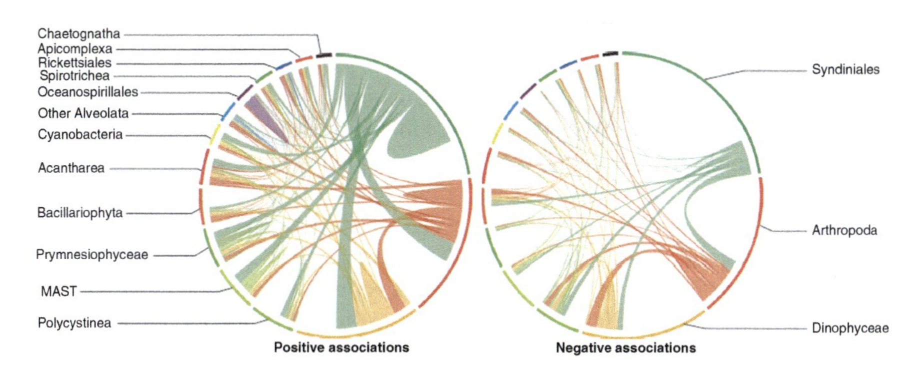


**Fig. 13.3** A visualization (CIRCOS plot) showing interactions (positive and negative) between major bacterial, protistan, and animal taxonomic groups in surface waters from around the world ocean, as derived from the TARA Oceans dataset. Connectivity (ribbons between groups) may indicate a variety of types of interactions, including various symbiotic interactions (commensalism, mutualism, parasitism). Syndiniales (presumed parasites) constituted an important fraction of the interactions (Fig. 2A of Lima-Mendez et al. 2015; see for additional details)

samples have documented a much richer diversity and geographical distribution of protist-protist relationships than previously recognized. Early sequence surveys of this type revealed the presence of large numbers of genetic signatures of alveolate protists for which few if any cultured representatives existed (Moon-van der Staay et al. 2001). These sequences were most closely related to several dinoflagellate-like parasites that had been well-characterized from cultures and field studies (Coats and Park 2002). Gene sequencing studies since then have documented a considerable diversity and widespread occurrence of many previously unknown, dinoflagellate-like alveolates (the novel marine alveolates: MALV groups), provided refinements of their phylogeny and described relatives, suggested and applied approaches for identifying their hosts, and begun to characterize their ecological impacts in planktonic food webs (Groisillier et al. 2006; Guillou et al. 2008).

One benchmark study implicating the degree to which we may have massively underestimated the importance of protistan parasitic symbioses involving alveolates was provided by an analysis of the TARA Oceans expedition's genetic database of samples collected in surface waters throughout the world ocean (Lima-Mendez et al. 2015). A substantial fraction of the sequences examined in that study were identified as putative parasitic taxa, particularly marine alveolates among the Syndiniales (Fig. 13.3). That study employed network analysis and microscopy to begin to link the parasitic taxa to possible hosts, many of which apparently infect dinoflagellates (i.e., other alveolate taxa). An analysis of the "plankton protist interactome" also concluded that a substantial portion (18%) of all protist-protist interactions in nature may be parasitic, based on network analysis of information collected in a Protist Interaction DAtabase (Bjorbækmo et al. 2020).

Chytrids are microscopic, flagellated species within the Fungi that have been repeatedly observed as parasites of marine phytoplankton. A review by Kagami et al. (2007) summarized numerous reports of freshwater and marine host-chytrid parasitisms reported in the literature (see Table 1 in Kagami et al. 2007). Chytrids play important but still largely unquantified roles in altering phytoplankton community composition and in population demise. Conspicuous chytrid infections are particularly common during phytoplankton blooms, presumably a consequence of efficient transmission of the parasite at high prey abundance and perhaps coinciding with times when phytoplankton growth conditions are deteriorating.

Parasitic lifestyles also occur among several clades of stramenopile (heterokont) and cercozoan protists. Oomycetes within the stramenopiles ("water molds," not true Fungi) are common parasites that infect and kill a variety of diatom species (Garvetto et al. 2018; Hanic et al. 2009). Similarly, species of the genus *Pirsonia* are tiny heterotrophic flagellated protists that also infect and kill a variety of diatoms. The phylogeny of these protists has been confused in part due to their minute size and therefore limited morphological features. Based on sequence information, some species of *Pirsonia* appear to be Cercozoa while other species are closely related to stramenopile taxa. The genus continues to undergo phylogenetic revision (Kim et al. 2017; Kühn et al. 2004). *Cryothecomonas* is a cercozoan genus containing a few well-known predators/parasites. Large-scale infections of phytoplankton and algae occurring in sea ice have been reported for species of this genus (Stoecker et al. 1993; Tillmann et al. 1999).

## 13.3 Mutualistic Nutritional Symbioses: $N_2$ Fixation

An important and intriguing group of microbial mutualistic symbioses are based on nitrogen ($N_2$) fixation (Fig. 13.4). $N_2$-fixing symbioses are common and best known in terrestrial systems involving bacteria or cyanobacteria and multicellular plants (Davies-Barnard and Friedlingstein 2020; Elmerich 2007; Fisher and Newton 2002; Rai et al. 2003; Valentine et al. 2018). In marine systems, there are known $N_2$-fixing symbioses in virtually every habitat (Zehr and Capone 2020). $N_2$-fixing symbioses are found in a variety of benthic organisms (Fiore et al. 2010; Petersen et al. 2017) such as corals (Benavides et al. 2017; Davy et al. 2012) clams (Cardini et al. 2019) and shipworms (Carpenter and Culliney 1975; Distel et al. 1991). There are also associations with benthic macroalgae and plants that may be symbiotic (Capone 1983; Cardini et al. 2018; Head and Carpenter 1975). From a global perspective, some of the most important marine $N_2$-fixing symbioses are those involving planktonic protists and cyanobacteria (Foster and Zehr 2019; Thompson and Zehr 2013), which are responsible for a large fraction of the N supply fueling net community production and vertical export in well-lit, nutrient-poor surface waters (Böttjer et al. 2017; Karl et al. 1997; Zehr and Capone 2020), such as the open ocean gyres which we focus on here.

$N_2$ fixation is energetically expensive (requiring ATP and reductant) and the enzyme nitrogenase, which catalyzes the reduction of $N_2$ to $NH_3$, is extremely sensitive to oxygen inactivation (Fay 1992; Postgate 1998; Zehr and Capone 2020) and is found only in Bacteria and some Archaea. $N_2$-fixing symbiotic microorganisms, like their free-living counterparts, must deal with the challenges

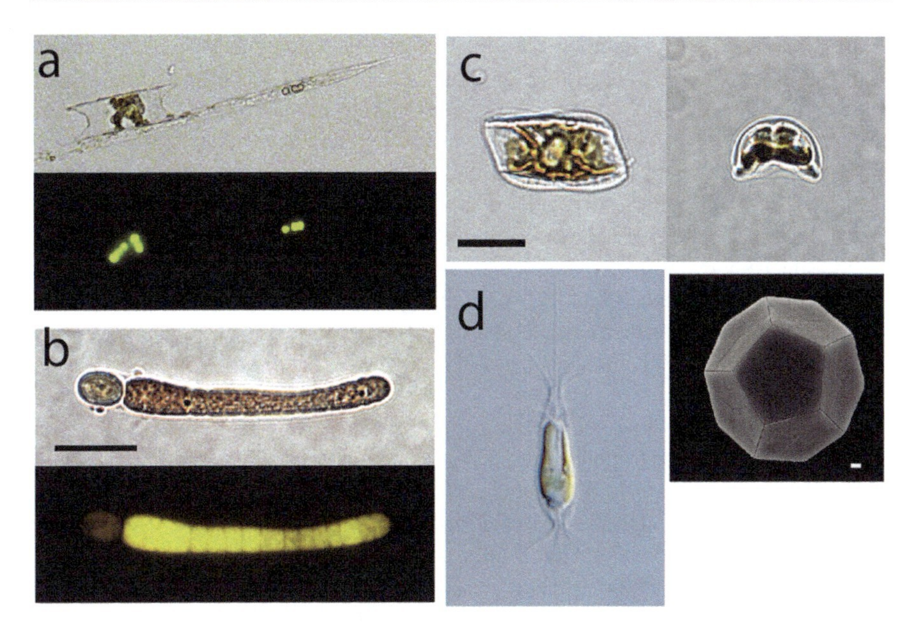


**Fig. 13.4** Open ocean $N_2$-fixing cyanobacterial symbioses. a. Light and epifluorescence micrographs of the symbiosis between the diatom *Hemiaulus* and the heterocyst-forming cyanobacterium *Richelia* and *Rhizosolenia-Richelia* (**a**). Light and epifluorescence micrographs of *Calothrix*, a frequent epibiont on *Chaetoceros* diatoms (**b**). Scale bar 10 mm. Light micrographs of symbiotic marine rhopalodiacean diatoms *Epithemia catenata* sp. nov. (left) and *Epithemia pelagica* sp. nov. (right) (**c**). The cyanobacterial endosymbionts are visible as coccoid cells. Scale bar 10 μm (images courtesy of C. Schvarcz). Light micrograph of flagellated and calcified forms of the *Braarudosphaera bigelowii* (flagellated form previously identified as *Chrysochromulina*) symbiosis with UCYN-A cyanobacteria (*Candidatus* Atelocyanobacterium thalassa) (**d**). The flagellated form (left) courtesy of K. Hagino and Y. Takiano). Images in **a**, **b**, and calcified form of *B. bigelowii* in **d** from Zehr and Capone (2021), with permission

of obtaining energy and avoiding oxygen damage, either from photosynthetically produced oxygen, or oxygen in the environment. Cyanobacteria have evolved several strategies for avoiding oxygen inhibition of $N_2$ fixation by photosynthetically evolved oxygen, including temporal and spatial separation of photosynthetic and fixation processes (Berman-Frank et al. 2003; Fay 1992). Symbiotic cyanobacteria use variations of these strategies to avoid oxygen inactivation, including symbiosis with photosynthetic protists.

Photosymbiotic $N_2$-fixing symbioses are common and diverse in the open ocean. Common $N_2$-fixing symbionts involve cyanobacteria, ranging from heterocyst-forming filamentous to unicellular taxa (Fig. 13.4) with a wide variety of protistan algae including diatoms and haptophytes (Foster and Zehr 2019). These are different than most photosymbioses because sometimes both partners are photosynthetic. Single-celled $N_2$-fixing symbioses are of evolutionary interest since they are analogous to the symbiotic events leading to the evolution of organelles, in this case leading to a $N_2$-fixing organelle ("nitroplast") (Zehr 2015).

In contrast to photosymbioses involving cyanobacteria, $N_2$-fixing non-cyanobacterial diazotroph (NCD) symbioses (presumed Bacteria and Archaea that are chemoheterotrophic or photoheterotrophic) in marine protists are much less commonly known, or possibly just not yet discovered. There are reports of NCD nitrogenase (*nifH*) gene sequences associated with phytoplankton cells (Bombar et al. 2013; Farnelid et al. 2010). Farnelid et al. (2010) showed that bacterial *nifH* genes were associated with heterotrophic dinoflagellates that had previously been noted as bearing cyanobacteria associates (Lucas 1991). Analysis of TARA Oceans metagenomic data showed that *nifH*-containing bacteria occurred in large plankton size-classes and were either particle-associated or associated with larger organisms (Gradoville et al. 2017; Karlusich et al. 2020). However, such associations are scarce (Farnelid et al. 2020).

The longest-known marine $N_2$-fixing symbioses involve diatoms and heterocyst-forming cyanobacteria (Caputo et al. 2019; Foster and Zehr 2019; Heninbokel 1986; Villareal 1989, 1990, 1991; White et al. 2007). Three common associations with the centric chain-forming diatom genera *Rhizosolenia*, *Hemiaulus*, and *Chaetoceros* (Fig. 13.4) have been observed in the environment, sometimes in blooms (Carpenter et al. 1999; Villareal 1994; Villareal et al. 2011). The filamentous cyanobacteria form short chains of vegetative cells with a terminal heterocyst and are epibiotic attached to the external surface (*Calothrix rhizosoleniae*), or endosymbiotic within the diatom frustule (*Richelia intracellularis*) (Villareal 1990, 1992). *R. intracellularis* in *Rhizosolenia* is between the frustule and the cell membrane (Caputo et al. 2019; Pyle et al. 2020) whereas in *Hemiaulus hauckii* they may be inside the diatom cell membrane (Caputo et al. 2019).

The two cyanobacterial genera (*Richelia* and *Calothrix*) are morphologically similar (Fig. 13.4) but there is genetic diversity among the strains and they are genetically distinct suggesting host-specificity (Caputo et al. 2019; Foster and Zehr 2006; Janson et al. 1999). No permanent stable cultures of these associations exist but cultures have been maintained sufficiently long in order to facilitate physiological experiments on growth and nutrient uptake (Pyle et al. 2020; Villareal 1989, 1990). The epibiont *Calothrix* sp. SC01 has been maintained in stable culture without the diatom symbiont (Foster et al. 2010). There is likely to be more uncultured diversity in the environment (Hilton et al. 2015) and it seems likely that there are strains of the diatom genera that do not harbor symbionts (Pyle et al. 2020). The diatom symbionts have been observed throughout the oceans in imaging and -omic data from TARA Oceans (Karlusich et al. 2020). It is unclear whether observed free *Richelia*-like filaments are truly free-living or lost from host diatom cells. It has not been determined whether the cyanobacterial symbionts are vertically transferred from generation to generation during cell division, or horizontally transferred via a free-filament stage in the environment. They have been observed in *Trichodesmium* aggregates (Momper et al. 2015). It is also unclear how they might be regained through the diatom frustule during the horizontal transfer from the environment. There have been reports of unattached cyanobacteria identified as *Richelia* (or unassociated diatoms, see Caputo et al. (2019)), but it is not clear whether or not they are genetically the same as those in symbiosis nor is it known how they would penetrate the silica frustule to be obtained from the environment to reinfect the diatoms.

The functions of the diatom symbioses are not entirely understood. It is clear that $N_2$ is fixed by the cyanobacterial heterocyst and that the N is rapidly transferred throughout the diatom (Foster et al. 2011). Vegetative cells are photosynthetic and likely fuel $N_2$ fixation in the heterocyst possibly along with photosynthesis by the host. $N_2$ fixation rates have been measured in natural populations and cultures (Carpenter et al. 1999; Foster et al. 2011; Villareal 1990) and are sufficient to support growth rates on the order of 1 $day^{-1}$ or even greater (Pyle et al. 2020). Metatranscriptomic analyses of natural populations suggest that cyclic phosphorylation might be important in *Richelia* associated with *Hemiaulus* and that a small antisense RNA may be involved in regulating cyclic photosynthesis by up- or down-regulating NADH dehydrogenase (Hilton et al. 2015).

Some genome reduction has been noted in the diatom symbionts compared to the epibiont *Calothrix* spp. SC01 (Hilton et al. 2013), perhaps an indication of movement towards organellogenesis. The genomic differences and presence or absence of genes coincide with physiological differences such as the response to nitrate (Pyle et al. 2020) and their intracellular location (Caputo et al. 2019). In nature, the growth of the symbionts needs to be coordinated with the growth of the host to maintain the partnership and metatranscriptomic studies show that there are coordinated rhythms of gene expression (Harke et al. 2019). However, in laboratory culture cell division of the diatom and cyanobacterium can become uncoupled leading to loss of the symbiont (Villareal 1989).

In freshwater, unicellular $N_2$-fixing cyanobacteria known as spheroid bodies are known to be associated with rhopalodian diatoms (Floener and Bothe 1980; Nakayama et al. 2011; Prechtl et al. 2004). These cyanobacteria are phylogenetically related to the coccoid cyanobacteria *Cyanothece* and *Crocosphaera* but have lost photosynthetic capability yet retained *nif* genes and fix $N_2$ (Bothe et al. 2010; Floener and Bothe 1980; Nakayama et al. 2014; Prechtl et al. 2004). The genome of the spheroid body of the diatom *Epithemia* has been sequenced and shows great genome reduction including loss of both photosystems and RuBisCO (Nakayama et al. 2014). Strains of a marine rhopalodian diatom have been isolated that also contain $N_2$-fixing spheroid bodies (Schvarcz et al. in press) (Fig. 13.4). These marine strains have likely been previously overlooked since the spheroid bodies lack pigments, but *nif* gene surveys show that they are widely distributed (Schvarcz et al. in press). They are similar in morphology and function to the UCYN-A symbionts discussed below.

A symbiosis between a unicellular cyanobacterium similar to the spheroid bodies in diatoms (now called UCYN-A or *Candidatus* Atelocyanobacterium thalassa) and a haptophyte (prymnesiophyte of the *Braarudosphaera bigelowii* group) was discovered in the last few decades (Hagino et al. 2013; Krupke et al. 2013, 2014; Thompson et al. 2012) following the initial report of a cyanobacterial *nifH* sequence from Station ALOHA in the North Pacific (Zehr et al. 1998). Similar to the spheroid bodies of freshwater *Rhopalodia*/*Epithemia*, the UCYN-A genome has been greatly reduced lacking even more enzymes than the *Epithemia* symbiont (Bothe et al. 2010;

Tripp et al. 2010). Tracer experiments demonstrated that N fixed by the symbiont was rapidly transferred to the haptophyte in exchange for fixed C (Thompson et al. 2012) and they fix $N_2$ in natural populations in a wide variety of locations including the Arctic (Cabello et al. 2016; Harding et al. 2018; Krupke et al. 2013, 2014; Martinez-Perez et al. 2016). The symbiotic relationship has led to physiological or genomic adaptations of the eukaryotic host such as the lack of ability to use exogenous nitrate (Mills et al. 2020). However, the full genome sequence of the haptophyte host has yet to be obtained. The UCYN-A symbioses exist as a cluster of closely related sublineages (termed UCYN-A1, UCYN-A2, etc.) that appear to have a specificity of host and symbiont (Cornejo-Castillo et al. 2019; Farnelid et al. 2016; Zehr et al. 2016).

The haptophyte host of one of the strains (the host of the strain UCYN-A2) is *Braarudosphaera bigelowii*, which has pentalith calcareous plates (Fig. 13.4) and has been found in sediments (Takayama 1972). Calcified cells collected in Japanese waters contain a spheroid body which was shown to be UCYN-A by nitrogenase (*nifH*) gene PCR (Hagino et al. 2013). The calcareous form appears to be only one life stage of the symbiotic cells in Japanese waters, as it also exists in a flagellated form with haptonema previously named *Chrysochromulina parkerae* (Hagino et al. 2013). Thus far in open ocean waters only the uncalcified forms have been reported by CARD-FISH observations (with the smaller UCYN-A1 form being the most abundant), but CARD-FISH procedures might destroy the calcareous plates.

Numerous other associations between cyanobacteria and protists have been observed microscopically that are possibly (but not proven) $N_2$-fixing symbioses (Foster and Zehr 2019). Among these are the diatom *Climacodium* and a unicellular cyanobacterium morphologically and phylogenetically similar to the *Crocosphaera/Cyanothece* group (Carpenter and Janson 2000). Cyanobacteria are associated with heterotrophic dinoflagellates, tintinnids, radiolarians, and amoebae (Foster and Zehr 2019). Some of such associations are suspected not to be $N_2$-fixing because the cyanobacterial cells are morphologically more similar to non-$N_2$-fixing cyanobacteria and are closely related to the non-$N_2$-fixing genera *Synechococcus* and *Prochlorococcus* (Foster et al. 2006a, b).

Knowledge of the UCYN-A $N_2$-fixing symbiosis came from targeted molecular (i.e., PCR-based) approaches. New technologies including metagenomic and metatranscriptomic approaches are providing new information on $N_2$-fixing symbioses. Data from the TARA Oceans dataset contained the complete sequence of the previously sequenced UCYN-A genome and was used to characterize UCYN-A gene expression (Cornejo-Castillo et al. 2016) as well as enabling application of FISH probes to characterize the distribution of UCYN-A symbioses across wide geographic areas (Cabello et al. 2016). Information on the host genome sequence is yet unpublished but metagenomic and metatranscriptomic studies are beginning to provide information on the host (Vorobev et al. 2020). The flagellated form of the symbiosis has been maintained in a nonaxenic culture finally facilitating biological experiments on symbiotic function.

## 13.4 Planktonic Rhizaria and Their Spectrum of Symbioses in the Ocean

The larger planktonic Rhizaria, specifically those species now encompassed by the clade Retaria containing the well-known Foraminifera and Radiolaria, are common and conspicuous protists that are found globally in the oceanic marine plankton. Planktonic Foraminifera contains a small number (roughly three dozen) of extant recognized species, whereas the Radiolaria are highly speciose. Radiolarian phylogeny has been the subject of considerable revision over the last few decades (Sierra et al. 2013), but the present-day Radiolaria includes the commonly encountered polycystine groups (Nassellaria, Spumellaria, Collodaria) and the Acantharia. Retaria includes some of the largest and most beautiful protistan taxa, with solitary adult specimens that can reach up to a centimeter or more in diameter and colonial forms that can form gelatinous ribbon-like structures exceeding one meter in length (Anderson 1983a; Hemleben et al. 1988).

Large planktonic Retaria have been a topic of fascination and biological research since the benchmark descriptive work and magnificent illustrations of Ernst Haeckel in the nineteenth century (Haeckel 1862, 1887). These organisms produce complex networks of pseudopodia, spider web-like extensions of their cellular cytoplasm that are used to entangle, capture, and digest prey and, in some species, harbor endosymbiotic algae (Fig. 13.5). All planktonic Foraminifera and Radiolaria are heterotrophic, feeding as generalists on a wide variety of planktonic organisms (Anderson 1983a, 1993; Hemleben et al. 1988; Swanberg 1983; Swanberg and Caron 1991). Additionally, many taxa within these groups form intricate skeletal structures of calcium carbonate (calcite) by the Foraminifera, silica (opal) in the nassellarian and spumellarian Radiolaria, or strontium sulfate (celestite) in the acantharian Radiolaria. The calcium carbonate and silica skeletons are fossilizable, making those specimens a mainstay of micropaleontological studies and paleoclimatological reconstructive work (Haq and Boersma 1998).

Pertinent to this chapter, planktonic Retaria take part in a variety of symbiotic interactions including commensalism, mutualism, and parasitism. Some species that participate in mutualistic associations harbor more or less intact photosymbiotic algae (an association that did not go unnoticed by Haeckel in the nineteenth century; Figs. 13.5 and 13.6), while others retain 'reduced' photosymbiotic algae or merely the chloroplasts of photosynthetic prey. As such, planktonic Retaria provide rich subject matter for research on the establishment and maintenance of symbiotic associations in the oceanic plankton, the physiological interactions that take place between hosts and symbionts, and the ecological consequences of those associations. However, the hosts are also extremely delicate creatures that have proven exceptionally difficult to culture in the laboratory although they can be hand-collected by SCUBA divers or gently collected in plankton nets and reared in the lab for days to a few weeks (Bé 1982; Kimoto 2015; Swanberg 1979). Beyond basic descriptions, simple experimental manipulations, and speculation regarding the nature of these associations, understanding of the roles of host and symbionts was limited until recently. This situation has begun to change due to the application of novel imaging

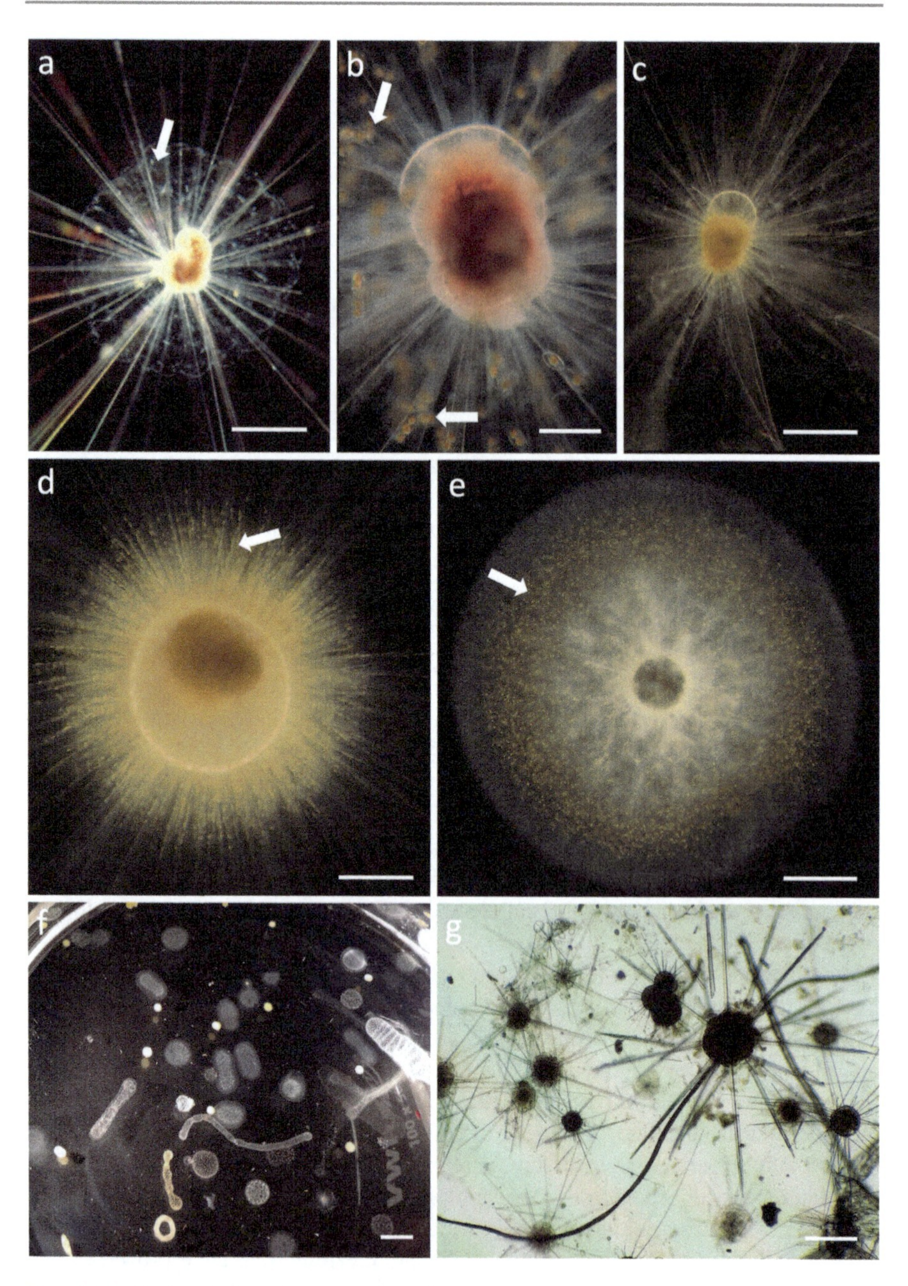


**Fig. 13.5** Planktonic Retaria from the oligotrophic ocean. The planktonic foraminiferan, *Hastigerina pelagica* produces long spines from its calcite skeleton on which it drapes sticky pseudopodia for capturing prey (**a**), and a bubble capsule to aid flotation (arrow in **a**). This species is often observed with commensal photosynthetic dinoflagellates in its pseudopodial network (**b**, arrows). *Globigerinella* (*aquilateralis*) *siphonifera* possesses a dense pseudopodial network and harbors either tiny endosymbiotic haptophytes (**c**, visible as yellow-brown color) or prasinophyte symbionts. The planktonic foraminiferan, *Orbulina universa*, with several thousand dinoflagellate endosymbionts (*Pelagodinium béii*) visible in its pseudopodia (**d**). The solitary radiolarian, *Thalassicolla nucleata*, with thousands of dinoflagellate endosymbionts, *Brandtodinium nutricula*

methodologies and genetic approaches. The examples below illustrate the breadth of retarian symbioses and how novel approaches are providing new insights and understanding into the functioning of these important and widespread symbioses.

### 13.4.1 Commensalistic and Mutualistic Photosymbioses Among Planktonic Retaria

Planktonic Retaria are voracious and highly efficient at capturing a tremendous array of sizes and types of planktonic prey ranging from bacteria to small or weak-swimming metazoa that are incapable of escaping the extensive and sticky pseudopodial networks of these specimens (Anderson 1993; Swanberg and Caron 1991) (Fig. 13.5a,c–e). Nonetheless, a few photosynthetic protistan species often found enmeshed within the pseudopodia of some planktonic Foraminifera appear to be immune to capture and digestion. The most commonly observed of these associates are photosynthetic dinoflagellates within the genus *Pyrocystis* and a few other dinoflagellate genera within the pseudopodial networks of the planktonic foraminiferan, *Hastigerina pelagica. H. pelagica* otherwise harbors no algal symbionts (unlike numerous other planktonic foraminiferal species). Other ectosymbiotic algae also have been observed occasionally associated with Foraminifera (Decelle et al. 2015). The photosynthetic dinoflagellates appear to suffer no harm from their host nor do they appear to provide it any benefit, therefore these associations have generally been thought to be commensal in nature (Fig. 13.5a,b; also see Figs. 5.3 and 5.4 in Hemleben et al. 1988). Physical protection from predation, acquired buoyancy, and nutritional supplementation have been proposed as possible benefits to the commensal algae although none of them have been substantiated.

Far more common and more extensively studied are photosymbioses that occur between numerous species of planktonic Retaria and a variety of photosynthetic protists (Figs. 13.5c–e and 13.6). The photosynthetic endosymbionts that occur in these innately heterotrophic species of Retaria often number in the thousands within an individual host. The associations are established by juvenile specimens each generation (i.e., horizontal transmission) and include situations that are probably the closest to true mutualisms among protistan symbioses (Figs. 13.5 and 13.6). However, more recent studies are also revealing that the relationship between symbionts and hosts varies among the many described associations and actually span a broad range of interactions from mutualism to almost complete organelle acquisition, as detailed below.

**Fig. 13.5** (continued) (**e**). Darkfield image of the contents of a plankton tow (1 mm mesh, 1 m diameter net) from the North Pacific Subtropical Gyre placed in a crystallizing dish showing the dominance of a variety of colonial Radiolaria (**f**). Silhouette micrograph of a 200 μm mesh plankton tow from the same location showing the dominance of Acantharia (one multi-chambered foraminiferan is also visible). Marker bars are 500 (**a**, **c**), 200 (**b**, **d**, **e**), 1000 (**f**) and 100 μm (**g**)

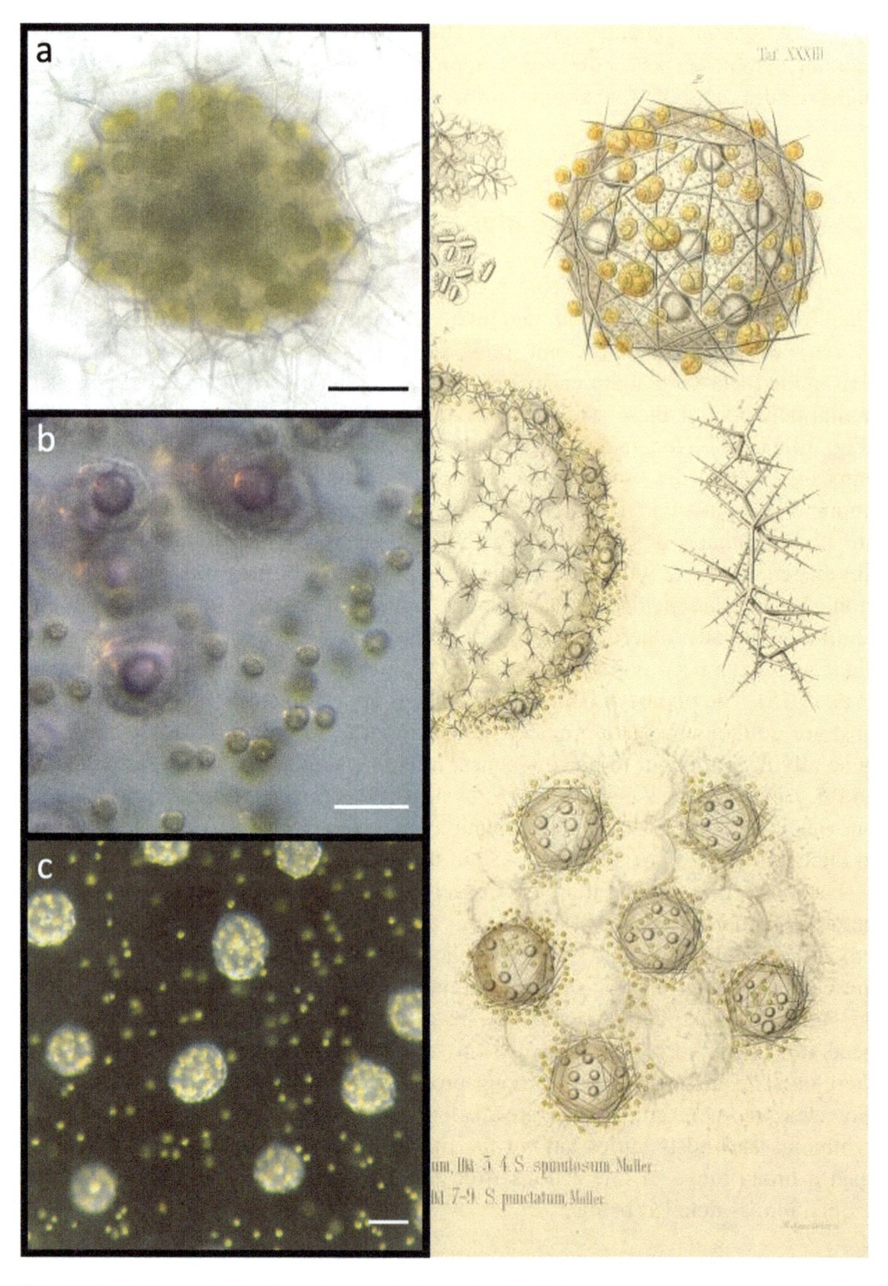


**Fig. 13.6** Endosymbiotic algae in association with Radiolaria were documented more than 150 years ago by Ernst Haeckel (1862). The right side of this figure is a portion of a plate from one of Haeckel's highly detailed monographs. The three insets on the left show what Haeckel must have seen through the microscope nearly 150 years ago. Contracted central capsules (not visible) of a spicule-producing colonial radiolarian (**a**) show numerous golden spheres of the dinoflagellate endosymbiont, *Brandtodinium nutricula*, embedded in the pseudopodial matrix around central capsules. Purple central capsules of the colonial radiolarian *Collosphaera* sp. (**b**) contrast with

Light and electron microscopy, and later gene sequencing, helped establish the taxonomy of the dominant algal types that are held intracellularly (i.e., endosymbiotically) within the pseudopodial matrices of planktonic Foraminifera and Radiolaria. A variety of photosymbionts have been documented as associates of Foraminifera (see Table 19.1 in Decelle et al. 2015, Table 1 in Takagi et al. 2019). The most common and best-studied photosymbionts among several spine-bearing foraminiferal species are a single species of dinoflagellate, *Pelagodinium (Gymnodinium) béii* (Siano et al. 2010; Spero 1987). Several non-spinose species of planktonic Foraminifera harbor non-dinoflagellate symbionts (Anderson 2014; Hemleben et al. 1988; Takagi et al. 2019) and at least one spinose species (*Globigerinella (aequilateralis) siphonifera*) harbors either haptophytes or prasinophytes in apparently mutually exclusive symbioses (Faber et al. 1988; Gast et al. 2000). Polycystine Radiolaria also form photosymbioses with a wide array of algal types including prasinophytes, haptophytes, and cyanobacteria (Anderson et al. 1983b; Decelle et al. 2015; Gast and Caron 2001) but the dominant and best-known relationships are formed with a single photosynthetic dinoflagellate species, *Brandtodinium (Scrippsiella) nutricula* (Anderson 1983a, b; Gast et al. 2000; Probert et al. 2014).

The two most common dinoflagellate photosymbionts are morphologically altered within the host cytoplasm. Both *B. nutricula* in Radiolaria and *P. béii* in Foraminifera lose their flagella and thecal plates in the cytoplasm of the host where they are held within perialgal vacuoles produced by the host (i.e., separate from direct cytoplasmic contact with the host, but completely engulfed within the host's cytoplasm). It is presumed although unsubstantiated that these changes are enacted by the host to facilitate molecular communication between host and symbiont and/or the translocation of photosynthate from symbiont to host. The cellular processes bringing about these morphological changes in the photosymbionts are unknown.

The dinoflagellate photosymbionts of planktonic Foraminifera and Radiolaria have been shown to contribute substantively and variously to the nutrition of their hosts. Symbiont abundances within the pseudopodial network remain more or less constant or even increase in number when the hosts are reared in the laboratory and fed prey, and fast repetition rate fluorometry (Fv/Fm) has revealed that the symbionts have high photosynthetic capacity during much of the host's life span (Takagi et al. 2016, 2019), even if a few symbionts are digested along the way (Anderson 1983a). Symbiont persistence has been taken as an indication that these relationships are mutualisms. Symbiont photosynthetic rates have been measured using microprobes to measure oxygen concentrations in response to light and dark (Jørgensen et al. 1985) and by traditional $^{14}C$-based measurements (Anderson 1978; Anderson et al. 1983a, 1989; Caron et al. 1995; Michaels 1991; Spero and Parker 1985). These studies have revealed extremely high rates of organic carbon production by the

**Fig. 13.6** (continued) the golden *B. nutricula* cells. Central capsules of the colonial radiolarian *Collozoum* sp. (**c**) have golden *B. nutricula* distributed throughout the pseudopodial matrix. Marker bars are all 100 μm

symbionts implying that symbiont primary production probably meets most if not all of the energetic demand of their hosts and that feeding by hosts may be as important for obtaining major nutrients, particularly nitrogen and phosphorus, as for organic carbon acquisition. One foraminiferan possessing *P. béii* symbionts, *Globigerinoides sacculifer*, reared in the light without prey remained alive for weeks, although symbiont number gradually dwindled and hosts eventually died without increasing in size (Bé et al. 1981). In contrast, specimens placed in continuous darkness or deprived of photosynthetic production by treatment with DCMU (3-(3,4-dichlorophenyl)-1,1-dimethylurea) died rapidly even in the presence of sufficient prey for foraminiferal growth indicating a strong dependency on the symbionts that might extend beyond energy/carbon requirements (Bé et al. 1982; Caron et al. 1982).

These "simple" experimental studies have revealed that different photosymbiont associates contribute uniquely to the life processes of their hosts. Some polycystine Radiolaria, for example, obtain so much nutrition from their symbionts that they have occasionally been considered phototrophic species because they have little apparent dependency on capturing prey although this generality does not apply to all symbiont-bearing species (Swanberg et al. 1986). In contrast, the foraminiferan *G. sacculifer* appears to be quite dependent on prey for growth as noted above (Bé et al. 1981; Caron et al. 1982; Takagi et al. 2016, 2018). A similar dependency on prey appears to be the case for the foraminiferan *Orbulina universa* despite the fact that it possesses large numbers of highly active dinoflagellate symbionts (Caron et al. 1987; Spero and Parker 1985) (Fig. 13.5d). Symbiont type for a given foraminiferal host also affects the nutritional state of the host. *G. siphonifera* establishes mutually exclusive photosymbioses with either prasinophyte or haptophyte algae, as noted above. The growth and longevity of the host differ depending on which symbiotic alga is present (Faber et al. 1989).

Knowledge of the exact nature of the metabolic coupling between hosts and photosymbionts in planktonic Retaria has been severely technique-limited until recent years, preventing a clearer understanding of the true nature of these symbioses and masking differences among them. Translocation of symbiont photosynthate to host was demonstrated decades ago (Anderson 1978; Anderson et al. 1983a) but progress beyond that basic tenet has been slow. The fragility and lack of amenability of these delicate organisms to laboratory culture have rendered them intractable for some traditional approaches commonly applied to large Cnidaria (Meyer and Weis 2012). The application of gene sequencing (and advanced imaging; see Sect. 13.4.2) to retarian photosymbioses, however, has begun to change that situation because such measurements can be made without the need for extensive handling or culture. The large genomes of these specimens, particularly those of the dinoflagellate symbionts, thus far still thwart the complete sequencing of genomes in most of these symbioses but transcriptomic analyses are providing novel insights into the metabolic interplay between host and symbionts in planktonic Retaria as well as many other ecologically relevant microbial eukaryotes lacking sequenced genomes (Caron et al. 2016).

Transcriptomics has begun to identify the genetic machinery that might be involved in establishing and maintaining retarian photosymbioses, and the impact of the associations on symbiotic algae. One study comparing gene expression of four rhizarian species (three symbiont-bearing Radiolaria and one aposymbiotic cercozoan rhizarian) concluded that c-type lectin-coding genes might be involved in establishing or maintaining photosymbiosis. This conclusion was based on differences in the expression levels of those genes in the symbiont-bearing versus aposymbiotic species (Balzano et al. 2015). Lectins play a role in cell-cell recognition in eukaryotes and therefore may hold clues as to how partners in rhizarian photosymbioses recognize each other.

Comparative studies of gene expression of *B. scrippsiella*, the dinoflagellate symbiont of the radiolarian *Thalassicolla nucleata* (Fig. 13.5e) in both the free-living and symbiotic state have implicated nitrogen transformations and amino acid production as key factors in the chemical interplay between host and symbiont (Liu et al. 2019). Symbiont genes that showed increased transcription in the symbiotic state included several genes involved in nitrogen transport and transformation, while genes involved in RNA and protein synthesis showed decreased transcription (Fig. 13.7). There was no evidence for increased carbohydrate or glycerol metabolism *in hospite* but pathways of amino acid synthesis were enhanced. Collectively, the findings implied a suppression of symbiont growth in the host's cytoplasm and amino acids as a potential form of organic carbon transfer from symbiont to host. Somewhat different biosynthetic pathways for symbiont photosynthate were identified from a study employing NanoSIMS with the planktonic foraminifer *O. universa* and its dinoflagellate symbiont *P. béii*. Starch appeared to be a major form of photosynthate produced by the symbionts during the day with translocation of photosynthate to the host mostly at night possibly in the form of lipids (LeKieffre et al. 2018).

These seminal applications of -omic approaches to retarian photosymbioses have begun to answer some questions regarding the establishment and function of retarian photosymbioses but have also raised other questions as to the presumed mutualistic nature of these relationships. It is clear that photosymbionts provide their hosts with substantial nutrition but a reciprocal benefit to the algae is difficult to argue in some situations. The ultimate fate of the symbionts amplifies this point. The ontogeny of planktonic Retaria involves development from swarmer cells that are a few micrometers in size (cells presumed to be haploid gametes) to adult specimens that may be macroscopic (Anderson 1976; Bé and Anderson 1976; Bé et al. 1983). Adult solitary forms or colonies at the time of reproduction resorb their pseudopodial networks and the entire cell undergoes multiple divisions into thousands to hundreds of thousands of minute flagellated swarmer cells. Symbionts are not transmitted vertically from adult to progeny. Swarmer formation has now been documented for numerous species but little is known and much is speculated regarding the early life stages of these species (Anderson 1983a; Hemleben et al. 1988). The ultimate fate of the symbiotic algae at the time of onset of the host's reproductive process is poorly understood for most species but their fate has great import for understanding whether Retarian photosymbioses are truly mutualistic.

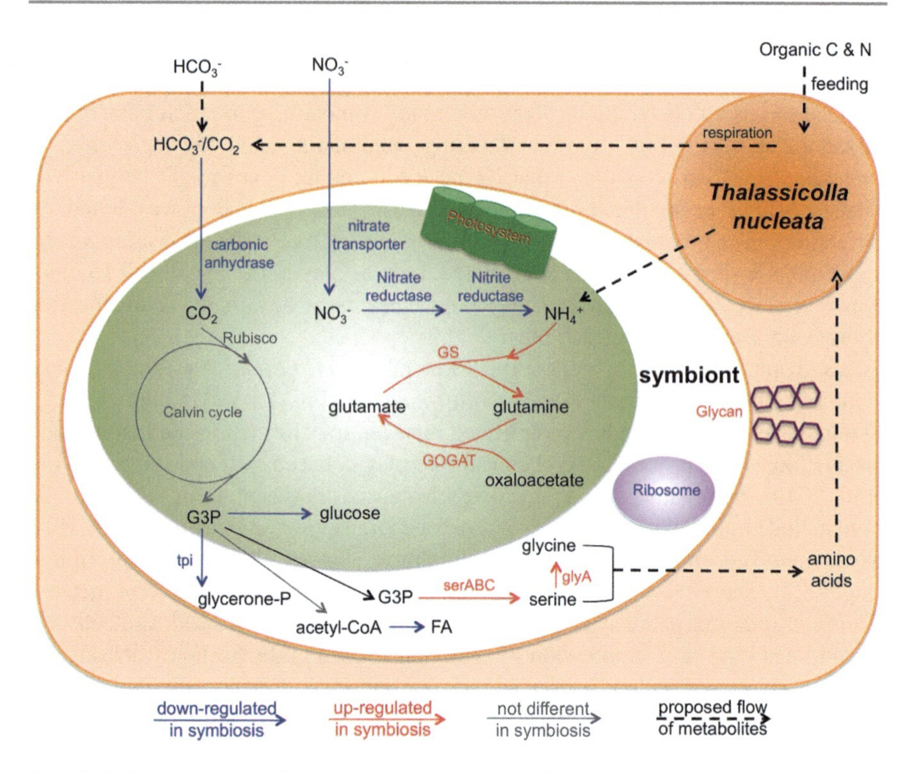


**Fig. 13.7** Proposed metabolic interactions between the radiolarian, *Thalassicolla nucleata* and its photosynthetic dinoflagellate endosymbionts, *Brandtodinium nutricula* inferred from transcriptomic data. Metabolic pathways that were up- and downregulated in the dinoflagellate in the symbiotic versus the free-living state are highlighted in red and blue, respectively (Fig. 5 from Liu et al. 2019; see for additional details)

Many Radiolaria appear to expel or discard their dinoflagellate symbiont, *B. nutricula*, at the onset of reproduction. The host's impending swarmer production thereby releases hundreds to thousands of symbionts into the environment, potentially a net "gain" for the dinoflagellate population in the free-living state if they have multiplied in the cytoplasm during the host's ontogeny. It could be argued that such a situation would constitute a mutualistic symbiosis. At least some of the symbionts released at the time of swarmer formation remain viable because they have repeatedly been cultured as free-living dinoflagellates from material discarded when swarmer cells are released. In contrast, thousands of the photosymbiont *P. béii* held in the cytoplasm of adult specimens of the foraminifer *G. sacculifer* are digested en masse at the onset of the host's reproductive cycle (Bé et al. 1983). A similar fate appears to be the case for several other Foraminifera that harbor *P. béii* and *G. siphonifera* which harbor non-dinoflagellate symbionts (Takagi et al. 2016). These findings imply an association that is less than mutualistic, and that the algae are merely 'farmed' to serve as an energy source for the energetically demanding process of cellular reorganization and division of the host's cytoplasm that characterizes reproduction in these Foraminifera. While symbionts are beneficial to the nutrition and life processes of the host, the alga ultimately gains nothing from the association if the symbionts are eventually digested.

The apparently disparate fates of symbionts in radiolarian and foraminiferal hosts raise questions with respect to whether it has been accurate to assume that all (or most) retarian photosymbioses are mutualistic relationships. Recent findings described below, obtained with cutting-edge imaging and -omic approaches, have begun to demonstrate a situation in the well-known Acantharia-*Phaeocystis* photosymbiosis that is far from mutualistic, where the relationship has clearly crossed the line from mutualism to symbiont "farming" or perhaps something even closer to organelle acquisition.

## 13.4.2 Photosymbioses, Organelle Acquisition, and the Acantharia–Phaeocystis Symbiosis

Early classifications of single-celled eukaryotes separated protists into two groups based on trophic modes: photosynthetic forms (algae), and heterotrophic forms (those that consume prey or other preformed organic matter: protozoa). There are, of course, examples of species that exhibit only phototrophy (e.g., many diatoms) or only heterotrophy (e.g., many ciliates), but modern protistologists (and their many phylogenetic revisions of the last few decades) recognize that trophic mode is a poor indicator of evolutionary relationships or ecology among these species. Many, perhaps most, planktonic protists exhibit a mixture of these fundamental trophic modes and the term "mixotrophy" has come into common use to describe individual species or pairs of species that exhibit a combination of photosynthetic and heterotrophic abilities (Flynn et al. 2019; Stoecker et al. 2017). The descriptor mixotrophy, therefore, encompasses many photosynthetic eukaryotes that possess their own chloroplasts (i.e., true eukaryotic algae) but also possess the ability to consume and digest prey as well as situations where heterotrophy and photosynthetic abilities originate from two different species. These mixotrophic capabilities have considerably changed the way we envision pelagic food web structure and biogeochemical cycles (Mitra et al. 2014; Stickney et al. 2000; Ward and Follows 2016).

The situation of heterotrophic hosts with photosynthetic symbionts is exemplified by many planktonic Retaria (Figs. 13.5 and 13.6). Such relationships cover a spectrum of interactions from the apparently mutualistic endosymbiosis detailed above for many Radiolaria, algal 'farming' noted for at least some planktonic Foraminifera, to the reduction and retention of specific organelles of photosynthetic prey ingested by heterotrophic protists (in the case of chloroplast retention this ability is often called kleptochloroplastidy). Numerous examples of the latter behavior exist across the heterotrophic protistan lineages most notably among planktonic retarians as described above, ciliates (Dolan 1992), and dinoflagellates (Stoecker 1999), but extending even to some flatworms (Stoecker et al. 1989) and molluscs (Hinde and Smith 1974). These interactions do not entail complete integration of the symbionts or their organelles into the host's metabolism, physiology, and cell cycle (a situation that characterizes true organelles) but they are also not mutualistic interactions as has often been assumed.

The relationship between some species of cryptophyte algae and species of the ciliate genus *Mesodinium (Myrionecta)* has become a well-studied system for investigating the one-sidedness that can exist among some protistan associations formerly assumed to be mutualistic photosymbioses (Johnson et al. 2017). Advanced imaging and genetic studies have begun to characterize the mechanism of algal capture, partial digestion, and organelle retention that results in the acquisition of functional chloroplasts for the heterotrophic/mixotrophic *Mesodinium*. Electron- and fluorescence microscopy (the latter enhanced with fluorescent in-situ hybridization (FISH) probing) has documented substantial reduction of the endosymbionts but retention of the cryptophyte's chloroplasts and nuclei, which remain transcriptionally active and serve to maintain chloroplast function in the host cytoplasm, endowing the host with a substantial photosynthetic ability (see Fig. 1 in Johnson et al. 2007).

Subsequent genomic and transcriptomic studies have detailed major changes in the transcriptional activity of the acquired cryptophyte components (Altenburger et al. 2021; Lasek-Nesselquist et al. 2015). Cryptophyte genes associated with photosynthesis, carbohydrate biosynthesis, and amino acid biosynthesis show enhanced transcription in *Mesodinium* containing sequestered and reduced cryptophytes, relative to free-living cryptophytes. The extensive metabolic rewiring of the cryptophyte organelles persists for weeks although chloroplast function eventually breaks down and new "symbionts" must be ingested.

The cryptophytes in the *Mesodinium*-cryptophyte interplay do not benefit nor do they survive these extraordinary alterations. As a consequence, the relationship is far from a mutualistic association, or even "farming" of algae as in the situation described above for the foraminifer *G. sacculifer* but appears to be rather far along an evolutionary pathway to permanent tertiary endosymbiosis (i.e., true organelle acquisition). The cercozoan *Paulinella chromatophore* and its cyanobacterial chromatophore noted above (Sect. 13.2.3) constitute a relationship that apparently is even farther along the path from mutualistic interaction between symbiotic partners to organellogenesis (see Fig. 1 in Nowack and Melkonian 2010). A greatly reduced genome of the cyanobacterium and substantive protein trafficking in *P. chromatophora* has been taken as evidence of an evolutionary progression towards photosynthetic organelle acquisition by a "heterotrophic" protist (Nowack and Grossman 2012; Singer et al. 2017).

The examples above illustrate a spectrum of symbiotic interactions ranging from truly mutualistic to nearly complete organelle acquisition. Nonetheless, there has been a persistent preconception that retarian photosymbioses are mutualisms. This notion has changed only slowly, in part as a result of studies of the photosymbiosis involving radiolarian Acantharia and their preferred symbiotic algae (species of the haptophyte genus *Phaeocystis*; Fig. 13.8) (Decelle et al. 2012). Recent studies have revealed similarities to aspects of the ciliate-cryptophyte association noted above through the application of advanced microscopy, isotopic elemental tracing, and

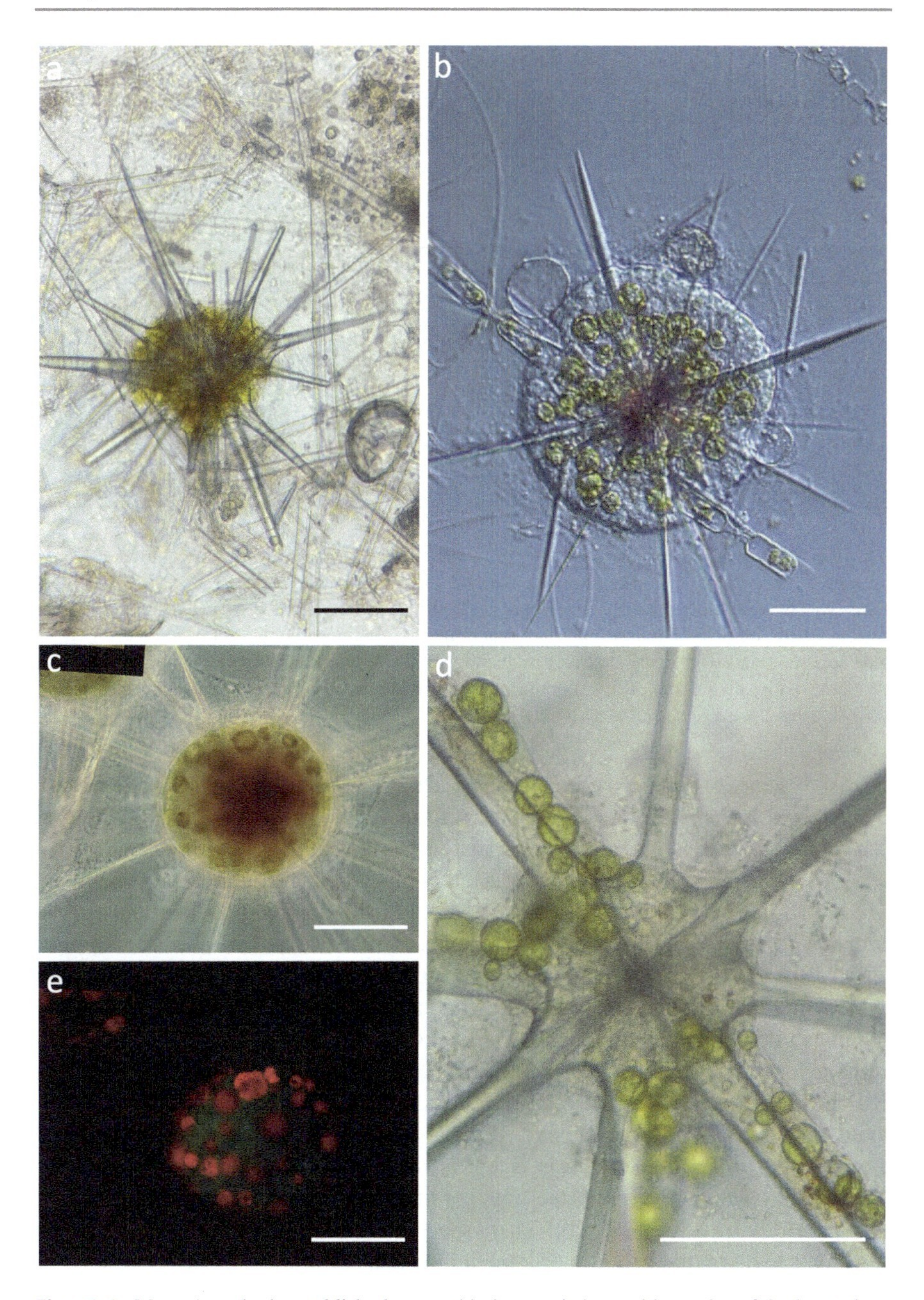


**Fig. 13.8** Many Acantharia establish photosymbiotic associations with species of the haptophyte genus *Phaeocystis*. Endosymbionts in these micrographs appear as yellowish spheres within the cytoplasm of the acantharian hosts. Light micrographs of compressed specimens of Acantharia reveal the presence of the closely-held symbionts (**a**, **b**). A light micrograph (**c**) and an epifluorescence micrograph of the same acantharian cell helps visualize the symbionts, the latter are autofluorescent with blue light excitation due to the presence of chlorophyll *a*. Algal cells of different sizes can be present in a single host (**d**), presumably an indication of the degree to which the algal architecture has been altered (see Decelle et al. 2019). Marker bars are all 100 μm

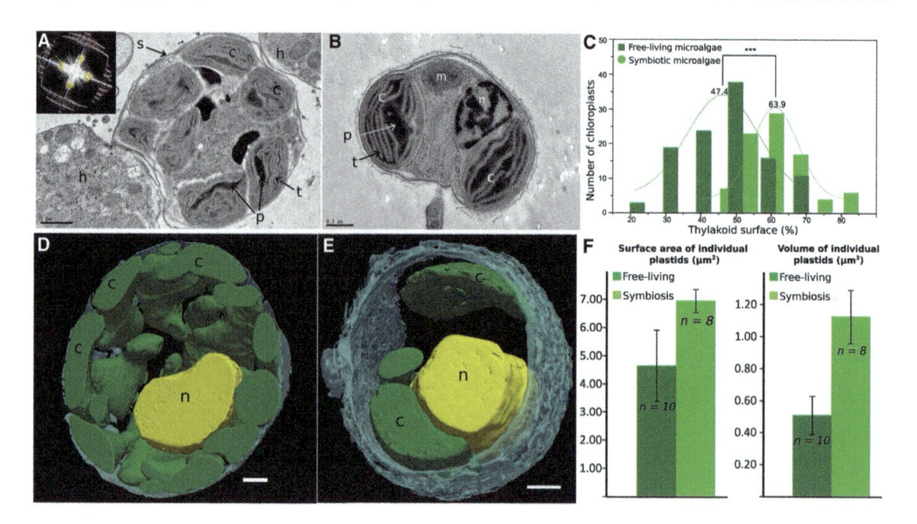


**Fig. 13.9** Transmission electron micrographs of *Phaeocystis cordata* in the symbiotic state (**a**) and cultured as free-living cells (**b**) showing marked differences in symbiont size and chloroplast number, with supporting data (**c**). Symbionts are highly modified in the symbiotic state to enhance photosynthetic capacity. Focused ion beam scanning electron microscopy of the same pairing (**d**, **e**) and supporting data (**f**) provide striking visualization of the differences in these features. The inset in (**a**) shows a low magnification micrograph of the acantharian host and its ornate skeleton (Fig. 1 from Decelle et al. 2019; see for additional details)

extensive genetic analyses (see Fig. 2 in Decelle et al. 2020) that have helped characterize the activities and ultimate fate of the photosymbionts.

Free-living *Phaeocystis* cells are transformed remarkably, and apparently irreversibly, when they are taken into the acantharian cytoplasm in this photosymbiosis. Decelle et al. (2019) reported that the volume of the haptophyte chloroplasts was dramatically increased (up to tenfold) and photosynthesis was enhanced accordingly (Figs. 13.8b,d and 13.9). Transcriptomic information was also obtained for this photosymbiosis. The analyses indicated that symbiont cell division was prevented, and expression of photosynthetic carbon fixation pathways was increased many-fold in the cytoplasm of the host (Uwizeye et al. 2020). Notably, the authors concluded that changes in the symbionts were so extensive that it is highly unlikely that they could reestablish themselves as free-living algae were they released from the host, meaning that this photosymbiosis is not at all a mutualism but rather a farming strategy, or "cytoklepty," with the host commandeering the photosymbiont for its own gain (Uwizeye et al. 2020). Their speculation of irreversible changes in the symbionts is supported by the anecdotal finding that "symbiotic" *Phaeocystis* has yet to be successfully cultured as free-living cells from the photosymbiosis.

These exciting new findings that retarian photosymbioses actually represent a wide spectrum of species-species interactions from mutualism to cytoklepty has opened interesting avenues of research regarding how and how fast 'acquired phototrophy' has evolved in planktonic Retaria, and other heterotrophic protists. The time frame for these transformations is the subject of much debate and research.

Similarly, the series of events that give rise to effective cross-talk between symbiont and host (and therefore host control of the symbionts) is still poorly constrained. Are the events that lead to eventual gene relocation and integration of the former-symbiont-now-plastid into the host genome rapid and fortuitous or slow and methodical? Both seem possible based on our current knowledge of horizontal gene transfer.

Questions also revolve around how algal species are chosen as photosymbionts. Are certain algal species more susceptible to establishing symbioses? A provisional answer to that question appears to be "yes" at least for some algae. *B. nutricula* establishes endosymbioses with numerous Radiolaria as well as some pelagic invertebrates, and *P. béii* is the preferred symbiont of several Foraminifera (Gast and Caron 1996, 2001). *Phaeocystis* species are commonly encountered as endosymbionts in several Acantharia, but retention of the plastids of these algae has also been observed in a common Antarctic heterotrophic dinoflagellate (Gast et al. 2007). These findings imply that some algae are more amenable to establishing and maintaining symbiotic associations (Gomaa et al. 2014), although the genetic bases for this ability are only slowly coming into focus (Nowack and Melkonian 2010). Photosymbioses in planktonic Retaria have proven informing yet challenging subjects with which to address these questions because they offer a wide range of endosymbiotic associations to explore.

### 13.4.3 Parasitic Symbioses Involving Planktonic Retaria

Morphological and physiological modifications of algal symbionts as described above represent a situation in which heterotrophic hosts commandeer their photosynthetic prey (or their organelles) but there are also situations in which protists infect and kill larger planktonic Retaria. Ectoparasites of planktonic Retaria have rarely been observed in part because of the difficulties of capturing and maintaining these specimens in the lab and in part because the pseudopodial networks of these specimens are exceedingly sticky, ensnaring and digesting most species that come in contact with them (Anderson 1993). Nonetheless, a few ectoparasites of planktonic Foraminifera have been documented and include a number of heterotrophic gymnodinoid and peridinoid dinoflagellates (see Fig. 5.4b in Hemleben et al. 1988). These species navigate the complex array of spines and pseudopodial networks of their hosts with apparent impunity, darting between pseudopodia-laden spines to occasionally remove and ingest bits of the host cytoplasm. Planktonic Retaria do not appear to be infected by many of the phylogenetically diverse endosymbiotic parasites common among phytoplankton taxa (e.g., chytrids, oomycetes; see Sect. 13.2.3).

Radiolaria and possibly Foraminifera are, however, hosts to dinoflagellate-like alveolate (MALV) groups (Coats 1999; Guillou et al. 2008) that possess heterotrophic parasitic lifestyles. These species "turn the tables" on their hosts relative to the outcomes of most of the photosymbioses described above. Life cycles of only a few of these parasites are well-known (e.g., *Amoebophrya*, *Parvilucifera*). They are commonly encountered in coastal plankton communities where it has been documented that they can considerably affect the abundances of their hosts (Alacid et al. 2015; Cachon and Cachon 1987; Chambouvet et al. 2008; Coats and Heisler 1989). The life cycle of these species involves entry into the host by motile zoospores, escape from the digestive processes of the host, growth and cell division inside the host by feeding on its cytoplasm, and subsequent release of the zoospores resulting in lysis and death of the host (see Fig. 1 in Alacid et al. 2015). The process is roughly analogous to viral infection, replication, and host lysis except that exploitation of the host's genetic machinery has not been documented among parasitic protists to our knowledge.

Awareness of parasitic protists in coastal ecosystems has been known for many years, including knowledge of some parasitic dinoflagellates associated with Radiolaria (Hertwig 1879). However, understanding the ubiquity and huge diversity of parasitic protists, and recognition that many have retarian hosts, has been enhanced greatly through gene sequencing surveys of environmental samples. Early genetic studies of parasitic symbioses between planktonic Retaria and alveolate parasites employed gene sequencing to draw the connection between the partners (Bråte et al. 2012; Dolven et al. 2007; Gast 2006; Guillou et al. 2008). Global surveys have used gene sequencing and network analysis to identify the diversity and putative links between parasites and retarian hosts (Lima-Mendez et al. 2015). Networking approaches have also been applied to understand linkages between specific retarian hosts, such as the planktonic foraminiferan *Neogloboquadrina pachyderma* and alveolate protists that may parasitize it (Greco et al. 2021). The transcriptomic study of (Liu et al. 2019) described above to characterize the photosymbiosis between the radiolarian *T. nucleata* and its dinoflagellate prey *B. nutricula* revealed the presence of a dinoflagellate-like, putative parasite alveolate in addition to the photosymbiont (Liu et al. 2019). Beyond these initial insights into parasitic symbioses involving planktonic Retaria largely open questions remain regarding host-parasite specificity, the contribution of alveolate parasites to host mortality, and the potential for these associations to alter retarian community structure in nature.

## 13.5 Concluding Remarks: Potential Scientific and Technological Benefits of Understanding Symbiosis

Innovations facilitated by symbiosis have played central roles in evolution enabling the evolution of eukaryotic respiration and photosynthesis and the subsequent enormous diversification of life. In addition, symbiotic associations continue to comprise major drivers of ecological responsiveness of species in the face of environmental change. The vast microbial diversity in the environment, including the marine environment, has been recognized in recent decades partially spawned by the revolution in molecular biology and genomics. New DNA sequence knowledge has dovetailed with classical and new observations to uncover previously unknown microorganisms, including interactions and symbioses. Surprising discoveries of novel microorganisms and the types of interactions among microorganisms suggest there are many more discoveries of symbiosis yet to be made in the ocean plankton.

The spectrum of symbiotic associations discussed in this review demonstrate a continuum of relationships ranging from independent free-living species to true organellogenesis are providing insight into how and why symbiotic relationships have evolved. New visualization, isotopic, and genetic approaches are facilitating a mechanistic understanding of how microbial symbioses function (Egan et al. 2020). In addition to understanding the diversity of metabolic interconnections between species, research on microbial symbiosis is poised to address even bigger questions with implications for evolution. What are the timescales of the evolution of new symbioses and organisms? What are the cellular and molecular steps leading from loose interactions, to cooperation and dependence in obligate symbioses and organelle evolution? What are the steps leading to gene exchange and protein translocations?

Understanding the processes and mechanisms involved in microbial symbioses is not just of significance to intellectual curiosity but has implications for understanding and predicting the real world at large scales. In addition to providing an understanding of the evolution of biological interactions, and evolutionary processes on Earth and perhaps elsewhere, symbioses play critical roles in biogeochemical cycles and food webs and need to be included accurately in predictive global biogeochemical models. Furthermore, microbial photosymbiosis and $N_2$-fixing symbioses provide key models that could lead to improvements in photosynthesis and $N_2$ fixation in agriculture.

It is an exciting and important time in microbial symbiosis research. This review provides only a glimpse of the exciting new understanding that has arisen in the last decade, and hopefully, will foster more research and appreciation of the complexities and importance of marine symbiosis.

**Acknowledgments** This work was partially supported by the Simons Foundation grants (SCOPE 724220 to JPZ, 824082 to JPZ, and P49802 to DAC) and NSF grants (OCE-1559165 to JPZ, and OCE-1737409 to DAC). We thank K. Hagino and C. Schvarcz for providing unpublished images, Andrea Dingeldein for graphic art, and M. R. Gradoville and K. Turk-Kubo for reviewing the manuscript.

## References

Alacid E, Reñé A, Garcés E (2015) New insights into the parasitoid *Parvilucifera sinerae* life cycle: the development and kinetics of infection of a bloom-forming dinoflagellate host. Protist 166: 677–699

Alegado RA, Brown LW, Cao S, Dermenjian RK, Zuzow R, Fairclough SR, Clardy J, King N (2012) A bacterial sulfonolipid triggers multicellular development in the closest living relatives of animals. eLife 1:e00013

Altenburger A, Cai H, Li Q, Drumm K, Kim M, Zhu Y, Garcia-Cuetos L, Zhan X, Hansen PJ, John U, Li S, Lundholm N (2021) Limits to the cellular control of sequestered cryptophyte prey in the marine ciliate *Mesodinium rubrum*. ISME J 15:1056–1072

Amin SA, Hmelo LR, van Tol HM, Durham BP, Carlson LT, Heal KR, Morales RL, Berthiaume CT, Parker MS, Djunaedi B, Ingalls AE, Parsek MR, Moran MA, Armbrust EV (2015)

Interaction and signalling between a cosmopolitan phytoplankton and associated bacteria. Nature 522:98–101
Anderson OR (1976) Fine structure of a Collodarian Radiolarian (*Sphaerozoum punctatum* Müller 1858) and cytoplasmic changes during reproduction. Mar Micropaleontol 1:287–297
Anderson OR (1977) The fine structure of a marine ameba associated with a blue-green alga in the Sargasso Sea. J Protozool 24:370–376
Anderson OR (1978) Fine structure of a symbiont-bearing colonial radiolarian, *Collosphaera globularis*, and $^{14}C$ isotopic evidence for assimilation of organic substances from its zooxanthellae. J Ultrastruct Res 62:181–189
Anderson OR (1983a) Radiolaria. Springer, New York, p 355
Anderson OR (1983b) The radiolarian symbiosis. In: Goff LJ (ed) Algal symbiosis. Cambridge University Press, Cambridge, pp 69–89
Anderson OR (1993) The trophic role of planktonic foraminifera and radiolaria. Mar Microb Food Webs 7:31–51
Anderson OR (2014) Living together in the plankton: a survey of marine protist symbioses. Acta Protozool 53:29–38
Anderson OR, Swanberg NR, Bennett P (1983a) Assimilation of symbiont-derived photosynthesis in some solitary and colonial radiolaria. Mar Biol 77:265–269
Anderson OR, Swanberg NR, Bennett P (1983b) Fine structure of yellow-brown symbionts (Prymnesiida) in solitary radiolaria and their comparison with similar acantharian symbionts. J Protozool 30:718–722
Anderson OR, Bennett P, Angel D, Bryan M (1989) Experimental and observational studies of radiolarian physiological ecology: 2. Trophic activity and symbiont primary productivity of *Spongaster tetras tetras* with comparative data on predatory activity of some Nassellarida. Mar Micropaleontol 14:267–273
Apprill A (2020) The role of symbioses in the adaptation and stress responses of marine organisms. Annu Rev Mar Sci 12:291–314
Archibald JM, Simpson AGB, Slamovits CH (2017) Handbook of the protists. Springer, Cham, p 1657
Balzano S, Corre E, Decelle J, Sierra R, Wincker P, Da Silva C, Poulain J, Pawlowski J, Not F (2015) Transcriptome analyses to investigate symbiotic relationships between marine protists. Front Microbiol. https://doi.org/10.3389/fmicb.2015.00098
Bauer-Nebelsick M, Bardele CF, Ott JA (1996) Redescription of *Zoothamnium niveum* (Hemprich & Ehrenberg, 1831), Eherenberg 1938 (Oligohymenophora, Peritrichida), a ciliate with ectosymbiotic, chemoautotrophic bacteria. Eur J Protistol 32:18–30
Bé AWH (1982) Biology of planktonic foraminifera. In: Broadhead TW (ed) Foraminifera: notes for a short course. University of Tennessee, Knoxville, pp 51–92
Bé AWH, Anderson OR (1976) Gametogenesis in planktonic foraminifera. Science 192:890–892
Bé AWH, Caron DA, Anderson OR (1981) Effects of feeding frequency on life processes of the planktonic foraminifer *Globigerinoides sacculifer* in laboratory culture. J Mar Biol Assoc UK 61:257–277
Bé AWH, Spero HJ, Anderson OR (1982) Effects of symbiont elimination and reinfection on the life processes of the planktonic foraminifer *Globigerinoides sacculifer*. Mar Biol 70:73–86
Bé AWH, Anderson OR, Faber WW Jr, Caron DA (1983) Sequence of morphological and cytoplasmic changes during gametogenesis in the planktonic foraminifer *Globigerinoides sacculifer* (Brady). Micropaleontology 29:310–325
Bell W, Mitchell R (1972) Chemotactic and growth responses of marine bacteria to algal extracellular products. Biol Bull 143:265–277
Benavides M, Bednarz VN, Ferrier-Pagès C (2017) Diazotrophs: overlooked key players within the coral symbiosis and tropical reef ecosystems? Front Mar Sci. https://doi.org/10.3389/fmars.2017.00010
Berman-Frank I, Lundgren P, Falkowski P (2003) Nitrogen fixation and photosynthetic oxygen evolution in cyanobacteria. Res Microbiol 154:157–164

Berney C, Romac S, Mahe F, Santini S, Siano R, Bass D (2013) Vampires in the oceans: predatory cercozoan amoebae in marine habitats. ISME J 7:2387–2399
Bjorbækmo MFM, Evenstad A, Røsæg LL, Krabberød AK, Logares R (2020) The planktonic protist interactome: where do we stand after a century of research? ISME J 14:544–559
Bombar D, Turk-Kubo KA, Robidart J, Carter BJ, Zehr JP (2013) Non-cyanobacterial nifH phylotypes in the North Pacific Subtropical Gyre detected by flow-cytometry cell sorting. Environ Microbiol Rep 5:705–715. https://doi.org/10.1111/1758-2229.12070
Bothe H, Tripp H, Zehr J (2010) Unicellular cyanobacteria with a new mode of life: the lack of photosynthetic oxygen evolution allows nitrogen fixation to proceed. Arch Microbiol 192:783–790
Böttjer D, Dore JE, Karl DM, Letelier RM, Mahaffey C, Wilson ST, Zehr J, Church MJ (2017) Temporal variability of nitrogen fixation and particulate nitrogen export at Station ALOHA. Limnol Oceanogr 62:200–216
Braga RM (2016) Microbial interactions: ecology in a molecular perspective. Brazil J Microbiol 47S:86–98
Bratbak G, Wilson W, Heldal M (1996) Viral control of *Emiliania huxleyi* blooms? J Mar Syst 9: 75–81
Bråte J, Krabberød AK, Dolven JK, Ose RF, Kristensen T, Bjørklund KR, Shalchian-Tabrizi K (2012) Radiolaria associated with large diversity of marine alveolates. Protist 163:767–777
Breitbart M (2012) Marine viruses: truth or dare. Ann Rev Mar Sci 4:425–448
Buchan A, LeCleir GR, Gulvik CA, Gonzalez JM (2014) Master recyclers: features and functions of bacteria associated with phytoplankton blooms. Nat Rev Microbiol 12:686–698
Cabello AM, Cornejo-Castillo FM, Raho N, Blasco D, Vidal M, Audic S, de Vargas C, Latasa M, Acinas SG, Massana R (2016) Global distribution and vertical patterns of a prymnesiophyte-cyanobacteria obligate symbiosis. ISME J 10:693–706
Cachon J, Cachon M (1987) Parasitic dinoflagellates. In: Taylor FJR (ed) The biology of dinoflagellates. Blackwell, Oxford, pp 571–610
Cantley AM, Woznica A, Beemelmanns C, King N, Clardy J (2016) Isolation and synthesis of a bacterially produced inhibitor of rosette development in choanoflagellates. J Am Chem Soc 138: 4326–4329
Capone DG (1983) $N_2$ fixation in seagrass communities. Mar Technol Soc J 17:32–37
Caputo A, Nylander JA, Foster RA (2019) The genetic diversity and evolution of diatom-diazotroph associations highlights traits favoring symbiont integration. FEMS Microbiol Lett 366:fny297
Cardini U, van Hoytema N, Bednarz VN, Al-Rshaidat MMD, Wild C (2018) $N_2$ fixation and primary productivity in a red sea *Halophila stipulacea* meadow exposed to seasonality. Limnol Oceanogr 63:786–798
Cardini U, Bartoli M, Lücker S, Mooshammer M, Polzin J, Lee RW, Micić V, Hofmann T, Weber M, Petersen JM (2019) Chemosymbiotic bivalves contribute to the nitrogen budget of seagrass ecosystems. ISME J 13:3131–3134. https://doi.org/10.1038/s41396-019-0486-9
Caron DA, Be AWH, Anderson OR (1982) Effects of variations in light intensity on life processes of the planktonic foraminifer *Globigerinoides sacculifer* in laboratory culture. J Mar Biol Assoc UK 62:435–451
Caron DA, Faber WW Jr, Bé AWH (1987) Growth of the spinose planktonic foraminifer *Orbulina universa* in laboratory culture and the effect of temperature on life processes. J Mar Biol Assoc UK 67:343–358
Caron DA, Michaels AF, Swanberg NR, Howse FA (1995) Primary productivity by symbiont-bearing planktonic sarcodines (Acantharia, Radiolaria, Foraminifera) in surface waters near Bermuda. J Plank Res 17:103–129
Caron DA, Alexander H, Allen AE, Archibald JM, Armbrust EV, Bachy C, Bharti A, Bell CJ, Dyhrman ST, Guida SM, Heidelberg KB, Kaye JZ, Metzner J, Smith SR, Worden AZ (2016) Probing the evolution, ecology and physiology of marine protists using transcriptomics. Nat Rev Microbiol 15:6–20

Carpenter EJ, Culliney JL (1975) Nitrogen fixation in marine shipworms. Science 187:551–552

Carpenter EJ, Janson S (2000) Intracellular cyanobacterial symbionts in the marine diatom *Climacodium frauenfeldianum* (Bacillariophyceae). J Phycol 36:540–544

Carpenter EJ, Montoya JP, Burns J, Mulholland MR, Subramaniam A, Capone DG (1999) Extensive bloom of a $N_2$-fixing diatom/cyanobacterial association in the tropical Atlantic Ocean. Mar Ecol Prog Ser 185:273–283

Chambouvet A, Morin P, Marie D, Guillou L (2008) Control of toxic marine dinoflagellate blooms by serial parasitic killers. Science 322:1254–1257

Childress JJ, Fisher CR (1992) The biology of hydrothermal vent animals—physiology, biochemistry, and autotrophic symbioses. Oceanogr Mar Biol 30:337–441

Coats DW (1999) Parasitic life styles of marine dinoflagellates. J Eukaryot Microbiol 46:402–409

Coats DW, Heisler JJ (1989) Spatial and temporal occurrence of the parasitic dinoflagellate *Duboscquella cachoni* and its tintinnine host *Eutintinnus pectinis* in Chesapeake Bay. Mar Biol 101:401–409

Coats DW, Park MG (2002) Parasitism of photosynthetic dinoflagellates by three strains of *Amoebophrya* (Dinophyta): parasite survival, infectivity, generation time, and host specificity. J Phycol 38:520–528

Cole JJ (1982) Interactions between bacteria and algae in aquatic ecosystems. Annu Rev Ecol Syst 13:291–314

Cordero OX, Datta MS (2016) Microbial interactions and community assembly at microscales. Curr Opin Microbiol 31:227–234

Cornejo-Castillo FM, Cabello AM, Salazar G, Sanchez-Baracaldo P, Lima-Mendez G, Hingamp P, Alberti A, Sunagawa S, Bork P, de Vargas C, Raes J, Bowler C, Wincker P, Zehr JP, Gasol JM, Massana R, Acinas SG (2016) Cyanobacterial symbionts diverged in the late Cretaceous towards lineage-specific nitrogen fixation factories in single-celled phytoplankton. Nat Commun 7:11071

Cornejo-Castillo FM, Munoz-Marin MDC, Turk-Kubo KA, Royo-Llonch M, Farnelid H, Acinas SG, Zehr JP (2019) UCYN-A3, a newly characterized open ocean sublineage of the symbiotic N2-fixing cyanobacterium *Candidatus* Atelocyanobacterium thalassa. Environ Microbiol 21: 111–124

Cruz-López R, Maske H, Yarimizu K, Holland NA (2018) The B-Vitamin mutualism between the dinoflagellate *Lingulodinium polyedrum* and the bacterium *Dinoroseobacter shibae*. Front Mar Sci. https://doi.org/10.3389/fmars.2018.00274

Davies-Barnard T, Friedlingstein P (2020) The global distribution of biological nitrogen fixation in terrestrial natural ecosystems. Global Biogeochem Cycles 34:e2019GB006387

Davy SK, Allemand D, Weis VM (2012) Cell biology of cnidarian-dinoflagellate symbiosis. Microbiol Mol Biol Rev 76:229–261

DeBary A (1879) Die Erscheinung der Symbiose. Vortrag auf der Versammlung der Naturforschung und Ärtze zu Kassel. Trübner KJ (ed) Verlag von Karl J Trübner, Strassburg, pp 1–30

Decelle J, Probert I, Bittner L, Desdevises Y, Colin S, de Vargas C, Galí M, Simó R, Not F (2012) An original mode of symbiosis in open ocean plankton. Proc Natl Acad Sci USA 109:18000–18005

Decelle J, Colin S, Forster RA (2015) Photosymbiosis in marine planktonic protists. In: Ohtsuka S, Suzaki T, Horiguchi T, Suzuki N, Not F (eds) Marine protists: diversity and dynamics. Springer, New York, pp 465–500

Decelle J, Stryhanyuk H, Gallet B, Veronesi G, Schmidt M, Balzano S, Marro S, Uwizeye C, Jouneau P-H, Lupette J, Jouhet J, Maréchal E, Schwab Y, Schieber NL, Tucoulou R, Richnow H, Finazzi G, Musat N (2019) Algal remodeling in a ubiquitous planktonic photosymbiosis. Curr Biol 29:968–978.e964

Decelle J, Veronesi G, Gallet B, Stryhanyuk H, Benettoni P, Schmidt M, Tucoulou R, Passarelli M, Bohic S, Clode P, Musat N (2020) Subcellular chemical imaging: new avenues in cell biology. Trends Cell Biol 30:173–188

Decelle J, Veronesi G, LeKieffre C, Gallet B, Chevalier F, Stryhanyuk H, Marro S, Ravanel S, Tucoulou R, Schieber N, Finazzi G, Schwab Y, Musat N (2021) Subcellular architecture and metabolic connection in the planktonic photosymbiosis between Collodaria (radiolarians) and their microalgae. bioRxiv. https://doi.org/10.1101/2021.03.13.435225:2021.2003.2013.435225

Distel DL, DeLong EF, Waterbury JB (1991) Phylogenetic characterization and in situ localization of the bacterial symbiont of shipworms (Teredinidae: Bivalvia) by usiing 16S rRNA sequence analysis and oligodeoxynucleotide probe hybridization. Appl Environ Microbiol 57:2376–2382

Dolan JR (1992) Mixotrophy in ciliates: a review of *Chlorella* symbiosis and chloroplast retention. Mar Microb Food Webs 6:115–132

Dolven JK, Lindqvist C, Albert VA, Bjørklund KR, Yuasa T, Takahashi O, Mayama S (2007) Molecular diversity of alveolates associated with neritic North Atlantic radiolarians. Protist 158: 65–76

Douglas AE, Raven JA (2003) Genomes at the interface between bacteria and organelles. Philos Trans R Soc Lond B Biol Sci 358:5–17. Discussion 517–518

Durham BP, Sharma S, Luo H, Smith CB, Amin SA, Bender SJ, Dearth SP, Van Mooy BA, Campagna SR, Kujawinski EB, Armbrust EV, Moran MA (2015) Cryptic carbon and sulfur cycling between surface ocean plankton. Proc Natl Acad Sci U S A 112:453–457

Egan S, Fukatsu T, Francino MP (2020) Opportunities and challenges to microbial symbiosis research in the microbiome era. Front Microbiol 11:1150

Elmerich C (2007) Historical perspective: from bacterization to endophytes. In: Elmerich C, Newton WE (eds) Associative and endophytic nitrogen-fixing bacteria and cyanobacterial associations. Springer, Dordrecht, pp 1–20

Faber WW Jr, Anderson OR, Lindsey JL, Caron DA (1988) Algal-foraminiferal symbiosis in the planktonic foraminifer *Globigerinella aequilateralis*: I. Occurrence and stability of two mutually exclusive chrysophyte endosymbionts and their ultrastructure. J Foram Res 18:334–343

Faber WW Jr, Anderson OR, Caron DA (1989) Algal-foraminiferal symbiosis in the planktonic foraminifer *Globigerinella aequilateralis*: II. Effects of two symbiont species on foraminiferal growth and longevity. J Foram Res 19:185–193

Farnelid H, Tarangkoon W, Hansen G, Hansen PJ, Riemann L (2010) Putative $N_2$-fixing heterotrophic bacteria associated with dinoflagellate–cyanobacteria consortia in the low-nitrogen Indian Ocean. Aquat Microb Ecol 61:105–117

Farnelid H, Turk-Kubo K, Munoz-Marin MD, Zehr JP (2016) New insights into the ecology of the globally significant uncultured nitrogen-fixing symbiont UCYN-A. Aquat Microb Ecol 77:125–138

Farnelid H, Turk-Kubo K, Zehr JP (2020) Cell sorting reveals few novel prokaryote and photosynthetic picoeukaryote associations in the oligotrophic ocean. Environ Microbiol 23:1469–1480. https://doi.org/10.1111/1462-2920.15351

Fay P (1992) Oxygen relations of nitrogen fixation in cyanobacteria. Microbiol Rev 56:340–373

Fiore CL, Jarett JK, Olson ND, Lesser MP (2010) Nitrogen fixation and nitrogen transformations in marine symbioses. Trends Microbiol 18:455–463

Fisher K, Newton WE (2002) Chapter 1—Nitrogen fixation—a general overview. In: Leigh GJ (ed) Nitrogen fixation at the millennium. Elsevier, Amsterdam, pp 1–34

Floener L, Bothe H (1980) Nitrogen fixation in *Rhopalodia gibba*, a diatom containing blue-greenish inclusions symbiotically. In: Schwemmler W, Schenk HEA (eds) Endocytobiology, endosymbiosis and cell biology. Walter de Gruyter, New York, pp 541–552

Flynn KJ, Mitra A, Anestis K, Anschütz AA, Calbet A, Ferreira GD, Gypens N, Hansen PJ, John U, Martin JL, Mansour JS, Maselli M, Medić N, Norlin A, Not F, Pitta P, Romano F, Saiz E, Schneider LK, Stolte W, Traboni C (2019) Mixotrophic protists and a new paradigm for marine ecology: where does plankton research go now? J Plank Res 41:375–391

Foster RA, Zehr JP (2006) Characterization of diatom-cyanobacteria symbioses on the basis of *nifH*, *hetR*, and 16S rRNA sequences. Environ Microbiol 8:1913–1925

Foster RA, Zehr JP (2019) Diversity, genomics, and distribution of phytoplankton-cyanobacterium single-cell symbiotic associations. Annu Rev Microbiol 73:435–456

Foster RA, Carpenter EJ, Bergman B (2006a) Unicellular cyanobionts in open ocean dinoflagellates, radiolarians, and tintinnids: ultrastructural characterization and immuno-localization of phycoerythrin and nitrogenase. J Phycol 42:453–463

Foster RA, Collier JL, Carpenter EJ (2006b) Reverse transcription PCR amplification of cyanobacterial symbiont 16S rRNA sequences from single non-photosynthetic eukaryotic marine planktonic host cells. J Phycol 42:243–250

Foster RA, Goebel NL, Zehr JP (2010) Isolation of *Calothrix rhizosoleniae* (Cyanobacteria) strain SC01from *Chaetoceros* (Bacillariophyta) spp. diatoms of the subtropical north Pacific Ocean. J Phycol 46:1028–1037

Foster RA, Kuypers MMM, Vagner T, Paerl RW, Musat N, Zehr JP (2011) Nitrogen fixation and transfer in open ocean diatom-cyanobacterial symbioses. ISME J 5:1484–1493

Frank AB (1877) Über die biologischen Verhältnisse des Thallus einiger Krustenflechten. Beitr Biol Pflanz 2:123–200

Frischkorn KR, Rouco M, Van Mooy BAS, Dyhrman ST (2017) Epibionts dominate metabolic functional potential of *Trichodesmium* colonies from the oligotrophic ocean. ISME J 11:2090–2101

Fuhrman JA, Suttle CA (1993) Viruses in marine planktonic systems. Oceanography 6:51–63

Garvetto A, Nézan E, Badis Y, Bilien G, Arce P, Bresnan E, Gachon CMM, Siano R (2018) Novel widespread marine oomycetes parasitising diatoms, including the toxic genus *Pseudo-nitzschia*: genetic, morphological, and ecological characterisation. Front Microbiol. https://doi.org/10.3389/fmicb.2018.02918

Gast RJ (2006) Molecular phylogeny of a potentially parasitic dinoflagellate isolated from the solitary radiolarian, *Thalassicolla nucleata*. J Euk Microbiol 53:43–45

Gast RJ, Caron DA (1996) Molecular phylogeny of symbiotic dinoflagellates from Foraminifera and Radiolaria. Mol Biol Evol 13:1192–1197

Gast RJ, Caron DA (2001) Photosymbiotic associations in planktonic foraminifera and radiolaria. Hydrobiologia 461:1–7

Gast RJ, McDonnell TA, Caron DA (2000) srDNA-based taxonomic affinities of algal symbionts from a planktonic foraminifer and a solitary radiolarian. J Phycol 36:172–177

Gast RJ, Moran DM, Dennett MR, Caron DA (2007) Kleptoplasty in an Antarctic dinoflagellate: caught in evolutionary transition? Environ Microbiol 9:39–45

Gomaa F, Kosakyan A, Heger T, Corsaro D, Mitchell E, Lara E (2014) One alga to rule them all: unrelated mixotrophic testate amoebae (Amoebozoa, Rhizaria and stramenopiles) share the same symbiont (Trebouxiophyceae). Protist 165(2):161–176

Gradoville MR, Crump BC, Letelier RM, Church MJ, White AE (2017) Microbiome of *Trichodesmium* Colonies from the North Pacific Subtropical Gyre. Front Microbiol 8:1122

Greco M, Morard R, Kucera M (2021) Single-cell metabarcoding reveals biotic interactions of the Arctic calcifier *Neogloboquadrina pachyderma* with the eukaryotic pelagic community. J Plank Res 43:113–125

Groisillier A, Massana R, Valentin K, Vaulot D, Guillou L (2006) Genetic diversity and habitats of two enigmatic marine alveolate lineages. Aquat Microb Ecol 42:277–291

Guerrero R, Pedros-Alio C, Esteve I, Mas J, Chase D, Margulis L (1986) Predatory prokaryotes: predation and primary consumption evolved in bacteria. Proc Natl Acad Sci U S A 83:2138–2142

Guillou L, Viprey M, Chambouvet A, Welsh RM, Kirkham AR, Massana R, Scanlan DJ, Worden AZ (2008) Widespread occurrence and genetic diversity of marine parasitoids belonging to Syndiniales (Alveolata). Environ Microbiol 10:3349–3365

Haeckel E (1862) Die Radiolarien (Rhizopoda Radiaria). Eine Monographie, Reimer, Berlin

Haeckel E (1887) Report on Radiolaria collected by H.M.S. Challenger during the 1873-1876. In: Thompson CW, Murray J (eds) The voyage of the HMS Challenger. Her Majesty's Stationary Office, London, pp 1–1760

Hagino K, Onuma R, Kawachi M, Horiguchi T (2013) Discovery of an endosymbiotic nitrogen-fixing cyanobacterium UCYN-A in *Braarudosphaera bigelowii* (Prymnesiophyceae). PLoS One 8:e81749

Hanic LA, Sekimoto S, Bates SS (2009) Oomycete and chytrid infections of the marine diatom *Pseudo-nitzschia pungens* (Bacillariophyceae) from Prince Edward Island, Canada. Botany 87: 1096–1105

Haq B, Boersma A (eds) (1998) Introduction to marine micropaleontology. Elsevier, Amsterdam, 376p

Harding K, Turk-Kubo KA, Sipler RE, Mills MM, Bronk DA, Zehr JP (2018) Symbiotic unicellular cyanobacteria fix nitrogen in the Arctic Ocean. Proc Natl Acad Sci U S A 115:13371–13375

Harke MJ, Frischkorn KR, Haley ST, Aylward FO, Zehr JP, Dyhrman ST (2019) Periodic and coordinated gene expression between a diazotroph and its diatom host. ISME J 13:118–131

Head WD, Carpenter EJ (1975) Nitrogen fixation associated with the marine macroalga *Codium fragile*. Limnol Oceanogr 20:815–823

Heide T, Govers L, de Fouw J, Olff H, Van der Geest M, Katwijk M, Piersma T, van de Koppel J, Silliman B, Smolders A, Van Gils J (2012) A three-stage symbiosis forms the foundation of seagrass ecosystems. Science 336:1432–1434

Hemleben C, Spindler M, Anderson OR (1988) Modern planktonic foraminifera. Springer, New York, p 363

Heninbokel JF (1986) Occurrence of *Richelia intracellularis* (Cyanophyta) within diatoms *Hemiaulus hauckii* and *H. membranaceus* off Hawaii. J Phycol 22:399–403

Hertwig R (1879) Der Organismus der Radiolarien. Jena Denksch 2:129–277

Hess S (2017) Hunting for agile prey: trophic specialisation in leptophryid amoebae (Vampyrellida, Rhizaria) revealed by two novel predators of planktonic algae. FEMS Microbiol Ecol 93. https://doi.org/10.1093/femsec/fix104

Hewson I, Poretsky RS, Dyhrman ST, Zielinkski B, White AE, Tripp HJ, Montoya JP, Zehr JP (2009) Microbial community gene expression within colonies of the diazotroph, *Trichodesmium,* from the Southwest Pacific Ocean. ISME J 3:1286–1300

Hilton JA, Foster RA, Tripp HJ, Carter BJ, Zehr JP, Villareal TA (2013) Genomic deletions disrupt nitrogen metabolism pathways of a cyanobacterial diatom symbiont. Nat Commun 4:1767

Hilton JA, Satinsky BM, Doherty M, Zielinski B, Zehr JP (2015) Metatranscriptomics of $N_2$-fixing cyanobacteria in the Amazon River plume. ISME J 9:1557–1569

Hinde R, Smith DC (1974) "Chloroplast symbiosis" and the extent to which it occurs in *Saccoglossa* (Gastropoda: Mollusca). Biol J Linnean Soc 6:349–356

Janson S, Wouters J, Bergman B, Carpenter EJ (1999) Host specificity in the *Richelia*-diatom symbiosis revealed by *hetR* gene sequence analysis. Environ Microbiol 1:431–438

Johansson ON, Pinder MIM, Ohlsson F, Egardt J, Töpel M, Clarke AK (2019) Friends with benefits: exploring the phycosphere of the marine diatom *Skeletonema marinoi*. Front Microbiol. https://doi.org/10.3389/fmicb.2019.01828

Johnson MD, Oldach D, Delwiche CF, Stoecker DK (2007) Retention of transcriptionally active cryptophyte nuclei by the ciliate *Myrionecta rubra*. Nature 445:426–428

Johnson MD, Lasek-Nesselquist E, Moeller HV, Altenburger A, Lundholm N, Kim M, Drumm K, Moestrup Ø, Hansen PJ (2017) *Mesodinium rubrum*: the symbiosis that wasn't. Proc Natl Acad Sci USA 114:E1040–E1042

Jørgensen BB, Erez J, Revsbech NP, Cohen Y (1985) Symbiotic photosynthesis in a planktonic foraminiferan, *Globigerinoides sacculifer* (Brady), studied with microelectrodes. Limnol Oceanogr 30:1253–1267

Kagami M, de Bruin A, Ibelings BW, Van Donk E (2007) Parasitic chytrids: their effects on phytoplankton communities and food-web dynamics. Hydrobiologia 578:113–129

Karl D, Letelier R, Tupas L, Dore J, Christian J, Hebel D (1997) The role of nitrogen fixation in biogeochemical cycling in the subtropical North Pacific Ocean. Nature 388:533–538

Karlusich JJP, Pelletier E, Carsique M, Dvorak E, Colin S, Picheral M, Pepperkok R, Karsenti E, Vargas Cd, Lombard F, Wincker P, Bowler C, Foster RA (2020) Global distribution patterns of marine nitrogen-fixers by imaging and molecular methods. bioRxiv. https://doi.org/10.1101/2020.10.17.343731:2020.2010.2017.343731

Keeling PJ, McCutcheon JP, Doolittle WF (2015) Symbiosis becoming permanent: survival of the luckiest. Proc Natl Acad Sci USA 112:10101–10103

Kiers ET, West SA (2015) Evolutionary biology. Evolving new organisms via symbiosis. Science 348:392–394

Kim S, Jeon CB, Park MG (2017) Morphological observations and phylogenetic position of the parasitoid nanoflagellate *Pseudopirsonia* sp. (Cercozoa) infecting the marine diatom *Coscinodiscus wailesii* (Bacillariophyta). Algae 32:181–187

Kimoto K (2015) Planktonic foraminifera. In: Ohtsuka S, Suzaki T, Horiguchi T, Suzuki N, Not F (eds) Marine protists: diversity and dynamics. Springer, New York, pp 129–178

King N, Westbrook MJ, Young SL, Kuo A, Abedin M, Chapman J, Fairclough S, Hellsten U, Isogai Y, Letunic I, Marr M, Pincus D, Putnam N, Rokas A, Wright KJ, Zuzow R, Dirks W, Good M, Goodstein D, Lemon D, Wanqing L, Lyons JB, Morris A, Nichols S, Richter DJ, Salamov A, Bork P, Lim WA, Manning F, Miller WT, McGinnis W, Shapiro H, Tijian R, Grigoriev IV, Rokhsar D (2008) The genome of the choanoflagellate *Monosiga brevicollis* and the origin of metazoans. Nature 451:783–788

Krupke A, Musat N, LaRoche J, Mohr W, Fuchs BM, Amann RI, Kuypers MMM, Foster RA (2013) In situ identification and $N_2$ and C fixation rates of uncultivated cyanobacteria populations. Syst Appl Microbiol 36:259–271

Krupke A, Lavik G, Halm H, Fuchs BM, Amann RI, Kuypers MMM (2014) Distribution of a consortium between unicellular algae and the $N_2$ fixing cyanobacterium UCYN-A in the North Atlantic Ocean. Environ Microbiol 16:3153–3167

Kühn SF (1998) Infection of *Coscinodiscus* spp. by the parasitoid nanoflagellate *Pirsonia diadema*: II. Selective infection behaviour for host species and individual host cells. J Plank Res 20:443–454

Kühn S, Medlin L, Eller G (2004) Phylogenetic position of the parasitoid nanoflagellate *Pirsonia* inferred from nuclear-encoded small subunit ribosomal DNA and a description of *Pseudopirsonia* n. gen. and *Pseudopirsonia mucosa* (Drebes) comb. nov. Protist 155:143–156

Lasek-Nesselquist E, Wisecaver JH, Hackett JD, Johnson MD (2015) Insights into transcriptional changes that accompany organelle sequestration from the stolen nucleus of *Mesodinium rubrum*. BMC Genomics 16:805

Lee MD, Webb EA, Walworth NG, Fu F-X, Held NA, Saito MA, Hutchins DA (2018) Transcriptional activities of the microbial consortium living with the marine nitrogen-fixing cyanobacterium *Trichodesmium* reveal potential roles in community-level nitrogen cycling. Appl Environ Microbiol 84:e02026–e02017

LeKieffre C, Spero HJ, Russell AD, Fehrenbacher JS, Geslin E, Meibom A (2018) Assimilation, translocation, and utilization of carbon between photosynthetic symbiotic dinoflagellates and their planktic foraminifera host. Mar Biol 165:104

Lima-Mendez G, Faust K, Henry N, Decelle J, Colin S, Carcillo F, Chaffron S, Ignacio-Espinosa JC, Roux S, Vincent F, Bittner L, Darzi Y, Wang J, Audic S, Berline L, Bontempi G, Cabello AM, Coppola L, Cornejo-Castillo FM, d'Ovidio F, De Meester L, Ferrera I, Garet-Delmas M-J, Guidi L, Lara E, Pesant S, Royo-Llonch M, Salazar G, Sánchez P, Sebastian M, Souffreau C, Dimier C, Picheral M, Searson S, Kandels-Lewis S, Tara Oceans Coordinators, Gorsky G, Not F, Ogata H, Speich S, Stemmann L, Weissenbach J, Wincker P, Acinas SG, Sunagawa S, Bork P, Sullivan MB, Karsenti E, Bowler C, de Vargas C, Raes J (2015) Determinants of community structure in the global plankton interactome. Science 348:1262073-1262071–1262073-1262079

Liu Z, Mesrop LY, Hu SK, Caron DA (2019) Transcriptome of *Thalassicolla nucleata* holobiont reveals details of a radiolarian symbiotic relationship. Front Mar Sci. https://doi.org/10.3389/fmars.2019.00284

Lucas IAN (1991) Symbionts of the tropical Dinophysiales (Dinophyceae). Ophelia 33:213–224

Mackiewicz P, Bodył A, Gagat P (2012) Protein import into the photosynthetic organelles of *Paulinella chromatophora* and its implications for primary plastid endosymbiosis. Symbiosis 58:99–107
Martinez-Perez C, Mohr W, Loscher CR, Dekaezemacker J, Littmann S, Yilmaz P, Lehnen N, Fuchs BM, Lavik G, Schmitz RA, LaRoche J, Kuypers MM (2016) The small unicellular diazotrophic symbiont, UCYN-A, is a key player in the marine nitrogen cycle. Nat Microbiol 1: 16163
Mayali X, Azam F (2004) Algicidal bacteria in the sea and their impact on algal blooms. J Eukaryot Microbiol 51:139–144
McFall-Ngai M (2014) Divining the essence of symbiosis: insights from the squid-*Vibrio* model. PLoS Biol 12:e1001783
Meheust R, Zelzion E, Bhattacharya D, Lopez P, Bapteste E (2016) Protein networks identify novel symbiogenetic genes resulting from plastid endosymbiosis. Proc Natl Acad Sci U S A 113: 3579–3584
Meyer E, Weis VM (2012) Study of cnidarian-algal symbiosis in the "omics" age. Biol Bull 223: 44–65
Michaels AF (1991) Acantharian abundance and symbiont productivity at the VERTEX seasonal station. J Plank Res 13:399–418
Mills MM, Turk-Kubo KA, van Dijken GL, Henke BA, Harding K, Wilson ST, Arrigo KR, Zehr JP (2020) Unusual marine cyanobacteria/haptophyte symbiosis relies on $N_2$ fixation even in N-rich environments. ISME J 14:2395–2406
Mitra A, Flynn KJ, Burkholder JM, Berge T, Calbet A, Raven JA, Granéli E, Glibert PM, Hansen PJ, Stoecker DK, Thingstad F, Tillmann U, Våge S, Wilken S, Zubkov MV (2014) The role of mixotrophic protists in the biological carbon pump. Biogeosciences 11:995–1005
Momper LM, Reese BK, Carvalho G, Lee P, Webb EA (2015) A novel cohabitation between two diazotrophic cyanobacteria in the oligotrophic ocean. ISME J 9:882–893
Monteil CL, Vallenet D, Menguy N, Benzerara K, Barbe V, Fouteau S, Cruaud C, Floriani M, Viollier E, Adryanczyk G, Leonhardt N, Faivre D, Pignol D, López-García P, Weld RJ, Lefevre CT (2019) Ectosymbiotic bacteria at the origin of magnetoreception in a marine protist. Nat Microbiol 4:1088–1095
Moon-van der Staay SY, De Wachter R, Vaulot D (2001) Oceanic 18S rDNA sequences from picoplankton reveal unsuspected eukaryotic diversity. Nature 409:607–610
Nakayama T, Ikegami Y, Nakayama T, Ishida K-i, Inagaki Y, Inouye I (2011) Spheroid bodies in rhopalodiacean diatoms were derived from a single endosymbiotic cyanobacterium. J Plant Res 124:93–97
Nakayama T, Kamikawa R, Tanifuji G, Kashiyama Y, Ohkouchi N, Archibald JM, Inagaki Y (2014) Complete genome of a nonphotosynthetic cyanobacterium in a diatom reveals recent adaptations to an intracellular lifestyle. Proc Natl Acad Sci USA 111(31):11407–11412
Norris RD (1996) Symbiosis as an evolutionary innovation in the radiation of Paleocene planktic foraminifera. Paleobiology 22:461–480
Not F, Probert I, Ribiero CG, Crenn K, Guillou L, Jeanthon C, Vaulot D (2016) Photosymbiosis in marine pelagic environments. In: Stal LJ, Cretoiu MS (eds) The marine microbiome. Springer, Cham, pp 305–330
Nowack EC, Grossman AR (2012) Trafficking of protein into the recently established photosynthetic organelles of *Paulinella chromatophora*. Proc Natl Acad Sci U S A 109:5340–5345
Nowack EC, Melkonian M (2010) Endosymbiotic associations within protists. Philos Trans R Soc Lond B Biol Sci 365:699–712
Ochman H, Moran NA (2001) Genes lost and genes found: evolution of bacterial pathogenesis and symbiosis. Science 292:1096–1099
O'Malley MA (2015) Endosymbiosis and its implications for evolutionary theory. Proc Natl Acad Sci U S A 112:10270–10277
Pasternak Z, Pietrokovski S, Rotem O, Gophna U, Lurie-Weinberger MN, Jurkevitch E (2013) By their genes ye shall know them: genomic signatures of predatory bacteria. ISME J 7:756–769

Paul C, Mausz MA, Pohnert G (2013) A co-culturing/metabolomics approach to investigate chemically mediated interactions of planktonic organisms reveals influence of bacteria on diatom metabolism. Metabolomics 9:349–359

Petersen JM, Kemper A, Gruber-Vodicka H, Cardini U, Van Der Geest M, Kleiner M, Bulgheresi S, Mußmann M, Herbold C, Seah BK (2017) Chemosynthetic symbionts of marine invertebrate animals are capable of nitrogen fixation. Nat Microbiol 2:16195

Postgate JR (1998) Nitrogen fixation. Cambridge University Press, Cambridge, UK, p 112

Prechtl J, Kneip C, Lockhart P, Wenderoth K, Maier UG (2004) Intracellular spheroid bodies of *Rhopalodia gibba* have nitrogen-fixing apparatus of cyanobacterial origin. Mol Biol Evol 21: 1477–1481

Probert I, Siano R, Poirier C, Decelle J, Biard T, Tuji A, Suzuki N, Not F (2014) *Brandtodinium* gen. nov. and *B. nutricula* comb. Nov. (Dinophyceae), a dinoflagellate commonly found in symbiosis with polycystine radiolarians. J Phycol 50:388–399

Pyle AE, Johnson AM, Villareal TA (2020) Isolation, growth, and nitrogen fixation rates of the *Hemiaulus-Richelia* (diatom-cyanobacterium) symbiosis in culture. PeerJ 8:e10115

Rai AN, Bergman B, Rasmussen U (2003) Cyanobacteria in symbiosis. Kluwer, New York, p 355

Ramanan R, Kim BH, Cho DH, Oh HM, Kim HS (2016) Algae-bacteria interactions: evolution, ecology and emerging applications. Biotechnol Adv 34:14–29

Rosset SL, Oakley CA, Ferrier-Pagès C, Suggett DJ, Weis VM, Davy SK (2020) The molecular language of the cnidarian–dinoflagellate symbiosis. Trends Microbiol 29:320–333. https://doi.org/10.1016/j.tim.2020.08.005

Sapp J (2004) The dynamics of symbiosis: an historical overview. Can J Bot 82:1046–1056

Sapp M, Schwaderer AS, Wiltshire KH, Hoppe HG, Gerdts G, Wichels A (2007) Species-specific bacterial communities in the phycosphere of microalgae? Microb Ecol 53:683–699

Schäfer H, Abbas B, Witte H, Muyzer G (2002) Genetic diversity of 'satellite' bacteria present in cultures of marine diatoms. FEMS Microbiol Ecol 42:25–35

Scholz B, Guillou L, Marano AV, Neuhauser S, Sullivan BK, Karsten U, Küpper FC, Gleason FH (2016) Zoosporic parasites infecting marine diatoms—a black box that needs to be opened. Fungal Ecol 19:59–76

Schvarcz CR, Wilson ST, Caffin M, Stancheva R, Li Q, Turk-Kubo KA, White AE, Karl DM, Zehr JP, Steward GF (in press) Overlooked and widespread pennate diatom-diazotroph symbioses in the sea. Nat Commun

Seymour JR, Amin SA, Raina JB, Stocker R (2017) Zooming in on the phycosphere: the ecological interface for phytoplankton-bacteria relationships. Nat Microbiol 2:17065

Shaked Y, de Vargas C (2006) Pelagic photosymbiosis: rDNA assessment of diversity and evolution of dinoflagellate symbionts and planktonic foraminiferal hosts. Mar Ecol Prog Ser 325:59–71

Sheridan CC, Steinberg DK, Kling GW (2002) The microbial and metazoan community associated with colonies of *Trichodesmium* spp.: a quantitative survey. J Plank Res 24:913–922

Siano R, Montresor M, Probert I, Not F, de Vargas C (2010) *Pelagodinium* gen. nov. and *P. bèii* comb. nov., a dinoflagellate symbiont of planktonic foraminifera. Protist 161:385–399

Siddiqui PJA, Bergman B, Carpenter EJ (1992) Filamentous cyanobacterial associates of the marine planktonic cyanobacterium *Trichodesmium*. Phycologia 31:326–337

Sierra R, Matz MV, Aglyamova G, Pillet L, Decelle J, Not F, de Vargas C, Pawlowski J (2013) Deep relationships of Rhizaria revealed by phylogenomics: a farewell to Haeckel's Radiolaria. Mol Phyl Evol 67:53–59

Singer A, Poschmann G, Mühlich C, Valadez-Cano C, Hänsch S, Hüren V, Rensing SA, Stühler K, Nowack ECM (2017) Massive protein import into the early-evolutionary-stage photosynthetic organelle of the amoeba *Paulinella chromatophora*. Curr Biol 27:2763–2773.e2765

Smriga S, Fernandez VI, Mitchell JG, Stocker R (2016) Chemotaxis toward phytoplankton drives organic matter partitioning among marine bacteria. Proc Natl Acad Sci U S A 113:1576–1581

Sockett RE (2009) Predatory lifestyle of Bdellovibrio bacteriovorus. Annu Rev Microbiol 63:523–539

Spero HJ (1987) Symbiosis in the planktonic foraminfer, *Orbulina universa*, and the isolation of its symbiotic dinoflagellate, *Gymnodinium beii* sp. nov. J Phycol 23:307–317
Spero HJ, Parker SL (1985) Photosynthesis in the symbiotic planktonic foraminifer *Orbulina universa*, and its potential contribution to oceanic primary productivity. J Foram Res 15:273–281
Stickney HL, Hood RR, Stoecker DK (2000) The impact of mixotrophy on planktonic marine ecosystems. Ecol Model 125:203–230
Stocker R (2012) Marine microbes see a sea of gradients. Science 338:628–633
Stoecker DK (1999) Mixotrophy among dinoflagellates. J Eukaryot Microbiol 46:397–401
Stoecker DK, Swanberg N, Tyler S (1989) Oceanic mixotrophic flatworms. Mar Ecol Prog Ser 58: 41–51
Stoecker DK, Buck KR, Putt M (1993) Changes in the sea-ice brine community during the spring-summer transition, McMurdo Sound, Antarctica. 2. Phagotrophic protists. Mar Ecol Prog Ser 95:103–113
Stoecker DK, Hansen PJ, Caron DA, Mitra A (2017) Mixotrophy in the marine plankton. Annu Rev Mar Sci 9:331–335
Swanberg NR (1979) The ecology of colonial radiolarians: their colony morphology, trophic interactions and associations, behavior, distribution and the photosynthesis of their symbionts. Ph.D., Woods Hole Oceanographic Institution and Massachusetts Institute of Technology
Swanberg NR (1983) The trophic role of colonial Radiolaria in oligotrophic oceanic environments. Limnol Oceanogr 28:655–666
Swanberg NR, Caron DA (1991) Patterns of sarcodine feeding in epipelagic oceanic plankton. J Plank Res 13:287–312
Swanberg NR, Anderson OR, Lindsey JL, Bennett P (1986) The biology of *Physematium muelleri*: trophic activity. Deep-Sea Res 33:913–922
Takagi H, Kimoto K, Fujiki T, Kurasawa A, Moriya K, Hirano H (2016) Ontogenetic dynamics of photosymbiosis in cultured planktic foraminifers revealed by fast repetition rate fluorometry. Mar Micropaleontol 122:44–52
Takagi H, Kimoto K, Fujiki T, Moriya K (2018) Effect of nutritional condition on photosymbiotic consortium of cultured Globigerinoides sacculifer (Rhizaria, Foraminifera). Symbiosis 76:25–39
Takagi H, Kimoto K, Fujiki T, Saito H, Schmidt C, Kucera M, Moriya K (2019) Characterizing photosymbiosis in modern planktonic foraminifera. Biogeosciences 16:3377–3396
Takayama T (1972) A note on the distribution of *Braarudosphaera bigelowii* (Gran and Braarud) Deflandre in the bottom sediments of Sendai Bay, Japan. Trans Proc Palaeont Soc Jpn 87:429–435
Tamm SL (1982) Flagellated ectosymbiotic bacteria propel a eukaryotic cell. J Cell Biol 94:697–709
Taylor FJR (1982) Symbioses in marine microplankton. Ann Inst Océanogr Paris 58(S):61–90
Thompson AW, Zehr JP (2013) Cellular interactions: lessons from the nitrogen-fixing cyanobacteria. J Phycol 49:1024–1035
Thompson AW, Foster RA, Krupke A, Carter BJ, Musat N, Vaulot D, Kuypers MMM, Zehr JP (2012) Unicellular cyanobacterium symbiotic with a single-celled eukaryotic alga. Science 337: 1546–1550
Tillmann U, Hesse K-J, Tillmann A (1999) Large-scale parasitic infection of diatoms in the Northfrisian Wadden Sea. J Sea Res 42:255–261
Tomaru Y, Kimura K, Nagasaki K (2015) Marine protist viruses. In: Ohtsuka S, Suzaki T, Horiguchi T, Suzuki N, Not F (eds) Marine protists: diversity and dynamics. Springer, Tokyo, pp 501–517
Trapp EM, Adler S, Zauner S, Maier U-G (2012) *Rhopalodia gibba* and its endosymbionts as a model for early steps in a cyanobacterial primary endosymbiosis. J Endocyt Cell Res 23:21–24
Tripp HJ, Bench SR, Turk KA, Foster RA, Desany BA, Niazi F, Affourtit JP, Zehr JP (2010) Metabolic streamlining in an open-ocean nitrogen-fixing cyanobacterium. Nature 464:90–94

Ushida K (2018) Symbiotic methanogens and rumen ciliates. In: Hackstein JHP (ed) (Endo)-symbiotic methanogenic archaea. Springer, Cham, pp 25–35

Uwizeye C, Mars Brisbin M, Gallet B, Chevalier F, LeKieffre C, Schieber N, Denis F, Wangpraseurt D, Schertel L, Stryhanyuk H, Musat N, Mitarai S, Schwab Y, Finazzi G, Decelle J (2020) Cytoklepty in the plankton: a host strategy to optimize the bioenergetic machinery of endosymbiotic algae. bioRxiv. https://doi.org/10.1101/2020.12.08.416644: 2020.2012.2008.416644

Valentine AJ, Benedito VA, Kang Y (2018) Legume nitrogen fixation and soil abiotic stress: from physiology to genomics and beyond. Annu Plant Rev 42:207–248

Villareal T (1989) Division cycles in the nitrogen-fixing *Rhizosolenia* (Bacillariophyceae)-*Richelia* (Nostocaceae) symbiosis. Br Phycol J 24:357–365

Villareal TA (1990) Laboratory cultivation and preliminary characterization of the nitrogen - fixing *Rhizosolenia - Richelia* symbiosis. Mar Ecol 11:117–132

Villareal TA (1991) Nitrogen-fixation by the cyanobacterial symbiont of the diatom genus *Hemiaulus*. Mar Ecol Prog Ser 76:201–204

Villareal TA (1992) Marine nitrogen-fixing diatom - cyanobacteria symbioses. In: Carpenter EJ, Capone DG, Rueter JG (eds) Marine pelagic cyanobacteria: *Trichodesmium* and other diazotrophs. Kluwer, Dordrecht, pp 163–175

Villareal TA (1994) Widespread occurrence of the *Hemiaulus*-cyanobacterial symbiosis in the southwest North Atlantic ocean. Bull Mar Sci 54:1–7

Villareal T, Adornato L, Wilson C, Shoenbachler C (2011) Summer blooms of diatom-diazotroph assemblages (DDAs) and surface chlorophyll in the N. Pacific gyre—a disconnect. J Geophys Res Oceans 116(C3):e6268

Vogels GD, Hoppe WF, Stumm CK (1980) Association of methanogenic bacteria with rumen ciliates. Appl Environ Microbiol 40:608–612

Vorobev A, Dupouy M, Carradec Q, Delmont TO, Annamalé A, Wincker P, Pelletier E (2020) Transcriptome reconstruction and functional analysis of eukaryotic marine plankton communities via high-throughput metagenomics and metatranscriptomics. Genome Res 30: 647–659

Wang Z, Wu M (2014) Phylogenomic reconstruction indicates mitochondrial ancestor was an energy parasite. PLoS One 9:e110685

Ward BA, Follows MJ (2016) Marine mixotrophy increases trophic transfer efficiency, mean organism size, and vertical carbon flux. Proc Natl Acad Sci USA 113:2958–2963

Weinbauer MG (2004) Ecology of prokaryotic viruses. FEMS Microbiol Rev 28:127–181

Wernegreen JJ (2017) In it for the long haul: evolutionary consequences of persistent endosymbiosis. Curr Opin Genet Dev 47:83–90

White AE, Prahl FG, Letelier RM, Popp BN (2007) Summer surface waters in the Gulf of California: prime habitat for biological $N_2$ fixation. Glob Biogeochem Cycle 21:GB2017. https://doi.org/10.1029/2006gb002779

Woznica A, Cantley AM, Beemelmanns C, Freinkman E, Clardy J, King N (2016) Bacterial lipids activate, synergize, and inhibit a developmental switch in choanoflagellates. Proc Natl Acad Sci USA 113:7894–7899

Wrede C, Dreier A, Kokoschka S, Hoppert M (2012) Archaea in symbioses. Archaea (Vancouver, BC) 2012:596846

Yarimizu K, Cruz-López R, Carrano CJ (2018) Iron and harmful algae blooms: potential algal-bacterial mutualism between *Lingulodinium polyedrum* and *Marinobacter algicola*. Front Mar Sci. https://doi.org/10.3389/fmars.2018.00180

Zehr JP (2015) How single cells work together. Science 349:1163–1164

Zehr JP, Capone DG (2020) Changing perspectives in marine nitrogen fixation. Science 368: eaay9514

Zehr JP, Capone DG (2021) Marine nitrogen fixation. Springer, Cham, p 186
Zehr J, Mellon M, Zani S (1998) New nitrogen-fixing microorganisms detected in oligotrophic oceans by amplification of nitrogenase (*nifH*) genes. Appl Environ Microbiol 64:34443450
Zehr JP, Shilova IN, Farnelid HM, Muñoz-Marín MD, Turk-Kubo KA (2016) Unusual marine unicellular symbiosis with the nitrogen-fixing cyanobacterium UCYN-A. Nat Microbiol 2: 16214
Zehr JP, Weitz JS, Joint I (2017) How microbes survive in the open ocean. Science 357:646–647

# Marine Extreme Habitats

# 14

Maria Pachiadaki and Virginia Edgcomb

**Abstract**

Extreme environments, habitats at the edge of survivability, can be found in most marine systems. These include habitats subjected to high radiation, high pressure, high or low temperatures, limited nutrient availability, or that contain high concentrations of salts, petroleum, or other toxic substances. Using this definition, the majority of the deep ocean and the marine deep subsurface—systems that harbor the most extensive microbiomes on Earth—would be classified as extreme. Because the microbial inhabitants of the deep sea, the subsurface, and the oceanic crust are discussed elsewhere in this book, this chapter will focus on hydrothermal vents and deep hypersaline anoxic basins, which have attracted the attention of the scientific community in recent decades due to their potential implications for astrobiology and biotechnology. Each of these two systems is characterized by the coexistence of *multiple* stressors (i.e., physicochemical parameters close to the limit of supporting life on Earth). The microorganisms inhabiting hydrothermal vents and deep hypersaline anoxic basins are called "polyextremophiles" and have attracted the attention of researchers who wish to gain knowledge about the adaptations to multiple extremes and the underlying mechanisms of the evolution of these marine microorganisms.

M. Pachiadaki (✉)
Biology Department, Woods Hole Oceanographic Institution, Woods Hole, MA, USA
e-mail: mpachiadaki@whoi.edu

V. Edgcomb
Geology & Geophysics Department, Woods Hole Oceanographic Institution, Woods Hole, MA, USA
e-mail: vedgcomb@whoi.edu



L. J. Stal, M. S. Cretoiu (eds.), *The Marine Microbiome*, The Microbiomes of Humans, Animals, Plants, and the Environment 3,
https://doi.org/10.1007/978-3-030-90383-1_14

**Keywords**

Astrobiology · Candidate phyla · Chemoautotrophy · Multiple stressors · Sulfur cycle · Symbiosis

## 14.1 Hydrothermal Vents

The discovery of the deep-sea hydrothermal vents in the mid-1970s on the Galapagos Rift (Weiss et al. 1977; Williams et al. 1974) profoundly changed our view of life on Earth. This discovery unveiled an entire ecosystem independent of sunlight and photosynthesis (Lonsdale 1977). Deep-sea hydrothermal vents have been found in several locations around the world in a variety of geological settings including mid-ocean ridges, back-arc spreading centers, volcanic arcs, and seamounts. Hydrothermal fluids, formed through seawater–rock interactions, are enriched in reduced chemical species such as hydrogen ($H_2$), hydrogen sulfide ($H_2S$), ferrous iron (Fe(II)), or methane ($CH_4$), with the actual composition and concentration depending on geological factors (Tivey 2007). The mixing of these reduced compounds with chemical oxidants from cold, oxygenated, deep-sea water either above or below the seafloor creates chemical disequilibria harnessed by metabolically versatile chemoautotrophic Bacteria and Archaea that fix $CO_2$ into biomass (Corliss et al. 1979; Jannasch and Wirsen 1979). The type of host rock and the rate of seafloor spreading as well as the depth, size, and shape of the heat source determine the hydrothermal geochemistry and the physical (e.g., temperature) characteristics of the vents. This influences the microbial community composition, specific metabolisms, and overall activity on both local and global scales (Amend et al. 2011). Vent microbial communities are also differentiated on a local scale through the creation of a variety of different niches along the steep thermal and chemical gradients within a vent site or even within individual vent structures such as the hydrothermal chimneys (e.g., Dick 2019 and references therein; Fortunato et al. 2018; Luther et al. 2001; Meier et al. 2017; Perner et al. 2013; Reysenbach et al. 2000a). The production of chemoautotrophic organic matter supports a diverse and abundant heterotrophic microbial community with high abundances[1] of Bacteria and Archaea. The organic carbon produced by chemoautotrophy at hydrothermal vents is transferred to higher trophic levels through grazing by microbial eukaryotes, small metazoans, and ultimately by invertebrate and vertebrate animals, or transferred through symbioses between different taxa. This makes these habitats the most productive ecosystems in the deep sea (Govenar 2012; Van Dover et al. 1988).

[1]Initial estimation of the abundance of Bacteria and Archaea in Galapagos Rift reached $10^9$ cells $ml^{-1}$ (Corliss et al. 1979). Subsequent studies at the same system reported abundances $10^5$–$10^6$ cells $ml^{-1}$ (Jannasch and Wirsen 1979; Karl et al. 1980). These abundances are 2–3-fold higher compared to the deep sea.

### 14.1.1 Processes and Microorganisms

#### 14.1.1.1 Sulfur Cycling

The initial hypothesis that explains the mechanism that sustains such high biomass in hydrothermal vent systems was the oxidation of sulfide as a source of energy (Corliss et al. 1979; Lonsdale 1977). Mafic host rocks (high MgO and FeO; Si>45%), typically emplaced along mid-ocean ridge spreading centers that exhibit fast to superfast spreading rates, expel high temperature acidic fluids enriched in sulfide and metals. Sulfide can serve as an electron donor for sulfur-oxidizing bacteria, while partially oxidized inorganic sulfur compounds (e.g., sulfite, thiosulfate, tetrathionate, elemental sulfur, and polysulfide) produced biotically and abiotically can be used as electron donors or acceptors in a variety of energy transformations. During the first expeditions to survey the Galapagos Rift hydrothermal vents more than 200 strains of microorganisms involved in sulfur cycling were isolated from water samples. The majority of these cultures oxidized hydrogen sulfide or thiosulfate as their preferred source of energy and were classified as *Thiobacillus*-like (Jannasch and Wirsen 1979). Detailed morphological and physiological characterization of strains from surface (biofilm) samples from the same system indicated their affiliation to the genus *Thiomicrospira* (Ruby and Jannasch 1982), which is now reclassified as *Hydrogenovibrio* (Boden et al. 2017; Jiang et al. 2017). Since then, several strains of *Thiomicrospira-Hydrogenovibrio* have been isolated from other vent systems reflecting the cosmopolitan abundance of the genus (e.g., Brinkhoff and Muyzer 1997; Jannasch et al. 1985; Muyzer et al. 1995; Wirsen et al. 1998). Several culture-independent surveys have also shown the high abundance of sulfur-oxidizing taxa belonging to Gammaproteobacteria (class of the bacterial phylum Proteobacteria) of the SUP05 lineage (e.g., Anderson et al. 2013; Fortunato et al. 2018).

Campylobacterota[2] (former class of Proteobacteria, Epsilonproteobacteria (Waite et al. 2017, 2018)) is another important phylum comprising sulfur-oxidizing bacteria that is found in hydrothermal vents. Members of this phylum have been detected and isolated from various habitats including vent fluids, walls of black-smoker chimneys, and biofilms (e.g., Huber et al. 2003, 2007; Longnecker and Reysenbach 2001; Reysenbach et al. 2000b; Takai et al. 2003a). Initial surveys of 16S rRNA gene sequences showed that the Campylobacterota inhabitants of hydrothermal vents were not among the cultured representatives of this phylum (Corre et al. 2001) but since then strains from all abundant genera have been isolated. Several of the Campylobacterota strains from hydrothermal vents appear to be sulfur and/or thiosulfate oxidizers including the mesophilic aerobic/microaerophilic *Sulfurimonas autotrophica* (Inagaki et al. 2003) and *Sulfurovum lithotrophicum* (Inagaki et al.

[2]For the taxonomy of Bacteria and Archaea in this chapter primarily the taxonomy used by the authors in the original papers is followed. For groups that have been reclassified such as the former class of Proteobacteria that are now a new phylum Campylobacterota the appropriate citation is provided and the newer classification is used.

2004) while other strains can oxidize both hydrogen and sulfur, e.g. *Sulfurimonas paralvinellae* (Takai et al. 2006) and *Sulfurimonas indica* (Hu et al. 2021a).

Several of the genes involved in sulfur oxidation have been surveyed in order to better understand the niche partitioning and ecophysiology of sulfur oxidizers in hydrothermal vents (e.g., Akerman et al. 2013). Bacteria use two main pathways for the oxidation of reduced sulfur compounds: (1) the reverse sulfate reduction (rDSR), and (2) the Sox (Dahl et al. 2008). The first pathway involves the same gene families as for sulfate reduction, i.e., the (reverse) dissimilatory sulfite reductase (*rdsr*) that encodes the enzyme for conversion of sulfide to sulfite and the adenylphosphosulfate reductase (*apr*) and sulfate adenylyltransferase (*sat*) that leads to the formation of sulfate. The complete Sox system (all genes of the cluster present: *soxY*, *soxZ*, *soxA*, *soxB*, *soxC*, and *soxD*) can be used to catalyze the oxidation of elemental sulfur, sulfide, and thiosulfate. The genomes of several organisms lack the *soxC* and *soxD* genes but these organisms can still use sulfite and the sulfone group in thiosulfate (Friedrich et al. 2005). Several other genes have been reported to participate in the oxidation of sulfur compounds such as sulfide:quinone oxidoreductase (*sqr*) and flavocytochrome c (*fcc*), which catalyze the oxidation of sulfide to elemental sulfur (Fukumori and Yamanaka 1979; Shahak et al. 1992).

Metatranscriptomic surveys have shown that the aforementioned genes are among the most highly transcribed in samples from the Axial Seamount hydrothermal vent (especially in the plume) and the Mid-Cayman Rise field, as well as in *Alviniconcha* snails living symbiotically with sulfur-oxidizing bacteria (Fortunato et al. 2018; Galambos et al. 2019; Sanders et al. 2013). The prevalence of expressed sulfur and sulfide oxidizing genes highlights the importance of sulfide oxidation in these systems. Although molecular signatures of Campylobacterota and Gammaproteobacteria SUP05 often overlap in environmental surveys, differences in abundance and/or activity (using messenger RNA as a proxy for activity) indicate that these two groups show niche partitioning. Generally, the SUP05 lineage appears to preferentially occupy cold, low sulfide conditions (e.g., plumes), while *Sulfurovum*-related and *Sulfurimonas*-related sulfur oxidizers dominate fluids with moderate temperatures and elevated sulfide concentrations. Vent location and a combination of conditions that determines the local geochemistry play a role in the distribution of specific taxa (Akerman et al. 2013; Fortunato et al. 2018; Meier et al. 2017).

Partially oxidized sulfur compounds, particularly sulfur and sulfate, can be reduced during respiration. Several hydrogen oxidizers (discussed in detail below) can use elemental sulfur as a terminal electron acceptor. Sulfur respiration requires at least two enzymes, a hydrogenase and a sulfur or polysulfide reductase, in the chemoautotrophic hyperthermophilic archaeon *Pyrodictium abyssi*, isolated from the Mid-Atlantic Ridge (Pley et al. 1991). In the hyperthermophilic hydrogenotrophic bacterium *Aquifex aeolicus* (Deckert et al. 1998) a multiprotein sulfur-reducing membrane-bound supercomplex is used coupling hydrogen oxidation to sulfur reduction in the presence of quinone (Guiral et al. 2005). Sulfur metabolism of *Thermovibrio ammonificans* appears to require (abiotic) transition from bulk elemental sulfur to polysulfide to nanoparticulate sulfur at an acidic pH. Sulfur reduction is coupled to biological hydrogen oxidation and mediated by an NADH-dependent sulfur reductase as the terminal reductase (Jelen et al. 2018). Several thermophilic heterotrophic taxa can also respire sulfur (discussed in detail in Bonch-Osmolovskaya 1994).

The other respiration pathway, sulfate reduction, has been well-studied in terms of microbial diversity and rates of the process. Several sulfate-reducing Bacteria and Archaea have been isolated from hydrothermal vents including the mesophilic *Desulfovibrio hydrothermalis* (Alazard et al. 2003), the thermophilic *Archaeoglobus profundus* (Burggraf et al. 1990), *Thermodesulfobacterium hydrogeniphilum* (Jeanthon et al. 2002), and *Thermodesulfatator indicus* (Moussard et al. 2004) as well as strains that had not previously been characterized (Stetter et al. 1987). Culture-independent approaches have unveiled diverse active sulfate-reducing microbial taxa belonging to Thermodesulfobacteria, Deltaproteobacteria, Archaeoglobi, Nitrospirae, and others in hydrothermal vent chimneys and sediments (e.g., Dhillon et al. 2003; Frank et al. 2013; Nakagawa et al. 2004; Nercessian et al. 2005). Sulfate reduction occurs above 100 °C in deep-sea hydrothermal vent sediments (Jørgensen et al. 1992), while temperature and pH appeared to be the key factors determining the rate of sulfate reduction (Frank et al. 2013, 2015).

#### 14.1.1.2 Hydrogen Oxidation

Ultramafic systems (ultraslow to slow spreading rates with high MgO and FeO, and Si<45%) are characterized by a temperature of 50–90 °C and alkaline fluids enriched in hydrogen, methane, and other hydrocarbons. Hydrogen oxidation is one of the most important inorganic energy sources in these systems providing large amounts of energy (237 kJ $mol^{-1}$) for ATP synthesis and autotrophic $CO_2$ fixation. Although oxidation of sulfide or thiosulfate yields approximately 3 times the amount of energy (797 kJ $mol^{-1}$) hydrogen oxidation is more favorable for autotrophic carbon fixation (Adam and Perner 2018a and references therein). Only a third of the energy is required for fixing 1 mol of carbon when oxidizing hydrogen since the redox potential of hydrogen is more negative than that of the reducing equivalent NAD(P)/H and a reverse electron transport is not required. The actual energy gain depends on the terminal electron acceptor with oxygen leading to the highest yields but several other compounds can be also be used including nitrate, sulfate, thiosulfate, sulfur, formate, $CO_2$, as well as various metals, e.g. Fe(III), Mn (III/IV), U (VI), Cr (VI), Co (III), and Tc (VII) (Lin et al. 2016; Liu et al. 2002).

Several aerobic and anaerobic hydrogen oxidizers have been isolated from hydrothermal vents. Many of the Camplylobacterota isolates can grow chemoautotrophically using hydrogen as energy source such as several species of the genus *Nautilia*, e.g., *Nautilia lithotrophica* (Miroshnichenko et al. 2002), *Nautilia abyssi* (Alain et al. 2009), and *Nautilia nitratireducens* (Pérez-Rodríguez et al. 2010), *Caminibacter hydrogeniphilus* (Alain et al. 2002), *Lebetimonas acidiphila* (Takai et al. 2005), *Cetia pacifica* (Grosche et al. 2015), *Hydrogenimonas thermophila* (Takai et al. 2004a), and *Sulfurovum aggregans* (Mino et al. 2014). These isolates use elemental sulfur as the primary electron acceptor producing hydrogen sulfide while some can also use nitrate leading to the formation of ammonium. Several hydrogenotrophic thermophilic Desulfobacterota (former class of Proteobacteria, Deltaproteobacteria (Waite et al. 2020)) have been identified from hydrothermal vents including the autotrophic sulfate-reducing *Desulfacinum hydrothermalis* (Sievert and Kuever 2000). Many autotrophic hydrogen-oxidizing Aquificae have also been isolated from hydrothermal vents such as *Desulfurobacterium thermolithotrophum* (L'Haridon et al. 1998), *Thermovibrio ruber* (Huber et al. 2002a) and *Thermovibrio ammonificans* (Vetriani et al. 2004), *Persephonella marina* and *Persephonella guaymasensis* (Götz et al. 2002), *Balnearium lithotrophicum* (Takai et al. 2003b), *Phorcysia thermohydrogeniphila* (Pérez-Rodríguez et al. 2012), and *Hydrogenothermus marinus* (Stohr et al. 2001), all of which are thermophilic or extreme thermophilic organisms.

Hydrogenases are enzymes involved in both the production and the consumption of molecular hydrogen. Genes encoding hydrogenases are widely distributed among Bacteria, Archaea, and unicellular Eukaryotes. The functional diversity of hydrogenases is reflected in their metal-binding motifs, phylogenetic clustering, and gene organization (Greening et al. 2016). Based on their structure, hydrogenases are categorized in three different classes (Vignais and Billoud 2007). Two of these classes, the [NiFe]-hydrogenases and the [FeFe]-hydrogenases, catalyze the reversible oxidation of hydrogen to protons and electrons. The [NiFe]-hydrogenases are involved in hydrogen sensing and consumption and have been subdivided into four subgroups (Greening et al., 2016; Vignais 2008). The [FeFe]-hydrogenases are subdivided into three groups. They are primarily involved in hydrogen evolution, i.e. hydrogen production and hydrogen sensing (Greening et al. 2016; Søndergaard et al. 2016). The third class, the [Fe]-hydrogenases, catalyzes the reduction of methenyltetrahydromethanopterin in methanogenic Archaea (Thauer 1998).

Hydrogenase genes have been frequently detected in hydrothermal vent metagenomes and in genomes from vent isolates (e.g., Brazelton et al. 2012; Fortunato et al. 2018; Han and Perner 2014; Hansen and Perner 2016). The high abundance of [NiFe]-hydrogenases involved in $H_2$ uptake and oxidation in the metagenomes from the vent chimneys of the ultramafic system Lost City indicates the potential importance of aerobic or facultatively anaerobic Betaproteobacteria belonging to the order Burkholderiales in the $H_2$-fueled (or $H_2$-controlled) primary production at that site (Brazelton et al. 2012). While, biofilm metatranscriptomes from Loki's Castle chimneys in the Arctic Mid-Ocean Ridge revealed high abundance of [NiFe]-hydrogenase transcripts belonging to Campylobacterota (Dahle et al. 2013). The analysis of RNA after stable isotope probing (RNA-SIP) using $^{13}$C-bicarbonate to target chemoautotrophs in the Axial Seamount also showed that Campylobacterota using hydrogen oxidation coupled to nitrate reduction dominated the active microbial community at moderate temperatures (Fortunato and Huber 2016). Activity-based screening applied to metagenomic fosmid libraries from three different hydrothermal vents recovered $H_2$-uptake and $H_2$-oxidizing hydrogenases affiliated to Gammaproteobacteria, Betaproteobacteria, Campylobacterota, Firmicutes, Planctomycetes, Actinobacteria, and Aquificae (Adam and Perner 2018b). The high levels of enzymatic $H_2$-uptake, the in vivo hydrogen consumption of engineered cells containing the metagenomically-identified hydrogenase gene clusters, and their low similarity to known hydrogenases indicate the large hydrogen-converting potential of the yet-uncultured hydrothermal vent microbes (Adam and Perner 2018b).

#### 14.1.1.3 Methanogenesis and Anaerobic Oxidation of Methane

The high concentrations of hydrogen in ultramafic systems can also be used by hydrogenotrophic methanogens, which thrive especially in high temperature regimes. Several thermophilic and hyperthermophilic methanogenic Archaea have been isolated from hydrothermal vents. *Methanocaldococcus*[3] *vulcanius* (Jeanthon et al. 1999), *Methanocaldococcus*[3] *infernus* (Jeanthon et al. 2002), *Methanocaldococcus indicus* (L'Haridon et al. 2003), and *Methanofervidicoccus abyssi* (Sakai et al. 2019) are strict hydrogenotrophic methanogens while *Methanocaldococcus*[3] *jannaschii* (Jones et al. 1983), the first extremophile that has been sequenced, *Methanothermococcus okinawensis* (Takai et al. 2002), *Methanotorris formicicus* (Takai et al. 2004b) can all use $H_2$ as well as formate for methanogenesis.

Amplicon sequencing surveys using as marker gene either 16S rRNA and/or methyl-coenzyme M reductase (subunit A; *mcrA*), a diagnostic gene for the last step of methanogenesis and the first step of anaerobic methane oxidation, have detected high abundance of hydrogen and formate utilizing methanogens of the Methanococcales and the Methanomicrobiales families in several hydrothermal vents (Dhillon et al. 2005; Huber et al. 2002b; Lever and Teske 2015). Archaeal lineages that can perform anaerobic methane oxidation have also been reported (Dhillon et al. 2005; Huber et al. 2002b; Lever and Teske 2015; Reed et al. 2009). The bimodal distribution of the archaeal taxa in Guaymas Basin suggests local peaks in hydrogenotrophic methanogenesis and potentially different $H_2$ sources, e.g., microbial diagenesis near the surface and thermal degradation of organic matter in deeper layers (Lever and Teske 2015). The detection of molecular signatures affiliated to ANME-2 in the same samples indicated that anaerobic oxidation of methane might occur concurrently with methanogenesis (Lever and Teske 2015). The archaeal microbial communities in the actively venting carbonate chimneys of Lost City showed low archaeal diversity. The higher-temperature samples were dominated by anaerobic methanogens from the Methanosarcinales order while taxa related to ANME-1 occupied the cooler areas (Brazelton et al. 2006; Kelley et al. 2005; Schrenk et al. 2004). The ANME-1 Archaea had been originally assumed to perform anaerobic oxidation of methane but other evidence indicated that they are involved in methanogenesis (Lloyd et al. 2011).

#### 14.1.1.4 Other Chemoautotrophic Processes

Several other sources of energy can be used for chemoautotrophic growth in hydrothermal vents. For instance, the elevated ammonium concentrations in hydrothermal vents can fuel aerobic ammonium-oxidizing Bacteria and Archaea. The high

[3]Former *Methanococcus*

rate of aerobic ammonium oxidation measured in the plume directly over Main Endeavour Field on the Juan de Fuca Ridge was projected to yield chemoautotrophic organic carbon production at least 4 times the amount of the sinking photosynthetically produced organic carbon that reach plume depths (Lam et al. 2004). *Nitrosomonas*-like Bacteria appeared to be the most abundant ammonium oxidizers. Bacteria dominate the ammonium-oxidizing community at a hydrothermal vent site at the Mid-Atlantic Ridge in the south Atlantic Ocean (Xu et al. 2014) but signatures of Marine Group I (MG-I) archaeal ammonium oxidizers have been recovered from several hydrothermal systems (e.g., Huber et al. 2002b; Takai and Horikoshi 1999; Wang et al. 2009). Anaerobic ammonium oxidation (anammox) occurs in chimney samples from hydrothermal vents. This has been demonstrated by using isotope pairing experiments while 16S rRNA gene sequences indicate that novel deep-branching taxa within the Planctomycetes phylum are likely the anammoxers in these hot habitats (Byrne et al. 2009; Russ et al. 2013).

Aerobic methane oxidation and ammonium oxidation appeared to be an important source of chemoautotrophically produced organic carbon in the plumes of the Endeavour Field (de Angelis et al. 1993). Amplicon surveys of the 16S rRNA gene and the methane monooxygenase gene (subunit A; pmoA) as well as metagenomic and metatranscriptomic studies have shown the presence of Type I methanotrophs belonging to the Gammaproteobacteria at various hydrothermal systems including the Lau Basin, Guaymas Basin, the Rainbow Vent fields, the Islands of Japan at the Mid-Okinawa Trough, and the Trans-Atlantic Geotraverse (e.g., Elsaied et al. 2004; Hirayama et al. 2007; Reed et al. 2009; Sylvan et al. 2013). Additionally, Type I methanotrophs such as *Methylomarinum vadi* (Hirayama et al. 2013) and *Methylomarinovum caldicuralii* (Hirayama et al. 2014) have been isolated from hydrothermal vent systems. Genomic reconstructions of Gammaproteobacterial methanotrophs from vent metagenomes indicated the existence of multiple adaptations to oxygen limitation within the highly dynamic vent environment and the potential to utilize hydrocarbons and organic sulfur in addition to C1 compounds (Skennerton et al. 2015; Zhou et al. 2020).

Ferrous iron (Fe(II)) is the most dominant metal in many vent fluids reaching concentrations of several mM (Charlou et al. 2002). The biological oxidation of Fe (II), a reaction that yields a minimal amount of energy for growth (109 kJ $mol^{-1}$), can also occur chemically in the presence of high concentrations of oxygen and in neutral pH. Thus, the oxygen-depleted and Fe(II)-rich fluids provide a favorable environment for iron-oxidizing bacteria. The first cultured isolate of a Fe(II) oxidizer from metalliferous deposits at Loihi Seamount, *Mariprofundus ferrooxydans* (Emerson et al. 2007), represents a new class of Proteobacteria, the Zetaproteobacteria. Molecular signatures of Zetaproteobacteria have been recovered from several hydrothermal vents (e.g., Emerson and Moyer 2002; Jesser et al. 2015; Johannessen et al. 2017; Kato et al. 2009; McAllister et al. 2011; Vander Roost et al. 2018) but the contribution of iron oxidizers to primary production has not been fully assessed.

#### 14.1.1.5 Carbon Fixation

One of the first activity measurements made on samples from the Galapagos Rift hydrothermal system soon after its discovery was the incorporation of inorganic carbon into microbial biomass and this demonstrated the importance of chemosynthesis in hydrothermal vents (Jannasch and Wirsen 1979). In situ incubation experiments showed high rates of $CO_2$ fixation (Jannasch and Wirsen 1979; Karl et al. 1980; Tuttle et al. 1983; Wirsen et al. 1986) comparable to the rates measured in oxic/anoxic interfaces of sulfidic systems such as the Cariaco Basin and the Black Sea (Gupta and Jannasch 1973; Tuttle and Jannasch 1979). Some of these earlier studies also detected high enzymatic activities of ribulose-1,5-bisphosphate carboxylase-oxygenase[4] (RuBisCO) the key enzyme of the Calvin–Benson–Bassham (CBB) carbon fixation cycle (e.g., Hügler and Sievert 2010 and references therein). The CBB cycle is probably the most prevalent pathway of $CO_2$ fixation because it is used by most photoautotrophs (e.g., Cyanobacteria and plants) but it is also found in several chemoautotrophs inhabiting hydrothermal vents including the free-living sulfur-oxidizing Gammaproteobacteria (Mußmann et al. 2007; Scott et al. 2006) and the iron-oxidizing Zetaproteobacteria (Emerson et al. 2007).

Campylobacterota and Aquificales assimilate inorganic carbon through the reductive tricarboxylic acid (rTCA) cycle (Beh et al. 1993; Hügler et al. 2005, 2007). The citrate cleaving enzymes, ATP-citrate lyase (*acl*) and citryl-CoA synthetase (*ccs*), diagnostic for the rTCA cycle have been used to investigate the distribution of carbon-fixating microorganisms in hydrothermal vents (e.g., Campbell and Cary 2004; Hügler et al. 2010; Perner et al. 2007). Genomic and biochemical evidence show that MG-1 archaeal ammonium oxidizers assimilate inorganic carbon via a modified version of the 3-hydroxypropionate/4-hydroxybutyrate cycle (3-HP/4-HB) (Berg et al. 2007; Hallam et al. 2006; Könneke et al. 2014) while methanogenic Archaea use the reductive acetyl coenzyme A (acetyl-CoA) pathway also known as the Wood–Ljungdahl (WL) pathway (Berg et al. 2010; Meuer et al. 2002).

The oxygen sensitivity of the enzymes involved and the energy costs required to fix $CO_2$ differ among pathways potentially affecting the distribution of chemoautotrophic lineages. The enzymes of the CBB and 3-HP/4-HB cycles are oxygen tolerant. The chemoautotrophs that use CBB and 3-HP/4-HB cycles at vents, the Gammaproteobacteria and MG-I Archaea, respectively, mostly inhabit the oxic or microoxic peripheral zones of hydrothermal vents while lineages using the oxygen-sensitive rTCA cycle, Campylobacterota and Aquificales, occur in habitats with more reduced conditions and temperatures between 20 and 90 °C (Nakagawa and Takai 2008 and references therein). The CBB and the modified 3-HP/4-HB pathways are energetically expensive as they require the investment of 7 and 5 ATP, respectively, for the synthesis of one molecule of pyruvate (Bar-Even

[4]Several forms of RuBisCO genes have been identified; only forms I and II have been enzymatically shown to fixate $CO_2$. The following discussion only refers to these two forms. The form III RuBisCO detected in hyperthermophilic Euryarchaeota inhabiting vent systems is involved in AMP metabolism (Sato et al. 2007).

et al. 2012; Könneke et al. 2014) whereas the rTCA cycle costs only 2 ATP. The acetyl-CoA pathway used by methanogenic Archaea, which are mostly found in the high temperature areas of hydrothermal vents, requires less than one ATP but has a high demand for metals, cofactors, and substrates with low reducing potential (Berg et al. 2010).

Estimates of chemoautotrophic productivity of hydrothermal vents range from 1 to 5 TgC per year (Jannasch 1995; McCollom and Shock 1997; McNichol et al. 2018). $CO_2$ fixation appears to be stimulated by supplementing vent emissions with reduced sulfur compounds (e.g., Tuttle et al. 1983) or $H_2$ (McNichol et al. 2018) that can be used as an energy source as well as the addition of terminal electron acceptors such as $O_2$ or $NO_3^-$ (McNichol et al. 2018). Single cell activity measurements in combination with 16S rRNA gene analysis show that in the Crab Spa hydrothermal vent of the East Pacific Rise Campylobacterota are the dominant carbon fixators with oxygen concentrations and temperature driving the niche partitioning of closely related taxa (McNichol et al. 2018).

### 14.1.2 Symbiosis

Soon after the discovery of hydrothermal vents the existence of symbiosis between invertebrates and bacteria has been suggested. Histologic analysis of the vestimentiferan tubeworm *Riftia pachyptila*, one of the most abundant and charismatic animals found at hydrothermal vents, showed the existence of a specialized organ, the trophosome, containing densely packed microbial cells (Cavanaugh et al. 1981). The presence of sulfur crystals and the high activity of enzymes involved in the sulfur cycle and in carbon fixation within the trophosomal tissue indicated that these microorganisms are likely chemoautotrophic sulfur-oxidizing bacteria that nourish their mouthless and gutless host through the release of nutrients (Cavanaugh et al. 1981). Since then, the existence of endobiotic or ectobiotic symbioses has been demonstrated in a variety of vent animals including mussels, bivalves, shrimps, and snails. Early surveys of chemosynthetic symbiotic associations indicated that they are highly specific (Distel et al. 1988; Polz et al. 1994) but subsequent studies have revealed that the diversity of symbionts is much higher than originally assessed and often cryptic (e.g., Beinart et al. 2012; Dubilier et al. 2008; Petersen et al. 2010).

Most of the bacteria involved in these symbioses are likely sulfur and/or hydrogen oxidizers belonging to the Gammaproteobacteria but Campylobacterota symbionts are also common especially as ectobionts (e.g., Beinart et al. 2012; Dubilier et al. 2008; Miyazaki et al. 2020; Petersen et al. 2010). Phylogenetic analysis of symbiotic and free-living sulfur-oxidizing chemoautotrophic Gammaproteobacteria suggests that symbioses have evolved on multiple independent occasions (Dubilier et al. 2008). The niche utilization of the host-symbiont systems (holobionts) appears to be determined by the function and activity of the bacterial symbionts corresponding to differences in vent geochemistry (Beinart et al. 2012; Sanders et al. 2013). A high degree of symbiont heterogeneity appears to exist within single holobionts. For example, the symbiotic consortia of the tubeworm

*Riftia pachyptila* consist of a polyclonal population of the gammaproteobacterial *Candidatus* Endoriftia persephone (Polzin et al. 2019) that exhibits remarkable metabolic versatility and redundancy including the expression of enzymes of two $CO_2$ fixation pathways, the CBB and rTCA pathways (Markert et al. 2007). Metaproteomics, microscopy, and flow cytometry indicate that symbiotic cells of different sizes represent metabolically dissimilar stages of a physiological differentiation process. The larger cells appear to be dedicated to carbon fixation whereas the smaller cells are actively dividing and potentially establishing symbiont–host interactions (Hinzke et al. 2021).

Methanotrophic symbiosis has also been discovered in deep-sea hydrothermal vents (e.g., Petersen and Dubilier 2010 and references therein). Mussels of the genus *Bathymodiolus* contain endosymbiotic aerobic methane-oxidizing Gammaproteobacteria related to the free-living methanotrophs of the genera *Methylobacter* and *Methylomicrobium* in several vent systems (Dubilier et al. 2008; Petersen and Dubilier 2009). The intracellular coexistence of methanotrophic and thiotrophic bacteria has also been observed in mussels from hydrothermal vents in the Mid-Atlantic Ridge and the Gulf of Mexico (Distel et al. 1988; Fisher et al. 1993).

### 14.1.3 Microbial Eukaryotes

Protists and fungi play important roles in deep-sea hydrothermal habitats as grazers of bacterial and archaeal prey, of parasites of other protists, and of Metazoa, in symbioses with bacteria and archaea and hydrothermal vent bivalves, and, in the case of fungi, as saprophytes living on particulate organic matter (Edgcomb 2016; Sauvadet et al. 2010). Heterotrophic protists are a critical energy link in hydrothermal food webs between microbial prey and metazoa. On the basis of microscopic and culture-based studies, ciliated protists and various flagellates were first documented to be present in hydrothermal vent sediments and waters from the Eastern Pacific and Northern Pacific Juan de Fuca Ridge (Atkins et al. 1998, 2000; Kouris et al. 2007; Small and Gross 1985). These studies were followed by marker gene-based studies of hydrothermal vent sediment protists at Guaymas Basin, Mexico (Edgcomb et al. 2002), Western Pacific Mariana Arc (Murdock and Juniper 2019), Mid-Atlantic Ridge and Lost City (López-García et al. 2003, 2007) that unveiled diverse assemblages of alveolates, stramenopiles, and Rhizaria. Microscopy and molecular marker gene surveys of six hydrothermal sites in the Pacific Ocean revealed differences between deep-pelagic water protist communities and those found in the immediate proximity of hydrothermal vent chimneys (Sauvadet et al. 2010). While deep-sea pelagic protist communities were dominated by dinoflagellates, radiolarians, and Syndiniales parasites, hydrothermal habitats were dominated by opportunistic detritivores and grazers belonging to Stramenopiles and Cercozoa. Additionally, protist communities found in association with the giant hydrothermal bivalves *Bathymodiolus thermophilus* and *Calyptogena magnifica* were primarily members of Ciliophora and Cercozoa. Microcolonizer devices deployed in sediments at the Mid-Atlantic Ridge were used to provide evidence of active protists in hydrothermal habitats (López-García et al. 2003). After 15 days of deployment an impressive diversity of protist colonizers appeared including kinetoplastids (predominantly bodonids) and alveolates (predominantly ciliates). Sauvadet et al. (2010) found that bivalve-associated protist communities were complex with differences between the hydrothermal fields. In some cases, these protists appeared to be parasites while in other cases they were grazers of food filtered by the bivalve, were symbionts of the bivalve, or were detritivores. While many hydrothermal vent habitats have yet to be explored for protist diversity the degree of endemism is still debated. The apparently endemic protist taxa to particular vent sites appear to be among the less frequently encountered and often are novel lineages at the phylum level as well as within defined clades of Rhizaria and Stramenopila that may serve as a reservoir ready to respond to the typically changing environmental conditions at hydrothermal vents (Murdock and Juniper 2019).

Grazing by protists on prey in hydrothermal fluids and on sediment microbiota is suspected to importantly impact carbon turnover in hydrothermal habitats within the temperature limits for eukaryotes. Protists transfer energy stored in this bacterial and archaeal biomass to higher trophic levels and they exhibit prey preferences and varied feeding strategies that can place selective pressure on prey species community composition. These processes make important contributions to the deep-sea microbial loop at hydrothermal sites. The first study of protist grazing was conducted using hydrothermal fluids collected from Gorda Ridge in the North East Pacific Ocean (Hu et al. 2021b). This study combined grazing experiments with a molecular survey of protist diversity to gauge grazing impacts at Gorda Ridge and to characterize the protist community in low temperature (10–80 °C) diffusely venting fluids collected at Sea Cliff and Apollo vent fields at this site. Consistent with higher (double) prey abundance in nutrient-rich vent fluids relative to nutrient-poorer deep seawater at Gorda Ridge, grazing pressure was higher in these vent fluids relative to background deep seawater. The protist community in Gorda Ridge vent fluids was dominated by ciliates, dinoflagellates, Syndiniales (a parasitic group of protists that may represent an important source of mortality to the protists themselves and possibly to other small metazoa), rhizaria, and stramenopiles. Network analyses suggest that the preferred prey groups included some of the most abundant bacterial groups: Alphaproteobacteria, Gammaproteobacteria, *Nitrososphaera*, and *Sulfurimonas*. Protists at this site grazed 28–62% of the daily stock of bacteria in the hydrothermal fluids that were used for these experiments, namely between 700 and 1828 cells $ml^{-1}$ $h^{-1}$. In nearby non-hydrothermally-influenced deep seawater the grazing rate was 255 cells $ml^{-1}$ $h^{-1}$. This shows that the vent microbial community was under greater top-down pressure relative to communities in background deep seawater. It is often thought that in the deep-sea protists are not important as grazers due to low predicted prey encounter rates. It now becomes evident that protists can at least partially overcome the challenge of low prey encounters by associating with sinking particles that contain abundant prey (Fenchel 1987; Sherr and Sherr 1994). Deep-sea chemocline habitats such as interfaces between hydrothermal fluids and deep-seawater host abundant bacterial and archaeal communities for reasons described above. Therefore, these habitats serve as oases for protists where feeding on dense populations of prey is possible. Because new technology enables grazing experiments in hydrothermal vents to be conducted in situ the results of such investigations may reveal that the grazing impact can be even greater than hitherto known.

## 14.2 Deep Hypersaline Anoxic Basins

Deep Hypersaline Anoxic Basins (DHABs) were discovered earlier than hydrothermal vents but the study of their microbial inhabitants were delayed for several years. In 1964, two brine-filled depressions with elevated temperatures were discovered in the Red Sea: the Atlantis II Deep and Discovery Deep (Charnock 1964; Miller 1964; Swallow and Crease 1965). Follow-up explorations have detected many more brine-filled basins in the central Red Sea (Backer and Schoell 1972). In 1977, Orca Basin, a 400 $km^2$ depression containing anoxic hypersaline water was found along the continental slope of the northern Gulf of Mexico (Shokes et al. 1977). The first Mediterranean DHAB, Tyro, was discovered in 1983 (Jongsma et al. 1983). Eight more have since been found in the Eastern Mediterranean Sea: L'Atalante, Bannock, Discovery, Hephaestus, Kryos, Medee, Thetis, and Urania (La Cono et al. 2011, 2019; Yakimov et al. 2015). The Mediterranean DHABs are the deepest (up to ~3500 m), the Orca Basin is ~2200 m deep, while the depth of the Red Sea DHABs varies from ~350 m (Afifi) to ~2800 m (Suakin). The basins in the Eastern Mediterranean Sea, the Gulf of Mexico, and the Red Sea differ in their geological settings but they all originate from ancient evaporites deposited during periods of extremely low sea levels. The evaporites in the Orca Basin are of the Jurassic age while those of the Red Sea and Mediterranean Sea were formed during the late Miocene Messinian salinity crisis 5590–5330 million years ago (Camerlenghi 1990; Ross et al. 1973). Deposition of seawater evaporites is a stepwise process of precipitation of different minerals determined by solubility and temperature. It begins with formation and sedimentation of calcium carbonates and sulfates followed by precipitation of halite (NaCl), kieserite ($MgSO_4 \cdot 3H_2O$), carnallite ($KMgCl_3 \cdot H_2O$), kainite ($MgSO_4 \cdot KCl \cdot 3H_2O$) and finally with the formation of bischofite ($MgCl_2 \cdot 6H_2O$). As the result, evaporitic deposits resemble a layer cake. Modern-day dissolution of certain exposed strata leads to formation of brines that accumulate in seafloor depressions and have distinct chemical compositions. These brines may have ion contents similar to that of seawater (thalassohaline brines) or drastically different (athalassohaline brines). In the Mediterranean Sea, Discovery, Kryos, and Hephaestus basins are filled with $MgCl_2$ brine formed by the dissolution of bischofite and represent the saltiest and densest athalassohaline deep-sea formations while the rest of the basins are NaCl dominated. Some of the NaCl dominated basins such as Bannock and Thetis are likely filled with ancient interstitial evaporated sea water rather than brine formed de novo by dissolution of halite (La Cono et al. 2011). Each hypersaline basin is geochemically distinct and reflects the different depositional stages of the evaporite of origin (Table 14.1). For example, the thalassohaline basin

**Table 14.1** Composition of major ions of DHABs mentioned in this chapter

| | $Na^+$ | $Cl^-$ | $Mg^{2+}$ | $K^+$ | $Ca^{2+}$ | $SO_4^{2-}$ | $HS^-$ |
|---|---|---|---|---|---|---|---|
| *Red Sea* | | | | | | | |
| Afifi[a] | 3133 | 3646 | 353 | n.a. | 69 | 189 | n.a. |
| Atlantis II[b,i] | 4646 | 5170 | 35 | 86 | 149 | 11 | n.a. |
| Discovery Deep[b] | 4636 | 5022 | 37 | 85 | 145 | 11 | n.a. |
| Erba[b] | 2622 | 2678 | 71 | 31 | 30 | 47 | n.a. |
| Kerbit[b] | 4806 | 5136 | 121 | 37 | 57 | 20 | 1.13 |
| Nereus[b] | 3557 | 4131 | 64 | 77 | 231 | 11 | n.a. |
| Shaban[b,i] | 4815 | 4936 | 91 | 49 | 21 | 50 | n.a. |
| Sea water[b] | 547 | 584 | 66 | 9 | 12 | 33 | n.a. |
| *Gulf of Mexico* | | | | | | | |
| Orca Basin[c] | 4240 | 4450 | 42 | 17 | 29 | 20 | 0.025 |
| Sea water[d] | 462 | 564 | 11 | 43 | 11 | 29 | 0 |
| *East Mediterranean* | | | | | | | |
| Bannock[e] | 4200 | 5378 | 644 | 126 | 16 | 135 | 2.9 |
| L'Atalante[f] | 4662 | 5296 | 596 | 334 | 7 | 328 | 2.86 |
| Medee[g] | 4178 | 5259 | 788 | 471 | 3 | 201 | 1.64 |
| Thetis[f] | 4760 | 5300 | 604 | 230 | 9 | 265 | 2.12 |
| Tyro[e] | 5300 | 5350 | 71 | 19 | 35 | 53 | 2.1 |
| Urania[f] | 3505 | 3730 | 315 | 122 | 32 | 107 | 15 |
| Discovery[h] | 84 | 10150 | 5150 | 20 | 1 | 110 | 0.85 |
| Hephaestus[h] | 93 | 9120 | 4720 | 28 | 2 | 203 | n.a. |
| Kryos[h] | 125 | 9043 | 4280 | 80 | 1 | 320 | 1.2 |
| Sea water[h] | 540 | 630 | 61 | 12 | 12 | 33 | n.a |

All values are in mmol $l^{-1}$
[a]Duarte et al. (2020)
[b]Schmidt et al. (2015)
[c]Van Cappellen et al. (1998)
[d]Joye et al. (2005)
[e]De Lange et al. (1990)
[f]La Cono et al. (2011)
[g]Yakimov (2013)
[h]La Cono et al. (2019)
[i]Average value

with the highest sodium concentration is Tyro (5300 mmol $l^{-1}$) while the Urania basin brine has a much lower sodium concentration (3505 mmol $l^{-1}$). The overall salinity of Urania is also lower. However, concentrations of methane and sulfide are much higher than the rest of the Eastern Mediterranean Sea basins (van der Wielen et al. 2005) because of the presence of hydrothermalism in this basin.

The high densities of these brine bodies act as a barrier that prevents mixing with the overlying seawater and hence limits the exchange of minerals and oxygen. These brine bodies are, therefore, physically isolated from other marine habitats much like

**Fig. 14.1** Photo of Eastern Mediterranean DHAB Discovery taken by *ROV Jason*. The light brown color (left) is the image of the sea floor; white zone in the middle pictures the transition (interface) from the seawater to the dense brine (right)

submersible islands and select for organisms adapted to their unique and multiple extremes. A narrow interface[5] layer (1–3 m thickness for the Mediterranean Sea DHABs) separates the hypersaline brine body from the overlying oxygenated sea water (Fig. 14.1). The interface of hot Red Sea DHABs is much broader. The stratification of these DHABs is destabilized by heating from below resulting in the formation of several distinct stratified layers of uniform temperature and salinity and with stepwise changes between them. Along the interface the redox potential decreases from +200 mV (in the oxic seawater) to –400 mV (in the anoxic reduced brines). The concentrations of salts increase 5–10 times and the oxygen becomes completely depleted. The physicochemical gradient formed along the interface and the associated chemical disequilibria form distinct niches to be occupied by different microorganisms.

Initially, the polyextreme conditions of brine cores in the Red Sea DHABs Atlantis II and Discovery Deep were considered to be too hostile for life since initial culturing efforts targeting bacteria were not successful (Watson and Waterbury 1969). But sulfate-reducing bacteria were detected and isolated from the less extreme interface (Trüper 1969; Watson and Waterbury 1969). Over the following decades isotope data, ATP measurements, and culturing efforts provided strong evidence for the existence of biogenic processes in Red Sea DHABs and Orca Basin (e.g., Blum and Puchelt 1991; Faber et al. 1998; Fiala et al. 1990; Heitzer and Ottow 1976; LaRock et al. 1979). The most important breakthrough discoveries

[5]The interface layer between the normoxic sea water and the brine is also called halocline, chemocline, redoxycline, or oxycline. The terms "interface" or "halocline" are more commonly used in the literature.

of DHAB microbiology took place in the last 20 years. The DNA sequencing revolution changed the field of Microbial Ecology and led to astonishing discoveries regarding the diversity, function, and evolution of microorganisms in diverse habitats and especially in extreme environments. The first 16S rRNA gene survey of sediment samples from the Kerbit Deep in the Red Sea detected a novel deep-branching lineage KB1 (Eder et al. 1999). Since then, several novel taxa have been detected in DHABs and several halophilic microorganisms have been isolated.

Although DHABs could be used as model systems for studying environmental selection and allopatric speciation of microorganisms (Stock et al. 2013) logistic challenges have hindered the study of these habitats in a systematic way. The differences in the methodology (clone libraries vs high-throughput libraries) and the primers used for molecular surveys of diversity prevent direct quantitative comparisons of the microbial community composition among the different basins. But there are some striking similarities and differences in the lineages recovered that will be discussed in detail below. In all cases, the microbial communities inhabiting DHABs are distinct from the overlying sea water (e.g., Antunes et al. 2011a; Merlino et al. 2018 and references therein). This lends support to the notion that DHAB communities are unique and do not only reflect microorganisms transported into these habitats via sinking organic particles. DHAB interfaces are highly productive ecosystems. Cell numbers of Bacteria and Archaea increase several folds in the interface compared to overlying seawater due to the high nutrient availability (from organic matter trapped by the change in density) and the variety of available terminal electron acceptors.

### 14.2.1 Microbial Diversity of the Red Sea DHABs

The geology and geochemistry of Red Sea DHABs have been studied thoroughly but microbiological investigations focused only on a few of them, namely Atlantis II, Discovery Deep, Erba, Kerbit, Nereus, Shaban, and more recently Afifi. A pyrosequencing survey of 16S rRNA genes from Atlantis II and Discovery Deep showed that the bacterial Marinimicrobia phylum (former SAR406 clade) was abundant in the deep sea and interface layers but declined with depth while the relative abundance of sequences affiliated to Bacteroidetes increased and Proteobacteria (of the class Gammaproteobacteria) dominated the brine cores (Bougouffa et al. 2013). The interfaces of the two basins had distinct microbial compositions but the brine cores appeared to be similar despite the striking differences in physicochemical parameters—especially in temperature and heavy metal composition (Bougouffa et al. 2013). The high bacterial diversity within the interfaces of Atlantis II, Discovery Deep, Erba, Kerbit, and Nereus was detected in a subsequent clone survey (Guan et al. 2015) showing that the majority of the sequences belong to Proteobacteria, Bacteroidetes, Marinimicrobia, and Chloroflexi. Deltaproteobacteria was the most abundant bacterial class in the cold brines (Erba, Kerbit, and Nereus). Analysis of the key gene for sulfate reduction, dissimilatory sulfide reductase (subunit A; *dsrA*), indicated that the Deltaproteobacteria in the cold brines are probably sulfate reducers. Sequences of the *dsrA* gene were also recovered from the hot DHABs Atlantis II and Discovery Deep but could not be affiliated to known phyla (Guan et al. 2015). Similarly, analysis of the gene encoding the particulate monooxygenase (subunitA; *pmoA*) revealed the presence of a diverse community of aerobic methane oxidizers in the interfaces of Atlantis II and Kebrit (Abdallah et al. 2014). Another study showed the presence of Bacteria capable of degrading aromatic compounds in Atlantis II while pathways related to the metabolism of sugars were more enriched in Discovery Deep (Wang et al. 2011).

Afifi has been surveyed using Illumina sequencing and revealed high relative abundances of Proteobacteria, Bacteroidetes, Marinimicrobia, and Chloroflexi as was the case in the other Red Sea DHABs. The deep-branching KB1 lineage was also prevalent (Duarte et al. 2020). This group has been found in a wide range of hypersaline systems including the interface of Kerbit and Shaban (Eder et al. 2001, 2002), sediments of the Orca Basin (Nigro et al. 2016, 2020), and the interfaces and the brine cores of the Mediterranean Sea DHABs (La Cono et al. 2011; van der Wielen et al. 2005; Yakimov et al. 2007a, b; 2013). Stable isotope experiments and single cell genomics indicated that amino acid utilization by KB1 shows adaptation to hypersalinity. This was concluded from the utilization of glycine betaine, which is a common compatible compound that is used as osmoprotectant, as a carbon and energy source (Nigro et al. 2016; Yakimov et al. 2013). This lineage still escapes culturing but its limited dispersal and its genomic characteristics indicate that it is a halophilic clade that dominates most brine cores. Besides the recovery of the KB1 lineage another novel lineage SB1, distantly related to Chlorobia, was detected in the Shabah interface and upper brine. Other sequences from the same DHAB were affiliated to Candidate Division OD1, Bacteroidetes, Marinimicrobia, and Proteobacteria (class Deltaproteobacteria).

Estimation of sequence abundances using qPCR indicates that Bacteria predominate over Archaea in all interfaces and brine bodies of the Red Sea DHABs with the exception of the Afifi brine (Duarte et al. 2020) and the Kebrit and Atlantis II interfaces (Guan et al. 2015). The high oxygen concentrations of Kebrit and Atlantis II interfaces (4–9 times higher than those at the other DHABs) are thought to support the proliferation of Ammonia Oxidizing Archaea (AOA). A 16S rRNA gene pyrosequencing survey suggested that in the interfaces of Red Sea DHABs sequences of the Thaumarchaeota phylum comprise 64% (Erba) to 99% (Atlantis II and Discovery Deeps) of total archaeal sequences (Ngugi et al. 2015). Single cell genomics unveiled that the most abundant AOA represents a novel species highly divergent from the bathypelagic *Nitrosopumilus* genomes. A putative proline-glutamate switch has been hypothesized to be an adaptation to osmotolerance (Ngugi et al. 2015).

Besides the Marine Group I (MG-I) Thaumarchaeota (i.e., AOA) that have been recovered from all DHAB interfaces (Bougouffa et al. 2013; Duarte et al. 2020; Guan et al. 2015) several groups of Euryarchaeota and the archaeal Candidate Divisions have been detected in the Red Sea DHABs including Halobacteria, Methanomicrobia, Archaeglobi, Mediterranean Sea Brine Lakes 1 (MSBL-1), Marine Benthic Group E (MBG-E), Deep Sea Euryarchaeota Group (DSEG), and

Methanobacteria (Bougouffa et al. 2013; Duarte et al. 2020; Eder et al. 2002; Guan et al. 2015). Most of the recovered archaeal lineages do not have cultured representatives thus their physiology and ecological role(s) in DHABs cannot yet be resolved. The sequence diversity of the alpha subunit of methyl-coenzyme A reductase genes (*mcrA*) encoding the key enzyme of methanogenesis and/or anaerobic methane oxidation has been investigated in several Red Sea DHABs (Guan et al. 2015). Several sequences are phylogenetically clustered with isolates of the genera *Methanohalophilus* and *Methanococcoides*, which have the potential to utilize methylated compounds as substrate to produce methane. It has been hypothesized that in hypersaline environments methylotrophic methanogenesis, due to its higher yield, is more favorable than acetoclastic or hydrogenotrophic methanogenesis because the methanogenic inhabitants of hypersaline environments have to spend more energy on the biosynthesis or the uptake of osmoprotecting compounds (McGenity 2010; Oren 1999). Additionally, in sulfate-rich habitats sulfate reducers would outcompete acetoclastic or hydrogenotrophic methanogens for substrates while methylated compounds derived from organic matter decomposition are preferentially consumed by methylotrophic methanogenesis (Oremland et al. 1982).

### 14.2.2 Microbial Diversity of the Orca DHAB

Shortly after the discovery of Orca Basin measurements of adenosine 5′-triphosphate (ATP) and uridine uptake indicated the presence of active microbial communities in its upper interface but both microbial biomass and activity decreased with depth (LaRock et al. 1979; Tuovila et al. 1987). Based on a high abundance of archaeal isoprenoid lipids it has been hypothesized that Archaea play an important role in the upper layers of the brine. However, a culture-dependent study showed the existence of manganese-oxidizing and iron-oxidizing bacteria in the upper interface. Iron-reducing and manganese-reducing bacteria increased in the lower depths where oxygen becomes depleted (Van Cappellen et al. 1998).

The importance of methylotrophic methanogenesis in hypersaline environments (mentioned above) became evident from a study of the hypersaline sediments of Orca Basin where high concentrations of methane and sulfate coexist (Zhuang et al. 2016). The biological source of this methane was evidenced by its strong $^{13}C$ depletion while the high abundance of methylated compounds such as methanol, dimethylsulfoniopropionate, trimethylamine, and dimethyl sulfide supported the methylotrophic origin of the methane. Stable and radio-isotope incubation experiments in combination with 16S rRNA gene sequence and lipid analyses verified that methylotrophic methanogenesis, likely conducted by methanogens related to the genus *Methanohalophilus*, was the major source of methane in these sediments (Zhuang et al. 2016).

The stable isotope composition of dissolved inorganic carbon and methane in porewater as well as sulfate concentrations decreasing downcore indicated that both methanogenesis and sulfate reduction occur in the anoxic sediments of Orca Basin (Nigro et al. 2020). Analysis of full length 16S rRNA gene sequences revealed the presence of the typical microbial key players of DHABs in the hematite-rich Orca sediments and the overlaying brine including the taxa belonging to the Proteobacteria (of potential sulfate reducers of the class Deltaproteobacteria as well as Alpha- and Gammaproteobacteria), Bacteroidetes, KB1, and methylotrophic methanogens affiliated to *Methanohalophilus* (Nigro et al. 2020).

### 14.2.3 Microbial Diversity of the Mediterranean DHABs

The microbiology of Mediterranean DHABs has been intensively studied. The first comparative microbial survey of four Mediterranean DHABs—the thalassohaline brines of L'Atalante, Bannock, and Urania, and the athalassohaline Discovery—showed distinct community compositions in each basin, with Discovery being the most dissimilar. A new lineage of Archaea, MSBL-1, dominated all brines (van der Wielen et al. 2005). Since then, MSBL-1 has also been recovered from Orca, Red Sea DHABs, and from other Mediterranean DHABs. The high abundance of MSBL-1 sequences in DHABs in which biogenic methane is present as well as the phylogenetic placement of this taxonomic group initially led to speculations that they are halophilic methanogens (Borin et al. 2009; van der Wielen et al. 2005; Yakimov et al. 2013). However, the metabolic reconstruction of genomes of members of MSBL-1 showed that a methanogenic lifestyle is not likely for this lineage. MSBL-1 possesses the potential to ferment sugars via the Embden–Meyerhof–Parnas pathway (Mwirichia et al. 2016).

Daffonchio et al. (2006) performed high-precision sampling of the Bannock DHAB interface. an approach that was subsequently adopted by several other research groups (e.g., Hallsworth et al. 2007; La Cono et al. 2011; Yakimov et al. 2007a, b). This sampling technique allowed high-resolution studies of the microbial communities along the steep interfaces of the Mediterranean DHABs. The abundance of Bacteria and Archaea in the interface increased 10–100 times (reaching up $10^6$ cells $ml^{-1}$) relative to overlying deep seawater. ATP concentrations as well as the activity of extracellular aminopeptidase and alkaline phosphatase were also significantly higher in the interface while sulfate reduction rates were elevated in the anoxic brine core. It was hypothesized that the density barrier leads to accumulation of sinking organic matter and thus to higher availability of nutrients and reactive surfaces, which stimulate the growth of microorganisms. Additionally, methane (and ammonium) originating from the brine can be oxidized under more energetically favorable conditions found in the interface supporting the proliferation of microbes that can use these substrates (Daffonchio et al. 2006). DNA and RNA 16S rRNA clones unveiled several new bacterial candidate divisions including MSBL2—phylogenetically related to the Red Sea SB1 lineage—which dominates Bannock's upper interface. In the deeper part of the interface right above the brine (22.2–25.0% salinity) the relative abundance of MSBL2 decreased while abundance of signatures of Sphingobacteria and KB1 increased. Signatures of the rest of the newly reported divisions MSBL3, MSBL4, MSBL5, and MSBL6 were found exclusively in the deeper interface of Bannock basin suggesting that they are adapted to anoxia and high salinity. Deltaproteobacterial clones related to sulfate reducers were also abundant with different classes detected in the upper and lower interface. Archaeal sequences affiliated to MSBL1, a ubiquitous lineage in DHABs, and ANME-1, a lineage of uncultured anaerobic methane oxidizers, were found in the cDNA libraries from the deeper interface where methane and sulfate coexist (Daffonchio et al. 2006).

Similar trends of microbial abundance and activity were observed in Urania Basin (Borin et al. 2009), a DHAB affected by hydrothermalism (Yakimov et al. 2007a). Urania is less salty than the other DHABs but contains high concentrations of methane (5.56 mM) and sulfide (up to 16 mM) (Charlou et al. 2003; Karisiddaiah 2000). Urania's water column has two chemoclines: a steep one in contact with the overlying oxic seawater and the other between the anoxic brines of different density (Borin et al. 2009). The ATP concentration and cell numbers in Urania's upper interface are higher (Borin et al. 2009) compared to values reported for Bannock (Daffonchio et al. 2006). Analysis of 16S rRNA gene sequences revealed that both interfaces are dominated by Proteobacteria of the classes Deltaproteobacteria and Campylobacterota, putative sulfate reducers and sulfur oxidizers, respectively. The peak in relative abundance of the uncultured novel deltaproteobacterial lineage MSBL8 coincided with a peak in rates of sulfate reduction in Urania's upper sharp interface indicating a major role of MSLB8 in this process. In the brine core methanogenesis greatly exceeded sulfate reduction (Borin et al. 2009).

The importance of autotrophic carbon fixation has been shown for the interfaces of L'Atalante and Thetis basins. Dark primary production in the upper interface of these two basins reached 538 nmol C $l^{-1}$ $day^{-1}$ and 315 nmol C $l^{-1}$ $day^{-1}$, respectively (La Cono et al. 2011; Yakimov et al. 2007b) while autotrophic production in the interface of the largest Mediterranean DHAB, Medee, was much lower (102 nmol C $l^{-1}$ $day^{-1}$) (Yakimov et al. 2013). The carbon fixation rates in the interfaces of L'Atalante and Thetis were 8–10 times higher than those in the overlying oxygenated deep-sea water. Inorganic carbon fixation was not detected in the anoxic lower interfaces or the brine cores of those two basins. The production rates measured by La Cono et al. (2011) and Yakimov et al. (2007b; 2013) are at least twice as high as the photosynthesis-driven biomass production rate in the euphotic layer of the East Mediterranean Sea (Vidussi et al. 2001) thus these local sources of abundant freshly-produced organic matter might be of high importance for Mediterranean deep-sea food webs. Aerobic and/or microaerophilic oxidation of sulfur or nitrogen compounds in the upper interface is probably the energy source fueling carbon fixation. The high abundance of 16S rRNA gene sequences affiliated to MG-I Thaumarchaeota, a chemoautotrophic lineage performing aerobic ammonium oxidation in the interface, encouraged the further investigation of their involvement in inorganic carbon fixation. Transcripts of the key gene for aerobic ammonium oxidation, ammonia monooxygenase (subunit A; *amoA*), were detected but their quantification showed that *amoA* was not expressed significantly more in the upper interface compared to the overlying sea water (Yakimov et al. 2007b). Transcripts of the key enzymes of the 3-hydroxypropionate/4-hydroxybutyrate (3-HP/4-HB) cycle used by MG-I for inorganic carbon assimilation, acetyl-CoA carboxylase (*accA*), and 4-hydroxybutyryl-CoA dehydratase were not detected (La Cono et al. 2011). Thus, aside from MG-I a variety of autotrophic lineages might be involved in inorganic carbon fixation in these DHAB interfaces. Indeed, the dark $CO_2$ uptake peak in the Thetis interface coincided with the recovery of mRNA of the two key genes involved in inorganic carbon fixation through the Calvin and rTCA cycles, the large subunit of RuBisCO (*cbbL*) and ATP-citrate lyase (subunit B; *aclB*), respectively. The taxonomic annotation of the recovered transcripts indicated that sulfur oxidizers belonging to Campylobacterota and *Thiomicrospira*-related Gammaproteobacteria were metabolically active and played a major role in $CO_2$ fixation in L'Atalante and Thetis (La Cono et al. 2011; Pachiadaki et al. 2014) whereas in Medee signatures of SUP05-related Gammaproteobacteria and Alphaproteobacteria sulfur oxidizers were recovered (Yakimov et al. 2013). Transcripts associated with sulfate reduction were detected along the entire interface of the Thetis basin while in the lower interface increased expression of genes associated with methane metabolism and osmoprotection was noted. In addition, in the lower interface nitrogenase transcripts affiliated to uncultured putative methanotrophic archaea were detected implying nitrogen fixation in this anoxic habitat and providing evidence of linked carbon, nitrogen, and sulfur cycles (Pachiadaki et al. 2014).

Microbial eukaryotes have also been documented in the haloclines of several east Mediterranean Sea DHABs. The adaptive mechanisms of Eukarya to hypersalinity have been explored to a lesser extent than those for Bacteria and Archaea, however they include maintenance of high guanine and cytosine ratios in DNA, high concentrations of acidic residues on exteriors of proteins, unique lipids, cellular pigments as well as typical architectures, physiologies, and metabolisms (discussed in Oren 2006). Heterotrophic nanoflagellates, amoebae, and ciliates have long been recognized as among the major groups of protists able to adapt to life in hypersaline waters (Post et al. 1983). Heterotrophic bicosoecids and non-pigmented chrysomonads affiliated with stramenopiles (e.g., *Halocafeteria*) were subsequently isolated from waters with 17.5% salinity (Park and Simpson 2010). Extremely halotolerant ciliates (e.g., *Trimyema koreanum* sp. nov.) were isolated from solar salterns with salinity of 29% (Cho et al. 2008) as well as a variety of other heterotrophs from nearly saturated brines (salinity >30%) that could not grow at salinities less than 7.5% (Cho et al. 2008; Park et al. 2007, 2009; Park and Simpson, 2010). Dinoflagellates were documented from environments with salinities of up to 30% (Feazel et al. 2008; Laybourn-Parry et al. 2002; Patterson and Simpson 1996). These and other studies have set the stage for justifying investigations of protist diversity in DHABs.

Abundant chemosynthetic bacterial communities in haloclines of DHABs provide a food source for active pelagic protist communities able to tolerate the extreme conditions of DHAB haloclines. The first investigations into protist diversity in DHABs were based on 18S rRNA gene analyses of the Bannock basin lower interface and brine (Edgcomb et al. 2009) as well as an 18S rRNA survey of L'Atalante upper and lower interface (Alexander et al. 2009). In Bannock, Alveolates dominated the recovered marker gene libraries, most of which were from Dinoflagellates and Ciliates (62 and 12%, respectively). Fungi were the third most abundant eukaryotic taxonomic group (17% of retrieved taxa) particularly in the deepest samples with the highest salinities. Being saprotrophs, it is not surprising to detect microbial fungi in interfaces and brines of DHABs because these places accumulate organic detritus on which fungi feed. Signatures of other groups included Stramenopiles, Euglenozoans, Cercozoans, and Kinetoplastids. While not necessarily reflective of living cells, ribosomal RNA in the L'Atalante and Thetis basins (Alexander et al. 2009; Stock et al. 2013) supports the idea of the presence of viable microbial eukaryotes. In Thetis, RNA-based marker gene analysis reveals likely viable fungi (38% of taxa), followed by ciliates and Stramenopiles (each accounting for ~20%) as well as Haptophytes, Choanoflagellates, and Jakobids. The ciliate taxa reported by Stock et al. (2012) were the most common group of alveolates in the upper and lower interface layers and were closely related to taxa detected in other east Mediterranean Sea DHABs suggesting adaptation of those groups to these habitats. Messenger RNA recovered from whole-community metatranscriptomes is a stronger indicator of cell viability or activity than ribosomal RNA or DNA. The analyses of metatranscriptomes collected from the Thetis interface and sediments from underneath the haloclines of Urania, Discovery, and Atalante DHABs revealed signatures of eukaryotic activities, particularly of fungi, as well as of microbe–microbe interactions, and of motility. Relative to a control site, increased expression of genes for osmolyte biosynthesis and heavy metal resistance and metabolism was observed. This gene expression indicates possible mechanisms for adaptation of protists and fungi to DHABs (Edgcomb et al. 2016; Pachiadaki et al. 2014). Moreover, grazing experiments using the Microbial Sampler-Submersible Incubation Device (MS-SID) that allows in situ studies of microbial activities showed enhanced rates of phagotrophy in Urania's interface than in the deep seawater above Urania DHAB. In fact, these rates of phagotrophy were higher than those observed in the euphotic zone of the same area (Pachiadaki et al. 2016). This reflects the high productivity of these deep-sea halocline habitats where elevated abundance of microbial prey exists along chemically-diverse (abundant sources of carbon and energy for microbial growth) haloclines providing a feeding oasis for protists.

While differences in salinity among DHABs and regions of individual haloclines may account for much of the variations in protist community composition also other factors are likely to contribute to the composition of the community including oxygen and ammonia the concentrations of which vary considerably among DHABs. This idea is supported by the broad comparison of eukaryotic communities of different DHAB haloclines and brines based on T-RFLP analyses (Filker et al. 2013). The high density of protists that DHAB interfaces can support is remarkable for these polyextreme deep-sea habitats. For example, $0.6 \times 10^4$ protist cells $l^{-1}$ were counted by microscopy in the anoxic Thetis basin brine (Stock et al. 2012). Visualization of intact protist cells using scanning electron microscopy (SEM) and fluorescence in situ hybridization (FISH) revealed that many ciliates and flagellates host bacterial epibionts (Edgcomb and Bernhard 2013) whose role is still poorly understood.

### 14.2.4 Microbial Diversity of the "Bittern"[6] Mediterranean Sea DHABs

Some of the most interesting polyextreme DHABs regarding limits of life are the athalassohaline Mg-rich Discovery, Kryos, and Hephaestus basins (Hallsworth et al. 2007; La Cono et al. 2019; Yakimov et al. 2015), which are known as the "bittern" DHABs. The habitability of these DHABs depends on the thermodynamic availability of water, i.e. water activity. The water activity in the brine cores of these basins is less than 0.400 (Hallsworth et al. 2007; La Cono et al. 2019; Yakimov et al. 2015). Values below 0.585 (Stevenson et al. 2015, 2017) are considered to be not supportive for life. Additionally, Mg-rich solutes are considered chaotropic, i.e. solutes that weaken electrostatic interactions and destabilize biological macromolecules. Chaotropic solutes such as, e.g., urea and phenol are extremely hostile for life (see Hallsworth et al. 2007 and references therein). The Discovery brine $MgCl_2$ concentration is 5.05 M (and close to purity, i.e., not containing substances that can stabilize macromolecules) while in the Kryos brine the $MgCl_2$ concentration is lower and contains kosmotropic substances such as sulfate that stabilize macromolecules mitigating the chaotropic effect of $MgCl_2$ (Yakimov et al. 2013). The youngest (700-year-old) DHAB, Hephaestus, contains $MgCl_2$ that reaches up to 4.77 M (La Cono et al. 2019).

The existence of life in the Discovery brine (which has among the harshest conditions for life) and the rest of the Mg-rich brines is debated. Two studies have recorded high rates of sulfate reduction in the brine cores (Steinle et al. 2018; van der Wielen et al. 2005) but several lines of evidence including the aforementioned biophysical constraints and the absence of mRNA suggest that the brine cores of the athalassohaline basins are sterile (Hallsworth et al. 2007; La Cono et al. 2019; Oren 2013; Yakimov et al. 2013). However, the interfaces of these athalassohaline basins appear to have active microbial communities that are adapted to a habitat where the water activity and chaotropicity are close to the limits of life. As is the case in the thalassohaline DHABs 16S rRNA sequences of MG-I and transcripts of *amoA* dominate the upper interface of the Discovery, Kryos, and Hephaestus basins (Hallsworth et al. 2007; La Cono et al. 2019; Yakimov et al. 2015). In Discovery, signatures of the 16S rRNA sequences belonging to halophilic methanogenic Archaea were found in the lower interface and brine but transcripts of *mcrA* were only detectable up to 2.3 M $MgCl_2$. Similarly, for Bacteria phylogenetic analysis of sequences amplified from *dsrAB* mRNA revealed two distinct lineages of sulfate reducers along the interface: Desulfobacteraceae were predominating in the upper layer and Desulfohalobiaceae, a family of halophilic Deltaproteobacteria, were abundant in the lower layer with $MgCl_2$ concentrations between 1.60 and 2.23 M (Hallsworth et al. 2007). Because of the presence of stabilizing substrates signals of

[6]In coastal ponds and solar salterns characterized by high rates of evaporation and sequential precipitation of the constituent salts of seawater the brine remaining after precipitation of the major salt NaCl—called a "bittern" owing to its bitter taste—is highly enriched in $MgCl_2$.

life in Kryos were retrieved from areas with higher concentrations of $MgCl_2$ (up to 3.03 M). MSBL1 and halophilic cluster 1 were the dominant Archaea while KSB1 and Desulfobacterota (related to *Desulfohalobium*) were the dominant Bacteria (Yakimov et al. 2015). Transcripts of *mcrA* from methanogenic Euryarchaeota belonging to *Methanohalophilus* and transcripts of *dsrA* were detected in Hephaestus up to 2.97 M $MgCl_2$ (La Cono et al. 2019).

The Discovery DHAB was also investigated for the presence of microbial eukaryotes. Jaccard indices supported the notion that unique chemistries in thalassohaline Bannock and athalassohaline Discovery DHABs select for unique protist communities (Edgcomb et al. 2009). Kinetoplastid signatures were recovered from the interface of several DHABs with contrasting chemistries and were used to develop FISH probes targeting a novel genus-level clade that dominates the Discovery interface library (Edgcomb et al. 2011). Kinetoplastids have been reported previously from hypersaline and anoxic habitats (Hauer and Rogerson 2005). Signatures of this novel clade were detected in samples from three out of six east Mediterranean Sea DHABs and represented up to 10% of the total protist community in the Discovery basin interface (Fig. 14.2) reaching up to $6.4 \times 10^3$ cells $l^{-1}$ (Edgcomb et al. 2011). The unique chemistries of different DHABs determine eukaryotic diversity in their haloclines and brines and these habitats harbor novel taxa that may aid to understand how certain eukaryotes are able to adapt to hypersaline conditions and the role of symbioses in these adaptations.

### 14.2.5 Culturing Efforts

Since the beginning of their discovery several halophilic or halotolerant strains have been isolated from the Red Sea and east Mediterranean Sea DHABs including the first cultured representative of the phylum Deferribacteres, *Flexistipes sinusarabici* (Fiala et al. 1990), the first representative of a novel bacterial phylum (formerly Haloplasmatales), *Haloplasma contractile* (Antunes et al. 2008a; 2011a), a sulfur-reducing, acetate-utilizing haloarchaeon *Halanaeroarchaeum sulfurireducens* (Sorokin et al. 2016), and a polysaccharide-degrader *Halorhabdus tiamatea* (Antunes et al. 2008b). A detailed description of all these isolates is out of the scope of this chapter but more information on the cultured strains from DHABs can be found in the reviews of Antunes et al. (2011b; 2020) and references therein.

## 14.3 Importance of Polyextreme Environments in Biotechnology

Polyextremophilic microorganisms are considered an underutilized and innovative source of novel compounds including pharmaceuticals. Extracts from DHAB isolates appeared to induce cytotoxic and apoptotic effects in human cancer lines (Esau et al. 2019; Sagar et al. 2013a, b). Several strains from hydrothermal vents encode antibiotic resistance and heavy metal resistance genes (Farias et al. 2015).

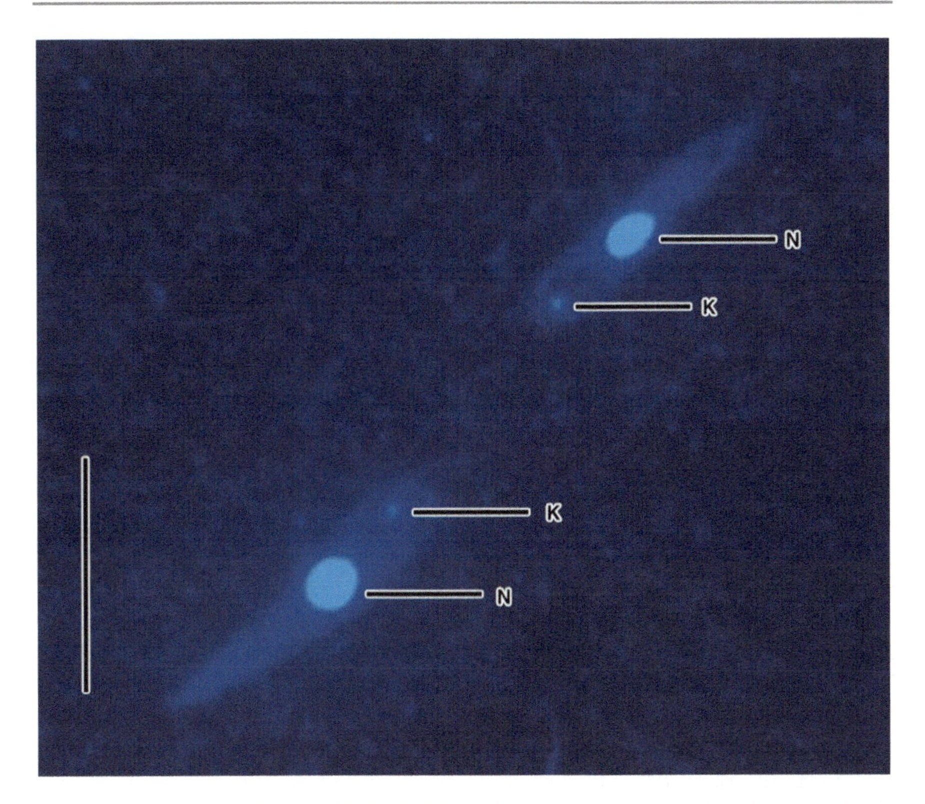


**Fig. 14.2** DAPI image of an "unidentified clade" of kinetoplastids on filters from Discovery basin halocline. The nucleus (N) and kinetoplast (K) are labeled on representative cells. Scale of image is 50 μm

Extremophiles have also been intensively studied for the production of enzymes that can be employed in industrial application enabling biocatalysis under unfavorable conditions (Sarmiento et al. 2015). Thermophilic and halophilic enzymes (including amylases, cellulases, esterases, and proteases) exhibit increased stability and biological activity at low water activity or/and high temperature (Vieille and Zeikus 2001). Such enzymes isolated from vent and DHAB microorganisms (e.g., Legin et al. 1997; Stepnov et al. 2019; Vuoristo et al. 2019) have been commercialized for industrial use (discussed in detail in Varrella et al. 2020) while several types of polymerases have been used for scientific purposes (e.g., Kong et al. 1993). Several more promising biocatalysts have been bioinformatically identified in metagenomes from Urania and Atlantis II (Ferrer et al. 2005; Mohamed et al. 2013). Novel in situ incubation devices containing different types of recalcitrant substrates have been deployed in the Jan Mayen hydrothermal Vent Field on the Arctic Mid-Ocean Ridge in order to enrich for taxa of bioprospecting interest (Stokke et al. 2020). Other potential applications of DHAB and vent microorganisms include the bioremediation of hydrocarbons and $CO_2$ sequestration (Minic and Thongbam 2011 and references therein; Varrella et al. 2020).

## 14.4 Relevance of Vents and Deep Hypersaline Anoxic Basins for Astrobiology

The origin of life still remains enigmatic but serpentinizing systems such as the alkaline hydrothermal vents appear to be an ideal habitat for the emergence of life (Martin et al. 2008). Serpentinization is a geological process leading to the formation of $H_2$ as well as methane and other short-chain hydrocarbons in the presence of $CO_2$. The methane- and acetate-producing geochemistry of alkaline hydrothermal vents such as the Lost City is considered to be the abiogenic precursor of microbial methanogenesis and acetogenesis (Martin et al. 2008). The mixing of alkaline fluids saturated in $H_2$ with acidic Hadean ocean waters through thin interconnected micropores rich in catalytic Fe(Ni)S minerals could have created an electrochemical flow reactor like an abiotic analog of a microbial cell. The difference in pH produced by these natural proton gradients is equivalent to the proton-motive force required for ATP production in modern microbes (Sojo et al. 2016). Thus, these alkaline hydrothermal vents have the conditions that might have led to the creation of protocells. For the same reason, methanogenesis and acetogenesis have been considered as "ancient" metabolisms.

The brine bodies of the athalassohaline DHABs are devoid of life (Hallsworth et al. 2007; La Cono et al. 2019; Yakimov et al. 2015) and their interfaces represent ideal systems for the study of the limits of life a research area of critical importance for the identification of other habitable planetary objects. The icy moons of the outer solar system have been considered as promising targets for humanity's quest to locate extraterrestrial life. The subsurface oceans of icy moons such as Enceladus and Europa are expected to be brines of variable chemistry, exposed to high pressure, and perhaps hydrothermally active and therefore may resemble DHABs.

**Disclaimer** The bibliography on the microbiology of vents and deep hypersaline anoxic basins is extensive including several books chapters and review articles. Important work might have been unintentionally omitted.

## References

Abdallah RZ, Adel M, Ouf A, Sayed A, Ghazy MA, Alam I, Essack M, Lafi FF, Bajic VB, El-Dorry H, Siam R (2014) Aerobic methanotrophic communities at the Red Sea brine-seawater interface. Front Microbiol 5:487

Adam N, Perner M (2018a) Microbially mediated hydrogen cycling in deep-sea hydrothermal vents. Front Microbiol 9:2873

Adam N, Perner M (2018b) Novel hydrogenases from deep-sea hydrothermal vent metagenomes identified by a recently developed activity-based screen. ISME J 12:1225–1236

Akerman N, Butterfield D, Huber J (2013) Phylogenetic diversity and functional gene patterns of sulfur-oxidizing subseafloor Epsilonproteobacteria in diffuse hydrothermal vent fluids. Front Microbiol 4:185

Alain K, Querellou J, Lesongeur F, Pignet P, Crassous P, Raguénès G, Cueff V, Cambon-Bonavita M-A (2002) *Caminibacter hydrogeniphilus* gen. nov., sp. nov., a novel thermophilic, hydrogen-oxidizing bacterium isolated from an East Pacific Rise hydrothermal vent. Int J Syst Evol Microbiol 52:1317–1323

Alain K, Callac N, Guégan M, Lesongeur F, Crassous P, Cambon-Bonavita M-A, Querellou J, Prieur D (2009) *Nautilia abyssi* sp. nov., a thermophilic, chemolithoautotrophic, sulfur-reducing bacterium isolated from an East Pacific Rise hydrothermal vent. Int J Syst Evol Microbiol 59: 1310–1315

Alazard D, Dukan S, Urios A, Verhé F, Bouabida N, Morel F, Thomas P, Garcia J-L, Ollivier B (2003) *Desulfovibrio hydrothermalis* sp. nov., a novel sulfate-reducing bacterium isolated from hydrothermal vents. Int J Syst Evol Microbiol 53:173–178

Alexander E, Stock A, Breiner H-W, Behnke A, Bunge J, Yakimov MM, Stoeck T (2009) Microbial eukaryotes in the hypersaline anoxic L'Atalante deep-sea basin. Environ Microbiol 11:360–381

Amend JP, McCollom TM, Hentscher M, Bach W (2011) Catabolic and anabolic energy for chemolithoautotrophs in deep-sea hydrothermal systems hosted in different rock types. Geochim Cosmochim Acta 75:5736–5748

Anderson RE, Beltrán MT, Hallam SJ, Baross JA (2013) Microbial community structure across fluid gradients in the Juan de Fuca Ridge hydrothermal system. FEMS Microbiol Ecol 83:324–339

Antunes A, Rainey FA, Wanner G, Taborda M, Pätzold J, Nobre MF, da Costa MS, Huber R (2008a) A new lineage of halophilic, wall-less, contractile bacteria from a brine-filled deep of the Red Sea. J Bacteriol 190:3580–3587

Antunes A, Taborda M, Huber R, Moissl C, Nobre MF, da Costa MS (2008b) *Halorhabdus tiamatea* sp. nov., a non-pigmented, extremely halophilic archaeon from a deep-sea, hypersaline anoxic basin of the Red Sea, and emended description of the genus *Halorhabdus*. Int J Syst Evol Microbiol 58:215–220

Antunes A, Alam I, El Dorry H, Siam R, Robertson A, Bajic VB, Stingl U (2011a) Genome sequence of *Haloplasma contractile*, an unusual contractile bacterium from a deep-sea anoxic brine lake. J Bacteriol 193:4551–4552

Antunes A, Ngugi DK, Stingl U (2011b) Microbiology of the Red Sea (and other) deep-sea anoxic brine lakes. Environ Microbiol Rep 3:416–433

Antunes A, Olsson-Francis K, McGenity TJ (2020) Exploring deep-sea brines as potential terrestrial analogues of oceans in the icy moons of the outer solar system. Curr Issues Mol Biol 38: 123–162

Atkins MS, Anderson OR, Wirsen CO (1998) Effect of hydrostatic pressure on the growth rates and encystment of flagellated protozoa isolated from a deep-sea hydrothermal vent and a deep shelf region. Mar Ecol Prog Ser 171:85–95

Atkins MS, Teske A, Anderson OR (2000) A survey of flagellate diversity at four deep-sea hydrothermal vents in the Eastern Pacific Ocean using structural and molecular approaches. J Eukaryot Microbiol 47:400–411

Backer H, Schoell M (1972) New Deeps with Brines and Metalliferous Sediments in the Red Sea. Nat Phys Sci 240:153–158

Bar-Even A, Flamholz A, Noor E, Milo R (2012) Thermodynamic constraints shape the structure of carbon fixation pathways. Biochim Biophys Acta Bioenerg 1817:1646–1659

Beh M, Strauss G, Huber R, Stetter K-O, Fuchs G (1993) Enzymes of the reductive citric acid cycle in the autotrophic eubacterium *Aquifex pyrophilus* and in the archaebacterium *Thermoproteus neutrophilus*. Arch Microbiol 160:306–311

Beinart RA, Sanders JG, Faure B, Sylva SP, Lee RW, Becker EL, Gartman A, Luther GW, Seewald JS, Fisher CR, Girguis PR (2012) Evidence for the role of endosymbionts in regional-scale habitat partitioning by hydrothermal vent symbioses. Proc Natl Acad Sci USA 109:E3241–E3250

Berg IA, Kockelkorn D, Buckel W, Fuchs G (2007) A 3-Hydroxypropionate/4-Hydroxybutyrate autotrophic carbon dioxide assimilation pathway in Archaea. Science 318:1782

Berg IA, Kockelkorn D, Ramos-Vera WH, Say RF, Zarzycki J, Hügler M, Alber BE, Fuchs G (2010) Autotrophic carbon fixation in archaea. Nat Rev Microbiol 8:447–460

Blum N, Puchelt H (1991) Sedimentary-hosted polymetallic massive sulfide deposits of the Kebrit and Shaban Deeps, Red Sea. Miner Depos 26:217–227

Boden R, Scott KM, Williams J, Russel S, Antonen K, Rae AW, Hutt LP (2017) An evaluation of *Thiomicrospira*, *Hydrogenovibrio* and *Thioalkalimicrobium*: reclassification of four species of *Thiomicrospira* to each *Thiomicrorhabdus* gen. nov. and *Hydrogenovibrio*, and reclassification of all four species of *Thioalkalimicrobium* to *Thiomicrospira*. Int J Syst Evol Microbiol 67: 1140–1151

Bonch-Osmolovskaya EA (1994) Bacterial sulfur reduction in hot vents. FEMS Microbiol Rev 15: 65–77

Borin S, Brusetti L, Mapelli F, D'Auria G, Brusa T, Marzorati M, Rizzi A, Yakimov M, Marty D, De Lange GJ, Van der Wielen P, Bolhuis H, McGenity TJ, Polymenakou PN, Malinverno E, Giuliano L, Corselli C, Daffonchio D (2009) Sulfur cycling and methanogenesis primarily drive microbial colonization of the highly sulfidic Urania deep hypersaline basin. Proc Natl Acad Sci USA 106:9151–9156

Bougouffa S, Yang JK, Lee OO, Wang Y, Batang Z, Al-Suwailem A, Qian PY (2013) Distinctive microbial community structure in highly stratified deep-sea brine water columns. Appl Environ Microbiol 79:3425–3437

Brazelton WJ, Schrenk MO, Kelley DS, Baross JA (2006) Methane- and sulfur-metabolizing microbial communities dominate the Lost City hydrothermal field ecosystem. Appl Environ Microbiol 72:6257–6270

Brazelton W, Nelson B, Schrenk M (2012) Metagenomic evidence for $H_2$ oxidation and $H_2$ production by serpentinite-hosted subsurface microbial communities. Front Microbiol 2:268

Brinkhoff T, Muyzer G (1997) Increased species diversity and extended habitat range of sulfur-oxidizing *Thiomicrospira* spp. Appl Environ Microbiol 63:3789–3796

Burggraf S, Jannasch HW, Nicolaus B, Stetter KO (1990) *Archaeoglobus profundus* sp. nov., represents a new species within the sulfate-reducing Archaebacteria. Syst Appl Microbiol 13: 24–28

Byrne N, Strous M, Crépeau V, Kartal B, Birrien J-L, Schmid M, Lesongeur F, Schouten S, Jaeschke A, Jetten M, Prieur D, Godfroy A (2009) Presence and activity of anaerobic ammonium-oxidizing bacteria at deep-sea hydrothermal vents. ISME J 3:117–123

Camerlenghi A (1990) Anoxic basins of the eastern Mediterranean: geological framework. Mar Chem 31:1–19

Campbell BJ, Cary SC (2004) Abundance of reverse tricarboxylic acid cycle genes in free-living microorganisms at deep-sea hydrothermal vents. Appl Environ Microbiol 70:6282–6289

Cavanaugh CM, Gardiner SL, Jones ML, Jannasch HW, Waterbury JB (1981) Prokaryotic cells in the hydrothermal vent tube worm *Riftia pachyptila* Jones: possible chemoautotrophic symbionts. Science 213:340–342

Charlou JL, Donval JP, Fouquet Y, Jean-Baptiste P, Holm N (2002) Geochemistry of high $H_2$ and $CH_4$ vent fluids issuing from ultramafic rocks at the Rainbow hydrothermal field (36°14′N, MAR). Chem Geol 191:345–359

Charlou JL, Donval JP, Zitter T, Roy N, Jean-Baptiste P, Foucher JP, Woodside J (2003) Evidence of methane venting and geochemistry of brines on mud volcanoes of the eastern Mediterranean Sea. Deep Sea Res Part I Oceanogr Res Pap 50:941–958

Charnock H (1964) Anomalous bottom water in the Red Sea. Nature 203:591–591

Cho BC, Park JS, Xu K, Choi JK (2008) Morphology and molecular phylogeny of *Trimyema koreanum* n. sp., a ciliate from the hypersaline water of a solar saltern. J Eukaryot Microbiol 55: 417–426

Corliss JB, Dymond J, Gordon LI, Edmond JM, von Herzen RP, Ballard RD, Green K, Williams D, Bainbridge A, Crane K, van Andel TH (1979) Submarine thermal springs on the Galápagos rift. Science 203:1073–1083

Corre E, Reysenbach A-L, Prieur D (2001) ε-Proteobacterial diversity from a deep-sea hydrothermal vent on the Mid-Atlantic Ridge. FEMS Microbiol Lett 205:329–335
Daffonchio D, Borin S, Brusa T, Brusetti L, van der Wielen PWJJ, Bolhuis H, Yakimov MM, D'Auria G, Giuliano L, Marty D, Tamburini C, McGenity TJ, Hallsworth JE, Sass AM, Timmis KN, Tselepides A, de Lange GJ, Hübner A, Thomson J, Varnavas SP, Gasparoni F, Gerber HW, Malinverno E, Corselli C, Garcin J, McKew B, Golyshin PN, Lampadariou N, Polymenakou P, Calore D, Cenedese S, Zanon F, Hoog S, Party BS (2006) Stratified prokaryote network in the oxic–anoxic transition of a deep-sea halocline. Nature 440:203–220
Dahl C, Friedrich C, Kletzin A (2008) Sulfur oxidation in prokaryotes. eLS. Wiley, Chichester
Dahle H, Roalkvam I, Thorseth IH, Pedersen RB, Steen IH (2013) The versatile in situ gene expression of an Epsilonproteobacteria-dominated biofilm from a hydrothermal chimney. Environ Microbiol Rep 5:282–290
de Angelis MA, Lilley MD, Baross JA (1993) Methane oxidation in deep-sea hydrothermal plumes of the endeavour segment of the Juan de Fuca Ridge. Deep Sea Res Part I Oceanogr Res Pap 40: 1169–1186
Deckert G, Warren PV, Gaasterland T, Young WG, Lenox AL, Graham DE, Overbeek R, Snead MA, Keller M, Aujay M, Huber R, Feldman RA, Short JM, Olsen GJ, Swanson RV (1998) The complete genome of the hyperthermophilic bacterium *Aquifex aeolicus*. Nature 392:353–358
De Lange GJ, Middelburg JJ, Van der Weijden CH, Catalano G, Luther GW, Hydes DJ, Woittiez JRW, Klinkhammer GP (1990) Composition of anoxic hypersaline brines in the Tyro and Bannock Basins, eastern Mediterranean. Mar Chem 31:63–88
Dhillon A, Teske A, Dillon J, Stahl DA, Sogin ML (2003) Molecular characterization of sulfate-reducing bacteria in the Guaymas Basin. Appl Environ Microbiol 69:2765–2772
Dhillon A, Lever M, Lloyd KG, Albert DB, Sogin ML, Teske A (2005) Methanogen diversity evidenced by molecular characterization of methyl coenzyme M reductase A (mcrA) genes in hydrothermal sediments of the Guaymas Basin. Appl Environ Microbiol 71:4592–4601
Dick GJ (2019) The microbiomes of deep-sea hydrothermal vents: distributed globally, shaped locally. Nat Rev Microbiol 17:271–283
Distel DL, Lane DJ, Olsen GJ, Giovannoni SJ, Pace B, Pace NR, Stahl DA, Felbeck H (1988) Sulfur-oxidizing bacterial endosymbionts: analysis of phylogeny and specificity by 16S rRNA sequences. J Bacteriol 170:2506–2510
Duarte CM, Røstad A, Michoud G, Barozzi A, Merlino G, Delgado-Huertas A, Hession BC, Mallon FL, Afifi AM, Daffonchio D (2020) Discovery of Afifi, the shallowest and southernmost brine pool reported in the Red Sea. Sci Rep 10:910
Dubilier N, Bergin C, Lott C (2008) Symbiotic diversity in marine animals: the art of harnessing chemosynthesis. Nat Rev Microbiol 6:725–740
Eder W, Ludwig W, Huber R (1999) Novel 16S rRNA gene sequences retrieved from highly saline brine sediments of Kebrit Deep, Red Sea. Arch Microbiol 172:213–218
Eder W, Jahnke LL, Schmidt M, Huber R (2001) Microbial diversity of the brine-seawater interface of the Kebrit Deep, Red Sea, studied via 16S rRNA gene sequences and cultivation methods. Appl Environ Microbiol 67:3077–3085
Eder W, Schmidt M, Koch M, Garbe-Schönberg D, Huber R (2002) Prokaryotic phylogenetic diversity and corresponding geochemical data of the brine–seawater interface of the Shaban Deep, Red Sea. Environ Microbiol 4:758–763
Edgcomb V (2016) Marine protist associations and environmental impacts across trophic levels in the twilight zone and below. Curr Opin Microbiol 31:169–175
Edgcomb VP, Bernhard JM (2013) Heterotrophic protists in hypersaline microbial mats and deep hypersaline basin water columns. Life (Basel) 3:346–362
Edgcomb VP, Kysela DT, Teske A, de Vera GA, Sogin ML (2002) Benthic eukaryotic diversity in the Guaymas Basin hydrothermal vent environment. Proc Natl Acad Sci USA 99:7658–7662
Edgcomb V, Orsi W, Leslin C, Epstein SS, Bunge J, Jeon S, Yakimov MM, Behnke A, Stoeck T (2009) Protistan community patterns within the brine and halocline of deep hypersaline anoxic basins in the eastern Mediterranean Sea. Extremophiles 13:151–167

Edgcomb VP, Orsi W, Breiner H-W, Stock A, Filker S, Yakimov MM, Stoeck T (2011) Novel active kinetoplastids associated with hypersaline anoxic basins in the Eastern Mediterranean deep-sea. Deep Sea Res Part I Oceanogr Res Pap 58:1040–1048

Edgcomb VP, Pachiadaki MG, Mara P, Kormas KA, Leadbetter ER, Bernhard JM (2016) Gene expression profiling of microbial activities and interactions in sediments under haloclines of E. Mediterranean deep hypersaline anoxic basins. ISME J 10:2643–2657

Elsaied HE, Hayashi T, Naganuma T (2004) Molecular Analysis of deep-sea hydrothermal vent aerobic methanotrophs by targeting genes of 16S rRNA and particulate methane monooxygenase. Mar Biotechnol 6:503–509

Emerson D, Moyer CL (2002) Neutrophilic Fe-oxidizing bacteria are abundant at the Loihi Seamount hydrothermal vents and play a major role in Fe oxide deposition. Appl Environ Microbiol 68:3085–3093

Emerson D, Rentz JA, Lilburn TG, Davis RE, Aldrich H, Chan C, Moyer CL (2007) A novel lineage of Proteobacteria involved in formation of marine Fe-oxidizing Microbial mat communities. PLoS One 2:e667

Esau L, Zhang G, Sagar S, Stingl U, Bajic VB, Kaur M (2019) Mining the deep Red-Sea brine pool microbial community for anticancer therapeutics. BMC Complement Altern Med 19:142

Faber E, Botz R, Poggenburg J, Schmidt M, Stoffers P, Hartmann M (1998) Methane in Red Sea brines. Org Geochem 29:363–379

Farias P, Espírito Santo C, Branco R, Francisco R, Santos S, Hansen L, Sorensen S, Morais PV (2015) Natural hot spots for gain of multiple resistances: arsenic and antibiotic resistances in heterotrophic, aerobic bacteria from marine hydrothermal vent fields. Appl Environ Microbiol 81:2534–2543

Feazel LM, Spear JR, Berger AB, Harris JK, Frank DN, Ley RE, Pace NR (2008) Eucaryotic diversity in a hypersaline microbial mat. Appl Environ Microbiol 74:329–332

Fenchel T (1987) Ecology of protozoa: the biology of free-living phagotrophic protists. Science Tech./Springer, Madison, WI

Ferrer M, Golyshina OV, Chernikova TN, Khachane AN, Martins dos Santos VAP, Yakimov MM, Timmis KN, Golyshin PN (2005) Microbial enzymes mined from the Urania deep-sea hypersaline anoxic basin. Chem Biol 12:895–904

Fiala G, Woese CR, Langworthy TA, Stetter KO (1990) *Flexistipes sinusarabici*, a novel genus and species of eubacteria occurring in the Atlantis II Deep brines of the Red Sea. Arch Microbiol 154:120–126

Filker S, Stock A, Breiner H-W, Edgcomb V, Orsi W, Yakimov MM, Stoeck T (2013) Environmental selection of protistan plankton communities in hypersaline anoxic deep-sea basins, Eastern Mediterranean Sea. Microbiologyopen 2:54–63

Fisher CR, Brooks JM, Vodenichar JS, Zande JM, Childress JJ, Burke RA Jr (1993) The co-occurrence of methanotrophic and chemoautotrophic sulfur-oxidizing bacterial symbionts in a deep-sea mussel. Mar Ecol 14:277–289

Fortunato CS, Huber JA (2016) Coupled RNA-SIP and metatranscriptomics of active chemolithoautotrophic communities at a deep-sea hydrothermal vent. ISME J 10:1925–1938

Fortunato CS, Larson B, Butterfield DA, Huber JA (2018) Spatially distinct, temporally stable microbial populations mediate biogeochemical cycling at and below the seafloor in hydrothermal vent fluids. Environ Microbiol 20:769–784

Frank KL, Rogers DR, Olins HC, Vidoudez C, Girguis PR (2013) Characterizing the distribution and rates of microbial sulfate reduction at Middle Valley hydrothermal vents. ISME J 7:1391–1401

Frank KL, Rogers KL, Rogers DR, Johnston DT, Girguis PR (2015) Key factors influencing rates of heterotrophic sulfate reduction in active seafloor hydrothermal massive sulfide deposits. Front Microbiol 6:1449

Friedrich CG, Bardischewsky F, Rother D, Quentmeier A, Fischer J (2005) Prokaryotic sulfur oxidation. Curr Opin Microbiol 8:253–259

Fukumori Y, Yamanaka T (1979) Flavocytochrome c of *Chromatium vinosum*: some enzymatic properties and subunit structure. J Biochem 8:1405–1414
Galambos D, Anderson RE, Reveillaud J, Huber JA (2019) Genome-resolved metagenomics and metatranscriptomics reveal niche differentiation in functionally redundant microbial communities at deep-sea hydrothermal vents. Environ Microbiol 21:4395–4410
Götz D, Banta A, Beveridge TJ, Rushdi AI, Simoneit BRT, Reysenbach AL (2002) *Persephonella marina* gen. nov., sp. nov. and *Persephonella guaymasensis* sp. nov., two novel, thermophilic, hydrogen-oxidizing microaerophiles from deep-sea hydrothermal vents. Int J Syst Evol Microbiol 52:1349–1359
Govenar B (2012) Energy transfer through food webs at hydrothermal vents. Oceanography 25: 246–255
Greening C, Biswas A, Carere CR, Jackson CJ, Taylor MC, Stott MB, Cook GM, Morales SE (2016) Genomic and metagenomic surveys of hydrogenase distribution indicate $H_2$ is a widely utilised energy source for microbial growth and survival. ISME J 10:761–777
Grosche A, Sekaran H, Pérez-Rodríguez I, Starovoytov V, Vetriani C (2015) *Cetia pacifica* gen. nov., sp. nov., a chemolithoautotrophic, thermophilic, nitrate-ammonifying bacterium from a deep-sea hydrothermal vent. Int J Syst Evol Microbiol 65:1144–1150
Guan Y, Hikmawan T, Antunes A, Ngugi D, Stingl U (2015) Diversity of methanogens and sulfate-reducing bacteria in the interfaces of five deep-sea anoxic brines of the Red Sea. Res Microbiol 166:688–699
Guiral M, Tron P, Aubert C, Gloter A, Iobbi-Nivol C, Giudici-Orticoni M-T (2005) A Membrane-bound multienzyme, hydrogen-oxidizing, and sulfur-reducing complex from the hyperthermophilic bacterium *Aquifex aeolicus*. J Biol Chem 280:42004–42015
Gupta RS, Jannasch HW (1973) Photosynthetic production and dark-assimilation of $CO_2$ in the Black Sea. Internationale Revue der gesamten Hydrobiologie und Hydrographie 58:625–632
Hallam SJ, Mincer TJ, Schleper C, Preston CM, Roberts K, Richardson PM, DeLong EF (2006) Pathways of carbon assimilation and ammonia oxidation suggested by environmental genomic analyses of marine Crenarchaeota. PLoS Biol 4:e95
Hallsworth JE, Yakimov MM, Golyshin PN, Gillion JLM, D'Auria G, De Lima AF, La Cono V, Genovese M, McKew BA, Hayes SL, Harris G, Giuliano L, Timmis KN, McGenity TJ (2007) Limits of life in $MgCl_2$-containing environments: chaotropicity defines the window. Environ Microbiol 9:801–813
Han Y, Perner M (2014) The role of hydrogen for *Sulfurimonas denitrificans*' metabolism. PLoS One 9:e106218
Hansen M, Perner M (2016) Hydrogenase gene distribution and $H_2$ consumption ability within the *Thiomicrospira* lineage. Front Microbiol 7:99
Hauer G, Rogerson A (2005) Heterotrophic Protozoa from Hypersaline Environments. In: Gunde-Cimerman N, Oren A, Plemenitaš A (eds) Adaptation to life at high salt concentrations in Archaea, Bacteria, and Eukarya. Springer, Dordrecht, pp 519–539
Heitzer RD, Ottow JCG (1976) New denitrifying bacteria isolated from Red Sea sediments. Mar Biol 37:1–10
Hinzke T, Kleiner M, Meister M, Schlüter R, Hentschker C, Pané-Farré J, Hildebrandt P, Felbeck H, Sievert SM, Bonn F, Völker U, Becher D, Schweder T, Markert S (2021) Bacterial symbiont subpopulations have different roles in a deep-sea symbiosis. eLife 10:e58371
Hirayama H, Sunamura M, Takai K, Nunoura T, Noguchi T, Oida H, Furushima Y, Yamamoto H, Oomori T, Horikoshi K (2007) Culture-dependent and -independent characterization of microbial communities associated with a shallow submarine hydrothermal system occurring within a coral reef off Taketomi Island, Japan. Appl Environ Microbiol 73:7642–7656
Hirayama H, Fuse H, Abe M, Miyazaki M, Nakamura T, Nunoura T, Furushima Y, Yamamoto H, Takai K (2013) *Methylomarinum vadi* gen. nov., sp. nov., a methanotroph isolated from two distinct marine environments. Int J Syst Evol Microbiol 63:1073–1082
Hirayama H, Abe M, Miyazaki M, Nunoura T, Furushima Y, Yamamoto H, Takai K (2014) *Methylomarinovum caldicuralii* gen. nov., sp. nov., a moderately thermophilic methanotroph

isolated from a shallow submarine hydrothermal system, and proposal of the family Methylothermaceae fam. nov. Int J Syst Evol Microbiol 64:989–999
Hu Q, Wang S, Lai Q, Shao Z, Jiang L (2021a) *Sulfurimonas indica* sp. nov., a hydrogen- and sulfur-oxidizing chemolithoautotroph isolated from a hydrothermal sulfide chimney in the Northwest Indian Ocean. Int J Syst Evol Microbiol 71:1466–5026
Hu SK, Herrera EL, Smith AR, Pachiadaki MG, Edgcomb VP, Sylva SP, Chan EW, Seewald JS, German CR, Huber JA (2021b) Protistan grazing impacts microbial communities and carbon cycling at deep-sea hydrothermal vents. PNAS 118 (29) e2102674118; https://doi.org/10.1073/pnas.2102674118
Huber H, Diller S, Horn C, Rachel R (2002a) *Thermovibrio ruber* gen. nov., sp. nov., an extremely thermophilic, chemolithoautotrophic, nitrate-reducing bacterium that forms a deep branch within the phylum Aquificae. Int J Syst Evol Microbiol 52:1859–1865
Huber JA, Butterfield DA, Baross JA (2002b) Temporal changes in archaeal diversity and chemistry in a Mid-Ocean Ridge subseafloor habitat. Appl Environ Microbiol 68:1585–1594
Huber JA, Butterfield DA, Baross JA (2003) Bacterial diversity in a subseafloor habitat following a deep-sea volcanic eruption. FEMS Microbiol Ecol 43:393–409
Huber JA, Mark Welch DB, Morrison HG, Huse SM, Neal PR, Butterfield DA, Sogin ML (2007) Microbial population structures in the deep marine biosphere. Science 318:97–100
Hügler M, Sievert SM (2010) Beyond the Calvin Cycle: autotrophic carbon fixation in the ocean. Ann Rev Mar Sci 3:261–289
Hügler M, Wirsen CO, Fuchs G, Taylor CD, Sievert SM (2005) Evidence for autotrophic $co_2$ fixation via the reductive tricarboxylic acid cycle by members of the ε subdivision of Proteobacteria. J Bacteriol 187:3020–3027
Hügler M, Huber H, Molyneaux SJ, Vetriani C, Sievert SM (2007) Autotrophic CO2 fixation via the reductive tricarboxylic acid cycle in different lineages within the phylum Aquificae: evidence for two ways of citrate cleavage. Environ Microbiol 9:81–92
Hügler M, Gärtner A, Imhoff JF (2010) Functional genes as markers for sulfur cycling and CO2 fixation in microbial communities of hydrothermal vents of the Logatchev field. FEMS Microbiol Ecol 73:526–537
Inagaki F, Takai K, Kobayashi H, Nealson KH, Horikoshi K (2003) *Sulfurimonas autotrophica* gen. nov., sp. nov., a novel sulfur-oxidizing ε-proteobacterium isolated from hydrothermal sediments in the Mid-Okinawa Trough. Int J Syst Evol Microbiol 53:1801–1805
Inagaki F, Takai K, Nealson KH, Horikoshi K (2004) *Sulfurovum lithotrophicum* gen. nov., sp. nov., a novel sulfur-oxidizing chemolithoautotroph within the ε-Proteobacteria isolated from Okinawa Trough hydrothermal sediments. Int J Syst Evol Microbiol 54:1477–1482
Jannasch HW (1995) Microbial interactions with hydrothermal fluids. In: Seafloor hydrothermal systems: physical, chemical, biological, and geological interactions. American Geophysical Union (AGU), Washington DC, pp 273–296
Jannasch HW, Wirsen CO (1979) Chemosynthetic primary production at East Pacific sea floor spreading centers. Bioscience 29:592–598
Jannasch HW, Wirsen CO, Nelson DC, Robertson LA (1985) *Thiomicrospira crunogena* sp. nov., a colorless, sulfur-oxidizing bacterium from a deep-sea hydrothermal vent. Int J Syst Evol Microbiol 35:422–424
Jeanthon C, L'Haridon S, Reysenbach A-L, Corre E, Vernet M, Messner P, Sleytr UB, Prieur D (1999) *Methanococcus vulcanius* sp. nov., a novel hyperthermophilic methanogen isolated from East Pacific Rise, and identification of *Methanococcus* sp. DSM 4213Tas *Methanococcus fervens* sp. nov. Int J Syst Evol Microbiol 49:583–589
Jeanthon C, L'Haridon S, Cueff V, Banta A, Reysenbach A-L, Prieur D (2002) *Thermodesulfobacterium hydrogeniphilum* sp. nov., a thermophilic, chemolithoautotrophic, sulfate-reducing bacterium isolated from a deep-sea hydrothermal vent at Guaymas Basin, and emendation of the genus *Thermodesulfobacterium*. Int J Syst Evol Microbiol 52:765–772
Jelen B, Giovannelli D, Falkowski PG, Vetriani C (2018) Elemental sulfur reduction in the deep-sea vent thermophile, *Thermovibrio ammonificans*. Environ Microbiol 20:2301–2316

Jesser KJ, Fullerton H, Hager KW, Moyer CL (2015) Quantitative PCR analysis of functional genes in iron-rich microbial mats at an active hydrothermal vent system (Lō'ihi Seamount, Hawai'i). Appl Environ Microbiol 81:2976–2984

Jiang L, Lyu J, Shao Z (2017) Sulfur metabolism of *Hydrogenovibrio thermophilus* strain s5 and its adaptations to deep-sea hydrothermal vent environment. Front Microbiol 8:2513

Johannessen KC, Vander Roost J, Dahle H, Dundas SH, Pedersen RB, Thorseth IH (2017) Environmental controls on biomineralization and Fe-mound formation in a low-temperature hydrothermal system at the Jan Mayen Vent Fields. Geochim Cosmochim Acta 202:101–123

Jones WJ, Leigh JA, Mayer F, Woese CR, Wolfe RS (1983) *Methanococcus jannaschii* sp. nov., an extremely thermophilic methanogen from a submarine hydrothermal vent. Arch Microbiol 136: 254–261

Jongsma D, Fortuin AR, Huson W, Troelstra SR, Klaver GT, Peters JM, van Harten D, de Lange GJ, ten Haven L (1983) Discovery of an anoxic basin within the Strabo Trench, eastern Mediterranean. Nature 305:795–797

Jørgensen BB, Isaksen MF, Jannasch HW (1992) Bacterial sulfate reduction above 100°C in deep-sea hydrothermal vent sediments. Science 258:1756–1757

Joye SB, MacDonald IR, Montoya JP, Peccini M (2005) Geophysical and geochemical signatures of Gulf of Mexico seafloor brines. Biogeosciences 2:295–309

Karisiddaiah SM (2000) Diverse methane concentrations in anoxic brines and underlying sediments, eastern Mediterranean Sea. Deep Sea Res Part I Oceanogr Res Pap 47:1999–2008

Karl DM, Wirsen CO, Jannasch HW (1980) Deep-sea primary production at the Galapagos hydrothermal vents. Science 207:1345–1347

Kato S, Yanagawa K, Sunamura M, Takano Y, Ishibashi J, Kakegawa T, Utsumi M, Yamanaka T, Toki T, Noguchi T, Kobayashi K, Moroi A, Kimura H, Kawarabayasi Y, Marumo K, Urabe T, Yamagishi A (2009) Abundance of Zetaproteobacteria within crustal fluids in back-arc hydrothermal fields of the Southern Mariana Trough. Environ Microbiol 11:3210–3222

Kelley DS, Karson JA, Früh-Green GL, Yoerger DR, Shank TM, Butterfield DA, Hayes JM, Schrenk MO, Olson EJ, Proskurowski G, Jakuba M, Bradley A, Larson B, Ludwig K, Glickson D, Buckman K, Bradley AS, Brazelton WJ, Roe K, Elend MJ, Delacour A, Bernasconi SM, Lilley MD, Baross JA, Summons RE, Sylva SP (2005) A serpentinite-hosted ecosystem: the Lost City hydrothermal field. Science 307:1428–1434

Kong H, Kucera RB, Jack WE (1993) Characterization of a DNA polymerase from the hyperthermophile archaea *Thermococcus litoralis*. Vent DNA polymerase, steady state kinetics, thermal stability, processivity, strand displacement, and exonuclease activities. J Biol Chem 268:1965–1975

Könneke M, Schubert DM, Brown PC, Hügler M, Standfest S, Schwander T, Schada von Borzyskowski L, Erb TJ, Stahl DA, Berg IA (2014) Ammonia-oxidizing archaea use the most energy-efficient aerobic pathway for $CO_2$ fixation. Proc Natl Acad Sci USA 111:8239–8244

Kouris A, Kim Juniper S, Frébourg G, Gaill F (2007) Protozoan–bacterial symbiosis in a deep-sea hydrothermal vent folliculinid ciliate (*Folliculinopsis* sp.) from the Juan de Fuca Ridge. Mar Ecol 28:63–71

La Cono V, Smedile F, Bortoluzzi G, Arcadi E, Maimone G, Messina E, Borghini M, Oliveri E, Mazzola S, L'Haridon S, Toffin L, Genovese L, Ferrer M, Giuliano L, Golyshin PN, Yakimov MM (2011) Unveiling microbial life in new deep-sea hypersaline Lake Thetis. Part I: Prokaryotes and environmental settings. Environ Microbiol 13:2250–2268

La Cono V, Bortoluzzi G, Messina E, La Spada G, Smedile F, Giuliano L, Borghini M, Stumpp C, Schmitt-Kopplin P, Harir M, O'Neill WK, Hallsworth JE, Yakimov M (2019) The discovery of Lake Hephaestus, the youngest athalassohaline deep-sea formation on Earth. Sci Rep 9:1679

Lam P, Cowen JP, Jones RD (2004) Autotrophic ammonia oxidation in a deep-sea hydrothermal plume. FEMS Microbiol Ecol 47:191–206

LaRock PA, Lauer RD, Schwarz JR, Watanabe KK, Wiesenburg DA (1979) Microbial biomass and activity distribution in an anoxic, hypersaline basin. Appl Environ Microbiol 37:466–470

Laybourn-Parry J, Quayle W, Henshaw T (2002) The biology and evolution of Antarctic saline lakes in relation to salinity and trophy. Polar Biol 25:542–552

Legin E, Ladrat C, Godfroy A, Barbier G, Duchiron F (1997) Thermostable amylolytic enzymes of thermophilic microorganisms from deep-sea hydrothermal vents. Comptes Rendus de l'Académie des Sciences - Series III - Sciences de la Vie 320:893–898

Lever MA, Teske AP (2015) Diversity of methane-cycling archaea in hydrothermal sediment investigated by general and group-specific PCR primers. Appl Environ Microbiol 81:1426–1441

L'Haridon S, Cilia V, Messner P, Raguénès G, Gambacorta A, Sleytr UB, Prieur D, Jeanthon C (1998) *Desulfurobacterium thermolithotrophum* gen. nov., sp. nov., a novel autotrophic, sulphur-reducing bacterium isolated from a deep-sea hydrothermal vent. Int J Syst Evol Microbiol 48:701–711

L'Haridon S, Reysenbach A-L, Banta A, Messner P, Schumann P, Stackebrandt E, Jeanthon C (2003) *Methanocaldococcus indicus* sp. nov., a novel hyperthermophilic methanogen isolated from the Central Indian Ridge. Int J Syst Evol Microbiol 53:1931–1935

Lin TJ, El Sebae G, Jung J-H, Jung D-H, Park C-S, Holden JF (2016) *Pyrodictium delaneyi* sp. nov., a hyperthermophilic autotrophic archaeon that reduces Fe(III) oxide and nitrate. Int J Syst Evol Microbiol 66:3372–3376

Liu C, Gorby YA, Zachara JM, Fredrickson JK, Brown CF (2002) Reduction kinetics of Fe(III), Co (III), U(VI), Cr(VI), and Tc(VII) in cultures of dissimilatory metal-reducing bacteria. Biotechnol Bioeng 80:637–649

Lloyd KG, Alperin MJ, Teske A (2011) Environmental evidence for net methane production and oxidation in putative ANaerobic MEthanotrophic (ANME) archaea. Environ Microbiol 13: 2548–2564

Longnecker K, Reysenbach A-L (2001) Expansion of the geographic distribution of a novel lineage of ε-Proteobacteria to a hydrothermal vent site on the Southern East Pacific Rise. FEMS Microbiol Ecol 35:287–293

Lonsdale P (1977) Clustering of suspension-feeding macrobenthos near abyssal hydrothermal vents at oceanic spreading centers. Deep Sea Res 24:857–863

López-García P, Philippe H, Gail F, Moreira D (2003) Autochthonous eukaryotic diversity in hydrothermal sediment and experimental microcolonizers at the Mid-Atlantic Ridge. Proc Natl Acad Sci USA 100:697–702

López-García P, Vereshchaka A, Moreira D (2007) Eukaryotic diversity associated with carbonates and fluid–seawater interface in Lost City hydrothermal field. Environ Microbiol 9:546–554

Luther GW, Rozan TF, Taillefert M, Nuzzio DB, Di Meo C, Shank TM, Lutz RA, Cary SC (2001) Chemical speciation drives hydrothermal vent ecology. Nature 410:813–816

Markert S, Arndt C, Felbeck H, Becher D, Sievert SM, Hügler M, Albrecht D, Robidart J, Bench S, Feldman RA, Hecker M, Schweder T (2007) Physiological proteomics of the uncultured endosymbiont of *Riftia pachyptila*. Science 315:247–250

Martin W, Baross J, Kelley D, Russell MJ (2008) Hydrothermal vents and the origin of life. Nat Rev Microbiol 6:805–814

McAllister SM, Davis RE, McBeth JM, Tebo BM, Emerson D, Moyer CL (2011) Biodiversity and emerging biogeography of the neutrophilic iron-oxidizing Zetaproteobacteria. Appl Environ Microbiol 77:5445–5457

McCollom TM, Shock EL (1997) Geochemical constraints on chemolithoautotrophic metabolism by microorganisms in seafloor hydrothermal systems. Geochim Cosmochim Acta 61:4375–4391

McGenity TJ (2010) Methanogens and methanogenesis in hypersaline environments. In: Timmis KN (ed) Handbook of hydrocarbon and lipid microbiology. Springer, Berlin, pp 665–680

McNichol J, Stryhanyuk H, Sylva SP, Thomas F, Musat N, Seewald JS, Sievert SM (2018) Primary productivity below the seafloor at deep-sea hot springs. Proc Natl Acad Sci USA 115:6756–6761

Meier DV, Pjevac P, Bach W, Hourdez S, Girguis PR, Vidoudez C, Amann R, Meyerdierks A (2017) Niche partitioning of diverse sulfur-oxidizing bacteria at hydrothermal vents. ISME J 11: 1545–1558

Merlino G, Barozzi A, Michoud G, Ngugi DK, Daffonchio D (2018) Microbial ecology of deep-sea hypersaline anoxic basins. FEMS Microbiol Ecol 94:fiy085

Meuer J, Kuettner HC, Zhang JK, Hedderich R, Metcalf WW (2002) Genetic analysis of the archaeon *Methanosarcina barkeri* Fusaro reveals a central role for Ech hydrogenase and ferredoxin in methanogenesis and carbon fixation. Proc Natl Acad Sci USA 99:5632–5637

Miller AR (1964) High salinity in sea water. Nature 203:590–591

Minic Z, Thongbam PD (2011) The biological deep sea hydrothermal vent as a model to study carbon dioxide capturing enzymes. Mar Drugs 9:719–738

Mino S, Kudo H, Arai T, Sawabe T, Takai K, Nakagawa S (2014) *Sulfurovum aggregans* sp. nov., a hydrogen-oxidizing, thiosulfate-reducing chemolithoautotroph within the Epsilonproteobacteria isolated from a deep-sea hydrothermal vent chimney, and an emended description of the genus *Sulfurovum*. Int J Syst Evol Microbiol 64:3195–3201

Miroshnichenko ML, Kostrikina NA, L'Haridon S, Jeanthon C, Hippe H, Stackebrandt E, Bonch-Osmolovskaya EA (2002) *Nautilia lithotrophica* gen. nov., sp. nov., a thermophilic sulfur-reducing epsilon-proteobacterium isolated from a deep-sea hydrothermal vent. Int J Syst Evol Microbiol 52:1299–1304

Miyazaki J, Ikuta T, Watsuji T, Abe M, Yamamoto M, Nakagawa S, Takaki Y, Nakamura K, Takai K (2020) Dual energy metabolism of the Campylobacterota endosymbiont in the chemosynthetic snail *Alviniconcha marisindica*. ISME J 14:1273–1289

Mohamed YM, Ghazy MA, Sayed A, Ouf A, El-Dorry H, Siam R (2013) Isolation and characterization of a heavy metal-resistant, thermophilic esterase from a Red Sea brine pool. Sci Rep 3: 3358

Moussard H, L'Haridon S, Tindall BJ, Banta A, Schumann P, Stackebrandt E, Reysenbach A-L, Jeanthon C (2004) *Thermodesulfatator indicus* gen. nov., sp. nov., a novel thermophilic chemolithoautotrophic sulfate-reducing bacterium isolated from the Central Indian Ridge. Int J Syst Evol Microbiol 54:227–233

Mußmann M, Hu FZ, Richter M, de Beer D, Preisler A, Jørgensen BB, Huntemann M, Glöckner FO, Amann R, Koopman WJH, Lasken RS, Janto B, Hogg J, Stoodley P, Boissy R, Ehrlich GD (2007) Insights into the genome of large sulfur bacteria revealed by analysis of single filaments. PLoS Biol 5:e230

Murdock SA, Juniper SK (2019) Hydrothermal vent protistan distribution along the Mariana arc suggests vent endemics may be rare and novel. Environ Microbiol 21:3796–3815

Muyzer G, Teske A, Wirsen CO, Jannasch HW (1995) Phylogenetic relationships of *Thiomicrospira* species and their identification in deep-sea hydrothermal vent samples by denaturing gradient gel electrophoresis of 16S rDNA fragments. Arch Microbiol 164:165–172

Mwirichia R, Alam I, Rashid M, Vinu M, Ba-Alawi W, Anthony Kamau A, Kamanda Ngugi D, Göker M, Klenk H-P, Bajic V, Stingl U (2016) Metabolic traits of an uncultured archaeal lineage -MSBL1- from brine pools of the Red Sea. Sci Rep 6:19181

Nakagawa S, Takai K (2008) Deep-sea vent chemoautotrophs: diversity, biochemistry and ecological significance. FEMS Microbiol Ecol 65:1–14

Nakagawa T, Nakagawa S, Inagaki F, Takai K, Horikoshi K (2004) Phylogenetic diversity of sulfate-reducing prokaryotes in active deep-sea hydrothermal vent chimney structures. FEMS Microbiol Lett 232:145–152

Nercessian O, Bienvenu N, Moreira D, Prieur D, Jeanthon C (2005) Diversity of functional genes of methanogens, methanotrophs and sulfate reducers in deep-sea hydrothermal environments. Environ Microbiol 7:118–132

Ngugi DK, Blom J, Alam I, Rashid M, Ba-Alawi W, Zhang G, Hikmawan T, Guan Y, Antunes A, Siam R, El Dorry H, Bajic V, Stingl U (2015) Comparative genomics reveals adaptations of a halotolerant thaumarchaeon in the interfaces of brine pools in the Red Sea. ISME J 9:396–411

Nigro LM, Hyde AS, MacGregor BJ, Teske A (2016) Phylogeography, salinity adaptations and metabolic potential of the Candidate Division KB1 Bacteria based on a partial single cell genome. Front Microbiol 7:1266

Nigro LM, Elling FJ, Hinrichs K-U, Joye SB, Teske A (2020) Microbial ecology and biogeochemistry of hypersaline sediments in Orca Basin. PLoS One 15:e0231676

Oremland RS, Marsh LM, Polcin S (1982) Methane production and simultaneous sulphate reduction in anoxic, salt marsh sediments. Nature 296:143–145

Oren A (1999) Bioenergetic aspects of halophilism. Microbiol Mol Biol Rev 63:334

Oren A (2006) Halophilic microorganisms and their environments. Springer, Dordrecht

Oren A (2013) Life in magnesium- and calcium-rich hypersaline environments: salt stress by chaotropic ions. In: Seckbach J, Oren A, Stan-Lotter H (eds) Polyextremophiles: life under multiple forms of stress. Springer, Dordrecht, pp 215–232

Pachiadaki MG, Yakimov MM, LaCono V, Leadbetter E, Edgcomb V (2014) Unveiling microbial activities along the halocline of Thetis, a deep-sea hypersaline anoxic basin. ISME J 8:2478–2489

Pachiadaki MG, Taylor C, Oikonomou A, Yakimov MM, Stoeck T, Edgcomb V (2016) In situ grazing experiments apply new technology to gain insights into deep-sea microbial food webs. Deep Sea Res Pt II Top Stud Oceanogr 129:223–231

Park JS, Simpson AGB (2010) Characterization of halotolerant Bicosoecida and Placididea (Stramenopila) that are distinct from marine forms, and the phylogenetic pattern of salinity preference in heterotrophic stramenopiles. Environ Microbiol 12:1173–1184

Park JS, Simpson AGB, Lee WJ, Cho BC (2007) Ultrastructure and phylogenetic placement within Heterolobosea of the previously unclassified, extremely halophilic heterotrophic flagellate *Pleurostomum flabellatum* (Ruinen 1938). Protist 158:397–413

Park JS, Simpson AGB, Brown S, Cho BC (2009) Ultrastructure and molecular phylogeny of two Heterolobosean Amoebae, *Euplaesiobystra hypersalinica* gen. et sp. nov. and *Tulamoeba peronaphora* gen. et sp. nov., isolated from an extremely hypersaline habitat. Protist 160: 265–283

Patterson DJ, Simpson AGB (1996) Heterotrophic flagellates from coastal marine and hypersaline sediments in Western Australia. Eur J Protistol 32:423–448

Pérez-Rodríguez I, Ricci J, Voordeckers JW, Starovoytov V, Vetriani C (2010) *Nautilia nitratireducens* sp. nov., a thermophilic, anaerobic, chemosynthetic, nitrate-ammonifying bacterium isolated from a deep-sea hydrothermal vent. Int J Syst Evol Microbiol 60:1182–1186

Pérez-Rodríguez I, Grosche A, Massenburg L, Starovoytov V, Lutz RA, Vetriani C (2012) *Phorcysia thermohydrogeniphila* gen. nov., sp. nov., a thermophilic, chemolithoautotrophic, nitrate-ammonifying bacterium from a deep-sea hydrothermal vent. Int J Syst Evol Microbiol 62:2388–2394

Perner M, Seifert R, Weber S, Koschinsky A, Schmidt K, Strauss H, Peters M, Haase K, Imhoff JF (2007) Microbial $CO_2$ fixation and sulfur cycling associated with low-temperature emissions at the Lilliput hydrothermal field, southern Mid-Atlantic Ridge (9°S). Environ Microbiol 9:1186–1201

Perner M, Gonnella G, Hourdez S, Böhnke S, Kurtz S, Girguis P (2013) In situ chemistry and microbial community compositions in five deep-sea hydrothermal fluid samples from Irina II in the Logatchev field. Environ Microbiol 15:1551–1560

Petersen JM, Dubilier N (2009) Methanotrophic symbioses in marine invertebrates. Environ Microbiol Rep 1:319–335

Petersen JM, Dubilier N (2010) Symbiotic methane oxidizers. In: Timmis KN (ed) Handbook of hydrocarbon and lipid microbiology. Springer, Berlin, pp 1977–1996

Petersen JM, Ramette A, Lott C, Cambon-Bonavita M-A, Zbinden M, Dubilier N (2010) Dual symbiosis of the vent shrimp *Rimicaris exoculata* with filamentous gamma- and epsilonproteobacteria at four Mid-Atlantic Ridge hydrothermal vent fields. Environ Microbiol 12:2204–2218

Pley U, Schipka J, Gambacorta A, Jannasch HW, Fricke H, Rachel R, Stetter KO (1991) *Pyrodictium abyssi* sp. nov. represents a novel heterotrophic marine archaeal hyperthermophile growing at 110°C. Syst Appl Microbiol 14:245–253
Polz MF, Distel DL, Zarda B, Amann R, Felbeck H, Ott JA, Cavanaugh CM (1994) Phylogenetic analysis of a highly specific association between ectosymbiotic, sulfur-oxidizing bacteria and a marine nematode. Appl Environ Microbiol 60:4461–4467
Polzin J, Arevalo P, Nussbaumer T, Polz MF, Bright M (2019) Polyclonal symbiont populations in hydrothermal vent tubeworms and the environment. Proc Royal Soc B 286:20181281
Post FJ, Borowitzka LJ, Borowitzka MA, Mackay B, Moulton T (1983) The protozoa of a Western Australian hypersaline lagoon. Hydrobiologia 105:95–113
Reed AJ, Dorn R, Van Dover CL, Lutz RA, Vetriani C (2009) Phylogenetic diversity of methanogenic, sulfate-reducing and methanotrophic prokaryotes from deep-sea hydrothermal vents and cold seeps. Deep Sea Res Pt II Top Stud Oceanogr 56:1665–1674
Reysenbach A-L, Banta AB, Boone DR, Cary SC, Luther GW (2000a) Microbial essentials at hydrothermal vents. Nature 404:835–835
Reysenbach A-L, Longnecker K, Kirshtein J (2000b) Novel bacterial and archaeal lineages from an in situ growth chamber deployed at a Mid-Atlantic Ridge hydrothermal vent. Appl Environ Microbiol 66:3798
Ross DA, Whitmarsh RB, Ali SA, Boudreaux JE, Coleman R, Fleisher RL, Girdler R, Manheim F, Matter A, Nigrini C, Stoffers P, Supko PR (1973) Red Sea drillings. Science 179:377–380
Ruby EG, Jannasch HW (1982) Physiological characteristics of *Thiomicrospira* sp. Strain L-12 isolated from deep-sea hydrothermal vents. J Bacteriol 149:161–165
Russ L, Kartal B, Op Den Camp H, Sollai M, Le Bruchec J, Caprais J-C, Godfroy A, Sinninghe Damsté J, Jetten M (2013) Presence and diversity of anammox bacteria in cold hydrocarbon-rich seeps and hydrothermal vent sediments of the Guaymas Basin. Front Microbiol 4:219
Sagar S, Esau L, Hikmawan T, Antunes A, Holtermann K, Stingl U, Bajic VB, Kaur M (2013a) Cytotoxic and apoptotic evaluations of marine bacteria isolated from brine-seawater interface of the Red Sea. BMC Complement Altern Med 13:29
Sagar S, Esau L, Holtermann K, Hikmawan T, Zhang G, Stingl U, Bajic VB, Kaur M (2013b) Induction of apoptosis in cancer cell lines by the Red Sea brine pool bacterial extracts. BMC Complement Altern Med 13:344
Sakai S, Takaki Y, Miyazaki M, Ogawara M, Yanagawa K, Miyazaki J, Takai K (2019) *Methanofervidicoccus abyssi* gen. nov., sp. nov., a hydrogenotrophic methanogen, isolated from a hydrothermal vent chimney in the Mid-Cayman Spreading Center, the Caribbean Sea. Int J Syst Evol Microbiol 69:1225–1230
Sanders JG, Beinart RA, Stewart FJ, Delong EF, Girguis PR (2013) Metatranscriptomics reveal differences in in situ energy and nitrogen metabolism among hydrothermal vent snail symbionts. ISME J 7:1556–1567
Sarmiento F, Peralta R, Blamey JM (2015) Cold and hot extremozymes: industrial relevance and current trends. Front Bioeng Biotechnol 3:148
Sato T, Atomi H, Imanaka T (2007) Archaeal type III RuBisCOs function in a pathway for amp metabolism. Science 315:1003
Sauvadet A-L, Gobet A, Guillou L (2010) Comparative analysis between protist communities from the deep-sea pelagic ecosystem and specific deep hydrothermal habitats. Environ Microbiol 12: 2946–2964
Schmidt M, Al-Farawati R, Botz R (2015) Geochemical classification of brine-filled Red Sea deeps. In: Rasul NMA, Stewart ICF (eds) The Red Sea: the formation, morphology, oceanography and environment of a young ocean basin. Springer, Berlin, pp 219–233
Schrenk MO, Kelley DS, Bolton SA, Baross JA (2004) Low archaeal diversity linked to subseafloor geochemical processes at the Lost City Hydrothermal Field, Mid-Atlantic Ridge. Environ Microbiol 6:1086–1095
Scott KM, Sievert SM, Abril FN, Ball LA, Barrett CJ, Blake RA, Boller AJ, Chain PSG, Clark JA, Davis CR, Detter C, Do KF, Dobrinski KP, Faza BI, Fitzpatrick KA, Freyermuth SK, Harmer

TL, Hauser LJ, Hügler M, Kerfeld CA, Klotz MG, Kong WW, Land M, Lapidus A, Larimer FW, Longo DL, Lucas S, Malfatti SA, Massey SE, Martin DD, McCuddin Z, Meyer F, Moore JL, Ocampo LH Jr, Paul JH, Paulsen IT, Reep DK, Ren Q, Ross RL, Sato PY, Thomas P, Tinkham LE, Zeruth GT (2006) The genome of deep-sea vent chemolithoautotroph *Thiomicrospira crunogena* XCL-2. PLoS Biol 4:e383

Shahak Y, Arieli B, Padan E, Hauska G (1992) Sulfide quinone reductase (SQR) activity in *Chlorobium*. FEBS Lett 299:127–130

Sherr EB, Sherr BF (1994) Bacterivory and herbivory: key roles of phagotrophic protists in pelagic food webs. Microb Ecol 28:223–235

Shokes RF, Trabant PK, Presley BJ, Reid DF (1977) Anoxic, hypersaline basin in the Northern Gulf of Mexico. Science 196:1443–1446

Sievert SM, Kuever J (2000) *Desulfacinum hydrothermale* sp. nov., a thermophilic, sulfate-reducing bacterium from geothermally heated sediments near Milos Island (Greece). Int J Syst Evol Microbiol 50:1239–1246

Skennerton CT, Ward LM, Michel A, Metcalfe K, Valiente C, Mullin S, Chan KY, Gradinaru V, Orphan VJ (2015) Genomic reconstruction of an uncultured hydrothermal vent gammaproteobacterial methanotroph (Family Methylothermaceae) indicates multiple adaptations to oxygen limitation. Front Microbiol 6:1425

Small EB, Gross ME (1985) Preliminary observations of protistan organisms, especially ciliates, from the 21°N hydrothermal vent site. Bull Biol Soc Wash 6:401–410

Sojo V, Herschy B, Whicher A, Camprubí E, Lane N (2016) The origin of life in alkaline hydrothermal vents. Astrobiology 16:181–197

Søndergaard D, Pedersen CNS, Greening C (2016) HydDB: a web tool for hydrogenase classification and analysis. Sci Rep 6:34212

Sorokin DY, Kublanov IV, Yakimov MM, Rijpstra WIC, Sinninghe Damsté JS (2016) Halobacteriaceae—click to open names for life widget. Int J Syst Evol Microbiol 66:2377–2381

Steinle L, Knittel K, Felber N, Casalino C, de Lange G, Tessarolo C, Stadnitskaia A, Sinninghe Damsté JS, Zopfi J, Lehmann MF, Treude T, Niemann H (2018) Life on the edge: active microbial communities in the Kryos $MgCl_2$-brine basin at very low water activity. ISME J 12: 1414–1426

Stepnov AA, Fredriksen L, Steen IH, Stokke R, Eijsink VGH (2019) Identification and characterization of a hyperthermophilic GH9 cellulase from the Arctic Mid-Ocean Ridge vent field. PLoS One 14:e0222216

Stetter KO, Lauerer G, Thomm M, Neuner A (1987) Isolation of extremely thermophilic sulfate reducers: evidence for a novel branch of Archaebacteria. Science 236:822

Stevenson A, Cray JA, Williams JP, Santos R, Sahay R, Neuenkirchen N, McClure CD, Grant IR, Houghton JD, Quinn JP, Timson DJ, Patil SV, Singhal RS, Antón J, Dijksterhuis J, Hocking AD, Lievens B, Rangel DEN, Voytek MA, Gunde-Cimerman N, Oren A, Timmis KN, McGenity TJ, Hallsworth JE (2015) Is there a common water-activity limit for the three domains of life? ISME J 9:1333–1351

Stevenson A, Hamill PG, O'Kane CJ, Kminek G, Rummel JD, Voytek MA, Dijksterhuis J, Hallsworth JE (2017) *Aspergillus penicillioides* differentiation and cell division at 0.585 water activity. Environ Microbiol 19:687–697

Stock A, Breiner H-W, Pachiadaki M, Edgcomb V, Filker S, La Cono V, Yakimov MM, Stoeck T (2012) Microbial eukaryote life in the new hypersaline deep-sea basin Thetis. Extremophiles 16: 21–34

Stock A, Filker S, Yakimov M, Stoeck T (2013) Deep hypersaline anoxic basins as model systems for environmental selection of microbial plankton. In: Seckbach J, Oren A, Stan-Lotter H (eds) Polyextremophiles: life under multiple forms of stress. Springer, Dordrecht, pp 499–515

Stohr R, Waberski A, Völker H, Tindall BJ, Thomm M (2001) *Hydrogenothermus marinus* gen. nov., sp. nov., a novel thermophilic hydrogen-oxidizing bacterium, recognition of *Calderobacterium hydrogenophilum* as a member of the genus *Hydrogenobacter* and proposal of the reclassification of *Hydrogenobacter acidophilus* as *Hydrogenobaculum acidophilum* gen.

nov., comb. nov., in the phylum 'Hydrogenobacter/Aquifex'. Int J Syst Evol Microbiol 51: 1853–1862

Stokke R, Reeves EP, Dahle H, Fedøy A-E, Viflot T, Lie Onstad S, Vulcano F, Pedersen RB, Eijsink VGH, Steen IH (2020) Tailoring hydrothermal vent biodiversity toward improved biodiscovery using a novel in situ enrichment strategy. Front Microbiol 11:249

Swallow JC, Crease J (1965) Hot salty water at the bottom of the Red Sea. Nature 205:165–166

Sylvan J, Sia T, Haddad A, Briscoe L, Toner B, Girguis P, Edwards K (2013) Low temperature geomicrobiology follows host rock composition along a geochemical gradient in Lau Basin. Front Microbiol 4:61

Takai K, Horikoshi K (1999) Genetic diversity of archaea in deep-sea hydrothermal vent environments. Genetics 152:1285–1297

Takai K, Inoue A, Horikoshi K (2002) *Methanothermococcus okinawensis* sp. nov., a thermophilic, methane-producing archaeon isolated from a Western Pacific deep-sea hydrothermal vent system. Int J Syst Evol Microbiol 52:1089–1095

Takai K, Inagaki F, Nakagawa S, Hirayama H, Nunoura T, Sako Y, Nealson KH, Horikoshi K (2003a) Isolation and phylogenetic diversity of members of previously uncultivated ε-Proteobacteria in deep-sea hydrothermal fields. FEMS Microbiol Lett 218:167–174

Takai K, Nakagawa S, Sako Y, Horikoshi K (2003b) *Balnearium lithotrophicum* gen. nov., sp. nov., a novel thermophilic, strictly anaerobic, hydrogen-oxidizing chemolithoautotroph isolated from a black smoker chimney in the Suiyo Seamount hydrothermal system. Int J Syst Evol Microbiol 53:1947–1954

Takai K, Nealson KH, Horikoshi K (2004a) *Methanotorris formicicus* sp. nov., a novel extremely thermophilic, methane-producing archaeon isolated from a black smoker chimney in the Central Indian Ridge. Int J Syst Evol Microbiol 54:1095–1100

Takai K, Nealson KH, Horikoshi K (2004b) *Hydrogenimonas thermophila* gen. nov., sp. nov., a novel thermophilic, hydrogen-oxidizing chemolithoautotroph within the ε-Proteobacteria, isolated from a black smoker in a Central Indian Ridge hydrothermal field. Int J Syst Evol Microbiol 54:25–32

Takai K, Hirayama H, Nakagawa T, Suzuki Y, Nealson KH, Horikoshi K (2005) *Lebetimonas acidiphila* gen. nov., sp. nov., a novel thermophilic, acidophilic, hydrogen-oxidizing chemolithoautotroph within the 'Epsilonproteobacteria', isolated from a deep-sea hydrothermal fumarole in the Mariana Arc. Int J Syst Evol Microbiol 55:183–189

Takai K, Suzuki M, Nakagawa S, Miyazaki M, Suzuki Y, Inagaki F, Horikoshi K (2006) *Sulfurimonas paralvinellae* sp. nov., a novel mesophilic, hydrogen- and sulfur-oxidizing chemolithoautotroph within the Epsilonproteobacteria isolated from a deep-sea hydrothermal vent polychaete nest, reclassification of *Thiomicrospira denitrificans* as *Sulfurimonas denitrificans* comb. nov. and emended description of the genus *Sulfurimonas*. Int J Syst Evol Microbiol 56:1725–1733

Thauer RK (1998) Biochemistry of methanogenesis: a tribute to Marjory Stephenson:1998 Marjory Stephenson Prize Lecture. Microbiology 144:2377–2406

Tivey MK (2007) Generation of seafloor hydrothermal vent fluids and associated mineral deposits. Oceanography 20:50–65

Trüper HG (1969) Bacterial sulfate reduction in the Red Sea hot brines. In: Degens ET, Ross DA (eds) Hot brines and recent heavy metal deposits in the Red Sea: a geochemical and geophysical account. Springer, Berlin, pp 263–271

Tuovila BJ, Dobbs FC, LaRock PA, Siegel BZ (1987) Preservation of ATP in hypersaline environments. Appl Environ Microbiol 53:2749

Tuttle JH, Jannasch HW (1979) Microbial dark assimilation of $CO_2$ in the Cariaco Trench. Limnol Oceanogr 24:746–753

Tuttle JH, Wirsen CO, Jannasch HW (1983) Microbial activities in the emitted hydrothermal waters of the Galápagos rift vents. Mar Biol 73:293–299

Van Cappellen P, Viollier E, Roychoudhury A, Clark L, Ingall E, Lowe K, Dichristina T (1998) Biogeochemical cycles of manganese and iron at the oxic–anoxic transition of a stratified marine basin (Orca Basin, Gulf of Mexico). Environ Sci Technol 32:2931–2939

Vander Roost J, Daae FL, Steen IH, Thorseth IH, Dahle H (2018) Distribution patterns of iron-oxidizing zeta- and beta-proteobacteria from different environmental settings at the Jan Mayen Vent Fields. Front Microbiol 9:3008

van der Wielen PWJJ, Bolhuis H, Borin S, Daffonchio D, Corselli C, Giuliano L, D'Auria G, de Lange GJ, Huebner A, Varnavas SP, Thomson J, Tamburini C, Marty D, McGenity TJ, Timmis KN, Party BS (2005) The enigma of prokaryotic life in deep hypersaline anoxic basins. Science 307:121–123

Van Dover CL, Fry B, Grassle JF, Humphris S, Rona PA (1988) Feeding biology of the shrimp Rimicaris exoculata at hydrothermal vents on the Mid-Atlantic Ridge. Mar Biol 98:209–216

Varrella S, Tangherlini M, Corinaldesi C (2020) Deep hypersaline anoxic basins as untapped reservoir of polyextremophilic prokaryotes of biotechnological interest. Mar Drugs 18:91

Vetriani C, Speck MD, Ellor SV, Lutz RA, Starovoytov V (2004) Thermovibrio ammonificans sp. nov., a thermophilic, chemolithotrophic, nitrate-ammonifying bacterium from deep-sea hydrothermal vents. Int J Syst Evol Microbiol 54:175–181

Vidussi F, Claustre H, Manca BB, Luchetta A, Marty J-C (2001) Phytoplankton pigment distribution in relation to upper thermocline circulation in the eastern Mediterranean Sea during winter. J Geophys Res Oceans 106:19939–19956

Vieille C, Zeikus GJ (2001) Hyperthermophilic enzymes: sources, uses, and molecular mechanisms for thermostability. Microbiol Mol Biol Rev 65:1–43

Vignais PM (2008) Hydrogenases and $H^+$-reduction in primary energy conservation. In: Schäfer G, Penefsky HS (eds) Bioenergetics: energy conservation and conversion. Springer, Berlin, pp 223–252

Vignais PM, Billoud B (2007) Occurrence, classification, and biological function of hydrogenases: an overview. Chem Rev 107:4206–4272

Vuoristo KS, Fredriksen L, Oftebro M, Arntzen MØ, Aarstad OA, Stokke R, Steen IH, Hansen LD, Schüller RB, Aachmann FL, Horn SJ, Eijsink VGH (2019) Production, characterization, and application of an alginate lyase, AMOR_PL7A, from hot vents in the Arctic Mid-Ocean Ridge. J Agric Food Chem 67:2936–2945

Waite DW, Vanwonterghem I, Rinke C, Parks DH, Zhang Y, Takai K, Sievert SM, Simon J, Campbell BJ, Hanson TE, Woyke T, Klotz MG, Hugenholtz P (2017) Comparative genomic analysis of the Class Epsilonproteobacteria and proposed reclassification to Epsilonbacteraeota (phyl. nov.). Front Microbiol 8:682

Waite DW, Vanwonterghem I, Rinke C, Parks DH, Zhang Y, Takai K, Sievert SM, Simon J, Campbell BJ, Hanson TE, Woyke T, Klotz MG, Hugenholtz P (2018) Addendum: comparative genomic analysis of the Class Epsilonproteobacteria and proposed reclassification to Epsilonbacteraeota (phyl. nov.). Front Microbiol 9:772

Waite DW, Chuvochina M, Pelikan C, Parks DH, Yilmaz P, Wagner M, Loy A, Naganuma T, Nakai R, Whitman WB, Hahn MW, Kuever J, Hugenholtz P (2020) Proposal to reclassify the proteobacterial classes Deltaproteobacteria and Oligoflexia, and the phylum Thermodesulfobacteria into four phyla reflecting major functional capabilities. Int J Syst Evol Microbiol 70:5972–6016

Wang S, Xiao X, Jiang L, Peng X, Zhou H, Meng J, Wang F (2009) Diversity and abundance of ammonia-oxidizing archaea in hydrothermal vent chimneys of the Juan de Fuca Ridge. Appl Environ Microbiol 75:4216–4220

Wang Y, Yang J, Lee OO, Dash S, Lau SCK, Al-Suwailem A, Wong TYH, Danchin A, Qian P-Y (2011) Hydrothermally generated aromatic compounds are consumed by bacteria colonizing in Atlantis II Deep of the Red Sea. ISME J 5:1652–1659

Watson SW, Waterbury JB (1969) The sterile hot brines of the Red Sea. In: Degens ET, Ross DA (eds) Hot brines and recent heavy metal deposits in the Red Sea: a geochemical and geophysical account. Springer, Berlin, pp 272–281

Weiss RF, Lonsdale P, Lupton JE, Bainbridge AE, Craig H (1977) Hydrothermal plumes in the Galapagos Rift. Nature 267:600–603

Williams DL, Von Herzen RP, Sclater JG, Anderson RN (1974) The Galapagos spreading centre: lithospheric cooling and hydrothermal circulation. Geophys J Int 38:587–608

Wirsen CO, Tuttle JH, Jannasch HW (1986) Activities of sulfur-oxidizing bacteria at the 21°N East Pacific Rise vent site. Mar Biol 92:449–456

Wirsen CO, Brinkhoff T, Kuever J, Muyzer G, Molyneaux S, Jannasch HW (1998) Comparison of a new *Thiomicrospira* strain from the Mid-Atlantic Ridge with known hydrothermal vent isolates. Appl Environ Microbiol 64:4057–4059

Xu W, Li M, Ding J-F, Gu J-D, Luo Z-H (2014) Bacteria dominate the ammonia-oxidizing community in a hydrothermal vent site at the Mid-Atlantic Ridge of the South Atlantic Ocean. Appl Microbiol and Biotechnol 98:7993–8004

Yakimov MM, Giuliano L, Cappello S, Denaro R, Golyshin PN (2007a) Microbial community of a hydrothermal mud vent underneath the deep-sea anoxic brine lake Urania (Eastern Mediterranean). Orig Life Evol Biosph 37:177–188

Yakimov MM, La Cono V, Denaro R, D'Auria G, Decembrini F, Timmis KN, Golyshin PN, Giuliano L (2007b) Primary producing prokaryotic communities of brine, interface and seawater above the halocline of deep anoxic lake L'Atalante, Eastern Mediterranean Sea. ISME J 1:743–755

Yakimov MM, La Cono V, Slepak VZ, La Spada G, Arcadi E, Messina E, Borghini M, Monticelli LS, Rojo D, Barbas C, Golyshina OV, Ferrer M, Golyshin PN, Giuliano L (2013) Microbial life in the Lake Medee, the largest deep-sea salt-saturated formation. Sci Rep 3:3554

Yakimov MM, La Cono V, Spada GL, Bortoluzzi G, Messina E, Smedile F, Arcadi E, Borghini M, Ferrer M, Schmitt-Kopplin P, Hertkorn N, Cray JA, Hallsworth JE, Golyshin PN, Giuliano L (2015) Microbial community of the deep-sea brine Lake Kryos seawater–brine interface is active below the chaotropicity limit of life as revealed by recovery of mRNA. Environ Microbiol 17:364–382

Zhou Z, Liu Y, Pan J, Cron BR, Toner BM, Anantharaman K, Breier JA, Dick GJ, Li M (2020) Gammaproteobacteria mediating utilization of methyl-, sulfur- and petroleum organic compounds in deep ocean hydrothermal plumes. ISME J 14:3136–3148

Zhuang G-C, Elling FJ, Nigro LM, Samarkin V, Joye SB, Teske A, Hinrichs K-U (2016) Multiple evidence for methylotrophic methanogenesis as the dominant methanogenic pathway in hypersaline sediments from the Orca Basin, Gulf of Mexico. Geochim Cosmochim Acta 187:1–20

# Part III

# Marine Microbiome from Genomes to Phenomes: Biogeochemical Cycles, Networks, Fluxes, and Interaction

# Marine Biogeochemical Cycles

# 15

Samantha B. Joye, Marshall W. Bowles, and Kai Ziervogel

**Abstract**

Microorganisms catalyze elemental cycling in the ocean and regulate geochemical processes that are crucial for the maintenance of planetary habitability. This chapter provides an overview of marine carbon, oxygen, nitrogen, phosphorus, sulfur, and trace gas cycling with an emphasis on modern processes in pelagic and sedimentary habitats. Microorganisms couple oxidation and reduction reactions to extract energy from chemical disequilibria via thermodynamics. The ocean microbiome represents a reservoir of genetic diversity and metabolic potential. It performs the basic functions of this living planet and responds to chronic and acute external forcing in order to maintain these functions. This chapter reviews the biogeochemical cycles at local and regional scales and assesses how ocean biogeochemistry may be altered by global change. Climate change is having substantial impacts on the ocean. Ocean warming may alter the ocean's microbiome. Biological interactions and distribution of plants, animals, and microorganisms will change to accommodate the new temperature and mixing regimes. Plankton moves with the water, and warming waters will exert a selective pressure on the microbiome, promoting the proliferation of

S. B. Joye (✉)
Department of Marine Sciences, University of Georgia, Athens, GA, USA
e-mail: mjoye@uga.edu

M. W. Bowles
Louisiana Universities Marine Consortium (LUMCON), Chauvin, LA, USA
e-mail: mbowles@lumcon.edu

K. Ziervogel
Institute for the Study of Earth, Oceans, and Space, University of New Hampshire, Durham, NH, USA
e-mail: Kai.Ziervogel@unh.edu

 
L. J. Stal, M. S. Cretoiu (eds.), *The Marine Microbiome*, The Microbiomes of Humans, Animals, Plants, and the Environment 3,
https://doi.org/10.1007/978-3-030-90383-1_15

microorganisms that are able to thrive under local conditions. Such changes in the microbiome may alter local and regional biogeochemistry.

**Keywords**

Carbon · Climate change · Elemental cycling · Microbial metabolism · Nitrogen · Phosphorus

## 15.1 Biogeochemistry in the Ocean

Biogeochemistry is an inherently interdisciplinary science that explores the chemical, biological, and geological processing of materials in an ecosystem in the context of the physical environment. The ocean is a dynamic biogeochemical reactor that enables planetary habitability and helps to keep climate in equilibrium by feedback mechanisms and by storing heat and carbon. The ocean acts as a source of atmospheric oxygen and as a sink for atmospheric carbon dioxide. It provides numerous critical ecosystem services such as nutrient cycling and detoxification and removal of pollutants. Ocean biogeochemistry is enabled by the tremendous biodiversity present at the microscale. Ocean biogeochemical reactions are mediated by microorganisms that couple oxidation and reduction reactions in order to extract power from chemical disequilibria via thermodynamics. Microorganisms use that energy to grow and generate organic biomass (Falkowski et al. 2008; Kleidon 2012). The tremendous metabolic plasticity and functional diversity coupled with the ability of ocean microbes to exhibit rapid growth rates allows the ocean's microbiome to respond quickly to changing environmental conditions (Sunagawa et al. 2015).

Ocean biogeochemical cycles and their spatial and temporal variations are tightly coupled to physical processes that are responsible for the large-scale movement of water within and between ocean basins. Wind-driven surface currents transfer water and heat from the equatorial regions to the polar regions. In polar regions, the cooling of salty water initiates thermohaline circulation that drives the global vertical water currents across the deep ocean. Together these surface and deep currents drive the "ocean conveyer belt" (Broecker 1987, 1991), a circulation system that transports oxygen, carbon, and other elements across the ocean and enable upwelling of nutrient-rich deep water. This phenomenon impacts nutrient availability and modulates ocean productivity patterns in certain ocean regions (Sarmiento and Toggweiler 1984; Toggweiler 1999). Ocean physics drives highly productive upwelling zones along the eastern side of ocean basins and along the equator while at the same time creating oligotrophic, nutrient-limited basins in the center of ocean gyres. Climate-driven changes in the veracity of the ocean conveyer belt are likely to alter the pattern and magnitude of biological processes because of changes in mixing regimes (Groeskamp et al. 2019).

Within the context of the dynamic physical system, autochthonous and allochthonous inputs of organic matter fuel elemental cycling at rates that impact biogeochemical cycling at a global scale. Ocean biogeochemical processes cycle the major elements of life—carbon, nitrogen, oxygen, phosphorus, and sulfur—along with key micronutrients and vitamins. These biogeochemical cycles act on a variety of scales, both temporal and spatial (Jackson 1994), and give rise to a suite of emergent properties in the marine environment (Levin 1998). The ocean's microbiome represents an enormous reservoir of metabolic potential that performs the basic biogeochemical functions and is able to quickly respond to chronic and acute environmental forcing (Baltar and Herndl 2019). This chapter provides an overview of ocean biogeochemistry of the major elements for life and of bioactive trace gases and highlight the linkages between the microbiome and functional dynamics at local and regional scales and assess how ocean biogeochemistry may be altered by global change. The processes that this chapter addresses includes primary production, which occur across the ocean but exhibits local and regional variations in the spatial-temporal patterns and magnitude of activity.

The biogeochemical cycles of the elements are inherently connected through the currency of carbon (Fig. 15.1), the primary structural building block of biological biomass. Carbon cycling is driven by the assimilation of inorganic carbon into biomass and the subsequent recycling of biomass-derived complex organic carbon to high molecular weight dissolved organic carbon and subsequently into labile intermediates of which a certain fraction is returned back to carbon dioxide while another portion forms labile intermediate metabolites and refractory dissolved organic compounds. Assimilation of inorganic carbon into biomass occurs concomitantly with the uptake of nutrients such as nitrogen, phosphorus, silica, and iron, as well as certain trace elements and vitamins. The stoichiometry of major nutrient elements drives profound patterns of ecology (Elser et al. 2008; Galbraith and Martiny 2015; Sterner 2015). In contrast, oxidation of organic matter recycles these materials to their mineral forms (Welti et al. 2017). A detailed review of the energetics and ecological stoichiometry of ocean ecosystems is beyond the scope of this chapter, but the reader is referred to Galbraith and Martiny (2015), Moreno and Martiny (2018), and Vallino and Algar (2016) for in-depth consideration of these topics.

## 15.2 Biogeochemical Cycles

The ocean's biogeochemical cycles are inherently interlinked. Microbial processes mediate sophisticated redox cycling of carbon, nitrogen, phosphorus, oxygen, and sulfur moving elements between oxidized and reduced reservoirs on short time scales and facilitating burial of solid phases that cycle on longer time scales. Tectonic processes influence the phosphorus and sulfur cycles disproportionately, while ocean–atmosphere dynamics play key roles in the carbon, nitrogen, and oxygen cycles. The key biological processes driving the biogeochemistry of these elements is summarized here, and references for critical reviews are provided for more in-depth exploration of specific sub-topics.

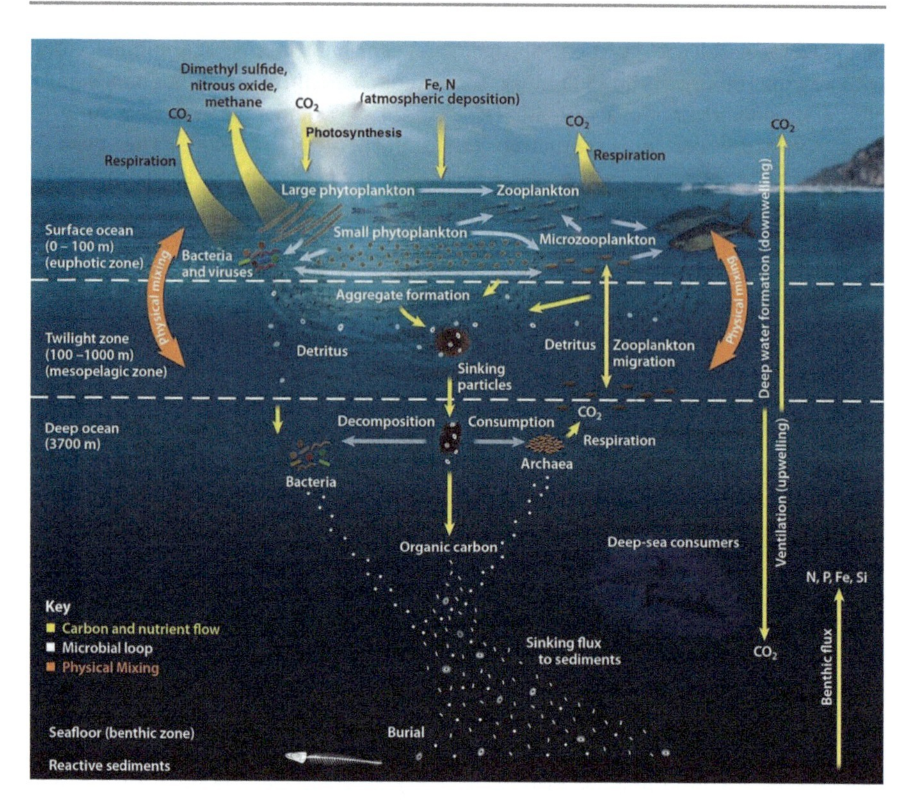


**Fig. 15.1** Biogeochemical cycling in the ocean. Microorganisms dominate transformations of carbon at the base of the food web, consuming inorganic carbon and producing particulate and dissolved organic matter. Particulate material can be oxidized or form aggregates and sink to the deep ocean or even to the seafloor. Dissolved organic matter partitions into discrete pools of that cycle on short (labile) or long (refractory) time scales. Figure adapted from one originally produced by Oak Ridge National Laboratory

### 15.2.1 Carbon

Primary production by phytoplankton in the ocean represents one of the largest fluxes of carbon on the planet. Phytoplankton use pigments to exploit chemical disequilibria—they convert light energy into chemical energy that is used to transform inorganic carbon into organic biomass. Oxygenic- and many anoxygenic phototrophic microorganisms are autotrophs, i.e., they fix inorganic carbon, and these organisms perform these processes in the surface ocean when sufficient sunlight and nutrients are present. Some anoxygenic phototrophs are heterotrophs, so do not fix carbon dioxide. Other organisms fix carbon dioxide into biomass in the absence of light energy (they are called chemoautotrophs). For example, microbes extract energy from the oxidation of ammonium, reduced sulfur compounds, or methane to fix inorganic carbon into biomass; this occurs throughout the ocean. Heterotrophic processes are fueled by organic carbon delivered to the ocean via terrestrial runoff (allochthonous organic matter) or from seafloor discharge of dissolved organic carbon, oil, or alkanes into the deep-water column and by organic carbon produced in situ (autochthonous organic matter). Autotrophic and heterotrophic processes occur contemporaneously, and while areas of net autotrophy versus net heterotrophy are typically spatially and/or temporally distinct, the differences are often small. Nonetheless, the ocean serves as a net sink for carbon dioxide overall.

Global estimates of primary production have been derived from mathematical models using satellite data describing ocean color (e.g., translated into chlorophyll concentration), global surface ocean temperature data from satellites, and eco-physiological data for phytoplankton that incorporates temperature-driven variability in photosynthetic rates and differences in photosynthetic efficiency (Behrenfeld and Falkowski 1997). Chlorophyll concentration derived from satellite data and algorithms are a robust estimate for mixed layer concentrations. The general rate of marine primary production is considered to be roughly 50 Gt C per year (Field et al. 1998), and approximately 30% of that production (16 gigatons of C per year) is exported to the ocean's interior (Behrenfeld et al. 2005; Falkowski et al. 1998; Sigman and Hain 2012). Estimates of phytoplankton production over time series between 1998 and 2018 varied between 38.8 and 42.1 Gt C per year (Kulk et al. 2020), which is similar in magnitude to the rate determined from models (~50 Gt C per year; Behrenfeld et al. 2005).

The average range of surface ocean primary production is 75–150 g C $m^{-2}$ $year^{-1}$. The concentration of chlorophyll and the rates of primary production are highest near continents (i.e., estuarine mean primary production is 252 g C $m^{-2}$ $year^{-1}$; Cloern et al. 2014) and in upwelling zones, such as the California coast (200–300 g C $m^{-2}$ $year^{-1}$), Peruvian coast (200–400 g C $m^{-2}$ $year^{-1}$), and in the Southern Ocean (200–400 g C $m^{-2}$ $year^{-1}$) (Webb 2020) where nutrient fluxes and nutrient concentrations tend to be higher (Fig. 15.2). High productivity areas account for a small proportion of ocean area (estuaries and shelves account for 8% of oceanic surface area; 92% of the area is open ocean). Rates of primary production are much lower on an aerial basis in oligotrophic ocean gyres (<50 g C $m^{-2}$ $year^{-1}$), but since gyres account for such a large proportion of ocean area, they account for a substantial amount of primary production (Fig. 15.2). Primary production in the ocean is controlled by the availability of nutrients, light, temperature, biological interactions (grazing or viral lysis), and by physical processes (Cloern et al. 2014). At any given location, one or more of these factors interact and thereby modulate the rate of primary production.

Inorganic carbon is also fixed into biomass by so-called "dark processes," namely by chemoautotrophic microorganisms or by anaplerotic metabolism in a variety of microorganisms (Baltar and Herndl 2019) that do not require light energy. Most chemoautotrophs fix carbon dioxide via the Calvin-Benson-Bassham cycle or via the reverse citric acid cycle (Berg 2011; Hügler and Sievert 2011). Notably, the ammonia-oxidizing archaea utilize a modified version of the hydroxypropionate/hydroxybutyrate cycle (Berg et al. 2007); this is the most efficient pathway of carbon dioxide fixation known (Könneke et al. 2014). Anaplerotic pathways involve the chemical reactions that replenish the intermediates of metabolic pathways, such as

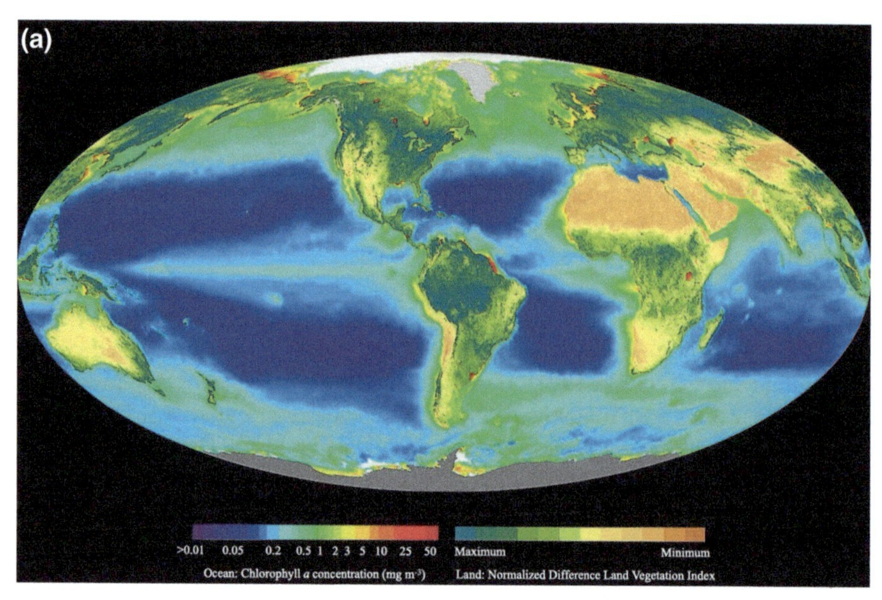


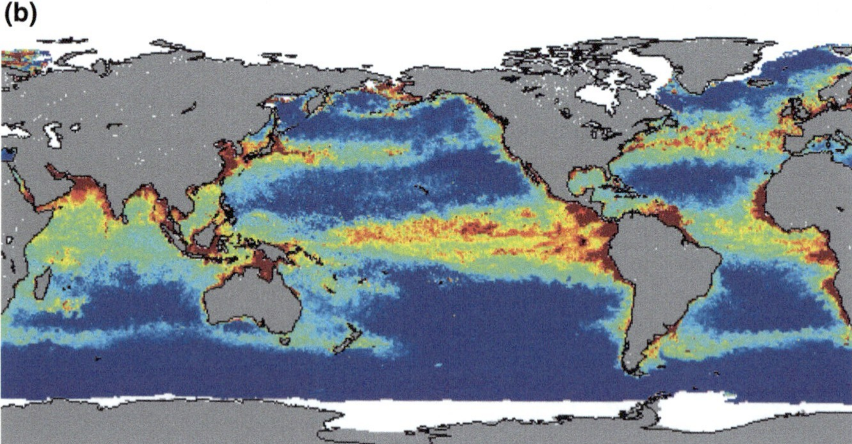


**Fig. 15.2** (**a**) Distribution of chlorophyll concentration in the surface ocean from the SeaWIFS project (NASA Goddard Space Flight Center, Wikimedia Commons; https://upload.wikimedia.org/wikipedia/commons/4/44/Seawifs_global_biosphere.jpg) and (**b**) Estimated rates of oceanic primary production from a vertically generalized production model using the temperature-dependence of primary production described by Eppley (1972) (red = high rates, blue = low rates) (from: http://sites.science.oregonstate.edu/ocean.productivity/custom.php)

the tricarboxylic acid cycle, to account for the shunting of intermediates into biosynthetic processes. Generally speaking, five major anaplerotic reactions can be distinguished. These are the conversion of pyruvate to oxaloacetate, aspartate to oxaloacetate, glutamate to alpha-ketoglutarate, adenylosuccinate to fumarate, and the beta-oxidation of fatty acids to succinyl-CoA.

Dark carbon fixation has long been recognized as an important source of organic carbon to the ocean (Nielsen 1960), but the relative importance of light-driven versus dark-driven primary carbon fixation remains poorly constrained to this day. Current estimates suggest that dark carbon fixation introduces an additional 1.2–11 Gt C per year of organic carbon into the ocean (Baltar and Herndl 2019), representing 5–22% of the annual global oceanic productivity. The rate of dark carbon fixation follows that of bacterial production in the deep ocean (Reinthaler et al. 2010). The bacterial production rate in the deep ocean is also similar to the organic carbon flux in the deep ocean (Baltar et al. 2010). While measurements of dark carbon fixation in the ocean are rare, available evidence suggests that this process contributes substantially to regional and global carbon cycles (Dyksma et al. 2016; Pachiadaki et al. 2017; Swan et al. 2011).

The presence of bioavailable autochthonous and allochthonous organic matter in the ocean supports heterotrophic microbial activity, which is reflected by the fact that net community production, which ranges between 5 and 12 Gt C per year, is, on average, less than 20% of gross primary production (see below; Sarmiento and Gruber 2006; Siegel et al. 2016; Sigman and Hain 2012). The net community production rate reflects the difference of autotrophic production and heterotrophic respiration of primary organic matter. This difference represents the amount of organic matter that is available for export from the euphotic zone to the deep ocean (Laws and Maiti 2019). Tight coupling between production and respiration of organic matter characterizes the open ocean environment, but some respiration—especially along continental margins—is fueled by allochthonous organic matter. Ocean margins are generally net heterotrophic—i.e., respiration of organic matter exceeds the production of organic matter (Smith and Mackenzie 1987) due to extensive inputs of terrigenous organic matter. The metabolic status of the offshore oligotrophic ocean is debated (Ducklow and Doney 2013) and depending on the evidence that is taken into account, its status is considered to be net autotrophic (Le et al. 2013) or net heterotrophic (Duarte et al. 2013). For the purpose of this chapter, the focus is more on the mechanisms of dissolved carbon cycling within the ocean than on the net flux of carbon in the system.

Dissolved organic carbon introduced into the ocean is metabolized and/or assimilated by heterotrophic bacteria (Moran et al. 2016; Zhuang et al. 2018, 2019b). Most of the terrigenous particulate organic matter introduced along the margins of the ocean via riverine and groundwater inputs is buried in river delta and shelf sediments (see Sect. 15.4 "Sediments"), while some material is respired and some is transported off the shelf (Burdige 2005; Hedges et al. 1997). Another source of organic matter to the ocean is the flux of dissolved organic matter from groundwater, which may account for as much as 18% of the total flux of dissolved organic carbon in some regions, e.g., the Arctic, and which may be increasing as a result of climate change (Connolly et al. 2020). The pool of carbon in dissolved organic matter in the ocean is similar to the amount of carbon present as carbon dioxide in the atmosphere (Hansell 2013; Hansell and Carlson 2015). Most of the dissolved organic matter, ~680 Gt, is recalcitrant (Hansell 2013) and roughly 6000 years old (Bauer et al. 1992). For more discussion regarding the nature and dynamics

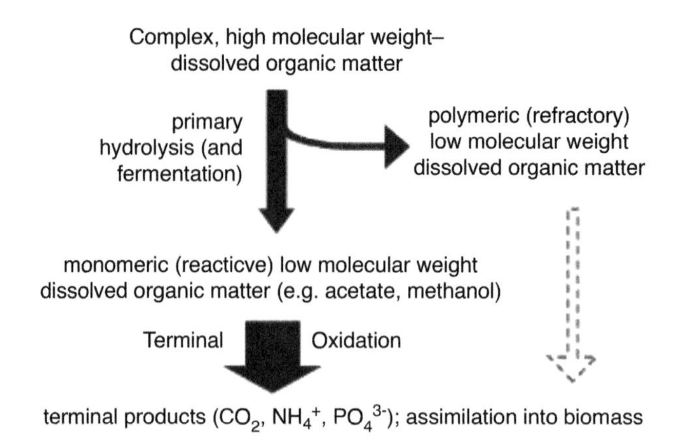


**Fig. 15.3** Scheme illustrating the processing of complex, high molecular weight dissolved organic matter to terminal products, including the production of heterotrophic bacterial biomass

of the refractory pool of dissolved organic matter, consult Arrieta et al. (2015), Jiao et al. (2010a), and Landry et al. (2017). The role of particles and particle-associated bacteria is discussed in Sect. 15.3.

Dissolved organic matter is generated through sloppy zooplankton grazing (Hygum and Peterson 1997) and by viral lysis of cells (Fuhrman 1999; Suttle 2007; Wilhelm and Suttle 1999) and by phytoplankton exudation. High molecular weight dissolved organic matter must be degraded via extracellular enzymes into monomers and small oligomers before it can be taken up by bacteria (Arnosti 2011; Cottrell and Kirchman 2000; Elifantz et al. 2007; Moran et al. 2016) (Fig. 15.3). However, some microorganisms, e.g., members of the *Bacteroidetes*, accumulate complex polysaccharides in the periplasmic space where they are degraded without diffusive loss of the products before they are further transported into the cell (Reintjes et al. 2017, 2019). This so-called 'selfish' (Cuskin et al. 2015; Reintjes et al. 2019) strategy of concentrating complex high molecular weight polysaccharides into the periplasmic space of cells appears to be widely distributed in the ocean and warrants further examination. Still, conventional knowledge stipulates that much of the high molecular weight dissolved organic matter is degraded extracellularly. This means that microorganisms that produce the exoenzymes may have to share the labile low molecular weight organic matter with those that do not invest in these enzymes and simply benefit from the metabolic investment of others (for reviews, see Arnosti 2011; Arnosti et al. 2011; Reintjes et al. 2019).

Labile low molecular weight organic compounds like methanol and acetate are important substrates for heterotrophic metabolism in the ocean (Mincer and Aicher 2016; Zhuang et al. 2018). Such labile compounds are often present at low concentration, but they cycle rapidly and support a substantial fraction of bacterial production (Zhuang et al. 2018). Methanol was long suspected to play an important role in the ocean's carbon cycle based on the abundance of methylotrophic bacteria commonly observed in sequence libraries (Giovannoni et al. 2008). Additionally, rapid turnover times in oligotrophic surface waters (Lidstrom 2006) pointed to a broad capacity for its metabolism. Mincer and Aicher (2016) showed that a wide variety of phytoplankton, including cyanobacteria, diatoms, coccoliths, and cryptophytes, produced methanol up at to 0.3% of the total cellular carbon production rate. The highest rate of methanol production occurred when the phytoplankton entered the stationary growth phase. The fate of methanol—oxidation to carbon dioxide or assimilation into biomass—depends on nitrogen availability (Zhuang et al. 2018). Methylamine, a byproduct of the metabolism of proteins and glycine betaine, an important osmolyte in phytoplankton, is also rapidly cycled by heterotrophic bacteria (Cirri and Pohnert 2019; Neufeld et al. 2007; Zhuang et al. 2018). The ammonium that is liberated from the decomposition of methylamine is recycled within the phytoplankton community (Suleiman et al. 2016). Acetate is another key intermediate of organic matter degradation, and like methanol and methylamine, it is rapidly oxidized or assimilated depending on local nutrient regimes (Zhuang et al. 2018). In some areas, acetate supported up to 50% of bacterial production (Zhuang et al. 2018).

Advances in organic geochemistry provide an unparalleled insight into the complexity of the dissolved organic matter pool (Moran et al. 2016). In parallel, modern genomics approaches have enhanced the understanding of the ability of microorganisms to process organic matter, the role of signaling molecules in microbial dynamics (Cirri and Pohnert 2019), algal-bacterial interactions in the phycosphere (Amin et al. 2012; Bell and Mitchell 1972; Seymour et al. 2010), and phycosphere metabolite exchange (Durham et al. 2015, 2017). Microbial genomics has also made it possible to understand how microbial populations and their activity respond to changing environmental conditions and perturbations (Joye and Kostka 2020). Linking geochemistry to genomics (Joye and Kostka 2020; Moran et al. 2016; Sunagawa et al. 2015) and single-cell techniques (Reintjes et al. 2019) holds great promise for advancing the understanding of organic carbon cycling in the ocean.

Photoheterotrophs are abundant in the ocean. For instance, aerobic anoxygenic photosynthetic bacteria (AAPB) constitute on average ~10% of the oceanic microbial biomass (Kolber et al. 2001). Anoxygenic phototrophs are particularly abundant in the oligotrophic ocean, where they may account for almost a quarter of microbial biomass (Lami et al. 2007). The AAPB are affiliated with the *Roseobacter* clade (Brinkhoff et al. 2008) in the alphaproteobacteria and the OM60/NOR5 clade of the gammaproteobacteria (Fuchs et al. 2007; Yan et al. 2009). It is estimated that bacteriochlorophyll *a*-mediated photophosphorylation is responsible for up to 6% of the total phototrophic energy flow in the surface ocean (Jiao et al. 2010b; Kolber et al. 2001). The high abundance of AAPBs is thought to originate from the selective advantage that photoheterotrophs enjoy by the utilization of abundant and relatively unlimited light energy compared to chemoheterotrophs, which require the availability of organic substrates. Also, the use of light energy by these organisms does not suffer from photoinhibition because AAPB carry out anoxygenic photosynthesis

(Spring and Riedel 2013). The regulation of activity by gammaproteobacterial and alphaproteobacterial AAPBs appears to vary substantially (Spring and Riedel 2013).

### 15.2.2 Oxygen

The oxygen inventory of the global ocean has declined substantially—by 2.7%—since 1960 (Schmidtko et al. 2017), which is largely caused by warming of the ocean, decreased oxygen solubility, increased stratification (Keeling et al. 2010), and increased biological consumption (Breitburg et al. 2018; Robinson 2019). Ocean models predict an additional decline in oxygen inventory by up to 7% by the end of this century (Keeling et al. 2010), assuming continued warming and intensifying stratification in the ocean. Ocean deoxygenation is most profound in pelagic oxygen minimum zones (Stramma et al. 2008) and in nearshore "dead zones" (Breitburg et al. 2018). Both types of low oxygen regions are expanding, which makes it important to document and understand the processes that lead to this situation.

Oxygen is produced during oxygenic photosynthesis and consumed by respiration, the Mehler reaction, and photorespiration. Cryptic oxygen cycling (Sutherland et al. 2020) is also important in some areas, especially in anoxic marine zones when photosynthetic cyanobacteria live atop the anoxic layer and produce oxygen that is then consumed rapidly by aerobic processes, limiting its accumulation (Garcia-Robledo et al. 2017). In these anoxic waters, the rates of oxygen production through cryptic cycling can be substantial and this process provides an additional source of carbon to fuel heterotrophy (Garcia-Robledo et al. 2017). In more ordinary pelagic habitats, microbial respiration plays a pivotal role in the oxygen cycle while at the same time controlling the storage of organic carbon in the ocean (Robinson 2019). Not much data is available to constrain the magnitude and variability of aerobic respiration rates in the ocean (Breitburg et al. 2018). Given the decline in oxygen concentration and depth-integrated inventories across the ocean (Schmidtko et al. 2017), more data describing the magnitude and variability in respiration rates is needed (Robinson 2019). In particular, the contribution of different microbial groups to respiration must be constrained, and the factors that control heterotrophic and autotrophic respiration, such as temperature, nutrient availability and element stoichiometry, must be elucidated.

Although microbial respiration is a major sink for oxygen in the ocean, it is not the only one. Photorespiration is the result of the oxygenase activity of the enzyme ribulose-1,5-bisphosphate carboxylase/oxygenase (rubisco) (Beardall et al. 2003), which oxygenates ribulose-bisphosphate, thereby short-circuiting $CO_2$ fixation. Photorespiration is essential for the recycling of 2-phosphoglycolate and 3-phosphoglycerate intermediates (Eisenhut et al. 2019). In order to prevent this loss of fixed carbon, some organisms limit photorespiration by an inorganic carbon concentration mechanism that favors the carboxylase activity of rubisco (Hagemann et al. 2016). Another sink for oxygen during photosynthetic metabolism is the Mehler reaction, which reduces oxygen to hydrogen peroxide (Strizh 2008). The Mehler reaction helps phytoplankton enhance the yield of ATP and reductant via the light-driven transfer of electrons generated by photosystem II back to water using ferredoxin to create a trans-thylakoid proton gradient that can drive ATP synthesis (Behrenfeld et al. 2008).

The biological production of reactive oxygen species such as superoxide via the extracellular reduction of oxygen by phototrophs and heterotrophic bacteria is another important process. Reactive oxygen may impact the physiology of an organism in various ways, such as by exacerbating oxidative stress, by serving as signaling molecules, and in some cases by playing a role in defense. Marine phototrophic microorganisms produce substantial quantities of superoxide (Hansel et al. 2016). Diaz et al. (2019) showed that the diatom *Thalassiosira* modulates the oxidation state of its internal $NADP^+/NADPH$ pool by producing superoxide. Marine bacteria are also an important source of reactive oxygen species, and this source does not depend on light. The formation of reactive oxygen dominates in the aphotic ocean, which represents the vast majority of the ocean volume (Diaz et al. 2013). It is estimated that the dark production of superoxide by marine microorganisms could consume as much as 15–50% of gross oxygen production in the marine environment (Sutherland et al. 2020). The processes that drive the release of extracellular superoxide are largely unknown. The influence of superoxide on the global oxygen and carbon budgets, on trace metal cycling, or on biological interactions is an important and open question.

### 15.2.3 Nitrogen

Nitrogen is an essential biological building block for organisms. It is necessary for the synthesis of amino acids, proteins, nucleic acids, enzymes, and vitamins. Dinitrogen gas ($N_2$) is the main component of the atmosphere, accounting for 78% of the composition of air. While nitrogen is required by all organisms, few organisms the so-called diazotrophs, can utilize atmospheric dinitrogen. Most of the organisms in the ocean rely on fixed nitrogen—e.g., ammonium, nitrate, nitrite, or organic nitrogen—to support their assimilatory demand. The relative proportion of carbon to nitrogen to phosphorus (C:N:P) in marine biomass in the modern ocean is roughly considered to be 106:16:1 (Redfield 1934, 1958), although deviations from this ratio are common (Moreno and Martiny 2018). Notably, different types of phytoplankton exhibit unique and flexible C:N:P stoichiometry (Weber and Deutsch 2010; White et al. 2006) and bacteria contain more N and have lower C:N ratios than phytoplankton (Bratbak 1985; Elser et al. 2008). Elemental stoichiometry controls biogeochemical cycling in the sea, and bulk stoichiometry has likely changed over geologic time (Planavsky 2014).

Certain microorganisms reduce dinitrogen gas to ammonia (=*nitrogen fixation*), which is subsequently assimilated into biomass (=*assimilation*) (Fig. 15.4). Biomass-nitrogen is recycled after the death of an organism and returned to the environment as ammonia (=*ammonification*). Ammonia can be oxidized to nitrite and then to nitrate (=*nitrification*), and that nitrate can be reduced to dinitrogen using organic carbon (mainly) as reductant (=*denitrification*), or ammonia is

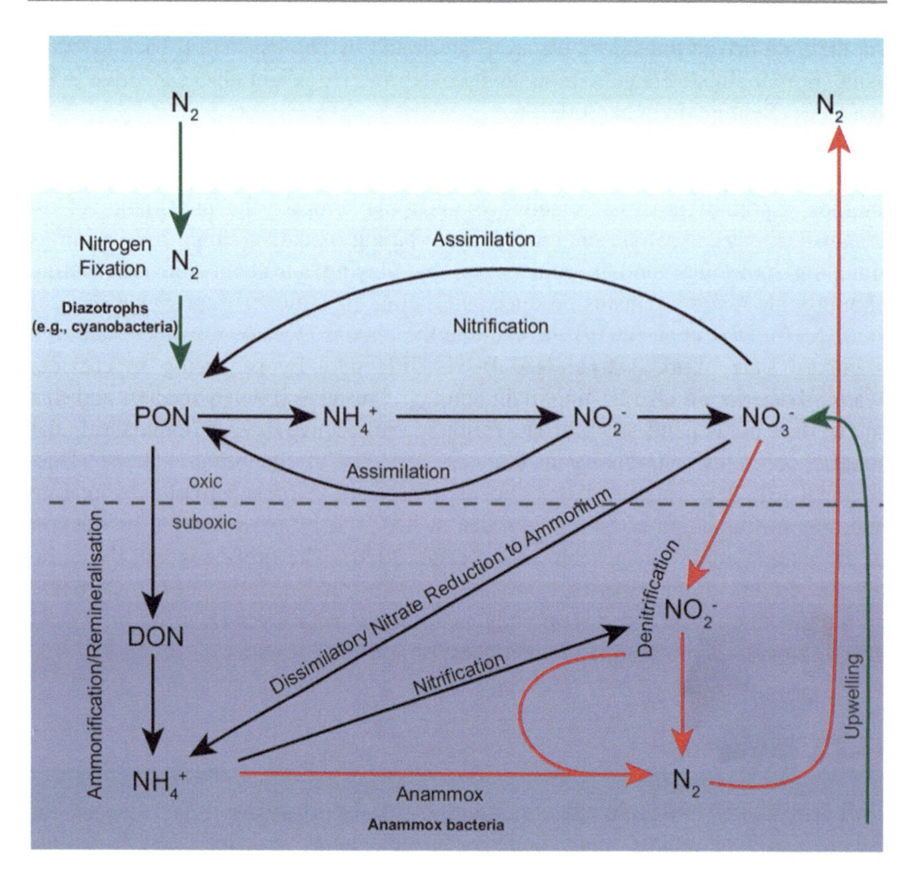


**Fig. 15.4** Key biogeochemical processes in the marine nitrogen cycle in oxic (upper portion) and suboxic/anoxic (lower portion) waters (adapted from Rush and Sinninghe-Damste 2017)

oxidized coupled to the reduction of nitrite (=*anaerobic ammonium oxidation; anammox*). The microorganisms mediating nitrification and anaerobic ammonium oxidation are chemoautotrophs, while organic carbon degrading denitrifiers are heterotrophs. Some autotrophic denitrifiers—microbes that couple nitrate reduction to the oxidation of Fe(II), Mn(III), hydrogen sulfide, or hydrogen while fixing inorganic carbon into biomass—proliferate in extreme environments (sludge digestors, deep-sea vents; Nakagawa and Takai 2008; Zhang et al. 2019a), but are not common in the open ocean. Nitrogen fixation can be mediated by photoautotrophic (i.e., cyanobacteria like *Trichodesmium*) and by heterotrophic microorganisms (i.e., sulfate-reducing microbes). A comprehensive overview of the microbial nitrogen cycle that includes an in-depth assessment of organisms and processes involved is provided by Kuypers et al. (2018).

In the ocean, nitrogen fixation introduces around 160 Tg of N per year across all ocean basins (Wang et al. 2019). Nitrogen fixation is mediated by microorganisms that possess the enzyme complex nitrogenase. Nitrogenase comes in three forms that differ in their cofactors (i.e., iron-iron, vanadium-iron, and molybdenum-iron; Eady 1996) but are functionally identical. Nitrogen fixation is an energy-demanding process. The reduction of one molecule of dinitrogen to ammonia requires 16 molecules of ATP and 8 low-potential electrons (via ferredoxin) (Bothe et al. 2010). Nitrogenase is inactivated by exposure to oxygen, which means that diazotrophs must protect nitrogenase from oxygen. Oxygenic phototrophic diazotrophs like *Trichodesmium* separate oxygen production from dinitrogen reduction (Berman-Frank et al. 2003), which comes at substantial energetic costs (Inomura et al. 2019). Potential adaptations to enable nitrogen fixation in the presence of oxygen include increased respiration rates to avoid inactivation of nitrogenase by oxygen, temporal separation of oxygenic photosynthesis (during the day) and nitrogen fixation (at night), and/or segregation/separation of the enzymatic machinery for the two processes. In the pelagic ocean, nitrogen fixation is mediated by free-living microorganisms such as *Trichodesmium* and *Crocosphaera* as well as by microbes that live associated with, e.g., diatoms (diatom-diazotroph associations; DDAs) (Inomura et al. 2020) or in symbiosis with eukaryotes (e.g., unicellular cyanobacterium "*Candidatus* Atelocyanobacterium thalassa" with the haptophyte *Braarudosphaera bigelowii*; Thompson et al. 2012). Heterotrophic nitrogen-fixing bacteria are widespread, and although their contribution to oceanic nitrogen fixation can be substantial (Wu et al. 2019), their importance appears to vary geographically and temporally (Sohm et al. 2011). Oceanic nitrogen fixation is often limited by phosphorus or iron availability (Sohm et al. 2011), and activity is also constrained by oxygen. Low oxygen locales such as aggregates and particles or oxygen minimum zones may be hotspots of nitrogen fixation activity.

Nitrification in the ocean is mediated by nitrifying bacteria and archaea (see Kuypers et al. 2018 and Ward 2011 for reviews). Nitrification is a key process of the ocean's nitrogen cycle that redistributes inorganic nitrogen via the sequential oxidation of ammonium to nitrite and subsequently by the oxidation of nitrite to nitrate. Nitrification rates are influenced by the availability of N substrate and oxygen as well as by the seawater pH (Beman et al. 2011). Nitrification also couples the nitrogen and carbon cycles because nitrifiers are chemoautotrophs that fix carbon dioxide into biomass. The two steps of nitrification are most commonly carried out by different groups of microbes. For a long time, nitrifying bacteria were presumed to be the sole nitrifiers in the ocean. Then, in the early twenty-first century, ammonia-oxidizing archaea were discovered (Könneke et al. 2005; Wuchter et al. 2006). Subsequently, many important discoveries were made in the microbiology and biogeochemistry of nitrification (reviewed by Kuypers et al. 2018). For example, a number of nitrifying archaea utilize organic nitrogen substrates, e.g., urea and cyanate (Kitzinger et al. 2019), and some nitrite-oxidizing *Nitrospira* are now known also to oxidize ammonia (Daims et al. 2015). The recognized diversity of nitrite oxidizers has expanded substantially. One of the most impactful papers reported the discovery of a bacterium, a *Nitrospira*, that could carry out both steps of nitrification and oxidize ammonium to nitrite and then nitrite to nitrate (Daims et al. 2015).

Reduction of dissolved nitrogen oxides such as nitrite and nitrate to dinitrogen gas completes the nitrogen cycle. There are three recognized mechanisms of nitrogen oxide reduction—denitrification, anaerobic ammonium oxidation, and dissimilatory nitrate reduction to ammonium (DNRA). The latter process, DNRA, occurs primarily in organic-rich sediments where it can be coupled to organic carbon or sulfide oxidation. This process conserves nitrogen in an easily accessible form, i.e., ammonium, retaining nitrogen in the system. The relative importance of dissimilatory nitrate reduction to ammonium in the pelagic ocean is unclear. However, this process could be important in anoxic zones where sulfide and nitrate co-occur (Lam et al. 2009) and perhaps in anoxic micro-niches on particles.

Denitrification involves the sequential reduction of nitrate to nitrite, nitric oxide, nitrous oxide, and ultimately dinitrogen, which is only available to diazotrophic microorganisms. Denitrification occurs in the water column as well as in sediments. The reductant for denitrification is usually dissolved organic carbon, and most denitrifying bacteria are heterotrophs. However, in anoxic sediments or in fluids from extreme environments like hydrothermal vents where inorganic reductants (i.e., hydrogen sulfide, hydrogen) are abundant, denitrification can be coupled to the oxidation of energy-rich inorganic compounds like hydrogen sulfide (Bourbonnais et al. 2012). Some denitrifiers produce substantial quantities of nitrous oxide, a potent greenhouse gas, during denitrification. Nitrous oxide production is exacerbated under fluctuating oxygen conditions and in the presence of sulfide (Joye 2002; Joye and Anderson 2008). Each year about 200 Tg of N is removed from the bioavailable N pool via nitrate reduction (Gruber 2016). At least half of that amount is removed via denitrification, which makes this process a key component of the ocean nitrogen cycle that can modulate oceanic N inventories and impact the relative importance of N versus P limitation of primary production. The organic carbon flux influences the rate of denitrification, and outside of oxygen minimum zones, denitrification can occur on particles where the suboxic or anoxic conditions promote denitrification (Smriga et al. 2021).

Anaerobic ammonium oxidation involves the reduction of nitrite to dinitrogen gas. The so-called anammox bacteria were discovered by Strous et al. (1999a, b) and mediate a comproportionation reaction where ammonium and nitrite are converted to dinitrogen and water (Jetten et al. 2009). Anammox bacteria are unique in having a novel organelle. This compartment is called the anammoxosome, and it is bounded by novel ladderane lipids (Boumann et al. 2009; Fuerst 2005). In the anammoxosome, hydrazine is produced as a metabolic intermediate of anaerobic ammonium oxidation (Kartal et al. 2011; Strous et al. 1999a). Anammox is now known to be widespread in the ocean and is an important component of the nitrogen cycle (Dalsgaard et al. 2005). Anammox accounts for up to half of the fixed nitrogen removal from the oceans (Dalsgaard et al. 2005). In contrast to denitrifiers, anammox bacteria are chemoautotrophs, and therefore their growth and metabolic activity are controlled by different environmental factors. Babbin et al. (2014) made a convincing argument and concluded that the relative importance of denitrification versus anammox is determined by local variations of organic matter quality and quantity.

Nitrogen-rich organic matter (Babbin et al. 2014) and ammonium excretion by vertical migrators (Bianchi et al. 2014) favor anammox over denitrification.

### 15.2.4 Phosphorus

Phosphorus is also an essential building block for organisms. It constitutes the third most abundant element in microbial biomass (Redfield 1934). Phosphorus is a critical component of nucleic acids, the adenosine derivatives utilized in energy transfer (adenosine diphosphate, adenosine triphosphate) and cell signaling (cyclic adenosine monophosphate), and in the phospholipids that constitute the key structural components of cell membranes. Phosphorus is also a key component of bone in vertebrates. On a global scale and over geologic time scales, the phosphorus cycle is controlled by tectonic processes. In stark contrast to nitrogen and carbon, the phosphorus cycle is not influenced by atmospheric processes. The main reservoir of phosphorus is sedimentary, and the phosphorus cycle is affected mostly by processes that are not biologically driven.

Tectonic processes thereby play a gate-keeping role in the global phosphorus cycle (Fig. 15.5). Phosphorus is rapidly cycled by organisms but very slowly by solid-earth processes in rocks, soils, and other sedimentary deposits. The authigenic mineral apatite, which consists of phosphorus that is bound to iron-oxyhydroxide, detrital-phosphate, and organic phosphorus, is the main form in which phosphorus occurs in rocks and sediments (Ruttenberg 2014). Tectonic uplift of rocks and subsequent weathering mobilizes and transfers phosphate from the solid phase to

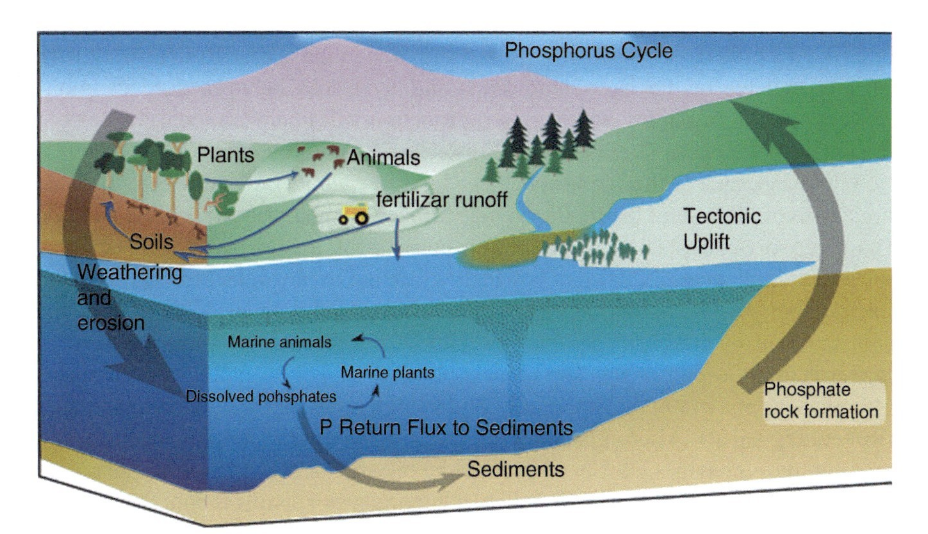


**Fig. 15.5** The global phosphorus cycle, underscoring the role of tectonic uplift and weathering on land as major controllers of the oceanic phosphorus inventory (adapted from original, Wikimedia, Creative Commons, open access)

its dissolved form, ortho-phosphate. Ortho-phosphate is a surface active in soils and is often associated with iron and aluminum oxyhydroxides. Some ortho-phosphate moves into the aqueous phase, where it is transported and rapidly assimilated by organisms at the base of the food web. In the aquatic realm, phosphate is recycled many times before ultimately being buried again in sediments. Comprehensive reviews of the ocean phosphorus cycle are provided in Duhamel et al. (2021), Paytan and McGlaughlin (2007) and Ruttenberg (2014).

The availability of inorganic ortho-phosphate may be a controlling factor for primary productivity and ecosystem composition and function in the ocean (Karl 2014; Martiny et al. 2019; Ryther and Dunstan 1971; Tyrrell 1999; Vallina et al. 2014). There is a continuing debate on whether nitrogen or phosphorus ultimately limits primary production in the ocean (see Tyrrell 1999). Because phosphorus inputs to the ocean are controlled by the long-term processes of tectonic uplift and weathering, geochemists generally consider phosphorus to be the ultimate limiting nutrient in the ocean (Tyrrell 1999). The atmosphere provides a virtually unlimited reservoir of nitrogen, and diazotrophic microorganisms increase nitrogen inputs locally and regionally largely as a function of phosphorus and iron availability. Hence, bound nitrogen is introduced biologically, while this is not the case for ortho-phosphate. However, experimental evidence (Ryther and Dunstan 1971) shows that sometimes nitrate limits phytoplankton productivity locally. Global ocean data sets show that in the vast majority of cases, nitrate is depleted before phosphate, which provides additional evidence for nitrogen limitation of production (Tyrrell 1999). Improvements in the analytical detection limits for phosphorus have provided additional insight into global phosphorus dynamics and revealed a disconnect between the relationships between nutrient availability and biogeochemical process dynamics (Martiny et al. 2019). Moreover, new phosphorus data revealed a tremendous degree of heterogeneity across space and time in the ocean (Ruttenberg 2014). This knowledge of the interrelationships and feedbacks between nutrient pools, biogeochemical dynamics, and ecosystem function will improve models of nutrient dynamics in the ocean (Vallina et al. 2014).

When inorganic (ortho) phosphate is limiting, autotrophic and heterotrophic organisms at the base of the food web may utilize phosphorus from organic sources. Dissolved organic phosphorus is a key component of the oceanic phosphorus inventory (Ruttenberg 2014) that is widely available to phytoplankton and to other microorganisms (Björkman and Karl 1994; Karl and Björkman 2002, 2015). The concentration of dissolved organic phosphorus in the ocean is often equal to or higher than ortho-phosphate. The organic phosphorus pool contains phosphonates, phosphate esters, as well phosphorus bound to high molecular weight organic matter (Ruttenberg 2014). Microorganisms utilize dissolved organic phosphorus using extracellular enzymes that cleave and release inorganic ortho-phosphate (Ammerman and Azam 1985). The relative importance of dissolved inorganic versus organic phosphorus cycling and how they vary over space and time is still unknown (Mather et al. 2008).

Phosphorus cycling within the ocean has long been considered to be driven largely by biological uptake and recycling. Indeed, heterotrophic bacteria use a

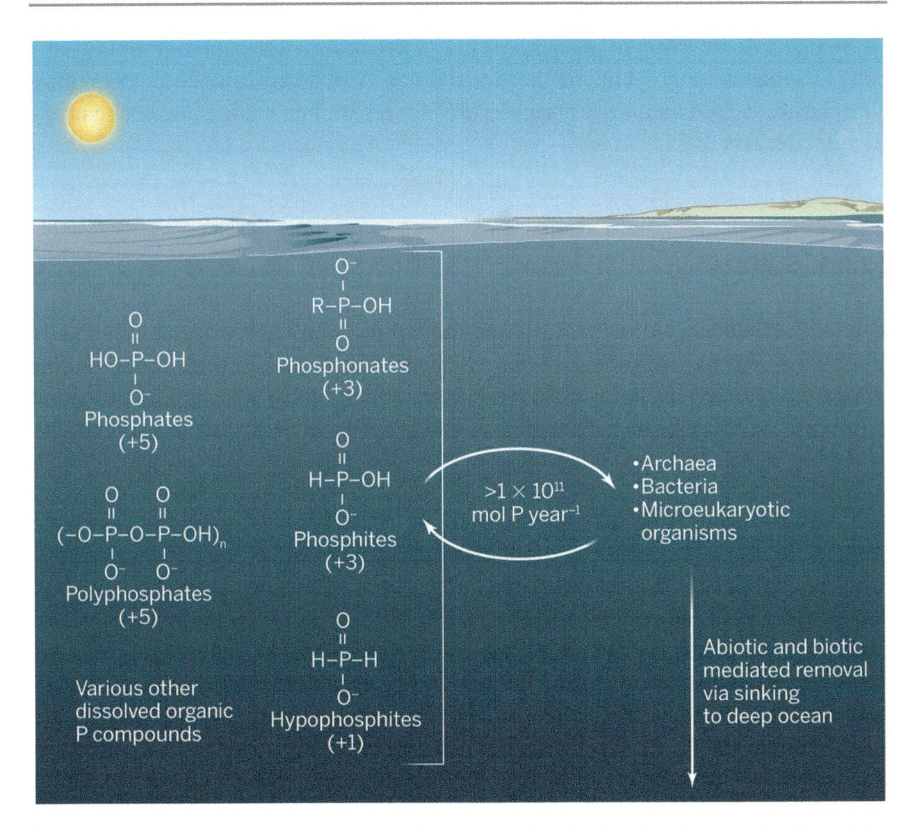


**Fig. 15.6** Recognition of the importance of redox chemistry in the phosphorus cycle changed the view of phosphorus cycling in the ocean (adapted from Benitez-Nelson 2015)

number of tricks to access phosphorus, ranging from secreting ligands to secreting redox-active antibiotics (Duhamel et al. 2021). Recently, Pasek et al. (2014) and Van Mooy et al. (2015) provided convincing evidence that redox chemistry also plays an important role in phosphorus cycling (Fig. 15.6). Phosphorus is present predominantly in an oxidized +5 valence state. However, reduced phosphorus forms, including phosphonates, phosphine, phosphite, and hypophosphite, are also present in the environment (Clark et al. 1998) and are utilized by microorganisms (Stone and White 2012; Van Mooy et al. 2015). Many bacteria—especially cyanobacteria—generate phosphonates, and in some organisms like *Trichodesmium,* up to 10% of cellular P is in the form of polyphosphate (Diaz et al. 2008; Dyhrman et al. 2009). Up to 25% of dissolved organic phosphorus may be present as phosphonates which link the oceanic phosphorus and methane cycles (Karl et al. 2008; Metcalf et al. 2012; Sosa et al. 2020; see Sect. 15.2.6 below). Other reduced forms of phosphorus, like phosphite, can be utilized by the majority of cultured bacteria that were screened for it, including *Prochlorococcus*, the predominant primary producer in the ocean (Martínez et al. 2012). This observation suggests that reduced phosphorus is an often overlooked but important source of phosphorus in the ocean (Karl 2014).

Given that redox cycling of phosphorus may account for a substantial (roughly half) part of the phosphorus cycling in the global ocean (Van Mooy et al. 2015), exploring phosphorus redox dynamics is a frontier area for research in marine biogeochemistry (Benitez-Nelson 2015).

### 15.2.5 Sulfur

The biogeochemical sulfur cycle shares similarities with the nitrogen, carbon, and phosphorus cycles. Microorganisms mediate redox transformations of sulfur converting sulfate into reduced forms that are incorporated into biomass (organic sulfur) via assimilatory mechanisms or utilizing sulfate as an electron acceptor under anaerobic or microaerobic conditions. Eventually, sulfate is reduced to hydrogen sulfide. Sedimentary processes strongly influence the sulfur cycle. Reduced (pyrite) and oxidized (anhydrite and gypsum) forms of sulfur are stored in sediments (Fig. 15.7). As is the case for phosphorus cycling, uplift of bedrock and weathering of pyrite and evaporites return sulfate to the ocean. The oceanic sulfur cycle was considered to be quite stable since atmospheric oxygen concentrations stabilized some ~600 million years ago. However, sulfate concentrations have varied substantially, and the mechanisms behind this variability were unclear. Using a non-steady state modeling approach Wortmann and Paytan (2012) showed that seawater sulfate concentrations increased sharply ~50 million years ago. These authors attributed this increase to the dissolution of evaporites driven by the expansion of the South Atlantic Ocean. Halevy et al. (2012) provided an estimate of changes in sulfate burial over 500 million years. Their modeling effort revealed highly variable burial rates of sulfate-containing evaporites and showed a strong impact on sulfur mineral weathering on atmospheric oxygen dynamics (Halevy et al. 2012). Together these

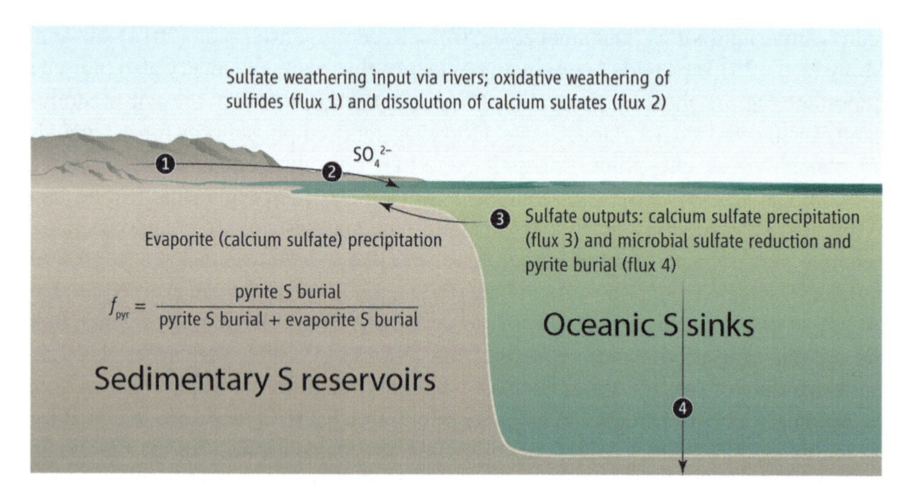


**Fig. 15.7** Sources and sinks of sulfate in the marine sulfur cycle. Adapted from Hurtgen (2012)

two studies provide convincing evidence that sulfate concentrations vary substantially and rapidly in ways that impact ocean biogeochemistry and global climate.

Sulfate is the dominant sulfur pool in the open ocean, and it is assimilated by phytoplankton; they reduce sulfate and assimilate it into biomass. Upon degradation of phytoplankton cells, reduced sulfur compounds are released and cycled via microbial processes. One of the major sulfur-containing molecules in marine phytoplankton is dimethylsulfoniopropionate (DMSP). Massive quantities of DMSP—more than a billion tons per year—are produced by marine phytoplankton. The intracellular DMSP concentration can reach up to 400 mM in some haptophytes (Curson et al. 2018). DMSP serves as an osmolyte, as a cryoprotectant, and can deter herbivory. It is released into the environment by grazing, viral lysis, or senescence of phytoplankton. Once in seawater, it is degraded into dimethyl sulfide (DMS) or dimethyl sulfoxide (DMSO). DMS and DMSP serve as signaling molecules that help connect organisms across trophic levels (Nevitt et al. 2008; Owen et al. 2021; Seymour et al. 2010). The release of DMS from the ocean to the atmosphere impacts the Earth's radiative balance. DMS is oxidized to sulfate, which serves as cloud condensation nuclei stimulating cloud formation (Chin et al. 1996; Cropp et al. 2007; Lovelock et al. 1972; Fig. 15.8); however, there is some debate as to the degree to which the oceanic DMS flux impacts global climate (Quin and Bates 2011). The fate of DMSP released into oceanic surface waters may be dictated by the microbial community's reduced sulfur demand (Vila-Costa et al. 2014). In fact, the bacterial sulfur demand may control the fate of DMSP in terms of DMS production and efflux to the atmosphere versus production of intermediate metabolites that are assimilated by the heterotrophic microbial population (Simó et al. 2009).

In low oxygen regions, sulfur is cycled in subtle but important ways (Callbeck et al. 2021; Canfield et al. 2010). Large regions of the ocean—such as the eastern tropical Pacific, upwelling zones off California, Namibia, and Peru, in the Arabian Sea, and in nearshore regions impacted by eutrophication—are characterized by low (hypoxic) or zero (anoxic) oxygen conditions. In these areas, high rates of surface productivity drive high rates of respiration in the midwater (offshore regions) or in bottom water (nearshore regions). In such situations, oxygen depletion rates can exceed oxygen supply rates leading to hypoxia or anoxia. Most previous studies of low oxygen waters have focused on nitrogen metabolism, assessing rates, patterns, and controls of denitrification and anammox (Kuypers et al. 2005; Thamdrup et al. 2006; Ward et al. 2009).

While assimilatory sulfate reduction (described above) is an important process in the oxic water column, especially in photosynthetically active surface waters, dissimilatory sulfate reduction is restricted to oxygen-limited waters. Hypoxic and anoxic zones were not considered to be hotspots of dissimilatory sulfate reduction since the areas are not sulfidic. Starting in 2008 a series of discoveries reset the understanding of sulfur biogeochemistry in low oxygen waters. A number of papers presented evidence for microbial sulfate reducers and/or sulfide oxidizers from oxygen minimum zones off Chile (Stevens and Ulloa 2008) and in Saanich Inlet (Walsh et al. 2009). Then, Canfield et al. (2010) reported a startling discovery in the waters of the oxygen minimum zone off Chile. They documented a rich microbial

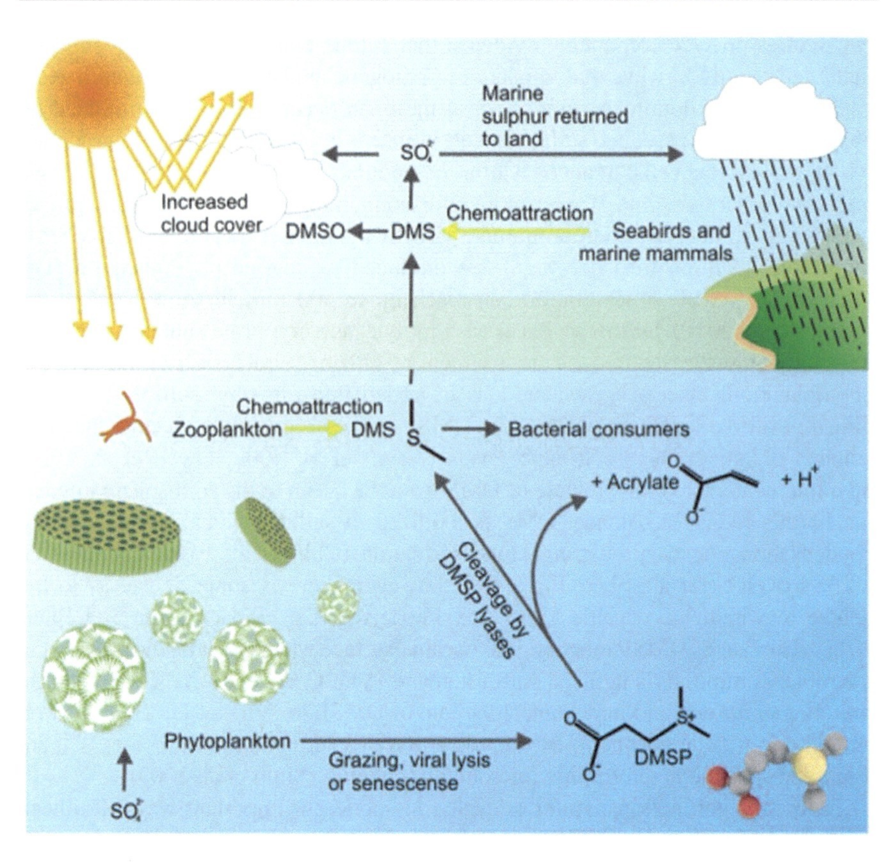


**Fig. 15.8** The role of dimethylsulfide in the ocean's sulfur cycle. Phytoplankton produces dimethylsulfoniopropionate, and upon release from cells, it is metabolized by microorganisms to dimethylsulfide (DMS) or to acrylate. In the atmosphere, DMS is oxidized to sulfate or dimethyl sulfoxide (DMSO). Atmospheric sulfate serves as cloud condensation nuclei. Adapted from Hehemann et al. (2014)

community of sulfate reducers and sulfide oxidizers, high rates of dinitrogen production via anaerobic ammonium oxidation, and high rates of sulfate reduction and sulfide oxidation. Sulfate reduction provided up to 22% of the ammonium demand of anaerobic ammonium oxidation (Canfield et al. 2010). Moreover, sulfide oxidation appeared to be coupled to the reduction of nitrate to nitrite, showing the intimate linking of the sulfur, carbon, and nitrogen cycles in the oxygen minimum zone. Callbeck et al. (2021) expanded on this work considerably by providing a comprehensive assessment of sulfur cycling in the context of carbon and nitrogen cycling in oxygen minimum zones. They confirmed and advanced previous reports showing that dissolved organic sulfur cycling is an underappreciated process in oxygen minimum zones. Organosulfur sulfidogenesis generates sulfide and other reduced intermediates that fuel oxidative sulfur metabolism. Carbon, nitrogen, and sulfur cycling are tightly coupled in oxygen minimum zones and the processes occurring in offshore areas are linked to processes occurring closer to the shore. Given the likely expansion of oxygen-limited regions in the future, understanding how elemental cycles in these areas are entwined will gain importance in marine biogeochemistry.

### 15.2.6 Trace Gases

#### 15.2.6.1 Methane

Methane is a powerful greenhouse gas with a warming potential much greater than carbon dioxide, and its atmospheric concentration is increasing. In the marine environment, methane is produced primarily in anoxic sediments where it can exist as a free or dissolved gas or as solid gas hydrate, when appropriate temperature and pressure conditions are met (i.e., within the gas hydrate stability zone). Methane can escape sediments via diffusion, advection, or bubbles (i.e., ebullition). The amount of methane that escapes from the sediment into the overlying water column is hard to quantify and thus poorly constrained. Estimates range from 1 to 200 Tmol year$^{-1}$. The concentration of dissolved methane in seawater is ~2 nM when it is in equilibrium with the atmosphere (Kirschke et al. 2013; Reeburgh 2007; Ruppel and Kessler 2017). Once in the water column, methane is subject to oxidation mediated by a diverse group of microorganisms collectively termed aerobic methanotrophs (Hanson and Hanson 1996; Knief 2015). Observations of microbial methane oxidation have been made at sites of natural gas seepage, oil spills, hydrothermal plumes, productive shelf regions, high latitude regions, as well as one report from hydrate containing sediments (Crespo-Medina et al. 2014; De Angelis et al. 1993; Kelley 2003; Mau et al. 2013; Pack et al. 2011, 2015; Reeburgh et al. 1991; Rogener et al. 2018; Valentine et al. 2001, 2010; Ward 1992; Ward and Kilpatrick 1990, 1993). Methanotrophs oxidize methane at rates ranging from ~0.001 to 1000 nM day$^{-1}$. Metal cofactors (e.g., copper, iron, and lanthanum) are key components of the enzymes involved in methanotrophy, and this raises questions as to whether the availability of metals limits methane oxidation (Crespo-Medina et al. 2014; Huang et al. 2019; Picone and Op den Camp 2019, and references therein). Although methane oxidation in the water column is a predominantly aerobic process, there are exceptions such as in anoxic basins (e.g., the Black Sea), where anaerobic methanotrophs couple the oxidation of methane to sulfate reduction (Reeburgh et al. 1991).

Paradoxically, methane production occurs in the fully oxygenated oceanic water column leading to methane concentrations that exceed those expected from atmospheric saturation in the tropical (Karl et al. 2008) and polar (Damm et al. 2010) ocean. Diverse explanations have been offered as the potential mechanism of aerobic methane production, e.g., methanogenesis in the anoxic microzones in particles or in the gut of zooplankton, but the most likely explanation is the microbially-mediated demethylation of methylphosphonates (Kamat et al. 2013). This mechanism is remarkable as the bacteria generate methane by cleaving the C-P bond of methylphosphonate (see Kamat et al. 2013 for a detailed explanation). Methylphosphonate is a natural component of dissolved organic matter (Repeta et al. 2016) and is produced by a wide variety of marine microorganisms (Metcalf et al. 2012; Ulrich et al. 2018). Using pure cultures, Carini et al. (2014) showed that *Pelagibacter* sp. strain HTCC7211, a member of the SAR11 clade, produced methane from methylphosphonate. The production of methane from methylphosphonate is thought to serve as a source of inorganic phosphorus under P-limited oceanic conditions (Sosa et al. 2019a, 2020). However, Sosa et al. (2019b) showed that *Prochlorococcus* oxidizes methylphosphonate to formate, thereby also releasing P for assimilatory uptake. Hence, it seems that demethylation with the consequence of methane formation is not the only way to generate inorganic phosphate from methylphosphonate. Bižić et al. (2020) reported that cyanobacteria from a range of different environments, including marine, freshwater, and terrestrial habitats, produce methane in the course of photosynthesis independent of and in the absence of methylphosphonates. Wang et al. (2021) documented the biologically-mediated aerobic production of methane from methyl amine; this mechanism has not been documented in the ocean but warrants further investigation. Hence, much remains to be learned about aerobic methane production in the ocean.

#### 15.2.6.2 Nitrous Oxide

Nitrous oxide is an important trace gas in the atmosphere that has 300 times the greenhouse warming potential of carbon dioxide. Currently, nitrous oxide accounts for ~6% of the greenhouse-gas-driven tropospheric warming (IPCC 2013). Nitrous oxide also contributes to stratospheric ozone depletion. The concentration of nitrous oxide in the ocean varies substantially in magnitude over space and time. The measured concentrations of nitrous oxide range from less than 1 nM to over 1500 nM. Some oceanic regions serve as a source, while others serve as a sink of nitrous oxide (Bange et al. 2019; Rogener et al. 2021). On average, the ocean accounts for about a quarter of the total global nitrous oxide emissions to the atmosphere (Ciais et al. 2013; Yang et al. 2020). However, nearshore low oxygen zones may be an unrecognized and globally important source of nitrous oxide (Rogener et al. 2021). In the ocean, nitrous oxide is produced during nitrification and denitrification, and some of the nitrous oxides may be consumed via denitrification (Bange et al. 2010; Ji et al. 2015). The present understanding of nitrous oxide sources, sinks, and cycling is incomplete (Rees et al. 2021). The different potential oceanic sources of nitrous oxide are discussed in detail by Babbin et al. (2015).

Oxygen minimum zones are hotspots of nitrous oxide cycling (Babbin et al. 2015). These areas are characterized by high nitrous oxide production rates, high nitrous oxide concentration, and a substantial potential for nitrous oxide efflux to the atmosphere. Babbin et al. (2015) showed that rates of incomplete denitrification resulting in nitrous oxide production were about an order of magnitude higher than previously thought. They also showed that the turnover rates of nitrous oxide were ~20 times higher than the net flux to the atmosphere, documenting turnover times as low as 1 day. These findings imply that a substantial increase in atmospheric efflux of nitrous oxide can be expected in the future.

Denitrification is considered the primary sink of nitrous oxide, but Rees et al. (2021) showed that nitrous oxide could be consumed under fully oxic conditions.

These authors performed incubation experiments using samples collected from the North Atlantic Ocean and observed that nitrous oxide was consumed in the upper 100 m of the water column. Furthermore, they documented the presence of microorganisms, e.g., *Pseudomonas stutzeri* and *P. fluorescens*, that are capable of nitrous oxide reduction, as was confirmed by the presence of the gene encoding for nitrous oxide reductase (*nosZ*) (Rees et al. 2021). Surprisingly, there was little evidence for denitrification (i.e., they did not detect the *nirS* gene) in their samples, which suggests that the nitrous oxide consumption was not related to denitrification. Rather, it appears that some nitrous oxide reducing microorganisms thrive in oxygenated waters and that these microorganisms may contribute substantially to oceanic nitrous oxide cycling.

## 15.3 Aggregates and Particles

### 15.3.1 Particulate Organic Matter

Marine particulate organic matter is part of a size continuum spanning from networks of high molecular weight biopolymers, collectively called exopolymeric substances (EPS), to suspended particles in the millimeter to centimeter size range (Verdugo et al. 2004). Exopolymeric substances are actively secreted by marine microbes (bacteria, archaea, and microeukaryotes) following a myriad of environmental cues such as nutrient and thermal stress as well as protection against potentially toxic substances (Xiao and Zheng 2016). Exopolymeric substances are generally comprised of polysaccharides, and to a lesser extent proteins, lipids, and nucleic acids (Quigg et al. 2016). Polysaccharides and proteins can form covalently cross-linked polymer chains resulting in the formation of hydrogels that coalesce to form larger gels and porous networks, many of which become visible as μm to mm sized gel-like particles or particulate EPS (Decho and Gutierrez 2017; Verdugo et al. 2004).

Transparent exopolymeric particles are operationally defined as polysaccharide-containing gels that are stained with the cationic copper phthalocyanine dye, Alcian Blue, and retained on filters with a pore size of 0.4 μm (Engel 2009; Passow and Alldredge 1995a). Particulate EPS of proteinaceous origin are referred to as Coomassie Brilliant Blue stainable particles, which is a dye specific for amino acids (Long and Azam 1996). Both transparent and Coomassie stainable exopolymeric particles are ubiquitous in the ocean (Engel et al. 2020; Long and Azam 1996) and fulfill a fundamental role in carbon cycling. About 10% of the annual oceanic primary productivity is channeled into transparent exopolymeric particles (Mari et al. 2017), with similar proportions expected for Coomassie stainable particles (Engel et al. 2020). Transparent exopolymeric particles and Coomassie stainable particles represent an important energy source for heterotrophic microbes (Busch et al. 2017; Passow and Alldredge 1994).

Due to their surface-active ("sticky") matrices, marine gels play a decisive role in the formation of sinking aggregates, also known as marine snow (Passow 2002;

Simon et al. 2002; Thornton 2002). Marine snow is defined as particles of >500 μm in diameter comprised of organic and inorganic matter, including prokaryotic and eukaryotic microbial cells (alive and dead) as well as fecal material and discarded mucus feeding structures. Owing to high concentrations of labile organic matter, marine snow is densely colonized by specialized heterotrophic bacteria present at several orders of magnitude higher abundance compared to that found in a similar volume of aggregate-free seawater (Smith et al. 1992; Ziervogel et al. 2011). Differences between particle-attached and free-living bacterial communities (DeLong et al. 1993; Fontanez et al. 2015; Moeseneder et al. 2001) may shape biogeochemical processes in sinking marine snow. In the growth of microorganisms on sinking, aggregates can shift the phylogenetic composition observed between marine snow and the ambient water. Aggregate-associated microbial communities change with time and season depending on the nutritional demands of the microbial community (Steiner et al. 2020) and with depth as the result of ingrowth, detachment, viral lysis, and grazing while the marine snow is sinking (Kiorboe et al. 2003; Thiele et al. 2015). This could partly explain the inconsistent relationships between community composition of free-living and aggregate-associated microbes in the ocean.

Fecal pellets represent another important component of the sinking particulate organic matter pool in the ocean (Jefferson and Turner 2002). These pellets occur in various sizes and shapes ranging from micrometer-sized “minipellets” produced by microzooplankton (e.g., dinoflagellates, ciliates, and copepod nauplii), cylindrical or ovoid pellets produced by copepods, larger fecal strings from euphausiids, tabular fecal flakes produced by salps and tunicates, and millimeter-sized fecal pellets produced by pelagic fish (Saba et al. 2021; Saba and Steinberg 2012). Similar to marine snow, fecal pellets often harbor dense bacterial populations attached in abundances that are orders of magnitude higher than found in ambient seawater. Therefore, fecal pellets represent “hotspots” for active microbial populations in the water column (Jefferson and Turner 2002).

### 15.3.2 Particulate Inorganic Matter

Particulate inorganic matter in the ocean is involved in the life cycles of many marine microbes. Inorganic particles, such as clay minerals, act as surfaces for microbial cells and may also trap metabolites (Ziervogel et al. 2007) and serve as transport vehicles for nutrients (C, N, P) and trace metals (e.g., Fe) from land to the sea (Zhang et al. 2017). A major source of particulate inorganic matter is of terrestrial origin and enters the coastal waters by riverine runoff. This export has been estimated 20 Pg of suspended sediments per year on a global scale (Beusen et al. 2005). Most of the land-derived particulate matter (inorganics and organics) is deposited on the continental shelf, where it is laterally transported to the deep slope bottom waters (Bianchi et al. 1997; Ziervogel et al. 2016). Major inputs of particulate inorganic matter to open ocean surface waters come from atmospheric deposition of dust, which is the primary source of iron in the open ocean. Iron is an essential micronutrient and may be limiting to photosynthetic and diazotrophic microorganisms (Jickells et al. 2005). In fact, the importance of atmospheric iron deposition and its potential use as fertilizer for open ocean primary production has developed into one of the most controversially discussed aspects in oceanography over the last four decades (Emerson 2019; Martin et al. 1990).

### 15.3.3 Remineralization of Organic Material

Most dissolved organic matter produced by phytoplankton in the surface ocean undergoes rapid microbial remineralization to $CO_2$. The fraction of primary productivity that escapes rapid remineralization accumulates in the surface ocean and, in the case of surface-active compounds, adds to the pool of particulate EPS (see above). Microbial decomposition of EPS and other high molecular weight biopolymers requires extracellular enzymes that cleave organic substrates outside the cell to sizes small enough for uptake (<600–1000 Da; Arnosti 2011; Azam and Malfatti 2007). Marine snow-associated microbial communities generally exhibit high rates of their extracellular enzymes, particularly proteolytic and carbohydrate-hydrolyzing enzymes (Ploug and Grossart 1999; Simon et al. 2002; Smith et al. 1992; Ziervogel et al. 2011). Because there seems to be a weak coupling between the rate of hydrolysis of polymers and the rate of uptake of the hydrolysis products, marine snow may release substantial amounts of dissolved organic matter to the water column (Smith et al. 1992). Marine snow also releases inorganic nutrients and hydrolytic enzymes (Baltar 2018; Zhao et al. 2020; Ziervogel and Arnosti 2008). Enzymes released from sinking marine snow and thus dissociated from microbial cells may remain active for a considerable time which has consequences for the biogeochemical cycles in the ocean (Azam and Long 2001; Thomson et al. 2019).

High rates of organic matter oxidation in sinking marine snow can drive the formation of anoxic microzones in otherwise oxygenated waters (Bianchi et al. 2018; Ploug et al. 1997). These small-scale physicochemical gradients in marine snow dictate microbial community composition and function in sinking aggregates (Fig. 15.9). The anoxic microzones in the water column enhance the use of alternative electron acceptors like nitrate, ferric iron, and sulfate, which are used to respire organic matter in the absence of oxygen (Wright et al. 2012). Sulfate is energetically less favorable than oxygen as a terminal electron acceptor for respiration. However, substantial rates of microbial sulfate reduction have been detected in sinking marine snow, producing S-containing organic compounds that may be resistant to hydrolysis (Raven et al. 2021). This biologically driven sulfurization of the organic matter could increase the efficiency of the biological pump with respect to carbon export and sequestration in the deep ocean.

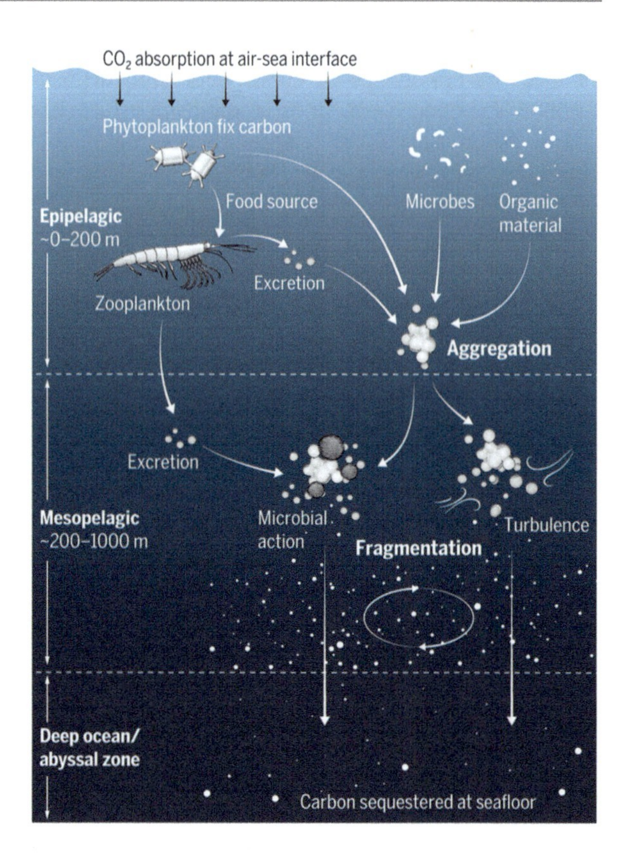


**Fig. 15.9** The role of aggregate formation and particle flux in the ocean biogeochemical cycling. Adapted from Nayak and Twardowski (2020)

### 15.3.4 POM Sedimentation and Associated Elemental Fluxes

Massive sedimentation events of marine snow have often been observed as phytoplankton blooms decline (Trudnowska et al. 2021) and EPS production increases in response to nutrient stress (Passow and Alldredge 1995b). The flux of primary production from sunlit surface layers into the ocean's interior represents a major energy source for carbon-limited microbial communities in the deep ocean. The small fraction of carbon that evades microbial remineralization and is buried in sediments represents the sequestration of $CO_2$ from the atmosphere and is a sink of reduced carbon, thus balancing atmospheric oxygen levels on geologic timescales (Hedges 1992). Numerous oceanographic programs and field campaigns over the past decades have been devoted to the quantification of the fluxes and fate of marine snow-associated carbon (Siegel et al. 2016). Quantifying vertical fluxes of particulate organic matter in the ocean's interior is complicated by the fact that marine snow formation and sedimentation follow seasonal and interannual variations mainly driven by surface water productivity (Lampitt et al. 2010; Smith et al. 2013). The quality and quantity of particulate organic matter reaching the deep-sea depend in part on the aggregates' settling rates, which are a function of the size, form, and density of marine snow (Burd and Jackson 2009). Particulate inorganic matter, including clay minerals from terrestrial sources as well as carbonate sheathes in the form of calcite foraminifera shells, coccolith plates, and aragonite pteropod shells and shell fragments may act as ballast for marine snow, increasing its density and thus altering its sinking rate (Armstrong et al. 2002; Passow and la Rocha 2006). Additional factors affecting the fate of sinking marine snow include microbial decomposition and grazing and physical fragmentation during transit through the water column (Briggs et al. 2020; Collins et al. 2015) (Fig. 15.9).

In addition to providing carbon and energy sources to deep ocean communities, sedimentation of marine snow may also have detrimental impacts on pelagic and benthic ecosystems. For instance, sedimentation of marine snow comprised of detritus from potentially toxic cyanobacteria coincided with fish kills in some areas (McInnes and Quigg 2010). Marine snow also "attracts" hydrophobic substances such as petroleum hydrocarbons and form marine oil snow (Passow 2016). The phenomenon of marine oil snow formation and sedimentation as a pathway to move weathered oil from the surface ocean into the ocean's interior and into the marine food web received great attention in the aftermath of the 2010 Deepwater Horizon oil spill in the Gulf of Mexico (Chanton et al. 2015; Daly et al. 2016; Passow and Ziervogel 2016). In addition to oil contaminants, the fate of (micro-)plastics in the water column is also linked to aggregation with particulate exopolymeric substances (Michels et al. 2018; Summers et al. 2018) and sinking particulate organic matter (Kvale et al. 2020; Porter et al. 2018), enhancing the downward transport through the water column of otherwise buoyant marine plastics.

## 15.4 Sediments/Benthic Habitats

Marine sediments are form as particles are deposited from the water column, accumulating atop oceanic basement rocks. These sediments are classified as lithogenous, biogenous, hydrogenous, or cosmogenous. Lithogenous and biogenous sediments are amongst the most abundant, with lithogenous sediments dominating near continents and biogenous sediments in open ocean settings. Many of the bottom sediments along the continental shelf (60–70%) are sands, but on a global scale, sands are rare (4%) (Emery 1968; Middelburg 2019). Regardless of sediment type or composition, sediment accumulation continues until subduction or uplift of the sediment takes place. Currently, there is an estimated $3.37 \times 10^8$ km$^3$ of sediment overlying the oceanic basement rock. Sediments are more concentrated along continental margins where the mean sediment thickness is 3044 m versus only 404 m for the deep ocean (Straume et al. 2019). Most of the oceanic sediments are anoxic and cold (0–20 °C) (Bowles et al. 2014; Bradley et al. 2020; LaRowe et al. 2017). Particles that settle on the seafloor contain carbon which has accumulated to 2322 Pg C in the upper 1 m of the sediment alone (Atwood et al. 2020). This carbon fuels microbial activity, and indeed marine sediments host a substantial population of microorganisms ($2.9 \times 10^{29}$), endospores ($0.25–1.9 \times 10^{29}$), and viruses ($3.3–16.2 \times 10^{31}$) (Kallmeyer et al. 2012; Pan et al. 2019; Wörmer et al. 2019).

### 15.4.1 Gradients-Depth, Near Versus Offshore

Carbon that has reached the seafloor serves as a reductant or as an energy source fueling many microbial processes (e.g., oxygen respiration, denitrification, iron reduction, sulfate reduction, and methanogenesis). The relationship between carbon input and microbial activity explains many of the lateral gradients that are observed across marine sediments and determine how sediment biogeochemical cycling is modeled and understood. The most carbon-rich sediments are located along the continental margins and beneath upwelling regions, as evidenced by global maps of seafloor surficial particulate organic carbon content (Seiter et al. 2005). Most empirical models of microbial abundance, microbial activity, and physical processes (e.g., sedimentation rates) have relied on water depth or near versus offshore approximations (Bowles et al. 2014; Egger et al. 2018; Middelburg et al. 1997). Middelburg et al. (1997) were among the first to establish empirical models using a large compilation of data to approximate sedimentation rates and microbial activity. In their work, water depth was a crucial explanatory variable, which roughly corresponds to near versus offshore sites. Empirical relationships have now been improved by including distance to shore as an explanatory variable, which directly resolves nearshore versus offshore (Bowles et al. 2014; Egger et al. 2018; Kallmeyer et al. 2012). Distance to shore most likely serves as a proxy for particulate and nutrient inputs from the continents and provides additional context that is lacking when only using water depth. Finally, the distribution of carbon on the seafloor is roughly mirrored in satellite-based sea surface observations of particulate organic carbon and chlorophyll *a* concentration, which have been used to predict sedimentation rates as well as microbial process rates with a high degree of spatial resolution (Bowles et al. 2014; Egger et al. 2018).

Microbial activity is linked to sedimentation rate as sinking particles supply carbon to the seafloor surface. Sedimentation rate varies from near shore to offshore, with higher rates typically found nearshore. Sedimentation rate ranges from <0.1 cm per 1000 years to several cm per year, with the higher rates found near the coast (Aller 2013). Although sedimentation rates indicate how quickly material accumulates at the seafloor, they do not necessarily predict the carbon deposition rate. Carbon content in sinking particles changes depending on many factors. The carbon flux has been measured throughout the water column at numerous sites globally. These fluxes ranged from <10 to 1238 mg C $m^{-2}$ $day^{-1}$. The flux of particulate organic carbon decreases with water depth. The carbon delivery to the seafloor typically decreases with increasing water depth for sites with similar carbon flux at the sea surface (Mouw et al. 2016). The total flux of carbon to the seafloor has been estimated using a variety of different methods and extrapolations as ranging from 12 to 191 Tmol C $year^{-1}$ (Berner 1982; Dunne et al. 2007; Hedges and Keil 1995; Jahnke 1996; Regnier et al. 2013; Seiter et al. 2005; Wallmann et al. 2012).

The lateral gradients of carbon supply result in changes in microbial abundance and activity, as well as in the dominant microbial processes. Across the ocean, oxygen consumption occurs in surface sediments. In coastal regions where carbon inputs are high oxygen penetration will be on the scale of mm versus sediments below the open ocean where oxygen penetration and respiration can extend meters below the seafloor (D'Hondt et al. 2015; Glud 2008). The rapid removal of oxygen in coastal and nearshore sediments promotes sulfate reduction (Bowles et al. 2014). Proceeding from the continental shelf towards the open ocean, the drop in carbon supply favors denitrification and metal oxide reduction (D'Hondt et al. 2004). Methanogenesis takes place in sediments where sulfate is depleted; such sediments are largely found in nearshore and coastal regions or in extreme environments. It is estimated that $10^8$ $km^3$, representing approximately one-third of the global sediment volume, is methanogenic and/or fermentative (Bowles et al. 2014).

## 15.4.2 General Biogeochemical Patterns in Sediments

Lateral microbial biogeochemical gradients resulting from high carbon loads and related dynamics drive vertical gradients in marine sediments. Carbon that is deposited at the seafloor is degraded by microbes, and the recalcitrant part that remains is buried. Labile carbon is degraded first, and, generally speaking, the buried portion represents more refractory material. A series of electron acceptors with decreasing electrochemical reduction potential is used to oxidize the deposited organic carbon. The most prominent electron acceptors in marine sediments are oxygen and sulfate, while nitrate, metals, and carbon dioxide are less available, and thus, less important (see Canfield et al. 2005, and references therein).

Oxygen respiration and sulfate reduction are responsible for most of the global marine carbon mineralization. Oxygen respiration rates range from <0.001 to ~10 μmol $cm^{-3}$ $day^{-1}$, with substantial variability over space and time (Thullner et al. 2009). Global oxygen respiration rates are estimated as ~3.1 Tmol $year^{-1}$ (Thullner et al. 2009). Sulfate reduction rates in marine sediments range from below detection to 100s of nmol $cm^{-3}$ $day^{-1}$. Depending on the method used, global estimates of sulfate reduction range from 11.8 to 75 Tmol $year^{-1}$ (Bowles et al. 2014; Kasten and Jørgensen 2000; Thullner et al. 2009). This means that 12–40% of the carbon flux to sediments is mineralized by sulfate reduction (Bowles et al. 2014; Dunne et al. 2007; Regnier et al. 2013; Thullner et al. 2009). It is important to note that within the sulfate reduction zone, a substantial fraction of sulfate (7–47%) can serve as the electron acceptor for the anaerobic oxidation of methane (Egger et al. 2018).

Denitrification and metal reduction typically occur in the sediment at depths between the zone of oxygen respiration and sulfate reduction, according to their potential thermodynamic energy yield. Denitrification ramps up immediately after oxygen is consumed, but due to the low concentration of nitrate in bottom water (20–30 μM), the overall contribution of denitrification to carbon mineralization is low. Global estimates for sediment denitrification range from 1.3 to 20.3 Tmol $year^{-1}$ (Bianchi et al. 2012; Bohlen et al. 2012; DeVries et al. 2013; Middelburg et al. 1996; Seitzinger et al. 2006; Thullner et al. 2009).

Metals (e.g., iron and manganese) also contribute to mineralization, but their process rates are often inferred from increases in concentrations of the reduced metal products (e.g., Fe(II), for example, see Canfield et al. 1993). Iron reduction in marine sediments has been approximated at 0.6 Tmol year$^{-1}$ (Thullner et al. 2009). Beneath the zone of sulfate reduction, methanogenesis is the most prominent pathway. It uses carbon dioxide as electron acceptor and reduces it to methane; methane production can also proceed through acetoclastic and methylotrophic mechanisms. Approximations of global marine sediment methanogenesis rates are rare. Reeburgh (2007) conflated freshwater and oceanic methanogenesis and calculated gross methane production rates of 5.3 Tmol $CH_4$ year$^{-1}$. Although the methods may not be directly comparable, Egger et al. (2018) estimated rates of methanogenesis to be 5.7–7.6 Tmol $CH_4$ year$^{-1}$ in diffusion-dominated marine sediments alone.

### 15.4.3 Hot Spots

Although much of the distribution and activity of microorganisms in the seafloor largely depends on the deposition of particulate carbon from the water column, certain hot spots exist where the carbon and other energy-rich substrates are supplied from subsurface reservoirs in the sediment (e.g., cold seeps and vents) and in bulk from whale falls. Cold seeps occur when deeply sourced oil, gas, and/or brine migrate through the sediment and fuel vast ecosystems at the seawater-sediment interface. Hydrothermal vents are areas where seawater heated by magma is expelled as hydrothermal fluids, creating chemical disequilibria. These disequilibria generate a thermodynamic situation in which many microorganisms can thrive. In contrast, whale falls receive carbon and other energetic substrates from whale carcasses that land on the seafloor. The whale carcasses are rich in oils and fatty acids, which stimulate microbial activity.

***Cold Seeps*** Cold seeps are found globally and often along continental margins where substantial carbon burial or evaporitic basins have formed over the geologic past (Joye 2019). These areas of seepage manifest as oil-stained sediments, brine pools and flows, gas hydrates, gas bubbles, bacterial mats, as well as a variety of macrofauna (e.g., tubeworms, clams, mussels, urchins) (Joye 2019). The reduced substrate supplied from deep reservoirs at these sites fuels robust microbial activity. Specifically, rates of oxygen consumption, denitrification, sulfate reduction, and methane oxidation are particularly elevated in these sediments (Boetius and Wenzhöfer 2013; Bowles and Joye 2011; Bowles et al. 2011, 2019; Joye 2019). Indeed, some of the highest ex situ rates of sulfate reduction measured (>20 μmol cm$^{-3}$ day$^{-1}$) were from a cold seep setting (Bowles et al. 2011). Although oil and endogenous dissolved organic carbon fuel these sulfate reduction rates, part of it can be attributed to the anaerobic oxidation of methane. When sediments are subjected to in situ pressure with realistic methane concentrations, the rate of anaerobic oxidation of methane can reach 5 μmol cm$^{-3}$ day$^{-1}$ (Bowles et al. 2019). At cold seeps alone, anaerobic methane oxidation rates are thought to be responsible for oxidizing 0.22–4.9 Tmol $CH_4$ year$^{-1}$ (Bowles et al. 2011; Hinrichs and Boetius 2002). For a more thorough review of rates of anaerobic methane oxidation in different seep environments, the reader is referred to Joye and Bowles (2022).

***Hydrothermal Vents*** Hydrothermal vents are found dispersed along tectonic plate boundaries and associated features (e.g., oceanic ridges and back-arc basins). Microbial communities associated with vents are phylogenetically and functionally diverse because the microorganisms inhabit multiple dynamic interfaces from hot to cold (hyperthermophiles to psychrophiles) and oxic to anoxic. Reduced compounds (e.g., sulfide and sulfur) in the hydrothermal fluid serve as a substrate for many microorganisms. A prominent sulfur-oxidizing microorganism is *Sulfurimonas*, which is commonly found in hydrothermal plumes (Mino et al. 2017). Chemosynthetic microbes, like *Sulfurimonas*, are thought to fix from 110 to 230 mmol carbon globally at vent sites, which underscores the importance of microbes in elemental cycling at hydrothermal vents (McNichol et al. 2018). Hyperthermophiles associated with these vents are often methanogens, particularly those that use hydrogen, which is enriched in vent fluids. The abundances of hydrothermal methanogens in the vent fluids are estimated as $10^6$–$10^8$ cells $mL^{-1}$ (Stewart et al. 2019). Detailed observations of microbial communities and their activities within the fractured rock circulating hydrothermal fluids are complicated. It is clear that hydrothermal environments play a critical role in the chemistry of the ocean (Dick 2019) and that these areas support novel and diverse microbial communities (see Dick et al. 2013 and Dick 2019 for reviews).

***Whale Falls*** Whale falls supply a large amount of carbon to the seafloor and thereby stimulate extensive microbial activity. However, whale falls do not constitute a major input of carbon to the seabed on a global scale; they account for about 0.1% of the global particulate organic carbon flux (Treude et al. 2009). Treude et al. (2009) monitored and documented a whale carcass over several years and found that bacterial mats and black sediments formed similar to those observed at cold seeps. Rates of sulfate reduction rivaled the highest rates observed at cold seeps (33 μmol $cm^{-3}$ $day^{-1}$), and methane production rates were also elevated. In another study of multiple whale falls, Goffredi et al. (2008) observed highly elevated total organic carbon (3.5%) in sediments near the whale fall. The same study reported diverse archaea, particularly species known to be methanogens and anaerobic methanotrophs. These studies show that a single whale fall has a multiyear impact on the local microbial community and on sediment biogeochemistry.

### 15.4.4 Deep Biosphere

The deep subsurface represents an expansive ecosystem that is important for global biogeochemical cycles. The number of microbes living in marine subsurface sediments is similar to that of the water column (Kallmeyer et al. 2012; Whitman et al. 1998). The subsurface harbors highly diverse communities of bacteria and archaea, and the community composition is determined by oxygen (Hoshino et al.

2020). In general, archaea outnumber bacteria in subsurface sediments (Bar-On et al. 2018; Biddle et al. 2006; Lipp et al. 2008). Bacteria appear to be more diverse than archaea in the subsurface sediment, but the diversity indices of both domains decrease with sediment depth (Hoshino et al. 2020). Another observation is that the number of endospores exceeds those of vegetative cells in subsurface sediments, which suggests the presence of considerable dormant populations (Wörmer et al. 2019). Spore formation is a survival strategy that allows an organism to overcome extended periods of unfavorable growth conditions, but their long-term viability is unclear. The processes carried out in deep subsurface microbes, and the factors limiting microbial activity have been explored by several research groups (Heuer et al. 2020; Inagaki et al. 2015; Trembath-Reichert et al. 2017). Notably, in deep sediments (1.5–2.5 km) of the Shimokita Peninsula, Japan, cell numbers rapidly decreased to $<10–10^4$ cells $mL^{-1}$. The terrigenous microbial communities that were found in this deep biosphere retained their typical characteristics even after being buried for millions of years. Microbial communities that originated from a coal environment were still capable of growth on methylated compounds (Inagaki et al. 2015; Trembath-Reichert et al. 2017). This work also emphasized that the dramatic decline in cell numbers coincided with modeled increases in abiotic amino acid racemization and DNA depurination, which could be interpreted as a mechanism that limits life at these depths. Abiotic racemization and depurination are driven by temperature, and observations from the Nankai Trough show that cells were undetectable in large sections of sediments above 45 °C. However, it is important to note that methanogenesis and acetate degradation have been observed in some sediments at the Nankai Trough, where the temperature was higher (Heuer et al. 2020). Although microbial process rates and biomass are low at depth, the activities are robust in more shallow sediments and contribute importantly to global elemental cycles by serving as a sink for sulfur and carbon.

## 15.5 Ocean Biogeochemical Cycling in a Changing World

The ocean has changed substantially in the past several decades due to global climate change: surface water has become warmer, seawater has become more acidic, and the water column has become more stratified. Global change-induced alterations in the oceans have been discussed in quite some detail in the Intergovernmental Panel on Climate Change's "Special Report on Ocean and Cryosphere in a Changing Climate" (IPCC 2020). The reader is referred to this report for an in-depth discussion and for references. Only a brief overview is provided here. Considerable and widespread changes are expected in the global climate in the coming decades, and these changes will certainly impact the ocean's microbiome and its functionality, and thereby the ocean biogeochemical cycles. Change is already evident across the ocean, but it is most notable in the polar regions where multiyear ice has declined substantially, and glacial melt from land has increased the flux of materials from the land to the sea. Changes in the intensity of ocean currents like the Gulf Stream are related to the increased freshwater inputs from the Arctic. These changes in ocean currents may have important impacts on the rate of deep-water formation, which could slow down thermohaline circulation. Across the globe, the ocean is warming, which leads to changes in physical circulation, in some cases leads to more stratification and less mixing, and elsewhere the expansion of areas with low oxygen conditions. Altered hydrological patterns on the continents may change the discharge regimes, which will alter coastal and nearshore biogeochemistry substantially.

Climate change and associated acidification and warming of the surface ocean affect microbial metabolic rates, including the release of particulate organic compounds that form particulate EPS and gels (Endres et al. 2014; Piontek et al. 2009). As a consequence, the accumulation of particulate EPS may accelerate organic matter sedimentation and carbon sequestration at depth in the future ocean (Arrigo 2007). Accelerated microbial oxidation rates of EPS in a warming ocean, in contrast, may counteract carbon export fluxes to the deep-sea, preserving more organic matter in the surface ocean (Endres et al. 2014; Piontek et al. 2010). However, microbial processes, including dynamics of EPS in a warmer and more acidic ocean, have been found to further depend on other environmental factors such as nutrient availability (Otero and Vincenzini 2004; Passow and Laws 2015), complicating future predictions of particulate organic matter sedimentation and thus the efficiency of the biological pump under future climate scenarios. Baseline data describing ocean biogeochemistry across space and time is needed to enable to monitor the impacts of climate change on ocean biogeochemistry.

## References

Aller RC (2013) Sedimentary diagenesis, depositional environments, and benthic fluxes. Treatise on geochemistry, 2nd ed, vol 8. Elsevier, Oxford, pp 293–334

Amin SA, Parker MS, Armbrust EV (2012) Interactions between diatoms and bacteria. Microbiol Mol Biol Rev 76:667–684

Ammerman JW, Azam F (1985) Bacterial 5′-nucleotidase activity in estuarine and coastal marine waters: role in phosphorus regeneration. Limnol Oceanogr 36:1437–1447

Armstrong RA, Lee C, Hedges JI, Honjo S, Wakeham SG (2002) A new, mechanistic model for organic carbon fluxes in the ocean based on the quantitative association of POC with ballast minerals. Deep-Sea Res Pt II 49:219–236

Arnosti C (2011) Microbial extracellular enzymes and the marine carbon cycle. Annu Rev Mar Sci 3:401–425

Arnosti C, Steen AD, Ziervogel K, Ghobrial S, Jeffrey WH (2011) Latitudinal gradients in degradation of marine dissolved organic carbon. PLoS One 6:e28900

Arrieta JM, Mayol E, Hansman RL, Herndl GJ, Dittmar T, Duarte CM (2015) Dilution limits dissolved organic carbon utilization in the deep ocean. Science 348:331–333

Arrigo KR (2007) Carbon cycle—marine manipulations. Nature 450:491–492

Atwood TB, Witt A, Mayorga J, Hammill E, Sala E (2020) Global patterns in marine sediment carbon stocks. Front Mar Sci 7:165

Azam F, Long RA (2001) Oceanography—sea snow microcosms. Nature 414:495–498

Azam F, Malfatti F (2007) Microbial structuring of marine ecosystems. Nat Rev Microbiol 5: 782–791

Babbin AR, Keil RK, Devol AH, Ward BB (2014) Organic matter stoichiometry, flux, and oxygen control nitrogen loss in the ocean. Science 344:406–408. https://doi.org/10.1126/science.1248364

Babbin AR, Bianchi D, Jayakumar A, Ward BB (2015) Rapid nitrous oxide cycling in the suboxic ocean. Science 348:1127–1129. https://doi.org/10.1126/science.aaa8380

Baltar F (2018) Watch out for the "Living Dead": cell-free enzymes and their fate. Front Microbiol 8:2438

Baltar F, Herndl GJ (2019) Ideas and perspectives: is dark carbon fixation relevant for oceanic primary production estimates? Biogeosciences 16:3793–3799. https://doi.org/10.5194/bg-16-3793-2019

Baltar F, Arístegui J, Sintes E, Gasol JM, Reinthaler T, Herndl GJ (2010) Significance of non-sinking particulate organic carbon and dark $CO_2$ fixation to heterotrophic carbon demand in the mesopelagic northeast Atlantic. Geophys Res Lett 37:L09602. https://doi.org/10.1029/2010GL043105

Bange HW, Freing A, Kock A, Löscher CR (2010) Marine pathways to nitrous oxide. In: Smith K (ed) Nitrous oxide and climate change. Earthscan, London, pp 36–62

Bange HW, Arévalo-Martínez DL, de la Paz M, Farías L et al (2019) A harmonized nitrous oxide ($N_2O$) ocean observation network for the 21$^{st}$ century. Front Mar Sci 6:157. https://doi.org/10.3389/fmars.2019.00157

Bar-On YM, Phillips R, Milo R (2018) The biomass distribution on Earth. Proc Natl Acad Sci USA 115:6506–6511

Bauer JE, Williams PM Druffel ERM (1992) $^{14}C$ activity of dissolved organic carbon fractions in the north-central Pacific and Sargasso Sea. Nature 357:667–670

Beardall J, Quigg A, Raven JA (2003) Oxygen consumption: photorespiration and chlororespiration. In: Larkum AWD, Douglas SE, Raven JA (eds) Photosynthesis in algae. Advances in photosynthesis and respiration, vol 14. Springer, Dordrecht, pp 157–181. https://doi.org/10.1007/978-94-007-1038-2_8

Behrenfeld MJ, Falkowski PG (1997) A consumer's guide to phytoplankton primary productivity models. Limnol Oceanogr 42:1479–1491

Behrenfeld MJ, Boss E, Siegel DA, Shea DM (2005) Carbon-based ocean productivity and phytoplankton physiology from space. Global Biogeochem Cycles 19:GB1006. https://doi.org/10.1029/2004GB002299

Behrenfeld MJ, Halsey KH, Milligan AJ (2008) Evolved physiological responses of phytoplankton to their integrated growth environment. Philos Trans Roy Soc Lond B 363:2687–2703. https://doi.org/10.1098/rstb.2008.0019

Bell W, Mitchell R (1972) Chemotactic and growth responses of marine bacteria to algal extracellular products. Biol Bull 143:265–277

Beman JM, Chow CE, King AL, Feng Y, Fuhrman JA et al (2011) Global declines in oceanic nitrification rates as a consequence of ocean acidification. Proc Natl Acad Sci USA 108:208–213. https://doi.org/10.1073/pnas.1011053108

Benitez-Nelson C (2015) The missing link in oceanic phosphorus cycling? Science 348:759–760. https://doi.org/10.1126/science.aab2801

Berg IA (2011) Ecological aspects of the distribution of different autotrophic $CO_2$ fixation pathways. Appl Environ Microbiol 77:1925–1936

Berg IA, Kockelkorh D, Buckel W, Fuchs G (2007) A 3-hydroxyproprionate/4-hydroxybutyrate autotrophic carbon dioxide assimilation pathway in Archaea. Science 318:1782–1786

Berman-Frank I, Lundgren P, Falkowski PG (2003) Nitrogen fixation and photosynthetic oxygen evolution in cyanobacteria. Res Microbiol 154:157–164

Berner RA (1982) Burial of organic carbon and pyrite sulfur in the modern ocean: its geochemical and environmental significance. Am J Sci 282:451–473. https://doi.org/10.2475/ajs.282.4.451

Beusen AHW, Dekkers ALM, Bouwman AF, Ludwig W, Harrison J (2005) Estimation of global river transport of sediments and associated particulate C, N, and P. Global Biogeochem Cycles 19:GB4S0. https://doi.org/10.1029/2005GB002453

Bianchi TS, Lambert CD, Santschi PH, Guo LD (1997) Sources and transport of land-derived particulate and dissolved organic matter in the Gulf of Mexico (Texas shelf/slope): the use of lignin-phenols and loliolides as biomarkers. Org Geochem 27:65–78

Bianchi D, Dunne JP, Sarmiento JL, Galbraith ED (2012) Data-based estimates of suboxia, denitrification, and $N_2O$ production in the ocean and their sensitivities to dissolved $O_2$. Global Biogeochem Cycles 26:GB2009. https://doi.org/10.1029/2011GB004209

Bianchi D, Babin AR, Galbriath ER (2014) Enhancement of anammox by the excretion of diel vertical migrators. Proc Natl Acad Sci USA 111:15653–15658. https://doi.org/10.1073/pnas.1410790111

Bianchi D, Weber TS, Kiko R, Deutsch C (2018) Global niche of marine anaerobic metabolisms expanded by particle microenvironments. Nat Geosci 11:263–268

Biddle JF, Lipp JS, Lever MA, Lloyd KG, Sørensen KB et al (2006) Heterotrophic Archaea dominate sedimentary subsurface ecosystems off Peru. Proc Natl Acad Sci USA 103:3846–3851

Bižić M, Klintzsch T, Ionescu D, Hindiyeh MY et al (2020) Aquatic and terrestrial cyanobacteria produce methane. Sci Adv 6:eaax5343. https://doi.org/10.1126/sciadv.aax5343

Björkman K, Karl DM (1994) Bioavailability of inorganic and organic P compounds to natural assemblages of microorganisms in Hawaiian coastal waters. Mar Ecol Progr Ser 111:265–273

Boetius A, Wenzhöfer F (2013) Seafloor oxygen consumption fueled by methane from cold seeps. Nat Geosci 6:725–734

Bohlen L, Dale AW, Wallmann K (2012) Simple transfer functions for calculating benthic fixed nitrogen losses and C:N:P regeneration ratios in global biogeochemical models. Global Biogeochem Cycles 26:GB3029. https://doi.org/10.1029/2011GB004198

Bothe H, Schmitz O, Yates MG, Newton WE (2010) Nitrogen fixation and hydrogen metabolism in cyanobacteria. Microbiol Mol Biol Rev 74:529–551

Boumann HA, Longo ML, Stroeve P, Poolman B et al (2009) Biophysical properties of membrane lipids of anammox bacteria: I. Ladderane phospholipids form highly organized fluid membranes. Biochim Biophys Acta 1788:1444–1451. https://doi.org/10.1016/j.bbamem.2009.04.008

Bourbonnais A, Lehmann MF, Butterfield DA, Juniper SK (2012) Subseafloor nitrogen transformations in diffuse hydrothermal vent fluids of the Juan de Fuca Ridge evidenced by the isotopic composition of nitrate and ammonium. Geochem Geophys Geosyst 13:1–23. https://doi.org/10.1029/2011gc003863

Bowles M, Joye SB (2011) High rates of denitrification and nitrate removal in cold seep sediments. ISME J 5:565–567

Bowles MW, Samarkin VA, Bowles KM, Joye SB (2011) Weak coupling between sulfate reduction and the anaerobic oxidation of methane in methane-rich seafloor sediments during ex situ incubation. Geochim Cosmochim Acta 75:500–519

Bowles MW, Mogollón JM, Kasten S, Zabel M, Hinrichs KU (2014) Global rates of marine sulfate reduction and implications for sub-sea-floor metabolic activities. Science 344:889–891

Bowles MW, Samarkin VA, Hunter KS, Finke N, Teske AP, Girguis PR, Joye SB (2019) Remarkable capacity for anaerobic oxidation of methane at high methane concentration. Geophys Res Lett 46:12192–12201

Bradley JA, Arndt S, Amend JP, Burwicz E, Dale AW et al (2020) Widespread energy limitation to life in global subseafloor sediments. Sci Adv 6:eaba0697

Bratbak G (1985) Bacterial biovolume and biomass estimations. Appl Environ Microbiol 49:1488–1493

Breitburg D, Levin LA, Oschlies A, Grégoire M et al (2018) Declining oxygen in the global ocean and coastal waters. Science 59:eaam7240. https://doi.org/10.1126/science.aam7240

Briggs N, Dall'Olmo G, Claustre H (2020) Major role of particle fragmentation in regulating biological sequestration of $CO_2$ by the oceans. Science 367:791

Brinkhoff T, Giebel H-A, Simon M (2008) Diversity, ecology, and genomics of the *Roseobacter* clade: a short overview. Arch Microbiol 189:531–539. https://doi.org/10.1007/s00203-008-0353-y

Broecker WS (1987) The biggest chill. Nat Hist Mag 97:74–82

Broecker WS (1991) The Great Ocean Conveyer. Oceanography 4:79–89

Burd AB, Jackson GA (2009) Particle aggregation. Annu Rev Mar Sci 1:65–90

Burdige DJ (2005) Burial of terrestrial organic matter in marine sediments: a re-assessment. Global Biogeochem Cycles 19:GB4011. https://doi.org/10.1029/2004GB002368

Busch K, Endre S, Iversen MH, Michels J et al (2017) Bacterial colonization and vertical distribution of marine gel particles (TEP and CSP) in the Arctic Fram Strait. Front Mar Sci 4:166

Callbeck CM, Canfield DE, Kuypers MMM, Yilmaz P et al (2021) Sulfur cycling in oceanic oxygen minimum zones. Limnol Oceanogr 66:2360–2392. https://doi.org/10.1002/lno.11759

Canfield DE, Thamdrup B, Hansen JW (1993) The anaerobic degradation of organic matter in Danish coastal sediments: iron reduction, manganese reduction, and sulfate reduction. Geochim Cosmochim Acta 57:3867–3883. https://doi.org/10.1016/0016-7037(93)90340-3

Canfield D, Kristensen E, Thamdrup B (2005) Aquatic geomicrobiology. Elsevier Science and Technology, 656 p, ISBN-13 9780121583408

Canfield DE, Stewart FJ, Thamdrup B, De Brabandere L et al (2010) A cryptic sulfur cycle in oxygen-minimum–zone waters off the Chilean coast. Science 330:1375–1378. https://doi.org/10.1126/science.1196889

Carini P, White AE, Campbell EO, Giovannoni SJ (2014) Methane production by phosphate-starved SAR11 chemoheterotrophic marine bacteria. Nat Commun 5:1–7

Chanton J, Zhao T, Rosenheim BE, Joye S, Bosman S et al (2015) Using natural abundance radiocarbon to trace the flux of petrocarbon to the seafloor following the deepwater horizon oil spill. Environ Sci Technol 49:847–854

Chin M, Jacob DJ, Gardner GM, Foreman-Fowler MS et al (1996) A global three-dimensional model of tropospheric sulfate. J Geophys Res 101:18667–18690

Ciais, P, Sabine C, Bala G, Bopp L et al (2013) Carbon and other biogeochemical cycles. In: Stocker TF, Qin D, Plattner G-K, Tignor M et al (eds) Climate change 2013: the physical science basis. Contribution of Working Group I to the fifth assessment report of the Intergovernmental Panel on Climate Change. Cambridge University Press, pp 465–570

Cirri E, Pohnert G (2019) Algae-bacteria interactions that balance the planktonic microbiome. New Phytol 223:100–106. https://doi.org/10.1111/nph.15765

Clark LL, Ingall ED, Benner R (1998) Marine phosphorus is selectively remineralized. Nature 393: 426–429. https://doi.org/10.1038/30881

Cloern JE, Foster SQ, Kleckner AE (2014) Phytoplankton primary production in the world's estuarine-coastal ecosystems. Biogeosciences 11:2477–2501

Collins JR, Edwards BR, Thamatrakoln K, Ossolinski JE et al (2015) The multiple fates of sinking particles in the North Atlantic Ocean. Global Biogeochem Cycles 29:1471–1494

Connolly CT, Cardenas MB, Burkart GA, Spencer RGM, McClelland JW (2020) Groundwater as a major source of dissolved organic matter to Arctic coastal waters. Nat Commun. https://doi.org/10.1038/s41467-020-15250-8

Cottrell MT, Kirchman DL (2000) Natural assemblages of marine Proteobacteria and members of the Cytophaga-Flavobacter cluster consuming low- and high-molecular-weight dissolved organic matter. Appl Environ Microbiol 66:1692–1697

Crespo-Medina M, Meile CD, Hunter KS, Diercks AR, Asper VL et al (2014) The rise and fall of methanotrophy following a deepwater oil-well blowout. Nat Geosci 7:423–427

Cropp R, Norbury J, Braddock R (2007) Dimethylsuphide, clouds, and phytoplankton: insights from a simple plankton ecosystem feedback model. Global Biogeochem Cycles 21. https://doi.org/10.1029/2006GB002812

Curson ARJ, Williams BT, Pinchbeck BJ, Sims LP et al (2018) DSYB catalyses the key step of dimethylsulfoniopropionate biosynthesis in many phytoplankton. Nat Microbiol 3:430–439. https://doi.org/10.1038/s41564-018-0119-5

Cuskin F, Lowe EC, Temple MJ, Zhu Y, Cameron EA et al (2015) Human gut *Bacteroidetes* can utilize yeast mannan through a selfish mechanism. Nature 517:165–169. https://doi.org/10.1038/nature13995

Daims H, Lebedeva E, Pjevac P, Han P et al (2015) Complete nitrification by *Nitrospira* bacteria. Nature 528:504–509. https://doi.org/10.1038/nature16461

Dalsgaard T, Thamdrup B, Canfield DE (2005) Anaerobic ammonium oxidation (anammox) in the marine environment. Res Microbiol 156:457–464. https://doi.org/10.1016/j.resmic.2005.01.011

Daly KL, Passow U, Chanton J, Hollander D (2016) Assessing the impacts of oil-associated marine snow formation and sedimentation during and after the Deepwater Horizon oil spill. Anthropocene 13:18–33

Damm E, Helmke E, Thoms S, Shauer U et al (2010) Methane production in aerobic oligotrophic surface water in the central Arctic Ocean. Biogeosciences 7:1099–1108

De Angelis MA, Lilley MD, Baross JA (1993) Methane oxidation in deep-sea hydrothermal plumes of the Endeavour Segment of the Juan de Fuca Ridge. Deep Sea Res Pt I Oceanogr Res Papers 40:1169–1186

Decho AW, Gutierrez T (2017) Microbial extracellular polymeric substances (EPSs) in ocean systems. Front Microbiol 8:922

DeLong EF, Franks DG, Alldredge AL (1993) Phylogenetic diversity of aggregate-attached- vs. free-living marine bacterial assemblages. Limnol Oceanogr 38:924–934

DeVries T, Deutsch C, Rafter PA, Primeau F (2013) Marine denitrification rates determined from a global 3-D inverse model. Biogeosciences 10:2481–2496

D'Hondt S, Jørgensen BB, Miller DJ, Batzke A et al (2004) Distributions of microbial activities in deep subseafloor sediments. Science 306:2216–2221

D'Hondt S, Inagaki F, Zarikian CA, Abrams LJ, Dubois N et al (2015) Presence of oxygen and aerobic communities from sea floor to basement in deep-sea sediments. Nat Geosci 8:299–304

Diaz JM, Ingall E, Benitez-Nelson C, Patterson D (2008) Marine polyphosphate: a key player in geologic phosphorus sequestration. Science 320:652–655. https://doi.org/10.1126/science.1151751

Diaz JM, Hansel CM, Volker BM, Mendes CM et al (2013) Widespread production of extracellular superoxide by heterotrophic bacteria. Science 340:1223–1226

Diaz JM, Plummer S, Hansel CM, Andeer PF et al (2019) NADPH-dependent extracellular superoxide production is vital to photophysiology in the marine diatom *Thalassiosira oceanica*. Proc Natl Acad Sci USA 116:16448–16453. https://doi.org/10.1073/pnas.1821233116

Dick GJ (2019) The microbiomes of deep-sea hydrothermal vents: distributed globally, shaped locally. Nat Rev Microbiol 17:271–283

Dick GJ, Anantharaman K, Baker BJ, Li M et al (2013) The microbiology of deep-sea hydrothermal vent plumes: ecological and biogeographic linkages to seafloor and water column habitats. Front Microbiol 4:124

Duarte CM, Regaudie-de-Gioux A, Arrieta JM, Delgado-Huertas A, Augsti S (2013) The oligotrophic ocean is heterotrophic. Annu Rev Mar Sci 5:551–569. https://doi.org/10.1146/annurev-marine-121211-172337

Ducklow HW, Doney SC (2013) What is the metabolic state of the oligotrophic ocean? A debate. Annu Rev Mar Sci 5:525–533. https://doi.org/10.1146/annurev-marine-121211-172331

Duhamel S, Diaz JM, Adams JC et al (2021) Phosphorus as an integral component of global marine biogeochemistry. Nat Geosci 14:359–368. https://doi.org/10.1038/s41561-021-00755-8

Dunne JP, Sarmiento JL, Gnanadesikan A (2007) A synthesis of global particle export from the surface ocean and cycling through the ocean interior and on the seafloor. Global Biogeochem Cycles 21:GB002907

Durham BP, Sharma S, Luo J, Smith CB et al (2015) Cryptic carbon and sulfur cycling between surface ocean plankton. Proc Natl Acad Sci USA 112:453–457

Durham BP, Dearth SP, Sharma S, Amin SA et al (2017) Recognition cascade and metabolite transfer in a marine bacteria-phytoplankton model system. Environ Microbiol 19:3500–3513

Dyhrman ST, Benitez-Nelson CR, Orchard ED, Haley ST, Pellechia PJ (2009) A microbial source of phosphonates in oligotrophic marine systems. Nat Geosci 2:696–699. https://doi.org/10.1038/ngeo639
Dyksma S, Bischof K, Fuchs BM, Hoffman K et al (2016) Ubiquitous Gammaproteobacteria dominate dark carbon fixation in coastal sediments. ISME J 10:1939–1953
Eady RR (1996) Structure-function relationships of alternative nitrogenases. Chem Rev 96:3013–3030
Egger M, Riedinger N, Mogollón JM, Jørgensen BB (2018) Global diffusive fluxes of methane in marine sediments. Nat Geosci 11:421–425
Eisenhut M, Roell M-S, Weber APM (2019) Mechanistic understanding of photorespiration paves the way to a new green revolution. New Phytol 223:1763–1769. https://doi.org/10.1111/nph.15872
Elifantz H, Dittel AI, Cottrell MT, Kirchman DL (2007) Dissolved organic matter assimilation by heterotrophic bacterial groups in the western Arctic Ocean. Aqut Microb Ecol 50:39–49
Elser JJ, Sterner RW, Gorokhova E, Fagan WF et al (2008) Biological stoichiometry from genes to ecosystems. Ecol Lett 3:540–550. https://doi.org/10.1111/j.1461-0248.2000.00185.x
Emerson D (2019) Biogenic iron dust: a novel approach to ocean iron fertilization as a means of large scale removal of carbon dioxide from the atmosphere. Front Mar Sci 6:22
Emery KO (1968) Relict sediments on continental shelves of world. AAPG Bull 52:445–464
Endres S, Galgani L, Riebesell U, Schulz K-G, Engel A (2014) Stimulated bacterial growth under elevated $pCO_2$: results from an off-shore mesocosm study. PLoS One 9:1–8
Engel A (2009) Determination of marine gel particles. In: Wurl O (ed) Practical guidelines for the analysis of seawater. CRC, Boca Raton, FL, pp 125–142
Engel A, Endres S, Galgani L, Schartau M (2020) Marvelous marine microgels: on the distribution and impact of gel-like particles in the oceanic water-column. Front Mar Sci 7:405
Eppley RW (1972) Temperature and phytoplankton growth in the sea. Fish Bull 70:1063–1085
Falkowski PG, Barber RT, Smetacek V (1998) Biogeochemical controls and feedbacks on ocean primary production. Science 281:200–206
Falkowski PG, Fenchel T, DeLong EF (2008) The microbial engines that drive earth's biogeochemical cycles. Science 320:1034–1039. https://doi.org/10.1126/science.1153213
Field CB, Behrenfeld MJ, Randerson JT, Falkowski P (1998) Primary production in the biosphere: integrating terrestrial and oceanic components. Science 281:237–240
Fontanez KM, Eppley JM, Samo TJ, Karl DM, DeLong EF (2015) Microbial community structure and function on sinking particles in the North Pacific Subtropical Gyre. Front Microbiol 6:469
Fuchs BM, Spring S, Teeling H, Quast C et al (2007) Characterization of a marine gammaproteobacterium capable of aerobic anoxygenic photosynthesis. Proc Natl Acad Sci USA 104:2891–2896. https://doi.org/10.1073/pnas.0608046104
Fuerst JA (2005) Intracellular compartmentation in planctomycetes. Annu Rev Microbiol 59:299–328. https://doi.org/10.1146/annurev.micro.59.030804.121258
Fuhrman JA (1999) Marine viruses and their biogeochemical and ecological effects. Nature 399:541–548
Galbraith ED, Martiny AC (2015) A simple nutrient-dependence mechanism for predicting the stoichiometry of marine ecosystems. Proc Natl Acad Sci USA 112:8199–8204. https://doi.org/10.1073/pnas.1423917112
Garcia-Robledo E, Padilla CC, Aldunte M, Stewart FJ et al (2017) Cryptic oxygen cycling in anoxic marine zones. Proc Natl Acad Sci USA 114:8319–8324. https://doi.org/10.1073/pnas.1619844114
Giovannoni SJ, Hayakawa DH, Tripp HJ et al (2008) The small genome of an abundant coastal ocean methylotroph. Environ Microbiol 10:1771–1782. https://doi.org/10.1111/j.1462-2920.2008.01598.x
Glud RN (2008) Oxygen dynamics of marine sediments. Mar Biol Res 4:243–289

Goffredi SK, Wilpiszeski R, Lee R, Orphan VJ (2008) Temporal evolution of methane cycling and phylogenetic diversity of archaea in sediments from a deep-sea whale-fall in Monterey Canyon, California. ISME J 2:204–220
Groeskamp S, Griffies SM, Iudicone D, Marsh R, Nurser AJG, Zika JD (2019) The water mass transformation framework for ocean physics and biogeochemistry. Annu Rev Mar Sci 11:271–305
Gruber N (2016) Elusive marine nitrogen fixation. Proc Natl Acad Sci USA 113:4246–4248
Hagemann M, Kern R, Maurino VG et al (2016) Evolution of photorespiration from cyanobacteria to land plants, considering protein phylogenies and acquisition of carbon concentrating mechanisms. J Exp Bot 67:2963–2976. https://doi.org/10.1093/jxb/erw063
Halevy I, Peters SE, Fischer WW (2012) Sulfate burial constraints on the Phanerozoic sulfur cycle. Science 227:331–334. https://doi.org/10.1126/science.1220224
Hansel CM, Buchwald C, Diaz JM, Ossolinski JE et al (2016) Dynamics of extracellular superoxide production by *Trichodesmium* colonies from the Sargasso Sea. Limnol Oceanogr 61:1188–1200. https://doi.org/10.1002/lno.10266
Hansell DA (2013) Recalcitrant dissolved organic carbon fractions. Annu Rev Mar Sci 5:421–445
Hansell DA, Carlson CA (2015) Dissolved organic matter in the ocean carbon cycle. EOS Trans Am Geophys Union 96:EO033011. https://doi.org/10.1029/2015EO033011
Hanson RS, Hanson TE (1996) Methanotrophic bacteria. Microbiol Rev 60:439–471
Hedges JI (1992) Global biogeochemical cycles: progress and problems. Mar Chem 39:67–93
Hedges JI, Keil RG (1995) Sedimentary organic matter preservation: an assessment and speculative synthesis. Mar Chem 49:81–115
Hedges JI, Keil RG, Benner R (1997) What happens to terrestrial organic matter in the ocean. Org Geochem 27:195–212
Hehemann J-H, Law A, Redecke L, Boraston AB (2014) The structure of RdDddP from Roseobacter denitrificans reveals that DMSP lyases in the DddP-family are metalloenzymes. PLoS One 9(7):e103128. https://doi.org/10.1371/journal.pone.0103128
Heuer VB, Inagaki F, Morono Y, Kubo Y et al (2020) Temperature limits to deep subseafloor life in the Nankai Trough subduction zone. Science 370:1230–1234
Hinrichs KU, Boetius A (2002) The anaerobic oxidation of methane: new insights in microbial ecology and biogeochemistry. In: Wefer G (ed) Ocean margin systems. Springer, pp 457–477
Hoshino T, Doi H, Uramoto GI, Wörmer L et al (2020) Global diversity of microbial communities in marine sediment. Proc Natl Acad Sci USA 117(44):27587–27597
Huang J, Yu Z, Groom J, Cheng JF, Tarver A, Yoshikuni Y, Chistoserdova L (2019) Rare earth element alcohol dehydrogenases widely occur among globally distributed, numerically abundant and environmentally important microbes. ISME J 13(8):2005–2017
Hügler M, Sievert SM (2011) Beyond the Calvin cycle: autotrophic carbon fixation in the ocean. Annu Rev Mar Sci 3:261–289
Hurtgen MT (2012) The marine sulfur cycle, revisited. Science 337:305–306. https://doi.org/10.1126/science.1225461
Hygum BH, Peterson JW (1997) Dissolved organic carbon released by zooplankton grazing activity—a high-quality substrate pool for bacteria. J Plankt Res 19:97–111
Inagaki F, Hinrichs K-U, Kubo Y, Bowles MW et al (2015) Exploring deep microbial life in coal-bearing sediment down to ~2.5 km below the ocean floor. Science 349:420–424
Inomura K, Wilson ST, Deutsch C (2019) Mechanistic model for the coexistence of nitrogen fixation and photosynthesis in marine *Trichodesmium*. mSystems 4:e00210-19. https://doi.org/10.1128/mSystems.00210-19
Inomura K, Follett CL, Masuda T, Eichner M et al (2020) Carbon transfer from the host diatom enables fast growth and high rate of $N_2$ fixation by symbiotic heterocystous cyanobacteria. Plants (Basel) 9:192. https://doi.org/10.3390/plants9020192
Intergovernmental Panel on Climate Change (IPCC) (2013) Climate change 2013: the physical science basis. Contribution of Working Group I to the fifth assessment report of the Intergovernmental Panel on Climate Change. Cambridge University Press, Cambridge

Intergovernmental Panel on Climate Change (IPCC) (2020) Summary for policymakers. In: Pörtner H-O, Roberts DC, Masson-Delmotte V, Zhai P, Tignor M, Poloczanska E, Mintenbeck K, Nicolai M, Okem A, Petzold J, Rama B, Weyer N (eds) IPCC special report on the ocean and cryosphere in a changing climate. https://doi.org/10.1002/essoar.10502454.1

Jackson JBC (1994) Constancy and change of life in the sea. Phil Trans Royal Soc B 344:55–60

Jahnke RA (1996) The global ocean flux of particulate organic carbon: areal distribution and magnitude. Global Biogeochem Cycles 10:71–88

Jefferson T, Turner J (2002) Zooplankton fecal pellets, marine snow and sinking phytoplankton blooms. Aquat Microb Ecol 27:57–102

Jetten MS, van Niftrik L, Strous M, Kartal B et al (2009) Biochemistry and molecular biology of anammox bacteria. Crit Rev Biochem Mol Biol 44:65–84. https://doi.org/10.1080/10409230902722783

Ji Q, Babbin AR, Jayakumar A, Oleynik S, Ward BB (2015) Nitrous oxide production by nitrification and denitrification in the Eastern Tropical South Pacific oxygen minimum zone. Geophys Res Lett 42:10755–10764. https://doi.org/10.1002/2015GL066853

Jiao N, Herndl G, Hansell D et al (2010a) Microbial production of recalcitrant dissolved organic matter: long-term carbon storage in the global ocean. Nat Rev Microbiol 8:593–599. https://doi.org/10.1038/nrmicro2386

Jiao N, Zhang F, Hong N (2010b) Significant roles of bacteriochlorophyll *a* supplemental to chlorophyll *a* in the ocean. ISME J 4:595–597. https://doi.org/10.1038/ismej.2009.135

Jickells TD, An ZS, Andersen KK, Baker AR et al (2005) Global iron connections between desert dust, ocean biogeochemistry, and climate. Science 308:67–71

Joye SB (2002) Denitrification in the marine environment. In: Collins G (ed) Encyclopedia of environmental microbiology. Wiley, New York, pp 1010–1019. https://doi.org/10.1002/0471263397.env141

Joye SB (2019) The geology and geobiology of hydrocarbon seeps. Annu Rev Earth Planet Sci 48: 205–231. https://doi.org/10.1146/annurev-earth-063016-020052

Joye SB, Anderson I (2008) Nitrogen cycling in coastal sediments. In: Capone D, Bronk D, Carpenter E, Mulholland M (eds) Nitrogen in the marine environment. Elsevier, pp 867–915

Joye SB, Bowles MW (2022) Microbial dynamics at cold seeps. In: Giovannelli D, Vertriani C (eds) Deep-sea microbiology. Springer, in press

Joye SB, Kostka JE (2020) Microbial genomics of the global ocean system. Am Acad Microbiol Colloq Rep. https://doi.org/10.1128/AAMCol.Apr.2019

Kallmeyer J, Pockalny R, Adhikari RR, Smith DC, D'Hondt S (2012) Global distribution of microbial abundance and biomass in subseafloor sediment. Proc Natl Acad Sci USA 109: 16213–16216

Kamat SS, Williams HJ, Dangott LJ, Chakrabarti M, Raushel FM (2013) The catalytic mechanism for aerobic formation of methane by bacteria. Nature 497:132–136. https://doi.org/10.1038/nature12061

Karl DM (2014) Microbially mediated transformations of phosphorus in the sea: new views of an old cycle. Annu Rev Mar Sci 6:279–337

Karl DM, Björkman KM (2002) Dynamics of DOP. In: Hansell D, Carlson C (eds) Biogeochemistry of marine organic matter. Academic, San Diego, CA, pp 249–366

Karl DM, Björkman KM (2015) Dynamics of dissolved organic phosphorus. In: Hansell D, Carlson C (eds) Biogeochemistry of marine organic matter, 2nd ed. Academic, San Diego, CA, pp 249–366. https://doi.org/10.1016/B978-0-12-405940-5.00005-4

Karl DM, Beversdorf L, Björkman KM, Church MJ et al (2008) Aerobic production of methane in the sea. Nat Geosci 1:473–478. https://doi.org/10.1038/ngeo234

Kartal B, Maalcke WJ, de Almeida NM, Cirpus I et al (2011) Molecular mechanism of anaerobic ammonium oxidation. Nature 479:127–130. https://doi.org/10.1038/nature10453

Kasten S, Jørgensen BB (2000) Sulfate reduction in marine sediments. In: Schulz HD, Zabel M (eds) Marine geochemistry. Springer, Berlin, pp 263–281

Keeling RF, Körtzinger A, Gruber N (2010) Ocean deoxygenation in a warming world. Annu Rev Mar Sci 2:199–229. https://doi.org/10.1146/annurev.marine.010908.163855
Kelley C (2003) Methane oxidation potential in the water column of two diverse coastal marine sites. Biogeochemistry 65:105–120
Kiorboe T, Tang K, Grossart HP, Ploug H (2003) Dynamics of microbial communities on marine snow aggregates: colonization, growth, detachment, and grazing mortality of attached bacteria. Appl Environ Microbiol 69:3036–3047
Kirschke S, Bousquet P, Ciais P, Saunois M et al (2013) Three decades of global methane sources and sinks. Nat Geosci 6:813–823
Kitzinger K, Padilla CC, Marchant HK, Hach PF et al (2019) Cyanate and urea are substrates for nitrification by Thaumarchaeota in the marine environment. Nature Microbiol 4:234–243. https://doi.org/10.1038/s41564-018-0316-2
Kleidon A (2012) How does the Earth system generate and maintain thermodynamic disequilibrium and what does it imply for the future of the planet? Philos Trans Royal Soc A 370:1012–1040. https://doi.org/10.1098/rsta.2011.0316
Knief C (2015) Diversity and habitat preferences of cultivated and uncultivated aerobic methanotrophic bacteria evaluated based on *pmoA* as molecular marker. Front Microbiol 6:1346
Kolber ZS, Plumley FG, Lang AS, Beatty JT et al (2001) Contribution of aerobic photoheterotrophic bacteria to the carbon cycle in the ocean. Science 29:2492–2495
Könneke M, Bernhard AE, de la Torre JR, Walker CB et al (2005) Isolation of an autotrophic ammonia-oxidizing marine archaeon. Nature 437:543–546. https://doi.org/10.1038/nature03911
Könneke M, Schubert DM, Brown PC, Hügler M et al (2014) Ammonia-oxidizing archaea use the most efficient aerobic pathway for $CO_2$ fixation. Proc Natl Acad Sci USA 22:8239–8244
Kulk G, Platt T, Dingle J, Jackson T, Jönsson BF et al (2020) Primary production, an index of climate change in the ocean: satellite-based estimates over two decades. Remote Sensing 12: 826–835. https://doi.org/10.3390/rs12050826
Kuypers MMM, Lavik G, Woebken D, Schmid M et al (2005) Massive nitrogen loss from the Benguela upwelling system through anaerobic ammonium oxidation. Proc Natl Acad Sci USA 102:6478–6483. https://doi.org/10.1073/pnas.0502088102
Kuypers MMM, Marchant HK, Kartal B (2018) The microbial nitrogen-cycling network. Nat Rev Microbiol 16:263–277. https://doi.org/10.1038/nrmicro.2018.9
Kvale KF, Friederike Prowe AE, Oschlies A (2020) A critical examination of the role of marine snow and zooplankton fecal pellets in removing ocean surface microplastic. Front Mar Sci 6:808
Lam P, Lavik G, Jensen MM, van de Vossenberg J et al (2009) Revising the nitrogen cycle in the Peruvian oxygen minimum zone. Proc Natl Acad Sci USA 106:4752–4757. https://doi.org/10.1073/pnas.0812444106
Lami R, Cottrell MT, Ras J, Ulloa O, Obernosterer I et al (2007) High abundances of aerobic anoxygenic photosynthetic bacteria in the South Pacific Ocean. Appl Environ Microbiol 73: 4198–4205. https://doi.org/10.1128/AEM.02652-06
Lampitt RS, Salter I, de Cuevas BA, Hartman S et al (2010) Long-term variability of downward particle flux in the deep northeast Atlantic: causes and trends. Deep Sea Res Pt II Top Stud Oceanogr 57:1346–1361
Landry Z, Swan BK, Herndl GJ, Stepanauskas R, Giovannoni SJ (2017) SAR202 genomes from the dark ocean predict pathways for the oxidation of recalcitrant dissolved organic matter. mBio 8: e00413-17. https://doi.org/10.1128/mBio.00413-17
LaRowe DE, Burwicz E, Arndt S, Dale AW, Amend JP (2017) Temperature and volume of global marine sediments. Geology 45:275–278
Laws EA, Maiti K (2019) The relationship between primary production and export production in the ocean: effects of time lags and temporal variability. Deep Sea Res Pt I 148:100–107
Le B, Williams PJ, Quay PD, Westberry TK, Behrenfeld MJ (2013) The oligotrophic ocean is autotrophic. Annu Rev Mar Sci. https://doi.org/10.1146/annurev-marine-121211-172335

Levin SA (1998) Ecosystems and the biosphere as complex adaptive systems. Ecosystems 1:431–436
Lidstrom ME (2006) Aerobic methylotrophic prokaryotes. Prokaryotes 2:618–634. https://doi.org/10.1007/0-387-30742-7_20
Lipp JS, Morono Y, Inagaki F, Hinrichs KU (2008) Significant contribution of Archaea to extant biomass in marine subsurface sediments. Nature 454:991–994
Long RA, Azam F (1996) Abundant protein-containing particles in the sea. Aquat Microb Ecol 10: 213–221
Lovelock JE, Maggs RJ, Rasmussen RA (1972) Atmospheric dimethyl sulphide and natural sulfur cycle. Nature 237:452–453. https://doi.org/10.1038/237452a0
Mari X, Passow U, Migon C, Burd AB, Legendre L (2017) Transparent exopolymer particles: effects on carbon cycling in the ocean. Prog Oceanogr 151:13–37
Martin JH, Gordon RM, Fitzwater SE (1990) Iron in Antarctic waters. Nature 345:156–158
Martínez A, Osburne MS, Sharma AK, DeLong EF, Chisholm SW (2012) Phosphite utilization by the marine picocyanobacterium *Prochlorococcus* MIT9301. Environ Microbiol 14:1363–1377
Martiny AC, Lomas MW, Fu W, Boyd PW et al (2019) Biogeochemical controls of surface ocean phosphate. Sci Adv 5:eaax0341. https://doi.org/10.1126/sciadv.aax0341
Mather RL, Reynolds SE, Wolff GA, Williams RG et al (2008) Phosphorus cycling in the North and South Atlantic Ocean subtropical gyres. Nat Geosci 1:439–443. https://doi.org/10.1038/ngeo232
Mau S, Blees J, Helmke E, Niemann H, Damm E (2013) Vertical distribution of methane oxidation and methanotrophic response to elevated methane concentrations in stratified waters of the Arctic fjord Storfjorden (Svalbard, Norway). Biogeosciences 10:6267–6278
McInnes AS, Quigg A (2010) Near-annual fish kills in small embayments: casual vs. causal factors. J Coast Res 26:957–966
McNichol J, Stryhanyuk H, Sylva SP, Thomas F et al (2018) Primary productivity below the seafloor at deep-sea hot springs. Proc Natl Acad Sci USA 115:6756–6761
Metcalf WM, Griffin BM, Cicchillo RM, Gao J et al (2012) Synthesis of methylphosphonic acid by marine microbes: a source for methane in the aerobic ocean. Science 337:1104–1107. https://doi.org/10.1126/science.1219875
Michels J, Stippkugel A, Lenz M, Wirtz K, Engel A (2018) Rapid aggregation of biofilm-covered microplastics with marine biogenic particles. Proc Roy Soc B 285:20181203
Middelburg JJ (2019) Carbon processing at the seafloor. In: Middelburg JJ (ed) Marine carbon biogeochemistry. Springer, Cham, pp 57–75
Middelburg JJ, Soetaert K, Herman PM, Heip CH (1996) Denitrification in marine sediments: a model study. Global Biogeochem Cycles 10:661–673
Middelburg JJ, Soetaert K, Herman PM (1997) Empirical relationships for use in global diagenetic models. Deep Sea Res Pt I Oceanogr Res Pap 44:327–344
Mincer TJ, Aicher AC (2016) Methanol production by a broad phylogenetic array of marine phytoplankton. PLoS One 11:e0150820. https://doi.org/10.1371/journal.pone.0150820
Mino S, Nakagawa S, Makita H, Toki T, Miyazaki J, Sievert SM et al (2017) Endemicity of the cosmopolitan mesophilic chemolithoautotroph *Sulfurimonas* at deep-sea hydrothermal vents. ISME J 11:909–919
Moeseneder MM, Winter C, Herndl GJ (2001) Horizontal and vertical complexity of attached and free-living bacteria of the eastern Mediterranean Sea, determined by 16S rDNA and 16S rRNA fingerprints. Limnol Oceanogr 46:95–107
Moran MA, Kujawinski EB, Stubbins A, Fatland R et al (2016) Deciphering ocean carbon in a changing world. Proc Natl Acad Sci USA 113:3143–3151. https://doi.org/10.1073/pnas.1514645113
Moreno AR, Martiny AC (2018) Ecological stoichiometry of ocean plankton. Annu Rev Mar Sci 10:43–69. https://doi.org/10.1146/annurev-marine-121916-063126
Mouw CB, Barnett A, McKinley GA, Gloege L, Pilcher D (2016) Global ocean particulate organic carbon flux merged with satellite parameters. Earth Syst Sci Data 8:531–541

Nakagawa S, Takai K (2008) Deep-sea vent chemoautotrophs: diversity, biochemistry and ecological significance. FEMS Microbiol Ecol 65:1–14. https://doi.org/10.1111/j.1574-6941.2008.00502.x

Nayak AR, Twardowski MS (2020) "Breaking" news for the ocean's carbon budget. Science 367 (6479):738–739. https://doi.org/10.1126/science.aba7109

Neufeld J, Schäfer H, Cox M, Boden R et al (2007) Stable-isotope probing implicates *Methylophaga* spp. and novel *Gammaproteobacteria* in marine methanol and methylamine metabolism. ISME J 1:480–491. https://doi.org/10.1038/ismej.2007.65

Nevitt GA, Losekoot M, Weimerskirch H (2008) Evidence for olfactory search in wandering albatross *Diomedea exulans*. Proc Natl Acad Sci USA 105:4576–4581. https://doi.org/10.1073/pnas.0709047105

Nielsen ES (1960) Dark fixation of $CO_2$ and measurements of organic productivity, with remarks on chemo-synthesis. Physiol Plant 13:348–357

Otero A, Vincenzini M (2004) Nostoc (Cyanophyceae) goes nude: extracellular polysaccharides serve as a sink for reducing power under unbalanced C/N metabolism. J Phycol 40:74–81

Owen K, Saeki K, Warren JD, Bocconcelli A et al (2021) Natural dimethyl sulfide gradients would lead marine predators to higher prey biomass. Commun Biol 4:149. https://doi.org/10.1038/s42003-021-01668-3

Pachiadaki MG, Sintes E, Bergauer K, Brown JM et al (2017) Major role of nitrite-oxidizing bacteria in dark ocean carbon fixation. Science 358:1046–1051. https://doi.org/10.1126/science.aan8260

Pack MA, Heintz MB, Reeburgh WS, Trumbore SE et al (2011) A method for measuring methane oxidation rates using low levels of $^{14}C$-labeled methane and accelerator mass spectrometry. Limnol Oceanogr Methods 9:245–260

Pack MA, Heintz MB, Reeburgh WS, Trumbore S et al (2015) Methane oxidation in the eastern tropical North Pacific Ocean water column. J Geophys Res Biogeosci 120:1078–1092

Pan D, Morono Y, Inagaki F, Takai K (2019) An improved method for extracting viruses from sediment: detection of far more viruses in the subseafloor than previously reported. Front Microbiol 10:878

Pasek MA, Sampson JM, Atlas Z (2014) Redox chemistry in the phosphorus biogeochemical cycle. Proc Natl Acad Sci USA 111:15468–15473. https://doi.org/10.1073/pnas.1408134111

Passow U (2002) Transparent exopolymer particles (TEP) in aquatic environments. Progr Oceanogr 55:287–333

Passow U (2016) Formation of rapidly-sinking, oil-associated marine snow. Deep Sea Res Pt II Top Stud Oceanogr 129:232–240

Passow U, Alldredge AL (1994) Distribution, size and bacterial colonization of Transparent Exopolymer Particles (TEP) in the ocean. Mar Ecol Progr Ser 113:185–198

Passow U, Alldredge AL (1995a) A dye-binding assay for the spectrophotometric measurement of transparent exopolymer particles (TEP). Limnol Oceanogr 40:1326–1335

Passow U, Alldredge AL (1995b) Aggregation of a diatom bloom in a mesocosm: the role of transparent exopolymer particles (TEP). Deep Sea Res Pt II Top Stud Oceanogr 42:99–109

Passow U, la Rocha CLD (2006) Accumulation of mineral ballast on organic aggregates. Global Biogeochem Cycles 20:GB1013

Passow U, Laws EA (2015) Ocean acidification as one of multiple stressors: growth response of *Thalassiosira weissflogii* (diatom) under temperature and light stress. Mar Ecol Prog Ser 541: 75–90

Passow U, Ziervogel K (2016) Marine snow sedimented oil released during the Deepwater Horizon spill. Oceanography 29:118–125

Paytan A, McGlaughlin K (2007) The oceanic phosphorus cycle. Chem Rev 107:563–576

Picone N, Op den Camp HJ (2019) Role of rare earth elements in methanol oxidation. Curr Opin Chem Biol 49:39–44

Piontek J, Händel N, Langer G, Wohlers J, Riebesell U, Engel A (2009) Effects of rising temperature on the formation and microbial degradation of marine diatom aggregates. Aquat Microb Ecol 54:305–318
Piontek J, Lunau M, Händel N, Borchard C, Wurst M, Engel A (2010) Acidification increases microbial polysaccharide degradation in the ocean. Biogeosciences 7:1615–1624. https://doi.org/10.5194/bg-7-1615-2010
Planavsky NJ (2014) The elements of marine life. Nat Geosci 7:855–856. https://doi.org/10.1038/ngeo2307
Ploug H, Grossart HP (1999) Bacterial production and respiration in suspended aggregates—a matter of the incubation method. Aquat Microb Ecol 20:21–29
Ploug H, Kühl M, Buchholz-Cleven B, Jørgensen BB (1997) Anoxic aggregates—an ephemeral phenomenon in the pelagic environment? Aquat Microb Ecol 13:285–294
Porter A, Lyons BP, Galloway TS, Lewis C (2018) Role of marine snows in microplastic fate and bioavailability. Environ Sci Technol 52:7111–7119
Quigg A, Passow U, Chin WC, Xu C et al (2016) The role of microbial exopolymers in determining the fate of oil and chemical dispersants in the ocean. Limnol Oceanogr Lett 1:3–26
Quin PK, Bates TS (2011) The case against climate regulation via oceanic phytoplankton sulphur emissions. Nature 480:51–56. https://doi.org/10.1038/nature10580
Raven MR, Keil RG, Webb SM (2021) Microbial sulfate reduction and organic sulfur formation in sinking marine particles. Science 371:178–181
Redfield AC (1934) On the proportions of organic derivations in sea water and their relation to the composition of plankton. James Johnstone Memorial Volume, University Press of Liverpool, pp 176–192
Redfield AC (1958) The biological control of chemical factors in the environment. Am Sci 46:205–222
Reeburgh WS (2007) Oceanic methane biogeochemistry. Chem Rev 107:486–513
Reeburgh WS, Ward BB, Whalen SC, Sandbeck KA et al (1991) Black Sea methane geochemistry. Deep Sea Res Pt A. Oceanogr Res Pap 38:S1189–S1210
Rees AP, Brown IJ, Jayakumar A, Lessin G et al (2021) Biological nitrous oxide consumption in oxygenated waters of the high latitude Atlantic Ocean. Commun Earth Environ 2:36. https://doi.org/10.1038/s43247-021-00104-y
Regnier P, Friedlingstein P, Ciais P, Mackenzie FT, Gruber N et al (2013) Anthropogenic perturbation of the carbon fluxes from land to ocean. Nat Geosci 6:597–607
Reinthaler T, Van Aken HM, Herndl GJ (2010) Major contribution of autotrophy to microbial carbon cycling in the deep North Atlantic's interior. Deep Sea Res Pt II 57:1572–1580
Reintjes G, Arnosti C, Fuchs B, Amann R (2017) An alternative polysaccharide uptake mechanism of marine bacteria. ISME J 11:1640–1650. https://doi.org/10.1038/ismej.2017.26
Reintjes G, Arnosti C, Fuchs B, Amann R (2019) Selfish, sharing and scavenging bacteria in the Atlantic Ocean: a biogeographical study of bacterial substrate utilisation. ISME J 13:1119–1132. https://doi.org/10.1038/s41396-018-0326-3
Repeta D, Ferrón S, Sosa O, Johnson CG et al (2016) Marine methane paradox explained by bacterial degradation of dissolved organic matter. Nat Geosci 9:884–887. https://doi.org/10.1038/ngeo2837
Robinson C (2019) Microbial respiration, the engine of ocean deoxygenation. Front Mar Sci 5:533. https://doi.org/10.3389/fmars.2018.00533
Rogener MK, Bracco A, Hunter KS, Saxton M, Joye SB (2018) Impact of the Deepwater Horizon oil well blowout on methane oxidation dynamics in the Northern Gulf of Mexico. Elementa Sci Anthropocene 6:73. https://doi.org/10.1525/elementa.332
Rogener MK, Hunter KS, Rabalais NN, Bracco A, Stewart FJ, Joye SB (2021) Pelagic denitrification and methane oxidation in oxygen-depleted waters of the Louisiana shelf. Biogeochemistry 154. https://doi.org/10.1007/s10533-021-00778-8
Ruppel CD, Kessler JD (2017) The interaction of climate change and methane hydrates. Rev Geophys 55:126–168

Rush D, Sinninghe-Damste JS (2017) Lipids as paleomarkers to constrain the marine nitrogen cycle. Environ Microbiol 19:2119–2132. https://doi.org/10.1111/1462-2920.13682
Ruttenberg KC (2014) The global phosphorus cycle. In: Holland HD, Turekian KK (eds) Treatise on geochemistr, vol 10. Elsevier, pp 499–558. https://doi.org/10.1016/b978-0-08-095975-7.00813-5
Ryther JG, Dunstan WM (1971) Nitrogen, phosphorus and eutrophication in the coastal marine environment. Science 171:1008–1013
Saba GK, Steinberg DK (2012) Abundance, composition and sinking rates of fish fecal pellets in the Santa Barbara Channel. Sci Rep 2:716
Saba GK, Burd AB, Dunne JP, Hernández-León S et al (2021) Toward a better understanding of fish-based contribution to ocean carbon flux. Limnol Oceanogr 66:1639–1664
Sarmiento JL, Gruber N (2006) Ocean biogeochemical cycles. Princeton University Press, Princeton, NJ
Sarmiento JL, Toggweiler JR (1984) A new model for the role of the oceans in determining atmospheric $pCO_2$. Nature 308:621–624
Schmidtko S, Stramma L, Visbeck M (2017) Decline in global oceanic oxygen content during the past five decades. Nature 542:335–339. https://doi.org/10.1038/nature21399
Seiter K, Hensen C, Zabel M (2005) Benthic carbon mineralization on a global scale. Global Biogeochem Cycles 19:GB1010
Seitzinger S, Bohlke JK, Bouwman AF, Lowrance R et al (2006) Denitrification across landscapes and waterscapes: a synthesis. Ecol Appl 16:2064–2090
Seymour JR, Simo R, Ahmed T, Stocker R (2010) Chemoattraction to dimethyl-sulfoniopropionate throughout the marine microbial food web. Science 329:342–345
Siegel DA, Buesseler KO, Behrenfeld MJ, Benitez-Nelson CR et al (2016) Prediction of the export and fate of global ocean net primary production: the EXPORTS Science Plan. Front Mar Sci 3:22. https://doi.org/10.3389/fmars.2016.00022
Sigman DM, Hain MP (2012) The biological productivity of the ocean. Nat Educ Knowledge 3:1–16
Simó R, Vila-Costa M, Alonso-Saez L, Cardelus C et al (2009) Annual DMSP contribution to S and C fluxes through phytoplankton and bacterioplankton in a NW Mediterranean coastal site. Aquat Microb Ecol 57:43–55. https://doi.org/10.3354/ame01325
Simon M, Grossart HP, Schweitzer B, Ploug H (2002) Microbial ecology of organic aggregates in aquatic ecosystems. Aquat Microb Ecol 28:175–211
Smith SV, Mackenzie FT (1987) The ocean as a net heterotrophic system: implications from the carbon biogeochemical cycle. Global Biogeochem Cycles 1:187–198. https://doi.org/10.1029/GB001i003p00187
Smith DC, Simon M, Alldredge AL, Azam F (1992) Intense hydrolytic enzyme activity on marine aggregates and implications for rapid particle dissolution. Nature 359:139–142
Smith KL, Ruhl HA, Kahru M, Huffard CL, Sherman AD (2013) Deep ocean communities impacted by changing climate over 24 y in the abyssal northeast Pacific Ocean. Proc Natl Acad Sci USA 110:19838–19841
Smriga S, Ciccarese D, Babbin AR (2021) Denitrifying bacteria respond to and shape microscale gradients within particulate matrices. Commun Biol 4:570. https://doi.org/10.1038/s42003-021-02102-4
Sohm JA, Webb EA, Capone DG (2011) Emerging patterns of marine nitrogen fixation. Nat Rev Microbiol 9:499–508. https://doi.org/10.1038/nrmicro2594
Sosa OA, Repeta DJ, DeLong EF, Ashkezari MD, Karl DM (2019a) Phosphate-limited ocean regions select for bacterial populations enriched in the carbon–phosphorus lyase pathway for phosphonate degradation. Environ Microbiol 21:2402–2414. https://doi.org/10.1111/1462-2920.14628
Sosa OA, Casey JR, Karl DM (2019b) Methylphosphonate oxidation in *Prochlorococcus* strain MIT9301 supports phosphate acquisition, formate excretion, and carbon assimilation into purines. Appl Environ Microbiol 85:e00289–e00219. https://doi.org/10.1128/AEM.00289-19

Sosa OA, Burrell TJ, Wilson ST, Foreman RK et al (2020) Phosphonate cycling supports methane and ethylene supersaturation in the phosphate-depleted western North Atlantic Ocean. Limnol Oceanogr 65:2443–2459. https://doi.org/10.1002/lno.11463
Spring S, Riedel T (2013) Mixotrophic growth of bacteriochlorophyll *a*-containing members of the OM60/NOR5 clade of marine gammaproteobacteria is carbon-starvation independent and correlates with the type of carbon source and oxygen availability. BMC Microbiol 13:117. https://doi.org/10.1186/1471-2180-13-117
Steiner PA, Geijo J, Fadeev E, Obiol A et al (2020) Functional seasonality of free-living and particle-associated prokaryotic communities in the coastal Adriatic Sea. Front Microbiol 11: 2875
Sterner RW (2015) Ocean stoichiometry, global carbon, and climate. Proc Natl Acad Sci USA 112: 8162–8163. https://doi.org/10.1073/pnas.1510331112
Stevens H, Ulloa O (2008) Bacterial diversity in the oxygen minimum zone of the eastern tropical South Pacific. Environ Microbiol 10:1244–1259. https://doi.org/10.1111/j.1462-2920.2007.01539.x. pmid:18294206
Stewart LC, Algar CK, Fortunato CS, Larson BI et al (2019) Fluid geochemistry, local hydrology, and metabolic activity define methanogen community size and composition in deep-sea hydrothermal vents. ISME J 13:1711–1721
Stone BL, White AK (2012) Most probable number quantification of hypophosphite and phosphite oxidizing bacteria in natural aquatic and terrestrial environments. Arch Microbiol 194:223–228. https://doi.org/10.1007/s00203-011-0775-9
Stramma L, Johnson GC, Sprintall J, Mohrholz V (2008) Expanding oxygen-minimum zones in the tropical oceans. Science 320:655–658. https://doi.org/10.1126/science.1153847
Straume EO, Gaina C, Medvedev S, Hochmuth K et al (2019) GlobSed: updated total sediment thickness in the world's oceans. Geochem Geophys Geosyst 20:1756–1772. https://doi.org/10.1029/2018GC008115
Strizh I (2008) The Mehler reaction as an essential link between environmental stress and chloroplast redox signaling. In: Allen JF, Gantt E, Golbeck JH, Osmond B (eds) Photosynthesis. Energy from the Sun. Springer, Dordrecht, pp 1343–1346. https://doi.org/10.1007/978-1-4020-6709-9_289
Strous M, Fuerst JA, Kramer EH, Logemann S et al (1999a) Missing lithotroph identified as new planctomycete. Nature 400:446–449. https://doi.org/10.1038/22749
Strous M, Kuenen JG, Jetten MSM (1999b) Key physiology of anaerobic ammonium oxidation. Appl Environ Microbiol 65:3248–3250
Suleiman M, Zecher K, Yucel O, Jagmann N, Philipp B (2016) Interkingdom cross-feeding of ammonium from marine methylamine-degrading bacteria to the diatom *Phaeodactylum tricornutum*. Appl Environ Microbiol 82:7113–7122
Summers S, Henry T, Gutierrez T (2018) Agglomeration of nano- and microplastic particles in seawater by autochthonous and de novo-produced sources of exopolymeric substances. Mar Poll Bull 130:258–267
Sunagawa S, Coelho LP, Chaffron S, Kultima JR et al (2015) Structure and function of the global ocean microbiome. Science 348:1261359. https://doi.org/10.1126/science.1261359
Sutherland KM, Wankel SD, Hansel CM (2020) Dark biological superoxide production as a significant flux and sink of marine dissolved oxygen. Proc Natl Acad Sci USA 117:3433–3439. https://doi.org/10.1073/pnas.1912313117
Suttle CA (2007) Marine viruses—major players in the global ecosystem. Nat Rev Microbiol 5: 801–812. https://doi.org/10.1038/nrmicro1750
Swan BK, Martinez-Garcia M, Preston CM, Sczyrba A et al (2011) Potential for chemolithoautotrophy among ubiquitous bacteria lineages in the dark ocean. Science 333: 1296–1300
Thamdrup B, Dalsgaard T, Jensen MM, Ulloa O et al (2006) Anaerobic ammonium oxidation in the oxygen-deficient waters off northern Chile. Limnol Oceanogr 51:2145–2156. https://doi.org/10.4319/lo.2006.51.5.2145

Thiele S, Fuchs BM, Amann R, Iversen MH (2015) Colonization in the photic zone and subsequent changes during sinking determine bacterial community composition in marine snow. Appl Environ Microbiol 81:1463–1471

Thompson AW, Foster R, Krupke A, Carter BJ et al (2012) Unicellular cyanobacterium symbiotic with a single-celled eukaryotic alga. Science 337:1546–1550. https://doi.org/10.1126/science.1222700

Thomson B, Wenley J, Currie K, Hepburn C, Herndl GJ, Baltar F (2019) Resolving the paradox: continuous cell-free alkaline phosphatase activity despite high phosphate concentrations. Mar Chem 214:103671

Thornton DCO (2002) Diatom aggregation in the sea: mechanisms and ecological implications. Eur J Phycol 37:149–161

Thullner M, Dale AW, Regnier P (2009) Global-scale quantification of mineralization pathways in marine sediments: a reaction-transport modeling approach. Geochem Geophys Geosyst 10: Q10012

Toggweiler JR (1999) Variation of atmospheric $CO_2$ by ventilation of the ocean's deepest water. Paleooceanography 14:571–588

Trembath-Reichert E, Morono Y, Ijiri A, Hoshino T et al (2017) Methyl-compound use and slow growth characterize microbial life in 2-km-deep subseafloor coal and shale beds. Proc Natl Acad Sci USA 114:E9206–E9215

Treude T, Smith CR, Wenzhöfer F, Carney E, Bernardino AF et al (2009) Biogeochemistry of a deep-sea whale fall: sulfate reduction, sulfide efflux and methanogenesis. Mar Ecol Prog Ser 382:1–21

Trudnowska E, Lacour L, Ardyna M, Rogge A et al (2021) Marine snow morphology illuminates the evolution of phytoplankton blooms and determines their subsequent vertical export. Nat Commun 12:2816

Tyrrell T (1999) The relative influences of nitrogen and phosphorus on oceanic primary production. Nature 400:525–531. https://doi.org/10.1038/22941

Ulrich EC, Kamat SS, Hove-Jensen B, Zechel DL (2018) Methylphosphonic acid biosynthesis and catabolism in pelagic archaea and bacteria. In: Moore BS (ed) marine enzymes and specialized metabolisms—Part B, Methods in enzymology, vol 605. Elsevier, pp 351–426. https://doi.org/10.1016/bs.mie.2018.01.039

Valentine DL, Blanton DC, Reeburgh WS, Kastner M (2001) Water column methane oxidation adjacent to an area of active hydrate dissociation, Eel River basin. Geochim Cosmochim Acta 65:2633–2640

Valentine DL, Kessler JD, Redmond MC, Mendes SD et al (2010) Propane respiration jump-starts microbial response to a deep oil spill. Science 330:208–211

Vallina S, Follows M, Dutkiewicz S, Montoya JM et al (2014) Global relationship between phytoplankton diversity and productivity in the ocean. Nat Commun 5:4299. https://doi.org/10.1038/ncomms5299

Vallino JJ, Algar CK (2016) The thermodynamics of marine biogeochemical cycles: Lotka revisited. Annu Rev Mar Sci 8:333–356. https://doi.org/10.1146/annurev-marine-010814-015843

Van Mooy BAS, Krupke A, Dyhrman ST, Fredericks HF et al (2015) Major role of planktonic phosphonate reduction in the marine phosphorus redox cycle. Science 348:783–785. https://doi.org/10.1126/science.aaa8181

Verdugo P, Alldredge AL, Azam F, Kirchman DL et al (2004) The oceanic gel phase: a bridge in the DOM-POM continuum. Mar Chem 92:67–85

Vila-Costa M, Rinta-Kanto JM, Poretsky RS, Sun S et al (2014) Microbial controls on DMSP degradation and DMS formation in the Sargasso Sea. Biogeochemistry 120:295–305. https://doi.org/10.1007/s10533-014-9996-8

Wallmann K, Pinero E, Burwicz E, Haeckel M et al (2012) The global inventory of methane hydrate in marine sediments: a theoretical approach. Energies 5:2449–2498

Walsh DA, Zaikova E, Howes CG, Cong YC et al (2009) Metagenome of a versatile chemolithoautotroph from expanding oceanic dead zones. Science 326:578–582. https://doi.org/10.1126/science.1175309pmid:19900896

Wang Q, Alowaifeer A, Kerner P, Balasubramanian N, Patterson A, Christian W, Tarver A, Dore JE, Hatzenpichler R, Bothner B, McDermott TR (2021) Aerobic bacterial methane synthesis. Proceedings of the National Academy of Sciences U.S.A.,118 (27) e2019229118; DOI: 10.1073/pnas.2019229118

Wang WL, Moore JK, Martiny AC, Primeau FW (2019) Convergent estimates of marine nitrogen fixation. Nature 566:205–211. https://doi.org/10.1038/s41586-019-0911-2

Ward BB (1992) The subsurface methane maximum in the Southern California Bight. Continent Shelf Res 12:735–752

Ward BB (2011) Nitrification in the ocean. In: Ward BB, Arp DJ, Klotz MG (eds) Nitrification. Online ISBN: 9781683671169. https://doi.org/10.1128/9781555817145

Ward BB, Kilpatrick KA (1990) Relationship between substrate concentration and oxidation of ammonium and methane in a stratified water column. Continent Shelf Res 10:1193–1208

Ward BB, Kilpatrick KA (1993) Methane oxidation associated with mid-depth methane maxima in the Southern California Bight. Continent Shelf Res 13:1111–1122

Ward BB, Devol AJ, Rich JJ, Chang BX et al (2009) Denitrification as the dominant nitrogen loss process in the Arabian Sea. Nature 461:78–81. https://doi.org/10.1038/nature08276

Webb P (2020) Introduction to oceanography. Creative commons attribution 4.0 international license. https://rwu.pressbooks.pub/webboceanography/

Weber T, Deutsch C (2010) Ocean nutrient ratios governed by plankton biogeography. Nature 467: 550–554. https://doi.org/10.1038/nature09403

Welti N, Striebel M, Ulseth AJ, Cross WF et al (2017) Bridging food webs, ecosystem metabolism, and biogeochemistry using ecological stoichiometry theory. Front Microbiol 8:1298. https://doi.org/10.3389/fmicb.2017.01298

White AE, Spitz YH, Karl DM, Letelier RM (2006) Flexible elemental stoichiometry in *Trichodesmium* spp. and its ecological implications. Limnol Oceanogr 51:1777–1790. https://doi.org/10.4319/lo.2006.51.4.1777

Whitman WB, Coleman DC, Wiebe WJ (1998) Prokaryotes: the unseen majority. Proc Natl Acad Sci USA 95:6578–6583

Wilhelm SW, Suttle CA (1999) Viruses and nutrient cycles in the sea: viruses play critical roles in the structure and function of aquatic food webs. BioScience 49:781–788. https://doi.org/10.2307/1313569

Wörmer L, Hoshino T, Bowles MW, Viehweger B et al (2019) Microbial dormancy in the marine subsurface: global endospore abundance and response to burial. Sci Adv 5:eaav1024

Wortmann UG, Paytan A (2012) Rapid variability of seawater chemistry over the past 130 million years. Science 337:334–336. https://doi.org/10.1126/science.1220656

Wright JJ, Konwar KM, Hallam SJ (2012) Microbial ecology of expanding oxygen minimum zones. Nat Rev Microbiol 10:381–394

Wu C, Kan J, Liu H, Pujari L et al (2019) Heterotrophic bacteria dominate the diazotrophic community in the Eastern Indian Ocean during pre-southwest monsoon. Microb Ecol 78:804–819. https://doi.org/10.1007/s00248-019-01355-1

Wuchter C, Abbas B, Coolen MJ, Herfort L, van Bleijswijk J, Timmers P, Strous M, Teira E, Herndl GJ, Middelburg JJ, Schouten S, Sinninghe Damsté JS (2006) Archaeal nitrification in the ocean. Proc Natl Acad Sci U S A 103(33):12317–12322

Xiao R, Zheng Y (2016) Overview of microalgal extracellular polymeric substances (EPS) and their applications. Biotechnol Adv 34:1225–1244

Yan S, Fuchs BM, Lenk S, Harder J et al (2009) Biogeography and phylogeny of the NOR5/OM60 clade of *Gammaproteobacteria*. Syst Appl Microbiol 32:124–139. https://doi.org/10.1016/j.syapm.2008.12.001

Yang S, Chang BX, Warner MJ, Weber TS et al (2020) Global reconstruction reduces the uncertainty of oceanic nitrous oxide emissions and reveals a vigorous seasonal cycle. Proc Natl Acad Sci USA 117:11954–11960. https://doi.org/10.1073/pnas.1921914117

Zhang X, Stavn RH, Falster AU, Rick JJ et al (2017) Size distributions of coastal ocean suspended particulate inorganic matter: amorphous silica and clay minerals and their dynamics. Est Coast Shelf Sci 189:243–251

Zhang M, Zhangshu G, Wen S, Lu H et al (2018) Chemolithotrophic denitrification by nitrate-dependent anaerobic iron oxidizing (NAIO) process: insights into the evaluation of seeding sludge. Chem Eng J 345:345–352. https://doi.org/10.1016/j.cej.2018.03.156

Zhao Z, Baltar F, Herndl GJ (2020) Linking extracellular enzymes to phylogeny indicates a predominantly particle-associated lifestyle of deep-sea prokaryotes. Sci Adv 6:eaaz4354.

Zhuang G-C, Peña-Montenegro TD, Montgomery A, Joye SB (2018) Microbial metabolism of methanol and methylamine in the Gulf of Mexico: insight into marine carbon and nitrogen cycling. Environ Microbiol 20:4543–4554. https://doi.org/10.1111/1462-2920.14406

Zhuang G-C, Montgomery A, Samarkin VA, Song M, Liu J, Schubotz F, Teske AP, Hinrichs KU, Joye SB (2019a) Generation and utilization of volatile fatty acids and alcohols in hydrothermally altered sediments in the Guaymas Basin, Gulf of California. Geophys Res Lett 46 (5):2637–2646. https://doi.org/10.1029/2018GL081284

Zhuang G-C, Peña-Montenegro TD, Montgomery A, Joye SB (2019b) Significance of acetate as microbial carbon and energy source in the water column of Gulf of Mexico: implications for marine carbon cycling. Global Biogeochem Cycles 33:223–235. https://doi.org/10.1029/2018GB006129

Ziervogel K, Arnosti C (2008) Polysaccharide hydrolysis in aggregates and free enzyme activity in aggregate-free seawater from the north-eastern Gulf of Mexico. Environ Microbiol 10:289–299

Ziervogel K, Karlsson E, Arnosti C (2007) Surface associations of enzymes and of organic matter: consequences for hydrolytic activity and organic matter remineralization in marine systems. Mar Chem 104:241–252

Ziervogel K, Steen AD, Arnosti C (2011) Changes in the spectrum and rates of extracellular enzyme activities in seawater following aggregate formation. Biogeosciences 7:1007–1015

Ziervogel K, Dike C, Asper V, Montoya J, Battles J et al (2016) Enhanced particle fluxes and heterotrophic bacterial activities in Gulf of Mexico bottom waters following storm-induced sediment resuspension. Deep Sea Res Pt II Top Stud Oceanogr 129:77–88

# A Holistic Approach for Understanding the Role of Microorganisms in Marine Ecosystems

16

Gerard Muyzer and Mariana Silvia Cretoiu

**Abstract**

Marine ecosystems are among the largest aquatic ecosystems on Earth. They harbor a wealth of biodiversity, provide essential ecosystem services, and are a rich source of bioproducts. Within these marine ecosystems, microorganisms play a crucial role in recycling chemical elements such as carbon, nitrogen, and sulfur. It has become increasingly recognized that microorganisms interact with marine macroorganisms, such as fish, corals, sponges, and seagrasses. Therefore, in order to obtain a comprehensive understanding of the role of microorganisms in marine ecosystems, a holistic scientific research approach is needed. This chapter will provide examples of studies that used systems biology to study the role of microorganisms in the cycling of chemical elements and nutrients.

**Keywords**

Bio-applications · Fluxes · Marine ecosystems · Metabolic interactions · Multi-omics toolbox · Ocean restoration

G. Muyzer (✉)
Microbial Systems Ecology, Department of Freshwater and Marine Ecology, Institute for Biodiversity and Ecosystem Dynamics, University of Amsterdam, Amsterdam, The Netherlands
e-mail: g.muijzer@uva.nl

M. S. Cretoiu
University of Utrecht, Utrecht, The Netherlands



L. J. Stal, M. S. Cretoiu (eds.), *The Marine Microbiome*, The Microbiomes of Humans, Animals, Plants, and the Environment 3,
https://doi.org/10.1007/978-3-030-90383-1_16

## 16.1 Introduction

Marine ecosystems, such as the open ocean, estuaries, coral reefs, mangrove forests, and seagrass meadows harbor an enormous diversity of organisms. They provide essential ecosystem services such as fishery, recreation, transportation, and carbon sequestration (Barbier 2017). Microorganisms, i.e., archaea, bacteria, protists, and biological entities such as viruses, play an essential role in the recycling of carbon and nutrients in marine ecosystems (Falkowski et al. 2008). These microorganisms are involved in a variety of different ecological processes, such as among others the "microbial loop," in which dissolved organic matter (DOM) is processed and consumed and eventually returned to the marine food web where it becomes available to higher trophic levels (Azam et al. 1983). An associated process is the so-called "viral shunt" by which microbial cells are infected and lysed by viruses, thereby releasing organic compounds and nutrients into the environment that might stimulate microbial growth (Wilhelm and Suttle 1999). In addition, the viral shunt stimulates the "biological pump" that transports organic matter to the seafloor, where its further diagenesis follows geological timescales. Although much research has been done on the degradation of organic matter, a comprehensive understanding of the role of microorganisms in the conversion of chemical elements and the recycling of nutrients is still missing. A holistic or systems biology approach is required in which the knowledge and technologies from different scientific disciplines such as biology, geology, physics, mathematics, and computer science are combined (Karsenti et al. 2011). The so-called "top-down" systems biology approach, in which innovative "state-of-the-art" omic techniques are used, will generate the data that are required to create the conceptual and mathematical models that will generate the important questions and hypotheses that can be answered in subsequent experiments (Shahzad and Loor 2012). Moreover, the mathematical models will help to predict the effects of changing environmental conditions such as climate change on the activities and proliferation of microorganisms and the marine ecosystems they thrive in (Raes and Bork 2008). Repetitive rounds of experiments and theoretical modeling will lead to a comprehensive understanding of the role of microorganisms in marine ecosystems (Fig. 16.1).

## 16.2 Multi-Omics as a Toolbox to Study Diversity and Function of Microbial Communities

### 16.2.1 Marine Microbiome Analysis Using rRNA Gene Amplicon Sequencing

Sequencing of hypervariable regions of the rRNA genes contributed greatly to the knowledge of the composition of the marine microbiome and to a certain extent, also of the ecological function of its component microorganisms. Analysis of one or multiple 16S or 18S rRNA gene hypervariable regions became the preferred method for many marine microbiologists. Different primer pairs, next generation sequencing

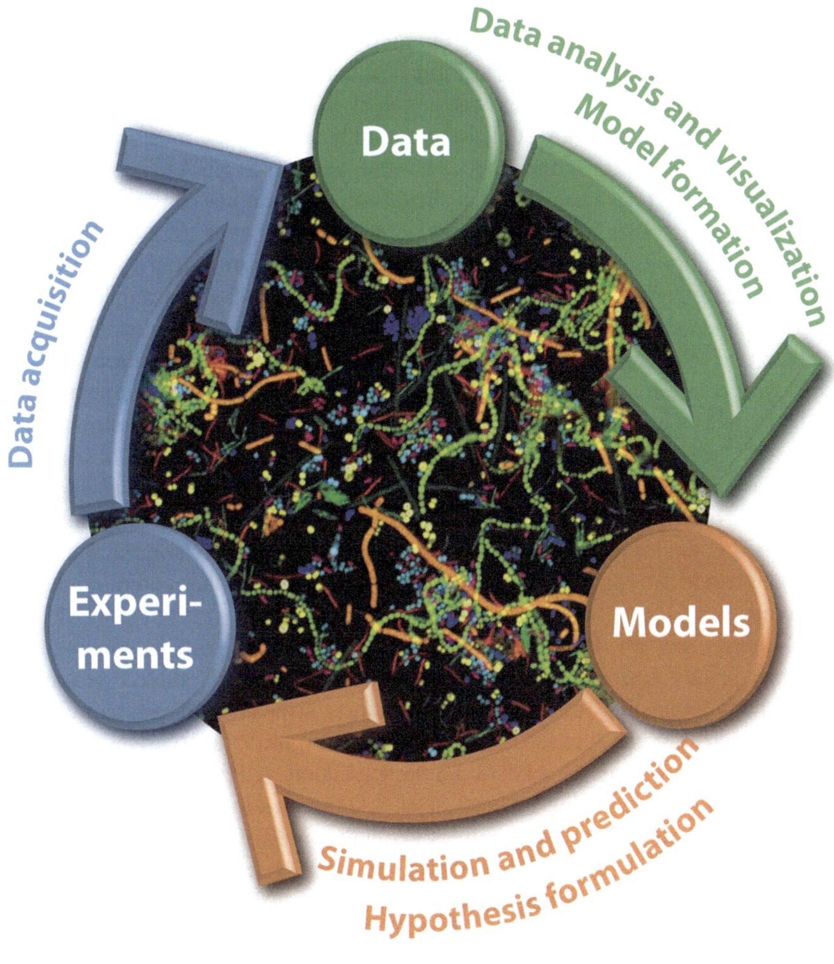

Fig. 16.1 Different steps in a holistic or systems biology approach to study microbial communities. Experiments including meta-omic analyses will generate data that can be used for the formation of conceptual and mathematical models. Subsequently, these models will generate novel questions and hypotheses that can be answered or tested in new experiments. Repetitive rounds of experimental and theoretical activities will lead to a comprehensive understanding of the role of microorganisms in the various processes in marine ecosystems. The micrograph of the microbial community was kindly provided by Alex Valm (University at Albany, SUNY, USA). The photo shows the enormous diversity of microorganisms that can be present in microbial communities

platforms, and data analysis protocols were used in order to accommodate the specific features of the investigated marine samples. A literature search with "16S rRNA gene methods" and "marine samples" as search terms returns more than 200 hits of scientific articles published in 2020. High-throughput sequencing of

16S rRNA genes offered a first snapshot on the composition and diversity of previously underexplored marine microbiomes. Deep-sea microbial communities (Queiroz et al. 2020), symbiotic associations in marine sponges (Waterworth et al. 2020), novel branches of marine archaea (Zou et al. 2020), microbial colonizers of marine plastic debris (Davidov et al. 2020) are only a few of the most recent topics. Analysis of copy numbers and sequence variation of the 18S rRNA gene and plastid-derived conserved genes (also known as metabarcoding) improved the characterization of marine microbial eukaryotes. Fungi-specific primer pairs were designed, and novel marine fungi were presented as promising sources of bioactive agents (Banos et al. 2018; Ettinger et al. 2021). The abundance of nematodes was monitored, and migration patterns between aquatic (marine and freshwater) and terrestrial habitats were recorded (Holterman et al. 2019). Ribosomal and cytochrome oxidase subunit I gene fragments (declared as molecular operational taxonomic units) extracted from total DNA of the water column (some researchers prefer the term "environmental DNA" or "e-DNA") were tested as markers for monitoring of benthic microeukaryotes (Antich et al. 2020). Although these authors identified a variety of organisms belonging to the Archaeplastida and Stramenopiles and they confirmed them as being similar to those of benthic origin, the study concluded that a direct sampling method would have a higher precision when complex communities are considered.

As emphasized by Eloe-Fadrosh et al. (2016) amplicon sequencing has limitations. After comparing 6000 MAGs (metagenome-assembled genomes), these authors found that mismatches in the target sequences of commonly used primer set such as, for instance, 515F-806R failed to amplify the 16S rRNA genes of several members of the Candidate Phyla Radiation (CPR) nor of the so-far uncharacterized archaea. Eloe-Fadrosh et al. (2016) therefore, argued for the use of metagenomics and metatranscriptomics to obtain a better description of microbial diversity. Figure 16.2 shows an overview of the different approaches that are currently used to study microbial communities.

### 16.2.2 Metagenomics

The next step in the characterization of microbial communities is the metagenomics approach, whereby all genomes of the organisms in the community are sequenced. Besides revealing information on the identity of the organisms in the community, metagenomics also gives insight into their metabolic potential. Although metagenomic sequencing is more expensive and the bioinformatic analysis of the sequence data more computer intensive than for 16S rRNA gene (fragment) amplicon sequencing it is currently an established and popular instrument in the molecular toolkit of microbial ecologists (Quince et al. 2017). Many scientists have used metagenomics to study microbial communities in a variety of ecosystems. Nayfach et al. (2021) published a genomic catalog of the Earth's microbiomes consisting of 52,515 MAGs reconstructed from metagenomes of microbial communities from a variety of different habitats, including those that are host-

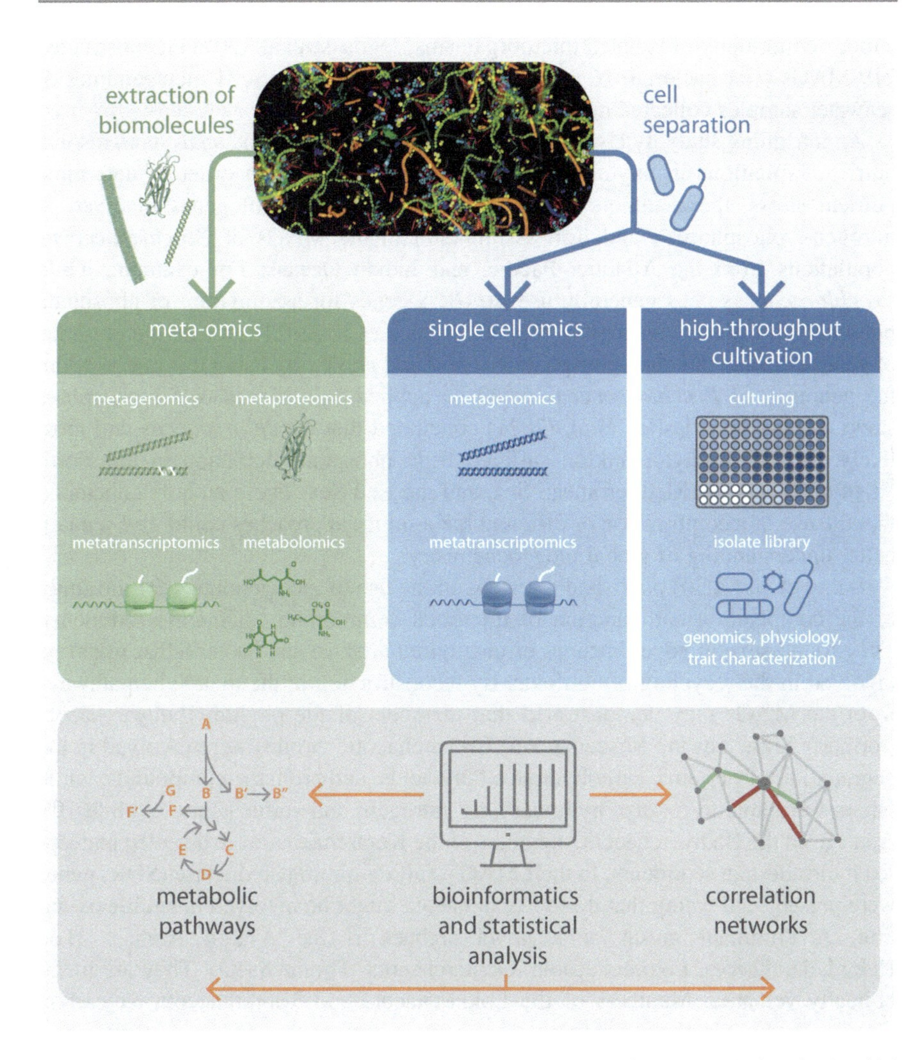


**Fig. 16.2** Schematic overview of the "toolbox" that is used for the analysis of microbial communities. The micrograph shows an example of a microbial community. This community can be studied with a variety of meta-omics techniques. Moreover, individual cells from the community can be physically isolated (e.g., by flow cytometric cell sorting) and further analyzed by single-cell omics or by high-throughput cultivation. The results can be used for the construction of correlation networks and/or for the reconstruction of the metabolic pathway

associated, aquatic, engineered, terrestrial, and atmospheric. This enormous number of MAGs represented 12,556 novel operational taxonomic units from 135 phyla.

Duarte et al. (2020) used metagenomics to discover novel genes in Red Sea plankton and did not find evidence of saturation. They concluded that there are still many novel genes to be discovered in the marine environment and that this can only be achieved by massive sequencing. By using a co-assembly approach supported by a mock community of isolated microorganisms, Jégousse et al. (2021) reconstructed 219 MAGs (191 bacterial, 26 archaeal and 2 eukaryotic) from 31 metagenomes of seawater samples collected near Iceland.

An intriguing study by Ustick et al. (2021) used metagenomic analysis to resolve nutrient limitations in phytoplankton from several oceans. In order to determine nutrient stress, these authors quantified the gain and loss of genes involved in nitrogen-, phosphorus-, and iron assimilation in the MAGs of *Prochlorococcus* populations from the Atlantic, Pacific, and Indian Oceans. For example, while *Prochlorococcus* cells generally use *pstABCS* genes for assimilation of phosphate but when phosphate was heavily depleted, Ustick et al. (2021) noticed the presence of genes encoding alkaline phosphatase (*phoA* and *phoX*). By using this information, the genomes of *Prochlorococcus* could be applied as a biosensor for phosphate stress. In this way, Ustick et al. (2021) concluded that *Prochlorococcus* and most likely also other phytoplankton suffered from phosphate depletion in the North Atlantic Ocean, the Mediterranean Sea, and the Red Sea. These authors concluded that the use of a combination of different meta-omics approaches could give a much better understanding of global biogeochemistry.

Baker et al. (2020) published a review on the use of metagenomics for the study of the composition and function of microbial communities in marine sediments. They found several novel lineages of uncultured archaea and bacteria that might be involved in the recycling of nutrients. By reconstruction of the metabolic pathways from the MAGs they demonstrated that members of the phylum Bathyarchaeota (formerly known as the Miscellaneous Crenarchaeotic Group) were involved in the degradation of proteins, carbohydrates, fatty acids, and aromatic compounds, while others were able to oxidize hydrocarbons, nitrogen, and sulfur (Zhou et al. 2019). Apart from the Bathyarchaeota, members of the Korarchaeota were detected in deep-sea hydrothermal sediments. In their MAGs, sulfide-quinone reductase (SQR) genes were present, indicating that these Korarchaeota might be involved in sulfide oxidation. A dominant group of a novel archaea is the Asgard Archaea (i.e., Heimdallarchaeota, Lokiarchaeaota, Odinarchaeota, Thorarchaeta). They are metabolically versatile. Members of the Lokiarchaeota and Thorarchaeota fix carbon dioxide, although they can also use organic substrates and can even grow syntrophically like the first isolate of the phylum Lokiarchaeota *Prometheoarchaeum syntrophicum* that was enriched in a coculture with a methanogen (Imachi et al. 2020). Members of the Heimdallarchaeota, the closest relatives of the eukaryotes, are growing heterotrophically via fermentation. Metagenomic analysis has also revealed novel marine lineages within the Euryarchaeota, such as the Hadesarchaea (Baker et al. 2016). Reconstruction of the metabolic pathways of these organisms indicated an adaptation to a nutrient-limited environment.

Apart from novel archaea, the metagenomic analysis also recovered many novel groups of bacteria present in marine sediment. For instance, Dombrowski et al. (2018) were able to reconstruct 304 bacterial MAGs from hydrothermal sediments of the Guaymas Basin, including three new phyla (GB-BP1, GB-BP2, and GB-BP3).

By using metagenomics, Kerou et al. (2021) reconstructed 11 MAGs from deep-sea sediments of the Atlantic- and Pacific Ocean that formed three novel lineages of ammonia-oxidizing archaea (AOA) within the order *Nitrosopumilales* (phylum Thaumarchaeota). Apart from known genes involved in ammonia oxidation and autotrophic carbon fixation, Kerou et al. (2021) also found genes involved in fermentation, amino acid uptake and in DNA repair indicating special adaptations to the harsh conditions of energy limitation and high pressure in the deep sea.

### 16.2.3 Metatranscriptomics and Metaproteomics

While metagenomics is well established in microbial ecology, this is not the case with metatranscriptomics and metaproteomics, which are still in their infancy. Some studies using metatranscriptomics and metaproteomics were reviewed by Muyzer (2016), and Shakya et al. (2019) summarized the advances and challenges in metatranscriptomics. The latter authors stated that even though reference genomes might be lacking, metagenome-assembled genomes (MAGs) from the same sample are necessary. Furthermore, Shakya et al. (2019) advocated that the submission of these metagenomic sequences to the public repositories must be accompanied by a detailed description of the metadata of the sampled environment in order to understand the context of the dataset and its interpretation. Metadata file containing "Minimum Information about any (x) Sequence (MIxS)" established by Genomics Standard Consortium (Yilmaz et al. 2011) is required by almost any scientific journal. As an excellent example, Salazar et al. (2019) published a study in which they used both metagenomics and metatranscriptomics to determine genes and transcripts of ocean microbial communities collected from 126 sampling stations around the world. They demonstrated that community turnover as a response to global warming was the strongest in the polar regions.

Saito et al. (2019) wrote a review on the progress and challenges in ocean metaproteomics. The authors gave a variety of examples of successful studies that used metaproteomics to characterize microbial communities in the marine environment. Saito et al. (2019) also gave insight into the challenges that scientists face in each of the steps of the metaproteomics approach they carry out, i.e., from the collection and analysis of the samples to the interpretation and sharing of the data. The main challenges are expected in the enormous diversity of (often closely related) microorganisms and the absence of (annotated) sequences in the databases. Saito et al. (2019) offer several suggestions in order to meet these challenges. The combination of high-resolution methods for peptide identification and the decrease of false discovery rate must be prioritized. Computational tools and data collecting approaches must be improved and should be synchronized across different research programs in order to maximize the taxonomic interpretation of peptide identification. It is necessary to implement the usage of pre-defined metadata files containing information regarding sampling expedition, contextual measurements (geographical location, physicochemical parameters) and sample acquisition. Providing details about software or algorithm used for taxonomy and functional analysis will prevent incorrect assignment of peptides from microorganisms that are not or only partly characterized. This proteomics approach is also applied to the orderly investigation of marine microorganisms that cause harmful algal blooms. Many shellfish diseases are caused by algal toxins and are monitored by measuring the most recurrent and abundant toxins derived from blooms of cyanobacteria and dinoflagellates. Accumulation of dinoflagellates belonging to genera *Dinophysis* and *Prorocentrum* in shellfish in the North Atlantic Ocean has caused diarrhetic shellfish poisoning in humans (Campos et al. 2020; Swan et al. 2018). These toxins form a broad group of polyketides with lipophilic properties and accumulate in adipose tissues, thus posing concern to food and public health safety (Davidson and Bresnan 2009). Using omics, it was confirmed that the mechanism of action of diarrhetic shellfish toxins involves the inhibition of protein phosphatases generating deregulation of many intracellular processes (Yadav et al. 2017). In the context of marine biotoxins and seafood safety, the use of proteomics based on high-throughput and high-resolution technologies (e.g., LC-MS/MS coupled with isotope-coded affinity tags, NMR) is considered essential for the field of ecotoxicology.

### 16.2.4 Single-Cell Omics

The metagenomic approach is undoubtedly powerful, but it also has its biases and limitations. These include, for instance, the difficulty in getting the full genome of rare microorganisms as well as the presence of chimera sequences from closely related populations (Sczyrba et al. 2017). A new and intriguing development in the analysis of microbial communities is the analysis of single cells using a variety of omics techniques (Kaster and Sobol 2020). Isolation and omics of single (individual) cells emerged in the last 10 years as among the most wanted methods to be applied to any molecular environmental survey of microorganisms. Many of the revolutionary instruments such as Fluorescence-Activated Cell Sorting (FACS) and digital high-resolution microscopes required for the isolation of individual cells from water, sediment, or soil were developed and improved during numerous projects located at marine and oceanographic research institutes (Olson et al. 1993; Sieracki et al. 1998; Yentsch and Yentsch 2008). Coupled with culturing and metagenomics, single-cell methods started to provide a better sizeable snapshot of the phylogenetic and functional diversity of the three domains of life (Fig. 16.2). It opened a new avenue for the deep analysis of the marine microbiome and will improve the prediction of the loss of microbial taxa, which goes unnoticed today. This may lead to the development of innovative solutions for ocean restoration (Levy et al. 2021). Several technologies exist for the isolation of single microbial cells from the marine environment. Fluorescence-activated cell sorting (FACS) has been used since 1970s, and its application in environmental microbiology was recognized early (Kamentsky 1973). Yet, it took until 2007 when the microfluidic techniques were coupled with the separation of microbial cells (Kvist et al. 2007). Cells could be separated, lysed, and nucleic acids extracted and amplified in situ. Maximization of the yield of DNA amplification allowed further progress of high-throughput sequencing (HTS) in single-cell omics.

A single-cell omics analysis involves two major steps: (1) isolation of cells in a multi-well plate and (2) amplification of their genomes. Using FACS and a water sample treated with a nucleic acid fluorescent stain, any number of cells belonging to a pre-declared (pre-defined) group based on cell size or nucleic acid content can be isolated simultaneously. After cell lysis, the genome is amplified by a Multiple Displacement Amplification (MDA) reactions carried out by a high-fidelity DNA polymerase (e.g., phage phi 29 DNA polymerase). The single amplified genomes can be directly sequenced and treated as representatives of the "whole genome" (whole-genome sequencing or WGS) and/or screened using specific marker genes (e.g., 18S rRNA gene for eukaryotes and 16S rRNA gene for bacteria and archaea) to resolve their taxonomical affiliation (Landry et al. 2017; Stepanauskas 2012; Woyke et al. 2017). Once a single-cell genome is accepted as curated according to the standards required by the "minimum information about a single amplified genome (SAG) of bacteria and archaea" (MISAG), it can be placed into a larger evolutionary and ecological context (Bowers et al. 2017). Phylogenetically informative genes such as those in the rRNA operon or conserved protein marker genes (such as those involved in the transport and metabolism of carbohydrates) have been used to assess the level of relationship of microbial lineages from which these genes were recovered. SAGs also proved to be of great use as provisory reference genomes for many of the yet uncultured microbes. Very often, metagenomic reads, assembled fragments, and metagenome-assembled genomes (MAGs) from marine samples are aligned to SAGs. This recruitment process is used for quantifying abundance across physicochemical gradients and for the reconstruction of metabolic pathways. Hence, SAGs are particularly powerful in deciphering the molecular organization of the marine microbiome.

One of the first studies that revealed the power of single-cell genomics for the analysis of in situ interactions from the marine environment was performed by Yoon et al. (2011). Individual cells isolated from a 50 ml coastal seawater sample served as material for DNA isolation for whole-genome sequencing. The cells were identified by fluorescent ultramicroscopy as the uncultured marine group Picobiliphyta and therefore difficult to place them in a phylogenetic tree of marine protists. WGS and analysis or the rRNA genes indicated that the cells represented three divergent clades. None of these clades showed genomic evidence of plastid DNA or nuclear-encoded plastid-targeted proteins. Yoon et al. (2011) suggested that the newly discovered picobiliphytes are heterotrophs. Most importantly, their study provided an optimized protocol to generate substantial amounts of genome data from single cells, which supported the conclusion that picobiliphytes lack light-harvesting proteins. Staining the food vacuoles in these picobiliphytes prior to cell sorting and the subsequent analysis of 16S- and 18S rRNA genes present in SAGs of the sorted cells resulted in the identification of micro-eukaryotic hosts as Chrysophytes, Apicomplexa, and Basidiomycota containing symbiotic Bacteroidetes and Actinobacteria (Martinez-Garcia et al. 2012). This study proved the physical association and co-existence of the bacterivorous flagellate MAST-4 and one of the most abundant bacterial taxa in the ocean—SAR11.

SAGs have shown their value in the analysis of genetic heterogeneity and recombination frequencies within discrete populations *Prochlorococcus* and SAR11. The high abundance of these groups of marine microorganisms in the surface waters and their extensive contribution to marine productivity was confirmed by numerous studies (Haro-Moreno et al. 2020; Kashtan et al. 2017; Thompson et al. 2019). *Prochlorococcus* cells recovered from water samples taken at ALOHA and at the Bermuda Atlantic Time Series (BATS) stations were analyzed at the level of SAGs (Kashtan et al. 2017). Comparative genomics analysis of co-occurring SAGs indicated that the two ocean sites were inhabited by non-overlapping and distinct subpopulations of *Prochlorococcus*. The Atlantic Ocean *Prochlorococcus* populations were composed of hundreds of subpopulations with distinct genomes. The Pacific Ocean *Prochlorococcus* subpopulations were more diverse than those of the Atlantic Ocean, but none of them were dominant. The genomic differences were attributed to the spatial separation at the level of micro-habitats. SAGs originated in co-occurring bacterial cells from the ALOHA and BATS stations allowed the differentiation of SAR11 ecotypes. Thrash et al. (2014) used 11 SAGs and a deep metagenomics approach to investigate the spatiotemporal distribution of SAR11 subclade 11 from surface waters to 250 m depth. Recruitment of metagenomic sequences against SAGs with the high level of genome completeness (up to 85%) allowed the recovery of many genes important for metabolic pathways essential for surviving in the deep ocean as well as provided evidence for their genomic context (e.g., sulfite oxidase genes located adjacent to cytochrome). A large number of metagenomic sequences showed high similarity to hypervariable genome regions and to regions involved in the defense against viral infection. Exploration of single cells offered a new perspective about the SAR11 clade orthologs. While presenting subtle genomic differences such as large intergenic regions and preferential amino acid substitutions, deep ocean SAR11 has a metabolism similar to those living in surface waters.

The incorporation of single-cell-based methods in molecular environmental surveys has greatly expanded the knowledge of microbial diversity by the discovery of new bacterial and archaeal taxa. Molecular environmental surveys using SAGs and their 16S rRNA genes increased the quality of many Candidatus groups as well (Eloe-Fadrosh et al. 2016; Eloe-Fadrosh 2019). Taxa lineages considered as "rare" or "unknown" were identified, characterized, and their metabolic features linked to the nutrient cycles in their habitat. Among these lineages were those of the uncultured Gammaproteobacteria from marine sediments. A study on $^{14}$C-bicarbonate assimilation in coastal sediments across Europe and Australia showed that these uncultured Gammaproteobacteria were responsible for 80% of the total dark carbon fixation in these sediments (Dyksma et al. 2016). Isolation of individual cells and omics analysis confirmed the hypothesis that chemolithoautotrophy in marine surface sediments is driven by sulfur oxidation and offered new evidence of genes and metabolic pathways involved in dark carbon fixation. Novel marine archaea belonging to the Thaumarchaeota were described using MAGs and SAGs, and the mechanisms for genetic variation were inferred (Aylward and Santoro 2020). Similarly, Pachiadaki et al. (2019) used SAGs to study the microbial diversity in 28 water samples from the Atlantic and Pacific Oceans. Using a randomized cell selection strategy, Pachiadaki et al. (2019) obtained more than 12,715 good-quality SAGs from planktonic bacteria and archaea and concluded that each of these SAGs was unique. Furthermore, Pachiadaki et al. (2019) detected a dominance of genes involved in light harvesting, $CO_2$ fixation, and the synthesis of secondary metabolites. They reported a large number of genes involved in energy metabolism and secondary metabolites production. This impressive number of single-cell genomes was termed Global Ocean Reference Genomes Tropics (GORG-Tropics), and it was shown that it outperformed the existing marine bacterial and archaeal reference genomes. To improve the utility of their dataset, Pachiadaki et al. (2019) developed a new computational pipeline (GORG Classifier) that facilitates the interpretation of meta-omics data using SAGs as a reference.

Single-cell methods also proved useful for the advancement of research on marine viruses. Linking viruses to their host cells in marine surface bacterioplankton was possible through SAGs (Labonté et al. 2015; Martinez-Hernandez et al. 2017). Virus types and stages of infections of the first known viruses of Thaumarchaeota, Marinimicrobia, Verrucomicrobia, and the Gammaproteobacterial clusters SAR86 and SAR92 were reported. Novel types of marine phages from oxygen minimum zones were first reported using SAGs (Roux et al. 2014). A wide diversity of novel nucleocytoplasmic large DNA viruses (NCLDVs) of marine protists were discovered using single-cell and single virus approaches (Martínez Martínez et al. 2020; Wilson et al. 2017). Fragment recruitment of viral metagenomic reads coupled with observations based on SAGs and single virus genomes (SVGs) are valuable tools for studying virus-cell interactions in complex microbial communities. Reference databases are now populated with single virus genomes (SVGs) from marine samples, and dedicated sequencing programs were made accessible to the scientific community (Paez-Espino et al. 2016, 2017). Isolation of single virus particles and their hosts increased the understanding of virus-driven molecular interactions in marine ecosystems.

There is no doubt that single-cell methodology has led to a high number (thousands) of individual genomes for uncultured microbes and viruses. The success of this method is highly depending on stringent wet-lab conditions and high-performance computers for data analysis. In the last 10 years, specific algorithms and pipelines have been developed to address the limitations of assembling genomes of single cells, thereby decreasing the number of artifacts and improving the construction of draft genomes.

### 16.2.5 Single-Cell Transcriptomics

Single-cell methods are often used to supplement microbiome studies at the population level, microscopic analysis, or remote sensing of primary production. Great progress in understanding life strategies and core metabolism of microbial inhabitants of various marine ecosystems has been achieved by using DNA-based molecular methods. Traditionally, studies were conducted under the assumption that individual cells have a metabolic status similar to that observed at the population level. However, whole-genome sequencing of single cells and of bulk populations have shown that individual behavior can be different from that of the population as a whole. Eventually, gene-expression heterogeneity among individual cells determines the fate of the whole microbial system. In the last decade, great progress in mammalian research has stimulated the usage of single-cell RNA-sequencing for the analysis of cell-to-cell heterogeneity in microbial communities. Microbial systems researchers faced some major challenges. First, transcriptomic profiling of a cell requires an average of 10 pg of total RNA. Compared to eukaryotic cells, this amount is not easily obtained from microorganisms. Second, eukaryotic single-cell transcriptomics employed oligo dT as primers for cDNA synthesis. However, oligo dT primers cannot be used directly to bacterial transcriptomics because bacterial RNA does not have these non-unique poly(A) structures. In addition to the special requirements for sample preservation prior to RNA extraction, these factors restricted the use of RNA-seq to only a few microbial taxa. Although not a marine species, the cyanobacterium *Synechocystis* sp. PCC 6803 has been used to combine single-cell transcriptomics with RNA-seq (Wang et al. 2015). Single cells of *Synechocystis* cultured under standard growth conditions were compared to those grown under nitrogen starvation. The cells were isolated using a single-cell manipulator coupled to an inverted microscope. RNA transcripts were amplified using single primer isothermal amplification. The results revealed a considerable and even increasing heterogeneity among isogenic (single strain) populations under stress. Wang et al. (2015) claimed the development of an integrated workflow that could be applied to any other taxa as well.

Aiming at method improvement and extension, Liu et al. (2017) used two marine microbial eukaryotes in an RNA-based comparative analysis. RNA isolated from batch cultures and single cells of the dinoflagellate *Karlodinium veneficum* and the haptophyte *Prymnesium parvum* was sequenced and assembled. For both taxa, the number of transcripts recovered from batch cultures was higher than those from single cells. Many known housekeeping genes were not recovered from single cells. Among transcripts detected in multiple cells, their differential abundance varied between different cells. Only the transcripts with high abundance levels in cultured-based transcriptomes were detected in all single cells. The major metabolic pathways and physiological functions were assembled from both types of transcriptomes, cultured-based, and single-cell. These observations suggest that single-cell transcriptomes reflect the level of high stochasticity of individual genes rather than the physiological differences among cells.

Single-cell transcriptomics was also used for the reconstruction of the origin and evolutionary interrelationship of the marine heterotrophic flagellate *Abedinium* (Cooney et al. 2020). Transcriptome sequencing of manually isolated *Abedinium*-like cells revealed distinct cellular types. Extensive protein-based phylogenetic analysis indicated *Abedinium* as a distinct lineage among dinoflagellates. Evidence of specific organellar features showed that RNA-seq improved the characterization of rare marine microorganisms. Integration of RNA-seq (next to other single-cell methods) to the geographic survey of rare or intractable microeukaryotes offers a high level of detail that may aid to unravel the evolution of many marine species.

## 16.3 Integration of Omics and Culturing

Although it is not an easy task to isolate bacteria from nature, generally referred to as the "Great Plate Count Anomaly" (Staley and Konopka 1985), it is certainly not impossible. Lewis et al. (2021) discussed the possible reasons for the failure to isolate microorganisms. These include the lack of knowledge of the substrates and growth conditions, symbiotic interdependencies, or dormancy of the target organisms (Zhang et al. 2021). They described innovative methods that improved the success rate of the isolation and culturing of microorganisms, such as membrane diffusion-based, cell sorting-based and microfluidics-based culturing. Droplet microfluidics was used by Hu et al. (2021) as a high-throughput single-cell method for the isolation and culturing of marine microorganisms. Liu et al. (2021) used FACS-iChip, which is a combination of cell sorting-based and membrane diffusion-based culturing and was deployed for the isolation of microorganisms. To briefly explain this approach, cells were labeled with fluorescent dyes, separated by flow cytometric FACS, and the individual separated cells were incubated in the wells of a 384-well microtiter plate that was sealed with two membranes (the "iChip"). The iChip was incubated in situ for several days, after which bacteria were isolated in pure culture. Henson et al. (2020) used large-scale dilution-to-extinction (DTE) culture and managed to obtain 328 isolates, which represented 5% of the bacterioplankton community. The first member of the Lokiarchaeaota, *Candidatus* Prometheoarchaeum synthrophicum, was isolated as a coculture with the methanogen *Methanogenium* (Imachi et al. 2020). From these reports, it is clear that innovative methods enhance the success rate of isolation of microorganisms, particularly from the marine environment. However, in many cases, efforts to isolate and culture hitherto unknown microorganisms can still be tedious and often require a lot of patience. For instance, it took more than 12 years to enrichment *Candidatus* Prometheoarchaeum synthrophicum.

Isolation of members of the microbial community provides an opportunity to perform a so-called "bottom-up" systems biology approach in which synthetic communities can be designed using mixed cultures of well-characterized microorganisms (Grosskopf and Soyer 2014). Well-defined synthetic communities of various levels of complexity are ideal to test hypotheses resulting from "top-down" research under perfectly controlled laboratory conditions. Advantages of these synthetic microbial communities are that any level of complexity can be introduced, the dynamics and gene expression of the individual members can be followed and compared to pure cultures or to other synthetic community compositions or to different external conditions, and modeling can be used to choose between the many options of synthetic community composition and external conditions, which will enhance to outcomes. All these approaches are far more difficult in natural communities because they are so much more complex than synthetic communities, and the external forcing cannot be easily controlled. Moreover, synthetic microbial communities can be used to monitor their behavior after perturbations, which is difficult or even impossible to study in natural communities.

## 16.4 Marine Microbial Ecosystems Beyond Genes and Genomes

Microbiomes are essential for ocean biogeochemical processes. Thousands of single-cell- and metagenome-assembled genomes of bacteria, archaea, eukarya, and their associated viruses have been reconstructed from an unprecedented amount of sequence data generated by global ocean expeditions (e.g., Tara Oceans, Malaspina, Oceana, Global Reef) and other more local studies. These genomes, together with those of reference (cultured) representatives, are essential for understanding the interactions between marine organisms and for an intelligent exploration of marine resources. All positive (e.g., cross-feeding, biofilm formation) and negative (e.g., competition for nutrients) interactions occur at the metabolic level. Some interactions and processes (such as a fast depletion of nutrients or the production of antimicrobial compounds) have a bigger impact on local ecosystem functioning, and therefore identification of critical or key species is important (Decho and Gutierrez 2017). Genome-scale metabolic networks have been proposed that can bridge the gap between environmental records of system variability that do not have empirical confirmation. A mechanistic phytoplankton cell model was used to test if physiological traits inferred from metatranscriptome data could help to identify physiological traits that provide adaptation to environmental conditions. For example, using a phytoplankton genome-derived cell model for simulation of cellular concentration of ribosomes and their RNAs it was shown how growth and cellular allocation strategies of nitrogen and phosphorous can be integrated into simulations involving patterns of light, temperature, and nutrients in the surface of the marine environment (Mock et al. 2017). Metabolic modeling of microbial marine species is still in its embryonic phase but begins to show promising results. Actively growing and steady-state equilibrium populations of the marine filamentous diazotrophic (nitrogen-fixing) cyanobacterium *Trichodesmium erythraeum* was assessed with the help of genome-scale metabolic networks (Gardner and Boyle 2017). Their model predicts an optimal growth rate in accordance with observed measurements: nitrogen-fixing cells conserve carbon while carbon-fixing cells conserve energy. Simulations under unconstrained carbon-nitrogen uptake indicate that non-nitrogen-fixing photoautotrophic cells have a lower optimal growth than those that grow diazotrophically. It was also shown that cell differentiation is driven by metabolite fluxes and that colony shape and diffusion characteristics can limit the cell's ability to grow optimally. While aiming at providing a system-level understanding of marine metabolism, models of marine microbes are mostly constructed for the development of application purposes (Fondi and Fani 2017). Extraction of bioactive molecules, identification of pathways for improving bioremediation processes, and the discovery of novel enzymes are among the most desired applications. Flux-balance and flux-variability analysis of multi-species models of marine microbial communities are promising methods to investigate the equilibrium of trophic chains and to predict response to contaminants. Ecological and metabolic models are connected by "interface reactions" and integrated into metabolism-food web networks (van Oevelen et al. 2010). This by-level approach has been applied for testing bioremediation capabilities of communities occurring at the interface of marine oil spills and detritus (Murawski et al. 2020; Perkovic et al. 2016). The current perspective indicates that metabolic models of marine bacteria and archaea will improve inferences about acclimation processes under temperature and pH shifts (Czajka et al. 2018; Li et al. 2018). However, because metabolic reconstruction totally depends on genome annotation and biochemical characterization of compounds, metabolic pathways, and transporters, essential improvements and developments in marine microbiology need to be achieved. The genomic catalog of Earth's microbiomes was announced (Nayfach et al. 2021), and the move towards a comprehensive and highly curated microbiome-oriented database is made. The need to translate genome information of the marine microbiome for a better understanding and manipulation is urgent. An "Ocean Virtual Microbiome" (similar to "Human Virtual Microbiome") that accommodates data and virtual space for monitoring and simulation experiments is under construction (Tara Ocean Foundation). Genome and biochemical data repositories, computational tools, and real-time interactions between researchers are going to advance ocean/marine research and development in the near future.

The importance of the microbiome extends from the organism to the ecosystem level and even directly affects the Ocean as a whole and thereby the entire planet. Irrespective of the location (surface waters or the deep sea) and temporal status (transient or persistent), the marine microbiome has an important impact on the fluxes of energy and materials in the ocean. The microbiome represents the fundament of biological activities. Recent advances in the field of marine systems microbiology led to many new discoveries. For instance, many marine protists have incorporated new biochemical functions such as denitrification thanks to their microbiomes (Graf et al. 2021). Microbiomes are also responsible for developing new metabolic pathways that are capable of degrading polymeric chemicals derived from plastics. Acquisition of novel symbionts has been shown to increase the phenotypic plasticity of coastal species (Herrera et al. 2020). Advances in marine microbiome research improved prediction models about responses of species and communities to stressors driven by human activity (Cullen et al. 2020). The study of marine microbiomes will be at the heart of sustainable marine development policies as the United Nations declared the present decade to Ocean Science for Sustainable Development (Reuver et al., Chap. 18). "The science we need for the Ocean we want" is the motto of the Atlantic Ocean Research Alliance between Canada, the European Union and the USA that will carry out the program to decelerate ocean degradation. There is no doubt that an exciting future for marine microbial systems research and valorization lies ahead of us.

**Acknowledgments** Gerard Muyzer was financially supported by the Research Priority Area *Systems Biology* of the University of Amsterdam. We thank Alex Valm for the micrograph of a microbial community that was used in the figures.

## References

Antich A, Palacín C, Cebrian E, Golo R, Wangensteen OS, Turon X (2020) Marine biomonitoring with eDNA: can metabarcoding of water samples cut it as a tool for surveying benthic communities? Mol Ecol. https://doi.org/10.1111/mec.15641

Aylward FO, Santoro AE (2020) Heterotrophic Thaumarchaea with small genomes are widespread in the dark ocean. mSystems 5:e00415–20

Azam A, Fenchel T, Field JG, Gray JS, Meyer-Reil LA, Thingstad F (1983) The ecological role of water-column microbes in the sea. Mar Ecol 10:257–263

Baker BJ, Saw JH, Lind AE et al (2016) Genomic inference of the metabolism of cosmopolitan subsurface Archaea, Hadesarchaea. Nat Microbiol 1:16002

Baker BJ, Appler KE, Gong X (2020) New microbial biodiversity in marine sediments. Ann Rev Mar Sci 13:161–175

Banos S, Lentendu G, Kopf A et al (2018) A comprehensive fungi-specific 18S rRNA gene sequence primer toolkit suited for diverse research issues and sequencing platforms. BMC Microbiol 18:190

Barbier EB (2017) Marine ecosystems services. Curr Biol 27:R507–R510

Bowers RM, Kyrpides NC, Stepanauskas R, Harmon-Smith M, Doud D, Reddy TBK et al (2017) Minimum information about a single amplified genome (MISAG) and a metagenome-assembled genome (MIMAG) of bacteria and archaea. Nat Biotechnol 35:725–731

Campos A, Freitas M, de Almeida AM, Martins JC, Domínguez-Pérez D, Osório H, Vasconcelos V, Reis Costa P (2020) OMICs approaches in diarrhetic shellfish toxins research. Toxins 12:493

Cooney EC, Okamoto N, Cho A, Hehenberger E, Richards TA, Santoro AE, Worden AZ, Leander BS, Keelin PJ (2020) Single-cell transcriptomics of *Abedinium* reveals a new early-branching dinoflagellate lineage. Genome Biol Evol 12:2417–2428

Cullen CM, Aneja KK, Beyhan S, Cho CE, Woloszynek S, Convertino M, McCoy SJ, Zhang Y, Anderson MZ, Alvarez-Ponce D, Smirnova E, Karstens L, Dorrestein PC, Li H, Gupta AS, Cheung K, Gloeckner Powers J, Zhao Z, Rosen GL (2020) Emerging priorities for microbiome research. Front Microbiol 11:136

Czajka JJ, Abernathy MH, Benites VT, Baidoo EEK, Deming JW, Tang YJ (2018) Model metabolic strategy for heterotrophic bacteria in the cold ocean based on *Colwellia psychrerythraea* 34H. Proc Natl Acad Sci USA 11:12507–12512

Davidov K, Iankelevich-Kounio E, Yakovenko I et al (2020) Identification of plastic-associated species in the Mediterranean Sea using DNA metabarcoding with Nanopore MinION. Sci Rep 10:17533

Davidson K, Bresnan E (2009) Shellfish toxicity in UK waters: a threat to human health? Environ Health 8(S12):1–4

Decho AW, Gutierrez T (2017) Microbial extracellular polymeric substances (EPSs) in Ocean systems. Front Microbiol 8:922

Dombrowski N, Teske AP, Baker BJ (2018) Expansive microbial metabolic versatility and biodiversity in dynamic Guaymas Basin hydrothermal sediments. Nat Commun 9:4999

Duarte CM, Ngugi DK, Alam I, Pearman J, Kamau A, Eguiluz VM, Gojobori T, Acinas SG, Gasol JM, Bajic V, Irigoien X (2020) Sequencing effort dictates gene discovery in marine microbial metagenomes. Environ Microbiol 22:4589–4603

Dyksma S, Bischof K, Fuchs BM et al (2016) Ubiquitous Gammaproteobacteria dominate dark carbon fixation in coastal sediments. ISME J 8:1939–1953

Eloe-Fadrosh EA (2019) Genome gazing in ammonia-oxidizing archaea. Nat Rev Microbiol 17:531

Eloe-Fadrosh EA, Ivanova NN, Woyke T, Kyrpides NC (2016) Metagenomics uncovers gaps in amplicon-based detection of microbial diversity. Nat Microbiol 1:15032
Ettinger CL, Vann LE, Eisen JA (2021) The global diversity of the *Zostera marina* microbiome. Appl Environ Microbiol. https://doi.org/10.1128/AEM.02795-20
Falkowski PG, Fenchel T, DeLong EF (2008) The microbial engines that drive the Earth's biogeochemical cycles. Science 320:1034–1039
Fondi M, Fani R (2017) Constraint-based metabolic modelling of marine microbes and communities. Mar Genom 34:1–10
Gardner JJ, Boyle NR (2017) The use of genome-scale metabolic network reconstruction to predict fluxes and equilibrium composition of N-fixing versus C-fixing cells in a diazotrophic cyanobacterium, *Trichodesmium erythraeum*. BMC Syst Biol 11:4
Graf JS, Schorn S, Kitzonger K, Ahmerkamp S, Woehle C, Huettel B, Schubert CJ, Kuypers MMM, Milucka J (2021) Anaerobic endosymbiont generates energy for ciliate host by denitrification. Nature 591:445–465
Grosskopf T, Soyer OS (2014) Synthetic microbial communities. Curr Op Microbiol 18:72–77
Haro-Moreno JM, Rodriguez-Valera F, Rosselli R, Martinez-Hernandez F, Roda-Garcia JJ, Gomez ML, Fornas O, Martinez-Garcia M, López-Pérez M (2020) Ecogenomics of the SAR11 clade. Environ Microbiol 22:1748–1763
Henson MW, Lanclos VC, Pitre DM, Weckhorst JL, Lucchesi AM, Cheng C, Temperton B, Thrash JC (2020) Expanding the diversity of bacterioplankton isolates and modeling isolation efficacy with large-scale dilution-to-extinction cultivation. Appl Environ Microbiol 86:e00943–e00920
Herrera M, Klein SG, Schmidt-Roach E, Campana S, Cziesielski MJ, Chen JC, Duarte CM, Aranda M (2020) Unfamiliar partnership limit cnidarian holobiont acclimation to warming. Global Change Biol 26:5539–5553
Holterman M, Schratzberger M, Helder J (2019) Nematodes as evolutionary commuters between marine, freshwater and terrestrial habitats. Biol J Linnean Soc 128:756–767
Hu B, Xu P, Ma L, Chen D, Wang J, Dai X, Huang L, Du W (2021) One cell at a time: droplet-based microbial cultivation, screening and sequencing. Mar Life Sci Technol 3:169–188
Imachi H, Nobu MK, Nakahara N et al (2020) Isolation of an archaeon at the prokaryote-eukaryote interface. Nature 577:519–525
Jégousse C, Vannier P, Groben R, Glöckner FO, Marteinsson V (2021) A total of 219 metagenome-assembled genomes of microorganisms from Icelandic marine waters. PeerJ 2021:9:e11112. Published 2021 Apr 2. https://doi.org/10.7717/peerj.11112
Kamentsky LA (1973) "Cytology automation". Adv Biol Med Phys 14:93–161. https://doi.org/10.1016/B978-0-12-005214-1.50007-8. ISBN 9780120052141. PMID 4579761
Karsenti E, Acinas SG, Bork P et al (2011) A holistic approach to marine eco-system biology. PLoS Biol 9:e1001177
Kashtan N, Roggensack SE, Berta-Thompson JW, Grinberg M, Stepanauskas R, Chisholm SW (2017) Fundamental differences in diversity and genomic population structure between Atlantic and Pacific *Prochlorococcus*. ISME J 11:1997–2011
Kaster A-K, Sobol MS (2020) Microbial single-cell omics: the crux of the matter. Appl Microbiol Biotechnol 104:8209–8220
Kerou M, Ponce-Toledo R, Zhao R, Abby SS, Hirai M, Nomaki H, Takaki Y, Nunoura T, Jøorgensen SL, Schleper C (2021) Genomes of Thaumarchaeota from deep sea sediments reveal specific adaptations of three independently evolved lineages. ISME J. https://doi.org/10.1038/s41396-021-00962-6
Kvist T, Ahring BK, Lasken RS, Westermann P (2007) Specific single-cell isolation and genomic amplification of uncultured microorganisms. Appl Microbiol Biotechnol 74:926–935
Labonté JM, Swan BK, Poulos B, Luo H, Koren S, Hallam SJ, Sullivan MB, Woyke T, Wommack KE, Stepanauskas R (2015) Single-cell genomics-based analysis of virus–host interactions in marine surface bacterioplankton. ISME J 9:2386–2399

Landry Z, Swan BK, Herndl GJ, Stepanauskas R, Giovannoni SJ (2017) SAR202 genomes from the dark ocean predict pathways for the oxidation of recalcitrant dissolved organic matter. mBio 8: e00413-17

Levy S, Elek A, Grau-Bové X, Menéndez-Bravo S, Iglesias M, Tanay A, Mass T, Sebé-Pedrós A (2021) A stony coral cell atlas illuminates the molecular and cellular basis of coral symbiosis, calcification, and immunity. Cell 184:1–15

Lewis WH, Tahon G, Geesink P, Sousa DZ, Ettema TJG (2021) Innovations to culturing the uncultured microbial majority. Nat Rev Microbiol 19:225–240

Li F, Xie W, Yuan Q, Luo H, Li P, Chen T, Zhao X, Wang Z, Ma H (2018) Genome-scale metabolic model analysis indicates low energy production efficiency in marine ammonia-oxidizing archaea. AMB Express 27:106

Liu Z, Hu SK, Campbell V, Tatters AO, Heidelberg KB, Caron DA (2017) Single-cell transcriptomics of small microbial eukaryotes: limitations and potential. ISME J 11:1282–1285

Liu H, Xue Y, Stirling E, Ye S, Xu J, Ma B (2021) FACS-iChip: a high-efficiency iChip system for microbial 'dark matter' mining. Mar Life Sci Technol 3:162–168

Martinez-Garcia M, Brazel D, Poulton NJ, Swan BK, Gomez ML, Masland D, Sieracki ME, Stepanauskas R (2012) Unveiling in situ interactions between marine protists and bacteria through single cell sequencing. ISME J 6:703–707

Martinez-Hernandez F, Fornas O, Lluesma Gomez M, Bolduc B, de la Cruz Peña MJ, Martínez Martínez J, Anton J, Gasol JM, Rosselli R, Rodriguez-Valera F, Sullivan MB, Acinas SG, Martinez-Garcia M (2017) Single-virus genomics reveals hidden cosmopolitan and abundant viruses. Nat Commun 8:15892

Martínez Martínez J, Martinez-Hernandez F, Martinez-Garcia M (2020) Single-virus genomics and beyond. Nat Rev Microbiol 18:705–716

Mock T, Daines SJ, Geider R, Collins S, Metodiev M, Millar AJ, Moulton V, Lenton TM (2017) Bridging the gap between omics and earth system science to better understand how environmental change impacts marine microbes. Global Change Biol 22:61–67

Murawski SA, Ainsworth CH, Gilbert S, Hollander DJ, Paris CB, Schlüter M, Wetzel DL (2020) Scenarios and responses to future deep oil spills: fighting the next war. Springer, Berlin, 881 p

Muyzer G (2016) Marine microbial systems ecology: microbial networks in the sea. In: Stal LJ, Cretoiu MS (eds) The marine microbiome. Springer, Cham, pp 335–344

Nayfach S, Roux S et al (2021) A genomic catalog of Earth's microbiomes. Nat Biotechnol 39:499–509

Olson RJ, Zettler ER, DuRand MD (1993) Phytoplankton analysis using flow cytometry. In: Kemp PF, Sherr BF, Sherr EB, Cole JJ (eds) Handbook of methods in aquatic microbial ecology. Lewis, Boca Raton, pp 175–186

Pachiadaki MG, Brown JM, Brown J, Bezuidt O, Berube PM, Biller SJ, Poulton NJ, Burkart MD, La Clair JJ, Chisholm SW, Stephanauskas R (2019) Charting the complexity of the marine microbiome through single-cell genomics. Cell 179:1623–1635

Paez-Espino D, Eloe-Fadrosh E, Pavlopoulos G et al (2016) Uncovering Earth's virome. Nature 536:425–430

Paez-Espino D, Chen IA, Palaniappan K, Ratner A, Chu K, Szeto E, Pillay M, Huang J, Markowitz VM, Nielsen T, Huntemann M, Reddy TBK, Pavlopoulos GA, Sullivan MB, Campbell BJ, Chen F, McMahon K, Hallam SJ, Denef V, Cavicchioli R, Caffrey SM, Streit WR, Webster J, Handley KM, Salekdeh GH, Tsesmetzis N, Setubal JC, Pope PB, Liu WT, Rivers AR, Ivanova NN, Kyrpides NC (2017) IMG/VR: a database of cultured and uncultured DNA viruses and retroviruses. Nucleic Acids Res 45:D457–D465

Perkovic M, Harsch R, Ferraro G (2016) Oil spills in the Adriatic Sea. In: Carpenter A, Kostianoy A (eds) Oil pollution in the Mediterranean Sea: Part II. The handbook of environmental chemistry, vol 84. Springer, Cham

Queiroz LL, Bendia AG, Duarte RTD et al (2020) Bacterial diversity in deep-sea sediments under influence of asphalt seep at the São Paulo Plateau. Ant Leeuwenhoek 113:707–717

Quince C, Walker AW, Simpson JT, Loman NJ, Segata N (2017) Shotgun metagenomics, from sampling to analysis. Nat Biotechnol 35:833–844
Raes J, Bork P (2008) Molecular eco-systems biology: towards an understanding of community function. Nat Rev Microbiol 6:693–699
Roux S, Hawley AK, Torres Beltran M, Scofield M, Schwientek P, Stepanauskas R, Woyke T, Hallam SJ, Sullivan MB (2014) Ecology and evolution of viruses infecting uncultivated SUP05 bacteria as revealed by single cell- and metagenomics. Elife 3:e03125
Saito MA, Bertrand EM, Duffy ME, Gaylord DA, Held NA, Hervey WJ, Hettich RL, Jagtap PD, Janech MG, Kinkade DB, Leary DH, McIlvin MR, Moore EK, Morris RM, Neely BA, Nunn BL, Saunders JK, Shepherd AI, Symmonds NI, Walsh DA (2019) Progress and challenges in ocean metaproteomics and proposed best practices for data sharing. J Proteome Res 18:1461–1476
Salazar G, Paoli L, Alberti A et al (2019) Gene expression changes and community turnover differentially shape the global ocean metatranscriptome. Cell 179:1068–1083
Sczyrba A, Hofmann P, Belmann P et al (2017) Critical assessment of metagenome interpretation—a benchmark of metagenomics software. Nat Methods 14:1063–1071
Shahzad K, Loor JJ (2012) Application of top-down and bottom-up systems approaches in ruminant physiology and metabolism. Curr Genomics 13:379–394
Shakya M, Lo C-C, Chain PSG (2019) Advances and challenges in metatranscriptomic analysis. Front Genet 10:904
Sieracki CK, Sieracki ME, Yentsch CS (1998) An imaging-in-flow system for automated analysis of marine microplankton. Mar Ecol Prog Ser 168:285–296
Staley JT, Konopka A (1985) Measurements of *in situ* activities of non-photosynthetic microorganisms in aquatic and terrestrial habitats. Ann Rev Microbiol 39:321–346
Stepanauskas R (2012) Single cell genomics: an individual look at microbes. Curr Opin Microbiol 15:613–620
Swan SC, Turner AD, Bresnan E, Whyte C, Paterson RF, McNeill S, Mitchell E, Davidson K (2018) *Dinophysis acuta* in Scottish coastal waters and its influence on diarrhetic shellfish toxin profiles. Toxins 28:399
Thompson LR, Haroon MF, Shibl AA, Cahill MJ, Ngugi DK, Williams GJ, Morton JT, Knight R, Goodwin KD, Stingl U (2019) Red Sea SAR11 and *Prochlorococcus* single-cell genomes reflect globally distributed pangenomes. Appl Environ Microbiol 85:e00369–e00319
Thrash JC, Temperton B, Swan BK, Landry ZC, Woyke T, DeLong EF, Stepanauskas R, Giovannoni SJ (2014) Single-cell enables comparative genomics of a deep ocean SAR11 bathtype. ISMEJ 8:1440–1451
Ustick LJ, Larkin AA, Garcia CA, Garcia NS, Brock ML, Lee JA, Wiseman NA, Moore JK, Martiny AC (2021) Metagenomic analysis reveals global-scale patterns of oceanic nutrient limitation. Science 372:287–291
van Oevelen D, Van den Meersche K, Meysman FJR, Soetaert K, Middelburg JJ, Vezina A (2010) Quantifying food web flows using linear inverse models. Ecosystems 13:32–45
Wang J, Chen L, Chen Z, Zhang W (2015) RNA-seq based transcriptomic analysis of single bacterial cells. Integr Biol (Camb) 7:1466–1476
Waterworth SC, Kalinski J-CJ, Madonsela LS, Parker-Nance S, Kwan JC, Dorrington RA (2020) Family matters: the genomes of conserved bacterial symbionts provide insight into specialized metabolic relationships with their sponge host. bioRxiv. https://doi.org/10.1101/2020.12.09.417808
Wilhelm SW, Suttle CA (1999) Viruses and nutrient cycles in the sea: viruses play critical roles in the structure and function of aquatic food webs. BioScience 49:781–788
Wilson WH, Gilg IC, Moniruzzaman M, Field EK, Koren S, LeCleir GR, Martínez Martínez J, Poulton NJ, Swan BK, Stepanauskas R, Wilhelm SW (2017) Genomic exploration of individual giant ocean viruses. ISME J 11:1736–1745
Woyke T, Doud DFR, Schulz F (2017) The trajectory of microbial single-cell sequencing. Nat Meth 14:1045–1054. https://doi.org/10.1038/nmeth.4469

Yadav L, Tamene F, Göös H, Van Drogen A, Katainen R, Aebersold R, Gstaiger M, Varjosalo M (2017) Systematic analysis of human protein phosphatase interactions and dynamics. Cell Syst 4:430–444

Yentsch CS, Yentsch CM (2008) Single cell analysis in biological oceanography and its evolutionary implications. J Plankton Res 30:107–117

Yilmaz P, Kottman R, Field D et al (2011) Minimum information about a marker gene sequence (MiMARKS) and minimum information about any (x) sequence (MIxS) specifications. Nat Biotechnol 29:415

Yoon HS, Price DC, Stepanauskas R, Rajah VD, Sieracki ME, Wilson WH, Yang EC, Duffy S, Bhattacharya D (2011) Single-cell genomics reveals organismal interactions in uncultivated marine protists. Science 332:714–717

Zhang X-H, Ahmad W, Zhu X-Y, Chen J, Austin B (2021) Viable but nonculturable bacteria and their resuscitation: implications for cultivating uncultured marine microorganisms. Mar Life Sci Technol 3:189–203

Zhou Z, Liu Y, Lloyd K, Pan J, Yang Y, Gu J-D, Li M (2019) Genomic and transcriptomic insights into the ecology and metabolism of benthic archaeal cosmopolitan, Thermoprofundales (MBG-D archaea). ISME J 13:885–901

Zou D, Liu H, Li M (2020) Community, distribution and ecological roles of estuarine Archaea. Front Microbiol 11:2060

# The Hidden Treasure: Marine Microbiome as Repository of Bioactive Compounds

17

Bathini Thissera, Ahmed M. Sayed, Hossam M. Hassan, Usama R. Abdelmohsen, Rainer Ebel, Marcel Jaspars, and Mostafa E. Rateb

**Abstract**

Marine-associated microbiome is known as a hub for novel chemistry and biology by producing interesting pharmacophores. Thus, in the area of natural product drug discovery, contribution and attention toward marine natural product investigation is a growing trend. The rapid swift in exploring the sea for harvest untapped plethora of marine resources to investigate associated microorganisms such as bacteria, fungi, and cyanobacteria are facilitated by technological advances. This chapter discusses the importance of chemical diversity of the marine microbiome in the natural product drug discovery pipeline giving specific reported examples of promising marine-derived bioactive candidates, as well as intriguing strategies to ramp up the discovery of pharmacologically inspiring secondary metabolites out of the marine microbial biosynthesis process.

**Keywords**

Marine bacteria · Marine cyanobacteria · Marine fungi · Marine natural products · Secondary metabolites elicitation

B. Thissera · M. E. Rateb (✉)
School of Computing, Engineering & Physical Sciences, University of the West of Scotland, Paisley, Scotland, UK
e-mail: m.rateb11@aberdeen.ac.uk

A. M. Sayed · H. M. Hassan
Department of Pharmacognosy, Faculty of Pharmacy, Nahda University, Beni-Suef, Egypt

U. R. Abdelmohsen
Department of Pharmacognosy, Faculty of Pharmacy, Deraya University, New Minia City, Egypt

R. Ebel · M. Jaspars
Marine Biodiscovery Centre, Department of Chemistry, University of Aberdeen, Aberdeen, Scotland, UK



L. J. Stal, M. S. Cretoiu (eds.), *The Marine Microbiome*, The Microbiomes of Humans, Animals, Plants, and the Environment 3,
https://doi.org/10.1007/978-3-030-90383-1_17

## 17.1 Introduction

Natural Products (NPs) continue to be one of the most inspiring sources for the development of new drugs due to their impressive chemical diversity and potent and selective biological activity. The structure of more than 450,000 NPs is available from a variety of different databases such as PubChem, REAXYS, ChEMBL, ZINC, NaprAlert, Natural Product Atlas, and SuperNatural II (Pereira 2019). The higher success rate of marine natural compounds (MNPs) (1 in 3500 MNPs against the industry-based average for synthetic compounds of (1 in 5–10,000 compounds) has led to the rejuvenation of interest in NP-like scaffolds for drug discovery campaigns. New approaches are required to combat the perceived disadvantages of NPs compared to synthetic drugs, such as the difficulty of access and supply, the complexity of NP chemistry and structure elucidation, and the slowness of working with NPs (Pereira and Aires-de-Sousa 2018).

MNP research is developing continuously with more new compounds added every year (Fig. 17.1). This rapid development in MNP discovery is associated with various technological advances (e.g., iChip, co-culture, OSMAC, and epigenetic manipulations) that have facilitated the exploration of this huge mine of chemical entities. To date, the global pharmaceutical pipeline from marine sources consists of thirteen approved drugs, ten of which are anticancer drugs (Table 17.1). Currently, there are about 23 marine natural products or antibody–drug conjugates in Phase I to Phase III clinical trials mainly in the area of cancer therapy (Jaspars et al. 2016; https://www.midwestern.edu/departments/marinepharmacology/clinical-pipeline.xml). There are four marine natural products currently in Phase III clinical trials

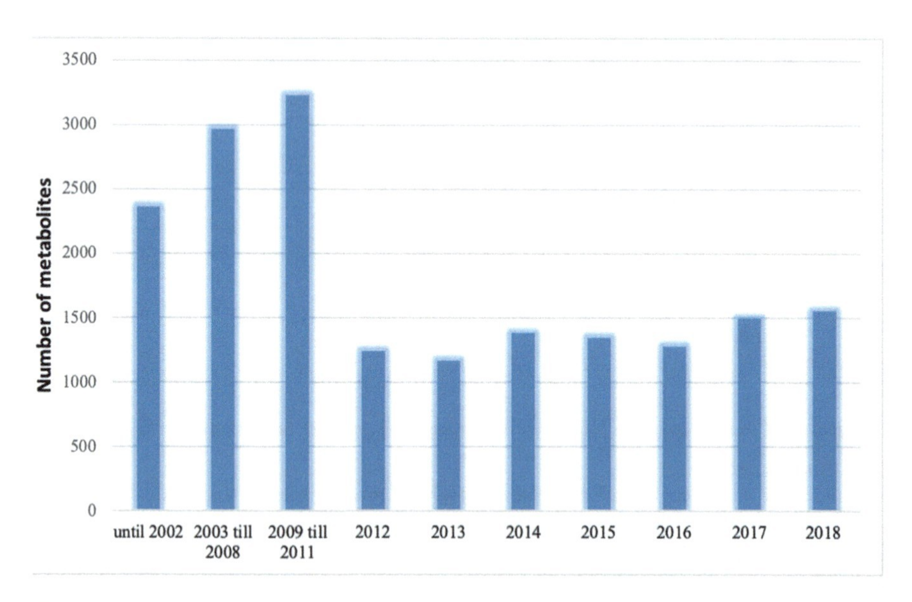


**Fig. 17.1** Marine natural products discoveries till 2018 extracted from Marine Natural Products periodic reviews—Natural Product Reports—Royal Society of Chemistry

**Table 17.1** Approved drugs from MNPs and derivatives (https://www.midwestern.edu/departments/marinepharmacology/clinical-pipeline.xml)

| Compound name | Marine organism | Chemical class | Molecular target | Disease associated |
|---|---|---|---|---|
| Lurbinectedin | Tunicate | Alkaloid | RNA Polymerase II | Cancer: metastatic small cell lung cancer |
| Belantamab Mafodotin-blmf | Mollusk/ cyanobacterium | ADC (MMAE) | BCMA | Relapsed/refractory multiple myeloma |
| Enfortumab vedotin | Mollusk/ cyanobacterium | ADC (MMAE) | Nectin-4 | Cancer: metastatic urothelial cancer |
| Polatuzumab vedotin | Mollusk/ cyanobacterium | ADC (MMAE) | CD76b & microtubules | Cancer: non-Hodgkin lymphoma, chronic lymphocytic leukemia, lymphoma, B-cell lymphoma |
| Plitidepsin | Tunicate | Depsipeptide | eEF1A2 | Cancer: multiple myeloma, leukemia, lymphoma |
| Trabectedin (ET-743) | Tunicate | Alkaloid | Minor groove of DNA | Cancer: soft tissue sarcoma and ovarian cancer |
| Brentuximab vedotin | Mollusk/ cyanobacterium | ADC (MMAE) | CD30 & microtubules | Cancer: anaplastic large T-cell systemic malignant lymphoma, Hodgkin's disease |
| Eribulin mesylate (E7389) | Sponge | Macrolide | Microtubules | Cancer: metastatic breast cancer |
| Omega-3-acid ethyl ester | Fish | Omega-3-fatty acids | Triglyceride-synthesizing enzymes (TSE) | Hypertriglyceridemia |
| Eicosapentaenoic acid ethyl ester | Fish | Omega-3-fatty acids | TSE | Hypertriglyceridemia |
| Omega-3-carboxylic acid | Fish | Omega-3-fatty acids | TSE | Hypertriglyceridemia |
| Ziconotide | Cone snail | Peptide | N-Type Ca channel | Pain: severe chronic pain |
| Vidarabine (Ara-A) | Sponge | Nucleoside | Viral DNA polymerase | Antiviral: Herpes simplex virus |
| Cytarabine (Ara-C) | Sponge | Nucleoside | DNA polymerase | Cancer: Leukemia |

which include three anticancer compounds (plinabulin, lurbinectedin, and salinosporamide A) and one analgesic (tetrodotoxin). In phase II there are 12 MNPs and derivatives, from those three are small molecules: one anticancer (plocabulin) and two against Alzheimer's disease (bryostatin and DMXBA). All the remaining nine MNPs and their derivatives in the clinical trials (phase I) are recognized as anticancer antibody–drug conjugates.

It is worth noting that the availability of funding has had a great influence on the biological activity space of MNP, e.g., 10 out of the 13 approved MNPs drugs and 19 out of the 23 MNPs and derivatives in all clinical trial phases have anticancer activity. This is correlated with the National Institutes of Health (NIH)/National Cancer Institute (NCI) being the leading funding agency in the USA for MNP research over many years (Newman and Cragg 2016).

In this chapter, the importance of marine natural products from different sources with examples of potential bioactive molecules as well as advances in discovery strategies will be discussed.

## 17.2 The Current Status of Marine Microbe-Derived Drug Discovery

### 17.2.1 Marine Bacteria

Like other marine microorganisms, marine bacteria have evolved unique metabolic pathways that enable them to survive in harsh environments and to biosynthesize their own specialized metabolites that terrestrial bacteria may lack. The number of new metabolites originating from marine bacteria increased exponentially after the

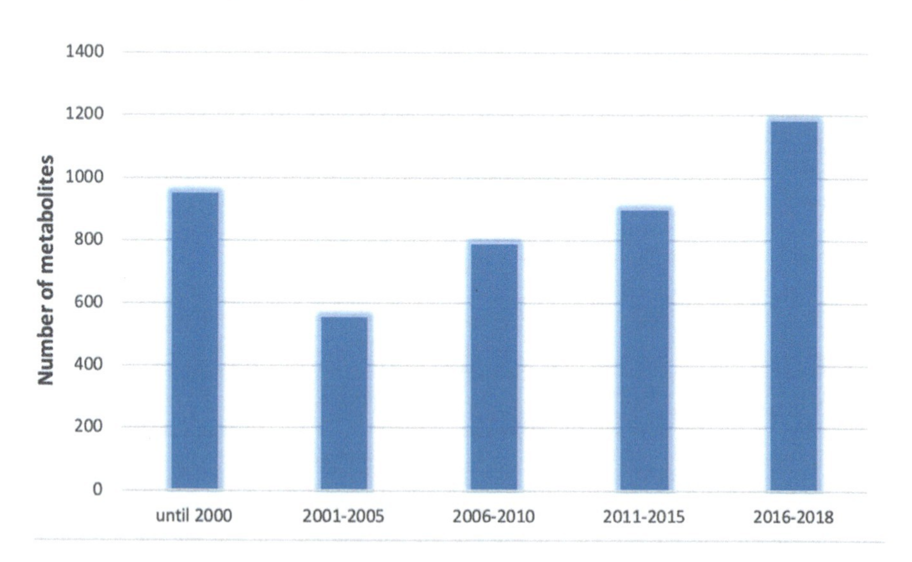


**Fig. 17.2** Numbers of new marine bacteria-derived natural products from 2000 to 2018

year 2000 (Fig. 17.2) (Blunt et al. 2003, 2004, 2005, 2006, 2007, 2008, 2009, 2010, 2011, 2012, 2013, 2014, 2015, 2016, 2017, 2018; Carroll et al. 2019, 2020). Until the end of the year 2018, the continued interest in discovering novel bioactive compounds from marine bacteria led to the isolation of around 4500 metabolites, many of which showed promising broad-spectrum bioactivities, particularly anticancer and antimicrobial properties.

Marine actinomycetes are considered to be the major source of new chemical entities, notably *Streptomyces* that have accounted for about 38% of newly discovered marine bacteria-derived natural products. Marine bacteria are often isolated from marine sediments and from marine macro-organisms (e.g., sponges, corals, or algae) but also from extreme habitats such as the deep-sea and hypersaline lakes (Jones et al. 2019).

#### 17.2.1.1 Early Discoveries of Marine Bacterial Natural Products

Isatin (**1**) was one of the earliest reported antimicrobial and anticancer metabolites. This compound is produced by several bacterial strains that colonize the surface of embryos of the shrimp *Palaemon macrodactylus*, which protects these embryos against the pathogenic fungus *Lagenidium callinectes* (Gil-Turnes et al. 1989). The *Alteromonas luteoviolaceus*-derived metabolite possesses broad-spectrum antimicrobial activity and its biosynthetic pathway has been fully elucidated (Laatsch 2017). The research interest in marine actinomycetes started to expand from the beginning of the 1990s when the unusual bicyclic depsipeptides salinamide A and B (**2** and **3**) were reported from a *Streptomyces* sp. that was isolated from the jellyfish *Cassiopeia xamachana*. Both depsipeptides exhibited moderate antibiotic activity but they had potent in vivo anti-inflammatory potential (Moore et al. 1999; Trischman et al. 1994). In 1994, the highly brominated pyrrole antibiotic pentabromopseudiline (**4**) was among the first reported marine bacteria-derived metabolites. The growing interest in marine actinomycetes as producers of bioactive compounds continued into this century and only in the period 2000–2002 nearly 250 new compounds have been reported from this group of microorganisms. During the same period, the number of reported new metabolites produced by terrestrial actinomycetes did not exceed 150 (Laatsch 2017). Although the exploration of marine actinomycetes as a source for new bioactive metabolites was at an early stage, numerous interesting compounds have been isolated during the period of 2000–2005. For example, the novel polycyclic polyketide antibiotic abyssomicin C (**5**) was reported from a marine *Verrucosispora* (Riedlinger et al. 2004). This unusual compound interferes with the biosynthesis of *p*-aminobenzoic acid and inhibits the biosynthesis of folic acid at an earlier stage than do the traditional sulfa drugs (Bister et al. 2004). As a result, abyssomicin C and its analogs showed antibacterial activity toward a broad spectrum of pathogenic bacteria including those that are multiple antibiotic resistant (Rath et al. 2005). Diazepinomicin (**6**) is another example of a unique farnesylated dibenzodiazepinone isolated from a marine-derived *Micromonospora* (Charan et al. 2004). This compound exhibits a wide spectrum of biological activities ranging from antibacterial to anticancer. In 2006, diazepinomicin (*aka* ECO-4601) has been submitted for clinical trials as an anticancer agent for many types of tumors and has successfully completed phase 1 trials (https://go.drugbank.com). Furthermore, a novel β-lactone-γ-lactam metabolite, salinosporamide A (NPI-0052, **7**), was reported from a new obligate marine actinomycete, *Salinispora tropica* (Feling et al. 2003). This halogenated metabolite is an orally active proteasome inhibitor and can induce apoptosis in multiple myeloma cancer cells with a unique mode of action (Chauhan et al. 2005). In 2007, NPI-0052 (**7**) was submitted for clinical trials as an anticancer agent for multiple myeloma and its evolution through phase 2 clinical trials is ongoing at the time of writing of this chapter (https://clinicaltrials.gov/ct2/show/NCT00461045).

#### 17.2.1.2 Recently Discovered Marine Bacterial Natural Products

With the advances in structural biology and fermentation processes marine bacteria have gained much attention in the fields of bioremediation and biotechnology (Andryukov et al. 2019). Moreover, extensive investigation of the biosynthetic pathways of marine-derived bacteria revealed that more than 70% of the secondary metabolites they produced were non-ribosomal peptides (NRPs), polyketides (PKs), and mixed PKS-NRPS (Pinu et al. 2017; Wang and Lei 2018). Most of these classes of metabolites have antimicrobial and anticancer potential. Terrestrial actinomycete-derived metabolites also produced a large number of NRPs and PKs but marine bacteria produce a greater chemical diversity of these molecules. Hence, the focus on exploring metabolic pathways of NRPs and PKs in marine bacteria, particularly in actinomycetes, has dramatically increased in recent years (Andryukov et al. 2019).

Over the last decade, lipopeptides were amongst the most frequently reported marine bacteria-derived metabolites with promising antimicrobial potential. Halobacillin (**8**) and methylhalobacillin (**9**) are examples of two cyclic lipopeptides obtained from bacteria isolated from deep-sea sediments (Zhou and Guo 2012). Both metabolites showed high efficacy against the growth of human colon tumor cells ($IC_{50}$ 0.98 μg/mL) (Mondol et al. 2013). Polyketides are another large class of microbial natural products that have provided many successful pharmaceutical products. Marine bacteria have further extended the chemical space of polyketides with several novel compounds (Tareq et al. 2012). The antibiotics ieodoglucomides A and B (**10** and **11**) show broad-spectrum antibacterial activity (MIC ~8 μg/mL) and selective cytotoxicity against human lung cancer cells ($IC_{50}$ 17 and 25 μg/mL, respectively). Both polyketides were recovered from the marine sediment-derived *Bacillus licheniformis* (Tareq et al. 2012). The antibiotic TP-1161 (**12**) is another thiopeptide-polyketide that was isolated from the marine actinomycete *Nocardiopsis* sp. (Engelhardt et al. 2010). This unusual metabolite shows in vitro antibacterial activity against a panel of Gram-positive bacteria (with MICs varying from 0.25 to 4 μg/mL). Additionally, TP-1161 is able to inhibit the growth of vancomycin-resistant bacterial strains, including *Enterococcus faecalis* and *Enterococcus faecium* at MIC = 1 μg/mL (Engelhardt et al. 2010).

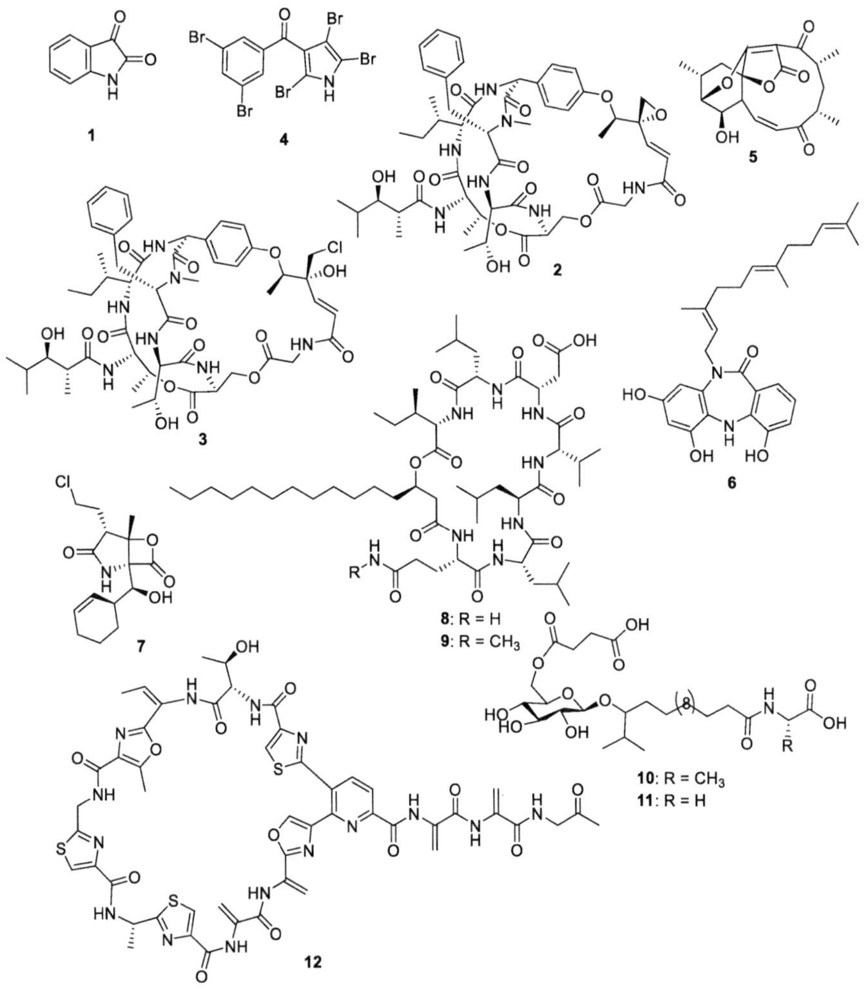

## 17.2.2 Marine Fungi

Fungi form part of marine microbiological communities and are present as saprotrophs, parasites, or symbionts in all ecosystems. There is considerable interest in marine fungi due to the structural diversity of their natural products. Despite the discovery of several interesting bioactive compounds from marine fungi, the number of these products that were isolated has increased only slowly over a long period of time. To date, only ~1100 species of fungi have been described from marine environments, although estimates of the total number of fungal species range from 1.5 to 5 million (Jones et al. 2019). Researchers discovered that the same marine fungal species obtained from distant geographical locations produce different

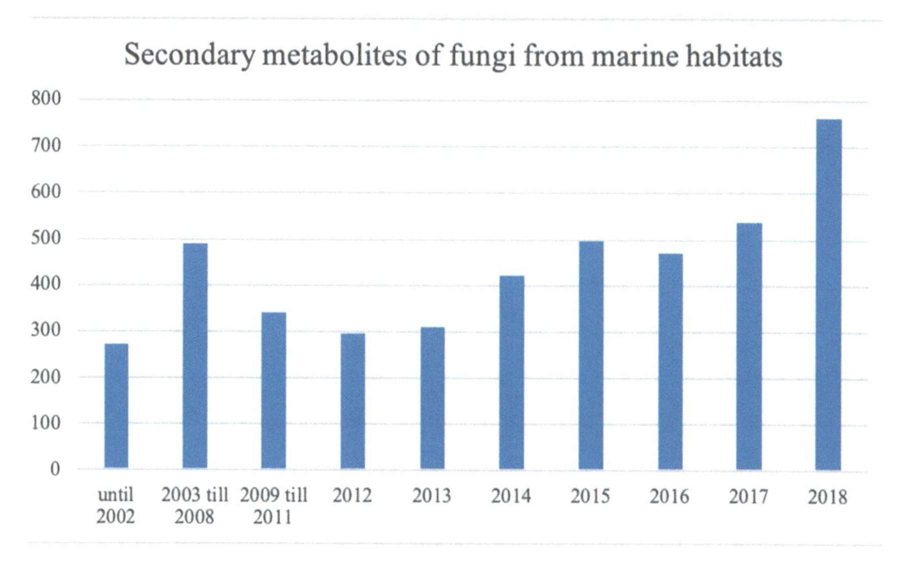


**Fig. 17.3** Secondary metabolites discovered from marine fungi

metabolites. Until 2002, only 270 natural products were reported from marine fungi. Thereafter, this number increased and reached 1120 by the end of 2010 (Rateb and Ebel 2011). The discovery of more new marine fungal bioactive compounds continued to increase and in the year 2018 alone 760 new compounds reported (Fig. 17.3). This data indicates that by the end of 2018 the total number of newly discovered natural marine fungal products reached approximately 4400.

The broad-spectrum antibiotic cephalosporin C (**13**) has been known since 1955 and has been obtained from the fungus *Acremonium chrysogenum* that was isolated from seawater sampled near a sewage outfall of the Sardinian coast (Abraham 1979). The tubulin depolymerizing agent diketopiperazine halimide (**14**) was isolated from the fungus *Aspergillus* sp. (Fenical et al. 1998). This molecule was later used for the development of the closely related synthetic analog plinabulin (NPI-2358) (**15**) which is currently in the clinical trial phase. Plinabulin is the lead asset of BeyondSpring Pharmaceuticals and is currently in the late stage III clinical phase and its intended use will be to avoid chemotherapy-induced neutropenia in non-small-cell lung- and brain tumors (https://clinicaltrials.gov/ct2/results?term=plinabulin&pg=1).

The small contribution that marine fungi have made up-to-date to the discovery of new drug leads may be attributed to the fact that the chemical investigation of these microorganisms for the production of promising bioactive metabolites has been virtually neglected between 1980 and 1992. By the end of the 1980s, only 15 secondary metabolites had been reported as being derived from marine fungi (Bugni and Ireland 2004).

#### 17.2.2.1 Anti-infective Marine Fungal Natural Products

Neoechinulin B (**16**), an indole prenylated diketopiperazine isolated from the marine fungus *Eurotium rubrum*, exhibits a strong inhibitory effect against the H1N1 virus in infected cells of the MDCK cell line and inhibited clinical isolates of amantadine-, oseltamivir-, and ribavirin-resistant influenza. The limited toxicity of Neoechinulin B, its wide-spectrum activity against drug-resistant clinical viral isolates, and decreased drug resistance induction made it a strong candidate for its possible use for the treatment of clinically resistant viral isolates (Chen et al. 2015). Chemical investigation of the gorgonian coral-derived fungus *Aspergillus terreus* SCSGAF0162 led to the isolation of the cyclic tetrapeptide asperterrestide A (**17**) that inhibits the replication of M2-resistant influenza strain A/WSN/33 H1N1 in MDCK cells (He et al. 2013). Screening of the *Stachybotrys chartarum* MXH-X73 marine sponge-associated fungus led to the isolation of phenylspirodrimane stachybotrin D (**18**), which inhibits HIV-1 replication by inhibiting reverse transcriptase without being toxic for humans. In addition, evaluation of phenylspirodrimane stachybotrin D revealed similar inhibitory effects on the replication of wild and multiple non-nucleoside reverse transcriptase inhibitor (NNRTI)-resistant HIV-1 strains to HIV-1 (Ma et al. 2013).

Chemical analysis of the marine-derived fungus *Stagonosporopsis cucurbitacearum* resulted in the isolation of the pyridone alkaloid didymellamide A (**19**), which showed promising antifungal activity against the azole-resistant and sensitive *Candida albicans*, *C. glabrata*, and *C. neoformans* (Haga et al. 2013). Sesquiterpene penicibilaene B (**20**), isolated from *Penicillium bilaiae* MA-267 recovered from mangrove rhizospheric soil, exhibited selective action against the plant pathogenic fungus *Colletotrichum gloeosporioides* (Meng et al. 2014).

Chemical analysis of the sponge-derived fungus *Penicillium adametzioides* AS-53 resulted in the discovery of the peniciadametizine A derivative of dithiodiketopiperazine (**21**), which exhibited selective antifungal activity against plant pathogenic fungus *Alternaria brassicae* (Liu et al. 2015). The alkaloid varioxepine A (**22**) was isolated from the marine alga-derived endophytic fungus *Paecilomyces variotii* and showed a potent inhibitory effect against the plant pathogen *Fusarium graminearum* (Zhang et al. 2014).

The chemical characterization of *Penicillium brocae* MA-231 isolated from a mangrove plant resulted in the isolation of pyranonigrin A (**23**) with a clear antimicrobial activity against several Gram-positive and -negative pathogenic bacteria (Meng et al. 2015). The bisthiodiketopiperazine derivative adametizine A (**24**) isolated from *Penicillium adametzioides* AS-53, a marine sponge-derived fungus, showed strong inhibitory activity against *Staphylococcus aureus*, *Aeromonas hydrophila*, *Vibrio* spp. *V. harveyi*, and *V. parahaemolyticus* (Liu et al. 2015). Aspergillusene A (**25**), a sesquiterpene isolated from the sponge-associated fungus *Aspergillus sydowii* ZSDS1-F6 displayed promising antimicrobial activity against *Klebsiella pneumoniae* and *Aeromonas hydrophila* (Wang et al. 2014). The isocoumarin derivative penicisimpin A (**26**) isolated from the mangrove plant-derived fungus *Penicillium simplicissimum* MA-332 exhibited strong activity against *Escherichia coli, Pseudomonas aeruginosa, V. parahaemolyticus*, and *V. harveyi* (Xu et al. 2016). Diaporthalasin (**27**), a pentacyclic cytochalasin isolated from the marine-derived fungus *Diaporthaceae* sp. PSU-SP2/4 displayed strong antibacterial activity against both *S. aureus* and methicillin-resistant *S. aureus* (Khamthong et al. 2014). The aminolipopeptide trichoderin A (**28**) isolated from the marine sponge-derived *Trichoderma* sp. exhibited potent anti-mycobacterial activity against *Mycobacterium smegmatis*, *M. bovis* BCG, and *M. tuberculosis* H37Rv (Pruksakorn et al. 2010).

#### 17.2.2.2 Anticancer Marine Fungal Natural Products

Diaporthalasin (**27**), a pentacyclic cytochalasin isolated from *Diaporthaceae* sp., a marine fungus (SP-SP2/4 PSU) demonstrated substantial antibacterial activity against both *S. aureus* and methicillin resistant *S. aureus* (Khamthong et al. 2014). Trichoderin A (**28**) aminolipopeptide was isolated from a marine sponge-derived *Trichoderma* sp. and showed potent anti-mycobacterial activity against *M. smegmatis*, *M. bovis* BCG, and *M. tuberculosis* H37Rv (Pruksakorn et al. 2010). Chemical investigation of the sponge-derived fungus *Stachylidium* sp. resulted in the isolation of a phthalimidine derivative mariline A1 (**30**) with potent inhibitory activity against the human leukocyte elastase (Almeida et al. 2012). Chloropreussomerin A (**31**) was the first chlorinated metabolite in the preussomerin family and was obtained from the fungus *Lasiodiplodia theobromae* ZJ-HQ1, an endophyte isolated from a mangrove plant. Chloropreussomerin A showed a potent in vitro cytotoxicity against several human cancer cell lines (Chen et al. 2016a). Chemical screening of the marine fungus *Aspergillus ochraceus* Jcma1F17 resulted in the isolation of a member of an unusual class of nitrobenzoyl sesquiterpenoid: 6β,9α-dihydroxy-14-p-nitrobenzoylcinnamolide (**32**), which exhibited strong cytotoxicity against 10 cancer cell lines (Fang et al. 2014). The discovery of 20, structurally diverse, complex indole-diterpene compounds resulted from genome mining of the fungus *Mucor irregularis* QEN-189, which was isolated from mangrove plants. Among them, rhizovarin B (**33**) showed good activity against the human A-549 and HL-60 cancer cell lines (Gao et al. 2016). A mangrove-derived endophytic fungus *Pestalotiopsis microspora* led to the isolation of the macrolides pestalotioprolides E (**34**) and F (**35**), which show strong cytotoxicity against the murine lymphoma cell line L5178Y while in addition, pestalotioprolide F shows potent activity against the human ovarian cancer cell line A2780 (Liu et al. 2016). Chaunolidone A (**36**), a pyridinone derivative isolated from the marine fungus *Chaunopycnis* sp. (CMB-MF028) showed selective and potent inhibition of human non-small-cell lung carcinoma cells (NCI-H460) (Shang et al. 2015).

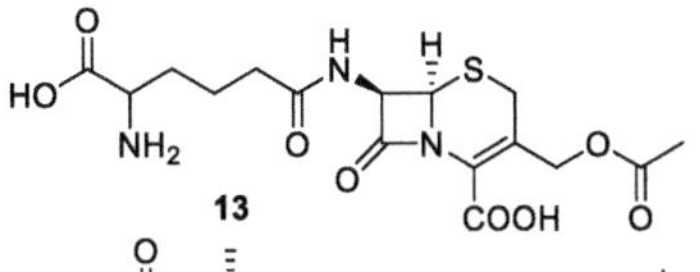

**13**

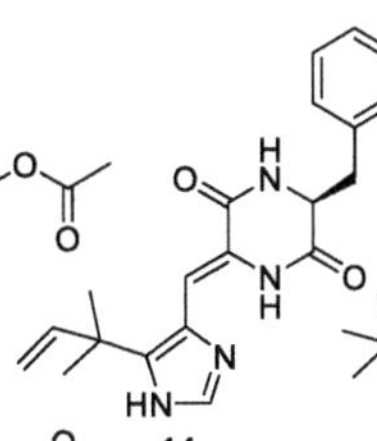

**14**

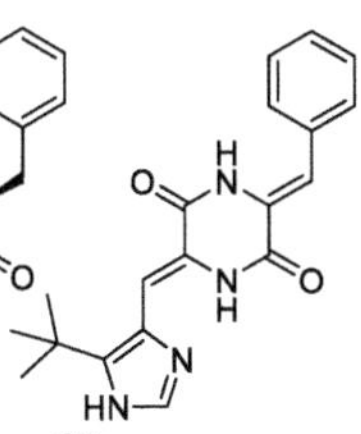

**15**

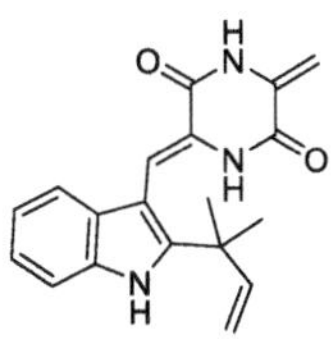

**16**

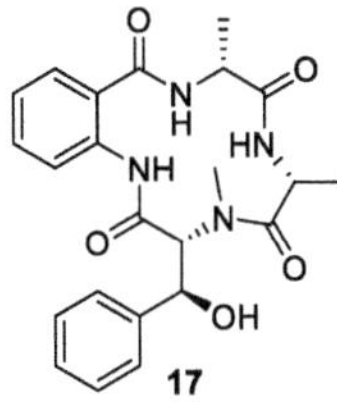

**17**

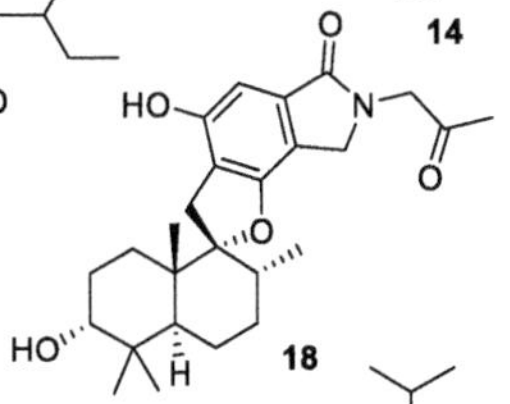

**18**

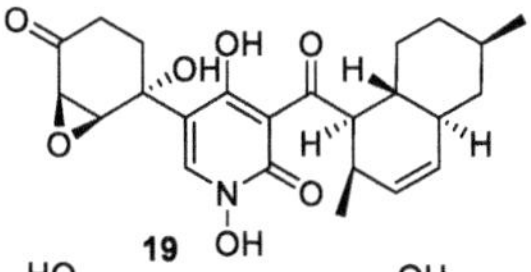

**19**

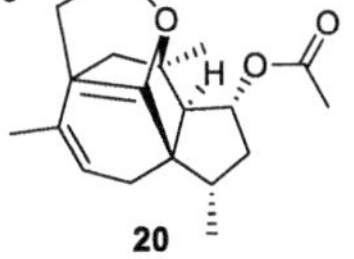

**20**

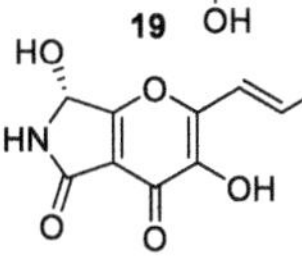

**23**

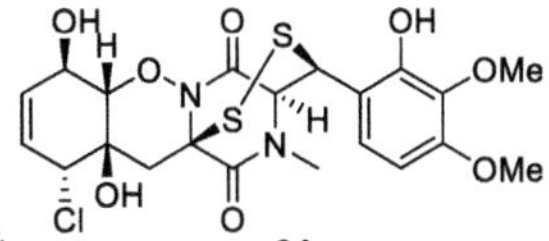

**24**

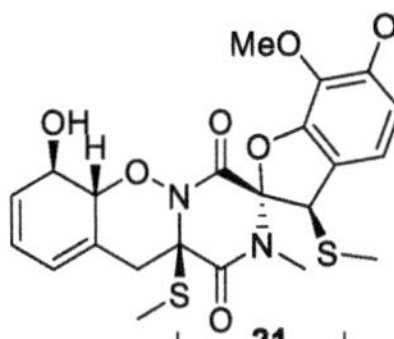

**21**

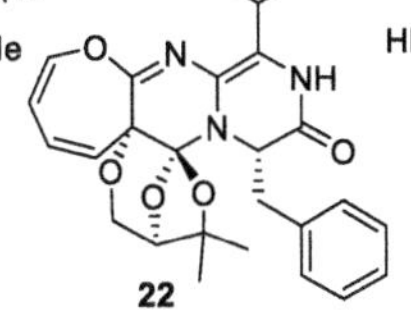

**22**

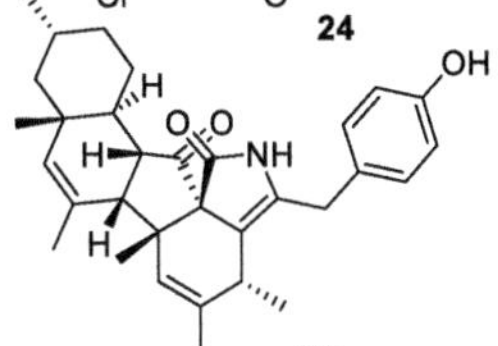

**27**

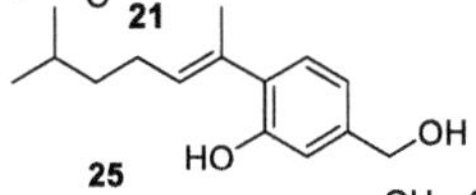

**25**

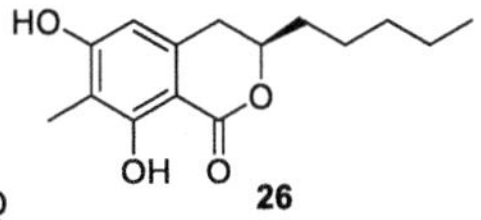

**26**

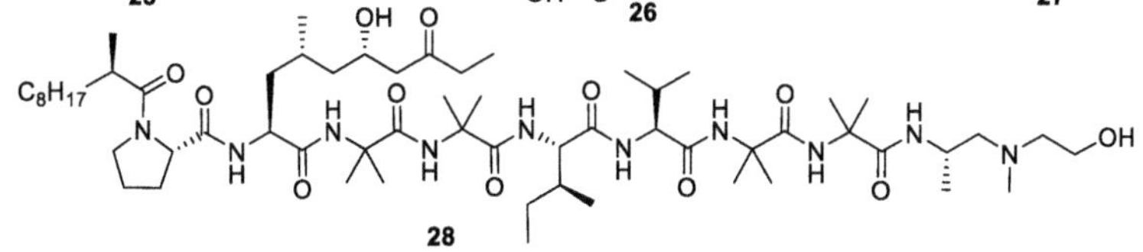

**28**

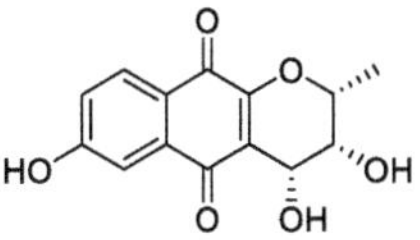

**29**

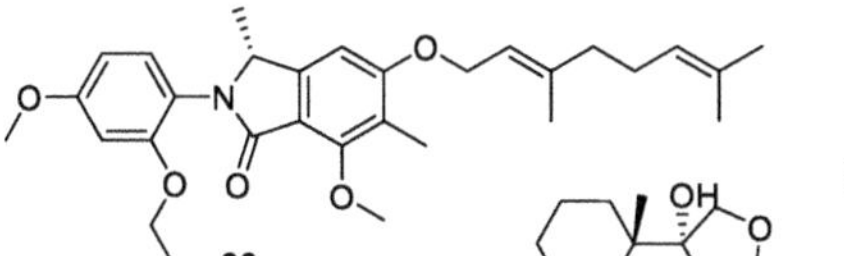

**30**

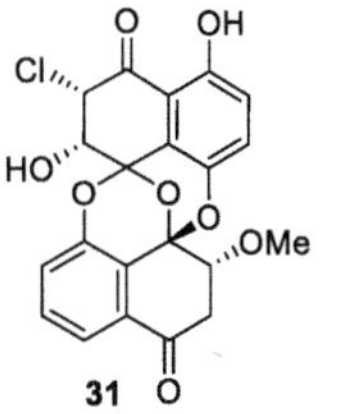

**31**

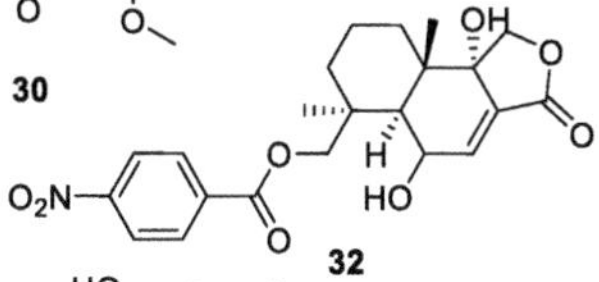

**32**

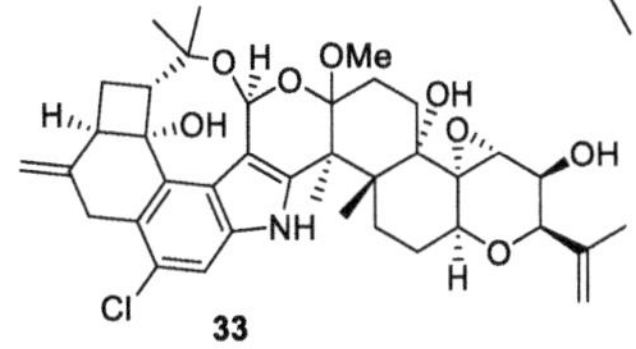

**33**

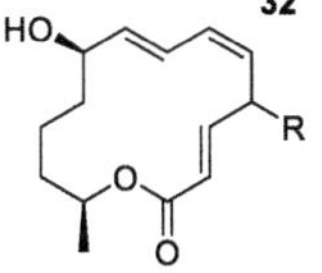

**34** R = α-OH
**35** R = β-OH

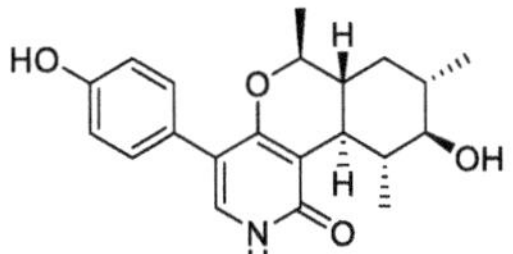

**36**

### 17.2.3 Marine Cyanobacteria

Cyanobacteria are an ancient group of oxygenic phototrophs equipped with a wide range of cellular strategies, physiological capacities, as well as other adaptations that allow their global colonization of a wide range of habitats. They thrive under diverse and often extreme range of environmental conditions (e.g., in marine environments, hypersaline lakes, terrestrial environments, freshwater lakes, and thermal springs) (Kurmayer et al. 2016; Mazard et al. 2016; Whitton and Potts 2012). More than 90 genera of cyanobacteria produce compounds with potential bioactivities, most of which belong to the orders Oscillatoriales, Nostocales, Chroococcales, and Synechococcales. In terms of their molecular diversity and relative bioactivity, the majority of the cyanobacterial orders remain poorly explored. The metabolites of cyanobacteria with potential bioactivity belong to about 10 different chemical classes (Demay et al. 2019).

In the marine environment, cyanobacteria occur as free-living organisms but also live associated with a variety of hosts (e.g., fungi, ascidians, corals, and protists). Up to the end of 2019, about 550 secondary metabolites were reported from diverse genera of marine cyanobacteria such as *Lyngbya*, *Moorea*, *Symploca*, and *Oscillatoria*. The biosynthetic pathways of cyanobacteria show unusual mechanistic and enzymatic features that result in the production of bioactive compounds with a variety of chemical structures (Tan and Phyo 2020). Several pharmacological trends have been observed amongst the various marine cyanobacterial secondary metabolites. An important number of molecules possess either potent cytotoxic, neuromodulating, or anti-infective properties (Aráoz et al. 2010; Costa et al. 2012; Niedermeyer 2015; Rivas and Rojas 2019). These compounds show potency and selectivity against human drug targets, including cancer, inflammation, and neuro-degenerative disorders. As such, these cyanobacterial secondary metabolites are considered prolific drug leads for drug discovery and development. For instance, cyanobacteria-derived compounds and their synthetic analogs have been reported as Antibody–Drug Conjugates (ADCs) and include dolastatin 10, auristatin E, and OKI-179. These ADCs have undergone or are currently undergoing clinical trials for the treatment of cancer diseases (Newman and Cragg 2014, 2017). Here we will discuss some of the mechanisms of the actions of ADCs.

Largazole (**37**), a cyclic depsipeptide, is a highly potent inhibitor of histone deacetylase (HDAC) class I, originally discovered as a secondary metabolite of the marine cyanobacterium *Symploca* sp. from Key Largo, USA. Largazole possesses a variety of unusual structural features, including the attachment of a 4-methylthiazoline unit to a thiazole and a 3-hydroxy-7-mercaptohept-4-enoic acid unit. In comparison with paclitaxel, actinomycin D and doxorubicin exhibited potent inhibition of growth of transformed human mammary epithelial cells (MDA-MB-231) with a $GI_{50}$ of 7.7 nM and exhibited exquisite antiproliferative activity against transformed fibroblastic osteosarcoma U2OS ($GI_{50}$ 55 nM) over non-transformed fibroblasts NIH3T3 ($GI_{50}$ 480 nM) (Taori et al. 2008). Carmaphycins A (**38**) and B (**39**) are potent novel proteasome inhibitors isolated in low yield from organic extracts of *Symploca* sp. from Carmabi beach, Curaçao. They were evaluated against

*Saccharomyces cerevisiae* 20S proteasome and have comparable $IC_{50}$ of ~2.5 nM. Such inhibitory effect is similar to that of epoxomicin and salinosporamide A (Pereira 2012).

The largamide D derivative (**41**) is generated by intramolecular largamide D condensation (**40**). This molecule, compared to largamide D, demonstrated an 11- and 33-fold decrease in activity against chymotrypsin and elastase, respectively. For serine protease inhibition, the Ahp moiety is necessary and any structural or conformational changes to this unit will influence the activity (Luo et al. 2016). Gallinamide A (**42**) is one of the most potent and selective marine cyanobacterial antimalarial compounds reported to date, with an $EC_{50}$ of 74 nM when tested against *Plasmodium falciparum* strain 3D7. *Plasmodium falciparum*-infected red blood cells treated with nanomolar concentrations of (**42**) have a swollen food vacuole phenotype. Using fluorescent probes based on rhodamine fluorophore-tagged molecules, it was discovered that (**42**) is a specific inhibitor of plasmodial cysteine proteases, falcipains 2, 2′, and 3. Moreover, for antimalarial activity, the methoxypyrrolinone unit in gallinamide A is critical (Stolze et al. 2012).

Grassystatins A (**43**) and B (**44**) displayed potent inhibitory activity against aspartyl proteases cathepsins D and E with an average $IC_{50}$ of 16.9 nM and 0.62 nM, respectively. Moreover, tasiamide B (**45**) is a statin-containing linear depsipeptide that displayed a potent activity against cathepsins D and E, with $IC_{50}$ of 50 nM and 9.0 nM, respectively (Al-Awadhi et al. 2017; Kwan et al. 2009; Tan et al. 2013; Turk 2006). One of the earliest examples of potent microtubule inhibitors reported from marine cyanobacteria is the dolastatin class of molecules, especially dolastatin 10 (**46**) (Poncet 1999). The apratoxins are a novel class of potent cytotoxic cyclodepsipeptides reported from several *Lyngbya* spp. When tested against a panel of cancer cell lines, including HT29, HeLa, and U2OS, nanomolar concentrations of apratoxins had a major anticancer effect. A total of nine compounds associated with apratoxin A has been identified to date with apratoxin A (**47**) being the most cytotoxic. Further research on apratoxin has shown that this molecule has a strong antiangiogenic activity by inhibiting the activation of retinal endothelial cells and pericytes by mediating multiple angiogenic pathways (Chen et al. 2011). Coibamide A (**48**) is a structurally novel cyclic depsipeptide with potent antiproliferative properties reported from a marine cyanobacterium of Panama. Coibamide A showed strong cytotoxicity against NCI-H460 lung cancer cells and mouse neuro-2a cells, with an $LC_{50}$ less than 23 nM. The compound was tested in the NCI 60-cell line and it exhibited activities against MDA-MB-231, LOX IMVI, HL-60(TB), and SNB-75 with $IC_{50}$ of 2.8 nM, 7.4 nM, 7.4 nM, and 7.6 nM, respectively. Coibamide A specifically targets the trimeric Sec61 translocon's Sec61 alpha subunit. Binding of coibamide A to Sec61 resulted in the inhibition of substrate-non-selective ER protein import and conferred strong cytotoxicity against particular cancer cell lines (Hau et al. 2013; Serrill et al. 2016; Tranter et al. 2020; Wan 2018).

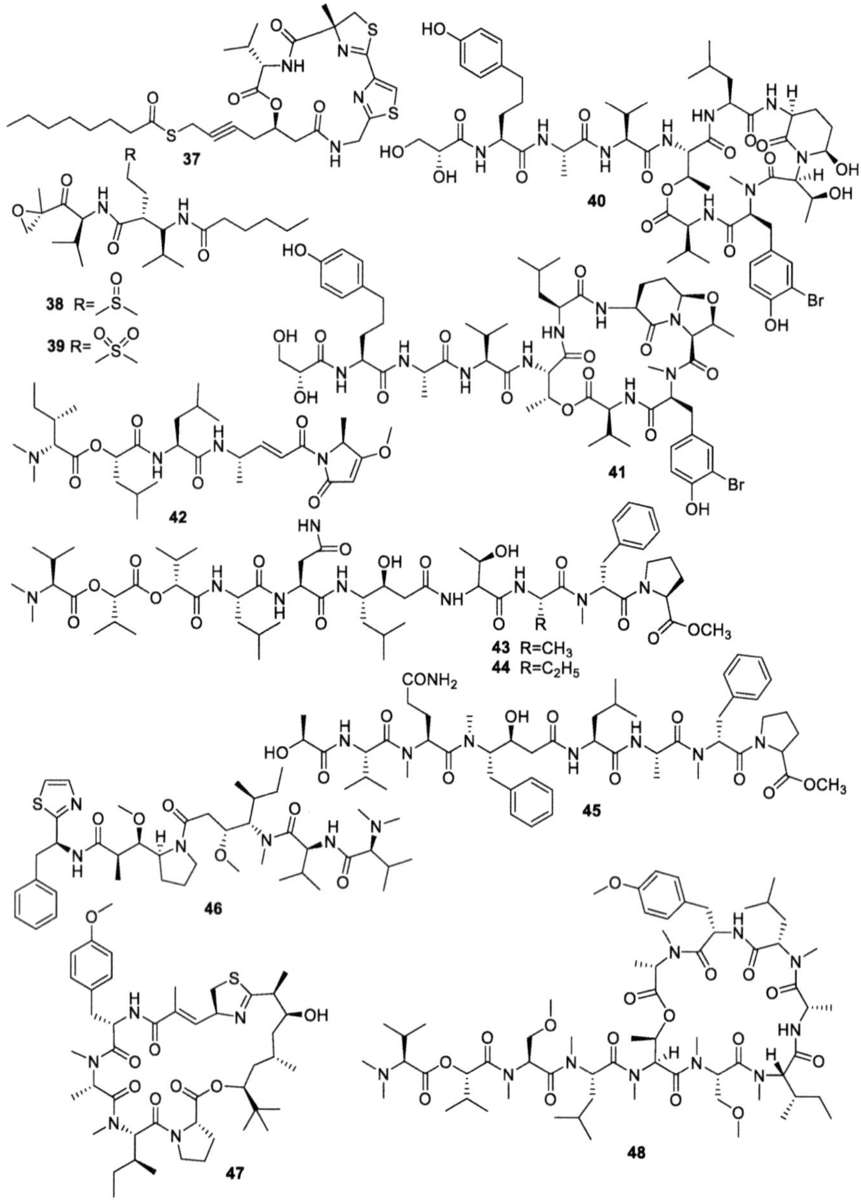
37
40
38 R=
39 R=
41
42
43 R=CH3
44 R=C2H5
45
46
47
48

## 17.3 Emerging Strategies for the Exploration of Marine Bioactive Compounds

### 17.3.1 In-Situ Isolation Technology

Thousands of microbial species remain un- or underexplored for their capacity to produce bioactive secondary metabolites as they cannot be grown in synthetic media (Nichols et al. 2010). The remarkable gap between the microbial richness in the biosphere and their often estimated as less than 1% culturability under laboratory conditions has been coined as 'The Great Plate Count Anomaly' (Staley and Konopka 1985). Accessing this missing microbial diversity almost certainly would lead to the discovery of a hitherto untapped mine of novel bioactive compounds (Epstein 2013). Recently, there has been an increase in efforts to isolate extremophiles with the use of cutting-edge equipment aiming to enhance the culturability of rare microbes. Uncovering marine rare actinomycetes has been attempted by focusing on deep-sea sediments sampled by using specialized remotely operated underwater vehicles (Bredholdt et al. 2007; Fenical et al. 1999; Pathom-Aree et al. 2006). Despite these efforts to obtain sediments and other materials, laboratory studies are still challenging as the current strategies of altering the nutritional composition and other physicochemical factors mimicking the natural habitat are painstakingly slow, emphasizing the need for radically new strategies (Berdy et al. 2017).

Kaeberlein et al. (2002) introduced the idea of moving the culturing into the natural habitat which led to the development of the in situ iChip. The use of a diffusion chamber (Fig. 17.4, 1) allowed naturally occurring growth factors to diffuse into the synthetic growth medium improving culturability (Berdy et al. 2017; Kaeberlein et al. 2002). Diffusion chambers are equipped on both sides with a membrane with 20–30 nm pore size. Using the in-situ diffusion chamber technique, marine microbes from intertidal sediment were serially diluted, mixed with warm agar made with sea salt, and the inoculated agar was placed in the diffusion chamber leaving a thin layer of air between agar and the top membrane of the diffusion chamber. Finally, the incubated diffusion chambers were transferred to an aquarium in which the natural environment was simulated by filling it with sediments collected from tidal flat and placing the diffusion chamber on the surface of the sediment and filling thin layer of air with seawater (Kaeberlein et al. 2002). The semipermeable membrane allows the exchange of nutrients and other chemicals between cells and the environment but retains the cells in their confined space. This resulted in a 300-fold improved recovery of microorganisms compared to conventional Petri dishes (Berdy et al. 2017; Kaeberlein et al. 2002). Subsequent studies have demonstrated that one to several incubations in a diffusion chamber leads to an increase in the number and diversity of environmental isolates and the ability to grow them in vitro (Bollmann et al. 2007).

A miniaturized in situ diffusion chamber has been developed by Ben-Dov et al. (2009) using a double encapsulation method for in situ culturing of microorganisms (Fig. 17.4, 2). This method cultures microorganisms in droplets of agar which are subsequently encapsulated by a polysulfonic polymeric membrane (PPM) forming a

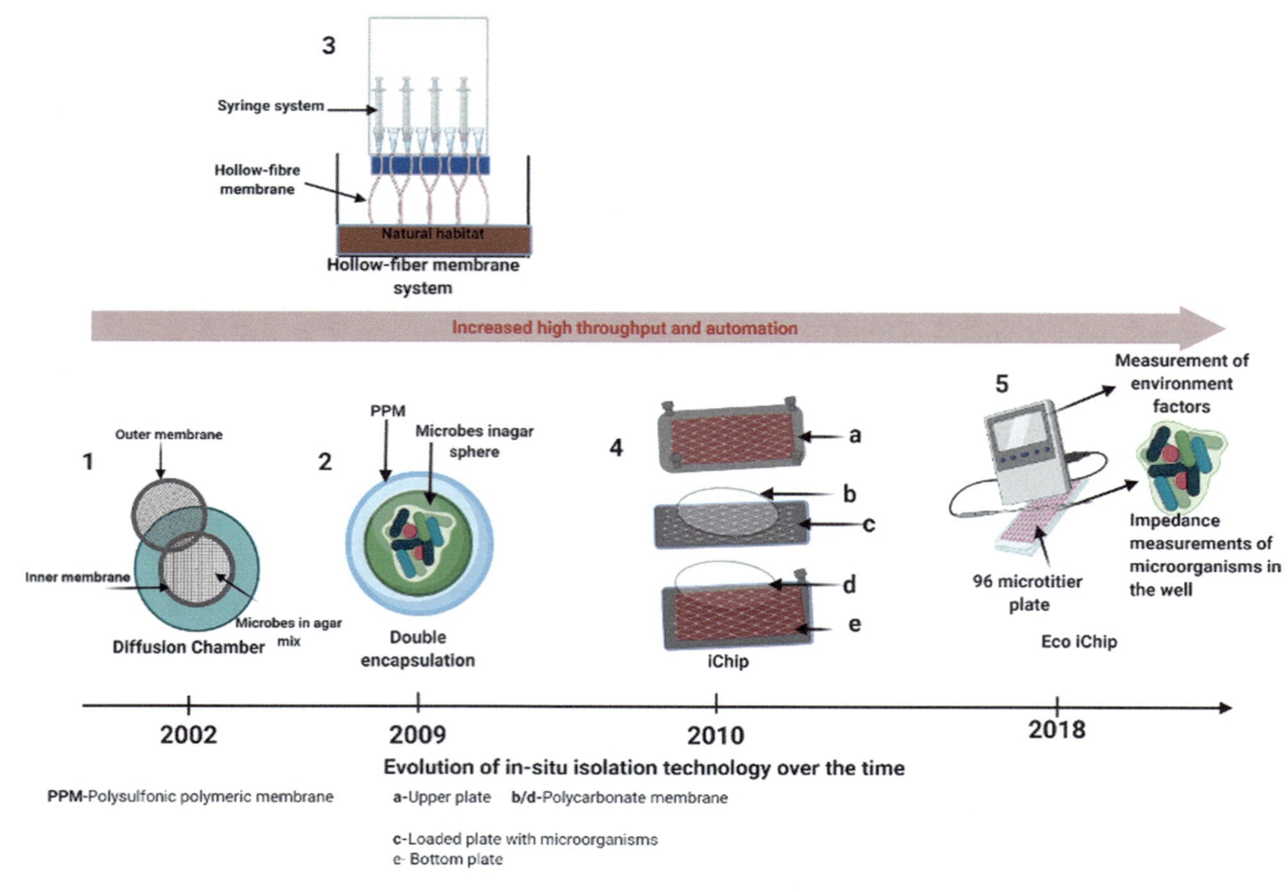


**Fig. 17.4** Evolution of in situ isolation and high throughput over time (figure was produced by licensed biorender.com software)

bilayer surrounding the capsule containing the microorganism. These capsules are then incubated under simulated or in natural environments such as the mucous surface of fungia coral or sediment. After incubation until a few generations of growth of the microorganism they acquired the ability to grow in conventional laboratory setups, presumably as a result of gradual adaptation to the growth conditions (Ben-Dov et al. 2009).

The hollow fiber membrane system is another invention that facilitates in situ isolation of microorganisms from simulated or natural environments (Aoi et al. 2009). The system consists of 48–96 hollow fiber chambers into which diluted environmental samples can be fed using syringes. One chamber unit consists of a porous hollow fiber polyvinylidene fluoride membrane (0.1 mm mean pore size, 67–70% porosity, 30 cm length, 1.2 mm outside diameter, 0.76 mm inside diameter) allowing an exchange of chemicals between microbes and the environment and also to exchange toxic by-products (Fig. 17.4, 3). The favorable feature of this system is the supply of various low concentration substrates, facilitating interspecies as well as intraspecies interactions through quorum sensing and other signaling pathways that are vital for many microorganisms for growth and survival (Aoi et al. 2009).

The above-mentioned diffusion chamber techniques for in situ isolation suffer from low throughput mainly because of the laborious isolation procedures needed for a single species. The diffusion chambers allow colonies containing different species to grow together, which limits their use for drug discovery (Berdy et al. 2017; Nichols et al. 2010). In order to improve the diffusion chamber technique, Nichols et al. (2010) invented the isolation chip or iChip (Fig. 17.4, 4). The iChip contains hundreds of miniature diffusion chambers each of which can accommodate a single microbial cell. During in situ incubation under simulated conditions or in a natural habitat, each miniature diffusion chamber allows the culturing of a single species in one step (Nichols et al. 2010). The iChip is a versatile concept that allows it to be applied in a variety of different situations such as in soil, in aquatic habitats, as well as for the human (and other) microbiomes (Berdy et al. 2017). Despite the high throughput nature of the modern in situ iChip concept, it still suffers from a few limitations which may require further modifications. Indeed, the isolation in one go of single colonies from environmental samples provides a non-laborious method for the isolation and identification of novel species. However, growing in isolation is not always ideal or even possible for some species as they may depend on symbiotic relationships or other types of association with other organisms. While the iChip seems to be an ideal device for an aquatic environment it works less well in a dry environment because the gel plaques in the micropores with microbes require moisture from its environment to prevent them from drying. Further modification may be possible when the microchambers are continuously supplied with water from the environment. Finally, long-term iChips fixed in position in an aquatic environment may run into anoxia. The thin layer of oxygen between sediment and the bottom surface of the iChip may represent an unnatural environment for the targeted microbes. It is therefore imperative to continuously aerate the water covering the sediment (Berdy et al. 2017).

Sylvain et al. (2018) developed the initial concept of the iChip of (Nichols et al. 2010) into an automated real time in situ microbial monitoring system called the Eco iChip (Fig. 17.4, 5) allowing the real-time measurement of growth conditions at the different growth phases of the life cycle eliminating the drawbacks associated with traditional non-automated iChips. This will be an important approach to allow these previously uncultured microorganisms to be domesticated in the laboratory by the continuous supply of their natural environmental factors.

The first successful application of the iChip in microbial drug research was the discovery of the new antibiotic teixobactin (**49**) (Hunter 2015). Teixobactin was isolated from an extract of a new species of β-proteobacteria provisionally named *Eleftheria terrae*. Another novel bacterium *Gallaecimonas mangrovi* HK-28, isolated from mangrove sediments using an in situ iChip, revealed antibacterial activity against the marine pathogen *Vibrio harveyi*. *Gallaecimonas mangrovi* HK-28 produces three new diketopiperazines gallaecimonamides A–C (**50–53**) (Ding et al. 2020; Zhang et al. 2018). A novel antibacterial *N*-acyltyrosine (**54**) was found in a new species of *Alteromonas* sp. RKMC-009 following application of an in situ iChip in the marine sponge *Xestospongia muta* (Macintyre et al. 2019).

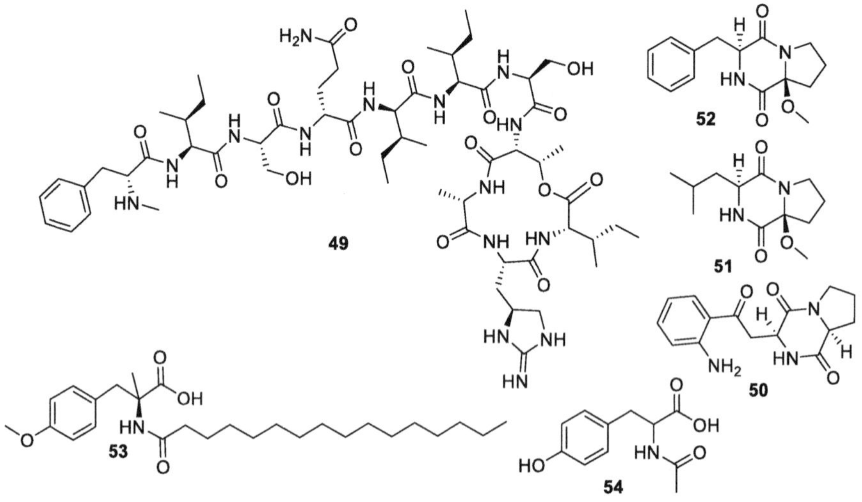

## 17.3.2 Microbial Co-culture

Culturing of two or more microbes as a strategy for the induction of new bioactive molecules has been applied since the first report of the serendipitous discovery of penicillin G (Fleming 1929). Growing two microbial species together has been repeatedly proven successful as a strategy of triggering silent secondary metabolite gene clusters (Rateb et al. 2013; Thissera et al. 2020). There are several theories that explain silent genetic perturbation by microbial co-culture in a synthetic environment (Seyedsayamdost et al. 2012). Co-existence in the natural environment comes with natural stresses such as competition for food and space or natural antagonism or

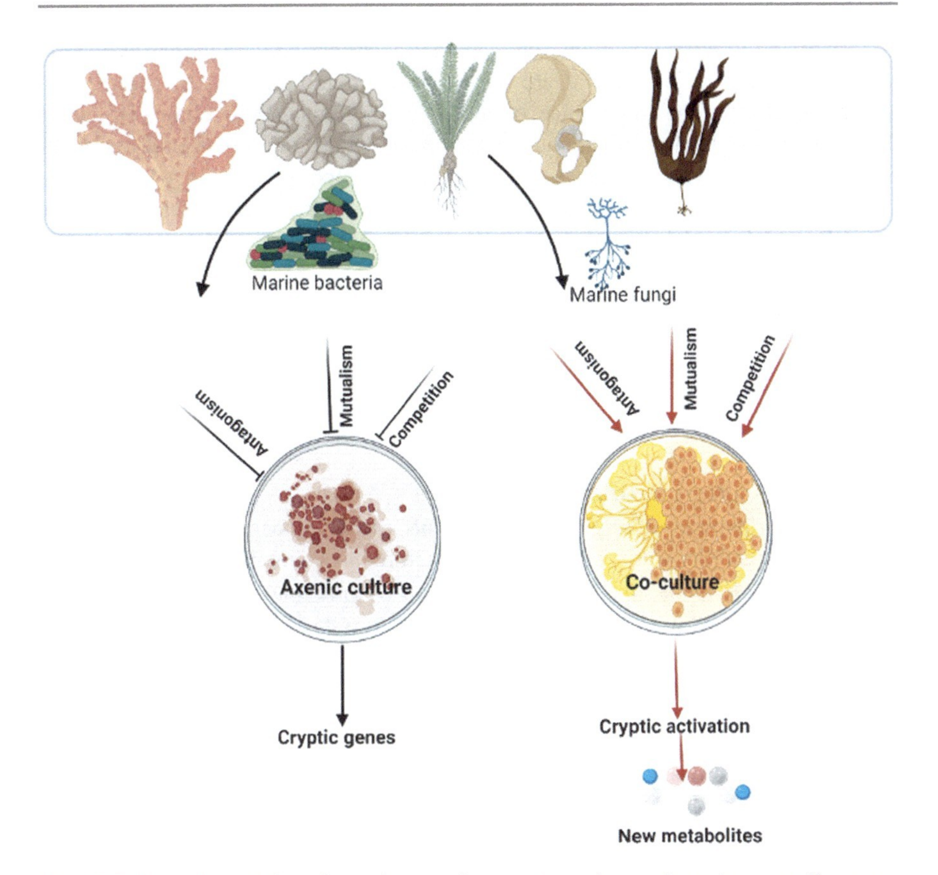


**Fig. 17.5** Natural simulation of co-culture environment over the axenic environment (figure was produced by using licensed biorender.com software)

mutualism which may affect gene silencing (Fig. 17.5) (Akone et al. 2019; Reen et al. 2015; Shin et al. 2018). Triggers relieving gene silencing included tight physical intimacy of the microorganisms (Cueto et al. 2001; Schroeckh et al. 2009; Wakefield et al. 2017), exchange of signaling molecules (Shi et al. 2020), or horizontal gene transfer (Kurosawa et al. 2010). The latter is especially the case when a pathogenic species is co-cultured with a non-pathogenic organism. In this section, we will discuss the most recent developments in the use of co-culture strategies to make the drug discovery pipeline more efficient for marine natural products.

#### 17.3.2.1 Marine Fungal-Bacterial Co-culture

The antibiotic agent pestalone (**55**) was reported as the first new secondary metabolite produced by a marine fungus *Pestalotia* sp., isolated from the surface of the brown alga *Rosenvingea* sp. that was collected in the Bahamas Islands and while being co-cultured with an unidentified marine bacterium (Cueto et al. 2001). Co-culturing of the marine sponge-derived actinomycete *S. rochei* MB037 with the coral reef-derived fungus *Rhinocladiella similis* 35 induced two new antibacterial fatty acids with a rare nitrile group: borrelidins J (**56**) and K (**57**) (Yu et al. 2019). The importance of the co-culture strategy to obtain structural diversification of bioactive compounds was illustrated by the isolation of ten novel prenylated 2,5 diketopiperazines 12b-hydroxy-13a-ethoxyverruculogen TR (**58**), 12b-hydroxy-13a-butoxyethoxyverruculogen TR-2 (**59**), hydrocycloprostatin A 9 (**60**), hydrocycloprostatin B (**61**), 25-hydroxyfumitremorgin B (**62**), 12b-hydroxy-13a-butoxyethoxyfumitremorgin B (**63**), 12b-hydroxy-13a-methoxyverruculogen (**64**), 26a-hydroxyfumitremorgin A (**65**), 25-hydroxyfumitremorgin A (**66**), and diprostatin A (**67**) with potent BRD4 inhibitory action. These structurally diverse metabolites were produced following co-culturing of the marine fungus *Penicillium* sp. DT-F29 and the marine bacterium *Bacillus* sp. B31 (Yu et al. 2017).

#### 17.3.2.2 Co-culturing of Marine Bacteria

A new depsipeptide dentigerumycin E (**68**) with moderate anticancer and antimetastatic activity was produced in a co-culture of two marine bacteria, *Streptomyces* sp. and *Bacillus* sp., both isolated from an intertidal mud flat in Wando, Republic of Korea (Shin et al. 2018). The interspecies bacterial interactions of a set of marine invertebrates associated *Micromonosporaceae* were studied using an innovative co-culture platform. While these bacteria exuded secondary metabolites, they did not as axenic cultures (Adnani et al. 2015). This co-culture platform was high throughput compared to conventional Petri dish approaches and it revealed that 12 species out of 65 *Micromonosporaceae* excreted different secondary metabolites in different combinations of species. The co-culturing of a *Rhodococcus* sp. and a *Micromonospora* sp. is another noteworthy case in which silent genes were induced. This led to a new antibacterial bis-nitroglycosylated anthracycline: keyicin (**69**), that possesses selective antibacterial activity against Gram-positive bacteria including *Rhodococcus* sp. as well as *Mycobacterium* sp.

#### 17.3.2.3 Co-culturing of Marine Fungi

Oppong-Danquah et al. (2020) used a systematic approach based on comparative metabolomics and bioactivity to select the best pair of fungi for co-culturing. Monoculture extracts of the marine fungi *Plenodomus influorescens*, *Penicillium bialowiezense*, *Sarocladium strictum*, *Helotiales* sp., two strains of *Pyrenochaeta* sp., and two strains of *Lentithecium* sp. were screened against a series of phytopathogens followed by ranking them through their antiphytopathogenic activities. Subsequently, species were paired as weak-weak, weak-strong, strong-strong on solid agar plates as the best visualizing modes for macro-interactions (Bertrand et al. 2013). All co-culture extracts were analyzed and compared with their monoculture extracts using metabolomics and screening for antiphytopathogenic activity in order to ascertain the deteriorated chemical profiles and enhanced bioactivities. *P. influorescens* (strong partner) with *Pyrenochaeta nobilis* (weak partner) was selected for a large-scale analysis. This resulted in the isolation of five polyketides of which dendrodolide N (**70**) and 8α-hydroxy-spiciferinone (**71**) were new (Oppong-Danquah et al. 2020). A co-culture of two fungi of the genus

*Aspergillus* (BM-05 and BM-05ML), isolated from a brown alga belonging to the genus *Sargassum* collected off the North Sea island Helgoland, produced a new cyclopeptide psychrophilin E (**72**), which has an antiproliferative effect against various cancer cell lines (Ebada et al. 2014). Two strains of *Aspergillus* sp., isolated from rotten fruit of the mangrove species *Avicennia marina*, produced a new antibacterial alkaloid aspergicin (**73**) during mixed fermentation (Zhu et al. 2011).

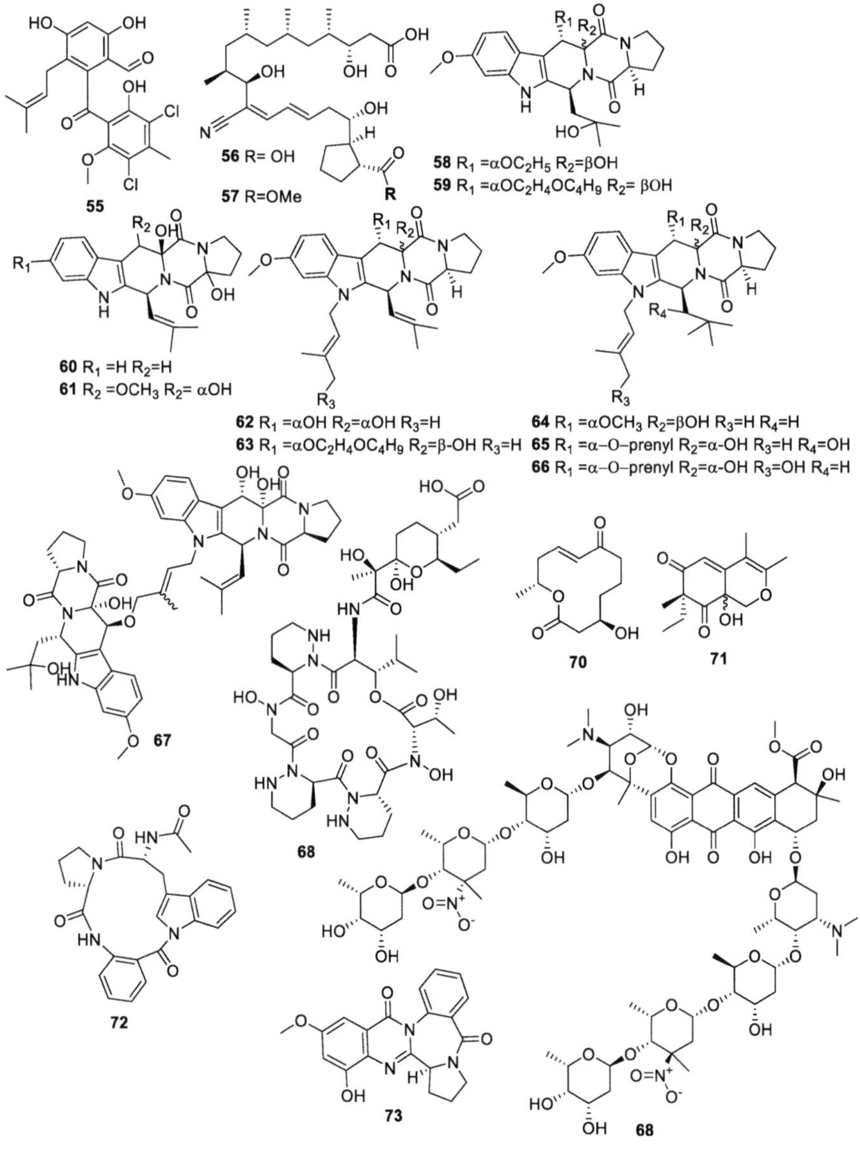

**55**

**56** R= OH
**57** R=OMe

**58** $R_1$ =$\alpha OC_2H_5$ $R_2$=$\beta$OH
**59** $R_1$ =$\alpha OC_2H_4OC_4H_9$ $R_2$= $\beta$OH

**60** $R_1$ =H $R_2$=H
**61** $R_2$ =$OCH_3$ $R_2$= $\alpha$OH

**62** $R_1$ =$\alpha$OH $R_2$=$\alpha$OH $R_3$=H
**63** $R_1$ =$\alpha OC_2H_4OC_4H_9$ $R_2$=$\beta$-OH $R_3$=H

**64** $R_1$ =$\alpha OCH_3$ $R_2$=$\beta$OH $R_3$=H $R_4$=H
**65** $R_1$ =$\alpha$–O–prenyl $R_2$=$\alpha$-OH $R_3$=H $R_4$=OH
**66** $R_1$ =$\alpha$–O–prenyl $R_2$=$\alpha$-OH $R_3$=OH $R_4$=H

**67**

**68**

**70**

**71**

**72**

**73**

**68**

### 17.3.3 The OSMAC (One Strain Many Compounds) Approach

In terms of the production of bioactive compounds, the genomes of many promising marine microorganisms reveal a much larger number of biosynthetic gene clusters than are expressed under normal culture conditions, hence genetic manipulation may be required in order to have them expressed (Romano et al. 2018). Approaches such as gene knock-out (Wang et al. 2006), heterologous or homologous expression, pathway-specific activation of regulatory or promotor genes, require a full and precise knowledge of the specific functions of the genes (transcriptional/regulatory functions). In addition, much biological information about the targeted organisms is needed, requiring advanced analytical equipment and study of the data that is generated. Bioinformatics allows the prediction of tentative structures of specialized metabolites produced by silent genes although it sometimes leads to wrong conclusions (Kim et al. 2017; Rutledge and Challis 2015). An example of an incorrect prediction is that for the gene cluster responsible for the biosynthesis of the pyrrolamides, a family of secondary metabolites known as DNA minor groove binders. This gene cluster was earlier reported as the gene cluster responsible for producing congocidine (netropsin) in *Streptomyces ambofaciens*. However, later studies demonstrated that there are two distinct pyrrolamide-like gene clusters in *Streptomyces netropsis* DSM40486 working reciprocally to produce three pyrrolamides: distamycin, congocidine, and a hybridized congocidine-distamycin called disgocidine (Vingadassalon et al. 2015). Without the second study, pathway-specific activation of only the pyrrolamide gene cluster would not have revealed the production of the other two pyrrolamides, distamycin and disgocidine as they are coded in other pyrrolamide-like gene clusters (Vingadassalon et al. 2015). Such pitfalls can be avoided by the global alteration of microbial physiology (Romano et al. 2018).

Global alteration of microbial physiology which aims at the whole biosynthetic network without targeting a specific one or two biosynthetic pathways to trigger different chemical profiles from single species is termed OSMAC (One Strain Many Compounds) by Zeeck and co-workers (Bode et al. 2002). However, similar experiments in which the culture conditions were varied date back to 1975 (Okazaki et al. 1975). In OSMAC, systematical alteration of culture conditions such as media composition, aeration, salinity, culture vessels, and temperature result in altered chemical profiles leading to the discovery of new secondary metabolites. The altered chemical profile that is obtained as a result of culturing under different conditions is assumed to originate from the provision of different environmental simulations causing activation of different biosynthetic pathways without the need for genetic manipulation (Bode et al. 2002; Romano et al. 2018). Changing environmental factors influence the biosynthesis of secondary metabolites at different levels such as at the transcriptome and proteome level (Fig. 17.6).

#### 17.3.3.1 OSMAC with Alteration of Food Source

The earliest example was reported by Okazaki et al. (1975) who used different nutrient conditions in order to trigger the production of the antibiotic SS-228 Y (**74**)

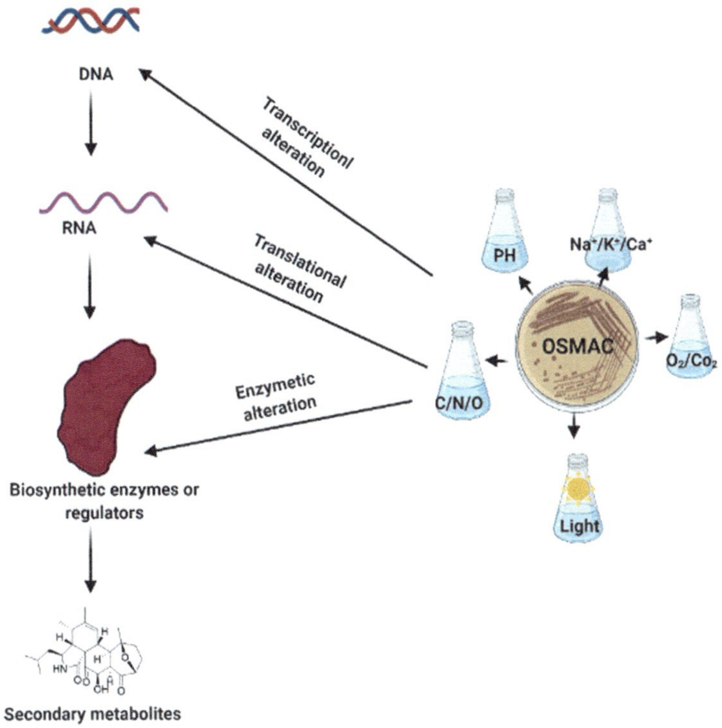

**Fig. 17.6** Different levels of influences by OSMAC in the biosynthesis of secondary metabolites (figure was produced using licensed biorender.com software)

by the marine bacterium *Chainia purperogena* that was isolated from shallow sea mud. After a series of optimizations using different media components, only the medium containing Kobu-Cha (powdered Laminaria seaweed) produced the new antibiotic SS-228-Y. The marine fungus *Spicaria elegans* KLA03 produced a new polyketide eleganketal A (**75**), possessing a rare and highly oxygenated spiro [isobenzofuran-1,3′-isochroman] ring, when cultured in a modified mannitol-based medium ($NH_4Cl$ as nitrogen source) (Luan et al. 2014). Growing the marine fungus *Scedosporium apiospermum* F41–1 in GPY medium boosted alkaloid production. Growth in GPY medium supplemented with L-tryptophan, L-phenylalanine, L-threonine, and L-methionine resulted in distinct chemical profiles and led to the isolation of 22 alkaloids, including 18 quinazoline-containing indole alkaloids, three formamides, and one isocyanide, as well as 14 new scedapins A–G (**76-82**) and scequinadolines A–G (**83-87**). Scedapin C (**88**) and scequinadolin D (**89**) have potential antiviral effects against hepatitis C (Huang et al. 2017).

### 17.3.3.2 OSMAC with Solid and Liquid Media

It has been shown that solidified media give rise to different metabolite profiles than their liquid version. Imanaka et al. (2010) showed that the use of a solid substrate for the fungus *Aspergillus oryzae* IAM 2706 led to a more diverse chemical profile than liquid shaking cultures with the same medium. Contrary to this work, Doran (2013) argued that the mobility of the nutrients, signaling molecules, oxygen, and foraging, are less well simulated in a solid medium compared to its liquid counterpart. However, the improved chemical profile may be understood as the lack of environmental stimulation that otherwise would produce these specialized metabolites in order to withstand environmental stresses. Another classical example of a possible chemical profile disparity and induction of new interesting metabolites when culturing was switched from liquid to solid medium is the marine *Penicillium* sp. F23-2. This fungus displayed an altered chemical profile when it was grown on a rice medium rather than in a liquid potato-based medium. Under these conditions *Penicillium* sp. F23-2 produced five new ambuic acid analogs: penicyclones A–E (**90-94**), with potential antibacterial activity against *Staphylococcus aureus* (Guo et al. 2015).

### 17.3.3.3 OSMAC with Changes in Physical Factors

Alteration of aeration, temperature, and pH have a strong effect on the production of microbial secondary metabolites. Aeration increases the metabolic rates and the conversion of the substrate(s) in the growth medium. This is usually achieved by shaking culture flasks or by bubbling air (Romano et al. 2018). The hypoxia-driven decrease in the production of the antibiotic napyradiomycin in the marine streptomycete strain CNQ-525 results in a spike of the intermediate 8-amino-flavioli. The conversion of 8-amino-flavioli into napyradiomycin is a redox reaction and a constant aeration is therefore important for the production and the yield of this antibiotic (Gallagher et al. 2018).

Considering the secondary metabolism of marine microbes, osmotic pressure, salinity, and pH are among the most relevant physical factors. For instance, habitats such as the deep sea, intertidal areas, shallow sea sediments, or mangroves all differ in osmotic pressure, salinity, and dissolved gases, and their microbiomes are adapted to that. Saha et al. (2005) showed the importance of mimicking the natural sea salinity conditions for the production of an antibacterial lipid from a marine-derived actinobacterium. Overy et al. (2017) published a comprehensive systematic study that confirmed the effect of osmotic pressure and salinity on the growth and production of secondary metabolites in fungi.

Hitherto, most studies involved tedious, laborious, and low throughput OSMAC applications in which one factor at time was optimized. Mathematical models are now used in order to avoid arduous and complicated optimization techniques during fermentation and improve the reproducibility of methods (Singh et al. 2017).

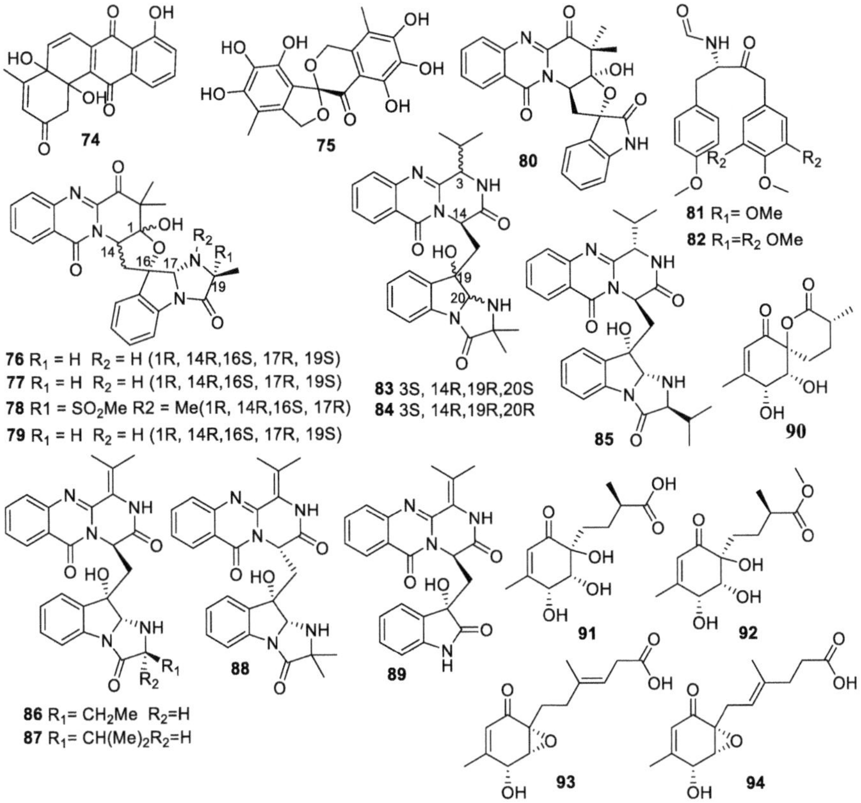

74 75 80

81 $R_1$= OMe
82 $R_1$=$R_2$ OMe

76 $R_1$ = H $R_2$ = H (1R, 14R,16S, 17R, 19S)
77 $R_1$ = H $R_2$ = H (1R, 14R,16S, 17R, 19S)
78 R1 = $SO_2Me$ R2 = Me(1R, 14R,16S, 17R)
79 $R_1$ = H $R_2$ = H (1R, 14R,16S, 17R, 19S)

83 3S, 14R,19R,20S
84 3S, 14R,19R,20R

85 90

86 $R_1$= $CH_2Me$ $R_2$=H
87 $R_1$= $CH(Me)_2R_2$=H

88 89 91 92 93 94

## 17.3.4 Chemical Elicitation

Additions to culture media were made in order to divert existing or induce new biochemical pathways in microorganisms. Alginate added to a culture of bifidobacteria (Akiyama et al. 1992) and chitosan to a culture of *Fusarium oxysporum* (El Ghaouth et al. 1994) enhanced growth and caused morphological and structural changes in the microorganisms. These studies indicated that small molecules reported as elicitors mainly served as signaling molecules (Eberhard et al. 1981; Nealson et al. 1970) or as epigenetic modulators (Shwab et al. 2007).

### 17.3.4.1 Quorum Sensing Elicitors

The chemical signaling network in Gram-negative bacteria is mainly mediated by AHLs (Acylated Homoserine Lactones) the release of which exhibits a population density-dependent regulation. The first report of an AHL involved in cell signaling was from the symbiotic marine bacterium *Vibrio fischeri* in which it induced luciferase synthesis (Eberhard et al. 1981). Four out of 43 marine snow-derived bacteria including *α-Proteobacteria* and *Roseobacter* spp. produced AHLs that induced phenotypic traits such as biofilm formation, co-enzyme synthesis, and antibiotic production (Gram et al. 2002). The use of quorum sensing (QS) molecules produced by distantly related species to introduce antagonism by disrupting the native QS system by altering the gene transcription is an intriguing strategy for the induction of QS-controlled antagonistic secondary metabolites. In Gram-negative bacteria, native AHL receptors are activated by native AHL as well as non-native AHL molecules that possess slightly changed structures. Thus, the exogeneous addition of non-native AHLs might initiate new biosynthetic pathways by interfering with native QS networks. This phenomenon is demonstrated by the inhibition of bioluminescence of the marine bacterium *Vibrio harveyi* by two phenethylamine metabolites produced by the marine bacterium *Halobacillus salinus* by competing with 3-oxo-hexanoyl-homoserine lactone (OHHL) which accounts for bioluminescence in *Vibrio harveyi* (Teasdale et al. 2009). *N*-acetyl-D-glucosamine (GlcNAc), a component of peptidoglycan in the bacterial cell wall and of chitin in fungal cell wall, was released into the environment during cell repairing and may play a role as a signaling molecule (Dashti et al. 2017; Konopka 2012). Incorporation of GlcNAc in the culture media of the sponge-derived actinobacteria *Rhodococcus* sp. RV157 and *Actinokineospora* sp. EG49 demonstrated the chemical elicitation of new secondary metabolites. These include the induction of the new siderophore bacillibactin (**95**) and the surfactin antibiotic (**96**) from *Rhodococcus* sp. RV157, and the amplification of the new metabolites actinosporins E–H (**97–100**) from *Actinokineospora* sp. EG49 (Dashti et al. 2017).

#### 17.3.4.2 Epigenetic Elicitation

Epigenetic elicitation or genetics-free manipulation allows upregulation of secondary metabolite pathways (Ganesan et al. 2019). Achieving epigenetic elicitation by means of small-molecule epigenetic modifiers is an efficient alternative for invasive genetic manipulations such as gene knockout, heterologous expression, homologous expression, and ribosome engineering. Epigenetics is the study of heritable phenotypes including secondary metabolites without the need to make changes to the DNA (Pfannenstiel and Keller 2019). In some instances, the biosynthetic gene cluster is not transcribed (so called "gene silencing") in eukaryotes, including fungi, because of the compact arrangement of the DNA strands around histones that form tightly packed nucleosomes. This is a reversible enzymatic process. The three main regulatory enzymes taking part in this process are histone acetyltransferases (HAT), histone deacetylases (HDAC), and DNA methyltransferase (DMT). These enzymes alter the nature of the packaging of DNA around histones by adding $-CH_3$ or $-C(=O)-CH_3$ (Ac) into DNA or histone tails. Incorporation of acyl moieties into histone tails driven by HAT, forming loosely packed chromatin (so called euchromatin regions), makes the conserved biosynthetic gene clusters (BGCs) available for

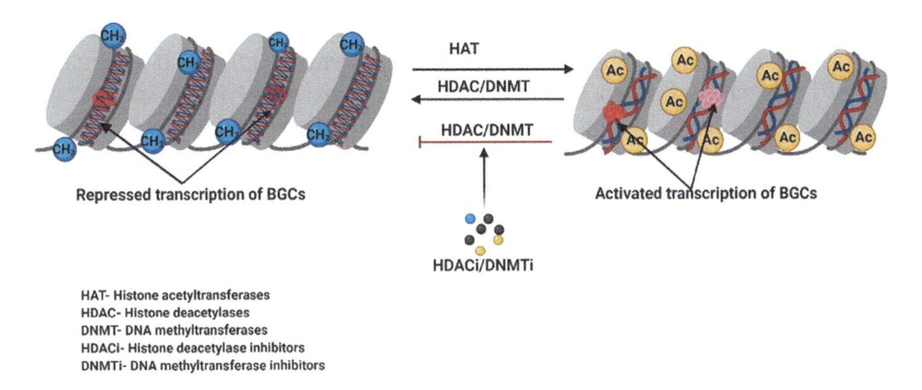


**Fig. 17.7** Cryptic gene induction by epigenetic elicitors (figure was produced using licensed biorender.com software)

transcription. While the incorporation of $-CH_3$ into DNA or histone tails driven by respectively DNMT and HDAC does the opposite action of acyl moieties making the chromatin packaging denser and tighter, forming so-called heterochromatins leaving them more conserved and transcriptionally inactive (Ganesan et al. 2019; Jin et al. 2011; Pfannenstiel and Keller 2019) (Fig. 17.7). The reverse action of HAT is driven by HDAC re-producing heterochomatins showing the importance of HDAC inhibitors (HDACi) thereby triggering cryptic gene activation. The use of DNMT inhibitors (DNMTi) discourages the formation of heterochromatin (Mao et al. 2018; Pfannenstiel and Keller 2019; Ramadan et al. 2015).

HDACi are small molecules and beneficial for use in oncology as promising cancer therapies. Their ability to revert abnormal epigenetic features associated with many types of cancers made them popular as anticancer medication (Hull et al. 2016; Sanaei and Kavoosi 2019). In eukaryotes, HDACi are known to regulate gene expression at different levels such as transcription factor activity, miRNA expression, and signal transduction pathways (Alao 2004; Fournel et al. 2008; Hull et al. 2016; Romano et al. 2018). These findings encouraged researchers to use HDACi for the induction of cryptic pathways in fungi to discover the hidden genome and hence the discovery of new secondary metabolites. The first application of HDACi was an induction of new metabolites from *Alternaria alternata* and *Penicillium expansum* with HDACi and the antifungal antibiotic trichostatin A (Shwab et al. 2007). This introduced the use of HDACi inhibitors into microbial natural product chemistry. The main mechanism of cryptic induction by HDACi is by inhibiting the histone deacetylases leaving the euchromatins accessible for gene transcription and initiating new cryptic biosynthetic pathways (Janssens et al. 2019; Pfannenstiel and Keller 2019). The marine microbiome has been reported to be a profound resource for producing HDAC inhibitors which are currently being used as promising anticancer therapies such as chromopeptide A (**101**), a depsipeptide isolated from the marine sediment-derived bacterium *Chromobacterium* sp. (Sun et al. 2017). Varghese et al. (2015) demonstrated the importance of marine actinomycetes for the production of HDACi. A study involving a variety of fungi, including several marine species, was conducted in order to assess the broad-spectrum elicitation effect on their chemical profiles by both HDACi and DNMTi (Williams et al. 2008). This demonstrated the importance of these molecules for cryptic induction. The study of Williams et al. (2008) showed that the marine fungus *Cladosporium cladosporioides*, treated with HDACi suberoylanilide hydroxamic acid, produced a complex series of perylenequinones, including the new metabolites cladochromes F (**102**) and G (**103**) along with four known cladochromes.

Basically, the mode of cryptic induction by DNMTi is by reversing the action of DNMT, which catalyzes the amalgamation of methyl groups from S-adenyl methionine to the fifth carbon of a cytosine residue to form 5-methylcytosine. Methylated DNA strands enhance the attractive forces with histones wrapping tightly around them making DNA transcriptionally quiescent (Moore et al. 2013). The marine fungus *Cochliobolus lunatus* treated with well-known DNMTi 5-azacytidine induced two new biologically active α-pyrones: cochliobopyrones A (**104**) and B (**105**) (Wu et al. 2019). In another study with the same species (strain TA26–46), the organism was cultured with exogenously incorporated 5-azacytidine. This study exemplifies the potential cryptic induction by producing seven new diethylene glycol phthalate esters: cochphthesters A –G (**106–111**) (Chen et al. 2016b).

The application of epigenetic elicitors in the cryptic induction of bacterial species was not encouraging due to the absence of highly organized DNA and histone complexes forming nucleosomes. This explains the failure of the epigenetic elicitation by HDACi and DNMTi. However, several studies have reported the possible epigenetic elicitation in bacterial cells by HDACi and DNMTi (Moore et al. 2012; Okada and Seyedsayamdost 2017). The identification of natural and synthetic elicitors at the microscale level, for instance, the ground-breaking invention of high throughput elicitor screening, (HiTES) (Seyedsayamdost 2014) and its recent advancement HiTES-IMS (Seyedsayamdost 2019) will be intriguing for future applications in the area of chemical elicitation.

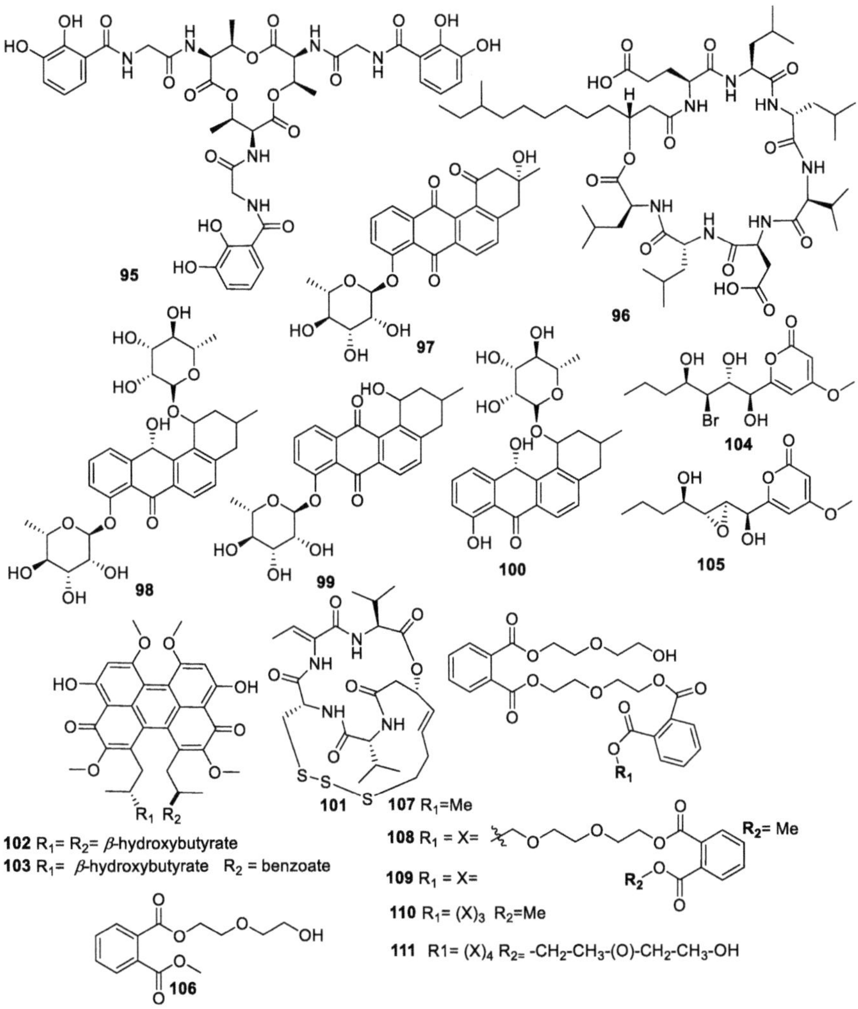

95 96 97 98 99 100 101 104 105 106

102 $R_1$= $R_2$= $\beta$-hydroxybutyrate
103 $R_1$= $\beta$-hydroxybutyrate $R_2$ = benzoate

107 $R_1$=Me
108 $R_1$ = X= $R_2$= Me
109 $R_1$ = X=
110 $R_1$= $(X)_3$ $R_2$=Me
111 R1= $(X)_4$ $R_2$= $-CH_2-CH_3-(O)-CH_2-CH_3-OH$

## References


Abraham EP (1979) A glimpse of the early history of the cephalosporins. Rev Infect Dis 1:99–105

Adnani N, Vazquez-Rivera E, Adibhatla SN et al (2015) Investigation of interspecies interactions within marine Micromonosporaceae using an improved co-culture approach. Mar Drugs 13: 6082–6098. https://doi.org/10.3390/md13106082

Akiyama H, Murata K, Endo T et al (1992) Effect of depolymerized alginates on the growth of Bifidobacteria. Biosci Biotechnol Biochem 56:355–356. https://doi.org/10.1271/bbb.56.355

Akone SH, Pham CD, Chen H et al (2019) Epigenetic modification, co-culture and genomic methods for natural product discovery. Phys Sci Rev 4:2018–0118. https://doi.org/10.1515/psr-2018-0118

Alao JP (2004) Histone deacetylase inhibitor trichostatin A represses estrogen receptor-dependent transcription and promotes proteasomal degradation of cyclin D1 in human breast carcinoma cell lines. Clin Cancer Res 10:8094–8104. https://doi.org/10.1158/1078-0432.CCR-04-1023

Al-Awadhi FH, Law BK, Paul VJ et al (2017) Grassystatins D-F, potent aspartic protease inhibitors from marine cyanobacteria as potential antimetastatic agents targeting invasive breast cancer. J Nat Prod 80:2969–2986

Almeida C, Hemberger Y, Schmitt SM et al (2012) Marilines A-C: novel phthalimidines from the sponge-derived fungus *Stachylidium* sp. Chem Eur J 18:8827–8834

Andryukov B, Mikhailov V, Besednova N (2019) The biotechnological potential of secondary metabolites from marine bacteria. J Mar Sci Eng 7:176. https://doi.org/10.3390/jmse7060176

Aoi Y, Kinoshita T, Hata T et al (2009) Hollow-fiber membrane chamber as a device for in situ environmental cultivation. Appl Environ Microbiol 75:3826–3833. https://doi.org/10.1128/AEM.02542-08

Aráoz R, Molgo J, Tandeau de Marsac N et al (2010) Neurotoxic cyanobacterial toxins. Toxicon 56: 813–828

Ben-Dov E, Kramarsky-Winter E, Kushmaro A (2009) An in-situ method for cultivating microorganisms using a double encapsulation technique. FEMS Microbiol Ecol 68:363–371. https://doi.org/10.1111/j.1574-6941.2009.00682.x

Berdy B, Spoering A, Ling L et al (2017) In situ cultivation of previously uncultivable microorganisms using the ichip. Nat Protocols 12:2232–2242. https://doi.org/10.1038/nprot.2017.074

Bertrand S, Schumpp O, Bohni N et al (2013) Detection of metabolite induction in fungal co-cultures on solid media by high-throughput differential ultra-high pressure liquid chromatography-time-of-flight mass spectrometry fingerprinting. J Chromatog A 1292:219–228. https://doi.org/10.1016/j.chroma.2013.01.098

Bister B, Bischoff D, Ströbele M et al (2004) Abyssomicin C—A polycyclic antibiotic from a marine *Verrucosispora* strain as an inhibitor of the p-aminobenzoic acid/tetrahydrofolate biosynthesis pathway. Angew Chem Int Ed 43:2574–2576

Blunt JW, Copp BR, Munro MH et al (2003) Marine natural products. Nat Prod Rep 20:1–48

Blunt JW, Copp BR, Munro MH, Northcote PT et al (2004) Marine natural products. Nat Prod Rep 21:1–49

Blunt JW, Copp BR, Munro MH, Northcote PT et al (2005) Marine natural products. Nat Prod Rep 22:15–61

Blunt JW, Copp BR, Munro MH, Northcote PT et al (2006) Marine natural products. Nat Prod Rep 23:26–78

Blunt JW, Copp BR, Munro MH, Northcote PT et al (2007) Marine natural products. Nat Prod Rep 24:31–86

Blunt JW, Copp BR, Munro MH, Northcote PT et al (2008) Marine natural products. Nat Prod Rep 25:35–94

Blunt JW, Copp BR, Munro MH, Northcote PT et al (2009) Marine natural products. Nat Prod Rep 26:170–244

Blunt JW, Copp BR, Munro MH, Northcote PT et al (2010) Marine natural products. Nat Prod Rep 27:165–237

Blunt JW, Copp BR, Munro MH, Northcote PT et al (2011) Marine natural products. Nat Prod Rep 28:196–268

Blunt JW, Copp BR, Munro MH, Northcote PT et al (2012) Marine natural products. Nat Prod Rep 29:144–222

Blunt JW, Copp BR, Munro MH, Northcote PT et al (2013) Marine natural products. Nat Prod Rep 30:237–323

Blunt JW, Copp BR, Keyzers RA et al (2014) Marine natural products. Nat Prod Rep 31:160–258

Blunt JW, Copp BR, Keyzers RA et al (2015) Marine natural products. Nat Prod Rep 32:116–211

Blunt JW, Copp BR, Keyzers RA et al (2016) Marine natural products. Nat Prod Rep 33:382–431

Blunt JW, Copp BR, Keyzers RA et al (2017) Marine natural products. Nat Prod Rep 34:235–294

Blunt JW, Carroll AR, Copp BR, Davis RA, Keyzers RA, Prinsep MR (2018) Marine natural products. Nat Prod Rep 35(1):8–53. https://doi.org/10.1039/C7NP00052A
Bode HB, Bethe B, Hofs R et al (2002) Big effects from small changes: possible ways to explore nature's chemical diversity. Chembiochem 3:619–627
Bollmann A, Lewis K, Epstein S (2007) Incubation of environmental samples in a diffusion chamber increases the diversity of recovered isolates. Appl Environ Microbiol 73:6386–6390. https://doi.org/10.1128/AEM.01309-07
Bredholdt H, Galatanko O, Engelhardt K et al (2007) Rare actinomycete bacteria from the shallow water sediments of the Trondheim fjord, Norway: isolation, diversity and biological activity. Environ Microbiol 9:2756–2764. https://doi.org/10.1111/j.1462-2920.2007.01387.x
Bugni TS, Ireland CM (2004) Marine-derived fungi: a chemically and biologically diverse group of microorganisms. Nat Prod Rep 21:143–163
Carroll AR, Copp BR, Davis RA et al (2019) Marine natural products. Nat Prod Rep 36:122–173
Carroll AR, Copp BR, Davis RA et al (2020) Marine natural products. Nat Prod Rep 37:175–223
Charan RD, Schlingmann G, Janso J et al (2004) Diazepinomicin, a new antimicrobial alkaloid from a marine *Micromonospora* sp. J Nat Prod 67:1431–1433
Chauhan D, Catley L, Li G et al (2005) A novel orally active proteasome inhibitor induces apoptosis in multiple myeloma cells with mechanisms distinct from Bortezomib. Cancer Cell 8:407–419
Chen QY, Liu Y, Luesch H (2011) Systematic chemical mutagenesis identifies a potent novel apratoxin A/E hybrid with improved in vivo antitumor activity. ACS Med Chem Lett 2:861–865
Chen X, Si L, Liu D et al (2015) Neoechinulin B and its analogues as potential entry inhibitors of influenza viruses, targeting viral hemagglutinin. Eur J Med Chem 93:182–195
Chen S, Chen D, Cai R, Cui H et al (2016a) Cytotoxic and antibacterial preussomerins from the mangrove endophytic fungus *Lasiodiplodia theobromae* ZJ-HQ1. J Nat Prod 79:2397–2402
Chen M, Zhang W, Shao W et al (2016b) DNA methyltransferase inhibitor induced fungal biosynthetic products: diethylene glycol phthalate ester oligomers from the marine-derived fungus *Cochliobolus lunatus*. Mar Biotechnol 18:409–417. https://doi.org/10.1007/s10126-016-9703-y
Costa M, Costa-Rodrigues J, Fernandes MH et al (2012) Marine cyanobacteria compounds with anticancer properties: a review on the implication of apoptosis. Mar Drugs 10:2181–2207
Cueto M, Jensen PR, Kauffan C et al (2001) Pestalone, a new antibiotic produced by a marine fungus in response to bacterial challenge. J Nat Prod 64:1444–1446
Dashti Y, Grkovic T, Abdelmohsen U et al (2017) Actinomycete metabolome induction/suppression with N-acetylglucosamine. J Nat Prod 80:828–836. https://doi.org/10.1021/acs.jnatprod.6b00673
Demay J, Bernard C, Reinhardt A et al (2019) Natural products from cyanobacteria focus on beneficial activities. Mar Drugs 17:320. https://doi.org/10.3390/md17060320
Ding L, Xu P, Zhang W et al (2020) Three new diketopiperazines from the previously uncultivable marine bacterium *Gallaecimonas mangrovi* HK-28 cultivated by iChip. Chem Biodiv 17: e2000221. https://doi.org/10.1002/cbdv.202000221
Doran P (2013) Mass transfer. In: Bioprocess engineering principles, 2nd edn. Academic, pp 379–444. https://doi.org/10.1021/ie50684a007
Ebada SS, Fischer T, Hamacher A et al (2014) Psychrophilin E a new cyclotripeptide from co-fermentation of two marine alga-derived fungi of the genus *Aspergillus*. Nat Prod Res 28: 776–781. https://doi.org/10.1080/14786419.2014.880911
Eberhard A, Burlingame A, Eberhard C et al (1981) Structural identification of autoinducer of *Photobacterium fischeri* luciferase. Biochemistry 20:2444–2449. https://doi.org/10.1021/bi00512a013
El Ghaouth A, Arul J, Wilson C et al (1994) Ultrastructural and cytochemical aspects of the effect of chitosan on decay of bell pepper fruit. Physiol Mol Plant Pathol 44:417–432
Engelhardt K, Degnes KF, Kemmler M et al (2010) Production of a new thiopeptide antibiotic TP-1161 by a marine *Nocardiopsis* species. Appl Environ Microbiol 76:4969–4976

Epstein S (2013) The phenomenon of microbial uncultivability. Curr Opin Microbiol 16:636–642. https://doi.org/10.1016/j.mib.2013.08.003

Fang W, Lin X, Zhou X et al (2014) Cytotoxic and antiviral nitrobenzoyl sesquiterpenoids from the marine derived fungus *Aspergillus ochraceus* Jcma1F17. Med Chem Comm 5:701–705

Feling RH, Buchanan GO, Mincer TJ (2003) Salinosporamide A: a highly cytotoxic proteasome inhibitor from a novel microbial source, a marine bacterium of the new genus *Salinospora*. Angew Chem Int Ed 42:355–357

Fenical W, Jensen PR, Cheng, XC (1998) US Pat. 6069146A. https://patents.google.com/patent/US6069146A/en. Last accessed Nov 2020

Fenical W, Baden D et al (1999) Marine derived pharmaceuticals and related bioactive compounds. In: Fenical W (ed) From monsoons to microbes: understanding the ocean's role in human health. National Academies Press, Washington, DC, pp 71–86

Fleming A (1929) On the antibacterial action of cultures of a penicillium, with special reference to their use in the isolation of *B. influenzae*. From the laboratory of the inoculation Department St Mary's Hospital, London, pp 226–234

Fournel M, Bonfils C, Hou Y et al (2008) MGCD0103, a novel isotype-selective histone deacetylase inhibitor has broad spectrum antitumor activity in vitro and in vivo. Mol Cancer Therap 7:759–768. https://doi.org/10.1158/1535-7163.MCT-07-2026

Gallagher KA, Wanger G, Henderson G et al (2018) Ecological implications of hypoxia-triggered shifts in secondary metabolism. Environ Microbiol 19:2182–2191. https://doi.org/10.1111/1462-2920.13700

Ganesan A, Arimondo PB, Rots MG et al (2019) The timeline of epigenetic drug discovery: from reality to dreams. Clin Epigenet 11:1–17. https://doi.org/10.1186/s13148-019-0776-0

Gao SS, Li XM, Williams K et al (2016) Rhizovarins A-F indole-diterpenes from the mangrove-derived endophytic fungus *Mucor irregularis* QEN-189. J Nat Prod 79:2066–2074

Gil-Turnes AS, Hay ME, Fenical W (1989) Symbiotic marine bacteria chemically defend crustacean embryos from a pathogenic fungus. Science 246:116–118

Gram L, Grossart HP, Schlingloff A et al (2002) Possible quorum sensing in marine snow bacteria: production of acylated homoserine lactones by *Roseobacter* strains isolated from marine snow. Appl Environ Microbiol 68:4111–4116

Guo W, Zhang Z, Zhu T et al (2015) Penicyclones A–E, antibacterial polyketides from the deep-sea-derived fungus *Penicillium* sp. F23-2. J Nat Prod 78:2699–2703. https://doi.org/10.1021/acs.jnatprod.5b00655

Haga A, Tamoto H, Ishino M et al (2013) Pyridone alkaloids from a marine-derived fungus *Stagonosporopsis cucurbitacearum* and their activities against azole-resistant *Candida albicans*. J Nat Prod 76:750–754

Hau AM, Greenwood JA, Lohr CV et al (2013) Coibamide A induces mTOR-independent autophagy and cell death in human glioblastoma cells. PLoS One 8:e65250

He F, Bao J, Zhang XY et al (2013) Asperterrestide A a cytotoxic cyclic tetrapeptide from the marine-derived fungus *Aspergillus terreus* SCSGAF0162. J Nat Prod 6:1182–1186

Huang L, Xu MY, Li HJ et al (2017) Amino acid-directed strategy for inducing the marine-derived fungus *Scedosporium apiospermum* F41−1 to maximize alkaloid diversity. Org Lett 19:4888–4891. https://doi.org/10.1021/acs.orglett.7b02238

Hull E, Montgomery MR, Leyva KJ (2016) HDAC inhibitors as epigenetic regulators of the immune system: impacts on cancer therapy and inflammatory diseases. BioMed Res Int 2016:8797206. https://doi.org/10.1155/2016/8797206

Hunter P (2015) Antibiotic discovery goes underground: the discovery of teixobactin could revitalise the search for new antibiotics based on the novel method the researchers used to identify the compound. EMBO Rep 16:563–565

Imanaka H, Tanaka S, Feng B et al (2010) Cultivation characteristics and gene expression profiles of *Aspergillus oryzae* by membrane-surface liquid culture, shaking-flask culture, and agar-plate culture. J Biosci Bioeng 109:267–273. https://doi.org/10.1016/j.jbiosc.2009.09.004

Janssens Y, Wynendaele E, Berghe WV et al (2019) Peptides as epigenetic modulators: therapeutic implications. Clin Epigenet 11:1–14. https://doi.org/10.1186/s13148-019-0700-7
Jaspars M, De Pascale D, Andersen JH et al (2016) The marine biodiscovery pipeline and ocean medicines of tomorrow. J Mar Biol Assoc UK 96:151–158
Jin B, Li Y, Robertson KD (2011) DNA methylation: superior or subordinate in the epigenetic hierarchy? Genes Cancer 2:607–617. https://doi.org/10.1177/1947601910393957
Jones EBG, Pang KL, Abdel-Wahab MA et al (2019) An online resource for marine fungi. Fungal Diver 96:347–433
Kaeberlein T, Lewis K, Epstein S (2002) Isolating "uncultivable" microorganisms in pure culture in a simulated natural environment. Science 296:1127–1129. https://doi.org/10.1126/science.1070633
Khamthong N, Rukachaisirikul V, Phongpaichit S et al (2014) An antibacterial cytochalasin derivative from the marine-derived fungus *Diaporthaceae* sp. PSU-SP2/4. Phytochem Lett 10:5–9
Kim HU, Blin K, Lee SY et al (2017) Recent development of computational resources for new antibiotics discovery. Curr Opin Microbiol 39:113–120. https://doi.org/10.1016/j.mib.2017.10.027
Konopka JB (2012) N-Acetylglucosamine functions in cell signaling. Scientifica 2012:489208. https://doi.org/10.6064/2012/489208
Kurmayer R, Deng L, Entfellner E (2016) Role of toxic and bioactive secondary metabolites in colonization and bloom formation by filamentous cyanobacteria *Planktothrix*. Harmful Algae 54:69–86
Kurosawa K, MacEachran DP, Sinskey AJ (2010) Antibiotic biosynthesis following horizontal gene transfer: new milestone for novel natural product discovery? Expert Opin Drug Discov 5: 819–825. https://doi.org/10.1517/17460441.2010.505599
Kwan JC, Eksioglu E, Liu C et al (2009) Grassystatins A-C from marine cyanobacteria, potent cathepsin E inhibitors that reduce antigen presentation. J Med Chem 52:5732–5747
Laatsch H (2017) AntiBase: the natural compound identifier. https://www.wiley.com/en-us/AntiBase%3A+The+Natural+Compound+Identifier-p-9783527343591. Last accessed Nov 2020
Liu Y, Li XM, Meng LH et al (2015) Bisthiodiketopiperazines and acorane sesquiterpenes produced by the marine-derived fungus *Penicillium adametzioides* AS-53 on different culture media. J Nat Prod 78:294–1299
Liu S, Dai H, Makhloufi G et al (2016) Cytotoxic 14-membered macrolides from a mangrove-derived endophytic fungus *Pestalotiopsis microspore*. J Nat Prod 79:2332–2340
Luan Y, Wei H, Zhang Z et al (2014) Eleganketal A a highly oxygenated dibenzospiroketal from the marine-derived fungus *Spicaria elegans* KLA03. J Nat Prod 77:1718–1723. https://doi.org/10.1021/np500458a
Luo D, Chen QY, Luesch H (2016) Total synthesis of the potent marine-derived elastase inhibitor lyngbyastatin 7 and in vitro biological evaluation in model systems for pulmonary diseases. J Org Chem 81:532–544
Ma X, Li L, Zhu T et al (2013) Phenylspirodrimanes with anti-HIV activity from the sponge-derived fungus *Stachybotrys chartarum* MXH-X73. J Nat Prod 76:2298–2306
Macintyre L, Charles M, Haltli B et al (2019) An ichip-domesticated sponge bacterium produces an N-acyltyrosine bearing an α-methyl substituent. Org Lett 21:7768–7771. https://doi.org/10.1021/acs.orglett.9b02710
Mao D, Okada BK, Wu Y et al (2018) Recent advances in activating silent biosynthetic gene clusters in bacteria. Curr Opin Microbiol 45:156–163. https://doi.org/10.1016/j.mib.2018.05.001
Mazard S, Penesyan A, Ostrowski M et al (2016) Tiny microbes with a big impact: the role of cyanobacteria and their metabolites in shaping our future. Mar Drugs 14:97

Meng LH, Li XM, Liu Y et al (2014) Penicibilaenes A and B sesquiterpenes with a tricyclo[6.3.1.0 (1,5)] dodecane skeleton from the marine isolate of *Penicillium bilaiae* MA-267. Org Lett 16: 6052–6055

Meng LH, Li XM, Liu YM (2015) Polyoxygenated dihydropyrano [2,3-c]pyrrole-4,5-dione derivatives from the marine mangrove-derived endophytic fungus *Penicillium brocae* MA-231 and their antimicrobial activity. Chin Chem Lett 26:610–612

Mondol M, Shin H, Islam M (2013) Diversity of secondary metabolites from marine *Bacillus* species: chemistry and biological activity. Mar Drugs 11:2846–2872

Moore BS, Trischman JA, Seng D, Kho D et al (1999) Salinamides, anti-inflammatory depsipeptides from a marine streptomycete. J Org Chem 64:1145–1150

Moore JM, Bradshaw E, Seipke R et al (2012) Use and discovery of chemical elicitors that stimulate biosynthetic gene clusters in *Streptomyces* bacteria. Meth Enzymol 517:367–385. https://doi.org/10.1016/B978-0-12-404634-4.00018-8

Moore LD, Le T, Fan G (2013) DNA methylation and its basic function. Neuropsychopharmacology 38:23–38. https://doi.org/10.1038/npp.2012.112

Nealson KH, Platt T, Hastings JW (1970) Cellular control of the synthesis and activity of the bacterial luminescent system. J Bacteriol 104:313–322. https://doi.org/10.1128/jb.104.1.313-322.1970

Newman DJ, Cragg GM (2014) Marine-sourced anti-cancer and cancer pain control agents in clinical and late preclinical development. Mar Drugs 12:255–278

Newman DJ, Cragg GM (2016) Drugs and drug candidates from marine sources: an assessment of the current "state of play". Planta Med 82:775–789

Newman DJ, Cragg GM (2017) Current status of marine-derived compounds as warheads in anti-tumor drug candidates. Mar Drugs 15:99. https://doi.org/10.3390/md15040099

Nichols D, Cahoon N, Trakhtenberg E et al (2010) Use of ichip for high-throughput in situ cultivation of "uncultivable" microbial species. Appl Environ Microbiol 76:2445–2450. https://doi.org/10.1128/AEM.01754-09

Niedermeyer TH (2015) Anti-infective natural products from cyanobacteria. Planta Med 81:1309–1325

Okada BK, Seyedsayamdost MR (2017) Antibiotic dialogues: induction of silent biosynthetic gene clusters by exogenous small molecules. FEMS Microbiol Rev 41:19–33. https://doi.org/10.1093/femsre/fuw035

Okazaki T, Kitahara T, Okami Y (1975) Studies on marine microorganisms. IV. A new antibiotic SS-228 Y produced by *Chainia* isolated from shallow sea mud. J Antibiot 28:176–184. https://doi.org/10.7164/antibiotics.28.176

Oppong-Danquah E, Budnicka P, Blümel M et al (2020) Design of fungal co-cultivation based on comparative metabolomics and bioactivity for discovery of marine fungal agrochemicals. Mar Drugs 18:73. https://doi.org/10.3390/md18020073

Overy D, Correa H, Roullier C (2017) Does osmotic stress affect natural product expression in fungi? Mar Drugs 15:254. https://doi.org/10.3390/md15080254

Pathom-Aree W, Nogi Y, Sutcliffe I et al (2006) *Dermacoccus abyssi* sp. nov., a piezotolerant actinomycete isolated from the Mariana Trench. Int J Syst Evol Microbiol 56:1233–1237. https://doi.org/10.1099/ijs.0.64133-0

Pereira R (2012) The carmaphycins: new proteasome inhibitors exhibiting an α,β-epoxyketone warhead from a marine cyanobacterium. ChemBioChem 13:810–817

Pereira F (2019) Have marine natural product drug discovery efforts been productive and how can we improve their efficiency? Expert Opin Drug Discov 14:717–722

Pereira F, Aires-de-Sousa J (2018) Computational methodologies in the exploration of marine natural product leads. Mar Drugs 16:236. https://doi.org/10.3390/md16070236

Pfannenstiel BT, Keller NP (2019) On top of biosynthetic gene clusters: how epigenetic machinery influences secondary metabolism in fungi. Biotechnol Adv 37:107345. https://doi.org/10.1016/j.biotechadv.2019.02.001

Pinu FR, Villas-Boas SG, Aggio R (2017) Analysis of intracellular metabolites from microorganisms: quenching and extraction protocols. Metabolites 7:53. https://doi.org/10.3390/metabo7040053

Poncet J (1999) The dolastatins, a family of promising antineoplastic agents. Curr Pharm Des 5: 139–162

Pruksakorn P, Arai M, Kotoku N et al (2010) Trichoderins, novel aminolipopeptides from a marine sponge-derived *Trichoderma* sp., are active against dormant mycobacteria. Bioorg Med Chem Lett 20:3658–3663

Ramadan U, Grkovic T, Balasubramanian S et al (2015) Elicitation of secondary metabolism in actinomycetes. Biotechnol Adv 33:798–811. https://doi.org/10.1016/j.biotechadv.2015.06.003

Rateb ME, Ebel R (2011) Secondary metabolites of fungi from marine habitats. Nat Prod Rep 28: 290–344

Rateb ME, Hallyburton I, Houssen WE et al (2013) Induction of diverse secondary metabolites in *Aspergillus fumigatus* by microbial co-culture. RSC Adv. 3:14444–14450. https://doi.org/10.1039/c3ra42378f

Rath P, Kinast S, Maier ME (2005) Synthesis of the fully functionalized core structure of the antibiotic abyssomicin C. Org Lett 7:3089–3092

Reen FJ, Romano S, Dobson ADW et al (2015) The sound of silence: activating silent biosynthetic gene clusters in marine microorganisms. Mar Drugs 13:4754–4783. https://doi.org/10.3390/md13084754

Riedlinger J, Reicke A, Zähner HANS et al (2004) Abyssomicins inhibitors of the para-aminobenzoic acid pathway produced by the marine *Verrucosispora* strain AB-18-032. J Antibiot 57:271–279

Rivas L, Rojas V (2019) Cyanobacterial peptides as a tour de force in the chemical space of antiparasitic agents. Arch Biochem Biophys 664:24–39

Romano S, Jackson SA, Party S et al (2018) Extending the "one strain many compounds" (OSMAC) principle to marine microorganisms. Mar Drugs 16:244. https://doi.org/10.3390/md16070244

Rutledge PJ, Challis GL (2015) Discovery of microbial natural products by activation of silent biosynthetic gene clusters. Nat Rev Microbiol 13:509–523. https://doi.org/10.1038/nrmicro3496

Saha M, Ghosh D, Garai D et al (2005) Studies on the production and purification of an antimicrobial compound and taxonomy of the producer isolated from the marine environment of the Sundarbans. Appl Microbiol Biotechnol 66:497–505

Sanaei M, Kavoosi F (2019) Histone deacetylases and histone deacetylase inhibitors: molecular mechanisms of action in various cancers. Adv Biomed Res 8:63. https://doi.org/10.4103/abr.abr_142_19

Schroeckh V, Scherlachb K, Nützman HW et al (2009) Intimate bacterial-fungal interaction triggers biosynthesis of archetypal polyketides in *Aspergillus nidulans*. Proc Natl Acad Sci USA 106: 14558–14563. https://doi.org/10.1073/pnas.0901870106

Serrill JD, Wan X, Hau AM et al (2016) Coibamide A, a natural lariat depsipeptide, inhibits VEGFA/VEGFR2 expression and suppresses tumour growth in glioblastoma xenografts. Invest New Drugs 34:24–40

Seyedsayamdost MR (2014) High-throughput platform for the discovery of elicitors of silent bacterial gene clusters. Proc Natl Acad Sci USA 111:7266–7271. https://doi.org/10.1073/pnas.1400019111

Seyedsayamdost MR (2019) Toward a global picture of bacterial secondary metabolism. J Ind Microbiol Biotechnol 46:301–311. https://doi.org/10.1007/s10295-019-02136-y

Seyedsayamdost MR, Traxler MF, Clardy J et al (2012) Old meets new: using interspecies interactions to detect secondary metabolite production in actinomycetes. Methods Enzymol 517:89–109

Shang Z, Li L, Espósito BP et al (2015) New PKS-NRPS tetramic acids and pyridinone from an Australian marine-derived fungus *Chaunopycnis* sp. Org Biomol Chem 13:7795–7802

Shi L, Wu YM, Yang XQ et al (2020) The cocultured *Nigrospora oryzae* and *Collectotrichum gloeosporioides*, *Irpex lacteus*, and the plant host *Dendrobium officinale* bidirectionally regulate the production of phytotoxins by anti-phytopathogenic metabolites. J Nat Prod 83:1374–1382. https://doi.org/10.1021/acs.jnatprod.0c00036

Shin D, Byun WS, Moon K et al (2018) Coculture of marine *Streptomyces* sp. with *Bacillus* sp. produces a new piperazic acid-bearing cyclic peptide. Front Chem 6:498. https://doi.org/10.3389/fchem.2018.00498

Shwab EK, Bok JW, Tribus M et al (2007) Histone deacetylase activity regulates chemical diversity in *Aspergillus*. Euk Cell 6:1656–1664. https://doi.org/10.1128/EC.00186-07

Singh V, Haque S, Niwas R et al (2017) Strategies for fermentation medium optimization: an in-depth review. Front Microbiol 7:2087. https://doi.org/10.3389/fmicb.2016.02087

Staley T, Konopka A (1985) Measurement of in situ activities of nonphotosynthetic microorganisms in aquatic and terrestrial habitats. Annu Rev Microbiol 39:321–346. https://doi.org/10.1146/annurev.mi.39.100185.001541

Stolze SC, Deu E, Kaschani F et al (2012) The antimalarial natural product symplostatin 4 is a nanomolar inhibitor of the food vacuole falcipains. Chem Biol 19:1546–1555

Sun JY, Wang JD, Wang X et al (2017) Marine-derived chromopeptide A, a novel class i HDAC inhibitor suppresses human prostate cancer cell proliferation and migration. Acta Pharmacol Sinica 38:551–560. https://doi.org/10.1038/aps.2016.139

Sylvain M, Lehoux F, Morency S et al (2018) The EcoChip: a wireless multi-sensor platform for comprehensive environmental monitoring. IEEE Trans Biomed Circuit Syst 12:1289–1300. https://doi.org/10.1109/TBCAS.2018.2878404

Tan LT, Phyo MY (2020) Marine cyanobacteria: a source of lead compounds and their clinically-relevant molecular targets. Molecules 25:2197. https://doi.org/10.3390/molecules25092197

Tan GJ, Peng ZK, Lu JP et al (2013) Cathepsins mediate tumour metastasis. World J Biol Chem 4: 91–101

Taori K, Paul VJ, Luesch H (2008) Structure and activity of largazole, a potent antiproliferative agent from the Floridian marine cyanobacterium *Symploca* sp. J Am Chem Soc 130:1806–1807

Tareq FS, Kim JH, Lee MA et al (2012) Ieodoglucomides A and B from a marine-derived bacterium *Bacillus licheniformis*. Org Lett 14:1464–1467

Teasdale ME, Liu J, Wallace J et al (2009) Secondary metabolites produced by the marine bacterium *Halobacillus salinus* that inhibit quorum sensing-controlled phenotypes in Gram-negative bacteria. Appl Environ Microbiol 75:567–572. https://doi.org/10.1128/AEM.00632-08

Thissera B, Alhadrami HA, Hassan MHA et al (2020) Induction of cryptic antifungal pulicatin derivatives from *Pantoea agglomerans* by microbial co-culture. Biomolecules 10:268. https://doi.org/10.3390/biom10020268

Tranter D, Paatero AO, Kawaguchi S et al (2020) Coibamide A targets Sec61 to prevent biogenesis of secretory and membrane proteins. ACS Chem Biol 15:2125–2136

Trischman JA, Tapiolas DM, Jensen PR (1994) Salinamides A and B: anti-inflammatory depsipeptides from a marine streptomycete. J Am Chem Soc 116:757–758

Turk B (2006) Targeting proteases: successes, failures and future prospects. Nat Rev Drug Discov 5:785–799

Varghese TA, Jayasri MA, Suthindhiran K (2015) Marine Actinomycetes as potential source for histone deacetylase inhibitors and epigenetic modulation. Lett Appl Microbiol 61:69–76. https://doi.org/10.1111/lam.12430

Vingadassalon A, Lorieux F, Juguet M et al (2015) Natural combinatorial biosynthesis involving two clusters for the synthesis of three pyrrolamides in *Streptomyces netropsis*. ACS Chem Biol 10:601–610. https://doi.org/10.1021/cb500652n

Wakefield J, Hassan HM, Jaspars M et al (2017) Dual induction of new microbial secondary metabolites by fungal bacterial co-cultivation. Front Microbiol 8:1284. https://doi.org/10.3389/fmicb.2017.01284

Wan X (2018) ATG5 promotes death signalling in response to the cyclic depsipeptides coibamide A and apratoxin A. Mar Drugs 16:77. https://doi.org/10.3390/md16030077

Wang YP, Lei QY (2018) Metabolite sensing and signalling in cell metabolism. Signal Transduction Targeted Therapy 3:30
Wang J, Soisson A, Young K et al (2006) Platensimycin is a selective FabF inhibitor with potent antibiotic properties. Nature 441:358–361. https://doi.org/10.1038/nature04784
Wang JF, Lin XP, Qin C et al (2014) Antimicrobial and antiviral sesquiterpenoids from sponge-associated fungus, *Aspergillus sydowii* ZSDS1-F6. J Antibiot 67:581–583
Whitton BA, Potts M (2012) Introduction to the cyanobacteria. In: Whitton BA (ed) Ecology of cyanobacteria II: their diversity in space and time. Kluwer, Dordrecht, pp 1–13
Williams RB, Henrikson JC, Hoover AR et al (2008) Epigenetic remodelling of the fungal secondary metabolome. Org Biomol Chem 6:1895–1997
Wu JS, Shi XH, Zhang YH et al (2019) Co-cultivation with 5-azacytidine induced new metabolites from the zoanthid-derived fungus *Cochliobolus lunatus*. Front Chem 7:763. https://doi.org/10.3389/fchem.2019.00763
Xu R, Li XM, Wang BG (2016) Penicisimpins A-C three new dihydroisocoumarins from *Penicillium simplicissimum* MA-332 a marine fungus derived from the rhizosphere of the mangrove plant *Bruguiera sexangula* var. Rhynchopetala. Phytochem Lett 17:114–118
Yu L, Ding W, Wang Q et al (2017) Induction of cryptic bioactive 2, 5-diketopiperazines in fungus *Penicillium* sp. DT-F29 by microbial co-culture. Tetrahedron 73:907–914. https://doi.org/10.1016/j.tet.2016.12.077
Yu M, Li Y, Banakar SP et al (2019) New metabolites from the Co-culture of marine-derived actinomycete *Streptomyces rochei* MB037 and fungus *Rhinocladiella similis* 35. Front Microbiol 10:915. https://doi.org/10.3389/fmicb.2019.00915
Zhang P, Mandi A, Li XM et al (2014) Varioxepine A a 3H-oxepine-containing alkaloid with a new oxa-cage from the marine algal-derived endophytic fungus *Paecilomyces variotii*. Org Lett 16: 4834–4837
Zhang W, Yuan Y, Su D et al (2018) *Gallaecimonas mangrovi* sp. nov., a novel bacterium isolated from mangrove sediment. Ant Leeuwenhoek 111:1855–1862
Zhou ZF, Guo YW (2012) Bioactive natural products from Chinese marine flora and fauna. Acta Pharmacol Sinica 33:1159–1169
Zhu F, Chen G, Choi X et al (2011) Aspergicin a new antibacterial alkaloid produced by mixed fermentation of two marine-derived mangrove epiphytic fungi. Chem Nat Comp 47:674–676

# Ocean Restoration and the Strategic Plan of the Marine Microbiome 18

Marieke Reuver, Jane Maher, and Annette M. Wilson

**Abstract**

The ocean is the largest ecosystem on Earth and is vital for human well-being. It is however greatly impacted by a range of pressures coming from human activities, such as pollution including oil spills, plastics, and endocrine-disrupting chemicals. Most critically, the ocean is being altered by climate change leading to ocean warming, sea-level rise, and ocean acidification. Their cumulative impact is putting further strain on many marine species and habitats. The health and resilience of marine ecosystems are severely compromised by all these pressures and large-scale ecological restoration is urgently needed to rebuild the ocean. New approaches are needed and a major aspect and focus of this chapter are giving an overview of some of the latest developments in the strategic use of the marine microbiome in ocean restoration. Thanks to scientific research there is a better understanding of the importance of the microbiome in maintaining ecosystem health and enhancing resilience. It is crucial to recognize the valuable role of the marine microbiome in ocean restoration. This book chapter aims to outline present state-of-the-art scientific, current policy and governance, and communication developments around the marine microbiome with the hope to inform and support the development of effective management strategies of the marine microbiome for sustainable ocean restoration.

**Keywords**

Bioremediation · Knowledge transfer · Marine management · Microbiome · Ocean literacy · Ocean restoration

M. Reuver (✉) · J. Maher · A. M. Wilson
ERINN Innovation Ltd, Dublin 8, Ireland
e-mail: marieke@erinn.eu; jane@erinn.eu; annette@erinn.eu



L. J. Stal, M. S. Cretoiu (eds.), *The Marine Microbiome*, The Microbiomes of Humans, Animals, Plants, and the Environment 3,
https://doi.org/10.1007/978-3-030-90383-1_18

## 18.1 Introduction

The ocean is the largest ecosystem covering over two-thirds of the Earth. It contains rich biodiverse habitats, regulates our climate, provides invaluable ecosystem services, plays a central role in global food security, and absorbs considerable amounts of heat and carbon dioxide (OECD 2020; Stocker 2015; Visbeck 2018). Hence, the ocean is vital for human well-being. The ocean also presents immense opportunities for economic growth, employment, and development. Today, more than 40% of the human global population lives in areas within 200 km of the ocean and there are 15 coastal megacities (Blackburn et al. 2019; Visbeck 2018). These coastal communities benefit from or rely directly on marine resources including coral reefs and kelp forests. These two ecosystems support cumulatively the highest levels of marine biodiversity. The ocean economy spans multiple sectors including oil and gas, fishing, aquaculture, shipping, tourism, offshore wind energy, mining, and marine biotechnology, and is growing rapidly. Prior to the COVID-19 pandemic, the OECD (2020) projected a doubling of the ocean economy from 2010 to 2030 to reach US$3 trillion and employ 40 million people.

The ocean is impacted greatly by our rapidly growing, affluent, and technologically advanced societies, and as a result, is in decline. Pressures come from a wide range of human activities including fishing (destructive demersal fisheries and bycatch), shipping and transport, offshore oil and gas, mining, pollution including plastics, nutrient pollution and coastal eutrophication, underwater noise, and contaminants such as lead or mercury as well as emerging pollutants such as endocrine disruptors, and introduction of invasive species. Most critically, the ocean is being altered by climate change leading to ocean warming, sea-level rise, and ocean acidification. The cumulative impacts of these changes are putting further strain on many marine species and habitats. The health and resilience of marine ecosystems are severely compromised by all these pressures and this will continue to increase unless bold action is taken at the global level to ensure the protection of natural resources for society (OECD 2020; Visbeck 2018). There is an urgent need to strengthen the ocean's resilience against environmental and climate stressors and the window of opportunity is closing rapidly. Rebuilding marine life is an ethical obligation and a smart economic objective to achieve a sustainable future (Duarte et al. 2020).

The year 2021 marks the start of the United Nations Decade of Ocean Science for Sustainable Development as well as the Decade on Ecosystem Restoration (2021–2030) and as such provides a timely opportunity to strengthen international collaboration in the science-based management of marine ecosystems (Allard et al. 2020). Underpinned by knowledge in the latest IPCC and IPBES reports large-scale ecological restoration is urgently needed to rebuild the ocean (Duarte et al. 2020). This urgency was recently reflected in the European Union's H2020 Funding Program Green Deal Call which included a topic on "Restoring biodiversity and ecosystem services" where largescale restoration of marine ecosystems will be a priority for which tens of millions of euros will be dedicated (EU Green Deal call topic LC-GD-7-1-2020). New approaches are needed to restore the ocean and one major aspect and focus of this chapter is giving an overview of some of the latest developments in the strategic use of the marine microbiome in ocean restoration. Until recently, there has been limited emphasis placed on the potential of microorganisms in ecosystem management plans and policy but thanks to ground-breaking research there is a better understanding of the importance of microorganisms in maintaining ecosystem health and enhancing resilience in the face of global change (Cavicchioli et al. 2019). It is crucial to recognize their valuable role in restoration (Allard et al. 2020). A consortium of microbiologists led by Cavicchioli et al. (2019) released a Consensus Statement in 2019 which highlighted that "an immediate, sustained and concerted effort is required to explicitly include microorganisms in research, technology development, and policy and management decisions" (Cavicchioli et al. 2019, p. 582). As a microbiologists' warning, the intent was to emphasize the importance of the microbial world and to prompt international microbiologists to engage in and integrate their research into the frameworks for addressing climate change and accomplishing the United Nations Sustainable Development Goals (Cavicchioli et al. 2019).

**This book chapter aims to outline present state-of-the-art scientific, current policy and governance, and communication developments around the marine microbiome, with the hope to inform and support the development of effective management strategies of the marine microbiome for sustainable ocean restoration.**

## 18.2 The Marine Microbiome in Ocean Restoration

### 18.2.1 Importance of the Marine Microbiome

The ocean is the largest habitat and microbes (i.e., viruses, bacteria, archaea, microeukaryotes) are its most abundant inhabitants that together are among one of the largest microbiomes (Ainsworth et al. 2017; Pita et al. 2018; Trevathan-Tackett et al. 2019). The Census of Marine Life (2018) estimated that 90% of marine biomass is microbial and emphasized that microbes can be found in all marine ecosystems ranging from coastal estuaries, mangroves, seagrass meadows, kelp forests, coral reefs to the open ocean (Cavicchioli et al. 2019) and deep-sea hydrothermal vents.

The marine microbiome plays a central role in many important processes such as harvesting solar energy, primary productivity, and driving biogeochemical cycles, including those that control the Earth's climate (Ainsworth et al. 2017; Trevathan-Tackett et al. 2019). Marine microorganisms form the basis of ocean food webs as they fix carbon and nitrogen and remineralize organic matter. Thus, they are the foundation of the global carbon and nutrient cycles (Azam and Malfatti 2007). Marine microbes produce 50% of the oxygen globally and remove approximately the same amount of carbon dioxide from the atmosphere through photosynthesis (Dubilier et al. 2016). They can also remove up to 90% of methane from the ocean

(Dubilier et al. 2016). Therefore, the marine microbiome is vital for carbon sequestration and combating the effects of climate change (Cavicchioli et al. 2019).

Marine microbes often exist in mutualistic symbiosis with macro-organisms such as animals, plants, and algae (Apprill 2017; Egan et al. 2013). There are many examples of marine mutualisms in which marine microbes enable hosts to utilize resources or substrates otherwise unavailable to the host alone. A well-known example of marine mutualism is seen in the coral reef system where the dinoflagellate alga from the family Symbiodiniaceae supplies the coral with glucose, glycerol, and amino acids while the coral provides the alga a protected environment and limiting compounds needed for photosynthesis (Wilkins et al. 2019). Through a growing body of research and increasing knowledge of the microbiome, it has been argued that all living organisms have, to varying degrees, a symbiotic relationship with microorganisms and that this contributes to their evolutionary success (Egan et al. 2020).

### 18.2.2 The Potential of the Marine Microbiome in Ocean Restoration

It is thanks to scientific developments such as low-cost, high-throughput sequencing, advances in sample preparation, improvements in computing power and imaging technologies, and the development of bioinformatic tools that allow to better understand the full extent and importance of the microbial world (Dubilier et al. 2016). Along with advances in genome sequencing and meta-omics, culture-independent analysis of microbiomes has considerably accelerated the progress of microbiome research and has made clearer how the composition of microbiomes can influence ecosystems (Foo et al. 2017). As a result, there has been a growing interest in engineering microbiomes for changing microbiota in order to alter ecosystems of interest (Foo et al. 2017), which has already been successfully used for terrestrial ecosystems (Rosado et al. 2019; Van Oppen et al. 2015). So far, microbiomes are largely underexploited in marine ecosystems (Rosado et al. 2019) but it has been stated that marine microbes may hold the key to addressing the impacts of climate change and pollution in the marine environment (Cavicchioli et al. 2019). Building upon microbiome manipulation developments in animal and plant systems there is increasing evidence of successfully applying this technique in marine environments, e.g., in coral reef ecosystems (Rosado et al. 2019). It is expected that the microbiome manipulation approach could help protect vulnerable marine habitats from the negative consequences of climate change (Egan et al. 2020). There has been a call to action to further develop innovative microbial technologies to support the mitigation of climate change impacts, reduce pollution, and eliminate reliance on fossil fuels (Cavicchioli et al. 2019).

One example of a microbiome manipulation approach is bioremediation, which uses living organisms or their metabolic products to break down contaminants into a less toxic or non-toxic form. It is considered to be a sustainable and ecologically friendly approach causing minimal or lesser impact to the environment (Villela et al. 2019) than traditional methods. It has been suggested that bioremediation offers a solution to many known and emerging contaminants in polluted marine waters e.g., aquaculture and fisheries effluents, trace metals, endocrine disrupters, mixed waste/ municipal wastewaters discharges, crude and refined oil pollution, as well as biological carbon sequestration (Paniagua-Michel and Rosales 2015).

### 18.2.3 State-of-the-Art in Ocean Restoration

The current understanding of the functional role and mechanisms of host interaction of microbial symbionts is still limited also because culturing marine microbial symbionts is often challenging and many have not been cultured yet. Modern technologies such as culture-independent methods, e.g., amplicon sequencing and shotgun metagenomics have only recently allowed biologists to get more insight into symbiont diversity and function at a much greater speed than ever before (Egan et al. 2020). Knowledge gaps regarding how to successfully harness bioremediation approaches are currently also being filled by combining -omics information from genomics, transcriptomics, proteomics, and metabolomics with advanced imaging techniques (Egan et al. 2020). These advances allow better insight into how the microbiome could potentially help in ocean restoration. The below paragraphs outline the state-of-the-art in ocean restoration for some of the current issues such as combating climate change and tackling pollution of oil, plastics, and Endocrine Disrupting Chemicals for which microbiome solutions are promising.

#### 18.2.3.1 Climate Change and Corals

In 2020, for the first time ever, the World Economic Forum ranked climate change and several related environmental issues as the top five risks to global economic stability in terms of likelihood (WEF 2020) recognizing the magnitude of the impact of climate change on society. Scientific evidence for the warming of the global climate system is unequivocal. Human activities such as the burning of fossil fuels have considerably raised atmospheric carbon dioxide ($CO_2$) levels in the last 150 years (IPCC 5th assessment report 2014) resulting in global warming and, after dissolving in surface seawater, ocean acidification (OA). The ocean has absorbed much of the increased heat with the top 100 m of ocean showing warming of more than 0.33 °C since 1969. Since the beginning of the Industrial Revolution, the acidity of surface ocean waters has increased by about 30% (NASA 2021).

The combined effects of OA with warming and other drivers such as increased UV exposure and deoxygenation are expected to affect and endanger many marine communities (Gao et al. 2019). The impacts of OA range from the reduced calcification of many calcifying organisms, reduced species diversity, and substantial modification of communities to lower ecosystem resilience (Gao et al. 2019). OA in the geological past has been associated with mass extinctions of marine life (Garbelli et al. 2016). It is therefore alarming to note that the current seawater pH and carbonate saturation state are declining faster than they have been in the last 300 million years (Gao et al. 2019). The ocean warming effects observed so far comprise, amongst others, of mass mortalities of marine invertebrates, loss of kelp forests, and coral bleaching (Straub et al. 2019). Coral bleaching is the breakdown of the critical symbiosis between the coral host and its algal endosymbionts and often leads to extensive coral mortality (Brown 1997).

At the microbiome level, climate change can directly and indirectly alter host-associated microbial communities which may negatively impact host health through the breakdown of symbiotic relationships or the loss of important microbial functions (Trevelline et al. 2019). For example, in sponges, it has been demonstrated that increased water temperatures lead to a loss of important microbial taxa and that any bacterial taxa that remain may lose important functions (Ramsby et al. 2018). As a result, corals that are stressed by ocean warming and acidification have microbiomes that are different from healthy hosts (Trevelline et al. 2019; Van Oppen and Blackall 2019). Higher ocean temperature and/or acidification can lead to a shift from mostly putatively beneficial bacterial taxa to opportunistic and potentially pathogenic groups (Vanwonterghem and Webster 2020). Changing ocean conditions as a result of climate change have already led to more frequent and more severe coral reef bleaching events and these are expected to increase importantly by the end of this century (van Hooidonk et al. 2016). Although coral reefs will regenerate through recolonization by thermal tolerant species this will take centuries. Current associated noncoral reef species will have no shelter resulting from the rapid collapse and it is expected that thousands of these species will be lost entirely (Novak et al. 2020). Therefore, the rapid loss of reef-forming scleractinian corals as a result of climate change would have severe consequences for entire coral reef ecosystems as well as for the coastal human populations depending on these coral reefs (Damjanovic et al. 2019).

While it is urgent to address the root causes of climate change it is also essential to explore the possibility of mitigating its effects. Current management methods of the ocean environment include restoration, maintenance (such as marine-protected areas and catchment management to improve water quality), and enforcement of national governmental agreements to decrease greenhouse gas emissions but the ocean is still rapidly declining and additional solutions are needed. Marine microorganisms are one such new solution that has the potential to mitigate some of the negative effects of climate change on coral and potentially other ocean ecosystems. Although microorganisms are crucial in regulating climate change, they have rarely been the focus of climate change studies (Cavicchioli et al. 2019). However, it has been postulated that the impact of climate change will depend heavily on the responses of microorganisms, which are essential for achieving an environmentally sustainable future (Cavicchioli et al. 2019).

Microbial inoculations have already been used in terrestrial plants, in humans, and in a variety of other host organisms. For example, as being explored in the EU H2020-funded SIMBA project (https://simbaproject.eu/) plant growth-promoting microorganisms (PGPM) are natural symbionts that colonize the rhizosphere, stimulate plant growth and development, and protect against biotic and abiotic stresses (Tabacchioni et al. 2021). The use of Beneficial Microorganisms for Corals (BMCs) has been proposed as a tool for the improvement of coral health (Peixoto et al. 2017) and although the field is still in its infancy knowledge in this research topic is advancing rapidly. Like PGPMs for crops BMCs are defined as consortia of microorganisms that can contribute to coral health through mechanisms that include promoting coral nutrition and growth, mitigating stress and impacts of toxic compounds, deterring pathogens, and benefiting early life-stage development (Peixoto et al. 2021). It is expected that similar to plants and other animals exposing corals to certain microbial communities may trigger a beneficial shift in the symbiosis and make the holobiont, a biological unit comprising a host and all the macro- and microorganisms that live in association with it (Peixoto et al. 2021), more resilient to external pressures including climate change (Damjanovic et al. 2019).

The use of BMCs to increase the resistance of corals to environmental stress has proven to be effective in laboratory trials (Assis et al. 2020). An example is a study by Damjanovic et al. (2019) who explored the feasibility of coral early life stage microbiome manipulation by repeatedly inoculating coral recruits with a bacterial cocktail generated in the laboratory. They co-cultured two species, namely *Acropora tenuis* and *Platygyra daedalea,* to simultaneously investigate the effect of host factors on the coral microbiome. Their initial results provide evidence to support the feasibility of coral microbiome manipulation at least in a laboratory setting (Damjanovic et al. 2019). Another example is the research performed by Rosado et al. (2019) who facilitated manipulation of the coral-associated microbiome through the addition of a consortium of native putatively Beneficial Microorganisms for Corals (pBMC) in a controlled aquarium experiment. They found that their pBMC consortium had the ability to partially mitigate coral bleaching. Their results indicate that the microbiome in corals can be manipulated to lessen the effect of bleaching thus helping to alleviate pathogen and temperature stresses. They concluded that the addition of pBMCs represents a promising novel approach for minimizing coral mortality in the face of increasing environmental impacts (Rosado et al. 2019). Damjanovic et al. (2017) showed that a single exposure of coral larvae to the mucus-associated microbes of four different coral species resulted in divergent microbial communities after four months of rearing in filter-sterilized seawater. Their experiment showed that coral-associated microbiomes can be influenced to develop in different directions following microbial treatment and that early coral life in particular may be suitable for targeted microbial inoculation (Damjanovic et al. 2019).

Hence, the use of Beneficial Microorganisms for Corals (BMCs) to increase the resistance of corals to environmental stress has proven to be effective in laboratory trials. However, delivery mechanisms for efficient and safe transmission of BMC consortia on a larger scale such as in the field is needed because direct inoculation can be challenging. Packaged delivery mechanisms have been successfully used to transmit probiotics to other organisms including humans, lobsters, and fish. In corals, Assis et al. (2020) tested a method for utilizing rotifers of the species *Brachionus plicatilis* for delivery of BMCs to corals of the species *Pocillopora damicornis*. It was found that the rotifers efficiently ingested BMCs and the BMC-enriched rotifers were actively ingested by *P. damicornis* corals indicating that this is a promising technique for administering coral probiotics in situ.

Peixoto et al. (2021) reviewed the current proposed BMC approach and outlined the studies that have proven its potential to increase coral resilience to stress. They have revisited and expanded the list of putative beneficial microorganisms associated with corals and their proposed mechanisms that facilitate improved host performance. They also discuss the caveats and bottlenecks affecting the efficacy of BMCs and include the next steps to facilitate application at larger scales that can improve outcomes for corals and reefs globally (Peixoto et al. 2021).

While the use of BMCs is a promising method to address climatic pressures on corals environmental degradation is probably occurring too fast for corals to be able to adapt through natural selection alone. Scientists are now also considering genomic technologies to facilitate adaptation and resilience to warmer water (Novak et al. 2020). An option that is being explored is using Assisted Evolution (AE) to enhance the environmental stress tolerance and resilience of corals and thereby the success of coral reef restoration efforts (Van Oppen et al. 2015, 2017). AE aims to accelerate the rate of naturally occurring evolutionary processes in order to develop corals that are better able to cope with current climate change trajectories (Van Oppen et al. 2015). AE encompasses selective breeding, assisted gene flow, conditioning or epigenetic programming, and the manipulation of the coral microbiome. For example, Chakravarti et al. (2017) proposed directed laboratory evolution in Symbiodinium as a strategy to enhance coral holobiont thermal tolerance. AE of the hosts' Symbiodiniaceae population can be facilitated through the maintenance and growth of cultures under future climate scenarios. Increasing the temperature tolerance range of specific species of Symbiodiniaceae has already been achieved in controlled experiments, ex situ, with *Cladocopium* increasing its thermotolerance after ~80 generations (2.5 years) combined with superior photophysiological performance and growth when reared at 31 °C compared to 27 °C (Chakravarti et al. 2017). It is likely that coral restoration will require a myriad of approaches and Peixoto et al. (2017) envisages that AE of coral could be developed in association with the use of BMC consortia to target specific coral under specific environmental conduction.

***Considerations for the Future*** Several studies have demonstrated the ability to manipulate the coral microbiome to enhance thermal tolerance and disease resistance at least in ex situ experiments. Considerations for the future include further research into an optimal inoculation regime, application to the field, the possibility of scaling up these efforts, as well as getting answers to questions on concerns around biosecurity of manipulating microbes in laboratories and releasing these into the environment (Peixoto et al. 2021; Sweet et al. 2017). Approaches that use non-native microorganisms are also something that could be considered in the future (Peixoto et al. 2019); however, as will be discussed below, a much greater understanding of all interactions and potential impacts will be required.

To gain more insights into the efficacy of probiotics for coral reef restoration and conservation an optimal inoculation regime should be developed through the determination of suitable inoculation frequency and bacterial cell density. Systematic experiments are required to determine optimal inoculation protocols by using the same bacterial taxa, coral species, and culture conditions while varying bacterial cell density and/or frequency of inoculum administration (Damjanovic et al. 2019).

In terms of further research into an application to the field industrial infrastructures to produce cultured bacterial consortia already exist but a suitable delivery approach needs further research (Damjanovic et al. 2019). Peixoto et al. (2017) suggested the preparation of inocula by encapsulating bacteria into microscopic feed particles. Damjanovic et al. (2019) also refer to an alternative in the form of administration of probiotic bacteria through a heterotrophic food source such as the brine shrimp *Artemia* something that has already been proven successful in spiny lobster larvae (Goulden et al. 2012). It is also essential to better understand how microorganisms contribute to the transgenerational acclimatization of reef organisms if we are to reliably predict the consequences of global change for reef ecosystems. The capability of sustained bacteria-coral associations after inoculation needs to be studied outside of laboratory-controlled conditions to understand the impact of several fluctuating factors including timescale, life stages, and numerous biotic and abiotic factors. (Damjanovic et al. 2019). Webster and Reusch (2017) highlighted the capacity and mechanisms for microbiome-mediated transgenerational acclimatization (MMTA) in reef species, proposed a modified Price equation as a framework for assessing MMTA and recommended future areas of research to better understand how microorganisms contribute to the transgenerational acclimatization of reef organisms.

In relation to biosecurity concerns, it has been argued that coral reef ecosystems might be inadvertently harmed through manipulations that we cannot entirely predict or control (Sweet et al. 2017). For example, there are concerns that manipulation of the corals' microbiome may result in decreased fitness of the meta-organism in other areas of life history (Peixoto et al. 2019). Also, pathogens or parasites could be transferred from captive systems to the natural environment and impact the native fauna (Sweet et al. 2017). Potential probiotic bacteria (e.g., *Vibrio*) could also possibly be converted into pathogens (Bruto et al. 2017). In general, an overabundance of certain bacteria in the wild could have unintended effects on the ecosystem (Damjanovic et al. 2019). Therefore, future research should focus on understanding the risks before releasing microbes into open ecosystems. It has already been advocated that rigorous scientific trials and risk/benefit analyses should be carried out prior to introducing any foreign microbial communities into the ocean (Damjanovic et al. 2019; Van Oppen et al. 2017).

However, business as usual cannot be an option as marine ecosystems are under immediate pressure and future climate modeling indicate that further decline is inevitable. There is an urgent need to develop and rigorously test new approaches that are safe and effective in building the resilience of corals to withstand climate change (Peixoto et al. 2019) as well as new approaches to protect and restore other sessile animals and ocean ecosystems affected by climate change. It is paramount that the marine science community makes rapid advances across multiple approaches considering the threats faced by our ocean.

#### 18.2.3.2 Oil Spills

Another major environmental problem is the contamination of aquatic ecosystems due to accidental spillage and chronic pollution of petroleum hydrocarbon compounds (Bordoloi and Boruah 2018; Mapelli et al. 2017). This pollution generates a great hazard to marine ecosystems including the deep ocean (Mahjoubi et al. 2017). Oil pollution has a well-documented negative effect on marine animals and birds (e.g., King et al. 2021; Troisi et al. 2016); however, the full impacts of oil pollution on ocean microbes are complex and not yet fully understood (Society for Applied Microbiology 2018). It has been observed that oil contamination on corals causes bleaching, decreases larvae's resistance to thermal stress, and disrupts microbial associated communities (Villela et al. 2019). Furthermore, oil remediation techniques commonly rely on chemical dispersants to remediate spills, which can be even more toxic to marine ecosystems than the oil contamination itself (Ados Santos et al. 2015; DeLeo et al. 2016). An example is the Deepwater Horizon catastrophe in 2010. This catastrophe is deemed the worst oil accident in history whereby more than 700,000 tons of crude oil were released into the Gulf of Mexico (Mapelli et al. 2017). Response teams attempted to encourage natural bacteria to remove the oil more quickly by applying a 'chemical dispersant' (Corexit 9500) to the spill. Over 7 million liters of this dispersant were used to separate the oil into small droplets making them more accessible to microbes for degradation. However, an unexpected side effect of this was that bacteria were growing that degraded the Corexit and most likely impeded the action of the oil-degrading bacteria (Society for Applied Microbiology 2018).

Physicochemical technologies are available for oil spill clean-up (e.g., skimmers, barriers, controlled in situ burning, chlorination, UV oxidation, and solvent extraction (Ghosal et al. 2016; Walker 2017) but natural oil-degrading bacteria play a fundamental role in the breakdown and removal of oil in the ocean. Hydrocarbons must ultimately be mineralized by microorganisms and their degradation therefore greatly rely on natural bacteria to help clear oil spills (Mapelli et al. 2017). This process, when controlled and optimized, is known as bioremediation (Society for Applied Microbiology 2018). Bioremediation is based on the catabolic activity of microorganisms and their ability to use pollutants as carbon and energy sources by transforming it into less toxic compounds. It is considered as an effective, safe, eco-friendly, and low-cost clean-up method for oil spills having several advantages compared to other methods (Bordoloi and Boruah 2018; Mahjoubi et al. 2017; Prince and Atlas 2016). Although a lot is still unknown bioremediation of oil spills is the earliest and one of the best-investigated methods for combating marine pollution and research activities are only increasing. In the past few years, an increasing trend has been observed in patenting activity in the field of bioremediation of petroleum hydrocarbon (Bordoloi and Boruah 2018).

Different bioremediation strategies can be adopted depending on several factors such as site characteristics and type and concentration of pollutants. Frequently used bioremediation approaches applied in marine environments impacted by an oil spill include bio-augmentation (adding oil-degrading bacteria to supplement or enrich the existing microbial biota) and bio-stimulation (applying fertilizers as nutrients to encourage and stimulate the growth of indigenous oil degraders) (Das and Chandran 2011). Bio-stimulation has been successfully employed many times but bio-augmentation has yet to be shown to be effective in real oil-spill situations and likely is unnecessary given the omnipresent distribution of oil-degrading bacteria in the sea (Prince and Atlas 2016).

Bioremediation efficiency depends mainly on the microbial community composition, the sites to be decontaminated, and the environmental conditions (Mahjoubi et al. 2017; Paniagua-Michel and Rosales 2015). The efficiency of bioremediation is limited by the knowledge of the microbial ecology of polluted marine systems. It is yet unknown how environmental factors such as hydrostatic pressure, temperature, and dispersant toxicity affect the assembly and activity of hydrocarbon-degrading microbial communities. This limits the application of microorganism-based clean-up using physicochemical remediation. Therefore, a research review has been conducted to integrate the results of microbial physiology, metabolism, and ecology to better understand how microbes can be exploited to create improved biotechnological solutions to clean up the marine surface and deep waters, sediments, and beaches (Mapelli et al. 2017). Mahjoubi et al. (2017) have presented a comprehensive overview of successful case studies of bioremediation strategies used for the decontamination of oil-polluted marine environments. They listed the type of bioremediation treatment, the microorganisms and nutrients added, the type of contaminants that were targeted, and the removal percentage in the number of days. A high number (over 200) and high diversity of potential hydrocarbon-degrading microorganisms were reported from several different habitats ranging from bacteria (including cyanobacteria), microalgae, to fungi (Villela et al. 2019). But bacteria represented the major class of microorganisms involved in the degradation of hydrocarbons (Mahjoubi et al. 2017). There are also microbial consortia that have dual functions, namely promoting oil degradation and improving ecosystems and plant health at the same time (Allard et al. 2020).

A major challenge for the bioremediation of oil-contaminated sites is the bioavailability of contaminants. The use of biosurfactants produced by bacteria is an effective way to address this as it decreases the surface tension and facilitates the contact between microorganisms and pollutants in a nontoxic biodegradable way. Studies dealing with the biotechnological use of biosurfactants have considerably increased and the use of biosurfactants is a potential alternative and promising tool for bioremediation (Mahjoubi et al. 2017). One example is the research that was undertaken by Ados Santos et al. (2015). They produced a probiotic bacterial consortium from the coral *Mussismilia harttii* that was trained to degrade water-soluble oil fractions (WSFs). The research found that the bacterial consortium was responsible for the highly efficient degradation of petroleum hydrocarbons while minimizing the effects of WSFs on coral health. Moreover, the impact of WSFs on the coral microbiome was diminished by the introduced bacterial consortium. Following its introduction, the bacterial consortium had a dual function, i.e., promoting oil WSF degradation and improving coral health with its probiotic features (Ados Santos et al. 2015).

Another novel approach is the integration of electrochemical methods and biological routes that employ microbes as catalysts. Several achievements were reported by researchers for the remediation of oil spills by using bioelectrochemical systems (Wang et al. 2015). Microbial fuel cells are another technique to convert chemical energy into electricity concurrent with contaminant degradation (Baniasadi and Mousavi 2018).

***Considerations for the Future*** Future research on the bioremediation of oil spills must consider all different aspects such as effects of environmental parameters, metabolic pathways, basis of hydrocarbon breakdown as substrate (dissimilation) from the genetic point of view, and effects of hydrocarbon contaminants on microorganisms (Baniasadi and Mousavi 2018). The ecological impact of pollution clean-up strategies must be understood down to the microbiological level as human efforts to manipulate may have unexpected consequences as shown by the Deepwater Horizon catastrophe (Society for Applied Microbiology 2018).

Novel fields of research for future bioremediation of oil contamination in the ocean include approaches adding innovative materials (such as novel biosurfactants) using genetically engineered microorganisms and integration of electrochemical strategies with biological methods (Baniasadi and Mousavi 2018). Evidently, the application of some of these approaches to the field (e.g., manipulating microorganisms with the aim of modification of enzymatic characteristic, metabolic pathway design, expansion of substrate rate) will need to be carefully considered because of the environmental and ethical issues that are associated with these approaches (Baniasadi and Mousavi 2018).

#### 18.2.3.3 Plastic Pollution

Over the last few decades, it has become clear that plastic pollution presents a global and major societal and environmental challenge given its increasing presence in our ocean (Jacquin et al. 2019).

It is estimated that up to 12 million ton of plastics enters the ocean annually (IUCN n.d.), which is likely to continue to increase by an order of magnitude within the next decade (Jambeck et al. 2015). Plastics are expected to persist in the environment for hundreds or even thousands of years (Barnes et al. 2009). Although current knowledge about the exact impact of plastic on marine microbial life and ecosystem functions is insufficient (Jacquin et al. 2019) plastic debris poses a considerable threat by choking, blocking digestive tracts, altering feedback behavior, and entangling wildlife (Barnes et al. 2009).

As plastic debris (of any size, including microplastics) enters the aquatic environment it rapidly becomes biofouled, i.e., colonized by microbial communities composed of diverse bacteria, single-celled algae, fungi, as well as macro-organisms such as barnacles, bryozoans, hydroids, or multicellular algae (Wright et al. 2020). The microbial biofilms that form on plastic have been coined the "Plastisphere" (Zettler et al. 2013) and its microbial composition is diverse and distinct from the surrounding planktonic communities (Oberbeckmann et al. 2014; Rogers et al. 2020; Zettler et al. 2013). The organisms that colonize buoyant marine plastic debris can be transported across the oceans and in some cases become invasive species of fragile ecosystems (Wright et al. 2020).

As plastic ages in the marine environment, it presents a potential chemical hazard due to the release of persistent organic pollutants (POPs) from the plastic surface, of chemical additives leaching out of the plastic, as well as of chemicals produced by the degradation of the plastic polymer itself (Gewert et al. 2015; Wright et al. 2020). Many of these chemicals are known for their toxicity to marine biota (Rochman et al. 2013; Teuten et al. 2009; Wright et al. 2020). Marine microorganisms that make up the plastisphere play a key role in the biogeochemical cycles of the ocean and therefore plastic pollution can potentially impact these cycles (Rogers et al. 2020). However, so far only one study has compared the heterotrophic production of bacteria living on plastic and in seawater (Dussud et al. 2018).

Literature on microbial ecotoxicology, the field dealing with the impact of microbial communities on pollutants and inversely including biodegradation, has grown considerably over the last 15 years (Jacquin et al. 2019). In parallel, there has been a tremendous increase in knowledge of the plastisphere (Wright et al. 2020). It has been suggested that the marine microbiome could play an important role in combating plastic pollution because microorganisms, including bacteria and fungi, possess the capabilities to degrade or deteriorate plastics using plastic as a carbon source and converting it (totally or partially) into biogas and biomass. Under laboratory conditions, the capability of microorganisms to biodegrade plastic usually by enzymatic hydrolysis or oxidation has been reported for numerous bacterial strains (Bhardwaj et al. 2013; Kale et al. 2015; Krueger et al. 2015; Oberbeckmann and Labrenz 2020).

Similarly, certain fungi also have the potential to tackle plastic pollution in the marine environment. For example, research done by Paço et al. (2017) investigated biodegradation of polyethylene (PE) microplastics by the naturally occurring marine fungus *Zalerion maritimum*. Results showed that *Z. maritimum* can break down PE pellets in both mass and size indicating that this fungus may actively contribute to the biodegradation of microplastics while requiring a minimum of nutrients (Paço et al. 2017).

**Case Study 1: State of the Art Project Generating New Knowledge: MycoPLAST**

A fungi-focused research project, MycoPLAST, is underway since 2020 aiming to assess the diversity, activity, and distribution of fungi associated with marine plastic debris samples and to evaluate and possibly unleash their ability to degrade complex plastic polymers. This will be achieved by providing detailed information on the identity and environmental importance of fungi associated with a wide variety of plastic samples (retrieved from different aquatic habitats) using molecular marker gene analyses, by establishing an extensive collection of fungal isolates, by determining their ability to efficiently degrade plastic polymers, by optimizing the utilization/degradation yield through microbial consortia, use of surfactants, and by constrained adaptation through iterative culturing. The project is based on the hypothesis that marine fungal communities represent an overlooked and untapped microbial component associated with marine plastic waste and that a better knowledge of their diversity, activity, function, and the exploration of their potential ability to degrade complex plastics could lead to efficient bioremediation applications. Final results from MycoPLAST are expected in 2023.

Website: https://anr.fr/Project-ANR-19-CE04-0001

Jacquin et al. (2019) presented an update of the current list of microorganisms proven to possess biodegradation capabilities for various types of polymers under laboratory conditions. However, so far even in optimized laboratory conditions rates of degradation of plastics by microorganisms are low (Krueger et al. 2015) and data on plastic mineralization in the ocean itself are still basically non-existent (Jacquin et al. 2019). Rogers et al. (2020) published a list of known plastic degraders (several bacterial and fungal species) in marine and terrestrial habitats. However, of these plastic degraders, no specific enzymatic pathway has been identified yet. The timescale of degradation of most conventional plastics is decades to centuries (Jacquin et al. 2019). Therefore, insufficient time has passed to know the most fundamental knowledge of the degradation process including its timescale and its use and effect on the marine microbiome, particularly in the ocean environment.

A review argued that "current standards and test methods are insufficient in their ability to realistically predict the biodegradability of carrier bags in aquatic environments, due to several shortcomings in experimental procedures and a paucity of information in the scientific literature" (Harrison et al. 2018, p. 14). Most studies have not been carried out under environmental conditions. Jacquin et al. (2019) presented the knowledge gaps on plastic biodegradation by marine microorganisms and attempted to identify possible directions for future research in this area. Summarizing, pathways of plastic degradation are still emerging, and it has not been possible yet to identify microbes that have proven to mineralize plastics in marine systems, especially within complex biofilms (Wright et al. 2020).

***Considerations for the Future*** Further research on biodegradation of plastics is needed in particular by identifying present gaps in knowledge and prioritizing the key aspects to address such as better physical characterization of marine biofilms and gaining a mechanistic and functional understanding of plastisphere communities (Wright et al. 2020). While the potential for plastics to be removed from the ocean by mineralization is evident this has not been shown outside the laboratory. As recommended by Wright et al. (2020) future work should focus on the identification of the mechanisms involved in degradation processes in order to be able to correctly interpret what is actually happening in the marine environment.

An added complication is that there are currently more than 5300 grades of synthetic polymers for plastics in commerce and they are commonly produced with a range of chemical additives such as plasticizers, flame retardants, antioxidants and other stabilizers, pro-oxidants, surfactants, inorganic fillers, or pigments (Wagner and Lambert 2018). The heterogeneous physical-chemical properties of plastics will likely result in heterogeneous metabolic pathways of biodegradation once they enter the environment, especially when considering the dynamic oceanic conditions and the large variety of microorganisms that may need to interact for the degradation of a single piece of plastic (Wagner and Lambert 2018). Therefore, treating plastic as a single compound does not make sense (Wagner and Lambert 2018) and any metabolic pathway of plastic biodegradation may ignore the complexity of the various processes that occur in the environment (Jacquin et al. 2019). Biodegradation of plastics in the marine environment depends on the composition as well as on the various ecosystems and environmental conditions encountered during the longevity of these xenobiotics (Jacquin et al. 2019).

Supporting the above findings, Roager and Sonnenschein (2019) emphasize the need for extended and reproducible collection of data to assess the presence of a core microbiome or functionalities of the plastisphere to confirm its capability for biodegradation of plastic. Furthermore, they also suggest next steps in research to elucidate the level of natural bioremediation and the exploitation of bacterial degradative mechanisms of plastic (Roager and Sonnenschein 2019). Jacquin et al. (2019) recommended that a complete study of the biodegradation of a polymer at sea must combine several monitoring parameters and especially be confirmed in the field with experiments in situ (Jacquin et al. 2019). Wright et al. (2020) suggested closing knowledge gaps on plastic–plastisphere–environment interactions through research on the physical characterization of marine biofilms, on the different successional stages, on the use of environmentally relevant concentrations of biofouled microplastics, and on the prioritization of gaining a mechanistic and functional understanding of plastisphere communities. Rogers et al. (2020) also recommended that future research must improve mechanistic understanding of the fate and environmental impacts of plastic litter, while also delivering much-needed information on microplastic exposure routes and levels in the environment to environmental managers. One suggested way forward is looking at the degradation pathway of naturally occurring polymers in the ocean that are also biodegradable such as chitin. Further exploring microorganisms that can degrade chitin might help to better understand the mechanisms behind polymer degradation and thereby learn about plastic decomposition (Wright et al. 2020). Substantial time and resources will be required to fully realize bioremediation as a strategy to combat the ocean plastic problem (Novak et al. 2020).

#### 18.2.3.4 Endocrine Disrupting Chemicals

The marine environment is exposed to the combined impacts of a cocktail of toxic chemicals and wastes entering the ocean. Many of these chemicals are Endocrine Disrupting Compounds (EDCs), which are contaminants that can interact with the endocrine system and negatively affect health, reproduction, and threaten survival. Understanding and concern about EDCs is a novel phenomenon developing only in the last three decades when an increasing number of omnipresent chemicals to which humans and wildlife are exposed have been identified to have endocrine-disrupting properties. Today, about 800 compounds are known or suspected to be able to affect the endocrine system, hormone receptors, hormone synthesis, or hormone conversion. Only a small fraction has been thoroughly investigated and most chemicals in commercial use have not been tested at all (United Nations Environment Programme and the World Health Organization 2013). These include well-known persistent organic pollutants that are banned in the western world such as PCBs and DDT as well as chemicals still in current use such as phthalates used in plastics and personal care products, brominated flame retardants, and perfluorinated compounds (Ingre-Khans et al. 2017). As research in this area has increased, evidence now points to exposure to EDCs playing an important role in the worldwide loss of species and decreased population numbers of amphibians, mammals, birds, reptiles, freshwater and marine fishes, and invertebrates (United Nations Environment Programme and the World Health Organization 2013). The exact degree of EDC impacts on fish and wildlife in nature is still uncertain but there is enough concern about EDCs that many developed nations are now requiring testing of new and existing chemicals for their endocrine-disrupting potential (McCormick and Romero 2017).

EDCs end up in the ocean through numerous ways including effluent discharges, agricultural runoff, urban emissions, and long-range transport via wind and ocean currents leading to worldwide exposure of humans and wildlife to EDCs (United Nations Environment Programme and the World Health Organization 2013). Examples of EDCs detected in the ocean are highly persistent water-soluble chemicals such as perfluorooctanesulfonate (PFOS), perfluorooctanoic acid (PFOA), and hexachlorocyclohexane (HCH) isomers, which are ingredients in insecticide. These EDCs have been found at great distances from where they were originally released transported by ocean currents (Armitage et al. 2009a, b; de Wit et al. 2004). It has been postulated that these assumed undegradable PFCs will be circulating in the ocean for many centuries after their emissions (United Nations Environment Programme and the World Health Organization 2013).

The presence of EDCs in the marine environment has only recently been recognized as a serious threat to biodiversity as well as human health but several studies have shown the high potential of marine microbes to naturally mitigate these harmful chemicals. EDCs that are potentially degradable through the use of marine microbiome approaches include the synthetic compounds estrogen 17a-ethinylestradiol (EE2), nonylphenol (NP), bisphenols (BPs), and mercury.

For example, Villela et al. (2019) have investigated the selection of putative beneficial microorganisms for corals (pBMCs) to degrade EE2. Their protocol describes how to isolate and test bacteria capable of degrading EE2 followed by a description of how to detect some putative beneficial characteristics of these associated microbes to their coral host. The methodologies described are cost-effective, easy to perform, and highly adaptable (Villela et al. 2019). Wang et al. (2019) investigated the intracellular absorption, extracellular adsorption, and bio-degradation of NP by four species of marine microalgae: *Phaeocystis globosa*, *Nannochloropsis oculata*, *Dunaliella salina*, and *Platymonas subcordiformis*. The results showed a sharp decrease of NP in the medium containing these four microalgal species which show strong biodegradation of this compound. Hence, it was concluded that these microalgae can improve the bioremediation of NP-contaminated water (Wang et al. 2019). Bisphenol A (BPA) is an important monomer in the manufacture of polycarbonate plastics, food cans, and other daily used chemicals. It has been identified as an environmental endocrine disruptor for its estrogenic and genotoxic activity. Although many factors affect the fate of BPA in the environment, BPA degradation is mainly dependent on the metabolism of bacteria. Many BPA-degrading bacteria have been identified from water, sediment/soil, and wastewater treatment plants. Zhang et al. (2013) reviewed and summarized BPA-degrading bacteria and the (proposed) BPA degradation pathway. BPS is the main replacement for BPA and may have similar effects as BPA. Danzl et al. (2009) investigated the biodegradation of bisphenol A, F, and S in seawater and did not find degradation of BPS. BPF is more biodegradable than BPA in seawater while BPS is more likely to accumulate in the aquatic environment. BPS poses a lower risk to human health and to the environment than BPA or BPF but since it does not seem to be biodegradable it might become an ecological burden when persistent (Danzl et al. 2009).

Mercury is one of the most serious pollutants threatening the ocean. It is a ubiquitous EDC and is a potent neurological poison for all life. In aquatic environments inorganic metallic mercury is converted by bacteria to the highly toxic methylmercury. Methylmercury has similar characteristics to Persistent Organic Pollutants (POPs) in terms of toxicity, persistence and bioaccumulation, and capacity for long-range transport. It is lipophilic and bioaccumulates in aquatic organisms biomagnifying and reaching high concentrations in top-order predators such as sharks, tuna, and swordfish. Methylmercury levels in some fish species can be up to a million times greater than the levels present in the surrounding water. Both the inorganic and organic forms of mercury (Hg) are powerful cytotoxic and neurotoxic agents in both humans and wildlife. Santos-Gandelman et al. (2014) analyzed the resistance profile and potential detoxification of inorganic and organic forms of Hg of bacteria isolated from marine sponges. Their results suggest a potential for mercury detoxification by marine sponge-associated resistant bacteria either through reduction or sequestration, as well as the possibility of bioremediation of toxic waste containing mercury (Santos-Gandelman et al. 2014).

**Case Study 2: State of the Art Project Generating New Knowledge: MER-CLUB**

Marine microorganisms also hold the genetic potential for Mercury (Hg) detoxification and represent an economical and highly efficient alternative for decontamination. The EU EASME-EMFF (Executive Agency for SMEs, European Maritime and Fisheries Fund) funded project MER-CLUB, underway since 2020, targets the largely unexplored diversity of marine Hg detoxifiers by isolating novel detoxifying marine strains and tests their application in a clean-up system. The project combines bio-augmentation and bio-stimulation strategies that will be evaluated and scaled up to a bioreactor study. The clean-up system will serve as a proof of concept for further development and commercial implementation facilitated by strong stakeholder involvement. As part of the project, a pilot study will be performed using dredged Hg contaminated sediments as a proof of concept of the developed MER-CLUB technology. The areas of expertise and experience of the partners will support the generation of valuable scientific knowledge including: (1) Disassembly, demolition, scrapping, and selective valorization of industrial facilities (such as chlor-alkali plants), (2) Research and treatment of contaminated soils, aquifers, and industrial ruins, (3) Integrated waste management and emergency response. This project, while in its early stages, has the potential to further broaden our insight into the potential of marine microorganisms for bioremediation and has the potential to contribute to large-scale restoration of our marine ecosystems.

Website: https://mer-club.eu/

Other major pollutants that have been shown to be potentially degradable through marine microbiome application are Polycyclic Aromatic Hydrocarbons (PAH), which are released in the environment due to the incomplete combustion of organic materials like coal, wood, and petroleum. Akhtar and Amin-ul Mannan (2020) state that a marine fungus, *Cochliobolus lunatus*, can be used to degrade PAH in soil and saline water. However, the study was not conducted in an area with high salinity and therefore future studies are required to evaluate the degradation potential of *C. lunatus* for PAHs in the marine environment.

***Considerations for the Future*** It is concluded that the bioremediation of EDCs is still in an early stage and few scientific articles have been published on this subject. However, the preliminary results are promising.

## 18.3 Policy and Governance. The Current State and Future Expectations

The "UN Decade on Ecosystem Restoration" (2021–2030) has just begun. This call to action has the purpose of recognizing the need to massively accelerate the prevention, halt, and reversion of the degradation of ecosystems worldwide. The Decade of Ecosystem Restoration declaration also coincides with the UN Decade of Ocean Science for Sustainable Development, which aims to provide a common framework to ensure that ocean science can fully support countries' actions to sustainably manage the ocean. The UN Decade of Ocean Science for Sustainable Development was successfully proposed by the Intergovernmental Oceanographic Commission (IOC) of UNESCO, a specialized organization of the United Nations system for ocean observations, data, services, and related capacity development. The timing of the UN Decade of Ocean Science for Sustainable Development was identified to align with the 2030 Agenda for Sustainable Development and corresponding Sustainable Development Goals (SDGs) adopted by the United Nations in 2015.

Both Decades together provide a powerful means to ensure restoration of the ocean, which is so urgently needed considering its rapid degradation and overuse. They open the way for the global ocean community to mobilize itself to achieve relevant SDGs, in particular SDG14: Life Below Water (14.2—Environmental restoration), with the aim to conserve and sustainably use the ocean, seas, and marine resources for sustainable development. The ambition is to use this gathering momentum to mobilize the scientific community, policymakers, business, civil society, and citizens around a program of joint research and technological innovation. To be successful in achieving ocean restoration there should be a focus on both deep disciplinary understanding of ocean processes and solution-oriented research to generate new knowledge (Visbeck 2018).

While at the European level there are various legal EU frameworks, strategies, and action plans that promote nature restoration, it has been said that these directives lack in most cases a specific prioritization framework for implementing conservation and restorative actions (Schoukens 2017). Further, still conceptual, regulatory, and implementation gaps have prevented existing legislation from achieving its objectives (van Tatenhove et al. 2020). To address this, the recent EU Biodiversity Strategy 2030 has been launched with its ambitious long-term plan for protecting nature and reversing the degradation of ecosystems. The Strategy is a key pillar of the European Green Deal (Europe's growth strategy) and sets out strategic plans to meet existing targets including transforming at least 30% of Europe's seas into effectively managed protected areas and, more broadly, the EU to become climate neutral by 2050 (EU 2020). The European Commission has proposed new legislation to ensure that these targets are met, e.g., the European Climate Law which intends to write into law the goals set out in the European Green Deal (EC 2020a) including those within the Biodiversity Strategy. The Climate Law would ensure all EU policies contribute to this goal and that all sectors of the economy and society play their part.

To date, microorganisms have been given limited consideration in ecosystem management policy but with the growing understanding of their importance in maintaining ecosystem health and enhancing resilience to global change so too grow the opportunities to utilize them in restoration (Allard et al. 2020). The Consensus Statement, published in Nature by Cavicchioli et al. (2019), calls for explicit considerations of microorganisms in policy and decision-making for ecosystem management. It is a call to action for microbiologists to become increasingly engaged in policy formation and for microbial research to become more integrated into the frameworks for addressing climate change and accomplishing the United Nations SDGs (Cavicchioli et al. 2019). In line with this, the Atlantic Ocean Research Alliance (AORA) published a Marine Microbiome Roadmap (2020) that states that "To ensure early coordination and interoperability guided by a shared vision, we need to bring together science, industry and policy makers to advance the Next Great Exploration of the Oceans" (Bolhuis et al. 2020). AORA scientists came together with policymakers to develop the Roadmap to help align and advance an Atlantic-wide approach to marine microbiome research and monitoring. One of their long-term aims is to communicate results to familiarize managers and policy makers about the microbiome and to develop management strategies that encompass ecosystems.

Moving into a post-COVID-19 world, the need for new policy decisions is evident. Existing policies might require changes to their trajectories and new policies will emerge as a response (Claudet et al. 2020). Considering the ocean's vastness there is a need to be context-specific and bottom-up approaches should be supported in order to meet binding targets (van Tatenhove et al. 2020). Further still, engaging in largescale ocean restoration will take more than good intentions and in the wake of COVID-19 there is an increasing awareness that we can no longer carry on with "business as usual." An example of this is the EU's Green Deal (EU 2019), which has now been supplemented by a Green Recovery Plan (EU 2020). In 2019, the EU assigned 1 billion € to the Horizon 2020 Green Deal call of which there was a call on *Restoring biodiversity and ecosystem services* (LC-GD-7-1-2020) that aimed to address the current barriers preventing the large-scale restoration of ecosystem (terrestrial and marine). Of the 1 billion € available, 80 million € is assigned for this specific call, highlighting the urgency and importance of restoration.

It is hoped that the newest policy and governance measures in parallel with post-COVID-19 sentiments will ensure a broad uptake and implementation and make it possible for marine restoration actions to accelerate the recovery, integrity, and resilience of degraded ecosystems, which will support biodiversity and improve human health and wellbeing. With continued advances marine ecosystem restoration can be elevated to a key rather than minor management intervention (Saunders et al. 2020).

## 18.4 Strategic Communication Around Usage of the Marine Microbiome in Ocean Restoration

There is an urgent need to restore the ocean and prevent further worsening of the conditions that are caused by human activities. Society as a whole must do more to preserve the ocean and its resources for current and future generations. Everyone, at all levels of society, needs to be aware of how important the ocean, including the marine microbiome, is and what can be done to protect, restore, and preserve it in a joint and determined effort. Effective and strategic communication is key to attaining this awareness and motivating mitigative behaviors thereby inducing successful marine restoration and conservation (see e.g., Kolandai-Matchett et al. 2020; Kolandai-Matchett and Armoudian 2020; Laffoley et al. 2020 Ockendon et al. 2018; Reuver et al. 2016).

Unfortunately, many people including high-level policy makers still have insufficient awareness of the importance of the ocean as central to human well-being. And even if they do, awareness does not necessarily lead to positive behavior change. There is a need for society to become more ocean and microbiome literate and to collectively engage in environmentally responsible behavior. This requires a better understanding of how our relationship with and knowledge of the ocean translates into a behavioral change that positively affects the ocean as well as what the most effective ways to achieve this are (Uyarra and Borja 2016). In general, marine conservation communication is still in emerging stages when compared to terrestrial-focused conservation communication (Kolandai-Matchett et al. 2020). Research-informed marine communication has been lagging behind particularly in conveying complex issues (such as the ocean microbiome) and so there is still limited knowledge on which approach(es) would work best (Kolandai-Matchett and Armoudian 2020). Here, the latest trends and scientific insights into ocean literacy, microbiome literacy, and effective and strategic communication are outlined that really motivates behavioral change to support academics, industry, policymakers, and society as a whole to restore and conserve the ocean.

## 18.5 Ocean Literacy

"Knowing and understanding the ocean's influence on us, and our influence on the ocean is crucial to living and acting sustainably. This is the essence of ocean literacy (OL)." (https://oceanliteracy.unesco.org). Being ocean literate is "being able to make informed and responsible decisions regarding the ocean and its resources" (Cava et al. 2005, p. 35). Increasing OL is an effective and essential part of the strategies that are needed to change people's individual behavior towards the ocean in a positive and constructive way (Brennan et al. 2019; Reuver et al. 2016) to help to restore and preserve this precious resource including its microbiome.

The OL movement began in the USA in the early 2000s with the National Marine Educators Association (NMEA) in the US at the forefront and the term Ocean Literacy being coined in 2005 (Cava et al. 2005). NMEA's efforts focused on defining OL centering around seven Ocean Literacy Principles (OLP), which represent a source of inspiration for those working toward achieving an ocean literate society. OL has become an international effort in recent times. In Europe, the European Marine Science Educators Association (EMSEA) was established in 2012 and the European Union committed to the development of OL within the EU amongst others by providing funding for two major OL projects: Sea Change (https://seachangeproject.eu/) and ResponSEAble (https://www.responseable.eu/). At the global level, the UN Decade of Ocean Science for Sustainable Development (2021–2030) has indicated ocean literacy as one of the priority research and development areas.

The main goal of OL is not only to educate but rather to empower individuals to make informed and responsible decisions related to marine resources and ocean sustainability (Santoro et al. 2017). The OL movement has developed a framework for the introduction of ocean science in schools' curricula globally and it is expected that as a result the younger generations will be the first in developing more sustainable behavioral patterns (Fernandez Otero et al. 2019). In the last decade, numerous educational and communication resources representing a valuable legacy for future OL activities have been developed of which many are available from the recently established Ocean Literacy Portal by the Intergovernmental Oceanographic Commission—UNESCO (https://oceanliteracy.unesco.org/about/). The Ocean Literacy Portal aims to support the creation of an ocean-literate society able to make informed and responsible decisions on ocean resources and ocean sustainability.

Until recently, OL focused mostly on the education sphere, including Massive Open Online Courses (MOOCS) (e.g., "Exploring our Oceans") (Fielding et al. 2019), formal and informal education (Barracosa et al. 2019), and elementary school students (Mogias et al. 2019). The UNESCO-IOC initiative "*Ocean for All: a global strategy to raise awareness for the conservation, restoration, and sustainable use of our ocean*" was launched in 2017 to increase all people's understanding and encourage wider participation from all stakeholders globally beyond the education realm. A UNESCO/IOC (2019) report states that "achieving the goal of having an ocean literate society requires the involvement of different actors and to use strategies that go far beyond the conventional policy making including new forms of communication. It would require the inclusion and the empowerment of communities and networks of business, universities, research centers, and civic groups to share the responsibility for reversing the current trend of ocean health degradation through the development of innovative solutions and the creation of partnerships." Already existing worldwide initiatives include the H2020-funded Atlantic Ocean Research Alliance (AORA), which coordinates marine science educators in Europe, the United States, and Canada to better inform and engage citizens about the ocean's influence on them and their influence on the ocean. Through AORA there is now a transatlantic strategy on OL, which aims to consolidate existing efforts toward a proof of concept for Transatlantic Ocean Literacy (TOL) (Ocean Literacy Working Group of the Atlantic Ocean Research Alliance 2020; Fernandez Otero et al. 2019).

While much progress has been made, building OL across society targeting all population groups from citizens to policy makers is a challenge for many different reasons. Many public marine outreach efforts for example carried out through aquaria and zoos have only local reach and are therefore limited in their effect and are not always affordable for everybody. Marine citizen science programs whereby scientific research is conducted by citizens as amateur scientists have the same issues and in addition tend to attract only those who are already interested and aware. McCauley et al. (2019) found that important actors, such as policy makers, teachers, and lecturers often have a lack of knowledge and understanding of the importance of the ocean in our cultural, social, and environmental heritage. There is a need for more marine outreach efforts that targets all stakeholders at all levels of society to bring about global ocean literacy. In addition, there is a need for effective communication approaches that achieve OL leading to actual ocean-related behavioral change. Numerous studies have been conducted on the links between environmental knowledge and attitude and behavior to protect the environment (including the ocean) finding that behavior change rarely occurs due to just simply providing information (e.g., Barreto et al. 2014; Stoll-Kleemann 2019). This is because at least 80% of the motives for pro-environmental or non-environmental behavior seem to be based on situational and other internal factors, which means that even when people receive information they may still give it a low priority within the context of everyday life (Stoll-Kleemann 2019).

There are not so many behavioral change studies focused on ocean conservation specifically. In one example, Stoll-Kleemann (2019) explored the barriers, opportunities, and incentives to encourage more ocean-friendly behavior, building on an interdisciplinary and multifactor approach, the existing pro-environmental behavior model (Kollmuss and Agyeman 2002), and theoretical frameworks on individual behavior (Darnton 2008). This publication reported that there are many factors that influence individual ocean-related behavior including personal factors (e.g., values, knowledge, emotions, personality traits), external factors (e.g., socio-cultural and politico-economic), and internal and external incentives (Stoll-Kleemann 2019). As behavior is determined by multiple variables, sometimes in interaction, the most effective behavior change programs often require combinations of intervention types (Stern 2000).

Intervention types include reputation-based incentives, social marketing, and successfully diffusing social change and many of these are not yet broadly applied to the field of ocean literacy/ocean related pro-environmental behavior. Stoll-Kleemann (2019) presents feasible options and incentives towards ocean-friendly behavior. For example, Lubchenco et al. (2016) showed that an approach based on increasing reputation and toward a positive self-image can create conditions that also incentivize companies and countries to engage in activities that support sustainability. Stoll-Kleemann (2019) also states that it is better to include more emotional and "feeling" elements in choices regarding ocean-relevant behavior. This author asserts feelings regarding responsibility, guilt, pride, hope, and gratitude as positive (moral) emotions. Altruism, ethical values, and reciprocity are powerful drivers of change. Community-Based Social-Marketing (CBSM) strategies can also influence ocean-related behavior based on the idea that creating new norms and commitment of individuals can result in social diffusion and aim at developing supportive and transformative social interaction (McKenzie-Mohr 2000). Furthermore, for marine conservation communication to be effective the messages need to be framed based on norms that are context and audience specific. The message needs to be easy to understand considering literacy levels and community language (Kolandai-Matchett and Armoudian 2020).

Mokos et al. (2020) researched how to increase Ocean Literacy through non-formal marine science education and whether it had a long-term effect. Their results indicate that there is a need for an integrated approach to the teaching of Ocean Literacy starting from the early grades by combining teacher's professional development, strengthening ocean-related topics in school curricula, and promoting non-formal educational activities.

## 18.6 Knowledge Transfer

Knowledge Transfer (KT) is another type of science communication. KT describes a two-way process through which a knowledge output moves from a knowledge source to a targeted potential user who then applies that knowledge ("*knowledge output*" is a unit of knowledge or learning generated by or through research activity) (Reuver et al. 2016). The reason that KT is described as a two-way process is because its core philosophy is to frame communication activities around target users' needs. Effective KT requires that a specific target user is profiled and bespoke materials are developed in a medium that is framed for and specific to that user's behaviors, needs, and interests. Argote and Fahrenkopf (2016) describe KT as the process through one social unit learns from or is affected by the experience (or knowledge gained) by another.

Within the academic world, there continues to be a different understanding and interpretation of the meaning of KT. There is undoubted agreement that KT seeks to organize, create, capture, or distribute knowledge and to ensure its availability for future users (Reuver et al. 2016). In organizational theory, KT is performed in response to the practical problem of losing knowledge often tacit knowledge when staff leaves an organization (Argote and Ingram 2000). Within the world of communication professionals a whole lexicon surrounds KT and is often overlapping in its meaning and commonly disputed and has not yet been entirely overcome (Reuver et al. 2016). In marine science, KT is more selfless in its manner with its primary purpose being to increase the likelihood that research evidence will be applied by policy, industry, science, and society (Fernandez Otero et al. 2019; Mitton et al. 2007). It is performed to spread any knowledge (explicit and tacit) more widely to those that may benefit from or exploit the knowledge via commercial and non-commercial settings (Thérin 2013) thus increasing its overall impact. Within the marine microbiome research field, KT has great potential to support the further development of the field particularly in case of the transfer of knowledge from other areas of microbiology fields such as ongoing methodologies and discoveries in soil and human microbiome research.

A well-established knowledge transfer methodology that is used in many European marine-related research projects is based on an approach originally developed by the Irish knowledge management organization AquaTT (https://aquatt.ie/) in the EC-funded FP7 MarineTT project (https://marinett.eu/). AquaTT was the coordinator of the pioneering MarineTT project which was recognized as an 'exemplar' project in the ex-post evaluation of FP7 to the EC. MarineTT was a European flagship project addressing the need for more effective KT in European publicly-funded marine research projects. This methodology was then adapted in the KT-focused H2020-funded COLUMBUS project (2015–2018), which focused on knowledge transfer to support European Blue Growth (https://www.columbusproject.eu/). The objective of the COLUMBUS project was to produce qualitative case studies on how the project activities successfully transferred marine knowledge to policy, industry, science, and society resulting in impacts that measurably contribute to "Blue Growth" (AquaTT 2015). Over the course of the project COLUMBUS identified 967 projects noted to have potentially produced solutions that responded to the needs of nine targeted marine and maritime activities. From this, a total of 1779 Knowledge Outputs was collected from these projects of which 246 KO were prioritized and 56 were transferred with 48 success stories released for public dissemination (Fernandez Otero et al. 2019).

Knowledge transfer was defined in COLUMBUS as the "overall process of moving knowledge between knowledge sources to the potential users of knowledge. Knowledge transfer consists of a range of activities which aim to capture, organize, assess and transmit knowledge, skills and competence from those who generate them to those who will utilize them." This is the methodology that several organizations working in the marine sector (e.g., AquaTT, ERINN Innovation) implement in a number of research projects today including numerous EU-funded projects, e.g., AQUAEXCEL$^{2020}$, AQUAEXCEL3.0, SIMBA, ERGO, ParaFishControl, PerformFish, SEAFOOD$^{TOMORROW}$, BIOGEARS, SEALIVE, MISSION ATLANTIC, MARBLES, SCORE, and others. For example, the H2020-funded SEALIVE project (https://sealive.eu/), which has the vision to decrease plastic waste and contamination on land and in the sea by boosting the use of biomaterials and contributing to the circular economy with cohesive bio-plastic strategies, implements the COLUMBUS KT methodology. The methodology ensures that the transfer of high potential knowledge produced in a project is strategic, coordinated, and effective. Furthermore, the methodology focuses on knowledge outputs rather than generalized information of the project. It is not limited to de novo or pioneering discoveries but may also include new methodologies, processes, adaptations, insights, alternative applications of prior know-how, and knowledge (ERINN Innovation 2020).

The intention of the methodology is to identify a high potential knowledge output's pathway to impact or "knowledge output pathway." A knowledge output pathway is a term coined by the COLUMBUS project and is used to represent the series of steps (and relevant actors) that connect a knowledge output with its end user who will apply the knowledge output and result in an eventual impact. Eventual impacts can vary widely depending on the knowledge and end user. Reuver et al. (2016) provided some examples of what an eventual impact might look like. These include:

- The development of Blue Economy —commercializing a product or service; improving existing business performance; creating new markets for an existing product or service; establishing of a new business or a strategic collaboration between businesses to market a new service; or, attracting inward investment, i.e., by finding new ways to exploit environmental resources and services optimally.
- Sustainable Blue Growth—applying knowledge to inform policy and regulation; improving environmental monitoring programs; or, enabling the development of ecosystem services.
- A Blue Society—enhancing public health and well-being; saving public sector money; or, enabling a resilient society (e.g., protecting vulnerable people, places and infrastructure; providing a secure supply of food, energy, water).

The European Commission has developed a platform, the Horizon Results Platform, where knowledge outputs can be uploaded so that EU research projects can share results to relevant stakeholders (such as policymakers, industry, or society) so they can "flourish into innovations that contribute to our society and economy, and to a sustainable future" (EC 2019a).

The place for knowledge transfer in EU-funded projects continues to grow as generating impact from projects is an increasing priority area for the European Commission. The continued emphasis on impact is seen most readily in the new Framework Program "Horizon Europe" that was recently launched (2021–2030). The 95.5bn € Program is considered the most ambitious research and innovation program in the world. Horizon Europe will maximize its impact and deliver on the EU's strategic priorities such as the recovery from the immediate economic and social damage brought about by the coronavirus pandemic, and green and digital transitions, and tackles global challenges to improve the quality of our daily lives (EC 2020b). Horizon Europe will be more impact focused from the start with a clear intervention logic including scientific impact and specific approaches targeting impact. (EC 2019b). There is a Mission Area on Oceans, Seas and Water that has the objective by 2030 to "*clean[ing] marine and fresh waters, restoring degraded ecosystems and habitats, decarbonizing the blue economy to sustainably harness the essential goods and services they provide*" (draft Mission Work Program 2021). In order to achieve this mission and to capture, measure, and ensure that impacts are realized for society and policy all funded projects will need strong knowledge management, knowledge transfer, and knowledge exploitation strategies in place from the outset. In ensuring the achievement of the Oceans, Seas and Water Mission Area objective, the marine microbiome has the potential to play a critical role.

## 18.7 Discussion and Conclusion

As can be seen from the increasing number and diversity of scientific publications on the marine microbiome, policy developments, and the emergence of large international collaborative networks, progress in the field of marine microbiome research is increasing exponentially. The opportunities to apply marine microbiome approaches to ocean restoration are promising but still there are also several challenges to be overcome ranging from scientific and technical limitations, ethical issues, policy barriers, and fragmentation in the field. In relation to the latter according to Dubilier et al. (2016), there are two major stumbling blocks to advancing understanding of microbes' role in the biosphere namely fragmentation of the life-sciences field and a lack of coordination among the different microbiome research efforts that are ongoing. Wilkins et al. (2019) also see great value in building a framework of broad collaborative networks including terrestrial microbiome links as they could support future progress in microbial symbiosis research overall and are more sustainable and ultimately more productive. Upcoming opportunities addressing this challenge include more collaboration at the European level through Horizon Europe—the next research and innovation framework program. Within this large funding program dedicated budgets will become available for large-scale collaborative research on a healthy ocean and on healthy seas, coastal-, and inland waters including finding systemic solutions for the prevention, decrease, mitigation, and removal of marine pollution, research on adaption to and mitigation of pollution and climate change in the ocean, and sustainable use and management of ocean resources. A collaboration effort is the implementation of a Microbial Resource Research Infrastructure in Europe: IS_MIRRI21, which will support research, development, and innovation in the use and preservation of microbial life for the purpose of basic and applied scientific research, preservation, and exploitation of microorganisms (https://ismirri21.mirri.org/). Another collaborative effort would be the creation of a central bank of microbial strains with potential probiotic characteristics and degradation capacity as proposed by Villela et al. (2019). This would be a crucial step for the progress of this field saving time and work and contributing to the assembly of new specific consortia worldwide. This is much needed because new bioremediation-probiotic consortia need to be assembled for each specific restoration situation (Villela et al. 2019).

Technological and methodological limitations and remaining research questions include challenges to obtain a better understanding of the function of the microbiome members. Advances in DNA-based molecular approaches have provided unparalleled insights into the taxonomic and metabolic diversity of the microbial world and the availability of sequencing data is vast. There is however a need to annotate functional genes within metagenomic and metatranscriptomic datasets (Trevathan-Tackett et al. 2019). In-depth functional analyses are needed to get a better insight into holobiont functioning and responses to environmental stress. This knowledge would also inform how microbiome functioning relate to modified biogeochemical fluxes within the ecosystem (Trevathan-Tackett et al. 2019). Another key issue is the interaction of host-associated microbes with their marine host species of which little is known for most marine host species (Wilkins et al. 2019). It is argued that such advances in research will help to predict responses of species, communities, and ecosystems to stressors driven by human activity and inform future management strategies (Wilkins et al. 2019).

Trevathan-Tackett et al. (2019) have outlined some remaining key research questions in relation to using the microbiome for restoration purposes. These include whether "a managed or manipulated microbiome can outlive its function, and if so, what impacts does this have on the microbiome, holobiont or ecosystem; whether there is there a way to 'stop' a microbial function or remove a community once a particular job has been done; and how resilient would a managed or manipulated microbiome be to disturbances and how would they be monitored to know if a desirable or undesirable outcome is achieved" (Trevathan-Tackett et al. 2019).

Policy challenges that will need to be addressed in the near future to further advance the field of marine microbiome ocean restoration are modifying a number of existing regulatory frameworks particularly those impacting biotechnological interventions. These include the 1972 Convention on International Trade of Endangered Species (CITES), the 1992 Convention on the Conservation of Biological Diversity (CBD), the 2000 Cartagena Protocol, and the 2014 Nagoya Protocol. It will be necessary to revise these policies to reflect the changing environment of conservation needs. The United Nations has begun drafting policy to rectify the lack of governance of sharing marine genetic resources beyond jurisdictional borders (Novak et al. 2020). Vanwonterghem and Webster (2020) advocate that microbes become an integral component of climate change research and decision-making in conservation and natural resource management given the critical role of microorganisms in driving global climate change. As has been suggested previously (Bourlat et al. 2013; Trevathan-Tackett et al. 2019) it would be beneficial to identify, engage, and communicate with stakeholders at all levels early in the research process in order to optimize research-policy transfer of knowledge that address real needs.

Associated with policy challenges are ethical issues in relation to biotechnological interventions specifically in the microbiome such as those in coral AE (Assisted Evolution). Ethical frameworks are necessary for effective and responsible practice and interventions are controversial but "for many marine areas the cost of inaction is too high to allow controversy to be a barrier to conserving viable ecosystems" (Novak et al. 2020).

Linked to its ability to restore the ocean several researchers suggest the microbiome could also be used as a biomarker or monitoring tool of ecosystem health using microbes as early warning indicators for stress at holobiont and ecosystem levels (Egan and Gardiner 2016; Pita et al. 2018; Vanwonterghem and Webster 2020). Used in combination with traditional monitoring programs microbiome monitoring would allow obtaining key data that enable intervention and restoration before crucial ecosystem functions are lost (Pita et al. 2018). Hutchins et al. (2019) state that robust predictive models should include microbial processes to interpolate and extrapolate observed microbial interactions with their environment to enable a predictive understanding of microbiome functions in diverse ecosystems.

To conclude, seeing current political will and advances together with developments in microbiome research and communication, there is a positive future outlook for using the microbiome in ocean restoration. Hopefully, the COVID-19 pandemic will spur on all actors to build a blue recovery through the use of the marine microbiome for a healthier planet for all.

## References

Ados Santos H, Duarte G, Rachid C, Chaloub RM, Calderon EN, de Barros Marangoni LF, Bianchini A, Nudi AH, Lima do Carmo F, van Elsas JD, Soares Rosado A, Barreira e Castro C, Silva Peixoto R (2015) Impact of oil spills on coral reefs can be reduced by bioremediation using probiotic microbiota. Sci Rep 5:18268. https://doi.org/10.1038/srep18268

Ainsworth TD, Fordyce AJ, Camp EF (2017) The other microeukaryotes of the coral reef microbiome. Trends Microbiol 25:980–991. https://doi.org/10.1016/j.tim.2017.06.007

Akhtar N, Amin-ul Mannan M (2020) Mycoremediation: expunging environmental pollutants. Biotechnol Rep 26:e00452. https://doi.org/10.1016/j.btre.2020.e00452

Allard SM, Costa MT, Bulseco AN, Helfer V, Wilkins LGE, Hassenrück C, Zengler K, Zimmer M, Natalia Erazo N, Rodrigues JLM, Duke N, Melo VMM, Vanwonterghem I, Junca H, Makonde HM, Jiménez DJ, Tavares TCL, Fusi M, Daffonchio D, Duarte CM, Peixoto RS, Rosado AS, Gilbert JA, Bowman J (2020) Introducing the mangrove microbiome initiative: identifying microbial research priorities and approaches to better understand, protect, and rehabilitate mangrove ecosystems. mSystems 5:5. https://doi.org/10.1128/mSystems.00658-20

Apprill A (2017) Marine animal microbiomes: toward understanding host–microbiome interactions in a changing ocean. Front Mar Sci 4:222. https://doi.org/10.3389/fmars.2021.643339

AquaTT (2015) Deliverable 2.2. Guidelines on carrying out COLUMBUS knowledge transfer and impact measurement. Columbus Website—Results. Available online at: http://www.columbusproject.eu/D2%202%20Guidelines%20on%20carrying%20out%20COLUMBUS%20KT%20v2%20Final%2030.11.15.pdf. Accessed 12.03.21

Argote L, Fahrenkopf E (2016) Knowledge transfer in organizations: the roles of members, tasks, tools, and networks. Organ Behav Hum Decis Process 136:146–159

Argote L, Ingram P (2000) Knowledge transfer: a basis for competitive advantage in firms. Organ Behav Hum Decis Process 82:150–169

Armitage J, MacLeod M, Cousins IT (2009a) Modeling the global fate and transport of perfluorooctanoic acid (PFOA) and perfluorooctanoate (PFO) emitted from direct sources using a multispecies mass balance model. Environ Sci Technol 43:1134–1140

Armitage JM, Schenker U, Scheringer M, Martin JW, MacLeod M, Cousins IT (2009b) Modeling the global fate and transport of perfluorooctane sulfonate (PFOS) and precursor compounds in relation to temporal trends in wildlife exposure. Environ Sci Technol 43:9274–9280

Assis JM, Abreu F, Villela HMD, Barno A, Valle RF, Vieira R, Taveira I, Duarte G, Bourne DG, Høj L, Peixoto RS (2020) Delivering beneficial microorganisms for corals: rotifers as carriers of probiotic bacteria. Front Microbiol 11:3243. https://doi.org/10.3389/fmicb.2020.608506

Azam F, Malfatti F (2007) Microbial structuring of marine ecosystems. Nat Rev Microbiol 5:782–791. https://doi.org/10.1038/nrmicro1747

Baniasadi M, Mousavi SM (2018) A comprehensive review on the bioremediation of oil spills. In: Kumar V, Kumar M, Prasad R (eds) Microbial action on hydrocarbons. Springer, Singapore. https://doi.org/10.1007/978-981-13-1840-5_10

Barnes DKA, Galgani F, Thompson RC, Barlaz M (2009) Accumulation and fragmentation of plastic debris in global environments. Philos Trans R Soc B 364:1985–1998. https://doi.org/10.1098/rstb.2008.0205

Barracosa H, de los Santos CB, Martins M, Freitas C, Santos R (2019) Ocean literacy to mainstream ecosystem services concept in formal and informal education: the example of coastal ecosystems of southern Portugal. Front Mar Sci 6:626. https://doi.org/10.3389/fmars.2019.00626

Barreto ML, Szóstek A, Karapanos E, Nunes NJ, Pereira L, Quintal F (2014) Understanding families' motivations for sustainable behaviors. Comput Hum Behav 40:6–15. https://doi.org/10.1016/j.chb.2014.07.042

Bhardwaj H, Gupta R, Tiwari A (2013) Communities of microbial enzymes associated with biodegradation of plastics. J Polym Environ 21:575–579. https://doi.org/10.1007/s10924-012-0456-z

Blackburn S, Pelling M, Marques C (2019) Megacities and the coast: global context and scope for transformation. Coasts Estuaries 661–669. https://doi.org/10.1016/b978-0-12-814003-1.00038-1

Bolhuis H, Buttigieg PL, Goodwin K, Groben R, Ludicone D, Lacoursière-Roussel A, Pesant S, Robinson S, Björnsson S, Erlendsson LS, Rae M (2020) Atlantic Ocean Research Alliance (AORA)—Marine Microbiome Roadmap. https://doi.org/10.5281/zenodo.3632526. https://www.atlanticresource.org/aora/site-area/publications/publications

Bordoloi J, Boruah HPD (2018) Analysis of recent patenting activities in the field of bioremediation of petroleum hydrocarbon pollutants present in the environment. Recent Pat Biotechnol 12:3–20. https://doi.org/10.2174/1872208311666170504111019

Bourlat SJ, Borja A, Gilbert J, Taylor MI, Davies N, Weisberg SB, Griffith JF, Lettieri T, Field D, Benzie J, Glöckner FO, Rodríguez-Ezpeleta N, Faith DP, Bean TP, Obst M (2013) Genomics in marine monitoring: new opportunities for assessing marine health status. Mar Pollut Bull 74:19–31. https://doi.org/10.1016/j.marpolbul.2013.05.042

Brennan C, Matthew A, Owen M (2019) A system dynamics approach to increasing ocean literacy. Front Mar Sci 6:360. https://doi.org/10.3389/fmars.2019.00360

Brown BE (1997) Coral bleaching: causes and consequences. Coral Reefs 16:S129–S138. https://doi.org/10.1007/s003380050249

Bruto M, James A, Petton B, Labreuche Y, Chenivesse S, Alunno-Bruscia M, Polz MF, le Roux F (2017) *Vibrio crassostreae*, a benign oyster colonizer turned into a pathogen after plasmid acquisition. ISME J 11:1043–1052. https://doi.org/10.1038/ismej.2016.162

Cava F, Schoedinger S, Strang C, Tuddenham P (2005) Science content and standards for ocean literacy: a report on ocean literacy. Available online at: http://coexploration.org/oceanliteracy/documents/OLit2004-05_Final_Report.pdf. Accessed 04.01.2021

Cavicchioli R, Ripple WJ, Timmis KN, Azam F, Bakken LR, Baylis M, Behrenfeld MJ, Boetius A, Boyd PW, Classen AT, Crowther TW, Danovaro R, Foreman CM, Huisman J, Hutchins DA, Jansson JK, Karl DM, Koskella B, Mark W, David B, Martiny JBH, Moran MA, Orphan VJ, Reay DS, Remais JV, Rich VI, Singh BK, Stein LY, Stewart FJ, Sullivan MB, van Oppen MJH, Weaver SC, Webb EA, Webster NS (2019) Scientists' warning to humanity: microorganisms and climate change. Nat Rev Microbiol 17:569–586. https://doi.org/10.1038/s41579-019-0222-5

Census of Marine Life (2018) Accessed 06.02.2021. https://ocean.si.edu/census-marine-life

Chakravarti LJ, Beltran VH, van Oppen MJH (2017) Rapid thermal adaptation in photosymbionts of reef-building corals. Glob Chang Biol 23:4675–4688. https://doi.org/10.1111/gcb.13702

Claudet J, Bopp L, Cheung W, Devillers R, Escobar-Briones E, Haugan P, Heymans JJ, Masson-Delmotte V, Matz-Lück N, Miloslavich P, Mullineaux L, Visbeck M, Watson R, Zivian AM, Ansorge I, Araujo M, Aricò S, Bailly D, Barbiere J, Barnerias C, Bowler C, Brun V, Cazenave A, Diver C (2020) A roadmap for using the UN Decade of Ocean Science for Sustainable Development in support of science, policy, and action. One Earth 2:34–42. https://doi.org/10.1016/j.oneear.2019.10.012

Damjanovic K, Blackall LL, Webster NS, van Oppen MJH (2017) The contribution of microbial biotechnology to mitigating coral reef degradation. Microb Biotechnol 10:1236–1243. https://doi.org/10.1111/1751-7915.12769

Damjanovic K, van Oppen MJH, Menéndez P, Blackall LL (2019) Experimental inoculation of coral recruits with marine bacteria indicates scope for microbiome manipulation in *Acropora tenuis* and *Platygyra daedalea*. Front Microbiol 10:1702. https://doi.org/10.3389/fmicb.2019.01702

Danzl E, Sei K, Soda S, Ike M, Fujita M (2009) Biodegradation of bisphenol A, bisphenol F and bisphenol S in seawater. Int J Environ Res Public Health 6:1472–1484. https://doi.org/10.3390/ijerph6041472

Darnton A (2008) Reference report: an overview of behaviour change models and their uses. Centre for Sustainable Development. University of Westminster

Das N, Chandran P (2011) Microbial degradation of petroleum hydrocarbon contaminants: an overview. Biotechnol Res Int 2011:941810. https://doi.org/10.4061/2011/941810

DeLeo DM, Ruiz-Ramos DV, Baums IB, Cordes EE (2016) Response of deep-water corals to oil and chemical dispersant exposure. Deep Sea Res Pt II Top Stud Oceanogr 129:137–147. https://doi.org/10.1016/j.dsr2.2015.02.028

de Wit C, Fisk A, Hobbs K, Muir D (2004) AMAP assessment 2002: persistent organic pollutants in the Arctic. Arctic Monitoring and Assessment Programme (AMAP), Oslo, Norway. https://www.amap.no/documents/doc/amap-assessment-2002-persistent-organic-pollutants-in-the-arctic/96

Duarte CM, Agusti S, Barbier E, Britten GL, Castilla JC, Gattuso J-P, Fulweiler RW, Hughes TP, Knowlton N, Lovelock CE, Lotze HK, Predragovic M, Poloczanska E, Roberts C, Worm B (2020) Rebuilding marine life. Nature 580:39–51. https://doi.org/10.1038/s41586-020-2146-7

Dubilier N, McFall-Ngai M, Zhao L (2016) Microbiology: create a global microbiome effort. Nature 526:631–634. https://doi.org/10.1038/526631a

Dussud C, Hudec C, George M, Fabre P, Higgs P, Bruzaud S, Delort A-M, Eyheraguibel B, Meistertzheim A-L, Jacquin J, Cheng J, Callac N, Odobel C, Rabouille S, Ghiglione J-F (2018) Colonization of non-biodegradable and biodegradable plastics by marine microorganisms. Front Microbiol 9:1571. https://doi.org/10.3389/fmicb.2018.01571

Egan S, Gardiner M (2016) Microbial dysbiosis: rethinking disease in marine ecosystems. Front Microbiol 7:991. https://doi.org/10.3389/fmicb.2016.00991

Egan S, Harder T, Burke C, Steinberg P, Kjelleberg S, Thomas T (2013) The seaweed holobiont: understanding seaweed-bacteria interactions. FEMS Microbiol Rev 37:462–476. https://doi.org/10.1111/1574-6976.12011

Egan S, Fukatsu T, Francino MP (2020) Opportunities and challenges to microbial symbiosis research in the microbiome era. Front Microbiol 7:1150. https://doi.org/10.3389/fmicb.2020.01150

ERINN Innovation (2020) SEALIVE deliverable 8.1 dissemination and exploitation plan

EU (2019) A European green deal: striving to be the first climate-neutral continent. Retrieved from https://ec.europa.eu/info/strategy/priorities-2019-2024/european-green-deal_en

EU (2020) The EU budget powering the recovery plan for Europe. Retrieved from https://ec.europa.eu/info/files/eu-budget-powering-recovery-plan-europe_en

European Commission (EC) (2019a) Horizon results platform. Available online at: https://ec.europa.eu/jrc/communities/en/community/tto-circle-community/news/horizon-results-platform. Accessed 12.03.21

European Commission (EC) (2019b) Horizon Europe presentation. Available online at: https://ec.europa.eu/info/sites/info/files/research_and_innovation/ec_rtd_he-presentation_062019_en.pdf. Accessed 12.03.21

European Commission (EC) (2020a) European Climate Law. Available online at: https://ec.europa.eu/clima/policies/eu-climate-action/law_en. Accessed 09.03.2021

European Commission (EC) (2020b) Horizon Europe factsheet. Available online at: https://op.europa.eu/en/publication-detail/-/publication/eef524e8-509e-11eb-b59f-01aa75ed71a1/. Accessed 12.03.2021

European Commission (EC) H2020 LC-GD-7-1-2020. Available online at: https://ec.europa.eu/info/funding-tenders/opportunities/portal/screen/opportunities/topic-details/lc-gd-7-1-2020. Accessed 30.12.2020

Fernandez Otero RM, Bayliss-Brown G, Papathanassiou M (2019) Ocean literacy and knowledge transfer synergies in support of a sustainable blue economy. Front Mar Sci 6:646. https://doi.org/10.3389/fmars.2019.00646

Fielding S, Copley JT, Mills RA (2019) Exploring our oceans: using the global classroom to develop ocean literacy. Front Mar Sci 6:340. https://doi.org/10.3389/fmars.2019.00340

Foo JL, Ling H, Lee YS (2017) Microbiome engineering: current applications and its future. Biotechnol J 12:3. https://doi.org/10.1002/biot.201600099

Gao K, Beardall J, Häder D-P, Hall-Spencer JM, Gao G, Hutchins DA (2019) Effects of ocean acidification on marine photosynthetic organisms under the concurrent influences of warming, UV radiation, and deoxygenation. Front Mar Sci 6:322. https://doi.org/10.3389/fmars.2019.00322

Garbelli C, Angiolini L, Shen SZ (2016) Biomineralization and global change: a new perspective for understanding the end-Permian extinction. Geology 45:19–22. https://doi.org/10.1130/G38430.1

Gewert B, Plassmann MM, MacLeod M (2015) Pathways for degradation of plastic polymers floating in the marine environment. Environ Sci Process Impacts 17:1513. https://doi.org/10.1039/C5EM00207A

Ghosal D, Ghosh S, Dutta TK, Ahn Y (2016) Current state of knowledge in microbial degradation of polycyclic aromatic hydrocarbons (PAHs): a review. Front Microbiol 7:1369

Goulden EF, Hall MR, Bourne DG, Pereg LL, Hoj L (2012) Pathogenicity and infection cycle of *Vibrio owensii* in larviculture of the ornate spiny lobster (*Panulirus ornatus*). Appl Environ Microbiol 78:2841–2849. https://doi.org/10.1128/AEM.07274-11

Harrison JP, Boardman C, O'Callaghan K, Delort A-M, Song J (2018) Biodegradability standards for carrier bags and plastic films in aquatic environments: a critical review. R Soc Open Sci 5: 171792. https://doi.org/10.1098/rsos.171792

Hutchins DA, Jansson JK, Remais JV, Rich VI, Singh BK, Trivedi P (2019) Climate change microbiology — problems and perspectives. Nat Rev Microbiol 17:391–396. https://doi.org/10.1038/s41579-019-0178-5

Ingre-Khans E, Ågerstrand M, Rudé C (2017) Endocrine disrupting chemicals in the marine environment. ACES report number 16. Department of Environmental Science and Analytical Chemistry, Stockholm University

IPCC (2014) Fifth Assessment Report from the Intergovernmental Panel on Climate Change (2014). https://climate.nasa.gov/evidence/. Accessed 05.01.2021

IUCN (n.d.) IUCN issues brief: marine plastics. https://www.iucn.org/resources/issues-briefs/marine-plastics. Accessed 08.03.2021

Jacquin J, Cheng J, Odobel C, Pandin C, Conan P, Pujo-Pay M, Barbe V, Meistertzheim A-L, Ghiglione J-F (2019) Microbial ecotoxicology of marine plastic debris: a review on colonization and biodegradation by the "plastisphere". Front Microbiol 10:865. https://doi.org/10.3389/fmicb.2019.00865

Jambeck J, Geyer R, Wilcox C, Siegler T, Perryman M, Andrady A, Narayan R, Law K (2015) Plastic waste inputs from land into the ocean. Science 347:768–771. https://doi.org/10.1126/science.1260352

Kale SK, Deshmukh AG, Dudhare MS, Patil VB (2015) Microbial degradation of plastic: a review. J Biochem Technol 6:952–961. https://doi.org/10.31838/ijpr/2021.13.01.245

King M, Elliott JE, Williams TD (2021) Effects of petroleum exposure on birds: a review. Sci Total Environ 755:142834. https://doi.org/10.1016/j.scitotenv.2020.142834

Kolandai-Matchett K, Armoudian M (2020) Message framing strategies for effective marine conservation communication. Aquat Conserv 30:2441–2463. https://doi.org/10.1002/aqc.3349

Kolandai-Matchett K, Armoudian M, Li E (2020) Communicating complex ocean issues: how strategically framed messages affect awareness and motivation when conveyed using narrative vs. expository language. Aquat Conserv 31:870–887. https://doi.org/10.1002/aqc.3484

Kollmuss A, Agyeman J (2002) Mind the gap: why do people act environmentally and what are the barriers to pro-environmental behavior? Environ Educ Res 8:239–260. https://doi.org/10.1080/13504620220145401

Krueger MC, Harms H, Schlosser D (2015) Prospects for microbiological solutions to environmental pollution with plastics. Appl Microbiol Biotechnol 99:8857–8874. https://doi.org/10.1007/s00253-015-6879-4

Laffoley D, Baxter JM, Amon DJ, Claudet J, Hall-Spencer JM, Grorud-Colvert K, Levin LA, Reid PC, Rogers AD, Taylor ML, Woodall LC, Andersen NF (2020) Evolving the narrative for protecting a rapidly changing ocean, post-COVID-19. Aquat Conserv Mar Freshw Ecosyst. https://doi.org/10.1002/aqc.3512

Lubchenco J, Cerny-Chipman EB, Reimer JN, Levin SA (2016) The right incentives enable ocean sustainability successes and provide hope for the future. Proc Natl Acad Sci USA 113:14507–14514. https://doi.org/10.1073/pnas.1604982113

Mahjoubi M, Cappello S, Souissi Y, Jaouani A, Cherif A (2017) Microbial bioremediation of petroleum hydrocarbon-contaminated marine environments. https://doi.org/10.5772/intechopen.72207

Mapelli F, Scoma A, Michoud G, Aulenta F, Boon N, Borin S, Kalogerakis N, Daffonchio D (2017) Biotechnologies for marine oil spill cleanup: indissoluble ties with microorganisms. Trends Biotechnol 35:860–870. https://doi.org/10.1016/j.tibtech.2017.04.003

McCauley V, McHugh P, Davison K, Domegan CT (2019) Collective intelligence for advancing ocean literacy. Environ Educ Res 25:280–291. https://doi.org/10.1080/13504622.2018.1553234

McCormick SD, Romero LM (2017) Conservation endocrinology. BioScience 67:429–442. https://doi.org/10.1093/biosci/bix026

McKenzie-Mohr D (2000) Promoting sustainable behavior: an introduction to community-based social marketing. J Soc Issues 56:543–554. https://doi.org/10.1111/0022-4537.00183

Mitton C, Adair CE, McKenzie E, Patten SB, Waye Perry B (2007) Knowledge transfer and exchange: review and synthesis of the literature. Milbank Q 85:729–768. https://doi.org/10.1111/j.1468-0009.2007.00506.x

Mogias A, Boubonari T, Realdon G, Previati M, Mokos M, Koulouri P, Cheimonopoulou MT (2019) Evaluating ocean literacy of elementary school students: preliminary results of a cross-cultural study in the Mediterranean region. Front Mar Sci 6:396. https://doi.org/10.3389/fmars.2019.00396

Mokos M, Realdon G, Cizmek IZ (2020) How to increase ocean literacy for future ocean sustainability? The influence of non-formal marine science education. Sustainability 12:10647. https://doi.org/10.3390/su122410647

NASA (2021) Global climate change. Available at: https://climate.nasa.gov/evidence. Accessed 06.02.2021

Novak BJ, Fraser D, Maloney TH (2020) Transforming ocean conservation: applying the genetic rescue toolkit. Genes 11:209. https://doi.org/10.3390/genes11020209

Oberbeckmann S, Labrenz M (2020) Marine microbial assemblages on microplastics: diversity, adaptation, and role in degradation. Ann Rev Mar Sci 12:209–232. https://doi.org/10.1146/annurev-marine-010419-010633

Oberbeckmann S, Loeder MGJ, Gerdts G, Osborn AM (2014) Spatial and seasonal variation in diversity and structure of microbial biofilms on marine plastics in Northern European waters. FEMS Microbiol Ecol 90:478–492. https://doi.org/10.1111/1574-6941.12409

Ocean Literacy Working Group of the Atlantic Ocean Research Alliance with support from Ciência Viva (2020). https://allatlanticocean.org/uploads/ficheiro/ficheiro_5fabcdc99e377.pdf

Ockendon N, Thomas DHL, Cortina J, Adams WM, Aykroyd T, Barov B et al (2018) One hundred priority questions for landscape restoration in Europe. Biol Conserv 221:198–208

OECD (2020) OECD work in support of a sustainable ocean. www.oecd.org/ocean. Accessed 24.12.2020

Paço A, Duarte K, da Costa JP, Santos PSM, Pereira R, Pereira ME, Freitas AC, Duarte AC, Rocha-Santos TAP (2017) Biodegradation of polyethylene microplastics by the marine fungus *Zalerion maritimum*. Sci Total Environ 586:10–15. https://doi.org/10.1016/j.scitotenv.2017.02.017

Paniagua-Michel J, Rosales A (2015) Marine bioremediation—a sustainable biotechnology of petroleum hydrocarbons biodegradation in coastal and marine environments. J Bioremediat Biodegrad 6:273. https://doi.org/10.4172/2155-6199.1000273

Peixoto RS, Rosado PM, Leite DC, Rosado AS, Bourne DG (2017) Beneficial microorganisms for corals (BMC): proposed mechanisms for coral health and resilience. Front Microbiol 8:341. https://doi.org/10.3389/fmicb.2017.00341

Peixoto RS, Sweet M, Bourne DG (2019) Customized medicine for corals. Front Mar Sci 6:686. https://doi.org/10.3389/fmars.2019.00686

Peixoto RS, Sweet M, Villela HDM, Cardoso P, Thomas T, Voolstra CR, Høj L, Bourne DG (2021) Coral probiotics: premise, promise, prospects. Annu Rev Anim Biosci 9:265–288. https://doi.org/10.1146/annurev-animal-090120-115444

Pita L, Rix L, Slaby BM, Franke A, Hentschel U (2018) The sponge holobiont in a changing ocean: from microbes to ecosystems. Microbiome 6:46. https://doi.org/10.1186/s40168-018-0428-1

Prince RC, Atlas RM (2016) Bioremediation of marine oil spills. In: Steffan R (ed) Consequences of microbial interactions with hydrocarbons, oils, and lipids: biodegradation and bioremediation. Handbook of hydrocarbon and lipid microbiology. Springer, Cham

Ramsby BD, Hoogenboom MO, Whalan S, Webster NS (2018) Elevated seawater temperature disrupts the microbiome of an ecologically important bioeroding sponge. Mol Ecol 27:2124–2137. https://doi.org/10.1111/mec.14544

Reuver M, Bayliss-Brown GA, Calis T, Cardillo P, NíCheallacháin C, Dornan N (2016) Outreach of the unseen majority. In: Stal LJ, Cretoiu MS (eds) The marine microbiome. An untapped source of biodiversity and biotechnological potential. Springer, Cham, pp 473–498. https://doi.org/10.1007/978-3-319-33000-6_18

Roager L, Sonnenschein EC (2019) Bacterial candidates for colonization and degradation of marine plastic debris. Environ Sci Technol 53:11636–11643. https://doi.org/10.1021/acs.est.9b02212

Rochman CM, Hoh E, Kurobe T, Teh SJ (2013) Ingested plastic transfers hazardous chemicals to fish and induces hepatic stress. Sci Rep 3:3263. https://doi.org/10.1038/srep03263

Rogers KL, Carreres-Calabuig JA, Gorokhova E, Posth NR (2020) Micro-by-micro interactions: how microorganisms influence the fate of marine microplastics. Limnol Oceanogr Lett 5:18–36. https://doi.org/10.1002/lol2.10136

Rosado PM, Leite DCA, Duarte GAS, Chaloub RM, Jospin G, Nunes da Rocha U, Saraiva JP, Dini-Andreote F, Eisen JA, Bourne DG, Peixoto RS (2019) Marine probiotics: increasing coral resistance to bleaching through microbiome manipulation. ISME J 13:921–936. https://doi.org/10.1038/s41396-018-0323-6

Santoro F, Scowcraft G, Santin S, Fauville G, Tuddenheim P (2017) Ocean literacy for all—a toolkit. IOC/UNESCO and UNESCO Venice Office (Paris, IOC Manuals and Guides 80)

Santos-Gandelman JF, Giambiagi-deMarval M, Muricy G, Barkay T, Laport MS (2014) Mercury and methylmercury detoxification potential by sponge-associated bacteria. Ant Leeuwenhoek 106:585–590. https://doi.org/10.1007/s10482-014-0224-2

Saunders MI, Doropoulos C, Bayraktarov E, Babcock RC, Gorman D, Eger AM, Vozzo ML, Gillies CL, Vanderklift MA, Steven ADL, Bustamante RH, Silliman BR (2020) Bright spots in coastal marine ecosystem restoration. Curr Biol 30:R1500–R1510. https://doi.org/10.1016/j.cub.2020.10.056

Schoukens H (2017) Ecological restoration as new environmental paradigm: a legal review of opportunities and challenges within the context of EU environmental law, with a particular focus on the EU nature directives. PhD dissertation, Ghent University

SIMBA (Sustainable Innovation of Microbiome Applications in the Food System). https://simbaproject.eu/. Accessed 23.12.2020

Society for Applied Microbiology (2018) The marine microbiome. Science notes for policy, November 2018. https://sfam.org.uk/knowledge/policy/priority-areas/preserving-protecting-our-oceans.html

Stern PC (2000) Toward a coherent theory of environmentally significant behavior. J Soc Issues 56: 407–424. https://doi.org/10.1111/0022-4537.00175

Stocker TF (2015) The silent services of the world ocean. Science 350:764–765. https://doi.org/10.1126/science.aac8720

Stoll-Kleemann S (2019) Feasible options for behavior change toward more effective ocean literacy: a systematic review. Front Mar Sci 6:273. https://doi.org/10.3389/fmars.2019.00273

Straub SC, Wernberg T, Thomsen MS, Moore PJ, Burrows MT, Harvey BP, Smale DA (2019) Resistance, extinction, and everything in between—the diverse responses of seaweeds to marine heatwaves. Front Mar Sci 6:763. https://doi.org/10.3389/fmars.2019.00763

Sweet M, Ramsey A, Bulling M (2017) Designer reefs and coral probiotics: great concepts but are they good practice? Biodiversity 18:19–22. https://doi.org/10.1080/14888386.2017.1307786

Tabacchioni S, Passato S, Ambrosino P, Huang L, Caldara M, Cantale C, Hett J, Del Fiore A, Fiore A, Schlüter A, Sczyrba A, Maestri E, Marmiroli N, Neuhoff D, Nesme J, Sørensen SJ, Aprea G, Nobili C, Presenti O, Giovannetti G, Giovannetti C, Pihlanto A, Brunori A, Bevivino A (2021) Identification of beneficial microbial consortia and bioactive compounds with potential as plant biostimulants for a sustainable agriculture. Microorganisms 9:426. https://doi.org/10.3390/microorganisms9020426

Teuten EL, Saquing JM, Knappe DRU, Barlaz MA, Jonsson S, Bjorn A, Rowland SJ, Thompson RC, Galloway TS, Yamashita R, Ochi D, Watanuki Y, Moore C, Viet PH, Tana TS, Prudente M, Boonyatumanond R, Zakaria MP, Akkhavong K, Ogata Y, Hirai H, Iwasa S, Mizukawa K, Hagino Y, Imamura A, Saha M, Takada H (2009) Transport and release of chemicals from plastics to the environment and to wildlife. Philos Trans R Soc B 364:2027–2045. https://doi.org/10.1098/rstb.2008.0284

Thérin F (2013) Handbook of research on techno-entrepreneurship, second edition: how technology and entrepreneurship are shaping the development of industries and companies. Edward Elgar, Cheltenham

Trevathan-Tackett SM, Sherman CD, Huggett MJ, Campbell AH, Laverock B, Hurtado-McCormick V, Seymour JR, Firl A, Messer LF, Ainsworth TD, Negandhi KL, Daffonchio D, Egan S, Engelen AH, Fusi M, Thomas T, Vann L, Hernandez-Agreda A, Gan HM, Marzinelli EM, Steinberg PD, Hardtke L, Macreadie PI (2019) A horizon scan of priorities for coastal marine microbiome research. Nat Ecol Evol 3:1509–1520. https://doi.org/10.1038/s41559-019-0999-7

Trevelline BK, Fontaine S, Hartup BK, Kohl KD (2019) Conservation biology needs a microbial renaissance: a call for the consideration of host-associated microbiota in wildlife management practices. Proc Roy Soc B 286:20182448. https://doi.org/10.1098/rspb.2018.2448

Troisi G, Barton S, Bexton S (2016) Impacts of oil spills on seabirds: unsustainable impacts of non-renewable energy. Int J Hydrog Energy 41:16549–16555. https://doi.org/10.1016/j.ijhydene.2016.04.011

UNESCO/IOC (2019) Ocean literacy for all initiative—summary of achievements (2017-2018). Venice, IOC/UNESCO & UNESCO Regional Bureau for Science and Culture in Europe (IOC Brochure 2019-2)

United Nations Environment Programme and the World Health Organization (2013) In: Bergman A, Heindel JJ, Jobling S, Kidd KA, Zoeller RT (eds) State of the science of endocrine disrupting chemicals 2012. http://www.who.int/ceh/publications/endocrine/en/

Uyarra MC, Borja A (2016) Ocean literacy: a 'new' socio-ecological concept for a sustainable use of the seas. Editorial. Mar Pollut Bull 104:1–2. https://doi.org/10.1016/j.marpolbul.2016.02.060

Van Hooidonk R, Maynard J, Tamelander J, Gove J, Ahmadia G, Raymundo L, Williams G, Heron SF, Planes S (2016) Local-scale projections of coral reef futures and implications of the Paris agreement. Sci Rep 6:39666. https://doi.org/10.1038/srep39666

Van Oppen MJH, Blackall LL (2019) Coral microbiome dynamics, functions and design in a changing world. Nat Rev Microbiol 17:557–567. https://doi.org/10.1038/s41579-019-0223-4

Van Oppen MJH, Oliver JK, Putnam HM, Gates RD (2015) Building coral reef resilience through assisted evolution. Proc Natl Acad Sci 112:2307–2313. https://doi.org/10.1073/pnas.1422301112

Van Oppen MJH, Gates RD, Blackall LL, Cantin N, Chakravarti LJ, Chan WY, Cormick C, Crean A, Damjanovic K, Epstein H, Harrison PL, Jones TA, Miller M, Pears RJ, Peplow LM, Raftos DA, Schaffelke B, Stewart K, Torda G, Wachenfeld D, Weeks AR, Putnam HM (2017) Shifting paradigms in restoration of the world's coral reefs. Glob Chang Biol 23:3437–3448. https://doi.org/10.1111/gcb.13647

Van Tatenhove JPM, Ramírez-Monsalve P, Carballo-Cárdenas E, Papadopoulou N, Smith CJ, Alferink L, Ounanian K, Long R (2020) The governance of marine restoration: insights from three cases in two European seas. Restor Ecol. https://doi.org/10.1111/rec.13288

Vanwonterghem I, Webster NS (2020) Coral reef microorganisms in a changing climate. iScience 23:100972. https://doi.org/10.1016/j.isci.2020.100972

Villela HD, Vilela CL, Assis JM, Varona N, Burke C, Coil DA, Eisen JA, Peixoto RS (2019) Prospecting microbial strains for bioremediation and probiotics development for metaorganism research and preservation. J Vis Exp 152:e60238. https://doi.org/10.3791/60238

Visbeck M (2018) Ocean science research is key for a sustainable future. Nat Commun 9:690. https://doi.org/10.1038/s41467-018-03158-3

Wagner M, Lambert S (eds) (2018) Freshwater microplastics: emerging environmental contaminants? Handb Environ Chem 58. https://doi.org/10.1007/978-3-319-61615-5

Walker AH (2017) Oil spills and risk perceptions. In: Oil spill science and technology. Elsevier, pp 1–70

Wang H, Luo H, Fallgren PH, Jin S, Ren ZJ (2015) Bioelectrochemical system platform for sustainable environmental remediation and energy generation. Biotechnol Adv 33:317–334. https://doi.org/10.1016/j.biotechadv.2015.04.003

Wang L, Xiao H, He N, Sun D (2019) Biosorption and biodegradation of the environmental hormone nonylphenol by four marine microalgae. Sci Rep 9:5277. https://doi.org/10.1038/s41598-019-41808-8

Webster NS, Reusch TBH (2017) Microbial contributions to the persistence of coral reefs. ISME J 11:2167–2174. https://doi.org/10.1038/ismej.2017.66

WEF (2020) The global risks report 2020. In: Insight Report, 15th edn. World Economic Forum in Partnership with Marsh & McLennan and Zurich Insurance Group, Cologny/Geneva, Switzerland. https://www.weforum.org/reports/the-global-risks-report-2020

Wilkins L, Leray M, O'Dea A, Yuen B, Peixoto RS, Pereira TJ, Bik HM, Coil DA, Duffy JE, Herre EA, Lessios HA, Lucey NM, Mejia LC, Rasher DB, Sharp KH, Sogin EM, Thacker RW, Vega Thurber R, Wcislo WT, Wilbanks EG, Eisen JA (2019) Host-associated microbiomes drive structure and function of marine ecosystems. PLoS Biol 17:e3000533. https://doi.org/10.1371/journal.pbio.3000533

Wright RJ, Erni-Cassola G, Zadjelovic V, Latva M, Christie-Oleza JA (2020) Marine plastic debris: a new surface for microbial colonization. Environ Sci Technol 54:11657–11672. https://doi.org/10.1021/acs.est.0c02305

Zettler ER, Mincer TJ, Amaral-Zettler LA (2013) Life in the "Plastisphere": microbial communities on plastic marine debris. Environ Sci Technol 47:7137–7146. https://doi.org/10.1021/es401288x

Zhang W, Yin K, Chen L (2013) Bacteria-mediated bisphenol A degradation. Appl Microbiol Biotechnol 97:5681–5689. https://doi.org/10.1007/s00253-013-4949-z

Printed by Printforce, the Netherlands